Understanding the development of communities, describing their structure and characterizing their function, are major objectives of ecologists worldwide. Such understanding is becoming increasingly important with the recognition that the conservation of biodiversity is best achieved through the management of natural communities. Progress remains slow, with community diversity still incompletely characterized, competitive interactions between species at only an early stage of understanding and the stability and resilience of any community the subject of debate. The study of Antarctic communities can provide a valuable step forward in investigating the controls on community development, the utilization of habitats and the interaction between species in both species-rich and species-poor communities. This book contains chapters characterizing the present approaches to both aquatic and terrestrial communities in the Antarctic. From biodiversity to trophic flows, from ecophysiological strategies to the impacts of environmental change and the effects of human disturbance, this volume provides an up-to-the-minute overview of community studies in terrestrial, marine and freshwater ecosystems of the Antarctic and the Southern Ocean.

Antarctic Communities

Species, Structure and Survival

Antarctic Communities

Species, Structure and Survival

EDITED BY B. BATTAGLIA, J. VALENCIA AND D. W. H. WALTON

CAMBRIDGE UNIVERSITY PRESS
Cambridge, New York, Melbourne, Madrid, Cape Town, Singapore, São Paulo, Delhi

Cambridge University Press
The Edinburgh Building, Cambridge CB2 8RU, UK

Published in the United States of America by Cambridge University Press, New York

www.cambridge.org
Information on this title: www.cambridge.org/9780521111799

First published 1997
This digitally printed version 2009

A catalogue record for this publication is available from the British Library

Library of Congress Cataloguing in Publication data

Antarctic communities : species, structure, and survival / edited by B. Battaglia, J. Valencia, and D.W.H. Walton.
p. cm.
'The meeting from which these papers were taken was the sixth biological symposium sponsored by the Scientific Committee on Antarctic Research (SCAR) . . . held at 'Le Zitelle' Centre in Venice, Italy from May 30 to June 3. 1994' – Pref.
Includes bibliographical references and index.
ISBN 0 521 48033 7 (hardbound)
1. Biological diversity – Antarctica – Congresses. 2. Ecology – Antarctica – Congresses. 3. Adaptation (Biology) – Antarctica – Congresses. 4. Nature – Effect of human beings on – Antarctica – Congresses. I. Battaglia, B. (Bruno), 1923– . II. Valencia, J. (José), 1935– . III. Walton, D. W. H. IV. International Council of Scientific Unions. Scientific Committee on Antarctic Research.
QH84.2.A57 1997
5777.8′2′09989–dc21 96–37973 CIP

ISBN 978-0-521-48033-8 hardback
ISBN 978-0-521-11179-9 paperback

Contents

III – SURVIVAL MECHANISMS

IV – ADAPTIVE MECHANISMS

Preface

For many disciplines Antarctic science has been traditionally viewed as of regional interest only, and thus far removed from the vanguard of important advances. Its practitioners were often considered as narrowly focused with data of little general relevance to the discipline. In 1995 these ideas are very far from reality with Antarctica contributing both to the investigation of global problems and to major scientific advances. This is especially true of biology, where the development of commercial exploitation in fishing and tourism and the effects of enhanced UV on the biota have produced a wide range of new research programmes investigating both fundamental and applied science problems. Despite all the pressures towards increasing specialization it has still remained possible to organize meetings at which the Antarctic ecosystem can be treated as a whole. This holistic approach has been evident for the last 30 years and this symposium in Venice marked a further milestone in progress.

The meeting from which these papers were taken was the sixth biological symposium sponsored by the Scientific Committee on Antarctic Research (SCAR). It was held at "Le Zitelle" Centre in Venice, Italy from May 30 to June 3, 1994 at the invitation of the SCAR Italian Committee and the Programma Nazionale di Ricerche in Antartide (PNRA). Previous symposia were held in Paris, France, 1962, Cambridge, England, 1968, Washington DC, USA, 1974, Wilderness, South Africa, 1983 and Hobart, Australia, 1988. The proceedings were edited by Carrick & Prevost (1964), Holdgate (1970), Llano (1977), Siegfried, Condy & Laws (1985) and Kerry & Hempel (1990), respectively. Each stands as a major landmark in the development of biology in the Antarctic and provides a clear indication of what were some of the major science preoccupations of the time.

The earlier symposia addressed very general subjects and published all the papers offered. This has now become impossible and present policy is to decide on a key area of interest and organize the meeting around it. The central focus of this symposium was at the level of the community. A clear understanding of the community species composition, structural relationships and survival strategies is essential for an understanding of ecological processes. Major programmes on the effects of global change, the need to predict and monitor impacts, the importance of conservation and environmental management measures, and the development of models have all made community research more important.

The meeting was attended by 334 registrants representing 25 nations. There were 53 oral papers, including 10 keynote papers. At the ten poster sessions 252 posters were displayed. From all of this material the Steering Committee selected 63 papers, chosen to illustrate in a coherent fashion the range of current research related to communities. All of these were peer reviewed and revised before inclusion in this volume.

At the end of the symposium in Hobart five people gave overviews of key groups and G. Hempel suggested what might constitute future research needs. In attempting to look forward Hempel suggested that, in the marine ecosystem, the benthos was in particular need of closer attention, long-term studies were required to 'observe the development of plankton blooms, grazing by krill swarms and to measure vertical fluxes' and sea ice should be a major focus for interdisciplinary studies utilizing shipborne work and remote sensing. He highlighted the gains to be expected from detailed comparisons between communities, their structure, dynamics and productivity, and recommended the value of ecophysiological studies on terrestrial communities as a useful contribution to global change studies. Noting the importance of taxonomy to all future studies on terrestrial and limnic systems he highlighted the need for basic descriptions of communities and better soils information as key constituents of any sound management and conservation policies.

How far have we come in these fields since Hobart? The structure of the symposium reflected some of Hempel's concerns. Sessions for oral presentations and posters were focused on: Biodiversity, Microbial Ecology, Planktonic Processes, Ecophysiology, Benthic Ecology, Bird and Mammal Population Dynamics, Biochemical Adaptations, Resources and Seasonality, Human Impact, and Environmental Change. There was strong emphasis on community research in each field and a strong showing for global change and human impact. Not presaged by Hempel was a significant number of contributions on molecular ecology and biology, clearly a developing field with a great deal to offer to evolution, population biology and adaptation. There was little evidence of a general increase in studies on soils, but the value of plants as indicators of climate change has been recognized. In the marine field there continued to be detailed studies on zooplankton and on phytoplankton with increased studies on benthos in shallow water. The lack of contributions to the meeting on birds and seals was not an indication of a decline in research on these groups but more a clash with other international meetings.

In preparing this volume we have selected important examples from the range of material presented in Venice to indicate progress in a wide variety of related fields. Each section is introduced by the Editors to indicate the way in which the papers answer key questions and how these many individual efforts take our understanding of community ecology forward.

REFERENCES

Carrick, R., Holdgate, M.W. & Prevost, J., eds.1964. *Biologie Antarctique*. Paris: Hermann.

Holdgate, M.W., ed. 1970. *Antarctic ecology*, 2 vols. London: Academic Press.

Llano, G.A., ed. 1977. *Adaptations within Antarctic ecosystems*. Washington, DC: Smithsonian Institution.

Siegfried, W.R. , Condy, P.R. & Laws, R.M. eds. 1985. *Antarctic nutrient cycles and food webs*. Berlin: Springer Verlag.

Kerry, K.R. & Hempel, G. 1990. *Antarctic ecosystems: ecological change and conservation*. Berlin: Springer Verlag.

November 1995

B. Battaglia
J. Valencia
D. W. H. Walton

Acknowledgements

The organization of the Symposium was undertaken by an International Steering Committee, constituted by Professor J. Valencia (Chile) Chairman, Dr P. Shaughnessy (Australia) Secretary, Professor B. Battaglia (Italy), Dr D.W.H. Walton (UK), Dr R.I. Lewis Smith (UK), Dr G.Hubold (Germany), Dr J.P. Croxall (UK), Dr W. Arntz (Germany), Dr D. Siniff (USA), G. di Prisco (Italy) and Dr R.M. Laws (UK). A local organizing committee under the leadership of B. Battaglia and including P.M. Bisol, L. Dalla Venezia, A. Libertini, T. Patarnello, V. Varotto and A. Zitelli provided crucial assistance in preparing the programme of events, printing and distributing information on the meeting and making arrangements in Venice. Of particular relevance were, in this context, the contributions of Luisella Dalla Venezia and Vittorio Varotto, greatly helped by the effective cooperation of Francesca Pirazzoli of ENEA, Caterina Dal Pra of 'Le Zitelle' Centre, and Stefania Marcato of the Biology Department, University of Padova. Special administrative and secretarial assistance was efficiently provided by Francesco Gratteri and Cecilia Zattara, Biology Department, University of Padova.

The meeting was organized under the auspices of the Scientific Committee on Antarctic Research and the patronage of the Accademia Nazionale dei Lincei, Istituto Veneto di Scienze, Lettere ed Arti, University of Venice, University of Padova, Regione Veneto, Provincia di Venezia and Comune di Venezia. Considerable financial support was provided by the Italian Antarctic Research Program (MURST, ENEA, CNR) The administrative aspects were taken care of by ENEA. Other financial support was received from: Consorzio Venezia Nuova, Tecnomare S.p.A., Ligabue Catering S.p.A., Societa' Italiana per il Gas p.a., Assicurazioni Generali S.p.A., Italteam Shipping s.r.l. and Datamat Ingegneria dei Sistemi S.p.A. Certain events were made possible thanks to contributions of the Universities of Venezia and Padova.

Without the assistance of a great many people this symposium would not have been possible. The Steering Committee and the Editors are grateful to them for their efforts in support of Antarctic science.

Producing a volume of this length and complexity is a time-consuming task in which the Editors need substantial help from authors and the community at large. The Editors wish to sincerely thank all those listed below who assisted in the refereeing process which has, as always, markedly improved the quality of the final contributions. It had been the Editors' expressed intention to publish the volume earlier but this depended on the timely co-operation of all the authors and reviewers involved. Unfortunately, there were significant delays in finding reviewers willing to undertake the task for many of the papers, many authors took considerable periods to provide the revisions of their manuscripts and many authors failed to follow the manuscript preparation instructions, greatly increasing the editorial work. We can only apologise for the delay to those who did make every effort to meet the original deadlines.

Referees: V. Ahmadjian, J. Anderson, J. Bale, S. Björck, M. Bölter, I. Boyd, P. Broady, A. Clarke, P. Convey, G. Cripps, N. Cronberg, B. Culik, W. Davidson, S. de Mora, G. di Prisco, J. Eastman, S. Eggington, E. El-Sayed, C. Ellis-Evans, Z. A. Eppley, G. Ernsting, S. Focardi, P. Franzmann, M. Fukuchi, P. Grime, W. Hagen, O. Holm-Hansen, G. Hosie, C. Howard-Williams, M. Hughes, A. Huiskes, B. Huntley, B. Jones, V. Jones, L. Kappen, D. Karentz, D. Keiller, J. King, M. Klages, G. Kooyman, R. Laslett, R. M. Laws, J. Laybourn-Parry, R.I. Lewis Smith, M. C. Malin, H. Marchant, D. Osborne, J. Overnell, D. O. Øvstedal, L. Peck, D. Peel, M. Perutz, J. Priddle, P. Rainbow, R. Ross, N. Russel, B. Schroeter, P. Shaughnessy, G. Somero, L. Sømme, M. Spindler, V. Spiridonov, W. Testa, P. Vernon, W. Vincent, J. Watkins, M. White, D. Wynn-Williams, G. P. Zauke.

Finally, we are especially grateful to Elizabeth Edwards and Sharon Cooke who undertook a great deal of the extensive secretarial work essential to bring this volume to publication.

I Biodiversity and evolution

INTRODUCTION

Biodiversity and its assessment is a field undergoing unexpected developments (Wilson 1988). This feature of all communities has assumed considerable political importance since the meeting in Rio de Janeiro. Unfortunately, before deciding 'what should be preserved' it is necessary to know what there is to preserve. Even in Antarctica, viewed nowadays as an immense natural reserve to be protected from human impact, biodiversity must be assessed. The traditional taxonomic approach is necessary but not sufficient to answer all the questions. Whilst the species concept remains at the centre of any classification, to understand better the processes and mechanisms which have led to the evolution of biodiversity in time and space, the problems need to be tackled at various levels of organisation, from communities or multispecies assemblages to the intraspecific levels.

The evaluation of intraspecific biodiversity, for instance at the levels of Mendelian populations, or of genes and their molecular constituents, may help in solving various problems of species taxonomy. This approach may also help with aspects of the functional role of biodiversity, provide criteria for the identification of 'key' species and reveal new mechanisms of adaptation and evolution (e.g. DeLong *et al.* 1994). Moreover, changes in biodiversity may also provide an effective tool for monitoring the effects of environmental impact, even those of global change (Solbrig *et al.* 1992).

Antarctic biodiversity, both at the species and community levels, varies from place to place and from group to group. Whilst in some cases we now have a positive indication of the factors causing this, in most cases a convincing picture is still lacking. New sampling methods, new analytical techniques and, above all, the adoption of conceptually and methodologically more advanced criteria to face new as well as old problems, are already taking us down new pathways. For many of those problems a multidisciplinary approach is absolutely essential. Understanding the evolutionary processes which have led to the present diversity relies upon the cooperative efforts of taxonomists, ecologists, geneticists, physiologists and molecular biologists. The contributions to this section of the Symposium can be divided into two groups – both having in common an evolutionary viewpoint. The first group deals with community and species diversity. Arntz *et al.* (Chapter 1), in their paper on biodiversity in the Antarctic marine ecosystem, stress the lack of agreement between various authors in estimates of species numbers. This may be due to insufficient taxonomic knowledge, to the multiplicity of techniques adopted for assessing biodiversity at the species level, to lack of standardized sampling, or to the misleading effect of older records on delimiting endemisms. Benthic biodiversity is clearly very considerable but many groups are still inadequately described, especially in the deep sea (Grassle 1991). There are continuing problems in providing accurate comparisons of community diversity from different areas.

On land there is also taxonomic confusion. The species status of the Antarctic lichen flora has been thoroughly revisited in the contribution of Castello and Nimis (Chapter 2), who consider valid only a few of the species previously described by Dodge (1973). In their view the known lichen flora is thus reduced from 415 to 260 species, with the percentage of endemic species falling from 91% to 38%, but the percentage of bipolar and cosmopolitan species increases greatly. Partly on the grounds of a high endemic component, previous workers have always concluded that the lichen flora is ancient but the present authors now suggest that the lichen flora is a young one, possibly due to long distance dispersal in the Quaternary.

An interesting example of taxonomic and bionomic bipolarity is provided in a study conducted by Svoboda *et al.* (Chapter 3) on a symbiotic relationship between a Hydractinia and a brittlestar. The Arctic and Antarctic counterparts of both symbionts are closely related, and the morphological features of this commensal symbiosis are remarkably similar. It appears that these unique relationships must have developed independently in both polar seas.

The first group of papers concludes with the contribution of Zimmerman *et al.* (Chapter 4) on the composition and community structure in the demersal fish fauna of the Lazarev Sea. In this area, abundance and biomass exhibit great variability and species diversity is higher than in any other region of the Southern Ocean. These results indicate distribution patterns characterised by pronounced small-scale heterogeneities.

The second group of papers deals with problems of variation at the gene level. It is at this level that studies on population characterization and microevolution need to be pursued. A very good example is provided by Adam *et al.* (Chapter 5) who studied the geographic and microgeographic patterns of genetic variation in the morphologically variable moss *Bryum argenteum.* Both isozyme and DNA (RAPD) analyses were employed, the latter revealing levels of variation much higher than the former. From these results the authors draw interesting inferences about the patterns of colonisation by this moss in Antarctica.

A stimulating ecogenetic study of fish and seal anisakid endoparasites in both polar regions by Bullini *et al.* (Chapter 6) used isozymes. As well as providing a description of colonization patterns they also draw conclusions about the host–parasite mechanisms of mutual adaptation and co-evolution. These anisakids exhibit a genetic variability which is higher in the Antarctic species than in the Boreal ones, a difference which is attributed to the lower habitat disturbance of the Antarctic area.

Molecular tools have also proved very useful for establishing phylogenetic relationships. The study by Bargelloni *et al.* (Chapter 7) on nineteen species of notothenioids used mitochondrial DNA genetic analysis, and allows them to infer phylogenetic relationships among and within five families of these fishes. This provides the first example of the use of molecular techniques to address phylogenetic issues in notothenioids, and offers new ways for determining the tempo and mode of their evolution.

These tools can also be effectively applied to the inheritance of specific gene sequences that may have significant evolutionary advantages. The disputed issue of whether myoglobin is expressed in any of the channichthyid fishes seems to be solved by the demonstration by Vayda *et al.* (Chapter 9) of the expression and accumulation of this protein in two icefish species. The problem is discussed in the light of the genetic events occurring during radiation of channichthyid species.

The use of molecular techniques has greatly increased our knowledge of prokaryotes diversity (Franzmann & Dobson 1993). The contribution of Franzmann *et al.* (Chapter 8) to the problem of adaptive speciation and evolution of this group is of considerable importance. Analysis of the 16 S RNA gene as a molecular clock shows that the Antarctic species diverged long before Antarctica established as a permanent cold environment. The problem is: are there any Antarctic prokaryotes? Whether or not individual species are unique to Antarctica remains an open question, which will perhaps find an answer with the development of adequate new methods.

REFERENCES

DeLong, E. F., Wu, K. Y., Prézelin, B. B. & Jovine, R. V. M. 1994. High abundance of Archaea in Antarctic marine picoplankton. *Nature*, **371**, 695–697.

Dodge, C. W. 1973. *Lichen flora of the Antarctic continent and adjacent islands*. Canaan, New Hampshire: Phoenix Publishing, 399 pp.

Grassle, J. F. 1991. Deep-sea benthic biodiversity. *Bioscience*, **41**, 464–469.

Solbrig, O. T., van Emden H.M. & van Oordt, eds. 1992. *Biodiversity and global change*. Paris: IUBS.

Wilson, E. O., ed. 1988. *Biodiversity*. Washington DC: National Academy Press.

1 Antarctic marine biodiversity: an overview

WOLF E. ARNTZ, JULIAN GUTT AND MICHAEL KLAGES
Alfred-Wegener-Institut für Polar- und Meeresforschung, Columbusstrasse, D-27568 Bremerhaven, Germany

ABSTRACT

The unique Antarctic marine environment, its evolutionary history, its biotic peculiarities and its (hitherto) comparatively low degree of human impact make a biodiversity approach and a comparison with other areas particularly worth while. Current knowledge seems to indicate that there is no common pattern for species richness in the various Antarctic subsystems (e.g. pelagic/benthic, shallow/deep) or for different taxonomic groups. Some assemblages appear to be fairly rich in species, others consist of only a few, and the same pattern applies to the various taxa at a higher taxonomic level. The Antarctic marine ecosystem as a whole seems to have a lower percentage of species known to date in most higher taxa than would be expected from its share of the area of the world's oceans. However, comparison with other marine ecosystems is difficult because of differences in area, environment, sampling and processing, and taxonomic knowledge. Comparison with the Arctic Ocean indicates that species numbers of most groups are much higher in Antarctic waters, but many more comparable data are needed to judge whether this also holds generally true for diversity, and whether large-scale latitudinal gradients exist for more than a few groups. Few authors have calculated diversity and evenness indices, and these, too, are often of very limited comparability. High species numbers do not necessarily imply high values of diversity and evenness.

Key words: biodiversity, Antarctica, species richness, evenness, benthos, plankton.

INTRODUCTION

The term 'biodiversity' is often used in a rather broad sense, which may sometimes lead to a confusion of terms. Many people believe, rather intuitively, that the 'richness' of a community is not only reflected by the numbers of species present and in the distribution of individuals among these species, but also by the total number of individuals, total biomass, and possibly also by a variety of trophic or other ecological functions. Furthermore, there is growing evidence that biological diversity may play a key role in protecting the global biosphere, which is increasingly affected by human influences. We need to measure biological diversity to identify those factors which govern it, and to arrive at a better understanding of the consequences of high or low biodiversity for different ecosystems. In this context, studies on Antarctic biodiversity are of special interest, since the Southern Ocean is still a rather pristine, and presumably a very sensitive, ecosystem.

An earlier paper (Arntz *et al.* 1994) looked at the Antarctic zoobenthos in general and it is useful to begin by considering some of the findings:

- Like other marine ecosystems, the South Polar Sea reveals distinct differences between its various subsystems in shallow water (<30 m), on the deeper shelf and slope, and in the deep sea, although these subsystems share a surprisingly high number of eurybathic species. The intertidal and upper sublittoral levels are heavily impacted by ice. The richest communities, mostly dominated by sessile suspension feeders, are found on the deeper shelf and the upper slope. Seemingly, the deep sea does not reveal great differences from other deep-sea areas of the world ocean; however, sampling in that area has been very limited to date, and further sampling may provide new evidence.
- Densities in most benthic communities, excluding those of

shallow water, are of the same order of magnitude as in similar marine communities in other areas, and are usually much below the numbers that are reached in temperate soft bottoms (see e.g. Linke 1939, Ziegelmeier 1970). Only in one single case (an infaunal community in McMurdo Sound; Dayton & Oliver 1977) have exceptionally high values been found.

- Biomass appears high at first glance in the epifaunal suspension-feeding communities (however, with an important share of silici- and calcimass), although it does not reach the peak values that can be found elsewhere, e.g. in temperate mussel banks (Thamdrup 1935) or intertidal clam beds of upwelling regions (Arntz *et al.* 1987). It seems comparatively low in most infaunal communities of the high Antarctic (Gerdes *et al.* 1992). However, according to Brey & Clarke (1993), average benthic biomass in the Antarctic is higher than that of temperate and subtropical communities. The distribution of both biomass and abundance values is highly patchy.
- Various kinds of life history strategies in the zoobenthos have been found in the Antarctic, from close coupling to the primary production cycle in the pelagic (which is considered to be the main ecological factor in the South Polar Sea, in terms of seasonal food limitation, rather than the low temperature; see Clarke 1988) to total uncoupling. However, meroplanktonic larvae seem to be scarce, despite the fact that a certain number of such larvae has been found recently, mostly belonging to larger organisms living in shallow water (Pearse *et al.* 1991).

This paper will be restricted to the narrower meaning of biodiversity, i.e. to species numbers and the distribution of individuals among the species. We will not be discussing the usefulness of different diversity indices (see, e.g. Hurlbert 1971, Margelef 1977). We think that with respect to the Southern Ocean the real problem is that technical progress has hampered comparative studies of biodiversity. People have been ingenious in inventing ever more perfect grabs, corers, plankton nets and trawls. There have also been major developments in sampling methods and treatment of samples. However, not even the most refined techniques can reduce data to a common denominator if some basic requirements of comparable sampling and processing of samples are neglected. If Antarctic researchers do manage to agree on a limited range of standardized equipment and procedures in the future, they should then be able to answer the basic questions of interest to them:

- Is biodiversity high or low in the Southern Ocean?
- What degree of variability in biodiversity is there among different Antarctic subsystems and assemblages?
- Which ecological factors (physical, biological) shape and characterize areas or assemblages of high and low diversity?
- How do Antarctic communities compare with those in other parts of the world ocean? In particular, is there any such thing as a general latitudinal gradient?

Despite the fact that the Southern Ocean, in terms of physical environmental factors, is a relatively homogenous ecosystem of enormous dimensions compared with the seas around other continents, a closer look reveals distinct differences among the different subsystems. This is true not only for the various zones or belts surrounding the continent, from the high Antarctic across the pack-ice belt, through the sub-Antarctic including parts of the Antarctic Peninsula, to the 'maritime' groups of islands (e.g. Hempel 1985). It is valid also in a vertical sense, from the barren Antarctic shores scoured by ice across the much richer sublittoral communities on the shelf and slope to the virtually unstudied vast areas of deep sea surrounding the continent. Talking about marine biodiversity in 'the Antarctic' we have to refer to a multitude of different sites and species assemblages. Life in the pelagial and a great part of life in the benthal has been described as being essentially circumpolar (Hedgpeth 1971), but in some benthic groups such as the asteroids (Voß 1988), molluscs (Hain 1990) and holothurians (Gutt 1991a), distinct differences have been observed in species composition between the subregions of the Southern Ocean.

Taxonomic knowledge of the Antarctic fauna may be better than one would suspect (see table 10.2 in Winston 1992); knowledge of the fauna of the pack-ice zone has improved considerably in recent time (Arntz *et al.* 1994), but much of the deep-sea fauna and the meio- and microfauna in general are almost unknown (Arnaud 1992, Dahms 1992). Unfortunately, many taxonomists work on timescales which are not particularly helpful to the ecologist who is in need of rapid species identification now, in order to be able to calculate species numbers and diversity indices. In this respect, limitations for Antarctic research resemble very much those for scientists working in the deep sea.

Finally, if we want to compare conditions in 'the Antarctic' with those in other marine areas (e.g. species numbers) what other area would make a valid comparison? All other continents have a wide range of climatic zones, and the extensions of most other relatively homogenous marine regions are much smaller than those of the Southern Ocean. The deep sea is an exception, and a comparison certainly is worth while, as was shown over a decade ago by Lipps & Hickmann (1982). Comparisons also appear reasonable with the Arctic Ocean, taking into account the great differences between the two systems (Dayton 1990, Dayton *et al.* 1994).

Thus, finding an answer to the questions as to what Antarctic marine biodiversity is like and how it compares to that of other areas is not simple. On the other hand, it is worth while making the attempt because Antarctica is unique, it has a number of abiotic and biotic peculiarities (Dayton 1990, Dayton *et al.* 1994, Arntz *et al.* 1994), it has a long evolutionary history (Clarke & Crame 1989, 1992), and, despite a growing human impact, it is still much less anthropogenically disturbed than any other marine ecosystem. We will have to consider these peculiarities, and also the difficulties outlined above, when looking at the available data. The focus will be on benthic species as the most numerous group, but other groups will be considered where data are available.

BIODIVERSITY

Species numbers

There have been several estimates of species worldwide (e.g. Wilson 1992), which also include terrestrial and limnic organisms. The recent estimate of species numbers in the major groups of marine organisms in the world's oceans by Winston (1992, table 10.1) is quite detailed, but may turn out to be an underestimate for many groups, in the light of a recent discussion on the subject (Grassle & Maciolek 1992, May 1992, Poore & Wilson 1993 and reply by May in that paper). Using estimates of species along spatial gradients, Grassle & Maciolek (1992) arrived at a few hundreds of millions of benthic invertebrates in the world ocean, which they scaled back to 10 million because they believed that species numbers should be lower on the floor of ocean basins than on continental shelves or slopes. May (1992) doubted their estimate, mainly because he thought that there is no linear increase of species numbers along global gradients, and he proposed an upper limit of half a million species. Poore & Wilson (1993) used isopod data to demonstrate that Grassle & Maciolek's final estimate might be rather on the low side since the ratio between known and unknown species (31% known) in their samples was unusually favourable. While May used a factor of two for the ratio of unknown to known deep-sea species, Poore & Wilson suggested a factor of 20. Furthermore, whereas Grassle & Maciolek assumed that shallow marine infaunal communities outside the tropics have generally lower species numbers, using as an example 200 species from comprehensive sampling on Georges Bank, Poore & Wilson mention values of 700 and 800 invertebrates from two SE Australian communities.

Without further information, the estimates in table 10.1 from Winston (1992) have to stand. We have added 'true' seabirds from Tuck & Heinzel (1980) and Harrison (1983), seals from Deimer (1987, based on the 'Marine Mammal Protection Act' of 1972) and Bonner (1989), and cetaceans from Gaskin (1982) to complete the list.

Probably the first person to call the attention to the fact that some groups in the Antarctic are 'rich' in species and others are 'poor' was Dearborn (1968). With improved taxonomic data, species numbers of Antarctic marine fauna were compiled by Dell (1972, table I) and White (1984, table III). The former author also provided figures on the percentage of endemic species, and the latter compared Antarctic species numbers with those in the Arctic, revealing in all cases higher species richness in the Southern Ocean. White's figures have been reused by Grebmeier & Barry (1991), so the Arctic species catalogue may not have changed much recently. Decapods do not fit the general picture in that they have higher species numbers in the Arctic (Dearborn 1968).

The Antarctic species numbers provided by White were in each individual case substantially higher than those given by Dell, which apparently again reflects taxonomic progress. White (1984, table VII) also cited endemism values on the species and genus levels from various sources, which in some cases compared well with Dell's figures but differed substantially in others. Based on extensive isopod material, White (1984, table VIII) showed that species and genera numbers differ considerably at different Antarctic localities, as do the percentages of endemists both on the species and genus level. This is a very important point in that it indicates that, as with the species numbers, most endemism figures found in the literature cannot simply be compared; they must be referred to specific regions of the Southern Ocean. Unfortunately, by no means all authors who have compiled the data of particular groups have indicated clearly whether and to what extent they include the area north of the Polar Front (in some cases even the Magellan Region is included). Obviously, both the species number and endemism values have to be lower than for the total Southern Ocean if they refer to subregions, but see Brandt (1991, table 2) for a different way of presentation. It is not surprising, considering the 7 years between the publication dates, that Brandt's (1991) data include more isopod species than those of White (1984), but the data (where they are comparable by regions) bear very little resemblance. This reflects the problems a reviewer has to face with Antarctic data.

Our recent compilation (Fig. 1.1A) suffers from the uncertainties just indicated. As can be seen from this illustration, there is a variety of species-rich and species-poor taxa. For example, stomatopods are totally absent, and reptant decapods are almost absent from the Antarctic fauna, natant decapod species are few, whereas groups such as sponges, bryozoans, mollusks, polychaetes, amphipods and isopods have a high number of species. Since White's (1984) compilation many new species of Antarctic amphipods, isopods, bryozoans and pycnogonids have been added to the list, which would increase the difference from the Arctic provided the figures there actually remained the same. Interestingly, those groups that are rich in species belong to quite different trophic guilds and also differ substantially in terms of motility. Some currently poor groups, such as the decapods and fish, used to be quite rich around the fragments of Gondwana in Cretaceous and Early Tertiary times, but may have been eliminated by glacial advances (Clarke 1990).

What do 'high' and 'low' species numbers mean, and how do they compare with other marine areas? Again, there is the problem of scales and depths, even if we agree to compare different latitudes and (thus) climatic zones. Clearly, we cannot compare species numbers from the total Southern Ocean with those from a single transect in another area, or compare data from a box corer with Agassiz trawl data. Gutt (1991b) has discussed these problems in detail. In some cases – e.g. most of the peracarids – the Southern Ocean will outcompete many marine areas in the world whereas in others, such as the decapods, it is obvious that they are under-represented under the present conditions. Interestingly, some of the taxa with few species in

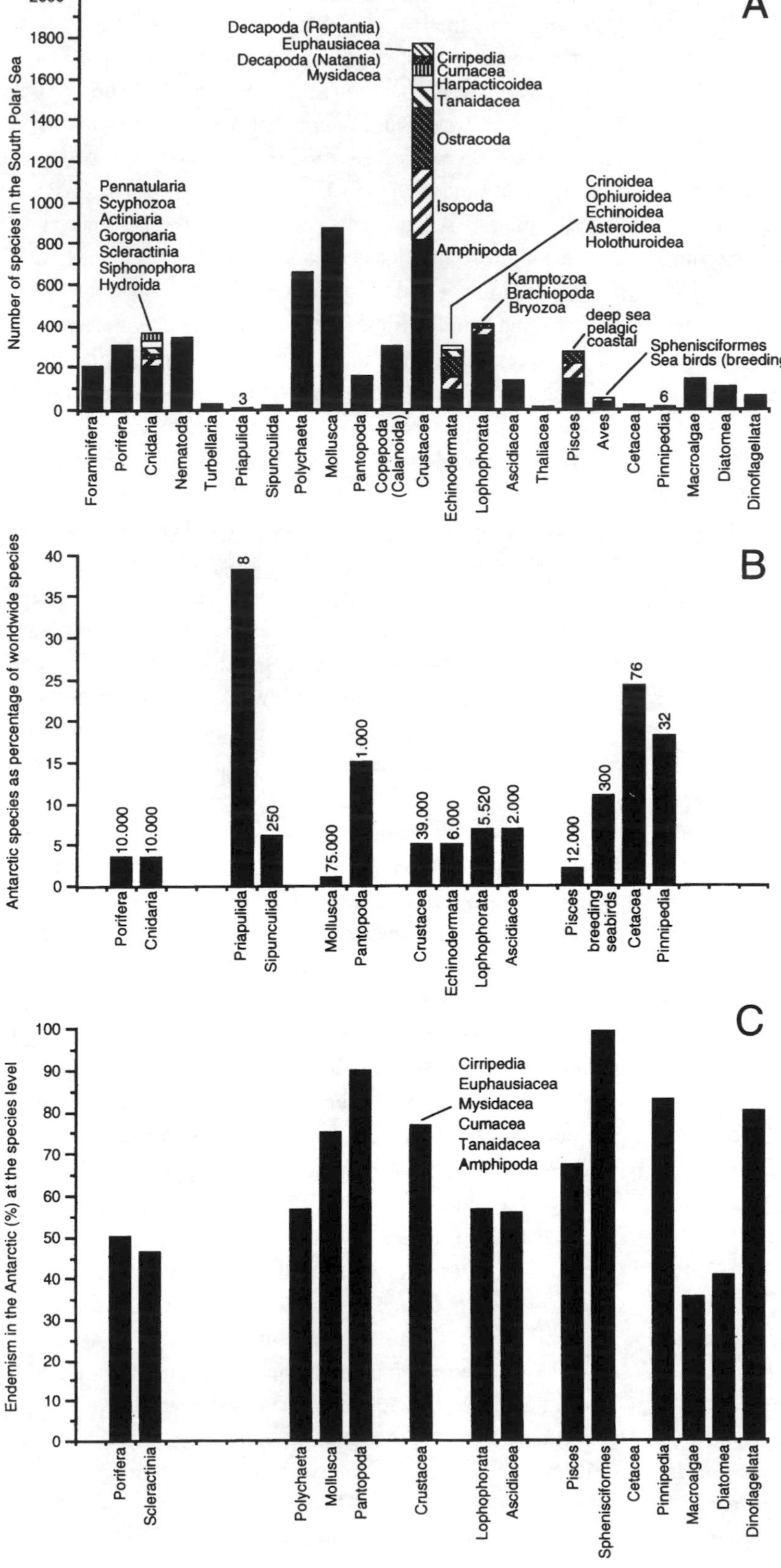

Fig. 1.1. A. Species numbers of different taxa in the Southern Ocean, present estimates. Data sources: Sieg & Wägele (1990); additional information: Porifera, Tendal (pers. commun.), Sarè *et al.* (1992); Hydroida, Svoboda (pers. commun.); Siphonophora & Scyphozoa, Pagés (pers. commun.), Larson (1986); Scleractinia, Cairns (1990); other Cnidaria, Turbellaria and Nematoda, Dell (1972); Copepoda (pelagic), Razouls (1992); Decapoda, Arntz & Gorny (1991); Euphausiacea, Kirkwood (1982); Cirripedia, Newman & Ross (1971); Cumacea, Ledoyer (1993); Isopoda, Brandt (1991); Amphipoda, de Broyer & Jazdzewski (1993); Asteroidea, Voβ (1988); Ophiuroidea, Dahm (pers. commun.); Echinoidea, Pawson (1969); Holothuroidea, Gutt (1988); Kamptozoa, Emschermann (1993); Thaliacea, Lohmann & Hentschel (1933); Pisces, Kock (1992); Aves, Odening (1984), Laws (1989); Pinnipedia, Bonner (1989); Cetacea, Gaskin (1982); Macroalgae, Wiencke (pers. commun.); Diatomea & Dinoflagellata, Balech (1970), Heywood & Whitaker (1984). B. Antarctic species numbers (as in A) as a percentage of species numbers in the world ocean (worldwide estimates derived from Winston 1992, table 10.1; see numbers above columns). C. Endemism at the species level of various taxa in the Southern Ocean (only south of Antarctic Convergence). Data sources as in A; in addition: Porifera, Koltun (1970); Polychaeta, Knox (1970); Pantopoda, Frey (1964); Ascidiacea, Kott (1969).

Antarctica, such as the euphausiids, natant decapods and brachiopods, include species with a high numerical or biomass dominance rank in their communities.

In terms of total species numbers within assemblages, the rich epifaunal suspension feeder communities of the Antarctic can be compared with tropical or subtropical seagrass (Ros *et al.* 1984, Gambi *et al.* 1992, Mazella *et al.* 1993) or even with coral communities (see Gutt 1991b), both of which are also three-dimensional. On the other hand, Antarctic intertidal assemblages have an extremely low species richness, much lower even than temperate or upwelling assemblages suffering from severe oxygen deficiency (Arntz 1981, Tarazona *et al.* 1988). Another example of very low species richness in the Antarctic is provided by the warm-blooded animal assemblages on sea ice; e.g. emperor penguin colonies host at most a handful of other species.

The total numbers of invertebrate species from the Southern Ocean, the overwhelming majority of which are benthic, are quite similar to those presented by Fredj & Laubier (1985, table 1) for the Mediterranean benthos. The echinoderm figure (144 species) is twice as high in the Antarctic, but in all other cases more species are known from the Mediterranean, which has been studied in more detail but has a much smaller extension. The 'total crustaceans' group has about the same number of species in both ecosystems, but decapods are quite rich in the Mediterranean and extremely poor in species in Antarctic waters.

Using our present best estimates for the Southern Ocean again and comparing these with the estimated marine species numbers in the world's oceans from Winston (1992, table 10.1), most of the Antarctic higher taxa provide between 3 and 7% of the worldwide marine species number of their respective group (Fig. 1.1B). Only pycnogonids and priapulids have much higher values, as have the warm-blooded animals, the majority of which, however, are migrant species. Three to seven per cent is clearly less than the share of the Southern Ocean in the world ocean; the area covered by pack ice in winter alone exceeds 10% of the world ocean surface (Laws 1989). However, the data presented by Winston (1992) have been derived from actual knowledge on shelf species, whereas shelf areas of the same depths are relatively scarce in the Antarctic under present geological conditions. The figures may change in either direction in the future depending on how many new species are detected in the deep sea and whether Grassle & Maciolek's (1992) estimate holds true.

For those species that live in Antarctic waters, our data confirm a high level of endemism in most groups (Fig. 1.1C). The problems that arise in calculating valid figures have been referred to above. Endemicity values of taxa may reflect environmental changes in the past and both duration and degree of isolation from other biogeographic zones. If marked environmental changes such as the advance and retreat of ice shelves coincide with isolation, as is suggested for the Antarctic, allopatric speciation may be favoured, leading to adaptive radiation into groups with many endemic species. Levels of endemism are thus

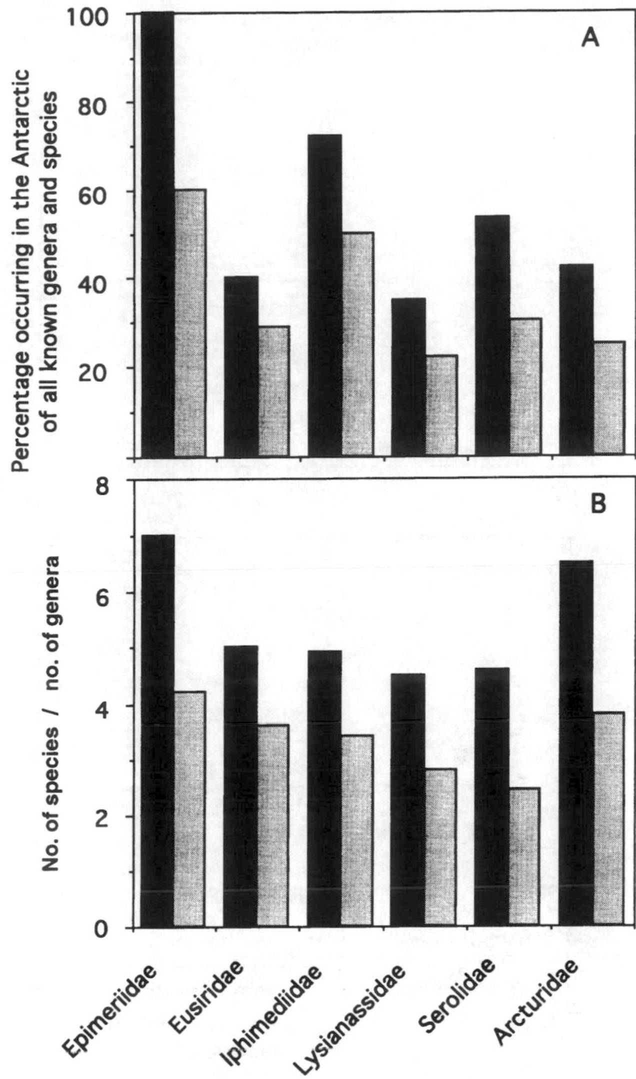

Fig. 1.2. A. Numbers of Antarctic genera (black columns) and species (shaded columns) of four amphipod and two isopod families as a percentage of the numbers of genera and species of these groups known worldwide. B. Taxonomic diversity. Ratio of species:genera numbers in four amphipod and two isopod families. Black: worldwide ratios, shaded: Antarctic ratios. Data are derived from de Broyer & Jazdzewski (1993), Klages (1991 and unpublished data), Brandt (1991) and Wägele (1994).

helpful in explaining the great differences in species richness found among Antarctic taxa.

On the other hand, the ice shelf processes which favour species formation must have caused extinctions of many species as well. This may be the reason why taxonomic diversity is not higher in most cases. For example, four common gammaridean amphipod and two common isopod families contribute a high share of their genera (35–100 %) and species (22–60 %) known in the world's oceans to the Antarctic ecosystem (Fig. 1.2A), but the species:genus number ratios, ranging between 2.8 and 4.2, are lower than on a worldwide level (Fig. 1.2 B). In the other groups referred to in Fig. 1.1 A this ratio always ranged between 1 and 4. However, a general trend (e.g. an increase from old to young groups), or the presence of any distinct taxonomic subgroups, were not recognizable. Another reason for the low ratios encountered (i.e. the high taxonomic diversity), which adds to

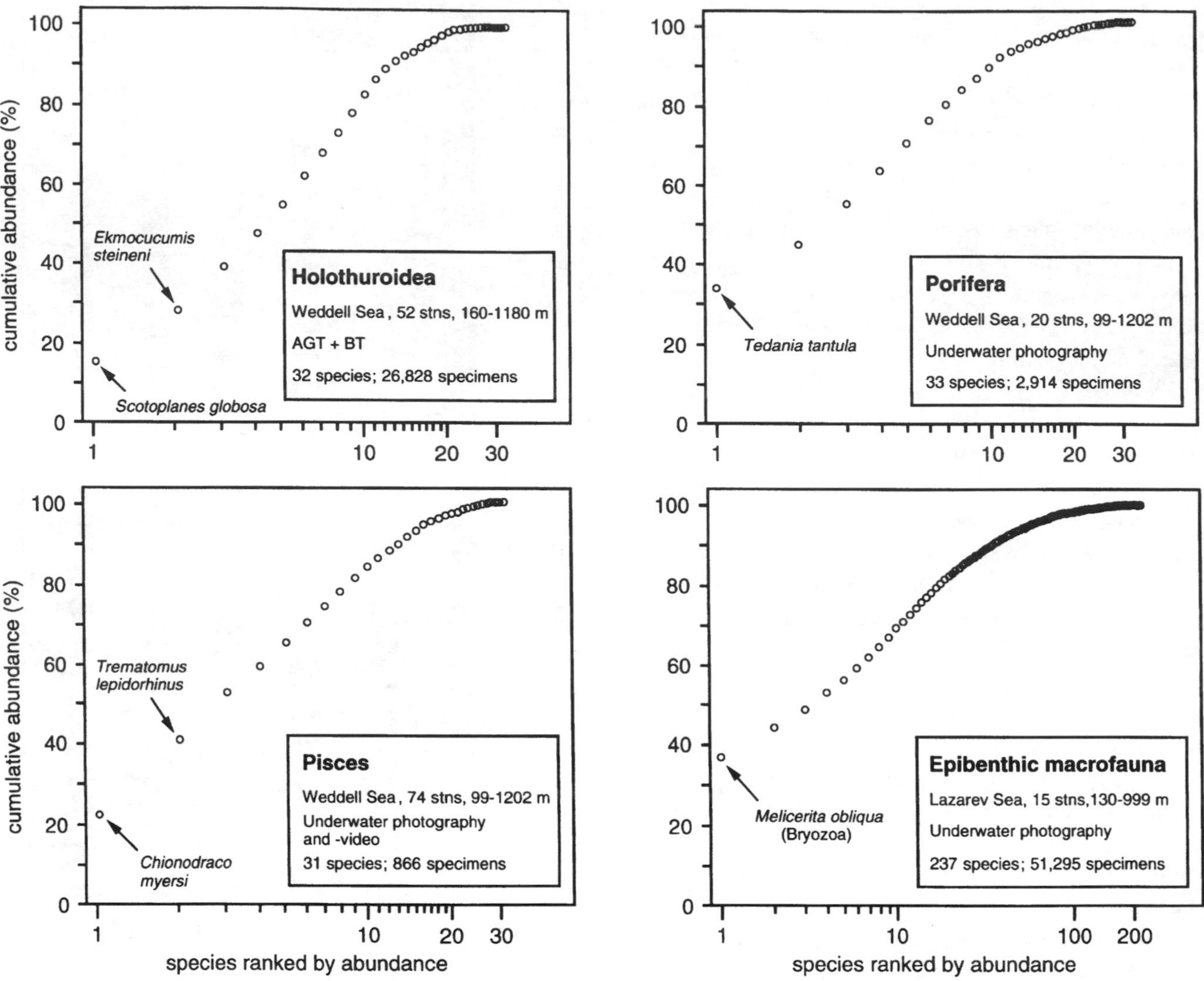

Fig. 1.3. Cumulative dominance plots of three benthic/demersal taxa in the Weddell Sea and total epibenthic macrofauna in the Lazarev Sea. Data derived from Gutt (1991), Barthel & Gutt (1992), Ekau & Gutt (1991), Gutt *et al.* (1994) and Gutt & Starmans in press AGT=Agassiz trawl, BT=bottom trawl.

the one discussed above, may be that within some families only a few genera have split up into numerous species, whereas many of the others are monospecific; an average then is of little use. For example, shelled gastropods and bivalves in the Weddell Sea reveal a remarkably high taxonomic diversity: the 145 gastropod species belong to at least 26 families and 69 genera, while the 43 bivalve species belong to 17 families and 25 genera. Many genera and some families are monospecific. High species numbers only occur in three families (Buccinidae, Turridae, Philobryidae). Some, elsewhere very successful, groups are missing in the Southern Ocean altogether: Cardoidea, Veneroidea, Tellinoidea, Mactroidea (Hain 1990). Holothurians (Gutt 1991a) also have a high taxonomic diversity in the Weddell Sea. All eight species of caridean shrimp in that area belong to different genera (Arntz & Gorny 1991).

Dominance curves: epibenthic macrofauna

If data are at hand that allow for a breakdown into species and include counts of individuals, cumulative dominance plots are the most illustrative way to demonstrate differences between assemblages and to show whether there is a certain balance (evenness) between the numbers of individuals of different species. Data, both for total epifaunal communities and for higher taxa such as holothurioids, most of which stem from surveys with imaging methods (UW camera and ROV) are available for the Weddell Sea and the Lazarev See (Fig. 1.3). From 1085 UW photos taken at 15 stations in the Lazarev Sea between 130 m and 1000 m depth (981 m^2 area), Gutt & Starmans (in press) counted 51 295 individuals of epibenthic macrofauna which could be assigned, on different taxonomic levels, to 237 taxa. Within-site species numbers as found by this method varied substantially (between 39 and 182) but were higher than those of epifaunal communities in temperate areas such as the Baltic (Arntz *et al.*, unpublished observations). Visual methods have proven to be very useful for the quantitative study of the epifaunal macrofauna and megafauna component, which is difficult to sample with grabs and cores (Hamada *et al.* 1986), or for comparison with trawl catch data (Brey & Gutt 1991, Ekau

& Gutt 1991, Gutt *et al.* 1991, 1994, Barthel & Gutt 1992, Gorny *et al.* 1993).

The resultant dominance curves are by no means flat; there are always a few dominant species, particularly in the three groups (holothurians, sponges and fish) presented in Fig. 1.3. On the whole, however, dominance appears to be lower than in boreal temperate communities (Arntz & Rumohr 1982). This is particularly true for the 'epibenthic macrofauna' curve (lower right in Fig. 1.3) where only the dominant bryozoan *Melicerita obliqua* reduces an otherwise very high evenness (Gutt & Starmans in press). A surprisingly high evenness – 11 species with populations > 2000 ind. m^{-2} – was also registered by Dayton & Oliver (1977), in a dense community in McMurdo Sound, which was, however, infaunal.

Diversity (Shannon–Wiener) and other indices

Macrobenthos

Almost all studies where diversity or evenness indices have been calculated in the Antarctic cannot be compared due to the use of different equipments and procedures (Clarke 1992). Even where the same persons worked at different localities, changes were made in the gear used (Gallardo 1992). In many other cases, either higher taxa only were presented due to taxonomic difficulties, or the breakdown to species was restricted to a few groups.

Using H' diversity and Margalef's species richness index SR, Richardson (1976) made an attempt to compare the soft-bottom macrofauna data from Arthur Harbor (Anvers Island) with data from similar environments elsewhere in the world's oceans. He concluded that the Arthur Harbor indices were quite high. However, Richardson & Hedgpeth (1977) noted the apparent lack of comparability of the then existing Antarctic macrofaunal data (Gallardo & Castillo 1968, 1969: Deception and Greenwich Islands, respectively; Lowry 1975: Arthur Harbor; Hardy 1972: Signy Island) and found only Boesch's (1972) investigations on the Virginia and N Carolina shelf to be comparable with their own data. The H' values did not differ between the two localities, and species richness (SR according to Margalef) was much higher and evenness much lower at Arthur Harbor – the latter probably due to recruitment of dominant species at the time of sampling. Independently, the other investigators mostly claimed that their diversity values from Antarctic sites were high; only Gallardo & Castillo (1969) reported lower values from Deception Island due to frequent volcanic activity.

There does not seem to have been much advance in the question of calculating diversity indices of infaunal macrobenthos communities in the Antarctic since Richardson & Hedgpeth (1977). Future, comparable, studies are necessary to reveal whether Antarctic macrobenthos – with the exception of the intertidal and upper sublittoral – is really more diverse than the benthic macrofauna in other areas.

Meiobenthos

To our knowledge, there has been only one diversity investigation of an Antarctic meiobenthos community to date. Herman & Dahms (pers. commun.) collected benthic meiofauna along a transect off Halley Bay (SE Weddell Sea) during the 'European Polarstern Study' in 1989. They could not distinguish the individual species but calculated different diversity indices for 13 major taxa. Consequently, their H' values were 'monotonously low' (Dahms 1992) due to the overwhelming dominance of nematodes. Much more, and more detailed, work on the meiofauna is needed to arrive at definite conclusions.

Fish

Hubold (1992) calculated diversity from trawl (bottom trawl and Agassiz trawl) catches taken by various investigators in different Antarctic regions and compared them with data from the North Sea and Arctic waters (S Greenland). Hubold himself admits that comparability is limited due to the use of different trawls and somewhat variable trawling duration and speed. However, it seems noteworthy that H' diversity and evenness values were consistently highest in the high Antarctic (Weddell Sea, Prydz Bay, and Ross Sea) compared with the Scotia Arc region, whereas the values for the latter were mostly in the same range as the northern hemisphere data. Within the Atlantic sector of the Southern Ocean the data indicate a steady latitudinal increase of H' diversity towards the high Antarctic (Hubold 1992, fig. 11). Towards the benthopelagial and the true pelagial, diversity and evenness of the respective fish communities seem to decrease (again, the values are not strictly comparable since different trawls were used).

Hubold (1992) stresses that the within-site species richness and diversity of Antarctic demersal fish communities is surprisingly high, despite a very low taxonomic diversity on the levels of orders and families (Kock & Kellermann 1991) and generally lower species numbers compared, for example, with the northwest Atlantic. He argues that the Antarctic shelf has been colonized by a mostly endemic (strongly eurybathic?) demersal fish fauna which has occupied many niches, whereas the development of species with a pelagic life cycle has been restricted by the advance of the ice shelves and distinct changes in hydrography during glacial periods.

Zooplankton

Boysen-Ennen (1987) and Boysen-Ennen & Piatkowski (1988) studied meso- and macrozooplankton communities in the southeastern Weddell Sea. Based on rectangular midwater trawl (RMT) catches, they distinguished three communities (with a number of regularly occurring species): the Southern Shelf (55), the Northeastern Shelf (64), and the Oceanic Community (61). They calculated mean H' diversity and evenness for each of the three communities. The Oceanic Community had the highest values followed by the Southern Shelf community; all values differed significantly. All three communities were much poorer at the surface (0–50 m) than at depths between 51 and 300 m.

Piatkowski (1987) also calculated mean H' diversity and evenness for four macrozooplankton communities in the Antarctic Peninsula area. As would be expected, species numbers for the macroplankton were generally lower because the mesoplankton was not considered. Again the oceanic community revealed the highest values. However, all figures were very low in this area and varied greatly with the consequence that no significant differences were detected. Piatkowski (1987) compares these data with macrozooplankton data from the three southeastern Weddell Sea communities mentioned above, taken by the same method, and shows the southeastern Weddell Sea values to be consistently higher. In this case the Northeast Shelf Community (which combines elements of the other two communities) was most diverse and had the most even distribution of individuals over species.

Using the same equipment again, Siegel *et al.* (1992, table 5) measured diversity of macro- and mesozooplankton in the open north Weddell Sea. The values were fairly high at the sea surface and increased to 200–300 m depth.

So, as with the fish communities, the available zooplankton diversity data seem to indicate highest diversity in the high Antarctic. There are, however, distinct differences, particularly between the oceanic communities of the two regions: whereas in the southeast Weddell Sea copepods contribute up to 85% of individuals, euphausiids may provide over 90% dominance off the Antarctic Peninsula and in the Scotia Sea and north Weddell Sea (Mujica & Torres 1982, Siegel *et al.* 1992), and at stations with a great preponderance of one group the other is almost excluded (Boysen-Ennen 1987).

Latitudinal clines

Within the Antarctic, regional differences in within-site diversity have already been discussed in the cases where data from different sites were comparable. This is clearly not the case for macrobenthos. However, a breakdown into species and their numbers of individuals is available for part of our material from the Weddell Sea, which may give an indication of what we might expect if we had comparable data.

The issue of latitudinal clines was first put forward by Thorson (1952), who showed that the average species number of three epifaunal groups (amphipods, nudibranchs and brachyuran crabs) increased strongly from the high Arctic towards the tropics, whereas there was no such change in the mean numbers of infaunal species. Surprisingly, this finding has been generalized by many scientists and extended in a way that would yield a bell-shaped curve for the world's oceans, with highest faunal species numbers on the equator and lowest numbers in the Arctic and Antarctic (note that the seaweed flora does not show highest species richness in the tropics, cf. Lüning 1990).

Clarke (1992) has shown that at this time the evidence for an overall latitudinal gradient in marine species richness is not convincing, although there are three cases where a bell-shaped curve does indeed apply: for gastropods (Fischer 1960), bivalves (Stehli *et al.* 1967) and planktonic foraminiferans (Stehli *et al.* 1972). As Clarke (1992) points out, these groups have in common a calcareous skeleton, and the metabolic cost of calcification is higher at low temperatures. However, even for these groups the database from the Southern Ocean is much smaller than from other latitudes. The evidence for a latitudinal cline in coastal waters remains weak (Gage & May 1993).

In the deep sea, latitudinal gradients seemed less likely than in coastal waters due to the large distance of the communities in this area from environmental impacts at the surface and the 'endless sameness of the deep-sea bed' (Gage & May 1993), although there has been much discussion on plankton aggregates reaching the seafloor at great velocity (Graf 1989). However, Rex *et al.* (1993) presented evidence from epibenthic sled data that shows continuous poleward declines in the Atlantic Ocean to be existent for deep-sea bivalves and gastropods, and also in the northern hemisphere for abyssal isopods. Again, the database is much weaker for the South Atlantic deep sea, and the authors lacked data south of 40° S. Rex *et al.* (1993) presented their data as 'normalized expected number of species', E (Sn), which is Hurlbert's (1971) modification of Sanders' rarefaction method. It normalizes species numbers to a common number of individuals and measures evenness, not species richness, although large values of this index will often indicate that more species are present (May 1992).

Brey *et al.* (1994, table 1) responded by providing E (Sn) values from the southeastern Weddell Sea which were normalized to $E\,(S_{100})$ for inter-taxon comparison, and the three groups that had been presented by Rex *et al.* (1993) were adapted to the normalized sample sizes used by these authors. The resultant values for the three groups in question were much higher than the Arctic values presented by Rex *et al.* and rather in the upper range of their values from the tropics, indicating that the trend of the bell-shaped curve may be reversed in the Antarctic in these cases. However, other benthic groups clearly have much lower $E\,(S_{100})$ values whereas amphipods had the highest value at all (same table).

Brey *et al.*'s (1994) paper certainly contributes to the discussion, but again there is a distinct question mark as to the comparability of the data. Agassiz trawls and epibenthic sleds have very different catch characteristics, the mesh size in the codend differed markedly, and the SE Weddell Sea data were clearly taken (on the average) from lower depths. There are also arguments in favour of a comparability of the datasets – the meshes are mostly clogged by sponge spicules or bryozoan debris which prevent smaller individuals from being washed out, and eurybathy is a characteristic property of Weddell Sea benthos. Whether increased numbers of species as taken by an epibenthic sled would increase or decrease the E (Sn) values would finally depend on the numbers of individuals. At any rate, the issue of latitudinal gradients has to be treated with great caution until better samples are available. At this time we cannot even be sure that the paradigm of exceptionally high diversity in the deep sea is true (Gray 1994).

DISCUSSION

The belief that Antarctic communities are rich and diverse, as has often been suggested in the literature, receives some support from our study but at the same time it has become obvious that this statement needs much qualification. To a person working in oxygen deficient areas of the Baltic or the Humboldt current, Antarctic benthos communities appear much richer than to someone studying tropical coral reefs. Comparing the Antarctic marine fauna with that of Australia, Australia will win; however, such a comparison is biased because Australia has warm and cold regions with totally different sets of species, whereas all the fauna living around Antarctica are part of the same immense cold-water system. Some communities in some subsystems are rich in species of some higher taxa, for example many of the suspension feeder communities on the deeper shelf, which may profit from increased habitat complexity (see Warwick 1993). However, the reverse is true for others, for example shallow-water communities where *Nacella concinna* may be almost the only macrofaunal species, areas impacted by iceberg scouring or the large colonies of emperor penguins where Weddell seals, skuas and snow petrels are the only other macrofauna. Some communities such as infaunal macrobenthos and meiobenthos do not seem to present any unusual trends. In the pelagic oceanic and transitional zooplankton communities seem to be richer in species and more diverse than strictly neritic communities, and those in slightly deeper water are apparently more diverse than those close to the surface. In all these cases the differences can be largely attributed to the degree of constancy and the relative harshness of environmental conditions. We have not found diversity calculations for the pack-ice community, but most scientists were surprised by the number of small floral and faunal groups inhabiting the channels of sea ice and the distinct distribution of sea-ice organisms depending on processes of ice formation and the age and structure of ice (Spindler 1990, Spindler & Dieckmann 1991). In the case of demersal fish, which are not rich in species compared, for example, with the North Atlantic (but which again, are much richer than in the Arctic), and which are of low taxonomic diversity on the higher levels, within-site diversity is comparatively high, whereas pelagic fish in the Antarctic are poor in species and of low diversity (Hubold 1992, Kock 1992).

It would be desirable for biodiversity studies in the Antarctic to be as much advanced as is the discussion on the possible reasons for the high species richness and high rates of endemism of the Antarctic fauna compared with that of the Arctic (the Antarctic flora is not particularly rich; Lüning 1990). The two systems – despite similarities in polar position and low temperatures – are strikingly different in geological history, impact from land and freshwater, recent interchange with adjacent oceans, age and variability of sea ice and many other factors (for a recent review see Dayton *et al.* 1994).

Among other factors, the much longer time of existence of the Antarctic ecosystem than of that of the Arctic is thought to be responsible for its greater species richness, and various ecological theories have been invoked to explain how this happened (see Lipps & Hickman 1982) and how latitudinal gradients may have developed (Gage & May 1993). Recent important contributions have shown that glaciation of the Antarctic reaches farther back (>36 Ma) than assumed hitherto (Hambrey *et al.* 1991, Ehrmann 1994). Shallow-water areas used to be much more common during some periods in the past than they are now (Clarke & Crame 1989). The drastic temperature gradient which hampers faunal and floral exchange between Antarctica and other parts of the world ocean is not typical of the past but is of relatively recent origin (< 16 Ma; Crame 1993a). Conditions in the deep sublittoral benthic system are not totally free of abiotic disturbance (Arntz *et al.* 1992) and the biotic disturbance is less than in the Arctic (Dayton 1990), but the role of disturbance for species development is somewhat equivocal (Dayton & Hessler 1972, Oliver & Slattery 1985). Most of the Antarctic marine fauna developed *in situ* (Clarke & Crame 1989) but several other ways may also have been realized (Sieg 1992). Recent geographic and hydrographic isolation of the Antarctic ecosystem may have been an important factor, but there has been much more exchange in the past than was assumed until recently, with examples of bipolarity continuing up to the present time (Crame 1992, 1993b). Repeated latitudinal range shifts may have increased high-latitude species richness in the southern hemisphere (Crame 1993a).

Whether diversity (and not only species richness) is actually higher in the Antarctic than in the Arctic remains to be demonstrated. So far, only the demersal fish data indicate such differences. There is no indication that high biodiversity might imply increased stability in the sense of resilience (Remmert 1992). As biological and physical processes are interactive, losses of biological diversity may cause environmental changes, which in turn may lead to even more impoverished biological systems, which are susceptible to collapse when faced with further changes (Thorne-Miller & Catena 1991).

Biodiversity studies have recently increased very much in importance. Although there is no immediate danger of species extinction in the Southern Ocean, biodiversity approaches can be helpful in the context of evaluating the effects of distinctly different environmental conditions on marine ecosystems. Also, in the context of conservation, an improved knowledge of the species richness in the Antarctic marine ecosystem and of the distribution of individuals over the species is required before human impact increases. At this time, the following suggestions can be made for future work:

- Within the forthcoming international EASIZ and GLOBEC programmes, an attempt should be made to use comparable sampling equipment and processing methods, to arrive at diversity data for different Antarctic communities that can be compared;
- an international sampling programme from the Arctic to the Antarctic should be established both for coastal and deep-sea macro- and meiobenthos, to obtain comparable data sets to judge on the question of latitudinal clines;
- new techniques such as imaging methods should be standardized and quantified, to be applied in difficult cases such as benthic hard bottoms and for comparison with results from other gear;

– taxonomists should be encouraged to cooperate with ecologists within the framework of international programmes, with the aim of arriving at biodiversity estimates in reasonable time scales.

ACKNOWLEDGEMENTS

The authors are indebted to a large number of colleagues who provided literature sources and data for this paper: J. M. Gili, F. Pagés (Barcelona), A. Clarke (Cambridge), A. Svoboda (Bochum), U. Piatkowski (Kiel), C. Monniot (Paris), O. Ristedt (Bonn), O. Tendal (Copenhagen), T. Brey, C. Dahm, C-P. Günther, C. Metz, S. Schiel, M. Stiller, C. Wiencke (all Bremerhaven). M. C. Gambi and A. Clarke reviewed the paper and made useful suggestions. We also thank SCAR for the invitation to the senior author to present this keynote in Venice. M. Chétioui-Romboy and C. Pichler-Dieckmann were helpful with the preparation of the manuscript and the literature search.

REFERENCES

Arnaud, P. M. 1992. The state of the art in Antarctic benthic research. In Gallardo, V. A., Ferretti, O. & Moyano, H. I., eds. *Oceanografia en Antártica*. Concepcion: Centro EULA, University of Concepción, 341–436.

Arntz, W. E. 1981. Biomass zonation and dynamics of macrobenthos in an area stressed by oxygen deficiency. In Barrett, G. & Rosenberg, R., eds. *Stress effects on natural ecosystems*. New York: John Wiley, 215–225.

Arntz, W. E. & Gorny, M. 1991. Shrimp (Decapoda, Natantia) occurrence and distribution in the eastern Weddell Sea, Antarctica. *Polar Biology*, **11**, 169–77.

Arntz, W. E. & Rumohr, H. 1982. An experimental study of macrobenthic colonization and succession, and the importance of seasonal variation in temperate latitudes. *Journal of Experimental Marine Biology and Ecology*, **64**, 17–45.

Arntz, W. E., Brey, T., Tarazona, J. & Robles, A. 1987. Changes in the structure of a shallow sandy-beach community in Peru during an El Niño event. In Payne, A. I. L., Gulland, J. A. & Brink, K. H., eds. *The Benguela and comparable ecosystems. South African Journal of Marine Science*, **5**, 645–658.

Arntz, W. E., Brey, T., Gerdes, D., Gorny, M., Gutt, J., Hain, S. & Klages, M. 1992. Patterns of life history and population dynamics of benthic invertebrates under the high Antarctic conditions of the Weddell Sea. In Colombo, G., Ferrari, I., Ceccherelli, V. U. & Rossi, R., eds. *Marine eutrophication and population dynamics*. Fredensborg: Olsen & Olsen, 221–230.

Arntz, W. E., Brey, T. & Gallardo, V. A. 1994. Antarctic zoobenthos. *Marine Biology Oceanographic Annual Review*, 241–303.

Balech, E. 1970. The distribution and endemism of some Antarctic microplankters. In Holdgate, M. W., ed. *Antarctic ecology*. Vol. 1. London: Academic Press, 143–147.

Barthel, D. & Gutt, J. 1992. Sponge associations in the eastern Weddell Sea. *Antarctic Science*, **4**, 137–150.

Boesch, D. F. 1972. Classification and community structure of macrobenthos in the Hampton Roads area, Virginia. *Marine Biology*, **21**, 226–244.

Bonner, W. N. 1989. *The natural history of seals*. London: Christopher Helm, 196 pp.

Boysen-Ennen, E. 1987. Zur Verbreitung des Meso- und Makrozooplanktons im Oberflächenwasser der Weddellsee (Antarktis). *Berichte zur Polarforschung*, **35**, 1–126.

Boysen-Ennen, E. & Piatkowski, U. 1988. Meso- and macrozooplankton communities in the Weddell Sea, Antarctica. *Polar Biology*, **9**, 17–35.

Brandt, A. 1991. Zur Besiedlungsgeschichte des antarktischen Schelfes am Beispiel der Isopoda (Crustacea, Malacostraca). *Berichte zur Polarforschung*, **98**, 1–240.

Brey, T. & Clarke, A. 1993. Population dynamics of marine benthic invertebrates in Antarctic and sub-Antarctic environments: are there unique adaptations? *Antarctic Science*, **5**, 253–266.

Brey, T. & Gutt, J. 1991. The genus *Sterechinus* (Echinodermata: Echinoidea) on the Weddell Sea shelf and slope (Antarctica): distribution, abundance and biomass. *Polar Biology*, **11**, 227–232.

Brey, T., Klages, M., Dahm, C., Gorny, M., Gutt, J., Hain, S., Stiller, M., Arntz, W. E., Wägele, J. W. & Zimmermann, A. 1994. Antarctic benthic diversity. *Nature*, **368**, 297.

Broyer, C. de & Jazdzewski, K. 1993. Contribution to the marine biodiversity index: a checklist of the Amphipoda (Crustacea) of the Southern Ocean. *Documents de Travail de l'Institut Royal des Sciences Naturelles de Belgique, Bruxelles*, **73**, 1–160.

Cairns, S. D. 1990. Scleractinia (Steinkorallen). In Sieg, J. & Wägele, J. W., eds. *Fauna der Antarktis*. Hamburg: Paul Parey, 26–29.

Clarke, A. 1988. Seasonality in the Antarctic marine environment. *Comparative Biochemistry and Physiology*, **90B**, 461–73.

Clarke, A. 1990. Temperature and evolution. Southern Ocean cooling and the Antarctic marine fauna. In Kerry, K. R. & Hempel, G., eds. *Antarctic ecosystems: ecological change and conservation*. Berlin: Springer, 9–22.

Clarke, A. 1992. Is there a diversity cline in the sea? *Trends in Ecology & Evolution*, **9**, 286–287.

Clarke, A. & Crame, J. A. 1989. The origin of the Southern Ocean marine fauna. In Crame, J. A., ed. *Origins and evolution of the Antarctic biota. Geological Society Special Publication*, **47**, 23–268.

Clarke, A. & Crame, J. A. 1992. The Southern Ocean benthic fauna and climatic change: a historical perspective. *Philosophical Transactions of the Royal Society*, **B 338**, 299–309.

Crame, J. A. 1992. Evolutionary history of the polar regions. *Historical Biology*, **6**, 37–60.

Crame, J. A. 1993a. Latitudinal range fluctuations in the marine realm through geological time. *Trends in Ecology & Evolution*, **8**, 162–166.

Crame, J. A. 1993b. Bipolar molluscs and their evolutionary implications. *Journal of Biogeography*, **20**, 145–161.

Dahms, H-U. 1992. Importance of zoosystematic research as demonstrated by the Antarctic meiofauna. *Verhandlungen der Deutschen Zoologischen Gesellschaft*, **85**, 277–284.

Dayton, P. K. 1990. Polar benthos. In Smith, W. O., ed. *Polar oceanography. Part B: Chemistry, biology, and geology*. London: Academic Press, 631–685.

Dayton, P. K. & Hessler, R. R. 1972. Role of biological disturbance in maintaining diversity in the deep sea. *Deep-Sea Research*, **19**, 199–208

Dayton, P. K., Mordida, B. J. & Bacon, F. 1994. Polar marine communities. *American Zoologist*, **34**, 90–99.

Dayton, P. K. & Oliver, J.S. 1977. Antarctic soft-bottom benthos in oligothophic and eutrophic environments. *Science*, **197**, 55–58.

Dearborn, J. H. 1968. Benthic invertebrates. *Australian Natural History*, Dec. 1968, 134–139.

Deimer, P. 1987. *Das Buch der Robben*. Hamburg: Rasch und Röhring, 184 pp.

Dell, R. K. ed. 1972. Antarctic benthos. *Advances in Marine Biology*, **10**, 1–216.

Ehrmann, W. U. 1994. Die känozoische Vereisungsgeschichte der Antarktis. *Berichte zur Polarforschung*, **137**, 1–152.

Ekau,W. & Gutt, J. 1991. Notothenioid fishes from the Weddell Sea and their habitat, observed by underwater photography and television. *Proceedings NIPR Symposium on Polar Biology*, **7**, 91–102.

Emschermann, P. 1993. On Antarctic Entoprocta: nematocyst-like organs in a loxosomatid, adaptive developmental strategies, host specificity, and bipolar occurrence of species. *The Biological Bulletin*, **184**, 153–185.

Fischer, A. G. 1960. Latitudinal variations in organic diversity. *Evolution*, **14**, 64–81.

Fredj, G. & Laubier, L. 1985. The deep Mediterranean benthos. In Moraitou-Apostolopoulou, M. & Kiortsis, V., eds. *Mediterranean marine ecosystems. Nato Conference Series. I. Ecology*, Vol. 8. New York: Plenum Press. 109–145.

Frey, W. G. 1964. The pycnogonid fauna of the Antarctic continental shelf. In Carrick, R., Holdgate, M. & Prévost, J., eds. *Biologie Antarctique*. Paris: Herman, 263–270.

Gage, J. D. & May, R. M. 1993. A dip into the deep seas. *Nature*, **365**, 609–610.

Gallardo, V. A. 1992. Estudios bentónicos en bahías someras antárticas del archipiélago de las Islas Shetland del Sur. In Gallardo, V. A., Ferretti, O. & Moyano, H. I., eds. *Oceanografía en Antártica*. Concepcion: Centro EULA, University of Concepción, 383–393.

Gallardo, V. A. & Castillo, J. G. 1968. Mass mortality in the benthic infauna of Port Foster resulting from the eruptions in Deception Island (South Shetland Is.). *INACH Publication*, **16**, 3–13.

Gallardo, V. A. & Castillo, J. G. 1969. Quantitative benthic survey of the infauna of Chile Bay (Greenwich I., South Shetland Is.). *Gayana*, **16**, 3–18.

Gambi, M. C., Lorenti, M., Russo, G. F., Scipione, M. B. & Zupo, V. 1992. Depth and spatial distribution of some groups of the vagile fauna of the *Posidonia oceania* leaf stratum: structural and trophical analyses. *Marine Ecology*, **13**, 17–39.

Gaskin, D. E. 1982. *The ecology of whales and dolphins*. London: Heinemann, 459 pp.

Gerdes, D., Klages, M., Arntz, W. E., Herman, R. L., Galéron, J. & Hain, S. 1992. Quantitative investigations on macrobenthos communities of the southeastern Weddell Sea shelf based on multibox corer samples. *Polar Biology*, **12**, 291–301.

Gorny, M., Brey, T., Arntz, W. E. & Bruns, T. 1993. Growth, development and productivity of *Chorismus antarcticus* (Crustacea: Decapoda: Natantia) in the eastern Weddell Sea, Antarctica. *Journal of Experimental Marine Biology and Ecology*, **174**, 261–275.

Graf, G. 1989. Benthic-pelagic coupling in a deep-sea benthic community. *Nature*, **341**, 437–439.

Grassle, J. F. & Maciolek, N. J. 1992. Deep sea species richness: regional and local diversity estimates from quantitative bottom samples. *American Naturalist*, **139**, 313–341.

Gray, J. S. 1994. Is deep-sea species diversity really so high? Species diversity of the Norwegian continental shelf. *Marine Ecology Progress Series*, **112**, 205–209.

Grebmeier, J. M. & Barry, J. P. 1991. The influence of oceanographic processes on pelagic-benthic coupling in polar regions: a benthic perspective. *Journal of Marine Systems*, **2**, 498–518.

Gutt, J. 1991a. On the distribution and ecology of holothurians in the Weddell Sea (Antarctica). *Polar Biology*, **11**, 145–155.

Gutt, J. 1991b. Are Weddell Sea holothurians typical representatives of the Antarctic benthos? *Meeresforschung*, **33**, 312–329.

Gutt, J., Gorny, M. & Arntz, W. E. 1991. Spatial distribution of Antarctic shrimps (Crustacea: Decapoda) by underwater photography. *Antarctic Science*, **3**, 363–369.

Gutt, J., Ekau, W. & Gorny, M. 1994. New results on the fish and shrimp fauna of the Weddell Sea and Lazarev Sea (Antarctic). *Proceedings NIPR Symposium on Polar Biology*, **7**, 91–102.

Gutt, J. & Starmans, A. in press. Biodiversity of the macrobenthos of the Lazarev Sea (Antarctic): indications of the ecological role of physical parameters and biological interactions. *Deep-Sea Research*.

Hain, S. 1990. Die beschalten benthischen Mollusken (Gastropoda und Bivalvia) des Weddellmeeres, Antarktis. *Berichte zur Polarforschung*, **70**, 1–181.

Hamada, E., Numanami, H., Naito, Y. & Taniguchi, A. 1986. Observation of the marine benthic organisms at Syowa Station in Antarctica using a remotely operated vehicle. *Memoirs of the National Institute for Polar Research*, Special Issue, **40**, 289–298.

Hambrey, M. J., Ehrmann, W. U. & Larsen, B. 1991. Cenozoic glacial record of the Prydz Bay continental shelf, East Antarctica. *Proceedings of the Ocean Drilling Program, Scientific Results 1991*, 77–132.

Hardy, P. 1972. Biomass estimates from some shallow-water infaunal communities at Signy Island, South Orkney Islands. *British Antarctic Survey Bulletin*, No. 31, 93–106.

Harrison, P. 1983. *Seabirds. An identification guide*. Beckenham: Croom Helm, 448 pp.

Hedgpeth, J. W. 1971. Perspectives of benthic ecology in Antarctica. In Quam, L., ed. *Research in the Antarctic*. Washington: American Association for the Advancement of Science, 93–136.

Hempel, G. 1985. Introductory theme paper: on the biology of polar seas, particularly the Southern Ocean. In Gray, J. S. & Christiansen, M. E., eds. *Marine biology of polar regions and effects of stress on marine organisms*. London: Wiley, 3–33.

Heywood, R. B. & Whitaker, R. M. 1984. The Antarctic marine flora. In Laws, R. M., ed. *Antarctic ecology*, Vol. 2. London: Academic Press, 373–419.

Hubold, G. 1992. Zur Ökologie der Fische im Weddellmeer. *Berichte zur Polarforschung*, **103**, 157 pp.

Hurlbert, S. N. 1971. The non-concept of species diversity: a critique and alternative parameters. *Ecology*, **52**, 577–586.

Kirkwood, J. M. 1982. A guide to the Euphausiacea of the Southern Ocean. *ANARE Research Notes*, **1**, 1–45.

Klages, M. 1991. *Biologische und populationsdynamische Untersuchungen an ausgewählten Gammariden (Crustacea: Amphipoda) des südöstlichen Weddellmeeres, Antarktis*. PhD Thesis, University of Bremen, 240 pp. [unpublished]

Knox, G. A. 1977. The Antarctic polychaete fauna: its characteristics, distribution patterns, and evolution. In Llano, G. A., ed. *Adaptations within Antarctic ecosystems*. Houston: Gulf Publishing Co., 1111–1127.

Kock, K-H. 1992. *Antarctic fish and fisheries*. Cambridge: Cambridge University Press, 359 pp.

Kock, K-H. & Kellermann, A. 1991. Reproduction in Antarctic notothenioid fish – a review. *Antarctic Science*, **3**, 125–150.

Koltun, V. M. 1976. Porifera-Part I: Antarctic sponges. In B.A.N.Z. Antarctic Research Expedition 1929–1931, *Reports Series* B (Zoology and Botany), **9**, 147–198.

Kott, P. 1969. Antarctic Ascidiacea. *Antarctic Research Series*, **13**, 239 pp.

Larson, R. J. 1986. Pelagic scyphomedusae (Scyphozoa: Coronatae and Semaeostomae) of the Southern Ocean. *Antarctic Research Series*, **41**, 59–165.

Laws, R. M. 1989. *Antarctica. The last frontier*. London: Boxtree, 208 pp.

Ledoyer, M. 1993. Cumacea (Crustacea) de la campagne EPOS 3 du R.V. *Polarstern* en mer de Weddell, Antarctique. *Journal of Natural History*, **27**, 1041–1096.

Linke, O. 1939. Die Biota des Jadebusenwattes. *Helgoländer wissenschaftliche Meeresuntersuchungen*, **1**, 201–348.

Lipps, J. H. & Hickmann, C. S. 1982. Origin, age, and evolution of Antarctic and deep-sea faunas. In Ernst, W. G. & Morin, J. G., eds. *The environment of the deep sea*. Eaglewood Cliffs, NJ: Prentice-Hall, 324–356.

Lohmann, H. & Hentschel, E. 1933. Die Appendicularien im Südatlantischen Ozean. *Wissenschaftliche Ergebnisse der Deutschen Atlantischen Expedition auf dem Forschungs-und Vermessungsschiff 'Meteor'*, **13**, 153–243.

Lowry, J. K. 1975. Soft bottom macrobenthic community of Arthur Harbor, Antarctica. *Antarctic Research Series*, **23**, 1–19.

Lüning, K. 1990. *Seaweeds. Their environment, biogeography, and ecophysiology*. New York: Wiley, 527 pp.

Margalef, R. 1977. Ecosystem diversity differences: poles and tropics. In Dunbar, M. J., ed. *Polar oceans*. Calgary: Arctic Institute of North America, 367–376.

May, R. M. 1992. Bottoms up for the oceans. *Nature*, **357**, 278–279.

Mazella, L., Scipione, M. B., Gambi, M. C., Buia, M. C., Lorenti, M., Zupo, V. & Canceuri, G. 1993. The Mediterranean sea grass *Posidonia oceanica* and *Cymodocea nodosa*. A comparative overview. In *The First International Conference on the Mediterranean Coastal Environment, MED COAST 93, Antalya, Turkey*. 103–116.

Mujica, A. R. & Torres, A. G. 1982. Qualitative and quantitative analysis of the Antarctic zooplankton. *Ser. Cient. INACH*, **28**, 165–174.

Newman, W. A. & Ross, A. 1971. Antarctic Cirripedia. *Antarctic Research Series*, **14**, 257 pp.

Odening, K. 1984. *Antarktische Tierwelt*. Leipzig: Urania-Verlag, 159 pp.

Oliver, J. S. & Slattery, P. N. 1985. Effects of crustacean predators on species composition and population structure of soft-bodied infauna from McMurdo Sound, Antarctica. *Ophelia*, **24**, 155–175.

Pawson, D. L. 1969. Echinoidea. *Antarctic Map Folio Series*, **11**, 38–41.

Pearse, J. S., McClintock, J. B. & Bosch, I. 1991. Reproduction of Antarctic benthic marine invertebrates: tempos, modes and timing. *American Zoologist*, **31**, 65–80.

Piatkowski, U. 1987. Zoogeographische Untersuchungen und Gemeinschaftsanalysen am antarktischen Makroplankton. *Berichte zur Polarforschung*, **34**, 1–150.

Poore, G. C. & Wilson, G. D. F. 1993. Marine species richness. *Nature*, **361**, 597–598.

Razouls, C. 1992. Inventaire des copépodes planctoniques marins antarctiques et sub-antarctiques. *Vie Milieu*, **42**, 337–343.

Remmert, H. 1992. *Ökologie*. Berlin: Springer-Verlag, 363 pp.

Rex, M. A., Stuart, C. T., Hessler, R. R., Allen, J. A., Sanders, H. L. & Wilson, G. D. F. 1993. Global-scale latitudinal patterns of species diversity in the deep-sea benthos. *Nature*, **365**, 636–639.

Richardson, M. D. 1976. *The classification and structure of marine macrobenthic assemblages at Arthur Harbour, Anvers Islands, Antarctica*. PhD thesis, Oregon State University. [unpublished]

Richardson, M. D. & Hedgpeth, J. W. 1977. Antarctic soft-bottom, macrobenthic community adaptations to a cold, stable, highly productive, glacially affected environment. In Llano, G. A., ed. *Adaptions within Antarctic ecosystem*. Washington DC: Smithsonian Institution, 181–195.

Ros, J., Olivella, I. & Gili, J. M. 1984. *Els sistemes naturals de les Iles Medes*. Barcelona: Institut d' Estudis Catalans, 828 pp.

Sarà, M., Balduzzi, A., Barbieri, M., Bavestrello, G. & Burlando, B. 1992. Biogeographic traits and checklist of Antarctic demosponges. *Polar Biology*, **12**, 559–585.

Sieg, J. 1992. On the origin and age of the Antarctic Tanaidacean fauna. In Gallardo, V. A., Ferretti, O. & Moyano, H. I., eds. *Oceanografia en Antártica*. Concepcion: Centro EULA, University of Concepción, 421–429.

Sieg, J. & Wägele, J. W. 1990. *Fauna der Antarktis*. Berlin: Paul Parey, 197 pp.

Siegel, V., Skibowski, A. & Harm, U. 1992. Community structure of the epipelagic zooplankton community under the sea-ice of the northern Weddell Sea. *Polar Biology*, **12**, 15–24.

Spindler, M. 1990. A comparison of Arctic and Antarctic sea ice and the effects of different properties on sea ice biota. In Bleil, U. & Thiede, J. eds. *Geological history of the polar oceans: Arctic versus Antarctic*. Dordrecht: Kluwer Academic Publishing, 173–186.

Spindler, M. & Dieckmann, G. S. 1991. Das Meereis als Lebensraum. *Spektrum der Wissenschaft*, **2**, 48–57.

Stehli, F. G., Douglas, R. & Kafescegliou, I. 1972. Models in palaeobiology. In Schopf, T. J. M., ed. *Models for the evolution of planktonic foraminifera*. San Francisco: W. H. Freeman & Co., 116–128.

Stehli, F. G., McAlester, A. L. & Helsley, C. E. 1967. Taxonomic diversity of recent bivalves and some implications for geology. *Bulletin Geological Society of America*, **78**, 455–466.

Tarazona, J., Salzwedel, H. & Arntz, W. E. 1988. Positive effects of 'El Niño' on macrozoobenthos inhabiting hypoxic areas of the Peruvian upwelling system. *Oecologia*, **76**, 184–190.

Thamdrup, H. M. 1935. Beiträge zur Ökologie der Wattenfauna auf experimenteller Grundlage. *Meddelser fra Kommissionen for Danmarks Fiskeri- og Havundersøgelser, Serie Fiskeri*, **10**, 1–125.

Thorne-Miller, B. & Catena, J. G. 1991. *The living ocean: understanding and protecting marine biodiversity*. Washington DC: Thorne-Miller, Island Press, 180 pp.

Thorson, G. 1952. Zur jetzigen Lage der marinen Bodentier-Ökologie. *Verhandlungen der Deutschen Zoologischen Gesellschaft*,1951, 276–327.

Tuck, G. & Heinzel, H. 1980. *Die Meeresvögel der Welt*. Hamburg: Paul Parey, 334 pp.

Voß, J. 1988. Zoogeographie und Gemeinschaftsanalyse des Makrozoobenthos des Weddellmeeres (Antarktis). *Berichte zur Polarforschung*, **45**, 1–145.

Wägele, J. W. 1994. Notes on Antarctic and South American Serolidae (Crustacea, Isopoda) with remarks on the phylogenetic biogeography and a description of new genera. *Zoologische Jahrbücher Abteilung für Systematik*, **121**, 3–69.

Warwick, R. 1993. Coastal marine biodiversity. *NERC News*, Oct. 1993, 4–5.

White, M. G. 1984. Marine benthos. In Laws, R. M., ed. *Antarctic ecology*, Vol.2. London: Academic Press, 421–461.

Wilson, E. O. 1992. Der gegenwärtige Stand der biologischen Vielfalt. In Wilson, E. O., ed. *Ende der biologischen Vielfalt? Der Verlust an Arten, Genen und Lebensräumen und die Chancen für eine Umkehr*. Heidelberg: Spektrum Akad. Verlag, 19–36.

Winston, J. E. 1992. Systematics and marine conservation. In Eldredge, N., ed. *Systematics, ecology and the biodiversity crisis*. New York: Columbia University Press, 144–168.

Ziegelmeier, E. 1970. Über Massenvorkommen verschiedener makrobenthaler Wirbelloser während der Wiederbesiedlungsphase nach Schädigungen durch 'katastrophale' Umwelteinflüsse. *Helgoländer wissenschaftliche Meeresuntersuchungen*, **21**, 9–20.

2 Diversity of lichens in Antarctica

MIRIS CASTELLO AND PIER LUIGI NIMIS
Department of Biology, University of Trieste, Via Giorgieri 10, I 34127 Trieste, Italy

ABSTRACT

The only extant lichen flora of Antarctica, that of C. W. Dodge, published in 1973, includes 415 species, 44.6% of which were described by Dodge as new to science. We have recently examined the types of 152 (of a total of 186) species described by Dodge and accept only 31 species (20%) as valid. All of Dodge's species were described as endemic to Antarctica; after revision Antarctica has lost 121 endemic lichens. On the other hand, recent lichenological research in Antarctica has added many lichen species to its lichen flora. This paper summarizes the results of the last 20 years of lichenological exploration of Antarctica, and considers the main floristic–phytogeographic elements in the flora. The known lichen flora of Antarctica is reduced from 415 to 260 species, the percentage of endemic species falls from 91% to 38%, and that of bipolar and cosmopolitan species increases from 2.4% to 41.5%. Although these figures are likely to change in the near future as a consequence of the present intensive lichenological research in Antarctica, they suggest that the lichen flora of the continent, and especially of continental Antarctica, is a young one, which mainly originated by long-distance dispersal in the Quaternary period.

Key words: Antarctica, biodiversity, flora, lichens, phytogeography.

INTRODUCTION

Although lichens, one of the most conspicuous elements of Antarctic terrestrial ecosystems, have been investigated since the early Antarctic expeditions, their taxonomy is still very confused, and knowledge of the Antarctic lichen flora is incomplete. According to some authors (e.g. Dodge 1973), Antarctica hosts more than 400 lichen species, while others (e.g. Hertel 1988, Galloway 1991) give much lower estimates of 160–200 species. Dodge (1973) and several earlier authors claimed that most Antarctic lichens are narrow-ranging endemics whereas, according to Hertel (1988), the endemic element does not exceed 20% of the total.

Such a confused situation is due to several reasons. Most collectors were not lichenologists, which resulted in incompletely representative collections. The identification of Antarctic material was often carried out by people who did not study the material in the field, with a consequent underestimation of phenotypic variation. Most of the early taxonomists working with Antarctic material started from the premise that the lichen flora of Antarctica, because of its isolation, had a very high degree of endemism, and rarely tried to compare their lichens with those already known from other parts of the world. Thus, many 'endemic' species were described, which later proved to be either morphotypes of the same species induced by different environmental conditions, or even synonyms of widespread species with a more-or-less cosmopolitan, or bipolar distribution (Smith 1984, Hertel 1987a, 1988, Nimis 1990). Finally, the type specimens of many species, and especially those of the two authors who were most prolific in describing new taxa from Antarctica, i.e. M. A. Hue and C. W. Dodge, were not available to specialists, either because they were lost (as was apparently the case for many of Hue's types), or because the author did not make them available for loan (as in the case of Dodge's types).

The most complete lichen flora of Antarctica, that of Dodge (1973), was reviewed with the following words by Almborn (1974): 'This author [Dodge] has caused untold damage to taxonomic lichenology. His publications unfortunately cannot be simply ignored. Future serious lichenologists will have to spend much time and trouble in evaluating and identifying all his many worthless taxa' (see also comments by Lindsay, 1974). We also became convinced that this flora was the major obstacle to any serious study of Antarctic lichens as we tried to identify the first collections from the surroundings of the Italian research station at Terra Nova Bay (Victoria Land). It was clear that little could be done without a revision of the many species described by Dodge and collaborators. The taxonomic criteria adopted by Dodge were heavily criticized by many lichenologists, but no serious revision of his species was carried out, mainly because of unavailability of the types. The lichen herbarium of C. W. Dodge was donated, before his death, to the Farlow Herbarium (FH, Cambridge, MA). Thanks to the curators of FH, we were recently able to carry out a critical revision of the types of 152 (of a total of 186) species described by Dodge (Castello & Nimis 1995).

The main purpose of this paper is to give a brief account of the results of this revision, and to deduce the floristic and phytogeographic elements of this flora from 20 years of post-Dodge lichenological research in Antarctica by various authors.

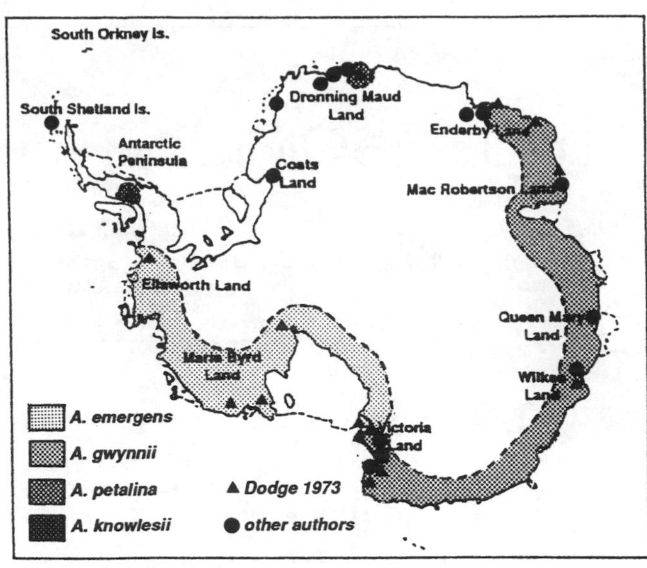

Fig. 2.1. Hitherto known distribution of *Acarospora gwynnii* and of three synonymous taxa in Antarctica. The four 'endemic' taxa proved to be a single circum-Antarctic species.

THE REVISION OF C. W. DODGE'S *FLORA*

In his *Lichen Flora of the Antarctic Continent* Dodge (1973) listed 415 species (excluding dubious records and infraspecific taxa); of these, nearly half (45%) were described by Dodge (alone, or with collaborators), 22% by Hue, 9% by Vainio, and 5% by Darbishire. In our revision (Castello & Nimis 1995) we considered only species described by Dodge, Dodge & Rudolph and Dodge & Baker which are reported in Dodge's *Flora* (Dodge 1973), also including the results of other lichenologists who analysed some of Dodge's types. Species described by Dodge from sub-Antarctic areas, and not cited in the flora, were not included.

When we started revising Dodge's material we discovered that some types were merely fragments of indeterminable sterile crusts, the same species was sometimes described several times under widely different generic names, the original descriptions do not comply with the characters of the types, and the characters given for some species are a mixture of those of different lichens growing together on the type sample.

To give an example of the taxonomic miasma created by Dodge, we summarize here the results of a recent revision of Antarctic yellow Acarosporaceae (Castello & Nimis 1994a). Four genera (*Acarospora*, *Biatorella*, *Biatorellopsis* and *Eklundia*) and ten species, all of them described as endemic to Antarctica, were reported by Dodge (1973). Our revision showed that in Antarctica *Acarospora* s.s. includes only two yellow species, *A. gwynnii* Dodge & Rudolph and *A. flavocordia* Castello & Nimis; *Biatorella* does not occur in Antarctica, the species described by Dodge belonging instead in *Candelariella*; the genus *Eklundia* is a synonym of *Candelariella*; *Biatorellopsis* does not occur in Antarctica either, and the type species of the genus belongs in *Pleopsidium*; *Pleopsidium* is represented in Antarctica by one species, *P. chlorophanum* (Wahlenb.) Zopf, which was treated under different generic and specific names by earlier authors. Altogether, the number of species of yellow Acarosporaceae known from Antarctica was reduced from ten to three, distributed in two genera, *Acarospora* and *Pleopsidium*. The two species of *Acarospora* are, so far as is known, endemic to Antarctica, while *P. chlorophanum* is a widespread bipolar–subcosmopolitan lichen. The case of *A. gwynnii* is particularly pertinent, and representative of the status of many Antarctic 'endemic' taxa: this species was described four times, as *A. gwynnii* Dodge & Rudolph, *A. emergens* Dodge, *A. knowlesii* Dodge and *A. petalina* Golubk. & Savicz; the four species were endemic to particular, more-or-less wide areas of the Antarctic continent. The recognition of their conspecificity shows that *A. gwynnii* is a widespread circum-Antarctic species (Fig. 2.1). Its status as an Antarctic endemic may be provisional, as *A. gwynnii* still has to be compared with the numerous similar species described from other parts of the world.

Similar results were obtained in a study on the genus *Candelariella* in Antarctica (Castello & Nimis 1994b). This genus was treated by Dodge (1973) under several genera–*Blastenia*, *Biatorella*, *Eklundia*, *Protoblastenia*, belonging to families which are quite unrelated to the Candelariaceae. The two species of *Candelariella* reported from Antarctica by Dodge (1973), *C. albovirens* Dodge & Baker and *C. rudolphi* Dodge, belong instead in *Lecanora* and *Xanthoria*, respectively. The six species examined by Castello & Nimis (1994b), all purported to be endemic to Antarctica, were found to be synonyms of three species, *Candelariella aurella* (Hoffm.) Zahlbr., *C. vitellina* (Hoffm.) Müll. Arg. and *C. flava* (Dodge & Baker) Castello & Nimis, two of them cosmopolitan and the latter endemic.

In the light of these results, it is not surprising that after the revision of Dodge's types only 31 species out of 152 (20%) were accepted as valid (Castello & Nimis 1995). This confirms the hypothesis by Hertel (1988) that *c.* 80% of the species created by Dodge would be reduced to synonyms. All of these species were considered by Dodge (1973) as endemic to Antarctica. Thus, Antarctica has lost 121 'endemic' lichens. This considerably alters the phytogeographic image of Antarctica as a refugial centre for old, Gondwanaland species, as postulated by Dodge (1973).

DATA AND METHODS

This paper is based on three species lists, covering the whole of the Antarctic Region.

(a) The 415 species included in *The Lichen Flora of the Antarctic Continent* by C. W. Dodge (1973).
(b) The same as (a), after the revision by Castello & Nimis (1995), and including synonymies proposed by several other authors which examined types of Antarctic species described by various authors (e.g. Lamb 1964, 1968, Brodo & Hawksworth 1977, Övstedal 1983, Filson 1984, 1987, Hertel 1984, Redon 1985, Walker 1985, Filson 1987, Hale 1987, Botnen & Övstedal 1988, Söchting & Övstedal 1991, Smith & Övstedal 1991, etc.), with a total of 241 species.
(c) A 'post-Dodge' check-list of Antarctic lichens including the records given by most authors who worked on Antarctic lichens after the publication of Dodge's flora (1973–1994), with a total of 260 species. This list, which is not included here, is available from the authors upon request.

The lists (b) and (c), although with a similar number of species, are widely different. The reason is that Dodge accepted, besides the many species described by him, also many other poorly known taxa described by earlier authors, and especially those described by Hue. Many of these taxa were not considered by post-Dodge authors, for two main reasons: (a) because his flora was largely unworkable, (b) because the descriptions of Hue, as those of Dodge, rarely give a clear idea of the species. However, these names cannot be simply ignored. They were validly published, and they could prove to have priority with respect to species described later. Many taxa described by Darbishire, Vainio, Zahlbruckner, and especially by Hue still await revision, and on these we are currently working. Until this revision is completed it would be premature to compile a general checklist of Antarctic lichens.

For the subdivision of Antarctica into regions we followed Smith (1984), who recognizes three main biological regions: sub-Antarctic, maritime Antarctic and continental Antarctic. In the present paper, the sub-Antarctic region is not considered. *Continental Antarctica* includes the coastal fringe of Greater Antarctica and Lesser Antarctica south of 70° S, and the east coast of the Antarctic Peninsula, south of *c.* 63° S. *Maritime Antarctica* includes the west coast of Antarctic Peninsula and offshore islands to 70° S, and also the northeast coast of the Peninsula to *c.* 63° S, plus South Shetland, South Orkney and South Sandwich Islands.

The three datasets provide the basis for determining some main phytogeographic groups of lichens in continental Antarctica, maritime Antarctica and in the whole Antarctic region. The following main phytogeographic elements are distinguished: *Endemic*: subdivided into endemic to continental Antarctica, endemic to maritime Antarctica and endemic to the Antarctic region (occurring both in continental and in maritime Antarctica). *Sub-Antarctic*: species occurring in the sub-Antarctic and Antarctic regions. *Cool-temperate*: species of the cool-temperate areas of southern South America (Fuegia and Falkland Islands), also occurring in Antarctica. *Austral*: species described from other southern continents, outside the tropical and subtropical zones, and also found in Antarctica. *Bipolar*: species distributed in the boreal and in the austral zones but absent from tropical lowlands, with or without intermediate populations in tropical mountain areas (Du Rietz 1940). *Cosmopolitan*: widespread species, occurring in both hemispheres, in more than one vegetation zone.

The comparison between the percentage occurrence of these phytogeographic groups in the three species lists indicates the drastic changes in the floristic–phytogeographic image of Antarctica brought about by the progress of lichenological research since 1973.

THE NEW FLORISTIC IMAGE OF ANTARCTICA

Table 2.1 presents the percentage occurrence of lichen species in the three species lists, subdivided into phytogeographic elements, in continental Antarctica, in maritime Antarctica and in the whole Antarctic region.

The data for continental Antarctica are shown in Fig. 2.2. The critical re-appraisal of many taxa reported in Dodge's flora led to a drastic decrease in the number of species, from 201 (Dodge 1973) to 81; the post-Dodge checklist includes 90 species. The present authors consider the lichen flora of continental Antarctica to be of approximately 100 species, i.e. half of the number given by Dodge (1973). It should be added, however, that this estimate is based on the hitherto available data, and that the number of species is likely to increase in the future, as several regional floras are under completion by different authors. The incidence of endemics in continental Antarctica reduces from 90% (Dodge 1973), to 65% (Dodge 1973 after revision), to 30% (post-Dodge checklist). This last figure is probably closer to reality, since in the flora of Dodge (1973) several poorly known, 'endemic' taxa described by other authors are accepted, many of which will certainly prove to be synonyms of other more widespread species. The decrease of continental endemics is paralleled by an increase in endemics for all of Antarctica (from 9 to 20%), mainly because many species known only from the type collection, and hence considered as 'narrow ranging endemic', proved to be identical with species occurring also in maritime Antarctica (see Fig. 2.1). Altogether, the endemic

Table 2.1. *Percentage occurrence of the main phytogeographic elements and total number of species in the lichen floras of continental Antarctica, maritime Antarctica and of the whole Antarctic region, according to three main sources: (a) the flora of Dodge (1973), (b) the same, after revision of many Antarctic taxa by post-Dodge authors, (c) a checklist of all lichen records by recent authors, excluding Dodge*

	Continental endemic		Maritime endemic		Antarctic endemic		Sub-Antarctic		Cool-temperate		Austral		Bipolar		Cosmopolitan		Total
	no.	%	no.	%	no.	%	no.	%	no.	%	no.	%	no.	%	no.	%	no.
Continental Antarctica																	
Dodge (1973)	180	89.5	—	—	19	9.4	—	—	1	0.5	1	0.5	—	—	—	—	201
Dodge (1973) after revisions	51	64.3	—	—	17	20.2	1	1.2	—	—	1	1.1	8	9.2	3	3.4	81
Recent authors excluding Dodge	27	30.0	—	—	18	20.0	5	5.5	—	—	1	1.0	28	31.1	11	12.2	90
Maritime Antarctica																	
Dodge (1973)	—	—	178	76.3	19	8.6	—	—	22	9.3	4	1.6	7	3.0	3	1.3	233
Dodge (1973) after revisions	—	—	77	45.6	20	11.8	19	11.3	19	11.2	5	3.0	20	11.8	9	5.3	169
Recent authors excluding Dodge	—	—	54	23.9	18	8.0	26	11.5	15	6.6	12	5.3	63	27.9	38	16.8	226
Antarctica																	
Dodge (1973)	180	43.4	178	42.9	19	4.6	—	—	23	5.5	5	1.1	7	1.7	3	0.7	415
Dodge (1973) after revisions	54	22.4	77	31.9	26	10.7	20	8.3	19	7.96	6	2.4	27	11.2	12	4.9	241
Recent authors excluding Dodge	27	10.4	54	20.8	18	6.9	26	10.0	15	5.7	12	4.5	6	26.5	39	15.0	260

element shows a substantial decrease, from 99% (Dodge 1973) to 85% (Dodge after revision), to 50% as deduced from all records of post-Dodge authors. The opposite trend characterizes cosmopolitan and bipolar species; these are not represented in Dodge's flora, but make up 13% of its revised version, and 43% of the post-Dodge checklist. Austral and sub-Antarctic species, practically absent in Dodge's flora, make up 7% of the post-Dodge checklist.

For the maritime Antarctic (Fig. 2.3), the trend is analogous to that of continental Antarctica, except for the number of species, which does not show such a drastic reduction. Dodge (1973) listed 233 species from maritime Antarctica; after revision the number of species drops to 169, but in the last 20 years many authors reported many new or additional species from this region, so that the post-Dodge checklist includes 226 species, only a few less than in Dodge's flora. The phytogeographic structure of maritime Antarctica, however, shows dramatic changes. Of all species listed by Dodge (1973) from maritime Antarctica, 76% were described as endemic to this region. After revision, the number of endemics is reduced to 46%, while the post-Dodge checklist includes only 24% of endemics.

Sub-Antarctic species, a category not present in Dodge's flora, are now *c.* 11–12% of the total, a number which is likely to increase in the future. Wide-ranging species (including cosmopolitan, bipolar, austral and cool-temperate), are less than 15% in Dodge's flora, and more than 50% in the post-Dodge checklist.

The previous information is summarized, for the entire Antarctic region, in Fig. 2.4. The total number of species given by Dodge (1973) for Antarctica as a whole was 415; the revision of many taxa reduced it to 241; the post-Dodge checklist includes 260 species. The endemic element is more than 90% in Dodge's flora, 65% in its revised version, and only 38% in the post-Dodge checklist. Wide-ranging species, which constitute a very minor part of Dodge's flora, make up more than 50% of the post-Dodge checklist.

The figures discussed above are not definitive. Most collections, especially from continental Antarctica, were made by non-lichenologists, who obviously tended to collect the most conspicuous species. There might still be several uncollected species, even in continental Antarctica, and some of them might even prove to be endemic to this region. Further work by lichenologists will continue to change our understanding of this flora. The new phytogeographic image of Antarctica, as it is emerging from the deep fog of past uncritical research, is very different from what is still found in many textbooks, and is worthy of brief discussion.

DISCUSSION – THE NEW PHYTOGEOGRAPHIC IMAGE OF ANTARCTICA

Early hypotheses on the origin of the Antarctic lichen flora, based on the assumption of a high degree of endemism, relied heavily on the theory of the migration of continents, which later developed into plate tectonics. According to some views the lichen flora of Antarctica should have been an old, relict flora with Gondwanaland affinities. Dodge (1973) claimed that continental Antarctica, with its many narrow-ranging endemics, hosts an ancient lichen flora, which survived on nunataks during the last glacial period; the flora of the Antarctic Peninsula, on the contrary, should be somewhat less well characterized, and might have derived from Tierra del Fuego.

It is now clear that the endemic element is less important in Antarctica than was supposed and that the percentage of

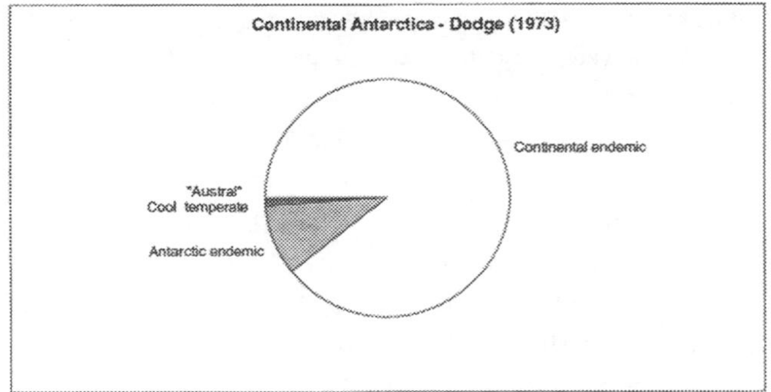

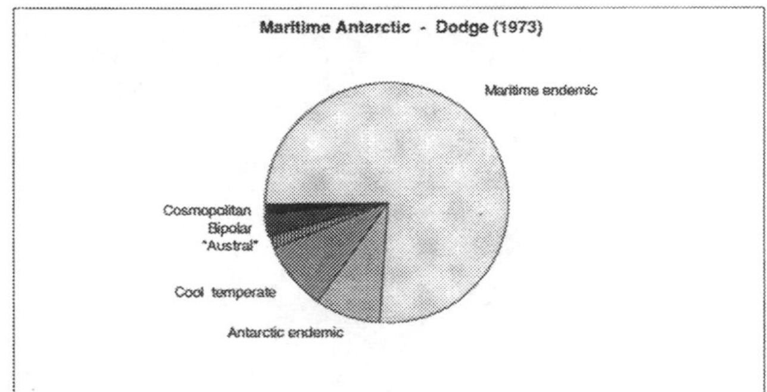

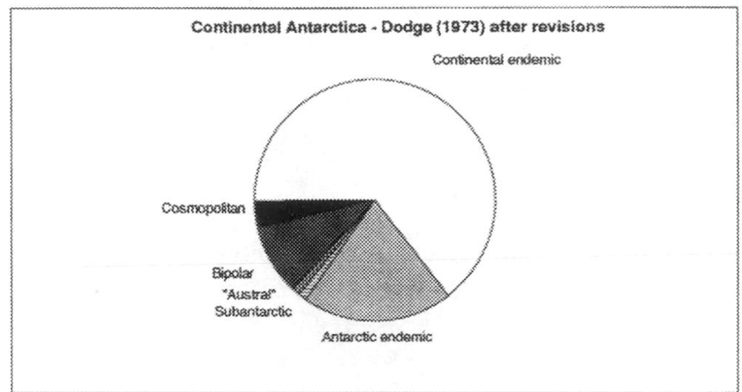

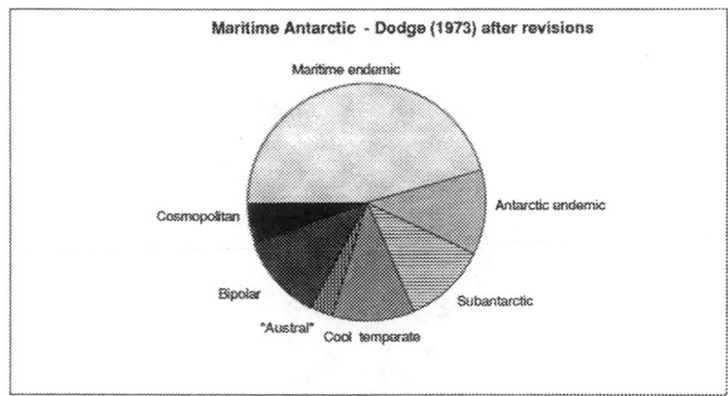

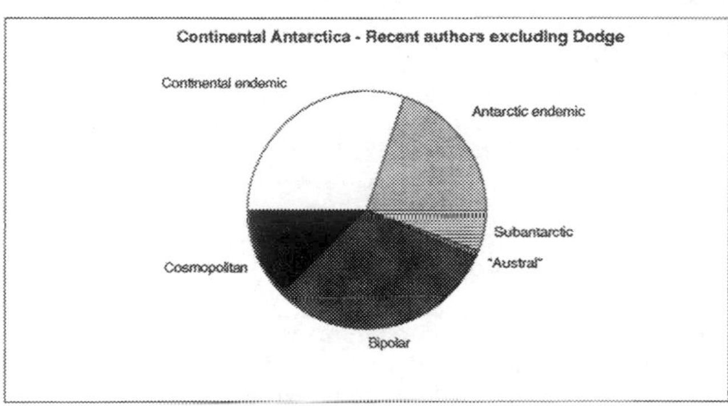

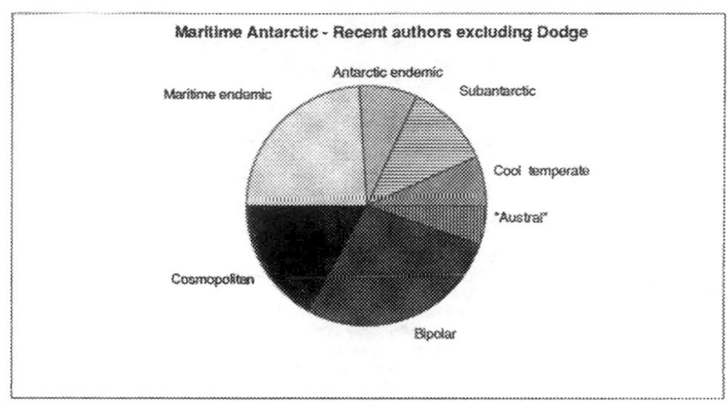

Fig. 2.2. Percentage occurrence of the main phytogeographic elements in the lichen flora of Continental Antarctica according to three main sources: (a) the flora of Dodge (1973), (b) the same, after revision of many Antarctic taxa by post-Dodge authors, (c) a checklist of all lichen records by recent authors, excluding Dodge.

Fig. 2.3. Percentage occurrence of the main phytogeographic elements in the lichen flora of Maritime Antarctica according to three main sources: (a) the flora of Dodge (1973), (b) the same, after revision of many Antarctic taxa by post-Dodge authors, (c) a checklist of all lichen records by recent authors, excluding Dodge.

endemics is likely to decrease even further in the future, as a consequence of the revision of the types described by pre-Dodge authors (especially by M. A. Hue), and of the comparison of Antarctic endemic species with taxa from other parts of the world, particularly from montane areas of southern continents. Several Antarctic endemics have closely related taxa in the Northern Hemisphere, e.g. *Caloplaca athallina* Darb. (Antarctica) and *C. tyroliensis* Zahlbr. (Antarctica and N Hemisphere), *Carbonea capsulata* (Dodge & Baker) Hale, and *C. vorticosa* (Flörke) Hertel, *Lecanora mons-nivis* Darb. and *L. hagenii s.l.*, *Lecanora fuscobrunnea* Dodge & Baker and *L. polytropa s.l.*, *Rhizocarpon flavum* Dodge & Baker and *Rh. geographicum s.l.*, *Xanthoria mawsonii* Dodge and *X. borealis* R.Sant. & Poelt. These taxa are only weakly differentiated, some of them are sympatric in Antarctica, and in many cases it is dubious whether they really deserve to be separated at species rank. These considerations apply to continental Antarctica and, to a lesser degree, to maritime Antarctica, whereas there is evidence that the sub-Antarctic region is relatively rich in endemics (also at generic level), and therefore, contrary to Dodge's (1973) theory, hosts an older flora (Hertel 1987a, b, 1988, Galloway 1991).

The presence of widespread and bipolar species in the Antarctic lichen flora has been explained by three main theories. The first is the 'migration hypothesis' of Stebbins (1950) and Dodge (1964). It postulates that bipolar species reached Antarctica from the north through the more-or-less uninterrupted mountain chains of North, Central and South America. Such a theory, however, is quite difficult to defend as many bipolar species are not known from the northern Andes in South America, or from the southern Cordillera in North America.

The second is the 'persistence theory' (see Lindsay 1977, Sheard 1977, Kärnefelt 1979), which postulates that bipolar

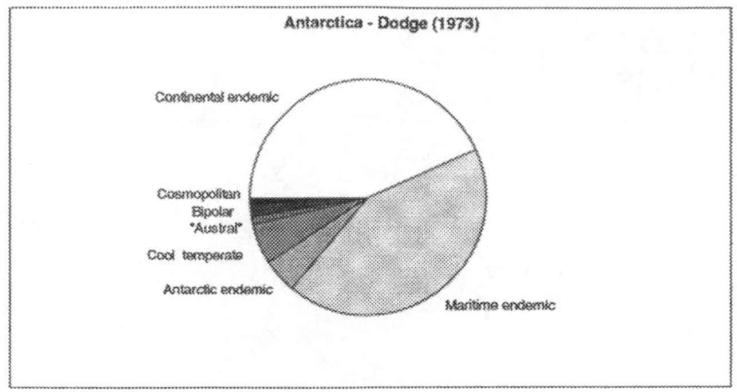

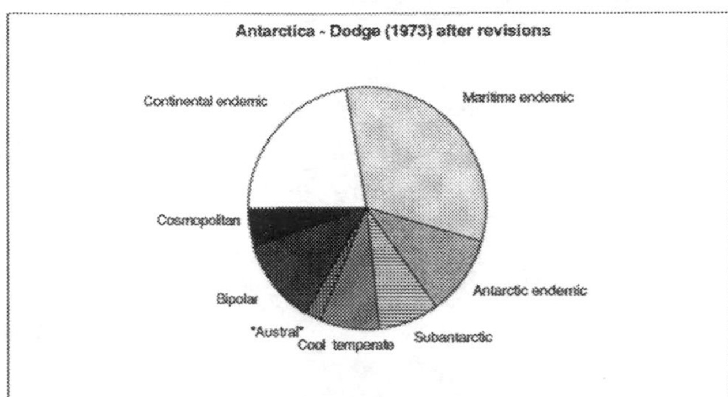

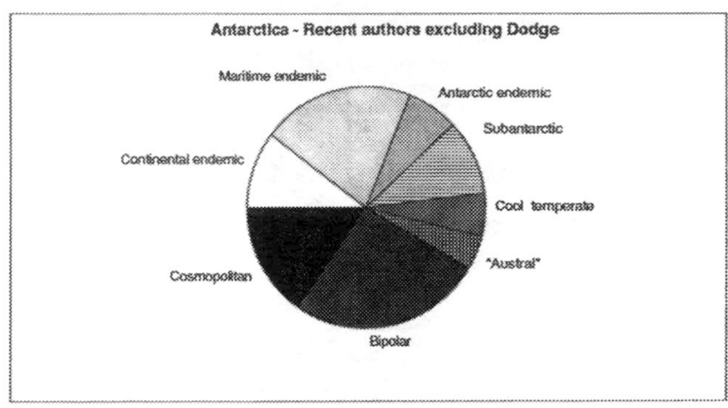

Fig. 2.4. Percentage occurrence of the main phytogeographic elements in the lichen flora of Antarctica according to three main sources: (a) the flora of Dodge (1973), (b) the same, after revision of many Antarctic taxa by post-Dodge authors, (c) a checklist of all lichen records by recent authors, excluding Dodge.

species, or closely related ancestors, were widespread already before the fragmentation of Pangea. The populations brought into the Arctic zone by continental drift, and those persisting in Antarctica, underwent only slight or no differentiation, due to the very slow evolutionary rates of lichens. This theory, which implies that Antarctica hosts a very old, relict flora, was previously supported by the presumed high incidence of endemic taxa, but cannot be maintained now that endemics appear to be much less numerous than hitherto assumed.

Finally, the third theory implies a recent, probably post-glacial recolonization by lichens through long-distance dispersal mechanisms. Contrary to the first two theories, which cannot be falsified, the long-distance dispersal hypothesis can be corroborated by field studies, as in the Antarctic region there are areas which were completely glaciated, and which, therefore, were only recently colonized by lichens. Informative results were obtained by Hertel (1984, 1987a) for Marion and Prince Edward Islands, and by Engelskjön & Jörgensen (1986) for Bouvetöya. These islands were subject to heavy glaciation, and their lichen floras have features which strongly support the long-distance dispersal hypothesis, such as a preponderance of wide-ranging and bipolar species, a low number of endemics, a high number of genera compared with that of species, and the absence of many other species which are otherwise common in the sub-Antarctic zone. Similar results were obtained by Schuster (1979) for the hepatics of South Shetland Islands.

In conclusion, the hitherto available information suggests that the lichen flora of Antarctica, and especially of continental Antarctica, is a young one, which mainly originated during the Quaternary period by long-distance dispersal.

ACKNOWLEDGEMENTS

This study was financed by the Italian National Programme for Antarctic Research (PNRA).

REFERENCES

Almborn, O. 1974. Review of Dodge, C. W. Lichen flora of the Antarctic Continent. *Botaniska Notiser*, **127**, 454–455.

Botnen, A. & Övstedal, D. O. 1988. Muscicolous *Lepraria* and other leprarioid lichens in the Antarctic. *Polar Research*, **6**, 129–133.

Brodo, I. M. & Hawksworth, D. L. 1977. *Alectoria* and allied genera in North America. *Opera Botanica*, **42**, 1–164.

Castello, M. & Nimis, P. L. 1994a. Critical notes on Antarctic yellow Acarosporaceae. *Lichenologist*, **26**, 283–294.

Castello, M. & Nimis, P. L. 1994b. Critical notes on the genus *Candelariella* (Lichenes) in Antarctica. *Acta Botanica Fennica*, **150**, 5–10.

Castello, M. & Nimis, P. L. 1995. A critical revision of Antarctic lichens described by C. W. Dodge. *Bibliotheca Lichenologica*, **57**, 71–92.

Dodge, C. W. 1964. Ecology and geographical distribution of Antarctic lichens. In Carrick, R., Holdgate, M. & Prévost, J., eds. *Biologie Antarctique*, Paris: Hermann, 165–171.

Dodge, C. W. 1973. *Lichen flora of the Antarctic continent and adjacent islands*. Canaan, New Hampshire: Phoenix Publishing, 399 pp.

Du Rietz, E. 1940. Problems of bipolar plant distribution. *Acta Phytogeographica Suecica*, **13**, 215–282.

Engelskjön, T. & Jörgensen, P. M. 1986. Phytogeographical relations of the cryptogamic flora of Bouvetöya. *Norsk Polarinstitutt Skrifter*, **185**, 71–79.

Filson, R. B. 1984. *Xanthoria elegans* and its synonyms in Antarctica. *Lichenologist*, **16**, 311–312.

Filson, R. B. 1987. Studies in Antarctic lichens VI. Further notes on *Umbilicaria*. *Muelleria*, **6**, 335–347.

Galloway, D. J. 1991. Phytogeography of southern hemisphere lichens. In Nimis, P. L. & Crovello, T. J., eds. *Quantitative approaches to phytogeography*. Dordrecht: Kluwer Academic Publishers, 233–262.

Hale, M. E. 1987. Epilithic lichens in the Beacon Sandstone formation, Victoria Land, Antarctica. *Lichenologist*, **19**, 269–287.

Hertel, H. 1984. Über saxicole, lecideoide Flechten der Subantarktis. *Beihefte Nova Hedwigia*, **79**, 399–500.

Hertel, H. 1987a. Progress and problems in taxonomy of Antarctic saxicolous lecideoid lichens. *Bibliotheca Lichenologica*, **25**, 219–242.

Hertel, H. 1987b. Bemerkenswerte Funde südhemisphärischer, saxicoler Arten der Sammelgattung. *Lecidea*. *Mitteilungen der Botanischer Staatssammlung München*, **23**, 321–340.

Hertel, H. 1988. Problems in monographing Antarctic crustose lichens. *Polarforschung*, **58**, 65–76.

Kärnefelt, I. 1979. The brown fruticose species of *Cetraria*. *Opera Botanica*, **46**, 1–150.

Lamb, I. M. 1964. Antarctic lichens. I. The genera *Usnea, Ramalina, Himantormia, Alectoria, Cornicularia*. *British Antarctic Survey Scientific Reports*, No. 34, 34 pp.

Lamb, I. M. 1968. Antarctic lichens II. The genera *Buellia* and *Rinodina*. *British Antarctic Survey Scientific Reports*, No. 61, 129 pp.

Lindsay, D. C. 1974. Review of Dodge, C. W. Lichen Flora of the Antarctic Continent and adjacent islands. *Lichenologist*, **6**, 130–131.

Lindsay, D. C. 1977. Lichens of cold deserts. In Seaward, M. R. D., ed. *Lichen ecology*, London: Academic Press, 183–209.

Nimis, P. L. 1990. Lo stato delle conoscenze lichenologiche sul continente Antartico. *Proceedings 1st Meeting Biology in Antarctica*. Rome: CNR, 5–54.

Övstedal, D. O. 1983. Some lichens from Vestfjella, Dronning Mauds Land, Antarctica. *Cryptogamie, Bryologie, Lichénologie*, **4**, 217–226.

Redon, J. 1985. *Liquenes antarcticos*. Santiago: Instituto Antarctico Chileno, 123 pp.

Schuster, R. M. 1979. On the persistence and dispersal of transantarctic Hepaticae. *Canadian Journal of Botany*, **57**, 2179–2225.

Sheard, J.W. 1977. Palaeography, chemistry and taxonomy in the lichenized ascomycetes *Dimelaena* and *Thamnolia*. *Bryologist*, **80**, 100–118.

Smith, R. I. L. 1984. Terrestrial plant biology of the Sub-antarctic and Antarctic. In Laws, R. M., ed. *Antarctic ecology*, Vol. 1. London: Academic Press, 61–162.

Smith, R. I. L. & Övstedal, D. O. 1991. The lichen genus *Stereocaulon* in Antarctica and South Georgia. *Polar Biology*, **11**, 91–102.

Söchting, U. & Övstedal, D. O. 1991. Contribution to the *Caloplaca* flora of the western Antarctic region. *Nordic Journal of Botany*, **12**, 121–134.

Stebbins, G. L. 1950. *Variation and Evolution in Plants*. New York: Columbia University Press.

Walker, F. J. 1985. The lichen genus *Usnea* subgenus *Neuropogon*. *Bulletin of the British Museum (Natural History), Botany series*, **13**, (1), 1–130.

3 Two polar *Hydractinia* species (Cnidaria), epibiotic on two closely related brittle stars (Echinodermata): an example for a taxonomic and ecological bipolarity

A. SVOBODA[1], S. STEPANJANTS[2] AND I. SMIRNOV[2]
[1]*Ruhr-Universität Bochum, Lehrstuhl für Spezielle Zoologie, D-44780 Bochum, Germany,* [2]*Zoological Institute, Russian Academy of Sciences, 199034 St. Petersburg, Russia*

ABSTRACT

The Antarctic hydroid species Hydractinia vallini *and the related Arctic species* H. ingolfi *both live on the epidermis of closely related ophiurids (Ophiolepididae). Both hydroid species are limited in distribution to polar sea temperatures. The polyps settle on the oral and aboral side of the ophiurid disc and the lateral sides of the arms. In both species the polyps of the oral disc side grow to a larger size but usually have reduced tentacles. The proboscis of these polyps is so much widened that the tentacles may even be hidden behind it. Numerous hypostomal digestive glands come into contact with the ground when the proboscis gets turned inside out. This suggests that this behaviour is used for extra-intestinal digestion of large prey which otherwise cannot be ingested. Everting of the hypostome of* H. vallini *could also be provoked for polyps which settled on the aboral surface of the host when the brittlestar is turned upside-down for a long period. All other polyps have long tentacles and a conical proboscis, as in other hydractiniids.*

The unique symbiotic relationships developed independently in both polar seas may be regarded as an example of taxonomic and bionomic bipolarity according to Andriashev's concept.

Key words: bipolar distribution, hydrozoa, *Hydractinia*, Ophiolepididae, symbiosis.

A bipolar distribution concept may either focus on the occurrence of one species limited strictly to both polar regions (Berg 1933) or on the bipolar distribution of closely related species, species groups or even biocoenoses. The latter bionomic (auto- and synecological) concept was proposed by Andriashev (1987).

Because of the species-specific symbiotic relationship of hydractiniids to distinct hosts (other Hydrozoa, Crustacea: hermit crabs, molluscs, echinoderms and even fish) these athecate hydroid genera are of special interest.

Two polar ocean hydroid species of the Hydractiniidae family, which live in symbiosis with brittle stars (Ophiolepididae, Ophiuroidea) have been reinvestigated (Svoboda *et al.* 1995). *Hydractinia vallini* Jaederholm, 1926, stolonizes on the epidermis of *Theodoria relegata* (Koehler, 1922, synonym of *T. wallini* Mortensen, 1925) in the Ross Sea (Discovery Inlet) at a depth of 550 m and in the southern Weddell Sea at depths of 256–470 m (Fig. 3.1a). According to Russian expedition material (1956–1958; 1953–1982; vessels *Knipovitch* (1981–1982) and *Skif* (1972–1975)) this species is circum-Antarctic spreading north to Kerguelen and South Georgia Islands at depths of between 39 and 2020 m. The similar *Hydractinia* (syn. *Stylactaria*) *ingolfi* (Kramp, 1932) is found on *Homophiura tesselata* Verrill, 1894 (Paterson, 1985) south of Greenland and Iceland at depths of 2137–3229 m (Fig. 3.1b) The Antarctic *H. vallini* lives at water temperatures of between -1.1 and 1.9 °C and the Arctic *H. ingolfi* between 1.4

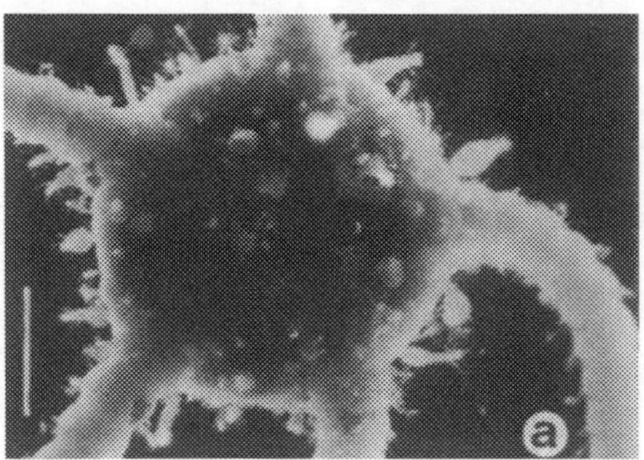

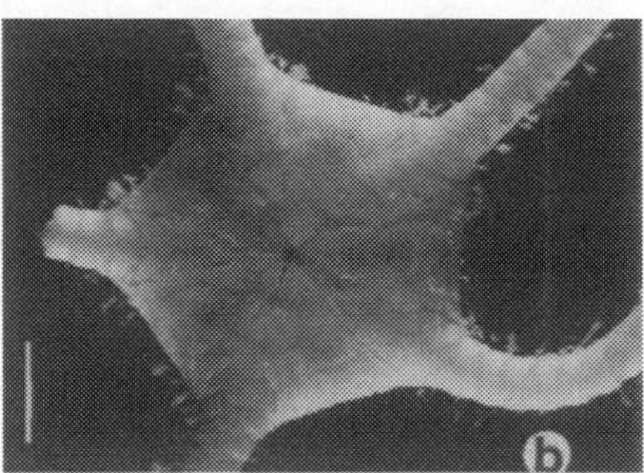

Fig. 3.1. (a) Living Antarctic *Hydractinia vallini* on its ophiurid host *Theodoria relegata* from the Weddell Sea. (b) Arctic *Hydractinia (Stylactaria) ingolfi* on *Homophiura tesselata*, type specimen from the Zoology Museum, Copenhagen. Scales: 2 mm

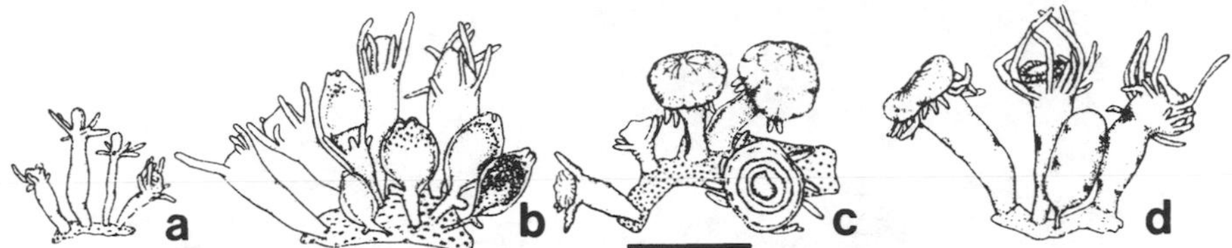

Fig. 3.2. Polyps of *Hydractinia vallini* from (a) tip of the arm, (b) aboral side of the disc and (c) oral side of its host, *Theodoria relegata*. (d) Polyps of *Hydractinia ingolfi* from the disc margin of its host, *Homophiura tesselata*. Scale 1 mm.

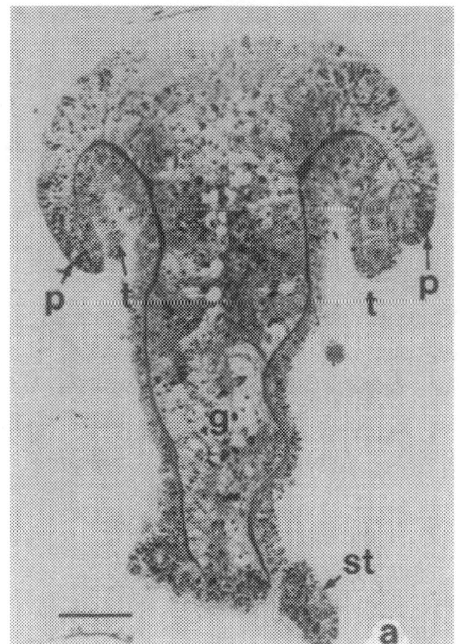

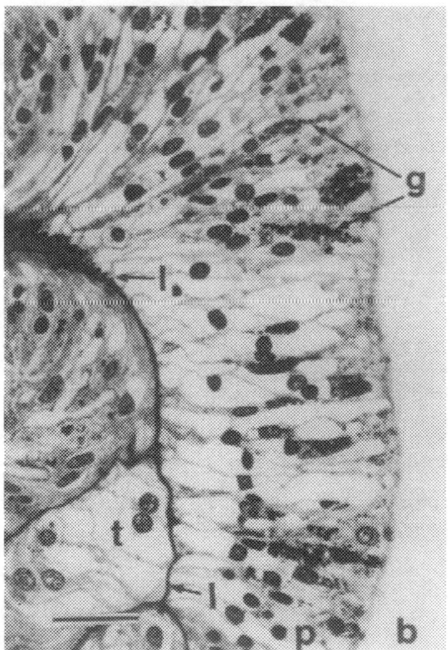

Fig. 3.3. (a) Polyp of *Hydractinia vallini* in longitudinal section with everted proboscis (p), hiding the tentacles (t); stolon (st). (b) Magnified part of everted proboscis with granular glands–g, tentacle endoderm cells (t), basal lamellal (l). (Semi–thin sections, stained with polychromatic methylene blue.) Scales: (a) 100 μm, (b) 30 μm.

and 3.0 °C. *H. tesselata* specimens found in warmer water (4.4–4.8 °C) contain no hydroid symbionts.

Polyps and gonophores of both hydractiniid species are common on the aboral side of the disc of their ophiurid hosts. Population density of the polyps is high along the margin of the disc and low on the oral plates. The lateral and aboral sides of the arms are covered nearly to the tips, but with smaller polyps than those of the central disc. In both species the hydranths on the aboral side of the brittle star have a rounded, conical proboscis (Fig. 3.2a, b). The polyps from the lower lateral region and from the oral side, which are able to contact the ground, show a trumpet-shaped or disc-shaped widened mouth (Fig. 3.2c, d). The disc-shaped mouth can be widened so much that the tentacles, being much shorter than those of the polyps with a conical proboscis, are hidden (Figs. 3.2c, 3.3a). Experiments with living *T. relegata* turned on to their aboral surface have shown that the mouth-widening in the polyps is provoked by their contact with the seabed. This behaviour is obviously used for extra-intestinal digestion. Thereby large prey (e.g. tissues of Bryozoa, sponges) or bacteria on the ground, which cannot be

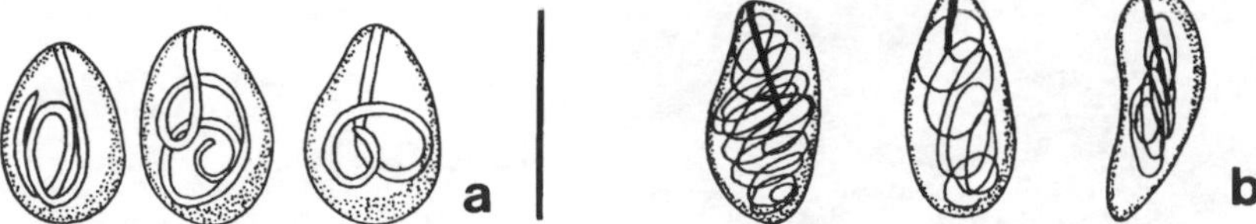

Fig. 3.4. Nematocysts of both hydractiniid species: (a) desmonems and (b) microbasic euryteles Scale 10 μm.

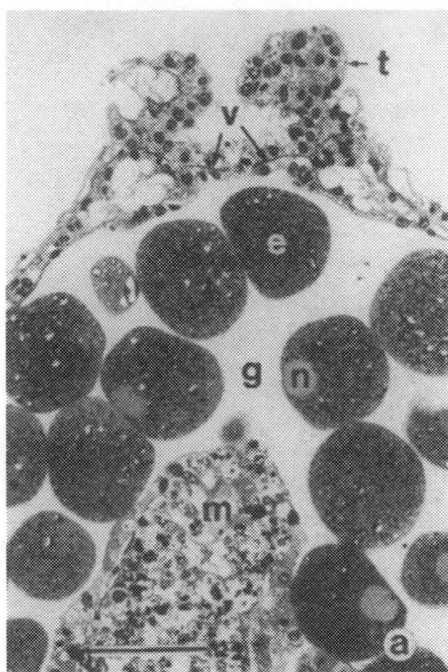

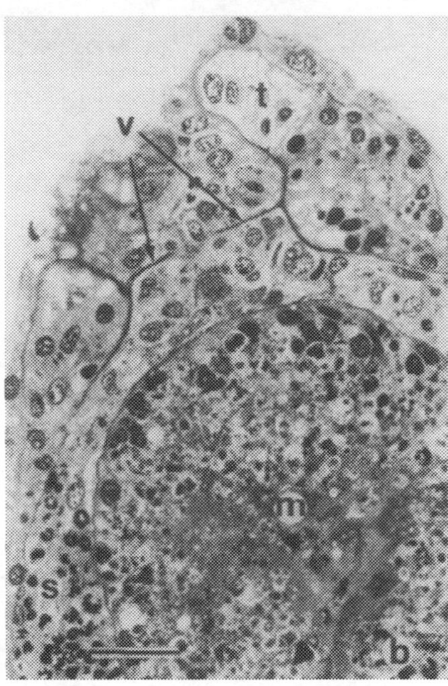

Fig. 3.5. Gonophores of *Hydractinia vallini* in longitudinal section. (a) Mature female gonophore; eggs (e), gastral cavity (g), manubrium (m), nucleus (n), tentacle (t), basal lamella of the velum (v). (b) Male gonophore: sperm (s), Scales: 30 μm.

ingested, may be utilized by direct contact with the numerous hypostomal granular glands (Fig. 3.3b). The change of the mouth form occurs very slowly, taking about one day at temperatures between 0 and 1 °C. Therefore, the proboscis eversion remains after fixation. The two distinctly different proboscis types in all colonies of both hydractiniids lead to the conclusion that their ophiurid hosts do not hide on the bottom like other species, otherwise the hydranths of the upper side of the brittle star would also have the disc-shaped proboscis.

Both hydroid species have many small, rounded desmoneme nematocysts (about 7.5 × μ4 m, Fig. 3.4a) and much fewer elongated microbasic euryteles (10 × 3 μm, Fig. 3.4b). The large gonophores of both species develop at the body of a rudimentary, tentacleless gonozoid, which is curved at the point of attachment. The gonophores of *H. ingolfi* are tentacleless cryptomedusoids, but those of *H. vallini* have four rudimentary tentacles and a velum with a small open bell cavity (Fig. 3.5a, b). In an early stage the gonophores of *H. vallini* show four separate gonads around the manubrium, which fuse later. It may be concluded from the absence of a bell jelly, ring muscles and a gastrovascular system that the medusoids do not release from the gonozoid, but shed their products freely in the water.

According to Kramp (1932) the netlike stolon of *H. ingolfi* should be covered by a perisarc which is free of spines. Therefore, *H. ingolfi* was referred to the genus *Stylactaria* Stechow (1921) by Namikawa (1991). Unfortunately, most *H. ingolfi* specimens from the Danish Ingolf Expedition are badly damaged by sediment abrasion from trawling and have been preserved and stored in alcohol for 100 years. Therefore, most stolons and polyp stalks lack even the ectoderm, but at lower magnification the thick and abrasion-resistent mesogloea simulates the presence of a perisarc on the upper side of the stolons and on the polyp bases. From selected specimens (Ingolf Stat. 35) embedded total stains or sections (1 and 10 μm) show a very thin perisarc sheet only between the stolon and the host tissue. Therefore, *H. ingolfi* should be retained in the genus *Hydractinia* van Beneden (1841). The stolons of *H. vallini* show numerous nematocysts grouped in patches (Fig. 3.2a b, c) and oriented towards the perisarc separating it from the host (Fig. 3.6a, b) whereas they seem to be more dense on the naked upperside of the stolons of *H. ingolfi*.

Undoubtedly, we are dealing with two closely related brittle stars of the family Ophiolepidae, which are restricted to the Antarctic and Arctic regions, living in symbiosis with two closely related hydroids of the same genus (*Hydractinia*). The benefit for the ophiurid host may be in the protection from predators by the dense forest of stinging tentacles, and such a protection function could also be proved for Cnidaria–Crustacea and Cnidaria–Mollusc symbioses. The benefit for the hydractiniid lies in transportation to new food

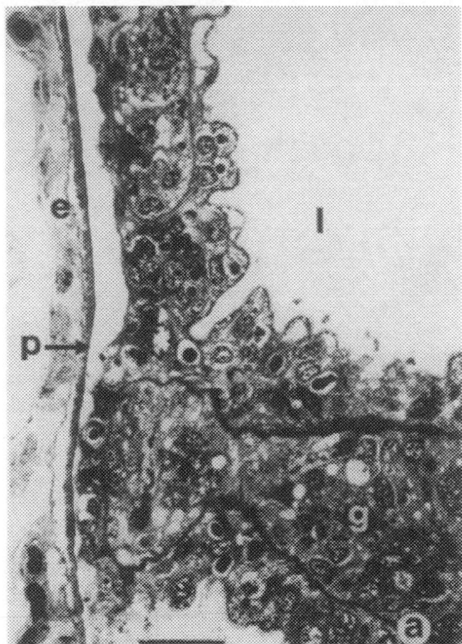

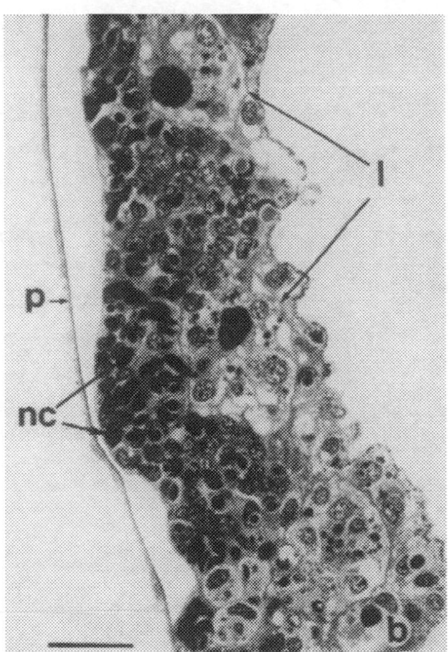

Fig. 3.6. (a) Stolon and pedal disc of *Hydractinia vallini* in longitudinal section; host epidermis (e), gastral cavity (g), basal lamella (l), perisarc (p). (b) Stolonal plate of *H. vallini* with aggregated nematocysts in cross section; stolon with perisarc lifted off the host; nematocyst (nc), Scales: 30 μm.

sources and protection from being buried in the sediment. Both commensal symbioses, most probably developed indepedently of each other, are established in polar seas of the Arctic and Antartic region and may be regarded as examples of taxonomic and bionomic bipolarity.

ACKNOWLEDGEMENTS

For the loan of the type and additional material of *Hydractinia ingolfi* we are indebted to K. W. Petersen and O. Tendal from the Zoological Museum, Copenhagen and Dr W. Vervoort, Zoological Museum, Leiden, for a loan of slides and kind suggestions and criticism of the manuscript.

REFERENCES

Andriashev, A. P. 1987. Development of Berg's concept of bipolarity of marine fauna. *Biologia Morya*, **2**, 60–66.

Berg, L. S. 1933. Die bipolare Verbreitung von Organismen und Eiszeit. *Zoogeographica*, **1**, 449–484.

Jäderholm, E. 1926. Über einige antarktische und subantarktische Hydroiden. *Arkiv für Zoologi*, **13A**(14), 1–7.

Kramp, P. L. 1932. Hydroids. In *The Godthaab Expedition 1928. Meddelelser om Grønland*, **79**, 1–86.

Namikawa, H. 1991. A new species of the genus *Stylactaria* (Cnidaria, Hydrozoa) from Hokkaido, Japan. *Zoological Science*, **8**, 805–812.

Paterson, G. L. J. 1985. The deep-sea Ophiuroidea of the North Atlantic Ocean. *Bulletin of the British Museum (Natural History), Zoology series*, No. 49 (1), 1–162.

Stechow, E. 1921. Neue Genera und Species von Hydrozoen und anderen Evertebraten. *Archiv für Naturgeschichte*, **87**, 249–265.

Svoboda, A., Stepanjants, S. & Smirnov. 1995. Zwei polare Hydractiniiden-Arten (Hydroida, Cnidaria) als Symbionten nahe verwandter Schlangensterne (Ophiolepididae, Echinodermata) – ein ökologisches Beispiel für Bipolarität. *Berichte zur Polarforschung*, **155**, 83–89.

4 On the demersal fish fauna of the Lazarev Sea (Antarctica): composition and community structure

CHRISTOPHER ZIMMERMANN

Institut für Polarökologie der Universität Kiel, Wischhofstr. 1-3, Geb. 12, D-24148 Kiel, Germany

ABSTRACT

The biogeography of Antarctic fishes from the Lazarev Sea (8° W–15° E) was investigated during two expeditions of RV Polarstern *in February–March 1991 and March–May 1992. The 21 catches with bottom and Agassiz trawl yielded 1842 fishes of 61 species and 11 families. Seventy-five per cent of the individuals belonged to the suborder Notothenioidei (Perciformes). The numerical dominant species were* Trematomus lepidorhinus *(22%),* Macrourus holotrachys *(17%) and* Chionodraco myersi *(12%).*

Three distinctly different communities could be separated by applying descriptive statistics (cluster analysis, multidimensional scaling). Characteristic species were Trematomus pennellii, Pleuragramma antarcticum, Trematomus eulepidotus *and* Prionodraco evansii *for cluster 'shallow',* T. scotti *for cluster 'shelf edge' and* Chionodraco myersi *for cluster 'slope', respectively.*

*The composition of the Lazarev Sea fish fauna is notably distinct from the adjacent Weddell Sea. In the Lazarev Sea, the percentage of benthopelagic nototheniids (*Trematomus lepidorhinus, Aethotaxis mitopteryx, Lepidonotothen kempi*) and non-notothenioid species is remarkably higher. Mean number of species per haul and evenness reach the highest values reported from investigations in other Antarctic seas, whilst diversity (*H′*) values exceed them. The high variability of abundance, biomass and species number per haul indicates small-scale patchiness of benthic and benthopelagic fish distribution patterns.*

Key words: Notothenioidei, zoogeography, high-Antarctic province, descriptive statistics.

INTRODUCTION

A large gap exists in our knowledge of the coastal fish fauna of East Antarctica. Previous research had focused on the high-Antarctic shelf seas south of 71 °S, namely the Weddell Sea and the Ross Sea, whereas the more northerly, narrow shelves between 70° and 65° S were virtually unexplored. A detailed study of the demersal communities in the Lazarev Sea was carried out for the first time during expeditions of the German RV *Polarstern* in 1991 and 1992. Its results were compared with the Weddell Sea ichthyofauna in order to clarify whether both areas are zoogeographically distinct or whether they belong to a uniform 'high-Antarctic province' as defined by Kock (1992) and Eastman (1993).

Investigated area

The Lazarev Sea is part of the Atlantic sector of the Antarctic Ocean, located between 8° W and 15° E to the east of the Weddell Sea (Canadian Hydrographic Service 1981). It is characterized by a narrow, rugged shelf with deep canyons perpendicular to the ice-shelf edge in the south (Dreyer *et al.* 1992), as well as small scale heterogeneities of biotic and abiotic para-

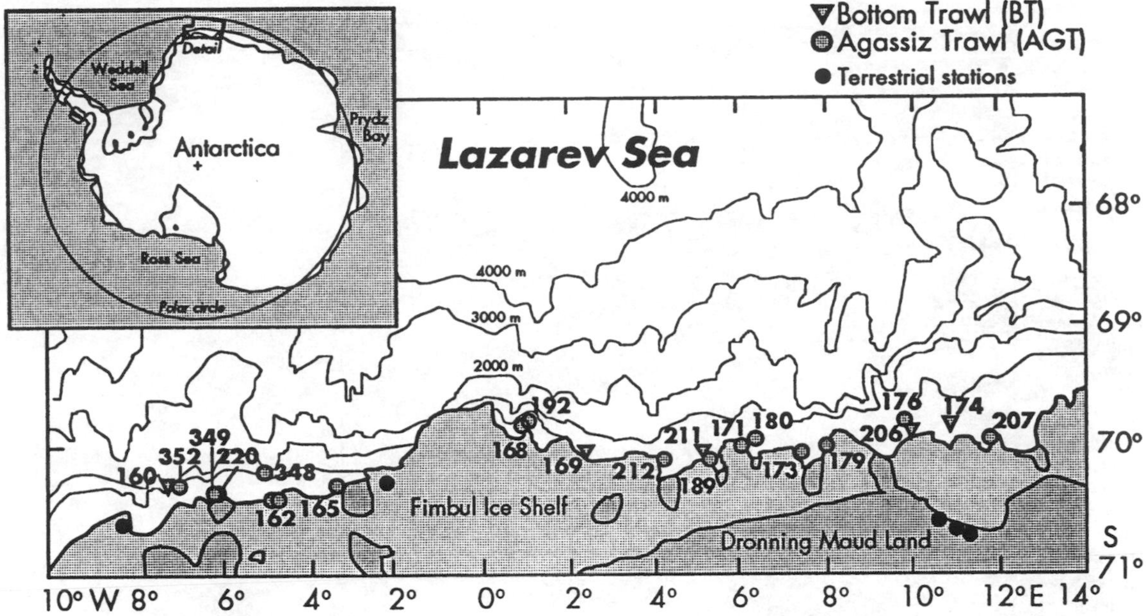

Fig. 4.1. Location of benthic stations during *Polarstern* cruises ANT IX/3 1991 (Sta. 160 to 212) and ANT X/3 1992 (Sta. 348 to 352) in the Lazarev Sea. Net type used is indicated.

meters. 'Innershelf depressions' as found on the Weddell Sea shelf are not present. A special hydrographic feature of the Lazarev Sea is a pronounced onshelf advection of relatively warm water originating at the Antarctic Divergence by the Weddell Gyre reflow. Bottom topography strongly affects circulation in this area, especially the Maud Rise in the north and the Astrid Ridge in the east (Schröder *et al.* 1992). Water temperature at the bottom was found to vary between –1.8 and 2 °C within some ten nautical miles. Considerable variation on a small spatial scale is also found in the plankton (Bathmann *et al.* 1992) and benthos (Gerdes *et al.* 1992, Gutt 1992). The Lazarev Sea is completely ice-covered in winter and usually ice free in summer. During the seasons investigated (February–May), even the fast-ice areas at the ice-shelf edge had gone.

MATERIAL AND METHODS

The investigated fish specimens were collected during two cruises of RV *Polarstern* in February–March 1991 and March–May 1992, using bottom trawls (BT, $n=5$; commercial 140 ft. trawl, mouth opening 22×5 m) and Agassiz trawls (AGT, $n=17$; epibenthic sledge, mouth opening 3×1 m). Hauls were carried out at 21 stations between ice shelf and continental slope (max. distance to the ice shelf edge *c.* 12 nautical miles, Fig. 4.1). The bottom depths were between 120 m and 1400 m. The swept area for all 22 net hauls totalled about 384 500 m^2 (Wöhrmann & Zimmermann 1992, di Prisco *et al.* 1993), three-quarters of which was BT sampling. Nearly half of the fished area was situated between 600 and 900 m depth. Fish processing and identification followed the routine procedures: immediately after the catch species were measured (total and standard length to the lower centimetre), weighed (fresh and gutted weight) and identified according to Fischer & Hureau (1985) and Gon & Heemstra (1990).

Explorative statistics

Station and species parameters were calculated from the raw dataset, including diversity, H' (Shannon & Weaver 1963), evenness, E (Pielou 1966), dominance and presence.

Both, classification and ordination techniques were applied to investigate multi-species distribution patterns. The faunistic resemblances between stations were compared by the Bray–Curtis coefficient on the basis of 4th root transformed abundances. The resulting resemblance matrix was subjected to a cluster analysis (UPGMA-classification), using the computer program COMM (Piepenburg & Piatkowski 1992), and to non-metric multidimensional scaling (MDS), using commercial statistic packages. Distinct station groups were identified from the cluster dendrogram and two-dimensional MDS plot. For details see Field *et al.* (1982). For the determination of characteristic species the degree of association regarding individuals (DAI) or regarding stations (DAS, Salzwedel *et. al.* 1985) was determined. Characteristic species are those showing a DAI or DAS of > 66%, a dominance within the cluster of > 5% and a presence of > 75% (Piepenburg & Piatkowski 1992).

RESULTS

The catches yielded 1842 fishes belonging to 61 species and eleven families. Seventy-five per cent of the individuals (68% of the species) belonged to the perciform suborder Notothenioidei. The numerically dominant species were *Trematomus lepidorhinus* (22%), *Macrourus holotrachys* (17%) and *Chionodraco myersi* (12%), which comprised more than 50% of all specimens.

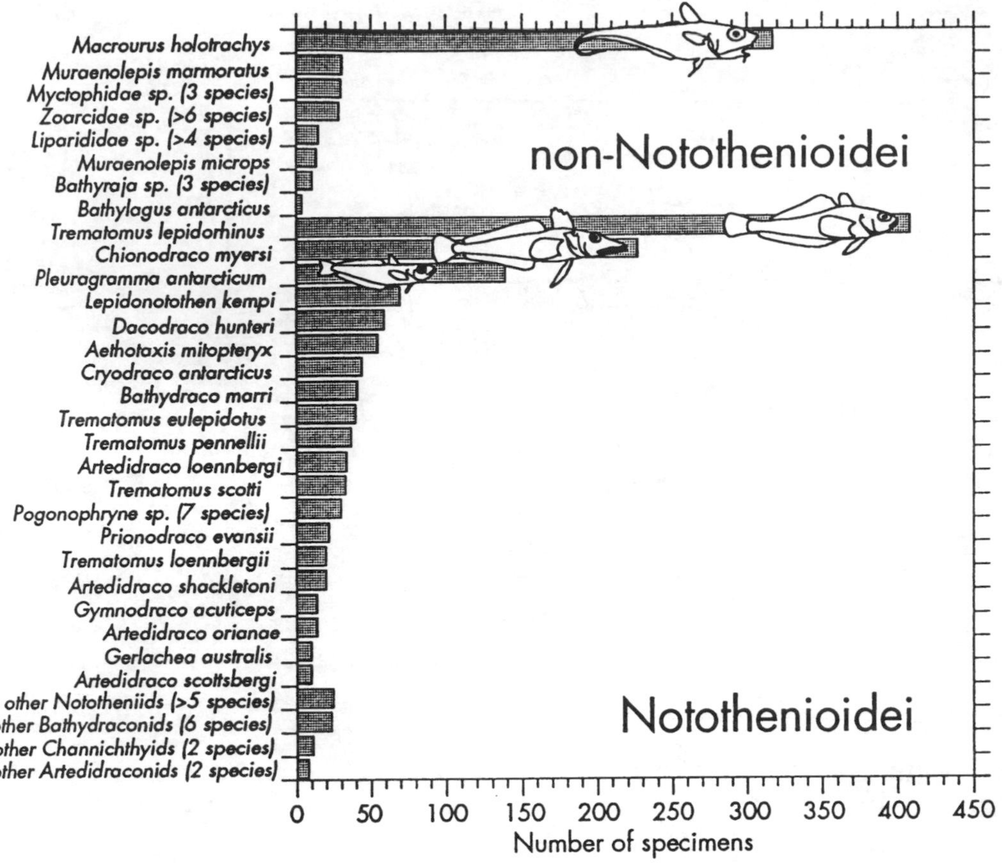

Fig. 4.2. Frequency distribution of demersal Lazarev Sea fishes. Notothenioid and non-notothenioid species are listed separately, some species occuring in very small numbers are combined.

Many species were caught in small numbers, the majority, including nearly all non–notothenioids, had fewer than 10 specimens (Fig. 4.2).

Cumulative species numbers indicated that 70% of the estimated total number of demersal fish species were represented in four bottom trawls. It is to be expected that nearly all Lazarev Sea's demersal species catchable with the sampling methodology are being recorded with the number of hauls employed.

Abundance and family composition varied markedly with depth and net type (Fig. 4.3). The shallowest AGT yielded the highest abundance (50 000 km^{-2}) and abundances decreased to 700 km^{-2} for the haul conducted at 1400 m depth. Mean densities were calculated to be 12 500 km^{-2} and 4700 km^{-2} for AGT and BT, respectively. Estimated fish biomass decreased with depth. Mean diversity H', evenness E and species numbers per haul attained comparatively high values (Table 4.1).

By taxa, a notable feature was that the relative abundance of artedidraconids decreased with depth, while that of non-notothenioids increased (Fig. 4.3). Nototheniids were dominant in all hauls down to 600 m; highest abundances of channichthyids were found between 300 and 600 m depth.

Descriptive statistics demonstrated that the distribution of fish species in the Lazarev Sea was mainly determined by water depth or associated factors (e.g. temperature). This zonation proved to be stable when other classification or ordination techniques were applied. The cluster analysis separated three distinct communities, called 'shallow', 'shelf edge' and 'deep' (Fig. 4.4). 'Shallow' stations are characterized by highest abundance and diversity, whereas 'shelf edge' encompassed mostly stations with low diversity and low mean species number per haul. The affiliation of stations to the clusters and their characteristic and dominant species are given in Fig. 4.4.

DISCUSSION

Comparing the Lazarev Sea with the neighbouring Weddell Sea, the overall similarities in species composition allows the Lazarev Sea to be included in the high-Antarctic province. Neyelov *et al.* (1982) proposed this on the basis of a Soviet study, dealing mainly with holopelagic and non-notothenioid deep-sea species. Furthermore, the vertical zonation of the demersal fish fauna in the Lazarev Sea corroborates the general pattern found in other Antarctic regions: a coastal fauna on the shelf and a deep-sea fauna on the slope, separated by a transitional zone (Kock 1992).

However, apart from the more general similarities, fish community patterns are remarkably different in the Lazarev Sea, when looking at the Weddell Sea fish fauna in more detail (as described by Hubold (1992) using the same sampling methodology):

- The total number of fish species is higher in the Lazarev Sea. Despite the low number of hauls conducted with only

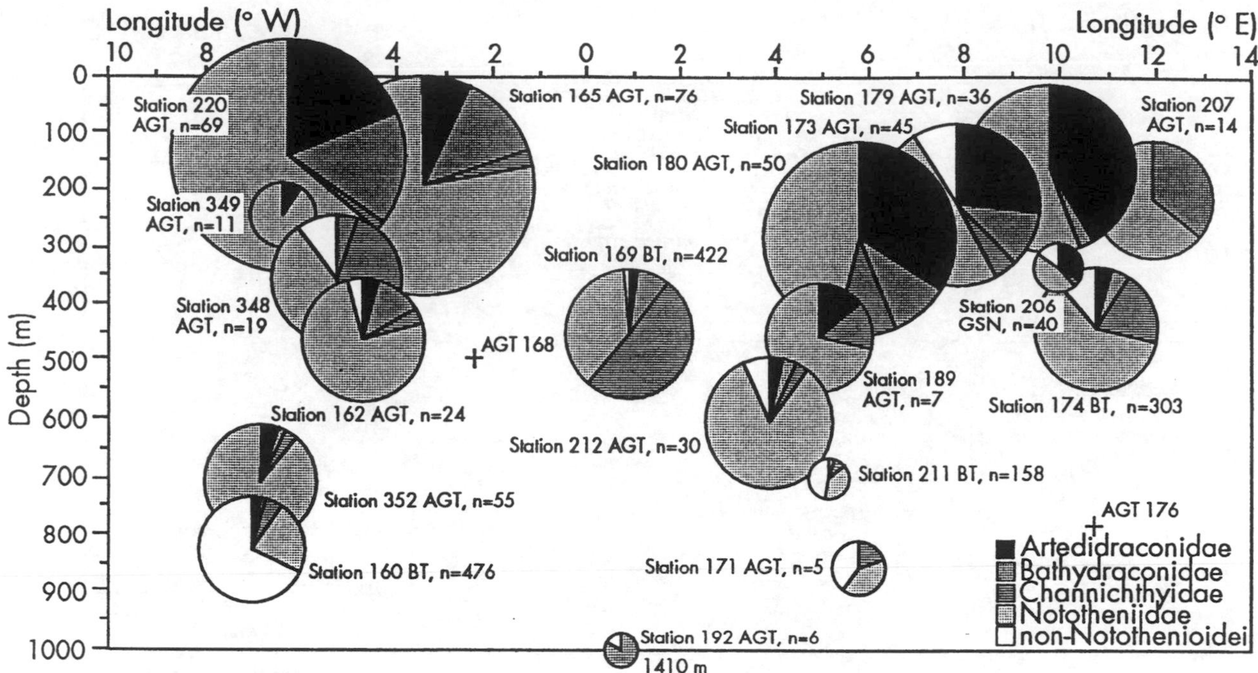

Fig. 4.3. Family composition and abundance (correlated to the logarithm of the circle diameter) of Lazarev Sea fishes from benthic hauls versus depth and longitude. The circle center indicates the position and depth. *n*=absolute catch number, AGT=Agassiz trawl, BT=bottom trawl. Sta. 168 and 176 omitted due to low number of individuals.

Table 4.1. *Comparison of ecological parameters for fish communities in different Antarctic areas*

Net type	Area	*n*	Diversity *H*	Evenness *E*	Species per haul
AGT	South Orkney Is.		0.702	0.293	
	Lazarev Sea	17	1.530	0.860	6
	Prydz Bay	10	1.590	0.862	7
	Weddell Sea	53	1.601	0.831	8
BT	South Orkney Is.	46	1.169	0.481	11
	South Shetland Is.	8	1.538	0.569	15
	Lazarev Sea	5	1.985	0.673	23
	Weddell Sea	27	1.789	0.624	17
	Ross Sea	4	1.677	0.786	9

Notes:
n=number of hauls.
Lazarev Sea: this study, all others taken from Hubold (1992, partly recalculated from data from Kock, Iwami & Abe, Ekau, Hubold, Schwarzbach, Targett and Williams).

two different gears in the Lazarev Sea, almost all demersal Weddell Sea species were found in our investigation area. Comparing the same net type and season only, 73 Weddell Sea hauls yielded fewer species (59) than the 22 hauls from the Lazarev Sea (61).

- Species number, abundance and biomass are highly variable in the Lazarev Sea, due to the small-scale heterogeneity of the habitat. This also increases the mean diversity and number of species per haul as compared with the more homogeneous Weddell Sea shelf.
- Different species are dominant and/or characteristic for the Lazarev Sea. The large-scale circulation, namely the Weddell Gyre reflow and the funneling effect of deep canyons carved in the steep slope, leads to a penetration of oceanic and deep-sea species onto the Lazarev shelf, and to a general softening of the border between deep-sea and coastal communities. This is illustrated by the much higher percentage of non-notothenioids in thc total catch of demersal species in the Lazarev Sea (25%), as compared with the Weddell Sea (1%).

Most of the differences may be explained by the sculptured topography and the corresponding complex current patterns on the Lazarev Sea shelf. However, except for the mentioned Soviet study (Neyelov *et al.* 1982, Trunov 1985), there are no additional data available from the Lazarev Sea to confirm the observed peculiarities.

Furthermore, distribution patterns for a number of notothenioid species show remarkable peculiarities which differ from other Antarctic shelf seas, e. g. the Weddell Sea (Hubold 1992). ***Pleuragramma antarcticum***. *Pleuragramma antarcticum* is the dominant fish species in the Weddell Sea, both in benthic and pelagic trawls. Based on pelagic trawls, Trunov (1985) described it as 'not rare' in the Lazarev Sea. In our benthic hauls *P. antarcticum* made up less than 10% of all specimens, whereas in comparable samples from the neighbouring Weddell Sea one month earlier it constituted almost half the catch. One explanation for the low presence of *Pleuragramma* in our Lazarev Sea catches may be the low catch efficiency of AGT and BT due to an epipelagic distribution of this species on the Lazarev shelf. This is indicated by the findings of *P. antarcticum* in Soviet pelagic trawls (Trunov 1985) and its

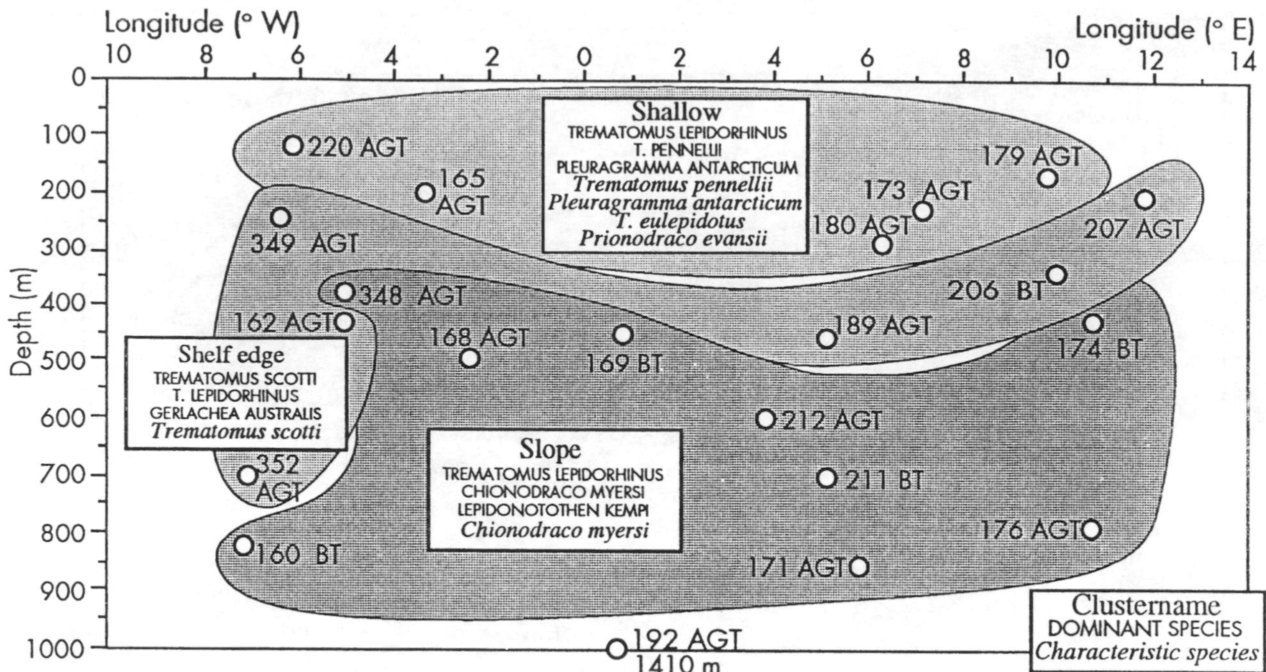

Fig. 4.4. Classification of stations, according to cluster analysis, depth and longitude of hauls. Sta. 192 could not be assigned to a cluster.

occurrence in stomachs of channichthyids (Zimmermann 1993), which make foraging trips in the water column (Daniels 1982). In the Lazarev Sea, *Pleuragramma* is also a characteristic species of shallow water stations, which contrasts with findings from the Weddell Sea (Hubold 1991). *Pleuragramma* could avoid warmer water at the bottom of the Lazarev Sea, or it could be shifted vertically by upwelling currents over the shelf. Furthermore, *P. antarcticum* could have been absent to a large extent in the Lazarev Sea in late summer. It is known to carry out extensive spawning migrations (Hubold 1992). The necessity of retention areas for larval development (Hubold 1992) and the presence of coastal polynyas favouring spawning success (Faleyeva & Gerasimchuk 1990) is discussed. Both features are absent in the Lazarev Sea.

Aethotaxis mitopteryx. Half of all *A. mitopteryx* specimens ever collected in the Antarctic were caught in benthic trawls in the Lazarev Sea (Kunzmann & Zimmermann 1992). In contrast, it rarely occurs in the southern shelf seas and at lower latitudes near the Antarctic peninsula. A preference for greater water depths, as stated by Ekau (1988) for *Aethotaxis* in the Weddell Sea, could not be confirmed for the Lazarev Sea. Abundance and depth distribution suggest that the Lazarev Sea and probably other narrow, northern shelves of East Antarctica are a main habitat of this species. *A. mitopteryx* and *P. antarcticum* are close relatives with a similar morphology, and to a certain degree a similar mode of life and food spectrum (summarized by Kunzmann & Zimmermann 1992). A main difference lies obviously in the vertical and horizontal distribution patterns. This spatial separation might be due to competitive exclusion, as it is discussed by White & Piatkowski (1993) for the larvae of both species in the Weddell Sea pelagial.

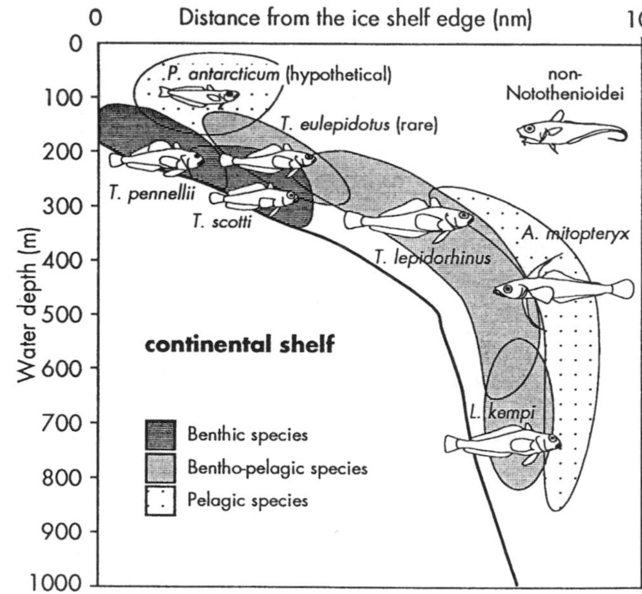

Fig. 4.5. Suggested vertical distribution of Lazarev Sea nototheniids.

***Trematomus* spp.** *Trematomus* species are far more common in the Lazarev Sea than in the Weddell Sea (30% versus 3% of total catch). Benthopelagic food generalists such as *T. lepidorhinus*, *T. eulepidotus* and – in terms of body shape and mode of life – the very similar *Lepidonotothen kempi* dominate this group, whereas *T. scotti*, the most frequent representative of this genus in the Weddell Sea, is comparatively rare (for species characterization see Ekau 1988, Schwarzbach 1988, DeWitt *et al.* 1990 and Ekau & Gutt 1991). An ecological equivalent (Kock 1992) to *T. scotti* is found only in the shallow waters of the Lazarev Sea, the strictly benthic *T. pennellii*. Fig. 4.5 gives a general distribution scheme of nototheniids in the Lazarev Sea.

Channichthyidae. This family had a seven-fold higher share of the total demersal catch in the Lazarev Sea than in the Weddell Sea (19% versus 3%). The species composition showed remarkable differences: *Pagetopsis maculatus* and *P. macropterus* were entirely absent in Lazarev Sea trawls, whereas every sixth channichthyid in the Weddell Sea belongs to this genus. The lack of *Pagetopsis* spp. is a corollary to the scarcity of *Pleuragramma antarcticum* in the Lazarev Sea and supports the discussed recruitment failure of the latter species in the area, for *Pagetopsis* larvae are known to depend on *Pleuragramma* larvae as major food (Hubold 1985, Hubold & Ekau 1990). Adults feed on *Euphausia crystallorophias* (Schwarzbach 1988), which appears to be rare in the Lazarev Sea. *Pagetopsis* might be replaced by *Dacodraco hunteri* and *Cryodraco antarcticus* (17% and 13% of channichthyid specimens, respectively), which both prey on fish. *C. antarcticus* seems to spawn in late summer on the shallower Lazarev Sea shelf (Zimmermann 1993). As discussed for *A. mitopteryx*, its main distribution is perhaps on the northern shelves of East Antarctica.

Artedidraconidae. The share of this family in the total catch and its diversity are notably higher in the Lazarev Sea than in the Weddell Sea. *Dolloidraco longedorsalis*, which is the dominant artedidraconid (56%) in the southern Weddell Sea, appears to be replaced by four *Artedidraco* species (59% of all artedidraconids in the Lazarev Sea).

CONCLUSIONS

Weddell Sea and Lazarev Sea ichthyofauna communities are more dissimilar than previously thought, probably due to differences in habitat structure, food availability and competition, leading to higher numbers of species per haul and higher evenness in the Lazarev Sea, and the highest fish diversity ever recorded in the Southern Ocean (Table 4.1). However, overall similarities in species composition justify the inclusion of the Lazarev Sea in the high-Antarctic province. The Weddell Sea is considered typical of high-Antarctic seas south of 71° S, with broad shelves and less affected by northern water masses. However, it remains to be shown whether the Lazarev Sea is representative of the narrow shelves around East Antarctica or whether the distinct demersal fish communities encountered here are a mere reflection of the complex current pattern in this particular area.

ACKNOWLEDGEMENTS

I would like to thank A. P. A. Wöhrmann and the captain and crew of RV *Polarstern* for support during the cruises; K-H. Kock, M. Stehmann and other specialists for the verification of species identification; W. Hagen, G. Hempel, G. Hubold, K-H. Kock and A. Kunzmann for valuable suggestions to this work, and C. Richter and an anonymous referee for improving the manuscript.

REFERENCES

Bathmann, U., Grossmann, S., Lochte, K., Scharek, R. & Schröder, M. 1992. The coastal current between the stations Georg-von-Neumayer and Georg-Forster. *Berichte zur Polarforschung*, **100**, 175–179.

Canadian Hydrographic Service. 1981. *General bathymetric chart of the oceans (GEBCO)*. 5th edn, Sheet 5.16. Ottawa: Canadian Hydrographic Service.

Daniels, R. A. 1982. Feeding ecology of some fishes of the Antarctic peninsula. *Fisheries Bulletin*, **80**, 575–588.

De Witt, H. H., Heemstra, P. C. & Gon, O. 1990. Nototheniidae. In Gon, O. & Heemstra, P. C., eds. *Fishes of the Southern Ocean*. Grahamstown: J. L. B. Smith Institute of Ichthyology, 279–331.

di Prisco, G., Tamburrini, M. & Kunzmann, A. 1993. Fisheries biology and structure and function of fish haemoglobin (ANT X/3, March–May 1992). *Berichte zur Polarforschung*, **121**, 92–98.

Dreyer, J., Hinze, H. & Munsch, I. 1992. Bathymetry and seafloor mapping by Hydrosweep. *Berichte zur Polarforschung*, **100**, 246–256.

Eastman, J.T. 1993. *Antarctic fish biology*. San Diego: Academic Press, 322 pp.

Ekau, W. 1988. Ökomorphologie nototheniider Fische aus dem Weddellmeer, Antarktis. *Berichte zur Polarforschung*, **51**, 1–140.

Ekau, W. & Gutt, J. 1991. Notothenioid fishes from the Weddell Sea and their habitat, observed by underwater photography and television. *Proceedings of the NIPR Symposium on Polar Biology*, **4**, 36–49.

Faleyeva, T. J. & Gerasimchuk, V. V. 1990. Features of reproduction in the Antarctic sidestripe, *Pleuragramma antarcticum* (Nototheniidae). *Journal of Ichthyology*, **30**, 67–79.

Field, J. G., Clarke, K. R. & Warwick, R. M. 1982. A practical strategy for analysing multispecies distribution patterns. *Marine Ecology Progress Series*, **8**, 37–52.

Fischer, W. & Hureau, J. C. 1985. FAO species identification sheets for fishery purposes. Southern Ocean (Areas 48, 58, 88). *FAO species identification sheets*, **1** (2). Rome: FAO, 232 pp.

Gerdes, D., Schanz, A. & Steinmetz, R. 1992. Quantitative macrofauna sampling and in-situ observation with the multibox corer. *Berichte zur Polarforschung*, **100**, 192–193.

Gon, O. & Heemstra, P.C., eds. 1990. *Fishes of the Southern Ocean*. Grahamstown: J. L. B. Smith Institute of Ichthyology, 462 pp.

Gutt, J. 1992. Faunistic-ecological investigations by means of underwater photography. *Berichte zur Polarforschung*, **100**, 189–192.

Hubold, G. 1985. The early life-history of the high-Antarctic silverfish *Pleuragramma antarcticum*. In Siegfried, W. R., Condy, P. R. & Laws, R. M., eds. *Antarctic nutrient cycles and food webs*. Berlin: Springer, 445–451.

Hubold, G. 1991. Ecology of notothenioid fish in the Weddell Sea. In di Prisco, G., Maresca, B. & Tota, B., eds. *Biology of Antarctic fish*. Heidelberg: Springer, 3–22.

Hubold, G. 1992. Zur Ökologie der Fische im Weddellmeer. *Berichte zur Polarforschung*, **103**, 1–157.

Hubold, G. & Ekau, W. 1990. Feeding patterns of post-larval and juvenile Notothenioids in the southern Weddell Sea (Antarctica). *Polar Biology*, **10**, 255–260.

Kock, K-H. 1992. *Antarctic fish and fisheries*. Cambridge: Cambridge University Press, 359 pp.

Kunzmann, A. & Zimmermann, C. 1992. *Aethotaxis mitopteryx*, a high-Antarctic fish with benthopelagic mode of life. *Marine Ecology Progress Series*, **88**, 33–40.

Neyelov, A. V., Permitin, Y. Y. & Trunov, I. A. 1982. The fish fauna of the Lazarev Sea (Eastern Antarctic) and its relation to neighbouring faunae. II. *All-Union Congress of Oceanologists: Biology of the Oceans*, **6**. Moscow: Oceanological Academy of Sciences of the USSR, 60–61. [In Russian]

Pielou, E. C. 1966. Shannon's formula as a measure of specific diversity: its use and disuse. *American Naturalist*, **100**, 463–465.

Piepenburg, D. & Piatkowski, U. 1992. A program for computer-aided analyses of ecological field data. *CABIOS*, **8**, 587–590.

Salzwedel, H., Rachor, E. & Gerdes, D. 1985. Benthic macrofauna communities in the German Bight. *Veröffentlichungen des Instituts für Meeresforschung Bremerhaven*, **20**, 199–267.

Schröder, M., Wisotzki, A., Witte, H., Griffith, S., Rau, I. & Bönisch, G. 1992. Physical oceanography. *Berichte zur Polarforschung*, **100**, 125–141.

Schwarzbach, W. 1988. Die Fischfauna des östlichen und südlichen Weddellmeeres: Geographische Verbreitung, Nahrung und trophische Stellung der Fischarten. *Berichte zur Polarforschung*, **54**, 93 pp.

Shannon, C. E. & Weaver, W. 1963. *The mathematical theory of communication*. Urbana: University of Illinois Press, 125 pp.

Trunov, I. A. 1985. Species composition and biology of fishes from shelf and slope of the Antarctic continent (in Lazarev Sea). In Yakovlev, V. N., ed. *Complex study of the bioproductivity of the Southern Ocean*, Vol. II. Moscow: Trudy VNIRO, 316–324. [In Russian]

White, M. G. & Piatkowski, U. 1993. Abundance, horizontal and vertical distribution of fish in Eastern Weddell Sea micronecton. *Polar Biology*, **13**, 41–54.

Wöhrmann, A. P. A. & Zimmermann, C. 1992. Comparative investigations on fishes of the Weddell Sea and the Lazarev Sea. *Berichte zur Polarforschung*, **100**, 208–222.

Zimmermann, C. 1993. *Beiträge zur Fischfauna des Lasarewmeeres (Antarktis)*. Master's thesis, University of Kiel , 97 pp. [unpublished]

5 Genetic variation in populations of the moss *Bryum argenteum* in East Antarctica

K. D. ADAM[1], P. M. SELKIRK[2*], M. B. CONNETT[1] AND S. M. WALSH[1]
[1] *University of Waikato, Department of Biological Sciences, Private Bag 3105, Hamilton, New Zealand,* [2] *Macquarie University, School of Biological Sciences, Sydney, N.S.W. 2109, Australia*

ABSTRACT

This paper reports on a study of the geographic and microgeographic patterns of variation in genetic marker loci in Bryum argenteum *populations from Ross Island (Cape Bird, Cape Royds) and southern Victoria Land (Granite Harbour, Eastern Taylor Valley, Garwood Valley) in East Antarctica. Approximately 300 samples were analysed for four isozymes (GPI, MDH, PGD and PGM); each had moderate levels of polymorphism. The geographic distribution of multi-isozyme banding patterns showed that there is a common genotype, occurring at high frequency in all five areas studied, while other, less frequent genotypes have a more restricted distribution. Using Randomly Amplified Polymorphic DNA (RAPD) markers, DNA-level variation was investigated in a subset of these samples, and found to be considerably higher than the isozyme variation. The genetic subdivision, and levels of genetic variability found in the populations sampled, suggest that present populations of* Bryum argenteum *in East Antarctica may either be remnants of genetically variable indigenous Antarctic populations, or the result of immigration by a range of genetically different propagules, or a combination of these.*

Key words: Antarctica, *Bryum argenteum*, DNA, genetic variation, isozymes, moss.

INTRODUCTION

Investigations of genetic diversity at marker loci have made a major contribution to assessing biodiversity, as well as to the study of colonization processes and population subdivision in many plant species (Barrett & Husband 1989, Barrett & Shore 1989, Hamrick & Godt 1989). In the past most of the markers studied were allozyme loci, which are highly suitable but limited in number and in the amount of genetic diversity they reveal. Recent developments in molecular genetic methods have made possible the study of genetic variation at the DNA level using genetic markers that are inherited in Mendelian fashion. This has vastly increased the number of loci for which the genetic composition of organisms and populations can be studied relatively quickly and at reasonable cost. Molecular markers such as RFLPs (Restriction Fragment Length Polymorphisms) and RAPDs (Randomly Amplified Polymorphic DNA) are available in large numbers spread throughout the genome, and show high levels of polymorphism and allelic diversity. More easily assayed than other DNA markers, RAPDs, first described in 1990 (Williams *et al.* 1990), are beginning to play a major role as genetic markers in population studies.

In bryophytes comparatively few population genetic studies have been published. Enzyme electrophoretic studies have shown genetic variability in temperate moss populations to be comparable with those in seed plants (Akiyama 1994, Derda & Wyatt 1990, Wyatt *et al.* 1989a) but to be higher than those in liverworts (Wyatt *et al.* 1989b). A comparison of isozyme and RAPD variability in temperate liverworts (*Porella* spp.) has confirmed the very low levels of isozyme variability in liverworts and shown levels of RAPD variability also to be lower than in many seed plants (Boisselier-Dubayle & Bischler 1994).

Mosses are important components of the terrestrial vegetation in Antarctica. On Ross Island and in Southern Victoria Land in East Antarctica mosses generally occur sparsely, in small clumps widely scattered in suitable habitats (Longton 1988), but are locally abundant in particularly favourable sites (Schwartz *et al.* 1992). In this region, two moss species, *Bryum argenteum* Hedw., and *Pottia heimii* (Hedw.) Hampe, are widespread, while several others including *Bryum pseudotriquetrum* (Hedw.) Gaern., Meyer et Scherb. and *Sarconeurum glaciale* (C. Muell.) Card. & Bryhn have a more restricted distribution.

Bryum argenteum is a cosmopolitan moss, known from all continents. Longton & MacIver (1977) compared growth in culture of material from Antarctic and North American sites. Between-population differences in size were not maintained in culture under controlled conditions, and were ascribed to phenotypic plasticity rather than to genetic differentiation. In a further study of *Bryum argenteum* from Antarctica, North America, Hawaii and Costa Rica, Longton (1981) found that clones from different locations remained morphologically distinct from one another in culture under controlled conditions. For some characters there was correlation between variation observed in field populations and in culture, but for other characters differences observed between populations in the field were not maintained in culture.

There has been little investigation of genetic variability within populations of mosses in Antarctica: to our knowledge there is one previous study of protein-level variability (Melick *et al.* 1994), and none of DNA-level variability.

The vegetation history of mosses in the ice-free areas of East Antarctica is unknown, but one view is that present floras comprise mostly post-glacial immigrants (Longton 1988), i.e. that colonization by propagules from outside the continent has taken place during the last 18 000 or so years. Other possibilities are that East Antarctic moss populations have been recruited from elements that have been present in ice-free areas (e.g. nunataks) for a long time (Kappen & Straka 1988), or are remnants of much more extensive bryophyte vegetation in times when the continental Antarctic climate was less extreme, or that the populations are derived from propagules that arrived in Antarctica a very long time ago.

Opinions differ as to whether the climate in southern Victoria Land has been hyper-arid cold desert, as now, and the Dry Valleys ice free continuously during the last 13–15 million years (Denton *et al.* 1993), or whether there was a warm interval with extensive deglaciation and associated expansion of the East Antarctic ice sheet about 3 million years ago (Barrett *et al.* 1992). Either of these views implies the presence of ice-free land in Southern Victoria Land for several millions of years, and hence allows the possibility of colonization by mosses over that time.

Bryum argenteum is not known to reproduce sexually in Antarctica. Antheridia have not been recorded there, but it is not known whether this is because male plants are absent or sterile (Longton 1988). *Bryum argenteum* in Antarctica reproduces vegetatively via bulbils and by deciduous shoot tips which, following detachment from the parent plants, are distributed by wind and water. Establishment may involve sprouting directly from an apical meristem, or formation of secondary protonema and subsequent shoot development.

We have used isozymes and RAPDs to study genetic variability in populations of *Bryum argenteum* from Ross Island and southern Victoria Land. Our aim was first to document the extent of genetic variability in Antarctic populations of this cosmopolitan moss, and second to interpret the geographic distribution of the variation in terms of probable colonization history of this region of Antarctica by this moss.

If present populations have arisen from few relatively recent introductions, we would expect a single or at best very few genotypes to be geographically widespread. If, on the other hand, present populations of *Bryum argenteum* were derived either from past extensive Antarctic populations, or from multiple introductions from outside Antarctica that completely colonized the area, we would expect present populations to show a homogeneous geographic distribution of multilocus haplotypes. (This assumes that the former widespread Antarctic populations, or the source populations for the multiple introduced propagules, were polymorphic.) Under this assumption, regional or local differentiation in haplotype frequencies could have occurred by (i) newly arriving propagules colonizing limited areas only, (ii) a drift-like effect of random elimination of haplotypes during periods of population decline, and/or (iii) adaptation to the local environment. In the absence of recombination, selection acting on genes coding for adaptive traits will be reflected at neutral marker loci in the same chromosomal linkage group.

MATERIAL AND METHODS

Plant material

Some 300 samples of *Bryum argenteum* were collected from five areas on Ross Island and southern Victoria Land, Cape Bird (CB), Cape Royds (CR), Garwood Valley (GV), the vicinity of Lake Fryxell (LF) in Eastern Taylor Valley and Granite Harbour (GH) (Fig. 5.1). *Bryum argenteum* occurs in places in which liquid water is available during most of the summer growing season (November to February). The species grows in clumps along drainage lines of meltwater from ice caps, glaciers and persistent snow banks. For our sampling we adopted a hierarchical sampling strategy subdividing each of the five areas (CB, CR, GV, LF, GH) into sites, each site consisting of one drainage line or a number of lines where these were clustered in an area of no more than *c.* 10 000 m^2. Samples of moss of between 0.5 cm^2 and 4 cm^2 were then collected from several clumps along a drainage line, usually one per clump. The number of sites per area ranged from one to 18, the number of clumps sampled per site ranged from one to 13, and the distance between adjacent clumps samples varied from 5 to 30 m. Each sample received a sample number plus a unique code for area,

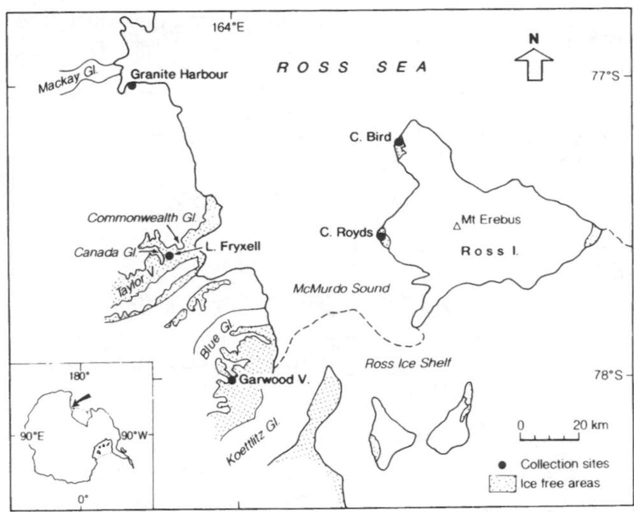

Fig. 5.1. Ross Island and Southern Victoria Land, East Antarctica, showing sites from which *Bryum argenteum* samples were collected. Arrow on inset map of Antarctica indicates location of main map.

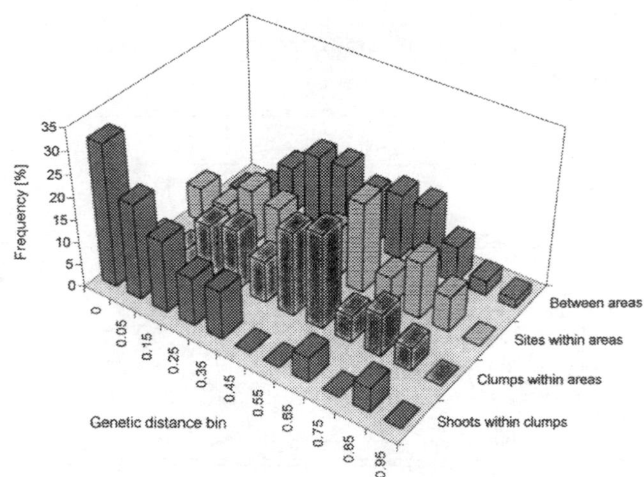

Fig. 5.2. Genetic distances estimated from RAPD data for *Bryum argenteum* populations in East Antarctica, breakdown according to geographic categories.

site within area, and clump number within site. A sample from the first clump in site 7 from the Cape Bird area, for example, would be coded CB7a.

Samples were collected into brown paper bags, air dried and stored at ambient temperatures (max. 5 °C) for up to eight days. Upon return to Scott Base, samples were frozen to −18 °C and kept at that temperature until genetic analysis at Waikato University in New Zealand. Of the samples collected, 277 have been analysed for isozyme polymorphism and 80 for RAPDs.

Isozymes

For isozyme analysis, three to five individual shoots of moss were incubated for 12 h at ambient temperature with 0.5 ml of Bold's Basal Medium (Bold 1967) in 1.5 ml centrifuge tubes. For isozyme extraction, 100 μl of extraction buffer (Triest 1991) was added to each tube and the plant material crushed with a rotating glass rod. Analyses for the following isozymes were conducted: AAT (EC 2.6.1.1), AH (=ACON=ACN, EC 4.2.1.3), APH (=ACP, EC 3.1.3.2), EST (EC 3.1.1.x), G6PDH (EC 1.1.1.49), GPI (=PGI, EC 5.3.1.9), GLUD (=GDH, EC 1.4.1.2) G3PDH (EC 1.2.1.12), IDH (EC 1.1.1.42), LAP (EC 3.4.11.x), MDH (EC 1.1.1.37), PER (EC 1.11.1.7), PGD (EC 1.1.1.44), PGM (EC 2.7.5.1), SDH (1.1.1.25) and TPI (EC 5.3.1.1).

Electrophoretic separation of isozymes was performed in 11% (w/v) starch gels, using the following buffer systems: Tris citrate/Lithium borate pH 8.1/8.5, L-Histidine citrate pH 5.7, and L-Histidine-HCl/trisodium citrate pH. 7.0, as described by Wendel & Weeden (1989). Enzyme-specific staining followed the methods decribed by the same authors. To ensure comparability of banding patterns between gels, extracts from a standard genotype of *Bryum argenteum* were applied with every 10 samples on all gels. In the absence of sexual reproduction in the material studied, no data on segregation at isozyme loci were available. Genetic interpretation of isozyme banding patterns was, of necessity, based on 'informed guesses'.

The genetic composition of samples from different areas was tested for homogeneity by pairwise comparisons of multilocus genotype frequency distributions using G tests. Genetic distances between groups of individuals were estimated using a multilocus band-sharing index as described for RAPDs below.

RAPDs

For RAPD analysis, DNA extractions using the CTAB method (Doyle & Doyle 1990) were initially made from three to five shoots of moss taken directly from the frozen clump samples. Later, DNA from individual shoots of moss was extracted following Wang *et al.* (1993), using 6 μl of 0.5M NaOH for grinding the plant material and adding 244 μl of TAE buffer immediately after grinding. For both extraction methods the DNA extracts were frozen immediately and stored at −18 °C until amplification.

DNA amplification was performed in a standard reaction mix; 25 μl mix contained 0.5 g genomic DNA, 2.5 μl 10×PCR buffer (500 mM KCl, 100 mM tris-HCl(pH 8.3), 15 mM $MgCl_2$, 0.1%(w/v) gelatin, 1% Triton X-100), 1.25 μl 10×dNTP mix (2 mM each dATP, dCTP, dGTP, dTTP (Boehringer Mannheim, Mannheim, Germany)), 0.25 μl primer stock (20 mM (Operon Technologies Inc., Alameda, California)), 0.4 μl Taq DNA polymerase (1 unit (Boehringer Mannheim)) and the balance of the volume being made up with water. The reaction mix was overlain with one drop of mineral oil (Perkin Elmer Co., Norwalk, Connecticut) and the following amplification protocol carried out using an Ericomp SingleBlock™ System (Ericomp Inc.,San Diego, California). Forty cycles of (1 min. 94 °C, 1 min. 35 °C, 2 min. 72 °C) were followed by 8 min. extension at 72 °C. Five microlitres of gel loading buffer (0.005%(w/v) bromophenol blue, 0.005%(w/v) xylene cyanol, 50%(v/v) glycerol, 0.05M $EDTA\text{-}Na_2$) was added to each sample and they were stored at −18 °C prior to electrophoretic analysis.

Table 5.1. *Occurrence of* Bryum argenteum *haplotypes in five sampled areas*

	Haplotypes								
	A	B	C	D	E	K	M	U	
Area	c	s	s	s	s	d	d	t	*n*
C. Bird	59	7	0	12	0	1	0	1	80
C. Royds	25	9	5	7	0	1	4	0	51
Garwood V.	34	10	0	1	0	0	0	0	45
L. Fryxell	49	1	4	7	0	0	8	0	69
Granite H.	18	2	0	5	1	0	6	0	32
Totals	185	29	9	32	1	2	18	1	277

Notes:

Haplotypes: c=most commonly occurring; s=differs from c at one locus; d=differs from c at two loci; t=differs from c at three loci; A–U, each letter designates a particular, different haplotype. *n*=number of samples.

Electrophoresis

Samples were run on horizontal agarose slab gels 25 cm long, 15 cm wide and with a total volume of 200 ml. Gels were made in 1×TAE buffer (40mM Tris-Acetate, 1mM EDTA) using 1%(w/v) NuSieve agarose (FMC Bioproducts, Rockland, Maine) and 1%(w/v) SeaKem LE agarose (FMC Bioproducts). Gels were run for a total of 400 volt hours (4 h at 4 V cm^{-1}) and subsequently stained with Ethidium Bromide prior to photographic recording.

Pairwise genetic distances between individuals were calculated using a band-sharing index based on Jaccard's similiarity index:

$$D=1-(2\,nxy/(nx+ny)) \qquad (5.1)$$

where *nxy* is the number of bands shared by two individuals compared, and *nx* and *ny* are the numbers of bands present in individual *x* and *y* only (Lynch 1990). For the comparison of groups of individuals, composite estimates of genetic distance based on Equation (5.1) can be derived as described by Clark & Lanigan (1993), since the plants we investigated were in their haploid state.

RESULTS

Isozymes

Of 16 enzyme systems studied, four (GPI, MDH, PGD and PGM) proved to be polymorphic, while the following 12 enzymes either were monomorphic (AAT, APH, AH, G6PDH, IDH, LAP, PER, SDH), or did not show sufficient enzyme activity or separation to allow reliable scoring (EST, GLUD, G3PDH, TPI). For each of the four polymorphic systems, two alleles were found, i.e. 16 possible multilocus genotypes or haplotypes, eight of which we encountered in our samples. At each of the four polymorphic loci one of the two alleles occurred at a much higher frequency than the other. The frequency distributions for haplotypes in the five sampling areas show the occurrence of a common haplotype in which the four common alleles are combined and which occurs at an overall frequency of $p=0.661$. The other haplotypes occur at a total frequency of $q=0.339$.

Table 5.2. *Genetic distances based on isozyme multilocus bandsharing data (Clark & Lanigan 1993) for populations of* Bryum argenteum *in East Antarctica*

	Cape Bird	Cape Royds	Granite Harbour	Gardwood Valley
Cape Royds	0.0648			
Granite Harbour	0.0673			
Garwood Valley	0.0406	0.0692	0.0708	
Lake Fryxell	0.0519	0.0702		09.0528

Of the possible haplotypes which differ from the common type at one locus four were found, whereas only two out of the six possible two-locus 'mutants' and only one out of the possible three-locus 'mutants' were found in our samples (Table 5.1).

Pairwise comparisons between the haplotype frequency distributions for the five sampling areas showed that the genetic composition of all areas is different at the 35% level or above, with the exceptions of comparisons between Granite Harbour and Cape Royds, and Granite Harbour and Lake Fryxell, which showed no significant differences. Genetic distances between pairs of populations showing significant differences in their genetic composition are given in Table 5.2.

RAPDs

For the initial primer screening we used DNA extracted from a subset of samples by the CTAB method, using three to five shoots per sample. DNA amplification with random oligonucleotide primers no. 1 to 20 from the Operon Technologies' A set revealed high levels of polymorphism for primer numbers 1, 3, 7, 8, 13, and 20, yielding a total of 36 polymorphic loci.

In the course of the primer screening, repeat samples taken from the same clump were analysed, and in some cases these repeat samples did not produce identical banding patterns. This gave rise to the concern that some or all of the amplified DNA observed came from template DNA of foreign origin, i.e. non-moss DNA or moss template carried over from previous extractions. However, after careful cleaning and sterilization of all relevant equipment and ingredients, as well as a number of control experiments, including amplification without added moss template DNA and repeated amplification with moss DNA from the same extract we conclude that the observed genetic differences between DNA extracts from the same clump sample must reflect genetic differences between shoots within single clumps.

As a consequence, all further DNA extractions were made from single shoots, and the subsequent statistical analysis includes only data from single shoot extractions. To estimate the degree of genetic heterogeneity within clumps repeat samples of between two and three individual shoots per clump were analysed from 34 clump samples. Multilocus analysis involving the data from the six primers listed above revealed that 19 of these 34 clumps, or 56%, were genetically homogeneous, whereas the remaining 15 clumps contained two or more different multilocus RAPD genotypes.

For this study pairwise genetic distances were calculated for 79 individual shoots from 38 different clumps representing all five sampling areas and including shoots from the same clump which had proved to be genetically identical, for which the genetic distance estimate is equal to zero. The frequency distribution for these genetic differences shows that while individual shoots taken from the same clump on average are more genetically similar than shoots taken from different clumps, there is no obvious correlation between genetic distance and spatial or geographic distance when shoots from different clumps are considered (Fig. 5.2).

DISCUSSION

In interpreting the results of this genetic analysis of Antarctic populations of *Bryum argenteum,* it is necessary to consider some aspects of its biology. First, in this environment, *Bryum argenteum* is exclusively vegetatively reproducing, and hence, an entirely haploid plant.

Second, the concept of individuals is difficult to apply to *Bryum argenteum* at least in Antarctica. We expected to be able to regard individual clumps as single-genotype colonies started by a single propagule, but the results clearly show that in a significant number of cases this is not true. Similarly, using isozyme markers, microscale genetic variation was found by Wyatt *et al.* (1989b) within clumps of *Plagiomnium ciliare*, and Cummins & Wyatt (1981) within clumps of *Atrichum angustatum* in the southeastern United States.

Third, the concept of generations and generation turnover is difficult to apply to mosses in the absence of sexual reproduction. It is not known how long individuals or colonies can persist, but we believe that their lifespan may be measured in decades rather than years.

The results of these isozyme analyses show moderate levels of genetic variability between samples from different areas. The genetic composition of samples from the five areas is sufficiently different to show that genetic exchange between these areas (presumably via windborne propagules) occurs, but is limited. Given the strong winds in this region of East Antarctica transport of propagules from one area to another is very likely to happen, but the genetic data show that these migrant propagules have not become established in sufficient quantities to overcome the effects of genetic subdivision. Estimates of genetic distances between populations in the five areas sampled fall within the range reported for other mosses in similar studies (Wyatt *et al.* 1989b, Derda & Wyatt 1990), even though they were derived by different methods. Because of the overall much lower levels of genetic variation for isozymes than for RAPDs, no estimates of genetic distances were derived for lower geographic levels, such as between sites within a sampling area.

The RAPD data indicate levels of genetic diversity considerably higher than those revealed by isozymes. This is in accordance with results from other plant studies (Liu & Furnier 1993) and our own observations in other species such as *Muehlenbeckia astonii*, a shrub endemic to New Zealand.

To look for genetic subdivision pairwise genetic distances between individuals were broken down into various geographic levels, and while shoots sampled from the same clump on average were more genetically similar than shoots taken from different clumps, there were no clear indications of genetic subdivision at higher geographic levels.

A detailed study of the RAPD results raises three interesting points. First, there is an apparent contradiction between the isozyme data, which suggest significant genetic subdivision between most sampling areas, and the RAPD data, which do not. This can be explained by the much higher resolution, compared with isozymes, with which RAPD analyses screen the genome for variability. Genetic subdivision is assumed to exist where genetic variation between (sub)populations is significantly greater than within (sub)populations. Hence, it is quite possible that indications of genetic subdivision are found for one type of genetic marker, but not the other.

The second point is the smaller genetic distance, on average, between non-identical shoots sampled from the same clump, than between shoots from different tufts. This could be explained by genetic differentiation through mutation of individuals of clonal origin within a single clump, but such an explanation would have to assume either that mutation rates were very high, or that these clumps have existed for a very long time.

Thirdly, there is the question as to why there was only one case of shoots with identical genotypes sampled from different areas. This can be explained by the high levels of RAPD variation found. The expected frequency of any particular genotype will therefore be very low, so the presence of a particular migrant genotype in different areas may well be missed due to sampling error.

In summary, this first intensive study of genetic variation in terrestrial Antarctic plants has shown high levels of variability, especially at RAPD loci. This indicates that present populations of *Bryum argenteum* in East Antarctica may either be remnants of genetically variable indigenous Antarctic populations, or the result of immigration by a range of genetically different propagules, or a combination of these. This study demonstrates how population genetic studies can contribute towards understanding the vegetation history and past colonization processes of mosses in Antarctica.

ACKNOWLEDGEMENTS

Thanks to Margaret Auger and Mike Moodie for help with the genetic analyses, to Allan Green for valuable advice, to Rod

Seppelt for help with the taxonomic identification of specimens, and to Frank Bailey for assistance in preparation of Fig. 5.1. The authors gratefully acknowledge the logistic support provided by the New Zealand Antarctic Programme and its staff, financial support from the University of Waikato, from the Macquarie University Research Grants Scheme, and from the Australian Antarctic Division.

REFERENCES

Akiyama, H. 1994. Allozyme variability within and among populations of the epiphytic moss *Leucodon* (Leucodontaceae: Musci). *American Journal of Botany*, **81**, 1280–1287.

Barrett, S. C. H. & Husband, B. C. 1989. The genetics of plant migration and colonization. In Brown, A. H. D., Clegg, M. T., Kahler, A. L. & Weir, B. S., eds. *Plant population genetics, breeding and genetic resources*. Sunderland, MA: Sinauer, 254–277.

Barrett, S. C. H. & Shore, J. S. 1989. Isozyme variation in colonising plants. In Soltis, D. & Soltis, P., eds. *Isozymes in plant biology*. London: Chapman & Hall, 106–126.

Barrett, P. J., Adams, C. J., McIntosh, W. C., Swisher III, C. C. & Wilson, G. C. 1992. Geochronological evidence supporting deglaciation three million years ago. *Nature*, **359**, 816–818.

Boisselier-Dubayle, M. C. & Bischler, H. 1994. A combination of molecular and morphological characters for the delimitation of taxa in European *Porella*. *Journal of Bryology*, **18**, 1–11.

Bold, H. C. 1967. *A laboratory manual for plant morphology*. New York: Harper & Row, 123 pp.

Clark, A. G. C. & Lanigan, C. M. S. 1993. Prospects for estimating nucleotide divergence with RAPDs. *Molecular Biology and Evolution*, **10**, 1096–1111.

Cummins, H. & Wyatt, R. 1981. Genetic variability in natural populations of the moss *Atrichum angustatum*. *Bryologist*, **84**, 30–38.

Denton, G. H., Sugden, D. E., Marchant, D. R., Hall, B. L. & Wilch, T. I. 1993. East Antarctic ice sheet sensivity to Pliocene climatic change from a Dry Valleys perspective. *Geografiska Annaler*, **75A**, 155–204.

Derda, G. S. & Wyatt, R. 1990. Genetic variation in the common haircap moss, *Polytrichum commune*. *Systematic Botany*, **15**, 592–605.

Doyle, J. J. & Doyle, J. L. 1990. Isolation of plant DNA from fresh tissue. *Focus*, **12**, 13–15.

Hamrick, J. L. & Godt, J. W. 1989. Allozyme diversity in plant species. In Brown, A. H. D, Clegg, M. T., Kahler, A. L. & Weir, B. S., eds. *Plant population genetics, breeding and genetic resources*. Sunderland, MA: Sinauer, 43–63.

Kappen, L. & Straka, H. 1988. Pollen and spores transport into the Antarctic. *Polar Biology*, **8**, 173–180.

Liu, Z. & Furnier, G. R. 1993. Comparison of allozyme, RFLP and RAPD markers for revealing genetic variation within and between trembling aspen and bigtooth aspen. *Theoretical and Applied Genetics*, **87**, 97–105.

Longton, R. E. 1981. Inter-population variation in morphology and physiology in the cosmopolitan moss *Bryum argenteum* Hedw. *Journal of Bryology*, **11**, 501–520.

Longton, R. E. 1988. *Polar bryophytes and lichens*. Cambridge: Cambridge University Press, 391 pp.

Longton, R. E. & MacIver, M. A. 1977. Climatic relationships in Antarctic and Northern Hemisphere populations of a cosmopolitan moss, *Bryum argenteum* Hedw. In Llano, G. A., ed. *Adaptations within Antarctic ecosystems*. Washington: Smithsonian Institution, 899–919.

Lynch, M. 1990. The similarity index and DNA fingerprinting. *Molecular Biology and Evolution*, **7**, 478–484.

Melick, D. R., Tarnawski, M. G., Adam, K. D. & Seppelt, R. D. 1994. Isozyme variation in three mosses from the Windmill Islands oasis, Antarctica: a preliminary study. *Biodiversity Letters*, **2**, 21–27.

Schwartz, A. M. J., Green, T. G. A. & Seppelt, R. D. 1992. Terrestrial vegetation at Canada Glacier, Southern Victoria Land. *Polar Biology*, **12**, 397–404.

Triest, L. 1991. Isozymes in water plants. National Botanic Gardens of Belgium. *Opera Belgica*, **4**, 167–192.

Wendel, J. F. & Weeden, N. F. 1989. Visualization and interpretation of plant isozymes. In Soltis, D. E. & Soltis, P.S., eds. *Isozymes in plant biology*. London: Chapman & Hall, 5–45.

Wang, H., Qi, M. & Cutler, A. J. 1993. A simple method of preparing plant samples for PCR. *Nucleic Acids Research*, **21**, 4153–4154.

Williams, J. G. K., Kubelik, A. R., Livak, K. J., Rafalski, J. A. & Tingey, S. V. 1990. DNA polymorphisms amplified by arbitrary primers are useful as genetic markers. *Nucleic Acids Research*, **18**, 6531–6535.

Wyatt, R., Stoneburner, A. & Ordrzykoski, I.J . 1989a. Bryophyte enzymes: systematic and evolutionary implications. In Soltis, D. E. & Soltis, P. S., eds. *Isozymes in plant biology*. London: Chapman & Hall, 221–240.

Wyatt, R., Stoneburner, A. & Ordrzykoski, I. J. 1989b. High levels of genetic variability in the haploid moss *Plagiomnium ciliare*. *Evolution*, **43**, 1085–1096.

6 Genetic and ecological research on Anisakid endoparasites of fish and marine mammals in the Antarctic and Arctic-Boreal regions

L. BULLINI[1], *P. ARDUINO*[2], *R. CIANCHI*[1], *G. NASCETTI*[2], *S. D'AMELIO*[3], *S. MATTIUCCI*[3], *L. PAGGI*[3], *P. ORECCHIA*[4], *J. PLÖTZ*[5], *B. BERLAND*[6], *J. W. SMITH*[7] *AND J. BRATTEY*[8]

[1]*Department of Genetics and Molecular Biology, University of Rome 'La Sapienza', Via Lancisi 29, I-00161 Rome, Italy,* [2]*Department of Environmental Sciences, Tuscia University, Via F. De Lellis, 100100 Viterbo, Italy,* [3]*Institute of Parasitology, University of Rome 'La Sapienza', Piazzale A. Moro 5, I-00185 Rome, Italy,* [4]*Department for Public Health and Cellular Biology, University of Rome 'Tor Vergata', Via O. Raimondo 8, I-00173 Rome, Italy,* [5]*Alfred-Wegener-Institute for Polar and Marine Research, Columbusstrasse, D-27515 Bremerhaven, Germany,* [6]*Zoological Laboratory, University of Bergen, Allègt. 41, N-5007, Bergen, Norway,* [7]*Scottish Office of Agriculture and Fisheries Department, Marine Laboratory, P.O.Box 101, Victoria Road, Aberdeen AB9 8DB, Scotland, UK,* [8]*Science Branch, Department of Fisheries and Oceans, P.O. Box 5667, St. John's, Newfoundland, Canada A1C 5X1*

ABSTRACT

Data are presented on the genetics and ecology of fish and marine mammal anisakid parasites of the genera Contracaecum, Pseudoterranova *and* Anisakis *from the Antarctic and Arctic-Boreal regions. The three morphospecies* C. osculatum, P. decipiens *and* A. simplex, *considered cosmopolitan and euriecious, were each shown by isozyme analysis to include a number of sibling species, differentiated genetically and ecologically. The reproductive isolation of* C. radiatum, *an Antarctic species often confused with* C. osculatum *s.l., was also shown. The* C. osculatum–radiatum, P. decipiens *and* A. simplex *complexes achieved a bipolar distribution at different times, from 5–6 to about 1 million years ago, through distinct colonizations of the Antarctic region. The more ancient bipolar distribution (*C. radiatum, P. decipiens *E) coincides with that of the first colonization of the Antarctic by seals; the more recent one (*C. osculatum *E) occurred in the Pleistocene. In the three anisakid complexes, Antarctic species show a higher genetic variability than the Boreal ones (average* H_e *0.21 and 0.14, respectively). This is apparently related to a lower habitat disturbance of the Antarctic region, allowing species to reach higher population sizes, with a lower probability of genetic drift phenomena. In both the Arctic-Boreal and Antarctic regions, differences in host preferences were seen which could be related both to differential host–parasite coadaptation and coevolution and to interspecific competition.*

Key words: anisakid nematodes, genetic markers, genetic diversity, peripatric speciation, parasite–host systems, bipolar distribution.

INTRODUCTION

Data are presented on the genetics and ecology of fish and marine mammal anisakid parasites of the genera *Contracaecum*, *Pseudoterranova* and *Anisakis* from the Antarctic, sub-Antarctic and Arctic-Boreal regions. Anisakids are ascaridoid nematodes dependent upon aquatic hosts for the completion of their life cycle, which generally involves an array of invertebrates and fish as intermediate or paratenic hosts, and marine mammals or fish-eating birds as definitive hosts. Morphological characters of taxonomic significance so far available in anisakid parasites are very few (i.e. morphology of excretory system, number and distribution of caudal papillae), and often applicable only to adults.

Molecular studies have contributed significantly to the study of parasite systematics, evolution and ecology. Such studies include: (*i*) isozyme analysis, (*ii*) restriction fragment length polymorphisms (RFLPs), (*iii*) random amplified polymorphic DNA analysis (RAPDs), and (*iv*) mini and microsatellite DNA polymorphisms. Among these approaches, electrophoretic analysis of gene–enzyme systems, widely used since the late 1960s, has so far yielded most data allowing, *inter alia*: analysis of the patterns of genetic variation in populations and species, and relating these to different life history and ecological variables; estimates of the amount of genetic divergence between conspecific and heterospecific populations; detection of sibling species (that is biological species virtually identical at the morphological level) in both sexes and at any life stage; quantification of gene flow within species; detection of reproductive isolation between both sympatric and allopatric populations; analysis of hybridization phenomena, by recognizing F_1 hybrids, backcrosses, recombinants, and introgressed genotypes; and evaluation, by means of an alternative method, of phylogenetic relationships between taxa.

Some of the main results obtained using multilocus electrophoresis in anisakid nematodes are reviewed and some perspectives, both at theoretical and practical levels, are examined.

DETECTION OF SIBLING SPECIES AND SPECIATION OF PARASITE–HOST SYSTEMS

The most important application of multilocus electrophoresis to parasite taxonomy concerns the detection of biological species, especially when speciation processes have occurred without morphological differentiation (*sibling species*), a case particularly frequent in endoparasites (Bullini 1985). Sibling species are morphologically very similar (or even identical), but they are differentiated at the genetic, ecological and behavioural level, and are reproductively isolated in the field. Speciation without morphological differentiation covers different cases, often unrelated (Mayr 1970), such as: (*i*) recent completion of reproductive isolation, with not enough time for morphological differentiation to be attained; (*ii*) pre-mating reproductive isolation mechanisms not involving visual cues (which are easily recognized by humans) but by acoustical or chemical signals; the last ones are very important in endoparasites; (*iii*) selection against phenotypic changes, as the realized phenotype represents an optimum in certain environmental conditions, such as endoparasitic life in a stable environment (e.g. the digestive tract in a homeothermic host); and (*iv*) species whose taxonomy is based on a small number of characters, often reflecting phenomena of adaptive convergence or parallelism (Bullini 1983, 1985).

Table 6.1. *Values of intraspecific gene flow estimated from the standardized variance of allele frequencies,* F_{ST} *(Crow and Aoki, 1984; Slatkin & Barton, 1989);* m *is the fraction of immigrant individuals in a population of effective size* N

Species	Gene flow (*Nm*)
P. decipiens A	4.38
P. decipiens B	3.66
P. decipiens C	9.36
C. osculatum A	5.20
C. osculatum B	4.90
C. osculatum C	3.90
C. osculatum D	4.60
C. osculatum E	6.10
A. simplex	5.80
A. pegreffii	6.30

In the case of sympatric or partially sympatric sibling species, a significant deficiency or complete lack of some heterozygote classes at polymorphic loci strongly suggests that the sample being dealt with comprises distinct gene pools. Often, 'diagnostic' loci occur, with alternative allozymes in the different sympatric forms (Bullini & Sbordoni 1980). Also in the case of allopatric populations, multilocus electrophoresis may provide reliable information on specific status. As the most common mode of speciation in animals, including endoparasites, has been shown to be the allopatric one (and in particular the 'peripatric' model, Mayr, 1963, 1970, see below), the build up of genetic divergence is generally a better predictor than conventional morphology of whether two allopatric populations will interbreed upon recontact. At least in the case of anisakid nematodes, when Nei's (1972) genetic distance between populations reaches values of about 0.2, gene exchange is interrupted by intrinsic reproductive isolating mechanisms (RIMs). On the other hand, in anisakid nematodes conspecific populations generally show similar allele frequencies even when located thousands of kilometres apart; accordingly, their values of D are quite low (0.001–0.02). For example, $D \sim 0.005$ was found between *Anisakis pegreffii* from the Mediterranean Sea and the Falkland Islands. Isozyme data indicate that genetic variation of anisakids is generally not structured in geographical races and subspecies (the rule in polytypic species), and high levels of interpopulation gene flow, indirectly estimated from allele frequencies, were detected in these ascaridoids (Table 6.1), which therefore are mostly monotypic species. Such high gene flow is apparently allowed by the high vagility of a number of hosts (e.g. squid, fish, seals, dolphins, aquatic birds).

Some controversial taxa recorded in this group have proved to be, when genetically tested, either synonyms or good species; the latter is the case in the Antarctic *Contracaecum radiatum* (often included within *C. osculatum* s.l.), the reproductive isolation of which has been demonstrated in the field (Arduino *et al.* 1994).

By the use of genetic markers, it has been shown that many anisakid morphospecies, considered cosmopolitan and euriecious, include a number of sibling species, reproductively isolated and differentiated both genetically and ecologically, for instance with respect to geographic distribution and host preferences. This is, for example, the case in *Contracaecum osculatum* (Bullini *et al.* 1992, Nascetti *et al.* 1984, 1990, 1993, Orecchia *et al.* 1994), *Pseudoterranova decipiens* (Orecchia *et al.* 1988, Paggi *et al.* 1985, 1988, 1990, 1991, D'Amelio *et al.* 1992) and *Anisakis simplex* (D'Amelio *et al.* 1993, Mattiucci *et al.* 1986, 1990, Nascetti *et al.* 1981, 1983, 1986 and unpublished). *C. osculatum* (Rudolphi 1802), which has been reported from at least 16 pinniped species, principally phocids, from both the southern and northern hemisphere (Delyamure 1955, Fagerholm & Gibson 1987), has been shown to include three Arctic-Boreal species (provisionally designated *C. osculatum* A, B, C) and two Antarctic ones (*C. osculatum* D and E), easy to recognize by diagnostic allozymic keys.

Five biological species were detected also within the morphospecies *P. decipiens* Krabbe 1878, a parasite recorded from various definitive hosts, pinnipeds and cetaceans, from the Arctic-Boreal region (Scandinavia, North America, Hudson Bay, Alaska), the Atlantic and Pacific Oceans and the Antarctic Ocean. Three members of the *P. decipiens* complex have so far been detected in the Arctic-Boreal region (*P. decipiens* A, B and C), one from Japan (*P. decipiens* D) and one from the Antarctic (*P. decipiens* E).

In addition, *Anisakis simplex,* the causal agent as a larva of human anisakiasis, which in the adult stage occurs in the stomach and intestine of various cetacean hosts, was found to include at least five sibling species: *A. simplex* s.s. (from the North East and North West Atlantic, and from Pacific Canada), *A. pegreffii* (from the Mediterranean Sea and Falkland Islands), *A. simplex* C (from Pacific Canada), *A. simplex* D (from the Falkland Islands), and *A. simplex* E (from the Tasman Sea).

Genetic relationships and times of evolutionary divergence estimated within the three complexes are shown in the UPGMA dendrograms in Fig. 6.1. Striking parallelisms are apparent, possibly owing to cospeciation phenomena. The *C. osculatum* s.l., *P. decipiens* and *A. simplex* complexes achieved a bipolar distribution in different times, from 5–6 to about 1 million years ago, through distinct colonizations of the Antarctic region (Table 6.2). The more ancient dating (*C. radiatum*, *P. decipiens* E) coincides with that of the estimated first colonization of the Antarctic by seals; the more recent one (*C. osculatum* E) occurred during the Pleistocene (Fig. 6.1, Table 6.2). These data support the hypothesis that a number of taxa from the Arctic-Boreal region have reached the Southern Ocean (or vice versa) in the comparatively recent past (Bullini *et al.* 1994, and unpublished).

Table 6.2. *Different colonizations of the Antarctic region by ascaridoid nematodes of the* Contracaecum osculatum *s.l.,* Pseudoterranova decipiens *and* Anisakis simplex *complexes, estimated on the basis of average genetic distance values of Antarctic members from their northern relatives*

Average Nei's *D*	Million years	Epoch	Antarctic species
0.9–1	5–6	Upper Miocene/ Lower Pliocene	*P. decipiens* E *C. radiatum*
0.3	1.5–2	Pleistocene	*C. osculatum* D *A. simplex* D
0.2	1	Pleistocene	*C. osculatum* E

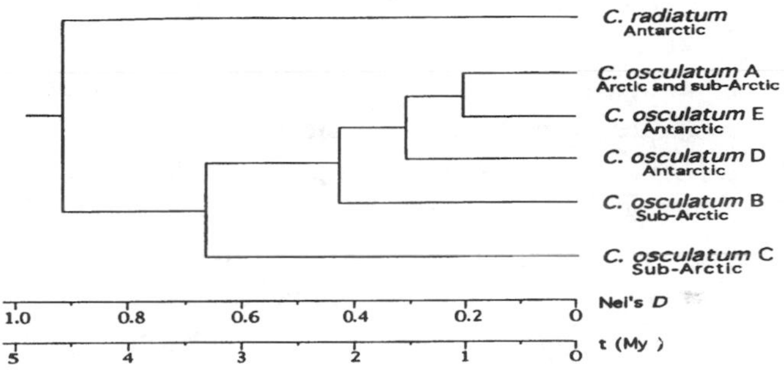

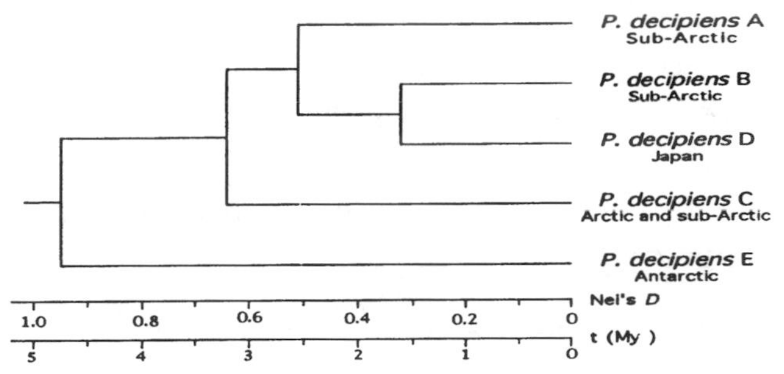

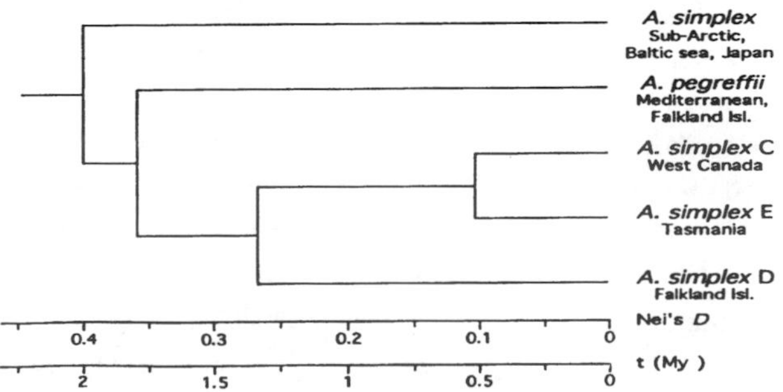

Fig. 6.1. Genetic relationships and times of evolutionary divergence among the members of the *Contracaecum osculatum–radiatum*, *Pseudoterranova decipiens* and *Anisakis simplex* complexes.

As to the mode of speciation of these parasites, the favourite model for many parasitologists was a sympatric one, involving the parasite's shift to a new host and its subsequent coadaptation, up to the rise of a new species. However, this hypothesis does not take into account that in sympatric conditions even limited levels of gene flow would homogenize population gene pools in distinct hosts, either preventing or disrupting ongoing adaptive processes.

A mode of speciation which apparently fits better for endoparasites is the *peripatric model* proposed by Mayr (1963, 1970). This involves geographical isolation of small populations, the genetic structure of which begins to differ from that of the parental population since the beginning of the process, by genetic drift phenomena. In the case of the *C. osculatum-radiatum*, *P. decipiens* and *A. simplex* complexes, genetic data strongly suggest that adaptation to different hosts and speciation is strictly related to geographic isolation of the hosts. Such processes apparently occurred in different times, from lower Miocene to Pliocene–Pleistocene, when extreme climatic variations took place. During glacial maxima (periods of highest reduction of sea level) lower-sized populations of hosts and their parasites would have remained isolated in marine refuges, promoting genetic divergence and coadaptation; during interglacial periods, geographic ranges might expand, favouring host shift. Similar coevolutionary processes have been proposed by Hoberg (1992), on a morphological basis, for other host–parasite interactions, involving holarctic cestodes (dilepidids and tetrabothriids) and their definitive hosts (fish-eating birds and pinnipeds).

GENETIC VARIATION AND ENVIRONMENTAL HETEROGENEITY

Allozyme studies carried out on 17 species of ascaridoid nematodes having different life-cycles have shown that their genetic variability is positively related to the degree of environmental heterogeneity they experience (Bullini *et al.*, 1986). It was found that species whose life-cycle is carried out in homeothermic hosts (genera *Parascaris*, *Ascaris*, *Baylisascaris*, *Toxascaris*, *Toxocara*) show a significantly lower genetic variability (average expected mean heterozygosity, H_e=0.04) than do those needing both poikilothermic and homeothermic hosts (genera *Anisakis*, *Pseudoterranova*, *Contracaecum*, *Phocascaris*) (average H_e= 0.15). A major role of natural selection in this phenomenon was suggested (Bullini *et al.* 1986). These and other data on helminth genetic variability were critically reviewed by Nadler (1990). The study of ten more ascaridoid species (several of which were from the Antarctic and sub-Antarctic regions) has confirmed the general trend observed: mean H_e=0.05 in the single-host group, versus 0.16 in the multiple-host group (Bullini *et al.* 1991). The overall picture confirms a major role of natural selection in determining different levels of genetic variability in these parasites. Among the various hypotheses proposed, our data appear to be in good agreement with the 'niche-width variation' hypothesis (Van Valen 1965), according to which great genetic variability is an adaptive strategy conferring higher fitness in spatio-temporally variable conditions. Interestingly, the average values of variability found in ascaridoid worms parasitizing only homeothermic vertebrates is similar to that of their hosts (H_e=0.04, from Nevo *et al.* 1984). On the other hand, the variability of ascaridoids parasitizing both poikilothermic and homeothermic species is similar to that found in their poikilothermic hosts (H_e=0.12, from Nevo *et al.* 1984).

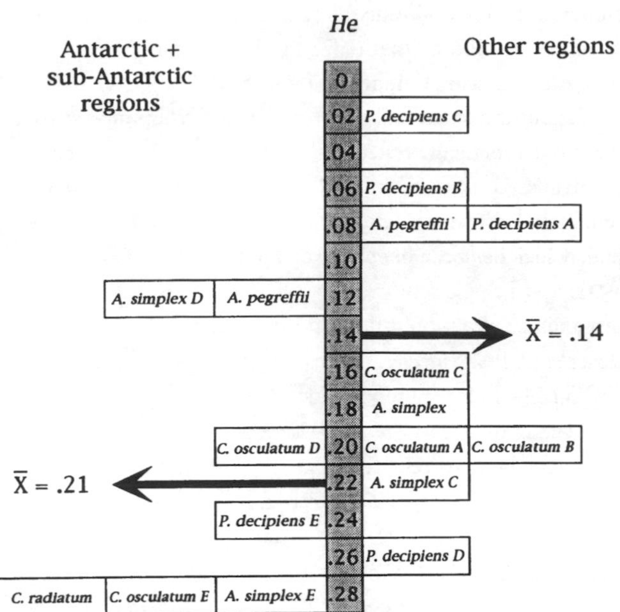

Fig. 6.2. Distribution of expected mean heterozygosity per locus (H_e) among ascaridoid species of the *Anisakis simplex*, *Contracaecum osculatum* s.l. and *Pseudoterranova decipiens* complexes from different geographic areas.

Antarctic and sub-Antarctic members of the genera *Anisakis*, *Contracaecum* and *Pseudoterranova* show a significantly higher genetic variability (average H_e=0.21) than those from other regions (H_e=0.14, Fig. 6.2).The same pattern is observed when Antarctic and sub-Antarctic anisakid species are compared only to the Arctic and sub-Arctic members of the three complexes. Accordingly, the observed values cannot be related to the extreme latitudes, a parameter often considered as relevant (cf. Nevo *et al.* 1984). The significantly higher genetic variability found in Antarctic and sub-Antarctic anisakid species can be related to a lower degree of habitat disturbance (e.g. by fishing, hunting and pollution) in the Antarctic region, allowing species to reach higher population sizes, with a reduced probability of genetic drift phenomena. Accordingly, much higher intensities of seal infection by *Contracaecum* species have been found in Antarctic waters (up to a hundred thousand individuals per host) than in the northern hemisphere (from several hundred to a few thousand individuals per host) (Klöser *et al.* 1992).

ECOLOGICAL NICHE AND COMPETITION

In both the Arctic-Boreal and Antarctic regions, the various members of the *C. osculatum*, *P. decipiens* and *A. simplex* complexes showed differences in host preferences. These can be due both to differential host–parasite coadaptation and coevolution, and to interspecific competition, which acts either by reducing the range of potential hosts, or by promoting a differential use of resources from a single individual host in sympatric conditions. For instance, Antarctic *C. radiatum* and *C. osculatum* D+E (which share the same definitive host, the Weddell seal *Leptonychotes weddellii*) appear to be adapted to benthic and pelagic food webs respectively; competition in the intermediate hosts would thus be avoided by spatial separation (Klöser *et al.* 1992).

Another example of niche subdivision was found between *P. decipiens* A and B as to their definitive hosts. Both species parasitize the common seal, *Phoca vitulina* and the grey seal, *Halichoerus grypus*. In the northeast Atlantic, *P. decipiens* A outnumbers *P. decipiens* B by about tenfold in the grey seal, while the opposite is true for the common seal, where *P. decipiens* B prevails by about tenfold. This could reflect differential host adaptation of the two parasites, which evolved by allopatry. However, in Canadian Atlantic waters, where *P. decipiens* A is lacking, *P. decipiens* B is found to parasitize equally both grey and common seals. Niche subdivision found in the northeast Atlantic appears therefore a consequence of the sympatric occurrence of the two *Pseudoterranova* species, due to secondary contact, which has promoted character displacement (Paggi *et al.* 1991).

Niche subdivision may also occur within the same host individual. For example, when members of the *P. decipiens* and *C. osculatum* complexes infect the same seal, a differential use of resources takes place, as the former invades the seal intestine if the stomach (the elective target for both) is highly infected by *Contracaecum* (Berland 1964, McClelland 1980, Klöser *et al.* 1992). *C. osculatum* s.l. appears to be the better competitor also when co-occurring with *Phocascaris phocae* and/or *Ph. cystophorae* in the same seal. Also in this case, *C. osculatum* is found almost exclusively in the stomach, while *Phocascaris* prevails in the upper section of the intestine; when *Phocascaris* is alone, it mostly occupies the stomach (Nascetti 1993 and unpublished). These phenomena of niche subdivision, evolving by character displacement, reduce the level of interspecific competition for trophic and spatial resources.

CONCLUDING REMARKS

It is expected that further studies on host-parasite systems using genetic markers will provide relevant data on their systematics, ecology and evolution, allowing inter alia:

- the use of anisakid and other endoparasites as markers to track and understand marine food webs, which are still poorly known especially in the Antarctic and sub-Antarctic regions;
- the evaluation by biodiversity estimates (at the levels of both species richness and intraspecific genetic variation) of the degree and impact of habitat disturbance in different geographic areas, with particular regard to the Southern Ocean;
- light to be thrown on the evolutionary relationships of Antarctic and sub-Antarctic taxa one with another and with relatives from northern latitudes, as well as on their patterns of speciation and cospeciation.

REFERENCES

Arduino, P., Nascetti, G., Cianchi, R., Plötz, J., Mattiucci, S., D'Amelio, S., Paggi, L., Orecchia, P. & Bullini, L. 1995. Isozyme variation and taxonomic rank of *Contracaecum radiatum* (v. Linstow, 1907) from the Antarctic Ocean (Nematoda, Ascaridoidea). Systematic Parasitology, 30, 1–9.

Berland, B. 1964. *Phocascaris cystophorae* sp. nov. (Nematoda) from the hooded seal, with an emendation of the genus. *Årbok*, **17**, 1–21.

Bullini, L. 1983. L'approccio elettroforetico all'identificazione delle specie gemelle negli insetti: risultati, prospettive e limiti. *In Proceedings 13th National Italian Congress Entomology, Sestriere, Italy*, 461–467.

Bullini, L. 1985. The electrophoretic approach to the study of parasites and vectors. *Parassitologia*, **27**, 1–11.

Bullini, L. & Sbordoni, V. 1980. Electrophoretic studies of gene-enzyme systems: microevolutionary processes and phylogenetic inference. *Bollettino di Zoologia*, **47**, 95–112.

Bullini, L., Nascetti, G., Paggi, L., Orecchia, P., Mattiucci, S. & Berland, B. 1986. Genetic variation of ascaridoid worms with different life cycles. *Evolution*, **40**, 437–440.

Bullini, L., Cianchi, R., Nascetti, G., D'Amelio, S., Mattiucci, S., Orecchia, P. & Paggi, L. 1991. Studies on the genetic structure of endoparasite nematodes and their hosts in the Antarctic region: adaptive processes and mechanims of genetic divergence and speciation. *Proceedings 1st Meeting Biology in Antarctica 1989, Roma*. Padova: Edizioni Universitarie Patavine, 177–198.

Bullini, L., Nascetti, G., Cianchi, R., Paggi, L., Mattiucci, S., D'Amelio, S., Orecchia, P., Brattey, J., Berland, B., Smith, J. W. & Plötz, J. 1992. Biochemical taxonomy of the *Contracaecum osculatum* complex (Nematoda, Ascaridoidea). In *Proceedings 6th European Multicolloquium Parasitology, The Hague, the Netherlands*, **95**.

Bullini, L., Arduino, P., Cianchi, R., Nascetti, G., D'Amelio, S., Mattiucci, S., Paggi, L. & Orecchia, P. 1994. Genetic and ecological studies on nematode endoparasites of the genera *Contracaecum* and *Pseudoterranova* in the Antarctic and Arctic-Boreal regions. *Proceedings 2nd Meeting Biology in Antarctica, Padova, 1992*. Padova: Aldo Martello, Edizioni Universitarie Patavine, 131–146.

Crow, J. F. & Aoki, K. 1984. Group selection for a polygenic behavioural trait: estimating the degree of population subdivision. *Proceedings of the National Academy of Sciences of the United States of America*, **81**, 8512–8514.

D'Amelio, S., Mattiucci, S., Paggi, L., Orecchia, P., Nascetti, G., Cianchi, R., Sato, N., Kikuchi, K., Ishikura, H. & Bullini, L. 1992. Genetic variation of a population of the *Pseudoterranova decipiens* complex from Japan. *Proceedings 6th European Multicolloquium Parasitology, The Hague, The Netherlands*, **45**.

D'Amelio, S., Mattiucci, S., Paggi, L., Orecchia, P., Cianchi, R., Nascetti, G., Arduino, P. & Bullini, L. 1993. New data on the *Anisakis simplex* complex (Nematoda, Ascaridida, Ascaridoidea), parasite of cetaceans. *European Cetacean Society Annual Conference, Cetacean Social Organization, Inverness, Scotland*, **42**.

Delyamure, S. L. 1955. Helminthofauna of marine mammals (ecology and phylogeny). Jerusalem: Israel Program for Scientific Translation.

Fagerholm, H. P. & Gibson, D. I. 1987. A redescription of the pinniped parasite *Contracaecum ogmorhini* (Nematoda, Ascaridoidea), with an assessment of its antiboreal circumpolar distribution. *Zoologica Scripta*, **18**, 33–41.

Hoberg, E.P. 1992. Congruent and synchronic patterns in biogeography and speciation among seabirds, pinnipeds, and cestodes. *Journal of Parasitology*, **78**, 601–615.

Klöser, H., Plötz, J., Palm, H., Bartsch, A. & Hubold, G. 1992. Adjustment of anisakid nematode life cycles to the high antarctic food web as shown by *Contracaecum radiatum* and *C. osculatum* in the Weddell Sea. *Antarctic Science*, **4**, 171–178.

Mattiucci, S., Nascetti, G., Bullini, L., Orecchia, P. & Paggi, L. 1986. Genetic structure of *Anisakis physeteris*, and its differentiation from the *Anisakis simplex* complex (Ascaridida: Anisakidae). *Parasitology*, **93**, 383–387.

Mattiucci, S., Nascetti, G., D'Amelio, S., Cianchi, R., Orecchia, P., Paggi, L., Berland, B. & Bullini, L. 1990. Nuovi dati sulla divergenza genetica tra nematodi dei generi *Contracaecum* e *Phocascaris* (Ascaridida: Anisakidae). *Parassitologia*, **32** (suppl. 1), 181.

Mayr, E. 1963. *Animal species and evolution*. Cambridge, Massachusetts: Belknap Press.

Mayr, E. 1970. *Populations, species and evolution*. Cambridge, Massachusetts: Belknap Press.

McClelland, G. 1980. *Phocanema decipiens*: pathology in seals. *Experimental Parasitology*, **49**, 405–419.

Nadler, S. A. 1990. Molecular approaches to studying helminth population genetics and phylogeny. *International Journal for Parasitology*, **20**, 11–29.

Nascetti, G. 1993. Recognition of sibling species in endoparasitic nematodes: data on genetic variation, reproductive isolation, ecological niche and competition. In Marchetti, R. & Cotta Ramusino, M., eds. *Ecologia. Proceedings 5th National Congress Italian Society Ecology, Milano, Italy 1992*, S.It.E./ Atti 15, 287–312.

Nascetti, G., Paggi, L., Orecchia, P., Mattiucci, S. & Bullini, L. 1981. Divergenza genetica in popolazioni del genere *Anisakis* del Mediterraneo. *Parassitologia*, **23**, 208–210.

Nascetti, G., Paggi., L., Orecchia, P., Mattiucci, S. & Bullini, L. 1983 . Two sibling species within *Anisakis simplex* (Ascaridida: Anisakidae). *Parassitologia*, **25**, 239–241.

Nascetti, G., Berland, B., Bullini, L., Mattiucci, S., Orecchia, P. & Paggi, L. 1984. Due specie gemelle in *Contracaecum osculatum* (Ascaridida, Anisakidae): isolamento riproduttivo e caratteri diagnostici a livello elettroforetico. *Annali Istituto Superiore di Sanita'*, **22**, 349–352.

Nascetti, G., Paggi, L., Orecchia, P., Smith, J. W., Mattiucci, S. & Bullini, L. 1986. Electrophoretic studies on the *Anisakis simplex* complex (Ascaridida: Anisakidae) from the Mediterranean and North-East Atlantic. *International Journal for Parasitology*, **16**, 633–640.

Nascetti, G., Cianchi, R., Mattiucci, S., D'Amelio, S., Orecchia, P., Paggi, L., Berland, B. & Bullini, L. 1990. Struttura genetica di specie antartiche del genere *Contracaecum* (Ascaridida: Anisakidae). *Parassitologia*, **32**, 187.

Nascetti, G., Cianchi, R., Mattiucci, S., D'Amelio, S., Orecchia, P., Paggi, L., Brattey, J., Berland, B., Smith, J. W. & Bullini, L. 1993. Three sibling species within *Contracaecum osculatum* (Nematoda, Ascaridida, Ascaridoidea) from the Atlantic Arctic-Boreal region: reproductive isolation and host preferences. *International Journal for Parasitology*, **23**, 105–120.

Nei, M. 1972. Genetic distance between populations. *American Naturalist*, **106**, 283-292.

Nevo, E., Beiles, A. & Ben Schlomo, R. 1984. The evolutionary significance of genetic diversity: ecological, demographic and life-history correlates. In Mani, G.S., ed. *Evolutionary dynamics of genetic diversity*. *Lecture notes in Biomathematics*, **53**, 13–213.

Orecchia, P., Berland, B., Bullini, L., D'Amelio, S., Mattiucci, S., Nascetti, G. & Paggi, L. 1988. Studi di tassonomia biochimica su *Pseudoterranova decipiens* Krabbe, 1878 (Ascaridida: Anisakidae). *Parassitologia*, **30**, 124–125.

Orecchia, P., Mattiucci, S., D'Amelio, S., Paggi, L., Plötz, J., Cianchi, R., Nascetti, G., Arduino, P. & Bullini, L. 1994. Two new members in the *Contracaecum osculatum* complex (Nematoda, Ascaridoidea) from the Antarctic. *International Journal for Parasitology*, **24**, 367–377.

Paggi, L., Nascetti, G., Orecchia, P., Mattiucci, S. & Bullini, L. 1985. Biochemical taxonomy of ascaridoid nematodes. *Parassitologia*, **27**, 105–112.

Paggi, L., Berland, B., Bullini, L., D'Amelio, S., Mattiucci, S., Nascetti, G. & Orecchia, P. 1988. Biochemical taxonomy of *Pseudoterranova decipiens* Krabbe, 1878 (Ascaridida: Anisakidae). *Proceedings 5th European Multicolloquium Parasitology, Budapest, Hungary, 1988*. Budapest: Statistical Publishing House, 100.

Paggi, L., Nascetti, G., Cianchi, R., Orecchia, P., Mattiucci, S., D'Amelio, S., Berland, B., Brattey, J., Smith, J. W. & Bullini, L. 1990. Further studies on sealworms of the *Pseudoterranova decipiens* complex. *Bulletin Societé Francaise Parasitologie*, **8** (suppl.), 721.

Paggi, L., Nascetti, G., Cianchi, R., Orecchia, P., Mattiucci, S., D'Amelio, S., Berland, B., Brattey, J., Smith, J. W. & Bullini, L. 1991. Genetic evidence for three species within *Pseudoterranova decipiens* (Nematoda, Ascaridida, Ascaridoidea) in the North Atlantic and Norwegian and Barents seas. *International Journal for Parasitology*, **21**, 195–212.

Slatkin, M. & Barton, N. H. 1989. A comparison of three indirect methods for estimating average levels of gene flow. *Evolution*, **43**, 1349–1368.

Van Valen, L. 1965. Morphological variation and the width of the ecological niche. *American Naturalist*, **99**, 377–389.

7 Molecular phylogeny and evolution of Notothenioid fish based on partial sequences of 12S and 16S ribosomal RNA mitochondrial genes

LUCA BARGELLONI[1], TOMASO PATARNELLO[1], PETER A. RITCHIE[2,3], BRUNO BATTAGLIA[1] AND AXEL MEYER[2]

[1]Department of Biology, University of Padova, 35121 Via Trieste, 75 Padova, Italy, [2]Department of Ecology and Evolution, State University of New York, Stony Brook, New York 11794, USA, [3]Ecology and Evolution, School of Biological Sciences, University of Auckland, Private Bag 92019, Auckland, New Zealand.

ABSTRACT

The perciform suborder Notothenioidei is a highly diversified group of fish inhabiting the Southern Ocean. In terms of numbers of species and biomass, it represents an important element in the Antarctic marine environment. The present study investigates the phylogenetic relationships of 18 species of notothenioids from five families (Bovichtidae, Nototheniidae, Artedidraconidae, Bathydraconidae and Channichthyidae). Two Antarctic species from the family Zoarcidae (Perciformes) were included in the analysis as an outgroup. Phylogenetic analyses were based on partial sequences of the 12S and 16S mitochondrial ribosomal RNA genes. A total of 928 base pairs (bp) were sequenced for each of the 20 taxa.

Both distance and maximum parsimony-based methods were used to infer the phylogenetic relationships among and within families of notothenioids. The topology of the trees obtained from the molecular data did not significantly differ from that proposed on the basis of morphological data. The Bovichtidae appear to be the sister group to all other notothenioid families. The DNA analysis suggests that the families Nototheniidae and Bathydraconidae are paraphyletic. The short branch lengths displayed by the neighbour-joining tree might account for a radiation-like mode of evolution. The time of divergence among notothenioids was estimated on the basis of their nucleotide divergence.

Key words: Notothenioid evolution, molecular phylogeny, 12S and 16S rRNA, mitochondrial DNA.

INTRODUCTION

The perciform suborder Notothenioidei is a highly diversified group of fish inhabiting the waters around the Antarctic continent. In terms of numbers of species and biomass it represents the dominant element of the Antarctic fish fauna (Gon & Heemstra 1990, Eastman 1991, Miller 1993). Notothenioids show a remarkable degree of endemism (97% of the species), and are distributed nearly exclusively in the Antarctic marine waters (Eastman 1993, p. 55). This is a unique environment, characterized, especially in the high-Antarctic Zone, by temperatures as low as −2 °C, presence of sea ice, and large seasonal fluctuations of the primary production. The presence of a circum-Antarctic current, the Antarctic Convergence, reduces the exchange of surface water, thus partially isolating the Antarctic Ocean. These features, together with a decreased availability of shelf habitats, seem to have reduced the opportunity for the fish fauna to diversify and colonize different habitats. Notothenioids have evolved several, remarkable, adaptations to the Antarctic conditions, enabling them to occupy successfully a large variety of ecological niches (Kock 1993, Eastman 1993, p.67). Among the physiological and molecular adaptations displayed by notothenioid fish, the most striking is certainly the presence of antifreeze glycopeptides (AFGPs), which are responsible for depressing the freezing point of the blood, thus ensuring that the fish do not freeze under Antarctic conditions (DeVries 1988, Cheng & DeVries 1991).

Also other relevant physiological features, such as blood viscosity, oxygen transport and the cardiovascular system, differ significantly in these fish from those in temperate species (Wells 1987, Macdonald & Wells 1991, Tota *et al.* 1991). In some Antarctic species haemoglobins have reduced affinity for oxygen; the extreme case is represented by the icefish (family Channichthyidae), which lack haemoglobin (Wells *et al.* 1990, di Prisco *et al.* 1990, 1991) and which have therefore been called 'white-blooded fish'. This major change in the blood biochemistry of Channichthyidae is allowed by the oxygen saturation of the Antarctic sea water, but whether it represents an adaptation or it is simply due to relaxed selection is difficult to say. On the contrary, neutral buoyancy, which is typical of some pelagic nototheniods (i.e. *Dissostichus* spp.), evolved from an ancestral benthic condition characterized by the absence of a swim bladder; this can probably be considered an adaptation to the pelagic niche (Eastman 1985).

The suborder Notothenioidei consists of six families (Bovichtidae, Nototheniidae, Artedidraconidae, Harpagiferidae, Bathydraconidae and Channichthyidae) with over 120 species (Eastman 1993, p. 59). The phylogenetic relationships among and within these families have been established, so far, solely on morphological characters (Iwami 1985). However, the complete lack of fossil records for these fish has made definition of their time and mode of evolution very difficult.

In the present study we report the molecular phylogeny of notothenioids based on the partial sequence of 12S and 16S rRNA mitochondrial genes (Bargelloni *et al.* 1994). This molecular approach provides an alternative hypothesis for the evolution of these Antarctic fish as well allowing comparison with phylogenies based on other characters. In addition, the DNA data permit (if the homogeneity of the nucleotide substitution rate is proven to hold by specific tests) timing of the evolutionary events of notothenioids, therefore providing insights into the mode of evolution, possibly related to paleoenvironmental events in Antarctica.

MATERIALS AND METHODS

Total genomic DNA was extracted from ethanol-preserved samples, as described elsewhere (Kocher *et al.* 1989). The species examined were the following: suborder Notothenioidei (family Bovichtidae) *Bovichtus variegatus* (family Nototheniidae) *Trematomus eulepidotus, T. hansoni, T. nicolai, T. pennellii, Pagothenia borchgrevinki, Notothenia coriiceps neglecta, Gobionotothen gibberifrons, Dissostichus mawsoni*; (family Artedidraconidae) *Histiodraco velifer, Pogonophryne scotti*; (family Bathydraconidae) *Gymnodraco acuticeps, Cygnodraco mawsoni, Parachaenichtys charcoti*; (family Channichthyidae) *Chaenocephalus aceratus, Chionodraco hamatus, Cryodraco antarcticus, Pagetopsis macropterus*. Two Antarctic species of Zoarcidae (Perciformes), *Lycodichthys dearborni* and *Pachycara brachycephalum*, were included in the study as outgroup.

The polymerase chain reaction (PCR) (Saiki *et al.*, 1988) was employed to amplify two segments of the mitochondrial DNA (mtDNA), from the 16S and 12S ribosomal RNA genes respectively. Double- and single-strand amplifications and direct sequencing were performed according to Patarnello *et al.* (1994). The PCR-primers used were 12Sa and 12Sb (Kocher *et al.* 1989) and 16Sa and 16Sb (Palumbi *et al.* 1991). Both strands of the amplified segments were sequenced; 375 base pairs (bp) of the 12S gene and 553 bp of the 16S gene were examined for each of the 18 taxa plus the two outgroups (GenBank Accession numbers Z32702-Z32739 and Z32747-Z32748). DNA sequences were aligned using a multiple-sequence alignment software (CLUSTAL: Higgins & Sharp 1988), with default settings. The alignment was unambiguous with the exception of a highly variable region of 45 bp in the 16S sequence (positions 223–267 in the dataset). The divergence times estimate among taxa was carried out both including and excluding this highly variable region of the 16S rRNA.

Phylogenetic analyses were performed by maximum parsimony (MP) implemented in PAUP (Swofford 1993). Due to the large number of taxa, heuristic search procedures were necessary to search for the most parsimonious tree(s). The reliability of the heuristic searches was improved by using the option 'random addition of taxa' with 100 replications in PAUP. Several different character-weighting schemes were used (Bargelloni *et al.* 1994).

Neighbour-joining analyses (NJ) (Saitou & Nei 1987) were carried out using the software package MEGA (Kumar *et al.* 1993). Different methods were used to estimate evolutionary distances accounting for multiple substitutions: Jukes–Cantor (Jukes & Cantor 1969), Kimura's two-parameter model (Kimura 1980), and Tamura–Nei's method (Tamura 1992).

Statistical confidence of MP and NJ evolutionary trees was assessed using bootstrapping (Felsenstein 1985), with 400 replications each.

The homogeneity of the rate was estimated using the relative rate test (Wilson *et al.* 1977); this approach tests whether the nucleotide substitution rates differ in two lineages using a third, 'reference' taxon; the advantage of such a test is that the knowledge of the divergence times between lineages is not required. The number of transversional substitutions per site was estimated using Kimura's formulas (Kimura 1980); the values calculated for each pairwise comparisons are reported in Table 7.1. Variances and covariances were determined according

Table 7.1. *Distance matrix. Genetic distances for pairwise comparisons of notothenioid mtDNA sequences. Percentage divergences based on all mutations are reported above the diagonal and those based on trasversions alone are presented below the diagonal. All distances were calculated using Kimura's two-parameter model (1980)*

	1	2	3	4	5	6	7	8	9	10	11	12	13	14	15	16	17	18
1 *Bovicthus variegatus*		17.64	18.25	17.80	18.52	17.48	16.85	17.44	18.07	16.57	16.57	16.52	18.05	17.91	16.70	16.39	16.55	16.69
2 *Trematomus eulepidotus*	4.96		0.93	0.46	1.16	0.93	3.54	4.04	4.77	4.28	4.04	4.28	5.90	5.52	4.39	4.52	4.27	4.89
3 *Trematomus hansoni*	5.09	0.11		1.16	1.39	1.39	3.90	4.16	4.65	4.78	4.53	4.52	6.02	5.77	4.63	4.76	4.51	5.13
4 *Trematomus nicolai*	4.96	0.00	0.11		1.16	0.93	3.54	3.79	4.52	4.28	4.04	4.03	5.90	5.52	4.51	4.64	4.39	4.76
5 *Trematomus pennellii*	5.47	0.46	0.35	0.46		1.39	3.90	4.52	5.01	4.89	4.65	4.64	5.76	5.88	5.00	5.13	4.88	5.50
6 *Pagothenia borchgrevinki*	4.96	0.23	0.35	0.23	0.69		3.53	3.78	4.27	4.28	4.03	4.02	5.63	5.39	4.63	4.26	4.51	4.88
7 *Notothenia coriiceps neglecta*	5.09	1.04	1.16	1.04	1.52	1.28		1.98	2.82	2.34	2.10	2.10	3.90	3.78	2.93	2.81	2.81	3.05
8 *Gobionotothen gibberifrons*	5.35	0.81	0.93	0.81	1.28	1.04	0.69		2.22	2.58	2.34	2.22	4.52	4.40	3.05	2.93	2.93	3.30
9 *Dissostichus mawsoni*	5.35	1.04	1.16	1.04	1.52	1.28	0.69	0.46		3.18	2.94	2.82	4.40	4.78	3.54	3.30	3.54	3.91
10 *Histiodraco velifer*	4.83	0.81	0.93	0.81	1.28	1.04	0.69	0.69	0.69		0.46	1.63	3.30	3.42	1.98	2.10	2.10	2.45
11 *Pogonophryne scotti*	4.96	0.69	0.81	0.69	1.16	0.93	0.58	0.58	0.58	0.11		1.39	2.93	3.30	1.98	1.86	1.86	2.22
12 *Gymnodraco acuticeps*	5.35	1.04	1.16	1.04	1.52	1.28	0.69	0.69	0.69	0.69	0.58		2.94	3.18	1.51	1.39	1.39	1.74
13 *Cygnodraco mawsoni*	5.47	1.04	1.52	1.40	1.63	1.63	1.28	1.04	1.04	1.04	0.93	0.81		3.30	3.53	2.93	3.42	3.78
14 *Parachaenicthys charcoti*	5.35	1.28	1.40	1.28	1.75	1.52	1.16	0.93	0.93	0.93	0.81	0.69	0.81		3.66	3.06	3.42	3.79
15 *Chaenocephalus aceratus*	5.09	1.52	1.63	1.52	1.99	1.75	1.40	1.16	1.16	0.93	1.04	0.93	1.28	0.93		0.81	0.35	1.16
16 *Chionodraco hamatus*	5.09	1.28	1.40	1.28	1.75	1.52	1.16	0.93	0.93	0.93	0.81	0.69	1.04	0.69	0.23		0.46	0.81
17 *Cryodraco antarticus*	5.09	1.28	1.40	1.28	1.75	1.52	1.16	0.93	0.93	0.93	0.81	0.69	1.04	0.69	0.23	0.00		0.81
18 *Pagetopsis macropterus*	5.22	1.40	1.52	1.40	1.87	1.63	1.28	1.04	1.04	1.04	0.93	0.81	1.16	0.81	0.35	0.11	0.11	

to Wu & Li (1985). Using *Bovichtus variegatus* as reference taxon, the homogeneity of the rate was statistically tested for all possible pairs of the remaining notothenioids; levels of significance were calculated using a standardized normal test (t test), with infinite degrees of freedom. Divergence times between taxa were calculated as the ratio of genetic distance (or mean genetic distance, when comparing clades with more than one taxon) and divergence rate.

RESULTS AND DISCUSSION

Phylogenetic evidence

Both parsimony and distance-based phylogenetic analyses yield an overall topology in good agreement with the phylogenetic pattern derived from cladistic analysis of morphological characters (Eastman 1993); the relationships among families inferred from molecular data are consistent with previous evidence. The family Bovichtidae appears to be the sister group to the other families of notothenioids, being highly divergent from them.

The taxonomical status of two families is questionable since both methods (PAUP and neighbour joining) indicate the families Nototheniidae and Bathydraconidae to be paraphyletic (Table 7.1, Fig. 7.1). Similarly, the subfamily Nototheniinae is an unnatural group because *G. gibberifrons* is more closely related to *D. mawsoni* than to *N. coriiceps* (Table 7.1, Fig. 7.1). This is not surprising, since nototheniid systematic relationships based on morphology were considered unsatisfactory, with revisions proposed recently (Balushkin 1990, 1991). Our molecular data support the splitting of the genus *Gymnodraco* from the rest of Bathydraconidae, previously proposed on the basis of morphological evidence by Hastings (1993, in Miller 1993). The neighbour-joining tree suggests that *Gymnodraco* is the sister group of the Channichthyidae (Fig. 7.1B). If this true, *Gymnodraco*, the only teleost with a single haemoglobin lacking the Bohr effect, should be regarded as an evolutionarily intermediate step toward the loss of haemoglobins in the Channichthyidae (Eastman 1993).

The short branch lengths at the inter-familial level displayed by the NJ tree (Fig. 7.1B), compared with the great distance separating *Bovichtus* from all the other notothenioids, suggest that rapid cladogenesis occurred long after the split of Bovichtidae, thus offering little time for mutations to accumulate.

Estimating time of divergence from DNA sequence data

Genetic distances between species can be used as a measure of time since their divergence; this is possible when the rate of genetic divergence is linear with time and homogeneous between taxa (molecular clock hypothesis). If the clock is valid, molecular data provide a valuable tool in timing evolutionary events.

However, even accepting the clock hypothesis, a problem still remains: calibrating the clock with absolute timing. This calibration can be accomplished when geological records for cladogenetic events are available; dates are usually provided by good fossil records of the taxa under investigation and/or by specific, well-established, geological events associated with separation of taxa (Caccone *et al.* 1994).

With regard to our data, evolutionary distances were calculated using transversions only, since transversions seem to evolve linearly with time (Miyamoto & Boyle 1989, Mindell &

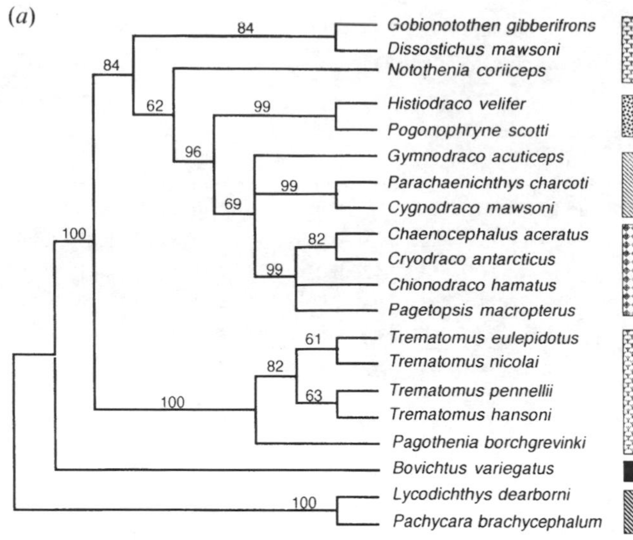

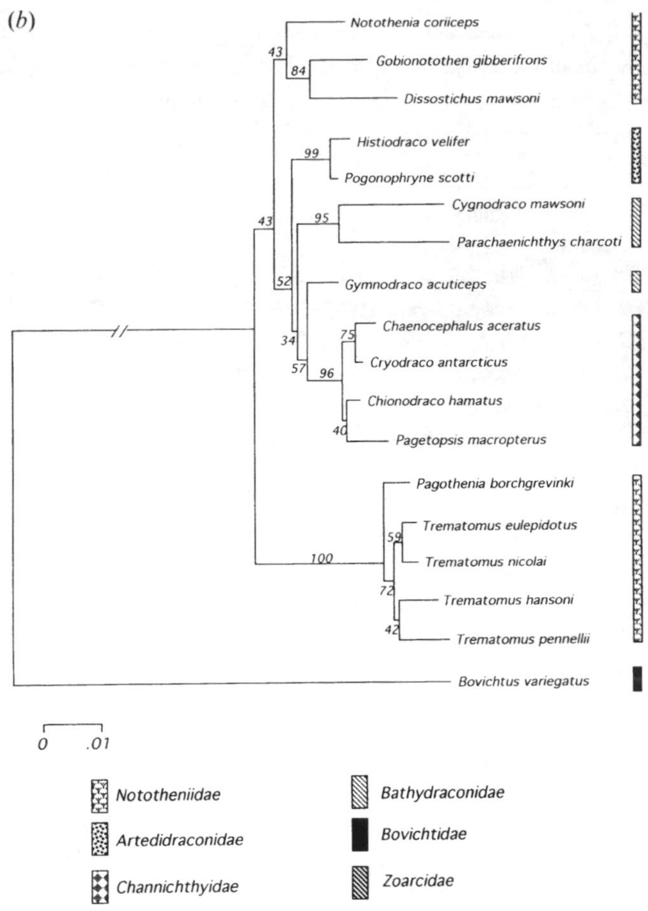

Fig. 7.1 (a) 50% majority rule consensus MP bootstrap tree based on a priori weighting scheme (see text). A MP analysis based on all characters unweighted, resulted in 15 shortest trees (344 steps) with a consistency index (CI)=0.791 (rescaled CI=0.635). (b) Neighbour-joining tree based on pairwise distances calculated on all mutations, according to Kimura (1980). The scale bar indicates 1% genetic distance, the branch-lengths are drawn according to the calculated distances. Numbers refer to bootstrap values (Felsenstein 1985) and vertical bars indicate the current taxonomy of notothenioid families, as reviewed in Eastman (1993).

Honeycutt 1990). Few minor differences in the rate were detected by the relative rate test, which involved the two Artedidraconidae spp. and *Cygnodraco mawsoni*. Therefore, rates were considered sufficiently homogeneous to be used in estimating time of divergence.

Two different estimates of divergence rates were applied to our data in order to calculate time of divergence between notothenioid species. A rate of 0.14% transversion (TV) per million years (My) was proposed for mountain newts (Caccone *et al.* 1994) and bovids (Allard *et al.* 1992); the substitution rate in newts was estimated using the same dataset (about 910 bp from 12S and 16S rRNA mitochondrial genes) as ours, whereas in bovids the complete sequence (2700 bp) of the 12S and 16S rRNA genes was used. Despite this difference in the length of the DNA segment considered, it is interesting that the rate proposed for newts and bovids are identical.

Using our data, we performed a transversion rate calibration with the complete separation of New Zealand from Antarctica (57 million years ago (Ma)) as the estimated time of divergence between *Bovichtus variegatus* and the rest of notothenioid taxa investigated. When our complete dataset was considered (928 bp), the mean nucleotide divergence between *Bovichtus* and the remaining species was 6.2% TV, yielding a rate of 0.11% TV/My which is very similar to that proposed for newts and bovids. When the reduced dataset (883 bp) was taken into account (excluding the highly variable sequence), the mean divergence of *Bovichtus variegatus* was 4.85% and the substitution rate was 0.085% TV/My, which represents a more conservative estimate. Due to the presence of a 45 bp unalignable and possibly hypervariable region, the complete dataset was regarded as less reliable and therefore the reduced dataset was used to calculate divergence times. Using the estimates proposed for newts and bovids (0.14% TV/My) as well as our more conservative estimates of 0.085% TV/My, times of divergence were calculated in relation to major events in notothenioid evolution (Table 7.2).

Tempo and mode of evolution of notothenioids

According to the distance-based phylogeny (NJ tree) *Bovichtus variegatus* diverged very early in the notothenioid evolution, probably after the separation of New Zealand from Antarctica (45–57 Ma).

Our estimate of the divergence time between *Dissostichus* and *Notothenia* ranges between 6–8 Ma (Table 7.2). It suggests that the northward expansion and successive regression of the Polar Front (6.5–5 Ma) possibly played a role as a vicariant factor in the distribution of species belonging to these two genera both outside and inside the Southern Ocean. Diversification of the Trematominae and Channichthyidae falls in a period (Mid Pliocene, 3.5–2.5 Ma) when partial deglaciation might have favoured speciation by providing new coastal and shelf- water habitats.

The short branching pattern displayed by the NJ tree for all the notothenioid taxa (except for *Bovichtus variegatus*) suggests

Table 7.2. *Major events in the evolution of notothenioids and their association with paleoclimatic and/or geological events. The estimate of the timing in notothenioid evolution was performed using a calibration of (a) 0.14% TV/My or (b) 0.085% TV/My (see text)*

Evolutionary event(s)	Mean divergence (a) (Ma)	Mean divergence (b) (Ma)	Associated paleoclimatic And/or geological events
Separation of *Bovichtus* from the rest of notothenioids	45	57	Complete separation of New Zealand from Antarctica (57 Ma)
Major divergence of notothenioid families (Trematominae versus remaining taxa, excluding Bovichtidae)	11	15	Development of East Antarctic sheet, continued drop in water temperatures, significant sea-ice formation (14–12 Ma)
Separation between *Notothenia* and *Dissostichus*	6	8	Northward expansion of Antarctic Polar Front (6.5–5 Ma)
Trematominae diversification	2.5–1.5	4.5–3	Partial deglaciation of Antarctica with relatively warm (2–6 °C) marine embayments; increase in shallow-water shelf habitat
Channichthyidae diversification	2	2.5	

rapid diversification of this group of fish, long after the separation of the Bovichtidae. On the basis of our estimates, this radiation-like mode of evolution possibly occurred 15–10 Ma, when a consistent sea-ice formation, drop in water temperatures and expansion of ice sheet played an important role in modifying the Antarctic climate (Eastman, 1993). These major changes in the Antarctic environment might have created empty niches to which very few non-notothenioid fish were able to adapt. If antifreeze glycopeptides (AFGPs) evolved only once in notothenioid evolution, long after the separation of Bovichtidae and before diversification of all the other notothenioids, as proposed by Bargelloni *et al.* (1994), AFGPs might have played a key role, among other features, in the adaptive radiation of notothenioid fishes.

ACKNOWLEDGEMENTS

We thank A. L. DeVries, E. Pisano, L. Camardella, G. diPrisco, and B. Dickson for providing us with samples. The present research was supported by the Italian PNRA (Programma Nazionale di Ricerche in Antartide) with grants to BB, TP and LB, by the Auckland University Research Committee (AURC) and the New Zealand Lottery Board with grants to PAR, and by US National Science Foundation with grant No. BSR-09107838 and BSR-9119867 to AM.

REFERENCES

Allard, M. W., Miyamoto, M. M., Jarecki, L., Kraus, F. & Tennant, M. R. 1992. DNA systematics and evolution of the artiodactyl family Bovidae. *Proceedings of National Academy of Sciences*, **89**, 3972–3976.

Balushkin, A. V. 1990. Review of blue notothenia of the genus *Paranotothenia* Balushkin (Nototheniidae) with description of a new species. *Journal of Ichthyology*, **30**, 132–147.

Balushkin, A. V. 1991. Review of a green notothenia, *Gobionotothen*, Balushkin (Nototheniidae) of the Antarctic and Subantarctic. *Journal of Ichthyology*, **31**, 42–55.

Bargelloni, L., Ritchie, P. A., Patarnello, T., Battaglia, B., Lambert, D. M. & Meyer, A. 1994. Molecular evolution at subzero temperatures: mitochondrial and nuclear phylogenies of fishes from Antarctica (suborder Notothenioidei) and the evolution of antifreeze glycopeptides. *Molecular Biology and Evolution*, **11**, 854–863.

Caccone, A., Milinkovitch, M. C., Sbordoni, V. & Powell, J. R. 1994. Molecular biogeography: using the Corsica–Sardinia microplate disjunction to calibrate mitochondrial rDNA evolutionary rates in mountain newts (*Euplotes*). *Journal of Evolutionary Biology*, **7**, 227–245.

Cheng, C. C. & DeVries, A. L. 1991. The role of antifreeze glycopeptides and peptides in the freezing avoidance of cold water fish. In di Prisco, G., ed. *Life under extreme conditions: biochemical adaptation*. Berlin: Springer-Verlag, 1–14.

DeVries, A. L. 1988. The role of antifreeze glycopeptides and peptides in the freezing avoidance of Antarctic fishes. *Comparative Biochemistry and Physiology*, **90B**, 611–621.

di Prisco, G., D'Avino, R., Camardella, L., Caruso, C., Romano, M. & Rutigliano, B. 1990. Structure and function of haemoglobin in Antarctic fishes and evolutionary implications. *Polar Biology*, **10**, 269–274.

di Prisco, G., D'Avino, R., Caruso, C., Tamburrini, M., Camardella, L., Rutigliano, B., Carratore, V. & Romano, M. 1991. The biochemistry of oxygen transport in red blooded Antarctic fish. In di Prisco, G., Maresca, B. & Tota, B., eds. *Biology of Antarctic fish*. Berlin: Springer-Verlag, 263–281.

Eastman, J. T. 1985. The evolution of neutrally buoyant notothenioid fishes: their specialization and potential interactions in the Antarctic marine food web. In Siegfried, W. R., Condy, P. R. & Laws, R. M., eds. *Antarctic nutrient cycles and food webs*. Berlin and Heidelberg: Springer-Verlag, 430–436.

Eastman, J. T. 1991. Evolution and diversification of Antarctic notothenioid fishes. *American Zoologist*, **31**, 93–109.

Eastman, J. T. 1993. *Antarctic fish biology: evolution in a unique environment*. San Diego: Academic Press, 322 pp.

Felsenstein, J. 1985. Confidence limits on phylogenies: an approach using the bootstrap. *Evolution*, **39**, 783–791.

Gon, O. & Heemstra, P. C., eds. 1990. *Fishes of the Southern Ocean*. Grahamstown: J. L. B. Smith Institute of Ichthyology, 462 pp.

Higgins, D. G. & Sharp, P. M. 1988. CLUSTAL: A package for performing multiple sequence alignments on a microcomputer. *Gene*, **73**, 237–244.

Iwami, T. 1985. Osteology and relationships of the family Channichthyidae. *Memoirs of National Institute of Polar Research, Tokyo, Series E*, **36**, 1–69.

Jukes, T. H. & Cantor, C. R. 1969. Evolution of protein molecules. In Munro H. N. ed. *Mammalian protein metabolism*. New York: Academic Press, 21–132.

Kimura, M. 1980. A simple method for estimating evolutionary rate of base substitutions through comparative studies of nucleotide sequences. *Journal of Molecular Evolution*, **16**, 111–120.

Kocher, T. D., Thomas, W. K., Meyer, A., Edwards, S. V., Paabo, S., Villablanca, F. X. & Wilson, A. C. 1989. Dynamics of mitochondrial DNA evolution in animals: amplification and sequencing with conserved primers. *Proceedings of National Academy of Sciences*, **86**, 6196–6200.

Kock, K.-H. 1993. *Antarctic fish and fisheries*. Cambridge: Cambridge University Press, 359 pp.

Kumar, S., Tamura, K. & Nei, M. 1993. *Molecular evolutionary genetics analysis (MEGA)*. University Park, PA: The Pennsylvania State University.

Macdonald, J. A. & Wells, R. M. G. 1991. Viscosity of body fluids from Antarctic notothenioid fish. In di Prisco, G., Maresca, B. & Tota, B., eds. *Biology of Antarctic fish*. Berlin: Springer-Verlag, 163–178.

Miller, R. G. 1993. *A history and atlas of the fishes from the Antarctic oceans*. Nevada: Foresta Institute, 792 pp.

Mindell, D. & Honeycutt, R. L. 1990. Ribosomal RNA: evolution and phylogenetic applications. *Annual Review of Ecology and Systematics*, **21**, 541–566.

Miyamoto, M. M. & Boyle, S. M. 1989. The potential importance of mitochondrial DNA sequence data to eutherian mammal phylogeny. In Fernholm, B., Bremer, K. & Jornvall, H., eds. *The hierarchy of life*. Amsterdam: Elsevier Science, 437–450.

Palumbi, S. R., Martin, A., Romano, S., McMillan, W. O., Stice, L. & Grabowski, G. 1991. *The simple fool's guide to PCR*. Honolulu: University of Hawaii Press.

Patarnello, T., Bargelloni, L., Caldara, F. & Colombo, L. 1994. Cytochrome *b* and 16S rRNA sequence variation in the *Salmo trutta* (Salmonidae, Teleostei) species complex. *Molecular Phylogenetics and Evolution*, **1**, 69–74.

Saiki, R. K., Gelfand, D. H., Stoffel, S., Scharf, S., Higuchi, R., Horn, R., Mullis, K. B. & Erlich, H. A. 1988. Primer-directed enzymatic amplification of DNA with a thermostable DNA-polymerase. *Science*, **239**, 487–491.

Saitou, H. & Nei, M. 1987. The neighbour-joining method: a new method for reconstructing phylogenetic trees. *Molecular Biology and Evolution*, **4**, 406–425.

Swofford, D. L. 1993. *Phylogenetic analysis using parsimony (PAUP)*. Champaign, IL: Illinois Natural History Survey.

Tamura, K. 1992. Estimation of the number of nucleotide substitutions when there are strong transition-transvertion and G+C content biases. *Molecular Biology and Evolution*, **9**, 678–687.

Tota, B., Agnisola, C., Schioppa, M., Acierno, R., Harrison, P. & Zummo, G. 1991. Structural and mechanical characteristic of the heart of the icefish *Chionodraco hamatus* (Lonnberg). In di Prisco, G., Maresca, B. & Tota, B., eds. *Biology of Antarctic fish*. Berlin: Springer-Verlag, 204–219.

Wells, R. M. G. 1987. Respiration of Antarctic fish from McMurdo Sound. *Comparative Biochemistry and Physiology*, **88A**, 417–424.

Wells, R. G. M., Macdonald, J. A. & di Prisco, G. 1990. Thin-blooded Antarctic fishes: a rheological comparison of the haemoglobin-free icefishes *Chionodraco kathleenae* and *Cryodraco antarticus* with a red blooded nototheniid, *Pagothenia bernacchii*. *Journal of Fish Biology*, **36**, 595–609.

Wilson, A. C., Carlson, S. S. & White, T. J. 1977. Biochemical evolution. *Annual Review of Biochemistry*, **46**, 573–639.

Wu, C. I. & Li, W. H. 1985. Evidence for higher rates of nucleotide substitution in rodents than in men. *Proceedings of National Academy of Sciences*, **82**, 1741–1745.

8 Prokaryotic Antarctic biodiversity

P. D. FRANZMANN[1], S. J. DOBSON[2], P. D. NICHOLS[3] AND T. A. MCMEEKIN[2]
[1]Current address: CSIRO, Division of Water Resources, Underwood Avenue, Floreat Park WA 6014, Australia (E-Mail: Peter.Franzmann@per.dwr.csiro.au), [2]Antarctic-CRC & Department of Agricultural Science, University of Tasmania, Box 252C Hobart 7001, Australia, [3]CSIRO Oceanography, GPO Box 1538, Hobart 7001, Australia

ABSTRACT

Prokaryotes dominate many Antarctic ecosystems. They play essential roles in food chains, in biogeochemical cycles, and as mineralizers of anthropogenic pollutants. Since the late 1970s our understanding of prokaryotic biodiversity has been greatly enhanced by the use of molecular techniques, such as 16S rRNA sequence comparisons. To date, all Antarctic prokaryotes that have been characterized by phenotypic and molecular techniques have represented new species. The prokaryotes of Antarctica are of broad phylogenetic origin with representatives from both domains, Archaea and Bacteria. Use of the 16S rRNA gene as a molecular clock shows that the Antarctic species diverged from their nearest known non-Antarctic relatives long before Antarctica established as a permanently cold environment. Most Antarctic bacterial species do not function optimally at the temperature of the environment from which they were isolated. Whether or not Antarctic environments contain prokaryotes that are exclusively Antarctic remains an open question.

Key words: prokaryotes, biodiversity, Antarctica, 16S rRNA, phylogeny.

INTRODUCTION

Prokaryotes, the simplest of life forms that independently self-replicate, dominate many Antarctic ecosystems. They play essential roles in food chains in both continental (Freckman & Virginia 1990, Friedmann *et al.* 1993) and oceanic systems (Tanoue & Hara 1986), and in biogeochemical cycles (Burton & Barker 1979, Matsumoto *et al.* 1989). Cyanobacteria are one of the most abundant and conspicuous groups of prokaryotes in non-marine Antarctic environments and play important roles in the fixation of carbon and energy (Vincent *et al.* 1993). Prokaryotes impact on human activities in Antarctica, as mineralizers of anthropogenic pollutants (Cahet *et al.* 1986), and as the group of organisms most likely to contaminate or colonize Antarctic environments as introduced species (Meyer *et al.* 1963). The diversity of microorganisms that inhabit Antarctic ecosystems may yield direct economic benefits through their exploitation for biotechnological applications (Kobori *et al.* 1984, Feller *et al.* 1992). The importance of microbial communities to Antarctic systems, and human endeavours within them, necessitates an understanding of Antarctic microbial biodiversity and processes.

The biota of continental Antarctica is restricted to refugia, and is largely confined to the less than 2 % of the continent that is ice free. Extensive microbial communities develop where liquid water is available and temperature fluctuations are ameliorated sufficiently to allow some growth, even though growth periods may be brief (Friedmann *et al.* 1987). Major Antarctic microbial communities occur in Southern Ocean waters and associated sea ice. Terrestrial microbial communities occur in endolithic and sublithic communities, Antarctic lakes, soils, and rivers. Microbial communities also occur as colonizers of other biota (Sieburth 1965).

The microbiota that colonize these niches have been examined from time to time, and bacterial isolates have been identified to genus but rarely to species level. Lists of genera inhabiting Antarctic environments have been published in many articles, including those by Hirsch *et al.* (1988), Vincent (1988),

Vishniac (1993) and Wynn-Williams (1990). Identification of prokaryotes was usually based on the application of a few tests to determine key attributes, for example Gram reaction, oxidative–fermentative tests, oxidase, catalase, morphology. When isolates have been characterized in detail, they have either been described as new species (Shivaji *et al.* 1992) or identified as 'atypical' representatives of a species (Shivaji *et al.* 1991). With the use of molecular techniques, such as 16S rRNA cataloguing and sequencing, to investigate prokaryotic phylogeny, a revolution has occurred in our understanding of microbial taxonomy. The recognition of two domains within the prokaryotes, the Archaea and Bacteria (previously designated Archaebacteria and Eubacteria) was pivotal in our appreciation of biological diversity and evolution (Woese & Fox 1977). Modern approaches to the classification of microorganisms have at last allowed a classification based on evolutionary comparisons that requires a polyphasic approach utilizing phenotypic, chemotaxonomic and molecular techniques. Classification of prokaryotes still requires functional or phenotypic descriptions to integrate the phylogenetic taxonomy with an organism's biogeochemical or ecological role.

The first study that examined nucleic acid sequence data for Antarctic microorganisms compared the sequence of 5S rRNA extracted directly from rocks of the Dry Valleys with a database of 5S rRNA sequences from over 500 organisms (Colwell *et al.* 1989). This study suggested that an organism most closely related to *Vibrio natriegens* inhabited this environment. Since that time, extensive databases of 16S rRNA (and 16S rDNA gene) sequences have been developed (Genbank; Ribosomal Database Project (RDP), Larsen *et al.* 1993) and a number of Antarctic prokaryotes are now represented in these databases. To obtain a clearer understanding of Antarctic prokaryotic diversity and the evolutionary origins of Antarctic microorganisms, comparisons were made between the available 16S rRNA sequences of Antarctic organisms and those of their nearest non-Antarctic relatives.

METHODS

The 16S rRNA sequence data used were obtained from Genbank, W. Ludwig of the Technical University of Munich, the Ribosomal Database Project (RDP, on the anonymous ftp server at the University of Illinios in Urbana, Illinios updated on 24 October, 1993, release version 3.1.) or derived in our own laboratory, in which 16S rDNA genes were directly sequenced from polymerase chain-reaction amplified fragments as described by Dobson *et al.* (1993a). All the available 16S rRNA sequences for Antarctic prokaryotes are from organisms isolated from the Vestfold Hills, except for 'Vesiculatum antarcticum', which was isolated from sea ice in McMurdo Sound (Staley *et al.* 1989, Larsen *et al.* 1993). A list of the Antarctic strains used in this study is given in Table 8.1. Nearest known phylogenetic relatives to the Antarctic isolates were obtained from the complete phylogenetic tree of prokaryotes for which 16S rRNA sequences are available (available through the ftp server of the Ribosomal Database, Larsen *et al.* 1993). The nearest phylogenetic relative to the 'wall-less spirochete' from Ace Lake was a strain 'PL12MY' about which no information other than its rRNA sequence is currently published. As a result, the next closest organism, *Spirochaeta bajacaliforniensis*, was included in the phylogenetic analysis.

Table 8.1. *Site of isolation of Antarctic prokaryotes for which 16S rRNA sequences are available, and temperature of the site at the time of isolation*

Species	Site	Temperature (°C)
Methanococcoides burtonii	Ace Lake bottom water	+1.7
Halobacterium lacusprofundi	Deep Lake sediment	+5.2
Carnobacterium funditum	Ace Lake bottom water	+1.7
Carnobacterium alterfunditum	Ace Lake bottom water	+1.7
'Wall-less spirochete'	Ace Lake bottom water	+1.7
Halomonas subglaciescola	Organic Lake (depth 2 m)	−7.3
Halomonas meridiana	Birch Lake (depth 6 m)	−3.9
'Vesiculatum antarcticum'	Below sea-ice (marine)	+0.6 to −1.25
Flectobacillus glomeratus	Burton Lake (depth 11.4 m)	−1.6
Favobacterium gondwanense	Organic Lake (depth 5 m)	−5.3

Percentage similarities and evolutionary distances (Jukes–Cantor formula), and a phylogenetic tree (using FITCH in PHYLIP, Felsenstein 1989) were determined for sequences from the Antarctic bacteria and their nearest known phylogenetic relatives after alignment to the RDP format as described by Dobson *et al.* (1993a).

RESULTS AND DISCUSSION

The diversity of Antarctic prokaryotes

A phylogenetic tree showing the relationship of Antarctic prokaryotes to their nearest known, non-Antarctic relatives is given in Fig. 8.1. The percentage divergence in 16S rRNA sequence between the Antarctic prokaryotes and their nearest known non-Antarctic relatives, and the optimum temperature for growth of each species, is shown in Table 8.2.

Representatives of both domains of the prokaryotes, Archaea and Bacteria, have been isolated from Antarctic environments. Archaean representatives include *Halobacterium*

Table 8.2. *A comparison of temperature characteristics of Antarctic bacteria and the similarity of their 16S rRNA gene sequence with the non-Antarctic bacteria to which they are phylogenetically most similar*

Antarctic species	T_{OPT}[1]	Non-Antarctic species	T_{OPT}[1]	% Dissimilarity in 16S rRNA
Methanococcoides burtonii	23	*Methanococcoides methylutens*	35	3
Halobacterium lacusprofundi	33	*Halobacterium saccharovorum*	*c*.50	3
Carnobacterium funditum	23	*Carnobacterium mobile*	*c*.30	4
Carnobacterium alterfunditum	23	*Carnobacterium mobile*	*c*.30	4
'Wall-less spirochete'	12	*Spirochaeta bajacaliforniensis*	36	12
Halomonas subglaciescola	22	*Deleya aquamarina*	*c*.35	5
Halomonas meridiana	34	*Deleya aquamarina*	*c*.35	0
'Vesiculatum antarcticum'	<10	*Flexibacter maritimus*	30	7
Flectobacillus glomeratus	17	*Flexibacter maritimus*	30	6
Favobacterium gondwanense	23	*Cytophaga marinoflava*	30	8

Note:

[1] T_{OPT} = temperatures in °C. Many temperatures are given as approximate values as descriptions often list a range of temperatures as T_{OPT}.

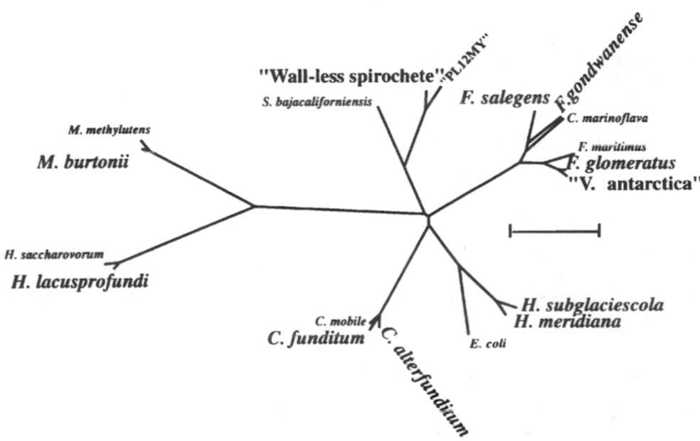

Fig. 8.1. The phylogeny of Antarctic bacteria for which 16S rRNA sequences are available, by comparison of these sequences with their nearest known relatives. Antarctic bacteria are in larger type. The bar equates to a Jukes–Cantor similarity measure of 0.1.

lacusprofundi, an extreme halophile from Deep Lake, and *Methanococcoides burtonii*, a methanogenic archaeon from Ace Lake. A range of 'phyla' from the Bacteria are represented, including Gram positives (*Carnobacterium funditum, Carnobacterium alterfunditum*), members of the Bacteroides/Flavobacteria group (*Flectobacillus glomeratus, Flavobacterium gondwanense, Flavobacterium salegans,* Vesiculata antarctica), spirochetes (wall-less spirochete), and the γ subgroup of the class Proteobacteria (*Halomonas subglaciescola* and *Halomonas meridiana*). The nearest non-Antarctic relative to the two *Halomonas* spp., *Deleya aquamarina*, is not included in the tree shown in Fig. 8.1 as its sequence is identical to that of *Halomonas meridiana*. The genera *Halomonas* and *Deleya* should be combined on the evidence of all available phylogenetic, phenotypic and chemotaxonomic data (Dobson *et al.* 1993b, Skerratt *et al.* 1991). It is interesting to note that the nearest known relatives of most of the Antarctic organisms from the present study and listed in Table 8.2 have been isolated from marine environments, the exceptions being *Halobacterium saccharovorum*, which was isolated from the Dead Sea, and *Carnobacterium mobile*, which was isolated from chilled foods.

Although not in the tree, representatives of most other 'phyla' of the domain Bacteria have been observed or isolated from Antarctic environments, including members of the planctomyces group (Hirsch *et al.* 1988), the cyanobacteria (Love *et al.* 1982), the green sulfur bacteria (Vincent 1988), *Deinococcus* spp. (Hirsch *et al.* 1988), and other members of the Proteobacteria. It is likely that sequences of phylogenetically conserved genes (genes that show considererable homology over all evolutionary groups) of Antarctic representatives of these groups will be added to the database in the future.

Molecular divergence and the 16S rRNA clock

Conserved genes, such as the small subunit rDNA gene, show an evolutionary branching pattern on which a 'molecular clock' can be loosely superimposed to estimate timing of the evolutionary branching points or nodes (Ochman & Wilson 1987). The small subunit rDNA gene is common to all independently self-replicating organisms and many cellular structures. In Prokaryotes, mitochondria and chloroplasts, the small subunit rRNA is the 16S rRNA. To our knowledge, divergence in the 16S rDNA gene has been calibrated as a molecular clock on two separate occasions (Ochman & Wilson 1987, Moran *et al.* 1993). Ochman & Wilson (1987) calibrated the clock using divergence in deep branches that must have occurred at about the time of geochemical change (eg. the divergence of microaerophiles and the appearance of oxygen < 2000 million years ago) as well as recent divergence (divergence of the salmonellae from *E. coli* with the appearance of the mammals < 150 million years ago). They estimated that a 1% dissimilarity in 16S rRNA sequence corresponded to an evolutionary separation of about 50 million years. Moran *et al.* (1993) examined the divergence in 16S rRNA sequence of maternally inherited endosymbionts of aphid host species. They found that the bacterial phylogeny was concordant with the phylogeny of the aphid hosts, and a fossil record was available to estimate dates of nodes in the aphid phylogenetic tree. The calibration of the clock spans only the period of aphid evolution (160–280 million years) but suggests a divergence rate of 1–2% per 50 million years; a rate perhaps slightly faster than but comparable with Ochman's & Wilson's calibration.

Molecular divergence, expressed as percentage dissimilarity of 16S rRNA sequence, of the Antarctic species from their nearest known non-Antarctic relatives is given in Table 8.2. Applying the molecular clock calibration to this data suggests that the majority of modern Antarctic prokaryotes diverged from their nearest non-Antarctic relatives in excess of 75–150 million years ago. This divergence is recent in terms of prokaryotic evolution, which has been in progress for about 3.6 billion years, but occurred much earlier than the establishment of a permanently cold Antarctic environment. The positioning of the nodes of divergence of Antarctic from non-Antarctic prokaryotes is limited by the incompleteness of the database. It is not known how representative of Antarctic environments the examined species are. Many organisms are yet to be discovered in both Antarctic and non-Antarctic environments, and it is likely that branches in the tree will shorten considerably. Two species, the non-Antarctic *Deleya aquamarina* and the Antarctic *Halomonas meridiana*, show no dissimilarity in 16S rRNA sequence and little difference in optimum growth temperatures (Table 8.2). This perhaps suggests recent divergence, and poor adaptation of the Antarctic species to cold environments.

Are there Antarctic prokaryotes?

Given the recent divergence of Antarctic isolates from non-Antarctic species the question arises as to whether there is such a thing as an Antarctic prokaryote. It is clear that the prokaryotes from Antarctica generally have decreased cardinal (maximum, optimal and minimal) temperatures for growth when compared with their nearest known non-Antarctic relatives (see Table 8.2 for a comparison of optimal growth temperatures). A number of laboratory findings have shown that there is a continuous selection for bacteria whose activity is greatest near the environmental temperature (Morita 1975). Although psychrophilic bacteria can be readily obtained from permanantly cold environments (Morita 1975, Staley *et al.* 1989), the greater proportion of organisms from cold environments are not psychrophilic (Gounot 1991). Indeed, many of the individual isolates of Antarctic bacteria for which phylogenetic data are available have not aligned their optimal temperatures for growth closely to the usual *in situ* temperatures experienced in their environment. This may be due to a number of factors, including the relatively recent development of Antarctica as a stable cold environment, the instability of temperatures of some Antarctic environments today or the recent colonization of Antarctic environments by organisms that have had only a short period for adaptation to low temperatures. Psychrotrophic bacteria, capable of growth at 0 °C, have been isolated from many non-Antarctic environments, including lakes and soils in the USA, plants, refrigerated foods and air (Gounot 1991).

The deep sea has been the most stable cold environment over time. The psychrosphere, a cold-water bottom layer in the world ocean, is thought to have formed in the early Oligocene, 36 million years ago (Kennett & Shackleton 1976). The Antarctic prokaryotes exhibiting the lowest optimum growth temperatures (Table 8.2) were 'Vesiculata antarctica' from the Southern Ocean and the 'wall-less spirochete' from an Antarctic saline lake that was probably colonized by marine microorganisms as it is of essentially marine salinity.

But there is evidence that the Antarctic continent was not a stable cold environment even in relatively recent times. The discovery of 2.5 to 3.0 million year old fossil *Nothofagus* wood in the Sirius Formation (Webb & Harwood 1987) suggests mean summer temperatures of approximately 5 °C (Hill *et al.* 1991) and probably much higher maximum temperatures. Even today, some Antarctic environments exhibit quite unstable temperature regimes. Temperatures in soils can reach as high as 30 °C and the microbiota must be able to survive this temperature (Bolter 1992).

In the last 730 000 years, ten major glaciations have occurred, perhaps a result of obliquity maxima in the Earth's orbit or 'Milankovitch cycles'. The current Antarctic prokaryotic biodiversity may be largely composed of quite recent immigrants that colonized suitable microbial refugia, like the lakes of the Vestfold Hills, which were exposed after the last major glaciation. It has been suggested that the concept of spatial isolation is not tenable to microbes (Zavarzin 1993), and that particles are constantly transported to the Antarctic continent (Burckle *et al.* 1988, Schlichting *et al.* 1978).

If Antarctic prokaryotes are endemic to Antarctica alone, then many types of organisms of quite separate evolutionary origins must be undergoing parallel evolutionary adaptation to life at cold temperatures. One mechanism by which these organisms achieve growth at low temperatures may be to increase the content of branched and unsaturated fatty acids in lipids in order maintain membrane fluidity at cold temperatures (Rotert *et al.* 1993). Early results indicate that the proportion of Antarctic strains that produce EPA (eicosapentaenoic acid, 20:5w3) is considerably higher than occurs in temperate marine bacteria; with EPA levels up to nearly 20% of the total fatty acids in some Antarctic bacteria (Nichols *et al.* 1993, and unpublished data). Similarly, a number of strains that produce up to 3% DHA (docosahexaenoic acid, 22:6w3) have been isolated.

Few Antarctic microorganisms have been examined for enzyme adaptation to cold temperatures but some studies have suggested that Antarctic microbes increase enzyme production at low temperatures in order to increase metabolic rates rather than alter the activity of the enzyme at low temperature (Reichardt 1988, Ray *et al.* 1992). Some proteins, cold-shock proteins and cold-acclimation proteins, that may be involved in membrane transport, freeze protection or transcription and translation are produced by some psychrophilic bacteria under appropriate conditions (Roberts & Inniss 1992).

It is clear that virtually all evolutionary branches of extant prokaryotes contribute to the diversity in Antarctica. Whether or not individual species are unique to Antarctica remains an open question. With the development of a database of rDNA sequences, sensitive detection techniques such as PCR can be used in other appropriate environments to examine for the signatures of known Antarctic organisms. A better understanding of Antarctic prokaryotic biodiversity will contribute to our knowledge of biogeochemical processes in the region, and is a prerequisite for the possible biotechnological exploitation of the resident microbiota.

ACKNOWLEDGEMENTS

This work was funded by grants from the Australian Research Council and Antarctic Science Advisory Committee and the CSIRO – University of Tasmania grants scheme. PDF and SJD are supported by ARC fellowships. We thank Harry Burton for his continued interest and encouragement and fellow ACAM colleagues for their support. We thank Andrew McMinn, Warwick Vincent, Roy Harden Jones and an anonymous referee for their comments on the manuscript.

REFERENCES

Bolter, M. 1992. Environmental conditions and microbiological properties from soils and lichens from Antarctica (Casey Station, Wilkes Land). *Polar Biology*, **11**, 591–599.

Burckle, L. H., Gayley, R. I., Ram, M. & Petit, J-R. 1988. Biogenic particles in Antarctic ice cores and the source of Antarctic dust. *Antarctic Journal of the United States*, **23** (5), 71–72.

Burton, H. R. & Baker, R. J. 1979. Sulfur chemistry and microbiological fractionation of sulfur isotopes in a saline Antarctic lake. *Geomicrobiology Journal*, **1**, 329–341.

Cahet, G., Delille, D. & Vaillant, N. 1986. Evolution a court terme des microfolres utilisant les hydrocarbures en milieu subantarctique. Comparison des systemes d'incubation naturels et artificiels. In *Deuxieme Colloque International de Bacteriologie marine – CNRS, Brest 1–5 October 1984*, 655–662.

Colwell, R. R., Macdonell, M. T. & Swartz, D. 1989. Identification of an Antarctic endolithic microorganism by 5S rRNA sequence analysis. *Systematic and Applied Microbiology*, **11**, 182–186.

Dobson, S. J., Colwell, R. R., Franzmann, P. D. & McMeekin, T. A. 1993a. Direct sequencing of the PCR-amplified 16S rRNA gene of *Flavobacterium gondwanense* sp.nov. (ACAM 44T=DSM 5425T), and *Flavobacterium salegens* sp. nov. (ACAM 48T=DSM 5424T), new species from a hypersaline Antarctic lake. *International Journal of Systematic Bacteriology*, **43**, 77–83.

Dobson, S. J., McMeekin, T. A. & Franzmann, P. D. 1993b. Phylogenetic relationships between some members of the genera Deleya, Halomonas, and Halovibrio. *International Journal of Systematic Bacteriology*, **43**, 665–673.

Feller, G., Lonhienne, T., Deroanne, C., Libioulle, C., Beeumen, J. & Gerday, C. 1992. Purification, characterization, and nucleotide sequence of the thermolabile a-amylase from the Antarctic psychrotroph *Altermonas haloplanctis* A23. *Journal of Biological Chemistry*, **267**, 5217–5221.

Felsenstein, J. 1989. PHYLIP-phylogeny inference package (version 3.2). *Cladistics*, **5**, 164–166.

Freckman, D. W. & Virgina, R. A. 1990. Nematode ecology of the McMurdo Dry Valley ecosystems. *Antarctic Journal of the United States*, **25**(4), 229–230.

Friedmann, E. I., Kappen, L., Meyer, M. A. & Nienow, J. A. 1993 . Long-term productivity in the cryptoendolithic microbial communitiy of the Ross Desert, Antarctica. *Microbial Ecology*, **25**, 51–69.

Friedmann, E. I., McKay, P. & Nienow, J. A. 1987. The cryptoendolithic microbial environment in the Ross Desert of Antarctica: satellite-transmitted continuous nanoclimate data 1984 to 1986. *Polar Biology*, **7**, 273–287.

Gounot, A-M. 1991. Bacterial life at low temperature: physiological aspects and biotechnological implications. *Journal of Applied Bacteriology*, **71**, 386–397.

Hill, R. S., Harwood, D. M. & Webb, P. N. 1991. Last remnant of Antarctica's Cenozoic flora: Nothofagus of the Sirius Group, Transantarctic Mountains. *Abstracts of the Eighth International Symposium on Gondwana, June 24–28, Hobart, Tasmania*. **43**.

Hirsch, P., Hoffmann, B., Gallikowski, C. C., Mevs, U., Siebert, J. & Sittig, M. 1988. Diversity and identification of heterotrophs from Antarctic rocks of the McMurdo Dry Valleys (Ross Desert). *Polarforschung*, **58**, 261–269.

Kennet, J. P. & Shackleton, N. C. 1976. Oxygen isotope evidence for the development of the psychrosphere 38 Myr ago. *Nature*, **260**, 513–515.

Kobori, H., Sullivan, C. & Shizuya, H. 1984. Heat-labile alkaline phosphatase from Antarctic bacteria: rapid 5' end-labelling of nucleic acids. *Proceedings of the National Academy of Sciences*, **81**, 6691–6695.

Larsen, N., Olsen G. J., Maidak, B. L., McCaughey, M. J., Overbeek, R., Macke, T. J., Marsh, T. L. & Woese, C. R. 1993. The ribosomal database project. *Nucleic Acids Research*, **21** (Supplement), 3021–3023.

Love, F. G., Simmons Jr., G. M., Parker, B. C., Wharton, R. A. & Seaburg, K. G. 1982. Modern Conophyton-like microbial mats discovered in Lake Vanda, Antarctica. *Geomicrobiology Journal*, **3**, 33–48.

Matsumoto, G. I., Watanuki, K. & Torrii, T. 1989. Vertical distribution of organic constituents in an Antarctic lake: Lake Fryxell. *Hydrobiologia*, **172**, 291–303.

Meyer, G. H., Morrow, M. B. & Wyss, O. 1963. Viable organisms from feces and foodstuffs from early Antarctic expeditions. *Canadian Journal of Microbiology*, **9**, 163–167.

Moran, N. A., Munson, M. A., Baumann, P. & Ishikawa, H. 1993. A molecular clock in endosymbiotic bacteria is calibrated using insect hosts. *Proceedings of the Royal Society London*, **B253**, 167–171.

Morita, R. Y. 1975. Psychrophilic bacteria. *Bacteriological Reviews*, **39**, 144–167.

Nichols, D. S., Nichols, P. D. & McMeekin, T. A. 1993. Polyunsaturated fatty acids in Antarctic bacteria. *Antarctic Science*, **5**, 149–160.

Ochman, H. & Wilson, A. C. 1987. Evolution in bacteria: evidence for a universal substitution rate in cellular genomes. *Journal of Molecular Evolution*, **26**, 74–86.

Ray, M. K., Devi, K. U., Kumar, G. S. & Shivaji, S. 1992. Extracellular protease from the Antarctic yeast *Candida humicola*. *Applied and Environmental Microbiology*, **58**, 1918–1923.

Reichardt, W. 1988. Impact of the Antarctic benthic fauna on the enrichment of biopolymer degrading psychrophilic bacteria. *Microbial Ecology*, **15**, 311–321.

Roberts, M. E. & Inniss, W. E. 1992. The synthesis of cold shock proteins and cold acclimation proteins in the psychrophilic bacterium *Aquaspirillum articum*. *Current Microbiology*, **25**, 275–278.

Rotert, K. R., Toste, A. P. & Steiert, J. G. 1993. Membrane fatty acid analysis of Antarctic bacteria. *FEMS Microbiology Letters*, **114**, 253–258.

Schlichting, J. R., Speziale, B. J. & Zink, R. M. 1978. Dispersal of algae and protozoa by Antarctic flying birds. *Antarctic Journal of the United States*, **13**(4), 147–149.

Sieburth, J. McN. 1965. Microbiology of Antarctica. In Van Oye, P. & Van Mieghem, J., eds. *Biogeography and ecology in Antarctica*. The Hague: Junk, 267–295.

Shivaji, S., Ray, M. K., Kumar, G. S., Reddy, G. S. N., Saisree, L. & Wynn-Williams, D. D. 1991. Identification of *Janthinobacterium lividum* from the soils of the islands of Scotia Ridge and from Antarctic Peninsula. *Polar Biology*, **11**, 267–271.

Shivaji, S., Ray, M. K., Rao, N. S., Saistree, L., Jagannadham, M. V., Kumar, G. S., Reddy, G. S. N. & Bhargava, M. P. 1992. *Sphingobacterium antarcticus* sp. nov., a psychrotrophic bacterium from the soils of Schirmacher oasis, Antarctica. *International Journal of Systematic Bacteriology*, **42**, 102–106.

Skerratt, J. H., Nichols, P. D., Mancuso, C. A., James, S. R., Dobson, S. J., McMeekin, T. A. & Burton, H. R. 1991. The phospholipid ester-linked fatty acid composition of members of the family Halomonadaceae and genus Flavobacterium. A chemotaxonomic guide. *Systematic and Applied Microbiology*, **14**, 8–13.

Staley, J. T., Irgens, R. L. & Herwig, R. P. 1989. Gas vacuolate bacteria from the sea ice of Antarctica. *Applied and Environmental Microbiology*, **55**, 1033–1036.

Tanoue, E. & Hara, S. 1986. Ecological implications of fecal pellets produced by the Antarctic krill *Euphausia superba* in the Antarctic Ocean. *Marine Biology*, **91**, 359–369.

Vincent, W. F. 1988. *Microbial ecosystems of Antarctica*. Cambridge: Cambridge University Press, 304 pp.

Vincent, W. F., Castenholtz, R. W., Downes, M. T. & Howard-Williams, C. 1993. Antarctic cyanobacteria: light, nutrients, and photosynthesis in the microbial mat environment. *Journal of Phycology*, **29**, 745–755.

Vishniac, H. S. 1993. The microbiology of Antarctic soils. In Friedman, E.I., ed. *Antarctic microbiology*. New York: Wiley-Liss, 297-341.

Webb, P. N. & Harwood, D. M. 1987. Terrestrial flora of the Sirius Formation: its significance for Late Cenozoic glacial history. *Antarctic Journal of the United States*, **22**, 7–11.

Woese, C. R. & Fox, G. E. 1977. Phylogenetic structure of the prokaryotic domain: the primary kingdoms. *Proceedings of the National Academy of Sciences of the USA*, **74**, 5088–5090.

Wynn-Williams, D. D. 1990. Ecological aspects of Antarctic microbiology. *Advances in Microbial Ecology*, **11**, 71–146.

Zavarzin, G. A. 1993. An ecological approach to the systematics of prokaryotes. In Guerrero, R. & Pedros-Alio, C., eds. *Trends in microbial ecology*. Barcelona: Spanish Society for Microbiology, 555–558.

9 Expression of the myoglobin gene in Antarctic channichthyid fishes

MICHAEL E. VAYDA[1*], *DEENA J. SMALL*[1] *AND BRUCE D. SIDELL*[2]

[1]*Department of Biochemistry, Microbiology and Molecular Biology,* [2]*Department of Zoology University of Maine, Orono, ME 04469-5735, USA*

ABSTRACT

Two channichthyid (Antarctic icefish) species, Chionodraco rastrospinosus *and* Pseudochaenichthys georgianus, *express and accumulate myoglobin (Mb) in heart ventricle tissue. This observation unequivocally resolves the hitherto disputed issue of whether this intracellular oxygen-binding protein is expressed in any of the channichthyid icefishes. Two other icefish species,* Chaenocephalus aceratus *and* Champsocephalus gunnari, *lack detectable Mb polypeptide in immunoblot assays. The absence of Mb in these two species must result from distinct mutational events: mRNA encoding Mb is present in the heart ventricle of* C. gunnari *and is associated with polyribosome complexes, whereas Mb mRNA is not detected in any of the tissues of* C. aceratus. *In contrast to a temperate zone teleost,* Morone saxatilis, *and other vertebrates, Mb expression in* C. rastrospinosus *and* P. georgianus *appears to be limited to the heart ventricle; Mb mRNA and polypeptide are not found in an aerobic skeletal muscle, the pectoral adductor profundus. This unusual pattern of Mb expression was also observed in three red-blooded Antarctic fish species,* Notothenia coriiceps, Gobionotothen gibberifrons *and* Trematomus newnesi. *This result suggests that a single mutation occurred early in the radiation of the Antarctic notothenioid fishes that abolished Mb expression in aerobic skeletal muscle, and independent mutations resulted in loss of Mb expression by heart muscle.*

Key words: Antarctic fish, myoglobin, cDNA cloning, immunoblot, RNA gel blot hybridization.

INTRODUCTION

The notothenioid fishes of coastal Antarctic waters provide an unique opportunity to study a closely related group of organisms that have radiated to fill niches of an unusual environment. These fishes inhabit waters that are thermally stable and very cold, ranging only between +0.3 and −1.87 °C throughout the year (Littlepage 1965, DeWitt 1971). The water column is vertically mixed, allowing an oxygen- and nutrient-rich environment for benthic organisms. Notothenioids are the dominant fishes in this ecosystem, representing almost 50% of all species and 50 to >90% of the coastal fish captures (DeWitt 1971, Anderson 1990, Eastman 1993). Glacial submergence and chronically cold temperatures are thought to have eliminated the diversity of coastal teleosts evident in the geological record of Antarctica, prior to the geographic isolation of the continent some 25 to 40 million years ago (Anderson 1990, Eastman 1991). Most extant species appear to have radiated from a narrow ancestral lineage that survived establishment of the polar Antarctic environment. With only a few exceptions, notothenioid fishes are found only in coastal Antarctic waters, suggesting that they have evolved in isolation during that period. Species migration from elsewhere in the world is minimized by strong circumpolar currents and a very deep bottom surrounding the continent (Kennett 1977, 1980). The unique physical characteristics of the Antarctic environment appear to have allowed the accumulation of mutations that would be disadvantageous or lethal in other environments.

Such mutations are particularly apparent in the Channichthyidae, or icefishes, one of the six families within the suborder Notothenioidei. Channichthyids are the only vertebrate organisms known to lack the circulating oxygen-carrier haemoglobin as adults. Although a decreased hematocrit is advantageous in reducing the high viscosity of blood at cold temperatures, icefish exhibit several adaptations to compensate for the low oxygen-carrying capacity of their blood. In comparison with their red-blooded notothenioid relatives, channichthyids have larger hearts with greater cardiac output, larger blood volumes and blood vessels of wider diameter (Hemmingsen & Douglas 1970, Fitch *et al.* 1984, Hemmingsen 1991). Absence of the intracellular oxygen-binding protein Mb has been assumed (Hamoir 1988, Eastman 1990) based upon the creamy-pale colouration of icefish tissues. Characteristics of Mb isolated from vertebrates, specifically the kinetic off-constant for dissociation of oxygen (Stevens & Carey 1981, Sato *et al.* 1990), suggest that Mb does not function at the physiological temperature of icefish. Whether the icefish possess modifications in cellular architecture, or in the Mb polypeptide, to facilitate intracellular diffusion of oxygen in aerobic muscles is not known.

We encountered two channichthyid species in our 1991 and 1993 field seasons that exhibited red-pigmented hearts: *Chionodraco rastrospinosus* and *Pseudochaenichthys georgianus*. These species were collected off the Antarctic Peninsula near the south shore of Low Island (63° 15′ S, 62° 05′ W) and Brabant Island near Astrolabe Needle (64° 11′ S, 62° 45′ W). The deep red-colored hearts of these species are in stark contrast to hearts of *Chaenocephalus aceratus* and *Champsocephalus gunnari,* which are pale greenish-brown, a colouration resulting from the high density of mitochondria present. Douglas *et al.* (1985) reported a haeme-containing protein in the high-speed supernatants of *P. georgianus* and *C. aceratus* heart homogenates, which they concluded was Mb. There was no direct evidence to support this conclusion. To the contrary, the high density of mitochondria in icefish heart tissue (Johnston *et al.* 1983, Tota *et al.* 1988) and the ionic strength of the buffer employed by Douglas *et al.* (1985) suggest that the haeme protein they detected was cytochrome c, not Mb. Further, the dramatic difference that we observe in heart pigmentation between *P. georgianus* and *C. aceratus* is inconsistent with the nearly equivalent amounts of haemochromagen reported by Douglas *et al.* (1985). The objective of our study presented here was to determine whether the red pigment observed in heart tissue of *C. rastrospinosus* and *P. georgianus* is Mb, by direct detection of the Mb polypeptide and Mb mRNA.

Mb POLYPEPTIDE IS PRESENT IN THE HEART VENTRICLE OF AT LEAST TWO CHANNICHTHYID ICEFISH SPECIES

We prepared a high speed supernatant (40 000*g*) from the ventricle of a freshly dissected *P. georgianus* heart. The absorption spectrum of this supernatant exhibited maxima at 530 and 580 nm, which are consistent with oxyMb, and which were not evident in preparations of *C. aceratus* heart ventricle. However, to prove conclusively that this absorption is due to Mb and not some other haeme-containing protein, we performed immunoblots to demonstrate the presence of the Mb polypeptide. Proteins in the high-speed extracts were resolved by electrophoresis through a 15% acrylamide gel and either stained with Coomassie Blue (Fig. 9.1A) or electroblotted to nitrocellulose. Blots were incubated with a commercially available (Sigma) rabbit anti-human Mb antibody, and subsequently a goat anti-rabbit IgG second antibody conjugated with horseradish peroxidase. Fig. 9.1B demonstrates that the anti-Mb antibody clearly identifies Mb from a temperate zone teleost, striped bass *Morone saxatilis*, and ventricular extracts of the red-blooded Antarctic fish *Trematomus newnesi*. An immunoreactive polypeptide of the same electrophorectic mobility is apparent in a ventricular extract of *P. georgianus* (Fig. 9.1). No trace of an immunoreactive polypeptide was present in the extracts of *C. aceratus* or *C. gunnari*, icefish species with pale-coloured hearts. Subsequent analyses have demonstrated the presence of an immunoreactive polypeptide of similar mobility in ventricular extracts of *C. rastrospinosus*, the other icefish with a pigmented heart, and two other red-blooded notothenioid fish, *Notothenia coriiceps* and *Gobionotothen gibberifrons*. The abundance of this polypeptide in the *C. rastrospinosus* extract appears similar to that of the red-blooded notothenioid fishes, whereas that of *P. georgianus* appears approximately two-fold less. These results suggest that the pigment in the two red-coloured icefish hearts is Mb.

ISOLATION OF A Mb cDNA CLONE FROM THE ANTARCTIC NOTOTHENIOID FISH, *N. CORIICEPS*

We attempted to confirm that the polypeptide detected by immunoblot analysis is indeed Mb by demonstrating the presence of Mb mRNA in northern blots. However, our initial efforts established only that cloned DNAs encoding mammalian Mb (Weller *et al.* 1984, Akaboshi 1985) do not cross-hybridize with *M. saxatilis, N. coriiceps* or other teleost Mb mRNAs to a reliable degree. Thus, it was necessary for us to clone a teleost Mb cDNA by the reverse transcriptase-polymerase chain reaction (RT-PCR) method (Saiki *et al.* 1988). mRNA was isolated from *N. coriiceps* heart ventricle by extraction with phenol:chloroform, essentially as described by Butler *et al.* (1990). Approximately 200 to 700 μg of total RNA was recovered per gram of ventricular tissue. Mb-specific, degenerate

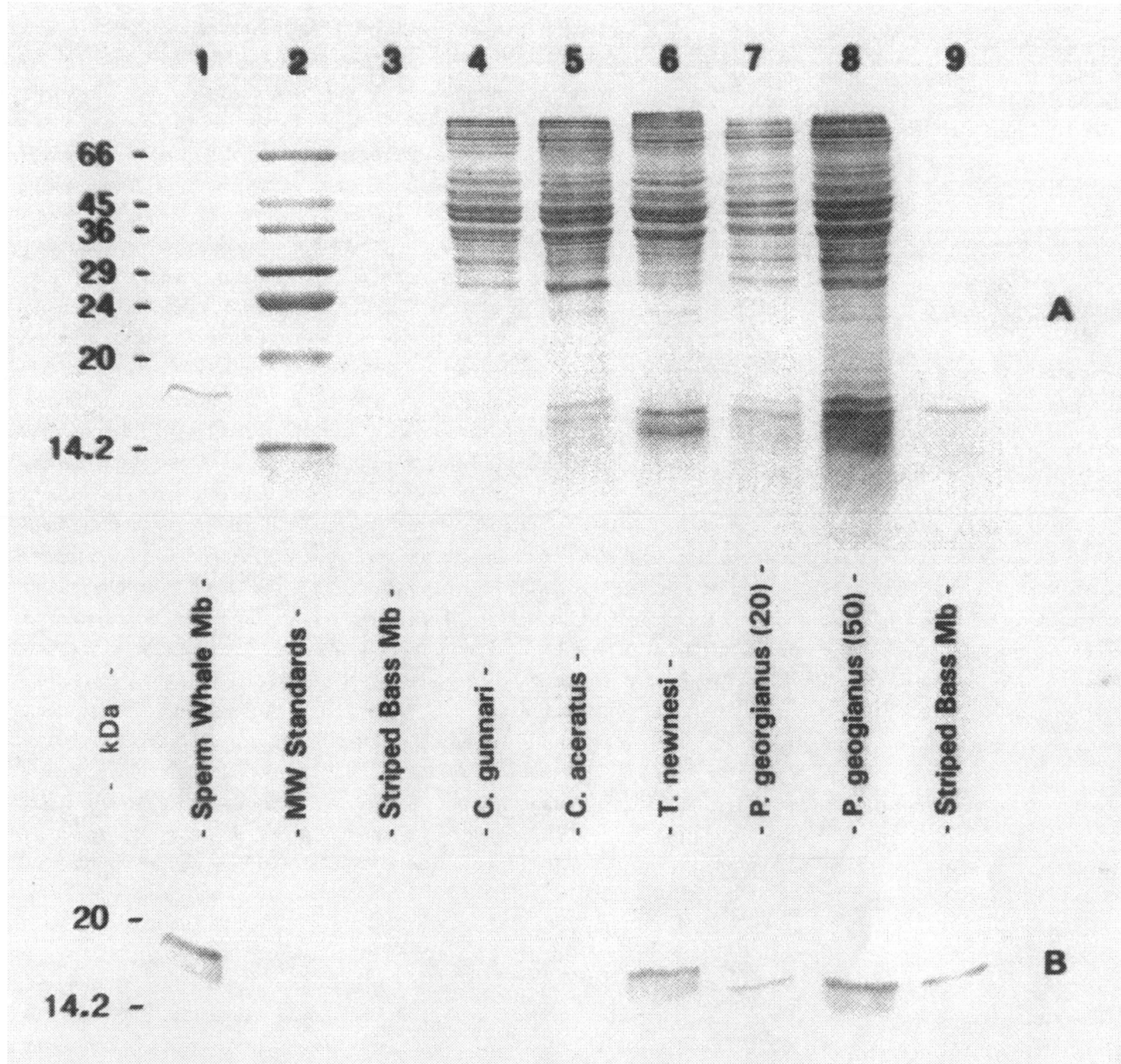

Fig. 9.1. Presence of myoglobin polypeptide in the heart ventricle of a channichthyid icefish. Polypeptides in a homogenate of heart ventricle supernatant were resolved by SDS-PAGE (A), electroblotted to nitrocellulose and the presence of myoglobin detected by incubation with a commercially available rabbit anti-human Mb antibody (B). Lane 1 – 0.5 μg purified sperm whale Mb; Lane 2 – mol. wt standards; Lane 3 – 0.5 μg purified striped bass (*M. saxatilis*) Mb; Lanes 4, 5, 6 – 20 μg each of supernatants of channichthyid icefishes *C. gunnari* and *C. aceratus*, and from the red-blooded notothenioid fish *T. newnesi*, respectively; Lanes 7, 8–20 μg and 50 μg crude supernatant of another channichthyid icefish, *P. georgianus*, respectively; Lane 9–1.0 μg of purified striped bass Mb.

oligonucleotide primers were synthesized based on the polypeptide sequences of Mb obtained from yellowfin tuna (*Thunnus albacares*, GenBank accession #P02205; Watts *et al.* 1980) and common carp (*Cyprinus carpio*, GenBank accession #P02204). A 329 bp PCR product was produced from *N. coriiceps* cDNA using a sense primer corresponding to codons 6 through 11 and an antisense primer spanning codons 115 through 110. This product was cloned into the TA cloning vector pCRII (Invitrogen) and propagated in *Escherichia coli*. Two independent clones were sequenced manually by the dideoxy method using Sequenase 2.0 (US Biochemical). Subsequently, the PCR product has been sequenced directly using the Cycle Sequencing Method (ABI) and an Automated DNA Sequencer (ABI). These sequences leave no doubt that the amplified product encodes *N. coriiceps* Mb; the cDNA sequence and deduced amino acid sequence of this clone are shown in Fig. 9.2. The predicted amino acid sequence of the *N. coriiceps* Mb segment bears >78% identity to the equivalent segments of the tuna and carp Mb polypeptide.

DETECTION OF Mb mRNA

Hybridization of radiolabelled *N. coriiceps* cDNA to RNA gel blots demonstrates that Mb is expressed in heart ventricular

```
A.
1   GTG CTG AAG TGC TGG GGT CCA ATG GAG GCG GAC TAC GCA ACC CAC GGG
49  GGG CTG GTG CTG ACC CGT TTA TTC ACA GAG CAC CCA GAA ACC CTG AAG
97  TTA TTC CCC AAG TTT GCT GGC ATC GCC CAT GGG GAC CTG GCC GGG GAT
145 GCA GGT GTT TCT GCC CAC GGT GCC ACA GTG CTG AAT AAA CTG GGT GAT
193 CTG CTG AAG GCC AGA GGC GCC CAC GCT GCC CTC CTC AAA CCT CTG TCC
241 AGC AGC CAC GCC ACC AAG CAC AAG ATC CCC ATT ATT AAC TTC AAG CTG
289 ATT GCA GAG GTC ATT GGT AAA GTC ATG GAG GAG AAG GCG GG

B.
Ncor:      VLKCWGPMEADYATHGGLVLTRLFTEHPETLKLFPKFAGIAHGDLAGDAGVSAHGA
Tuna: ADFDAVLKCWGPVEADTYYMGGLVLTRLFKEHPETQKLFPKFAGIAHGDLAGNAAISAHGA
Carp: HDAELVLKCWGGVEADFEGTGGEVLTRLFKQHPETQKLFPKFVGIASNELAGNAAVKAHGA

Ncor: TVLNKLGDLLKARGAHAALLKPLSSSHATKHKIPIINFKLIAEVIGKVMEEKAG
Tuna: TVLKKLGELLKAKGSHAAILKPLANSHATKHKIPINNFKLISEVLVKVMHEKAGLDAGGQT
Carp: TVLKKLGELLKARGDHAAILKPLATTHANTHKIALNNFRLITEVLVKVMAEKAGLDAGGQS

Ncor:
Tuna: ALRNVMGIIIADLEANYKELGFSG
Carp: ALRRVMDVVIGDIDTYYKEIGFAG
```

Fig. 9.2. Partial cDNA sequence and deduced amino acid sequence of *N. coriiceps* Mb. (A) Sense strand nucleotide sequence of the PCR product amplified from *N. coriiceps* cDNA using degenerate, Mb-specific primers. (B) The partial amino acid sequence of *N. coriiceps* Mb (Ncor) deduced from the nucleotide sequence in A, and comparison to the mature Mb polypeptide sequences of *T. albacares* (Tuna) and *C. carpio* (Carp).

tissues of *C. rastrospinosus* and *P. georgianus* (Fig. 9.3). In addition, the blot hybridization shown in Fig. 9.3 confirms that *C. aceratus* does not express the Mb gene. By contrast, ventricular extracts of *C. gunnari* consistently exhibit mRNA that hybridizes to the Mb probe, despite an absence of Mb polypeptide in this tissue. This mRNA species exhibits the same mobility (approximately 0.9 to 1.1 kb) as *C. rastrospinosus*, *P. georgianus*, *N. coriiceps* and *G. gibberifrons* Mb mRNA but is much less abundant. Five *C. gunnari* individuals caught in three different field seasons all exhibit the presence of Mb mRNA in this same low steady-state abundance. We have subsequently isolated mRNAs associated with polyribosome complexes from *C. gunnari*, *C. rastrospinosus*, *C. aceratus* and *N. coriiceps* heart ventricle using the procedure described by Crosby & Vayda (1991). This analysis indicates that essentially all of the Mb mRNA present in the heart ventricle of *C. gunnari*, *C. rastrospinosus* and *N. coriiceps* is associated with ribosomes *in vivo* (data not shown). However, we were unable to detect the presence of an immunoreactive Mb polypeptide upon translation of *C. gunnari* heart ventricle mRNA *in vitro*. We conclude from these observations that the *C. gunnari* Mb gene has suffered a point mutation that either causes premature termination of the Mb polypeptide or results in a Mb polypeptide that lacks epitopes recognized by polyclonal anti-Mb antisera. Reduction in the steady-state level of an mRNA species is often observed upon introduction of a mutation that causes premature termination of the polypeptide product.

Quiescence of the *C. aceratus* Mb gene appears to result from a mutation distinct from that of *C. gunnari*. There is a complete absence of mRNA hybridizing to the Mb cDNA probe in both *C. aceratus* total ventricular RNA and polyribosome-associated mRNA preparations. Further, the currently accepted cladistic tree of the Channichthyidae based upon morphological characters (Iwami 1985), suggests that *C. aceratus* and *C. gunnari* are more distantly related than either *C. aceratus* and *C. rastrospinosus* or *C. gunnari* and *P. georgianus*, suggesting a distinct mutational event. However, *C. aceratus* genomic DNA contains sequences that hybridize efficiently to the *N. coriiceps* Mb cDNA probe (data not shown). Thus, the Mb gene has not been deleted

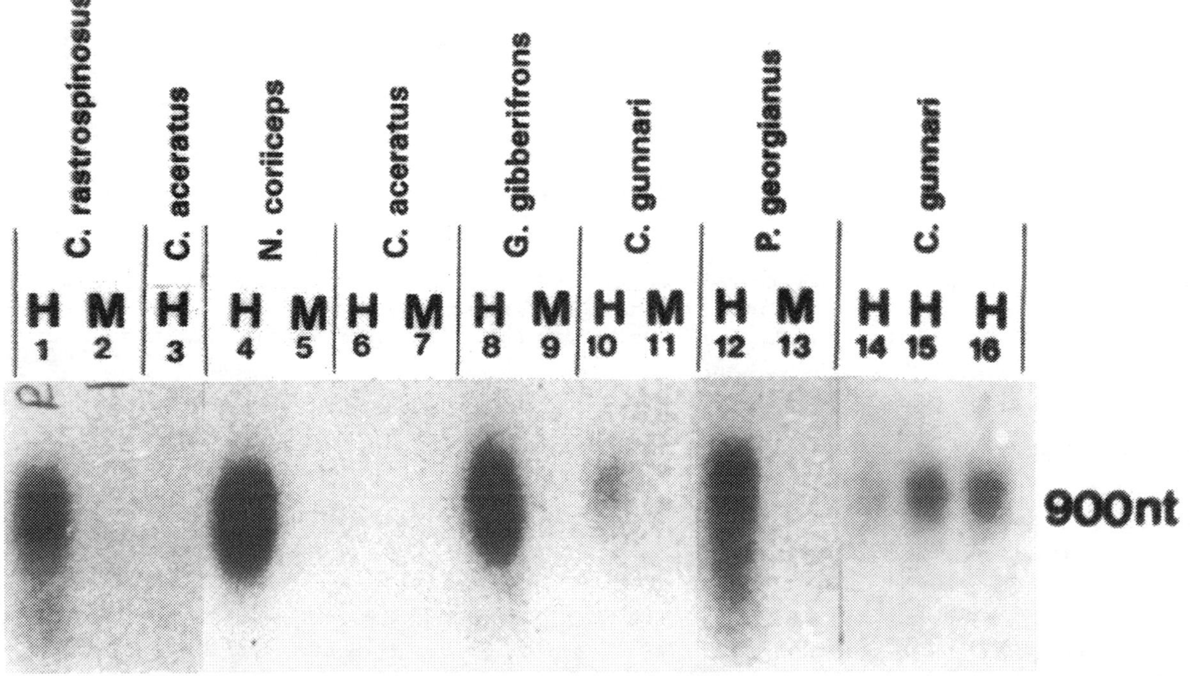

Fig. 9.3. Myoglobin mRNA is present in the heart ventricles of three channichthyid icefish species but absent from the aerobic skeletal muscle of all of the Antarctic notothenioid fishes examined. RNA was purified from heart ventricle (H) or aerobic pectoral adductor profundus muscle (M) of *C. rastrospinosus* (lanes 1,2), *N. coriiceps* (lanes 4,5), *C. aceratus* (lanes 3,6,7), *G. gibberifrons* (lanes 8,9), *C. gunnari* (lanes 10,11,14,15,16) and *P. georgianus* (lanes 12–13). The redundant samples of *C. aceratus* and *C. gunnari* were obtained from different individuals to confirm the absence or presence of Mb mRNA.

entirely from the *C. aceratus* genome. Loss of expression is due either to a promoter mutation, an RNA processing mutation, genomic rearrangement or partial deletion. We are currently engaged in nuclear run-on transcription analyses to determine whether the *C. aceratus* Mb sequence is expressed. Our efforts to sequence *C. aceratus* genomic clones hybridizing to the Mb probe should reveal the nature of the mutation in this species.

Fig. 9.3 also illustrates another striking feature of Mb expression in Antarctic fish: none of the notothenioid species we have tested to date accumulate Mb mRNA in the aerobic skeletal muscle, pectoral adductor profundus. By contrast, Mb is present in aerobic skeletal muscles of the temperate zone fishes *M. saxatilis, T. albacares* and *C. carpio*. The unusual pattern of Mb expression in two distant families of the Notothenioidei suggests that Mb is not an absolute requirement to support aerobic metabolism of skeletal muscle in the cold Antarctic environment, and further, that this mutation likely occurred in a common ancestor early in the radiation of the notothenioid fishes. Comparison of Mb promoter structure of temperate zone and notothenioid fishes may provide insight into mechanisms of tissue-specific gene expression.

CONCLUSIONS

We have demonstrated conclusively that at least two haemoglobin-less channichthyid icefish express Mb, *C. rastrospinosus* and *P. georgianus*. Individuals of a third icefish species, *C. gunnari*, consistently exhibit low steady-state levels of Mb mRNA: this mRNA is bound by ribosomes but Mb polypeptide is not detected by immunoblot either *in vivo* or when the polysome-associated mRNA is translated *in vitro*. Failure to detect Mb polypeptide in this species may be due either to the absence of the polypeptide or the lack of the epitopes recognized by the anti-human Mb antiserum. A fourth icefish species, *C. aceratus*, does not produce detectable levels of either Mb polypeptide or mRNA, but contains sequences in its genomic DNA that hybridize to the Mb coding sequence of *N. coriiceps*. All of the Antarctic notothenioid fishes examined to date lack Mb expression in aerobic skeletal muscle, in stark contrast to other teleosts such as tuna, carp, and striped bass. Thus, the icefish species examined to date constitute a closely related group of animals that exhibit control of Mb expression at three distinct levels: polypeptide accumulation, mRNA accumulation, and gene activation. We are currently attempting to determine whether the *C. aceratus* Mb sequences are transcribed. We plan to clone the Mb sequences from these species for comparison of coding sequences and promoter sequence elements, which may shed light on the failure to express Mb in specific species, or in a tissue-specific fashion.

ACKNOWLEDGEMENTS

The authors sincerely thank Drs. R. Londraville and K. Rodnick for their efforts in preparing the immunoblot shown. This work was supported by NSF grants DPP 88-19469 and DPP 92-20775 to BDS. The *Notothenia coriiceps* partial Mb cDNA is GenBank Accession #U68350.

REFERENCES

Akaboshi, E. 1985. Cloning and sequence analysis of porcine myoglobin cDNA. *Gene*, **40**, 137–140.

Anderson, M. E. 1990. The origin and evolution of the Antarctic ichthyofauna. In Gon, O. & Heemstra, P. C., eds. *Fishes of the Southern Ocean*. Grahamstown: JLB Smith Institute of Ichthyology, 26–33.

Butler, W., Cook, L. & Vayda, M. E. 1990. Hypoxic stress inhibits multiple aspects of the potato tuber wound response. *Plant Physiology*, **93**, 264–270.

Crosby, J. S. & Vayda, M. E. 1991. Stress-induced translational control in potato tubers may be mediated by polysome-associated proteins. *The Plant Cell*, **3**, 1013–1023.

DeWitt, H. H. 1971. Coastal and deep-water benthic fishes of the Antarctic. In Bushnell, V. C., ed. *Antarctic Map Folio Series*, Folio 15. New York: American Geographical Society, 1–10.

Douglas, E. L., Peterson, K. S., Gyso, J. R. & Chapman, D. J. 1985. Myoglobin in the heart tissue of fishes lacking hemoglobin. *Comparative Biochemistry and Physiology*, **81**A, 885–888.

Eastman, J. T. 1990. The biology and physiological ecology of Notothenioid fishes. In Gon, O. & Heemstra, P. C., eds. *Fishes of the Southern Ocean*. Grahamstown: J. L. B. Smith Institute of Ichthyology, 34–51.

Eastman, J. T. 1991. Evolution and diversification of Antarctic notothenioid fishes. *American Zoologist*, **31**, 93–109.

Eastman, J. T. 1993. *Antarctic fish biology: evolution in a unique environment*. San Diego: Academic Press, 322 pp.

Fitch, N. A., Johnston, I. A. & Wood, R. E. 1984. Skeletal muscle capillary supply in a fish that lacks respiratory pigments. *Respiration Physiology*, **57**, 201–211.

Hamoir, G. 1988. Biochemical adaptation of the muscle of the Channichthyidae to their lack of haemoglobin and myoglobin. *Comparative Biochemistry and Physiology*, **90B**, 557–559.

Hemmingsen, E. A. 1991. Respiratory and cardiovascular adaptation in hemoglobin-free fish: resolved and unresolved problems. In di Prisco, G., Maresca, B. & Tota, B., eds. *Biology of Antarctic fish*. New York: Springer-Verlag, 191–203.

Hemmingsen, E. A. & Douglas, E. L. 1970. Respiratory characteristics of the haemoglobin-free fish *Chaenocephalus aceratus*. *Comparative Biochemistry and Physiology*, **33**, 733–744.

Iwami, T. 1985. Osteology and relationships of the family Channichthyidae. *Memoirs of the National Institute of Polar Research, Series* **E36**, 1–69.

Johnston, I. A., Fitch, N., Zummo, G., Wood, R. E., Harrison, P. & Tota, B. 1983. Morphometric and ultrastructural features of the ventricular myocardium of the haemoglobin-less icefish *Chaenocephalus aceratus*. *Comparative Biochemistry and Physiology*, **76A**, 475–480.

Kennett, J. P. 1977. Cenozoic evolution of Antarctic glaciation, the circum-Antarctic Ocean, and their impact on global paleoceanography. *Journal of Geophysical Research*, **82**, 3843–3860.

Kennett, J. P. 1980. Paleoceanographic and biogeographic evolution of the southern ocean during the Cenozoic, and Cenozoic microfissil datums. *Palaeogeography, Palaeoclimatology and Palaeoecology*, **31**, 123–152.

Littlepage, J. L. 1965. Oceanographic investigation in McMurdo Sound, Antarctica. *Antarctic Research Series*, **2**, 1–37.

Saiki, R. K., Gelfand, D. H., Stroffel, S., Scharf, S. J., Higuchi, R., Horn, G. T., Mullis, K. B. & Erlich, H. A. 1988. Primer-directed enzymatic amplification of DNA with a thermostable DNA polymerase. *Science*, **239**, 487–491.

Sato, F., Shiro, Y., Sakaguchi, Y., Iizuka, T. & Hayashi, H. 1990. Thermodynamic study of protein dynamic structure in the oxygen binding reaction of myoglobin. *Journal of Biological Chemistry*, **265**, 18823–18828.

Stevens, E. D. & Carey, F. G. 1981. One why of the warmth of warm-bodied fish. *American Journal of Physiology*, **240**, R151–R155.

Tota, B., Farina, F. & Zummo, G. 1988. Ultrastructural aspects of functional interest in the ventricular myocardial wall of the Antarctic icefish *Chaenocephalus aceratus*. *Comparative Biochemistry and Physiology*, **90B**, 561–566.

Watts, D. A., Rice, R. H & Brown, W. D. 1980. The primary structure of myoglobin from yellowfin tuna *Thunnus albacares*. *Journal of Biological Chemistry*, **255**, 10 916–10 924.

Weller, P., Jeffreys, A. J., Wilson, V. & Blanchetot, A. 1984. Organization of the human myoglobin gene. *European Molecular Biology Organization Journal*, **3**, 439–446.

II Community structure and function

INTRODUCTION

Antarctic communities can be divided into two types – simple systems with few species in terrestrial and limnic habitats, and complex systems with many species in pelagic and benthic habitats. This difference does not necessarily reflect the ecological opportunities as far as niche utilization is concerned – for example, there are many inadequately exploited niches in terrestrial ecosystems – but is attributable to opportunities for dispersal, colonization and development. The continuity of the marine ecosystem around the Antarctic, and the comparative stability of the marine environment, when compared with the extreme seasonal changes in environment in terrestrial ecosystems, has made the buffered aquatic system an easier place for community assemblage and persistence.

The claims for simplicity in terrestrial ecosystems (Walton 1987) must be qualified. It is true that many trophic levels are missing and species diversity is low compared with temperate and tropical ecosystems. Nevertheless, in terms of understanding the functioning of communities these systems are still proving quite complex to analyse and model. In the benthic and pelagic communities there is generally too little information at the species level to do other than pick out interesting features and examine various indices of functionality. Linking the terrestrial and marine ecosystems are the birds and seals, providing the nutrient flow from the sea to the land.

In the Southern Ocean phytoplankton communities, and their grazers, have been the subject of a great deal of research. Previous observations have shown great variablity, both within and between years, in phytoplankton communities. Isolating the causes of this is proving difficult. Of particular interest is how primary production is related to nutrient availability and to oceanographic variables such as currents, stratification, sea-ice formation and winds. Research on this is a fundamental part of the US Long Term Ecological Research Program at Anvers Island. Moline *et al.* (Chapter 10) have investigated the way in which nutrient limitation is linked to stratification in surface waters by sea-ice and freshwater input. Their evidence shows that phytoplankton community structure may be greatly changed by depletion of nitrate and phosphate and advection of a new water mass into the area during a period of storms. A diatom-dominated community in a stable water column then becomes one dominated by *Phaeocycstis*. Comparing their data with those of previous authors suggests that this is a typical summer pattern at least for coastal areas on the west side of the Antarctic Peninsula.

Further north in the South Shetland Islands Villafane *et al.* (Chapter 11) sampled Drake Passage waters. Their data, from four summers, show considerable fluctuations in the floristic composition of the phytoplankton, with diatoms, dinoflagellates and flagellates all being dominant in different years. They suggest that variability in the abundance and species composition of the phytoplankton in this area may be attributable to grazing by salps, copepods, amphipods and euphausiids, but there is at present insufficient information to substantiate this.

Identifying the phytoplankton species involved in the blooms can be difficult, especially if some species are colonial which can be difficult to deal with by standard microscopical techniques. Skerratt *et al.* (Chapter 12) describe a technique using lipid signatures for identification, which was successfully applied to both ocean water and sea ice. Interestingly, *Phaeocystis* was found as a dominant species only once in five summers at their shelf site near Davis Station, but their data show that during that summer the pattern of floristic change was very similar to that shown for Anvers Island on the other side of the Antarctic.

Much further north Razouls *et al.* (Chapter 13) identified temporal variability in the sub- Antarctic bacterial, phytoplankton and zooplankton communities. One important result of this study is the importance of the sampling interval required to characterize key seasonal changes which may occur on a timescale of a few days. Their work again relates changes in community dominance to nutrient depletion and was undertaken on a shelf area at Kerguelen. Such shelf areas which support *Phaeocystis* are known to produce considerable amounts of dimethylsulphide (DMS) (Gibson *et al.* 1990). DMS is considered to account for over half of the biogenic global sulphur flux to the atmosphere and may be important for global change processes. At present its production and cycling are poorly understood but are clearly linked to phytoplankton communities. Studies by DiTullio & Smith (Chapter 14) have shown high levels of DMS in late summer in the Ross Sea, and have demonstrated that iron enrichment appears to increase growth in

Phaeocystis, which supports the suggestion that iron-limitation may be limiting DMS flux at present.

One of the greatest areas of interest at present is in the sea-ice zone. The physical structure of the niches to be occupied is not only complex but dynamic (in size, salinity and irradiance) in a way that few other ecosystems can match. Whilst the general features of sea-ice algae were described some time ago (e.g. Bunt & Wood 1963) it has recently become clear that a much more detailed picture is necessary. The seasonal dynamics of these communities are clearly fundamental to any understanding of their contribution to carbon flow into higher trophic levels. Fritsen & Sullivan (Chapter 15), working on first-year and multi-year ice, have provided a conceptual model of ice dynamics and the seasonal changes in the associated microbial communities. The generality of this model has yet to be tested, but it consitutes an important stage in marrying the physics, chemistry and biology of this unique ecosystem.

Zooplankton are important parts of the food chain, and none more so than krill. During the BIOMASS studies it became clear that an important key to understanding the dynamics of krill was to describe its location and activity during winter. The first suggestions of the relationship between krill and sea ice were made in Guzman(1983) but progress has been fairly slow since this, in part due to logistic difficulties of working in the Antarctic winter. Whilst the association between krill and ice is now not disputed, there are still major questions on how it moves from the water column into the forming sea ice, how and when aggregation occurs, and what life stages are represented in what numbers over the winter period. This last question is addressed here by Frazer *et al.* (Chapter 16) who found larval krill in areas west of the Antarctic Peninsula, which showed greater abundance and increasing aggregation during late winter.

Reproduction and development have been carefully investigated for both euphausiids and copepods. Egg production is a parameter both of estimates of population dynamics of the species (which clearly relates to frequency and adundance at the community level) but also of availability of food for larval fish. Kurbjeweit (Chapter 17), focusing on the dominant copepods *Calanus propinquus* and *Metridia gerlachei*, identified two modes of reproduction, which are both apparently food dependent and temperature independent but result in quite different temporal patterns of egg laying.

How far have we gone in understanding the relationships between the different community components for the benthic ecosystem? German work in the Weddell Sea has provided a conceptual model of energy flows at a large scale (Schalk *et al.* 1993), but at the smaller scale of shelf benthos accumulating data were waiting to be modelled. Jarre-Teichmann *et al.* (Chapter 18) have applied a steady-state model to the benthic communities of the southeastern Weddell Sea to characterize the transfers between the major trophic groups. Whilst the model identifies major contributors to carbon flow, for example Polychaetes and Holothuroideans, sensitivity analysis clearly indicates inadequate data on the role of protozoans and meiofauna, the population dynamics of dominant species and the physiology and biochemistry of high-Antarctic benthos.

At the more local level there have been an increasing number of papers on benthic species and communities. One approach has been to study those species which are important in terms of dominance or biomass, especially where their activities structure or change various aspects of the community. This can be especially valuable when attempting to model gross carbon transfers for coastal environments, generally believed to be highly productive (e.g. Gilbert 1991), and with rich benthic communities (e.g. see Knox 1994). However, low phytoplankton production can occur in areas with high benthic biomass, raising the question of where the carbon for the benthos is principally derived from. Schloss *et al.* (Chapter 19) have examined this question for a bay in the South Shetland Islands and concluded that there is no simple or general answer. Whilst planktonic production provides particulate input in spring, for the rest of the year resuspension of benthic material appears to be the key process. For coastal sites with significant meltwater input or which are especially shallow there are further difficulties for benthos feeding caused by suspended inorganic matter and by wind-driven mixing. Coupling physical and biological models for such sites is clearly going to be complex. A different approach was used by Ahn (Chapter 20), who chose to study the lamellibranch *Laternula elliptica,* which is widely distributed in shallow water and not infrequently the dominant species in benthic communities on soft substrates. She found that benthic micro-algae may be the principal food source for *Laternula* after the phytoplankton bloom is over, apparently confirming the conclusion of Dayton *et al.* (1986) that benthic invertebrate production is more closely related to benthic primary production than phytoplankton production.

Some rather different questions have been studied in terrestrial and freshwater communities. Community membership is believed to be determined by assemblage rules. The sub-Antarctic islands are seen by Chown (Chapter 21) as ideal sites for investigating such questions, in this instance using the weevils of Marion Island as the test case. He is able to show that whilst regional processes determine original assemblage membership, local interactions control ecological disposition. Applying his understanding of the rules to the effects of climate change on weevils he suggests that the key responses will need to be unravelled at the level of new interspecific interactions or changes in the strength of existing ones; looking for major changes at species distribution level is too crude a tool.

Microbial populations should offer ideal vehicles for investigations of community structure and function, in particular to examine the factors controlling species abundance and distribution. There have been various studies on individual parts of the terrestrial community and comparison of these indicates a mismatch in estimates of microbial biomass (Bölter 1994). To address this microscopy and vital staining were used by Bölter (Chapter 22) to identify the bacteria and autotrophs directly, rather than by culture or chemical assay, and characterise the microbial communities of a range of distinct habitats on the South Shetland Islands. In the aquatic communities there are a variety of unique Antarctic features which make research especially stimulating. In particular the pools on the McMurdo Ice Shelf by Bratina Island offer an amazing range of water types and diverse assemblages of algae and cyanobacteria. Hawes *et al.* (Chapter 23) chose to examine a protected tidal lagoon, a habitat type apparently quite rare in Antarctica. Their conclusions that the mud flat diatom community differed little in biomass or species composition from similar habitats in lower latitudes is unexpected since the habitat was considered to be probably the most extreme in the lagoon. Another extreme aquatic environment is Ace Lake, a saline meromictic water body near Davis Station. The occurrence of the marine picocyanobacterium *Synechococcus* in the lake in large numbers allowed Rankin *et al.* (Chapter 24) to use it to investigate why the species is present in such low numbers in Antarctic marine communities when it frequently dominates in more temperate oceans.

REFERENCES

Bölter, M. 1994. Estimations of microbial biomass by direct and indirect methods with special respect to monitoring programs. *Proceedings of the NIPR Symposium on Polar Biology*, **7**, 198–208.

Bunt, J. S. & Wood, E. J. F. 1963. Microalgae and Antarctic sea ice. *Nature*, **199**, 1254–1255.

Dayton, P. K., Watson, D., Palmisano, A., Barry, J. P., Oliver, J. S. & Rivera, D. 1986. Distribution patterns of benthic microalgal standing stock at McMurdo Sound, Antarctica. *Polar Biology*, **6**, 207–213.

Gibson, J. A. E., Garrick, R. C. Burton, H. R. & McTaggart, A. R. 1990. DMS and the alga *Phaeocycstis pouchetii* in Antarctic coastal waters. *Marine Biology*, **104**, 339–346.

Gilbert, N. S. 1991. Primary production by benthic microalgae in nearshore marine sediments of Signy Island, Antarctica. *Polar Biology*, **11**, 339–346.

Guzman, F. O. 1983. Distribution and abundance of Antarctic krill (*Euphausia superba*) in the Bransfield Strait. *Berichte zur Polarforschung Special Issue* No. 4, 169–190.

Knox, G. A. 1994. *The biology of the Southern Ocean*. Cambridge: Cambridge University Press, 444 pp.

Schalk, P. H., *et al.* 1993. Towards a conceptual model for the Weddell Sea ecosystem, Antarctica. In Christensen, V. & Pauly, D., eds. *Trophic models of aquatic ecosystems. ICLARM Conference Proceedings*, **26**, 323–337.

Walton, D. W. H. 1987. Antarctic terrestrial ecosystems. *Environment International*, **13**, 83–93.

10 Temporal dynamics of coastal Antarctic phytoplankton: environmental driving forces and impact of a 1991/92 summer diatom bloom on the nutrient regimes

MARK A. MOLINE[1], BARBARA B. PRÉZELIN[1], OSCAR SCHOFIELD[1] AND RAYMOND C. SMITH[2]

[1]*Marine Primary Productivity Group, Department of Biological Sciences and Marine Science Institute, University of California, Santa Barbara, CA 93106, USA,* [2]*Department of Geography, the CSL/Center for Remote Sensing and Environmental Optics, and the Marine Science Institute, University of California, Santa Barbara, CA 93106, USA*

ABSTRACT

Within the Palmer Long Term Ecological Research Program (PAL-LTER), a suite of environmental data sets were collected at a nearshore station throughout the 1991/92 austral summer. Seasonal changes are presented in the context of phytoplankton community ecology. Subseasonal fluctations in sea-ice coverage, freshwater inputs, as well as wind driven and advective processes disrupting stratified surface waters, appeared to be the major driving forces affecting the timing, duration and demise of local phytoplankton blooms. During a large diatom-dominated bloom (~30 mg chl a m^{-3}), macronutrients were depleted to detection limits (NO_3^- < 0.05 mmol m^{-3}, PO_4^{3-} < 0.03 mmol m^{-3}) and significant shifts in nutrient ratios were observed. Phytoplankton populations were light limited below ~5 m during the bloom, resulting from self-shading. The depth of light limitation deepened after the bloom was physically disrupted and removed from the region by strong advective processes.

Key words: phytoplankton, Antarctica, bloom, time series, nutrient limitation, mixed layer depths, photoadaptation, light limitation

INTRODUCTION

Recent studies have integrated physical, chemical and biological data in an attempt to understand the mechanisms controlling phytoplankton bloom dynamics in the Southern Ocean (Smith & Sakshaug 1990, Holm-Hansen & Mitchell 1991, Mitchell & Holm-Hansen 1991, Sakshaug *et al.* 1991). Theories and empirical models derived from such studies suggest that resource limitation and/or water column stability are the major factors governing phytoplankton bloom dynamics. Most studies have been conducted shipboard in pelagic regions and have focused on describing the short-term spatial variability of phytoplankton distribution, abundance, productivity and physiology. Temporal variations in bloom dynamics on the timescale of weeks are less documented (however, see Whitaker 1982, Krebs 1983, Priddle *et al.* 1986, Mitchell & Holm-Hansen 1991) and are generally based on a limited number of environmental parameters and/or insufficient information to track variations on the timescale of a few days over an entire season.

With the PAL-LTER data set from the summer 1991/92 season, we have just that opportunity. Here, we document the occurrence of a large bloom and present a detailed study of phytoplankton dynamics off the coast of Anvers Island on subseasonal timescales ranging from days to months within the summer season. By examining these variations on timescales of days, we hope to advance the understanding of the mechanisms regulating phytoplankton productivity and bloom development in Antarctic coastal waters.

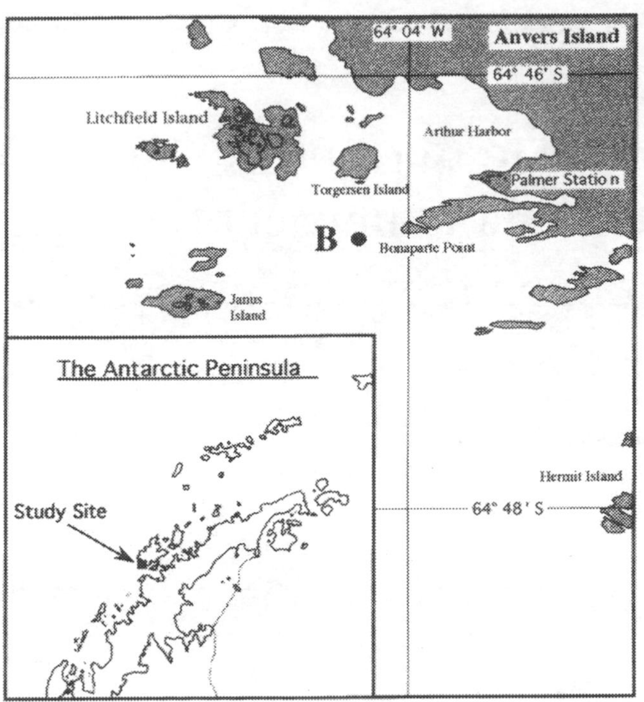

Fig. 10.1. Location of LTER sampling station B (64° 46.45' S, 64° 03.27' W) with respect to Palmer Station and (inset) the Antarctic Peninsula.

MATERIALS AND METHODS

Between November 1991 and February 1992, a total of 257 discrete water samples were collected repeatedly from Station B (Sta B) in the nearshore waters adjacent to Palmer Station, Antarctica (Fig. 10.1). Daily precipitation, snow cover and average wind speed and direction measurements were made at Palmer Station during the study period. These measurements were part of a longterm database collected by the US National Science Foundation. Prior to sample collection, in-water irradiances of photosynthetically available irradiation (Q_{par}) were made with a Biospherical Scalar Irradiance Meter (QSR-170DT) equipped with a QSP-100DT underwater sensor, deployed from an inflatable Zodiac boat. Samples were collected in 5l GoFlo bottle, transferred to dark carboys and immediately returned to Palmer Station for analyses.

Replicate subsamples for nutrient determination were filtered within an hour of collection through a 0.2 μm Nuclepore membrane, and the 20 ml filtrate for each sample was stored in polycarbonate scintillation vials (acid washed) at −70 °C. Samples were later transported at −20 °C to the Marine Science Analytical Laboratories, University of California, Santa Barbara for nutrient analyses. Methods for determination of the dissolved inorganic NO_3^-, PO_4^{3-}, and $Si(OH)_4$ concentrations were those of Johnson *et al.* (1985).

Reverse-phase HPLC procedures of Bidigare *et al.* (1989) were followed to determine the abundance of 17 phytoplankton pigments. Replicate 1 litre samples were filtered on 0.4 μm nylon 47 mm Nuclepore filters and extracted in 3 ml 90% acetone for 24 h in the dark (–20 °C). Pigment separation was carried out with a Hitachi L-6200A liquid chromatograph (436 nm). Peak identities of algal extracts were determined by comparing their retention times with pure pigment standards.

Blue-green photosynthetron methods described by Prézelin *et al.* (1994) were used to determine photosynthesis–irradiance (P–I) relationships for 77 of the collected samples. Non- linear curve fits for P–I data were calculated using the simplex method of Caceci & Cacheris (1984). Curve fitting provided estimates of P_{max} (the light saturated rate of photosynthesis), α (the affinity for photosynthesis at light-limited irradiances) and $I_k = P_{max}/\alpha$ (an estimate of the minimum irradiance required to light saturate photosynthesis).

Physical data were collected with instrumentation on a second inflatable Zodiac boat described by Smith *et al.* (1992). A total of 21 conductivity and temperature profiles were collected at Sta B using a SeaBird CTD. Here, we make a preliminary estimate of the upper mixed layer (UML) based on Sigma-t (σ_t) plots derived from these profiles, using the formula

$$\max\left|\frac{d\sigma_t}{dz}\right|$$

which assumes the UML depth to be equal to the depth where the gradient in σ_t is maximal. Like Mitchell & Holm-Hansen (1991), we assume that if the maximal σ_t gradient was less than 0.05 per metre, then the water column is essentially well mixed to the bottom (*c.* 80 m at Sta B).

There were several occasions when a freshwater lens was clearly evident on top of the marine layer. In these instances, an estimate of the mixing depth of the freshwater lens (FW-MLD) was made in addition to the UML depth.

Contour Plots were generated using the Delaunay triangulation method (DeltaGraph Pro3, DeltaPoint Inc., Monterey, CA, USA).

RESULTS

Until mid-December, the water column at Sta B was routinely covered with ice, buffered from relatively weak local winds, isothermal (–1.3 °C) and mixed to the bottom (Fig. 10.2). Glacier calving and a major wind event combined to break up and blow out local fast ice in mid-December. Due to solar insolation, ice-free conditions, major snow and glacier melt and precipitation, a freshwater lens (σ_t=26–26.6) developed and persisted throughout the summer season (Fig. 10.2). In addition to the effects of freshwater, relatively low wind speeds (<5 m s^{-1}) recorded through the first week in January allowed the water column to stratify and the UML shallowed to 20 m. Surface water temperature within the FW-MLD had warmed to +1.3 °C during this time, while within the UML, below the FW-MLD, the temperature was *c.* −0.2 °C (data not shown). Storm activity, beginning the second week of January, produced high precipitation and maintained strong northerly winds (av. ~15 m s^{-1}) for the following two weeks (Fig. 10.2). Water stratification broke down and within four days, the UML depth deepened to 55 m. Temperature–salinity relationships over the season indicate that

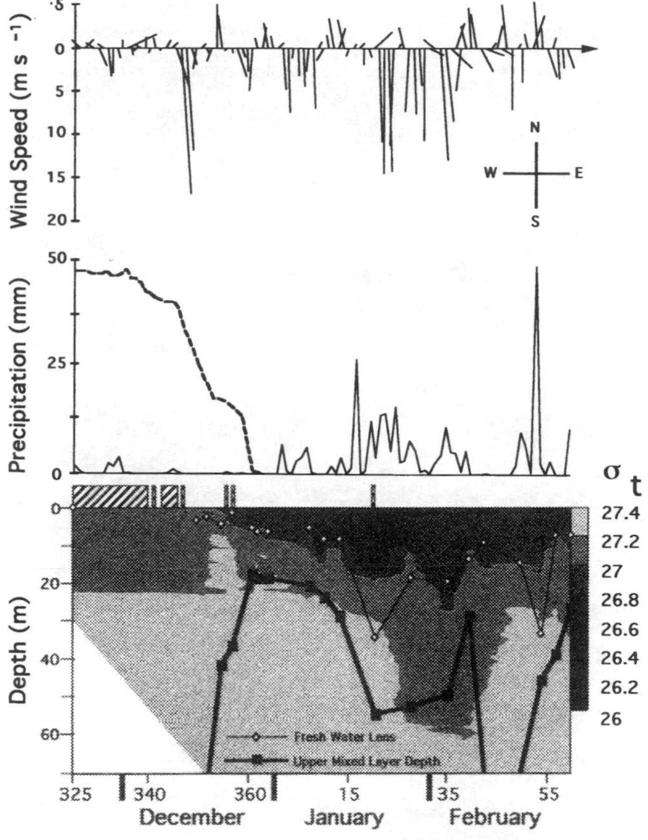

Fig. 10.2. (upper) Seasonal changes in daily average wind speed and direction at Palmer Station between November 21, 1991 and February 27, 1992. Inserted compass provides direction toward which the wind is blowing. (middle) Seasonal patterns in daily precipitation (solid line) and relative snow cover (dashed line) at Palmer Station. (lower) Seasonal contour plot of sigma-t (σ_t) at Station B, indicated by shading; see scale on the right. Seasonal changes in the depth of the upper mixed layer (black squares) and the freshwater lens (open diamonds) overlay the σ_t contour. Daily presence or absence of pack ice is indicated by the hatched bars.

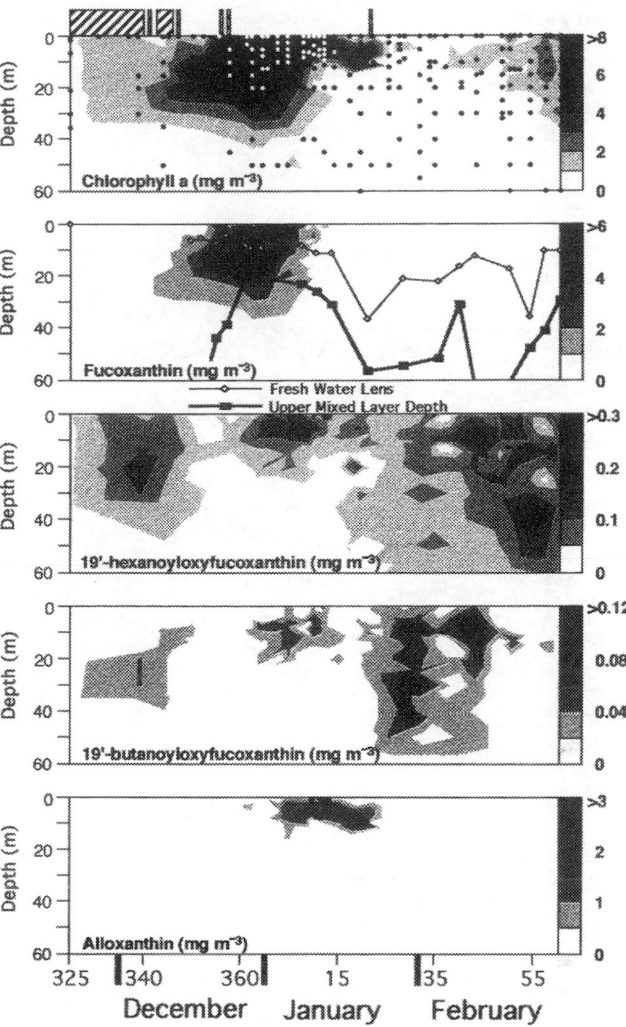

Fig. 10.3. Comparison of the seasonal changes in the depth distribution of chlorophyll *a* and of key phytoplankton pigments at Station B between November 21, 1991 and February 27, 1992. Pigments shown are indicators for diatoms (fucoxanthin), prymnesiophytes (19'-hexanoyloxyfucoxanthin), chrysophytes (19' butanoyloxyfucoxanthin), and cryptophytes (alloxanthin). Pigment concentrations are shown by shading; see scale on the right. Distribution of discrete samples is shown with closed circles in the upper panel. Seasonal changes in the depth of the upper mixed layer (black squares) and the freshwater lens (open diamonds) are shown in the second panel. The daily presence or absence of pack ice is indicated by the hatched bars.

a different water mass was advected into the area during this period (data not shown). With variable winds, the UML depth fluctuated and was not stable until mid February.

Concentrations of chlorophyll *a* (mg chl *a* m^{-3}) at Sta B varied 340-fold over the summer period, ranging from 0.086 to 29.2 mg m^{-3} (Fig. 10.3). Integrated water column values ranged from 59 to 612 mg chl *a* m^{-2}, with up to 70% of the biomass below the UML (Fig. 10.4). Early summer communities, present under the ice, were dominated by prymnesiophytes (indicated by 19'-hexanoyloxyfucoxanthin) and chrysophytes (19'-butanoyloxyfucoxanthin) (Fig. 10.3). In mid to late December, coincident with the shallowing of the UML depth to *c*. 20 m, a large diatom-dominated (fucoxanthin) bloom developed at Sta B (Fig. 10.3). For three weeks, the bloom intensified with the highest chl *a* concentrations being observed between 0–10 m. Taxonomic identification indicated bloom samples were dominated by a centric diatom, *Coscinodiscus* spp. (D. Karentz, pers. commun.). Within the last week of the bloom, there was a ten-fold increase in the abundance of prymnesiophyte pigmentation and a corresponding decrease in diatom abundance. The presence of single-celled *Phaeocystis* spp. was confirmed (D. Karentz, pers. commun.). During the second week of 1992 and corresponding to the advection event described above, chl a and all other pigments showed a rapid decrease, with the exception of significant surface concentrations of cryptophytes (alloxanthin) present for a week after the bloom disappearance. Chrysophyte communities were found until the beginning of February when *Phaeocystis* became dominant. As the UML depth began to shallow at the end of the summer monitoring period, diatoms were once again evident.

NO_3^-, PO_4^{3-} and $Si(OH)_4$ concentrations and the corresponding molar ratios of $Si(OH)_4$:NO_3^- and NO_3^-:PO_4^{3-} showed dramatic changes over the season (Fig. 10.5). Associated

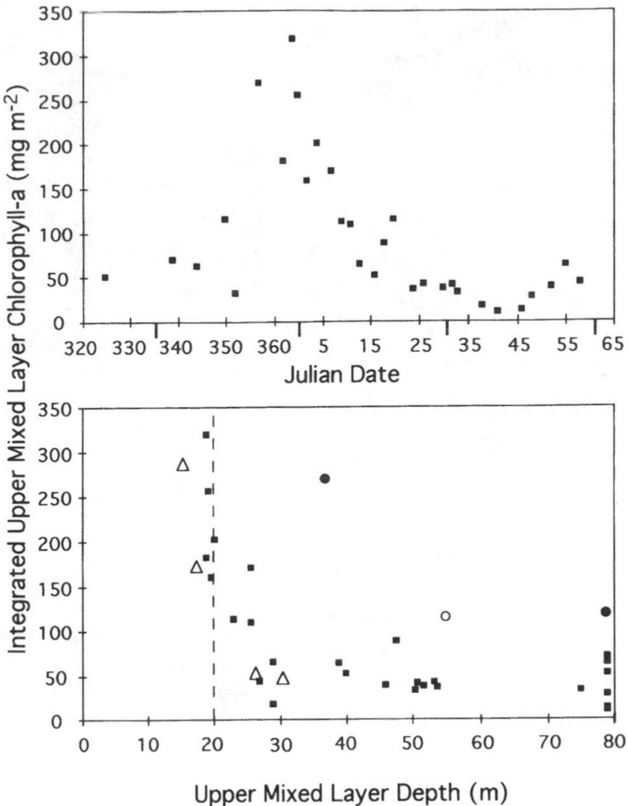

Fig. 10.4. Variations in integrated chl *a* above the upper mixed layer depth (UML) displayed as a function of time over the austral summer (upper) and UML (lower) for Station B, sampled between November 21, 1991 and February 27, 1992. In the lower panel, discrete data points within the PAL-LTER dataset have been differentially labelled (by closed squares) to indicate integrated chl *a* values present prior to the major shallowing of the UML (closed circles) noted in December, 1991 (Fig. 10. 2) and one value during a major advection event in late January, 1992 (open circle). For comparison, monthly mean values (open triangles), derived from chl *a* and σ_t profiles collected along the Antarctic Peninsula between December 1986 and March 1987, are presented (Mitchell & Holm-Hansen 1991). Dashed line represents UML depth in late Dec./early Jan.

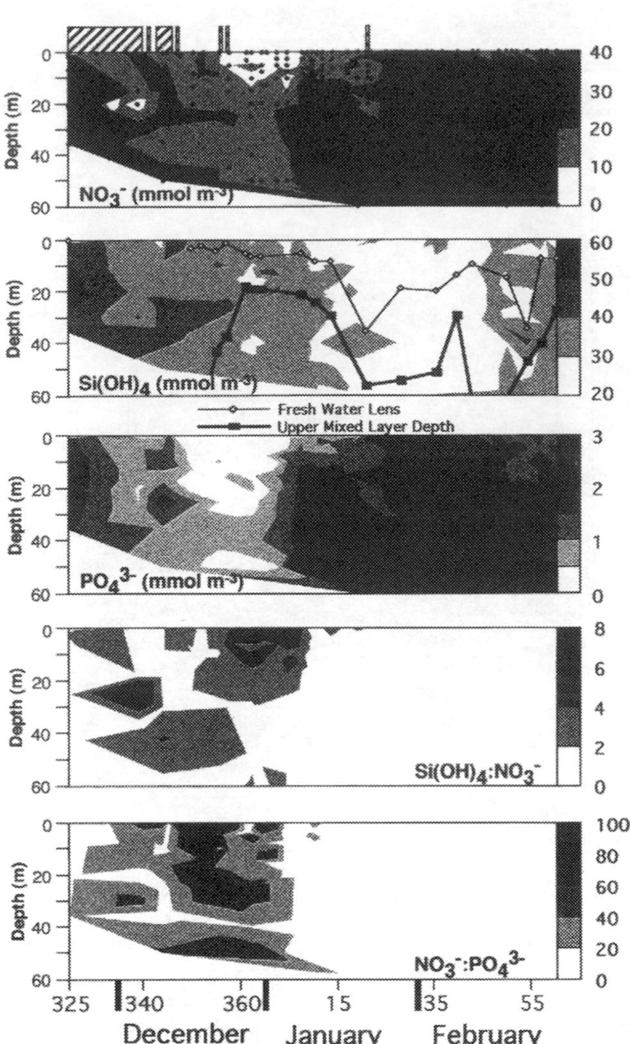

Fig. 10.5. Comparison of the seasonal changes in the depth distribution of the major macronutrients NO_3^-, $Si(OH)_4$ and PO_4^{3-}, and the derived molar ratios for $Si(OH)_4:NO_3^-$ and $NO_3^-:PO_4^{3-}$ determined for discrete samples (closed circles) collected at Station B between November 21, 1991 and February 27, 1992. Seasonal changes in the depth of the upper mixed layer (black squares) and the freshwater lens (open diamonds) are shown in the second contour plot. The daily presence or absence of pack ice is indicated by the hatched bars.

with the bloom was a depletion of NO_3^- and PO_4^{3-} to detection levels ($PO_4^{3-}<0.03$ mmol m^{-3}, $NO_3^-<0.05$ mmol m^{-3}). The ratio of $NO^{-3}:PO_4^{3-}$ also increased during the bloom and the difference in this ratio between 'nonbloom' ($x=14.14\pm2.99$, $n=166$) and bloom waters ($x=48.65\pm28.66$, $n=51$) was found to be significant at $p<0.01$ (Fig. 10.6). The large increase in the $NO_3^-:PO_4^{3-}$ ratio was due to the disproportionately large uptake of PO_4^{3-} by phytoplankton. After the bloom and coincident with the advection event, PO_4^{3-} and NO_3^- concentrations returned to pre-bloom levels.

The early summer period was characterized by high levels of $Si(OH)_4$ throughout the water column below the ice (Fig. 10.5) when diatoms were not abundant (Fig. 10.3). During the diatom bloom, $Si(OH)_4$ concentrations throughout the water column were reduced from >40 mmol m^{-3} to <30 mmol m^{-3}. High $Si(OH)_4:NO_3^-$ ratios during the bloom were primarily due to the large reduction of NO_3^- in the highly stratified UML. Following the bloom and advection of different water masses into the region, $Si(OH)_4$ concentrations thoughout the water column showed a further reduction (Fig. 10.5). Only in late February did $Si(OH)_4$ concentrations begin to rise.

Fig. 10.7 shows the phytoplankton's photoadaptive state over the season. Phytoplankton were light limited ($Q_{par}/I_k>1$) below 20 m for the entire season; however, the depth of light limitation shallowed to ~5 m during the bloom.

DISCUSSION

Results from this study document phytoplankton dynamics on timescales of days and illustrate the linkages and feedback mechanisms between the biological, physical and chemical environments. Seasonal freshwater inputs and decreased local winds caused stratification of the water column, which has been found in other studies to be a major controlling factor initiating biomass growth (Whitaker 1982, Smith & Sakshaug 1990,

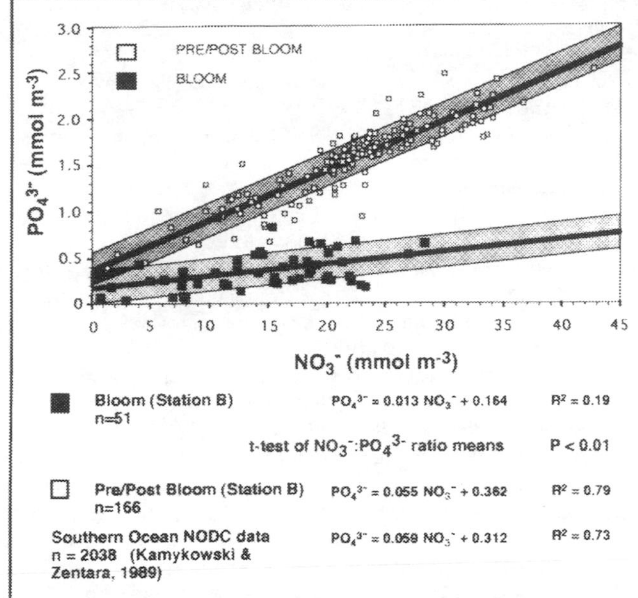

Fig. 10.6. Comparison of regression plots for changes in the abundance of inorganic NO_3^- and PO_4^{3-} for all depths sampled at Station B between November 21, 1991 and February 27, 1992. One regression line represents the PO_4^{3-} to NO_3^- relationship (open squares) for both the pre- and post-bloom period. The other regression represents the PO_4^{3-} to NO_3^- relationship (closed squares) during the diatom-dominated bloom occurring between December 10, 1991 to January 9, 1992 (Fig. 10.3). The shaded areas bordering each regression line represent ± one standard deviation.

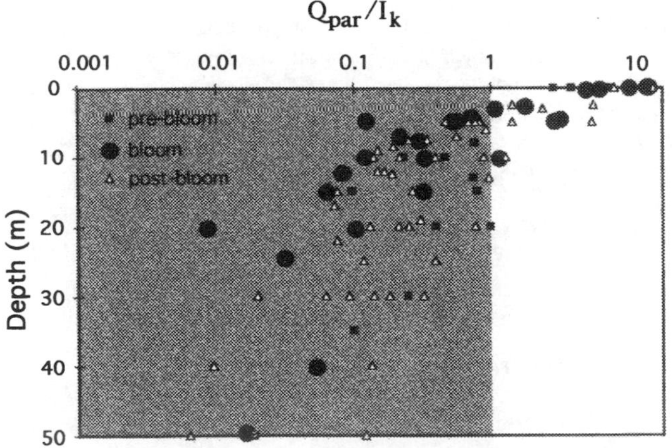

Fig. 10.7. Depth distribution of the photophysiological parameter Q_{par}/I_k determined for discrete samples collected from Station B between November 27, 1991 and February 27, 1992. Productivity measurements made prior (closed squares), during (closed circles) and after (open diamonds) the major diatom bloom are indicated. The shaded area indicates data where light limitation was evident.

Holm-Hansen & Mitchell 1991, Mitchell & Holm-Hansen 1991, Sakshaug *et al.* 1991). Water column stability in December allowed the phytoplankton to photoacclimate and overcome light limitation of growth. Biomass then accumulated in the upper 20 m and resulted in a major diatom bloom. As predicted by Kiefer & Kremer (1981), the zone of maximal growth and biomass shallowed as the bloom progressed. With the increase in biomass and light attenuation, the depth at which the phytoplankton were light limited also shallowed to ~5 m during the bloom (Fig. 10.7). With the shallowing of the UML, there was a shift in the phytoplankton composition from *Phaeocystis* (~1–3 m) to a larger diatom species (*Coscinodiscus* spp. ~40–80 m). Previous studies in the Antarctic have found an identical pattern of transition to larger phytoplankton as water column stability increases (Whitaker 1982, Rivkin 1991).

With the diatom biomass approaching concentrations thought maximal for Antarctic waters (Holm-Hansen *et al.* 1989), the chemical environment was significantly altered. The relationship between PO_4^{3-} and NO_3^- concentrations before and after the bloom was nearly identical to the Southern Ocean NODC (National Oceanographic Data Center) data given by Kamykowski & Zentara (1989) (Fig. 10.6). However, during the bloom, a significant shift in the relationship of NO_3^- to PO_4^{3-} showed the preferential uptake of PO_4^{3-}. Without a vertical nutrient flux caused by the strong stratification, this trend continued to detection limits, and therefore, PO_4^{3-} could have become the limiting resource. When PO_4^{3-} concentrations dropped to detection limits in the upper 10 m there was a change in the phytoplankton community from diatoms to *Phaeocystis.* Antarctic diatoms have shown competitive interactions for limiting nutrients, suggesting that algal assemblages may vary with limited resources (Sommer 1988, 1991), but this is the first study to present evidence that a shift in the nutrient field may result in changes in a natural Antarctic phytoplankton population. The change in the phytoplankton assemblage may have taken place as a result of PO_4^{3-} limiting the growth of *Coscinodiscus* spp., despite relatively high $Si(OH)_4$ concentrations, and the ability of *Phaeocystis* to utilize low concentrations of both PO_4^{3-} and NO_3^-. The nutrient absorption and storage capacity of the mucilage surrounding *Phaeocystis* colonies has been demonstrated in a range of both PO_4^{3-} (Veldhuis *et al.* 1991) and NO_3^- (Verity *et al.* 1988) concentrations and is a possible mechanism to explain the transition from diatoms to *Phaeocystis* in the low macronutrient environment at Sta B.

The dramatic macronutrient shifts in the water column during the second week of January were due to the advection of a different water mass into the area. The PO_4^{3-}-rich, NO_3^--rich and $Si(OH)_4$-poor water mass observed in the latter half of this study could have been a result of water with high PO_4^{3-}, NO_3^- and $Si(OH)_4$ concentrations mixing with coastal glacial meltwater and island runoff with high PO_4^{3-} and NO_3^- concentrations and no $Si(OH)_4$. Such mixing processes would have a diluting affect for $Si(OH)_4$ while not effecting the PO_4^{3-} and NO_3^- concentrations. Supporting evidence for the presence of glacial flour at Sta B is revealed in the σ_t data (Fig. 10.2), indicating a large freshwater input and high light attenuation in the top 20 m in the absence of high biomass (data not shown).

The presence of flagellates (cryptophytes (alloxanthin) and chrysophytes (19'-butanoyloxy-fucoxanthin)) during the month following the advection event may indicate a better adaptability to this period of increased vertical mixing, which has been documented in pelagic regions off the Antarctic Peninsula

(Kopczynska 1992). The final two weeks of the study are of particular interest because it was during this period that a second diatom bloom began to develop. Conditions were very similar to those found earlier in November and December; decreased winds, a shallowing UML and a transition from *Phaeocystis* to diatoms (fucoxanthin increased from <0.1 to 0.5 mg m^{-3}). These conditions initiated a second bloom that continued through the third week in March (Haberman, pers. commun.). Krebs (1983) showed that during a sampling period from December 1971 to January 1974 there were major blooms occurring in December–January and secondary less-intense blooms during the March–April time period. These findings, along with data from this study and from Holm-Hansen *et al.* (1989), suggest that near Palmer Station phytoplankton dynamics and the processes controlling them are similar between years. Physical factors, thought to partially initiate ice-edge blooms, such as a meltwater lens, also appear to be important in coastal systems, providing the stability needed for biomass accumulation. The depth of the UML, as defined in this study, was found to be an important factor in the development of phytoplankton biomass (Fig. 10.4). As the UML shallowed (~20 m), integrated chlorophyll *a* within the UML increased an order of magnitude. Mitchell & Holm-Hansen (1991) also showed the same relationship of increasing biomass with shallowing of the UML (Fig. 10.4). The degree of water column stability appears to also select for certain species: specifically, diatoms in a stratified high-nutrient condition, and *Phaeocystis* and flagellates in a mixed variable nutrient environment prior to stabilization. Although continued analyses of seasonal and interannual data is needed, results from this study give preliminary indications that temporal variations in phytoplankton biomass and community structure result from changes in water column stability and in the nutrient regime, and occur on the equivalent timescales as the transitional events (i.e. mixing, advection, stratification) seen in the physical dynamics.

ACKNOWLEDGEMENTS

K. Seydell, K. Scheppe, P. Handley and T. Newberger are thanked for their assistance in sample collection and analyses. Thanks to J. Priddle and O. Holm-Hansen for helpful comments. We also acknowledge R. Bidigare and M. Ondrusek for providing the HPLC training and pigment standards. Research was supported by National Science Foundation grant DPP 90–901127 to B. B. Prézelin. This is Palmer LTER publication number 47.

REFERENCES

Bidigare, R. R., Schofield, O. & Prézelin, B. B. 1989. Influence of zeaxanthin on quantum yield of photosynthesis of *Synechococcus* clone WH7803 (DC2). *Marine Ecology Progress Series*, **56**, 177–188.

Caceci, M. S. & Cacheris, W. P. 1984. Fitting curves to data. *Byte*, **9**, 340–362.

Holm-Hansen, O., Mitchell, B. G., Hewes, C. D. & Karl, D. M. 1989. Phytoplankton blooms in the vicinity of Palmer Station, Antarctica. *Polar Biology*, **10**, 49–57.

Holm-Hansen, O. & Mitchell, B. G. 1991. Spatial and temporal distribution of phytoplankton and primary productivity in the western Bransfield Strait region. *Deep-Sea Research*, **38**, 961–980.

Johnson, K. S., Petty, R. L. & Thomsen, J. 1985. Flow injection analysis for seawater micronutrients. In Zirino, A., ed. *Mapping Strategies in Chemical Oceanography. Advances in Chemistry Series*, **209**, 7–30.

Kamykowski, D. & Zentara, S. 1989. Circumpolar plant nutrient covariation in the Southern Ocean: patterns and processes. *Marine Ecology Progress Series*, **58**, 101–111.

Kiefer, D. A. & Kremer, J. N. 1981. Origins of vertical patterns of phytoplankton and nutrients in the temperate, open ocean: a stratigraphic hypothesis. *Deep-Sea Research*, **28**, 1087–1105.

Kopczynska, E. E. 1992. Dominance of microflagellates over diatoms in the antarctic areas of deep vertical mixing and krill concentrations. *Journal of Plankton Research*, **14**, 1031–1054.

Krebs, W. N. 1983. Ecology of neritic marine diatoms, Arthur Harbor, Antarctica. *Micropaleontology*, **29**, 267–297.

Mitchell, B. G. & Holm-Hansen, O. 1991. Observations and modeling of the Antarctic phytoplankton crop in relation to mixing depth. *Deep-Sea Research*, **38**, 981–1007.

Prézelin, B. B., Boucher, N. P. & Smith, R.C. 1994. Marine primary production under the influence of the Antarctic ozone hole: Icecolors '90. *Antarctic Research Series*, **62**, 159–186.

Priddle, J., Heywood, R. B. & Theriot, E. 1986. Some environmental factors influencing phytoplankton in the Southern Ocean around South Georgia. *Polar Biology*, **5**, 65–79.

Rivkin, R. B. 1991. Seasonal patterns of planktonic production in McMurdo Sound, Antarctica. *American Zoology*, **31**, 5–16.

Sakshaug, E., Slagstad, D. & Holm-Hansen, O. 1991. Factors controlling the development of phytoplankton blooms in the Antarctic Ocean – a mathematical model. *Marine Chemistry*, **35**, 259–271.

Smith, R. C., Baker, K. S., Handley, P. & Newberger, T. 1992. Palmer LTER program: hydrography and optics within the Peninsula grid, Zodiac sampling grid during the 1991–1992 field season. *Antarctic Journal of the United States*, **27**, 253–255.

Smith Jr., W. O. & Sakshaug, E. 1990. Polar Phytoplankton. In Smith, W.O. ed. *Polar oceanography; Part A: Physical science. Part B: Chemistry, biology, geology.* San Diego: Academic Press, 477–526.

Sommer, U. 1988. The species composition of Antarctic phytoplankton interpreted in terms of Tilman's competition theory. *Oecologia*, **77**, 464–467.

Sommer, U. 1991. Comparative nutrient status and competitive interactions of two Antarctic diatoms (*Corethron criophilum* and *Thalassiosira antarctica*). *Journal of Plankton Research*, **13**, 61–75.

Veldhuis, M. J. W., Colijn, F. & Admirall, W. 1991. Phosphate utilization in *Phaeocystis pouchetii* (Haptophyceae). *Marine Ecology*, **12**, 53–62.

Verity, P. G., Villareal, T. A. & Smayda, T. J. 1988. Ecological investigations of blooms of colonial *Phaeocystis pouchetii*. I. Abundance, biochemical composition, and metabolic rates. *Journal of Plankton Research*, **10**, 219–248.

Whitaker, T. M. 1982. Primary production of phytoplankton off Signy Island, South Orkneys, the Antarctic. *Proceedings of the Royal Society London*, **B214**, 169–189.

11 Distribution of phytoplankton organic carbon in the vicinity of Elephant Island, Antarctica, during the summers of 1990–1993

VIRGINIA E. VILLAFAÑE[1], *E. WALTER HELBLING*[1] *AND OSMUND HOLM-HANSEN*

Polar Research Program, Scripps Institution of Oceanography, University of California San Diego, La Jolla, CA, 92093-0202, USA, [1]Present address: Laboratorio de Fotobiologia y Productividad Primaria, Universidad Nacional de Pategonia, Casilla de Correos N 153, 9100 Trelew, Chubut, Argentina.

ABSTRACT

Phytoplankton studies have been conducted in a standard multi-station sampling grid around Elephant Island, South Shetland Islands, during four successive summers (early January to mid-March) of 1990–93, as part of the US Antarctic Marine Living Resources (AMLR) program. Results indicate that there is considerable spatial and temporal variability in distributions of total particulate organic carbon and phytoplankton carbon, as well as in the taxonomic composition of the crop during the time periods of our studies. Relatively low phytoplankton carbon concentrations were found in Drake Passage waters, with the highest carbon concentrations being in the zone of confluence between Drake Passage and Bransfield Strait waters. In three of the four years, phytoplankton biomass increased significantly from Leg I (January–February) to Leg II (February–March). The interannual variability in particulate organic carbon concentrations was more pronounced, with the years 1990–91 showing considerably greater concentrations of organic carbon throughout most of the sampling grid than the years 1992–93. The floristic composition of the phytoplankton also varied considerably, with diatoms being dominant in terms of total cellular organic carbon in 1990–91, but with flagellates being dominant in 1992–93. Leg II of 1993, which was the only year in which phytoplankton biomass decreased compared with Leg I, was also unique in that dinoflagellates dominated the crop.

Key words: phytoplankton, particulate organic carbon, biomass, variability, Elephant Island, Antarctica.

INTRODUCTION

The Elephant Island area, the study site of the US Antarctic Marine Living Resources (AMLR) program, has received special attention during the past decade (George 1984) since it has been one of the most important areas for commercial harvesting of the Antarctic krill, *Euphausia superba* Dana, (Macaulay *et al.* 1984, Everson 1988). Krill generally account for more than 50% of the total zooplankton biomass in many areas and occupy a key position within the Antarctic food chain (Marr 1962, Laws 1985). As krill are actively feeding in waters around Elephant Island, even during periods of intense swarming (Antezana & Ray 1984), it is important to assess the distribution, quantity, and quality of food available for grazing in this area in order to see if krill abundance is directly related to availability of food reservoirs.

Previous papers resulting from work in the AMLR study area have reported (i) data on chlorophyll *a* (chl *a*) concentrations and floristic composition during the austral summer of 1991 (Villafañe *et al.* 1993), (ii) the relationship between phytoplank-

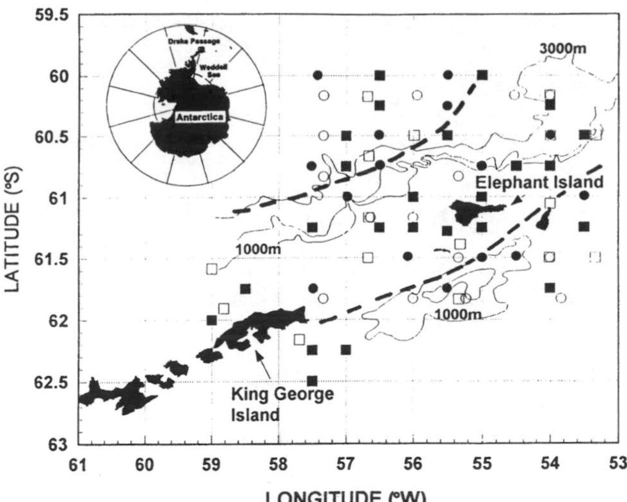

Fig. 11.1. Map showing the study area for the United States Antarctic Marine Living Resources program in the Antarctic. Depth contours are shown in metres. The symbols indicate the positions of the 'complete' stations during four consecutive years: (○) 1990; (□) 1991; (●) 1992; (■) 1993. The two heavy dashed lines indicate the approximate zones of demarcation between three major water masses as determined by the temperature–salinity characteristics of the upper water column for the four years of study (Drake Passage waters to the northwest, Bransfield Strait waters to the south and southeast, and Drake–Bransfield Confluence waters in the central region). The inset shows a polar view of Antarctica, with the small square close to the northern end of the Antarctic Peninsula indicating the location of the sampling grid.

ton species distribution and hydrographic conditions (Villafañe *et al.* 1994), and (iii) depth-integrated chl *a* concentrations (Helbling *et al.* 1994). In this paper we describe the spatial and temporal variability of food resources available for grazing zooplankton in the area around Elephant Island during the four austral summers of 1990–93. Emphasis is placed on the total organic carbon concentrations as well as on the carbon content of the major taxonomic groups of the phytoplankton.

MATERIALS AND METHODS

During the austral summers (early January–mid March) of 1990–93, phytoplankton studies were conducted in an area of approximately 50×103 km^2 in the vicinity of Elephant Island, South Shetland Islands, (Fig. 11.1), on board NOAA Ship *Surveyor*. The sampling grid consisted of 36, 50, 72 and 91 stations in 1990, 1991, 1992 and 1993, respectively. Each station was occupied twice each year, once during Leg I (early January–mid February) and once during Leg II (mid February–mid March).

The physical characteristics of the upper water column (0–750 m) were obtained at all stations with a rosette (General Oceanics) equipped with sensors for conductivity, temperature and depth (CTD, model SBE 9, Sea Bird Electronics). A sensor for Photosynthetic Active Radiation (PAR, 400–700 nm, Biospherical Instruments Inc. model QCP-200L), a 25 cm pathlength transmissometer (Sea Tech), a pulsed fluorometer (Sea Tech) and eleven 10 litre Niskin bottles were also attached to the rosette.

Water samples were taken at every station from the Niskin bottles at 11 standard depths (from 5 to 750 m, or within 10 m of the bottom at shallow stations) for measurement of chl *a*. At selected 'complete' stations (Fig. 11.1) water samples were taken at eight depths (5–75 m) for primary production measurements and also from 5, 20, 50 and 100 m for floristic analysis and determination of total particulate organic carbon (POC).

For chl *a* determinations, an aliquot of 100 ml was filtered through a Whatman GF/F glass fibre filter (25 mm) and the pigments were extracted in 10 ml of absolute methanol (Holm-Hansen & Riemann 1978). Fluorometric techniques were used to calculate chl *a* concentrations in the sample (Holm-Hansen *et al.* 1965). The chl *a* content in the nanoplankton fraction (cells $<20\mu$m) was obtained routinely during 1991, 1992 and 1993 studies (at 5, 20, 50 and 100 m) by prefiltering a subsample through a 20 µm Nitex® mesh, and the filtrate treated as described above for total chl *a* concentrations.

Measurements of POC were done by filtering one litre of sample through a Whatman GF/F filter (25 mm) which had been precombusted for 24 hours at 450 °C. Samples were kept dry until further analysis, which was carried out with a Perkin Elmer 2400 CHN analyser following the technique of Sharp (1974). The data from these chemical analyses were compared with beam attenuation data obtained with the transmissometer mounted on the rosette (Villafañe *et al.* 1993). As a good correlation (r^2=0.78, n=157) was found to exist between measured POC and the particulate beam coefficient (cp) during the four years of study, cp values were used to generate maps showing the distribution of estimated particulate organic carbon (POC_{est}) throughout the study area.

Samples for floristic analysis (125 ml) were poured into brown glass bottles and preserved with borate-buffered formalin to a final concentration in the sample of 0.4% of formaldehyde. The quantitative analysis of phytoplankton cells larger than 2 µm was done by inverted microscope techniques (Utermöhl 1958). Cell measurements were taken in order to get their volumes, so that phytoplankton carbon (in mg C m^{-3}) could be obtained, applying the equations of Strathmann (1967). The organisms were placed in one of the following three main taxonomic categories: diatoms, dinoflagellates (naked and thecates) and flagellates (includes chlorophytes, cryptophytes, chrysophytes, prymnesiophytes and unidentified flagellates).

RESULTS AND DISCUSSION

In a study area such as the AMLR grid, a large amount of data is necessary in order to obtain reliable estimates regarding the distribution of particulate organic carbon (either by microscopical or chemical determinations) throughout the entire upper water column. Both the chemical determination of POC and the estimation of phytoplankton organic carbon are labour and/or cost intensive. We have therefore used data from the transmissometer as an alternative method to estimate POC concentrations in surface waters. As the upper mixed layer (UML) in the

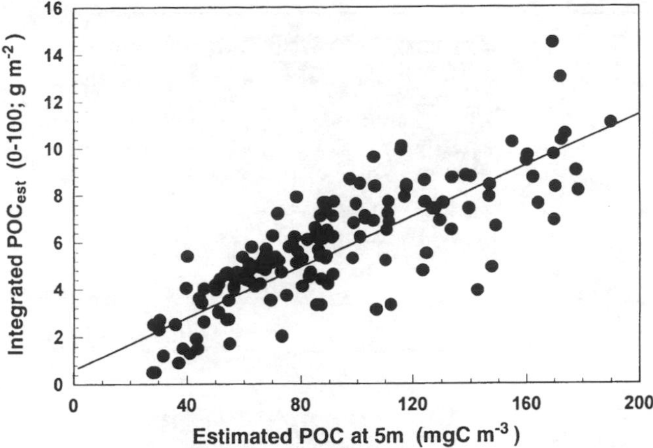

Fig. 11.2. Scatter plot of POC_{est} at 5 m depth and POC_{est} integrated from 0 to100 m. The line indicates the linear least square fit that corresponds to the equation $POC_{int}=0.05\times POC_{(5\,m)}+1.24$ ($r^2=0.70$, $n=157$). Note that POC_{int} is expressed in g C m^{-2} while $POC_{(5\,m)}$ is in mg C m^{-3}.

Antarctic is generally deep (approximately 50 m), surface values usually provide a good representation of the entire euphotic zone, as has been pointed out by studies using chl *a* as an estimator of biomass (Holm-Hansen & Mitchell 1991, Helbling *et al.* 1994). For the Elephant Island area, our data indicate that there is a good correlation ($r^2=0.70$, $n=157$) between POC_{est} at 5 m depth and POC_{est} integrated (0–100 m), as shown in Fig. 11.2. The fractionated chl *a* analyses and microscopic observations from four depths also indicated that there were no significant differences, either in size-fraction distribution or in species assemblages at different depths, within the euphotic zone. For the above reasons, we concluded that data from 5 m could be considered representative for the UML and hence most of the data presented in this paper are from 5 m depth.

As can be seen from the data presented in Fig. 11.3, there was much variability in distribution of POC_{est} within the AMLR area between Legs I and II of each year, but even more dramatic differences on the interannual basis (Table 11.1). A rather uniform distribution of POC_{est} was obtained in Leg I of 1990 (Fig. 11.3A), with the exception of high values (125–150 mg C m^{-3}) measured in the most northern and southern portions of the sampling grid and also in the southwestern corner. The distribution was more patchy during Leg II (Fig. 11.3B), with values up to 125–150 mg C m^{-3} to the north of Elephant Island, and values up to 200 mg C m^{-3} to the southwest of Elephant Island. POC_{est} values were slightly lower for most of the study area during Leg I of 1991, although some high values were associated with the shelf-slope north of Elephant Island and to the east of King George Island (Fig. 11.3C). An increase in the concentration of POC_{est} was noticed from Leg I to Leg II (Fig. 11.3D), with high POC_{est} values (up to 200 mg C m^{-3}) being associated with the continental shelf-slope north of King George Island and Elephant Island. South of this area the concentrations of POC_{est} were low, less than 100 mg C m^{-3}, as well as in the northwestern corner of the grid. During 1992, POC_{est} values were much lower than in the previous years, with concentrations of less than 75 mg C m^{-3} (Fig. 11.3E and F) throughout most of the study area. Small patches of about 175 mg C m^{-3} were observed in some areas north of Elephant Island area during Leg II (Fig. 11.3F). The POC_{est} concentrations during January–February of 1993 (Fig. 11.3G) were rather low throughout the area, with the exception of the region to the northeast of Elephant Island. During Leg II (Fig. 11.3H), there was a general decrease in POC_{est} concentrations throughout the study area. From the above data it is seen that the most dramatic change in particulate organic carbon concentrations in the AMLR study area occurred between the years 1990–91, when POC concentrations were relatively high, and the years 1992–93, when concentrations were much lower throughout the entire area.

In a recent paper (Silva *et al.* 1994) the AMLR study area was divided into three regions, based on the temperature–salinity characteristics of the water column. As shown by the dashed lines in Fig. 11.1, one region corresponded to Drake Passage waters (the northwestern portion of the sampling area), the second to Bransfield Strait waters (the south and southeastern portion of the sampling area) and the third was the confluence zone of Drake Passage and Bransfield Strait waters (the central region of the sampling grid). When looking at the spatial distribution of POC_{est} (Fig. 11.3, Table 11.1) in relation to these regions, it is seen that during 1990–91 the concentration of POC_{est} was quite high throughout most of the study area. During 1992–93 the Drake Passage and Bransfield Strait waters were characterized by low particulate carbon contents. However, during these two years, the Drake–Bransfield Confluence waters still showed high organic carbon concentrations, although they were lower than during 1990–91. These observations are in agreement with previous studies (El-Sayed 1988) in which pelagic Drake Passage waters had low biomass as compared with more coastal areas. The low biomass in these waters has been postulated to be due to iron deficiency (Martin *et al.* 1990, Holm-Hansen *et al.* 1994).

The zonal differences in distribution of organic carbon noted above are also reflected in differences in the species composition of the phytoplankton in the three regions. The species composition in Drake Passage waters was mainly dominated by small nanoplanktonic diatoms such as *Fragilariopsis pseudonana* and *F. cylindrus* in the 1990, 1991 and 1993 studies; during 1992, small flagellates (<10 m) and cryptophytes were dominant (Villafañe *et al.* 1994). The region of the confluence of Drake–Bransfield waters, which generally showed high carbon concentrations (Fig. 11.3, Table 11.1), includes a persistent frontal system that seems to be responsible for the enhanced phytoplankton biomass (Jacobs 1991, Helbling *et al.* 1993). In this area the phytoplankton crop was dominated by microplanktonic chain-forming diatoms, mainly *Pseudonitzschia* species (ranging from 40 μm to 50 μm in the longest dimension), as well as *Proboscia alata* and *Rhizosolenia antennata* f. *semispina* during 1990–91, and by small flagellates and cryptophytes during 1992–93. Bransfield Strait waters showed in general

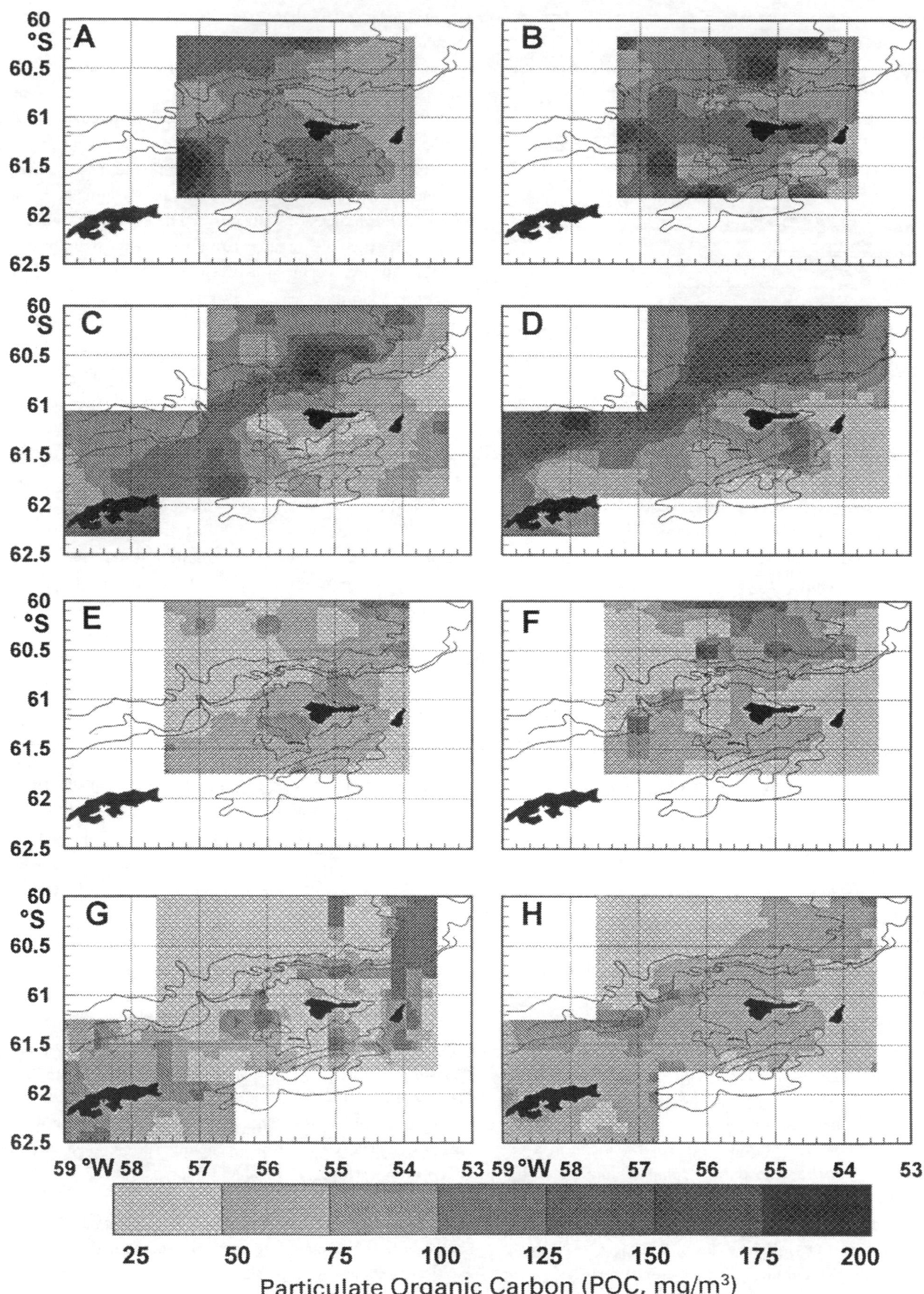

Fig. 11.3. Contour maps showing distribution of POC_{est} values (in mg C m^{-3}, as shown by scale at the bottom) at 5 m depth for all stations within the AMLR study area. A, Leg I 1990; B, Leg II 1990; C, Leg I 1991; D, Leg II 1991; E, Leg I 1992; F, Leg II 1992; G, Leg I 1993; H, Leg II 1993.

Table 11.1. *Mean and standard deviation of POC_{est} concentration values (in mg C m^{-3}) at 5 m depth for Legs I and II and for the four years of study (1990–93). Data are presented for each hydrographic region (see Fig. 11.1 and explanation in text). The inset for each region shows the results of a non-parametric Kruskal–Wallis test and an* a posteriori *test, Nemenyi procedure (Zar 1984). The lines under the years and Legs (roman numerals) indicate groups of data that were not significantly different*

Region A (Drake Passage waters)

	Leg I			Leg II		
Year	Mean	SD	*n*	Mean	SD	*n*
1990	73.3	12.7	7	93.3	22.7	7
1991	83.3	25.0	10	84.7	19.3	12
1992	27.7	12.2	24	36.7	23.3	26
1993	13.7	4.7	25	27.3	16.7	30

$P<0.0005$

93I 92I 93II 92II 90I 90II 91I 91II

Region B (Drake Passage – Bransfield Strait Confluence waters)

	Leg I			Leg II		
Year	Mean	SD	*n*	Mean	SD	*n*
1990	86.7	25.3	10	97.3	21.3	18
1991	93.3	18.7	29	104	19.7	24
1992	68.3	11	37	59.3	16	37
1993	83.3	28.7	46	63.3	10	34

$P<0.001$

92II 93II 92I 91I 90I 93I 91II 90II

Region C (Bransfield Strait waters)

	Leg I			Leg II		
Year	Mean	SD	*n*	Mean	SD	*n*
1990	80	30	4	86.7	33.3	8
1991	56.7	13.3	10	76.7	16.7	13
1992	21.3	5	3	22.3	3	8
1993	28.3	28	20	23.7	11	23

$P<0.006$

92II 92I 93II 93I 91II 91I 90I 90II

intermediate values of POC_{est}, as well as intermediate chl *a* concentrations and rates of primary production (Helbling *et al.* 1994). The phytoplankton in this region was generally dominated by flagellates.

Phytoplankton carbon, as determined by microscopical observations, represented between 41% to 59% of the total POC_{est}; these percentages are quite similar to those reported by Hewes *et al.* (1990) for diverse areas in Antarctic waters. When considering the variation in the distribution of phytoplankton

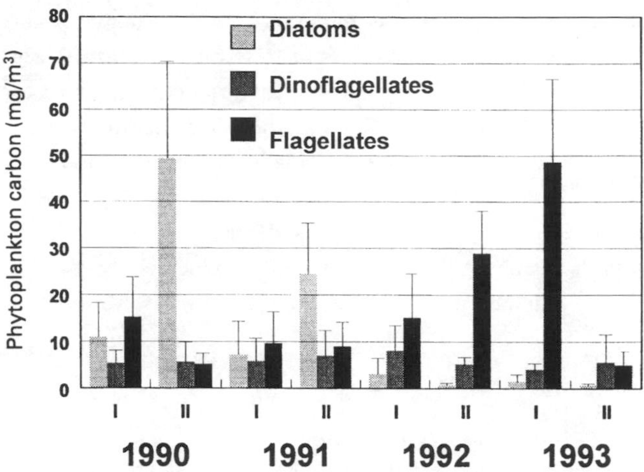

Fig. 11.4. Mean phytoplankton carbon content (in mg C m^{-3}) at 5 m depth in the main taxonomic groups as determined by microscopical analysis. The thin lines represent one standard deviation. The roman numerals I and II indicate the two legs of each yearly cruise.

carbon within the different taxonomic groups (Fig. 11.4), it is seen that diatoms were dominant during 1990–91. However, their numbers and biomass during 1992–93 were very small, with flagellates being the dominant group during these two years. This change in phytoplankton abundance and in dominance by different taxonomic groups during the four years of our study could be related in part to the fact that relatively high krill biomass occurs throughout the study area, and that krill concentrations apparently increased from 1990–91 to 1992–93 (Macaulay *et al.* 1990, Hewitt & Demer 1993). As krill have a preference for microplanktonic cells (Pavlov 1971, Meyer & El-Sayed 1983), especially diatoms, it is to be expected that intense krill grazing may cause a change in cell size and composition of the phytoplankton from a microplanktonic to a nanoplanktonic assemblage.

The results from this four-year study show that the shelf-break area to the north of Elephant and King George Islands, which is within the confluence zone of Drake Passage and Bransfield Strait waters, generally contains relatively high phytoplankton biomass, and that it often is dominated by chain-forming diatoms. This area should thus be capable of supporting relatively high concentrations of krill, and in fact, this has been corroborated by acoustic surveys of krill performed during the same time periods as our studies (Hewitt & Demer 1993). It should be noted, however, that during 1992–93 this area had lower phytoplankton biomass than during 1990–91 and that microplanktonic cells, especially diatoms, were in relatively low numbers. This may possibly be explained by results from previous studies in this area which have indicated that the preferential grazing impact of krill swarms may result in a nanoplankton-dominated phytoplankton crop (Holm-Hansen & Huntley 1984). In order to interpret variability in the abundance and species composition of the phytoplankton, it is necessary to have data on all the major factors which result in differential loss of phytoplankton biomass. Other studies of ours (Holm-Hansen *et al.* 1994) indicate that the marked temporal variability in phytoplankton noted above apparently is not due to physical mixing processes or to settling of cells to deep water. The most likely source of such variability is related to overall grazing effects, and it should be noted that the AMLR study area often has abundant biomass of salps, in addition to copepods, amphipods, and other Euphausiids (Nast 1986, Nordhausen 1991). Future analysis of phytoplankton variability in the Elephant Island area will have to include the grazing impact of all these major zooplankton groups.

ACKNOWLEDGEMENTS

This research was supported by the National Oceanic and Atmospheric Administration through the US Antarctic Marine Living Resources Program. We thank the Officers and Crew of NOAA Ship *Surveyor* for excellent support during field operations, and all the people that worked in our group during cruises. Grateful acknowledgment is also made to F. M. H. Reid for help in floristics analysis.

REFERENCES

Antezana, T. & Ray, K. 1984. Active feeding of *Euphausia superba* in a swarm north of Elephant Island. *Journal of Crustacean Biology,* **4,** 142–155.

El-Sayed, S. Z. 1988. Seasonal and interannual variabilities in Antarctic phytoplankton with reference to krill distribution. In Sahrhage, D., ed. *Antarctic ocean and resources variability*. Berlin: Springer-Verlag, 101–119.

Everson, I. 1988. Can we satisfactorily estimate variation in krill abundance? In Sahrhage, D., ed. *Antarctic ocean and resources variability*. Berlin: Springer-Verlag, 199–208.

George, R. Y., ed. 1984. The biology of the Antarctic krill *Euphausia superba*. *Journal of Crustacean Biology*, **4,** 1–337.

Helbling, E. W., Amos, A. F., Silva, S. N., Villafañe, V. & Holm-Hansen, O. 1993. Phytoplankton distribution and abundance as related to a frontal system north of Elephant Island, Antarctica. *Antarctic Science,* **5,** 25–36.

Helbling, E. W., Villafañe, V. & Holm-Hansen, O. 1994. Variability of phytoplankton distribution and primary production around Elephant Island, Antarctica, during 1990–1993. *Polar Biology,* **15,** 233–246.

Hewes, C. D., Sakshaug, E., Reid, F. M. H. & Holm-Hansen, O. 1990. Microbial autotrophic and heterotrophic eucaryotes in Antarctic waters: relationships between biomass and chlorophyll, adenosine triphosphate and particulate organic carbon. *Marine Ecology Progress Series,* **63,** 27–35.

Hewitt, R. P. & Demer, D. A. 1993. Dispersion and abundance of Antarctic krill in the vicinity of Elephant Island in the 1992 austral summer. *Marine Ecology Progress Series,* **99,** 29–39.

Holm-Hansen, O. & Huntley, M. 1984. Feeding requirements of krill in relation to food sources. *Journal of Crustacean Biology,* **4,** 156–173.

Holm-Hansen, O. & Mitchell, B. G. 1991. Spatial and temporal distribution of phytoplankton and primary production in the western Bransfield Strait region. *Deep-Sea Research,* **38,** 961–980.

Holm-Hansen, O. & Riemann, B. 1978. Chlorophyll *a* determination: improvements in methodology. *Oikos,* **30,** 438–447.

Holm-Hansen, O., Amos, A. F., Silva, N., Villafañe, V.E. & Helbling, E.W. 1994. *In situ* evidence for a nutrient limitation of phytoplankton growth in pelagic Antarctic waters. *Antarctic Science,* **6,** 315–324.

Holm-Hansen, O., Lorenzen, C. J., Holmes, R. W. & Strickland, J. D. H. 1965. Fluorometric determination of chlorophyll. *Journal du Conseil*, **30,** 3–15.

Jacobs, S. S. 1991. On the nature and significance of the Antarctic Slope Front. *Marine Chemistry*, **35,** 9–24.

Laws, R. M. 1985. The ecology of the Southern Ocean. *American Scientist*, **73,** 26–40.

Macaulay, M. C., English, T. S. & Mathisen, O. A. 1984. Acoustic characterization of swarms of Antarctic krill *(Euphausia superba)* from Elephant Island and Bransfield Strait. *Journal of Crustacean Biology,* **4,** 16–44.

Macaulay, M. C., Madriolas, A., Daly, K. & Morrison, P. 1990. Hydroacoustic survey for prey organisms. In AERG staff, eds. *AMLR 1989/90 field season report. Objectives, accomplishments and tentative conclusions*. Administrative Report LJ-90-11, 34–46.

Marr, J. W. S. 1962. The natural history and geography of the Antarctic krill (*Euphausia superba* Dana). *Discovery Report*, **32,** 33–464.

Martin, J. H., Gordon, R. M. & Fitzwater, S. E. 1990. Iron in Antarctic waters. *Nature*, **345,** 156–158.

Meyer, M. A. & El-Sayed, S. Z. 1983. Grazing of *Euphausia superba* Dana on natural phytoplankton populations. *Polar Biology,* **1,** 193–197.

Nast, F. 1986. Changes in krill abundance and in other zooplankton relative to the Weddell– Scotia Confluence around Elephant Island in November 1983, November 1984 and March 1985. *Archiv für Fischereiwissenschaft*, **37,** 73–94.

Nordhausen, W. 1991. AMLR program: horizontal separation of larval and adult *Thysanoessa macrura* around Elephant Island, Antarctica, during the 1991 austral summer. *Antarctic Journal of the United States*, **26** (5), 190–193.

Pavlov, V. Ya. 1971. On the physiology of feeding in *Euphausia superba*. *Proceedings Academiia nauk USSR*, **196,** 147–150.

Sharp, J. H. 1974. Improved analysis for 'particulate' organic carbon and nitrogen from seawater. *Limnology and Oceanography,* **19,** 984–989.

Silva, S. N., Helbling, E. W., Villafañe, V., Amos, A. F. & Holm-Hansen, O. 1994. Variability in nutrient concentrations around Elephant Island, Antarctica, during 1991–1993. *Polar Research*, **14,** 69–82.

Strathmann, R. R. 1967. Estimating the organic carbon content of phytoplankton from cell volume or plasma. *Limnology and Oceanography,* **12,** 411–418.

Utermöhl, H. 1958. Zur Vervollkommung der quantitativen Phytoplankton-Methodik. *Mitteilungen der Internationaler Vereins der Theoretisehen und Angewandten Limnologie*, **9,** 1–38.

Villafañe, V. E., Helbling, E. W. & Holm-Hansen, O. 1993. Phytoplankton around Elephant Island: distribution, biomass and composition. *Polar Biology,* **13,** 183–191.

Villafañe, V. E., Helbling, E. W. & Holm-Hansen, O. 1995. Spatial and temporal variability of phytoplankton biomass and taxonomic composition around Elephant Island, Antarctica, during summers of 1990-1993. *Marine Biology*, **123,** 4, 677–686.

Zar, J. H. 1984. *Biostatistical analysis.* Englewood Cliffs, NJ: Prentice Hall, 718 pp.

12 Identification of dominant taxa in coastal Antarctic water and ice core samples using lipid signatures

JENNIFER. H. SKERRATT[1], P. D. NICHOLS[1,2], T. A. MCMEEKIN[1,4] AND H. R. BURTON[3]
[1]Antarctic Co-operative Research Centre, GPO Box 252–80, Hobart, Tasmania, 7001, Australia, [2]CSIRO Division of Marine Research, Box 1538, Hobart, Tasmania, 7001, Australia, [3]Australian Antarctic Division, Channel Highway, Kingston, Tasmania, 7050, Australia, [4]Department of Agricultural Science, University of Tasmania, GPO Box 252–54, Hobart, Tasmania, 7001, Australia

ABSTRACT

Diatoms (predominantly Nitzschia*) and the prymnesiophyte* Phaeocystis *sp. form major algal blooms in Antarctic waters. Lipid components and their ratios were used to differentiate between these two algal groups. Lipid profiles were compiled for water-column particulate and ice algal communities collected near Davis station in Eastern Antarctica over five summer seasons (1988 to 1993).* Phaeocystis *sp. exhibited high levels of the fatty acid 14:0 and brassicasterol, low levels of polyunsaturated fatty acids (PUFAs) and a low 16:1ω7c to 16:0 ratio. In contrast, during periods of diatom blooms the dominant sterol and fatty acid were trans-22-dehydrocholesterol and 20:5ω3, which were accompanied by a high 16:1ω7c to 16:0 ratio. Sea-ice diatom blooms in the same area reflected lower levels of PUFAs than observed in the water column and had high concentrations of the sterol 24-ethylcholesterol. Branched-chain fatty acids were used as indicators of bacterial biomass and increased in concentration after the summer algal blooms each year.* Phaeocystis *sp. can be difficult to identify and enumerate using conventional methods, hence, during a bloom in 1989, the lipid signatures facilitated an estimate of the abundance of this alga.*

Key words: *Phaeocystis*, lipids, diatoms, fatty acids, sterols.

INTRODUCTION

In Antarctica, phytoplankton blooms occur during the period of total incident radiation from December to February. Diatom blooms in particular are prolific in the Antarctic sea ice and water column and are a major source of nutrients for higher trophic levels. The major fast-ice algal blooms occur in November (Garrison *et al.* 1987, Perrin *et al.* 1987) with smaller ice-algal blooms occurring in the austral autumn (Watanabe & Satoh 1987). Studies of phytoplankton in the water column at Davis station have shown similar algal successions over the summer season (Everitt & Thomas 1986, Perrin *et al.* 1987, Davidson & Marchant 1992a); however, the concentration of these blooms can vary markedly between years (Skerratt *et al.* 1995). The prymnesiophyte, *Phaeocystis* sp., can also be a major alga in Antarctic waters (Davidson & Marchant 1992a, b). However, due to its colonial nature it can be difficult to enumerate (Perrin *et al.* 1987, Davidson & Marchant 1992a). It can also be difficult to identify the flagellate form in field samples to species level with classical microscopic procedures.

Many species, including some Antarctic microorganisms (White 1983, Mancuso *et al.* 1990), contain a lipid signature. The lipid signature can provide a fingerprint to differentiate between taxa. The analysis of pure cultures establishes the concentration and profile of lipids within a particular species. The same analysis can then be applied to the determination of microbial biomass and community structure of samples collected from the field. Many bacteria contain branched-chain fatty acids that are not found in algae and other marine organisms. Therefore, the bacterial component of a field sample can be estimated by measuring the abundance of branched-chain fatty acids. Some bacteria contain fatty acids other than branched-chain fatty acids; however, analyses of mixed bacterial cultures indicate that branched-chain fatty acids can account for up to 70% of the total fatty acids (Gillan *et al.* 1983).

Although variation in lipid profiles may occur due to significant changes in culture conditions there is as yet little published evidence for such large shifts in nature. The use of lipid signatures in environmental studies has been validated in many environmental studies (Balkwill *et al.* 1988, Nichols *et al.* 1987, Mancuso *et al.* 1990). While variation in culture conditions are often dramatic the natural environment would be expected to restrict severely the survival of specific microbial strains to much narrower conditions of growth (White & Findlay 1988).

In this study we report on the application of lipid signature techniques on particulate matter from water samples and ice cores collected from a coastal site near Davis station in Eastern Antarctica. This study was performed to enable a broader understanding of the Antarctic near-shore environment using lipids to identify water-column and ice-algal species at this coastal region, with particular emphasis on the role of key algal groups such as diatoms and *Phaeocystis* sp.

SAMPLING AND EXPERIMENTAL METHODS

Study site

The Australian Antarctic station of Davis Base is situated in the Vestfold Hills in Eastern Antarctica (68°35′ S, 77°57′ E). During five summer seasons (1988–93) water-column particulate matter was collected regularly at a depth of 10 m, 500 m offshore from Magnetic Island, an island 4 km NNW of Davis Base. During collection in 1990/91 triplicate ice cores samples were also collected from the same site.

Field collection

Three water column particulate samples were collected for lipid analyses in an initial pilot season in 1988/89. This was followed by two more detailed, weekly studies in the summers of 1989/90 and 1990/91. Four samples was also analysed from the 1991/92 summer season and five from 1992/93.

Thirty to forty litres of water were sampled for lipid analyses; these were filtered *in situ* with a water sampler (Seastar or Infiltrex water pump) through a 14.2 cm diameter (Schleicher and Schuell) glass fibre filter (nominally 0.8 μm) over a 3 to 4 h period. Triplicate ice core- samples were obtained from the lower 20 cm of the sea ice with a 7 cm diameter (SPIRE) ice-coring auger. The cores were melted separately in the dark at 4 °C in 1.2 litres of filtered seawater and subsamples (50 ml) were taken for microscopy. Both water-column particulate and ice-core filters were frozen at −20 °C for later analysis.

Samples for microscopy were taken concurrently from all sites in all seasons. The January 1989 microscopy samples used for *Phaeocystis* sp. cell counts were taken at a location close (250 m) to where the lipid samples were collected. The *Phaeocystis* sp. bloom was prolific and widespread throughout the area at this time. The microscopy samples were fixed with Lugol's solution and stored at 4 °C. Species composition and phytoplankton counts were made using an inverted light microscope.

Analytical procedures

Water-column particulate and sea-ice filters were quantitatively extracted by the modified one-phase chloroform–methanol–water Bligh–Dyer method (Bligh & Dyer 1959, White *et al.* 1979). After phase separation, lipids were recovered from the lower chloroform layer, concentrated, sealed under nitrogen and stored at −20 °C.

A portion of the total lipid extract was analysed with an Iatroscan Mk. III TH-10 thin layer chromatography-flame ionization detector analyser (Volkman *et al.* 1989, Volkman & Nichols 1991). Sterol and free fatty acid fractions were obtained by saponification of 30% of the total lipid extract (Nichols *et al.* 1991). The sterols were converted to trimethyl silyl ethers by reaction with bis(trimethylsilyl)trifluoroacetamide. The fatty acid fraction was treated with methanol-hydrochloric acid-chloroform (10:1:1, v/v/v; 3 ml) to produce the corresponding fatty acid methyl esters.

Analyses were performed on a Hewlett Packard 5890 gas chromatograph using conditions as described by Nichols *et al.* (1993) and selected fatty acid samples were analysed on a polar phase (Supelcowax ™ 60×0.32 id (Supelco)) fused silica column. Identifications were confirmed by gas chromatography – mass spectrometry (GC–MS) using conditions as described by Volkman *et al.* (1989) and by the comparison of retention time and mass spectral data of authentic and laboratory standards.

Precision and accuracy

The logistics of sampling large volumes of water for lipid analysis did not allow replicate filters to be taken. The water column samples taken for lipid analyses were integrated samples taken over a 4 to 5 hour period. A nearby site was sampled weekly to estimate spatial variation (data not shown). The relative standard deviations of replicate samples for gas chromatographic analyses were 5% or less in replicates. The variation in the analyses of replicate samples was at least one order of magnitude less than observed environmental variation (see Table 12.2).

Nomenclature

Fatty acids are designated as total number of carbon atoms: number of double bonds followed by the position of the double bond from the ω (aliphatic) end. The suffixes *c* and *t* indicate *cis* and *trans* geometry.

Many of the trivial or common names do not reflect the stereochemistry of the sterol. The following list contains the common sterol name used followed by the systematic name in brackets: trans-22-dehydrocholesterol (cholesta-5,22E-dien-3ß-ol), cholesterol (cholest-5-en-3ß-ol), brassicasterol

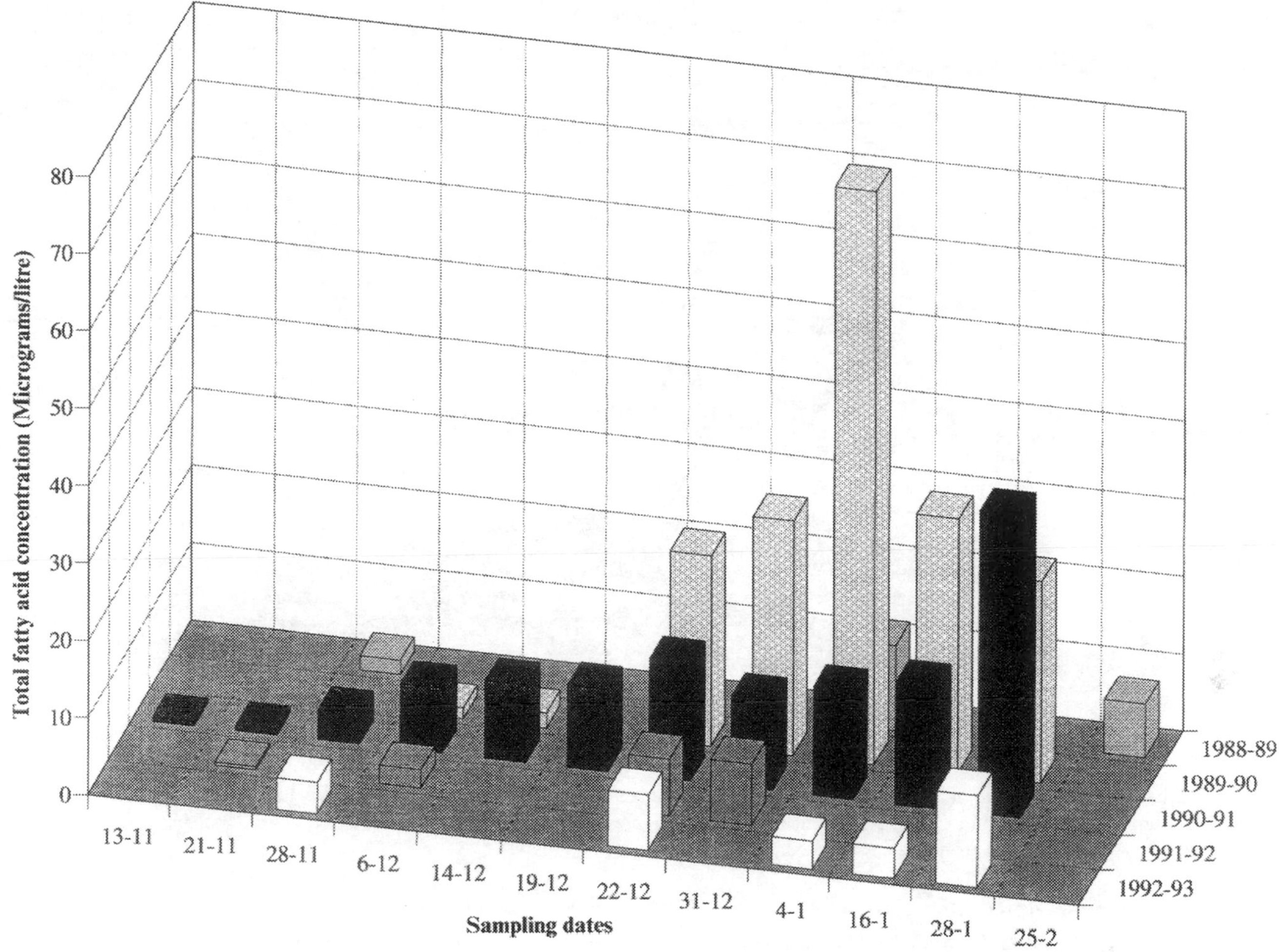

Fig. 12.1. Total fatty acid concentration of the water column particulates collected at Davis station through five summer seasons.

(24-methylcholesta-5,22E-dien-3ß-ol), 24-ethylcholesterol (24-ethylcholest-5-en-3ß-ol).

RESULTS AND DISCUSSION

Maxima in lipid concentration in the water column particulates occurred in samples collected in December and January (Fig. 12.1) and corresponded with peaks in algal biomass. Peaks in the lipid concentration of the algal communities in the ice cores were observed during the major sea-ice diatom blooms that occurred in November before the ice melt.

Phaeocystis sp.

Pure cultures of Antarctic *Phaeocystis* sp. strains are high in the sterol brassicasterol (97–100% of total sterols) and the concentration of this sterol in a singular cell was estimated (Nichols *et al.* 1991). The fatty acid profile also showed that the Antarctic strains of *Phaeocystis* sp., unlike many other algal species in Antarctica, are low in essential polyunsaturated fatty acids (PUFAs) (Nichols *et al.* 1991).

During the field study it was found that when *Phaeocystis* sp. was present, brassicasterol increased and the abundance of PUFAs was low (Fig. 12.2; Table 12.1). Increased levels of the fatty acid 14:0 and a low 16:1 to 16:0 ratio were also noted (Table 12.1). This combined lipid profile was used to identify the occurrence of this algae over the five summers. A low biomass is required to ascertain water-column community structure and this has the potential to enable measurement of low concentrations of *Phaeocystis* sp. which can not always be easily identified in its flagellate form. In January 1989, *Phaeocystis* sp. was estimated at 2.6×10^6 cells per litre based on the concentration of brassicasterol in particulate matter. Microscopic observations for water samples from a site 250 metres away estimated the bloom at 6×10^7 cells per litre (Davidson & Marchant 1992a). The differences in *Phaeocystis* sp. abundance at the two locations may reflect real variation between sites, as blooms of this species are not always homogeneous. An alternate, although we believe less likely, explanation is that the variation may be due to differences in biochemical composition between cultured and field samples. Combining brassicasterol with the other dominant *Phaeocystis* sp. markers provides a useful approach for identifying *Phaeocystis* sp. and estimating this alga's biomass, despite the possible underestimate in cell numbers.

The *Phaeocystis* sp. blooms at Davis station showed considerable variation in concentration over the five summer seasons (Skerratt *et al.* 1995). After the major bloom in January 1989,

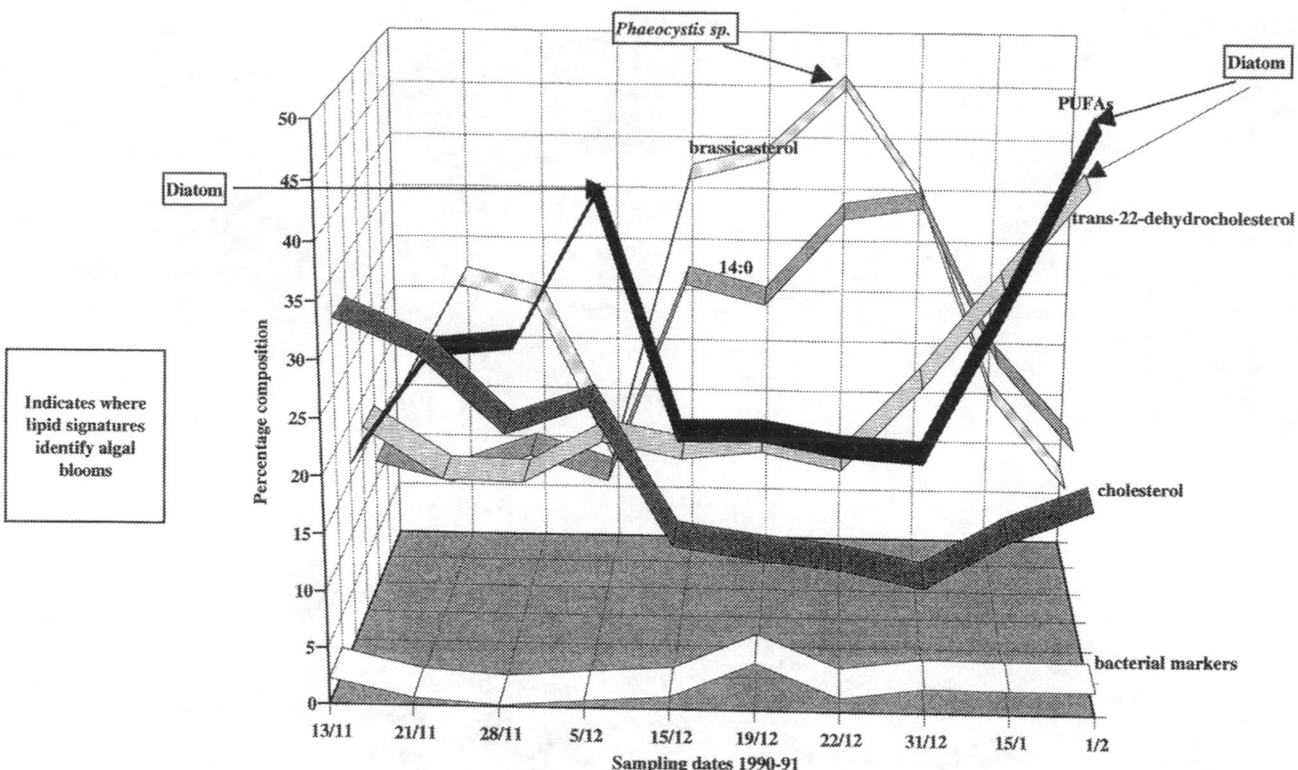

Fig. 12.2. Variation in the major algal, bacterial and heterotrophic lipid profiles for the water column particulates collected at Davis station during the 1990/91 season. Brassicasterol and trans-22-dehydrocholesterol are represented as percentage of total sterols. PUFAs, bacterial markers (branched-chain fatty acids) and 14:0 are represented as percentage of total fatty acids.

Table 12.1. *Sterol and fatty acid profiles of the water-column particulates that reflect* Phaeocystis *and* Nitzschia *spp. blooms near Magnetic Island for the 1988/89, 1989/90 and 1990/91 summer seasons*

	Percentage composition				
	Phaeocystis sp.			*Nitzschia* spp.	
Summer season	1988/89	1989/90	1990/91	1989/90	1990/91
Date	Jan.	Dec.	Dec.	Jan.	Feb.
Fatty acids					
14:0	34.1	20.9	35.4	14.4	13.8
20:5ω3	4.7	4.5	6.6	34.9	23.2
Branched chain fatty acids	0.9	0.4	1.2	0.7	2.0
Monounsaturated fatty acids	21.2	25.2	18.4	6.3	11.4
Other saturated fatty acids	29.4	35.5	27.5	22.1	26.2
Other PUFAs	7.8	13.1	10.9	21.6	23.5
Other fatty acids	1.9	0.4	—	0.1	—
Sum	100	100	100	100	100
16:1ω7*c*/16:0 ratio	0.8	0.5	0.9	2.8	1.6
Sterols					
Trans-22-deydrocholesterol	14.4	10.8	14.0	61.7	33.5
Brassicasterol	41.3	50.2	46.0	2.5	7.8
Other sterols	44.3	39.0	40.0	35.7	58.8
Sum	100	100	100	100	100

the following seasons did not contain a bloom of *Phaeocystis* sp. that was as abundant or as accessible as during the first season. In the following seasons however, lipid markers for *Phaeocystis* sp. were observed in December prior to the diatom blooms (Table 12.1) (see also Skerratt *et al.* 1995). This sequence of algal community variation is in agreement with other microscopy studies in the Antarctic, which have reported the common occurrence of *Phaeocystis* sp. before a diatom bloom in the water column (Everitt & Thomas 1986, Perrin *et al.* 1987, Davidson & Marchant 1992a). In this study, when classical microscopy identified the algal group as flagellates, lipid signatures were observed for *Phaeocystis* sp., thereby confirming that it is possible to identify and estimate the abundance of this alga using lipid signatures (Skerratt *et al.* 1995).

Diatoms in the water column

In the water column, Antarctic diatoms, in contrast to *Phaeocystis* sp., have a very high lipid and PUFA content (up to 50% of the total fatty acids, Table 12.1). These high levels of PUFAs can be used in the cell membrane to maintain cell fluidity (Parrish 1988). Other factors also contribute to cell fluidity such as cell structure and size. In the two major diatom blooms examined (1989/90 and 1990/91), the dominant algae in the water column were the diatoms *Nitzschia* spp. (*N. closterium* 1989/90; *N. seriata* and *N. kerguelensis* 1990/91) and the major sterol during both blooms was trans-22-dehydrocholesterol. As diatoms have a larger cell size (5–60 μm) than species such as *Phaeocystis* sp. (3–8 μm) higher total lipid concentrations were

Table 12.2. *Variation in lipid composition for averaged replicate 20 cm ice cores at Magnetic Island 1990/91 (variation in ice core replicates: ±10%)*

Date	Percentage composition (ice core replicate average)						
	13/11	21/11	26/11	5/12	15/12	22/12	31/12
Polar lipid	66	85	79	80	45	30	25
Triglyceride	16	3	25	18	55	62	65
Sterol	3	2	2	tr	nd	2	1
Free fatty acid	15	10	3	2	nd	6	9
Sum	100	100	100	100	100	100	100

Notes:
nd: not detected, tr: trace (<1%).

measured during the planktonic diatom blooms. Polar lipid was the predominant lipid in all water column samples analysed (Skerratt *et al.* 1995). High levels of polar lipid would be expected in a diatom bloom in the water column as the cells would generally be in logarithmic growth stage. The cells would not be storing or degrading the lipid as triglyceride or free fatty acid respectively, as this occurs when cells reach stationary phase growth.

Table 12.3. *Major sterol and fatty acid profiles in averaged replicate 20 cm ice cores at Magnetic Island 1990/91 (variation in ice core replicates: ±10%)*

Date	Percentage composition (ice core replicate average)				
	13/11	21/11	26/11	5/12	15/12
Fatty acids					
14:0	13.1	13.5	11.5	12.4	7.5
20:5ω3	11.7	5.5	7.5	0.3	3.7
Branched-chain fatty acids	1.4	0.8	0.7	0.2	0.5
Monounsaturated fatty acids	41.6	37.4	48.1	46.6	50.7
Other saturated fatty acids	18.9	36.3	27.4	39.4	31.4
Other PUFAs	13.4	6.6	4.8	1.4	3.7
Other fatty acids	—	—	—	—	2.5
Sum	100	100	100	100	100
16:1ω7*c*/16:0 ratio	2.1	0.9	1.7	1.1	1.6
Sterols					
Brassicasterol	40.5	32.7	38.3	49.0	63.0
24-ethylcholesterol	40.5	37.9	40.4	12.0	11.8
Other sterols	19.1	29.4	21.4	39.0	25.2
Sum	100	100	100	100	100

Ice algae

The dominant alga in the ice cores throughout the 1990/91 season was the diatom *Nitzschia stellata.* As the species composition remained relatively similar, the lipid profiles of the sea-ice diatom communities over the November to December period reflect physiological rather than species change.

The maxima in *Nitzschia stellata* abundance occurred at the beginning of November. The initial major lipid class of the ice algae was polar lipid (Table 12.2), but this decreased throughout November and December with a corresponding increase in the storage lipid, triglyceride. In December increasing sea-water temperatures and decreasing salinity at the ice–water interface accompanied a fall in algal biomass. This was reflected in the lipid concentration and composition. The rise in triglyceride in December corresponded to a decrease in PUFAs and an increase in saturated and monounsaturated fatty acids (Table 12.3). We believe there would be little benefit for the alga to use PUFAs as storage lipids when saturated and monounsaturated fatty acids require fewer biosynthetic steps. Therefore, PUFAs observed at the end of the season may be contained in the membrane component while the triglyceride fraction would contain predominantly saturated and monounsaturated components, as has been reported in previous studies (e.g. Henderson & Mackinlay 1992). The sea-ice diatoms therefore contained a lower abundance of PUFAs (Table 12.3) than observed in the water-column particulates (Table 12.1) Due to the decreasing salinity and increasing water temperatures, the ice-algal community was under greater natural stress than the underlying water column algae.

The sterol profiles of the ice cores also differed from the profiles observed in the water column. The major sterol in the *Nitzschia stellata* dominated ice cores was 24-ethylcholesterol (12–41% of total sterols; Table 12.3). This sterol has been found in previous Antarctic sea-ice diatoms studies near McMurdo station (Nichols *et al.* 1989, 1993). In the latter stages of the ice algal bloom in this study 24-ethylcholesterol decreased in abundance and brassicasterol increased to be the dominant sterol (Table 12.3). The presence of 24-ethylcholesterol at high levels adds further to the view that C_{29} sterols can be a major component in diatom species as well as traditionally being thought to derive from terrestrial or blue-green algal sources (Huang & Meischein 1979, Matsumoto *et al.* 1982).

Bacteria

Bacterial abundance was determined by measuring levels of branched-chain fatty acids in the field samples. In this study, the branched-chain fatty acids increased in concentration and relative abundance after all the summer algal blooms. Throughout most of the summer seasons the relative abundance was generally 2% of the total fatty acids in the water column and this increased to 4–7% after major phytoplankton blooms (Fig. 12.2). Branched-chain fatty acids were also low in the ice-algal communities and were highest in the initial sample (Table 12.3). Over the five seasons the bacterial biomarkers in the suspended particulates and ice cores were low in comparison with the high concentration of algal-derived biomarkers (Skerratt *et al.* 1995). These results therefore strongly illustrate that bacteria only contribute a small (<10%) proportion of the total biomass of water column and ice-associated microbial communities at this coastal site.

Heterotrophs

Algae from temperate and tropical waters, unlike heterotrophs, do not normally contain high concentrations of cholesterol, therefore, under such conditions this sterol can be used to analyse heterotrophic biomass. In this study Antarctic diatoms may also have contributed to the observed increase in cholesterol during the diatom blooms (Skerratt *et al.* 1995). Other studies of Antarctic diatoms show that some *Nitzschia* spp. are quite high in this sterol (*N. cylindrus*: 34%, Nichols *et al.* 1986).

Based on microscopic observation heterotrophic biomass increased during and after the summer diatom blooms (Skerratt *et al.* 1995). For most of the seasons there was an increase in the water-column particulates in the concentration of 22:6ω3 and C_{18} PUFAs at the end of the summer, which corresponded to a rise in the number of cryptomonads and tintinnids present (Skerratt *et al.* 1995). This finding points to a possible important role of these two groups in this Antarctic near-shore environment. There was also an increase in the abundance of these fatty acids and cholesterol in early December before the major phytoplankton blooms in each of the three seasons (e.g. Fig. 12.2; 1990) (Skerratt *et al.* 1995). The observed increase in these specific lipid signatures is consistent with an increase in heterotrophic organisms entering the water column after the November ice algal blooms.

CONCLUSIONS

Lipid profiles indicated intra- and inter-annual variation in both the concentration of algal biomass and the dominance of individual algal species. However, in the water column there were similarities in the succession of lipid profiles and phytoplankton over the five summers. For all summers the presence of *Phaeocystis* sp. was identified before the diatom blooms, which were also accompanied by an increase in markers for cryptomonads. Water-column diatom blooms were dominated by polar lipid and their sterol composition differed from that observed in the ice cores. Variations in fatty acid profiles of the ice-core diatoms were indicative of changes in physiological status. In times of stress during ice melt the ice algae produced high levels of triglyceride which were accompanied by a decrease in PUFAs.

Lipid profiles differentiated between *Phaeocystis* sp. and *Nitzschia* spp. blooms in the water column. Lipid profiles also reflected that the diatom *Nitzschia stellata,* which bloomed in the ice, was a different species and had a different lipid profile to the *Nitzschia* spp. (*N. seriata* and *N. kerguelensis*) that bloomed in the same area in the underlying water column. Significant differences occurred in the absolute and relative importance of *Phaeocystis* sp. between seasons. *Phaeocystis* sp. was only a major component of the phytoplankton population during one (1988/89) of the five summers. In this season lipid analysis quantified the concentration of *Phaeocystis* sp. to within a factor of 10 in comparison with microscopic data. In three of the following summer seasons *Phaeocystis* sp. was identified as a minor contributor to the marine particulates. Our results indicate that over the five summer seasons examined, the abundance of this alga was variable and generally low for this area of Antarctica.

ACKNOWLEDGEMENTS

This work was supported in part by funding from the Antarctic Science Advisory Committee and the CSIRO-University of Tasmania grants program. Thanks to Andrew Davidson, John van den Hoff, Joe Mancuso, Andrew McMinn, fellow Davis expeditioners and colleagues in the Marine Resources and Pollution programme of CSIRO Oceanography. We thank Geoff Cripps (BAS) and an anonymous referee for helpful review comments.

REFERENCES

Balkwill, D. L., Leach, F. R., Wilson, J. T., McNabb, J. F. & White, D. C. 1988. Equivalence of microbial biomass measures based on membrane lipid and cell wall components, adenosine triphosphate, and direct counts on subsurface aquifer sediments. *Microbial Ecology*, **16**, 73–84.

Bligh, E. H. & Dyer, W. J. 1959. A rapid method of total lipid extraction and purification. *Canadian Journal of Biochemistry and Physiology*, **37**, 911–917.

Davidson, A. T. & Marchant, H. J. 1992a. Protist abundance and carbon concentration during a *Phaeocystis* dominated bloom at an Antarctic coastal site. *Polar Biology*, **12**, 387–395.

Davidson, A. T. & Marchant, H. J. 1992b. The biology and ecology of *Phaeocystis* (Prymnesiophyceae). In Round, F. E. & Chapman, D. J., eds. *Progress in phycological research*, Vol. 8. Bristol: Biopress Ltd, 2–45.

Everitt, D. A. & Thomas, D. P. 1986. Observations of seasonal changes in diatoms at inshore localities near Davis Station, East Antarctica. *Hydrobiologia*, **139**, 3–12.

Garrison, D. L., Buck, K. R. & Fryxell, G. A. 1987. Algal assemblages in Antarctic pack ice and in ice-edge plankton. *Journal of Phycology*, **23**, 564–572.

Gillan, F. T., Johns, R. B., Verheyen, T. V., Nichols, P. D., Esdaile, R. J. & Bavor, H.J. 1983. Monounsaturated fatty acids as specific bacterial markers in marine sediments. In Bjøroy, M., ed. *Advances in organic geochemistry*. Amsterdam: Elsevier, 198–207.

Henderson, R. J. & Mackinlay, E. E. 1992. Radiolabelling studies of lipids in the marine cryptomonad *Chroomonas salina* in relation to fatty acid desaturation. *Plant Cell Physiology*, **33**, 395–406.

Huang, W. Y. & Meischein, W. G. 1979. Sterols as ecological indicators. *Geochimica et Cosmochimica Acta*, **43**, 739–745.

Mancuso, C. A., Franzmann, P. D., Burton, H. R. & Nichols, P. D. 1990. Microbial community structure and biomass estimates of a methanogenic Antarctic lake ecosystem as determined by phospholipid analyses. *Microbial Ecology*, **19**, 73–95.

Matsumoto, G., Torii, T. & Hanya, T. 1982. High abundance of algal 24–ethylcholesterol in Antarctic lake sediment. *Nature*, **299**, 52–54.

Nichols, D. S., Nichols, P. D. & Sullivan, C. W. 1993. Fatty acid, sterol and hydrocarbon composition of Antarctic sea-ice diatom communities from McMurdo Sound. *Antarctic Science*, **5**, 271–278.

Nichols, P. D., Mancuso, C. A. & White, D. C. 1987. Measurement of methanotroph and methanogen signature phospholipids for use in assessment of biomass and community structure in model systems. *Organic Geochemistry*, **11**, 451–461.

Nichols, P. D., Palmisano, A. C., Rayner, M. S., Smith, G. A. & White,

D. C. 1989. Changes in the lipid composition of Antarctic sea-ice diatom communities during a spring bloom: an indication of community physiological status. *Antarctic Science*, **1**, 133–140.

Nichols, P. D., Palmisano, A. C., Smith, G. A. & White, D. C. 1986. Lipids of the Antarctic sea-ice diatom *Nitzschia cylindrus*. *Phytochemistry*, **25**, 1649–1653.

Nichols, P. D., Skerratt, J. H., Davidson, A., Burton, H. & McMeekin, T. A. 1991. Lipids of cultured *Phaeocystis pouchetii*: signatures for food-web, biogeochemical and environmental studies in Antarctica and the Southern Ocean. *Phytochemistry*, **30**, 3209–3214.

Parrish, C. C. 1988. Dissolved and particulate marine lipid classes: a review. *Marine Chemistry*, **23**, 17–40.

Perrin, R. A., Lu, P. & Marchant, H. J. 1987. Seasonal variation in marine phytoplankton and ice algae at a shallow Antarctic coastal site. *Hydrobiologia*, **146**, 33–46.

Skerratt, J. H., Nichols, P. D., McMeekin, T. A. & Burton, H. 1995. Seasonal and inter- annual changes in planktonic biomass and community structure in eastern Antarctica using signature lipids. *Marine Chemistry*, **51**, 93–113.

Volkman, J. K. & Nichols, P. D. 1991. Applications of thin layer chromatography-flame ionisation detection to the analysis of lipids and pollutants in marine and environmental samples. *Journal of Planar Chromatography*, **4**, 19–25.

Volkman, J. K., Jeffery, S. W., Nichols, P. D., Rogers, G. I. & Garland, C. D. 1989. Fatty acid and lipid composition of 10 species of microalgae used in mariculture. *Journal of Experimental Marine Biology and Ecology*, **128**, 219–240.

Watanabe, K. & Satoh, H. 1987. Seasonal variation of ice algal standing crop near Syowa station, East Antarctica, in 1983/84. *Bulletin of the Plankton Society of Japan*, **34**, 131–150.

White, D. C. 1983. Analysis of microorganisms in terms of quantity and activity in sediments. In Slater, J. H., Whittenbury, R. & Wimpenny, J. W. T., eds. *Microbes in their natural environments, Society for General Microbiology, Symposium* Vol. 34. Cambridge: Cambridge University Press, 37–66.

White, D. C. & Findlay, R. H. 1988. Biochemical markers for measurement of predation effects on the biomass community structure, nutritional status and metabolic activity of microbial biofilms. *Hydrobiologia*, **159**, 119–132.

White, D. C., Davis, W. M., Nickels, J. S., King, J. D. & Bobbie, R. J. 1979. Determination of the sedimentary microbial biomass by extractible lipid phosphate. *Oecologia*, **40**, 51–62.

13 Temporal variability of bacteria, phytoplankton and zooplankton assemblages of the sub-Antarctic Morbihan Bay (Kerguelen Archipelago)

S. RAZOULS[1], F. DE BOVÉE[1], D. DELILLE[1], M. FIALA[1] AND P. MAYZAUD[2]

[1]*Observatoire Océanologique de Banyuls, Université P. et M. Curie, U.R.A. CNRS 2017, F- 66 650 Banyuls-sur-mer, France,*
[2]*Observatoire Océanologique de Villefranche, Université P. et M. Curie, E.P. CNRS 017, F- 06 230 Villefranche-sur-mer, France*

ABSTRACT

*Several year-round surveys in the Kerguelen Islands area have provided the opportunity to compare the temporal variations of pelagic living particulate matter with respect to environmental parameters. Analyses of bacterial, phytoplankton and zooplankton assemblages were carried out on different temporal scales, ranging from twice a week to once a month. Seasonal and pluri-annual changes in the biological parameters, including amino acids, carbohydrates, lipids, POC and chlorophyll a are emphasized in relation to the corresponding changes of environmental parameters. Changes in biochemical components mirrored those of phytoplankton. Chlorophyll a concentrations showed moderate values during winter and autumn. They increased sharply in summer, reaching a maximum value of about 27 mg m^{-3} corresponding to an important depletion of nutrients. Thus, phytoplankton blooms are correlated with optimal values of solar radiation. In contrast, bacteria showed maximal abundance during spring and autumn and do not seem directly linked to other parameters. During the phytoplankton bloom, the mesozooplankton biomass and the copepod density reached up to 1 g dry weight m^{-3} and 2×10^3 individuals m^{-3} respectively. A single species of copepod (*Drepanopus pectinatus*) dominated the zooplankton community. A pattern is drawn from the calculated chronological time series of biological communities.*

Key words: Kerguelen Islands, sub-Antarctic, bacteria, phytoplankton, zooplankton, temporal variability.

INTRODUCTION

In the high latitudes, the trophic environmental heterogeneity driven by seasonal influences constitutes a forcing parameter for the life history of pelagic communities (Hopkins *et al.* 1993). Numerous surveys along latitudinal transects in the western Indian Ocean have concerned oceanic areas located between the Polar Front and the pack-ice zone (Jacques 1978, 1982, Hoshiai *et al.* 1986, Fontugne & Fiala 1987, Mayzaud 1990). Although there are fewer investigations concerning living particulate matter in the sub-Antarctic area than in the Antarctic (see Vincent 1988, Smith 1990, Friedmannn 1993 for reviews), nevertheless sub-Antarctic data are necessary to model the Southern Ocean comprehensively. The vicinity of the following islands has been investigated: Macquarie Island (Bunt 1955), Prince Edward and Marion Islands (Kok & Grobbelaar 1978, Boden & Parker 1986, Boden 1988, Perissinotto 1989, Perissinotto & Duncombe Rae 1990, Perissinotto & McQuaid 1992), Heard Island (Woehler & Green 1992), and South Georgia (Ward 1989, Atkinson *et al.* 1990, 1992, Oresland & Ward 1993, Whitehouse

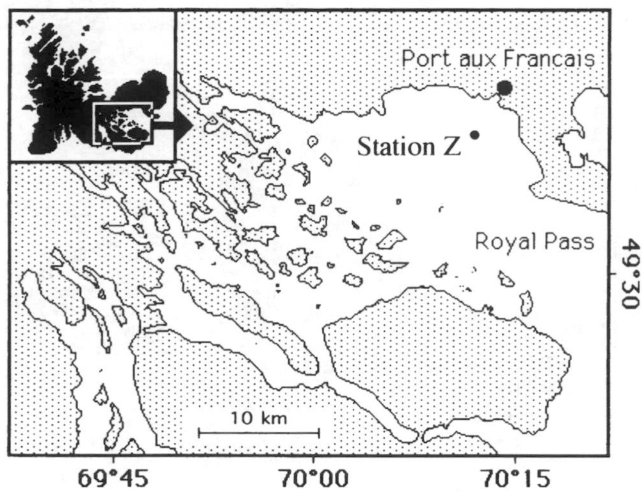

Fig. 13.1. Location of Morbihan Bay (Kerguelen Archipelago) and sampling station (Z).

Table 13.1. *Data set of measured parameters*

	1983	1984	1986	1987	1988	1990	1991
Temperature	*		*	*		*	*
Salinity			*			*	*
Nitrate			*				
Phosphate			*				
Silicate			*				
Bacterial biomass	*	*		*			
CFU	*	*		*			
Chlorophyll *a*			*	*	*	*	*
Carbohydrates				*	*		
Proteins				*	*		
Lipids				*	*		
POC and C:N				*		*	
Mesozooplankton biomass	*		*			*	*
Copepod numerations	*					*	

et al. 1993). However, these studies remain generally restricted to some summer and winter months. In contrast, the Morbihan Bay, in the Kerguelen Archipelago, offers the opportunity to perform year-round sampling, and to observe changes of hydrological and biological characteristics for short-term as well as for pluri-annual series (Delille 1977, 1990, Delille & Cahet, 1985, Bouvy *et al.* 1986, Bouvy & Delille 1988, Razouls & Razouls 1990, Mayzaud & Razouls 1992). The purpose of the present study, which is based on several years of observations undertaken in Morbihan Bay, was to describe a general pattern of temporal changes in physical and biochemical conditions in relation to bacterial, phytoplankton and zooplankton communities in a coastal sub-Antarctic ecosystem. It was an attempt to identify chronological periodicity among the different parameters and to synthesize their relationships.

MATERIAL AND METHODS

Located in the southeast of the archipelago, Morbihan Bay (about 600 km^2) opens to the Indian Ocean through Royal Pass, which is 12 km wide and about 40 m deep (Fig. 13.1). Hydrological and biological samples were taken in surface waters at station Z (40 m depth) located in the open zone of the bay at *c.* 3 km from the coast (49° 21' S, 70° 13' E). The study was carried out between January 1983 and December 1991. The initial set of data includes 45 annual time series of various parameters (Table 13.1). In so far as the meteorological and logistical conditions were fulfilled, hydrological and biological parameters (bacteria, phytoplankton and zooplankton) were sampled on different temporal scales, ranging from twice a week to once a month, but no parameters were collected continuously from 1983 to 1991. Surface water samples were collected with a Niskin bottle. Samples for bacteriological analysis were collected aseptically with sterile glass bottles. Analysis was performed in the base of Port aux Français within one hour of sample collection.

Temperature was measured by reversing thermometers mounted on a Niskin sampler and salinity was determined with an induction salinometer (Beckman Instrument Model RS7-C). Water samples for chlorophyll *a* analysis were prefiltered through a 200 μm mesh filter to remove larger detritic material and biota. After filtration through a Whatman GF/C glass-fibre, pigments were extracted in 90% acetone (Neveux & Panouse 1987). Chlorophyll *a* and phaeopigment were determined by measurements of fluorescence using a Turner Designs fluorometer. Nitrates, phosphates and silicates were analysed on filtered (0.45 μm) water samples using the AutoAnalyzer method (Treguer & Le Corre 1975).

Carbohydrates, proteins and lipids were extracted from aliquots of water samples, sieved on a 100 μm filter before filtering on GF/C glass-fibre filter. They were then estimated by the usual biochemical methods (Lowry *et al.* 1951, Dubois *et al.* 1956, Bligh & Dyer 1959, Myklestadt 1978).

Particulate organic carbon (POC) and particulate organic nitrogen (PON) were determined on pre-combusted GF/C glass fibre filter using a Perkin-Elmer 2400 CHN analyser.

Direct counts of total bacteria (AODC) were determined using the method of Hobbie *et al.* (1977). Viable heterotrophic bacteria (CFU) were enumerated on the 2216E medium (Oppenheimer & Zobell 1952).

Mesozooplankton (size from 200 μm to 2000 μm) was sampled with WPII nets by vertical hauls to the surface. The counts of animals and the dry weight (dw) biomass were estimated from duplicate samples.

The time series was studied with the help of the Systat (1992) and Mandrake (1991) packages. Cross correlations identify relationships between the different series and any time delays in the relationships.

RESULTS

Solar irradiation ranged from near 50 J cm^{-2} d^{-1} in May–June to more than 2500 J cm^{-2} d^{-1} in December and January, surface

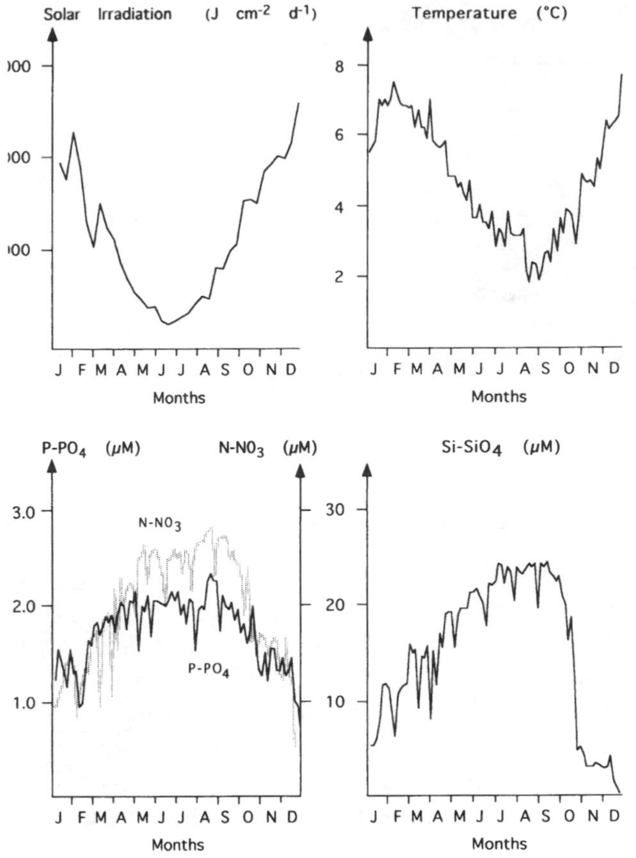

Fig. 13.2. Meteorological and hydrological parameters recorded in 1986. Solar irradiation and nutrients were measured only during this year. Surface water temperatures were collected during four other years (1983, 1987, 1990 and 1991). For clarity, all data do not appear on the drawings.

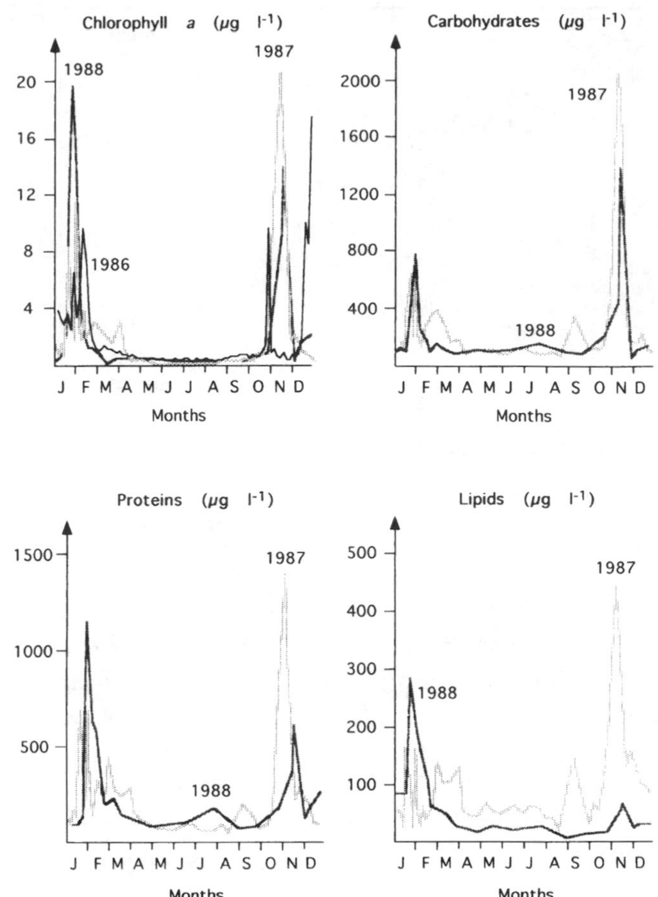

Fig. 13.3. Seasonal changes of biochemical data: chlorophyll *a*, total carbohydrates, proteins and lipids.

water temperatures rising from 2 °C during the winter to 7 °C during the summer (Fig. 13.2). Temperatures did not vary considerably from year to year but a small increase of annual mean (0.25 °C) was observed from 1983 to 1990. There were only very small seasonal changes in seawater salinity, which varied from 33.2‰ in winter to 33.6‰ in summer (data not shown). The surface waters are slightly less saline in Morbihan Bay (annual mean: 33.4‰) than in the offshore area (annual mean: 33.7‰; Murail *et al.* 1977).

Inorganic nutrients exhibited similar seasonal patterns. During winter, from May to September, the concentrations were high: 25, 2, and 24 µM for NO_3, PO_4 and SiO_4 respectively (Fig. 13.2). These concentrations decreased dramatically during October and remained at a very low level during the summer (5.5 µM for nitrate, 0.7 µM for phosphate and 0.2 µM for silicate). The summer depletion of inorganic nutrients corresponded to a sharp increase in phytoplankton biomass (Fig. 13.3). A first bloom appeared in spring (October–November) when the water temperature and the solar irradiation were optimal, and a second one in late summer (January–February). The peak concentrations varied from one year to another but were always higher than 8 µg chl *a* l^{-1}, the highest value recorded being 27 µg l^{-1} during November 1987. During other seasons, chlorophyll *a* concentration showed moderate values ranging from 0.4 to 0.6 µg l^{-1}.

The carbohydrates, proteins and lipids on average represented 29%, 53% and 18% (in weight) respectively of total sestonic material. Their mean annual amounts varied from 80 µg l^{-1} for lipids, to 129 µg l^{-1} for carbohydrates and 240 µg l^{-1} for proteins. They showed strong seasonal changes that mirrored those of phytoplankton (Fig. 13.3). Carbohydrates were dominant (35%) in spring and early summer, while protein and lipid proportions were highest in late summer and winter, reaching respectively, 59% and 20% of the sestonic amount. Chlorophyll *a* concentration is a good indicator of the total amount of organic material, as indicated by the significant correlations between the pigment content and total carbohydrates (r=0.83), proteins (r=0.95) and lipids (r=*82*) for the 80 sets of measurements from December 1986 to December 1988.

POC concentrations ranged from 138 µg l^{-1} in winter 1987 to a maximum value of 1800 µg l^{-1} in November 1987, with annual mean concentrations averaging 300 µg l^{-1} (data not shown). Large discrepancies in the mean POC level between years were observed (from 150 µg in 1984 to 402 µg l^{-1} in 1987). In contrast, a weak multi-year variation in the C:N ratio indicated simultaneous changes of POC and PON concentrations. The C:N ratio averaged 7 with lowest values (mean 5.5) corresponding to the spring and summer maxima of phytoplankton.

Total bacterial abundance ranged from less than 10^4 cells ml^{-1} in April and October 1984 to more than 10^6 cells ml^{-1} in

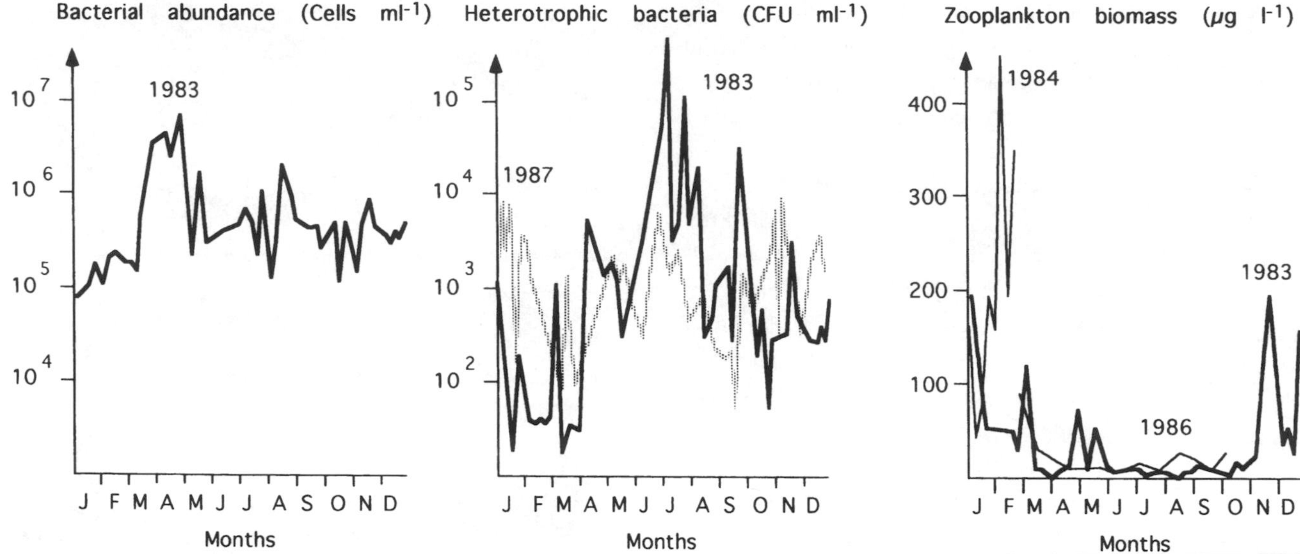

Fig. 13.4. Seasonal changes of total and heterotrophic bacterial abundance, and mesozooplankton biomasses. For clarity all data do not appear on the drawings.

March–April 1983. At the same time viable counts ranged from about 10 cells ml $^{-1}$ in early March 1983 to more than 10^5 cells ml^{-1} in June 1983. With one exception for high values observed in March–April 1983, direct counts were fairly constant throughout the year. Heterotrophic viable counts were generally low in late spring and summer and reached their highest values in the autumn, before a marked decline occurred in late winter (Fig. 13.4).

The mesozooplankton constitutes the major zooplanktonic fraction in Morbihan Bay, taking into account the small episodic amount of macroplankton individuals (<4 ind. 100 m^{-3}) of Euphausiacea, Amphipoda, larval stages of Decapoda or Medusa (Koubbi 1992). Copepods constituted the bulk of the zooplankton community. Their abundance (average from 200 ind. m^{-3} in winter to 6000 ind. m^{-3} in summer) was clearly correlated to the total biomass (r=0.61). One neritic and endemic species (*Drepanopus pectinatus*) represented 90% of the huge population of Copepods, whilst three others, two *Oithona* spp. (*O. frigida* and *O. similis*) and *Calanus simillimus,* accounted for 6% and 2% respectively. The overall mean biomass averaged 159 mg dw m^{-3} and displayed a double maximum period in November–December and in January–February with up to 450 mg dw m^{-3} (exceptionally 1000 mg dw m^{-3}). In contrast, during the winter period, the biomass decreased to values <10 mg m^{-3} (Fig. 13.4). The seasonal biomass may differ by a factor of two to eight years.

DISCUSSION

The recorded seasonal changes of inorganic nutrients in Morbihan Bay confirmed previous data obtained in the same area (Delille & Lagarde 1974). Neritic sub-Antarctic waters show some similarities with Antarctic waters (see review of El Sayed & Fryxell 1993) during autumn and winter: low phytoplanktonic biomass in spite of the high nutrient concentrations exceeding phytoplankton requirements. Marked increases in phytoplankton biomass between October and February occur in Antarctic coastal areas (Davidson & Marchant 1992) as well as sub-Antarctic neritic waters (Lipski 1987, Leakey *et al.* 1994). However, the double peak observed in the present study seems to be distinctive. Chlorophyll peaks were clearly associated with a depletion of dissolved nutrients, which indicates a high uptake rate (Jennings *et al.* 1984). The first peak in October–November is related to a sharp drop of silicate concentration, which decreased to 20 µM within few days. Such severe depletion seems to be an indirect indication of a diatom bloom. The high chlorophyll values observed are among the highest recorded, and are of the same magnitude as those observed in other sub-Antarctic and Antarctic islands (Perissinotto *et al.* 1992). The period of increase in the phytoplanktonic biomass corresponds to a water warming up and to a solar irradiation greater than 1500 J cm^{-2} d^{-1}. Thus, in the Kerguelen Archipelago, phytoplankton enhancement seems to be due to an increase in water temperature and solar irradiation during summer. However, the shallow inshore environment will also influence water-column stability and nutrient regeneration, and thus phytoplankton growth. This direct influence of temperature and solar irradiation contrast with the general idea of the island mass effect as the main factor responsible for an increase in productivity in the Southern Ocean (Planke 1977, Nast & Gieskes 1986, Perissinotto *et al.* 1992). Seasonal variabilities in particulate organic constituents were found to be closely related to that of chlorophyll and in good accordance with pigment enhancement. They reflected the important contribution of phytoplankton activity to carbon production. However, seasonal variations in the proportions of the biochemical constituents suggested changes in the physiological state and/or species of phytoplankton (Bedo 1987). Besides, the lowest estimations of the biochemical elements and POC in winter exceeded by far those of the sub-Antarctic oceanic zone (Copin-Montegut &

Copin-Montegut 1978, Bedo *et al.* 1987, 1990). Benthic derived POC could enhance levels of pelagic POC and biochemical constituents in shallow inshore waters. However, such an effect will be in contradiction with the close relationship observed between POC and phytoplankton.

Although there is increasing evidence that protozoans can contribute significantly to the diet of zooplankton (Gifford 1991), in Morbihan Bay the C:N values (5 to 7) evoked more a dominance of phytoplankton than of protozoans (C:N=4). Directly or indirectly, the episodic blooms of phytoplankton have to sustain the herbivorous zooplankton population. The time-series treatment showed the best correlation (r=0.43) between the zooplanktonic biomass, and the water chlorophyll content, with a time lag of 30 to 35 days. From late October to November and from late February to March the periods of maximum activity of the phytoplankton induce bursts of successive generations of copepods (Razouls & Razouls 1990).

The standing stock appears to be higher in the neritic sectors, for example 68 mg dw m^{-3} around Crozet Island (Razouls *et al.* 1994), than in other oceanic Indian sectors about 15–30 mg dw m^{-3} (Razouls & Razouls 1982, Voronina *et al.* 1994). A land effect could explain the high biomass values, observed too in other subantarctic or Antarctic bays of the Southern Ocean, which are often due to the dominance of *D. forcipatus* (Minoda & Hoshiai 1982, Ward 1989).

In contrast to the other biological parameters studied bacterial abundance was not directly related to temperature. This observation confirms previous studies in the Southern Ocean (Delille *et al.* 1988, Vincent 1988, Delille & Perret 1989, Fukunaga & Russell 1990, Vosjan & Olanczuk-Neyman 1991) which demonstrate that, despite the extreme cold, the temperature seems to have only a rather limited influence on Antarctic bacterioplanktonic populations. Trophic sources are probably the major regulating factor of the sub-Antarctic bacterial communities (Delille & Bouvy 1989, Delille & Perret 1991). However, on the time-scale of this study there were no direct significant correlations between bacterial parameters and any of the potential trophic sources studied (phytoplankton, POC). A decrease in predation pressure during the winter could also be involved.

The repetitive seasonal variations, even with differences in the timing of annual patterns, allow the calculation of annual mean values of hydrological and biological parameters (Fig. 13.5). However, attention needs to be drawn to the importance of the sampling interval. With a short time interval between samples rapid changes in biological parameters can be observed. For example, pulses of chlorophyll (every 1 to 3 days) would not be identified with long sampling intervals. Therefore, a short sampling interval appears to be essential for understanding the flux processes *in situ* and the precise temporal relationship between the biological parameters. Similarly, the quality of the time interpolation will depend on the frequency of data collection.

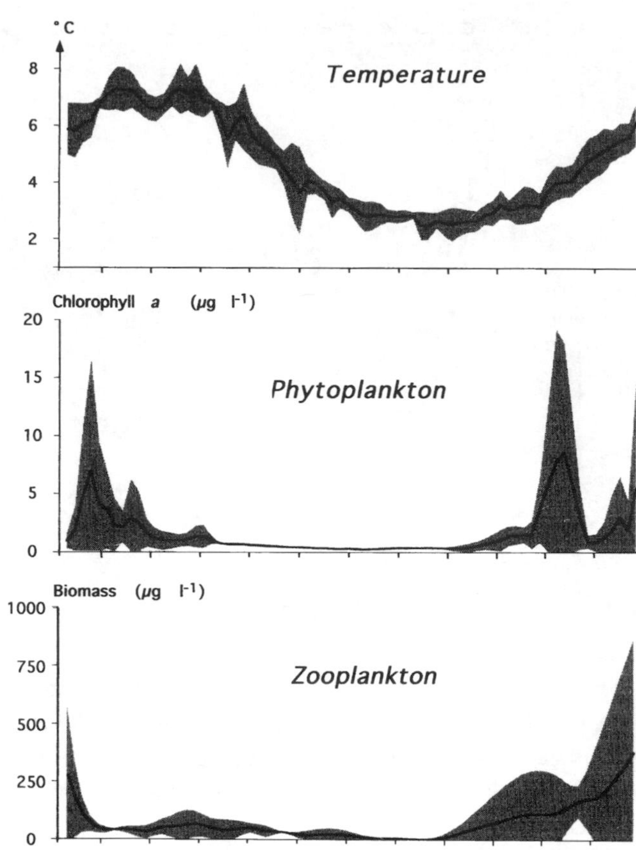

Fig. 13.5. Mean time series (mean ± standard deviation) of temperature, chlorophyll *a* and zooplankton biomass.

ACKNOWLEDGEMENTS

This research was supported by the Institut Français pour la Recherche et la Technologie Polaires.

REFERENCES

Atkinson, A., Ward, P., Peck, J. M. & Murray, A. W. A. 1990. Mesoscale distribution of zooplankton around South Georgia. *Deep-Sea Research*, **8,** 1213–1227.

Atkinson, A., Ward, P., Williams, R. & Poulet, S. A. 1992. Diel vertical migration and feeding of copepods at an oceanic site near South Georgia. *Marine Biology*, **113,** 583–593.

Bedo, A. 1987. *Caractérisation et relation du seston avec la structure hydrologique dans le secteur est-indien de l'Océan Antarctique*. Thèse Doctorat d'Etat. Université P. et M. Curie, 102 pp.

Bedo, A., Mayzaud, P. & Corre, M. C. 1987. Caractéristiques biochimiques de la matière organique particulaire dans le secteur sud-Indien de l'Océan Antarctique. *Rapports des campagnes à la mer, Mission de recherche des TAAF*, No. 84-01, 143–163.

Bedo, A., Mayzaud, P. & Corre, M. C. 1990. Définition de l'environnement trophique potentiel d'*Euphausia superba*: caractéristiques biochimiques et granulométriques du seston durant l'été austral dans le secteur sud Indien de l'Océan Antarctique. *Rapports des campagnes à la mer, Mission de recherche des TAAF*, No. 85–01, 71–86.

Bligh, E. G. & Dyer, W. J. 1959. A rapid method of total lipid extraction and purification. *Canadian Journal of Biochemistry and Physiology*, **37,** 911–917.

Boden, B. P. 1988. Observations of the island mass effect in the Prince Edward Archipelago. *Polar Biology*, **9,** 61–68.

Boden, B. P. & Parker, L. D. 1986. The plankton of the Prince Edward Islands. *Polar Biology*, **5,** 81–93.

Bouvy, M. & Delille, D. 1988. Spatial and temporal variations of Antarctic and sub-Antarctic bacterioplankton. *Netherlands Journal of Sea Research*, **22,** 139–147.

Bouvy, M., Le Romancer, M. & Delille, D. 1986. Significance of microheterotrophs in relation to the degradation process of sub-Antarctic kelp beds (*Macrocystis pyrifera*). *Polar Biology*, **5,** 249–253.

Bunt, J. S. 1955. The importance of bacteria and other microorganisms in the sea water at Macquarie Island. *Australian Journal of Marine and Freshwater Research*, **6,** 60–65.

Copin-Montegut, C. & Copin-Montegut, G. 1978. The chemistry of particulate matter from the south Indian and Antarctic oceans. *Deep-Sea Research*, **25,** 911–931.

Davidson, A. T. & Marchant, H. J. 1992. Protist abundance and carbon concentration during a *Phaeocystis*-dominated bloom at an Antarctic coastal site. *Polar Biology*, **12,** 387–392.

Delille, D. 1977. Bacterial sulfur and nitrogen cycles in sedimentary deposits of the fjord Bossière (Kerguelen Archipelago). In Llano, G.A., ed. *Adaptations within Antarctic ecosystems*. Washington DC: Smithsonian, 159–180.

Delille, D. 1990. Seasonal changes of sub-Antarctic heterotrophic bacterioplankton. *Archiv für Hydrobiologie*, **119,** 267–277.

Delille, D. & Bouvy, M. 1989. Bacterial responses to natural organic inputs in a marine sub- Antarctic area. *Hydrobiologia*, **182,** 225–238

Delille, D. & Cahet, G. 1985. Heterotrophic processes in a Kerguelen mussel-bed. In Siegfried W. R., Condy, P. R. & Laws, R. M., eds. *Antarctic nutrient cycles and food web*. Berlin: Springer-Verlag, 128–135.

Delille, D. & Lagarde, E. 1974. Cycles hydrologiques de 18 stations de la Baie du Morbihan (Archipel de Kerguelen). *Revue de l'Institut Pasteur de Lyon*, **7,** 321–338.

Delille, D. & Perret, E. 1989. Influence of temperature on the growth potential of southern polar bacteria. *Microbial Ecology*, **18,** 117–123.

Delille, D. & Perret, E. 1991. The influence of giant kelp *Macrocystis pyrifera* on the growth of sub-Antarctic marine bacteria. *Journal of Experimental Marine Biology and Ecology*, **153,** 227–239.

Delille, D., Bouvy, M. & Cahet, G. 1988. Short term variations of bacterio-plankton in Antarctic zone: Terre Adélie area. *Microbial Ecology*, **15,** 293–309.

Dubois, M., Gilles, K. A., Hamilton, P. A., Rebers, P. A. & Smith, F. 1956. Colorimetric method for determinations of sugars and related substances. *Annals of Chemistry*, **28,** 350–356.

El Sayed, S. Z. & Fryxell, G. A. 1993. Phytoplankton. In Friedmann, E.I., ed. *Antarctic microbiology*. New York: Wiley-Liss, 65–122.

Fontugne, M. & Fiala, M. 1987. MD38/APSARA II–ANTIPROD III à bord du '*Marion-Dufresne*', 16 janvier–22 février 1984. *Rapports des campagnes à la mer, Mission de Recherches des TAAF*, No. 84-01, 345 pp.

Friedmann, E. I., ed. 1993. *Antarctic microbiology*. New York: Wiley-Liss, 634 pp.

Fukunaga, N. & Russell, N. J. 1990. Membrane lipid composition and glucose uptake in two psychrotolerant bacteria from Antarctica. *Journal of General Microbiology*, **136,** 1669–1673.

Gifford, D. J. 1991. The protozoan–metazoan link in pelagic ecosystems. *Journal of Protozoology*, **38,** 81–86.

Hobbie, J. E., Daley, R. J. & Jasper, S. 1977. Use of nuclepore filters for counting bacteria by fluorescence microscopy. *Applied and Environmental Microbiology*, **33,** 1225–1228.

Hopkins,T. L., Lancraft, T. M., Torres, J. J. & Donnelly, J. 1993. Community structure and trophic ecology of zooplankton in the Scotia Sea marginal ice zone in winter (1988). *Deep-Sea Research*, **40,** 81–105.

Hoshiai, T., Nemoto, T. & Naito, Y., eds. 1986. Proceedings of the seventh symposium on Polar Biology. *Memoirs of National Institute of Polar Research*, Special Issue, **40,** 496.

Jacques, G. 1978. Campagne Antiprod I '*Marion-Dufresne*' 1–28 Mars 1977. *Résultats des campagnes à la mer, CNEXO*, No. 16, 149 pp.

Jacques, G. 1982. Campagne océanographique MD 21/Antiprod II (Mars 1980). *CNFRA*, No. 53, 141 pp.

Jennings Jr, J. C., Gordon, L. I. & Nelson, D. M. 1984. Nutrient depletion indicates high primary productivity in the Weddell–Scotia Confluence area. *Nature*, **309,** 51–54.

Kok, O. B. & Grobbelaar, J. U. 1978. Observations on the crustaceous zooplankton in some freshwater bodies of the sub-Antarctic island Marion. *Hydrobiologia*, **59,** 3–8.

Koubbi, P. 1992. *L'ichtyoplancton de la partie indienne de la province Kerguelenienne (Bassin de Crozet et plateau de Kerguelen): identification, distribution spatio-temporelle et stratégies de développement larvaire*. Thèse de Doctorat d'Etat, Université P. et M. Curie, 391 pp.

Leakey, R. J. G., Fenton, N. & Clarke, A. 1994. The annual cycle of planktonic ciliates in near shore waters at Signy Island, Antarctica. *Journal of Plankton Research*, **16,** 841–856.

Lipski, M. 1987. Variations of physical conditions, nutrients and chlorophyll *a* contents in Admiralty Bay (King George Island, South Shetland Islands, 1979). *Polish Polar Reasearch*, **8,** 307–332.

Lowry, O. M., Rosenbrough, N. J., Farr, A. L. & Randall, R. J. 1951. Protein measurement with the Folin phenol reagent. *Journal of Biological Chemistry*, **193,** 265–275.

MANDRAKE. 1991. *Un progiciel expert en analyse de séries temporelles*. Palaiseau (France): CEMS.

Mayzaud, P. 1990. MD 42/Sibex à bord du '*Marion-Dufresne*', 3 Janvier–18 Février 1985. *Rapports des campagnes à la mer, Mission de Recherches des TAAF*, No. 85–01, 210 pp.

Mayzaud, P. & Razouls, S. 1992. Degradation of gut pigment during feeding by a sub-Antarctic copepod: importance of feeding history and digestive acclimation. *Limnology and Oceanography*, **37,** 393–404.

Minoda, T. & Hoshiai, T. 1982. Zooplankton community in the cove of Cumberland Bay, South Georgia, in the Southern summer from January to February 1973. *Memoirs of the National Institute of Polar Research*, Special Issue, No.23, 32–37.

Murail, J. F., David, P. & Panouse, M. 1977. Hydrologie du plateau continental des Iles Kerguelen. *CNFRA*, No.42, 41–64.

Myklestadt, S. 1978. B-1,3 Glucans in diatoms and brown seaweeds. In Hellebust, J. A. & Craigie, J. S., eds. *Handbook of phycological methods*. Cambridge: Cambridge University Press, 512 pp.

Nast, F. & Gieskes, W. 1986. Phytoplankton observations relative to krill abundance around Elephant Island in November 1983. *Archiv für Fischereiwissenschaft*, **37,** 95–106.

Neveux, J. & Panouse, M. 1987. Spectrofluorometric determination of chlorophylls and pheophytins. *Archiv für Hydrobiologie*, **109,** 567–561.

Oppenheimer, C. H. & Zobell, C. E. 1952. The growth and viability of sixty-three species of marine bacteria as influenced by hydrostatic pressure. *Journal of Marine Research*, **11,** 10–18.

Oresland, V. & Ward, P. 1993. Summer and winter diet of four carnivorous copepod species around South Georgia. *Marine Ecology Progress Series*, **98,** 73–78.

Perissinotto, R. 1989. The structure and diurnal variations of the zooplankton of the Prince Edward Islands: implications for the biomass build-up of higher trophic levels. *Polar Biology*, **9,** 505–510.

Perissinotto, R. & Duncombe Rae, C. M. 1990. Occurence of anticyclonic eddies on the Prince Edward Plateau (Southern Ocean): effects on phytoplankton biomass and production. *Deep-Sea Research*, **37,** 777–793.

Perissinotto, R. & McQuaid, C. D. 1992. Land-based predator impact on vertically migrating zooplankton and micronekton advected to a Southern Ocean archipelago. *Marine Ecology Progress Series*, **80,** 15–27.

Perissinotto, R., Laubscher, R. K. & McQuaid, C. D. 1992. Marine productivity enhancement around Bouvet and the South Sandwich Islands (Southern Ocean). *Marine Ecology Progress Series*, **88,** 41–53.

Planke, J. 1977. Phytoplankton biomass and productivity in the subtropical convergence area and shelves of the western Indian sub-Antarctic islands. In Llano, G.A., eds. *Adaptations within Antarctic ecosystems*. Washington DC: Smithsonian, 51–73.

Razouls, C. & Razouls, S. 1982. Elements du bilan énergétique du mesozooplancton antarctique. Campagne océanographique MD 21/Antiprod II (mars 1980). *CNFRA*, No. 53, 131–141.

Razouls, C & Razouls, S. 1990. Biological cyle of a population of sub-Antarctic copepod, *Drepanopus pectinatus* (Clausocalanidae), Kerguelen Archipelago. *Polar Biology*, 10, 541–543.

Razouls, S., de Bovée, F., Razouls, C. & Panouse, M. 1994. Are size-spectra of mesozooplankton a good tag for characterizing pelagic ecosystem? *Vie et Milieu*, **44**, 59–68.

Smith Jr, W. O. 1990. *Polar oceanography. Part B: Chemistry, biology and geology*. San Diego: Academic Press, 760 pp.

SYSTAT. 1992. *Systat for windows: statistics; version 5*. Evanston (USA): Systat Inc.

Treguer, P. & Le Corre, P. 1975. *Manuel d'analyses des sels nutritifs dans l'eau de mer. Utilisation de l'AutoAnalyzer II Technicon*, 2nd edn. Brest: UBO, 110 pp.

Vincent, W. F. 1988. *Microbial ecosystems of Antarctica*. Cambridge: Cambridge University Press, 304 pp.

Voronina, N. M., Kosobova, K. N. & Pakhomov, E. A. 1994. Composition and biomass of summer metazoan plankton in the 0–200 m layer of the Atlantic sector of the Antarctic. *Polar Biology*, **14**, 91–95.

Vosjan, J. H. & Olanczuk-Neyman, K. M. 1991. Influence of temperature on respiratory ETS-activity of micro-organisms from Admiralty Bay, King George Island, Antarctica. *Netherlands Journal of Sea Research*, **28**, 221–225.

Ward, P. 1989. The distribution of zooplankton in an Antarctic fjord at South Georgia during summer and winter. *Antarctic Science*, **1**, 141–150.

Whitehouse, M. J., Symon, C. & Priddle, J. 1993. Variations in the distribution of chlorophyll *a* and inorganic nutrients around South Georgia, South Atlantic. *Antarctic Science*, **5**, 367–376.

Woehler, E. J. & Green, K. 1992. Consumption of marine resources by seabirds and seals at Heard Island and the McDonald Islands. *Polar Biology*, **12**, 659–665.

14 Studies on dimethyl sulphide in Antarctic coastal waters

GIACOMO R. DITULLIO[1] *AND WALKER O. SMITH, JR*

Dept. of Ecology and Evolutionary Biology, University of Tennessee, Knoxville, TN 37996, USA, [1] *Present address: University of Charleston, Grice Marine Lab, 205 Fort Johnson, Charleston, SC 29412, USA*

ABSTRACT

High (>90 nmol l^{-1}) dimethyl sulphide (DMS) concentrations were measured in the Ross Sea during the late summer (February, 1992). Phytoplankton pigment concentrations revealed the dominance of 19'-hexanoyloxyfucoxanthin (a proxy for prymensiophytes) in areas where high DMS concentrations were observed. Sediment trap fluxes of dimethylsulphoniopropionate (DMSP) in the Antarctic Peninsula region represented < 0.1% of the integrated water column concentration of DMS and particulate DMSP ($DMSP_p$). An Fe-enrichment experiment was performed in the northern Ross Sea, an area predominately inhabited by slow-growing (0.1 d^{-1}) diatom populations. $DMSP_p$ was 2.5 times higher relative to the control treatment following an 11-day enrichment experiment. Fucoxanthin concentrations (a proxy for diatoms) mirrored the increase in chlorophyll a *concentrations following Fe enrichment relative to the control. The chlorophyll* a*-specific DMSP production rate, however, was not significantly different than the control. Pigment-specific growth rates for* Phaeocystis antarctica, *however, were preferentially stimulated over other algal classes following Fe addition relative to their respective control populations. Thus,* P. antarctica *may represent a possible link between the iron and sulphur cycles in the Antarctic.*

Key words: DMS in the Antarctic.

INTRODUCTION

The volatile S gas dimethyl sulphide (DMS) accounts for over half of the global biogenic S flux to the atmosphere (Andreae 1990). Phytoplankton are the major source of DMS in the ocean. DMS is produced by the breakdown of its precursor, dimethylsulphoniopropionate (DMSP) (Challenger & Simpson 1948). However, the factors governing DMS production and fluxes are not well understood, especially in the Southern Ocean. It has been suggested that DMS may be important in regulating the Earth's radiation balance and hence may be involved in global climate change processes (Charlson *et al.* 1987). Martin (1992) has suggested that the close correlation between the DMS oxidation product, methanesulphonic acid (MSA) and Fe concentrations in Vostok ice cores (LeGrand *et al.* 1991) was evidence supporting the link between the Fe and S cycles (Zhuang 1992) and the Fe hypothesis (Martin 1990).

Antarctic coastal waters are frequently dominated by the colonial prymnesiophyte *Phaeocystis antarctica*, especially during the spring phytoplankton bloom (El-Sayed *et al.* 1983, Gibson *et al.* 1990). DMS concentrations in Antarctic coastal waters can be very high during the austral summer (Gibson *et al.* 1990, Fogelqvist 1991) compared with values in the Drake Passage (Berresheim 1987). Based on DMS production by *P. antarctica* in the austral summer it was suggested that Antarctic coastal waters could account for approximately 10% of the global DMS flux to the atmosphere (Gibson *et al.* 1990). DMS production from sea-ice algae (Kirst *et al.* 1991) during non-summer months, however, could increase the annual flux to the atmosphere, especially considering the wide expanse of sea-ice in the Southern Ocean (20×10^6 km^2).

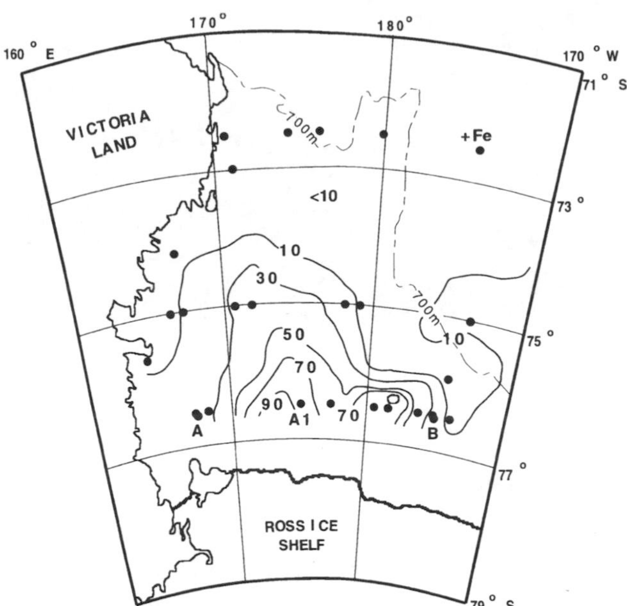

Fig. 14.1. Surface DMS concentrations in the Ross Sea, Antarctica during February 1992. Three transects were performed. Dots represent station locations along the transect line. Contour intervals are 20 nM. 700 m depth contour indicated by dashed line. A, A1 and B mark sites referred to in the text.

The goals of our study were two-fold. First, we investigated the magnitude of DMS concentrations in the Ross Sea and in coastal waters near the Antarctic Peninsula on two separate expeditions during the non-bloom conditions of late austral summer and early spring, respectively. In addition, we monitored the flux of DMSP into free-floating sediment traps deployed near the Antarctic Peninsula to assess the relative importance of sedimenting S out of the photic zone. Secondly, we monitored DMSP concentrations following an Fe-enrichment experiment in the Ross Sea to investigate the possible link between Fe and S cycles as postulated by Martin (1992).

MATERIALS AND METHODS

Three transects were sampled in the Ross Sea between 6 February and 28 February, 1992 (Fig. 14.1). In addition, stations at Charlotte Bay (64° 27' S, 61° 44' W) and Paradise Harbour (64° 51' S, 62° 54' W) along the Antarctic Peninsula were sampled during November 1992. Both sets of samples were obtained aboard the RV *Polar Duke*.

DMS measurements were performed using a liquid N_2 cryogenic purge and trap apparatus similar to the one described in Radford-Knoery & Cutter (1993). A Hewlett Packard 5890 Series II gas chromatograph equipped with a flame photometric detector (FPD) and a 3393 integrator were used to measure DMS. A Chromosil 330 column heated at 70 °C was used for the chromatographic separation. A permeation device was used for calibration and also as a doping gas to linearize the FPD response and to increase sensitivity. Samples were not filtered but comparative tests revealed only an approximate 10% difference in our studies. DMSP samples were frozen in liquid N_2 for later analyses. The methodology employed here for measuring DMS and DMSP has been described in detail (DiTullio & Smith 1995).

Phytoplankton pigments were used as a chemotaxonomic tool to identify the major classes of phytoplankton present (Weber & Wettern 1980). Algal pigments were separated by high-performance liquid chromatography (HPLC) using a C18 Spherisorb ODS II column and a solvent gradient (Solvent A: MeOH: 0.5N aq. ammonium acetate, 80:20; Solvent B: MeOH; Solvent C: acetone) similar to that reported by Goericke & Welschmeyer (1993). Concentrations of 19'-hexanoyloxyfucoxanthin (Hex) are associated with prymnesiophytes (Arpin *et al.* 1976). In the Antarctic, Hex is a good proxy for the presence of *Phaeocystis antarctica* (Buma *et al.* 1991) and fucoxanthin is assumed to be predominately associated with diatom cells. Concentrations of peridinin and 19'-butanoyloxyfucoxanthin are chemotaxonomic indicators for the presence of dinoflagellates and pelagophytes, respectively.

Samples from free-floating sediment traps were obtained to determine the flux of $DMSP_p$ out of the photic zone near the Antarctic Peninsula in November, 1992. Cylindrically shaped Multi-traps (VERTEX-style) were deployed at 60, 80, and 100 m for approximately 31 to 33 h in Charlotte Bay and Paradise Harbour (both in the Gerlache Strait region). Two sets of traps were deployed at each depth. One set of traps received no additions while the other set was poisoned with formalin.

Seawater for the Fe-enrichment experiment was collected at 72° 29' S, 172° 53' W in the Ross Sea on February 27, 1992. This station was located approximately 600 km east of Cape Adare in *c*. 2000 m of water, and hence was far removed from sedimentary and ice sources of iron (Martin *et al.* 1990). Seawater was collected into two acid-cleaned 20 l polycarbonate carboys from a rubber boat positioned approximately 3 km upwind from the ship. Seawater from the carboys was dispensed into 12 acid-cleaned 2.7 l polycarbonate bottles using trace-metal clean techniques. The control treatment received no additions. The Fe-amended bottles received a 2 nM addition with a 1:1 EDTA chelated Fe standard (Fe in HCl, atomic absorption standard). This small addition of EDTA does not affect the trace metal speciation as calculated by theTitrator program. Bottles were sealed using parafilm and tape and placed in three separate polyethylene bags to prevent contamination (DiTullio *et al.* 1993). Incubations were performed on-deck for 11 days at surface seawater temperature and irradiance. Two bottles were completely filtered at each time point to prevent contamination from sequential sampling.

Seawater was obtained from 72° 28' S, 172° 19.3' E to perform a ^{14}C-pigment labelling experiment. The experiment was started at dawn by inoculating 250 mCi of ^{14}C into a 2.7 l polycarbonate bottle. Light levels were simulated with neutral density screening. Following a 48 h incubation, subsamples were filtered onto GF/F filters for absolute growth rate determinations by the chl *a* labelling method (Redalje & Laws 1981). Pigments were collected following chromatography on both a reverse-phase C18 column

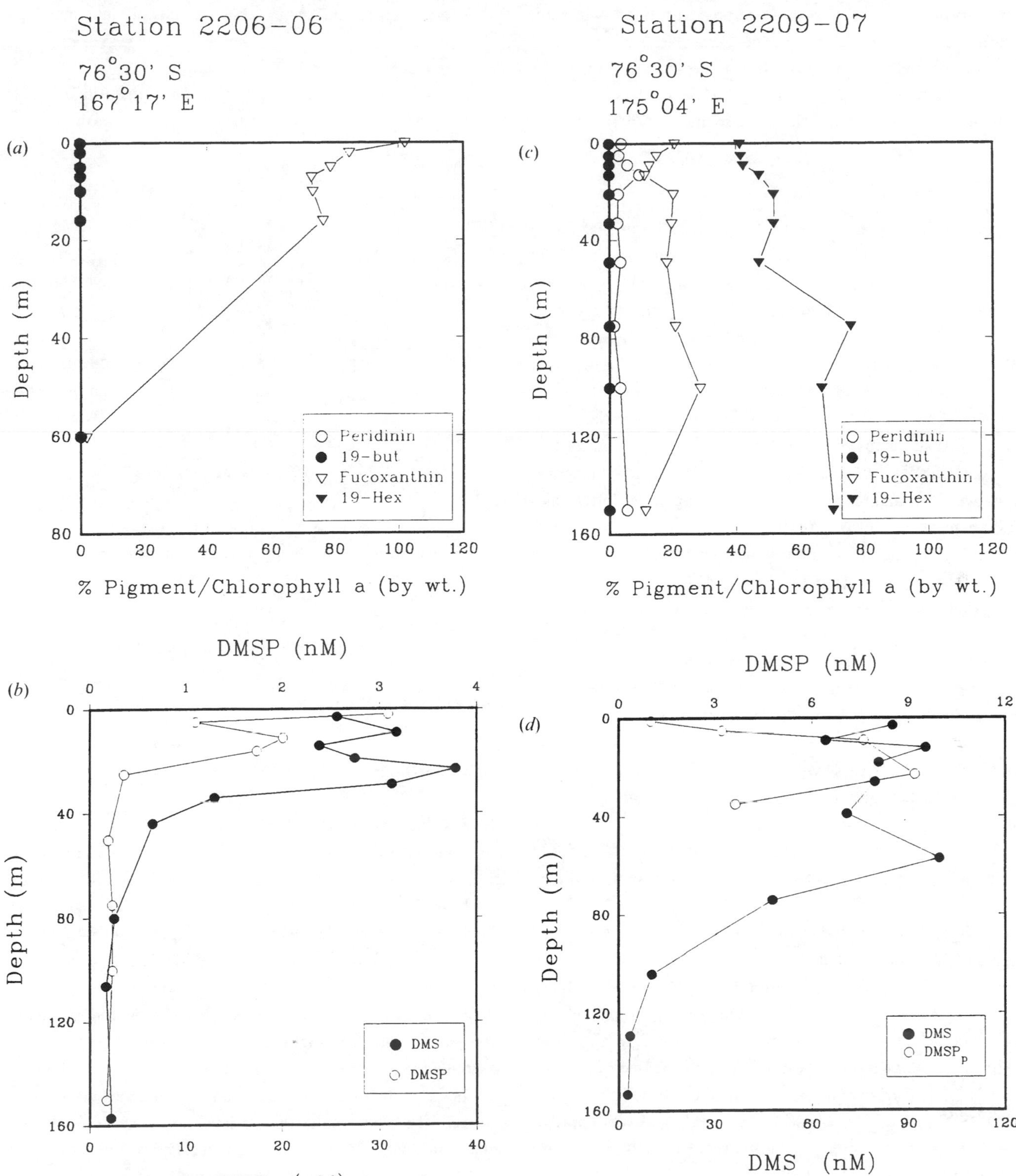

Fig. 14.2. (a) Phytoplankton pigment distributions as measured by high performance liquid chromatography (HPLC) in the southwestern Ross Sea at site A on Fig. 14.1 in late summer. (b) Vertical profiles of DMS and DMSP at site A. (c) HPLC pigment distributions at site A1, approximately 200 km east of site A. (d) Vertical profiles of DMS and DMSP at this site.

and a normal-phase silica column before radio assay on a liquid scintillation counter. The pigment C-specific growth rate of various phytoplankton groups was estimated from the incorporation of ^{14}C into xanthophylls (Gieskes & Kraay 1989) using the equations in Goericke & Welschmeyer (1993).

RESULTS

Surface DMS concentrations in the southern (along 76° 30' S) Ross Sea during February, 1992 were very high (range 11–97 nmol l^{-1}; Fig. 14.1) and were associated with a phytoplankton assemblage that was dominated by *Phaeocystis antarctica* as determined by HPLC pigment analyses (DiTullio & Smith 1995). In the southern Ross Sea, at locations closest to the coast (within 200 km), DMS concentrations were markedly lower (10–50 nmol l^{-1}) and the phytoplankton assemblage was dominated by diatoms, as indicated by the high concentrations of fucoxanthin (Fig. 14.2). Particulate DMSP ($DMSP_p$) concentrations were approximately an order of magnitude lower than DMS concentrations (Fig. 14.2).

Phytoplankton absolute growth rates in the northern Ross Sea in February 1992 were approximately 0.1 d^{-1} (Fig. 14.3) as estimated from the ^{14}C pigment labelling method. The highest specific growth rate observed was 0.21 d^{-1} for *Phaeocystis antarctica* populations incubated at 14% surface irradiance (Fig. 14.3). Although only the total population growth rate was available for the 30% light level it is probable that the highest growth rates of *P. antarctica* were at the 14% light level since the total phytoplankton community (chl *a*) also displayed the highest growth rate (0.15 d^{-1}) at that light level (Fig. 14.3).

The Fe-enrichment experiment resulted in a 2–2.5-fold increase in pigment concentrations (except peridinin) in the Fe treatment relative to the control after 11 days (Fig. 14.4). Based on the initial and final fucoxanthin:chl *a* ratios relative to other pigment:chl *a* ratios, diatoms contributed most to the total increase in chl *a* biomass following Fe enrichment (Fig. 14.5). The changes in pigment-specific growth rate ($P_{11} = P_6 e^{mt}$) relative to the control was displayed by *Phaeocystis antarctica* (Fig. 14.5). After the 11-day incubation $DMSP_p$ increased by approximately 2.5-fold relative to the control (Fig. 14.6). This increase in $DMSP_p$ production following Fe enrichment was approximately equal to the increase in *P. antarctica* growth rate relative to the control treatment.

During November, 1992 DMS and $DMSP_p$ integrated values were very similar to each other in Charlotte Bay near the Antarctic Peninsula (Fig. 14.7a). In contrast, integrated $DMSP_p$ in Paradise Harbour was approximately an order of magnitude higher than the integrated DMS value (Fig. 14.7c). Sediment trap fluxes of $DMSP_p$ at both sites were $<0.1\%$ of integrated water- column values. It is possible, however, that these fluxes represent a lower bound on the true $DMSP_p$ flux because of the potential artifact associated with frozen samples. Nevertheless,

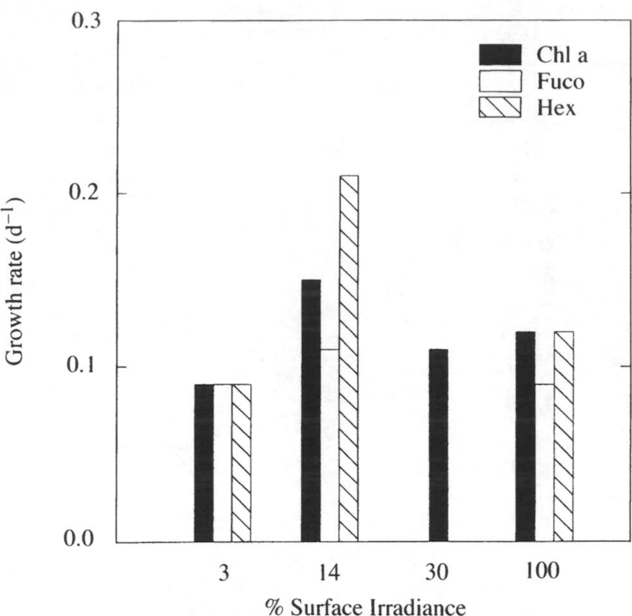

Fig. 14.3. A 24 h ^{14}C pigment labelling experiment performed in the northern Ross Sea on 28 February 1992. Algal specific growth rates were determined using the methodology of Goericke & Welschmeyer (1993).

even if our $DMSP_p$ values were underestimated by *c.* 40% (M. Keller, pers. commun. 1993) it is apparent that our $DMSP_p$ fluxes were small with respect to the DMS levels we observed during these pre-bloom conditions.

DISCUSSION

The possible link between the Fe and S cycles in the Antarctic was suggested to explain the correlation between MSA and Fe in ice samples from Vostok ice cores (Martin 1992). Although Fe enrichment resulted in an approximate 2.5-fold increase in $DMSP_p$ production relative to the control (Fig. 14.6), there was no significant difference in the chl *a*-specific $DMSP_p$ production rate. Hence, Fe addition did not appear to stimulate cellular production of $DMSP_p$. Nevertheless, Fe addition did stimulate *Phaeocystis antarctica* pigment-specific growth rates preferentially relative to those of diatoms (Fig. 14.5). Hence, it may be possible that dust deposition during glacial times could have changed the phytoplankton floristic structure to yield an enhanced DMS flux to the atmosphere. For instance, if lower irradiances favor *P. antarctica* growth relative to other phytoplankton classes in the Antarctic (e.g. Fig. 14.3), then a climatic change such as lower temperatures and irradiances during the last glacial maximum (as a result of Fe stimulated DMS production) could result in a positive-feedback loop whereby greater DMS production by *P. antarctica* could result in even lower irradiances and temperatures (see Charlson *et al.* 1987). Although an atmospheric connection between the Fe and S cycles has been postulated via the sulphur aerosol reduction of Fe^{3+} to Fe^{2+} (Zhuang *et al.* 1992), more evidence is required to

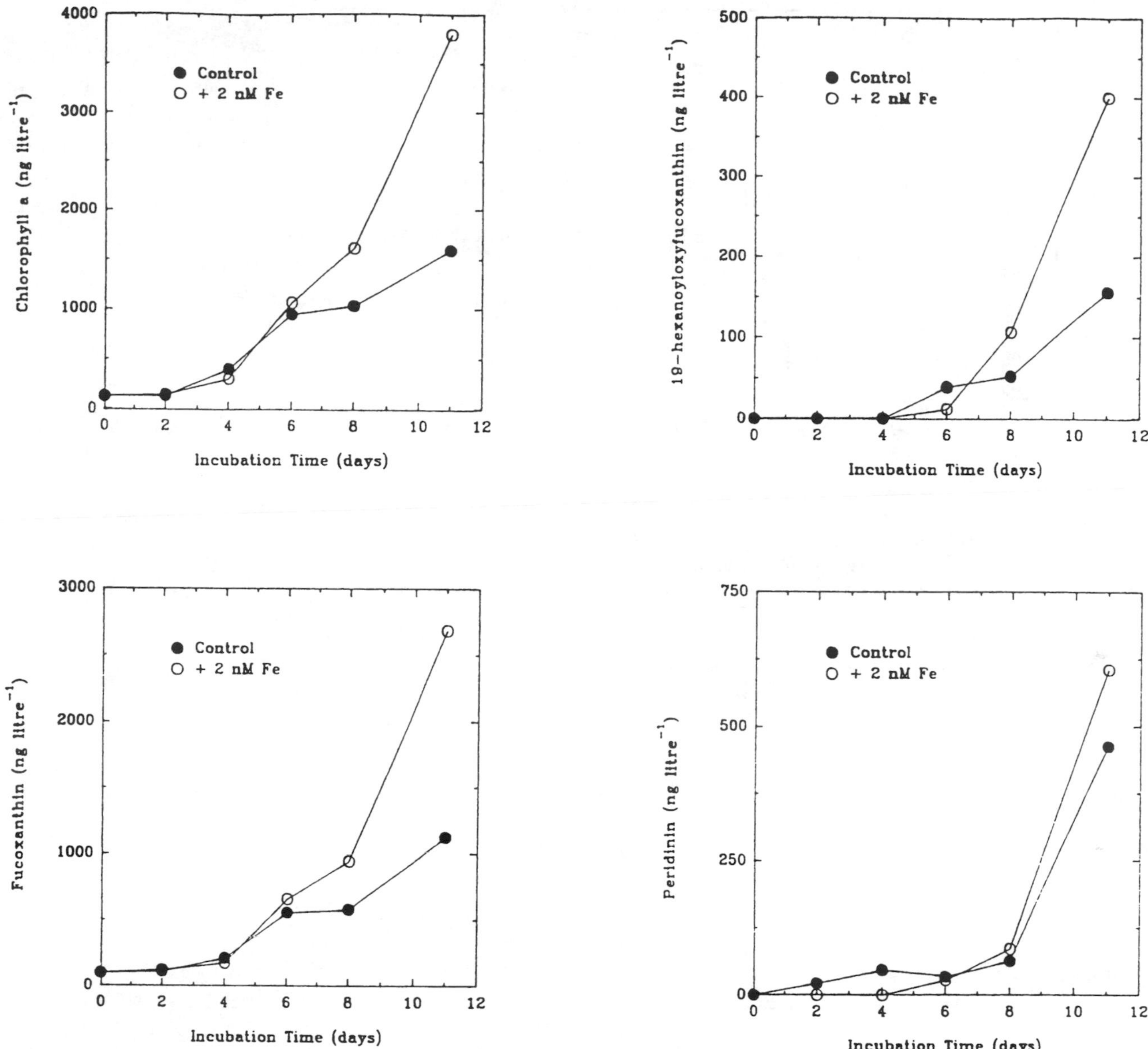

Fig. 14.4. HPLC pigment concentrations from an 11 day Fe enrichment experiment in the northern Ross Sea in February 1992 (see Materials and methods, in text, for details). Surface seawater was obtained by boat and incubated under simulated *in situ* conditions. Two nM Fe additions were made with an EDTA chelated solution.

support the Fe hypothesis and its possible link to glacial MSA levels in the Antarctic.

The dominance of diatoms in the southwestern Ross Sea (Site A; Fig. 14.2a) is consistent with biogenic Si production and fluxes at this site (DeMaster *et al.* 1992). To the east of this site (Sta. A1: Fig. 14.1) *Phaeocystis antarctica* was dominant as evidenced by the Hex concentrations (Fig. 14.2c). DMS concentrations mirrored this change in species composition (Fig. 14.2). Consistent with these findings, fucoxanthin dominated the flux into sediment traps near the coast at Site A in the Ross Sea (data not shown). In contrast, the flux at the eastern end of 76° 30' in the Ross Sea (Site B; Fig. 14.1) was dominated by the *P. antarctica* pigment, Hex (data not shown).

During the early spring, $DMSP_p$ concentrations can dominate over DMS production rates in Antarctic coastal waters (Fig. 14.7). During late summer, as the phytoplankton bloom in Antarctic coastal waters is remineralized, the concentrations of DMS can reach an order of magnitude higher than the $DMSP_p$ pool (e.g. Fig. 14.2). Fluxes of $DMSP_p$ into free-floating sediment traps were higher in live traps compared with formalin-preserved traps (Fig. 14.7). It is not clear, at this point, what factors were responsible for producing these results.

Fluxes of $DMSP_p$ out of the photic zone in early spring were negligible relative to integrated water-column concentrations near the Antarctic peninsula (e.g. Fig. 14.7). If other loss terms such as DMS microbial consumption (Kiene & Bates 1990) are also low in the Antarctic, as recent estimates of low bacterial biomass and metabolic rates have suggested (Karl

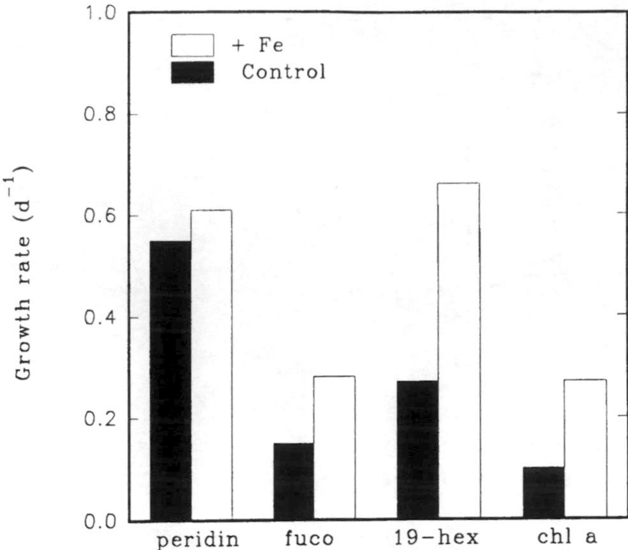

Fig. 14.5. Pigment-specific growth rates calculated by assuming exponential growth between days 6 and 11 of the grow-out experiment (see Fig.14. 4). It is assumed that after six days any photo-adaptive changes between treatments would be relatively small.

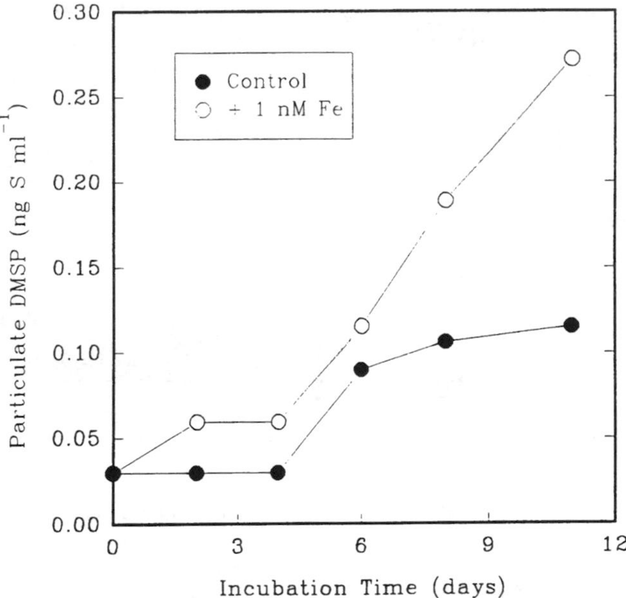

Fig. 14.6. Particulate DMSP concentrations in the control and Fe treatments. DMSP samples were subjected to base hydrolysis and the resultant DMS measured by gas chromatography.

1993), then DMS fluxes to the atmosphere may be greater than those estimated from just summer DMS concentrations (Gibson *et al.* 1990). Other S compounds such as dimethyl sulphoxide (DMSO) may also prove to be important in Antarctic coastal waters (Gibson *et al.* 1990), especially during high UV fluxes in the spring. Only by attempting S mass balances will it be possible to determine the relative importance of various components of the DMS cycle in Antarctic coastal waters.

ACKNOWLEDGEMENTS

We thank A. Close, S. Etnier, A. Jacobsen and S. Polk for assistance on the Ross Sea cruise and K. Daly, J. Goodlaxen, M. Pascual and A. Schauer for assistance on the Antarctic Peninsula cruise. Special thanks are extended to chief scientist D. M. Karl for providing sediment trap samples from the latter cruise. We express our sincere gratitude to the two referees, Professor S. J. de Mora and Dr G. Malin, whose comments improved the paper significantly. This research was supported by NSF grants DPP-88170070 and DPP-9116872.

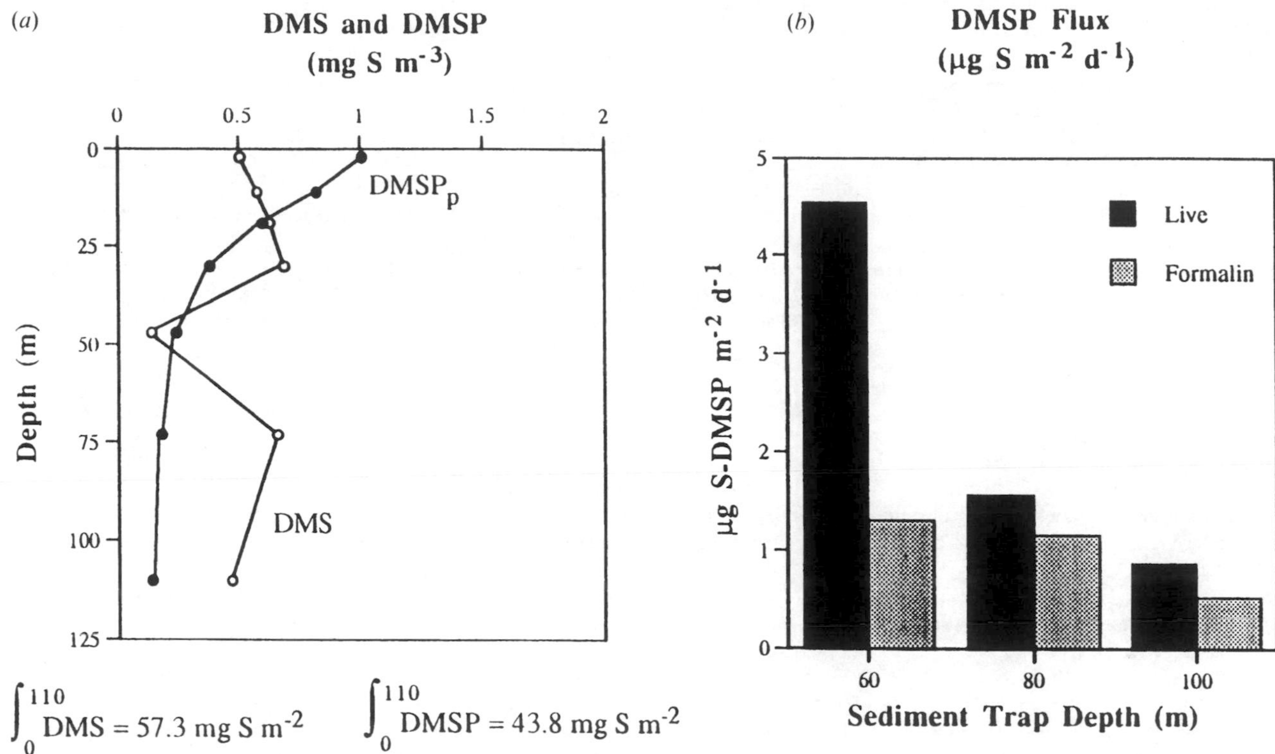

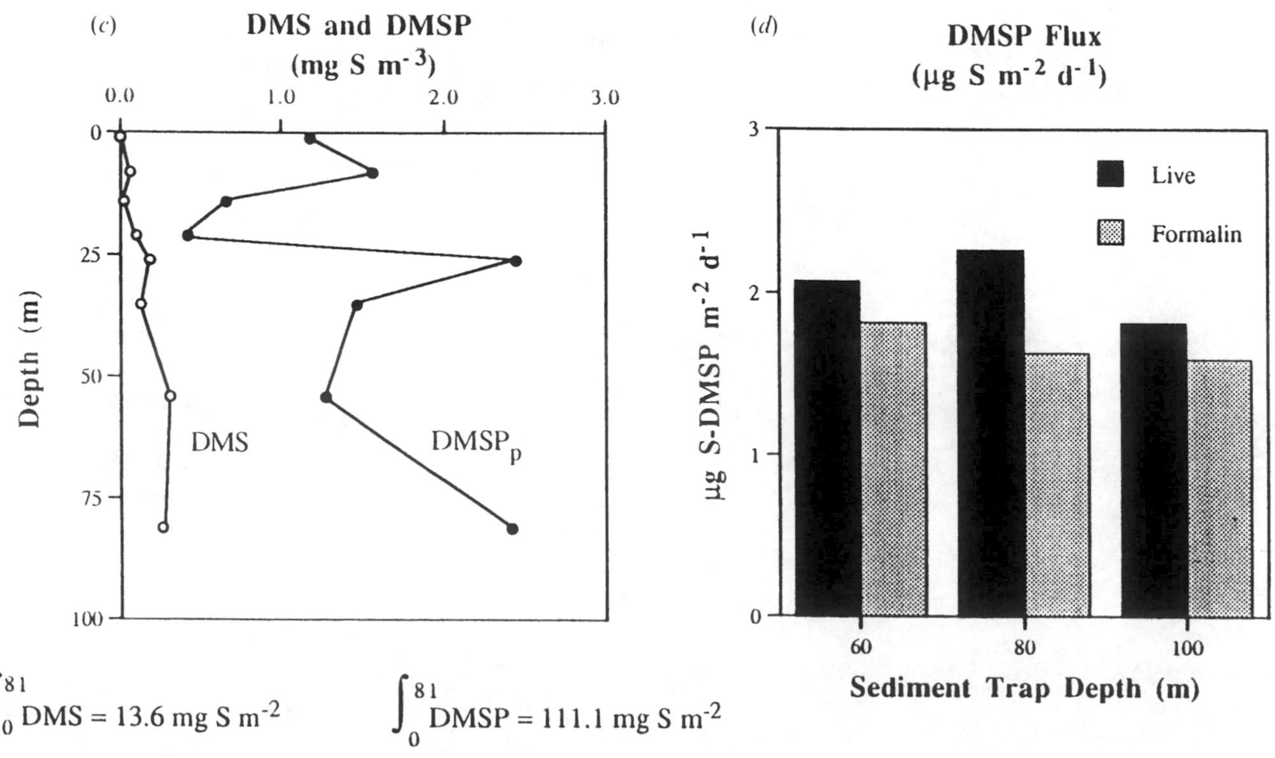

Fig. 14.7. (a) Vertical profiles and integrated values of DMS and DMSP in Charlotte Bay during November, 1992. (b) Sediment trap fluxes of DMSP in both live and formalin-treated traps. Trap deployment time was 31 h. Traps were provided by D.M. Karl. (c) DMS and DMSP vertical profiles and integrated values in Paradise Harbour. (d) DMSP fluxes associated with the water-column integrated values of DMS and DMSP. Trap deployment time was 33 h.

REFERENCES

Andreae, M. O. 1990. Ocean–atmosphere interactions in the global biogeochemical sulfur cycle. *Marine Chemistry*, **30**, 1–29.

Arpin, N., Svec, W. A. & Liaaen-Jensen S. 1976. New fucoxanthin-related carotenoids from *Coccolithus huxleyi*. *Phytochemistry*, **15**, 529–532.

Berresheim, H. 1987. Biogenic sulfur emissions from the sub-Antarctic and Antarctic oceans. *Journal of Geophysical Research*, **92**, 13 245–13 262.

Buma, A. G. J., Bano, N., Veldhuis, M. J. W. & Kraay G. W. 1991. Comparison of the pigmentation of two strains of the prymnesiophyte *Phaeocystis* sp. *Netherland Journal of Sea Research*, **27**, 173–182.

Challenger, F. & Simpson, M. I. 1948. Studies on biological methylation. Part XII. A precursor of the DMS evolved by *Polysiphonia fastigiata*. Dimethylcarboxyethyl-sulphonium hydroxide and its salts. *Journal of the Chemical Society*, **3**, 1591–1597.

Charlson R. J., Lovelock, J. E., Andrae, M. O. & Warren S. G. 1987. Oceanic phytoplankton, atmospheric sulphur, cloud albedo and climate. *Nature*, **326**, 655–661.

DeMaster, D. J., Dunbar, R. B., Gordon, L. I., Leventer, A. R., Morrison, J. M., Nelson, D. M., Nittrouer, C. A. & Smith, W. O., Jr 1992. Cycling and accumulation of biogenic silica and organic matter in high latitude environments: The Ross Sea. *Oceanography*, **5**, 146–153.

DiTullio, G. R. & Smith, W. O., Jr 1995. Relationship between DMS and phytoplankton pigment concentrations in the Ross Sea, Antarctica. *Deep-Sea Research*, Pt. 1, **42**, No. 6, 873–892.

DiTullio, G. R., Hutchins, D. A. & Bruland, K. W. 1993. Interaction of iron and major nutrients controls phytoplankton growth and species composition in the tropical North Pacific Ocean. *Limnology and Oceanography*, **38**, 495–508.

El-Sayed, S. Z., Biggs, D. C. & Holm-Hansen, O. 1983. Phytoplankton standing crop, primary productivity, and near-surface nitrogenous nutrient fields in the Ross Sea, Antarctica. *Deep-Sea Research*, **30**, 878–886.

Fogelqvist, E. 1991. Dimethyl sulfide (DMS) in the Weddell Sea surface and bottom water. *Marine Chemistry*, **35**, 169–177.

Gibson, J. A. E., Garrick, R. C., Burton, H. R. & McTaggart A. R. 1990. DMS and the alga *Phaeocystis pouchetti* in Antarctic coastal waters. *Marine Biology*, **104**, 339–346.

Gieskes, W. W. C. & Kraay, G. W. 1989. Estimating the carbon specific growth rate of major algal species in eastern Indonesian waters by ^{14}C labeling of taxon-specific carotenoids. *Deep-Sea Research*, **36**, 1127–1139.

Goericke, R. & Welschmeyer, N. A. 1993. The chlorophyll labeling method: measuring specific rates of chlorophyll *a* synthesis in cultures and in the open ocean. *Limnology and Oceanography*, **38**, 80–95.

Karl, D. M. 1993. Microbial processes in the southern ocean. In Freidmann, E. I., ed. *Antarctic microbiology*. Boca Raton, FL: Wiley-Liss, 1–63.

Kiene, R. P. & Bates, T. S. 1990. Biological removal of DMS from seawater. *Nature*, **345**, 702–705.

Kirst, G. O., Thiel, C., Wolff, H., Nothnagel, J., Wanzek, M. & Ulmke, R. 1991. Dimethylsulfoniopropionate (DMSP) in ice algae and its possible biological role. *Marine Chemistry*, **35**, 381–388.

LeGrand, M., Feniet-Saigne, C., Saltzman, E. S., Germain, C., Barkov, A. R. & Petrov, V. N. 1991. Ice core record measurements of oceanic emissions of dimethylsulfide during the last climate cycle. *Nature*, **350**, 144–146.

Martin, J. H. 1990. Glacial–interglacial CO_2 change: the iron hypothesis. *Paleoceanography*, **5**, 1–9.

Martin, J. H. 1992. Iron as a limiting factor in oceanic productivity. In Falkowski, P. G. & Woodhead, A. D., eds. *Primary productivity and biogeochemical cycles in the sea*. New York: Plenum Press, 123–138.

Martin, J. H., Fitzwater, S. E. & Gordon, R. M. 1990. Iron deficiency limits phytoplankton growth in Antarctic waters. *Global Biochemical Cycles*, **4**, 5–12.

Radford-Knoery, J. & Cutter, G. A. 1993. Determination of carbonyl sulfide and hydrogen sulfide species in natural waters using specialized collection procedures and gas chromatography with flame photometric detection. *Analytical Chemistry*, **65**, 976–982.

Redalje, D. G. & Laws, E. A. 1981. A new method for estimating phytoplankton growth rates and carbon biomass. *Marine Biology*, **62**, 73–79.

Weber, A. & Wettern, M. 1980. Some remarks on the usefulness of algal carotenoids as chemotaxonomic markers. In Czygan, F. C., ed. *Pigments in plants*. Stuttgart: Gustav Fischer Verlag, pp. 104–116.

Zhuang, G., Yi, Z., Duce, R. A. & P. R. Brown. 1992. Link between Fe and S cycles suggested by detection of Fe(II) in remote marine aerosols. *Nature*, **335**, 537–539.

15 Distributions and dynamics of microbial communities in the pack ice of the western Weddell Sea, Antarctica

CHRISTIAN H. FRITSEN[1,3] *AND CORNELIUS W. SULLIVAN*[2,4]

[1]*Department of Biological Sciences, University of Southern California, Los Angeles, CA 90089-0371, USA,* [2]*Hancock Institute for Marine Studies, University of Southern California, Los Angeles, CA 90089-0374, USA,* [3]*Present address: Department of Biology, Montana State University, Bozeman, MT 59717, USA,* [4]*Present address: National Science Foundation, Office of Polar Programs, Arlington, VA 22230, USA*

ABSTRACT

Antarctic pack ice contains diverse microbial communities in a variety of sea-ice habitats. A fundamental step towards understanding the processes which form and maintain these communities is the understanding of their seasonal cycles and population dynamics. We investigated the biological and physicochemical dynamics within first-year and second-year sea ice in the western Weddell Sea during the austral summer-to-winter transition which takes place from February to June. Both the multi-year ice and the first-year ice supported the net accumulation of both microalgae and bacteria. The microbes showed net growth near the surface of the second-year ice, whilst in the first-year ice net growth was concentrated near the ice/seawater interface. Therefore, the first time-series of Antarctic pack-ice microbial communities revealed concurrent, yet spatially separated, autumnal blooms that were previously unaccounted for in conceptual models of seasonal cycles of pack-ice production. This new information is used as the basis to develop a conceptual model of the dynamics of the pack ice in the western Weddell Sea and the associated microbial communities over seasonal to biannual cycles.

Key words: pack ice, algae, bacteria, Antarctica, Southern Ocean, ecosystems.

INTRODUCTION

Microbial communities can be found throughout the vertical profiles of Antarctic pack ice in a variety of physically distinct environments. Near the surface of the ice, microbial communities are often found in seawater pools caused by snow loading and deformation (e.g. rafting and ridging) (Horner *et al.* 1992). Within the interior of the ice, microbes often reside in the sea-ice matrices within microhabitats ranging from large porous areas, small brine pockets (millimetres in diameter) or larger brine drainage tubes and channels (Horner *et al.* 1992). At the ice/seawater interface, bottom-ice communities are found in direct contact with the underlying seawater or within the ice only centimetres distant from the seawater.

Accumulating evidence indicates that microbial communities throughout the range of pack-ice habitats are capable of active growth (e.g. Clarke & Ackley 1984, Kottmeier & Sullivan 1987, Garrison & Buck 1989, 1991, Lizotte & Sullivan 1991) and contribute to the productivity of the Southern Ocean (Legendre *et al.* 1992). Additional studies (Kottmeier & Sullivan 1987, Daly 1990, Smetacek *et al.* 1992, Franeker 1992) suggest that the sea-ice based production provides the necessary energy required by higher trophic levels such as krill, birds and mammals within the pack-ice regime, and sea-ice production augments the conventional water-column food chain (Legendre *et al.* 1992).

Despite the evidence for the occurrence and the ecological importance of these communities little is known about their seasonal dynamics within the Antarctic pack ice. The lack of information has been attributed primarily to the logistical difficulties of conducting long-term investigations in the drifting sea ice, especially the ability to obtain time-series information from contiguous sections of ice over appropriate timescales.

In the austral autumn of 1992, a US–Russian drifting ice station, Ice Station Weddell (ISW-1) was established in the southwestern Weddell Sea. Working from this platform allowed investigations of the pack-ice dynamics during a seasonal transition. The first part of this paper focuses on the distributions and dynamics of the microbial communities (comprised mainly of algae and bacteria but which also contain protozoa, Garrison *et al.* 1986) in first-year and multi- year pack ice during the transition from austral summer to winter. The information is then integrated into a conceptual model of the pack-ice dynamics in the Weddell Sea. The goal is to provide a general framework for understanding the biological and physico-chemical dynamics of the sea ice, which is a fundamental step towards understanding the structure and function of the pack-ice ecosystem.

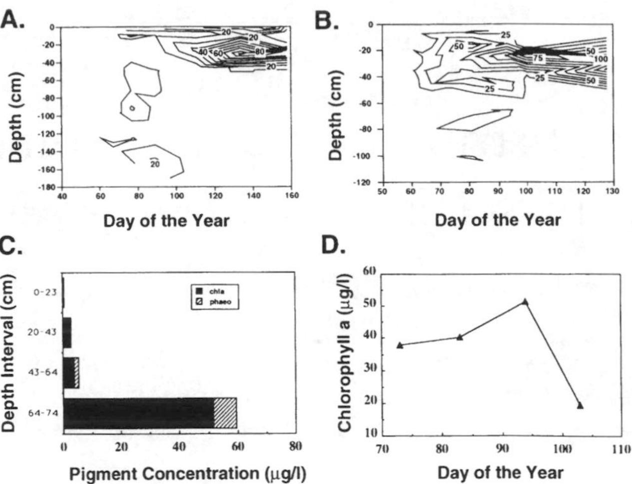

Fig. 15.1. (A) Contours of the time series of chlorophyll *a* concentrations (in µg l^{-1}) within the multi-year ice at site A and (B) within the multi-year ice at site B. Ice thicknesses at sites A and B were 1.2 to 2.2 m and 0.98 to 1.2 m respectively (C) Vertical profile of chlorophyll *a* (chla) and phaeopigments (phaeo) in the FY ice at site J on day 93. (D) Time series of chlorophyll *a* concentrations in the bottom 10 cm of FY ice at site J.

METHODS

Study areas

Ice Station Weddell was deployed on a multi-year (MY) ice floe (ice surviving at least one summer season) in the southwestern Weddell Sea on February 9, 1992. The floe drifted northward over several months along 53° W to a final latitude of 64.1° S on June 6 (Gordon *et al.* 1993). Three primary study sites were monitored throughout the drift. Sites A and B were in MY ice while Site J was in first-year (FY) ice that formed on a lead in early March. These three areas were instrumented with thermistors that recorded hourly temperatures at 5 to 20 cm spacings from the air, through the ice profile and into the water column (Ackley & Lytle 1992). Ice cores were repeatedly collected using 7 and 10 cm diameter SIPRE augers. Cores were placed in polyethylene bags and transported in dark containers back to the ISW-1 laboratories for further processing. Within an hour of collection, ice samples were cut into 10–20 cm sections and placed in darkened containers with filtered seawater (FSW) and allowed to melt at −1.8 to 0 °C. Melting in FSW takes four to six hours and minimizes osmotic stresses and cell lysis associated with salinity changes during the ice melting process (Garrison & Buck 1986). Subsamples for bacteria counts and sizing were fixed immediately upon melting in 1% formalin (final concentration, v/v) and subsamples for pigment analysis were also taken immediately, filtered through Whatman Gf/f filters and frozen at −20 to −30 °C prior to analysis. Parallel cores were also processed without adding FSW, and subsamples taken for chlorophyll *a* and phaeopigments were determined fluorometrically (Parsons *et al.* 1984) on a Turner 111 fluorometer calibrated with *Anacystis* chlorophyll *a* (Sigma Chemical Co.) as the standard. Fixed samples were transported back to the University of Southern California where bacteria were sized and counted via epifluorescence microscopy following staining with acridine orange (~ 0.04% final concentration) (Hobbie *et al.* 1979).

RESULTS AND DISCUSSION

Physical Environment

Although open leads were observed surrounding the ISW-1 floe at the beginning of the study, 80–90% of the sea's surface was covered by multi-year ice. Large areas of the MY ice were flooded with seawater, and porous layers, internal to the ice, were common (Fritsen *et al.* 1994).

In early February air temperatures were relatively warm, varying from 0 to −10 °C; and temperatures in the MY ice were relatively isothermal with seawater, ranging from −1.65 to −1.8 °C. By February 26, air temperatures had dropped and remained below −10 °C for several weeks. Sustained low temperatures propagated freezing fronts into the snow and ice and caused freezing of the seawater at or near the surface of the MY ice (Ackley & Lytle 1992, Fritsen *et al.* 1994). Open leads also began freezing between the MY ice floes at this time. By day 81 (March 21), freezing fronts had progressed through 50 to 70 cm of the porous MY ice, and by day 100 (April 11) they reached the bottom of the MY ice at several locations. First-year ice growing in the leads continued to thicken and reached 80 to 90 cm by day 103. Temperatures in the FY ice were approximately linear from the ice/seawater interface to the surface of the ice (Ackley & Lytle 1992).

Distribution and dynamics of algae and bacteria

Chlorophyll *a* accumulated in the MY ice over the course of the study with the majority of the net accumulation occurring in the upper reaches of the ice (Fig. 15.1A and B). In the FY ice, the majority (>90%) of the chlorophyll *a* was restricted to the lower 10 to 20 centimetres of the ice (Fig. 15.1C) where chlorophyll *a* concentrations initially increased (Fig. 15.1D) and then rapidly decreased following day 94. *Fragilariopsis* (mostly *F. cylindrus* and *F. curta*), *Nitzschia* spp. and chrysophytes were common in

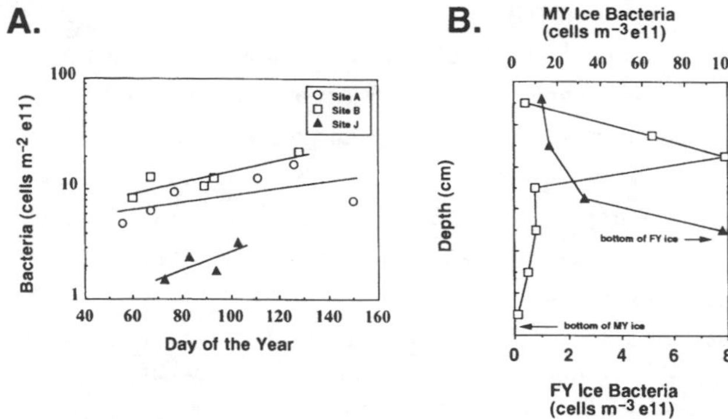

Fig. 15.2. (A) Time-series of the standing stocks of bacteria in the MY ice (sites A and B) and FY ice (site J) studied at ISW-1. Lines are least squares exponential regressions of the entire times series for each site. (B) Vertical profiles of bacteria concentrations in the MY ice (site B, day 128) (open squares) and FY ice (site J, day 94) (closed triangles) showing their contrasting accumulations and distributions in the different ice types.

the MY ice while *Fragilariopsis*, *Archeomonas* and *Corethron* spp. were common in the FY ice. Growth rates estimated from changes in pigment concentrations in the MY ice ranged from 0.07 d^{-1} early in the study (when the ice was relatively warm and daily irradiances were relatively high) to -0.019 d^{-1} in the lower portions of the ice after day 100 (when the ice was colder and irradiances were lower). Estimated algal growth rates in the FY ice ranged from 0.065 d^{-1} (day 73 to 94) to -0.106 d^{-1} (day 94 to day 103).

Bacterial populations also showed net increases within both the MY and FY ice throughout the study (Fig. 15.2A). The bacteria primarily accumulated in the same regions of the FY and MY ice that the microalgae were accumulating. This is evident in the vertical profiles of bacterial concentrations in the FY and MY ice determined near the end of the time-series (Fig. 15.2B). Bacterial growth rates estimated from changes in standing stocks ranged from 0.007 d^{-1} at site A (MY ice) to 0.02 d^{-1} in the bottom of the FY ice. Despite high concentrations of algae and bacteria within the ice, bacterioplankton and phytoplankton biomass in the waters beneath the ice remained characteristically low (Kottmeier & Sullivan 1990) throughout the entire study period (chlorophyll *a* <0.1 mg m^{-3}; bacteria, <1.4 e11 cells m^{-3}).

Population dynamics represent a balance between gains and losses that are caused by both biological (e.g. growth and grazing) and physical processes (e.g. advection, scavenging, dilution). If scavenging processes accounted for the observed increases in the sea-ice microbes, over 200 m^3 of seawater would have had to have been incorporated or scavenged into each square metre of ice to account for the net increases in algal and bacterial standing stocks in the MY ice. Fritsen *et al.* (1994) estimated that brine and seawater were exchanging within the MY ice during the study, but brine and seawater fluxes were estimated to be only 0.04 to 0.1 m^3 m^{-2} d^{-1} over the initial 21 days when the ice was relatively porous. Therefore, only a few cubic metres of seawater passed through the each square meter of the ice in the regions where the bacteria and algae were accumulating. Furthermore, brine–seawater exchanges would have been more likely to dilute the free-living sea-ice microbes within the ice because concentrations of bacteria and plant pigments in the brine were generally 5–10-fold higher in the brine than in the underlying seawater. *In situ* experiments within the MY ice also showed uptake of ^{14}C labelled bicarbonates at relatively high rates (Fritsen *et al.* 1993). Active metazoans (including copepods) and protozoans (including small flagellates, ciliates and foraminifera) were also observed in ice and brine samples taken from within the MY ice and possibly contributed to grazing losses. As a result of these observations, we suggest that accumulations in the MY ice were due primarily to *in situ* growth, and the net accumulation rates reflect only a portion of the growth and production of the sea-ice algae and bacteria.

Similar lines of reasoning suggest that the accumulations of algae and bacteria in the FY ice are primarily due to *in situ* growth as opposed to physical processes. Nitrate concentrations in the FY ice were shown to be nonconservative and depleted relative to concentrations of nitrate in the seawater from which the ice formed (Fritsen *et al.* 1993). This is interpreted as an indication of microalgal growth. The decrease in concentrations of the microalgae following day 94 is an additional indication that losses from bottom-ice communities occur while the ice is actively thickening, and by day 94 the loss rates exceeded the rates of *in situ* growth. Therefore, similar to the MY ice, net accumulations in FY ice are likely to underestimate *in situ* growth and primary production.

Seasonal, annual and biennial cycles

The time-series data on the sea-ice environments and the sea-ice microbial communities observed at ISW-1 provide new insights to the dynamics of seasonal production within the Weddell Sea. Based on these observations, previous descriptions on the distributions of sea-ice biota in the Weddell Sea (e.g. Kottmeier & Sullivan 1990, Garrison & Buck 1991, Kristianson *et al.* 1992, Ackley & Sullivan 1994) and the physical dynamics of Antarctic pack ice (Lange *et al.* 1989), we propose a conceptual model of the annual to biennial cycle of pack ice in the Weddell Sea and the succession of the associated microbial communities (Fig. 15.3). While portions of this model have been outlined previously (e.g. surface community development; Meguro 1962), the proposed scheme integrates several of these observations into a model that accommodates seasonal aspects of biological as well as physical processes.

In the austral autumn, pack ice forms via frazil-ice formation and the pancake cycle (Lange *et al.* 1989). Microbes are initially scavenged from the upper few metres of the water column and rapidly colonize newly forming ice (Garrison *et al.* 1983). Temperature gradients in FY ice are expected to be nearly linear from the ice/seawater interface and may restrict, but not eliminate (Garrison & Thomson 1993), the growth of the colonizing microbes to regions near the ice/seawater interface. If the ice forms early in the autumn, sufficient light may be available to

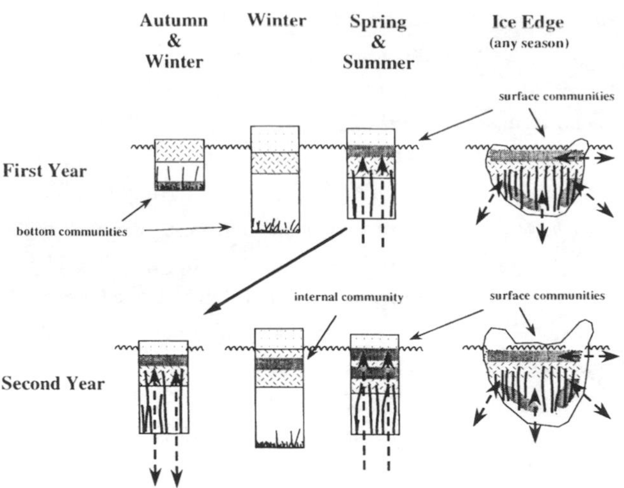

Fig. 15.3. Conceptual model of the annual to biennial dynamics of the Weddell Sea pack ice and the succession of the associated sea-ice microbial communities (see text for full description). Hatched areas indicate granular ice formed via frazil-ice or snow-ice formation. Shaded areas denote the *in situ* growth and net accumulation of dense microbial communities. Vertical lines illustrate large brine channels and porous ice, and dashed arrows illustrate potential movement and exchanges of water, nutrients and organisms.

enable net primary production, which allows bottom-ice communities to grow and accumulate additional biomass (this study and Hoshiai 1977).

During the first autumn and winter, the ice continues to thicken via thermodynamic growth (this study), if the conductive heat loss from the ice/seawater interface is higher than the oceanic heat flux (Maykut & Untersteiner 1971). Bottom-ice communities experience net losses as a result of grazing and physical processes which occur during the autumn and winter periods (this study and Hoshiai 1977). By the end of winter, remnants of bottom-ice blooms of microalgae which occurred in autumn are likely to be scarce. However, heterotrophic organisms (including bacteria) may benefit from the autumnal pulse of primary production and continue to proliferate throughout the winter as proposed by Kottmeier & Sullivan (1987).

In the early spring, daily insolations and atmospheric temperatures begin to increase and bottom-ice microalgae may once again bloom. The rates of microalgal and bacterial production are thought to be primarily controlled by the quantity and quality of photosynthetically active radiation reaching the lower ice (Grossi *et al.* 1987), which is largely determined by how much snow cover accumulates throughout the winter. As atmospheric temperatures rise throughout the spring, temperature gradients in the ice are expected to decrease and brine volumes increase (Weeks & Ackley 1982, Kottmeier & Sullivan 1988). As a result, communities at or near the bottom of the ice are likely to become more susceptible to grazing and ice ablation, which release the microbes into the water column (Grossi *et al.* 1987).

Ablation from the ice/seawater interface leads to ice thinning in the spring and summer, and coupled with sufficient accumulation of snow causes the ice surface to be depressed below sea level (Ackley *et al.* 1990). As ice temperatures increase, brine volumes continue to increase and the ice becomes more permeable and allows seawater to infiltrate and flood the surface of the ice. Lateral flooding also occurs (Kristianson *et al.* 1992), but the relative importance of each flooding mechanism is unknown. When flooding occurs, nutrient-rich seawater and colonizing organisms are transported into a relatively high irradiance regime and microbial populations bloom within the surface layers (Meguro 1962, Garrison & Buck 1991, Horner *et al.* 1992, Syversten & Kristianson 1993). Rates of primary and secondary production in the surface environments can exceed 1 g C m^{-3} day (e.g. Meguro 1962, Garrison & Buck 1991), and over time, nutrients become depleted to levels that are potentially limiting (Garrison & Buck 1991, Syversten & Kristianson 1993, Fritsen *et al.*, 1994). Although ice floes in the summer are highly porous ('rotten'), vertical nutrient exchange through the ice may become restricted by small-scale stratification caused by low-salinity brine, which results from internal melting of the snow and ice (this study, Garrison & Buck 1991).

Near the marginal ice-edge zones, ice floes break apart via enhanced exposure to deformation and wave action. During this process, sea-ice communities are increasingly exposed to nutrient exchanges, grazing pressures and variable light regimes. Eventually, ice floes completely deteriorate, and upon melting release their contents into the water column where they may act as inocula for marginal ice-edge blooms on a large scale (Sullivan *et al.* 1988) or contribute to the downward flux of particulate matter.

In the Western Weddell Sea, FY ice floes often survive the summer melt and are subjected to a second autumnal freezing season. Freezing fronts solidify partially deteriorated and flooded floes (Ackley & Lytle 1992, Fritsen *et al.* 1994) and initiate an exchange of brine and seawater (Fritsen *et al.* 1994). Surface and internal communities are replenished with new nutrients and bloom (Fritsen *et al.* 1994) if sufficient light is available for net photosynthesis and growth. These autumnal blooms in the MY ice occur at the same time that bottom-ice communities are developing in the FY ice (Fig. 15.1 and Fig. 15.3).

Surface and internal communities in the MY ice are frozen into the ice matrices during the winter (this study). Despite low irradiances and cold temperatures, microalgae have been shown to remain viable under conditions representative of this environment (Garrison & Thomson 1993). Continued bacterial growth in the winter may be sustained by degrading particulate and dissolved organic matter within the brine pockets (Arrigo *et al.* 1995). Low temperatures and high brine salinities, however, restrict microbial activity to varying degrees (Kottmeier & Sullivan 1988, Arrigo & Sullivan 1992).

Upon the second season of warming, MY ice floes may flood once again and successive blooms of surface microbial communities occur causing a layering or banded pattern through the vertical ice profile (Kottmeier & Sullivan 1990). Sea ice generally does not survive more than two summer seasons in the Weddell Sea (Ackley 1991), and second-year ice floes are advected generally to the northwestern region of the gyre where they encounter warmer waters, and release their contents to the water column. Sullivan *et al.* (1993) attributed the input of this inoculum as instrumental to the persistent blooms observed in the Scotia Ridge regions by CZCS (Coastal Zone Colour Scanner) imagery (also see Comiso *et al.* 1993).

The proposed model is intended to describe general processes that are believed to occur over seasonal timescales and over relatively large spatial scales in the pack ice of the western Weddell Sea and perhaps more extensively. Additional dynamics may prevail on shorter time and spatial scales and coastal ice has been shown to have additional interactions with the continent and continental shelves that are not included in this model (e.g. Smetacek *et al.* 1992). Deformation events in pack ice, for instance, are dynamic processes which produce surface flooding through rafting and ridging during all seasons and create additional habitats via surface loading. The temporal and spatial scales over which some of these processes occur are relatively unknown at the present time but likely are related to tidal or atmospheric forcing (Rowe *et al.* 1989), and therefore, can be expected to influence all of the pack ice.

ACKNOWLEDGEMENTS

We wish to thank S. F. Ackley, C. W. Mordy, S. Kristianson, D. Garrison, I. Melnikov and J. Kremer for helpful discussions, and all of the ISW-1 organizers and participants for the successful implementation of ISW-1. This research was supported by NSF grants OPP 90-23669 and DPP 90-9317380.

REFERENCES

Ackley, S. F. 1991. The growth, structure and properties of Antarctic sea ice. In *Glaciers, ocean, atmosphere interactions*. IAHS publication 208. Wallingford: Institute of Hydrology.

Ackley, S. F. & Lytle, V. I. 1992. Sea-ice investigations on Ice Station Weddell #1. II: ice thermodynamics. *Antarctic Journal of the United States*, **27**(5), 109–110.

Ackley, S. F. & Sullivan, C. W. 1994. Physical controls on the development and characteristics of Antarctic sea ice biological communities: a review and synthesis. *Deep-Sea Research*, **41**, 1583–1604.

Ackley, S. F., Lange, M. A. & Wadhams, P. 1990. Snow cover effects on Antarctica sea ice thickness. In Ackley, S. F. & Weeks, W. F., eds. *Sea ice properties and processes*. *CRREL Monographs*, **90–1**.

Arrigo, K. R. & Sullivan, C. W. 1992. The influence of salinity and temperature covariation on the photophysiological characteristics of Antarctic sea ice microalgae. *Journal of Phycology*, **28**, 746–756.

Arrigo, K. R., Dieckmann, G., Gosselin, M., Robinson, D. H., Fritsen, C. H. & Sullivan, C. W. 1995. A high resolution study of the platelet ice ecosystem in McMurdo Sound, Antarctica: Biomass, nutrient, and production profiles within a dense microalgal bloom. *Marine Ecology Progress Series*, **127**, 255–268.

Clarke, D. B. & Ackley, S. F. 1984. Sea ice structure and biological activity in the Antarctic marginal ice zone. *Journal of Geophysical Research*, **89** (C2), 2087–2095.

Comiso, J. C., McClain, C. R., Sullivan, C. W., Ryan, J. P. & Leonard, C. L. 1993. Coastal zone color scanner pigment concentrations in the Southern Ocean and relationships to geophysical features. *Journal of Geophysical Research*, **98** (C2), 2419–-2451.

Daly, K. 1990. Overwintering development, growth, and feeding of larval *Euphausia superba* in the Antarctic marginal ice zone. *Limnology and Oceanography*, **35**, 1564–1576.

Franeker, J. A. 1992. Top predators as indicators for ecosystem events in the confluence zone and marginal ice zone of the Weddell and Scotia Seas, Antarctica, November 1988 to January 1989 (EPOS, Leg 2). *Polar Biology*, **12**, 93–102.

Fritsen, C. H., Lytle, V. I., Ackley, S. F. & Sullivan, C. W. 1994. Autumn bloom of Antarctic pack-ice algae. *Science*, **266**, 782–784.

Fritsen, C. H., Mordy, C. W. & Sullivan, C. W. 1993. Primary production in the Weddell Sea pack ice during the austral autumn. *Antarctic Journal of the United States*, **28**(5), 124–126.

Garrison, D. L. & Buck, K. R. 1986. Organisms losses during ice melting: a serious bias in sea ice community studies. *Polar Biology*, **6**, 237–239.

Garrison, D. L. & Buck, K. R. 1989. The biota of the Antarctic sea ice in the Weddell Sea and Antarctic Peninsula regions. *Polar Biology*, **10**, 211–219.

Garrison, D. L. & Buck, K. R. 1991. Surface-layer sea ice assemblage in Antarctic pack ice during the austral spring: environmental conditions, primary production, and community structure. *Marine Ecology Progress Series*, **75**, 161–172.

Garrison, D. L. & Thomson, H. A. 1993. *Polarstern* 'Ant X/3' austral autumn in the ice 1992: sea-ice community studies. *Antarctic Journal of the United States*, **28**(5), 126–128.

Garrison, D. L., Ackley, S. F. & Buck, K. R. 1983. A physical mechanism for establishing algal populations in frazil ice. *Nature*, **306**, 363–365.

Garrison, D. L., Sullivan, C. W. & Ackley, S. F. 1986. Sea ice microbial communities in Antarctica. *Bioscience*, **36**, 243–250.

Gordon, A. L. & Ice Station Weddell Group of Principle Investigators and Chief Scientist. 1993. Weddell Sea exploration from Ice Station. *EOS*, **74**, 121 & 124–126.

Grossi, S. M., Kottmeier, S. T., Moe, R. L., Taylor, G. T. & Sullivan, C. W. 1987. Sea ice microbial communities. VI. Growth and primary production in bottom ice under graded snow cover. *Marine Ecology Progress Series*, **35**, 153–164.

Hobbie, J. E., Daley, R. J. & Jasper, S. 1979. Use of Nucleopore filters for counting bacteria by fluorescence microscopy. *Applied and Environmental Microbiology*, **33**, 122–1228.

Horner, R., Ackley, S. F., Dieckmann, G. S., Gulliksen, B., Hoshiai,T., Legendre, L., Melnikov, I. A., Reeburgh,W. S. , Spindler, M. & Sullivan, C. W. 1992. Ecology of sea ice biota. 1. Habitat, terminology, and methodology. *Polar Biology*, **12**, 417–427.

Hoshiai, T. 1977. Seasonal changes of ice communities in the sea ice near Swoya Station, Antarctica. In Dunbar, M. J., ed. *Polar oceans*. Calgary: Arctic Institute of North America, 307–317.

Kottmeier, S. T. & Sullivan, C. W. 1987. Late winter primary production and bacterial production in sea ice and seawater west of the Antarctic Peninsula. *Marine Ecology Progress Series*, **36**, 287–298.

Kottmeier, S. T. & Sullivan, C. W. 1988. Sea ice microbial communities (SIMCO). IX. Effects of temperature and salinity on rates of metabolism and growth of autotrophs and heterotrophs. *Polar Biology*, **8**, 293–304.

Kottmeier, S. T. & Sullivan, C. W. 1990. Bacterial biomass and production in pack ice of Antarctic marginal ice edge zones. *Deep-Sea Research*, **37**, 1311–1330.

Kristianson, S., Syversten, E. E. & Farbot, T. 1992. Nitrogen uptake in the Weddell Sea during late winter and spring. *Polar Biology*, **12**, 245–257.

Lange, M. A., Ackley, S. F., Wadhams, P., Dieckmann, G. S. & Eicken, H. 1989. Development of sea ice in the Weddell sea, Antarctica. *Annals of Glaciology*, **12**, 92–96.

Legendre, L., Ackley, S. F., Dieckmann, G. S., Gulliksen, B., Horner, R., Hoshiai, T., Melnikov, I. A., Reeburgh, W. S., Spindler, M. & Sullivan, C. W. 1992. Ecology of sea ice biota. 2. Global significance. *Polar Biology*, **12**, 429–444.

Lizotte, M. P. & Sullivan, C. W. 1991. Photosynthesis–irradiance relationships of microalgae associated with Antarctic pack ice: evidence for *in situ* activity. *Marine Ecology Progress Series*, **71**, 175–184.

Maykut, G. A. & Untersteiner, N. 1971. Some results from a time-dependent thermodynamic model of sea ice. *Journal of Geophysical Research*, **76**, 1550–1575.

Meguro, H. 1962. Plankton ice in the Antarctic Ocean. *Antarctic Record*, **14**, 192–199.

Parsons, T. R., Maita, Y. & Lalli, C. M. 1984. *A manual of chemical and biological methods for seawater analysis*. New York: Pergamon Press.

Rowe, M. A., Sear, C. B., Morrison, S. J., Wadhams, P., Limbert, D. W. S. & Crane, D. R. 1989. Periodic motions in the Weddell Sea pack ice. *Annals of Glaciology*, **12**, 145–151.

Smetacek, V., Scharek, R., Gordon, L. I., Eicken, H., Fahrbach, E., Rohardt, G. & Moore, S. 1992. Early spring phytoplankton blooms in ice platelet layers of the Southern Weddell Sea, Antarctica. *Deep-Sea Research*, **39**(2A), 153–168.

Sullivan, C. W., Arrigo, K. R., McClain, C. R., Comiso, J. C. & Firestone, J. 1993. Distributions of phytoplankton blooms in the Southern Ocean. *Science*, **262**, 1832–1837.

Sullivan, C. W., McClain, C. R., Comiso, J. C. & Smith, W. O. Jr 1988. Phytoplankton standing crops within an Antarctic ice edge assessed by satellite remote sensing. *Journal of Geophysical Research*, **93**, 12487–12498.

Syversten, E. E. & Kristianson, S. 1993. Ice algae during EPOS, leg 1: assemblages, biomass, origin, and nutrients. *Polar Biology*, **13**, 61–65.

Weeks, W. F. & Ackley, S. F. 1982. The growth, structure and properties of sea ice. *CRREL Monographs*, **82–1**, 1–130.

16 Abundance and distribution of larval krill, *Euphausia superba,* associated with annual sea ice in winter

THOMAS K. FRAZER [1], *LANGDON B. QUETIN* [2] *AND ROBIN M. ROSS* [2]

[1] *Department of Fisheries and Aquatic Sciences, University of Florida, Gainesville, Florida, FL 32653, USA,* [2] *Marine Science Institute, University of California at Santa Barbara, Santa Barbara, CA 93106, USA*

ABSTRACT

Larval krill, Euphausia superba, *associated with annual sea ice were censused visually using SCUBA during three winter cruises to a region west of the Antarctic Peninsula. Sampling during September 1991 and June 1993 was restricted to a small number of stations off Adelaide Island. A more extended, mesoscale survey was conducted in September 1993. Larval krill were observed feeding on ice-associated biota at all sampling locations (n = 36) on each of the three cruises. Mean numbers of larval krill (individuals per m^2) per station for September 1991, June 1993 and September 1993 were 24.60, 2.04 and 16.72, respectively. Larval abundances were significantly greater during late winter (September) sampling periods. A majority of larval krill censused during late winter occurred in large aggregations ($\geq 10^3$ individuals). Large aggregations of larvae were not found in early winter. Eighty per cent of larvae were observed under a complex habitat provided by over-rafted and/or eroded ice floes, and were generally associated with upward facing ice surfaces. Larval krill were rarely observed on the downward facing surfaces of unilayer floes, though ice-algae were often visible in these areas.*

Key words: Antarctic Peninsula, ecology, Long-Term Ecological Research (LTER).

INTRODUCTION

Qualitative observations of larval krill, *Euphausia superba*, feeding on the undersides of sea ice (e.g. Guzman 1983, Stretch *et al.* 1988, Marschall 1988, Daly 1990) are suggestive of an important ecological coupling. Quetin & Ross (1991) suggested that the association is based on the need to feed on the ice-associated community, although sea ice may also provide a refuge from predation. The two scenarios are not mutually exclusive, and it is clear that the intricacies of the linkage are not completely understood.

Smetacek *et al.* (1990) suggested 'that the bulk of the krill population moves into the ice habitat upon its formation', but quantitative data needed to evaluate this claim are lacking. Direct measures of the abundance of krill associated with sea ice are few (O'Brien 1987, Marschall 1988, Hamner *et al.* 1989), and no systematic surveys of larvae have been reported to date. As part of our ongoing study on the ecology and physiology of larval krill during winter, we have quantified the abundance and distribution of larvae in the sea ice habitat during three cruises in the austral winter.

We intend to combine small-scale distributional data and behavioral observations to increase our understanding of how krill larvae exploit the ice habitat. Quantitative census data from early and late winter sampling periods provides additional insight into temporal variablity of krill associated with sea ice. Our results provide essential detailed information for formulating hypotheses and investigating further the krill/ice interaction.

METHODS

Larval krill associated with annual sea ice were censused visually with SCUBA during three winter cruises to a region west of the

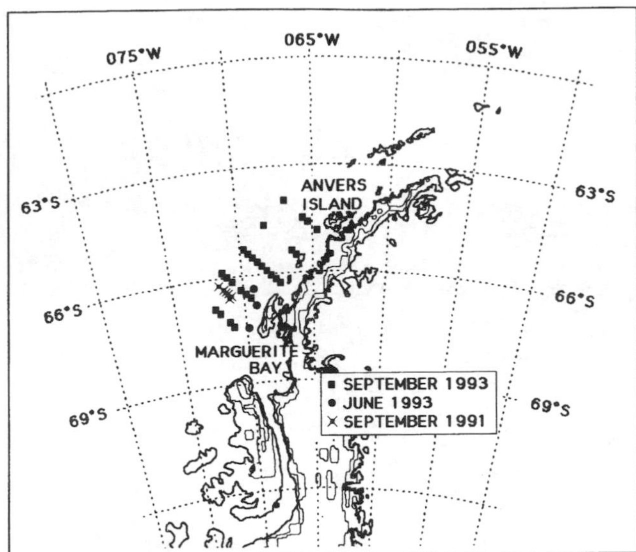

Fig. 16.1. Sampling stations occupied during each of the three winter cruises to the west of the Antarctic Peninsula in September 1991, June 1993 and September 1993. Adelaide Island is the large island north of Marguerite Bay.

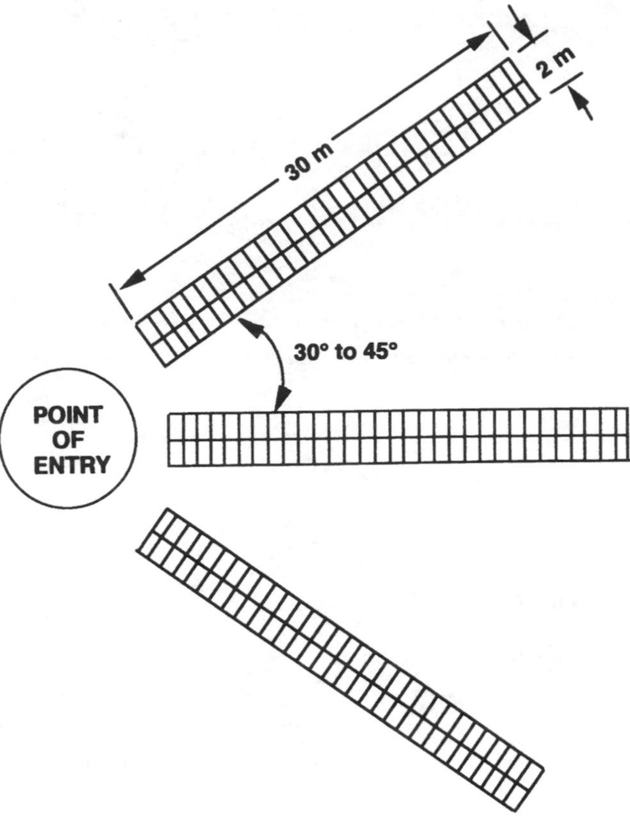

Fig. 16.2. Schematic depiction of diver transects at each sampling station. Krill larvae were censused continuously along replicate strip transects (30 m×2 m).

Antarctic Peninsula (Fig. 16.1). During September 1991 and June 1993, sampling was restricted to a relatively small area offshore of Adelaide Island. During September 1993, a mesoscale survey was conducted between Anvers Island and Marguerite Bay. The mesoscale survey of the under-ice habitat for larval krill was conducted within the confines of the Palmer Long-Term Ecological Research (LTER) site (Waters & Smith 1992) and encompassed the sampling areas of the two previous cruises.

At each sampling station krill were censused along 2 m by 30 m long transects originating at the point of entry of the divers (Fig. 16.2). Observations were made to a depth of 3 m below the nearest ice surface, although larval krill were rarely observed more than 0.5 m away from any ice surface. Three replicate transects between 30° and 45° apart were generally completed at each station (32 of 36 total) during a single 30 to 60 minute dive. At the conclusion of each dive additional observations were recorded, e.g. estimated percentage of over-rafted ice in the sampling area and other fauna present.

Estimates of animal abundance *in situ* are influenced by such variables as visibility, sampling by different individuals and animal behaviour. The censuses reported here were completed by a single individual to eliminate between-sampler variability. Krill were censused during daylight hours. Visibility was generally greater than 30 m so it was easy to see individual krill to either side of the transect line. Larval krill associated with sea ice generally remain site-attached during a dive and do not exhibit an escape response into the water column when approached slowly by a diver; likewise, larvae do not exhibit directed movements along ice surfaces over any appreciable distance (<1 m). Even when disturbed by a diver with a collecting net, the integrity of the aggregation is soon re-established, generally at the collection site. For small aggregations, of less than 50 individuals, absolute counts were possible. The total number of krill in larger aggregations was estimated by counting the number of volumes in an aggregation equivalent to a sub-volume with a counted number of individuals (10s, 20s, 50s).

Unless otherwise specified, larval abundances were calculated as the number of individuals per square metre and are reported as the mean of replicate transect counts at each sampling station. A Kruskal–Wallis non-parametric procedure with chi-square approximation (Sokal & Rolf 1981) was used to compare larval counts among sampling periods as variances were determined to be heterogeneous. For latitudinal comparisons the data from September 1993 were binned into one of three broad categories: (1) transect data collected at or north of 65° S, (2) transect data collected between 65° S and 66° S, and (3) transect data collected at or south of 66° S. Standard ANOVA procedures were used for the analysis.

RESULTS

Larval krill were observed in close association with sea ice at all under-ice stations during each of the three winter cruises where quantitative sampling was conducted (Fig. 16.3). The mean number of larvae per station did not differ between September 1991 and September 1993, but counts during the sampling period in June 1993 (early winter) were

Table 16.1. *Mean number of krill larvae per square metre (±standard error) for each of the three winter cruises. Sampling periods are ranked in order of increasing larval abundance (June 1993<September 1993=September 1991)*[1]

Sampling period	Larvae per m^2	Sampling stations	No. of transects
June 1993	2.04 (±1.32)	3	9
Sept. 1993	16.72 (±3.39)	26	73
Sept. 1991	24.60 (±12.42)	6[2]	21

Notes:

[1] The inequality is significant at the $p \leq 0.08$ level, Kruskal–Wallis test with chi-square approximation ($\chi^2=4.92$, df=2).

[2] One sampling station was occupied twice during September 1991.

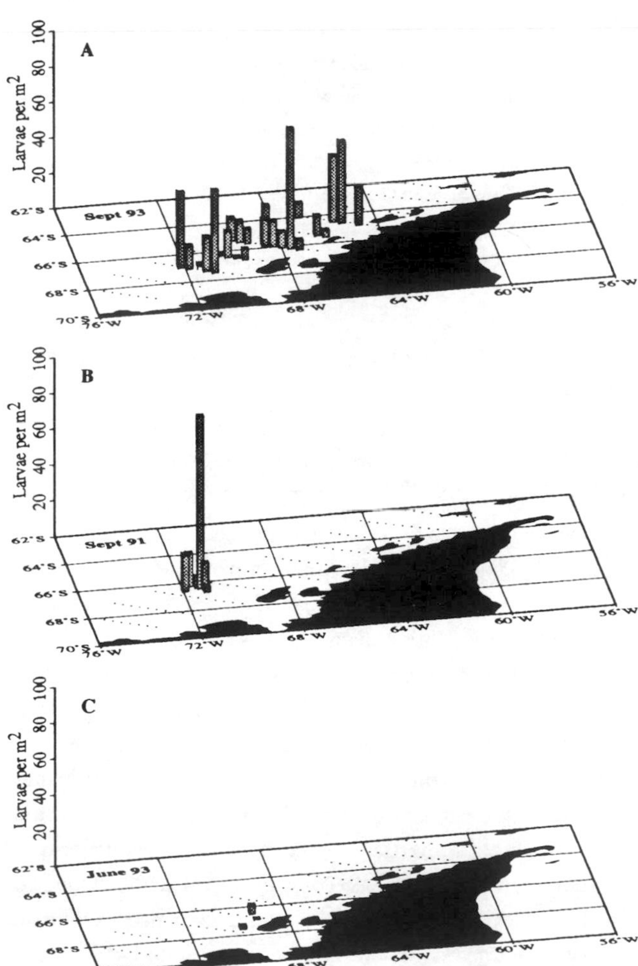

Fig. 16.3. Mean number of krill larvae per square metre at each sampling station for each of the three winter cruises: (A) September 1993, (B) September 1991 and (C) June 1993. All data are plotted with reference to established sampling locations within the Palmer LTER grid (Waters & Smith 1992).

significantly less than those from either late winter cruise (Table 16.1).

Large, distinct patches of larval krill (≥1000 individuals) accounted for less than 5% of the observations, but comprised 50% or more of the censused population (Fig.16.4). Large

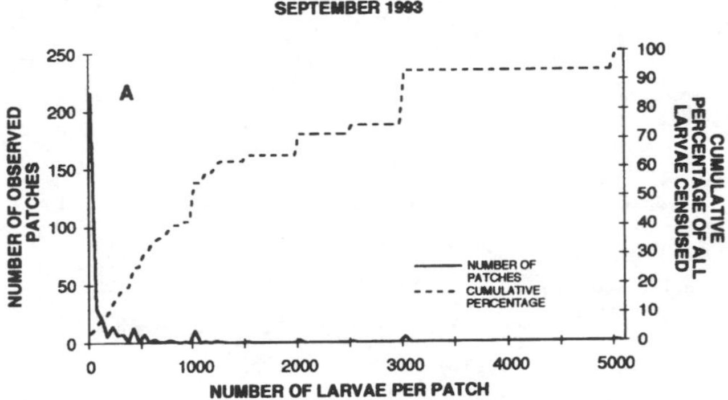

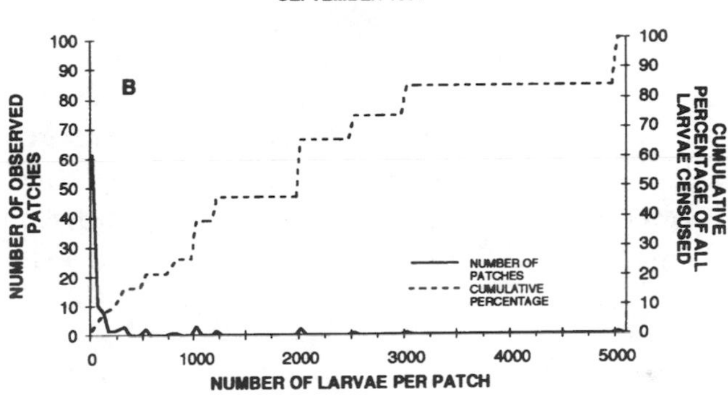

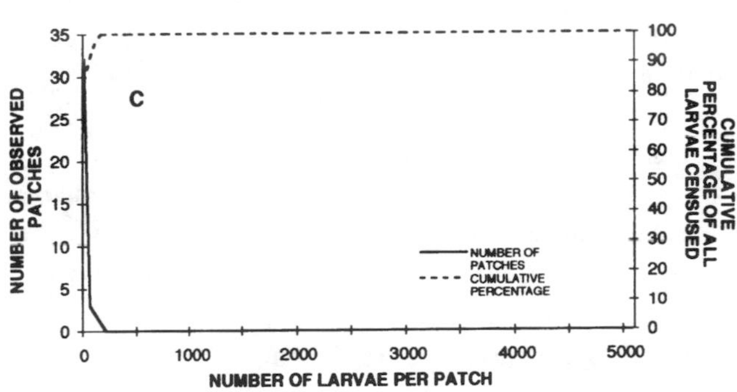

Fig. 16.4. Observed frequency of krill patches and cumulative abundance of individuals plotted against patch size for each of the three winter cruises: (A) September 1993, (B) September 1991 and (C) June 1993.

patches of larval krill were not observed on diver transects in June 1993. Early-winter sampling during both years was conducted in the same general geographic area west of Adelaide Island. Within the mesoscale sampling grid, there was no latitudinal gradient in krill abundance or distribution (ANOVA, df=2, $p \geq 0.05$). No cross-shelf gradients in abundance and distribution of larval krill were apparent (Fig. 16.3A).

Behavioural observations of larvae and the green coloration of their digestive gland indicated that most animals were feeding. However, larvae were not necessarily found in areas of highest plant pigment concentration. Greater than 80% of all larval krill were associated with over-rafted and/or eroded ice surfaces where plant pigment was rarely visible to the naked eye. Larval krill were less common on the downward facing surfaces of floes that were not over-rafted, although ice-algae were often visible in these areas.

DISCUSSION

Though sea ice had just recently formed in the northern portion of the mesoscale sampling grid in September 1993, larvae were observed at all stations with no statistical differences in abundance with latitude. This raises several questions: (1) do krill larvae inhabiting the water column stay with sea ice once it is formed, (2) did larvae sampled in the northern part of the LTER grid overwinter in ice-free water or (3) were they advected into the region prior to ice formation?

The distributional range of larval krill under sea ice during winter is not known. Nordhausen (1994) investigated the abundance and distribution of *E. superba* in ice-covered waters west of the Antarctic Peninsula. Although Nordhausen's study was restricted to the Gerlache Strait and inshore waters of Crystal Sound, no larvae were found during July 1992. This suggests that larval krill may, in general, overwinter in shelf and slope waters of the Antarctic Peninsula as observed and reported here. Comparative sampling in these distinctly different areas during the same year(s) would provide a test of the implicit hypothesis above.

Quantitative sampling of krill larvae is essential to evaluate spawning success and for subsequent predictions of year class strength and recruitment to the adult population. However, few quantitative studies of larval krill abundance and distribution have been conducted during the winter in the Southern Ocean, especially in the region covered by sea ice. One major difficulty is that in the complex ice habitat, larvae often occur on upward facing ice surfaces and net sampling is clearly not representative of larval abundance *in toto*. Quantitative surveys by divers are currently the best way to evaluate larval abundance and distribution in the sea-ice habitat.

The mesoscale survey during September 1993 covered a region large enough to allow estimation of the importance of sea ice to larval krill as a population. Larval krill were consistently found under the ice during September 1993, and a simple extrapolation of their mean abundance (16.72 larvae m^{-2}) yields a total estimate of 1.3×10^{12} larvae within the 200 km×400 km sampling region. Assuming that instantaneous rates of mortality (M) for all age-classes of post-larval krill are equal and ≤ 1.0 (cf. Priddle *et al.* 1988 and references therein, Siegel & Kalinowski 1994), and that recruitment is the same every year, then the larval concentrations in September 1993 are sufficient to generate an adult krill population on the same order of magnitude (Table 16.2). The projected number of adults within the sampling region would be 9.7–17.1 m^{-2}, with M values of 1.0 and 0.66, respectively (Table 16.2). These numbers compare favorably with Smetacek *et al.*'s (1990) estimate of 10–30 adults m^{-2} under sea ice at its maximal extent (based on population estimates by El-Sayed (1988)), and, if emigration is zero, suggest that larval concentrations immediately west of the Antarctic Penisula are not greater than necessary to maintain an average adult population. This would be counter to a generally held contention that the region west of the Antarctic Peninsula is an important nursery area for larval krill compared with other regions of the Southern Ocean.

Table 16.2. *Projected numbers of post-larval krill within the 200 km × 400 km sampling region based on larval abundance estimates from September 1993*[1]. *The projected numbers will vary depending on the assumed instantaneous rates of mortality (M). Two different scenarios are given*

Age-class (year)	Mean number of krill per m^2	Total number of krill
age-class 1[1]	16.72	1.3×10^{12}
Projection 1 ($M = 1.0$[2])		
age-class 2–7	9.70	7.5×10^{11}
Projection 2 ($M = 0.66$[3])		
age-class 2–7	17.10	1.3×10^{12}

Notes:

[1] The larval population present at the end of winter (September 1993) and the one-year-old age-class are, for the purposes of this exercise, assumed to be the same. It is assumed also that the numbers of krill in age-class 1 are constant from year to year. Rates of immigration and emigration of post-larval krill into and out of the 200 km × 400 km sampling region are assumed to be equal.

[2] $M = 1.0$ is the average value assumed by Priddle *et al.* (1988).

[3] $M = 0.66$ is the lowest value expected by Siegel & Kalinowski (1994).

We recognize that larval numbers observed in September 1993 may not be representative of those in other years, but larval abundances in September 1993 were remarkably similar to those estimated during September 1991 and suggest otherwise. Continued censuses of larval krill associated with sea ice in late winter need to be interpreted relative to the variability of winter ice cover and the spawning success of the previous season to understand the population dynamics of krill in this region.

Larval krill were found to be most concentrated in areas of over-rafted sea ice with upward-facing ice surfaces. Although Smetacek *et al.* (1990) suggested that ice surface area would be greater by about one-third because of ridging of cover, estimates of over-rafting by divers west of the Antarctic Peninsula are generally less ($\leq 5\%$), particularly in early winter. More work is needed to understand the ecology of larval krill relative to habitat complexity. A lack of habitat complexity may explain, in part, patch characteristics and low counts of larval krill during early winter (Fig. 16.4A, Table 16.1). Little is known about krill's early-winter transition from the water column to the ice, but the fact that large patches of larval krill observed during late winter were not common during early-winter sampling is consistent with the hypothesis that sea ice facilitates aggregation and formation of krill swarms (Hamner *et al.* 1989). Since krill numbers and the frequency of large patches appear to increase over winter in the sea-ice habitat, estimates of larval numbers based on diver observations in early winter may not be valid.

Contrary to several schematic depictions of krill feeding on downward facing ice surfaces during winter (e.g. Garrison *et al.*

1986, Smetacek *et al.* 1990), larvae associated with sea ice are generally oriented with feeding appendages down and scraping the upward facing and eroded ice surfaces they tend to occupy. Since larval krill are generally not found in areas where plant pigment is most concentrated, structural characteristics of sea ice appear to be a primary determinant of krill abundance and distribution during winter. Larval krill appear to have an affinity for areas of over-rafted ice and the refuge it might afford. Unfortunately, there is a paucity of information regarding predator–prey interactions in the annual sea-ice zone during winter and direct observations of predation on krill larvae are few. Hamner *et al.* (1989) observed ctenophores and an amphipod feeding on furcilia and further suggested that larval krill might be a significant component of the diet of fishes and migratory invertebrates, particularly in winter. Several alternative hypotheses can be suggested to explain the apparent affinity of larval krill for areas of over-rafted sea ice: (1) Feeding costs may be less for larvae on upward facing ice surfaces since feeding on downward facing surfaces will entail the additional cost of maintaining contact with the ice surface against a negative buoyancy. (2) Upward facing surfaces may also act as sediment traps which concentrate food resources and/or provide a substrate for prefered food items (O'Brien 1987). The relative importance of food resources and shelter to larval krill, as provided by annual sea ice, merits further investigation.

CONCLUSION

This is the first quantitative account of larval krill associated with sea ice during austral winter and observations using SCUBA have proved to be a requisite method for investigating the the krill/ice interaction. The data reported here are essential information for the formulation of hypotheses and further study of this ecologically important linkage.

ACKNOWLEDGEMENTS

We thank the numerous volunteers and scientific staff, Office of Polar Programs, Antarctic Support Associates and captains and crews of the RV *Polar Duke* for their support during each of our winter cruises. All contributed to the success of this project. Funding was provided by the United States National Science Foundation, Office of Polar Programs (grants DPP-8820589 and OPP-9117633 to L. B. Quetin and R. M. Ross and OPP-9011927 to R. M. Ross, L. B. Quetin, B. B. Prezelin and R. C Smith). This is Palmer LTER publication no. 49.

REFERENCES

Daly, K. L. 1990. Overwintering development, growth and feeding of larval *Euphausia superba* in the Antarctic marginal ice zone. *Limnology and Oceanography*, **35**,1546–1576.

El-Sayed, S. Z. 1988. The BIOMASS program. *Oceanus*, **31**,75–79.

Garrison, D. L., Sullivan, C. W. & Ackley, S. F. 1986. Sea ice microbial communities in Antarctica. *BioScience*, **36**, 243–250.

Guzman, O. 1983. Distribution and abundance of Antarctic krill (*Euphausia superba*) in the Bransfield Strait. In Schnack, S. B., ed. *On the biology of krill* Euphausia superba. *Berichte zur Polarforschung*, **4**, 169–190.

Hamner, W. M., Hamner, P. P., Obst B. S. & Carleton, J. H. 1989. Field observations on the ontogeny of schooling of *Euphausia superba* furcilia and its relationship to ice in Antarctic waters. *Limnology and Oceanography*, **34**, 451–456.

Marschall, H-P. 1988. The overwintering strategy of Antarctic krill under the pack-ice of the Weddell Sea. *Polar Biology*, **9**, 129–135.

Nordhausen, W. 1994. Winter abundance and distribution of *Euphausia superba, E. crystallorophias*, and *Thysanoessa macrura* in Gerlache Strait and Crystal Sound, Antarctica. *Marine Ecology Progress Series*, **109**, 131–142.

O'Brien, D. P. 1987. Direct observations of the behavior of *Euphausia superba* and *Euphausia crystallorophias* (Crustacea: Euphausiacea) under pack ice during the Antarctic spring of 1985. *Journal of Crustacean Biology*, **7**, 437–448.

Priddle, J., Croxall, J. P., Everson, I., Heywood, R. B., Murphy, E. J., Prince, P. A. & Sear, C. B. 1988. Large-scale fluctuations in distribution and abundance of krill – a discussion of possible causes. In Sahrhage, D., ed. *Antarctic ocean and resources variability*. Berlin: Springer-Verlag, 169–182.

Quetin, L. B. & Ross, R. M. 1991. Behavioural and physiological characterisitics of the Antarctic krill, *Euphausia superba. American Zoologist*, **31**, 49–63.

Siegel, V. & Kalinowski, J. 1994. Krill demography and small-scale processes: a review. In El-Sayed, S.Z., ed. *Southern Ocean ecology, the biomass perspective*. Cambridge: Cambridge University Press, 145–163.

Smetacek, V., Scharek, R. & Nothig, E-M. 1990. Seasonal and regional variation in the pelagial and its relationship to the life history cycle of krill. In Kerry, K. R. & Hempel, G., eds. *Antarctic ecosystems: ecological change and conservation*. Berlin: Springer-Verlag, 103–114.

Sokal, R. R. & Rohlf, F. J. 1981. *Biometry*, 2nd edn. San Francisco: W. H. Freeman, 859 pp.

Stretch, J. J., Hamner, P. P., Hamner, W. M., Michel, W. C., Cook, J. & Sullivan, C. W. 1988. Foraging behaviour of Antarctic krill, *Euphausia superba*, on sea ice microalgae. *Marine Ecology Progress Series*, **44**, 131–139.

Waters, K. J. & Smith, R.C. 1992. Palmer LTER: a sampling grid for Palmer LTER program. *Antarctic Journal of the United States*, **27**, 236–238.

17 Reproduction of *Calanus propinquus* and *Metridia gerlachei* in spring and summer in the Weddell Sea, Antarctica

F. KURBJEWEIT[1]

Alfred-Wegener-Institut für Polar- und Meeresforschung, PB 120161, 27515 Bremerhaven, Germany,[1]present address: Ernst-Moritz-Arndt University Greifswald, Department of Geography, Friedrich-Ludwig-Jahn-Str. 16, 17489 Greifswald, Germany

ABSTRACT

Egg production of the dominant calanoid copepod species Calanus propinquus *and* Metridia gerlachei *was studied in the Weddell Sea from November to March 1990/91 and December to January 1992/1993 under laboratory and* in situ *conditions.*

In the laboratory both species were fed with high food concentrations (15 μg chl a l^{-1}) of the diatom Thalassiosira antarctica *at two- and three-day intervals to investigate maximum production rates. In* Calanus propinquus *overall egg production decreased over the investigation period with increasing production rates during the third day after feeding, indicating that egg production was food dependent. The same was true for* Metridia gerlachei, *but no clear signal in egg production occurred due to feeding, although spawning intervals were 3.0 and 2.9 days for* C. propinquus *and* M. gerlachei, *respectively. Average daily egg production was 26.2 and 5.2 eggs female^{-1} d^{-1} for both species during the first nine days, equivalent to 2.9 and 1.6% body carbon female^{-1} d^{-1}.* In situ *experiments show that egg production of* C. propinquus *was restricted to the time period from December to January, while* M. gerlachei *was reproducing from at least the end of November until mid March in the same area. Analysis of variance for the* in situ *egg production experiments indicate that it is largely determined by food availability (chl* a, *POC) and ice coverage. These results are discussed and compared with investigations on reproduction of Antarctic and Arctic copepods.*

Key words: copepods, reproduction, Weddell Sea, *Metridia gerlachei*, *Calanus propinquus.*

INTRODUCTION

Investigating the rate of egg production by copepods can provide estimates of secondary production and the availability of food for many larval fish (Mullin 1993). This is in accordance with results from Sekiguchi *et al.* (1980) for *Acartia clausi hudsonica* and those of Kiørboe *et al.* (1985) for *A. tonsa.* Furthermore, it is of major interest for the understanding of marine ecosystems to know which parameters do have the greatest influence on egg production. Some authors believe that abiotic factors, e.g. currents or ice cover, influence the biology and thus the reproduction of Antarctic copepods (Smith 1990), whilst others have suggested that food limitation determines egg production (Checkley 1980). Some scientists believe that reproduction should be seen as more closely linked to lipid content and dry weight of copepods or parameters of the food (Cahoon 1981, Kiørboe 1989).

The two dominant Antarctic copepods, *Calanus propinquus*

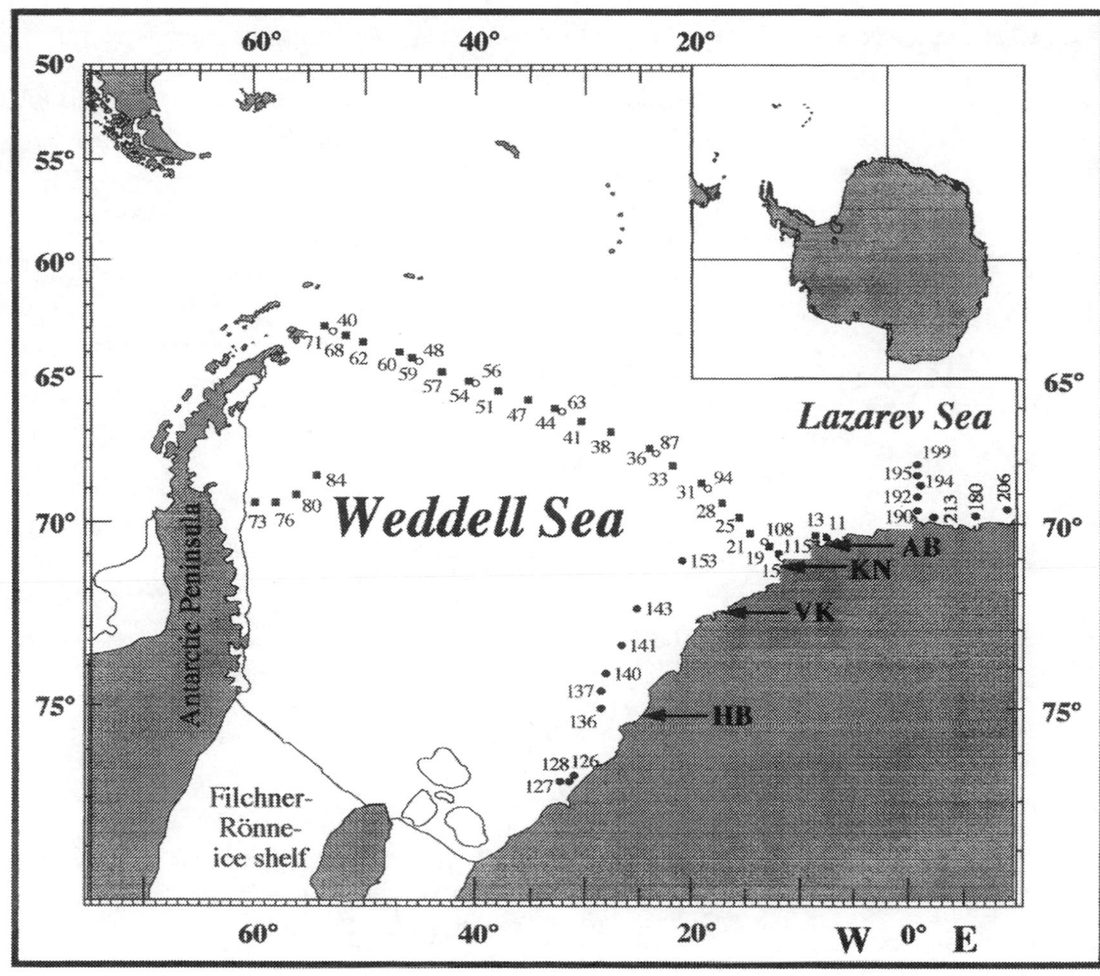

Fig. 17.1. Station map of ANT IX/2 (white circles: sta. 40–115) from 16 November to 30 December 1990, ANT IX/3 (black circles: sta. 126–213) from 3 January to 28 March 1991 and ANT X/7 (black squares: sta. 11–84) from 15 December 1992 to 15 January 1993 (AB=Atka Bay; HB=Halley Bay; KN=Kapp Norvegia; VK=Vestkapp).

and *Metridia gerlachei*, have been the subject of many investigations on distribution, life cycles and feeding strategies (Schnack 1983, 1985, Marin 1988, Huntley & Escritor 1992), but egg production has been more or less neglected. Both species seem to have different life-cycle strategies (Andrews 1966, Conover & Huntley 1991, Kurbjeweit 1993), and it is therefore of great importance to know more about their egg production and thus secondary production, since both belong to the most abundant macrozooplankton species and are major prey organisms for larval fish (Hubold 1992), squids and birds (Everson 1984), and contribute to the biogeochemical pathways via faecal pellet production (see González *et al.* 1994).

MATERIAL AND METHODS

Females of *Calanus propinquus* and *Metridia gerlachei* were collected at 51 stations with a vertically towed Bongo net (100 μm; 0.3 m s^{-1}) from the upper 300 to 500 m of the water column depending on the water depth at the station (Fig. 17.1; see Kurbjeweit 1993). Water temperature ranged between +0.8 and −1.9 °C. Immediately after the haul females of *C. propinquus* and *M. gerlachei*, respectively, were sorted either individually (laboratory experiments), or into groups of up to 10 individuals, into plexiglass cylinders having mesh (300 μm) false bottoms to separate eggs from females.

In the laboratory experiments in December/January small cylinders were used in which up to 48 single females were suspended in 250 ml TPX beakers containing cultures of *Thalassiosira antarctica* of 15 μg chl *a* l^{-1} (*c.* 450 μg C l^{-1}; C:chl *a* ratio of 30:1, Baumann pers. commun.). Usually every second or third day cylinders were transferred to new containers with fresh food and the eggs counted. The experiments were run for several weeks to estimate maximum egg production.

From November to March *in situ* egg production of groups of 10 females was measured in up to five replicates following the procedure described by Runge (1984, 1985) and Hirche & Bohrer (1987). The seawater for the experiments was taken from the surface of the stations. Light was provided by a daylight fluorescent bulb at about 8 μE m^{-2}. For each station the eggs were counted and removed over two days every 24 h.

For converting egg production into carbon the equation by Kiørboe *et al.* (1985) has been shown to give reliable results (Hirche 1990). Since the volume of an egg corresponds to 0.14×10^{-6} μg C μm^{-3}, eggs of *Calanus propinquus* and *Metridia*

Table 17.1. *Diameters, volumes and calculated C contents of eggs of* Calanus propinquus *and* Metridia gerlachei (*n=number of measurements*)

Species	Diameter μm (±SD)	*n*	Volume 10^6 μm^3 (±SD)	μg C egg^{-1} (±SD)
Calanus propinquus	175 (±5.8)	145	2.81 (±0.29)	0.39 (±0.04)
Metridia gerlachei	155 (±6.9)	120	1.95 (±0.27)	0.27 (±0.04)

Notes:
After Kiørboe *et al.* (1985).

Table 17.2. *Subdivision of chlorophyll* a, *ice cover and dry weight for the ANOVA*

Category	Low	Medium	High
Chl *a* (μg l^{-1})	0.0–0.2	0.2–1.0	>1.0
Ice cover (%)	0–10	20–50	60–100
Calanus propinquus			
Dry weight (μg)	<1000	1000–1500	>1500
Metridia gerlachei			
Dry weight (μg)	<200	200–300	>300

gerlachei are equivalent to 0.39 and 0.27 μg C, respectively (Table 17.1). To compare egg production rates with body mass (i.e. carbon), females from the experiments were first dried at 60 °C in aluminum dishes for 24 hours and subsequently analysed using a Carlo Erba CHN analyser.

Since *in situ* egg production rates of both copepod species did not show any clear correlation with parameters such as chlorophyll *a*, ice cover or dry weight and all parameters were independent variables from each other, an ANOVA was applied to test which parameters do influence egg production most. For the ANOVA chlorophyll *a*, ice cover and dry weight of females were subdivided in three categories – low, medium and high, characterized in Table 17.2.

RESULTS

Laboratory experiments

Calanus propinquus clearly responded to new food with an increase in egg production three days after feeding, while for *Metridia gerlachei* this signal was not evident (Fig. 17.2). For gaining a rough estimate of the maximum egg production rate for both species the first nine days have been averaged (see bar in Fig. 17.2). Table 17.3 indicates that the average clutch sizes of *C. propinquus* and *M. gerlachei* for this period differed significantly with about 47 and 29 eggs female^{-1} d^{-1}, respectively, while the spawning intervals were almost equal to *c.* 3 d. Considering the mean carbon content of the females for each station the average

Table 17.3. *Comparison of clutch size, spawning intervals, carbon content of females, average daily egg production, corresponding carbon content (μg C) and percentage of carbon content (%C) of females of* Calanus propinquus *and* Metridia gerlachei *respectively in experiments with high food concentrations of 15 μg chl* a *l^{-1}* (Thalassiosira antarctica)

	Calanus propinquus			*Metridia gerlachei*		
	M	SD	*n*	*M*	SD	*n*
Clutch size (eggs)	47.1	39.6	303	29.1	18.4	73
Spawning interval (d)	3.0	3.4	234	2.9	1.9	36
Females (μg C)	350	43.5	46	87.8	13.3	46
Eggs female^{-1} d^{-1}	26.1	10.2	48	5.2	4.0	46
μg C	10.2	4.0	48	1.4	1.1	46
% C	2.9	1.1	48	1.6	0.9	46

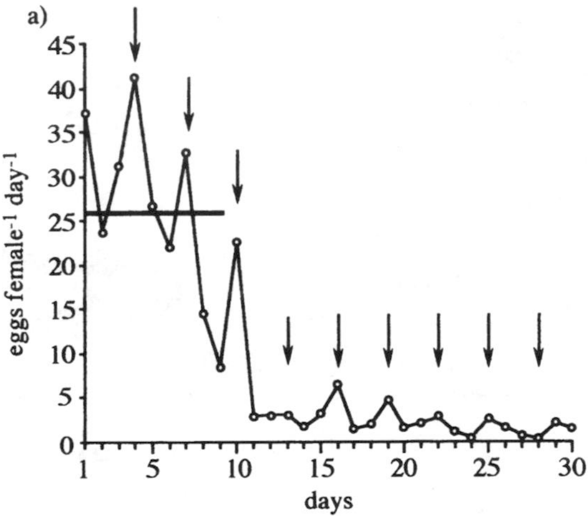

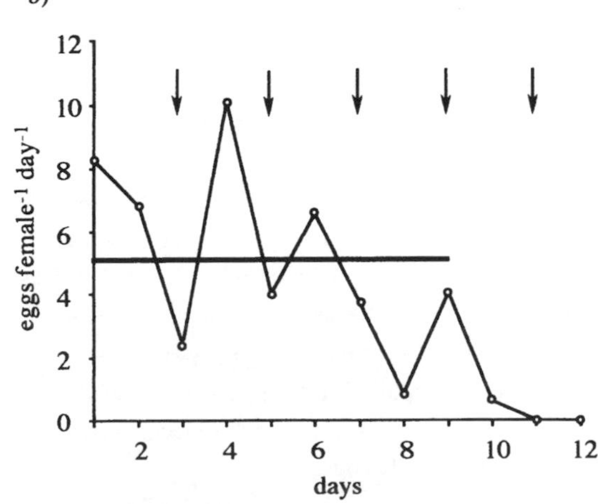

Fig. 17.2. Mean egg production of (a) *Calanus propinquus* and (b) *Metridia gerlachei* in long-term experiments. The bars indicate the average egg production for the first nine days. Arrows indicate days on which containers were exchanged and fresh food offered.

egg production rates in the laboratory under superabundant food of 26.1 and 5.2 eggs female^{-1} d^{-1} for *C. propinquus* and *M. gerlachei*, respectively, are equal to 2.9% and 1.6% body carbon d^{-1} (Table 17.3). Even under such very high food conditions the egg production rates dropped dramatically in both species after

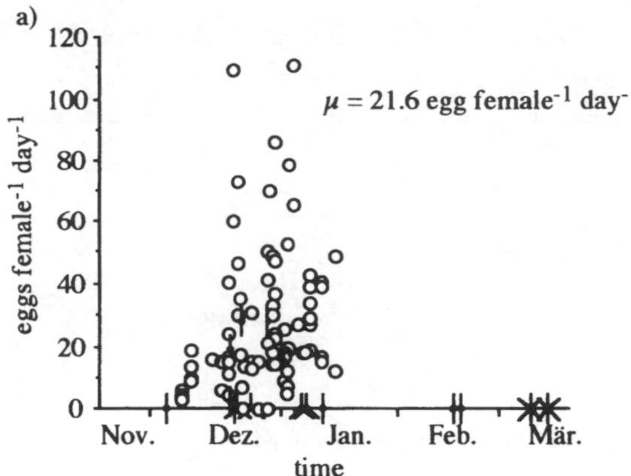

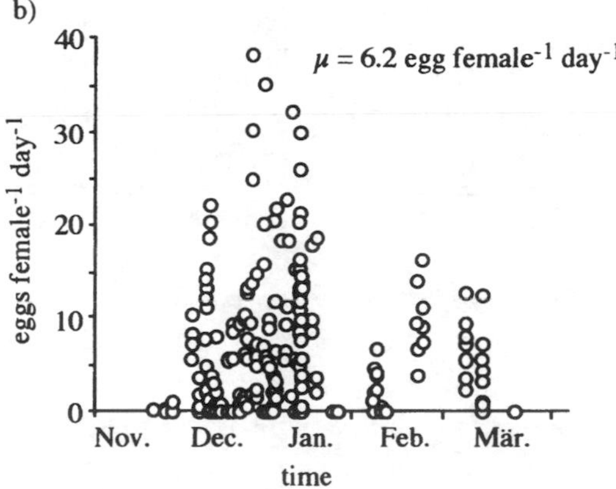

Fig. 17.3. *In situ* egg production of (a) *Calanus propinquus* and (b) *Metridia gerlachei* against time.μ=average daily egg production rate.

the ninth day (Fig. 17.2). In *C. propinquus* egg production continued at a low level for more than four weeks, but in *M. gerlachei* it dropped to zero on the eleventh day (Fig. 17.2).

In situ experiments

Although egg production rates were highly variable in the *in situ* experiments for *Calanus propinquus* and *Metridia gerlachei* highest rates occurred in late December and early January in the Weddell Gyre for both species with up to 120 and 40 eggs female^{-1} d^{-1}, respectively (Fig. 17.3). The average daily egg production rate during this period of time was 21.6 eggs female^{-1} d^{-1} (±23) for *C. propinquus* and 6.2 eggs female^{-1} d^{-1} (±7.4) for *M. gerlachei*, equivalent to 2.4% (±2.5%) and 1.4% (±1.9%) body carbon d^{-1}, respectively. While egg production in *C. propinquus* was observed from December to mid January and females failed to produce eggs in February and March, *M. gerlachei* reproduced during the whole period of 3.5 months (Fig. 17.3).

To investigate which of the environmental parameters do influence egg production most and to test hypotheses about food-limited egg production, the *in situ* egg production rates were plotted against food concentration (i.e. chl *a*, POC), ice cover and dry weight, but no correlation was apparent. A multiple regression analysis shows that the slopes are significantly different from zero. Thus, an ANOVA was applied showing that egg production in *C. propinquus* was at least in two cases significantly higher at high food concentrations (p=0.05), lower at high ice cover and lower at high dry weights (Table 17.4). In *M. gerlachei* the same trends are even more pronounced for food and ice concentration, but it is the inverse for the dry weight (Table 17.5).

Table 17.4a. Calanus propinquus*: Mean egg production,* M *(eggs female^{-1} d^{-1}), and standard deviation of* in situ *egg production for the in low, medium and high classified parameters (*n*=number of measurements)*

	chl *a*		Ice cover		Dry weight	
	n	*M* (±SD)	*n*	*M* (±SD)	*n*	*M* (±SD)
Low	33	11.4 (18.3)	79	20.2 (24.6)	48	17.1 (26.2)
Medium	70	16.8 (18.0)	14	25.1 (23.8)	45	25.1 (21.9)
High	24	29.7 (33.4)	34	9.3 (12.7)	34	9.3 (12.7)

Table 17.4b. *ANOVA for the average daily egg production rates in Table 17.3 by using Fisher PLSD test*

	Chl *a*	Ice cover	Dry weight
F-test	$p \leq 0.0001$	$p \leq 0.0001$	$p \leq 0.0001$
Low vs. medium	2.39*	2.45*	2.54*
Low vs. high	2.68*	2.31*	3.58*
Medium vs. high	2.21*	2.92*	2.98*

Note:
* significant at the 95% level

Table 17.5a. Metridia gerlachei*: Mean egg production,* M *(eggs female^{-1} d^{-1}), and standard deviation of* in situ *egg production for the low, medium and high classified parameters (*n*=number of measurements)*

	Chl *a*		Ice cover		Dry weight	
	n	M (±SD)	*n*	M (±SD)	*n*	M (±SD)
Low	48	3.1 (4.8)	133	6.4 (7.0)	38	2.1 (5.6)
Medium	113	5.6 (6.8)	41	10.6 (9.5)	158	6.6 (7.2)
High	61	9.9 (8.8)	48	2.1 (3.4)	34	10.0 (9.0)

Table 17.5b. *ANOVA for the average daily egg production rates in Table 17.3 by using Fisher PLSD test*

	Chl *a*	Ice cover	Dry weight
F-test	$p \leq 0.0079$	$p \leq 0.0256$	$p \leq 0.0073$
Low vs. medium	9.10	12.62	8.94
Low vs. high	11.56*	8.92*	9.65
Medium vs. high	10.19*	13.82*	9.79 *

Notes:
* significant at the 95% level.

DISCUSSION

Growth and reproduction of calanoid copepods are dependent on temperature and food conditions (e.g. Durbin *et al.* 1983). Although some authors (Corkett & McLaren 1978, Davis 1987) argue that reproduction is not food dependent, the most striking results are those of Checkley (1980), Runge (1980) and Tiselius *et al.* (1991), who found food-limited growth and reproduction rates under field conditions. They showed that both parameters were positively correlated with chlorophyll concentrations. In the present laboratory and *in situ* investigations evidence is given that egg production is mainly food dependent for the Antarctic copepod species *Calanus propinquus* and *Metridia gerlachei*, but it is also dependent on ice conditions and presumably other 'internal factors' such as dry weight or lipid content.

In both species spawning intervals were about three days. In *Calanus propinquus* the food limitation in terms of chlorophyll *a* seems to be more pronounced in comparison with *Metridia gerlachei*, since highest egg production is always observed between the second and third day after feeding. It seems that only permanent feeding – at least every 24 h – may maintain high production rates in the long run as was shown by Hirche (1989, 1990) for the Arctic copepods *C. glacialis* and *C. finmarchicus*. Since egg production occurred in all areas of the Weddell Sea and temperatures ranged in general between +1.0 and −1.9 °C it is believed that temperature plays a negligible role in this temperature range for both species. Normally, ice conditions determine indirectly the quantity of chlorophyll in the water column and thus play an important role in polar waters for reproduction of copepods (see Smith 1990). In the present investigations no positive correlation was found between chlorophyll and ice conditions (r^2=0.3), probably due to different velocities of ice and underlying water (Fahrbach, pers. commun.). An ANOVA was applied which shows that chlorophyll and ice cover as well as dry weight may have an influence on egg production. This is in accordance with the results of Hirche & Kattner (1993) who showed for *C. glacialis* two different kinds of egg production modes: a food-dependent and a food-independent one. Fransz (1988) and Huntley & Escritor (1992) propose that species like *C. propinquus* and *M. gerlachei* can fuel their egg production by use of lipids and are not dependent on phytoplankton, because nauplii of those species were found in late winter. This seems to be true for *C. hyperboreus* and *C. glacialis* in Arctic waters (Smith 1990), but for Antarctic waters it has not yet been demonstrated (Kurbjeweit 1993). Otherwise, results of Hirche & Bohrer (1987) for *C. glacialis* indicate that no egg production took place in areas with heavy pack ice and negligible phytoplankton stocks, whereas production was high in polynyas with comparatively high chlorophyll values.

The present investigations show that the time range of egg production is different in both species. While *M. gerlachei* can fuel its reproduction in spring although lipid contents are low, *C. propinquus* laid no eggs in February and March although lipid reserves were more than doubled in comparison with spring

Table 17.6. *Egg production of dominant calanoid copepod species in the Arctic and Antarctic as % body carbon day^{-1}*

Species	Area	% body C day^{-1}	Source
Calanus glacialis	Arctic	1.7–3.3	Tourangeau & Runge 1991
		3.3	Hirche & Bohrer 1987
		4.2	Smith 1990
		5	Hirche 1990
		5	Diel 1989
Calanus propinquus	Antarctic	2.4	Kosobokova 1994
		2.9	Kurbjeweit 1993
Calanoides acutus	Antarctic	2.9–4.0	Huntley *et al.* 1987
			Huntley & Escritor 1991
		2.1	Kurbjeweit unpublished
Metridia longa	Arctic	1.4–2.2	Diel 1989
Metridia gerlachei	Antarctic	1.6	Kurbjeweit 1993

(Kurbjeweit 1993), probably due to the preparation of this species for overwintering. In addition, both species do have different feeding modes: *C. propinquus* seems to prefer phytoplankton, but can switch to a more omnivorous, opportunistic mode like *M. gerlchei* at specific times of the year.

Although Kosobokova (in press) suggests a two to threefold higher egg production in Arctic calanoid copepods (namely *Calanus*) than in those in Antarctic areas, literature data do not indicate a significant difference in terms of body carbon day^{-1} (Table 17.6). At the present stage we do not have clear evidence for such differences in egg production for Arctic versus Antarctic copepods (and other invertebrates) and there is no obvious reason for such distinctive pattern, at least in planktonic animals.

In conclusion, *Metridia gerlachei* and *Calanus propinquus* do have different reproduction modes in relation to average daily egg production rates, clutch sizes and secondary production due to different life-cycle and feeding strategies. Furthermore, both species do not differ significantly in terms of production from their Arctic congeners. Finally, we need more sophisticated production measurements, if we want to gain more information on secondary production in justifiable time-spans to fill gaps in ecosystem flow diagrams.

ACKNOWLEDGEMENTS

I would like to thank the captain and crew of RV *Polarstern* for their professional assistance during the expeditions ANT IX/2+3 and ANT X/7. I am indebted to S. Schnack-Schiel for her critical comments on a former German draft of this manuscript. Dr M. Baumann helped with chl *a* sampling and phytoplankton cultures and K. Fahl, M. Gorny, and S. Günther took care of experiments on board. T. Brey gave good advice with the statistical analysis. This is publication number 814 of the Alfred Wegener Institute for Polar and Marine Research.

REFERENCES

Andrews, K. J. H. 1966. The distribution and life history of *Calanoides acutus* (Giesbrecht). *Discovery Reports*, **34,** 117–162.

Cahoon, L. B. 1981. Reproductive response of *Acartia tonsa* to variations in food ration and quality. *Deep-Sea Research*, **28A**, 1215–1221.

Checkley Jr, D. M. 1980. Food limitation of egg production by a marine, planktonic copepod in the sea off southern California. *Limnology and Oceanography*, **25**, 991–998.

Conover, R. J. & Huntley, M. 1991. Zooplankton and sea ice: distribution, adaptations to seasonally limited food, metabolism, growth patterns and life cycles strategies in polar seas. *Journal of Marine Systems*, **2**, 1–41.

Corkett, C. J. & McLaren, I. A. 1978. The biology of *Pseudocalanus*. *Advances in Marine Biology*, **15**, 1–231.

Davis, C. S. 1987. Components of the zooplankton production cycle in the temperate ocean. *Journal of Marine Research*, **45**, 947–983.

Diel, S. 1989. Zur Lebensgeschichte dominanter Copepodenarten (*Calanus finmarchicus, C. glacialis, C. hyperboreus, Metridia longa*) in der Framstraße. Dissertation, Universität Kiel, 140 pp. [Unpublished]

Durbin, E. G., Durbin, A. G., Smayda, T. J. & Verity, P. G. 1983. Food limitation of production by adult *Acartia tonsa* in Narragansett Bay, Rhode Island. *Limnology and Oceanography*, **28**, 1199–1213.

Everson, I. 1984. Zooplankton. In Laws, R. M., ed. *Antarctic ecology*. Vol. 2. London: Academic Press, 463–490.

Fransz, H. G. 1988. Vernal abundance, structure and development of epipelagic copepod populations of the eastern Weddell Sea, Antarctica. *Polar Biology*, **9**, 107–114.

González, H. E., Kurbjeweit, F. & Bathmann, U. 1994. Occurrence of cyclopoid copepods and faecal material in the Halley Bay region, Antarctica, during January–February 1991. *Polar Biology*, **14**, 331–342.

Hirche, H-J. 1989. Egg production of the Arctic copepod *Calanus glacialis* Jaschnov – laboratory experiments. *Marine Biology*, **103**, 311–318.

Hirche, H-J. 1990. Egg production of *Calanus finmarchicus* at low temperature. *Marine Biology*, **106**, 53–58.

Hirche, H-J. & Bohrer, R. N. 1987. Reproduction of the Arctic copepod *Calanus glacialis* in Fram Strait. *Marine Biology*, **94**, 11–17.

Hirche, H-J. & Kattner, G. 1993. Egg production and lipid of *Calanus glacialis* in spring: indication of a food-dependent and food-independent reproductive mode. *Marine Biology*, **117**, 615–622.

Hubold, G. 1992. Zur ökologie der Fische im Weddellmeer. *Berichte Polarforschung*, **103**, 157.

Huntley, M. & Escritor, F. 1991. Dynamics of *Calanoides acutus* (Copepoda: Calanoida) in Antarctic coastal waters. *Deep-Sea Research*, **38**, 1145–1167.

Huntley, M. & Escritor, F. 1992. Ecology of *Metridia gerlachei* Giesbrecht in the western Bransfield Strait, Antarctica. *Deep-Sea Research*, **39**, 1027–1055.

Huntley, M., Marin, V. & Resland, V. 1987. RACER: Feeding and egg production rates of some Antarctic copepods. *Antarctic Journal of the United States*, **22**, 158–160.

Kiørboe, T. 1989. Phytoplankton growth rate and nitrogen content: implications for feeding and fecundity in a herbivorous copepod. *Marine Ecology Progress Series*, **55**, 229–234.

Kiørboe, T., Møhlenberg, F. & Hamburger, K. 1985. Bioenergetics of the planktonic copepod *Acartia tonsa*: relation between feeding, egg production and respiration, and composition of specific dynamic action. *Marine Ecology Progress Series*, **26**, 85–97.

Kosobokova, K. N. 1994. Reproduction of the calanoid copepod *Calanus propinquus* in the southern Weddell Sea, Antarctica: observations in laboratory. *Hydrobiologia*, **293**, 219–227.

Kurbjeweit, F. 1993. Reproduction and life cycles of dominant copepod species from the Weddell Sea, Antarctica. *Report on Polar Research*, No. 129, 238.

Marin, V. 1988. Qualitative models of the life cycles of *Calanoides acutus, Calanus propinquus*, and *Rhincalanus gigas*. *Polar Biology*, **8**, 439–446.

Mullin, M. M. 1993. Reproduction by the oceanic copepod *Rhincalanus nasutus* off southern California, compared to that of *Calanus pacificus*. *CalCoFi Report 34.*

Runge, J. A. 1980. Effects of hunger and season on the feeding behaviour of *Calanus pacificus*. *Limnology and Oceanography*, **25**, 134–145.

Runge, J. A. 1984. Egg production of the marine, planktonic copepod, *Calanus pacificus* Brodsky: laboratory observations. *Journal of Experiments in Marine Biology and Ecology*, **74**, 53–66.

Runge, J. A. 1985. Relationship of egg production of *Calanus pacificus* to seasonal changes in phytoplankton availability in Puget Sound, Washington. *Limnology and Oceanography*, **30**, 382–396.

Schnack, S. B. 1983. Feeding of two Antarctic copepod species (*Calanus propinquus* and *Metridia gerlachei*) on a mixture of centric diatoms. *Polar Biology*, **2**, 63–68.

Schnack, S. B. 1985. Feeding by *Euphausia superba* and copepod species in response to varying concentrations of phytoplankton. In Siegfried, W. R. & Laws, R. M., eds. *Antarctic nutrient cycles and food webs*. Berlin: Springer-Verlag, 311–323.

Sekiguchi, H., McLaren, I. A. &Corkett, C. J. 1980. Relationship between growth and egg production in the copepod. *Marine Biology*, **58**, 133–138.

Smith, S. L. 1990. Egg production and feeding by copepods prior to the spring bloom of phytoplankton in Fram Strait, Greenland Sea. *Marine Biology*, **106**, 59–69.

Tiselius, P., Nielsen, T. G., Breuel, G., Jaanus, A., Korshenko, A. & Witek, Z. 1991. Copepod egg production in the Skagerrak during SKAGEX, May–June 1990. *Marine Biology*, **111**, 445–453.

Tourangeau, S. & Runge, J. A. 1991. Reproduction of *Calanus glacialis* under ice in spring in southeastern Hudson Bay, Canada. *Marine Biology*, **108**, 227–233.

18 Trophic flows in the benthic shelf community of the eastern Weddell Sea, Antarctica

A. JARRE-TEICHMANN[1], T. BREY, U. V. BATHMANN[1], C. DAHM, G. S. DIECKMANN, M. GORNY, M. KLAGES, F. PAGÉS, J. PLÖTZ, S. B. SCHNACK-SCHIEL, M. STILLER AND W.E. ARNTZ

Alfred-Wegener-Institut für Polar- und Meeresforschung, PB 120161, 27515 Bremerhaven, Germany, present address: Danish Institute for Fisheries Research, North Sea Centre, P.O.Box 101, 9850 Hirtshals, Denmark

ABSTRACT

Collaborative research in the Weddell Sea performed during the past decade has substantially increased insight into the different community components of this large ecosystem, and has led to a conceptual diagram of the biomass flows among its principal subsystems. In order to integrate the results of the various research efforts directed towards the shelf communities into a coherent whole, a static modelling approach was used to obtain the first balanced model of trophic flows between the dominant groups of the benthic shelf community of the eastern Weddell Sea. Polychaeta, Holothuroidea, Ophiuroidea and benthic Crustacea were found to be the most important groups in terms of carbon flow in the benthos. Preliminary results of network analysis performed on the model suggest a rather mature ecosystem in terms of community structure and carbon cycling, but yield ambiguous results with respect to the structure of the food web and overall system homeostasis. Further studies are suggested on the population dynamics and trophic requirements of the dominant groups in this High Antarctic community.

Key words: macrobenthos, static model, carbon flow budget, trophic interactions, network analysis, ecosystem maturity.

INTRODUCTION

In a conceptual model of energy flows in the Weddell Sea, Schalk *et al.* (1993) defined a number of different subsystems, setting apart a retention system at the centre of the Weddell gyre, and a pelagic export system at its border. The boundaries of the Weddell Sea are areas of high but very variable activity, i.e. the pelagic export system segregates into an 'ice-associated system' in summer and a 'deep system' in winter. However, not all biological activity is restricted to the ice-free period of the year (Spindler & Dieckmann 1991, Schnack-Schiel *et al.* 1991, Gorny *et al.* 1992). The benthos depends to a large extent on the sedimentation of organic matter from the pelagic (Arntz *et al.* 1992). On the Weddell Sea shelf, three major benthic communities have been classified, one on the eastern shelf, one on the southern shelf and a southern trench community (Voß 1988, Galéron *et al.* 1992). Gerdes *et al.* (1992), by means of cluster analysis, identified three different benthic assemblages on the southeastern Weddell Sea shelf (70–80 °S). Their 'Cluster B' forms the basis of this contribution (Fig. 18.1). It comprises the majority of stations on the shelf off Kapp Norvegia and Halley Bay, and is characterized by high benthic biomass and the dominance of echinoderms in terms of carbon.

The purpose of this study is to integrate the present knowledge of energy flow in a high Antarctic ecosystem. A further goal is to assess the special role of this shelf ecosystem in view of ongoing discussions on the fragility of the communities in the Antarctic Ocean. We emphasize the benthos because benthic communities on the shelf show considerable diversity (Brey *et*

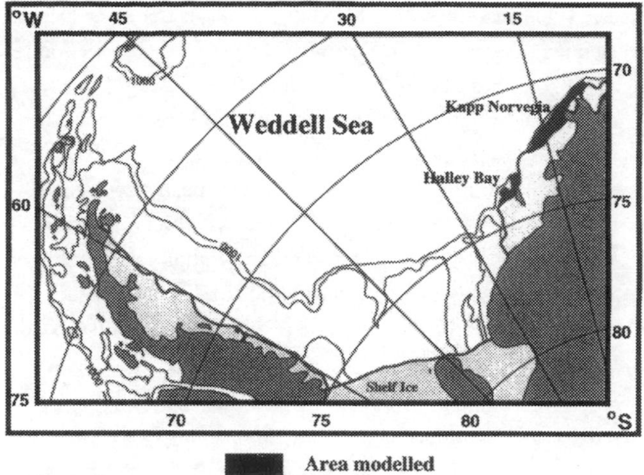

Fig. 18.1. Map of the Weddell Sea, Atlantic Sector, Southern Ocean, showing the shelf area represented in this study. The isobaths of 500 and 1000 m, as well as the shelf ice zone, are indicated.

al. 1994), and have higher biomasses than those from equal depths of temperate regions (Brey & Clarke 1993).

THE MODELLING APPROACH

The model was constructed using the ECOPATH II software (Christensen & Pauly 1992), based on the approach of J. J. Polovina and associates (Polovina & Ow 1983, Polovina 1984, 1985). Steady-state interactions of the components (species or species groups) in an ecosystem are described through a set of linear equations wherein the production of each component equals its withdrawal by other components in the system (predation mortality), its export from the system (fishing mortality and other exports), plus the baseline mortality, i.e.

Production by (i) = All predation on (i) + nonpredatory biomass losses of (i) + Fishery catches of (i) + other exports of (i) (18.1a)

The terms in this equation can be replaced by

$$\text{Production by } (i) = B_i - P/B_i$$
$$\text{Predatory losses of } (i) = M_2 = -_j (B_j - Q/B_j - Dc_{j,i}) \quad (18.2)$$
$$\text{Other losses of } (i) = (1 - EE_i) - B_i\, P/B_i$$

and this leads, for any component in the system, to

$$B_i - P/B_i - EE_i -_{-j} (B_j - Q/B_j - DC_{j,i}) - Ex_i = 0 \quad (18.1b)$$

where i=a component of the model, j=any of its predators, B=the biomass, P/B=the ratio of production of a population to its biomass (=total mortality under steady-state conditions), Q/B=the ratio of the consumption of a population to its biomass, DC=the average relative diet composition (in terms of weight), EE=the ecotrophic efficiency (the fraction of the total production consumed by predators or exported from the system), Ex=the export from the system (e.g. by emigration or advection, or fishery catch).

This structure defines the necessary inputs to the model. These are, for each component, an estimate of its

- biomass;
- production;
- total food consumption;
- assimilation efficiency;
- diet composition;
- exports from the system;
- ecotrophic efficiency.

It is a characteristic of systems of linear equations that, for each component, one of the above parameters (except diet composition) may be unknown, and is estimated when solving the system. If an acceptable result for each of the unknowns is achieved from the inputs, the model is regarded as balanced and may be subject to further study on the ecosystem level.
The energy balance of each component is given by

Consumption = Production + Respiration + Non-assimilated food (18.3)

where consumption is composed of consumption within the system and consumption of imports, and production may be consumed by predators, be exported from the system, or be a contribution to the detritus.

Besides various statistics that characterize the components, a number of statistics on the ecosystem level were computed that can be used for assessing the system as a whole, e.g. statistics with respect to maturity *sensu* Odum (1969), and statistics related to flow network theory as introduced by Ulanowicz and associates (Ulanovicz & Platt 1985, Ulanowicz 1986, in press, Ulanowicz & Norden 1990, Ulanowicz & Puccia 1990).

In this study, ecotrophic efficiencies were computed from the equations and used to balance the model, where, for obvious reasons, $0 \leq EE < 1$ served as a constraint for each component in the system. The biomass of the various benthic groups was obtained from the average biomass of these groups given in Gerdes *et al.* (1992, Cluster B, total of 21 stations). Where specific estimates of consumption per unit biomass were lacking, estimates of the gross efficiency (GE) of each group, i.e., the ratio of its production and consumption, were used to estimate consumption from a given production. If no specific estimates were available for the gross efficiency of a given group in the macrozoobenthos, we used the average of the values presently available for macrozoobenthos in the primary literature, i.e. 9% for herbivores and/or detritivores, and 30% for carnivores (Brey, unpublished data compilation). It should also be noted that the term 'detritus' in our model pertains to any form of non-living organisms, i.e., scavengers were considered as detritivores for the present purpose.

THE COMPONENTS OF THE MODEL

Primary producers

For the shelf of the southeastern Weddell Sea from Halley Bay to Kapp Norvegia, an average primary production of 557 mg C

m^{-2} day^{-1} was estimated during February/March 1983 (Bröckel 1985). Assuming the length of the primary productive phase as 150 days, this corresponds to an average annual production of 83 g C m^{-2} $year^{-1}$. Nöthig (1988) estimated primary production on the southeastern Weddell Sea shelf in January/February 1985 at about 380 mg C m^{-2} day^{-1}, corresponding to 57 g C m^{-2} $year^{-1}$ on the same length of the productive phase. We used an estimate of 70 g C m^{-2} $year^{-1}$ for our model.

Nöthig *et al.* (1991) gave an average phytoplankton biomass of 375 mg C m^{-2} in the upper 10 m of the water column. The depth profiles in Bröckel (1985) and Nöthig (1988) would suggest that this covers about half of the total phytoplankton biomass, and the biomass of phytoplankton during austral summer in this study was therefore calculated at 0.75 g C m^{-2}. Considering once more that phytoplankton is only present during 5 months of the year, a 'diluted' biomass of 0.313 g C m^{-2} and a corresponding P/B ratio of 224 $year^{-1}$, were used as input to the model.

Detritus import from ice algae

Algae inside and below fast ice maintain high stocks and may contribute a significant fraction to the primary production (Spindler & Dieckmann 1991). Arrigo *et al.* (1993) report a production rate of 0.38 g C m^{-2} day^{-1} under optimal conditions in the layer of platelet ice under the fast ice, a significant amount of which will sediment directly on the shelf during ice retreat (Smetacek *et al.* 1992). We think that the sedimentation of algae from the platelet ice is more important for the sedimentation on the shelf than sedimentation from inside the fast ice. Algae inside the fast ice maintain significant stocks (Table 18.1), however, the major fraction of the pack ice will melt beyond the shelf break and, accordingly, contribute to the sedimentation over the deeper parts of the slope and the central Weddell Sea, but not to the sedimentation on the shelf. As the dynamics of the ice-bound system are largely unknown, we opted to treat the production from this system as import to the detritus in our model. However, in view of the potentially high production given by Arrigo *et al.* (1993), the detritus import value of 5 g C m^{-2} $year^{-1}$ from ice algae, used here, should still be considered a conservative estimate of the total detrital fallout from the sea ice to the benthos.

Meso-zooplankton

Following Boysen-Ennen *et al.* (1991), copepods constitute 87.6% of the meso-zooplankton biomass in summer (Table 18.2) if euphausiids (which are modelled separately here) are excluded. Seasonal data on copepod biomass are given in Table 18.3. Significant differences in abundance and developmental stage composition of the dominant herbivorous copepod species were found in January and February. Therefore, the January data are considered to represent a spring situation within the copepod population while those from February represent a summer one (Schnack-Schiel *et al.* (1991). Average copepod biomass is 1.17 g C m^{-2}, using dry-weight data from Hagen (1988 and unpubl.), Mizdalski (1988), Schnack (1985) and Schnack *et al.* (1985). Dry weights of smaller copepodite stages were extrapolated from older stages using log-transformed length–weight relationships. Zooplankton body carbon was assumed to be equivalent to 40% of body dry weight (Drits *et al.* 1993). Assuming that the difference in mesh size used (100 μm and 320 μm for the studies of Schnack-Schiel and Boysen-Ennen, respectively) does not significantly change the relative contribution of the various taxa to the total zooplankton, an estimate of zooplankton biomass (excluding euphausiids) of 1.34 g C m^{-2} was obtained and used for the model.

Voronina *et al.* (1980 a,b) give *P*/*B* ratios of 4.5 $year^{-1}$ and 5.3 $year^{-1}$ for *Calanoides acutus* and *Rhincalanus gigas*, respectively. Taking into account that the biomass of *C. acutus* is more important in the zooplankton than that of *R. gigas* (Table 18.4), we used the lower estimate for the model.

Consumption and respiration rates of various copepod species, weighted by the fraction of the various taxa in the total biomass and taking into consideration their respective life strategies, are given in Table 18.5. Population food consumption averaged at about $Q/B = 110$ $year^{-1}$; however, we noted that this led to a very low a gross efficiency of 4%. An average of 65% of the ingested food is assimilated (Schnack *et al.* 1985). Accordingly, the seasonal average of the ratio of respiration to assimilation was computed as 66%.

Metridia gerlachei, *Oithona* sp. and *Oncaea* spp. are omnivores, the other major zooplankton groups are mainly herbivores/detritivores. The diet composition of the meso-zooplankton was assumed at 55% phytoplankton, 5% zooplankton and 40% detritus, taking into account the high fraction of detritus in the food during the winter months.

Studies on grazing impact suggest that grazers are not food limited, especially during bloom periods, and that losses due to vertical flux and phytoplankton respiration or excretion generally exceed those due to zooplankton grazing (Schnack *et al.* 1985, Hopkins 1987, Huntley *et al.* 1991). Schnack *et al.* (1985) found that during the early part of the bloom copepod grazing accounted for <1% primary production in the Bransfield Strait, while it accounted for 55% in the Drake Passage. Hopkins (1987) estimated daily total zooplankton consumption to be about 2% of phytoplankton standing stock in McMurdo Sound. Huntley *et al.* (1991) estimated that grazing could maximally account for 1–5% of production in Dec.–Jan. and 9–72% in Feb.–Mar. in the Bransfield Strait. The role of microzooplankton is presently unclear. Copepods are an integral part of the microbial network, feeding on it (Hopkins 1985, 1987) and resupplying it with faecal material (Smetacek *et al.* 1990).

Euphausiacea

Although benthopelagic aggregations of krill (*Euphausia superba*) have been observed on the deeper shelf of the eastern Weddell Sea (Gutt & Siegel 1994), they are less important on the shelf than in the oceanic area (Boysen-Ennen *et al.* 1991). The total biomass of euphausiids, of which *E. crystallorophias* was

Table 18.1. *Standing stocks of ice algae in the pack ice of the Weddell Sea*

Cruise	Year	Range of Julian days	No. of cores	Average core length (m)	Range of core lengths (m)	Mean integrated chlorophyll *a* ($\mu g\ m^{-2}$)	Standard deviation	Range of mean integrated chl *a* ($\mu g\ m^{-2}$)	Mean chlorophyll *a* ($\mu g\ l^{-1}$)	Standard deviation on mean	Range of mean chlorophyll *a* ($\mu g\ l^{-1}$)
AM 83	1983	318–333	12	1.37	0.94–2.11	8.62	14.11	0.32–47.60	5.31	7.22	0.3–22.56
AM86	1986	64–79	9	1.34	0.68–2.09	6.16	4.61	0.77–10.07	4.27	3.01	0.77–10.07
AM88	1988	162–180	55	0.67	0.11–1.49	4.70	6.02	0.05–29.18	6.74	6.37	0.12–23.42
ANT III/3	1985	19–54	13	2.29	0.75–3.55	131.66	159.43	5.90–435.3	62.35	73.40	7.84–202.47
ANT V/2	1986	201–249	79	0.74	0.23–4.00	1.58	3.17	0.02–16.10	1.91	2.95	0.04–14.63
ANT V/3	1986	278–339	55	0.77	0.20–2.20	5.48	8.11	0.01–35.43	5.90	6.20	0.02–26.68
ANT VII/2	1988	293–310	21	1.57	0.43–3.75	6.50	5.20	0.60–18.08	4.53	3.27	1.21–15.41
ANT VIII/2	1989	255–293	41	0.94	0.22–2.41	4.17	4.84	0.14–18.40	3.62	3.46	0.4–14.36
ANT X/3	1992	99–127	22	0.54	0.15–1.74	7.95	9.99	0.08–32.24	12.02	11.00	0.29–36.51
ANT X/4	1992	155–209	48	0.70	0.07–2.43	1.68	3.27	0.01–19.44	1.87	2.90	0.06–15.81
ANT XI/3	1993	29–53	29	2.24	0.81–5.93	34.22	26.28	5.55–99.12	17.79	13.32	1.17–58.54
WC 87	1987	163–189	44	0.56	0.17–3.92	3.36	7.99	0.07–52.7	4.25	3.40	0.29–18.54
Grand mean				1.14		18.01			10.88		
Mean (excl ANT III/3)				1.04		7.67			6.20		
Sum			428								

the most important species, was 0.49 g C m^{-2} during February/March 1983 (Boysen-Ennen *et al.* 1991). Lacking information on other euphausiids, we had to adopt production and consumption estimates for krill. The *P*/*B* ratio of *E. superba* ranges from 0.8 to 1.1 $year^{-1}$ (Siegel 1986). We are aware that production may be underestimated, as *E. crystallorophias* is smaller than krill and higher *P*/*B* ratios have been found for other euphausiids (Voronina *et al.* 1980c).

Table 18.2. *Zooplankton composition in the eastern Weddell Sea*

Taxon	Biomass (g C m^{-2})	Fraction (%)
Coelenterata	0.0081	0.60
Mollusca	0.0061	0.45
Polychaeta	0.0024	0.18
Ostracoda	0.0005	0.04
Copepoda	0.7561	55.69
Euphausiacea	0.4945	36.42
Decapoda	0.0006	0.04
Amphipoda	0.0008	0.06
Chaetognatha	0.0250	1.84
Tunicata	0.0609	4.49
Pisces (postlarvae)	0.0028	0.21
Total	1.3578	100.00

Note:
Based on Boysen-Ennen *et al.* (1991).

Clarke & Morris (1983) give a daily ration of 5.1% body mass for adult male krill, yielding a gross efficiency of 5.4%. The diet composition of krill was assumed at 85% phytoplankton, 10% detritus and 5% zooplankton.

Gelatinous zooplankton

Our knowledge of large gelatinous zooplankton does not yet allow the inclusion of this presumably important consumer group in the model. The first data on species composition and abundance of mesoplanktonic medusae and siphonophores off the eastern Weddell shelf are given in Table 18.6 and indicate a significant role in the pelagic off the shelf break. Macro- and megaplanktonic fractions were also collected during the same cruise, but are only now being quantified. However, on average, cnidarians contribute 52.6% to the biovolume of macrozooplankton in the eastern part of the Weddell Gyre in summer (Pagès *et al.* 1994), where they can exert a great impact on the ecosystem. We thus opted to restrict 10% of the zooplankton production for consumption of carnivorous gelatinous zooplankton, and to treat this fraction as exports from our system, since we do not know the fate of the energy that flows through this assemblage.

Table 18.3. *Biomass of copepods in the eastern Weddell Sea, sampled with a net of 100 μm mesh size. See text for classification of seasons. Seasonal mean is 1.17 g C m^{-2}*

Date	29 Jan. 1985	15 Feb. 1985	beg. May 1992	mid-Oct. 1986	beg. Nov. 1986	mid Nov. 1986
Season	Spring	Summer	Autumn	Winter	Winter	Winter
Sampling depth (m)	200	200	300	400	400	400
Total Copepoda biomass (g C m^{-2})	0.55	1.11	2.41	0.62	0.69	0.58

Note:
From Schnack-Schiel, unpublished.

Table 18.4. *Relative contribution (%) of major zooplankton taxa to the total biomass*

Date Season	29 Jan. 1985 Spring	15 Feb. 1985 Summer	beg. May 1992 Autumn	mid Oct. 1986 Winter	beg. Nov. 1986 Winter	mid Nov. 1986 Winter
Euchaeta spp.	11.56	2.75	1.45	9.67	14.02	9.52
Metridia gerlachei	10.08	3.50	6.32	24.68	15.98	15.69
Calanus propinquus	8.13	8.89	19.97	11.92	10.58	6.91
Calanoides acutus	35.99	21.36	6.31	7.81	5.66	7.04
Rhincalanus gigas	0.04	0.09	0.09	2.14	1.27	1.14
Microcalanus pygmaeus	0.04	0.02	0.41	5.80	8.83	4.88
Ctenocalanus citer	0.62	0.74	4.45	0.55	1.30	1.23
Stephos longipes	0.86	0.69	6.13	0.37	0.00	0.02
Oithona spp.	3.34	9.97	31.12	5.92	7.18	10.65
Oncaea spp.	18.25	41.76	17.41	27.26	31.18	37.52
Copepod nauplii	2.93	8.12	6.14	1.72	1.99	4.38

Holothuroidea

Sea cucumbers are the most important benthic group of the eastern Weddell Sea shelf. Two groups of holothurians with different distributions can be found. One group (Aspidochirotida and Elasipodida) lives on soft bottoms, whereas the second group (Dendrochirotida) lives on sand, hard bottoms and biogenic structures (Gutt 1991a). Holothurians feed on detritus, either in suspension or as deposit. They are rarely preyed upon (Gutt 1991b); however, they are prey to one amphipod species, *Bathypanoploea schellenbergi* (Coleman 1990).

As their ecology is unknown to us, we guesstimated P/B=0.1 year^{-1}, and Q/B=1.1 year^{-1} based on a gross efficiency of 9%.

Table 18.5. *Consumption and respiration rates of copepods on the eastern Weddell Sea shelf. The contribution of the various taxa was weighted by their fraction in the total copepod biomass by season as given in Table 18.2*

Month	Season	Consumption (g C m^{-2} d^{-1})	Respiration (g C m^{-2} d^{-1})	Q/B (quarter^{-1})
January	Spring	0.135	0.048	24.5
February	Summer	0.346	0.163	28.1
April	Autumn	0.930	0.442	35.1
October	Winter	0.154	0.068	22.6
Beg. November	Winter	0.223	0.084	29.4
Mid November	Winter	0.160	0.076	25.1

Asteroidea

Asteroidea are the second most important group in terms of benthic biomass (17%). The most important species on the southeastern Weddell shelf are *Odontaster validus*, *Acodontaster conspicuus* and *Notasterias armata*. The P/B ratio of *Odontaster validus* from the Ross Sea was estimated at 0.036–0.045 year^{-1} (Brey & Clarke 1993 citing Dayton *et al.* 1974 and McClintock *et al.* 1988), whereas the P/B ratio of *Acodontaster conspicuus* was slightly higher (0.069 year^{-1}) (Brey & Clarke 1993). The highest P/B ratio was found for *Perknaster fuscus*, i.e. 0.135 year^{-1} (Brey & Clarke 1993 based on Dayton *et al.* 1974). All studies were based on specimens from the Ross Sea. We used the biomass estimate of Gerdes *et al.* (1992) of 2.088 g C m^{-2} year^{-1}, along with a guesstimate for the P/B ratio of 0.08 year^{-1}.

Odontaster validus and several other sea star species are reported to be opportunistic feeders, predators and also scavengers (Jangoux 1982). *Acodontaster hodgsoni* feeds on sponges. We hence used a diet composition of 5% polychaetes, 5% brittle

Table 18.6. *Abundance of carnivorous gelatinous meso-zooplankton off the eastern Weddell Sea shelf. Values rounded to integers*

Station, Position	Depth (m)	Volume filtered (m^{-3})	Medusae n (1000 m^{-3})	Siphonophora n (1000 m^{-3})	Total Cnidaria n (1000 m^{-3})
452 (68.0° S, 12.0° W)	1000–330	167.5	1343	483	
	330–220	32.5	3704	2035	
	200–130	17.5	3369	1761	
	130–90	10.0	12800	1500	
	90–0	22.5	4040	44	
Total			25255	5823	31079
455 (68.7° S, 12.0° W)	1000–350	162.5	715	644	
	350–170	45.0	2819	1199	
	170–120	12.5	2480	1360	
	120–100	5.0	400	600	
	100–0	25.0	0	0	
Total			6414	3803	10217
461 (69.5° S, 12.0° W)	1000–250	187.5	69	293	
	250–160	22.5	222	932	
	160–100	15.0	2067	333	
	100–80	5.0	200	200	
	80–0	20.0	100	0	
Total			2658	1759	4417
467 (70.3° S, 12.0° W)	1000–330	167.5	472	846	
	330–280	12.5	2320	1520	
	280–180	25.0	2240	2520	
	180–100	20.0	0	800	
	100–0	25.0	80	0	
Total			5112	5686	10798

Table 18.7. *Production of the five dominant species of ophiuroids off Kapp Norvegia*

Species	P/B ($year^{-1}$)	B ($g\,C\,m^{-2}$)	P ($g\,C\,m^{-2}\,year^{-1}$)
Astrotoma agassizii	0.045	0.096	0.005
Ophionotus victoriae	0.180	0.384	0.069
Ophiurolepis gelida	0.156	0.197	0.031
Ophiurolepis brevirima	0.135	0.178	0.024
Ophioceres incipiens	0.212	0.466	0.099
Total		1.321	0.228

Note:
Based on Dahm (1995).

stars, 2% sponges, 2% tunicates, 1% molluscs, 1 % crustaceans and 84% detritus (including carcasses) for the model.

Ophiuroidea

Brittle stars are a very important predator group in the eastern Weddell Sea shelf ecosystem. The bulk of their biomass is composed by five species, *Ophionotus victoriae*, *Astrotoma agassizii*, *Ophiurolepis gelida*, *Ophiurolepis brevirima* and *Ophioceres incipiens*. Gerdes et al. (1992) estimated the total biomass to be 1.37 g C m^{-2}, corresponding to about 11% of the total benthic biomass. Following Dahm (1995), who combined data from box corers, underwater photography and agassiz trawls, brittle stars account for a fraction of about 20% of the total benthos in the area of interest here, corresponding to a biomass of 2.49 g C m^{-2}. As brittle stars are known to easily escape from grabs and corers, we used the higher biomass estimate in our model.

For *Ophionotus hexactis* from South Georgia, a (somatic) *P/B* ratio of 0.304 to 0.588 $year^{-1}$ was estimated (Brey & Clarke 1993, based on Morrison 1979). For the high Antarctic, productivity is expected to decrease markedly. The estimates of Dahm (1995) for the five dominant species (Table 18.7) average at P/B=0.173 $year^{-1}$. As this estimate stems directly from the area of interest here, we used it as model input.

The diet of brittle stars was found to be composed of approximately 40% benthic crustaceans, 45% detritus, 8.5% polychaetes, 5% phytoplankton, 1% bryozoans and 0.5% molluscs (Dahm 1995). Given the high fraction of detritus in their diet, but their otherwise predatory feeding habits, we used a gross efficiency of 20% to estimate their consumption. Brittle stars are preyed upon by sea stars, polychaetes and demersal fish.

Crinoidea

Gerdes *et al.* (1992) give an estimate of 0.62 g C m^{-2} for crinoids biomass. Lacking detailed knowledge on the ecology of this species group, we used a guesstimate of *P/B* of 0.1 $year^{-1}$ and a *Q/B* of 1.0 $year^{-1}$, based on a gross efficiency of 10%. The diet of this group was assumed to consist of 85% detritus, 10% phytoplankton and 5% zooplankton.

Echinoidea

Sea urchins are the smallest group among the echinoderms on the southeastern Weddell Sea shelf, with a biomass of 0.054 g C m^{-2} (Gerdes *et al.* 1992). On the southeastern Weddell Sea shelf, *Sterechinus neumayeri* is the predominant echinoid species (Brey & Gutt 1991). Its somatic *P/B* ratio was computed at 0.07 $year^{-1}$ (Brey 1991). Echinoids were assumed to be feeding on bryozoans (25%), polychaetes (6%), crustaceans (4%) and detritus (65%).

Polychaeta and other 'worms'

Polychaetes are rich in species and cover a wide range in terms of motility and trophic function (Arntz *et al.* 1994). With a biomass of 2.751 g C m^{-2}, they account for some 15% of the total benthic biomass (Gerdes *et al.* 1992). About 3% of polychaetes are polynoid (Stiller 1995), but the bulk of them are sedentary species. Another 5% are other 'worm' groups, e.g. Sipunculidae, Echiuridae, Priapulidae and Turbellaria. As information useful for this model is only available for polychaetes, we treated the other groups in the same way. A *P/B* ratio of 0.845 $year^{-1}$ was estimated for *Amphicteis gunneri* from the Kerguelen (Brey & Clarke 1993). Assuming that the *P/B* ratio will be lower in the high Antarctic, we used a *P/B* of 0.6 $year^{-1}$ for the model, and a *Q/B* ratio of 4.0 based on a gross efficiency of 15%. The diet of this group is largely unknown. As the major fraction are Sedentaria, most of the polychaetes were assumed to feed on detritus, as well as echiurids, sipunculids and priapulids. Polynoid polychaetes prey on other polychaetes, amphipods, brittle stars and sea cucumbers. *Laetmonice producta* (Aphroiditidae) feed on other polychaetes, amphipods and detritus. In turn, polychaetes and other 'worms' are preyed upon by several fish species, gastropods, and echinoderms, and some groups of predatory polychaetes. We hence used a diet composition for polychaetes of 92% detritus, 5% phytoplankton, 1% crustacea, 1% polychaetes and 1% echinoderms (holothurians and ophiuroids in equal fractions).

Benthic crustacea and chelicerata

This is a very heterogenous box, justified only by the relatively low biomass of its components (a total of less than 3% of the benthic biomass found by Gerdes *et al.* 1992). Isopods, amphipods and pantopods are by far the most important groups.

Amphipods are rich in species. They are part of the motile epibenthos, the biomass of which could have been underestimated. The most important species in terms of biomass are *Epimeria robusta*, *Gnathiphimedia mandibularis*, *Paraceradocus gibber* and *Orchomene plebs*. No production estimates are available for high-Antarctic amphipods, sub-Antarctic estimates of (somatic) *P/B* ratio range from 0.775 $year^{-1}$ to 1.2 $year^{-1}$ (Brey & Clarke 1993 based on Thurston 1968, 1970, Bone 1972 for *Bovallia gigantea*). We assumed that the productivity in the southeastern Weddell Sea is in the lower range of these sub-Antarctic values, and hence used P/B=0.7 $year^{-1}$ for the model.

P. gibber is an opportunistic predator mainly feeding on polychaetes and crustacea. Polychaetes, crustaceans and echinoderms were found in the stomachs of *E. robusta*, whereas *G. mandibularis* feeds on bryozoans (Klages & Gutt 1990). *P. gibber* and *E. robusta* were also observed to have ingested sediment. *O. plebs* is necrophagous (Rakusa-Suczczewski 1982).

Isopods occupy an intermediate level in species richness (Arntz *et al.* 1994). Arcturidae live as filter feeders on phytoplankton and detritus. Most isopods, however, are scavengers or predators and thus independent of the seasonality of primary production. Amphipods and polychaetes predominate in the diet of the opportunistic feeder *Serolis polita* (Arntz *et al.* 1994 citing Luxmoore 1985).

Decapods are scarce, being represented in the benthos only by five species of long-tailed shrimps (Decapoda: Natantia). The most common species on the shelf are *Notocrangon antarcticus* and *Chorismus antarcticus*. Shrimps have not been present in the multicorer samples of Gerdes *et al.* (1992). Their biomass was hence estimated from underwater photographs and trawl catches (Arntz & Gorny 1991, Gutt *et al.* 1994). However, at about 0.045 g C m^{-2}, biomass is rather low. *C. antarcticus* is one of the top predators of the system, preying on amphipods and isopods (Gorny 1992). Adult specimens have few predators. The somatic *P/B* ratio of *Chorismus antarcticus* has been estimated at 0.587 $year^{-1}$ (Gorny *et al.* 1993).

Little is published on the ecology of Antarctic pantopods, hence, no special attention could be paid to this group apart from adding their biomass to the overall crustacean biomass.

Q/B was estimated at 3.5 $year^{-1}$ based on a gross efficiency of 20% for this partly predatory and scavenging group. The diet was assumed to consist of 10% polychaetes, 3% crustaceans, 2% asteroids, 2% ophiuroids, 1% bryozoans and 82% detritus. Crustaceans are in general preyed upon by fish, echinoderms and other crustacean groups.

Lophophora and Cnidaria

Lophophorates (Bryozoa and Brachiopoda) and cnidarians (Hydrozoa and Anthozoa) are rich in species in the Weddell Sea and contribute slightly more than 7% to the benthic biomass. The major part of this fraction is made up by bryozoans. Unfortunately, very little is published on their ecology. Brachiopod *P/B* was estimated at 0.04 $year^{-1}$ (Brey *et al.* unpublished); we do, however, expect that bryozoan productivity is higher. Predators on bryozoans include amphipods (*Gnathiphimedia mandibularis*), sea urchins (*Sterechinus neumayeri*) and various species of gastropods.

We used the biomass of 0.749 g C m^{-2} of Gerdes *et al.* (1992), along with a *P/B* guesstimate of 0.1 $year^{-1}$ and an estimated gross efficiency of 0.1, hence, a *Q/B* of 1.0 $year^{-1}$.

Hemichordata

Gerdes *et al.* (1992) give an estimate of 0.626 g C m^{-2} for hemichordates. Lacking knowledge on their ecology, we used a guesstimate of *P/B* of 0.3 $year^{-1}$ and a *Q/B* of 2.0 $year^{-1}$, based on a gross efficiency of 15%. The diet of this group was assumed to consist entirely of detritus.

Porifera

The sponge fauna of the Weddell Sea is rich in species. Three eurybathic sponge associations were identified on the eastern Weddell Sea shelf (Barthel & Gutt 1992), however, with patchy distributions. Large sponges provide an additional substratum which is used by a variety of epizoic animals (Kunzmann 1992). Densely populated biogenic sediments are derived from sponge and bryozoan debris, forming thick mats (Dayton 1990). Production by sponges is very low, although sudden rises in population biomass may occur (Dayton 1989). Sea stars (*Odontaster meridionalis, O. validus, Perknaster fuscus antarcticus*) are major predators on sponges. Being filter feeders, the diet of sponges was assumed to consist of 95% detritus and 5% phytoplankton.

Sponges are dominant in wet mass; however, when converted to carbon, this group is no longer among the important ones. We used the biomass of 0.481 g C m^{-2} found by Gerdes *et al.* (1992) in the model, along with an assumed low *P/B* of 0.03 $year^{-1}$ and a gross efficiency of 0.05, hence a *Q/B* of 0.6 $year^{-1}$.

Tunicata

With an estimate of 0.280 g C m^{-2}, tunicates are not a major group on the shelf area under consideration. Given the limited knowledge on their ecology, we guesstimated *P/B* at 0.1 $year^{-1}$, similar to that of holothurians, and a *Q/B* at 1.0 $year^{-1}$, based on a gross efficiency of 10%. The diet of this group was assumed to consist of 90% detritus and 10% phytoplankton.

Benthic mollusca

The species richness of the benthic molluscs is not as high as that of other groups (Arntz *et al.* 1994, Brey *et al.* 1994), their biomass only accounts for slightly less than 2% of the total macrobenthic biomass. Abundant species are the bivalves *Limnopsis marionensis* (Limnopsidae), *Philobrya sublaevis, Lissarca notorcadensis* and *Adacnara nitens* (all Philobryidae), and the gastropods *Parmaphorella mawsoni* (Fissurellidae), *Margarella refulgens* (Trochidae), *Harpovoluta charcoti* (Volutidae) and *Ponthiothauma ergata* (Turridae) (Hain 1990). Production estimates of molluscs from the Weddell Sea are only available for *L. notorcadensis*, for which a (somatic) P/B of 0.305 $year^{-1}$ was computed by Brey & Hain (1992). In the absence of better information, we have used the *P/B* estimate for *L. notorcadensis*, as this is one of the dominant bivalves of the Weddell Sea.

A variety of different feeding modes is found in gastropods; however, only the feeding habits of the four most important species are considered here. *Margarella* and *Parmaphorella* are deposit feeders, *Harpavoluta* is a scavenger, *Ponthiothauma* is a predator (Hain 1990). With the filter-feeding bivalves, the diet composition of this group was assumed as 90% detritus, 5% phytoplankton and 5% polychates and other worms.

Octopods are preyed upon by Weddell Seals (Plötz 1986, Plötz *et al.* 1991a), but as their biomass is unknown and little is known on their ecology, they were omitted from consideration in the present contribution.

Pisces

Hubold (1992, table 19), based on Ekau (1990) and Hubold & Ekau (1987), gives a mean of 0.91 t km^{-2} wet mass for demersal, and 0.1 t km^{-2} wet mass for pelagic fish off Vestkapp, within our area of concern. *Chionodraco myersi* accounts for about half of the biomass of the demersal fish. The pelagic fish are dominated by *Pleuragramma antarcticum*. The biomass of this species increases by an order of magnitude towards the south, and it is assumed to migrate towards the eastern shelf during the spawning season. Assuming a wet weight/carbon ratio of 10, we used initial biomass estimates of 0.09 g C m^{-2} for demersal, and 0.01 g C m^{-2} for pelagic fish.

The *P/B* ratio of *Pleuragramma antarcticum* was estimated at 0.21 $year^{-1}$ (Hubold 1992, table 18). The total production by the demersal fish was estimated at as 0.415 g wet mass m^{-2} $year^{-1}$ (Hubold 1992, table 19). Accordingly, we used a mean *P/B* ratio of 0.46 $year^{-1}$ for the demersal fish in our model.

Demersal fish prey on a variety of benthic groups, as well as euphausiids (Schwarzbach 1988). *Chionodraco myersi* feeds exclusively on euphausiids and *Pleuragramma antarcticum*. We used a diet composition of 50% polychaetes, 10% crustaceans, 10% euphausiids, 7% brittle stars, 5% pelagic fish, 4% molluscs, 4% tunicates, 1% sea cucumbers, 1% starfish, 1% sea urchins, 1% bryozoans and 1% hemichordates in the model. Pelagic fish (mainly *Pleuragramma antarcticum*) feed on zooplankton and euphausiids (Hubold 1985, Hubold & Ekau 1990). As zooplankton was found to be by far the dominant prey, we used a diet composition of 75% zooplankton and 25% euphausiids.

Pelagic cephalopods

Cephalopods are important in the diet of top predators, i.e. fish (e.g. Kock 1992), Weddell seals (e.g. Plötz *et al.* 1991a), beaked whales (e.g. Laws 1989) and penguins (e.g. Klages 1989). *Psychroteuthis glacialis* is by far the most important squid species in the Weddell Sea. It is distributed over the shelf and slope in 300–700 m depth and feeds on euphausids (U. Piatkowski, pers. commun. 1994). However, as no biomass, production or consumption estimates exist to date for squids, they have not been included as a separate box in the model. Pütz (1994) indicates that the fraction of squids in the diet of emperor penguins has previously probably been overestimated based on the long residence time of beaks in their stomachs.

Marine mammals

Weddell seals (*Leptonychotes weddellii*) are top predators in high-Antarctic shelf waters and prefer coastal fast-ice areas throughout the year. A total adult population of 12 000 Weddell seals was estimated for a 1000 km coastline between Atka and Gould Bay (Hempel & Stonehouse 1987). With an average adult body mass of *c.* 300 kg, and a mean foraging radius of 100 km, the calculation of body (wet) mass is 0.036 t km^{-2}. Assuming a wet mass/carbon ratio of 10 led to an estimate 0.0036 g C m^{-2}, about half of the biomass estimated by Schalk *et al.* (1993, based on Franeker 1989, 1992) for the Scotia–Weddell Sea confluence zone. A tentative estimate for the average daily food intake of 10 kg per adult seal yielded a *Q/B* ratio of 12 $year^{-1}$. The *P/B* ratio was estimated at 0.07 $year^{-1}$, based on an assumed longevity of 15 years. Weddell seals forage in shelf waters and are opportunistic predators, feeding predominantly on pelagic fish (Plötz 1986) or, if not available, on bottom fish and squid (Plötz *et al.* 1991a), hence, sharing their food resource with other vertebrates of the area (e.g. Rau *et al.* 1992). As demersal fish are far more abundant on the shelf than pelagics, we used an initial diet compositon of 40% pelagic fish, 40% demersal fish and 20% squid treated as imports for the model.

Pinnipeds other than Weddell seals are summer guests in the high-Antarctic shelf areas of the eastern Weddell Sea and are neglected as they could not be quantified. The same applies to whales. To compensate for possible significant feeding by baleen whales, however, some two-thirds of the euphausiid production was assumed to be exported from the system.

Penguins

The emperor penguin (*Aptenodytes forsteri)* is also an important top predator in the coastal shelf system throughout the entire year. Sea birds other than emperor penguins are summer guests and can be neglected as they are comparatively rare in the study area and more abundant in the oceanic pack-ice areas of the northeastern Weddell Sea (Plötz *et al.* 1991b). An adult population of 120 000 emperor penguins has been counted on the 1000 km of coastline between Atka and Gould Bay (Hempel & Stonehouse 1987). Adult individuals have a body weight of about 30 kg. Their food consumption was estimtaed at about 1 kg per day, which left us with the same estimate of B=0.004 g C m^{-2} and Q/B=12 $year^{-1}$ as for the Weddell seals. The diet of emperor penguins on the eastern Weddell Sea coast mainly consists of pelagic fish, squid and krill (Klages 1989). The *P/B* ratio was also estimated at 0.07 $year^{-1}$, based on an assumed longevity of 15 years.

RESULTS

Balancing the models

While the flows between most of the groups were balanced without difficulties (Table 18.8), the meso-zooplankton, the benthic crustaceans and the pelagic fish could not be balanced with the initial estimates. The low gross efficiency of meso-zooplankton of 4% did not match the observed primary poduction, hence, it was raised to 7%, corresponding to a *Q/B* estimate of 64.3 $year^{-1}$. An average ingestion rate of 10% day^{-1} would have resulted in an annual consumption rate of 36.5 $year^{-1}$, and, with the *P/B* ratio of 4.5 $year^{-1}$ used, in a gross efficiency of 12.3%. The daily ration estimate of 17% we used is in line with results by

Table 18.8. *Assimilation coefficients entered and estimated parameters of the balanced model*

Component	Assimilation coefficient	Ecotrophic efficiency	Respiration assimilation
Phytoplankton	not applicable	0.79	not applicable
Zooplankton	0.65	0.90	0.89
Euphausiacea	0.65	0.79	0.91
Holothuroidea	0.60	0.23	0.83
Asteroidea	0.75	0.77	0.60
Ophiuroidea	0.70	0.60	0.71
Crinoidea	0.65	0.00	0.85
Echinoidea	0.70	0.16	0.93
Polychaeta etc.	0.70	0.38	0.79
Crustacea	0.70	0.95[1]	0.71
Bryozoa and Hydrozoa	0.65	0.84	0.85
Hemichordata	0.65	0.02	0.77
Porifera	0.65	0.93	0.92
Tunicata	0.65	0.79	0.85
Mollusca, benthic	0.66	0.27	0.85
Pisces, demersal	0.80	0.87	0.81
Pisces, pelagic	0.80	0.95[2]	0.81
Seals	0.80	0.00	0.99
Penguins	0.80	0.00	0.99
Detritus	not applicable	0.92	not applicable

Notes:
[1] EE=0.950 entered. The biomass of benthic crustaceans was estimated at 1.8 g C m^{-2}.
[2] EE=0.950 entered. The biomass of pelagic fish was estimated at 0.08 g C m^{-2}.

Huntley *et al.* (1991), who found zooplankton grazing rates of 5–20% body carbon mass per day (i.e. 18–73 $year^{-1}$), and by Schnack *et al.* (1985, two stations in the Bransfield Strait). Grazing rate estimates by Ward *et al.* (1995) around South Georgia (9–11% of zooplankton body carbon mass per day, corresponding to 33–40 $year^{-1}$) were slightly lower, whereas Schnack *et al.* (1985) estimated a daily grazing rate of 32% (117 $year^{-1}$) at a station in the Drake Passage.

The comparatively low biomass of benthic crustaceans could not be balanced given the high predation pressure mainly by ophiuroids. In order to meet the requirements of the predators, the ecotrophic efficiency of this group was set at 0.95 and the biomass estimated. The results yielded a biomass about five times higher than that found in the box-corers. This may, however, still be realistic given the extraordinarily high mobility of this group in the epifauna and a likely loss of biomass during the treatment of the samples on board.

The production of pelagic fish was too low to meet the consumption by seals. As seals are known to be opportunistic predators also diving to the shelf bottom, we increased the fraction of demersal fish in their diet to 75%, and lowered the fraction of pelagics to 5%. This appeared feasible in view of the biomass of the demersals exceeding that of the pelagics by one order of magnitude. We further increased the biomass of pelagic fish, based on the observations of seasonal migrations to the eastern shelf (Hubold 1992). This yielded a biomass estimate for pelagic fish (0.08 g C m^{-2}) almost as high as that of demersals with an ecotrophic efficiency of 0.95, in line with the indication by Hubold (1992) that the biomass of *Pleuragramma antarcticum* in the Weddell Sea is likely to be underestimated.

Sedimentation

Bathmann *et al.* (unpublished) estimated an export of 50 g C m^{-2} $year^{-1}$ from the pelagial in the Lazarev Sea, 25% of which can be expected to be available for consumption by the benthos, based on data in Martin *et al.* (1984) and a shelf depth of 500 m.

The highest daily sedimentation rates may be expected after sea-ice melting and throughout the short summer period. During January and February 1988, three distinct sedimentation pulses were recorded off Kapp Norvegia with daily maxima between 80 and 112 mg carbon m^{-2} day^{-1} (Bathmann *et al.* 1991). The first of these pulses (January) was characterized by high carbonate content, relatively high ^{13}C values (−24 PDB) and high content of amino acids, indicating sea-ice organisms and ice-associated diatoms to be the main source of sinking organic particles. The following pulses were caused by intense krill (*E. superba*) grazing and by a phytoplankton bloom (dominated by *Corethron criophilum*), respectively. Such sedimentation pulses must occur five to seven times throughout the summer period at the shelves of the eastern Weddell Sea as calculated from nutrient deficiencies found in late autumn (Bathmann *et al.* unpublished data).

The difference between annual phytoplankton production and its consumption by zooplankton and euphausiids in our model was 15.7 g C m^{-2} $year^{-1}$, and the difference between detritus consumption of the benthos and its contribution to the detritus amounts to 13 g C m^{-2} $year^{-1}$. Both figures correspond reasonably well with the above observations of sedimentation, and with the range given by Schalk *et al.* (1993) of 5–30 g C m^{-2} $year^{-1}$ for input of organic matter to the benthos. We expect that future studies will be able to incorporate the microbial food web, which we neglected in the present contribution, i.e. which is 'hidden' in the detritus box.

Flow diagram

The major flows in the ecosystem are illustrated in Fig. 18.2. The components of the system are arranged on the vertical axis according to their trophic level, which is defined as 1 for producers and detritus, and computed as 1 + the trophic level of their prey for consumers, weighted according to the fraction of the prey items in their diet. The size of the boxes is proportional to the biomass of the respective system components. It should, however, be kept in mind that the size of the detritus box is a rough guess only, based on the regression of Pauly *et al.* (1993). The major flows in the ecosystem occur in the plankton, as well as from detritus to the polychaetes. Other important flows comprise the crustaceans, sea cucumbers and brittle stars. Tunicates and hemichordates, as well as lophophorates and cnidarians

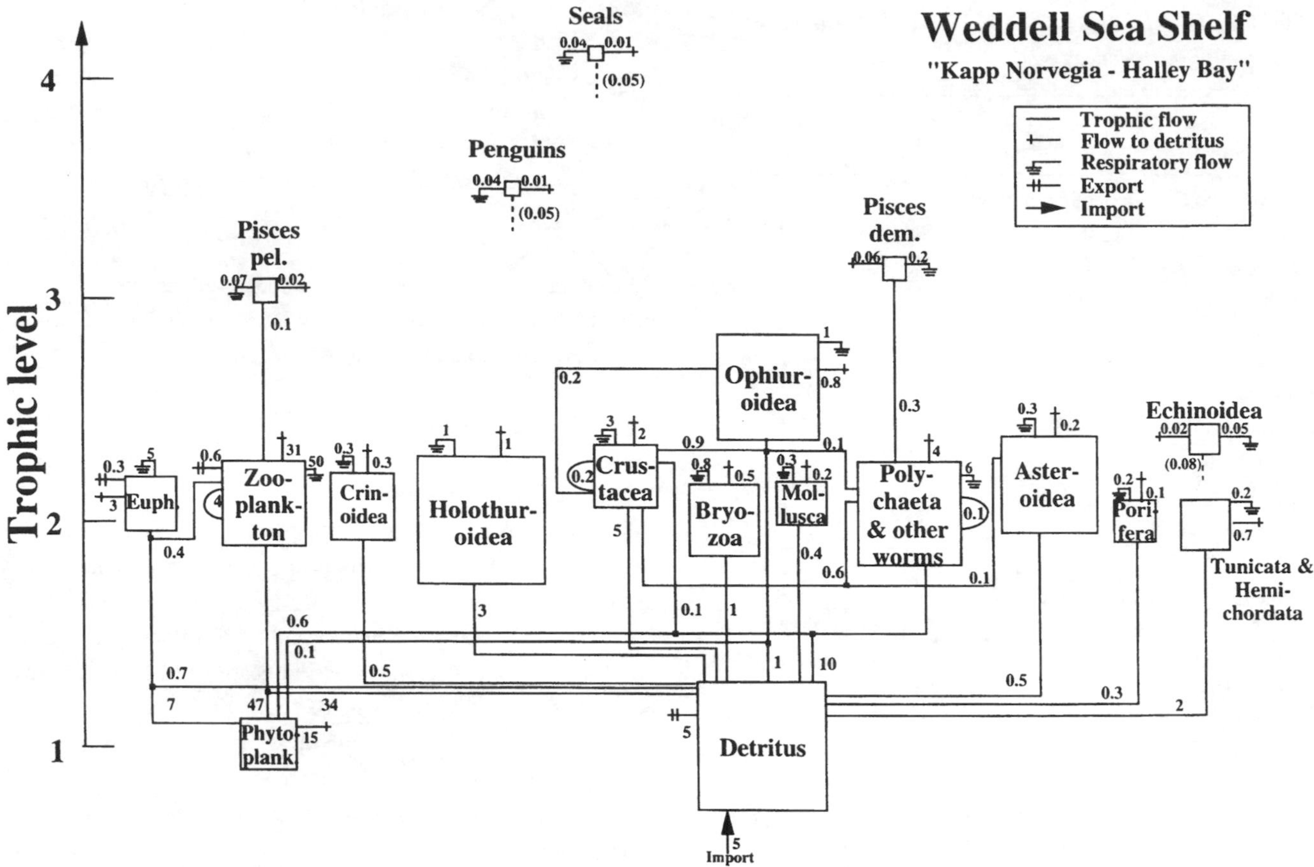

Fig. 18.2. Diagram of carbon flows of the eastern Weddell Sea shelf modelled in this study. The components of the model are arranged along the vertical axis according to their trophic level. The size of the boxes is proportional to the biomass of the components in the system if they are imagined as cubes instead of squares. Respiratory flows, backflows to the detritus, as well as exports from the system, are indicated. Flows enter boxes on the lower half, and leave them on the upper half. Flows of 1 g C m^{-2} $year^{-1}$ or more are rounded to integers, flows between 0.1 g and 1 g C m^{-2} $year^{-1}$ are rounded to one digit. Trophic flows of less than 0.1 g C were omitted for clarity.

(combined in the 'Bryozoa' box), contribute a noticeable share to the total flows as well, whereas each of the other groups by itself is of minor importance.

The flow network appears less web-like than it was actually computed, because the very small flows have been omitted for clarity's sake. About 19% of all possible (but not necessarily biologically realistic) connections between each two boxes in the system are realized in our model. The range of this connectance in the 37 models of aquatic ecosystems compared by Christensen & Pauly (1993b) was 4–61%. However, the structure of models compared and consequently the number of possible links between the boxes was rather heterogeneous.

The benthic groups are placed between trophic level 2.0 (herbivores/detritivores) and 2.5 (half herbivores/detritivores, half first-order carnivores). The top predators, i.e. fish, seals and penguins, are placed distinctly higher. The exact trophic level of the penguins remains to be determined, as they feed, in part, on squids, not included in our model. The flows to the top predators are very small in comparison with those on the lower trophic levels (0.2% of the total system througput), suggesting that they contribute little to the overall system homeostasy. In spite of the short ice-free season, roughly 20% more phytoplankton is consumed in the system (55 g C m^{-2} $year^{-1}$) than detritus (45 g C m^{-2} $year^{-1}$). The total backflows from all system components to the detritus amount to 64 g C m^{-2} $year^{-1}$, comprised mainly of zooplankton faeces (47.5%). The backflows to detritus from all benthic groups is 10 g C m^{-2} $year^{-1}$ (15.7%), and their total detritus consumption is 23 g C m^{-2} $year^{-1}$ (see also above). Respiratory flows in the system are 26.6% of the total throughput. General flow summary statistics are provided in Table 18.9.

Table 18.9. *Summary statistics for the balanced model*

Statistic	Value (g C m^{-2} y^{-1})
Sum of all consumption	121.8
Sum of all exports	5.8
Sum of all respiratory flows	69.4
Sum of all flows into detritus	63.5
Total system throughput	260.5
Sum of all production	80.9

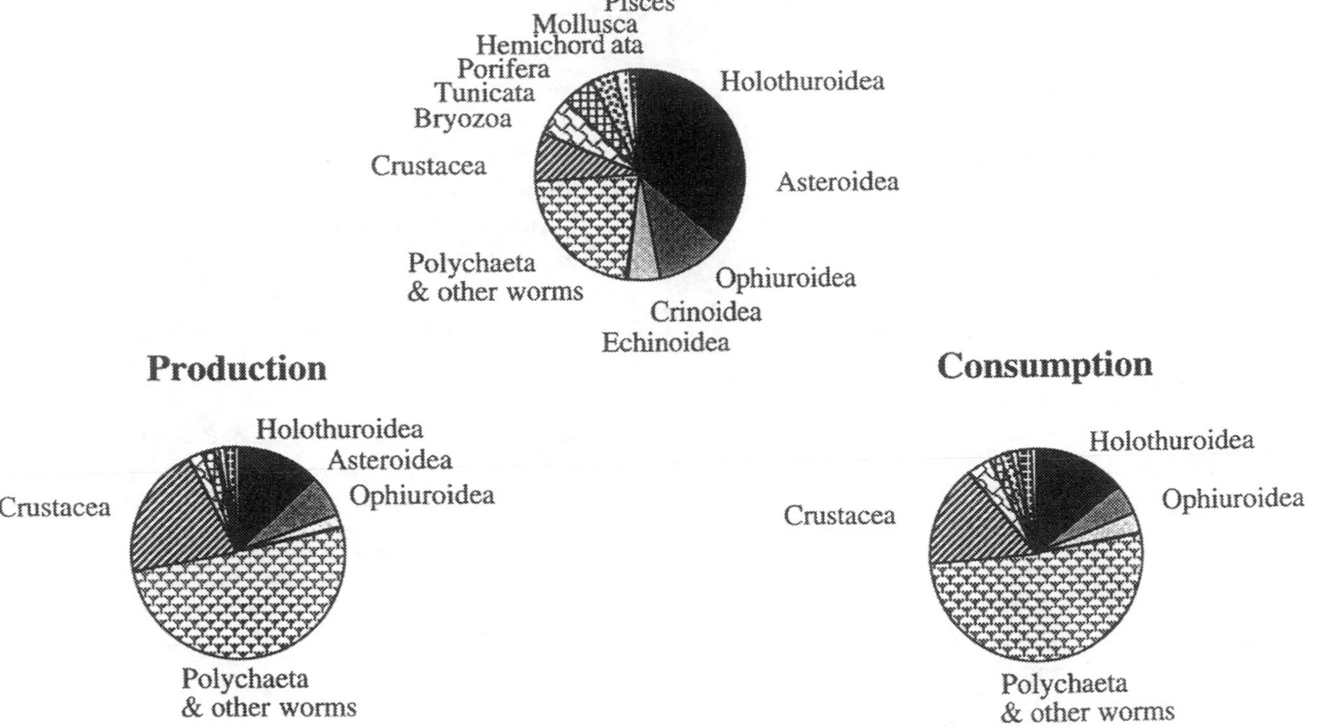

Fig. 18.3. Classification of benthic components according to biomass, production and consumption as used in this study.

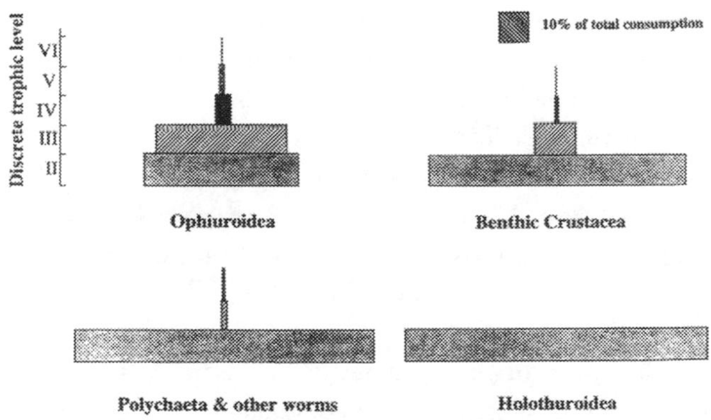

Fig. 18.4. Lindeman trophic pyramids for the four most important consumer groups in the system. The diagrams show the fraction of the total diet consumed on the discrete trophic levels II (herbivores/detritivores) to VI (4th-order carnivores).

Fig. 18.5. Transfer efficiencies of our Weddell Shelf model by discrete trophic level. The average transfer efficiencies (mean ± SD) of 37 trophic flow models ranging from ponds and freshwater lakes to open ocean systems, computed by Christensen & Pauly (1993b), are shown for comparison.

Partitioning of carbon flows in the benthos, trophic aggregation

The relative contribution of each benthic group to the total benthic biomass, production and consumption is given in Fig. 18.3. Although echinoderms contribute about half of the benthic biomass, they only account for about a quarter of the production and consumption. Polychaetes are the most important group in terms of both production and consumption, followed by benthic crustaceans, holothurians and ophiuroids. Breaking the trophic levels down into discrete trophic levels *sensu* Lindeman (1942) following Ulanowicz (in press) shows that the diet of these four major consumer groups is based on quite different trophic levels (Fig. 18.4). Sea cucumbers feed exclusively on trophic level one, whereas the brittle stars, feeding on five trophic levels simultaneously, are placed 'highest' in the benthic invertebrate food web. The trophic levels of benthic crustaceans and polychates take intermediate positions, both regarding fractionate levels as well as the numbers of discrete trophic levels they feed on.

The transfer efficiencies through the (discrete) trophic consumer levels are represented in Fig. 18.5. All transfer efficiencies of the discrete trophic levels III and above are higher than 11%.

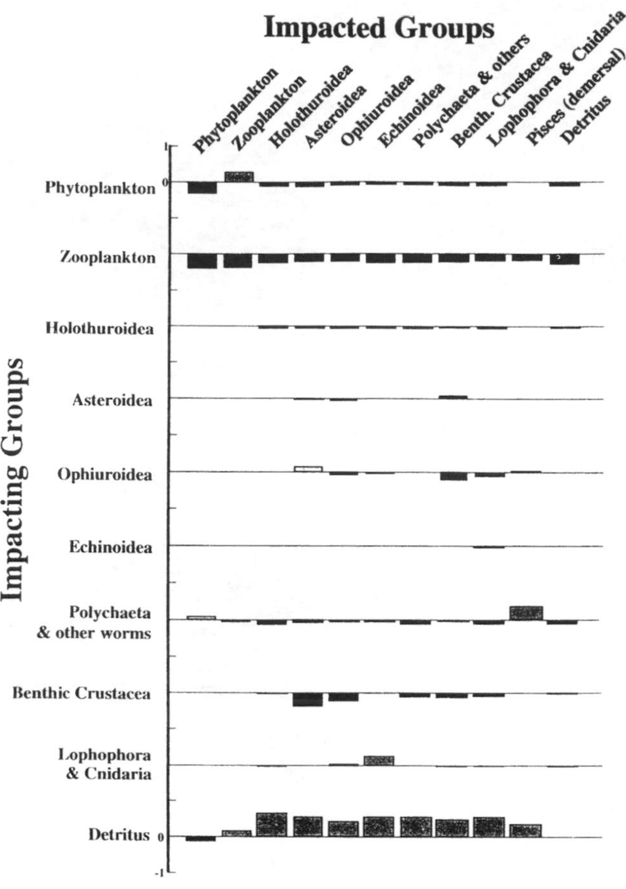

Fig. 18.6. Mixed trophic impacts on the shelf of the eastern Weddell Sea. The diagram shows the direct and indirect impacts that dominant components (vertical axis) would have on each other due to an increase in their biomass. Positive impacts are shown above the line, negative below. The impacts are relative, but comparable between groups.

The mean transfer efficiency (after arcsine-transformation) of all trophic levels is 10.3%, still distinctly higher than the corresponding value calculated from the models presented in Christensen & Pauly (1993a) with an average of 9.2%.

Given the high fraction of detritus in the diet of the majority of the components in our model, it is not surprising that detritus is an important component of the system. In spite of the higher amount of phytoplankton directly consumed in the system, it should be noted that sightly more than half of the total flows in the system (51%) are based on detritus. The respective range of the models in Christensen and Pauly (1993a) is 18–78%.

Mixed trophic impacts

The mixed trophic impacts between the various components in the model were computed following Ulanowicz & Puccia (1990), the results are presented in Fig. 18.6. The idea behind this approach is to assess the direct and the indirect impact that changes in the biomass of a given group would have on the biomass of the other groups in the system, assuming that the trophic structure remained the same. An increase in primary production positively impacts all pelagic components and the top predators, whereas the benthic community strongly depends on the presence of detritus, i.e. on the detrital input from the euphotic zone. Zooplankton is hence a major competitor for food to the benthos. Ophiuroids are important prey for asteroids and fish, but impact crustaceans and bryozoans as predators. Polychaetes are crucial prey to demersal fish, but rather act as competitors for the other groups. This group, but probably also crustaceans and holothurians, could impact the availability of detritus for the other benthic components. Crustaceans compete with asteroids and ophiuroids for food, but in part also directly prey on them. Consequently, the impact of the crustaceans on these two groups is rather strong. Bryozoans are crucial prey only to sea urchins. Detritus has a major positive impact on almost all benthic consumers, as well as on the top predators feeding on groups depending on the availability of detritus. In general, however, the cross-impact between the benthic components is lower than expected, pointing to a relatively balanced food web in the benthos on the one hand, and, due to the pronounced positive impact of phytoplankton and detritus (i.e. the lowest trophic level), to a food-limited system on the other.

Ecosystem attributes and goal functions

To assess and compare ecosystems a number of attributes and goal functions have been proposed, starting with Odum's (1969) attributes of ecosystem maturity. For community structure, he suggested that the ratio between primary production and respiration approaches unity when systems mature. Developed systems tend to sustain high biomasses, but their productivity is low. The amount of biomass supported by an unit of energy flow, as well as the average size of the organisms in the system, increase during maturation, whereas the net system production decreases. A mature system *sensu* Odum (1969) has a diverse food web with long, efficient food chains, predominantly based on detritus. Mature systems show a high degree of energy cycling in long and slow cycles. Several goal functions have been proposed in order to assess the overall system homeostasis. Our present analysis focuses on the theory of ascendency (Ulanowicz 1986). In this concept, the size of an ecosystem, computed as the total system throughput, is multiplied by a measure of its internal structure called 'factor of average mutual information', computed from the distribution of flows in the system. The product is called ascendency. The difference between ascendency and its upper boundary, the system capacity, is called the system's overhead, and represents the ability of the system to meet unexpected perturbations. Christensen (1995) showed that maturity *sensu* Odum (1969) is positively correlated with relative system overheads.

The results of our model in terms of Odum's attributes of ecosystem maturity (Odum 1969) are given in Table 18.10. With respect to community structure and carbon cycling, our system on the eastern Weddell shelf appears to be rather mature, as would be expected from its long history. However, the statistics related to food web complexity, as well as those related to overall system homeostasis, yield ambiguous results. Whereas the residence times of the organisms and flows in the system are rather high, the food web is less complex than would be expected for a mature system.

Table 18.10. *Ecosystem attributes and goal functions following Odum's (1969) theory of ecosystem development. Boxed sections indicate a rather mature system, estimates pertaining to the unshaded sections are ambiguous*

Attribute	Quantification	Weddell Sea Shelf
Community structure		
Community P/R	Prim. production/respiration	1.0
Community P/B	Prim. production/biomass	4.1 y^{-1}
Biomass supported by unit energy flow	Biomass/throughput	0.065 y^{-1}
Net community production	Prim. production – respiration	0.6 g C m^{-2} y^{-1}
Total organic matter	Total biomass	16.9 g C m^{-2}
Organism size, production, growth form	Biomass/production	0.21 y
Food web		
Food web structure, flow diversity	Connectance Index; System Omnivory Index[1]	0.19; 0.08
Length of food chains	Average path length; no. of discrete trophic levels	3.5; 8
Food web efficiency	Transfer efficiency between discrete trophic levels	10.3%
Cycling		
Mineral cycles, nutrient conservation	Finn's cycling index[2]	18.8%
Nutrient regeneration	Overhead on exports/capacity[3]	2.7%
Role of detritus in nutrient regeneration	Fraction of flows based on detritus	51.0%
Residence time	Biomass/(respiration+export)	0.22 y
System homeostasis		
Stability (resistance to external preturbations)	Overhead/system capacity[3]	0.77
Entropy	Respiration/biomass	4.1 y^{-1}
Information content of flows	Information factor; Ascendency/system capacity[4]	0.90; 0.23

Notes:
[1] following Christensen & Pauly (1992), [2] Finn (1976), [3] Ulanowicz & Norden (1990), [4] Ulanowicz (1986)

DISCUSSION

Model parameterization

While our knowledge on the major groups and their dynamics in the Antarctic Ocean has greatly increased over recent years, we are aware that many gaps still exist. The model parameterization should accordingly be regarded as preliminary in many ways. Additionally, the data availability does not yet allow the estimation of interannual variability, which is well-known to be characteristic of the area (Hempel 1985). Results from underwater photography indicate that the biomass of several groups may have been underestimated, e.g. the fraction of brittle stars appears too small relative to sea cucumbers and sea stars if box-corers are compared with underwater photos. This may be a general problem of underestimation of the role of highly motile epibenthic fauna, which, however, cannot yet be quantified, and which may need to be reassessed in the future.

Also, the role of the highly motile shrimp fauna may have been underestimated. Whereas their mean biomass values are low compared with other groups, locally, patches of *Notocrangon antarcticus* exist, with densities of 3.3 g wet mass m^{-2} (=0.3 g C m^{-2}) and maxima up to 27.5 g wet mass m^{-2} (=2.5 g C m^{-2}) (e.g. Halley Bay near shore, see Gutt *et al.* 1994). Approximately equal biomasses of amphipods and shrimps have been observed in some agassiz trawl samples. However, in a general energy flow model like the one presented here, spatial generalization is essential to obtain a more holistic view of the system.

The ecotrophic efficiency of polychaetes (and other 'worms') appears very low considering their frequent occurrence as food items. A better separation of this very heterogeneous group (into polychates and non-polychaetes), specific production estimates of the dominant subgroups, as well as more detailed food studies on the predators may help to enlighten this situation.

Using only one box of detritus, comprising carcasses, as well as particulate and dissolved organic matter, has been rather unsatisfactory. Whereas it would have been relatively straightforward to split this one box into several components, their proper trophic level, currently defined as 1, may need to be computed in a more flexible way in the future. Such reassessment would, for example, allow for the possibility of addressing in a more explicit way the common notion of scavenging isopods as benthic 'top-predators'.

Model output

Our results indicate that sea cucumbers, brittle stars, polychaetes and crustacea are the dominant benthic groups expending flow of energy. While we begin to understand the population dynamics of ophiuroids in the Weddell Sea (Dahm 1995), the population dynamics of sea cucumbers, polychaetes and benthic crustacea, as well as their food requirements need to be studied in greater detail to obtain a better picture of the energy flow in the benthic community of the southeastern Weddell Sea shelf. This is in line with Arntz *et al.* (1994), who point out that descriptive biogeographical work as well as studies on community distribution have been the leading topics during the past 6 years, followed by papers on species interactions. Population dynamics and studies on reproduction and life histories have been increasing in importance but are still only a minority of the papers. Few studies have to date been focussed on physiology and biochemistry of High Antarctic benthos. However, the role of protozoans and meiofauna in the Antarctic benthos, as well as the importance of lateral advection in structuring communities, remains to be quantified. Ecological growth efficiencies are generally expected to be higher in Antarctic waters, as the cost for maintenance metabolism is lower (Clarke 1985, 1987). However, it seems as if the surplus energy is rather put towards maintaining high body masses, with correspondingly low gross efficiency, than towards higher production (Brey & Clarke 1993). Most Antarctic benthic species seem to correspond to the traditional view of low annual growth rates, large final size and enhanced longevity. Lower growth performance has been found for Antarctic benthic invertebrates compared with temperate species (Arntz *et al.* 1994).

In spite of the gaps discussed, we are confident that our model depicts the general structure of the system, and that the flows are at the right order of magnitude. It should also be noted that, in spite of the uncertainty concerning the production rates, our estimates lie well in the range of 0.3–7.5 g C m^{-2} $year^{-1}$ for annual macrobenthic production given by Schalk *et al.* (1993), based on annual sedimentation and average benthic gross efficiency. The model can, therefore, be regarded as an improvement over previous conceptual models of Antarctic food webs (e.g. Clarke 1985). Nevertheless, the summary statistics obtained from the model should be regarded with appropriate caution, and will need further clarification from future refinement of the model. The relative simplicity of the food web expressed by the low omnivory indices could well be an artifact of the model construction. Most of the components are regarded as omnivores, pointing to an extreme limitation in food supply, similar to the situation in the deep sea (see, e.g. Gage & Tyler 1991). Nevertheless, the development time of the Weddell Sea shelf community after ice-retreat from the shelf may have been long enough to establish a well-structured community in spite of the limitations in food supply. However, given the slow turnover rates in the system, there may not have been enough time for the evolution into an overall stable ecosystem, capable of resisting external perturbations. As our present knowledge does not allow for a further breakdown even of the dominant benthic groups, further research effort is essential for our understanding of the Weddell Shelf benthic community.

ACKNOWLEDGEMENTS

This is Alfred Wegener Institute Publication No. 865.

REFERENCES

Arntz, W. E. & Gorny, M. 1991. Shrimp (Decapoda, Natantia) occurrence and distribution in the eastern Weddell Sea, Antarctica. *Polar Biology*, **11**, 169–177.

Arntz, W. E., Brey, T. & Gallardo, V. 1994. Antarctic zoobenthos. *Oceanography and Marine Biology Annual Review*, **32**, 241–304.

Arntz, W. E., Brey, T., Gerdes, D., Gorny, M., Gutt, J., Hain, S. & Klages, M. 1992. Patterns of life history and population dynamics of benthic invertebrates under the high Antarctic conditions of the Weddell Sea. In Colombo G., Ferrary, I., Ceccherelli, V. U. & Rossi, R., eds. *Marine eutrophication and population dynamics*. Fredensborg: Olsen & Olsen, 221–230.

Arrigo, K. R., Kremer, J. N. & Sullivan, C. W. 1993. A simulated Antarctic fast ice ecosystem. *Journal of Geophysical Research*, **98**, 6929–6946.

Bathmann, U. V., Fischer, G., Müller, P. J. & Gerdes, D. 1991. Short-term variations in particulate matter sedimentation off Kapp Norvegia, Weddell Sea, Antarctica: relation to water mass advection, ice cover, plankton biomass and feeding activity. *Polar Biology*, **11**, 185–195.

Barthel, D. & Gutt, J. 1992. Sponge associations in the eastern Weddell Sea. *Antarctic Science*, **4**, 137–150.

Bone, D. G. 1972. Aspects of the biology of the Antarctic amphipod *Bovallia gigantea* Pfeffer at Signy Island, South Orkney Islands. *British Antarctic Survey Bulletin*, No. 27, 105–122.

Boysen-Ennen, E., Hagen, W., Huboldt, G. & Piatkowski, U. 1991. Zooplankton in the ice- covered Weddell Sea, Antarctica. *Marine Biology*, **11**, 227–235.

Brey, T. 1991. Population dynamics of *Sterechinus antarcticus* (Echinodermata: Echinoidea) on the Weddell Sea shelf and slope, Antarctica. *Antarctic Science*, **3**, 251–256.

Brey, T. & Gutt, J. 1991. The genus *Sterechinus* (Echinodermata: Echinoidea) on the Weddell Sea shelf and slope (Antarctica): distribution, abundance and biomass. *Polar Biology*, **11**, 227–232.

Brey, T. & Hain, S. 1992. Growth, reproduction and production of *Lissarca notorcadensis* (Bivalvia: Phyllobryidae) in the Weddell Sea, Antarctica. *Marine Ecology Progress Series*, **82**, 219–226.

Brey, T. & Clarke, A. 1993. Population dynamics of marine benthic invertebrates in Antarctic and subantarcic environments: are there unique adaptations? *Antarctic Science*, **5**, 253–266.

Brey, T., Klages, M., Dahm, C., Gorny, M., Gutt, J., Hain, S., Stiller, M., Arntz, W. E., Wägele J-W. & Zimmermann, A. 1994. Antarctic benthic diversity. *Nature*, **368**, 297.

Bröckel, K. von. 1985. Primary production data from the south-eastern Weddell Sea. *Polar Biology*, **4**, 75–80.

Christensen, V. 1995. Ecosystem maturity – towards quantification. *Ecological Modelling*, **77**, 3–32.

Christensen, V. & Pauly, D. 1992. *A guide to the ECOPATH II software system (Ver. 2.1). ICLARM Software 6*. Manila: International Center for Living Aquatic Resources Management, 72 pp.

Christensen, V. & Pauly, D. eds., 1993a. *Trophic models of aquatic ecosystems. ICLARM Conference Proceedings*, No. 26, 390 pp.

Christensen, V. & Pauly, D. 1993b. Flow characteristics of aquatic ecosystems. In Christensen, V. & Pauly, D., eds. *Trophic models of aquatic ecosystems. ICLARM Conference Proceedings*, No. 26, 338–352.

Clarke, A. 1985. Energy flow in the southern ocean food web. In Siegfried, W.R., Condy, P.R. & Laws, R.M., eds. *Antarctic nutrient cycles and food webs*. Berlin: Springer-Verlag, 573–580.

Clarke, A. 1987. The adaptations of aquatic animals to low temperatures. In Grout, B. W. W. & Morris, G. J., eds. *The effects of low temperature on biological systems*. London: Edward Arnold, 315–348.

Clarke, A. & Morris, D. 1983. A preliminary energy budget of krill. *Berichte zur Polarforschung, Sonderheft*, **4**, 102–110.

Coleman, C. O. 1990. *Bathypanoploea schellenbergi* Holman & Watling 1983, an Antarctic amphipod (Crustaceae) feeding on holothuroidea. *Ophelia*, **31**, 197–205.

Dahm, C. 1995. *Populationsdynamik ausgewählter antarktischer Ophiuroiden aus den Gebieten Weddellmeer, antarktische Halbinsel und Ross-Meer*. Doctoral thesis, University of Bremen, Germany.

Dayton, P. K., Robillard, G. A., Payne, R. T. & Dayton, L. B. 1974. Biological accommodation in the benthic community at McMurdo Sound, Antarctica. *Ecological Monographs*, **44**, 105–128.

Dayton, P. K. 1989. Interdecadal variation in an Antarctic sponge and its predators from oceanographic climate shifts. *Science*, **245**, 1484–1486.

Dayton, P. K. 1990. Polar benthos. In Smith, W. O., ed. *Polar oceanography*. Part B: *Chemistry, biology and geology*. London: Academic Press, 631–685.

Drits, A. V., Paternak, A. F. & Kosobokova, K. N. 1993. Feeding, metabolism and body composition of the Antarctic copepod *Calanus propinquus* Brady with special reference to its life cycle. *Polar Biology*, **13**, 13–21.

Ekau, W. 1990. Demersal fish fauna of the Weddell Sea, Antarctica. *Antarctic Science*, **2**, 129–137.

Finn, J. T. 1976. Measures of ecosystem structure and function derived from analysis of flows. *Journal of Theoretical Biology*, **56**, 363–380.

Franeker, J. A. van. 1989. Seabirds, seals and whales. In Hempel,G., Schalk, P. H. & Smetacek, V. S., eds. *The expedition ANTARKTIS VII/3 (EPOS Leg 3) of RV 'Polarstern' in 1988/89. Berichte zur Polarforschung*, **65**, 10–13.

Franeker, J. A. van. 1992. Top predators as indicators for ecosystem events in the confluence zone and marginal ice zone of the Weddell and Scotia Seas, Antarctica, November 1988 to January 1989 (EPOS Leg 2). *Polar Biology*, **12**, 93–102.

Gage, J. D. &. Tyler, P. A. 1991. *Deep-sea biology: a natural history of organisms at the deep sea-floor*. Cambridge: Cambridge University Press, 504 pp.

Galéron, J., Herman, R. L., Arnaud, P. M., Arntz, W. E., Hain, S. & Klages, M. 1992. Macrofaunal communities on the continental shelf and slope of the southeastern Weddell Sea, Antarctica. *Polar Biology*, **12**, 283–290.

Gerdes, D., Klages, M., Arntz, W. E., Herman, R. L., Galéron, J. & Hain, S. 1992. Quantitative investigations on macrobenthos communities of the southeastern Weddell Sea shelf based on multibox corer samples. *Polar Biology*, **12**, 291–301.

Gorny, M. 1992. *Untersuchungen zur Ökologie Antarktischer Garnelen (Decapoda, Natantia)*. Doctoral thesis, University of Bremen, 129 pp.

Gorny, M., Arntz, W. E., Clarke, A. & Gore, D. J. 1992. Reproductive biology of caridean decapods from the Weddell Sea. *Polar Biology*, **12**, 111–120.

Gorny, M., Brey, T., Arntz, W. E. & Bruns, T. 1993. Growth, development and productivity of *Chorismus antarcticus* (Pfeffer) (Crustacea: Decapoda: Natantia) in the eastern Weddell Sea, Antarctica. *Journal of Experimental Marine Biology and Ecology*, **174**, 261–275.

Gutt, J. 1991a. On the distribution and ecology of holothurians in the Weddell Sea (Antarctica). *Polar Biology*, **11**, 145–155.

Gutt, J. 1991b. Are Weddell Sea holophurians typical representatives of the Antarctic benthos? *Meeresforschung*, **33**, 312–329.

Gutt, J. & Siegel ,V. 1994. Benthopelagic aggregations of krill (*Euphausia superba*) on the deeper shelf of the Weddell Sea (Antarctica). *Deep-Sea Research*, **41**, 169–178.

Gutt, J., Ekau, W. & Gorny, M. 1994. New results on the fish and shrimp fauna of the Weddell Sea and Lazarev Sea (Antarctica). *Proceedings of the NIPR (Tokyo) Symposium on Polar Biology*, **7**, 91–102.

Hagen, W. 1988. Zur Bedeutung der Lipide im antarktischen Zooplankton. *Berichte Polarforschung*, **49**, 129 pp.

Hain, S. 1990. Die beschalten benthischen Mollusken (Gastropoda & Bivalvia) des Weddellmeeres, Antarktis. *Berichte zur Polarforschung*, **70**, 181 pp.

Hempel, G. 1985. Antarctic marine food webs. In Siegfried, W.R., Condy, P.R. & Laws, R.M., eds. *Antarctic nutrient cycles and food webs*. Berlin: Springer-Verlag, 266–270.

Hempel. G. & Stonehouse, B. 1987. Aerial counts of penguins, seals and whales in the eastern Weddell Sea. In Schnack-Schiel, S., ed. *The winter-expedition of RV 'Polarstern' to the Antarctic (ANT V/1-3). Berichte zur Polarforschung*, **39**, 227–230.

Hopkins, T. L. 1985. Food web of an Antarctic midwater ecosystem. *Marine Biology*, **89**, 197–212.

Hopkins, T. L. 1987. Midwater food web in McMurdo Sound, Ross Sea, Antarctica. *Marine Biology*, **96**, 93–106.

Hubold, G. 1985. Stomach contents of the Antarctic silverfish *Pleuragramma antarcticum* from the southern and eastern Weddell Sea. *Polar Biology*, **5**, 43–48.

Hubold, G. 1992. Zur Ökologie der Fische im Weddellmeer. *Berichte zur Polarforschung*, **103**, 157 pp.

Hubold, G. & Ekau, W. 1987. Midwater fish fauna of the Weddell Sea, Antarctica. *Proceedings Fifth Congress of European Ichthyologists, Stockholm 1985*, 391–396.

Hubold, G. & Ekau, W. 1990. Feeding patterns of postlarval and juvenile notothenioids in the southern Weddell Sea (Antarctica). *Polar Biology*, **10**, 255–260.

Huntley, M., Karl, D. M., Niilher, P. & Holm-Hansen, O. 1991. Research on Antarctic coastal ecosystem rates (RACER): an interdisciplinary field experiment. *Deep-Sea Research*, **38**, 911–941.

Jangoux, M. 1982. Food and feeding mechanisms: Asteroidea. In Jangoux, M. & Lawrence, J. M., eds. *Echinoderm nutrition*. Rotterdam: A. A. Balkema, 117–184.

Klages, N. 1989. Food and feeding ecology of emperor penguins in the eastern Weddell Sea. *Polar Biology*, **9**, 385–390.

Klages, M. & Gutt, J. 1990. Comparative studies on the feeding behaviour of high Antarctic amphipods (Crustacea) in the laboratory. *Polar Biology*, **11**, 73–79.

Kock, K-H. 1992. *Antarctic fish and fisheries*. Cambridge: Cambridge University Press, 359 pp.

Kunzmann, K. 1992. *Die mit ausgewählten Schwämmen (Hexactinellida und Demospongiae) aus dem Weddellmeer, Antarktis, vergesellschaftete Fauna*. PhD thesis, University of Kiel, Germany. 121 pp.

Laws, R. M. 1989. *Antarctica: the last frontier*. London: Boxtree Ltd, 208 pp.

Lindeman, R. L. 1942. The trophic-dynamic aspect of ecology. *Ecology*, **23**, 399–418.

Luxmoore, R. A. 1985. The energy budget of a population of the Antarctic isopod *Serolis polita*. In Siegfried, W. R., Condy, P. R. & Laws, R. M., eds. *Antarctic nutrient cycles and food webs*. Berlin: Springer, 389–396.

Martin, J. H., Knauer, G. A., Karl, D. M. & Broenkow, W. W. 1984. VERTEX: carbon cycling in the northeast Pacific. *Deep-Sea Research*, **34**, 267–285.

McClintock, J. B., Pearse, J. S. & Bosch, I. 1988. Population structure and energetics of the shallow-water Antarctic sea star *Odontaster validus* in contrasting habitats. *Marine Biology*, **99**, 235–246.

Mizdalski, E. 1988. Weight and length data of zooplankton in the Weddell Sea in austral spring of 1986 (ANT V/3). *Berichte zur Polarforschung*, **55**, 1–72.

Morrison, G. W. 1979. *Studies on the ecology of the subantarctic ophiuroid* Ophionotus hexactis *(E.A. Smith)*. M Phil thesis, University of London, 213 pp.

Nöthig, E-M. 1988. Untersuchungen zur Ökologie des Phytoplanktons im südöstlichen Weddellmeer (Antarktis) im Januar/Februar 1985. *Berichte zur Polarforschung*, **53**, 1–118.

Nöthig, E-M., Bodungen, B.van & Sui, Q. 1991. Phyto- and protozooplankon biomass during austral summer in surface waters of the Weddell Sea and vicinity. *Polar Biology*, **11**, 293–304.

Odum, E. P. 1969. The strategy of ecosystem development. *Science*, **164**, 262–270.

Pagés, F., Pugh, P. R. & Gili, J-M. 1994. Macro- and megaplanktonic cnidarians collected in the eastern part of the Weddell Gyre during summer 1979. *Journal of the Marine Biological Association of the United Kingdom*, **74**, 873–894.

Pauly, D., Soriano-Bartz, M. L. & Palomares, M. L. D. 1993. Improved construction, parameterization and interpretation of steady-state ecosystem models. In Christensen, V. & Pauly, D., eds. *Trophic models of aquatic ecosystems. ICLARM Conference Proceedings*, No. 26, 1–13.

Plötz, J. 1986. Summer diet of Weddell seals (*Leptynchotes weddelli*) in the eastern and southern Weddell Sea, Antarctica. *Polar Biology*, **6**, 97–102.

Plötz, J., Ekau, W. & Reijnders, H. P. 1991a. Diet of Weddell seals *Leptonychotes weddellii* at Vestkapp, eastern Weddell Sea (Antarctica) in relation to food supply. *Marine Mammals Science*, **7**, 136–144.

Plötz, J., Weidel, H. & Bersch, M. 1991b. Winter aggregations of marine mammals and birds in the north eastern Weddell Sea pack ice. *Polar Biology*, **11**, 305–309.

Polovina, J. J. 1984. Model of a coral reef ecosystem. I. The ECOPATH model and its application to French Frigate Shoals. *Coral Reefs*, **3**, 1–11.

Polovina, J. J. 1985. An approach to estimating an ecosystem box model. *Fisheries Bulletin (US)*, **83**, 457–460.

Polovina, J. J. & Ow, M. D. 1983. ECOPATH: a user's manual and program listings. *National Marine Fisheries Service, US/NOAA, Honolulu, Administrative Report* H83-26, 46 pp.

Pütz, C. 1994. Untersuchungen zur Ernärungsökologie von Kaiserpinguinen (*Aptenodytes forsteri*) und Königspinguinen (*Aptenodytes patagonicus*). *Berichte zur Polarforschung*, **136**, 139.

Rakusa-Suczczewski, S. 1982. The biology and metabolism of *Orchomene plebs* (Hurley 1965) (Amphipoda: Gammaridea) from McMurdo Sound, Ross Sea, Antarctica. *Polar Biology*, **1**, 47–54.

Rau, G. H., Ainley, D. G., Bengtson, J. L., Torres, J. J. & Hopkins, T. L. 1992. $\partial^{15}N/\partial^{14}N$ and $\partial^{13}C/\partial^{12}C$ in Weddell Sea birds, seals and fish: implications for diet and trophic structure. *Marine Ecology Progress Series*, **84**, 1–8.

Schalk, P. H., Brey, T., Bathmann, U., Arntz, W. E., Gerdes, D., Diekmann, G. S., Ekau, W., Gradinger, R., Plötz, J., Nöthig, E-M., Schnack-Schiel, S. B., Siegel, V., Smetacek, V. S. & Franeker, J. A. van. 1993. Towards a conceptual model for the Weddell Sea, Antarctica. In Christensen, V. & Pauly, D., eds. *Trophic models of Antarctic ecosystems. ICLARM Conference Proceedings*, No. 26, 323–337.

Schnack, S. B. 1985. Feeding by *Euphausia superba* and copepod species in response to varying concentrations of phytoplankton. In Siegfried, W. R., Condy, P. R. & Laws, R. M., eds. *Antarctic nutrient cycles and food webs*. Berlin: Springer-Verlag, 311–323.

Schnack, S. B., Smetacek, V., Bodungen, B. van & Stegmann, P. 1985. Utilization of phytoplankton by copepods in Antarctic waters during spring. In Gray, J. S. & Christiansen, M. E., eds. *Marine biology of polar regions and effects of stress on marine organisms*. London: Wiley, 65–81.

Schnack-Schiel, S. B., Hagen, W. & Mizdalski, E. 1991. A seasonal comparison of *Calanoides acutus* and *Calanus propinquus* (Copepoda, Calanoida) in the eastern Weddell Sea, Antarctica. *Marine Ecology Progress Series*, **70**, 17–27.

Schwarzbach, W. 1988. Die Fischfauna des östlilchen und südlichen Weddellmeeres: geographische Verbreitung, Nahrung und trophische Stellung der Fischarten. *Berichte zur Polarforschung*, **54**, 1–93.

Siegel, V. 1986. Age and growth of Antarctic euphausiaceae (Crustacea) under natural conditions. *Marine Biology*, **96**, 483–495.

Smetacek, V. S., Scharek, R. & Nöthig, E-M. 1990. Seasonal and regional variation in the pelagial and its relationship to the life and history cycle of krill. In Kerry, K. R. & Hempel, G., eds. *Antarctic ecosystems: ecological change and conservation*. Berlin: Springer-Verlag, 103–114.

Smetacek, V. S. *et al.* 1992. Early spring phytoplankton blooms in ice platelet layers of the southern Weddell Sea, Antarctica. *Deep-Sea Research*, **39**, 153–168.

Spindler, M. & Diekmann, G. S. 1991. Das Meereis als Lebensraum. *Spektrum der Wissenschaft*, No. 2/1991, 48–57.

Stiller, M. 1995. *Untersuchungen zur Lebensweise antarktischer Polychaeten unter besonderer Berücksichtigung der Aphroditiden und Polynoiden*. Doctoral thesis, University of Bremen, Germany.

Thurston, M. H. 1968. Notes on the life history of *Bovallia gigantea* (Pfeffer) (Crustacea Amphipoda). *British Antarctic Survey Bulletin*, No. 16, 57–64.

Thurston, M. H. 1970. Growth in (Pfeffer) (Crustacea Amphipoda). In Holdgate, M. W., ed. *Antarctic ecology*, Vol. 1. London: Academic Press, 269–278.

Ulanowicz, R. E. 1986. *Growth and development: ecosystems phenomenology*. New York: Springer-Verlag, 203 pp.

Ulanowicz, R. E. Ecosystem trophic foundations: Lindeman exonerata. In Patten, B. C. & Jørgensen, S. E., eds. *Complex ecology*. Eaglewood Cliffs, New Jersey, USA: Prentice Hall. [In press].

Ulanowicz, R. E. & Platt, T. 1985 Ecosystem theory for biological oceanography. *Canadian Bulletin of Fisheries and Aquatic Science*, **213**, 260 pp.

Ulanowicz, R. E. & Norden, J. S. 1990. Symmetrical overhead in flow networks. *International Journal of Systems Science*, **21**, 429–437.

Ulanowicz, R. E. & Puccia, C.J. 1990. Mixed trophic impacts in ecosystems. *Coenoses*, **5**, 7–16.

Voronina, N. M., Menshutkin, V. V. & Tseytlin, V. B. 1980a. Production of the common species of Antarctic copepods, *Calanoides acutus*. *Oceanology*, **20**, 90–93.

Voronina, N. M., Menshutkin, V. V. & Tseytlin, V. B. 1980b. Model investigations of the annual population cycle of the abundant copepod species *Rhincalanus gigas* an estimate of its production in the Antarctic. *Oceanology*, **20**, 709–713.

Voronina, N. M., Menshutkin, V. V. & Tseytlin, V. B. 1980c. Secondary production of the Antarctic pelagic region. *Oceanology*, **20**, 714–715.

Voß, J. 1988. Zoogeographie und Gemeinschaftsanalyse des Makrozoobenthos des Weddellmeeres (Antarktis). *Berichte zur Polarforschung*, **45**, 1–145.

Ward, P., Atkinson, A., Murray, A. W. A., Wood, A. G., Williams, R. & Poulet, S. A. 1995. The summer zooplankton community at South Georgia: biomass, vertical migration and grazing. *Polar Biology*, **15**, 195–208.

19 Factors governing phytoplankton and particulate matter variation in Potter Cove, King George Island, Antarctica

IRENE SCHLOSS[1], HEINZ KLÖSER[2], GUSTAVO FERREYRA[1], ANTONIO CURTOSI[1], GUILLERMO MERCURI[1] AND EMILIO PINOLA[1]

[1]Instituto Antártico Argentino, Cerrito 1248, 1010 Buenos Aires, Argentina, [2]Alfred-Wegener- Institut für Polar- und Meeresforschung, PB 120161, 27515 Bremerhaven, Germany

ABSTRACT

Phytoplankton biomass and composition, and suspended particulate matter (organic and inorganic fractions) were related to light intensity, wind speed and direction, wind-driven currents, water temperature and salinity at two permanent stations in the inner and outer parts of Potter Cove, King George Island, South Shetland Islands, from November 1991 to February 1993. Results of both summer seasons showed that after seasonal ice retreat no phytoplankton bloom occurred. Local winds and wind-driven currents prevented biomass accumulation in early spring, before glacial melting. Later, creeks and possibly glacial subsurface drainage discharged suspended particles of mainly inorganic origin, which limited light available for photosynthesis in the cove. These effects were more evident in the inner than in the outer cove. Absolute values for particulate organic matter were relatively constant over the year, ranging between 0.35 and 3.3 mg l^{-1} and with no significant difference between the inner and outer cove. Typical Antarctic pelagic diatom species dominated the summer phytoplankton, but benthic diatoms were also abundant. Small flagellates dominated the autumn community. In winter water temperature and salinity were vertically homogeneous after ice consolidation and suspended particulate matter and chlorophyll a sharply decreased. In contrast to the summer situation, higher suspended particulate matter values occurred at greater depths and the percentage of organic matter was very high. Over the year, the carbon source for the benthic animals may change in origin and composition. Whereas in summer phytoplankton production, resuspension of benthic material and possibly input of terrigenous material constituted the main carbon source, resuspended material or secondary bacterial production would account for it during other seasons.

Key words: phytoplankton, particulate matter, coastal environment, hydrography, South Shetland Islands, Antarctica.

INTRODUCTION

While Antarctic coastal environments are generally believed to be highly productive, and indeed frequently are (Krebs 1977, 1983, Ferreyra & Tomo 1979, Clarke *et al.* 1988, Gilbert 1991a), poor phytoplankton development in coastal areas has also been recorded (Hart 1942, Warnke *et al.* 1973, Platt 1979, Hapter *et al.* 1983, Clarke *et al.* 1988, Yang 1990, Schloss *et al.* 1994). Despite low planktonic production, rich benthic animal communities may be found in these areas (Platt 1979, Klöser *et al.*

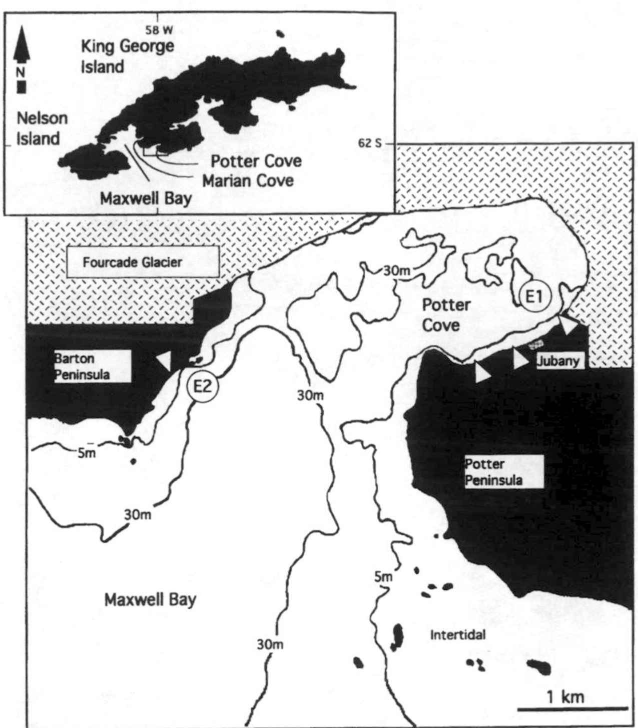

Fig. 19.1. Map of Potter Cove and its location on King George Island. White triangles show the positions of the mouths of the meltwater streams. Stations E1 and E2 are marked (corresponding to station S1 and N4, respectively: Klöser *et al.* 1994b). Also, the position of Jubany base is given, where meteorological data were collected.

1994a). Thus, an important question concerns the causes of the low microalgal biomass observed in the area (Yang 1990, Brandini 1993, Schloss *et al.* 1994). A second question is whether planktonic production provides the base for benthic animal nutrition, or whether alternative food sources exist. Some of the data analysed in this paper have been previously published (Klöser *et al.* 1993, 1994a, Schloss *et al.* 1994, Varela 1994), but are included to provide a longer temporal scale. This work is part of the multidisciplinary program RASCALS (Research on Antarctic Shallow Coastal and Littoral Systems).

MATERIAL AND METHODS

Two permanent stations were monitored from November 1991 to February 1993 at intervals of 7 to 15 days, depending on weather conditions, with the exception of November 1992, when a prolonged stormy period interrupted the investigation for 30 days. The stations, E1 and E2 (corresponding to station S1 and N4, respectively: Klöser *et al.*, 1993, 1994b) were chosen in the inner part and the mouth area of Potter Cove, King George Island, South Shetland Islands (62°14' S, 58°38' W: Fig. 19.1). Information on wind speed and direction was provided by the weather station of the National Meteorological Service of the Argentine Air Force. Water temperature and salinity were recorded with a 'M&E Ecosonde' type CTD. Underwater photon flux density was measured with a 'Li-Cor193SB' spherical quantum sensor, recording in the approximate range of 400 to 700 nm photosynthetically active radiation (PAR). Unfortunately, this was only possible during the two summer seasons. Water samples were taken with a 5 l Niskin bottle at 5 standard depths: 0, 5, 10, 20 and 30 m. Hydrographic measurements and water sampling were taken from a small rubber dinghy or, in periods with solid ice cover, through an artificial ice hole. Water samples were processed immediately in the laboratory. Chlorophyll *a* concentrations, of volumes between 0.5 and 2.0 l, were measured according to the method of Strickland & Parsons (1972). From equivalent volumes suspended particulate matter (SPM) concentration was determined following the procedure of Banse *et al.* (1963), modified after Ferreyra (1987): preweighed GF/F filters were dried at 40 °C for 24 h after filtering and then re- weighed to obtain the weight of SPM. After combustion of the particulate organic matter (POM) at 400 °C, the filters were weighed again to obtain the weight of the inorganic fraction (PIM). The weight of POM was obtained by subtraction. Selected phytoplankton and sea-ice samples, representative of different hydrographic situations, were examined using an inverted microscope according to Utermöhl (1958) and Hasle (1978a,b).

RESULTS

Wind

The wind field over Potter Cove was bimodal, and the prevalent wind directions corresponded to the E–W axis, close to the main axis of the cove. The mean wind speed for the whole study period was 8.55 m s^{-1} (Fig. 19.2a), but values exceeding 30 m s^{-1} were also measured. The median value for daily mean wind speed was around 10 m s^{-1} (Fig. 19.2b). The strongest winds were measured during winter and spring. A better resolution is available for the spring (Fig. 19.2c), where the maximum seasonal chlorophyll *a* value was recorded.

Sea-ice conditions

The start of this study coincided with the ice retreat in November 1991 (Figs. 19.3 and 19.4). The pack ice was broken in October, but due to strong west winds ice floes remained packed inside the cove. In 1992, pack-ice formation began by mid-May with thin ice floes a few centimetres thick. On May 25, the ice cover was firm. By September 16 the pack had broken again, and the ice fragments left the cove within 24 hours, due to strong northeasterly winds. Ice conditions can be highly variable: for example, in 1993 (i.e. outside the period of observation described in this paper) no persistent ice cover occurred at all.

Water temperature

In general, the water column at both permanent stations was thermally homogeneous for most of the time (Figs. 19.3a, 19.4a). Temperature rose from −1.3 °C in November 1991 to about +0.9 °C in February 1992. During this period, a shallow thermal stratification was observed in the upper 2 to 5 m, in which temperatures up to 2 °C were reached (January 17). This

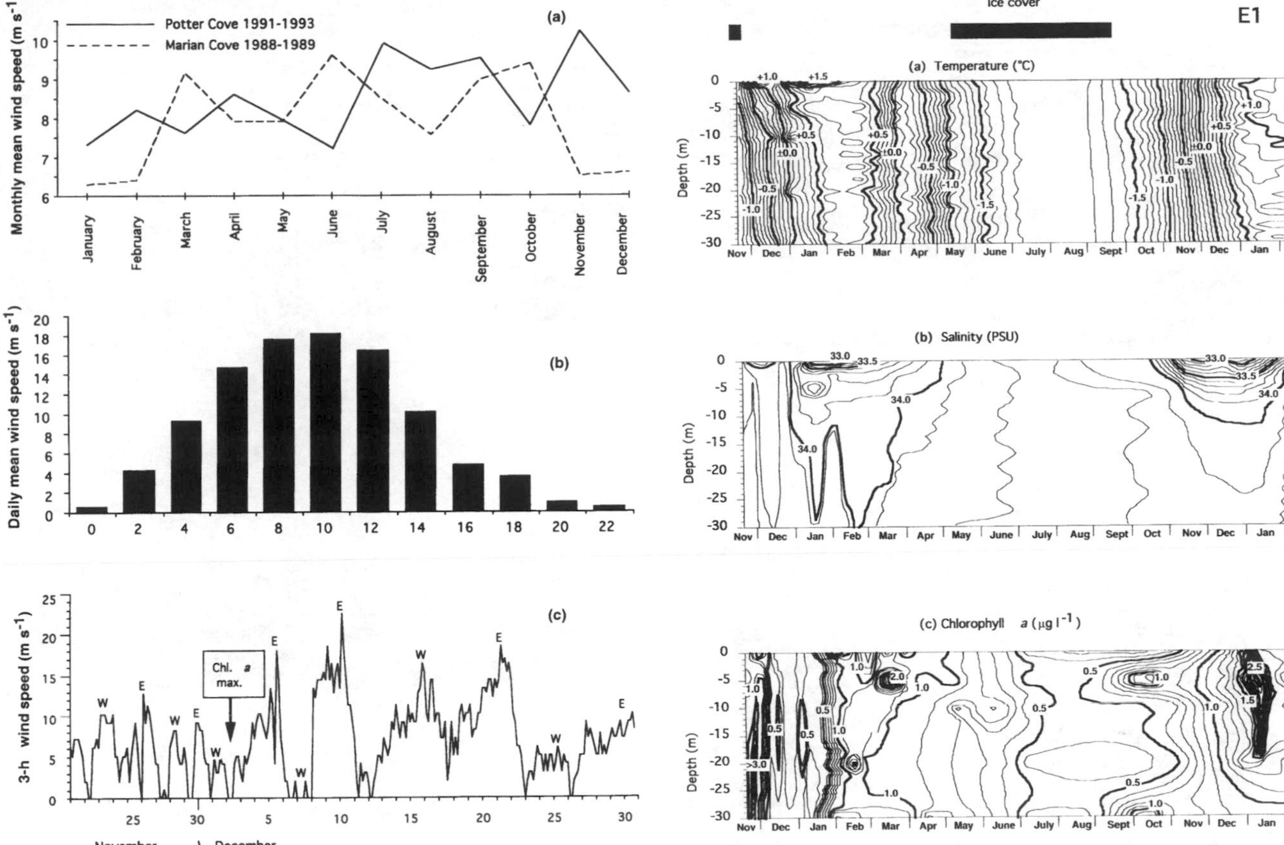

Fig. 19.2. (a) Monthly mean wind speed in Potter Cove for 1991–1993. For comparison, data from Marian Cove for 1988–89 (Chang, S.K. *et al.* 1990) are also given. (b) Frequency of wind speeds averaged over the day. Wind speed was recorded every 3 hours for 450 days. The strongest instantaneous wind speed exceeded 30 m s^{-1}. (c) Wind speed in the period between sea-ice retreat and onset of turbid meltwater discharge in November and December 1991. The date of the maximum chlorophyll *a* observed is indicated.

Fig. 19.3. Ice cover (black bars at top of figure), water temperature (a), salinity (b), and chlorophyll *a* (c) in a seasonal gradient at station E1, for a period of 15 months. Note that isolines are not substantiated by data in November 1992, when field work had to be interrupted for 30 days due to severe storms.

stratification disappeared at the end of February, and water column temperature started to decline. By the time of pack-ice consolidation, the water column temperature was around −1.2 °C, diminishing gradually further during the winter. Almost no change in water temperature occurred from June to September under the ice cover, when the lowest temperatures, close to the freezing point, were reached. After the ice break-up, temperature increased slowly and continuously, but did not reach the levels of the previous summer.

Salinity

Over most of the water column, salinity showed little variation over the year with values ranging between 34.2 PSU in winter and 33.9 PSU in February (Figs. 19.3b, 19.4b). In the upper 5 m of the water column, however, a low-salinity water mass was observed during the summer of 1991/92. In contrast to temperature gradients, the salinity gradient in the surface layer of station E1 (core value=32.13 PSU) was higher than at station E2 (core value=33.40 PSU). The salinity reduction also reached greater depths at station E1: while at E2 salinity increased to 33.9 PSU within 4.8 m, at E1 this salinity was reached at a depth of 10.8 m. From March 1992, the water column was uniform, with salinity values of 34.2 PSU. This homogeneity persisted even after the break up of the pack ice in spring. By the end of October 1992, however, strong freshwater input was evident, being again more important at E1 than at E2. This situation persisted through the summer of 1992/93.

Phytoplankton

Chlorophyll *a* concentrations were similar at both stations, with maximum and minimum values for E1 of 3.30 μg l^{-1} and 0.05 μg l^{-1}, respectively, whereas for E2 maximum and minimum values were 2.18 μg l^{-1} and 0.36 μg l^{-1} (Fig. 19.3c, 19.4c). Occasionally, higher but transient chlorophyll *a* concentrations have been observed near the beach, away from the permanent stations (6.85 μg l^{-1} in January 1991, 4.3 μg l^{-1} in January 1993), with clear evidence for resuspension of benthic material by strong northwesterly winds. Planktonic diatoms such as *Corethron criophilum*, *Eucampia antarctica*, *Odontella weissflogii* and *Thalassiosira antarctica* dominated the phytoplankton in November 1991 (*c*. 3μg chl *a* l^{-1}), but certain benthic diatoms such as *Licmophora* spp. were also abundant. Small phytoflagellates (<5 μm) were more abundant in a second peak (*c*. 1.4 μg chl *a* l^{-1}) in February 1992. A few *Corethron criophilum* cells were also present, but their physiological state, evidenced by

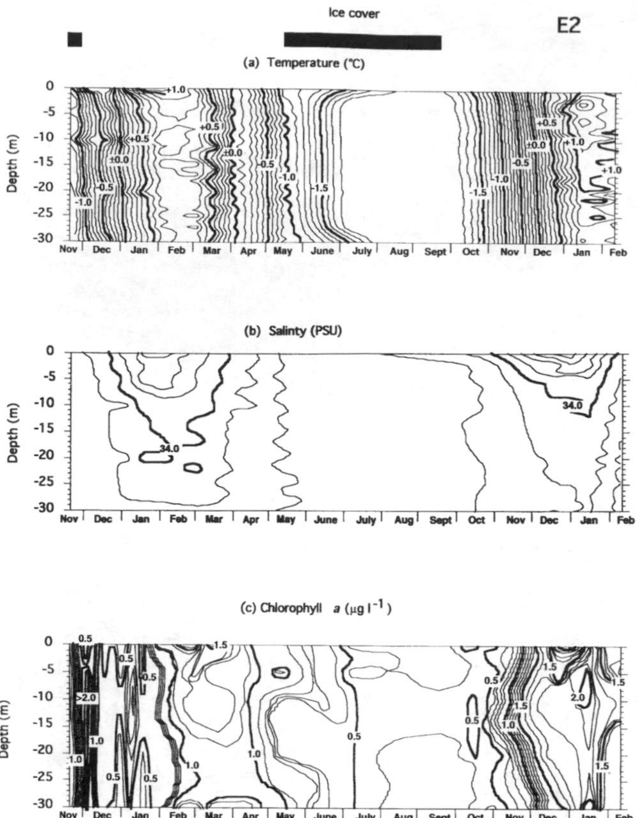

Fig. 19.4. Ice cover (black bars at top of figure), water temperature (a), salinity (b), and Chlorophyll *a* (c) in a seasonal gradient at station E2, for a period of 15 months. Note that isolines are not substantiated by data in November 1992, when field work had to be interrupted for 30 days due to severe storms.

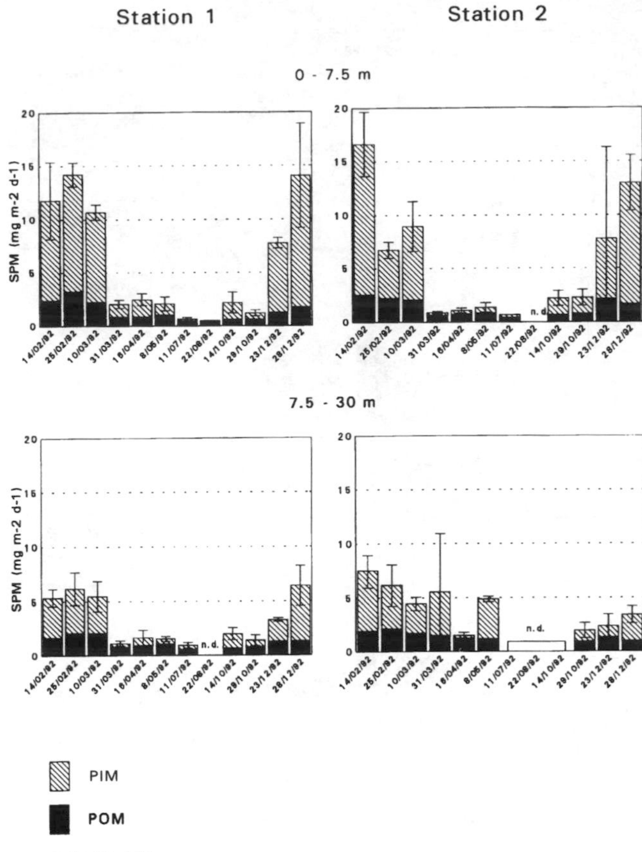

Fig. 19.5. Total suspended particulate matter (SPM) and the fractions of particulate inorganic and organic material (PIM and POM, respectively) for the upper (0–7.5 m) and lower (7.5–30 m) water column in each station. Error bars are ± 1 SD

their collapsed content, was poor. During winter, mainly benthic species of the genera *Cocconeis*, *Gyrosigma* and *Achnanthes* contributed to the meagre phytoplankton (*c*. 0.2 µg chl a l^{-1}). In the sea ice, an intense brown colour was noted shortly before the ice retreat. This was caused by diatoms of the genera *Navicula*, *Nitzschia* and *Pseudonitzschia*. Phytoplankton composition in the summer 1992/93 was very similar to that of November 1991.

Suspended particulate matter (SPM)

Averaged over the five depths analysed, total SPM amounts increased during summer with a strong predominance of the inorganic fraction (PIM). This was more evident in the upper water column (above 7.5 m, coinciding with the summer pycnocline) than below the pycnocline, and in late summer samples (February 25 and March 31) from E1 compared with those from E2 (Fig. 19.5). The proportion of the organic fraction (POM) nevertheless was quite high during the whole year, even in those samples where the total SPM was low. While vertical POM distribution was fairly uniform between depths, PIM was concentrated close to the surface in summer (Fig. 19.5). The latter was significantly correlated ($p<<0.01$) with the low surface layer salinity, exponentially at E1 and linearly at E2 (Fig. 19.6). PIM below the pycnocline at E2 was not so well correlated with the salinity field.

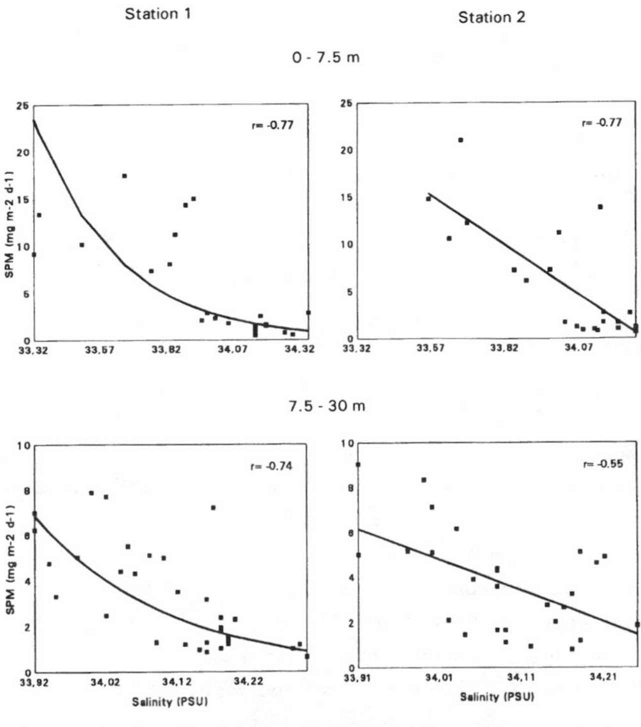

Fig. 19.6. Regression between total SPM and salinity for the upper (0–7.5 m) and lower (7.5–30 m) water column at each station. r=correlation coefficient. In all cases, $p<<0.001$.

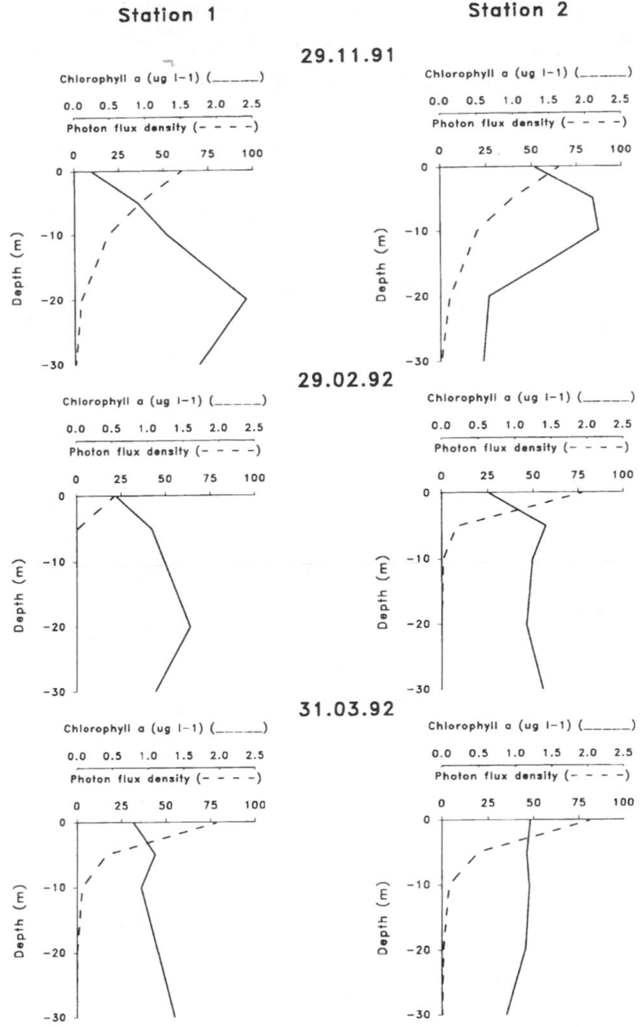

Fig. 19.7. Selected profiles of light attenuation and chlorophyll *a* concentrations at stations E1 and E2. In early summer (29 November 1991) profiles at both stations do not exhibit any unusual features. However, in late summer (25 February 1992), light is strongly reduced at E2 and almost absent at E1 below 5 m due to high SPM levels close to the surface (cf. Fig. 19.5). In late March 1992, conditions are similar to those in November 1991.

Light penetration

SPM concentrations heavily affected the penetration of light into the water column. Whereas in periods with low SPM concentrations, light levels of approximately 1% of surface irradiation could be measured at 25 to 30 m depth, light was strongly reduced at E2 and near to absent at E1 below 5 m due to the high SPM levels close to the surface (Fig. 19.7). Chlorophyll *a* was mainly concentrated around 20 m. This means that in samples from February 25 at E1, the biomass received less than 1% of the surface photon flux density. This affected the phytoplankton's physiology, as evidenced by the poor physiological state of the cells.

DISCUSSION

The hydrographical features of the data set are vertical homogeneity of the water column over the year and very shallow surface layers of low salinity during summer. The latter apparently were not strongly related to melting during the sea-ice retreat, as in other areas (Krebs 1983). Since the sea ice was rapidly driven out of the cove by the winds, no important melting could take place. Instead, melting of the glacier surface, which starts somewhat late due to its high elevation (Klöser *et al.* 1994b), is more likely to produce sufficient amounts of freshwater to establish the observed summer surface layers (Varela 1994). This meltwater modifies conditions in the cove drastically. The input of suspended sediments (Warnke *et al.* 1973, Jonasz 1983, Griffith & Anderson 1989) leads to the high concentrations of PIM observed in the upper water column, thereby shading the lower regions almost completely (Fig. 19.7; figure 9 in Klöser *et al.*, 1993). The most important effect on the phytoplankton is that, in this case, an increase in stratification does not imply good growth conditions, as is usually assumed. In turn, during spring and early summer the water column is fairly homogeneous. During this period, however, sufficient light penetrates down to at least 30 m, illuminating most of the inner cove water mass (greatest depth *c.* 50 m). Nevertheless, the cove was ice covered up to November during 1991 and from May to September 1992, while from the end of December 1991 to the end of February 1992 PIM of terrigenous origin was present. The period of favourable light conditions for phytoplankton growth appeared therefore to be very limited, to approximately 1.5 months that summer.

Even during this short period, phytoplankton growth may be prevented by physical dynamics: the hydrographic situation may change completely within one day following a change of winds (Klöser *et al.* 1994b). From current meter records in February 1992, Roese *et al.* (1993) calculated that, for the mean wind speed observed during that month (7.3 m s^{-1} for easterly winds, 6.6 m s^{-1} for westerly winds), the current speed driven by easterly winds is 13 cm s^{-1} at 15 m depth and 2 cm s^{-1} at 30 m. For currents driven by westerly winds the values are 11 cm s^{-1} and 1 cm s^{-1}, respectively. Easterly winds will drive surface water out of the cove, which will be replaced by bottom water in the inner part of the cove (Fig. 19.8). Westerly winds drive surface water into the cove. Therefore, surface outflow is blocked and meltwater will accumulate at the head of the cove (Klöser *et al.* 1994b). In contrast, the surface water imported from outside the cove will sink in the inner part of the cove and leave it at low speed close to the bottom (Fig. 19.8; Roese *et al.* 1993). Even if we do not consider that over the year stronger winds were more frequent than in February 1992, a particle will enter the cove at 9 km per day with westerly winds and leave it close to the bottom at *c.* 1 km per day. With easterly winds, a particle may enter the cove at the bottom with *c.* 1 km per day and leave it at *c.* 11 km per day at 15 m depth. Since the distance travelled would be approximately 4.5 km, a particle may stay in the cove for about four days under mean summer conditions, three days of which would be spent close to the bottom in more-or-less complete darkness. Even in spring and early summer, with more favourable light conditions, the residence time of a phytoplankton cell

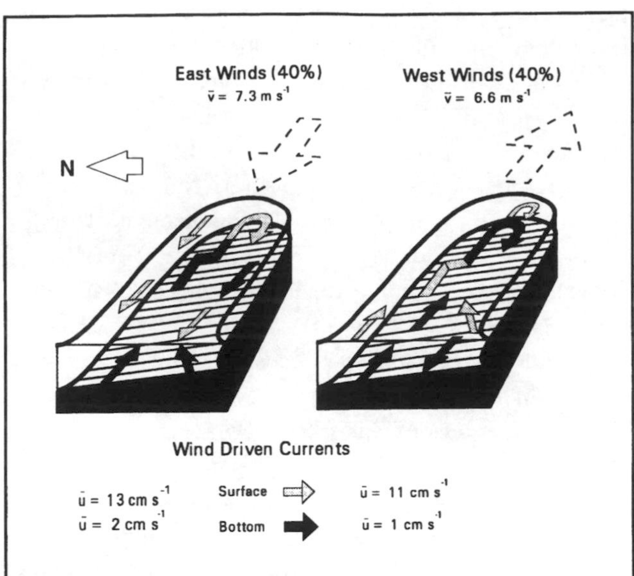

Fig. 19.8. Schematic view of currents under east and west wind regimes. Note that at the bottom a cyclonic circulation is maintained in both cases. Modified from Roese *et al.* (1993).

is not sufficiently long for substantial growth, if a doubling rate of 0.6 per day, mean growth rate for Antarctic phytoplankton (Jacques 1983), is assumed. Neither does the retention of melt-water create lasting stratification. After two or three days, a critical amount of water is reached, and the surface plume starts to flow out of the cove along its southern shore (Klöser *et al.*, 1994b).

At present, we do not know whether hydrographical processes observed in Potter Cove are representative of the other parts of Maxwell Bay. But as hydrographical data reported for the area by Hong *et al.* (1991) and Chang, K. I. *et al.* (1990), both for Marian Cove, were similar to our values, and sediment plumes from river mouths are common along the coast of King George Island, the situation in other parts of the Maxwell Bay coast might not differ too much. The circulation in the bay follows a cyclonic pattern and on this course Potter Cove is the last inlet to be passed before the water leaves Maxwell Bay again. Growth of phytoplankton would then be hindered during its travel along the coast, while grazing would continue (Brandini, 1993), leaving little phytoplankton to reach Potter Cove.

Phytoplankton has only a short time to grow between the ice retreat and the start of turbid meltwater discharge, providing that the wind is less than 4 to 5 m s^{-1}, at which velocity the wind-induced circulation is minimal (Pruszak 1980, Klöser *et al.* 1994b). This was indeed observed in November 1991, when 3.3 μg chl *a* l^{-1} (Fig. 19.2c) were accumulated at the start of a bloom of *Corethron criophilum*. These somewhat elevated pigment concentrations did not last, however, as the wind changed its direction and increased in speed.

POM was relatively constant over the study period with higher values during summer. In winter the POM maximum was generally found at a depth of 20 m at both stations. As chlorophyll *a* concentrations were negligible in the water column during this time, autotrophic organisms obviously did not contribute substantially to winter POM, in contrast to other observations (Berkman *et al.* 1986, Everitt & Thomas 1986). Rather, POM may have resulted from bacterial biomass in the water column or from resuspension of detritus (Dunbar *et al.* 1989). Although the absolute POM values are low in winter, this material may sustain the feeding of some benthic organisms, as has been described for other coastal areas (Demers *et al.* 1987). Therefore, the significance of POM as a food source during the summer could be relatively low compared with the winter, because of a higher dilution of these materials by PIM. This could be of ecological significance for filter feeders, as it would demand a higher filtration effort during summer (Ferreyra 1987).

The origin and composition of the carbon source for benthic animals may thus change with the seasons, and differ for the two sampling stations studied, E1 and E2. In summer, a part of this will be provided by planktonic production at both stations. Another part will be contributed by resuspension of benthic microalgae, as is shown by the high chlorophyll *a* concentrations during resuspension events and the abundance of benthic, mainly epiphytic diatoms such as *Licmophora* spp. in the summer plankton (Rat'kova 1993, and this study). Resuspension of benthic material has already been confirmed as an important process in Antarctic coastal waters (Krebs 1983, Everitt & Thomas 1986, Ferrario 1990, Gilbert 1991b). This process, together with the circulation pattern, could be responsible for the low correlation coefficient observed for the regression between SPM and salinity for the greater depths at E2 (Fig. 19.6). Detritus from the sediments, which originated from secondary bacterial production (Reichardt & Dieckmann 1985), and is resuspended by tidal currents at the sediment–water interface (Anderson & Mayer 1986, Hamblin *et al.* 1988), may be important in winter. Organic material of terrigenous origin must also be considered in future work as a potential food source (Platt 1979, Spolski & Ventura 1994).

ACKNOWLEDGEMENTS

We would like to thank Pablo Ljungberg, Oscar Picurú Alvarenga and Silvina Fernández Giuliano for their enthusiastic help in the field, Martin Roese and José Gallo for the utilization of some current data, and the personnel of Jubany Station for logistic support. The present work is part of the Argentinean–German cooperation program RASCALS (Research on Antarctic Shallow Coastal and Littoral Systems).

REFERENCES

Anderson, F. E. & Mayer, L. M. 1986. The interaction of tidal currents on a disturbed intertidal bottom with a resulting change in particulate matter quantity, texture and food quality. *Estuarine, Coastal and Shelf Sciences*, **22**, 19–29.

Banse, K., Falls, C. P. & Hobson, L. A. 1963. A gravimetric method for determining suspended matter in sea water using Millipore filters. *Deep-Sea Research*, **10**, 639–642.

Berkman, P. A., Marks, D. S. & Shreve, G. P. 1986. Winter sediment resuspension in McMurdo Sound, Antarctica, and its ecological implications. *Polar Biology*, **6**, 1–3.

Brandini, F. P. 1993. Phytoplankton biomass in an Antarctic coastal environment during stable water conditions – implications for the iron limitation theory. *Marine Ecology Progress Series*, **93**, 267–275.

Chang, K. I., Jun, H. K., Park, G. T. & Eo, Y. S. 1990. Oceanographic conditions of Maxwell Bay, King George Island, Antarctica (austral summer 1989). *Korean Journal of Polar Research*, **1**, 27–46.

Chang, S. K., Kim, D. Y., Lee, B. Y. & Chung, H. S. 1990. Environment around King Sejong Station, King George Island, Antarctica in 1988/89. *Korean Journal of Polar Research*, **1**, 59–65.

Clarke, A., Holmes, L. J. & White, M. G. 1988. The annual cycle of temperature, chlorophyll and major nutrients at Signy Island, South Orkney Islands, 1969–82. *British Antarctic Survey Bulletin*, No. 80, 65–86.

Demers, S., Therriault, J-C., Bourget, E. & Bah, A. 1987. Resuspension in the shallow sublittoral of a macrotidal estuarine environment: wind influence. *Limnology and Oceanography*, **32**, 271–287.

Dunbar, R. B., Leventer, A. R. & Stockton, W. L. 1989. Biogenic sedimentation in McMurdo Sound, Antarctica. *Marine Geology*, **85**, 155–179.

Everitt, D. A. & Thomas, D. P. 1986. Observations of seasonal changes in diatoms at inshore localities near Davis Station, East Antarctica. *Hydrobiologia*, **139**, 3–12.

Ferrario, M. 1990. RACER: phytoplankton of the northern Gerlache Strait. *Antarctic Journal of the United States*, **22** (5), 145–146.

Ferreyra, G. A. 1987. *Etude spatio-temporelle des transports particulaires vers une communauté benthique intertidale de substrat meuble (Estuaire maritime du Saint-Laurent)*. MSc Thesis, Université du Québec à Rimouski, 98 pp.

Ferreyra, G. A. & Tomo, A. P. 1979. Variación estacional de las diatomeas planctónicas en Puerto Paraiso-I. *Contribución del Instituto Antártico Argentino*, **264**, 149–184.

Gilbert, N. S. 1991a. Primary production by benthic microalgae in nearshore marine sediments of Signy Islands, Antarctica. *Polar Biology*, **11**, 339–346.

Gilbert, N. S. 1991b. Microphytobenthic seasonality in near-shore marine sediments at Signy Island, South Orkney Islands, Antarctica. *Estuarine, Coastal and Shelf Science*, **33**, 89–104.

Griffith, T. W. & Anderson, J. B. 1989. Climatic control of sedimentation in bays and fjords of the northern Antarctic Peninsula. *Marine Geology*, **85**, 181–204.

Hamblin, P. F., Lum, K. R., Comba, M. E. & Kaiser, K. L. E. 1988. Observations of suspended sediment flux over a tidal cycle in the region of the turbidity maximum of the upper St. Lawrence river estuary. In Aubrey, D. G. & Weishar, L., eds. *Hydrodynamics and sediment dynamics of tidal inlets. Lecture Notes on Coastal and Estuarine Studies*, Vol. 29. New York: Springer- Verlag, 244–256

Hapter, R., Wozniak, B. & Dobrowolski, K. 1983. Primary production in Ezcurra Inlet during the Antarctic summer of 1977/78. *Oceanologia*, **15**, 175–184.

Hart, T. J. 1942. Phytoplankton periodicity in Antarctic surface waters. *Discovery Reports*, **21**, 261–356.

Hasle, G. R. 1978a. Settling: the inverted-microscope method. In Sournia, A., ed. *Phytoplankton manual*. Paris:UNESCO, 88–96.

Hasle, G. R. 1978b. Using the inverted-microscope. In Sournia, A., ed. *Phytoplankton manual*. Paris:UNESCO, 191–196.

Hong, G. H., Kim, D. Y., Chung, H. & Pae, S. 1991. Coastal and inshore water interaction, mixing and primary productivity in the Bransfield Strait, Antarctica during austral summer 1989/1990. *Korean Journal of Polar Research*, **2**, 43–59.

Jacques, G. 1983. Some ecophysiological aspects of the Antarctic phytoplankton. *Polar Biology*, **2**, 27–33.

Jonasz, M. 1983. Particulate matter in the Ezcurra Inlet: concentration and size distributions. *Oceanologia*, **15**, 65–74.

Klöser, H., Ferreyra, G., Schloss, I., Mercuri, M., Laturnus, F. & Curtosi, A. 1993. Seasonal variation of algal growth conditions in sheltered Antarctic bays: the example of Potter Cove (King George Island, South Shetlands). *Journal of Marine Systems*, **4**, 289–301.

Klöser, H., Mercuri, G., Laturnus, F., Quartino, M. L. & Wiencke, C. 1994a. On the competitive balance of macroalgae at Potter Cove (King George Island, South Shetlands). *Polar Biology*, **14**, 11–16.

Klöser, H., Ferreyra, G., Schloss, I., Mercuri, M., Laturnus F. & Curtosi A. 1994b. Hydrography of Potter Cove, a small fjord-like inlet on King George Island (South Shetlands). *Estuarine, Coastal and Shelf Science*, **38**, 523–537.

Krebs, W. N. 1977. *Ecology and preservation of neritic marine diatoms, Arthur Harbor, Antarctica*. PhD thesis. University of California, 216 pp.

Krebs, W. N. 1983. Ecology of neritic marine diatoms, Arthur Harbor, Antarctica. *Micropalaeontology*, **29**, 267–297.

Platt, H. M. 1979. Ecology of the King Edward Cove, South Georgia: macro-benthos and the benthic environment. *British Antarctic Survey Bulletin*, No. 49, 231–238.

Pruszak, Z. 1980. Currents circulation in the waters of Admiralty Bay (region of Arctowsky Station on King George Island). *Polish Polar Research*, **1**, 55–74.

Rat'kova, T. N. 1993. The phytoplankton composition and distribution in the Ardley Harbour (King George Island, Antarctic). *Oceanologia*, **33**, 367–371 [In Russian with English summary].

Reichardt, W. & Dieckmann, G. 1985. Kinetics and trophic role of bacterial degradation on macro-algae in Antarctic coastal waters. In Siegfried, W. R., Condy, P. R. & Laws, R. M., eds. *Antarctic nutrient cycles and food webs*. Berlin: Springer-Verlag, 115–122.

Roese, M., Speroni, J., Drabble, M. & Pascucci, C. 1993. Medición de corrientes en Caleta Potter, Antártida. *Resúmenes de las Jornadas Nacionales de Ciencias del Mar, Puerto Madryn, 1993.*

Schloss, I., Ferreyra, G., Pinola, E., Mercuri, G. & Curtosi, A. 1994. Variación de la biomasa fitoplanctónica y del material particulado en suspensión en relación a algunos parametros ambientales en Caleta Potter. *Contribución del Instituto Antártico Argentino*, **419**, 17–30.

Spolski, B. & Ventura, A. 1994. Composición del material aportado por los chorillos – estudio preliminar. *Contribución del Instituto Antártico Argentino*, **419**, 36–41.

Strickland, J. D. H. & Parsons, T. R. 1972. A practical handbook of seawater analysis. *Bulletin of Fisheries Research Board Canada*, **167**, 1–310.

Utermöhl, H. 1958. Zur Vervollkommnung der quantitativen Phytoplankton-Methodik. *Mitteilungen des internationalen Vereins der theoretischen und angewandten Limnologie*, **9**, 1–38.

Varela, L. 1994. Estudio sobre el escurrimiento fluvial de arroyos de deshielo. *Contribución del Instituto Antártico Argentino*, **419**, 31–35.

Warnke, D., Richter, J. & Oppenheimer, C. H. 1973. Characteristics of the nearshore environment off the south coast of Anvers Island, Antarctic Peninsula. *Limnology and Oceanography*, **18**, 131–142.

Yang, J. S. 1990. Nutrients, chlorophyll *a* and primary productivity in Maxwell Bay, King George Island, Antarctica. *Korean Journal of Polar Research*, **1**, 11–18.

20 Feeding ecology of the Antarctic lamellibranch *Laternula elliptica* (Laternulidae) in Marian Cove and vicinity, King George Island, during one austral summer

IN-YOUNG AHN

Polar Research Center, Korea Ocean Research & Development Institute, Ansan P.O. Box 29, Seoul 425-600, Republic of Korea

ABSTRACT

The suspension-feeding bivalve Laternula elliptica *is one of the most common infauna in the Antarctic nearshore waters. Primary food items of* L. elliptica *and changes in faecal composition in response to a short-term variation in food quantity and quality were investigated at the South Shetland Islands. Epiphytic and epipelic diatoms predominated in ambient seawater, bottom and trap sediments and gut contents of* L. elliptica. *Chlorophyll* a *concentration in seawater was positively correlated with wind velocity, indicating that these benthic diatoms filtered by* L. elliptica *were apparently resuspended from a variety of benthic substrates by wind-generated waves. The low carbon contents of bottom surface sediment implied that benthic diatoms may be rapidly and efficiently utilized by benthic animals. In a flow-through culture system, faecal materials of* L. elliptica *were collected, and the carbon content and algal composition were compared with those of seston averaged during the faeces-collecting period. During the 19-day period, chl* a *concentration in the seawater flowing into the culture system fluctuated widely from 0.7 to 16* $\mu g\ l^{-1}$. *This variation affected the faecal materials. Organic carbon content and chl* a/*phaeopigment ratio of faecal materials, which were used as indices reflecting assimilation efficiency, increased with increase in chl* a *concentration within most of the range tested. In fact, more undigested cells were microscopically observed in faecal materials at higher chl* a *concentration, showing that* L. elliptica *was fed in excess of its need during most of the experimental period. The results of this study suggest that* L. elliptica *relies on benthic primary producers and is not food-limited even when phytoplankton production is low. The high biomass of* L. elliptica *seems to be sustained at least in part by a tight coupling between benthic primary production and secondary production during spring/summer.*

Key words: Antarctic, nearshore, *Laternula elliptica*, benthic diatoms, feeding ecology.

INTRODUCTION

The Antarctic marine benthic ecosystem is characterized by a stable and uniform physical environment, and highly seasonal but predictable food input from the overlying water column. Benthic communities show relatively high diversity and biomass comparable to the most productive areas of the world (White 1984). Benthic communities in the nearshore waters, however, are subject to variable environmental changes. Ice abrasion is a prevailing physical factor affecting the spatial distribution of benthic communities. Grounded icebergs (Shabica 1972, Gruzov 1977, Richardson & Hedgpeth 1977) and anchor ice (Dayton *et al.* 1969, Shabica 1972) effectively eliminate intertidal and shallow subtidal benthos. The ice-scoured substrates are virtually devoid of sessile epifauna (Picken 1985). Soft bottoms, however, support rich infaunal communities dominated by bivalve molluscs, polychaetes and amphipods (White 1984, Picken 1985).

The Antarctic lamellibranch *Laternula elliptica* (King and Broderip) is widely distributed in shallow waters around the Antarctic Continent and islands. It occurs in dense patches in shallow (<30 m) sheltered bays, on the order of tens of individuals per m² (Stout & Shabica 1970, Hardy 1972, Zamorano *et al.* 1986, Ahn 1993, 1994), dominating infaunal biomass (Picken 1985). This species is known to burrow deep (frequently >50 cm) into sediment (Hardy 1972). A pair of stout and highly extendable siphons are a morphological feature in adaptation to an environment mechanically affected by ice. With only the siphonal opening exposed at the surface, it feeds while the rest of the body stays deep in the sediment to avoid ice impacts (Ahn 1994).

High benthic biomass in the Antarctic nearshore waters has generally been explained as a result of slow growth, delayed maturation and longevity (White 1984). However, one of the largest bivalves in the Antarctic waters, *Laternula elliptica* grows to a shell length of approximately 90 mm in 12 or 13 years (Ralph & Maxwell 1977). Although the growth rate is lower than related species of similar ecological niche in temperate waters, it is much higher than those of other Antarctic bivalves (White 1984). In a recent experimental study, Ahn (1993) demonstrated that *L. elliptica* has high weight-specific biodeposition rates (0.26–2.17 mg dry wt g wet wt^{-1} d^{-1}), similar to those of a typical suspension-feeding bivalve, *Mytilus edulis,* in temperate water at an equivalent seston concentration. High rates of biodeposition imply high feeding activity, which in turn reflects a potential for rapid growth when food supply is sufficient. Clarke (1980, 1990) noted that growth of polar invertebrates was often markedly seasonal, suggesting that since temperatures fluctuate very little at high latitudes, food availability may be a major regulating factor.

In Antarctic open waters, phytoplankton is the primary food source for benthic fauna. In the nearshore waters, however, there are various food sources for benthic fauna other than phytoplankton. Ice-algal (Burkholder & Mandelli 1965, Andriashev 1968, Grossi *et al.* 1987) and benthic microalgal pro-

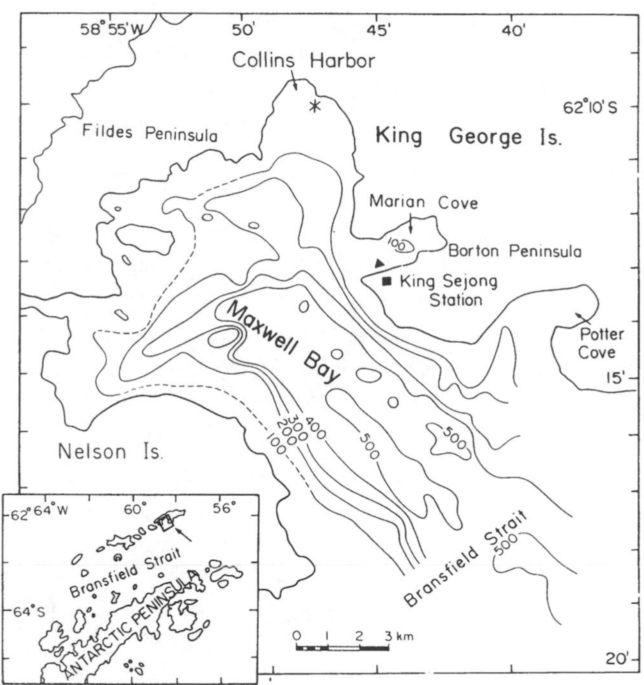

Fig. 20.1. The geographic location of Maxwell Bay. Bathymetric contours based on the information from *Atlas Hidrografico Chileno Antarctico* given by Instituto Hidrografico de la Armada, Chile (1982). *: sampling site (25–30 m depth) for *Laternula elliptica* and sediment; ▲: site for sediment traps (20 m depth).

duction (Palmisano *et al.* 1985, Dayton *et al.* 1986, Rivikin & Putt 1987, Gilbert 1991a), and horizontal advection of allochtonous food particles (Littlepage & Pearse 1962, Dayton & Oliver 1977, Fréchette & Bourget 1985, Barry & Dayton 1988, Dunbar *et al.* 1989) have been considered as other important processes providing organic matter for benthic organisms. Quantitative studies on the relative importance of these various resources, however, have rarely been conducted in Antarctic nearshore waters.

The primary food source for *Laternula elliptica*, and changes of faecal composition in response to a short-term variation in food quantity and quality were investigated during one austral summer as a first step to an understanding of the adaptive tactics of this species in an environment with variable food sources and availability. Algal composition and organic carbon content of *L. elliptica* gut content were analysed and compared with those of ambient seawater and habitat sediment. Faecal materials of *L. elliptica* were collected in a flow-through culture system and the variation of carbon content and algal compositions in the faecal materials were compared with those of suspended particulate matter (SPM) of the seawater provided as a primary food source.

MATERIALS AND METHODS

Environment

Maxwell Bay is a typical Antarctic bay with a central U-shaped deep basin. The geographic location and bathymetric contours of Maxwell Bay are shown in Fig. 20.1. Marian Cove (62°13' S,

58°45' W) and Collins Harbour (62°10' S, 58°47' W) are small embayments located at the northeastern part of the bay. Surface water freezes in winter and melts in summer. However, year-to-year variation is large, and a variable cover of drifting icebergs occurs during most of the year. Large icebergs are often grounded in shallow waters. Anchor-ice development was observed in shallow subtidal bottom during early winter (pers. commun. from S-K. Chang). These areas receive larger quantities of freshwater from land and submerged glaciers than the rest of the bay. Therefore, surface waters are colder and less saline, showing a gradual increase in temperature and salinity toward the southeastern sector and the mouth of the bay (Chang *et al.* 1990). Bottom sediments are poorly sorted and mixed with a considerable amount of terrigenous sand and gravel (Ahn & Kang 1991). During the past two seasons of the survey, a high occurrence of *Laternula elliptica* was observed at 10–20 m depth in Marian Cove (average 65 ind. m^{-2}) near the King Sejong Station (Ahn 1993) and at 25–30 m in Collins Harbour (average 86 ind. m^{-2}) (Ahn 1994).

Sample collection in field

Several dozen individuals of *Laternula elliptica* were collected at 25–30 m depth in Collins Harbour in early February 1993 by SCUBA divers. Sediment samples were taken using 28 cm^2 hand-held PVC corers (n=4) which sampled to a depth of 30 cm. After sampling, animals were transported in a few minutes to the lab at the King Sejong Station. Several of the collected animals were dissected to collect gut content, and the rest of them were kept in a flow-through culture tank for later use in the faeces-collecting experiment.

Collection of suspended particles and *L. elliptica* faecal pellets in a flow-through culture system

Seawater was pumped from 1 m depth beneath the low tide line in Marian Cove near the base (Ahn 1993). Pumped seawater was prefiltered through a 1 mm screen to remove zooplankton and other larger particles. For analysis of algal composition, seawater flowing into the system was collected every 12 hours in a 125 ml polyethylene bottle and preserved with 5% formalin. Chlorophyll *a* concentration and water temperature in the experimental tank (high-density polyethylene, 90×50×50 cm) were measured every 5 min. using a fluorometer equipped with a temperature probe (Turner Designs, 10-AU-005). To determine the amount and organic carbon content of seston, seawater was sampled every 4 to 6 hours and filtered on precombusted Whatman GF/C filters (Strickland & Parsons 1972). Salinity was measured to the nearest 0.01% by a Smart CTD every 12 hours.

Faecal materials of *Laternula elliptica* were collected from 20 animals (70–110 mm in a shell length of 86 mm and a weight of 113 g) in a stainless bath (50×30×15 cm) which was placed in the experimental tank. Between the seawater sampling, every two to three hours, animals were taken out of the tank and allowed to produce faecal materials in a container filled with 0.45 μm filtered seawater. Bulk samples of biodeposits were pipetted from the container and the chemical compositions were compared with those of suspended particles averaged over the period of faecal material collection, and weight-specific production rates at different food concentrations were determined. In addition, several batches of faeces and pseudofaeces were collected separately for microscopic analysis of algal composition.

Collection of naturally sedimenting materials using sediment traps

During the experimental period, a pair of sediment traps (14 cm in internal diameter and 75 cm in length) were deployed at 20 m water depth of Marian Cove to analyse the algal and chemical compositions of the naturally sedimenting particles, and also to estimate the downward flux of organic matter from overlying water. Traps were suspended 5 m above the bottom for 10 days. Traps were placed and removed by divers.

Microscopic analysis of cell composition and concentration

Membrane filter mounts with water-soluble embedding medium (HPMA, 2-hydroxyprophy methacrylate) (Crumpton 1987) were used for quantitative determination of cell composition, abundance and distribution. Subsamples of 50 to 100 ml were filtered onto membrane filters, and the HPMA slides were prepared for cell counting. At least 10 fields or 300 cells were enumerated using a Zeiss Axiophot microscope with a combination of light and epifluoresence microscopy (×400). A scanning electron microscope (Philips 515) was used to identify the species when identification was not possible under a light microscope. Phytoplankton were filtered, dehydrated and critical-point-dried according to the standard method of Dykstra (1992). Whole mounts coated with gold were used for the scanning electron microscopy. The number of phytoplankton cells per litre of sea water was obtained by counting numbers of cells per unit area, corrected for total area and volume used, calculated as in Kang & Fryxell (1991) from raw microscope counts. Cell volume of various diatom species was estimated from microscopic measurements of the linear dimensions of equivalent geometric shapes for the species. The estimated cell volumes were converted to cell carbon using an empirical equation (eq. 7 in Smayda 1978).

Chemical analysis of specimens

Subsamples of the collected SPM, sediment and faecal materials were taken, and chlorophyll *a* and phaeopigment contents were immediately measured by the spectrophotometric method (Strickland & Parsons 1972). Organic carbon contents were determined using a Carlo Erba NA-1500 Analyser after removing calcium carbonate with 8% sulphurous acid (Verardo *et al.* 1990).

RESULTS

***In situ* sediment, and *Laternula elliptica* gut content**

On the surface sediment of *Laternula elliptica* habitats, *Odontella litigiosa*, *Cocconeis* spp. and *Trachyneis aspera* were

Table 20.1. *The relative abundances (RA) of the most common diatoms by cell carbon in the top 1 cm layer of the sediment cores taken at 30 m depth in Collins Harbour. Each value represents the relative contribution of each species to total cell carbon*

Species	RA (%)
Odontella litigiosa	46
Cocconeis spp.	19
Trachyneis aspera	14
Pinnularia quatratareoides	5
Synedra spp.	3
Triceratium sp.	3
Amphora ovalis	2
Licmophora spp.	2
Pleurosigma spp.	1
Charcotia actiochilus	1
Thalassiosira antarctica	1
Pseudogomphonema kamtschaticum	1
Achnanthes brevipes var. *angustata*	1
Fragilariopsis kerguelensis	+
Fragilaria sp.	+
Navicula spp.	+
Melosira sol	+
Coscinodiscus sp.	+

Note:
+ = trace

Table 20.2. *The relative abundance (RA) of diatoms in the gut content of* Laternula elliptica *sampled from 25–30 m depth in Collins Harbour*

Species	RA(%)
Cocconeis spp.	30
Licmorphora spp.	18
Trachyneis aspera	8
Coscinodiscus spp.	8
Corethron criophilum	6
Pseudogomphonema kamtschaticum	4
Thalassiosira antarctica	4
Melosira sol	4
Pinnularia quadratareoides	4
Nitzschia sp.	4
Odontella litigiosa	2
Pinnularia sp.	2
Triceratium sp.	2
Pleurosigma sp.	2
Navicula sp.	2

pre-dominant (Table 20.1). These three diatoms species combined constituted *c.* 80% of total diatom biomass. In the gut content of *L. elliptica*, *Cocconeis* spp. were the most common. *Licmophora* spp., *Trachyneis aspera* and *Coscinodiscus* spp. were the next common (Table 20.2). Organic carbon content of the sediment cores varied between 0.2% and 0.8% with no particular trend along the vertical axis (Fig. 20.2).

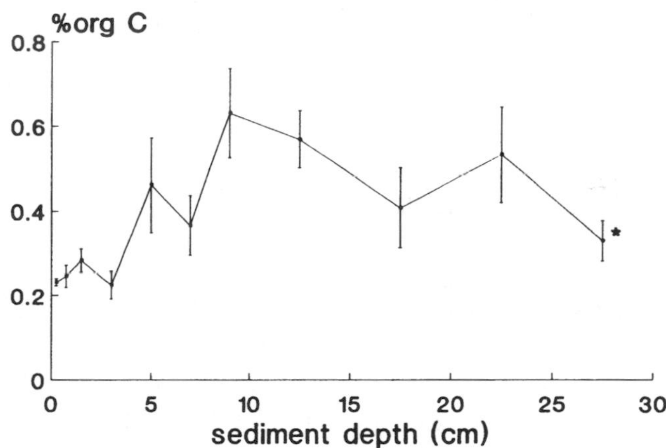

Fig. 20.2. Organic carbon contents in the sediment cores taken at *Laternula elliptica* sampling site at Collins Harbour. $n=5$ except * ($n=2$).

Experimental condition in the flow-through culture tank

During the 19-day experimental period, seawater temperature in the tank ranged from 0 to 2.2 °C (mean = 1.1 °C), and salinity ranged from 33.4 to 35.4‰ (mean = 34.2‰). Seston fluctuated between 2.5 and 20.2 mg l^{-1} dry wt with the mean of 6.1 mg l^{-1} dry wt and chl *a* concentration 0.7 to 16 μg l^{-1} (mean = 2.37 μg l^{-1}) (Fig. 20.3).

Suspended particles, and faecal materials in the flow-through culture tank

Total diatom cell concentration fluctuated from 0.86 to 153 μgC l^{-1} (mean = 25 μg C l^{-1}) (Fig. 20.4), showing a good correlation with the chl *a* values measured using a fluorometer (a product–moment correlation coefficient, $r=0.82$, $p<0.01$) (Fig. 20.5). Benthic forms, *Synedra* spp. (63±26%), *Licmophora* spp. (28±22%) and *Navicula* spp.(5.3±6.8%), were the pre- dominant microalgal species in seawater. Planktonic forms, such as *Thalassiosira antarctica*, constituted only a minor fraction (<4%) (Table 20.3).

Synedra spp. and *Licmophora* spp. were also the dominant species in *Laternula elliptica* faeces and pseudofaeces collected from the experiment tank (Table 20.4). Organic carbon contents and chl *a*/phaeopigments ratios of faecal materials increased with increase in chl *a* concentration in seawater within the range of 0.8 to <4 chl *a* l^{-1} with a tendency of a slight decrease at >10 μg chl *a* l^{-1} (Fig. 20.6). Weight-specific biodeposition rates ranged from 0.14 to 1.2 mg g dry tissue^{-1} h^{-1}, and did not show any relationship with chl *a* concentration in seawater within the range measured during the 2-week period (Fig. 20.7).

Trap sediment

The majority of the trap sediment particles was mineral and only a small fraction was organic detritus and benthic diatoms.

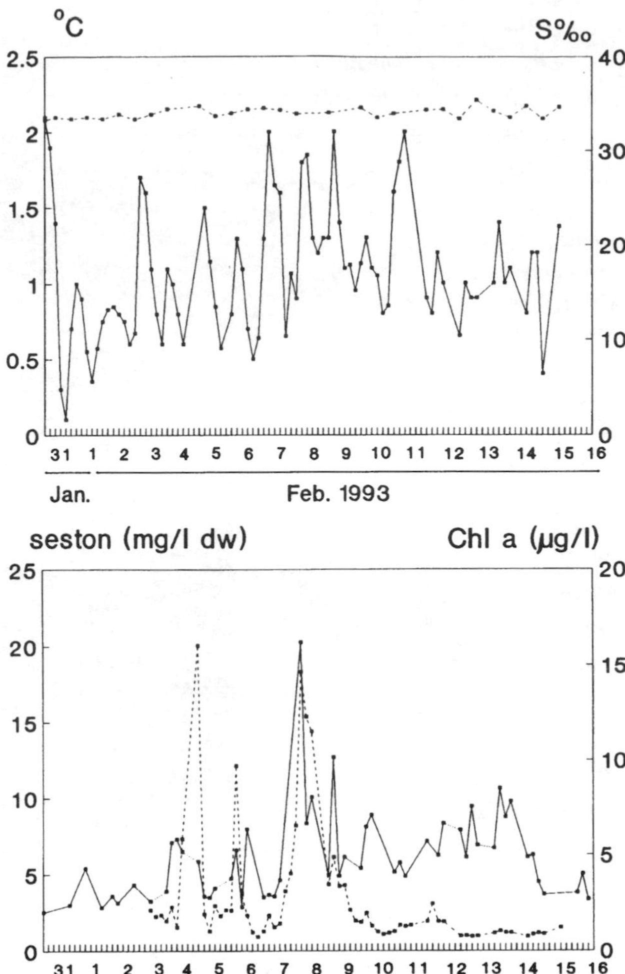

Fig. 20.3. Sea water temperature, salinity, and concentrations of seston and chlorophyll *a* in the flow-through culture system during the faecal matter collecting experiment at King Sejong Station.

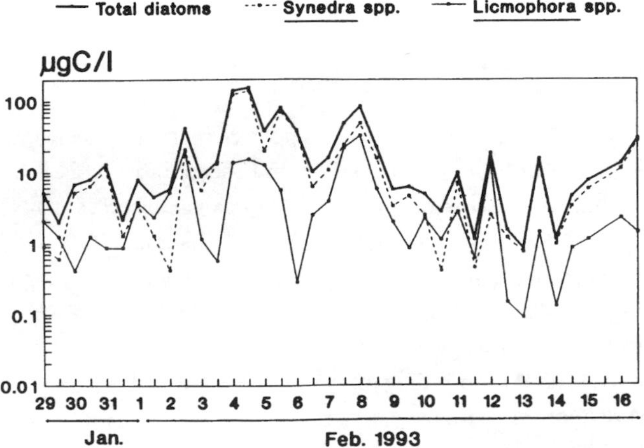

Fig. 20.4. Short-term variation of concentrations of total diatoms and of the most dominant diatoms in the flow-through culture system. Seawater was pumped from 1 m depth in Marian Cove and prefiltered onto a 1 mm screen to remove zooplankton and other large particles prior to use.

Table 20.3. *The relative abundances (RA) of the most common diatom species as a percentage of total cell carbon in the flow-through culture tank at King Sejong Station. Seawater was pumped from a 1 m depth in Marian Cove from 28 January to 16 February 1993. Seawater was sampled every 12 hours to collect data for diatom composition. Values are the means of the pooled data*

Species	RA (%)
Synedra spp.	63
Licmophora spp.	28
Navicula spp.	5
Thalassiosira antarctica	<4
Achnanthes brevipes var. *angustata*	+
Cocconeis spp.	+
Pseudogomphonema kamtschaticum	+
Cylindrotheca closterium	+

Note:
+ = <1%

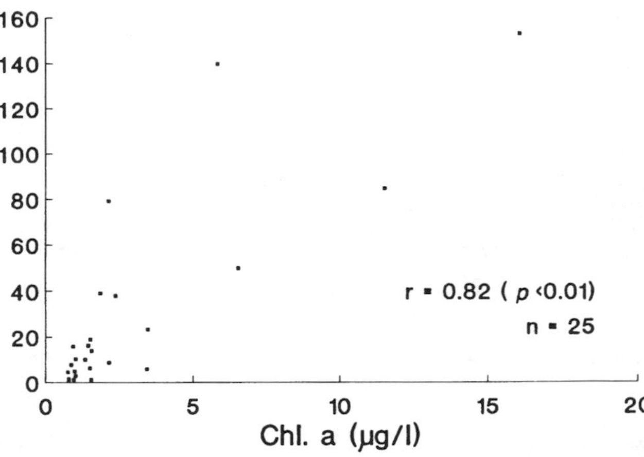

Fig. 20.5. A positive correlation between the carbon biomass of diatoms estimated from cell counts and the chl *a* values obtained from a fluorometer. *r*: product–moment correlation coefficient (Sokal & Rohlf, 1981).

The diatoms *Licmophora* spp. and *Synedra* spp. predominated in the trap sediment (Table 20.5). Organic carbon content of trap sediment was *c.* 0.52%. Organic carbon flux estimated from the deposits in the traps was 165 mg C m^{-2} d^{-1}.

DISCUSSION

Benthic diatoms predominated in SPM, bottom and trap sediments, and in the gut content and faecal materials of *Laternula elliptica*. The epiphytic diatoms *Synedra* spp. and *Licmophora* spp. predominated in seawater flowing into the culture system, which was pumped from 1 m depth in Marian Cove. In a companion study conducted during the same period in Marian

Table 20.4. *The relative abundances (means (standard deviations) in %) of the most common diatom species in the faeces and pseudofaeces collected in the flow-through culture tank on a three- or four-day interval between 3 and 14 February 1993. n=4*

Species	Faeces	Pseudofaeces
Synedra spp.	42.0(17)	40.0(7)
Licmophora spp.	46.0(18)	52.0(4)
Pseudogomphonema kamtschaticum	4.0(0.6)	4.0(2)
Cocconeis spp.	4.0(2.7)	2.0(0.9)
Thalassiosira antarctica	0.3(0.7)	1.4(1.6)
Achnanthes brevipes var. *angustata*	1.5(0.6)	0.9(0.7)
Navicula spp.	0.2(0.1)	0.2(0.4)

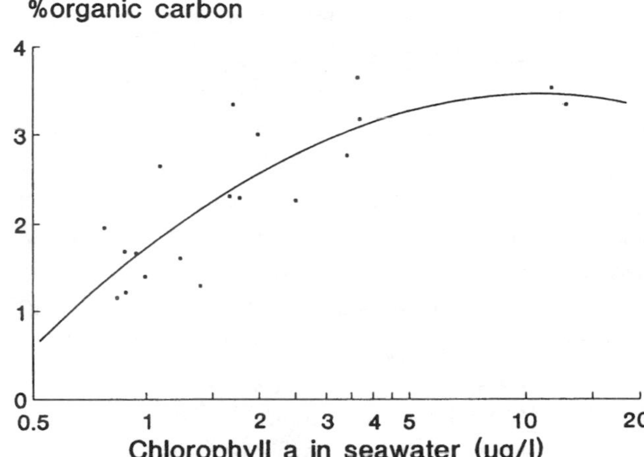

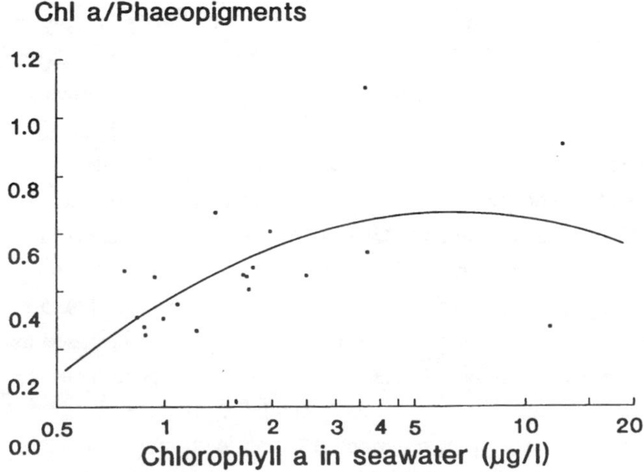

Fig. 20.6. Changes in the chemical composition in the faecal materials of *Laternula elliptica* at different algal concentrations in the flow-through culture tank. Regression lines fitted by polynomial regression analysis (Sokal & Rohlf, 1981). Regression lines: $Y_1=1.738+3.333 \log X - 1.603 (\log X)^2$, *Fs*: 17.15 with df 2 and 16, $p<0.001$; Percentage of organic carbon, chl *a*/phaeopigment ratios, $Y_2=0.366+0.749 \log X - 0.461(\log X)^2$, *Fs*: 3.712 with df 2 and 16, $0.025<p<0.05$, where X is chl *a* concentration in the seawater, and Y_1 and Y_2 are % organic carbon and chl *a*/phaeopigment ratios.

Table 20.5. *The relative abundance (RA) of the most common diatom species in sediment traps. Traps were deployed for 10 days (6 to 16 February 1993) at 5 m off the bottom at 20 m depth in Marian Cove near King Sejong Station*

Species	RA (%)
Licmophora spp.	51
Synedra kerguelensis	11
Fragilaria sp.	8
Pseudogomphonema kamtschaticum	8
Navicula spp.	5
Corethron criophilum	5
Pleurosigma spp.	3
Achnanthes brevipes var. *angustata*	3
Thalassiosira antarctica	3
Coscinodiscus sp.	3
Charcotia actiochilus	+
Cylindrotheca closterium	+
Cocconeis spp.	+
Pinnularia quatratareoides	+
Trachyneis aspera	+
Odontella litigiosa	+
Amphora ovalis	+
Triceratium sp.	+
Fragilariopsis kerguelensis	+

Note:
+=trace

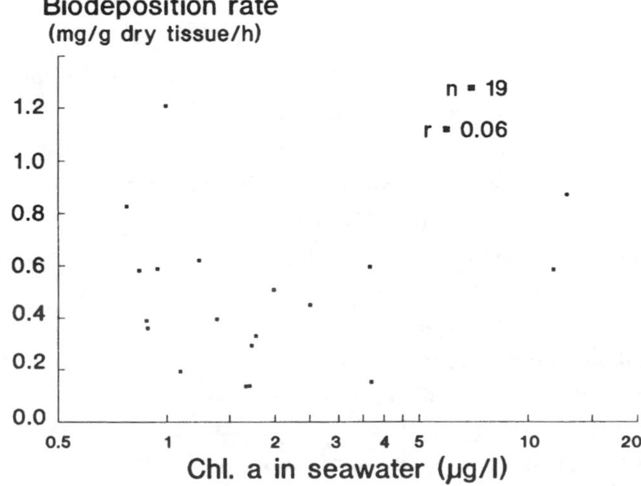

Fig. 20.7. Weight-specific biodeposition rates of *Laternula elliptica* at different food concentrations in the flow-through culture tank. Tissue dry weight: *c.* 8.5% of total wet weight including shell, *r*: product–moment correlation coefficient

Cove and vicinity, Ahn *et al.* (in press) reported that the epiphytic diatoms *Synedra* spp., *Achnanthes brevipes* var. *angustata* and *Licmophora* spp. dominated the water column algal populations down to 30 m depth and that densities of diatoms were several times higher in nearshore sites (2.4–14 µg C l^{-1})

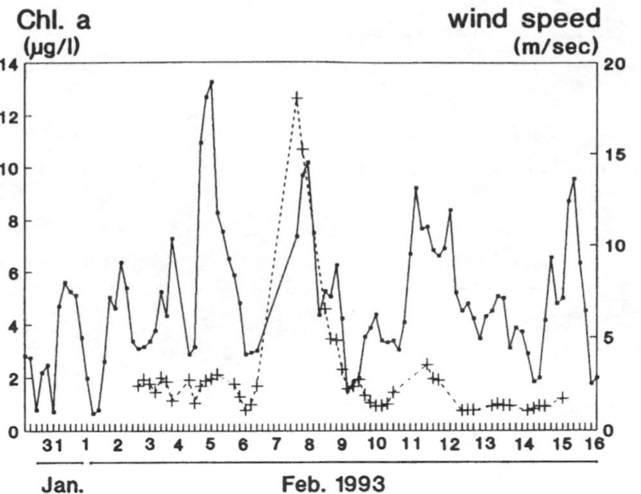

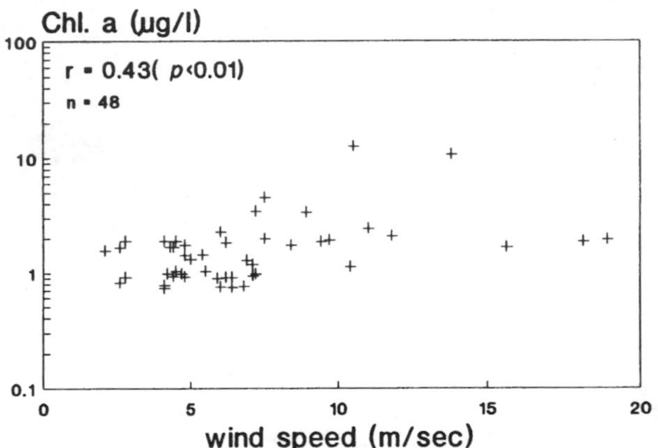

Fig.20.8. A positive correlation between chl *a* concentration (•) in the seawater and wind speed (+) during the flow-through culture experiment. Both chl *a* and wind are means of 4 h integrated values.

than offshore sites (1.2–3.2 µg C l^{-1}), and the proportion of the benthic diatoms sharply decreased from the nearshore (>98%) toward the bay mouth (20%). This suggests that benthic diatoms may be important as a food source for benthic fauna, particularly in nearshore waters. Many of the dominant epiphytic diatoms were observed attached to a variety of macroalgae in the intertidal and shallow subtidal benthic substrates.

Seawater chl *a* concentrations in the flow-through culture tank were positively correlated with wind speed (Fig. 20.8), indicating that these benthic diatoms were detached from benthic substrates by wind-generated waves. Resuspension of benthic microalgae was considered an important process for providing food to benthic herbivores (Everitt & Thomas 1986, Demers *et al.* 1987, Berkman 1991, Gilbert 1991a). Since the majority of microalgal cells sedimented in the traps were benthic diatoms, it is suggested that benthic diatoms are constantly resuspended and sedimented again.

In particular, *Synedra* spp. were found to attach only on a filamentous red alga, *Bangia* sp., which was forming a dense brown algal mat in the intertidal zone. It appears that *Synedra* spp. were washed out of the intertidal zone and transported to the subtidal waters by tides or wind-generated waves, and that as a result the proportion of *Synedra* spp. was higher in the flow-through seawater, which was pumped from 1 m depth of Marian Cove, than at the deeper water. Krebs (1983) also reported that *Synedra kerguelensis*, *Licmophora decora* and *Achnanthes brevipes* var. *angustata* were the most abundant diatoms in the intertidal zone of Arthur Harbour. Diatom composition on the surface sediment taken from the *Laternula elliptica* habitat of Collins Harbour was considerably different from those of the water column. These changes may be due to differences in substrate type and depth, and to differences in buoyancy related to cell morphologies. In the gut content of *L. elliptica* that had fed in the field the dominant diatom species from both the water column and the bottom sediment occurred together, implying that *L. elliptica* was filtering benthic diatoms detached from a variety of benthic substrates.

During the culture experiment, diatom composition changed slightly. The epiphytic diatoms *Synedra* spp. and *Licmophora* spp. predominated throughout the 19-day period, averaging 63% and 28% respectively of total diatom carbon biomass (Table 20.3). The composition of diatoms in the faeces and pseudofaeces reflects that of the suspended particles in seawater. However, there was a tendency for *Laternula elliptica* to reject *Licmophora* spp. The proportion of *Licmophora* spp. was higher in both faeces and pseudofaeces than in the algal populations in seawater. The cell volume of the large-sized species such as *L. luxuriosa* (*c.* 1.6×10^4 μm^3) is one or two orders of magnitude greater than other dominant species, *Synedra* spp. (1.5×10^3 μm^3) and *Navicula* spp. (2.8×10^2 μm^3). Selection of particle size has been reported for many suspension-feeding bivalves (Kiørboe & Møhlenberg 1981). Ahn (1993) reported that *L. elliptica* fed on algal particles in preference to inorganic particles, like many other suspension- feeding bivalves. Thus, *L. elliptica* seems to select particles based on size as well as food value.

Diatom concentration, however, fluctuated considerably in the culture tank, by two orders of magnitude. Cell numbers were positively correlated with chl *a* concentration in seawater. Therefore, chl *a* values were used as indices. Changes in algal concentration were strongly reflected in the faecal pellets. It is generally known that assimilation efficiencies of suspension-feeding bivalves tend to decrease at high food concentration (Bayne & Newell 1983). The assimilation efficiency of *Laternula elliptica* was not measured. Instead, organic carbon content and the chl *a*/phaeopigment ratio of faecal materials were used as indices of assimilation efficiency. It has been reported that at high algal concentrations, ingested algal cells were often partially digested and even egested as intact cells, and changes in faecal composition indicate differences in assimilation efficiency in bivalves (Morton 1983). Organic carbon contents and chl *a*/phaeopigments ratio of the faecal materials increased with an increase in chl *a* concentration in seawater within most of the range tested. More undigested cells were microscopically observed in faecal materials at higher chl *a* concentration. These results indicate that *L. elliptica* was feeding in excess of its need during most of the experimental period.

Table 20.6. *The relative abundances (RA) of the most common diatoms in the flocculent matter taken using a plastic syringe in* Laternula elliptica *habitat at 30 m depth in Collins Harbour*

Species	RA (%)
Odontella litigiosa	43
Cylindrotheca closterium[1]	19
Trachyneis aspera	11
Cocconeis spp.	11
Pleruosigma spp.	6
Licmophora luxuriosa	3
Pinnularia quatratareoides	2
Amphora ovalis	1
Charcotia actiochilus	1
Achnanthes brevipes var. *angustata*	1
Synedra spp.	1
Fragilaria sp.	1
Pseudogomphonema kamtchaticum	+
Thalassiosira antarctica	+
Tricertium arcticum	+
Fragilariopsis kerguelensis	+
Amphiprora sp.	+

Notes:

[1] By number, *C. closterium* constituted 77% of total diatom cells.

+ = trace

The organic carbon contents (0.2–0.8%) of the sediment cores collected from *Laternula elliptica* habitat were very low for nearshore sediment, but fall well within the ranges reported for other Antarctic nearshore sediments (Warnke *et al.* 1973, Mills & Hessler 1974, Richardson & Hedgpeth 1977, Schnack 1985, Ahn 1993). The carbon content was also low in the trap sediment (0.52%), but relatively high (1.23%) in flocculent matter which was sampled by a diver using a 50 ml plastic syringe. Low carbon values in both traps and bottom sediment indicate that the organic-rich flocculent matter was of benthic origin. Microscopic observation showed that the surface film consisted almost entirely of benthic diatoms and organic aggregates. It is interesting that the nanoplanktonic pennate *Cylindrotheca closterium* dominated by number (77%) the diatom populations, although it constituted only 19% of the diatom biomass due to its needle-like cell morphology (50–80 μm×2–3 μm, cell volume *c.* 63 m^3) (Table 20.6). *C. closterium* was rarely observed in core and water-column samples. *C. closterium* is known as an ice-related species that explosively blooms in the marginal sea-ice zones during austral summer (Stockwell *et al.* 1991, Kang *et al.* 1993). Stockwell *et al.* (1991) reported that *C. closterium* was not observed in sediment samples, whereas it dominated in the overlying water-column populations in Prydz Bay, and it is suggested that the lightly silicified *C. closterium* may not be well preserved in sediment. Since the flocculent matter tends to be easily resuspended, and is apparently winnowed while being sampled with the cores, *C. closterium* was found to be destroyed during the processing for electron microscope study, implying that this species is labile and thus has a higher food value for suspension feeders like *Laternula elliptica*. The low organic carbon values in sediment cores, in spite of the high value of the flocculent matter, imply that organic matter on the sediment surface is rapidly and efficiently utilized by benthic animals (Mills & Hessler 1974, Richardson & Hedgpeth 1977), and that there is a tight coupling between benthic primary production and secondary production.

Vertical flux of carbon estimated from the trap deposits was 165 mg C m^{-2} d^{-1}, comparable to values measured in other Antarctic continental margins using sediment traps during summer months (Schnack 1985, Wefer *et al.* 1988, Dunbar *et al.* 1989). However, the majority of the cells in the traps were benthic diatoms which were apparently resuspended by wind-generated waves and sedimented again. Therefore, the vertical flux of organic matter from the overlying water column was virtually nil.

The results of these studies imply that benthic microalgae can be an important food source for benthic fauna during a certain period of time in the year. Previous studies also revealed the potential importance of benthic microalgae as a major food source for benthic fauna during the austral summer and other seasons. Dayton *et al.* (1986) measured relatively high rates of benthic primary production at McMurdo Sound, and concluded that benthic invertebrate production was more closely related to benthic primary production than to water column production. Everitt & Thomas (1986) observed benthic diatoms predominating in water-column populations during winter, spring and summer in samples of inshore waters near Davis Station, East Antarctica. Gilbert (1991a) showed that benthic primary production clearly played an important part in seasonal production cycle in Signy island, and suggested that the benthic microalgae may assist in seeding the water-column bloom through wind- and wave-induced resuspension. Gilbert (1991b) also demonstrated that benthic microalgae were highly productive in the nearshore sediments before and after phytoplankton bloom during summer time.

CONCLUSIONS

Benthic diatoms may be the most important food source for *Laternula elliptica* at least during certain periods of the austral summer months in phytoplankton-impoverished waters. During this period, *L. elliptica* seems not to be food-limited. The high biomass may occur in part due to a tight coupling between benthic primary and secondary production, and rapid and efficient utilization of organic matter by *L. elliptica* during austral summer. However, seasonal variation of the relative importance of benthic diatoms to other food sources has to be further investigated.

ACKNOWLEDGEMENTS

I am grateful to Hosung Chung for sampling clams and Ja-Shin Kang for identifying diatoms. I also thank Sung-Ho Kang for helpful discussion on diatom ecology. I acknowledge the 6th Korean Antarctic Summer Research Team for their logistic support. This work was supported by Ministry of Science and Technology in Korea.

REFERENCES

Ahn, I-Y., Chung, H., Kang, J-S. & Kang, S-H. (in press). Diatom composition and biomass variability in nearshore waters of Maxwell Bay, Antarctica, during 1992/93 austral summer. *Polar Biology*.

Ahn, I-Y. 1993. Enhanced particle flux through the biodeposition by the Antarctic suspension- feeding bivalve *Laternula elliptica* in Marian Cove, King George Island. *Journal of Experimental Marine Biology and Ecology*, **171,** 75–90.

Ahn, I-Y. 1994. Ecology of the Antarctic bivalve *Laternula elliptica* (King & Broderip) in Collins Harbour, King George Island, benthic environment and an adaptive strategy. *Memoirs of the National Institute of Polar Research*, Special Issue **50,** 1–10.

Ahn, I-Y. & Kang, Y-C. 1991. Preliminary study on the macrobenthic community of Maxwell Bay, South Shetland Islands, Antarctica. *Korean Journal Polar Research*, **2,** 61–71.

Andriashev, A. P. 1968. The problem of the life community associated with the Antarctic fast ice. In *Symposium on Antarctic Oceanography, Santiago, Chile, 16, September 1966*. Cambridge: SCAR, 147–155.

Barry, J. P. & Dayton, P. K. 1988. Current patterns in McMurdo Sound, Antarctica and their relationship to local biotic communities. *Polar Biology*, **8,** 367–376.

Bayne, B. L. & Newell, R. C. 1983. Physiological energetics of marine molluscs. In Saleuddin, A. S. M. & Wilber, K. M., eds. *The Mollusca, vol. 4. Physiology, Part 1*. New York: Academic Press, 407–515.

Berkman, P. A. 1991. Holocene meltwater variations recorded in Antarctic coastal marine benthic assemblages. In Weller, G., Wilson, C. L. & Severin, B. A. B., eds. *International conference on the role of the polar regions in global changes.* Fairbanks: University of Alaska, 440–449.

Burkholder, P. R. & Mandelli, E. F. 1965. Productivity of microalgae in Antarctic sea ice. *Science*, **149,** 872–874.

Chang, K. I., Jun, H. K., Park, G. T. &. Eo, Y. S. 1990. Oceanographic conditions of Maxwell Bay, King George Island, Antarctica (austral summer 1989). *Korean Journal Polar Research*, **1,** 27–46.

Clarke, A. 1980. A reappraisal of the concept of metabolic cold adaptation in polar marine invertebrates. *Biological Journal of the Linnean Society*, **14,** 77–72.

Clarke, A. 1990. Temperature and evolution: Southern Ocean Cooling and the Antarctic marine fauna. In Kerry, K. R. & Hempel, G., eds. *Antarctic ecosystems, ecological change and conservation.* Berlin and Heidelberg: Springer-Verlag, 9–22.

Crumpton, W. G. 1987. A simple and reliable method for making permanent mounts of phytoplankton for light and fluorescence microscopy. *Limnology and Oceanography*, **32,** 1154–1159.

Dayton, P. K. & Oliver, J. S. 1977. Antarctic soft-bottom benthos in oligotrophic and eutrophic environments. *Science*, **197,** 55–58.

Dayton, P. K., Robilliard, G. A. & DeVries, A. L. 1969. Anchor ice formation in McMurdo Sound, Antarctica, and its biological effects. *Science*, **163,** 273–274.

Dayton, P. K., Watson, D., Palmisano, A., Barry, J. P., Oliver, J. S. & Rivera, D. 1986. Distribution patterns of benthic microalgal standing stock at McMurdo Sound, Antarctica. *Polar Biology*, **6,** 207–213.

Demers, S., Therriault, J. C., Bourget, E. & Bah, A. 1987. Resuspension in the shallow zone of a macrotidal estuarine environment: wind influence. *Limnology and Oceanography*, **32**, 327–339.

Dunbar, R. B., Leventer, A. R.& Stockton, W. L. 1989. Biogenic sedimentation in McMurdo Sound, Antarctica. *Marine Geology*, **85,** 155–179.

Dykstra, M. J. 1992. *Biological electron microscopy. Theory, techniques, and trouble shooting*. New York & London: Plenum Press, 360 pp.

Everitt, D. A. & Thomas, D. P. 1986. Observations of seasonal changes in diatoms at inshore localities near Davis Station, East Antarctica. *Hydrobiologia*, **139,** 3–12.

Fréchette, M. & Bourget, E. 1985. Energy flow between the pelagic and benthic zones: factors controlling particulate organic matter available to an intertidal mussel bed. *Canadian Journal of Fisheries and Aquatic Science*, **42,** 1158–1163.

Gilbert, N. S. 1991a. Microphytobenthic seasonality in near-shore marine sediments at Signy Island, South Orkney Islands, Antarctica. *Estuary and Coastal Shelf Science*, **33,** 89–104.

Gilbert, N.S. 1991b. Primary production by benthic microalgae in nearshore marine sediments of Signy Island, Antarctica. *Polar Biology*, **11,** 339–346.

Grossi, S. M., Kottmeier, S. T., Moe, R. L., Taylor G. T. & Sullivan C. W. 1987. Sea ice microbial communities. VI. Growth and primary production in bottom ice under graded snow cover. *Marine Ecology Progress Series*, **35,** 153–164.

Gruzov, E. N. 1977. Seasonal alterations in coastal communities in the Davis Sea. In Llano, G. A., ed. *Adaptations within Antarctic ecosystems.* Washington DC: Smithsonian Institution, 263–278.

Hardy, P. 1972. Biomass estimates from some shallow-water infaunal communities at Signy Island, South Orkney Island. *British Antarctic Survey Bulletin,* No. 31, 93–106.

Kang, S-H. & Fryxell, G. A. 1991. Most abundant diatom species in water column assemblages from five Leg 119 drill sites in Prydz Bay, Antarctica: distributional patterns. *Proceedings of the Ocean Drilling Program, Scientific Results*, **119,** 645–666.

Kang, S-H., Fryxell, G. A. & Roelke, D. L. 1993. *Fragilariopsis cylindrus* compared with other species of the diatom family Bacillariaceae in Antarctic marginal ice-edge zones. *Nova Hedwigia*, Beiheft, **106,** 335–352.

Kiørboe, T. F. & Møhlenberg, F. 1981. Particle selection in suspension-feeding bivalves. *Marine Ecology Progress Series*, **5,** 291–296.

Krebs, W. N. 1983. Ecology of neritic marine diatoms, Arthur Harbour, Antarctica. *Micropaleontology*, **29,** 267–297.

Littlepage, L. & Pearse, J. 1962. Biological and oceanographic observations under an Antarctic ice shelf. *Science*, **137,** 679–680.

Mills, E. & Hessler, R. R. 1974. Antarctic benthic communities: Hudson '70 expedition. *Antarctic Journal of the United States*, **9,** 312–316.

Morton, B. 1983. Feeding and digestion in bivalvia. In Saleuddin, A.S.M. & Wilbur, K.M., eds. *The Mollusca, Vol. 5. Physiology, Part II*. New York: Academic Press, 65–147.

Palmisano, A. C., SooHoo, J. B., White, D. C., Smith, G. A. & Stanton, G. R. 1985. Shade adapted benthic diatoms beneath Antarctic sea ice. *Journal of Phycology*, **21,** 664–667.

Picken, G. B. 1985. Marine habitats – benthos. In Bonner, W. N. & Walton, D. W. H., eds. *Key environments: Antarctica*. Oxford: Pergamon Press, 154–172.

Ralph, R. & Maxwell, J. G. H. 1977. Growth of two Antarctic lamellibranchs: *Adamussium colbecki* and *Laternula elliptica*. *Marine Biology*, **42,** 171–175.

Richardson, M. D. & Hedgpeth, J. W. 1977. Antarctic soft-bottom, macrobenthic community adaptations to a cold, stable, highly productive, glacially affected environment. In Llano, G.A., ed. *Adaptations within Antarctic ecosystems*. Washington DC: Smithsonian Institution, 181–196.

Rivikin, R. B. & Putt, M. 1987. Photosynthesis and cell division by antarctic microalgae: comparison of benthic, planktonic and ice algae. *Journal of Phycology*, **23,** 223–229.

Schnack, S. B. 1985. A note on the sedimentation of particulate matter in Antarctic waters during summer. *Meeresforschung*, **30,** 306–315.

Shabica, S. V. 1972. Tidal zone ecology at Palmer Station. *Antarctic Journal of the United States*, **7** (5), 185–186.

Sokal, R. R. & Rohlf, F. J. 1981. *Biometry*, 2nd edn. New York: W. H. Freeman & Company, 859 pp.

Stockwell, D. A., Kang, S-H. & Fryxell, G. A. 1991. Comparisons of diatom biocoenoses with holocene sediment assemblages in Prydz Bay, Antarctica. *Proceedings Ocean Drilling Program, Scientific Results*, **119,** 667–673.

Stout, W. E. & Shabica, S. V. 1970. Marine ecological studies at Palmer Station and vicinity. *Antarctic Journal of the United States*, **5** (4), 134–135.

Strickland, J. D. H. & Parsons, T. R. 1972. A practical handbook of seawater analysis, 2nd edn., *Bulletin, Fisheries Research Board of Canada* **167,** 1–310.

Verardo, D. J., Froelich, P. N. & McIntyre, A. 1990. Determination of organic carbon and nitrogen in marine sediments using the Carlo Erba NA-1500 Analyzer. *Deep-Sea Research*, **37,** 157–165.

Warnke, D. A., Richter, J. & Oppenheimer, C. H. 1973. Characteristics of the nearshore environment off the south coast of Anvers Island, Antarctic Peninsula. *Limnology and Oceanography,* **18,** 131–142.

Wefer, G. G., Fischer, D., Füetterer, D. & Gersonde, R. 1988. Seasonal particle flux in the Bransfield Strait, Antarctica. *Deep-Sea Research*, **35,** 891–898.

White, M. G. 1984. Marine benthos. In Laws, R.M., ed. *Antarctic ecology*, Vol. 2. London: Academic Press, 421–461

Zamorano, J. H., Durate, W. E. & Moreno, C. A. 1986. Predation upon *Laternula elliptica* (Bivalvia, Anatinidae): a field manipulation in South Bay, Antarctica. *Polar Biology,* **6,** 139–143.

21 Sub-Antarctic weevil assemblages: species, structure and survival

STEVEN L. CHOWN

Department of Zoology and Entomology, University of Pretoria, Pretoria 0002, South Africa

ABSTRACT

Sub-Antarctic islands are moderately harsh environments that are fairly species poor. They may serve as early warning stations for the prediction of global climate change and as model ecosystems that can be used to study such change, based on the relationships shown by weevil faunas in the trans-Antarctic track. In the sub-Antarctic weevil assemblages examined here (including the Falkland Islands and New Zealand sub-Antarctic islands), regional climate change has largely determined the species pool from which membership was drawn, whereas local processes such as interspecific interactions and adaptation have determined final assemblage structure. The relationship between physiological tolerances and assemblage structure is shown to be a complex one, although changes in these relationships may be most suitable for recognizing and assessing the effects of climate change. Changes in the strength and direction of interspecific interactions, particularly those between introduced predators and their prey (in this case feral house mice and Ectemnorhinus*-group weevils on Marion Island) are shown to be valuable indicators of climate change. This is because these changes are more readily discernible than direct, adaptive effects mediated through the physiological tolerances of species. It is suggested that sub-Antarctic research is now at the stage where it can be used for understanding and predicting the effects of future climate change. Hence, the current trend away from sub-Antarctic research towards more continental Antarctic research should be re-evaluated.*

Key words: biogeography, *Ectemnorhinus*-group, interspecific interactions, climate change, ecophysiology.

INTRODUCTION

Although reasons for the current changes in the planet's climate are far from clear (see e.g. Friis-Christensen & Lassen 1991), there is abundant evidence that global climates are changing (e.g. Hansen & Lebedeff 1987, Schneider 1992). General Circulation Models have provided reasonable forecasts of global, and to a lesser extent, regional changes in temperature and precipitation regimes (Goodess & Palutikof 1992), but these have yet to be accurately translated into predictable effects on ecological systems at local through to global scales (Schneider 1993). Research effort in this regard has largely focused on bioclimatic modelling (Cammel & Knight 1992), ecosystem ecology (Carpenter *et al.* 1993), individual-based ecophysiological models (Dunham 1993, Murdoch 1993), and palaeobiological analogies (Adams & Woodward 1992). Each of these research areas has delivered profound insights into the effects of environmental change (see Kingsolver *et al.* 1993), but each remains bedevilled by a unique set of problems (Dunham 1993, Schneider 1993). Quinn & Karr (1993) succinctly summarized these problems when they advised that the best that can be done in the face of change is to be on the lookout for ecological 'surprises'.

Given the rapidity of global climate change (Schneider 1992, 1993) and habitat fragmentation (Groom & Schumaker 1993), the urgency of implementing a research agenda focussed on understanding the ways in which current and past abiotic and biotic processes interact to produce the present patterns is clear. Carpenter *et al.* (1993) argued that the only way to achieve a clear understanding of the effects of global environmental change will be to embed small-scale, short-term experiments within large-scale experiments, sustained for long periods. Substantial manipulations will be required for these to succeed. On the other hand, Brown (1994) has argued that analysis of large databases and comparative, non-manipulative field studies are as important (see also Murdoch 1993). A reason for this difference in approach is that lakes are more 'closed' systems than most terrestrial habitats (Brooks & Wiley 1988). In closed systems, species are less likely to respond by tracking environmental change than in open systems (see Foster *et al.* 1990, Quinn & Karr 1993).

One way of integrating the large- and small-scale approaches, as well as those currently used to study environmental change, is to study terrestrial systems which more closely approximate a 'closed status' like that of freshwater lakes, and of which we have knowledge of both current conditions as well as the history of the systems and taxa involved (see also Schimel 1993). The only systems that approximate these conditions are isolated islands, where diverse, monophyletic taxa have been well studied. Most of the archipelagos that qualify for inclusion in this category (e.g. Hawaii, Galapagos) lie in tropical or temperate regions where the signal of present climate change is not liable to be as strong as on polar ones (see Parry 1992 for discussion; Adamson *et al.* 1988a, Smith & Steenkamp 1990 for data). The only isolated, high-latitude islands that harbour well-studied, fairly diverse monophyletic groups are those of the sub-Antarctic (Gressitt 1970, Chown 1990a). On these islands, weevils represent the only monophyletic taxon that is both reasonably diverse and well studied (Kuschel 1991, Chown 1994). In this paper, I therefore examine the past and present processes that have contributed to local (intra-archipelago) and regional (inter-archipelago) patterns in sub-Antarctic weevil diversity, and the possible effects continued environmental change is likely to have on weevil assemblages.

SUB-ANTARCTIC ISLANDS

The division of the Sub-Antarctic into biogeographic zones has been the focus of considerable attention (e.g. Skottsberg 1960, Gressitt 1970, Stonehouse 1982). The most recent treatment is that of Lewis Smith (1984), who considered Macquarie Island (South Pacific Province–SPP), South Georgia (South Atlantic Province–SAP), and the Prince Edward Islands, Crozet Archipelago, Kerguelen Archipelago and Heard and McDonald Islands (South Indian Province – SIP) to be 'true' sub-Antarctic Islands. Lewis Smith's (1984) study was based on differences in climate and vegetation structure which, although useful in categorizing islands based on the structure and functioning of their terrestrial systems, does not provide insight into their biogeographic relationships (see Myers & Giller 1988, Rosen 1988). To this end, presence/absence (and other, e.g. cladistic) data for a variety of taxa are usually subjected to various analyses (see Brundin 1988, Craw 1988, Humphries *et al.* 1988, Rosen 1988) to establish the biogeographic affinities of the landmasses they inhabit (see also Brooks & McLennan 1991). Although such a full analysis has not been completed for all sub-Antarctic Islands, three recent studies of sub-Antarctic insects suggest that the Falkland Islands and the sub-Antarctic Islands of New Zealand (the Auckland, Campbell and Snares Islands) should be included in the sub-Antarctic region of the South Polar 'biome' as was previously suggested by Skottsberg (1960). First, the genera *Heterexis* Broun and *Oclandius* Blanchard (Coleoptera: Curculionidae: Entiminae), from the Auckland and Campbell Islands have been shown to be the sister taxa of the *Ectemnorhinus* group of weevil genera, endemic to the South Indian Province (Chown 1994, Kuschel & Chown 1995). The former genera occur on the Auckland and Campbell Islands are closely related to genera from New Zealand (Kuschel 1964, 1971, 1991). Second, Serra-Tosio's (1986) cladistic analysis of the Podonomini (Diptera: Chironomidae) indicated clear relationships between the Crozet archipelago and New Zealand. Third, the ground beetle *Kenodactylus audouini* (Guérin) has populations on the Falkland Islands as well as on the Auckland, Campbell and Snares Islands (Johns 1974). Various other insect genera (e.g. *Halmaeusa, Crymus* and *Meropathus* (Coleoptera), and *Telmatogeton, Paractora* and *Apetaenus* (Diptera)) have distributional patterns which also support the inclusion of the Falkland Islands and the New Zealand sub-Antarctic Islands in the sub-Antarctic region (see also discussion in Kuschel 1971, Gressitt & Wise 1971, and particularly fig. 4 in Craw 1989).

In this scheme, ten archipelagos, including at least 17 islands (of which the 13 major islands are given in Table 21.1) may be considered sub-Antarctic (Fig. 21.1). These islands vary considerably in their extent, age and geological and climatological histories. Chown (1994) recently summarized this information for the SIP Islands, and information on the geology and glacial history of the other archipelagos can be found in Gressitt (1964), Gressitt & Wise (1971), Clapperton & Sugden (1976), Mercer (1983), Headland (1984), Adamson *et al.* (1988b), Williamson (1988) and Hall (1990). This information is summarized in Table 21.1. Because of their location in the vast Southern Ocean, the climates of these archipelagos can all be considered cold and oceanic, although they vary considerably (see Walton 1984 for review). The flora and vegetation structure of the islands considered here to be sub-Antarctic are also varied. Although the colder islands tend to have depauperate floras compared with their warmer counterparts, the sub-Antarctic islands share a remarkable number of both vascular and non-vascular species (Gremmen 1981). Although the

Table 21.1. *Geological and glacial histories of the major sub-Antarctic islands (see text for definition)*

Island	Age (Ma)	Origin	Glacial history
Falklands	>200	Continental	Cirque glaciers
South Georgia	100–140	Continental?	Heavily glaciated
Marion	0.45	Oceanic	Glaciated
Prince Edward	0.45	Oceanic	No glaciation
Cochons	0.4	Plateau–Continental?	No glaciation
Possession	8	Plateau–Continental?	No glaciation
Est	9	Plateau–Continental?	No glaciation
Kerguelen	40	Plateau–Continental?	Heavily glaciated
Heard	40	Plateau–Continental?	Heavily glaciated
Macquarie	10	Uplifted–plateau	No glaciation
Campbell	16	Plateau–Continental	Partly glaciated
Auckland	18	Plateau–Continental	Glaciated
Snares	10	Plateau–Continental	Glaciated?

Notes:
Data from Gressitt (1964), Gressitt & Wise (1971), Nur & Ben-Avraham (1982), Martin & Hartnady (1986) and Hall (1990).

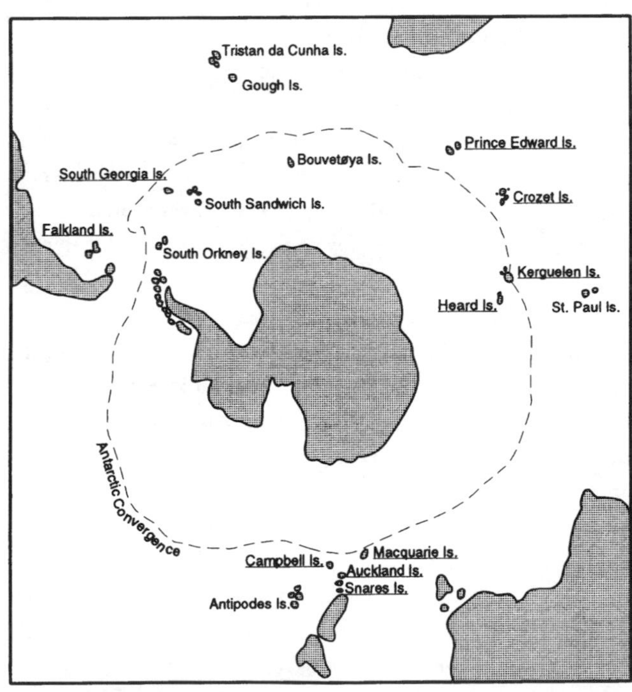

Fig. 21.1. The islands of the Southern Ocean, or trans-Antarctic track (*sensu* Craw 1989). Sub-Antarctic islands, as defined in this paper, are underlined.

vegetation varies between warmer and colder islands (Lewis Smith 1984), the predominant closed vegetation communities on all these islands can be considered tundra-like (see also Smith & French 1988). Closed vegetation communities predominate on the more northerly and/or warmer islands, but on the cooler islands, closed vegetation communities rapidly give way to open communities and fellfield with increasing altitude. Chown (1989) recognized two distinct biotopes on the SIP islands based on the presence/absence of angiosperms and vegetation structure. The vegetated biotope is dominated by angiosperms and supports closed vegetation communities, whereas the epilithic biotope is dominated by cryptogams which do not form closed communities. This classification can readily be applied to South Georgia, Macquarie, the Falkland Islands and the New Zealand sub-Antarctic islands, although the epilithic biotope is greatly reduced in extent on the latter islands because of the absence of true fellfield. Nonetheless, this distinction remains important when analysing the weevil faunas of these islands.

Table 21.2. *Weevil species richness and endemicity (to archipelago – A; to island – I, generic endemicity – G) on the sub-Antarctic islands. Where applicable, percentages are given in parentheses*

Island	Genera + species	Endemicity A	Endemicity I	Endemicity G
Falklands (SAP)	8+16	14 (88)	14 (88)	1 (13)
South Georgia (SAP)	0	0	0	0
Marion (SIP)	3+6	4 (67)	0	0
Prince Edward (SIP)	3+6	4 (67)	0	0
Cochons (SIP)	4+8	6 (75)	3 (38)	1 (25)
Possession (SIP)	4+15	13 (87)	4 (27)	1 (25)
Est (SIP)	4+11	9 (82)	0	1 (25)
Kerguelen (SIP)	5+10	5 (50)	5 (50)	1 (25)
Heard (SIP)	3+4	0	0	0
Macquarie (SPP)	0	0	0	0
Campbell (NZ)	8+14	3 (21)	3 (21)	0
Auckland (NZ)	12+25	9 (36)	9 (36)	1 (8)
Snares (NZ)	9	5 (56)	5 (56)	0

Notes:
SAP=South Atlantic Province, SIP=South Indian Ocean Province, SPP=South Pacific Province, NZ=New Zealand sub-Antarctic Islands.
Source: Data from Kuschel (1964, 1971, 1991), Greenslade (1990) and Chown (1994).

WEEVIL SPECIES

With the exception of South Georgia and Macquarie Island, all of the sub-Antarctic archipelagos support weevil faunas (Table 21.2). The fauna of the Falkland Islands has not been well studied, although some work is currently underway (J. Fuller, pers. commun. 1993). Seven or eight genera (depending on the taxonomic treatment) and 16 species are known from the islands, of which *Malvinius* is the only endemic genus (Kuschel 1991). Robinson's (1984, 1988) catalogue of the Falkland Islands insect fauna indicates, however, that more weevil species are known than Kuschel's (1991) analysis suggests, and that there are pronounced differences between East and West Falkland, which have rather different geological histories. The islets of the Falklands also seem set to deliver new species (Robinson 1988). One of the most noteworthy new records is that of *Pentarthum carmichaeli*, which is indigenous to the southern tip of New Zealand and the Snares and Auckland Islands, but which is thought to have drifted in *Nothofagus* logs to the Falkland Islands, Tristan da Cunha and Gough Island (Kuschel 1991). All of the non-endemic genera contain species from both the Falkland Islands and Tierra del Fuego (Kuschel 1971, 1991), which shows that the fauna has a South American origin, despite the fact that prior to 105 Ma the Falkland Islands lay adjacent to southern Africa (Martin & Hartnady 1986).

The weevil faunas of the sub-Antarctic islands of New Zealand were treated in detail by Kuschel (1964, 1971, 1991). A total of 17 genera (of which four are endemic) containing 35 species is known from these islands (see Table 21.2 for island endemicity). Kuschel (1964, 1971) argued that the non-endemic species (14 spp.) are largely of post-glacial origin (*c.* 16 000 y) and reached the islands as a result of recent dispersal, while those endemic to the islands (21 spp.) have a much earlier origin. That most of the 'pre-glacial' species have ectophytic larvae (i.e. larvae that do not feed within plant parts), which are more resistant to climate extremes than endophytic larvae, supports Kuschel's contention. Endophytic species tend to be much smaller (modal length=1.75 mm, n=15) than their ectophytic counterparts (modal length 6.6 mm, n=19), which is in keeping with their habits. The fauna of these islands is largely of New Zealand origin, although there are some relationships with the fauna of southern South America and the Falkland Islands.

The South Indian Province Islands are inhabited by a single, monophyletic weevil taxon, the *Ectemnorhinus* group (Dreux & Voisin 1989, Chown 1994). The group consists of six genera, of which three (*Canonopsis, Disker* and *Palirhoeus*) are monotypic. *Christensenia* holds two species, *Bothrometopus* 17 and *Ectemnorhinus* 14. The phylogenetically basal taxa are restricted to the oldest SIP islands. *Christensenia*'s two species are restricted to Îles Crozet, and *Canonopsis* occurs on Îles Kerguelen and Heard. *Palirhoeus* can be found on all the SIP islands, with the exception of Heard and McDonald Islands. *Bothrometopus* and *Ectemnorhinus* have species on all the SIP islands and reach their greatest species richness on Îles Crozet. The basal genera *Christensenia* and *Canonopsis* are morphologically and ecologically very similar to their adelphotaxon and are thought to have occurred on the islands since the Paleogene. *Disker* is likewise thought to have been on Îles Kerguelen since before the Neogene. Chown (1994) suggested that *Bothrometopus* diversified during the Neogene glaciations and that this may also have been the case in the *Ectemnorhinus vanhoeffenianus* species-group on Îles Crozet. The bulk of the diversification in *Ectemnorhinus* is thought to have taken place in response to post-glacial climate amelioration (see Chown 1990b, 1994).

The distributional patterns of sub-Antarctic weevils clearly show that most of the faunas have been derived from nearby continents, but that there are very close relationships between the faunas on all of the archipelagos. The demonstration of a phylogenetic relationship between weevils on the SIP archipelagos and those on the New Zealand sub-Antarctic islands supports previous contentions that the faunas of the sub-Antarctic islands have biogeographical affinities which link the major ocean basins through a generalized trans-Antarctic track (Craw 1988, 1989). This track may also extend to the faunas of Tristan da Cunha and Gough Island. The danger of restricting studies of the evolution and biogeography of southern island biotas to those islands considered 'sub-Antarctic' solely on climate and vegetational grounds, is highlighted by the long-standing debate concerning the relationships of the *Ectemnorhinus* group of genera. Previously, few authors thought to examine 'non-sub-Antarctic' islands for related genera, and restricted their investigations to continents, rather than ocean basins (e.g. Dreux & Voisin 1989). The above analysis suggests that Craw's (1988) call for a re-evaluation of both the philosophy underlying biogeographic analysis, and the affinities of the Southern Ocean basins, should be heeded. A re-appraisal of the 'trans-Antarctic' track is underway.

WEEVIL ASSEMBLAGE STRUCTURE

Community membership and structure are determined by processes operating at both regional and local scales (Brooks 1988, Ricklefs 1989, Brooks & McLennan 1991, Levin 1992). Based on the relative contributions of these processes, communities may be viewed as lying on a continuum which ranges from communities whose membership is unlimited and determined largely by the individualistic response of species to an unpredictable and variable environment (see e.g. Coope 1979, Foster *et al.* 1990, Harrison 1993), to those where strong linkages develop both within and between trophic levels (Southwood 1987, Pimm 1991, DeAngelis 1992; see Roughgarden 1989 for discussion). Bennett (1990) suggested, however, that during climate change, community membership is determined by the individualistic response of species to such changes, but that during climate stasis ecological processes such as biotic interactions assume more important roles in determining community structure and membership. This view would be in line with Drake's (1990, 1991) experimental studies of artificial laboratory communities and would explain many features of modern natural communities, such as priority effects (Drake 1991), the prevalence of interspecific interactions in moderately harsh environments (Connell 1980), and the predominance of interspecific interactions in old communities compared with younger ones (Arthur 1987). These 're-shuffling' effects of climate change, and the subsequent evolution of communities during climate stasis, should be readily discernible in the sub-Antarctic, where small changes in the position of the Antarctic Convergence Zone had large effects on the climates of these islands (Hall 1990).

Few investigations of insect community and/or assemblage structure have been undertaken on the sub-Antarctic islands (see Burger 1985). In the case of the New Zealand sub-Antarctic islands, the Falkland Islands, and Macquarie Island, the only data available are species lists which, in the case of the former islands, have been interpreted as indicative of the effects of Neogene glaciations (Kuschel 1964). On South Georgia, an investigation of the role of predation as a factor structuring beetle assemblages is underway (Ernsting 1993). Some work has also been undertaken on the Diptera of the SIP islands (Tréhen *et al.* 1985), but has focused largely on the roles of these species in terrestrial systems, rather than on the relative contribution of historical versus ecological factors, or abiotic versus biotic factors, to assemblage structure. Nonetheless, Tréhen *et al.* (1985) concluded that interspecific competition is an important factor contributing to the structure of modern dipteran assemblages on Île de la Possession. Likewise, Davies (1987) used indirect evidence to demonstrate the presence of interspecific competition between two *Amblystogenium* carabid beetles in the epilithic biotope on Île de la Possession. He also showed character release in *Amblystogenium pacificum* in the vegetated biotope on the same island.

The only detailed investigations of insect assemblage structure on sub-Antarctic islands are those of Chown (1990a, 1992, 1994), who investigated the possible effects of Quaternary glaciations on the SIP insect faunas as a whole, and the relative contribution of historical and ecological factors to the structure of weevil assemblages on these islands. Because island insects are unable to track changing climates like their continental counterparts (Chown & Scholtz 1989), the effects of Neogene climate change are pronounced on most of the sub-Antarctic islands. A regional analysis of geographic variation in species richness of all the SIP insects and the *Ectemnorhinus* group of weevils showed that species richness depends directly on the age and extent of glaciation of the archipelagos. Those islands that were most heavily glaciated tend to have the lowest species-richness. On the other hand, a local analysis of the Prince Edward Islands, of which only Marion Island has been glaciated (Hall 1990), indicated no difference between the weevil assemblages of the two islands. On these islands, interspecific competition contributed largely to assemblage structure in the moderately harsh, predator-free epilithic biotope. In the vegetated biotope no indications of interspecific competition were found, although large between-phenotype variation in the *Ectemnorhinus similis* species complex suggested that weevils have only recently colonized this habitat and are currently radiating therein (Chown 1990b). A similar pattern can be seen when all the sub-Antarctic islands are compared. Species richness tends to be greatest on the oldest archipelagos and those closest to continental areas, although in the latter case generic endemicity is lowest. Isolated archipelagos tend to have the highest endemicity at the species level (Table 21.2). Superimposed on this pattern is one of elevated species richness on multi-island archipelagos. In the SIP, these patterns have been interpreted as a consequence of inter-archipelago and inter-island allopatric speciation (Chown 1994), although intra-island allopatric and sympatric speciation may also have played important roles in the generation of species diversity on these islands (Chown 1990b, Chown 1994). Speciation has proceeded in much the same way on these islands as on more northerly archipelagos (Paulay 1985, Grant 1986, Otte 1989, Williamson 1981).

The study of weevil assemblages on the SIP islands highlighted the importance of scaling effects, and data from the other islands supports this conclusion. At a regional scale variation in disturbances due to climate change (e.g. glaciations) determines the pool of species available for assemblage membership, but the evolution of assemblage structure is very much dependent on local processes such as interspecific interactions. The importance of these interactions also varies with the age of the habitat concerned (which in turn is once again dependent on regional processes). The older the habitat (e.g. fellfield in the epilithic biotope) the more likely are interspecific interactions (see Arthur 1987, Chown 1992). Davies' (1987) data on carabid beetles suggests that this may also be the case in other taxa. The weevil assemblages therefore clearly show that during climate change a re-shuffling of communities takes place, followed during periods of stasis by local trajectories, leading finally to very different assemblage structures on the different SIP islands. When considering the apparently large influences of regional events and subsequent interactions on these assemblages, it should, however, be borne in mind that both survival of climate change and the effects of abiotic factors on processes such as interspecific interactions are always mediated through the physiologies of the organisms involved (Myers & Giller 1988: 150, Pimm 1991, Levin 1992).

SURVIVAL STRATEGIES

Survival strategies of the sub-Antarctic weevils have not been as well studied as those of the Perimylopidae from South Georgia. Much information is available on the resistance to cold and dry conditions of the latter at the biochemical and whole-organism levels (see e.g. Sømme *et al.* 1989, Ring *et al.* 1990, Haderspeck & Hoffmann 1991, Worland *et al.* 1992, 1993). Like other ecophysiological studies, these have demonstrated a suite of unique adaptations which enable the perimylopids to survive the relatively cold, fairly seasonal habitats on South Georgia (Block *et al.* 1988). In comparison, only two ecophysiological studies have been undertaken on the SIP weevils. Chown & Van Drimmelen (1992) showed that the larvae of *Palirhoeus eatoni*, which live in the relatively moist, supra-littoral zone, lose water more rapidly than those of *Ectemnorhinus similis* and *Bothrometopus randi*, which live in drier habitats. *B. randi* was, however, better able to replace lost water by drinking than *E. similis*, which lives in much moister habitats. In addition, larvae of the supra-littoral *P. eatoni* are capable of regulating their haemolymph osmolality and water content in both freshwater and seawater, whereas *B. randi* larvae do not have this ability. This adaptation to the supra-littoral habitat is unique amongst the Curculionidae.

Chown (1993) compared desiccation resistance and body water content in adults of all six weevil species inhabiting Marion Island to investigate the possible effects of differing tolerances on assemblage structure. The similarity in tolerance of these species to the perimylopid beetles from South Georgia can be considered adaptive convergence between these unrelated taxa (see also Crafford & Chown 1993). Weevil species inhabiting dry, rockface habitats had lower rates of water loss, took longer to reach the maximum water loss and had higher body water contents in the field than species from the moister vegetated biotope. These differences were regarded as being adaptations in these species to their different microhabitats. Furthermore, it was shown that body-size differences in certain weevil species pairs are unlikely to have evolved to promote coexistence via the elimination of competition for refugia, one of the major arguments in favour of competition between the epilithic species (Chown 1992). Chown (1993) also suggested that the very small body size of *B. elongatus* may have a large effect on its reponse to abiotic conditions, making it rather different from the other species that were examined (see Harrisson *et al.* 1991).

Bothrometopus elongatus's fellfield habitat may also carry an extra cost associated with these habitats at high altitude – that of enhanced cold hardiness, which may be energetically expensive (Block 1990). Investigations of cold tolerance strategies in the weevils from Marion Island are underway. In isolation, data on the supercooling points (SCP) of arthropods do not constitute a reliable guide to cold hardiness strategies (Worland *et al.* 1992, 1993), but so far only SCP data are available for Marion Island weevils. Higher-altitude weevil species and populations tended to have lower SCPs than those at lower altitude, as would be expected (unpublished data). However, *B. elongatus* does not follow this trend, which suggests that the moister fellfield sites, like the one where this species was collected for studies of desiccation resistance, may indeed pose a greater physiological challenge than drier ones. However, this may have more to do with the presence/absence of free water, which may inoculate freezing, than with absolute temperature. *Palirhoeus eatoni* has the lowest supercooling point of all the species, yet temperatures are least likely to drop below 0 °C in the supra-littoral zone. This cautions against the hasty interpretation of supercooling point data. The low SCP of this species may have more to do with substances involved in haemolymph osmoregulation than with cold tolerance (Chown & Van Drimmelen 1992).

Differences in the SCP of weevil populations occurring at high- versus low- altitude habitats suggest that many physiological responses in these organisms may vary as a result of the change of abiotic conditions with altitude, in much the same way that survival strategies may differ between species at different latitudes (Sømme & Block 1991). Of the physiological responses that vary with latitude, the latitudinal change in whole-organism oxygen uptake is surely one of the most polemical. Various authors have claimed that polar arthropods exhibit elevated respiration rates at low temperatures compared with their more temperate counterparts (e.g. Young 1979, Block 1984, Haderspeck & Hoffmann 1991). Sømme & Block (1991), *inter alia,* suggested that elevated oxygen uptake can be considered a low-temperature adaptation. This interpretation has recently been the subject of considerable criticism (e.g. Clarke 1991), and Clarke (1993) has argued that metabolic cold adaptation (or latitudinal compensation) does not exist. On Marion Island, the slope of the regression of metabolic rate on temperature in the six weevil species is lower in populations from high altitudes than in those from low altitudes (Chown, Van der Merwe & Smith, in press – data shown in Table 21.3). In addition, a strong inverse relationship between altitude and slope of the metabolic rate/temperature regression was found (Spearman rank correlation, $R_s=-0.822$, $P=0.014$, $n=10$). Because higher-altitude areas tend to be colder than lower ones (Schulze 1971, Blake & Smith unpublished data), the case for metabolic cold adaptation or low-temperature adaptation seems a strong one, particularly because activation energies of the species also tended to be lower at higher altitudes. Although lowered activation energy is also considered to be an adaptation to cold climates (Sømme & Block 1991), this interpretation appears to be simplistic in the case of the Marion Island weevils. Here, no significant relationship was found between activation energy and altitude ($R_s=-0.160$, $P=0.632$, $n=10$). However, there was a strong inverse relationship between body water content and activation energy in the lowland populations ($R_s=-0.943$, $P=0.035$, $n=6$). Because water content seems to be directly related to the dryness of the habitat (Chown 1993), the differences in activation energies and the slope of the lines could also be ascribed to metabolic adaptation to desiccation. During activity, insects can lose up to 65% of their water through the respiratory system (Wharton 1985). Thus, a lowered slope, and/or activation energy, may have more to do with the depression of metabolic rate at high temperatures, than with the elevation thereof at low temperatures. This interpretation is not counter to Clarke's (1993) argument that organisms should act to lower their ATP utilization (see also Sibly & Calow 1986). Kennedy (1994) has argued that adaptation to desiccation is of critical importance to Antarctic arthropods, and this also seems to be the case in sub-Antarctic weevils.

Table 21.3. *Relationship between metabolic rate (ml $O_2\,g^{-1}\,h^{-1}$) and temperature for the high- and low-altitude populations of weevil species on Marion Island The regression equations are in the form log metabolic rate*=b *log temperature*+a

Species	Slope	Intercept	r^2	*n*
Palirhoeus eatoni	0.924	−2.344	0.873	60
Bothrometopus randi	0.885	−2.991	0.842	60
B. parvulus low	0.921	−2.467	0.850	60
B. parvulus high	0.615	−1.951	0.875	60
B. elongatus low	0.698	−2.053	0.904	60
B. elongatus high	0.622	−1.752	0.845	60
Ectemnorhinus marioni low	0.734	−2.731	0.787	60
E. marioni high	0.701	−2.136	0.866	60
E. similis low	1.160	−3.998	0.850	60
E. similis high	0.824	−3.162	0.950	60

Source: From Chown, Van der Merwe & Smith, in press.

This information on the survival strategies of the Marion Island weevils shows that on the more northerly sub-Antarctic islands physiological responses may differ between different populations on the islands (compare the above data with that of Sømme *et al.* 1989), thus making the interpretation of survival under changing climatic conditions, and that of the role physiological responses play in structuring assemblages, somewhat more complex than has previously been assumed. Nonetheless, these findings highlight the role that physiological responses have to play not only in the survival of sub-Antarctic weevil species, but also in the way that their assemblages are structured. In addition, they suggest that it is necessary to consider ecophysiological responses when investigating assemblage structure on these islands (Chown 1992, 1994, Dunson & Travis 1991, Lubchenco *et al.* 1991, Kingsolver *et al.* 1993). This is likely to be of considerable importance when investigating the influence of current climate change on sub-Antarctic weevil assemblages.

THE EFFECTS OF ENVIRONMENTAL CHANGE

The knowledge we have of sub-Antarctic weevil assemblages, and the way that past climate change affected them, suggests that in this region climate change is liable to have a pronounced effect on both assemblage membership and structure. Sub-Antarctic weevils are usually polyphagous (Chown & Scholtz 1989) and are mostly eurytopic (Chown 1989), and appear therefore to be amongst those least likely to be affected by climate change (Holt 1990). Nonetheless, it has been shown here that abiotic factors have a pronounced influence on weevil assemblage structure via the differential responses of species (Chown 1993). Therefore the effects of climate change that should be most readily discerned on sub-Antarctic islands are those involving changes in the strength and/or direction of species interactions. Ives & Gilchrist (1993) came to similar conclusions concerning the effects of climate change on terrestrial systems as a result of theoretical analyses and simple laboratory experiments.

One of the strongest interspecific interactions on the sub-Antarctic islands is the effect of introduced predator species on the native biotas (Bonner 1984). House mice have been present on Marion Island for well over one century, but are absent from nearby Prince Edward Island (Crafford 1990). Although they are thought to have been responsible for between-island differences in the biomasses of their major prey species, *Pringleophaga marioni* Viette (Lepidoptera: Tineidae) and *Ectemnorhinus* weevils (Crafford & Scholtz 1987), it was originally suggested that the predators and their prey had reached an equilibrium (Gleeson & Van Rensburg 1982, Crafford 1990). Smith & Steenkamp (1990) suggested, however, that the impact of the mice could change as a result of an increase in their population density associated with climate amelioration on the islands. Matthewson (1993) argued that the summer population density of house mice had indeed increased in the 1991/92 season, but he failed to demonstrate this conclusively. Chown & Smith (1993) showed, however, that house mice had switched from feeding mainly on *Pringleophaga marioni* larvae to feeding mainly on weevils (see Murdoch & Oaten 1975 for discussion of prey switching). They also showed that mice are size-selective in their predation on weevils and have markedly affected body size in prey species compared to non-prey species. Likewise, mice were shown to be suppressing the spread of the sedge *Uncinia compacta* on Marion Island, which has come to dominate mire communities on Prince Edward Island. Based on these findings and other theoretical analyses (see Hassell *et al.* 1993, Ives & Gilchrist 1993), it seems safe to suggest that the first indications of climate change should be sought in novel changes in species interactions in the sub-Antarctic.

CONCLUSIONS

In moderately harsh, species-poor environments, both abiotic and biotic factors are liable to play equally important roles in community organization. Furthermore, these roles are liable to be more amenable to analysis in semi-closed, island-like systems than in more open, continental ones. Here, I have shown that sub-Antarctic islands are exemplars of such systems and that analysis of the weevil assemblages inhabiting these islands can contribute greatly to our understanding of the processes responsible for the geographical distribution of diversity in this region. In sub-Antarctic weevil assemblages, large-scale regional processes determine assemblage membership, whereas more local processes determine the trajectory that finally leads to current patterns in assemblage structure. In addition, the effects of current climate change are liable to be most readily discerned, not via their direct effects on the distribution of particular species, but rather through novel interspecific interactions, or changes in the strength of existing interactions.

The level of knowledge we currently have of sub-Antarctic weevils and the processes that have lead to their assemblage structure is largely a result of an interdisciplinary research effort that has made information on the geology, climate history and botany of the islands available for integration with modern, medium-term data on the faunas of these islands. In fact, sub-Antarctic research is now at the stage where we can not only explain present patterns, but where we also stand a reasonable chance of elucidating the effects that climate change is liable to have on these terrestrial ecosystems. Given that the signal of climate change is stronger on these high-latitude islands than in similar continental areas, the biotas may serve as early-warning systems, alerting us to changes we may expect in more complex continental situations. Unfortunately, abiotic factors predominate to a much greater extent in true Antarctic systems, making them far less suitable for this role. In this light, the current trend towards a reduction in sub-Antarctic research and an escalation in more continental activities should be re-evaluated.

ACKNOWLEDGEMENTS

Dr Robin Craw (formerly of the DSIR, New Zealand) and Dr G. Kuschel and T. Crosby (Landcare Research, New Zealand) provided information for this study. Prof. C. H. Scholtz, Dr M. McGeoch and Dr D. J. Browne (Department of Zoology and Entomology, University of Pretoria) commented on an earlier draft of the manuscript. M. McGeoch assisted with preparation of the manuscript. The constructive criticism of P. Vernon and an anonymous referee greatly improved the manuscript. Part of this work was written during a sabbatical with the British Antarctic Survey at Husvik, South Georgia. This research and travel to the SCAR VI Symposium on Antarctic Biology was funded by grants from the Foundation for Research Development (Pretoria) and the University of Pretoria.

REFERENCES

Adams, J. M. & Woodward, F. I. 1992. The past as a key to the future: the use of palaeoenvironmental understanding to predict the effects of man on the biosphere. *Advances in Ecological Research*, **22,** 257–314.

Adamson, D. A., Whetton, P. & Selkirk, P. M. 1988a. An analysis of air temperature records for Macquarie Island: decadal warming, ENSO cooling and southern hemisphere circulation patterns. *Papers and Proceedings of the Royal Society of Tasmania*, **122,** 107–112.

Adamson, D. A., Selkirk, P. M. & Colhoun, E. A. 1988b. Landforms of aeolian, tectonic and marine origin in the Bauer Bay – Sandy Bay region of sub-Antarctic Macquarie Island. *Papers and Proceedings of the Royal Society of Tasmania*, **122,** 65–82.

Arthur, W. 1987. *The niche in competition and evolution*. New York: Wiley, 175 pp.

Bennett, K. D. 1990. Milankovitch cycles and their effects on species in ecological and evolutionary time. *Paleobiology*, **16,** 11–21.

Block, W. 1984. Terrestrial microbiology, invertebrates and ecosystems. In Laws, R. M., ed. *Antarctic ecology*, Vol. 1. London: Academic Press, 163–236.

Block, W. 1990. Cold tolerance of insects and other arthropods. *Philosophical Transactions of the Royal Society of London*, **B326,** 613–633.

Block, W., Sømme, L., Ring, R., Ottesen, P. & Worland, M. R. 1988. Adaptations of arthropods to the sub-Antarctic environment. *British Antarctic Survey Bulletin*, No. 81, 65–67.

Bonner, W. N. 1984. Introduced mammals. In Laws, R. M., ed. *Antarctic ecology*, Vol. 1. London: Academic Press, 237–278.

Brooks, D. R. 1988. Scaling effects in historical biogeography: a new view of space, time and form. *Systematic Zoology*, **37,** 237–244.

Brooks, D. R. & McLennan, D. T. 1991. *Phylogeny, ecology and behaviour. A research programme in comparative biology*. Chicago: University of Chicago Press, 434 pp.

Brooks, D. R. & Wiley, E. O. 1988. *Evolution as entropy: towards a unified theory of biology*. 2nd edn. Chicago: University of Chicago Press, 415 pp.

Brown, J. H. 1994. The ecology of coexistence. *Science*, **263,** 995–996.

Brundin, L. Z. 1988. Phylogenetic biogeography. In Myers, A. A. & Giller, P. S., eds. *Analytical biogeography: an integrated approach to the study of animal and plant distributions*. London: Chapman & Hall, 343–369.

Burger, A. E. 1985. Terrestrial food webs in the sub-Antarctic: island effects. In Siegfried, W. R., Condy, P. R. & Laws, R. M., eds. *Antarctic nutrient cycles and food webs*. Berlin: Springer, 582–591.

Cammell, M. E. & Knight, J. D. 1992. Effects of climate change on the population dynamics of crop pests. *Advances in Ecological Research*, **22,** 117–162.

Carpenter, S. R., Frost, T. M., Kitchell, J. F. & Kratz, T. K. 1993. Species dynamics and global environmental change: a perspective from ecosystem experiments. In Karieva, P. M., Kingsolver, J. G. & Huey, R.B., eds. *Biotic interactions and global change*. Sunderland, MA: Sinauer Associates, 267–279.

Chown, S. L. 1989. Habitat use and diet as biogeographic indicators for sub-Antarctic Ectemnorhinini (Coleoptera: Curculionidae). *Antarctic Science*, **1,** 23–30.

Chown, S. L. 1990a. Possible effects of Quaternary climate change on the composition of insect communities of the South Indian Ocean Province Islands. *South African Journal of Science*, **86,** 386–391.

Chown, S. L. 1990b. Speciation in the sub-Antarctic weevil genus *Dusmoecetes* Jeannel (Coleoptera: Curculionidae). *Systematic Entomology*, **15,** 283–296.

Chown, S. L. 1992. A preliminary analysis of sub-Antarctic weevil assemblages: local and regional patterns. *Journal of Biogeography*, **19,** 87–98.

Chown, S. L. 1993. Desiccation resistance in six sub-Antarctic weevils (Coleoptera: Curculionidae): humidity as an abiotic factor influencing assemblage structure. *Functional Ecology*, **7,** 318–325.

Chown, S. L. 1994. Historical ecology of sub-Antarctic weevils (Coleoptera: Curculionidae): patterns and processes on isolated islands. *Journal of Natural History*, **28,** 411–433.

Chown, S. L. & Scholtz, C. H. 1989. Cryptogam herbivory in Curculionidae from the sub-Antarctic Prince Edward Islands. *Coleopterists Bulletin*, **43,** 165–169.

Chown, S. L. & Smith, V. R. 1993. Climate change and the short-term impact of feral house mice at the sub-Antarctic Prince Edward Islands. *Oecologia*, **96,** 508–516.

Chown, S. L. & Van Drimmelen, M. 1992. Water balance and osmoregulation in weevil larvae (Coleoptera: Curculionidae: Brachycerinae) from three different habitats on sub-Antarctic Marion Island. *Polar Biology*, **12,** 527–52.

Chown, S. L., Van der Merwe, M. & Smith, V. R. in press. The influence of habitat and altitude on oxygen uptake in sub-Antarctic weevils. *Physiological Zoology*.

Clapperton, C. M. & Sugden, D. E. 1976. Maximum extent of glaciers in part of West Falkland. *Journal of Glaciology*, **17,** 73–77.

Clarke, A. 1991. What is cold adaptation and how should we measure it? *American Zoologist*, **31,** 81–92.

Clarke, A. 1993. Seasonal acclimatization and latitudinal compensation in metabolism: do they exist? *Functional Ecology*, **7,** 139–149.

Connell, J. H. 1980. Diversity and the coevolution of competitors, or the ghost of competition past. *Oikos*, **35,** 131–138.

Coope, G. R. 1979. Late Cenozoic fossil Coleoptera: evolution, biogeography, and ecology. *Annual Review of Ecology and Systematics*, **10,** 247–267.

Crafford, J. E. 1990. The role of feral house mice in ecosystem functioning on Marion Island. In Kerry, K. R. & Hempel, G., eds. *Antarctic ecosystems, ecological change and conservation*. Berlin: Springer, 359–364.

Crafford, J .E. & Chown, S. L. 1993. Respiratory metabolism of sub-Antarctic insects from different habitats on Marion Island. *Polar Biology*, **13,** 411–415.

Crafford, J. E. & Scholtz, C. H. 1987. Quantitative differences between the insect faunas of sub-Antarctic Marion and Prince Edward Islands: a result of human intervention? *Biological Conservation*, **40,** 255–262.

Craw, R. 1988. Panbiogeography: method and synthesis in biogeography. In Myers, A. A. & Giller, P. S., eds. *Analytical biogeography: an integrated approach to the study of animal and plant distributions*. London: Chapman & Hall, 405–434.

Craw, R. 1989. New Zealand biogeography: a panbiogeographic approach. *New Zealand Journal of Zoology*, **16,** 527–547.

Davies, L. 1987. Long adult life, low reproduction and competition in two sub-Antarctic carabid beetles. *Ecological Entomology*, **12,** 149–162.

DeAngelis, D. L. 1992. *Dynamics of nutrient cycling and food webs*. London: Chapman & Hall, 270 pp.

Drake, J. A. 1990. Communities as assembled structures: do rules govern pattern? *Trends in Ecology and Evolution*, **5,** 159–164.

Drake, J. A. 1991. Community-assembly mechanics and the structure of an experimental species ensemble. *American Naturalist*, **137,** 1–26.

Dreux, P. & Voisin, J. F. 1989. Sur la systématique des genres de la sous-famille des Ectemnorrhininae (Coleoptera: Curculionidae). *Nouvelle Revue d'Entomologie (N.S.)*, **6,** 111–118.

Dunham, A. E. 1993. Population responses to environmental change: operative environments, physiologically structured models, and population dynamics. In Karieva, P. M., Kingsolver, J. G. & Huey, R. B., eds. *Biotic interactions and global change*. Sunderland, MA: Sinauer Associates, 95–119.

Dunson, W. A. & Travis, J. 1991. The role of abiotic factors in community organization. *American Naturalist*, **138,** 1067–1091.

Ernsting, G. 1993. Observations on the life cycle and feeding ecology of two recently introduced predatory beetle species at South Georgia, Sub-Antarctic. *Polar Biology*, **13,** 423–428.

Foster, D. R., Schoonmaker, P. K. & Pickett, S. T. A. 1990. Insights from paleoecology to community ecology. *Trends in Ecology and Evolution*, **5,** 5–21.

Friis-Christensen, E. & Lassen, K. 1991. Length of the solar cycle: an indicator of solar activity closely associated with climate. *Science*, **254,** 698–700.

Gleeson, J. P. & Van Rensburg, P. J. J. 1982. Feeding ecology of the house mouse *Mus musculus* Linnaeus on Marion Island. *South African Journal of Antarctic Research*, **12,** 34–39.

Goodess, C. M. & Palutikoff, J. P. 1992. The development of regional climate scenarios and the ecological impact of greenhouse gas warming. *Advances in Ecological Research*, **22,** 33–62.

Grant, P. R. 1986. *Ecology and evolution of Darwin's finches*. Princeton: Princeton University Press, 458 pp.

Greenslade, P. 1990. Notes on the biogeography of the free-living terrestrial invertebrate fauna of Macquarie Island with an annotated checklist. *Papers and Proceedings of the Royal Society of Tasmania*, **124,** 35–50.

Gremmen, N. J. M. 1981. *The vegetation of the sub-Antarctic islands Marion and Prince Edward*. The Hague: Dr. W. Junk Publishers, 149 pp.

Gressitt, J. L. 1964. Insects of Campbell Island: introduction. In Gressitt, J. L., ed. *Insects of Campbell Island. Pacific Insects Monograph*, **7,** 3–33

Gressitt, J. L. 1970. Sub-Antarctic entomology and biogeography. In Gressitt, J. L., ed. *Sub-Antarctic entomology, particularly of South Georgia and Heard Island. Pacific Insects Monograph*, **23,** 295–374.

Gressitt, J. L. & Wise, K. A. J. 1971. Entomology of the Aucklands and other islands south of New Zealand: introduction. In Gressitt, J. L., ed. *Entomology of the Aucklands and other islands south of New Zealand. Pacific Insects Monograph*, **27,** 1–45.

Groom, M. J. & Schumaker, N. 1993. Evaluating landscape change: Patterns of worldwide deforestation and local fragmentation. In Karieva, P. M., Kingsolver, J. G. & Huey, R. B., eds. *Biotic interactions and global change*. Sunderland: Sinauer Associates, 24–44.

Haderspeck, W. & Hoffmann, K. H. 1991. Thermal properties of digestive enzymes of a sub- Antarctic beetle, *Hydromedion sparsutum* (Coleoptera: Perimylopidae) compared to those in two thermophilic insects. *Comparative Biochemistry and Physiology*, **100A,** 595–598.

Hall, K. J. 1990. Quaternary glaciation in the southern ocean: Sector 0° long – 180° long. *Quaternary Science Reviews*, **9,** 217–228.

Hansen, J. & Lebedeff, S. 1987. Global trends of measured surface air-temperature. *Journal of Geophysical Research*, **92,** 3345–3372.

Harrison, S. 1993. Species diversity, spatial scale and global change. In Karieva, P. M., Kingsolver, J. G. & Huey, R. B., eds. *Biotic interactions and global change*. Sunderland, MA: Sinauer Associates, 388–401.

Harrisson, P. M., Rothery, P. & Block, W. 1991. Drying processes in the Antarctic collembolan *Cryptopygus antarcticus* (Willem). *Journal of Insect Physiology*, **37,** 883–890.

Hassell, M. P., Godfray, H. C. J. & Comins, H. N. 1993. Effects of global change on the dynamics of insect host–parasitoid interactions. In Karieva, P. M., Kingsolver, J. G. & Huey, R. B., eds. *Biotic interactions and global change*. Sunderland, MA: Sinauer Associates, 402–423.

Headland, R. 1984. *The island of South Georgia*. Cambridge: Cambridge University Press, 293 pp.

Holt, R. D. 1990. The microevolutionary consequences of climate change. *Trends in Ecology and Evolution*, **5,** 311–315.

Humphries, C. J., Ladiges, P. Y., Roos, M. & Zandee, M. 1988. Cladistic biogeography. In Myers, A. A. & Giller, P. S., eds. *Analytical biogeography: an integrated approach to the study of animal and plant distributions*. London: Chapman & Hall, 371–404.

Ives, A. R. & Gilchrist, G. 1993. Climate change and ecological interactions. In Karieva, P. M., Kingsolver, J. G. & Huey, R. B., eds. *Biotic interactions and global change*. Sunderland, MA: Sinauer Associates, 120–146.

Johns, P. M. 1974. Arthropoda of the sub-Antarctic islands of New Zealand (1) Coleoptera: Carabidae. Southern New Zealand, Patagonian, and Falkland Islands insular Carabidae. *Journal of the Royal Society of New Zealand*, **4,** 283–302.

Kennedy, A. D. 1994. Water as a limiting factor in the Antarctic terrestrial environment: a biogeographical synthesis. *Arctic and Alpine Research*, **25,** 308–315.

Kingsolver, J. G., Huey, R. B. & Karieva, P. M. 1993. An agenda for population and community research on global change. In Karieva, P. M., Kingsolver, J. G. & Huey, R. B., eds. *Biotic interactions and global change*. Sunderland, MA: Sinauer Associates, 480–486.

Kuschel, G. 1964. Insects of Campbell Island. Coleoptera: Curculionidae of the sub-Antarctic islands of New Zealand. In Gressitt, J. L., ed. *Insects of Campbell Island. Pacific Insects Monograph*, **7,** 416–493.

Kuschel, G. 1971. Entomology of the Aucklands and other islands south of New Zealand: Coleoptera: Curculionidae. In Gressitt, J. L., ed. *Entomology of the Aucklands and other islands south of New Zealand. Pacific Insects Monograph*, **27,** 225–259.

Kuschel, G. 1991. Biogeographic aspects of the subantarctic islands. In *International symposium on: biogeographical aspects of insularity, Rome, 18–22 May 1987. Atti dei Convegni Lincei*, **85,** 575–591.

Kuschel, G. & Chown, S. L. (1995) Phylogeny and systematics of the *Ectemnorhinus*-group of genera (Insecta: Coleoptera). *Invertebrate Taxonomy*, **8,** 841–863.

Levin, S. A. 1992. The problem of pattern and scale in ecology. *Ecology*, **73,** 1943–1967.

Lewis Smith, R. I. 1984. Terrestrial plant biology of the Sub-Antarctic and Antarctic. In Laws, R. M., ed. *Antarctic ecology*, Vol. 1. London: Academic Press, 61–162.

Lubchenco, J., Olson, A. M., Brubaker, L. B., Carpenter, S. R., Holland, M. M., Hubbell, S. P., Levin, S. A., MacMahhon, J. A., Matson, P. A., Melillo, J. M., Mooney, H. A., Peterson, C. H., Pulliam, H. R., Real, L. A., Regal, P. J. & Risser, P. G. 1991. The sustainable biosphere initiative: an ecological research agenda. *Ecology*, **72,** 371–412.

Martin, A. K. & Hartnady, C. J. H. 1986. Plate tectonic development of the south west Indian Ocean: a revised reconstruction of east Antarctica and Africa. *Journal of Geophysical Research*, **91,** 4767–4786.

Matthewson, D. C. 1993. *Population biology of the house mouse (*Mus musculus *Linnaeus) on Marion Island*. MSc thesis, University of Pretoria, 133 pp. [Unpublished.]

Mercer, J. H. 1983. Cenozoic glaciation in the southern hemisphere. *Annual Review of Earth and Planetary Sciences*, **11,** 99–132.

Murdoch, W. W. 1993. Individual-based models for predicting effects of global change. In Karieva, P. M., Kingsolver, J. G. & Huey, R. B., eds. *Biotic interactions and global change*. Sunderland, MA: Sinauer Associates, 147–162.

Murdoch, W. W. & Oaten, A. 1975. Predation and population stability. *Advances in Ecological Research*, **9,** 1–131.

Myers, A. A. & Giller, P .S., eds. 1988. *Analytical biogeography an integrated approach to the study of animal and plant distribution*. London: Chapman & Hall, 578 pp.

Nur, A. & Ben-Avraham, Z. 1982. Oceanic plateaus, the fragmentation of continents, and mountain building. *Journal of Geophysical Research*, **87,** 3644–3661.

Otte, D. 1989. Speciation in Hawaiian crickets. In Otte, D. & Endler, J.A., eds. *Speciation and its consequences*. Sunderland, MA: Sinauer Associates, 482–526.

Parry, M. 1992. The potential effect of climate change on agriculture and land use. *Advances in Ecological Research*, **22,** 63–91.

Paulay, G. 1985. Adaptive radiation on an isolated oceanic island: the Cryptorhynchinae (Curculionidae) of Rapa revisited. *Biological Journal of the Linnean Society*, **26,** 95–187.

Pimm, S. L. 1991. *The balance of nature? Ecological issues in the conservation of species and communities*. Chicago: University of Chicago Press, 434 pp.

Quinn, J. F. & Karr, J. R. 1993. Habitat fragmentation and global change. In Karieva, P. M., Kingsolver, J. G. & Huey, R. B., eds. *Biotic interactions and global change*. Sunderland, MA: Sinauer Associates, 451–463.

Ricklefs, R. E. 1989. Speciation and diversity: the integration of local and regional processes. In Otte, D. & Endler, J. A., eds. *Speciation and its consequences*. Sunderland, MA: Sinauer Associates, 599–622.

Ring, R. A., Block, W., Sømme, L. & Worland, M. R. 1990. Body water content and desiccation resistance in some arthropods from sub-Antarctic South Georgia. *Polar Biology*, **10,** 581–588.

Robinson, G. S. 1984. *Insects of the Falkland Islands: a checklist and bibliography*. London: British Museum (Natural History), 38 pp.

Robinson, G. S. 1988. *New records of insects from the Falkland Islands*. London: British Museum (Natural History), 7 pp.

Rosen, B. R. 1988. From fossils to earth history: applied historical biogeography. In Myers, A. A. & Giller, P. S., eds. *Analytical biogeography: an integrated approach to the study of animal and plant distributions*. London: Chapman & Hall, 437–478.

Roughgarden, J. 1989. The structure and assembly of communities. In Roughgarden, J., May, R. M. & Levin, S. A., eds. *Perspectives in ecological theory*. Princeton: Princeton University Press, 203–226.

Schimel, D. S. 1993. Population and community processes in the response of terrestrial ecosystems to global change. In Karieva, P. M., Kingsolver, J. G. & Huey, R. B., eds. *Biotic interactions and global change.* Sunderland, MA: Sinauer Associates, 45–54.

Schneider, S. H. 1992. The climate response to greenhouse gases. *Advances in Ecological Research*, **22,** 1–32.

Schneider, S. H. 1993. Scenarios of global warming. In Karieva, P. M., Kingsolver, J. G. & Huey, R. B., eds. *Biotic interactions and global change*. Sunderland, MA: Sinauer Associates, 9–23.

Schulze, B. R. 1971. The climate of Marion Island. In Van Zinderen Bakker, E. M., Winterbottom, J. M. & Dyer, R. A., eds. *Marion and Prince Edward Islands. Report on the South African biological and geological expedition 1965–1966.* Cape Town: A. A. Balkema, 16–31.

Serra-Tosio, B. 1986. Un nouveau chironomide Antarctique des Îles Crozet, *Parochlus crozetensis* n. sp. (Diptera, Nematocera). *Nouvelle Revue d'Entomologie (N.S.)*, **3,** 149–159.

Sibly, R.M. & Calow, P. 1986. *Physiological ecology of animals: an evolutionary approach*. Oxford: Blackwell Scientific Publications, 179 pp.

Skottsberg, C. 1960. Remarks on the plant geography of the southern cold temperate zone. *Proceedings of the Royal Society of London*, B **152,** 447–457.

Smith, V. R. & French, D. D. 1988. Patterns of variation in the climates, soils and vegetation of some sub-Antarctic and Antarctic islands. *South African Journal of Botany*, **54,** 35–46.

Smith, V. R. & Steenkamp, M. 1990. Climate change and its ecological implications at a sub- Antarctic island. *Oecologia*, **85,** 14–24.

Sømme, L. & Block, W. 1991. Adaptations to alpine and polar environments in insects and other terrestrial arthropods. In Lee, Jr., R. E. & Denlinger, D. L., eds. *Insects at low temperature*. London: Chapman and Hall, 318–359.

Sømme, L., Ring, R. A., Block, W. & Worland, M. R. 1989. Respiratory metabolism of *Hydromedion sparsutum* and *Perimylops antarcticus* (Col., Perimylopidae) from South Georgia. *Polar Biology*, **10,** 135–139.

Southwood, T. R. E. 1987. The concept and nature of the community. In Gee, J. H. R. & Giller, P. S., eds. *Organization of communities past and present*. Oxford: Blackwell Scientific Publications, 3–27.

Stonehouse, B. 1982. La zonation écologique sous les hautes latitudes australes. *Comité National Française des Recherches Antarctiques*, **51,** 531–537.

Tréhen, P., Bouche, M., Vernon, P. & Frenot, Y. 1985. Organization and dynamics of Oligochaeta and Diptera on Possession Island. In Siegfried, W. R., Condy, P. R. & Laws, R. M., eds. *Antarctic nutrient cycles and food webs*. Berlin: Springer, 606–613.

Walton, D. W. H. 1984. The terrestrial environment. In Laws, R. M., ed. *Antarctic ecology*, Vol. 1. London: Academic Press, 1–60.

Wharton, G. W. 1985. Water balance of insects. In Kerkut, G. A. & Gilbert, L. I., eds. *Comparative insect biochemistry, physiology and pharmacology*, Vol. 4. Oxford: Pergamon Press, 565–601.

Williamson, M. 1981. *Island populations*. Oxford: Oxford University Press, 286 pp.

Williamson, P. E. 1988. Origin, structural and tectonic history of the Macquarie Island region. *Papers and Proceedings of the Royal Society of Tasmania*, **122,** 27–43.

Worland, R., Block, W. & Rothery, P. 1992. Survival of sub-zero temperatures by two South Georgian beetles (Coleoptera, Perimylopidae). *Polar Biology*, **11,** 607–613.

Worland, R., Block, W. & Rothery, P. 1993. Ice nucleation studies of two beetles from sub-Antarctic South Georgia. *Polar Biology*, **13,** 105–112.

Young, S. R. 1979. Respiratory metabolism of *Alaskozetes antarcticus*. *Journal of Insect Physiology*, **25,** 361–369.

22 Microbial communities in soils and on plants from King George Island (Arctowski Station, Maritime Antarctica)

MANFRED BÖLTER
Institut für Polarökologie, Wischhofstraße 1-3, D-24148 Kiel, Germany

ABSTRACT

A microscopical study was performed on samples of soils and plants from King George Island, Maritime Antarctica. Counts were made of phototrophs (cyanobacteria and eukaryotic algae) and bacteria using epifluorescence microscopy with respect to their contributions on numbers and biomass of microorganisms in different habitats. Sampling was carried out during austral summer 1992/93 at sites with different plant cover in the area of Arctowski Station. Numbers and biomass of bacteria and phototrophs (algae and cyanobacteria) were found to be highest in surface layers of soils which contained high amounts of organic matter. Active shoots of Deschampsia antarctica *and* Colobanthus quitensis *were colonized by dense populations of eukaryotic algae, cyanobacteria and bacteria, whereas lichens (*Usnea antarctica, U. aurantiaco-atra*) were much less covered with bacteria and nearly no algae were found on these species.* D. antarctica, C. quitensis, *mosses and soil surfaces were mainly colonized by diatoms. Deeper layers of soils show high contributions of cyanobacteria. Biovolumes of these parts of phototrophs are similar to those of the concomitant bacterial flora or even exceed it considerably. Fungi and yeasts were seldom found.*

Key words: Maritime Antarctica, soils, bacteria, cyanobacteria, algae, pigments.

INTRODUCTION

Living biomass of soils of the Antarctic comprises mainly microorganisms, i.e. algae, fungi, yeasts, cyanobacteria and bacteria. Higher organisms, e.g. nematodes or microarthropods, are much rarer (Vishniac 1993, Block 1994). Eukaryotic algae and cyanobacteria represent important constituents of phototrophs in these soils (Wynn-Williams 1990), as well as microlichens. Microorganisms are thus of special interest for studies of cycles of organic matter. The occurrence of heterotrophic organisms is closely related to the availability of inorganic nutrients and organic matter and water supply. Low bioturbation and slow digestion of organic material determine cycling and storage of organic matter. Its degradation is mainly due to microorganisms (Block 1984, Smith 1985).

Close relationships between autotrophic and heterotrophic organisms have been shown during studies of the production of dissolved and particulate organic matter and its decomposition (Tearle 1987, Roser *et al.* 1992, Melick & Seppelt 1992, Bölter 1989, 1990a, Roser *et al.* 1994). These studies have shown the high contributions of different carbohydrates and polyols of either intact cells from plants or algae, or of leaching products. Such products serve as preferred substrates for bacteria, either directly (e.g. glucose, fructose) or after hydrolytic splitting of disaccharides via exoenzymes (Bölter 1993, Melick *et al.* 1994).

Further, high amounts of microbial biomass as estimated by indirect methods (e.g. ATP, lipid phosphate, CO_2-evolution, heat production) have shown that phototrophs can be regarded as major contributors to living biomass (Bölter 1989, 1990a,b), a

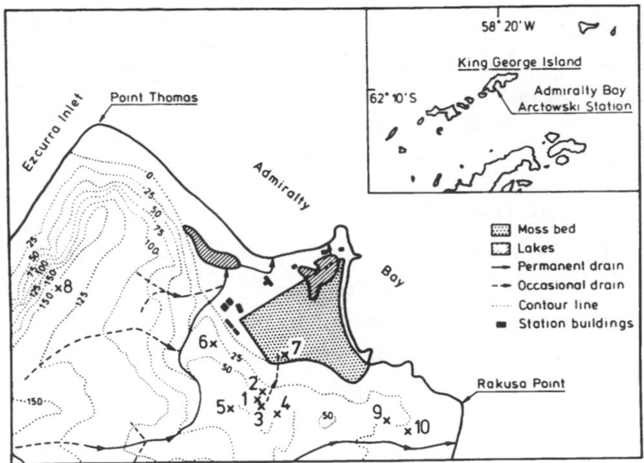

Fig. 22.1. Map of the sampling locations at King George Island, Antarctica. (after Rakusa-Suszczewski & Krzyszowska, 1991). Site descriptions and sample numbers are given in Table 22.1.

fact which often leads to mismatches between estimations of microbial biomass in terrestrial systems (Bölter 1994a, b).

Thus, it seemed meaningful to look into these habitats using direct microscopy for the contributions of individual parts of the microbial community. This study concerns microscopic estimates of numbers and biovolumes of bacteria, cyanobacteria and eukaryotic algae in different Antarctic soils and on plants.

The procedure uses epifluorescence microscopy for direct studies of microhabitats. Both acridine orange staining for bacterial counts and autofluorescence of phototrophs are used to identify organisms in the microbial community. This approach should give more insights into the actual microbiological habitats and the community structure in relation to individual soil environments. The aim is to show relationships between phototrophs and heterotrophs at the scale of bacteria and algae, and to investigate their communities in different soil and plant habitats.

MATERIALS AND METHODS

Sites

King George Island and the vicinity of Arctowski Station provides an area with different types of soils and plant cover; extended areas of meadows with higher plants can be found as well as large lichen heaths and barren soils in the fjells. Comprehensive descriptions of soils and vegetation patterns of the Admiralty Bay region have been published by Bölter (1992a), Olech (1993), Fabiszewski & Wojtuń (1993) and Zarzycki (1993).

Sampling

Samples were taken from various sites in the vicinity of Arctowski Station (Admiralty Bay) during December 1992 and January 1993. Sampling sites for this study are presented in Fig. 22.1. They were located in different environments of the Admiralty Bay area, representing the main habitats: stands of higher plants, lichens and mosses, fjell regions with sparse plant cover, and barren soils of wet and dry areas.

Soil surfaces (depths 0–0.5 cm and 0–4 cm) were sampled from sites 1–10 and on a patch (≈2 m^2) of diverse plant cover (site 5) which showed *Deschampsia antarctica, Colobanthus quitensis, Usnea antarctica, Polytrichum* sp., *Drepanocladus uncinatus* covered by *Ochrolechia frigida* and *Candelariella vitellina*, and *Placopsis contortuplicata*. Two depth profiles close to sites 1 (samples 64–67) and 2 (samples 52–57) were sampled in 1 cm steps. Green shoots of plants were sampled for analysis of the colonization of microorganisms. A short summary of the sampling sites is given in Table 22.1. A more detailed description of all sampling sites is given in Bölter (1995).

Rakusa-Suszczewski *et al.* (1993) present a meteorological compilation for this region. Bölter (1995) presents soil temperatures of depth profiles at sites 1–3 for 28 December 1992 to 16 January 1993 and for site 4 for 16 January 1993 to 22 January 1993. In brief, temperatures (°C, 1 cm depth) were:

site 1: min.: −2.0, max.: 21.6, mean: 6.4 (n=2625);
site 2: min.: −2.4, max.: 35.2, mean: 7.9 (n=2625);
site 3: min.: −2.4, max.: 35.2, mean: 7.4 (n=2625);
site 4: min.: −2.8, max.: 26.8, mean: 6.4 (n=864);
air (site 1, 60 cm above ground): min.: −3.2, max.: 12.0, mean: 2.8 (n=2404);
air (site 4, 80 cm above ground): min.: −2.8, max.: 8.0, mean: 1.6 (n=864).

METHODS

Water content was analysed on fresh samples in the field lab (drying at 105 °C for 24 h). Organic matter was analysed via loss on ignition (LOI) after combustion of dry samples at 550 °C for 24 h.

Plant pigments: Chlorophylls (*a, b, c*), carotenoids and phaeopigments were analysed according to the spectrophotometric method for algal pigments as described by Parsons *et al.* (1985).

Microscopic analyses: A Zeiss Standard Microscope equipped with a blue light illuminator for epifluorescence microscopy was used for bacterial (Plan 100/1.25) and for phototroph counts (Plan 63/1.2 and Neofluar NPL 25/0.8). Analyses were carried out immediately after sampling in the laboratory at Arctowski Station. Total bacterial numbers (TBN) were determined by epifluorescence microscopy (acridine orange staining) on polycarbonate membranes (pores 0.2 μm). Four hundred cells per sample were classified into sizes of cocci (<0.5 μm, 0.5–1 μm, >1 μm) and rods (length 0.5–1 μm, 1–2 μm, 2–3 μm, >3 μm) for further calculations on bacterial biovolume (BBV), bacterial biomass (TBB) and mean cell volumes (MCV). Methodological details are described by Bölter *et al.* (1993).

Phototrophs (TAN) were counted in aliquots by using the autofluorescence of the chlorophyll (green and blue excitation), on polycarbonate membranes (pores 3.0 μm); 50 microscopic

Table 22.1. *Description of sites and samples used for this study from Arctowski region. Sample numbers refer to soils if not otherwise stated*

Site	Description	Sample (no.:origin)
1	Meadow with dense plant cover of *Deschampsia antarctica* and *Colobanthus quitensis*, as well as some moss cushions (*Drepanocladus uncinatus, Polytrichum* sp.)	1: 0–4 cm 2: 4–8 cm 3: 8–12 cm 4: 12–16 cm 5: 16–20 cm 64: *D. antarctica* 65: 0–0.5 cm 66: 0.5–1.5 cm 67: 1.5–2.5 cm 78: *D. antarctica* shoot 79: *D. antarctica* root 83: *C. quitensis*, shoot 84: *C. quitensis*, root
2	Mineral soil lacking plant cover (Regosol)	8: 0–4 cm 51: 0–0.5cm 52: 0.5–1.5 cm 53: 1.5–2.5 cm 54: 2.5–3.5 cm 55: 3.5–4.5 cm 56: 4.5–5.5 cm 57: 5.5–6.5 cm 98: 0–0.5 cm
3	Moss cushions of *D. uncinatus*, covered with the epiphytic lichen *Ochrolechia frigida*	16: 0–4 cm 17: 4–8 cm 18: 8–12 cm 19: 14–18 cm 20: 20–24 cm 99: 0–0.5 cm
4	Meadow of *D. antarctica* and *C. quitensis*, with few cushions of *D. uncinatus*	24: 0–4 cm 25: 4–8 cm 26: 15–19 cm 27: 21–25 cm 90: moss with *O. frigida*
5	Barren soil inbetween some rocks and stones covered with *Usnea antarctica*, altitude. *c*. 50 m asl (Regosol)	29: 0–4 cm 30: 4–8 cm 80: *D. antarctica*, shoot 81: *D. antarctica*, root 85: *C. quitensis*, shoot 86: *C. quitensis*, root 87: *D. uncinatus* with microlichens 88: *Polytrichum* sp., shoots 92, 93: *U. antarctica* 97: *P. contortuplicata* 100: soil 0–0.5 cm 104: soil 0–0.5 cm
6	Barren muddy soil, no plant cover, altitude *c*. 40 m asl	32: 0–4 cm 101: 0–0.5 cm
7	Wet waterlogged plain with mainly *D. uncinatus* and *Polytrichum* sp., at sea level	35: 0–4 cm 36: 5–9 cm 37: 10–14 cm 38: 16–20 cm 91: *D. uncinatus*, shoots
8	Barren soil from a large plain at Panorama Ridge, altitude *c*. 150 m asl	102: 0–0.5 cm
9	Moraine near Ecology Glacier with a well-established cover of lichens (mainly *U. antarctica, O. frigida, Buellia caniops, B.* sp., *Rhizocarpon geographicum, Lepraria neglecta*) and mosses (mainly *D. uncinatus*), and higher plants (*D. antarctica* and *C. quitensis*), altitude *c*. 30 m asl	44: 0–4 cm 45: 7–11 cm 96: *O. frigida*
10	Moraine close to Ecology Glacier, no plant cover, altitude *c*. 20 m asl	48: 0–4 cm 103: 0–0.5 cm

fields were inspected. Small unicellular algae (<2 μm in diameter) may have been excluded from detection and counting. Cells were classified into the following morphological groups: cylinders, ellipsoids, coccoids and filaments. Biovolumes of phototrophs (TAV) were calculated from measurements of diameters (for coccoid cells), length and width (for cylinders and filaments) and the parameters of ellipsoids. No species determinations were carried out for this study.

RESULTS

Data on actual water contents and loss on ignition are presented in Table 22.2. More detailed results of analyses of soil properties (pH, stone content, C/N) and of the bacterial communities are given by Bölter (1995). In brief, increasing pH values are found with increasing depths, maximal contents of organic matter are found in surface layers with active plant covers, high C/N ratios (>20) are mainly associated with high amounts of organic matter, low C/N ratios (<10) are found in connection with low amounts of organic matter.

Pigments

The analyses of the photosynthetic pigments give some indications about the phototrophic communities of the samples. They are analyzed in samples of sites 1–10 (Table 22.2). The highest chlorophyll contents were found in the surface layers; they decreased towards deeper layers. Chlorophyll was present in significant amounts (>1 $\mu g\ g^{-1}$) only in soil horizons above 15 cm. Samples with plant cover contained more than 10 $\mu g\ g^{-1}$ of chlorophyll and at some places contained even more than 40 $\mu g\ g^{-1}$. Such high levels, however, were due to plant tissues. Elevated levels of other

Table 22.2. *Sample characteristics and contents of plant pigments (chlorophylls, carotenoids, phaeopigments, µg g⁻¹ dry wt) as found in samples from sites 1–10. Only those samples where at least one of the pigments was found are shown. 'Percentage phaeopigments' relates to total chlorophylls. Conversion from chlorophyll* a *to algal biomass was performed according to Roser* et al. *(1993b)*

Site	Sample no.	Depth (cm)	H_2O[1]	LOI[2]	Chl *a*	Chl *b*	Chl *c*	Carot.	Phaeop.	% Phaeop. of chl	Algal biomass (mg C g^{-1})
1	1	0–4[3]	30.2	24.5	142.9	103.4	5.5	82.2	3.1	1.2	2.86
1	2	4–8	23.7	6.3	2.5	1.3	1.0	0	0.5	10.1	0.05
1	3	8–12	20.9	5.6	2.4	1.7	0.3	0	0.2	5.1	0.05
1	4	12–16	23.6	5.1	2.3	1.4	0	0	0.5	11.3	0.05
1	5	16–20	19.5	5.0	0	0	0	0	0.1	100	0
1	65	0–0.5	36.5	21.7	0.9	0.4	0	0	0.2	11.2	0.02
1	66	0.5–1.5	28.4	11.5	0.9	0.7	0	0	0.03	2.3	0.02
2	8	0–4	14.0	3.2	0.5	0	0	0	0.1	9.2	0.01
2	51	0–0.5	10.5	3.4	14.9	10.2	5.8	0.7	0.2	0.7	0.30
2	52	0.5–1.5	11.0	3.3	8.1	4.0	3.3	0	0.2	1.1	0.16
2	53	1.5–2.5	14.2	3.4	6.3	4.7	0	0.3	0.3	2.5	0.13
2	54	2.5–3.5	15.7	3.6	2.1	1.3	0	0	0.2	6.7	0.04
2	55	3.5–4.5	16.1	3.6	2.6	1.0	0	0	0.1	3.8	0.05
2	56	4.5–5.5	17.0	3.3	1.4	0.6	0	0	0.1	5.7	0.03
2	57	5.5–6.5	16.3	3.2	1.8	0.6	1.1	0	0.1	4.3	0.04
3	16	0–4	24.5	15.3	8.1	5.9	2.9	0.1	0.7	3.9	0.16
3	17	4–8	22.1	6.4	3.1	2.5	0	2.5	0.2	2.6	0.06
3	18	8–12	19.1	6.4	2.0	1.2	0	0	0.2	6.1	0.04
3	19	14–18	15.9	4.2	0.5	0	0	0	0.2	29.6	0.01
3	20	20–24	14.9	3.5	0.5	0.2	0.3	0	0.2	14.0	0.01
4	24	0–4	24.7	10.7	63.1	44.8	1.9	18.7	3.4	3.0	1.26
4	25	4–8	32.5	13.1	6.4	4.5	0.9	0	0.4	3.4	0.13
4	26	15–19	31.9	12.3	5.3	4.3	1.9	0	0.6	4.9	0.11
4	27	21–25	33.0	3.0	0.5	0.2	0	0	0.2	16.7	0.01
5	29	0–4	11.2	5.3	0.5	0	0	0	0.1	11.7	0.01
5	30	4–8	22.9	4.9	0.4	0.6	0	0	0.1	10.1	0.01
6	32	0–4	23.1	6.8	0.8	0.5	0.1	0.1	0.2	13.4	0.02
7	35	0–4	67.1	51.1	69.2	46.1	1.2	2.4	3.8	3.1	1.38
7	36	5–9	66.4	15.7	38.8	32.1	1.2	0.7	3.1	4.1	0.78
7	37	10–14	26.6	6.0	8.9	5.4	0.9	0	1.2	7.0	0.18
7	38	16–20	19.3	2.4	0.1	0.2	0	0	0.1	15.0	>0.01
9	44	0–4	21.7	9.9	26.9	18.4	0.4	1.9	2.1	4.5	0.54
9	45	7–11	15.7	4.3	1.2	0.6	0.1	0	0.3	11.7	0.02
10	48	0–4	16.5	4.4	0.4	0.2	0	0	0.1	15.0	>0.01

Notes:

[1] H_2O: actual water content (% of fresh weight, 105 °C, 24 h). [2] LOI: loss on ignition (% of dry weight, 550 °C, 24 h).

[3] This sample includes active shoots of *D. antarctica*.

samples were due to algae or cyanobacteria, since no significant amounts of moss protonema or microlichens were found during the microscopic inspections; these levels of chlorophyll are generally below 20 µg g⁻¹.

After conversion into area-related data, maximal values of pigments can be found in samples 1, 24, 35 and 44, which contain 2.5, 1.6, 1.1 and 0.7 g chlorophyll m^{-2}, respectively.

Similar trends can be found with carotenoids and phaeopigments. The proportions of phaeopigments generally increased with depth (e.g. samples 51–54). Many surfaces contained less than 10% phaeopigment in relation to chlorophyll *a*, whereas other surfaces showed higher percentages. Carotenoids were found more seldom and only in few surface layers, i.e. at places with a plant cover of grass, mosses or lichens.

Ratios between chlorophyll *a* and *b* were in the range between 1 and 2 in all profiles. There are only few exceptions with values <1 (samples 30, 38) or >2 (samples 20, 48); no trend was detectable with respect to depth. Ratios between chlorophyll *a* and *c* show much greater variability. Low amounts of chlorophyll *c* led to high ratios (>10) of *a*/*c* in samples 1, 2, 35, 36 and 44. A

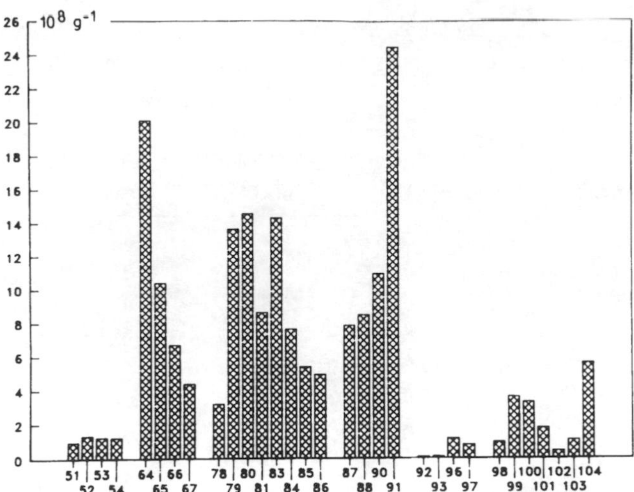

Fig. 22.2. Total bacterial counts (TBN, $n \times 10^8$ g^{-1} dry wt) of surface samples from soils, plants (*D. antarctica, C. quitensis*), mosses and lichens. (For origin of samples see Table 22.1.)

decreasing trend can be seen for this ratio with depth; values fall below 10 underneath the upper horizon.

Bacterial communities

Bacterial counts and biomass are presented in detail in Bölter (1995). Data for TBN of surface samples and plant material are displayed in Fig. 22.2. Highest values of TBN and BBM were measured in samples of soil surfaces, higher plants, shoots and roots, and mosses. Lichens show very low values, similar to those of barren soils.

The pattern of the mean cell volume (MCV) is more heterogeneous and does not show such obvious trends. Surfaces of *D. antarctica* and *C. quitensis* and mosses showed generally higher MCV than samples of related roots. High values of MCV were also found on a sample of *U. antarctica* and on *P. contortuplicata* and in the soil surface of samples 102 and 104. These data point to significantly different bacterial communities in these samples.

The size-class analysis shows that small cocci (<0.5 μm) were found only in low amounts in soil samples (samples 51–67, 98–104), samples of roots from *D. antarctica* and *C. quitensis*, samples of mosses and microlichens, and on samples of *Usnea*. Signficant amounts of large rods (>2 μm) were found mainly on plants or roots; this fraction was of minor importance for soils (except in that from sample 102 from site 8). Samples of plants also contained higher amounts of rods of 1–2 μm than soils (except samples from locations 8 and 10). Further details are given in Bölter (1995).

Phototrophs

Phototrophs were found only either with plant samples or in surface layers of soils to a maximal depth of 2.5 cm (Fig. 22.3, sample 67). A steady decrease of phototrophs with depth can be shown by samples of profiles at sites 1 and 2 (samples 51–54 and 64–67). Fruticose lichens (*U. antarctica, U. aurantiaco-atra*) were only rarely or not colonized by phototrophs (i.e. algae and/or cyanobacteria). Highest biovolumes of phototrophs were found

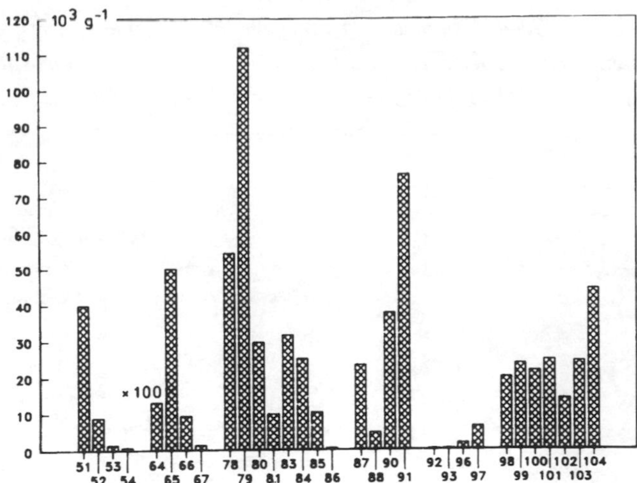

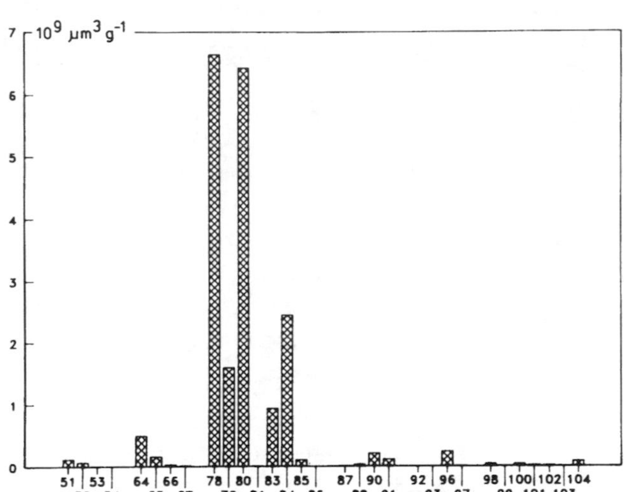

Fig. 22.3. (top) Total algal counts (TAN, $n \times 10^3$ g^{-1} dry wt), and (bottom) total algal biovolume (TAB, 10^9 μm^3) for soil and plant surfaces.

on plants (*D. antarctica, C. quitensis*, and mosses, Fig. 22.3). Shoots of *D. antarctica* and *C. quitensis* showed counts of phototrophs in the range between 10×10^3 and 54×10^3 g^{-1}, root systems in a range between 0.5×10^3 and 11.2×10^3 g^{-1}, mosses showed algal counts between 4.6×10^3 and 76×10^3 g^{-1}, and soil surfaces showed fairly constant values between 14×10^3 and 44×10^3 g^{-1}.

Greater variability in the distributions of phototrophs (using a morphological classification) was found for the individual samples than for the microscopic counts (Fig. 22.4). Although high numbers of small coccoid phototrophs can be found in soil samples (up to 90% of the total community, for the surface sample of site 8), these morphotypes did not show a significant contribution to the total biovolume of phototrophs. Cylindric and ellipsoid forms, probably diatoms according to their shape, represent the dominant community. Filamentous forms with high contributions (> 50%) to total biovolume were found only in soil samples of sites 3 and 10, as well as in the root system of *C. quitensis*, and on the lichens *P. contortuplicata* and *U. antarctica*.

In terms of biovolume, coccoid forms were of minor importance. This holds especially true for plant surfaces (shoots and

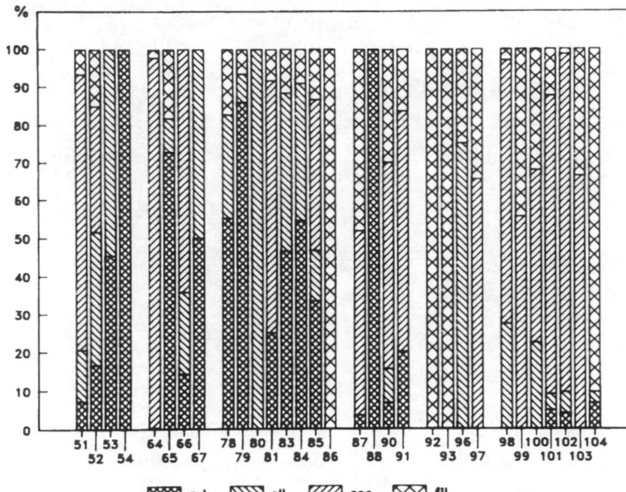

Fig. 22.4. Proportions of dominant algal forms as a percentage of the total algal number (TAN) for soil and plant surfaces (cyl=cylindric cells, ell=ellipsoid cells, coc=coccoid cells, fil=filamentous cells).

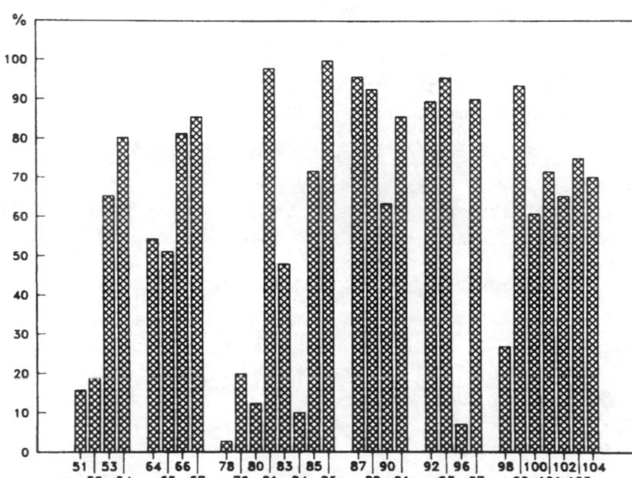

Fig. 22.5. Proportions of bacterial biovolume (TBV) as a percentage of the total microbial biovolume (bacterial biovolume+phototroph biovolume).

roots) where the algal biovolume was represented mainly by diatoms. Exceptions can be found only with one root of *C. quitensis* (100% filaments) and a moss cushion which was overgrown with *O. frigida*, where cylindric, coccoid and filaments shared the algal biovolume in equal parts. No cylindric or ellipsoid forms were found on fruticose lichens (*U. antarctica*) which indicates that these plants differ significantly from other plants with respect to their epiphytic community. Other morphological forms, such as *Cosmarium* sp. and *Calothrix* sp., were found only sporadically and they did not contribute significantly to total algal biovolume.

Phototroph biovolume versus bacterial biovolume showed the dominance of bacteria (>50% of the total biovolume, TBV) in most soil samples, except the surface layer of site 2, most root systems and most lichens (Fig. 22.5). Ratios between 0.25 and 1 were found in surface samples of most soils, and on shoots of *C. quitensis*. Comparisons of biovolumes from phototrophs (algae and cyanobacteria) and heterotrophs (bacteria) show that the biovolume of heterotrophs represents 15.7% of the total biovolume (phototrophs+heterotrophs) in sample 51, and 54.4% in sample 65. In deeper horizons (where algae were present) those figures rise to 80 and 85%, respectively. Shoots of *D. antarctica* had 3–12% bacterial biovolume, shoots of *C. quitensis* 48–52% bacterial biovolume of total biovolume (phototrophs+heterotrophs). Corresponding root systems had 10–99% bacterial biovolume, depending on depth. Mosses (*D. uncinatus and Polytrichum* sp.) showed bacterial abundance between 26 and 96%, whereas lichens had mainly more than 90% bacteria in the total epiphytic cover.

DISCUSSION

These data give a detailed picture of the microbial environment of Antarctic soils with respect to bacteria and phototrophic organisms. Both are important constituents for the description and understanding of processes within this environment.

Bacterial numbers were between 10^8 and 10^9 per gram of soil. These values are somewhat lower than reported from other studies in Antarctic soils (e.g. Ramsay 1983, Wynn-Williams 1985, Ramsay & Stannard 1986, Heatwole *et al.* 1989, Roser *et al.* 1993a, b), but they match well with data from my own earlier studies in this environment (Bölter 1990a, 1992a, b). The high values for mean cell volumes, which are generally related to soil horizons with high contents of organic matter and/or plant cover, show that significant divisons for the bacterial community can be related to edaphically rich and poor sites, similar to results from French & Smith (1986) from Marion Island. Hence, this descriptor becomes important in order to recognize nutrional stages of soils via this bacterial property.

Pigment analyses have shown that chlorophylls can be found only in the top layers, which are influenced either by higher plants or algae. Data in the ranges described here are similar to those from Ohtani *et al.* (1991) from Yukidori Valley close to Syowa Station in continental Antarctica. Area-related data (cf. Table 22.2) correspond to values from soil surfaces from Wilkes Land (Heatwole *et al.* 1989). Data from other surface samples from areas not covered by plants are similar to those from Signy Island (Davey 1988, 1991).

There are only a few studies which deal with phaeopigments in Antarctic soils. Davey (1988, 1991) shows some data for phaeophytin from Signy Island. He finds high amounts of such degraded chlorophyll mainly during the growth season (December–March), i.e. when algae are actively changing growth patterns. Also, he shows increasing concentrations of phaeophytin with an increasing distance from the polygons, and with increasing total plant pigments (Davey 1988). Such an effect is not found during this study: surface samples show generally low values for degraded chlorophyll, increasing in deeper layers.

Ratios between chlorophyll *a* and *c* have been described by Heatwole *et al.* (1989). They report ratios of 13.7 for soil from

Nelly Island (close to bird colonies) and 1.8 for soils from Casey Station. Elevated values for this ratio found during this study cannot be related to the influence of bird colonies. They are related to the plant cover, mainly to mosses and lichens. Elevated levels of chlorophyll *c* (>0.2 $\mu g\ g^{-1}$) found in the surface layers of certain sites indicate substantial amounts of diatoms and xanthophytes.

Counts of algae vary between 10^2 and 10^5 per gram dry weight. Algae are found in soil surfaces and on plants as epiphytes. Depth profiles show an occurrence of algal cells down to 2.5 cm, with their numbers decreasing strongly with depth. This has to be attributed to decreasing light levels in soils, since appropriate light conditions limits their living space with regard to depth. These results correspond to analyses of other authors who have also found highest numbers on plants (e.g. Broady 1979a, b, 1982, 1989; Ohtani & Kanda 1987; Ohtani *et al.* 1991) or in relation to soil particles (e.g. Davey & Clarke 1991, Davey *et al.* 1991) where algae and cyanobacteria play an important role in the stabilization of fellfield soils. Further, high algal counts are found below herbaceous vegetation and are highest at places with *D. antarctica* and *C. quitensis* rather than at places with mosses (e.g. *D. uncinatus*; Broady 1979a).

Analyses of algal communities in other Antarctic soils have shown that diatoms and cyanobacteria are dominant groups (Broady 1979a, b, Davey 1988). Those analyses, however, were mainly based on culture counts of algal propagules or other semiquantative techniques. This makes direct comparisons difficult, although trends similar to those found here are shown in Broady's studies from Signy Island (Broady 1979a).

The morphological grouping carried out during this study shows that soil surfaces (0–0.5 cm) are dominated by coccoid and filamentous forms. Filaments were similar to those presented by Davey *et al.* (1991) and Davey & Clarke (1991). Other filaments resemble *Pseudanabaena* sp. and *Phormidium* sp. The latter has been described as the most succesful phototroph organism in soils from Signy Island by Davey (1988). Coccoid forms found in the samples from Arctowski have shown only small diameters (~5–10 μm). Only a few had sizes of e.g. *Planktosphaerella* sp., and most of them are similar to *Synechococcus* sp. or *Gloeocapsa* sp. Deeper layers contained cylindric and ellipsoid forms in higher numbers, which are probably diatoms, due to their morphology: cylindric forms, as found in those samples, can be regarded as *Pinnularia* sp. (cf. figure 2d in Davey *et al.* 1991); ellipsoids can be described as forms of *Navicula* sp.

These preliminary identifications, however, have to be confirmed by taxonomic studies. Ohtani (pers. commun.) finds a great variety of both cyanobacteria and green algae (filamentous and unicellular forms) in surface soils and moss cushions around Great Wall Station (King George Island); diatoms are not an important constituent at that place.

A more detailed analysis of the taxonomic aspects is necessary. The data presented here have shown that such microscopic analysis provides substantial information for interpretations of the inter-microbial relationships in these habitats. Despite the problems of comparing data from direct observations and those from cultures, the method used here gives a first overview of the distribution pattern and the relationships between autotrophic and heterotrophic parts of the total microbial community. Further attempts to do this, however, need close cooperation with taxonomists, so that both qualitative and quantitative aspects can be used for better understanding of structure and function of soil systems.

ACKNOWLEDGEMENTS

I am greatly indepted to the Polish Academy of Science who enabled my visit to the Antarctic Station H. Arctowski at King George Island during December/January 1992/93. My thanks also to the 16th and 17th expedition at this base, especially to Dr. Maria Olech for the kind cooperation during my field work and for determinations of plant species. I have also to thank the crew of the M/S *Columbus Caravelle* for their logistic support. My thanks go to Drs P. Broady, R. Gradinger, S. Ohtani and D. D. Wynn-Williams for their constructive comments on this manuscript. The study was financially supported by the Deutsche Forschungsgemeinschaft (Bo 918/4-3).

REFERENCES

Block, W. 1984. Terrestrial microbiology, invertebrates and ecosystems. In Laws, R.M., ed. *Antarctic ecology*, Vol. 1. London: Academic Press, 163–236.

Block, W. 1994. Terrestrial ecosystems: Antarctica. *Polar Biology*, **14**, 293–300.

Bölter, M. 1989. Microbial activity in soils from Antarctica (Casey Station, Wilkes Land). *Proceedings of the NIPR Symposium on Polar Biology*, **2**, 146–153.

Bölter, M. 1990a. Microbial ecology of soils from Wilkes Land, Antarctica: I. The bacterial poulation and its activity in relation to dissolved organic matter. *Proceedings of the NIPR Symposium on Polar Biology*, **3**, 104–119.

Bölter, M. 1990b. Evaluation – by cluster analysis – of descriptors for the establishment of significant subunits in Antarctic soils. *Ecological Modelling*, **50**, 79–94.

Bölter, M. 1992a. *Vergleichende Untersuchungen zur mikrobiellen Aktivität in Böden und an Kryptogamen aus der kontinentalen und maritimen Antarktis*. Habilitationsschrift, University Kiel, 202 pp. [Unpublished.]

Bölter, M. 1992b. Environmental conditions and microbiological properties from soils and lichens from Antarctica (Casey Station, Wilkes Land). *Polar Biology*, **11**, 591–599.

Bölter, M. 1993. Effects of carbohydrates and leucine on growth of bacteria from Antarctic soils (Casey Station, Wilkes Land). *Polar Biology*, **13**, 297–306.

Bölter, M. 1994a. Estimations of microbial biomass by direct and indirect methods with special respect to monitoring programs. *Proceedings of the NIPR Symposium on Polar Biology*, **7**, 198–208.

Bölter, M. 1994b. Microcalorimetry and CO_2-evolution of soils and lichens from Antarctica. *Proceedings of the NIPR Symposium on Polar Biology*, **7**, 209–220.

Bölter, M. 1995. Distribution of bacterial numbers and biomass in soils and on plants from King George Island (Arctowski Station, Maritime Antarctic). *Polar Biology*, **15**, 115–124.

Bölter, M., Möller, R. & Dzomla, W. 1993. Determination of bacterial biovolume with epifluorescence microscopy: comparison of size distributions from image analysis and size classifications. *Micron*, **24**, 31–40.

Broady, P. A. 1979a. Quantitative studies on the terrestrial algae of Signy Island, South Orkney Islands. *British Antarctic Survey Bulletin*, No. 47, 31–41.

Broady, P.A. 1979b. A preliminary survey of the terrestrial algae of the Antarctic Peninsula and South Georgia. *British Antarctic Survey Bulletin*, No. 48, 47–70.

Broady, P. A. 1982. Ecology of non-marine algae at Mawson Rock, Antarctica. *Nova Hedwigia*, **36**, 209–229.

Broady, P. A. 1989. Survey of algae and other terrestrial biota at Edward VII Peninsula, Marie Byrd Land. *Antarctic Science*, **1**, 215–224.

Davey, M. C. 1988. Ecology of terrestrial algae of the fellfield ecosystems of Signy Island, South Orkney Islands. *British Antarctic Survey Bulletin*, No. 81, 69–74.

Davey, M. C. 1991. The seasonal periodicity of algae on Antarctic fellfield soils. *Holarctic Ecology*, **14**, 112–120.

Davey, M.C. & Clarke, K. J. 1991. The spatial distribution of microalgae on fellfield soils. *Antarctic Science*, **3**, 257–263.

Davey, M. C., Davidson, H. P. B., Richard, K. J. & Wynn-Williams, D. D. 1991. Attachment and growth of Anatrctic soil cyanobacteria and algae on natural and artificial substrata. *Soil Biology and Biochemistry*, **23**, 185–191.

Fabiszewski, J. & Wojtuń, B. 1993. Peat-forming vegetation. In Rakusa-Suszczewski, S., ed. *The Maritime Antarctic coastal ecosystem of Admiralty Bay*. Warsaw: Department of Antarctic Biology, Polish Academy of Sciences, 189–195.

French, D. D. & Smith, V. R. 1986. Bacterial populations in soils of a subantarctic island. *Polar Biology*, **6**, 75–82.

Heatwole, H., Saenger, P., Spain, A., Kerry, E. & Donelan, J. 1989. Biotic and chemical characteristics of some soils from Wilkes Land, Antarctica. *Antarctic Science*, **1**, 225–234.

Melick, D. R. & Seppelt, R. D. 1992. Loss of soluble carbohydrates and changes in freezing point of Antarctic bryophytes after leaching and repeated freeze–thaw cycles. *Antarctic Science*, **4**, 399–404.

Melick, D.R., Bölter, M. & Möller, R. 1994. Rates of soluble carbohydrate utilization in soils from the Windmill Islands Oasis, Wilkes Land, continental Antarctica. *Polar Biology*, **14**, 59–64.

Ohtani, S. & Kanda, H. 1987. Epiphytic algae on the moss community of *Grimmia lawiana* around Syowa Station, Antarctica. *Proceedings of the NIPR Symposium on Polar Biology*, **1**, 255–264.

Ohtani, S., Akiyama, M. & Kanda, H. 1991. Analysis of Antarctic soil algae by direct observation using the contact slide method. *Antarctic Record*, **35**, 285–295.

Olech, M. 1993. Lower plants. In Rakusa-Suszczewski, S., ed. *The Maritime Antarctic coastal ecosystem of Admiralty Bay*. Warsaw: Department of Antarctic Biology, Polish Academy of Sciences, 173–179.

Parsons, T. R., Maita, Y. & Lalli, C. M. 1985. *A manual of chemical and biological methods for seawater analysis.* Oxford: Pergamon Press, 173 pp.

Rakusa-Suszczewski, S. & Krzyszowska, A. 1991. Assessment of the environmental impact of the 'H. Arctowski' Polish Antarctic Station (Admiralty Bay, King George Island, South Shetland Islands). *Polish Polar Research*, **12**, 105–121.

Rakusa-Suszczewski, S., Mietus, M. & Piasecki, J. 1993. Weather and climate. In Rakusa-Suszczewski, S., ed. *The Maritime Antarctic coastal ecosystem of Admiralty Bay*. Warsaw: Department of Antarctic Biology, Polish Academy of Sciences, 19–25.

Ramsay, A. J. 1983. Bacterial biomass in ornithogenic soils of Antarctica. *Polar Biology*, **1**, 221–225.

Ramsay, A. J. & Stannard, R. E. 1986. Numbers and viability of bacteria in ornithogenic soils of Antarctica. *Polar Biology*, **5**, 195–198.

Roser, D. J., Melick, D. R., Ling, H. U. & Seppelt, R. D. 1992. Polyol and sugar content of terrestrial plants from continental Antarctica. *Antarctic Science*, **4**, 413–420.

Roser, D. J., Seppelt, R. D. & Ashbolt, N. 1993a. Microbiology of ornithogenic soils from the Windmill Islands, Budd Coast, continental Antarctica: some observations on methods for measuring soil biomass in ornithogenic soils. *Soil Biology and Biochemistry*, **25**, 177–183.

Roser, D. J., Seppelt, R. D. & Ashbolt, N. 1993b. Microbiology of ornithogenic soils from the Windmill Islands, Budd Coast, continental Antarctica: Microbial biomass distribution. *Soil Biology and Biochemistry*, **25**, 165–175.

Roser, D. J., Seppelt, R. D. & Nordstrom, O. 1994. Soluble carbohydrate and organic acid content of soil and associated microbiota from the Windmill Islands, Budd Coast, Antarctica. *Antarctic Science*, **6**, 53–59,

Smith, R. I. L. 1985. Nutrient cycling in relation to biological productivity in Antarctic and Sub-Antarctic terrestrial and freshwater ecosystems. In Siegfried, W. R., Condy, P. R. & Laws, R. M., eds. *Antarctic nutrient cycles and food webs*. Berlin: Springer-Verlag, 138–155.

Tearle, P. V. 1987. Cryptogamic carbohydrate release and microbial response during freeze–thaw cycles in Antarctic fellfield fines. *Soil Biology and Biochemistry*, **19**, 381–390.

Vishniac, H. S. 1993. The microbiology of Antarctic soils. In Friedmann, E. I., ed. *Antarctic microbiology.* New York: Wiley-Liss, 297–341.

Wynn-Williams, D. D. 1985. Photofading retardent for epifluorescence microscopy in soil micro-ecological studies. *Soil Biology and Biochemistry*, **17**, 739–746.

Wynn-Williams, D. D. 1990. Ecological aspects of Antarctic microbiology. *Advances in Microbial Ecology*, **11**, 71-146.

Zarzycki, K. 1993. Vascular plants and terrestrial biotopes. In Rakusa-Suszczewski, S. ed. *The Maritime Antarctic coastal ecosystem of Admiralty Bay*. Warsaw: Department of Antarctic Biology, Polish Academy of Sciences, pp. 181–187.

23 Environment and microbial communities in a tidal lagoon at Bratina Island, McMurdo Ice Shelf, Antarctica

I. HAWES, C. HOWARD-WILLIAMS, A-M. J. SCHWARZ AND M. T. DOWNES
National Institute of Water and Atmospheric Research, Ecosystems Division, P.O. Box 8602, Christchurch, New Zealand

ABSTRACT

The McMurdo Ice Shelf, with its 1500 km² of fresh and saline ponds has been described as Antarctica's biggest wetland and is the largest concentration of non-marine biota in the Ross Sea area. At rare intervals around the landward margins of this ice shelf, the semi-permanent ponds are replaced by areas which are inundated by the single daily tide. At Bratina Island (78° S, 166° E), one such lagoon was investigated during the 1992/93 summer. Low tide revealed a network of ponds, streams and mudflats and the flora and fauna of these was in many ways quite distinct from the non-tidal ponds in the area. The algal community in the driest areas was of low biomass (10 ± 1 mg chlorophyll a *m^{-2}) and dominated by motile diatoms (spp. of* Nitzschia *and* Navicula*). A biomass of 340 ± 20 mg chlorophyll* a *m^{-2} was recorded in shallow ponds and permanently moist sediments, where cyanobacteria reached maximum relative abundance. Pools greater than 30 cm deep (at low tide) were strongly stratified, with highly saline water underlying the brackish water which flushed the lagoon. The lower, saline layers were super-saturated with oxygen and contained very little dissolved inorganic carbon. As a result of solar heating, these layers reached stable temperatures of up to 12 °C, while shallower waters could fluctuate daily by up to 17 °C. Dense populations of benthic diatoms (580 ± 20 mg chlorophyll* a *m^{-2}) colonized the isolated saline layer of water though their photosynthetic rates were low. Within the stable water columns of these shallow ponds, dense, stratified populations of photosynthetic flagellates (*Pyramimonas, Chroomonas*) and ciliates (*Pseudobalanion, Askenasia*) developed.*

Key words: algal community, stratified pools, algal biomass.

INTRODUCTION

The McMurdo Ice Shelf is an ablation region in the northwest corner of the Ross Ice shelf. The surface of this part of the ice shelf is covered to varying degrees by moraine and marine sediments thought to be transported up from the seabed as material frozen into anchor ice, as well as windblown sediment (Debenham 1920, Swithinbank 1970). Ice movement throughout the McMurdo Ice Shelf is slow, of an order of 2 m y^{-1} between Brown Peninsula and Black Island (Swithinbank 1970, Kellogg *et al.*, 1990).

The surface of the McMurdo Ice Shelf carries networks of ponds and streams (Howard-Williams *et al.* 1990). The bottoms of many of these ponds are coated with thick mats of cyanobacteria, diatoms and other algae, making this area one of the largest concentrations of non-marine biota in the Ross Sea area (Howard-Williams *et al.* 1990, Hawes *et al.* 1993). In addition to the ponds on the surface of the ice shelf, tidal lagoons are occasionally noted at the landward margins of the McMurdo Ice Shelf. In these lagoons, extensive areas of sediment are regularly exposed and inundated by the single daily tide (Howard-Williams *et al.* 1990).

Preliminary observations suggested that these temporally and spatially variable lagoons contained diverse assemblages of algae and cyanobacteria. The biota of intertidal zones in Antarctica is poorly documented, as most shores are heavily impacted by wave and ice abrasion (Dayton & Robilliard 1974). The protected lagoons of the McMurdo Ice Shelf offer an opportunity to study intertidal biota, in an area where this disturbance is less apparent. The objective of this study, carried out in January 1993, was to characterize the broad ecosystem properties of a tidal lagoon at Bratina Island, to determine the diversity of the microbial communities within the lagoon complex, and to examine their biomass and productivity.

METHODS

Initial observations showed that water flows into and out of the lagoon via a tide crack. Within the lagoon, an anastomosing network of deep ponds (>1 m), shallow ponds (<1 m), streams and large areas of exposed sediment were revealed at low tide. A Stevens F-type level recorder was installed in this tide crack to monitor tidal height. The recorder was not surveyed to a bench mark, so the readings are relative, termed relative tidal height (RTH).

Temperature was recorded at 15 min intervals using Campbell 107 temperature probes connected to a Campbell CR10 datalogger at five locations in the lagoon. These were selected as a representative shallow pond and deep pond (surface and bottom temperature) and exposed mudflats (high and low tidal heights). Air temperature was also recorded.

Water samples were collected into acid-washed polythene bottles. All samples were filtered within a few hours of collection (Whatman GF/F), packed in ice for up to one week, then deep frozen for transport to New Zealand. Conductivity, pH, dissolved oxygen and temperature were determined *in situ* using Hannah, Yokogawa and YSI instruments respectively. Extinction of light through the water column was measured using a LiCor 2π cosine-corrected PAR sensor. Water samples from deeper ponds were collected using a depth-calibrated polythene tube connected to a vacuum flask and a hand-operated vacuum pump. The tube was lowered into the centre of the pond from a long rod. Conductivity of mudflat interstitial waters after air exposure was estimated by suspending a weighed sediment sample in a known volume of distilled water and measuring conductivity of the resulting suspension. The volume of water in the sediment sample was determined by reweighing the sample after drying, and conductivity of interstitial water was then estimated by correcting to the initial volume.

All water analyses were carried out using a Technicon II autoanalyser system. Dissolved reactive phosphorus (DRP), was measured by the method of Downes (1988), but without the arsenate reductant. Nitrate was measured by reduction to nitrite with cadmium followed by diazotization, and ammonium was analysed by the phenol-hypochlorite technique of Solorzano (1979).

Samples for planktonic chlorophyll *a* determination were filtered (Whatman GF/F) and filters frozen for return to New Zealand. Chlorophyll *a* was determined spectrophotometrically after grinding in 90% acetone. Phytoplankton samples for species identification were preserved with Lugol's iodine, sedimented and counted at ×100 and ×400 magnification.

Transects were established across representative parts of the lagoon for determination of algal distribution and biomass. Relative tidal height along these transects was determined at regular intervals by measuring water depth at a known (high) tidal height, on a calm day.

Benthic algal samples were collected using 5.3 or 2.5 cm^2 corers. Samples for chlorophyll *a* determination were extracted into boiling 90% ethanol for 10 min and chlorophyll determined spectrophotometrically. Samples for community analysis were examined at ×100 and ×400 magnification. Because of the variable and cohesive nature of many benthic mats, cell counts could not be made with confidence. Instead, taxa were quantified, by biovolume, on a scale of 0–5, where 0=not observed, 1=rare, 2=occasional, 3=frequent, 4=abundant and 5=dominant.

Measurements of benthic and planktonic productivity were undertaken *in situ* in one relatively deep pond (unofficially named Pancreas Pond). Planktonic productivity was measured by enclosing water in 125 ml polycarbonate bottles and incubating for 6 h with 2 μCi ^{14}C HCO_3. Benthic production was measured similarly, with subsamples of cores incubated in filtered pond water from the depth of collection. Where cohesive mats were present, three 2 mm diameter sub-cores were used in each bottle; where non-cohesive mats were found, subsamples of a slurry made from the surface 2 mm of the core were incubated. In all cases three light and two dark bottles were incubated. Incubations were terminated by filtration (Whatman GF/F), and the filters were fumed over concentrated HCl, frozen and returned to New Zealand for determination of ^{14}C uptake by liquid scintillation counting. Phytoplankton samples were counted directly in Brays scintillation cocktail, wheareas benthic mats were first digested in a NCS solubilizer (Amersham). Dissolved inorganic carbon (DIC) was measured on site using an infra-red gas analyser, after sparging acidified samples with CO_2-free air.

RESULTS

Tidal data

Over the study period, tidal inundation was daily, with a spring–neap pattern evident in the data (Fig. 23.1). The lagoon entrance was just below mid-tide level and on all days some water flowed into the lagoon. Within the lagoon, the range of tidal oscillation was restricted to 20 and 80 cm at neaps and springs respectively. A considerable proportion of the mud flats,

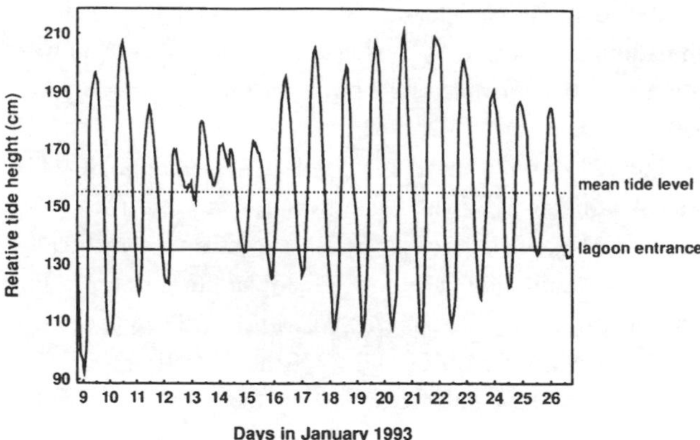

Fig. 23.1. Record of tidal oscillation during January 1993 at the tide crack through which the Bratina Island lagoon fills and drains. All heights referred to are relative to this gauge, and no absolute levels were obtained.

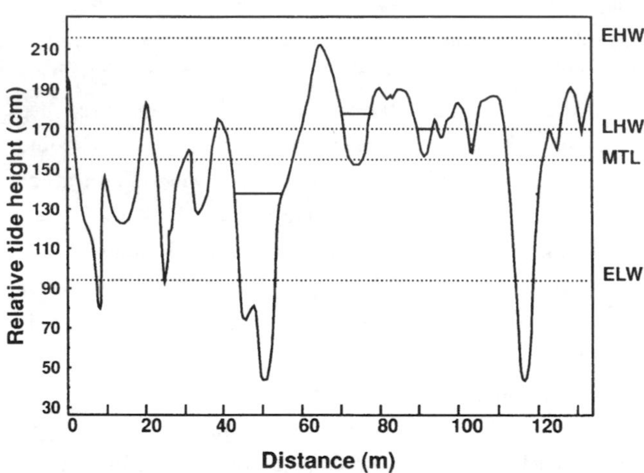

Fig. 23.2. A transect acrross the lagoon from South to North. The entrance is at the left hand of the figure. EHW – extreme high water, LHW – low high water, MTL – mean tide level, ELW extreme low water.

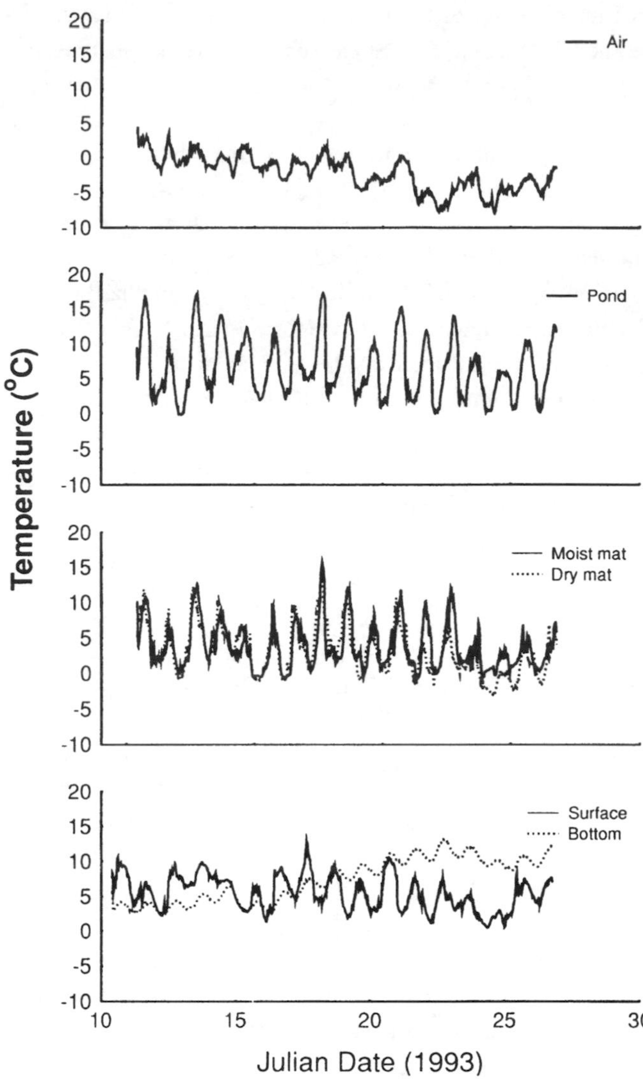

Fig. 23.3. Temperature records for the lagoon over the course of January. Temperatures of : air, a shallow pond, moist and dry mat communities, and surface and bottom in a deeper pond.

including some ponds, was not inundated on all tides (Fig. 23.2) and the zone between 190 and 210 cm RTH was not inundated for six days during the neap cycle.

Topography

Transects across the lagoon revealed a gently undulating mudflat, with deep (up to 2.4 m) and shallow pools (e.g. Fig. 23.2). The shallow ponds, stream channels and shallow margins of deeper ponds were covered by benthic algal mats ranging in colour from brown, through orange to yellow-green. Based on physical appearance, we recognized a number of zones within the transects.

Upper zone: bare sediment with no visible algal growth; mostly >190 cm RTH

Middle zone: a visible golden brown film of algae on the sediments; 180–190 cm RTH; not cohesive

Lower zone: thick (2–5 mm) orange-brown, cohesive cyanobacterial mats on sediments retaining moisture and in areas of standing water.

Temperature and conductivity

Most areas underwent diel temperature fluctuations (Fig. 23.3). These were most marked in the shallow ponds where daily ranges in excess of 15 °C occurred. Exposed flats also experienced large diel temperature fluctuations (Fig. 23.3c). In the deep pond (Fig. 23.3d), there was a clear difference in temperature between the bottom water, which increased from 3 to 11 °C over the course of January, and the surface layers, which fluctuated in a similar, though less extreme, diel pattern to that of the shallow ponds (Fig. 23.3b). In all cases pond and soil temperatures exceeded air temperatures. Evaporative concentrations resulted in elevated pore water conductivities of the upper layers of exposed mudflats. While the lower tidal zone retained a conductivity similar to that of the overlying water (up to 4 $\mu S\ cm^{-1}$), the upper and middle zones had pore water conductivities in the upper 2 mm of 26–28 $\mu S\ cm^{-1}$ after 12 h air exposure.

Gradients of increasing temperature and conductivity with depth were found in all ponds of depth greater than 30 cm (e.g. Fig. 23.4). Of 27 ponds examined, all 20 ponds with depths

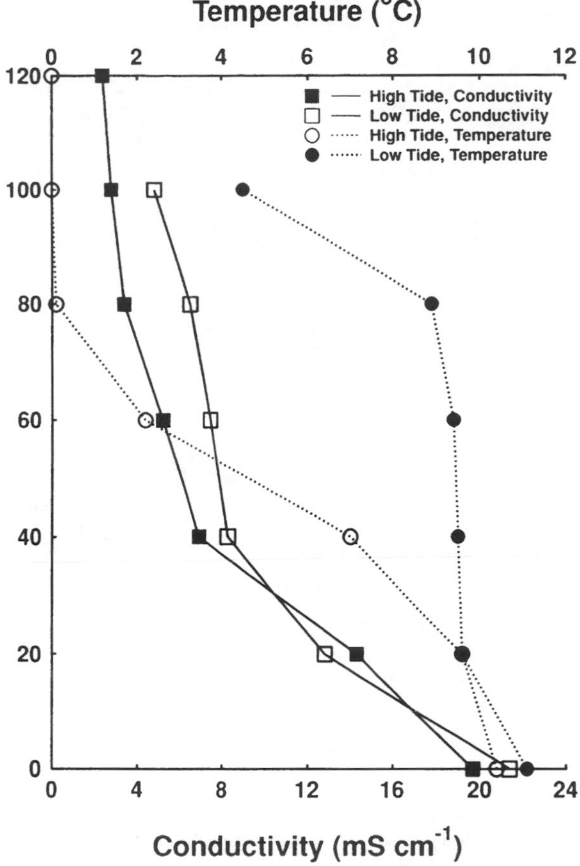

Fig. 23.4. Profiles of temperature and conductivity through Pancreas Pond at high and low tide on 25–26 January 1993.

greater than 30 cm were stratified, all seven ponds less than 30 cm deep were not. Despite the shallow depth of the ponds (the deepest pond measured was only 2.4 m at low tide), and the windy environment, the strong stratification persisted in all ponds investigated to the end of January, when ice was reforming at night.

Conductivity in the upper layer of the deep ponds and in the shallow ponds reflected the conductivity of water emerging from the tide crack which supplied water to the lagoon (2–3 mS cm^{-1}). Shallow ponds mixed daily with tidal water (when inundated), though conductivity rose over the low tide period by up to 100% (to 5 mS cm^{-1}) through evaporative concentration. The low conductivity of the tidal water indicates that it is largely derived from ice-shelf meltwater rather than seawater.

Dissolved oxygen

Within the freely mixing tidal water and the shallow ponds, oxygen concentrations were typically at or close to air saturation over the entire tidal cycle. Within the stratified ponds, the monimolimnion was supersaturated with oxygen. Our instrumentation could not measure oxygen concentrations above 20 mg l^{-1} so the degree of supersaturation was not determined. There were no detectable diel fluctuations in oxygen concentrations.

Nutrient concentrations

(i) Shallow ponds

At high tide, shallow ponds mixed with incoming tidal water. The composition of this water changed over the study period. At high tide on 11 January 1993, three shallow ponds, Adrenal, Heart and Spleen Ponds (unofficial names) contained 25–30 mg m^{-3} NO_3-N and 20–27 mg m^{-3} DRP. On 26 January 1993, the concentrations of DRP (23–30 mg m^{-3}) were similar to the 11 January values, while NO_3-N had been reduced to 4.1–5.6 mg m^{-3}. During low tide isolation of the ponds on 26 January both DRP and NO_3-N concentrations fell, to 13–18 and 0.2–2.6 mg m^{-3} respectively. At the same time conductivity increased by 50%, suggesting that there had been evaporative concentration and implying that there was biological incorporation of NO_3 and DRP.

(ii) Deep ponds

Seasonal and daily patterns of change in concentration of nutrients in the upper layers of Pancreas Pond were similar to those seen in shallow ponds consistent with mixing of these layers with incoming water. The monimolimnion of Pancreas Pond contained low concentrations of nutrients relative to the upper layer (2 mg m^{-3} NO_3-N and 5 mg m^{-3} DRP on 26 January).

TAXONOMIC DISTRIBUTION

Phytoplankton communities

The phytoplankton comprised an assemblage of flagellates, with a number of heterotrophic ciliates also present (Table 23.1). In tidally flushed waters of both shallow and deep ponds, *Chroomonas* was dominant. While *Chroomonas* dominated the surface layer, *Pyramimonas* dominated the 20–60 cm layer. Phytoplankton were sparse in the monimolimnion. Most other taxa were not common enough to be ascribed to clearly defined depth zones. Three heterotrophic ciliates, namely an *Askenasia* sp., a haptorid and a prostome (tentatively identified as a *Pseudobalanion* sp.), were present in samples from Pancreas Pond, but not in the monimolimnion.

Benthic communities

Benthic communities were dominated by diatoms and filamentous cyanobacteria, though the proportion of different taxa varied in different zones of the lagoon. A number of communities were recognized, which tended to grade into each other with no clear boundaries (Table 23.2).

Diatom films – Diatom films occurred in the upper and middle zones of the lagoon, with a film of motile diatoms (mostly species of *Navicula* and *Nitzschia*) and a few trichomes of the N-fixing cyanobacterium *Nodularia*.

Dry mat – At lower elevations, on relatively flat surfaces, a community comprised of a matrix of broad trichome oscillatoriales, *Nodularia* and naviculoid diatoms formed a relatively cohesive mat.

Table 23.1. *Composition of planktonic assemblages in water bodies of the Bratina Island lagoon at 1830 on 25 January 1993. Values (cells l^{-1}) are means of two replicate counts*

	Pond							
	Spleen	Heart	Pancreas					
Taxon	10 cm[1]	10 cm[1]	10 cm[1]	20 cm[1]	40 cm[1]	60 cm[1]	75 cm[1]	100 cm[1]
Phormidium (1 μm diameter)	0.8×10^3	1.5×10^3	1.2×10^3	4.2×10^3	2.3×10^3	2.6×10^3	6.8×10^3	0.9×10^3
Chroomonas	13.4×10^3	10.7×10^3	21.2×10^3	11.2×10^3	5.6×10^3	2.9×10^3	3.6×10^3	6.2×10^3
Chromulina	2.7×10^3	2.7×10^3	3.3×10^3	14.8×10^3	4.4×10^3	3.4×10^3	3.2×10^3	3.0×10^3
Chrysococcus	—	—	1×10^2	—	—	—	—	—
Chlamydomonas	—	—	3×10^2	1×10^2	1×10^2	1×10^2	2×10^2	—
Chlorogonium	—	0.3×10^2	1×10^2	—	—	—	—	—
Pyramimonas	0.5×10^3	0.4×10^3	2.6×10^3	21.5×10^3	26.5×10^3	19.2×10^3	4.7×10^3	0.7×10^3
Brachiomonas	—	2×10^2	—	2×10^2	—	1×10^2	5×10^2	3×10^2
Gymnodinium	1×10^2	—	—	2×10^2	—	—	—	—
Navicula	1.8×10^3	1.3×10^3	0.9×10^3	0.5×10^3	0.3×10^3	0.4×10^3	0.2×10^3	0.6×10^3
Askenasia[2]	—	—	1.4×10^2	3.3×10^2	2.7×10^2	0.8×10^2	1.7×10^2	—
?Pseudobalanion[2]	—	—	0.4×10^2	0.8×10^2	0.6×10^2	0.4×10^2	—	—
Litostome[2]	—	—	—	—	—	—	0.40×10^2	1.2×10^2

Notes:

[1] Depth of sample. [2] Heterotrophic ciliates.

Table 23.2. *Abundance scores and biomass as chlorophyll* a *of algae and cyanobacteria in typical benthic communities in the Bratina Island lagoon system. Taxa were scored as 0–5 (see text). For chlorophyll* a, *all values are mean ±SD of 12 samples, except for the deep pond where* n=5

Taxon	Diatom film	Dry mat	Moist mat	Shallow pond	Deep pond
Phormidium 1 μm diameter	0	0	4	5	0
Phormidium 3 μm diameter	0	4	4	4	1
Nodularia	2	2	1	0	0
Navicula sp. A *?muticopsis*	4	4	3	1	4
Navicula sp. B	1	0	0	0	0
Navicula sp. C	4	2	3	1	3
Nitzschia antarctica	3	1	0	0	0
Achnanthes sp. A	1	1	1	0	0
Achnanthes sp. B	0	0	1	0	0
Amphiprora ?kjellmanii	1	0	1	0	3
Pyramimonas sp.	1	0	0	0	0
Chlorophyll *a*	60±12	138±58	280±74	340±20	578±20

Moist mat – At lower tidal elevations and where sediments retained moisture, moist mat occurred which contained an increasing proportion of narrow oscillatorealean trichomes with decreasing tidal height.

Pond mat – Moist mat graded into pond mat which contained few diatoms and no *Nodularia,* and was dominated by narrow trichome oscillatoriales.

Deep-pond mat – Communities from within the deep, more saline parts of the ponds were quite distinct from shallow-pond mats, with a low abundance of cyanobacteria, but a dominance of diatoms, including genera previously found in diatom film and dry mat, as well as *Amphiprora* sp.

Biomass

Planktonic biomass, measured as chlorophyll *a,* conformed to the pattern seen in cell counts. In the upper, freely exchanging layer of water which was dominated by *Chroomonas,* chlorophyll *a* concentration was low and varied little over the tidal cycle other than what might be expected by evaporative concentration. In Pancreas Pond, maximum chlorophyll *a* concentration was observed immediately above the chemocline, where *Pyramimonas* was dominant, and this remained so over the tidal cycle. Planktonic chlorophyll *a* in the monimolimnion of Pancreas Pond was low (Fig. 23.5).

Transects across parts of the lagoon showed that biomass

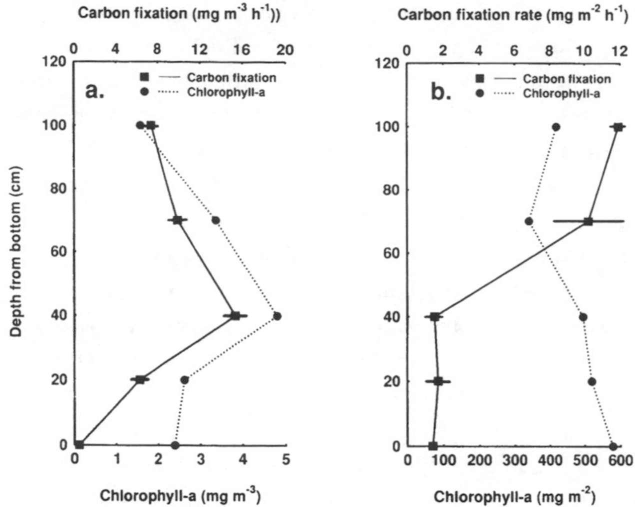

Fig. 23.5. Profiles of (a) planktonic and (b) benthic chlorophyll *a* and *in situ* carbon fixation in Pancreas Pond. Bars are 2 × standard error (n=3).

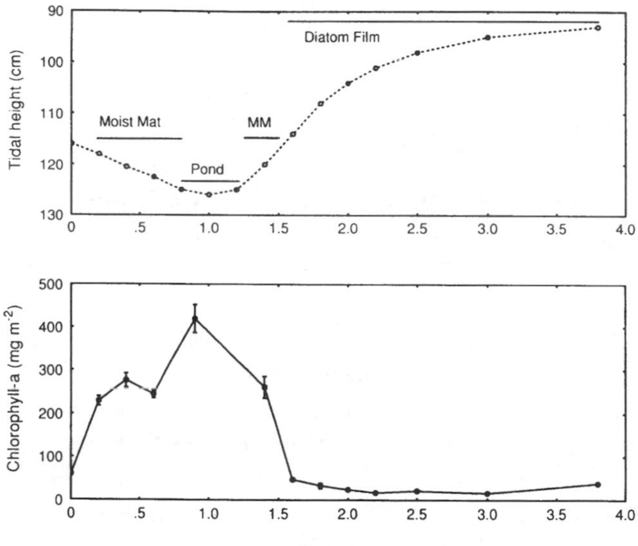

Fig. 23.6. Transect across a lagoon pond, showing (a) tidal height (dashed line) and community type (bars) and (b) biomass as chlorophyll *a*. Vertical bars are 2 × standard error (n=3).

increased with decreasing tidal height on mudflat areas, whereas shallow pond mats had a very high biomass (Fig. 23.6 and Table 23.2). Division of communities into types recognised by taxonomy and gross morphology showed that minimum biomass occurred in diatom films, increasing through dry mat, moist mat and pond mat. Maximum benthic biomass, however, was recorded in the diatom dominated monimolimnion of Pancreas Pond.

Photosynthesis rates *in situ*

Planktonic and benthic carbon fixation rates in Pancreas Pond were measured from 1200 to 1600 on 14 January 1993 and 16 January 1993 (Fig. 23.5). Photosynthetic rates per unit chlorophyll were higher in the planktonic communities than the benthos. Integrated areal photosynthesis of the planktonic community was 8.1 mg C $m^{-2} h^{-1}$, comparable to that of the benthos despite the much lower biomass of the former. The vertical extinction coefficient for PAR in Pancreas Pond on 18 January 1993 was 0.64 which meant that even at high tide (120 cm above the pond bottom), 46% of surface radiation penetrated to the base of the pond. Light is therefore unlikely to limit photosynthetic rate and, with the exception of the monimolimnion, photosynthesis tended to follow distribution of chlorophyll *a* in both benthic and planktonic assemblages. Dissolved inorganic carbon concentration was very low in the monimolimnion of Pancreas Pond (<1 g m^{-3}), compared with 4.5 g m^{-3} in the tidal component, and this may be the reason for the low observed photosynthetic rates.

DISCUSSION

Despite the distance of 30 km from Bratina Island to the open water of the McMurdo Sound, tidal movement of water underneath the McMurdo Ice Shelf is sufficient to displace water in the cracks between the blocks of ice adjacent to Bratina Island, such that the lagoon is flushed daily to some extent. Similar tidal movements at the margins of King George V Ice Shelf have been reported by Heywood (1977), resulting in tidal fluctuations of freshwater, which floats on seawater and is trapped between the ice shelf and the shore. The low conductivity of the tidal water at Bratina Island suggests that this water too is largely derived from meltwater generated from the ice shelf, which has flowed into the cracks, rather than from seawater.

The ponds on the surface of the McMurdo Ice Shelf, covering an area of 1500 km^2, are recognized as the site of one of the largest concentrations of non-marine organisms in the southern Ross Sea, supporting a variety of cyanobacteria- and diatom-dominated communities (Howard-Williams *et al.* 1990). The lagoon system created by the tidal oscillation provided a further group of contrasting habitats with differing physical, chemical and biological characteristics, most of which are not found elsewhere on the McMurdo Ice Shelf. Tidal influx of low salinity water makes the lagoon in some ways more analagous to an estuary or saltmarsh, than to the rest of the McMurdo Ice shelf. While lagoon habitats tended to grade into each other, they could be broadly categorized into tidally exposed mudflats, shallow ponds which freely exchanged with the tidal water, and relatively isolated bottom waters in the deeper ponds.

The diatom-dominated mudflat communities have not previously been reported from this area, though diatom-dominated communities are found to colonize rapidly unstable habitats of the ice shelf (Howard-Williams *et al.* 1990). They are analogous to the *Navicula*/*Nitzschia* assemblages which are typical of many temperate mudflats where they are tolerant of widely fluctuating salinities, as seen here (e.g. Round 1981). The diel fluctuations in temperature, which would have resulted in overnight freeze–thaw cycles later in the season, coupled with the fluctuating salt content associated with evaporation, probably made this the most extreme environment in the lagoon. *Nodularia* is also common on temperate mudflats and its presence in the lagoon may have been favoured by the low nitrate concentrations and low inorganic N:P ratios (<1), which would confer an advantage to N_2-fixing organisms.

The biomass of diatom films measured as chlorophyll *a* at 60 mg m^{-2} was similar to published values for temperate mudflats, for example 57–160 mg m^{-2} in a Mississippi saltmarsh (Sullivan & Moncrieff 1988) and 40–70 mg m^{-2} in a South Carolina saltmarsh (Pickney & Zingmark 1993). In both biomass and species composition, this Antarctic mudflat community differed little from similar habitats in lower latitudes.

Communities in the shallow ponds were the most similar of the lagoon communities to those in the ponds of the undulating ice at the McMurdo Ice Shelf proper. Average benthic biomass in three shallow ponds of the lagoon (340 mg m^{-2}) falls within the range of 80–641 given for six ponds on the undulating ice by Howard-Williams *et al.* (1990). Mat composition was also similar, both groups of mats being dominated by oscillatorialean cyanobacteria. *In-situ* production estimates of the cyanobacterial mats in the lagoon of 8–12 mg C m^{-2} h^{-1} fell within the range of other Antarctic ponds and streams, including the McMurdo Ice Shelf. Benthic production in streams in southern Victoria Land ranged between 5–22 mg C m^{-2} h^{-1} (Howard-Williams & Vincent 1989), McMurdo Ice shelf ponds 9–21 mg C m^{-2} h^{-1} (Howard-Williams *et al.* 1990) and mats from maritime Antarctic streams 30–36 mg C m^{-2} h^{-1} (Hawes 1993). The McMurdo Ice Shelf Ponds are thought to be nutrient limited, with the high biomass of cyanobacteria taking many years to accumulate (Hawes *et al.* 1993). The tidal ponds of the lagoon were no more productive than these, and growth and accumulation of biomass is likely to be slow in the lagoon as well. Low nutrient concentrations, particularly nitrate-N in the tidally flushed water, and the depletion of nitrogen:phosphorus to the ratio of 0.3–0.5 over the course of tidal isolation, suggest that nitrogen may be limiting. This ratio is much less than that typically required by algae and cyanobacteria. Flushing environments can be more productive than stagnant ones, where this results in an enhanced nutrient supply. This does not appear to be the case in the Bratina Island lagoon, as the dilute flushing water, which is trapped within the cracks of the ice shelf and does not exchange with seawater, is nitrogen poor.

Perhaps the most unusual environments on the Bratina Island Lagoon complex were the stratified ponds. This stratification seems likely to have persisted through to freeze-up, despite occasional high winds and the shallow depth of the ponds. Current data are insufficient to determine the origin of the deep layer, though it may be a result of freeze-out of salts during winter (cf. Schmidt *et al.* 1991). Heat gain by the bottom layer was continuous through January, with a superimposed daily oscillation. This heat gain is similar, though less intensive than in solar ponds of the Dead Sea (e.g. Dor & Paz 1989). Physical isolation of the warm, deep layers from the mixing zone is likely to be responsible for the depletion of nutrients and DIC, and supersaturation with oxygen in this layer as algal growth proceeds. Low rates of photosynthesis of both phytoplankton and phytobenthos measured in this study probably reflect the low concentration of DIC. The reasons for the absence of cyanobacteria from this layer is not clear. Salinity is unlikely to be a factor as cyanobacteria were tolerant of high salinities in other ponds of the McMurdo Ice Shelf (Howard-Williams *et al.* 1990).

Stratified water columns, stabilised by temperature and salinity gradients are found in many deep Antarctic lakes (Vincent 1988). Stable water columns allow motile organisms to select their optimal growth conditions. This appears to be how the *Pyramimonas* population is distributed in Pancreas Pond, concentrating at the base of the mixing zone, at slightly elevated conductivity. Since there is no flux of nutrients from the bottom water, unlike some deeper antarctic lakes where the monimolimnion is anoxic and a source of nutrients to a deep chlorophyll maximum (e.g. Vincent *et al.* 1981), this is most likely a response to a radiation or salinity preference. This flagellate has not been recorded from any other ponds of the McMurdo Ice Shelf, where *Chroomonas* and *Ochromonas* were dominant in the phytoplankton (M.R. James, pers. commun. and authors' observations). The assimilation numbers (mg C fixed mg^{-1} chlorophyll *a*) for the planktonic communities were much higher than for the benthic ones (approximately 5 compared with 0.01), suggesting that the planktonic community was relatively dynamic, though there was no evidence for an accumulation of chlorophyll *a* over tidal isolation or the sampling period.

Another interesting alga in the deep water layer was the diatom *Amphiprora*. While this is commonly found on tidal flats (e.g. Round 1981), in Antarctica it is commonly associated with sea- ice algal communities (Horner 1985a). Species of *Pyramimonas* have also been recorded as sea-ice algae (McFadden *et al.* 1982). Should the saline layer represent a remnant of winter brine formed by salt freeze-out it will be a similar habitat to that of ice algae which develop in brine channels in sea ice (Horner 1985b), and a similar flora may be produced.

It is not known how widespread similar tidal lagoons are in Antarctica. We surveyed the coast of Brown Peninsula and the mainland opposite and found only two other lagoons. There are almost certainly other similar lagoons around the continent. This habitat, which offers a range of exotic micro-environments in rarely found a small space elsewhere in Antarctica may, however, be an important local refuge of biodiversity. Within these specialized microhabitats the communities may attain levels of biomass and productivity which are comparable to mudflat communities at lower latitudes and show similar species composition.

ACKNOWLEDGEMENTS

We are grateful to Drs J. C. Ellis-Evans and N. J. Russell for useful comments on the original paper.

REFERENCES

Dayton, P. K. & Robilliard, G. A. 1974. Biological accommodation in the benthic community at McMurdo Sound, Antarctica. *Ecological Monographs*, **44**, 105–128.

Debenham, F. 1920. A new mode of transportation by ice: the raised marine muds of South Victoria Land (Antarctica). *Quarterly Journal of the Geological Society London*, **75**, 51–76.

Dillon, R. D. & Dierle, D. A. 1980. Microbiocoenoses in an Antarctic pond. In Giesy, J. P., ed. *Microcosms in ecological research*, DoE Symposium Series No. 52. Washington DC: Department of Energy. 446–457.

Dor, I. & Paz, N. 1989. Temporal and spatial distribution of mat microalgae in the experimental solar ponds, Dead Sea Area, Israel. In Cohen, Y. & Rosenberg, E., eds. *Microbial mats: physiological ecology of benthic microbial communities*. Washington, DC: American Society for Microbiology, 114–122.

Downes, M. T. 1988. Taupo research laboratory chemical methods manual. *Taupo Research Laboratory Report*, **102**, 1–70.

Hawes, I. 1993. Photosynthesis in thick cyanobacterial films: a comparison of annual and perennial Antarctic mat communities. *Hydrologia*, **252**, 203–209.

Hawes, I., Howard-Williams, C. & Pridmore, R. D. 1993. Environmental control of microbial biomass in the ponds of the McMurdo Ice Shelf, Antarctica. *Archiv für Hydrobiologie*, **127**, 271–287.

Heywood, R. B. 1977. Limnological survey of the Ablation Point area, Alexander Island, Antarctica. *Philosophical Transactions of the Royal Society of London*, **B279**, 39–54.

Horner, R. A. 1985a. Taxonomy of sea ice microalgae. In Horner, R. A., ed. *Sea ice biota*. Boca Raton, Florida: CRC Press, 147–157.

Horner, R. A. 1985b. Ecology of sea ice microalgae. In Horner, R. A., ed. *Sea ice biota*. Boca Raton, Florida: CRC Press, 83–103.

Howard-Williams, C., Pridmore, R. D., Broady, P. A. & Vincent, W. F. 1990. Environmental and biological variability in the McMurdo Ice Shelf ecosystem. In Kerry, K. & Hempel, G., eds. *Ecological change and the conservation of Antarctic ecosystems*. Berlin: Springer-Verlag, 23–31.

Kellogg, T. B., Kellogg, D. E. & Stuiver, M. 1990. Lake Quaternary history of the southwest Ross Sea: Evidence from debris bands on the McMurdo Ice Shelf, Antarctica. *Antarctic Research Series*, **20**, 25–26.

Mcfadden, G. I., Moestrup, Ø. & Whetherlase, R. 1982. *Pyramimonas gelidicola* sp. nov. (Prasinophyceae) a new species isolated from Antarctic sea ice. *Phycologia*, **21**, 103–111.

Pinckney, J. L. & Zingmark, R. G. 1993. Modelling the annual production of intertidal benthic microalgae in estuarine ecosystems. *Journal of Phycology*, **29**, 396–407.

Round, F. G. 1981. *The ecology of algae*. Cambridge: Cambridge University Press, 653 pp.

Schmidt, T., Moskal, W., DeMora, S. J., Howard-Williams, C. & Vincent, W. F. 1991. Limnological properties of Antarctic ponds during winter freezing. *Antarctic Science*, **3**, 379–388.

Solorzano, L. 1979. Determination of ammonia in natural waters by the phenolhypochlorite method. *Limnology and Oceanography*, **24**, 799–801.

Sullivan, M. J. & Moncrieff, C. A. 1988. Primary production of edaphic algal communities in a Mississippi salt marsh. *Journal of Phycology*, **24**, 49–58.

Swithinbank, C. 1970. Ice movement in the McMurdo Sound area of Antarctica. In *Proceedings of the International Symposium on Antarctic Glaciological Exploration, Hanover, NH, USA, 1970*, Cambridge: SCAR, 472–482.

Vincent, W. F. 1988. *Microbial ecosystems of Antarctica*. Cambridge: Cambridge University Press, 304 pp.

Vincent, W. F., Downes, M. T. & Vincent, C. L. 1981. Nitrous oxide cycling in Lake Vanda, Antarctica. *Nature*, **292**, 103–111.

24 Seasonal distribution of picocyanobacteria in Ace Lake, a marine-derived Antarctic lake

LYNNE M. RANKIN[1], P. D. FRANZMANN[2], T. A. MCMEEKIN[3] AND H. R. BURTON[1]
[1]*Australian Antarctic Division, Channel Highway, Kingston, Tasmania, 7052, Australia,* [2]*Antarctic CRC, University of Tasmania, GPO Box 252C, Hobart, Tasmania, 7001, Australia,* [3]*Department of Agricultural Science, University of Tasmania, GPO Box 252C, Hobart, Tasmania, 7001, Australia*

ABSTRACT

Recently, a Synechococcus *sp. was identified in Ace Lake, a marine-derived meromictic lake in the Vestfold Hills, Antarctica, where it occured in numbers 40 000 times greater than reported for southern polar waters. Ace Lake is 24 m deep and is usually ice covered for 11 months of the year. The lake can be divided into three regions: a mixed zone (top 7 m); an oxycline (between 7 m and 12 m); and an anaerobic zone (below 12 m). The greatest number of* Synechococcus *sp. cells occured at 11 m in December when numbers reached 8×10^6 ml^{-1}. To our knowledge this population density is higher than any previously reported for marine* Synechococcus *spp. At this depth the temperature (6 °C) and salinity (30 ppt) remained near constant throughout the year.* Synechococcus *sp. numbers and fluorescence were monitored by flow cytometry. The distinct orange fluorescence emitted from phycoerythrin and the small cell size (0.8 μm × 1.5 μm) distinguished the* Synechococcus *sp. population from populations of other photosynthetic cells. As temperature is considered to be an important control of* Synechococcus *spp. numbers in this region, the existence of a population of* Synechococcus *sp. in a marine-derived Antarctic lake may now provide an insight into controls of cell numbers in the Southern Ocean.*

Key words: *Synechococcus*, flow cytometry, Antarctic lakes, meromictic, picoplankton.

INTRODUCTION

Synechococcus spp. are small unicellular cyanobacteria that occur abundantly in surface waters of temperate and tropical oceans. The contribution of *Synechococcus* spp. to primary production in these regions is variable but often high (Waterbury *et al.* 1986). Cell numbers range between 10^2 cells ml^{-1} to almost 10^6 cells ml^{-1} in tropical and temperate oceans, with the contribution of *Synechococcus* spp. to primary productivity ranging from 5% to 65% of total primary productivity depending on location and assay techniques used (Waterbury *et al.* 1986).

Unlike temperate and tropical waters it has been reported that *Synechococcus* spp. occur in low numbers in polar oceans (Murphy & Haugen 1985, Marchant *et al.* 1987, Letelier & Karl 1989). It has been shown that there is a direct relationship between temperature and the abundance of *Synechococcus* spp. in polar oceans (Murphy & Haugen 1985, Gradinger & Lenz 1989, Walker & Marchant 1989), although there may be some other factor or factors other than temperature limiting the growth of the picocyanobacteria. In coastal Antarctic waters *Synechococcus* sp. numbers have been reported to be less than 200 cells ml^{-1} (Walker & Marchant 1989).

Although several lakes in Antarctica have deep-water chlorophyll maxima that may be dominated by picocyanobacteria (Spaulding *et al* 1994, Vincent 1988) the numbers of

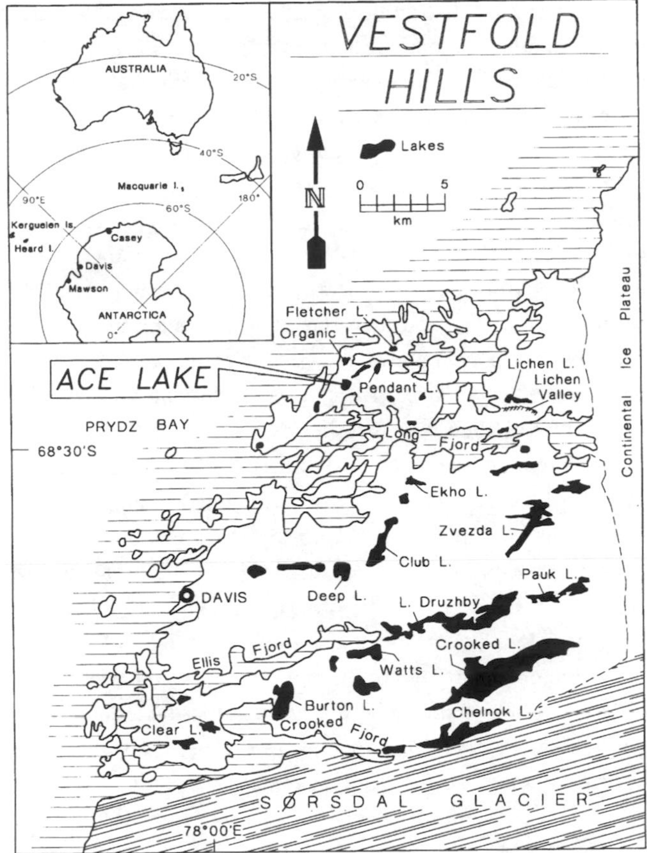

Fig. 24.1. Map of the Vestfold Hills showing Ace Lake on Long Peninsula.

Synechococcus sp. measured in Ace Lake, Vestfold Hills, Antarctica during the austral summer of 1992/93 are, to our knowledge, greater than any previously reported for marine *Synechococcus* spp. Ace Lake was isolated from the marine system less than 2000 years ago (Burton & Barker 1979). The ratio of major cations to chloride is typical for marine waters (Masuda *et al.* 1988) and the copepods and eukaryotic phytoplankton in Ace Lake are of marine origin (Bayly & Burton 1987, Burch 1988). The discovery of *Synechococcus* sp. in the marine-derived Ace Lake provides a unique opportunity to characterize an Antarctic population of marine *Synechococcus* sp. and to investigate environmental constraints on their distribution in polar regions.

MATERIALS AND METHODS

Ace Lake

Ace Lake (68°29' S, 78°10' E) is one of approximately 30 saline meromictic lakes and marine basins in the Vestfold Hills, Antarctica. Ace Lake covers an area of 1.3×10^5 m^2 and is situated on Long Peninsula (Fig. 24.1). The lake is 25 m deep at its deepest spot. The lake is ice covered for 11 months of the year and is usually ice free in February. In cold years, however, the lake remains permanently ice covered. Ace Lake has been investigated extensively since 1979 and is best known for its microbial processes in the anoxic zone. The upper 12 m of the lake are also interesting as there is a permanent stratification of phytoplankton and a relatively small number of marine-derived species constitute the plankton. It is therefore a simpler ecosystem than the nearby ocean itself.

Sample collection and storage

Water samples were collected using a Kemmerer Bottle and water sampling was always carried out at the deepest spot in Ace Lake. Samples were collected monthly between February 1992 and August 1992 and fortnightly between September 1992 and January 1993. Samples were collected at two metre intervals between 2 m and 12 m with an additional sample taken at 11 m. Samples were refrigerated at 4 °C in the dark and flow cytometric analysis was conducted within 4 h of sampling. Subsamples were stored in 0.2 μm filtered, pH adjusted, formalin (1.5%) for future analysis by epifluorescent microscopy. A submersible data logger (Platypus Engineering) was used for the collection of temperature and conductivity data. Conductivity data were converted to salinity using the formulas of Fofonoff & Millard (1983) and Gibson *et al.* (1990). Light readings were collected with a Digital Scalar Irradiance Meter and an underwater quantum sensor. Light readings were taken as close to solar noon as possible.

Flow cytometry

Samples were analysed using a Becton Dickinson FACScan fitted with an argon laser emitting light of 488 nm. Forward light scatter (FSC), side angle light scatter (SSC) and fluorescence emission from phycoerythrin (FL2, BP585/42) and from chlorophyll (FL3, LP650) were used to detect *Synechococcus* sp. populations. The instrument was set with a FL3 threshold and a low aquisition rate. A known concentration of 1.98 μm fluoresbrite calibration grade micro-spheres (Polysciences Inc.) was added to each sample to enable accurate counting. Flow cytometric data was analysed using the 'Lysis II' software on the FACScan.

Epifluorescent microscopy

Slides were prepared by filtering 5–10 ml (1.5%) formalin-preserved lake water samples onto 0.2 μm membranes, prestained with Irgalan black (Millipore). Samples were counted using a Zeiss Axioscope epifluorescence microscope with green light (filter block 12, excitation filter G546, dichromatic beam splitter FT 580, barrier filter LP 590). Cells were also observed with blue light (filter block 9, excitation filter BP450-490, dichromatic beam splitter FT 510, barrier filter LP 520).

Light photomicrographs

Photomicrographs were taken using a Leitz DMRBE epifluorescent microscope. Cells were applied to precoated agar slides (Noble Agar, Difco) and photographed under oil immersion 100× phase contrast and epifluorescence. Green light was used for the epifluorescence microscopy (Filter block M2, excitation filter BP 546/14, dichromatic mirror RKP 580, suppression filter LP 580).

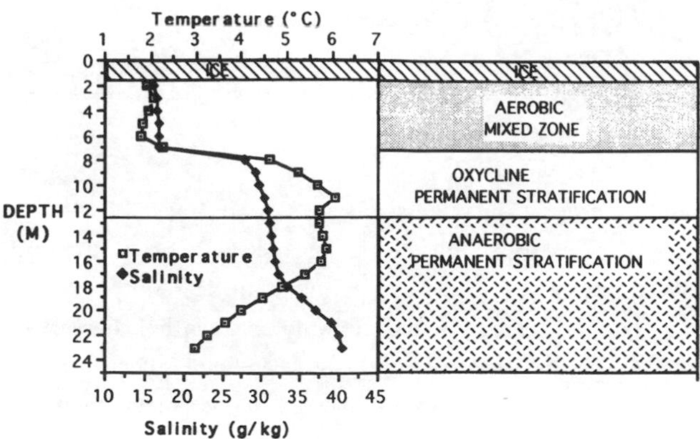

Fig. 24. 2. Summer (17 December 1992) temperature and salinity profile for Ace Lake, showing the aerobic zone and the stratification in Ace Lake. Below 8 m the temperature and salinity of Ace Lake remained relatively stable throughout the year. Temperature changed in the mixed zone, where it increased from a minimum of −1 °C in August to a maximum of 4 °C in January.

RESULTS AND DISCUSSION

Ace Lake can be divided into three zones, the mixed zone, the oxycline and the anaerobic zone or anoxylimnion (Fig. 24.2). The depth of the mixed zone can change slightly from year to year. During 1992/93 the top 7 m of the lake were mixed, with a summer temperature of 2 °C and a salinity of 16 ppt. Between 7 and 12 m, through the oxycline, there was a sharp increase in salinity and temperature. At 11 m the salinity was 30 ppt. Below 12 m the water was anaerobic. Salinity continued to increase with depth, and the temperature decreased. The lake remained relatively stable throughout the year, except in the mixed zone where the temperature in winter fell to a minimum of −1 °C in August, and in the summer the temperature of the mixed zone reached a maximum of 4 °C in January.

Although Ace Lake has been studied since 1979, single-celled cyanobacteria have not been reported in the lake until now. It was through the application of flow cytometric methods that the organisms were identified. The naturally fluorescing *Synechococcus* sp. cells were identified by the distinctive orange fluorescence (Fig. 24.3) emitted by the phycobiliproteins. As other organisms of this size do not emit light at this wavelength, *Synechococcus* sp. was clearly distinguished from other photosynthetic organisms on the flow cytometer.

Phase contrast and epifluorescence photomicrographs confirmed the *Synechococcus* sp. morphology in Ace Lake (Figs. 24.4 and 24.5). The cells were typically 1.5 μm in length and 0.9 μm in diameter. The photosynthetic pigments in *Synechococcus* sp. caused it to fluoresce red under a green (BP 546 nm) and orange under a blue (BP 450–490 nm) filter when excited by a mercury vapour lamp.

During winter (August) *Synechococcus* sp. cell numbers were less than 10^4 cells ml^{-1} throughout the aerobic zone (Fig. 24.6). During summer (December) cell numbers remained low in the mixed zone but rapidly increased through the oxycline. Maximum cell numbers (8×10^6 cells ml^{-1}) were recorded at 11 m in December. This distribution of *Synechococcus* spp. has been reported in other meromictic lakes (Venkateswaran *et al.* 1993), but these lakes were in tropical regions, water temperatures were higher and low temperature as a control of cell numbers was therefore not an issue in these lakes. The increase in *Synechococcus* sp. cell numbers over summer was obvious when a seasonal profile at 10 m in the lake was considered (Fig. 24.7). Over the winter months (May to August) cell numbers were below 10^4 cells ml^{-1}. Cell numbers increased through September, peaked in December and then started to fall again in January. From this data an estimate of growth rate could be made for the population of *Synechococcus* sp. in the field. A plot of log (cell number) versus time over October, November and December, gave a regression line with an r^2 value of 0.999. A generation time of 14.85 days at 6 °C was determined from the plot. Generation times for *Synechococcus* spp. in tropical and temperate environments have been estimated to be less than a day (Campbell & Carpenter 1986, Waterbury *et al.* 1986); thus this long generation time in Ace Lake is likely to be due, at least partially, to temperature. However, as the temperature at 10 m is near constant throughout the year (6 °C) other factors (such as light, grazing or nutrient availability) were also controlling the population growth of *Synechococcus* sp.. The rapid loss of 20% of the total population, which was in an exponential growth phase (Fig. 24.7), between December and January, would suggest grazing had an important controlling influence on cell numbers until May (Fig. 24.7). Potential grazers such as ciliates, heterotrophic nanoflagellates and small copepods occur in Ace Lake and both ciliates (Perris, pers. commun.) and copepods (Bayly & Burton 1987) have population peaks at approximately 11 m, where the *Synechococcus* sp. population also peaked at 8×10^6 cells ml^{-1}. This concentration of biota indicates that the microbial loop is active. It will be valuable to compare the growth rate of *Synechococcus* sp. in the lake with that obtained from cultures in which there is no loss from grazing and the physical conditions are optimal.

During winter (August) the light intensity at 10 m was 0.13 μmol m^{-2} s^{-1} (Fig. 24.8). Over summer (December) the light intensity at 10 m was 12.4 mol m^{-2} s^{-1}, a 100-fold increase in light compared with the August light intensity. It is possible that this increase in light intensity resulted in the *Synechococcus* sp. bloom. It has been reported that in the class *Cyanophyceae,* the minumum light intensity for growth is 5 mol m^{-2} s^{-1} (Richardson *et al.* 1983). *Synechococcus* sp. numbers peaked at 11 m, where there was less than 1% incident light. It is likely, however, that the *Synechococcus* sp. cells living at the high latitude, in Ace Lake, have adapted to survival at lower light intensities, as they must survive for four months with light intensities below 1 mol m^{-2} s^{-1}. No light was detected past 12 m, where a band of photosynthetic sulphur bacteria occurred.

The discovery of the single-celled cyanobacterium, *Synechococcus* sp., in a marine-derived Antarctic lake has now provided an unique opportunity to investigate factors that control the abundance and distribution of a natural population

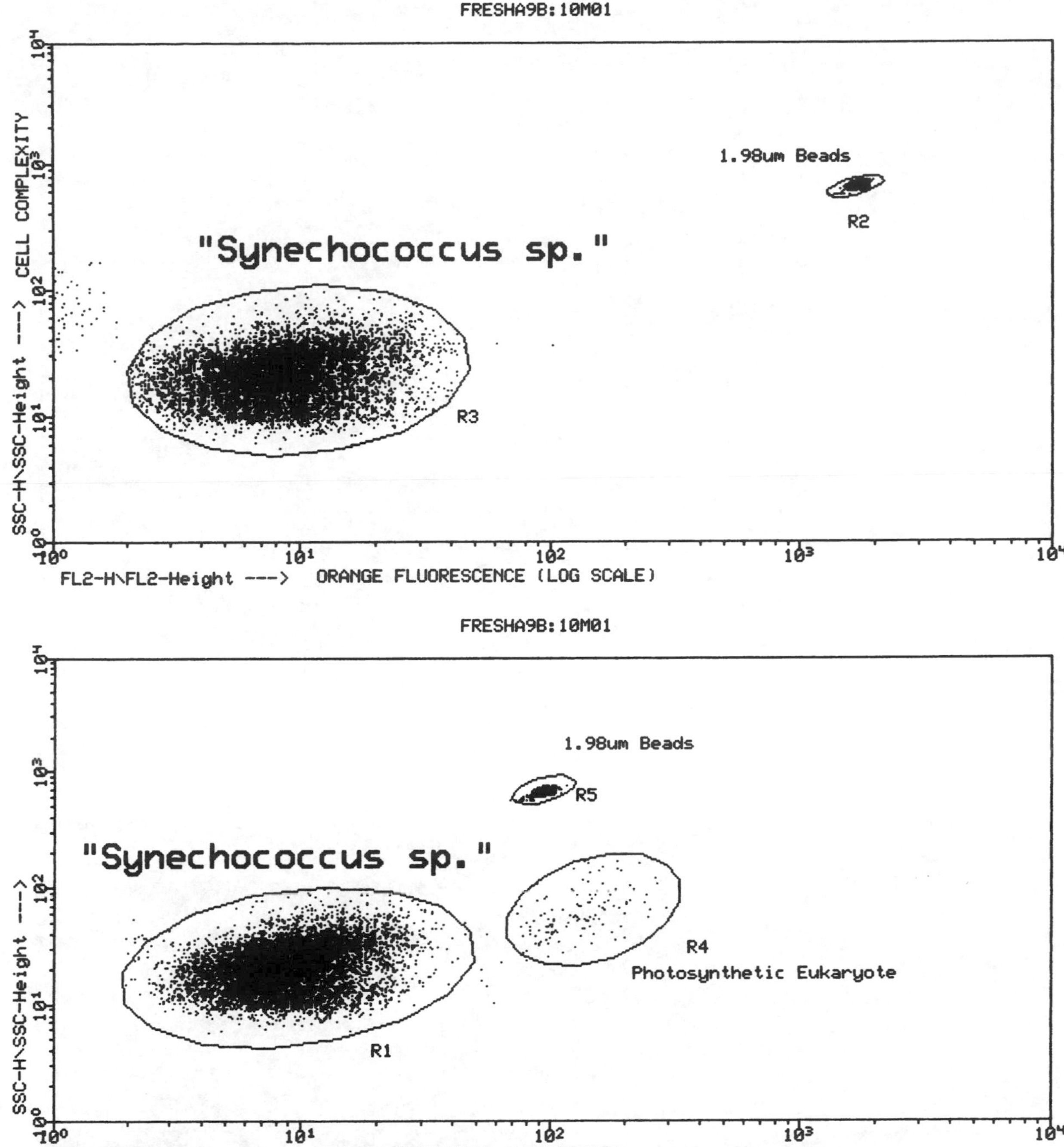

Fig. 24.3. Flow cytograms show the distinctive orange fluorescence (FL2) emitted from phycoerythrin pigments in *Synechococcus* sp. that were absent from other photosynthetic organisms of a similar size. Chlorophyll pigments (FL3) were detected in all photosynthetic organisms present in Ace Lake. The axes are in log units and show the increase in relative fluorescence (FL2/FL3) on the *x* axis and relative cell complexity (SSC) on the *y* axis.

of the picocyanobacterium in southern polar regions. One of the advantages of doing ecological studies in lake environments is that the population is captive, and thus not removed by currents. It is generally considered that temperature limits the abundance of *Synechococcus* spp. in southern polar waters (Gradinger & Lenz 1989). In Ace Lake, between 8 m and 11 m, where *Synechococcus* sp. was most abundant over summer and temperatures were relatively stable all year round, it was not temperature alone which controlled the populations abundance. A combination of factors such as light, nutrient availability and grazing were controlling the abundance of *Synechococcus* sp. at these depths. Further investigations are underway to determine the extent these factors play in controlling *Synechococcus* sp. abundance and distribution in Ace Lake. Understanding what is controlling the abundance of *Synechococcus* sp. in an Antarctic lake will then give more insight into the factors that are limiting the abundance of the species in the Southern Ocean. The relationship of the Ace Lake *Synechococcus* sp. to species from temperate and tropical regions is also being investigated.

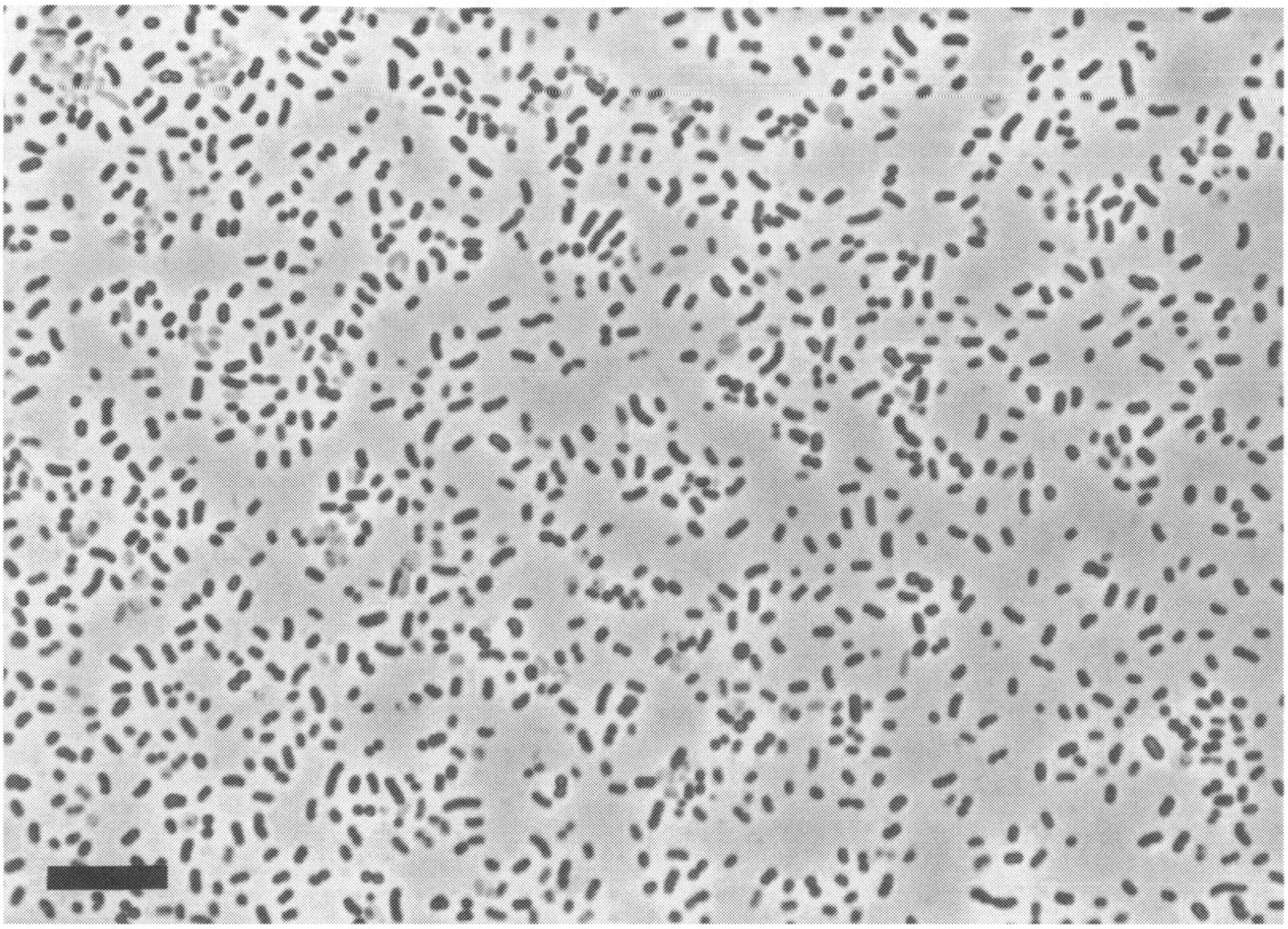

Fig. 24.4. Phase contrast micrograph of *Synechococcus* sp. Scale = 10 μm

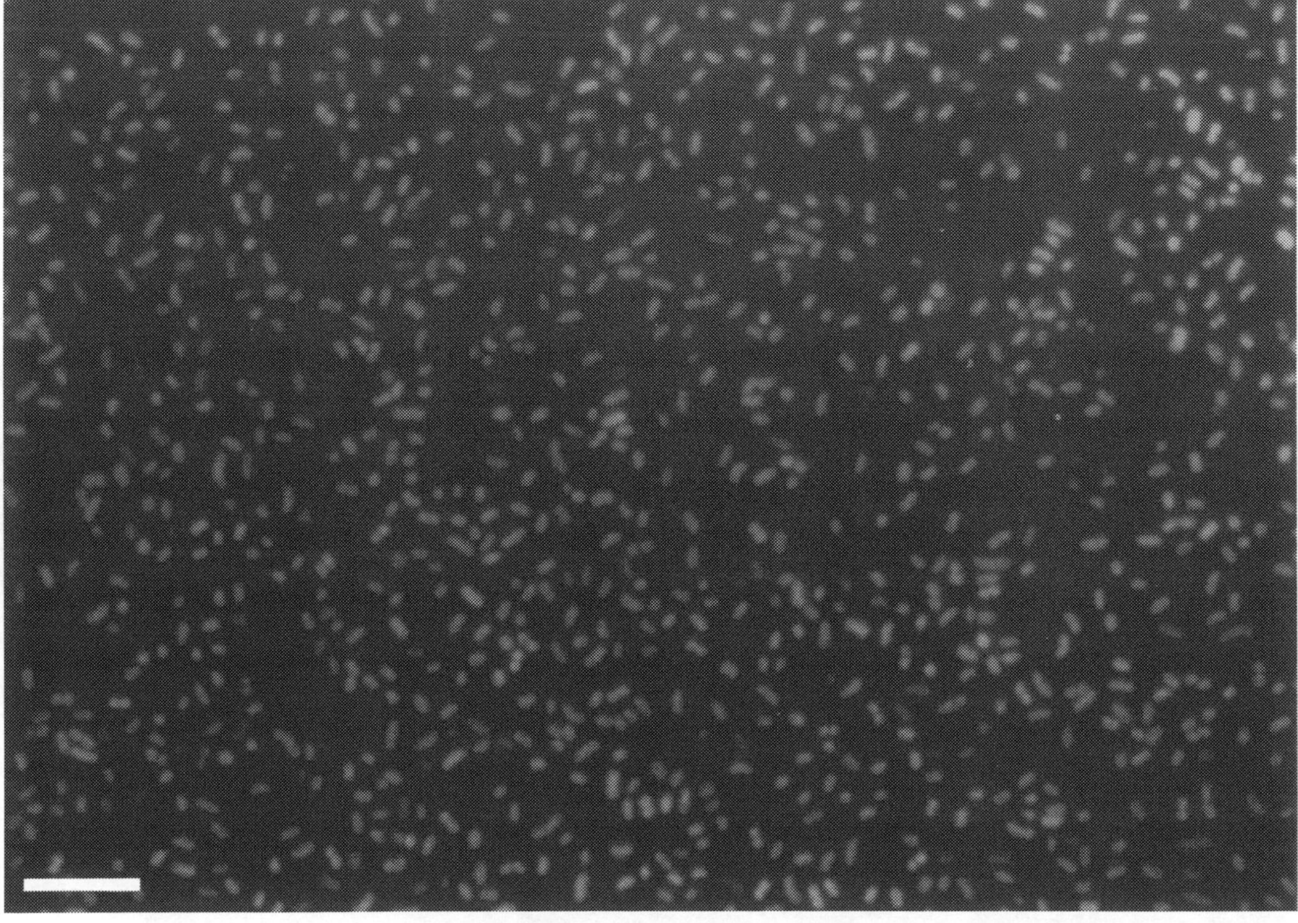

Fig. 24.5. Epifluorescence photomicrograph of *Synechococcus* sp. Under green light the cells fluoresce red and under blue light the cells fluoresce orange. Scale = 10 μm

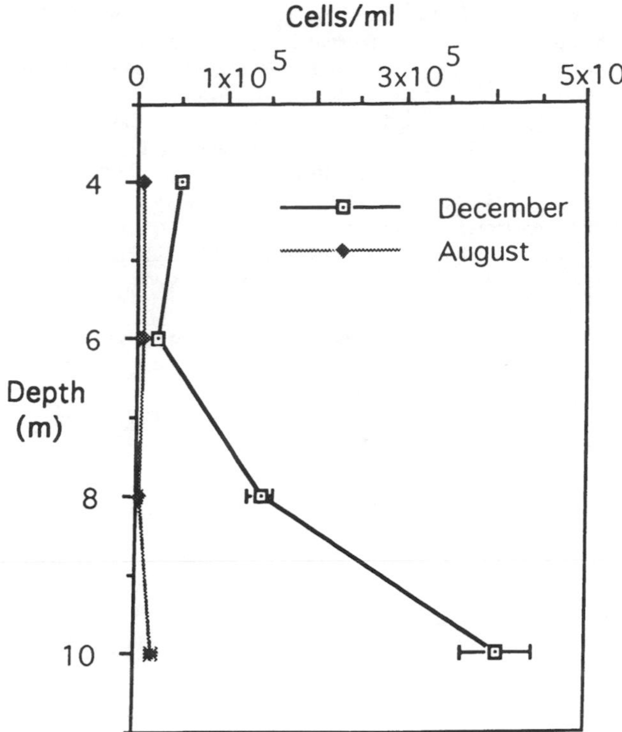

Fig. 24.6. A summer and winter profile through the aerobic zone in Ace lake showed that during winter *Synechococcus* sp. numbers were low at all depths (below 10^4 cells ml^{-1}) Throughout summer numbers remained low in the mixed zone of the lake but increased within the oxycline, (7–12 m).

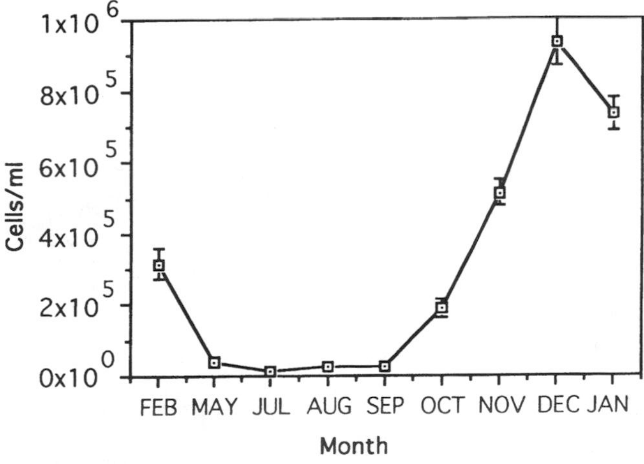

Fig. 24.7. Annual profile of *Synechococcus* sp. numbers at 10 m depth in Ace Lake. Over a one- year period *Synechococcus* sp. numbers peaked in December. The maximum population of *Synechococcus* sp. measured in Ace Lake was 8×10^6 cells ml^{-1} and occurred during mid December at a depth of 11 m.

ACKNOWLEDGEMENTS

We would like to thank Tracey Pitman for her invaluable assistance in Antarctica during the 1992/93 season, Sharee McCammon for assistance with microscopy and the support staff at Davis Station, Antarctica and the Australian Antarctic Divison. PDF is funded by the Australian Research Council. This work was supported by the Australian Scientific Antarctic Committee.

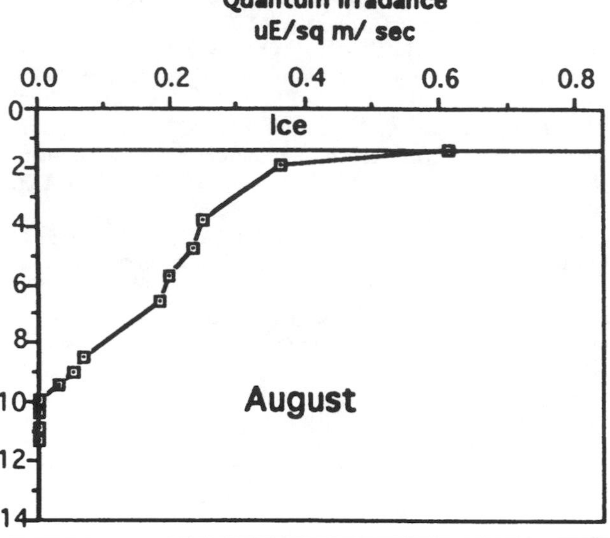

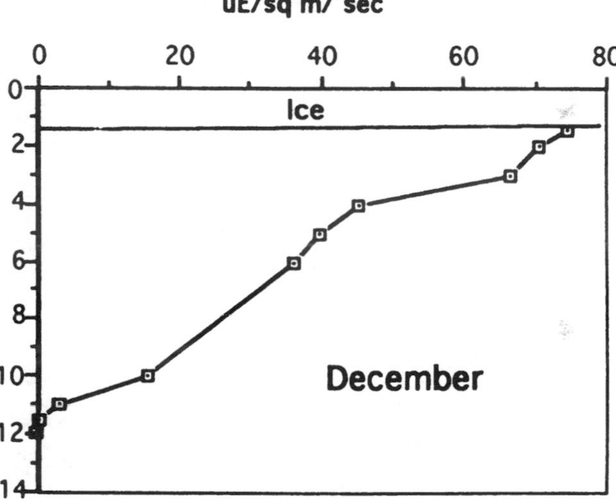

Fig. 24.8. Light penetration (μmol $m^{-2} s^{-1} = \mu E\, m^{-2} s^{-1}$) in Ace Lake. *Synechococcus* sp. numbers peaked at 11 m depth where there was less than 1% incident light. No light was detected past 12 m, where a band of photosynthetic sulphur bacteria occurred.

REFERENCES

Bayly, I. A. E. & Burton, H. R. 1987. Vertical distribution of *Paralabidocera antarctica* (Copepoda, Calanoida) in Ace Lake, Antarctica, in summer. *Australian Journal of Marine and Freshwater Research*, **38**, 537–543.

Burch, M. D. 1988. Annual cycle of phytoplankton in Ace Lake, an ice covered, saline meromictic lake. *Hydrobiologica*, **165**, 59–75.

Burton, H. R. & Barker, R. J. 1979. Sulfur chemistry and microbiological fractionation of sulfur isotopes in a saline Antarctic lake. *Geomicrobiology Journal*, **1**, 329–340.

Campbell, L. & Carpenter, E. J. 1986. Diel patterns of cell division in marine *Synechococcus* spp. (Cyanobacteria): use of the frequency of dividing cells technique to measure growth rate. *Marine Ecology Progress Series*, **32**, 139–148.

Fofonoff, N .P. & Millard, Jr, R. C. 1983. Algorithms for computation of fundamental properties of seawater. *UNESCO technical papers in marine sciences* No. 44.

Gibson, J., Ferris, J. & Burton, H. 1990. Temperature, density, temperature condictivity and conductivity–density relationships for marine derived saline lake waters. *ANARE Research Notes* 78, p. 31.

Gradinger, R. & Lenz, J. 1989. Picocyanobacteria in the high Arctic. *Marine Ecology Progress Series,* **52,** 99–101.

Letelier, R. M. & Karl, D. M. 1989. Phycoerythrin-containing cyanobacteria in surface waters of the Drake Passage during February 1987. *Antarctic Journal of the United States*, **24,** 185–188.

Marchant, H. J., Davidson, A. T. & Wright, S. W. 1987. The distribution and abundance of chroochoccoid cyanobacteria in the Southern Ocean. *Proceedings of the NIPR Symposium on Polar Biology*, **1,** 1–9.

Masuda, N., Nakaya, S., Burton, H. R. & Torii, T. 1988. Trace element distributions in some saline lakes of the Vestfold Hills, Antarctica. *Hydrobiologica*, **165,** 103–114.

Murphy, L. S. & Haugen, E. M. 1985. The distribution and abundance of phytotrophic ultraplankton in the North Atlantic. *Limnology and Oceanography*, **30,** 47–58.

Richardson, K., Beardall, J. & Raven, J. A. 1983. Adaptation of unicellular algae to irradiance: analysis of strategies. *New Phytologist*, **93,** 157–191.

Spaulding, S. A., McKnight, D. M., Smith, R. L. & Dufford, R. 1994 Phytoplankton population dynamics in perennially ice-covered Lake Fryxell, Antarctica. *Journal of Plankton Research*, **16,** 527–541.

Venkateswaran, K., Maruyama, A., Higashihara, T., Sakou, H. & Maruyama, T. 1993. Microbial characteristics of Palau Jellyfish Lake. *Canadian Journal of Microbiology,* **39,** 506–512.

Vincent, W. F. 1988 *Microbial ecosystems in Antarctica*. Cambridge: Cambridge University Press, 304 pp.

Walker, T. D. & Marchant, H. J. 1989 The seasonal occurrence of chrococcoid cyanobacteria at an Antarctic coastal site. *Polar Biology*, **9,** 193–196.

Waterbury, J. B., Watson, S. W., Valois, F. W. & Franks, D. G. 1986. Biological and ecological characterization of the marine unicellular cyanobacterium *Synechococcus*. In Platt, T. & Li, W. K. W., eds. Phytosynthetic picoplankton. *Canadian Bulletin of Fisheries and Aquatic Sciences*, **214,** 71–120.

III Survival mechanisms

INTRODUCTION

Antarctic organisms face varied and extreme environmental conditions that challenge their ability to perform all living functions. Colonization, establishment, growth and reproduction may all be seriously threatened by temperature, water or light availability, or extreme seasonality in the food supply. The study of survival mechanisms under such apparently harsh enviromental conditions is clearly a necessity if we wish to understand how Antarctic communities work.

In the aquatic environment the annual cycle of ice formation and melting imposes severe constraints on many organisms (Gray & Christiansen 1985). This is especially true for planktonic species, since the ice changes the structure of the habitat, the ionic gradient, the light penetration and the potential for stratification. In the marine ecosystem there is a particular interest in discovering if the sea ice effectively promotes a niche differentiation for zooplankton and how the ice dynamics control the communities. The ice-edge habitat has become an increasing focus for research (see Knox 1994) since it appears to play a major role in the ecological and biogeochemical cycles of the Southern Ocean.

Euphausia crystallorophias has evolved in habitats over the coastal shelf favoured by the regular formation of plankton-rich polynyas. The study of Pakomov & Perissinotto (Chapter 25) showed that the spawning success and abundance of *E. crystallorophias* is only possible because of increased food availability, but when present it appears to utilize the niche more effectively than *E. superba.*

On the other hand *E. superba* has to survive during winter under sea ice. Virtue *et al.* (Chapter 26) in careful 130-day starvation experiments, used lipid analysis to demonstrate that symbiotic bacteria in the gut produce fatty acids that are used by krill for routine metabolic maintenance, and that the summer reserves of lipid are kept to be used in reproduction and moulting. In this way, the reproductive cycle of *E. superba* becomes independent of the environmental cycle of ice formation and food availability.

Fish have a salt concentration in their body fluids which will only depress their freezing point by −0.7 °C. With seawater freezing at −1.8 °C fish face the risk of freezing in Antarctic waters. The presence of liver-synthesized glycoproteins prevents freezing of body fluids and tissues. Recent studies by De Vries (Chapter 27) show that the functional mechanism of the antifreeze utilizes the tridimensional structural match of the antifreeze molecule to the ice lattice during the initial ice crystal formation. The antifreeze (AFGP) production depends on a large gene family, responsible for the synthesisis of several kinds of molecules. The levels of circulating antifreeze are strongly correlated with the severity of the cold environment.

The work of Wöhrmann (Chapter 28) has shown for the first time the presence of AFGP in members of the Gadiforms, Scorpaeniforms and Mictophiforms collected in the Weddell Sea, with water temperature as the key to the induction of antifreeze synthesis by the liver. He also found that these molecules are not species specific, and discusses the evolutionary implications of this.

Survival of fish will be dependent on their ability to find their prey, and yet there has been relatively little work on how they actually do this. Montgomery (Chapter 29) studied the retina of young *Pagothenia borchgrevinki* and followed the ontogenetic changes in packing of retinal receptors. He found a gradual loss of visual acuity with increased age, apparently limiting the ability of young fish to capture prey visually. But these fish are exposed to a gradual reduction of light as winter aproaches and sea ice forms, imposing a constraint to visual foraging. Montgomery postulates an ontogenetic shift to using the lateral line system to capture prey at close range in a dark environment.

The high variability of water availability is one of the most important environmental constraints on terrestrial maritime ecosystems. Schroeter *et al.* (Chapter 30) studied photosynthetic activity of lichens under water stress. Photobionts in a single thallus are able to respond and modify their level of photosythetic activity depending on the presence of liquid water or water vapour, and this is affected by the microtopography of the substrate. In *Buellia* the green algal photobiont is activated by water vapour, whilst the cyanobacterial photobiont needs liquid water in *Placopsis.* The development of instrumentation for automated measurement of metabolic activity coupled with a gas exchange model now allows estimates of primary production of lichens in specific habitats.

As yet there has been little published on cryoprotection in Antarctic plants. The studies of Jackson & Seppelt (Chapter 31) on the photosynthetic activity of *Prasiola* revealed a seasonal accumulation of proline. This amino acid decreases during summer and when the alga is kept under favourable cultivation conditions. They postulate that proline is involved in cryoprotection of *Prasiola* against freezing. They also found that UV radiation causes photoinhibition of photosythesis during summer, which may prove to be a more widespread feature of algal response.

Morphological variability is a frequent feature of many species of Antarctic lichens. It is usually attributable to exposure of the habitat to high winds and abrasion. However, this did not seem to be the case for *Mastodia*. The stability of the lichen symbiosis in *Mastodia* under water stress was studied by Huiskes *et al.* (Chapter 32). Prolonged thallus hydration under illumination causes the phycobiont to photosythesize for longer periods leading to accumulation of photosynthates. The phycobiont then increases production of cells, and this causes an intermediate form between the lichen and the alga, destabilizing the lichen symbiosis.

The survival of marine mammals is related to population dynamics and complex environmental relationships affecting reproduction. Boveng & Bengtson (Chapter 33), using mathematical models and simulation techniques, attempt to interpret the observed variability in cohort strength with the intention of clarifying the relationship with environmental fluctuations. They found that there is no periodicity involved and that the relative cohort strengths are well determined and derive from demographic events.

REFERENCES

Gray, J. S. & Christiansen, M. E., eds. *Marine biology of polar regions and effects of stress on marine organisms.* Chichester: J. Wiley & Sons, 639 pp.

Knox, G. A. 1994. *The biology of the Southern Ocean.* Cambridge: Cambridge University Press, 444 pp.

25 Spawning success and grazing impact of *Euphausia crystallorophias* in the Antarctic shelf region

EVGENY A. PAKHOMOV[1] *AND RENZO PERISSINOTTO*[2]

[1]*Southern Ocean Group, Department of Zoology and Entomology, Rhodes University, P.O. Box 94, Grahamstown 6140, South Africa,* [2]*Department of Zoology, University of Fort Hare, PB X 1314, Alice 5700, South Africa*

ABSTRACT

A long-term investigation on the distribution and feeding dynamics of the neritic krill Euphausia crystallorophias *was undertaken in the Indian and Atlantic sectors of the Southern Ocean from 1977 to 1995. This study presents the main results obtained during 12 research cruises in the shelf region of the Cooperation Sea and two cruises in the Lazarev Sea. Both larvae and adult* E. crystallorophias *occurred in high numbers only in shelf waters, between 100 and 600 m depth. In the Lazarev Sea, at stations where swarming of adults occurred, average abundances attained levels of 470–775 ind. 1000 m^{-3}. Gut pigment levels, measured fluorometrically during the austral summer, ranged from 19 to 212 ng (pigm.) $ind.^{-1}$ in larvae/juveniles, and from 49 to 1000 ng (pigm.) $ind.^{-1}$ in adults. Ingestion rates varied between 0.25 and 6.84 μg (pigm.) $ind.^{-1}$ d^{-1}, or 0.1–3.8 μg (C) m^{-2} d^{-1}. Grazing impact on the phytoplankton stock generally did not exceed 1% of total daily production. However, in areas with swarms of adult* E. crystallorophias *daily consumption rates could potentially reach levels of c. 4% of total primary production. Due to the general absence of the Antarctic krill* Euphausia superba *and other large planktonic microphages such as salps on the Antarctic shelf, it is, thus, possible that* E. crystallorophias *constitutes the most important consumer of phytoplankton biomass in this environment. This is corroborated by a clear covariance between the spatial distribution of* E. crystallorophias, *its spawning success and the occurrence of phytoplankton-rich polynyas observed in the Prydz Bay region.*

Key words: East Antarctica, *Euphausia crystallorophias*, spawning success, grazing impact, polynyas.

INTRODUCTION

Euphausia crystallorophias, Holt & Tattersall, is an endemic circumpolar euphausiid inhabiting the Antarctic shelf region (John 1936, Lomakina 1964). It is the dominant euphausiid species in the shelf pelagic community (Pakhomov 1993, Hosie 1994) and is also a very important food source for many coastal predators (e.g. Hubold 1985, Bushuev 1986, Thomas & Green 1988, Whitehead *et al.* 1990). Nevertheless, its ecological importance has generally been ignored because of the priority given to the study of the larger Antarctic krill, *Euphausia superba* Dana.

The ice cover along the coast of the Antarctic continent breaks up irregularly, forming shore leads and polynyas. Coastal polynyas are particularly common in the seas surrounding the continent during late winter and early spring (Dmitrash 1975, Zwally *et al.* 1985). The Cooperation Sea (or Prydz Bay region) is an area where coastal polynyas can reach a size of up to 60 000 km^2 and may be observed every year (Romanov 1984, Smith *et al.* 1984). During spring, coastal polynyas have relatively high phytoplankton stocks, productivities and abundance of predators (Ainley 1985, Comiso *et al.* 1990, Sullivan *et al.* 1993). Therefore, the life history of *E. crystallorophias* is expected to

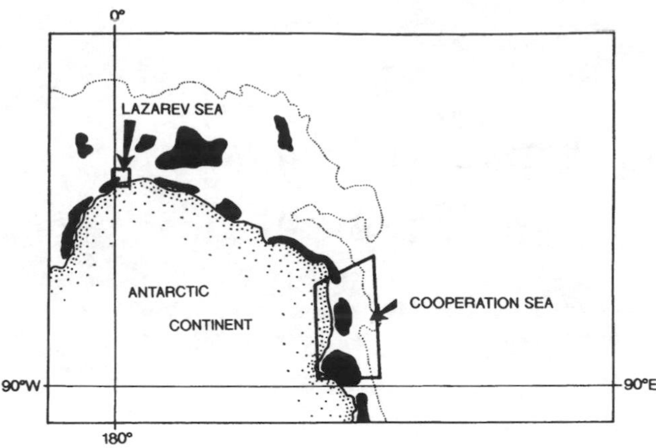

Fig. 25.1. Locations of the sampling areas (boxes), ice edge (dotted line) and seasonal polynyas (dark areas) in the Indian sector of the Southern Ocean during spring. Positions of the ice edge and polynyas are adapted from Dmitrash (1975) and Romanov (1984).

reflect the unique features of the Antarctic neritic biotope. The main aims of this study were: (a) to test the hypothesis that coastal polynyas affect the spawning success of *E. crystallorophias* in the Cooperation Sea, and, (b) to provide realistic estimates of the *in situ* grazing impact of this species on the phytoplankton stock in the shelf region of the Lazarev Sea.

METHODS

Abundance and distribution

Larvae were collected during a 12-year-long investigation on the pelagic ecosystem of the Cooperation Sea (south of 60° S, between 60 and 85° E, Fig. 25.1), from 1977 to 1989. Samples were collected with vertical hauls through the upper 500 m layer using a Juday Net, Oceanic Model (mouth area: 0.5 m^2, mesh size: 120 μm). Developmental stages of euphausiid larvae were identified according to Makarov *et al.* (1986) and counted. The timing of spawning of *E. crystallorophias* was calculated using data on the duration of the period of larval development obtained under laboratory conditions by Ikeda (1986) and the age composition of the various developmental stages (Pakhomov & Perissinotto 1996). The beginning of spawning was calculated using the last larval stages found in the samples and the duration of its developmental time. Similarly, peak spawning was derived from the most abundant developmental stages found in the samples.

Ingestion rates

Samples of subadult (total length 20–25 mm) and larval *E. crystallorophias* were collected in the Lazarev Sea during austral summer 1990/91. Vertical tows at 2–3 hours intervals were made using a WP-2200 μm mesh net during 26–27 December 1990 (0–300 m) and 2–3 February 1991 (0–400 m). Juveniles (10–15 mm) and adults (26–32 mm) were collected during austral summer 1994/95. A 500 μm mesh Bongo net towed obliquely from 300 m to the surface was used during 12 December 1994 and 16–17 January 1995. Gut pigment contents were measured fluorometrically and calculated according to Strickland & Parsons (1968) as modified by Perissinotto (1992). To estimate the gut evacuation rate constant, experimental animals were placed into containers filled with filtered seawater to which charcoal particles (<100 μm) were added. Gut pigment contents of *E. crystallorophias* were then monitored over time and ingestion rates were calculated as described by Perissinotto (1992). To estimate the gut pigment degradation efficiency, a two-compartment pigment budget approach was employed, by comparing the decrease in pigment content in the grazing bottles with the increase in gut pigment levels of animals incubated in these bottles (Lopez *et al.* 1988, Mayzaud & Razouls 1992).

RESULTS AND DISCUSSION

Spawning success

With few exceptions, in the Cooperation Sea *E. crystallorophias* larvae were found only in shelf waters (Pakhomov & Perissinotto 1996). The spatial distribution of larvae was very patchy and similar during all years investigated. Maximum concentrations (more than 2000 ind. 1000 m^{-3}) were always found within Prydz Bay, where the highest peak in abundance (up to 23 969 ind. 1000 m^{-3}) was recorded in 1988. Only once, during the 1985 survey, were high concentrations (>2000 ind. 1000 m^{-3}) of larvae observed in the shelf waters of the western part of the Cooperation Sea, while elsewhere in the shelf region larval abundances were usually <500 ind. 1000 m^{-3} (Pakhomov & Perissinotto 1996). The sharp spatial fluctuations in abundance and biomass of both larvae and adults *E. crystallorophias* are a typical feature of this species and are well documented in the literature (e.g. Lomakina 1964, Brinton & Townsend 1991, Hosie 1991, Pakhomov 1993).

In the different regions of the Southern Ocean, eggs and spawning females of *E. crystallorophias* have consistently been found in November and December (Harrington & Thomas 1987, Brinton & Townsend 1991). In the Cooperation Sea, in December–January most larvae were in the stage of metanauplius and calyptopis I, while February samples were usually dominated by calyptopis III and furcilia I. This developmental stage composition of *E. crystallorophias* larvae indicates that, in the Cooperation Sea, the spawning period varies from year to year, and may start as early as at the beginning of October and extend until mid January. During the period 1977–89, the peak of spawning varied from the beginning of November to the end of December (Fig. 25.2).

According to Whitehead *et al.* (1990), the breakout time in the inshore areas of Prydz Bay varied by up to seven weeks during a period of study from 1981 to 1988. In the summers of 1981/82, 1982/83, 1986/87 and 1987/88, breakout of fast ice occurred 2–4 weeks earlier than in the other years of the study. With the exception of the summer 1986/87, when Prydz Bay was substantially undersampled, our data on the timing (e.g. beginning and duration) of the spawning of *E. crystallorophias* show a remarkable coincidence with periods when earlier breakout was

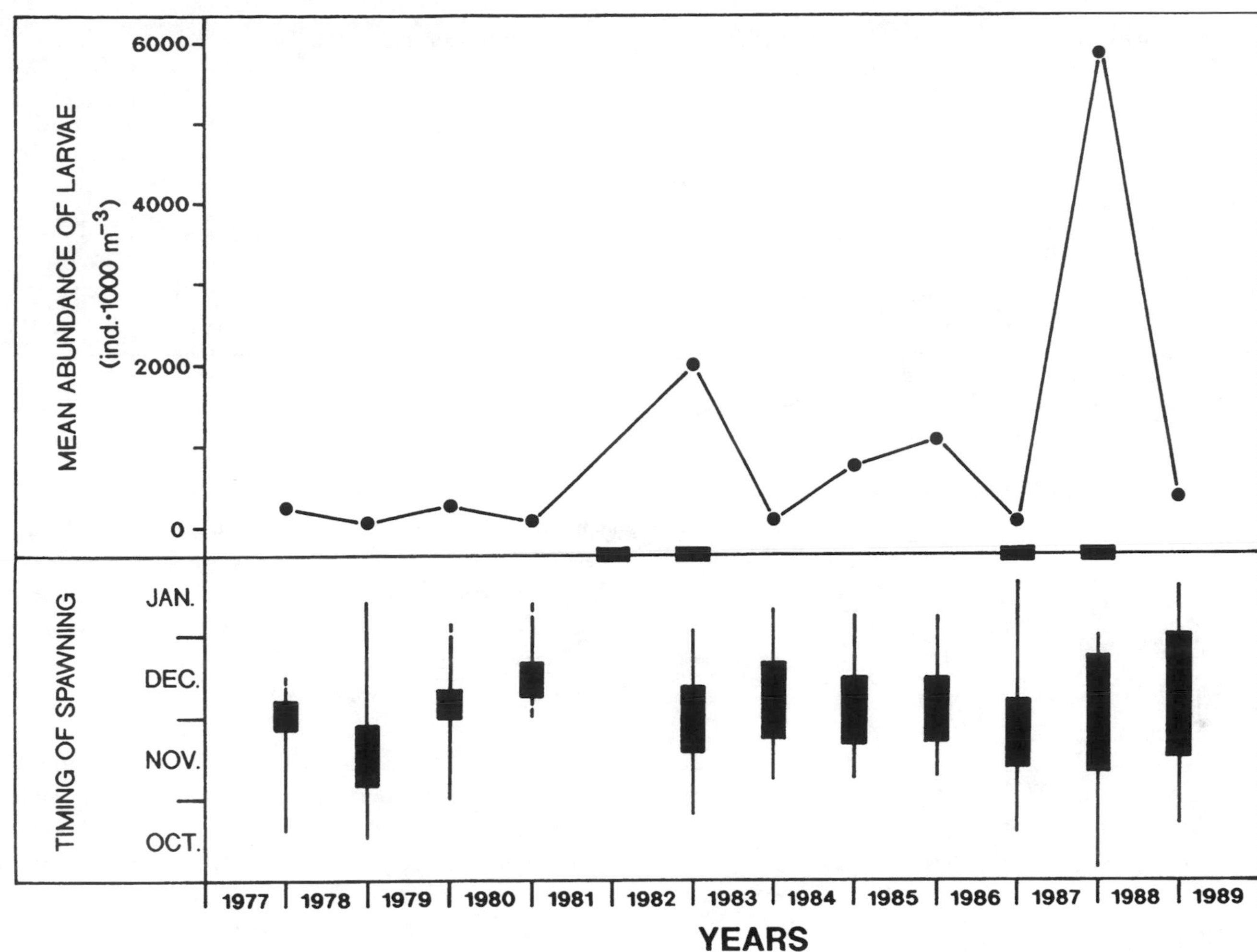

Fig. 25.2. The spawning time and mean abundances of *E. crystallorophias* larvae in the Cooperation Sea. Thick vertical bars indicate the peak of spawning. Mean values of larval abundance were calculated considering only stations where larvae were present in the net tows. Years when breakout of fast ice in Prydz Bay occurred earlier than in other years (from 1981 to 1988, according to Whitehead *et al.* 1990) are shown by thick bars on the horizontal axis.

observed. The summers of 1982/83 and 1987/88 are also when the highest concentrations of *E. crystallorophias* larvae were found within the Prydz Bay region (Fig. 25.2). Therefore, the position, size and time of breakout are important factors affecting the timing of spawning of *E. crystallorophias*, its success and spatio-temporal variability in abundance. The development of large polynyas within Prydz Bay causes an 'oasis effect' (Smith *et al.* 1984, Yamada & Kawamura 1986, Pakhomov & Karpenko 1988), while the ocean to the north is still extensively covered by pack ice (Whitehead *et al.* 1990). According to recent investigations (Harrington & Thomas 1987, Smith & Schnack-Schiel 1990), the spawning of *E. crystallorophias* may start under fast ice. However, it has been shown that its peak coincides with periods of development of coastal polynyas (Pakhomov & Perissinotto 1996). Also, the special conditions found within these polynyas seems to allow *E. crystallorophias* to start spawning about one month earlier than *Euphausia superba* (Hosie *et al.* 1988, Pakhomov & Karpenko 1988).

Grazing impact

Levels of gut pigment concentration at the 24 h stations varied between 0.8 and 84.1, 19 and 218 and 49 and 1000 ng (pigm.) ind.$^{-1}$ in larvae/juveniles, subadults and adults, respectively (Table 25.1). During austral summer 1990/91, subadults of *E. crystallorophias* exhibited highest gut pigment levels in the afternoon (1200–1700 GMT), while larvae fed most intensively in the early evening and morning. Conversely, during austral summer 1994/95 no clear diurnal variations in the ingestion rate of juveniles were found. However, a sharp increase in the feeding activity of adults during nighttime (00.00–01.00 GMT) was recorded.

A negative exponential model provided the best fit to the decline in gut pigment content of *E. crystallorophias* subadults and adults (Fig. 25.3). Gut evacuation rates (k) decreased according to body size, ranging from 2.25 h^{-1} in subadults to 0.32–0.50 h^{-1} in adults (Table 25.1). These results also suggest that the gut passage time ($1/k$) covaries negatively with surface chlorophyll *a* concentrations. Losses of pigment fluorescence during digestion in adult *E. crystallorophias* ranged from 0 to 43% (average 14.3±24.8% (±SD)) during December 1994 and

Table 25.1. *Gut pigment contents and ingestion rates of different size groups of* E. crystallorophias *in the Lazarev Sea*

	Total length (mm)	Gut pigment content (ng (pigm.) ind.$^{-1}$)	Ingestion rates (μg (pigm.) ind.$^{-1}$ d^{-1})	k (h^{-1})	Surface chl *a* concentration (mg m^{-3})
27 December 1990	3–5 (cal. III)	0.8–18.6	0.25	2.25[1]	0.9–2.1
	20–25 (subad.)	19–218	4.91	2.25	0.9–2.1
12 December 1994	26–30 (adults)	84–159	1.00	0.32	0.37
17 January 1995	10–15 (juv.)	8.3–84.1	1.621	2.25[1]	1.3
	29–32 (adults)	49–1000	6.84	0.50	1.3

Note:

[1] k-value taken from the gut evacuation experiment carried out with *E. crystallorophias* 20–25 mm long.

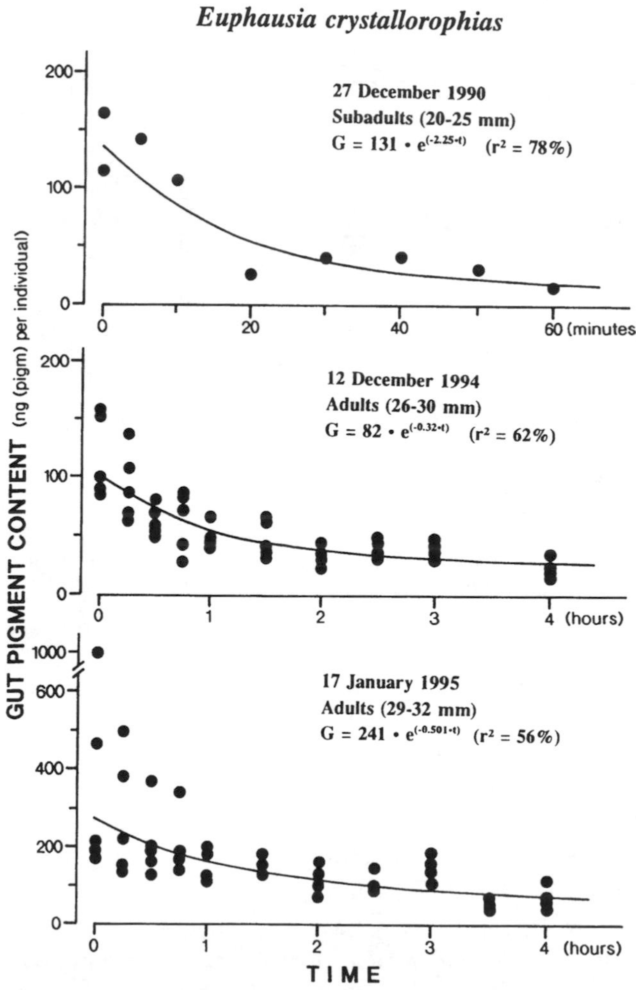

Fig. 25.3. Gut evacuation rates of *E. crystallorophias* adults in the Lazarev Sea, during December 1990, December 1994 and January 1995. *G*: total gut pigment at time *t* (in hours).

from 36 to 99% (average 72.3±23.5%) during January 1995. Thus, an average degradation efficiency factor of 0.86 for December and of 0.247 for January was used to correct the ingestion rates of adult *E. crystallorophias* (Perissinotto 1992).

Maximum ingestion rates for larvae, juveniles and subadults/adults were respectively 25.2, 67.5 and 285–471 ng (pigm.) ind.$^{-1}$ h^{-1}. Daily integrated consumption of phytoplankton increased from 0.25 μg (pigm.) ind.$^{-1}$ d^{-1} in larvae to 6.84 μg (pigm.) ind.$^{-1}$ d^{-1} in adults (Table 25.1). By pooling the results obtained in the experiments conducted at stations where chlorophyll *a* concentrations exceeded ~ 1 mg. m^{-3}, a multiplicative regression model of the form $I=0.025\times L^{1.653}$ was fitted to the variation in daily ingestion rate (I, μg (pigm.) ind.$^{-1}$d^{-1}) versus total body length, L (mm). The model accounted for over 99% of the total variance ($P<0.005$).

In the Lazarev Sea, typical abundances of *E. crystallorophias* in non–swarming conditions were ~6.4–86.1 ind. 1000 m^{-3} for juveniles and 3.1–100 ind. 1000 m^{-3} for subadults–adults. However, at stations where swarming of adults occurred average abundances reached levels of 469.7–774.9 ind. 1000 m^{-3}. Daily consumption rates of juveniles ranged, therefore, between 2.5 and 56.1 μg (pigm.) m^{-2} d^{-1}. Similarly, subadult and adult rates were of 19.6–196.4 and 2.1–12.5 μg (pigm.) m^{-2} d^{-1}, respectively. Estimation of daily population impact in the region where *E. crystallorophias* swarms were found attained maximum levels of 145 μg (pigm.) m^{-2} d^{-1} in December 1994 and 507 μg (pigm.) m^{-2} d^{-1} in January 1995.

Chlorophyll *a* concentrations in the shelf waters of the Lazarev Sea were generally low in mid-December (0.1–0.4 mg m^{-3}) but increased sharply during the end of December and in early January, ranging from 0.9 to 6.2 mg m^{-3} (Perissinotto *et al.* 1992, Laubscher *et al.* 1993, Table 25.1). Similarly, depth-integrated chlorophyll *a* levels were lowest (52.5 mg m^{-2}) during mid-December but increased to 395 mg m^{-2} at the beginning of February. In relation to the total phytoplankton stock available during the period of the investigation, the daily impact of non-swarming *E. crystallorophias* was very limited, ranging from <0.1 to 0.3% of the total chlorophyll *a*. In swarming regions, however, daily impact may have been equivalent to *c.* 0.4–1.2% of the total phytoplankton stock.

Using the C:chl *a* ratio derived from the equation of Hewes *et al.* (1990), larvae, juveniles and subadults/adults of *E. crystallorophias* would have consumed 139–257, 158–896 and 125–1901 μg C m^{-2} d^{-1}, respectively. The maximum impact due to adults during swarming conditions would have reached levels of ~1.5 mg C m^{-2} d^{-1} in December and of ~3.4 mg C m^{-2} d^{-1} in January. Daily primary production rates ranged from <200 mg

C m^{-2} d^{-1}, in mid- December, to 370 mg C m^{-2} d^{-1}, in early January (Laubscher unpublished). Thus, the population impact of *E. crystallorophias* would generally be only <1% of total daily production. Swarming behaviour could, however, result in the local removal of ~1–4% of daily production. The shelf region is characterized by the absence of large planktonic microphages, e.g. *E. superba* and tunicates (Pakhomov 1993, Hosie 1994) and the grazing impact of copepods appears to be very limited, usually <2% of total primary production (Drits & Semenova 1989, Drits & Pasternak 1993). *E. crystallorophias* may, therefore, be the major single pelagic consumer of phytoplankton stock in the Antarctic shelf region.

CONCLUSIONS

It is suggested that the spawning success of *E. crystallorophias* may strongly depend on the timing of ice breakout with formation of polynyas within the continental shelf of Antarctica. Indeed, coastal polynyas, which often exhibit an increase in the level of phytoplankton productivity, are areas of enhanced spawning success and grazing activity for *E.crystallorophias*. Thus, this euphausiid provides a good example of adaptation to the oscillating regime of the shelf community. The utilization of the unique features of the Antarctic coastal biotope allows this species to occupy a key position as an intermediate link between primary productivity and top predators. Also, due to their influence on the population size of *E.crystallorophias*, coastal polynyas may represent one of the most important phenomena controlling pelagic biological production and carbon flux to the higher trophic levels of the Antarctic shelf region.

ACKNOWLEDGEMENTS

We are very grateful to Rhodes University and the South African Department of Environment Affairs and Tourism for providing funds and facilities for this study.

REFERENCES

Ainley, D. G. 1985. The biomass of birds and mammals in the Ross Sea. In Siegfried, W. R., Condy P. R.& Laws, R. M., eds. *Antarctic nutrient cycles and food webs*. Berlin: Springer- Verlag, 498–515.

Brinton, E. & Townsend, A. W. 1991. Developmental rates and habitat shifts in the Antarctic neritic euphausiid *Euphausia crystallorophias*, 1986-87. *Deep-Sea Research*, **38,** 1195–1211.

Bushuev, S. G. 1986. Feeding of Minke whales, *Balaenoptera acutorostrata*, in the Antarctic. *Report of the International Whaling Commission*, **36,** 241–245.

Comiso, J. C., Maynard, N. G., Smith Jr., W. O. & Sullivan, C. W. 1990. Satellite ocean colour studies of Antarctic ice edges in summer and autumn. *Journal of Geophysical Research*, **95** (C6), 9481–9496.

Dmitrash, Zh. D. 1975. Antarctic polynyas from CZCS data. *Uchenye zapiski Leningradskogo Universiteta*, **24,** 168–179 [in Russian].

Drits, A. V. & Semenova, T. N. 1989. Trophic characteristics of major planktonic phytophagous from South Shetland Islands region during early spring. In Ponomareva, L. A., ed. *Complex investigations of the pelagic zone of the Southern Ocean*. Moscow: Shirshov Institute Oceanology Publishers, 66–78 [in Russian].

Drits, A. V. & Pasternak, A. F. 1993. Feeding of dominant species of the Antarctic herbivores zooplankton. In Voronina, N. M., ed. *Pelagic ecosystems of the Southern Ocean*. Moscow: Nauka Press, 250–259 [in Russian].

Harrington, S. A. & Thomas, P. G. 1987. Observations on spawning by *Euphausia crystallorophias* from waters adjacent to Enderby Land (East Antarctica) and speculations on the early ontogenetic ecology of neritic euphausiids. *Polar Biology*, **7,** 93–95.

Hewes, C. D., Sakshang, E., Reid, F. M. H. & Holm-Hansen, O. 1990. Microbial autotrophic and heterotrophic eucaryotes in antarctic waters: relationships between biomass and chlorophyll, adenosine tryphosphate and particulate organic carbon. *Marine Ecology Progress Series*, **63,** 27–35.

Hosie, G. W. 1991. Distribution and abundance of euphausiid larvae in the Prydz Bay region, Antarctica. *Antarctic Science*, **3,** 167–180.

Hosie, G. W. 1994. The macrozooplankton communities in the Prydz Bay region, Antarctica. In El-Sayed, S. Z., ed. *Southern Ocean ecology: the BIOMASS perspective*. Cambridge: Cambridge University Press, 93–123.

Hosie G. W., Ikeda, T. & Stolp, M. 1988. Distribution, abundance and population structure of the Antarctic krill (*Euphausia superba* Dana) in the Prydz Bay region, Antarctica. *Polar Biology*, **8,** 213–224.

Hubold, G. 1985. The early life-history of the high-Antarctic silverfish *Pleuragramma antarcticum*. In Siegfried, W. R., Condy, P. R. & Laws, R. M., eds. *Antarctic nutrient cycles and food webs*. Berlin: Springer-Verlag, 445–451.

Ikeda, T. 1986. Preliminary observations on the development of the larvae of *Euphausia crystallorophias* Holt & Tattersall in the laboratory. *Memoirs of the National Institute of Polar Research*, **40,** 183–186.

John, D. D. 1936. The southern species of the genus *Euphausia*. *Discovery Reports*, **14,** 193–324.

Laubscher, R. K., Perissinotto, R. & McQuaid, C. D. 1993. Phytoplankton production and biomass at frontal zones in the Atlantic sector of the Southern Ocean. *Polar Biology*, **13,** 471–481.

Lomakina, N. B. 1964. The euphausiid fauna of the Antarctic and notal regions. In *Biological Reports of the Soviet Antarctic Expedition 1955–1958*, Vol. 2. Jerusalem: Israel Programme for Scientific Translation, 254–334.

Lopez, M. D. G., Huntley, M. E. & Syles, P. F. 1988. Pigment destruction by *Calanus pacificus*: impact on the diel feeding patterns. *Journal of Experimental Marine Biology and Ecology*, **10,** 715–734.

Makarov, R. R., Menshenina, L. L. & Spiridonov, V. A. 1986. *Methodological recommendations for the investigation on the Antarctic euphausiid larvae*. Moscow: VNIRO Publishers, 1–104. [in Russian].

Mayzaud, P. & Razouls, S. 1992. Degradation of gut pigment during feeding by a subantarctic copepod: Importance of feeding history and digestive acclimation. *Limnology and Oceanography*, **37,** 393–404.

Pakhomov, E. A. 1993. The faunistic complexes of macroplankton in the Cooperation Sea (Antarctica). *Antarctica*, **32,** 94–110 [in Russian].

Pakhomov, E. A. & Karpenko, G. P. 1988. *Euphausia superba* and *E.crystallorophias* reproduction peculiarities in the Cooperation Sea. In *Stocks and biological bases of rational utilization of fishery invertebrates* (Collected papers), Vladivostok: TINRO Publishers, 38–39 [in Russian].

Pakhomov, E. A. & Perissinotto, R. 1996. Antarctic neritic krill *Euphausia crystallorophias*: spatio- temporal distribution, growth and grazing rates. *Deep-Sea Research*, **43,** 59–87.

Perissinotto, R. 1992. Mesozooplankton size-selectivity and grazing impact on the phytoplankton community of the Prince Edward Archipelago (Southern Ocean). *Marine Ecology Progress Series*, **79,** 243–258.

Perissinotto, R., Laubscher, R. K. & McQuaid, C. D. 1992. Marine productivity enhancement around Bouvet and the South Sandwich Islands (Southern Ocean). *Marine Ecology Progress Series*, **88,** 41–53.

Romanov, A. A. 1984. *Ice in the Southern Ocean*. Leningrad: Gidrometeoisdat Press, 1–88 [in Russian].

Smith, N. R., Dong, Z., Kerry, K. R. & Wright, S. 1984. Water masses and circulation in the region of Prydz Bay, Antarctica. *Deep-Sea Research*, **31,** 1121–1147.

Smith, S. L & Schnack-Schiel, S. B. 1990. Polar zooplankton. In Smith Jr., W. O. ed. *Polar oceanography. Part B: Chemistry, biology, and geology*. London: Academic Press, pp. 527–598.

Strickland, J. D. H. & Parsons, T. R. 1968. A practical handbook of sea-water analysis. *Bulletin of the Fisheries Research Board of Canada*, **167,** 1–311.

Sullivan, C. W., Arrigo, K. R., McClain,C. R., Comiso, J. C. & Firestone, J. 1993. Distributions of phytoplankton blooms in the Southern Ocean. *Science*, **262,** 1832–1837.

Thomas, P. G. & Green, K. 1988. Distribution of *Euphausia crystallorophias* within Prydz Bay and its importance to the inshore marine ecosystem. *Polar Biology*, **8,** 327–331.

Whitehead, M. D., Johnstone, G. W. & Burton, H. R. 1990. Annual fluctuations in productivity and breeding success of Adélie penguins and fulmarine petrels in Prydz Bay, East Antarctica. In Kerry, K. R. & Hempel, G., eds. *Antarctic ecosystems. Ecological change and conservation*. Berlin: Springer-Verlag, 214–223.

Zwally, H. J., Comiso, J. C. & Gordon, A. L. 1985. Antarctic offshore leads and polynyas and oceanographic effects. *Antarctic Research Series*, **43,** 203–226.

Yamada, S. & Kawamura, A. 1986. Some characteristics of the zooplankton distribution in the Prydz Bay region of the Indian sector of the Antarctic Ocean in the summer of 1983/84. *Memoirs of the National Institute of Polar Research*, **40,** 86–95.

26 Dietary-related mechanisms of survival in *Euphausia superba*: biochemical changes during long-term starvation and bacteria as a possible source of nutrition

P. VIRTUE[1], P. D. NICHOLS[1,2] AND S. NICOL[3]

[1]IASOS and CRC for the Antarctic and Southern Ocean Environment, University of Tasmania, GPO 252C, Hobart, Tasmania 7001, Australia, [2]CSIRO Division of Marine Research, Marine Laboratories, GPO Box 1538, Hobart, Tasmania 7001, Australia. [3]Australian Antarctic Division, Channel Highway, Kingston, Tasmania 7050, Australia

ABSTRACT

Lipid class, fatty acid content and composition of Euphausia superba *collected from the field and starved under laboratory conditions were determined. Krill collected from Prydz Bay, Antarctica, were subjected to long-term starvation (130 days) upon capture. The total lipid content of starved krill remained relatively constant (17–24 mg g^{-1}) throughout starvation. Body length decreased significantly with a reduction of up to 35% and lipid decreased from 19 to 4.1 mg $animal^{-1}$. No significant differences were detected in levels of triacylglycerol during starvation. There was a significant decrease in polar lipid levels between Day 0 and Day 5; however, levels did not change significantly throughout the remainder of the starvation period. Sterols, predominantly of algal origin, fell from 0.5–3.8 μg per animal to trace levels by Day 5 of starvation. Bacterial cultures isolated from the stomach and the hepatopancreas of krill were also analysed and found to contain strains able to produce polyunsaturated fatty acids including eicosapentaenoic acid (20:5ω3) from trace levels to 0.7% of the total fatty acids. These and related findings may partially explain the high levels (approximately 50% of total fatty acids) of essential fatty acids found in the hepatopancreas in this species and the fact that krill are able to utilize food sources deficient in essential lipid components (e.g.* Phaeocystis*).*

Key words: *Euphausia superba*, lipids, fatty acids, sterols, bacteria, starvation.

INTRODUCTION

Euphausia superba is the dominant species of crustacean in the Southern Ocean. The high biomass reflects krill's ability to adapt to extreme seasonal changes and their efficient use of available resources. *E. superba* are subjected to a marked seasonality in phytoplankton production which may result in limited food availability for up to eight months of the year in some areas. *E. superba* may adopt a number of different strategies to survive periods of reduced phytoplankton abundance. Ikeda & Dixon (1982) reported that starving krill utilize body protein, resulting in body shrinkage associated with a decrease in metabolic rate. Body shrinkage in winter in Arctic euphausiids is well documented (Båmstedt 1976, Falk-Petersen 1981). Negative growth rates have been reported in *E. superba* in summer samples caught and subsequently starved in the laboratory at sea (Nicol *et al.* 1992, Quetin & Ross 1991).

Krill feeding on detritus and switching to a more omnivorous diet during winter has been suggested (El-Sayed & McWhinnie 1979, Mauchline 1980). Krill also have been reported to feed on ice algae during the winter (Spiridonov *et al.* 1985, Kawaguchi *et al.* 1986). Rakusa- Suszczewski & Zdanowski (1989) reported a high concentration of bacteria in the stomach contents of *E. superba* and suggested bacteria may play an important role in the diet, particularly in the winter. In addition to being a component of the diet, Rakusa-Suszczewski & Zdanowski (1989) suggested that the bacteria found in the stomach may be living *in situ* and providing a source of amino acids, enzymes or vitamins for krill.

Bacterioplankton constitute a substantial portion of the planktonic biomass in the Southern Ocean. From September through to April biomass estimates for Antarctic polar waters of between 0.6×10^5 and 0.6×10^6 cells ml^{-1} from a variety of Antarctic oceanic areas have been reported (reviewed by Karl 1993). In the marginal ice edge zone, bacteria comprise up to 16% of the plankton biomass in the Weddell Sea region (Lancelot *et al.* 1993). Bacteria, in addition to heterotrophic nanoplankton, may be a viable component of the diet during periods of low phytoplanktonic production. One of the aims of this investigation was to determine the fatty acid composition and therefore the nutritional attractiveness of bacteria found in both the krill stomach and hepatopancreas.

Conflicting evidence of an overwintering storage of lipid has been reported in *E. superba*. Hagen (1988) and Quetin & Ross (1991) suggest lipids in the form of triacylglycerols are used as an energy source through the winter. Clarke (1980, 1984), however, reported that in the absence of substantial wax ester synthesis, lipid reserves were not an important energy source in the winter. Although lipids were not analysed, Ikeda & Dixon (1982), also deduced from the C:N ratio of krill subjected to starvation in the laboratory that protein rather than lipid was metabolized during starvation.

The physiological responses of *E. superba* to starvation are not fully understood. Reported physiological changes include decreased oxygen uptake, decreased ammonia and phosphate excretion, changes in inter moult period and sexual regression (Ikeda & Dixon 1982, Thomas & Ikeda 1987). Sexual regression was found to be associated with negative growth during starvation in *E. superba* (Ikeda & Dixon 1982, Siegel 1987) and this was followed by sexual re-maturation when growth resumed (Thomas & Ikeda 1987). This phenomena has not been well documented in males of this species.

To date the role of lipids in *E. superba* during periods of limited food supply is not clear. An aim of this study was to determine the lipid composition of krill under conditions of controlled starvation. The lipid class, fatty acid and sterol profiles of *E. superba* were determined immediately after capture and during a subsequent period of starvation. Physiological changes associated with starvation such as growth rate, sexual maturity and mortality were also monitored. The parallel analysis of lipid content and composition with these biological parameters will provide further insight into the survival strategy employed by this key species of the Southern Ocean ecosystem.

METHODS AND MATERIALS

Krill collection and experimental conditions

Several thousand specimens of *E. superba* were caught on 27 January 1993 in the Prydz Bay region of Antarctica 66° 30.83′ S and 67° 30.32′ E with a rectangular midwater trawl net (8×8 m) which was allowed to drift in the top 20 m of the water column. Krill were maintained in darkness in a cold room (0±0.5 °C) on board RSV *Aurora Australis* for a 40 day period until they were transferred to the Antarctic Division cold room facilities at Kingston, Tasmania where they were kept under similar conditions for the remainder of the experiment.

From this collection over 250 krill were selected, shortly after capture, according to size and sexual maturity, and maintained individually in 2 l plastic containers for up to 130 days. Krill used for this experiment were all of similar size with a total body length, from the tip of rostrum to the tip of uropod, of approximately 35–45 mm (Standard measurement 1, described by Kirkwood 1982). Animals used for analysis were, at the onset of the experimental period, either adult non-gravid females (3A or 3E) (Kirkwood 1982) or subadult males. Seawater was filtered through a 0.45 μ Millipore filter and changed weekly. Bottles were examined every 24 h for exuviae. Three krill of each sex were taken for lipid analysis every 5, 10 or 20 days throughout the starvation period. To isolate the effects of starvation from those of moulting, where practicable, animals were not taken within four days of ecdysis. Animals were frozen in liquid nitrogen for up to 2 months prior to extraction. Wet weights were converted to dry weights, where reported, using a mean value of 77.2% water content (Nicol *et al.* 1992). Animals dying naturally were not used for lipid analysis.

Sexual maturation was monitored using light microscopy. Throughout the experiment both exuviae and whole dead animals were examined. Maturation stages reported are those described by Kirkwood (1982). Mortality rates were determined using all animals dying naturally from the time of capture until the end of the experiment (n=211). Accidental deaths incurred were not included in this analysis.

Moults were collected throughout the experimental period for growth rate analysis. Both uropods from each exoskeleton were measured and the mean was taken. Upon death (either naturally or killed for biochemical analysis), the mean of both body uropods were taken from each animal (n=143). Growth was determined from the percentage size change in the mean uropod length at first moult during the experimental period to the mean uropod length at the time of death. Animals taken for analysis or dying naturally before their first moult were not included in this analysis. Uropod exopodite measurements and whole body lengths were taken using standard measurement 7 and 1 described by Kirkwood (1982).

Bacteria collection and culture

Krill used for analysis of stomach and hepatopancreas bacteria were immersed in a 1:1 sterile seawater to cryogenic solution (either 10% di-methyl sulphoxide or 10% glycerol) immediately after capture. Krill were frozen in liquid nitrogen and analysed within two months. Krill were dissected under sterile conditions and individual organs were cultured in 10 ml broths of Zobell's medium (Zobell 1946) and incubated at 10 °C until turbidity was apparent. Inoculations on agar plates were made with tenfold serial dilutions in to sterilized sea water and incubated at 10 °C for 2–4 days. Single colonies were spread on to new plates and allowed to incubate. Both mixed culture plates and single colony plates were harvested for lipid analysis.

Lipid extraction and analysis

Krill were analysed individually. Wet whole krill were weighed prior to grinding in $CHCl_3$. Krill and bacteria were extracted quantitatively by the modified one-phase $CHCl_3$-MeOH Bligh & Dyer (1959) method (White *et al.* 1979). After phase separation, the lipids were recovered in the lower $CHCl_3$ layer (solvents were removed *in vacuo*) and stored in 1.5 ml vials fitted with teflon-lined caps at −20 °C. A portion of the total lipid extract of krill samples were analysed for lipid content and lipid class composition with an Iatroscan MK III TH10 TLC-FID (Thin Layer Chromatography-Flame Ionisation Detector) analyser (Iatron Laboratories, Japan) (Volkman *et al.* 1986, Volkman & Nichols 1991). The response of the FID was calibrated using external standards covering the concentration range used in analysis of the samples (0.5–5.0 μg spotted). As different lipid classes have different FID responses, separate calibration curves were used for each class.

Fatty acid methyl esters (FAME) and free alcohols and sterols were formed by direct transesterification of an aliquot of the total lipids with methanol/$CHCl_3$/HCl (10:1:1; 100 °C; 60 min.). After cooling, 1 ml milli-Q-water was added and products were extracted into hexane/$CHCl_3$ (4:1). Solvents were removed under a stream of nitrogen and the alcohols and sterols were converted to their corresponding O-trimethylsilylethers by treatment with N,O-bis(trimethylsilyl)trifluoroacetamide.

Gas chromatographic analyses of fatty acid methyl esters were performed on a Hewlett Packard 5890 GC equipped with a methyl silicone fused-silica capillary column and a flame ionization detector (Nichols *et al.* 1989, Nichols *et al.* 1991). In addition, selected fatty acid samples were analysed using a polar BP 20 fused silica column (Volkman *et al.* 1989). GC-MS analysis of samples was performed on a HP 5890 GC and 5970 Mass Selective Detector fitted with a direct capillary inlet and a split/splitless injector (Nichols *et al.* 1989, Nichols *et al.* 1991). Fatty acids, sterol and isoprenoid C_{25} alkenes were identified by comparing retention time and mass spectral data with those obtained for authentic and laboratory standards.

Statistical analysis

A one-factor analysis of variance was performed for total lipid, lipid class and sterol data from the krill starvation experiment. Fisher's PLSD multiple comparison test was used and results were reported using a significance level of 99%. Regression lines were compared by Analysis of Covariance (ANCOVA). All data are reported using plus or minus standard deviation.

RESULTS

Growth, mortality and sexual maturation of krill

Growth decreased significantly throughout starvation. Individual growth of krill ranged between 13.5% and −35.2% (n=143) from the day of their first moult during the experimental period until they either died naturally or were taken for analyses. There were no significant differences between the calculated regressions of female and male growth (p<0.01) (Fig. 26.1). As no significant differences were found between sexes, an

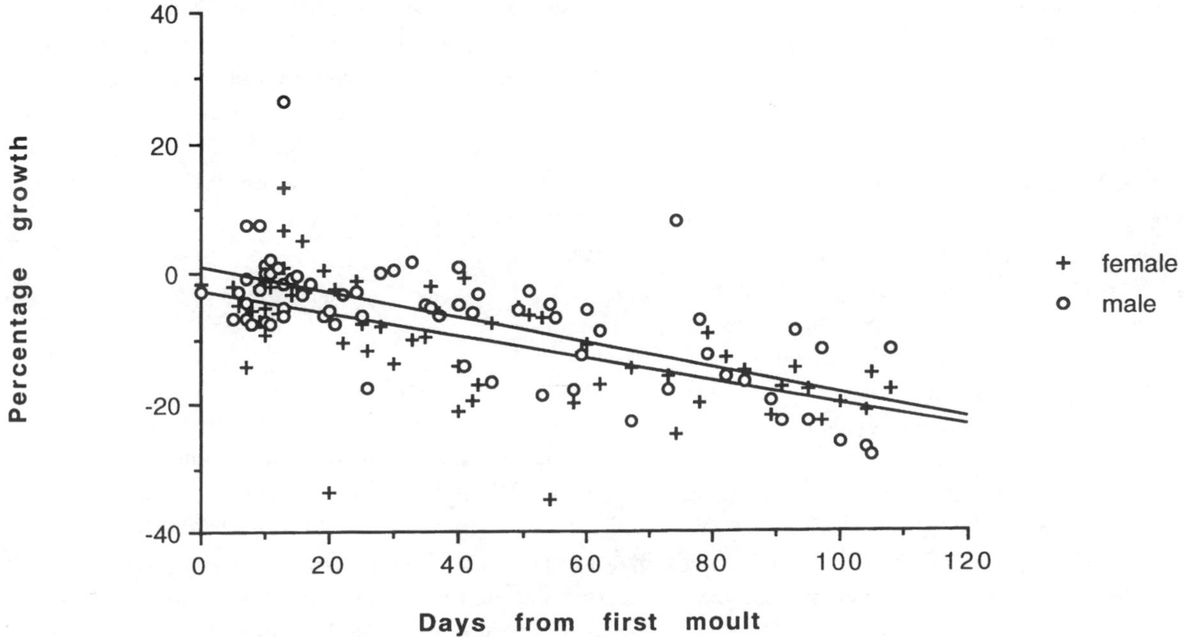

Fig. 26.1. Percentage growth of *Euphausia superba* during starvation, determined from the day of first moult until the end of the starvation period. Female: $y=-2.5094-0.17628x$, $R^2=0.396$. Male: $y=1.0483-0.19609x$, $R^2=0.458$.

Table 26.1. *Total lipid in* Euphausia superba *during starvation. Data are the mean of three analyses ±SD*

Day	0	5	15	25	35	55	7	95	115
Male									
Wet weight (g)	0.7±0.2	0.5±0.1	0.6±0.1	0.5±0.2	0.4±0.1	0.4±0.0	0.4±0.1	0.4±0.1	0.3±0.1
Percentage total lipid (dry wt)	11.9±6.1	11.3±2.0	7.7±3.5	9.9±2.0	7.1±0.3	8.8±1.8	7.7±0.5	9.3±0.9	9.7±0.5
Total lipid (mg animal^{-1})	17.9±5.9	12.2±0.7	10.2±1.8	10.4±2.5	6.8±1.6	7.1±0.7	7.2±1.6	9.3±0.8	7.8±0.5
Female									
Wet weight (g)	0.9±0.2	0.6±0.0	0.5±0.1	0.5±0.1	0.5±0.2	0.4±0.0	0.4±0.1	0.3±0.1	0.2±0.1
Percentage total lipid (dry wt)	8.8±1.6	8.1±2.2	10.4±3.6	10.3±4.6	7.8±1.7	7.3±2.0	8.9±0.4	9.6±2.3	7.5±0.8
Total lipid (mg animal^{-1})	19.0±8.4	10.3±2.1	12.0±1.1	10.2±2.7	9.5±4.2	6.8±2.8	8.5±1.4	5.5±2.5	4.1±1.4

average growth of the total population was determine to be −0.08% body length per day starvation.

Mortality over 130 days was 12% (25 individuals). There was no significant differences between sexes with respect to mortality ($p<0.01$). Mortality was particularly high during ecdysis among the experimental population (18 out of the 25 krill died during moulting).

All female krill used in this experiment were initially adults either post-spawning or non-gravid pre-spawning. Sexual regression among female krill was noted at the second moult. Loss of thelycum pigmentation was evident and colour changed from bright red observed in the first moult to pale pink. By the third moult, pigmentation was reduced further and the thelycum was colourless. The decrease in pigmentation was accompanied by a successive loss of form and a decrease in chitinization of the thelycum. Although not as obvious as in female krill, sexual regression was noted in subadult male *E. superba*. Upon capture subadult male ampullae were pale but visible to the naked eye. Single and bilobed petasmas were evident at first moult. By the second and third moults either the petasma was single lobed or endopods were undifferentiated and the ampullae were no longer visible.

Lipids

Considering both male and female data together, animal wet weight (single krill) decreased from 0.9 g on Day 0 to 0.2 g by day 115 (Table 26.1). No significant differences were found in total lipid per animal from Day 5 through to Day 115. There was however, a significant decrease in total lipid per animal between Day 0 and Day 5 ($p<0.01$) (Table 26.1). The same trend was seen in the polar lipid fraction; however there were no significant differences in the triacylglycerol fraction throughout starvation (Fig. 26.2). No significant differences in percentage composition of fatty acids were found between males and females and among samples throughout starvation (Table 26.2). Free fatty acids were either negligible or found in only trace amounts indicating lipid hydrolysis had not occurred in these samples. Total sterol content in krill throughout starvation ranged from 2.6– 5.8 mg g^{-1} dry weight. Sterols in order of elution included: cholesta-5,22E-dien-3ß-ol (22-dehydrocholesterol), cholest-5-en-3ß-ol (cholesterol), cholesta-5,24-dien-3ß-ol (desmosterol), 24-methylcholesta-5,22E-dien-3ß-ol (brassicasterol), 24-methylcholesta-5,24(28)-dien-3ß-ol (24-methylene cholesterol), 24-methylcholest-5-en-3ß-ol (campesterol), 24-ethylcholesta-5,22E-dien-3ß-ol (stigmasterol), 24-ethylcholest-5-en-3ß-ol (sitosterol), 24-ethylcholesta-5,24(28)Z-dien-3ß-ol (iso-fucosterol) and 4,23,24-trimethyl-5-cholest-22E-en-3ß-ol (dinosterol). There was no significant difference in the absolute levels of cholesterol and desmosterol throughout starvation. Cholesterol was the major sterol and ranged from 1.5 to 3.8 mg g^{-1} dry weight. Although total sterol per animal did not change throughout starvation, sterols of algal origin decreased significantly from Day 0 to Day 5. Levels of algal sterols ranged from 0.5–3.8 μg per animal on Day 0 and fell to trace and zero levels by Day 5 and Day 15 for dinosterol.

Fatty acid profiles of bacteria

Eight of the 11 bacterial species analysed produced EPA (eicosapentaenoic acid) at levels ranging from trace to 0.7%. EPA-producing bacteria were isolated from both the stomach and the hepatopancreas. Levels of 18:4ω3 and 18:2ω6 ranged from trace to 0.8% and 0.4% respectively. Levels of saturated, branched and monounsaturated fatty acids varied markedly in the cultures. Representative fatty acid profiles of bacteria isolated from the stomach and hepatopancreas are given in Table 26.3. Four of the cultures were tested for gram reactions and all were found negative.

DISCUSSION

Growth of krill during starvation

Siegel (1987), using seasonal age and growth parameters, reported that *E. superba* did not grow in winter. Although it is known that krill can feed on ice algae during winter (Spiridonov *et al.* 1985, Kawaguchi *et al.* 1986), krill measurements reported by Kawaguchi *et al.* (1986) indicate that no significant growth occurred during the austral winter. Body shrinkage of *E. superba* has been observed both in the field (Nicol *et al.* 1992, Quetin & Ross 1991) and in the laboratory (Mackintosh 1967, Ikeda & Dixon 1982). Under controlled laboratory conditions,

Table 26.2. *Percentage fatty acid of* Euphausia superba *during starvation. Data are the mean of six analyses ±SD*

Day	0	5	15	25	35	55	75	95	115
14:0	9.5±1.1	11.8±1.4	9.8±0.1	12.0±0.4	12.9±1.2	8.7±1.4	10.9±3.2	7.5±1.7	8.0±2.3
16:1ω7c	9.9±1.6	6.7±0.8	5.6±2.8	8.1±0.6	7.0±0.7	10.4±3.0	7.8±1.0	7.9±2.1	7.2±0.2
16:0	20.8±0.8	24.6±3.8	19.1±1.5	21.0±0.3	23.0±0.6	23.8±2.6	23.7±0.4	22.8±1.4	22.1±0.4
18:4ω3	0.8±0.0	1.6±0.5	2.2±1.2	2.0±0.2	1.7±0.1	0.2±0.2	0.9±0.8	0.7±0.1	0.4±0.1
18:2ω6	2.3±0.4	3.1±0.2	2.8±0.3	2.5±0.1	2.6±0.5	2.3±0.3	2.7±0.3	2.5±0.3	2.9±0.3
18:3ω3	0.6±0.1	1.0±0.3	1.4±0.5	1.1±0.0	0.9±0.2	0.7±0.3	0.8±0.1	0.9±0.1	0.9±0.0
18:1ω9c	10.5±1.7	12.7±1.5	10.5±0.9	10.3±0.3	11.9±0.5	14.0±0.7	13.7±1.2	13.3±0.3	12.8±1.0
18:1ω7c	10.3±0.7	10.2±0.9	11.5±1.2	9.6±0.4	8.6±0.9	14.1±0.6	10.6±2.5	12.8±0.3	13.5±0.3
18:0	0.9±0.2	1.1±0.3	0.7±0.0	0.7±0.0	1.1±0.5	0.7±0.0	1.0±0.0	0.8±0.3	0.9±0.3
20:5ω3	18.2±2.7	13.4±3.3	18.0±0.1	17.0±0.7	15.4±0.9	12.7±2.9	13.6±0.5	15.3±2.5	15.2±1.4
22:6ω3	9.5±1.3	5.8±2.9	9.9±1.0	8.1±0.4	8.7±2.2	8.1±0.8	7.3±0.7	9.3±2.9	11.2±1.5
Total ω3	28.5	20.8	30.2	27.0	25.9	21.1	21.7	25.4	26.8
Total ω7	20.2	16.9	17.1	17.7	15.6	24.5	18.4	20.7	20

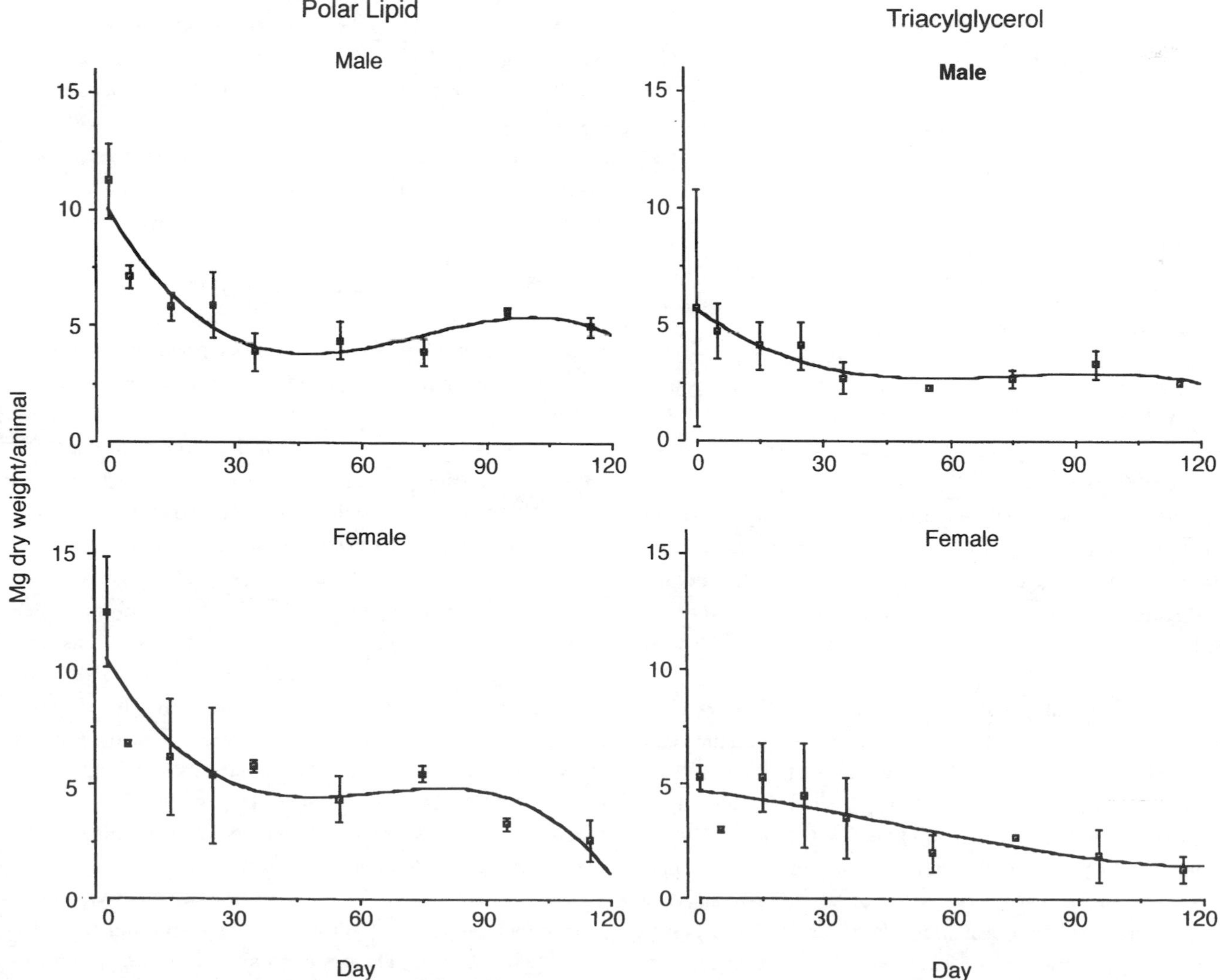

Fig. 26.2. Triacylglycerol and polar lipid levels in male and female *Euphausia superba* during starvation (n=3±SD). Regression equations are polar lipid – Male: y=9.9774–0.30615x+4.7048e-3x^2–2.1025e-5x^3, R^2=0.858. Female: y=10.439–0.30909x+5.1162e-3x^2–2.6571e-5x^3, R^2=0.803. Triacylglycerol – Male: y=5.6320–0.13025x+1.8311e-3x^2–8.0267e6x^3, R^2=0.906. Female: y=4.6668–1.9553e2x–s3.5031e-4x^2+2.4011e-6x^3, R^2=0.702.

Table 26.3. *Percentages of fatty acids in bacterial cultures from* Euphausia superba

Fatty Acid	Stomach	Hepatopancreas
13:0	0.4	0.4
i14:0	0.4	0.3
14:1ω7	0.9	0.3
14:1ω5	0.1	0.0
14:0	2.5	0.5
a15:1	13.4	0.9
i15:0	1.5	9.6
a15:0	2.1	2.0
15:0	6.0	9.9
i16:1	0.5	0.0
i16:0	0.0	0.8
16:1ω7c	37.3	36.5
16:1ω7t	0.0	5.2
16:0	9.2	11.6
i17:1	0.9	0.5
i17:0	0.0	0.5
17:1ω8	15.3	2.7
17:1ω6	1.6	3.2
17:0	2.3	5.6
18:4ω3	0.0	0.8
18:2ω6	0.0	0.4
18:1ω9c	2.2	5.7
18:1ω7c	2.9	0.4
18:0	0.6	1.1
i19:0	0.4	0.6
a19:0	0.3	0.6
20:5ω3	0.1	0.7

Ikeda & Dixon (1982) demonstrated the extremely high tolerance to starvation of this species, which survived for 211 days without food. In the present study both body size and weight decreased throughout starvation. Animals were not weighed on Day 0 so the individual reduction in body weight was not quantified. Negative growth, however, was demonstrated during starvation. An average decrease of 0.08% body length per day was exhibited from first moult until death during starvation, with animals decreasing in body length up to 35% over 130 days.

Sexual regression of krill during starvation

The observation of female sexual regression in this study confirms earlier suggestions (Makarov 1976, McWhinnie *et al.* 1979) and laboratory observations (Thomas & Ikeda 1987, Denys *et al.* 1981) of this characteristic for *E. superba*. Thomas & Ikeda (1987) reported that both starving and feeding female *E. superba* underwent sexual regression after spawning which was accompanied by a decrease in body length. Females were then able to reverse this process after five months in captivity, which was independent of nutrient conditions. It was suggested that this versatility would allow krill to reach maturity and spawn independently of environmental conditions and thereby ensure survival. Using length frequency data during different seasons, Siegel (1987) concluded that rejuvenation is common among Antarctic euphausiids and that rematuration and spawning can occur up to three times during a lifespan. Sexual regression in male *E. superba* is not well documented. All adult post-spawning males died within the first few days of capture and we assumed that once mating had taken place they were unable to regress. Poleck & Denys (1982) observed regression in the male petasma as late as Stage D (Bargmann 1937); however, at this stage spermatophores are not yet present in ampullae. Sexual regression was observed in subadult males in the present study. Regression from a developed petasma through to an undifferentiated endopod accompanied a successive regression of the ampullae until it was no longer visible. As observed for female krill, the degree of maturation in subadult male krill appears to be a function of body size.

Lipids

Euphausia superba are reported to utilize lipid reserves, mainly triacylglycerols, as an energy source in winter (Hagen 1988, Quetin *et al.* 1994). These findings were based on data comparing field samples taken in the late summer to those taken in the early spring. Lipid levels of krill caught at the end of January, on Day 0 in this study, were similar to that reported by Quetin *et al.* (1994) for krill analysed at the same time of the year. Although krill are actively feeding in late January, the lipid levels were lower than that reported for krill caught in late February to March and in the early spring (August and September) (Quetin *et al.* 1994). If lipids were utilized as an overwintering energy source then krill lipid levels would be at their lowest in early spring. These higher reported lipid levels found in krill sampled in spring than in summer samples would suggest that lipid is primarily used for reproduction rather than as an overwintering energy source.

Ikeda & Dixon (1982) found that the carbon, nitrogen and phosphorus content of starving krill resembled that of wild krill. Although lipid content and composition were not determined, Ikeda & Dixon (1982) concluded from this analysis that lipids were not catabolized during starvation. The absolute levels of total lipid (mg animal^{-1}) in the present study decreased throughout starvation; however, this decrease was not significant except between Day 0 and Day 5. This initial decrease indicates that lipid is an important energy source during short-term starvation. In the long term, however, lipid reserves are essentially being conserved. The primary role of lipid during starvation is for biosynthesis requirements or other essential functions such as moulting and maintaining membranes rather than for energy production.

Triacylglycerol is a storage lipid which is usually metabolized during food shortages. Krill in this study were not using triacylglycerol as such. There was no significant decrease in triacylglycerol level throughout starvation; however, levels were variable particularly in Day 0 males. Polar lipids, which make up the lipid bilayer of cell membranes, were also maintained during long-term starvation. Polyunsaturated fatty acids, which are major constituents of the polar lipid fraction in *E. superba* (Virtue *et al.* 1993b), remained relatively constant throughout starvation. The fatty acid composition of the total lipid, reflected in the lipid class composition, also did not change substantially throughout starvation.

Apart from levels of cholesterol which remained constant throughout starvation, minor sterols decreased. Sterols of algal origin in krill indicate recent dietary input. The rate of depletion of algal sterols reflects the rates of assimilation and/or conversion of these sterols to cholesterol. Crustaceans do not have the capacity for *de novo* sterol synthesis but are able to dealkylate dietary C_{28} and C_{29} sterols to cholesterol (Teshima 1982). It is difficult to detect this conversion by merely examining levels of cholesterol because the levels of these algal sterols found in krill are an order of magnitude smaller than levels of cholesterol. Algal sterols found in krill are not only useful indicators of dietary composition but they can potentially be used in the quantitative determination of various phytoplanktonic components in the diet. Similarly, unsaturated isoprenoid C_{25} alkenes ($ipC_{25:3}$, $ipC_{25:4}$ and $ipC_{25:5}$) were found to be the dominant components in the hydrocarbon fraction of krill collected from the South Georgia region (Fig. 26.3). As diatoms are to our knowledge the sole source of these components (e.g. Nichols *et al.* 1993a and references therein), it may be possible to use these novel biomarkers in krill food-chain studies. Further research with these novel biomarkers is recommended for future studies of krill–diatom trophodynamics.

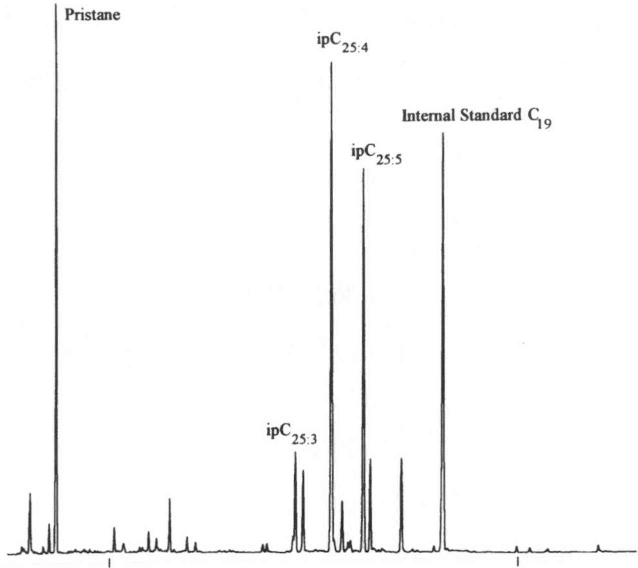

Fig. 26.3. Partial gas chromatograph showing isoprenoid hydrocarbons in *Euphausia superba*.

Bacteria

There is evidence that the Antarctic food chain is not as simple as previously proposed. The incorporation of particulate and dissolved organic material, bacteria and heterotrophs into the traditional diatom–krill–vertebrate food chain is warranted. Tanoue & Hara (1986) showed that krill fed on choanoflagellates, and Hewes *et al.* (1985) suggested that protozoans may be a major food source for krill. Rakusa-Suszczewski & Zdanowski (1989) found high concentrations of bacteria in krill stomachs. Krill have been shown to be able to filter particles as small as some bacteria (Kils 1983, Morris 1984). Some bacteria are probably too small to be filtered by krill from the water column; however, they may be ingested via heterotrophs such as choanoflagellates and ciliates, or associated with detrital material. Rakusa-Suszczewski & Zdanowski (1989) compared the bacterial concentration found in the krill stomach to that found in the water column. Taking into account the filtration rates of krill, they considered concentrations of bacteria found in the stomach too high to be solely of dietary origin and suggested that in addition to dietary input, bacteria may be living *in situ* in the stomach and may provide a source of amino acids, enzymes or vitamins for krill.

Some strains of bacteria found in Antarctica have been reported to produce polyunsaturated fatty acids (PUFA), including eicosapentaenoic acid (EPA) (Nichols *et al.* 1993a). The ability to produce PUFAs is not widespread among bacteria. PUFAs have a vital nutritional function for grazing zooplankton as they play a major physiological role. Crustaceans are reported to be incapable of synthesizing polyunsaturated fatty acids *de novo* (Clarke 1983) and dietary sources of the polyunsaturated fatty acids, eicosapentaenoic and docosahexaenoic are considered essential in many crustacean species (Kanazawa 1985).

The bacteria found in the hepatopancreas of krill analysed in this study may indicate that they are resident rather than dietary in nature. Only a small number of bacterial isolates were analysed in the present study; however, 70% of the strains produced EPA, albeit in small amounts. In companion studies, Nichols *et al.* (1993b) have reported that approaching 30% of Antarctic bacteria analysed (n=23) contained PUFA; the relative level of EPA was as high as 20% (Nichols unpublished data). Virtue *et al.* (1993a) found total ω3 fatty acids in the hepatopancreas of *E. superba* of up to 50% of the total fatty acids. Levels of ω3 fatty acids in algal species on which krill are reported to graze are substantially lower than those found in the hepatopancreas and hence dietary fatty acids cannot account for these high levels. Possible chain elongation and desaturation of dietary derived fatty acids may be occurring in this organ, or other dietary sources of PUFA such as bacteria may be partially responsible for these high levels. In another study krill fed on *Phaeocystis*, which is deficient in EPA, and krill fed on diatoms, which are high in EPA, showed no significant differences in their fatty acid profiles (Virtue *et al.* 1993b). In that study, biochemical capabilities such as chain elongation and desaturation of dietary fatty acids were suggested mechanisms to achieve this observation. However, at this stage we believe that the role of bacteria in krill nutrition cannot be dismissed.

ACKNOWLEDGEMENTS

Presentation at the Sixth SCAR Biology Conference was made possible through a CSIRO INRE postgraduate award. This research was supported by the CSIRO Division of Oceanography and the Antarctic Science Advisory Committee. David Nichols is acknowledged for his help with isolation and growth of bacteria. Geoffrey Cripps of the British Antarctic Survey kindly provided krill from South Georgia used for the hydrocarbon analysis. Stephanie Barrett of CSIRO is thanked for help during identification of the branched hydrocarbons. Members of the Marine Resources and Pollution Programme, CSIRO Division of Oceanography, are acknowledged for their assistance in this study. Jon Watkins, Fred Buchholz, David Walton and one anonymous reviewer are thanked for reviewing this paper.

REFERENCES

Båmstedt, U. 1976. Studies on the deep-water pelagic community of Korsfjorden, western Norway: changes in the size and biochemical composition of *Meganyctiphanes norvegica* (Euphausiacea) in relation to its life cycle. *Sarsia*, **61**, 15–30.

Bargmann, H. E. 1937. The reproductive system of *Euphausia superba*. *Discovery Reports*, **23**, 103–176.

Bligh, E. G. & Dyer, W. J. 1959. A rapid method of total lipid extraction and purification. *Canadian Journal of Biochemical Physiology*, **37**, 911–917.

Clarke, A. 1980. The biochemical composition of krill, *Euphausia superba* Dana, from South Georgia. *Journal of Experimental Marine Biology and Ecology*, **43**, 221–236.

Clarke, A. 1983. Life in cold water: the physiological ecology of polar marine ectotherms. In Barnes, M., ed. *Oceanography and marine biology; an annual review*, **21**, 341–453.

Clarke, A. 1984. Lipid content and composition of Antarctic krill, *Euphausia superba* Dana. *Journal of Crustacean Biology*, **4**, 285–294.

Denys, C. J., Poleck, T. P. & O'Leary, M. M. 1981. Biological studies of krill, austral summer 1979–80. *Antarctic Journal of the United States*, **15**, 146147.

El-Sayed, Z. Z. & McWhinnie, M. A. 1979. Protein of the last frontier. *Oceanus*, **22**, 13-20.

Falk-Petersen, S. 1981. Ecological investigations on the zooplankton community of Balsfjorden, northern Norway: seasonal changes in body weight and the main biochemical composition of *Thysanoessa inermis* (Kroyer), *T. raschii* (M. Sars) and *Meganyctiphanes norvegica* (M. Sars) in relation to environmental factors. *Journal of Experimental Marine Ecological Series*, **49**, 103-120.

Hagen, W. 1988. Zur Bedeutung der Lipide im antarktischen zooplankton [On the significance of lipids in Antarctic zooplankton.] *Berichte zur Polarforschung*, **49**, 1-129.

Hewes, C. D., Holm-Hansen, O. & Sakshaug, E. 1985. Alternate carbon pathways at lower trophic levels in the Antarctic food web. In Siegfried, W. R., Condy, P. R. & Laws, R. M., eds. *Antarctic nutrient cycles and food webs.* Berlin: Springer-Verlag, 227-283.

Ikeda, T. & Dixon, P. 1982. Body shrinkage as a possible over-wintering mechanism of the Antarctic krill, *Euphausia superba* Dana. *Journal of Experimental Marine Biology and Ecology*, **62**, 143-151.

Karl, D. M. 1993. Microbial processes in the Southern Oceans. In Friedmann, E. I., ed. *Antarctic microbial processes*. New York: Wiley-Liss, 1–63.

Kawaguchi, K. S., Ishikawa, S. & Matsuda, O. 1986. The overwintering strategy of Antarctic krill, (*Euphausia superba* Dana) under the coastal fast ice off the Ongul Island in Lutzow-Holm Bay, Antarctica. *Memoirs of the National Institute Polar Research*, **44**, 67–85.

Kanazawa, A. 1985. Nutrition of penaeid prawns and shrimps. *In Proceedings of the First International Conference on the Culture of Penaeid Prawn/Shrimp, Iloilo City, The Philippines, 1984.* SEAFDEC Aquaculture Department, 123–130.

Kils, U. 1983. Swimming and feeding of Antarctic krill, *Euphausia superba*, some outstanding energetics and dynamics, some unique morphological details. *Berichte zur Polarforschung*, **4**, 130–155.

Kirkwood, J. M. 1982. A guide to the Euphausiacea of the Southern Ocean. Australian National Antarctic Research Expeditions. *ANARE Research Notes*, **1**, 1–45.

Lancelot, C., Mathot, S., Becquevort, S., Dandois, J-M. & Billen, G. 1993. Carbon and nitrogen cycling through the microbial network of the marginal ice zone of the Southern Ocean with particular emphasis on the northwestern Weddell Sea. In Caschetto, S., ed. *Plankton ecology and marine biogeochemistry.* Brussels: Belgian Science Policy Office, **1**, 1–102.

Mackintosh, N. A. 1967. Maintenance of living *Euphausia superba* and frequency of moults. *Norsk Hvalfangst-Tidende*, **56**, 97–102.

Makarov, R. R. 1976. Reproduction of *Euphausia superba* (Dana) (Crustacean: Euphausiacea). *Vsesiuznyi nauchno-issledovatel'skii Institut morskogo rybnogo Khoziaistva i Okeanografii*, **110**, 85–89.

Mauchline, J. 1980. The biology of mysids and euphausiids. *Advances in Marine Biology*, **18**, 681 pp.

McWhinnie, M. A., Denys, C. J., Parkin, R. & Parkin, K. 1979. Biological investigations of *Euphausia superba* Dana. *Antarctic Journal of the United States*, **14**, 163–164.

Morris, D. J. 1984. Experimental investigations of the ecological physiology of *Euphausia superba*. *Berichte zur Polarforschung*, **4**, 111–120.

Nichols, D. S., Nichols, P. D. & McMeekin, T. A. 1993a. Polyunsaturated fatty acids in Antarctic bacteria. *Antarctic Science*, **5**, 149–160.

Nichols, D. S., Nichols, P. D. & Sullivan, C. W. 1993b. Seasonal studies of the fatty acid, sterol and hydrocarbon composition of Antarctic sea-ice diatom communities collected during the spring bloom of 1989 in McMurdo Sound. *Antarctic Science*, **5**, 271–279.

Nichols, P. D., Palmisano, A. C., Rayner, M. S., Smith, G. A. & White, D. C. 1989. Changes in the lipid composition of Antarctic sea-ice diatom communities during a spring bloom: an indication of community physiological status. *Antarctic Science*, **1**, 133–140.

Nichols, P. D., Skerratt, J. H., Davidson, A., Burton, H. & McMeekin, T. A. 1991. Lipids of cultured *Phaeocystis pouchetii:* signatures for food-web, biogeochemical and environmental studies in Antarctica and the Southern Ocean. *Phytochemistry*, **30**, 3209–3214.

Nicol, S., Stolp, M., Cochran, T., Geijsel, P. & Marshall, J. 1992. Summer growth and shrinkage of Antarctic krill *Euphausia superba* from the Indian Ocean sector of the Southern Ocean during summer. *Marine Ecology Progress Series*, **89**, 175–181.

Poleck, T. P. & Denys, C. J. 1982. Effects of temperature on the moulting, growth and maturation of the Antarctic krill *Euphausia superba* (Crustacea: Euphausiacea) under laboratory conditions. *Marine Biology*, **70**, 255–265.

Quetin, L. B. & Ross, R. M. 1991. Behavioural and physiological characteristics of the Antarctic krill, *Euphausia superba*. *American Zoologist*, **31**, 49–63.

Quetin, L. B., Ross, R. M. & Clarke, A. 1994. Krill energetics: seasonal and environmental aspects of the physiology of *Euphausia superba*. In El-Sayed, S. Z., ed. *Southern Ocean ecology, the BIOMASS prospective.* Cambridge: Cambridge University Press, 165–184.

Rakusa-Suszcewski, S. & Zdanowski, M. K. 1989. Bacteria in krill (*Euphausia superba* Dana) stomach. *Acta Protozoologica*, **28**, 87–90.

Siegel, V. 1987. Age and growth of Antarctic Euphausiacea (Crustacea). *Marine Biology*, **95**, 483–495.

Spiridonov, V. A., Gruzov, E. N. & Pushkin, A. 1985. Observations on schools of the Antarctic *Euphausia superba* (Crustacea: Euphausiacea) under ice. *Zoologicheskii Zhurnal*, **64**, 1655–1660. [In Russian]

Tanoue, E. & Hara, S. 1986. Ecological implications of fecal pellets produced by the Antarctic krill *Euphausia superba* in the Antarctic Ocean. *Marine Biology*, **91,** 359–369.

Teshima, S. 1982. Sterol metabolism. In Pruder, G. D., Langdon ,C. J. & Conklin, D. E., eds. *Proceedings of the second conference on aquaculture nutrition; biochemical and physiological approaches to shellfish nutrition.* Baton Rouge: Louisiana State University Press, 205–215.

Thomas, P. G. & Ikeda, T. 1987. Sexual regression, shrinkage, rematuration and growth of spent female *Euphausia superba* in the laboratory. *Marine Biology*, **95,** 357–363.

Virtue, P., Nicol, S. & Nichols, P. D. 1993a. Changes in the digestive gland of *Euphausia superba* during short-term starvation: lipid class, fatty acid and sterol content and composition. *Marine Biology*, **117,** 441–448.

Virtue, P., Nichols, P. D., Nicol, S., McMinn, A. & Sikes, E. L. 1993b. The lipid composition of *Euphausia superba* Dana in relation to the nutritional value of *Phaeocystis pouchetii* (Hariot) Lagerheim. *Antarctic Science*, **5,** 169–177.

Volkman, J. K. & Nichols, P. D. 1991. Applications of thin-layer chromatography-flame ionisation detection to the analysis of lipids and pollutants in marine and environmental samples. *Journal of Planar Chromatography*, **4,** 19–26.

Volkman, J.K., Everitt, D.A. & Allen, D.I. 1986. Some analysis of lipid classes in marine organisms, sediments and sea-water using thin-layer chromatography-flame ionisation detection. *Journal of Chromatography*, 6, 147–162.

Volkman, J. K., Jeffrey, S. W., Nichols, P. D., Rogers, G. I. & Garland, C. D. 1989. Fatty acid and lipid composition of 10 species of microalgae used in mariculture. *Journal of Experimental Marine Biology and Ecology*, **128,** 219–240.

White, D. C., Davis, W. M., Nickels, J. S., King, J. D. & Bobbie, R. J. 1979. Determination of the sedimentary microbial biomass by extractable lipid phosphate. *Oecologia*, **40,** 51–62.

Zobell, C. E. 1946. *Marine microbiology*. Waltham: Chronica Botanica, 240pp.

27 The role of antifreeze proteins in survival of Antarctic fishes in freezing environments

ARTHUR L. DEVRIES
Department of Physiology, University of Illinois, Urbana, Illinois 61801, USA

ABSTRACT

Freezing avoidance in Antarctic fishes is associated with the presence of high levels (3.5% W/V) of blood borne antifreeze proteins (AFs). Synthesized in the liver, they are secreted into the blood and distributed into most body fluids. They prevent freezing by adsorbing to, and inhibiting the growth of small ice crystals which have entered from the sea water. The notothenioid fish AFs are composed of repeats of Ala–Ala–Thr– with a disaccharide, attached to each Thr and range in size from 2.6 to 33 KDa. The zoarcid fish peptides with molecular weights of 7 and 14 KDa show no biased amino acid composition, show no repeats in sequence and are globular. Adsorption to ice involves a structural match between the AF molecules and the ice lattice, with both type of AFs binding to the prism planes. Shallow-water fishes have ice associated with their integument, intestinal fluid and spleen. The AF-free ocular fluid and urine remain supercooled throughout the life of the fish, and are protected from inoculation because the surrounding tissues contain AFs. Endogenous 'AF-inhibited' ice crystals appear to be cleared from the circulation and disposed of in the spleen. Levels of circulating AFs are correlated with severity of the environment. They are encoded by large gene families, some of which have multiple copies per gene, which provide the extremely high gene dosage necessary for survival.

Key words: antifreeze proteins, freezing avoidance, Antarctic fishes, adsorption inhibition, supercooling, freezing point.

INTRODUCTION

Fishes have been living in the cold Antarctic waters for approximately 25 million years and the fact that some reproduce and grow at relatively rapid rates indicates how well they are adapted to their cold, icy environment. Adaptations that have been identified are ones that have resulted in compensation for the rate-depressing effects of temperature on energy producing thermochemical reactions and the cold-induced solidification of membranes. These adaptations include the presence of efficient enzymes which rapidly catalyze reactions at −2 °C, producing the necessary energy for locomotion, growth and reproduction. At these low temperatures membrane function is maintained because membrane fluidity is preserved by replacement of membrane saturated fatty acids with unsaturated ones. Adaptations that ensure levels of activity and energy production commensurate with an active life are not unique to the cold Antarctic waters but are also found in organisms inhabiting other cold, deep oceans and cold terrestrial environments. One adaptation that is unique to fish in ice-laden freezing waters is the presence of a biological antifreeze that provides for freezing avoidance. The importance of this adaptation for the Antarctic marine ecosystem cannot be over-emphasized because if it were absent, fish would be poorly represented and the nature of the food web would be very different as fishes are a major prey for many marine birds and mammals. The nature of these unique antifreeze proteins and how they confer survival in the Antarctic fishes as well as their evolution in response to the ice-laden environment is beginning to be understood.

THE ICE-LADEN ANTARCTIC MARINE ENVIRONMENT AND FREEZING AVOIDANCE

The shallow waters (<100 m) of the Antarctic ocean are at the freezing temperature of seawater (−1.9 °C), and the upper 50 m of the water column is ice-laden during the winter and early summer (Littlepage 1965, Dayton *et al.* 1969). With the exception of ice platelets captured in a closing net at a depth of 250 m in the Weddell Sea near the Filchner Ice Shelf (Dieckman *et al.* 1986), the presence of ice in deep waters has only been inferred from oceanographic data (Foldvid & Kvinge 1977). The underside of thick (600 m) ice shelves are freezing (growing) and thus the cold seawater associated with this freezing interface can continue freezing as it flows out from under the shelves and rises. This water is probably the source of the minute ice crystals which eventually rise and grow or coalesce into macroscopic platelets, which are characteristic ice formations present in the shallow waters near ice shelves. These ice-laden waters paradoxically support extensive populations of fishes at all depths, some of which in fact live among the various ice formations (Andriashev 1970, DeVries 1974).

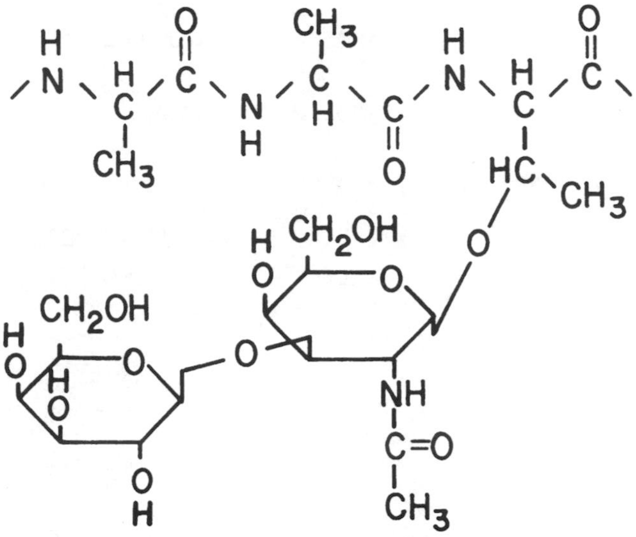

Ice-laden waters preclude supercooling as a mechanism of freezing avoidance

Marine teleosts contain only about one third the amount of salt present in seawater and are therefore strongly hypo-osmotic with a freezing point of about −0.8 °C (Black 1951, Prosser 1973). Freezing avoidance by supercooling in the presence of ice is impossible (Scholander *et al.* 1957) because this metastable state cannot exist in the presence of ice. The ability of Antarctic fishes to thrive in their icy, cold habitat is due to the presence of special biological antifreeze proteins (AFs) in their blood and most other body fluids, which with the normal levels of blood electrolytes, primarily NaCl, effectuate a combined freezing point (freezing temperature) of −2.2 °C, which is three tenths of a degree below the freezing seawater (DeVries, 1982). In contrast to fishes, Antarctic invertebrates are isosmotic to seawater (DeVries & Lin 1977b) and they are also osmoconformers. Thus, for the invertebrates there is no need for special freezing avoidance mechanisms because as osmoconformers their freezing points will never be higher than that of their seawater habitat.

STRUCTURES, HETEROGENEITY AND FUNCTION OF ANTIFREEZE PROTEINS IN ANTARCTIC FISHES

In the Antarctic fishes, the antifreeze proteins are either glycopeptides (AFGPs) or small proteins (AFPs). The AFGPs were first described in the Antarctic nototheniid fish, *Pagothenia borchgrevinki* as a heterogeneous population of at least eight sizes based on gel mobility and designated as AFGP 1 to 8. They are composed of the same basic glycotripeptide unit, (Ala–Ala–Thr–)$_n$, with the disaccharide galactosyl-N-acetylgalactosamine linked to the threonines (Fig 27.1) (DeVries *et al.* 1970, Komstsu *et al.* 1970, Shier *et. al*, 1972, 1975), with molecular weights ranging from 34 to 2.6 KDa. The smallest AFGPs, 7 and 8, are most abundant, constituting about 66% of AFGPs in the blood. With better gel resolution techniques, AFGPs 5, and 6 have recently been found to consist of multiple bands. Thus, as many as 16 sizes have been found in the large Antarctic cod *Dissostichus mawsoni*, but within a similar molecular weight range and abundance distribution. The small AFGPs 7 and 8 differ slightly from the larger AFGPs in that some of the alanines are replaced by prolines (Lin *et al.* 1972, Morris *et al.* 1978). AFGPs of similar sizes and identical compositions are found in most members of the suborder Notothenioidei inhabiting Antarctic waters south of the Antarctic convergence (Ahlgren & DeVries 1984, Wohrmann chapter 28 this volume), although the absolute amounts of the various sizes vary significantly with the thermal severity of the environment (DeVries & Lin 1977b). Interestingly, similar AFGPs are also found in several northern gadids (true cods) (Van Voorhies *et al.* 1978), but their composition and size is not as conserved within this one family as compared with the uniformity within the suborder Notothenioidei.

Fig. 27.1. The chemical structure of the basic repeating structural unit of antifreeze glycoproteins. The polypeptide backbone is composed of two alanyl residues followed by a threonine to which the disaccharide galactose-N-acetylgalactosamine is linked. The different sized AFGPs are made up of differing numbers of this unit and in the small ones (AFGPs 6,7 and 8) alanines are often replaced by a proline residue.

In contrast to the structurally conserved AFGPs, the known fish AFPs are highly varied in amino acid composition and structure. The AFPs can be grouped into three classes. The alanine-rich, α-helical AFPs of winter flounder, Alaskan plaice and short-horn sculpin (M.W. 4.5 KDa), the cysteine-rich, ß-structured AFPs of sea raven (M.W. 14 KDa) and the AFPs of both the northern and Antarctic zoarcid fishes (M.W. 7 KDa). For a review of the structures of these AFPs see Davies & Hew (1990) and Cheng & DeVries (1991). The AFPs found in two Antarctic zoarcids have no biased amino acid composition, no repeating sequence elements, and in contrast to the AFGPs are globular structures (Cheng & DeVries 1989).

```
                 1                                            ◇                        64
 7 kDa RD1:  NKASVVANQLIPINTALTLIMMKAEVVTPMGIPAEEIPKLVGMQVNRAVPLGTTLMPDMVKNYE

                                                           ◇  ◇◇◇
 7 kDa RD2:  NKASVVANQLIPINTALTLIMMKAEVVTPMGIPAEDIPRIIGMQVNRAVPLGTTLMPDMVKNYE

                                                                                  (connector)
14 kDa RD3:  NKASVVANQLIPINTALTLIMMKAEVVTPMGIPAEEIPNLVGMQVNRAVPLGTTLMPDMVKNYE - DGTTSPGLK -
               74                                                              134
              -SVVANQLIPINTALTLVMMKAEEVSPKGIPSEEISKLVGMQVNRAVYLDQTLMPDMVKNYE

                 1                                 ▼    ▼▼            ▼            63
 7 kDa AB2:   TKSVVANQLIPINTALTLVMMKAEEVSPKGIPAEEIPRLVGMQVNRAVYLDETLMPDMVKNYE
```

Fig. 27.2. The sequence of the AFPs from the two McMurdo Sound zoarcids fishes, *Lycodichthys dearborni* and *Pachycara brachycephalum*. *Lycodichthys* has two major 7 KDa variants (RD 1 and RD2) and one 14 KDa variant (RD3) which is made up of two 7KDa AFPs separated by a 9 residue linker. The two halves of RD3 differ by 10 residues which are unlined. The first half of RD3 differs from RD1 and 2 by one and four residues respectively and differences are indicated by diamonds above the RD1 and RD2 sequences. The second half of RD3 more closely resembles the sequence of the *Pachycara* AB2 isoform with only four residues being different, as indicated by the solid triangles over the AB2 sequence.

The McMurdo Sound zoarcid fish *Pachycara brachycephlum* (formerly *Austrolycicthys brachycephalus)* synthesizes only two AFPs, AB1 and AB2, which make up 95% and 5% respectively, of the total circulating AF. In the other McMurdo Sound zoarcid fish, *Lycodichthys dearborni* (formerly *Rhigophila dearborni),* seven AFPs have been isolated from the blood with three distinct isoforms accounting for 90% of the AFP. One of these isoforms is unusual in that it is composed of two similar but not identical 7 KDa domains in tandem. Within this large molecule, the 7 KDa isoforms are joined head to tail by a unique nine amino acid residue sequence (Fig. 27.2). The extreme heterogeneity in one of the McMurdo Sound species and its absence in the other is at present unexplained.

The AFs are synthesized by the liver (Hudson *et al.* 1979, O'Grady *et al.* 1982a) and secreted into the blood, from which they diffuse into most extracellular fluid compartments, where they confer freezing avoidance. Three notable exceptions are the urine, ocular fluid and endolymph (Ahlgren *et al.* 1988), which are formed by secretion and thus contain no AFs. They are metastably supercooled by 1 °C, and remain so because ice cannot propagate through the AF-fortified surrounding tissues (Turner *et al.* 1885). Although the fish skin is a barrier to ice propagation to 1 °C below the blood freezing point (Valerio *et al.* 1992), it is not an absolute barrier. The lack of AFs in the urine is explained by absence of filtration in urine formation; urine is formed by secretion by aglomerular nephrons in the notothenioid fishes (Dobbs *et al.* 1974), and non-functional glomerular nephrons in the eel pouts (Eastman *et al.* 1979).

The intestinal fluid of marine fishes is half the osmotic concentration of seawater and in the Antarctic fishes it is liable to freeze as ice crystals are present from the ingested seawater. To avoid freezing, the intestinal fluid of the Antarctic nototheniid fishes is fortified only with small AFGPs (7 and 8), which it receives via translocation from blood to bile (O'Grady *et al.* 1982b, 1983). Ice growth in the hypo-osmotic intestinal fluid is inhibited down to −2.2 °C. The AFs adsorbed to growth-inhibited ice are lost in the excretory waste, which is energetically costly but obligatory for survival.

Adsorption–inhibition as a non-colligative mechanism of antifreeze function

In the McMurdo Sound fishes the circulating levels of AFs are high enough so that in conjunction with sodium chloride the blood freezing point is lowered well below the freezing point of seawater. When dissolved in water AFGPs and AFPs lower the freezing point or temperature at which ice starts to grow by 1.5 °C for a 2 %W/V. But, they have little effect on the melting temperature of ice (DeVries 1971, 1986). Adsorption of the AFs onto ice, leading to inhibition of ice growth and therefore a depression of the freezing point has been the generally accepted mechanism of antifreeze action (Raymond & DeVries 1977). The adsorption–inhibition mechanism is consistent with the observation that ice occasionally enters the circulation of the Antarctic fishes and ice crystal growth is inhibited by the AFs.

Lattice matches – the molecular basis of adsorption of AFs to specific ice crystal planes

It has been proposed for quite some time that AF adsorption to ice involved a lattice match between the hydrogen-bonding groups in the AF molecules and the oxygens in the ice lattice (DeVries & Lin 1977a, DeVries 1984). Adsorption planes for several AFs have recently been determined and the alignment direction of the AF molecules deduced. For some of them a lattice match model is clearly in accord with the experimental observations (Knight *et al.* 1991, 1993). For the α-helical AFPs of flounder and plaice, the polar side chain of the four regularly repeated Thr residues are spaced 16.7 Å apart, a distance which matches the repeat spacing of the ice lattice along their alignment direction of <0112> on their adsorption plane of {2110} (Knight *et al.* 1991). For AFGPs 7 and 8, a 9.1 Å repeat spacing between the disaccharide moieties was identified, which matches twice the repeat spacing along the AFGPs alignment direction of <1120> (or a-axes) on their adsorption plane of {1010} (Knight *et al.* 1993). The alignment direction on the adsorption planes has yet to be determined for the other AFs, including the high molecular weight and the zoarcid fish AFPs. Recently obtained NMR solution structure of the AFP of *Macrozoarces americanus* (Sonnichsen *et al.* 1993) and *P. brachycephalum*

(DeVries unpublished) showed that this globular protein contains anti-parallel ß-sheets which may have appropriately spaced polar residues that can form a lattice match with the prism plane.

Inhibition of ice growth

The different AFs preferentially adsorb to specific crystallographic planes in ice, but the common result is an overall inhibition of ice growth along a-axes, the normal fast growth direction in water and ordinary solutions (Raymond & DeVries 1977; Raymond *et al.* 1989, Knight *et al.* 1991, 1993). Initially antifreezes were found to completely inhibit the growth of small polycrystalline ice 'seeds' (2 μm) within the hysteresis gap; no change in shape or size was visually detectable even after several days of holding it at 0.6 °C below its melting point (DeVries & Lin 1977b). At this temperature supercooling ice growth in pure water is about 1mm $min.^{-1}$ (Hobbs 1974). However, when large single ice crystals (0.5 mm^2) of known orientation were used, microscopic growth was observed on certain faces (Raymond *et al.* 1989). With AFGPs 1–5, growth restricted to a few hundred μm in the c-axis direction occurs on the basal plane, producing hexagonal pits. With the small AFGPs 7 and 8, and all AFPs, pits also form on the basal plane but fill in as c-axis growth continues and the ice crystal eventually becomes a hexagonal bipyramid (Raymond *et al.* 1989). Thus, AFs do allow some ice growth within the hysteresis gap, but completely stop it after the characteristic morphology is reached. When freezing does occur at temperatures lower than those within the hysteresis gap, rapid growth parallel to the thermodynamically non-preferred c-axis direction occurs, resulting in the formation of 0.1 to 5 μm thick hexagonal spicules. When Antarctic fishes are cooled below their freezing point, this rapid spicular growth leads to nearly instant death.

How adsorption leads to the inhibition of ice growth is not entirely clear. The hypothesis is that the Kelvin Effect is involved. AF molecules adsorbed on a growth step or growth surface are thought to force ice to grow in between them resulting in highly curved growth fronts (Fig. 27.3). These curved fronts have a high surface free energy and thus are unstable. That is, the water molecules in the growth front tend to leave the solid front and reside in the liquid phase. Although lowering the temperature probably results in sub-microscopic growth of these fronts within the hysteresis gap, there is no macroscopic growth. Thus, the freezing point of the bulk solution is lowered (Raymond & DeVries 1977, DeVries 1984, Knight *et al.* 1991). The occurrence of the postulated highly curved growth fronts has yet to be experimentally verified.

ENDOGENOUS ICE IN FISH AND ITS RELATIONSHIP TO ENVIRONMENTAL SEVERITY

Endogenous ice is present in most Antarctic fishes

The adsorption–inhibition mechanism (Raymond & DeVries 1977) implies that ice must be present somewhere in the live fish for AFs to be biologically useful. Recently, all shallow-water (< 30 m) McMurdo Sound fishes (*Trematomus* spp.) have been shown to have ice associated with their skin, gills, intestinal fluid, and surprisingly the spleen as well, during the austral winter and early summer. Since the spleen in fish receives venous blood, splenic ice implies that ice must somehow enter the circulation and becomes retained there, although thus far the blood has always tested negative for ice. During the warmest summer month (December), the environmental shallow-water fishes of McMurdo Sound lose their splenic ice despite the fact that the water only warms from the freezing temperature of –1.93 °C to an average of –1.8 °C. At this time ice is still observed to be associated with the skin; this is probably acquired from the periodic tidal fronts of freezing seawater that originate from beneath the Ross Ice Shelf and pass through the shallow waters. The disappearance of splenic ice at –1.8 °C, a temperature 0.8 °C below the melting temperature of ice in the blood suggests that the spleen is somehow able to dispose of its accumulated ice. The mechanism by which this occurs is unknown; however, since the spleen is implicated in the clearance of ice, it may be that splenic macrophages are involved. If so, explaining how macrophages dissolve (melt) ice crystals would require novel research approaches. It is perhaps useful to view the invasion of ice and the response to it as an immunological one similar to that shown by vertebrates when pathogens invade the body. Immunological elements (macrophages) are brought into play and often dispose of the foreign particles (virus, bacteria or particulate matter) by phagocytosis. Perhaps the response of Antarctic fishes to circulatory ice is yet another manifestation of the vertebrate immune systems way of handling foreign particles that have breached the barrier separating the internal milieu from the environment.

Endogenous ice in deep-water fishes?

It is unclear whether the deep-water (500 m) species in McMurdo Sound encounter ice on a regular basis or even at all during the course of their life. At depths of 600 m in McMurdo Sound the *in situ* freezing point of seawater is 0.5 °C lower than the ambient seawater temperature (−1.9 °C) because of the effect of hydrostatic pressure on the freezing point of seawater and thus ice nucleation cannot occur. It appears paradoxical then that the deep-water (500 m) *Trematomus* and zoarcid fishes are well fortified with AFs (Ahlgren & DeVries 1984). However, intermittent upward excursions of potentially supercooled water from the bottom of the ice shelf do occur and could cause ice formation as the water rises (Foldvid & Kvinge 1977). The underside of the Ross Ice Shelf is 400 m below sea level at its edge, and the water beneath it at times is very cold (−2.3 °C). Sporadic influx and upwelling of this cold water mass into McMurdo Sound occurs and at some point during its vertical displacement the transient shelf/Sound water mixture is colder than, i.e. supercooled with respect to, the *in situ* freezing point at that point. This supercooled water can nucleate and give rise to an abundance of small ice crystals in the water column. Thus, fishes swimming at depths shallower than the bottom of the shelf could periodically be exposed to ice. Thus far no ice has

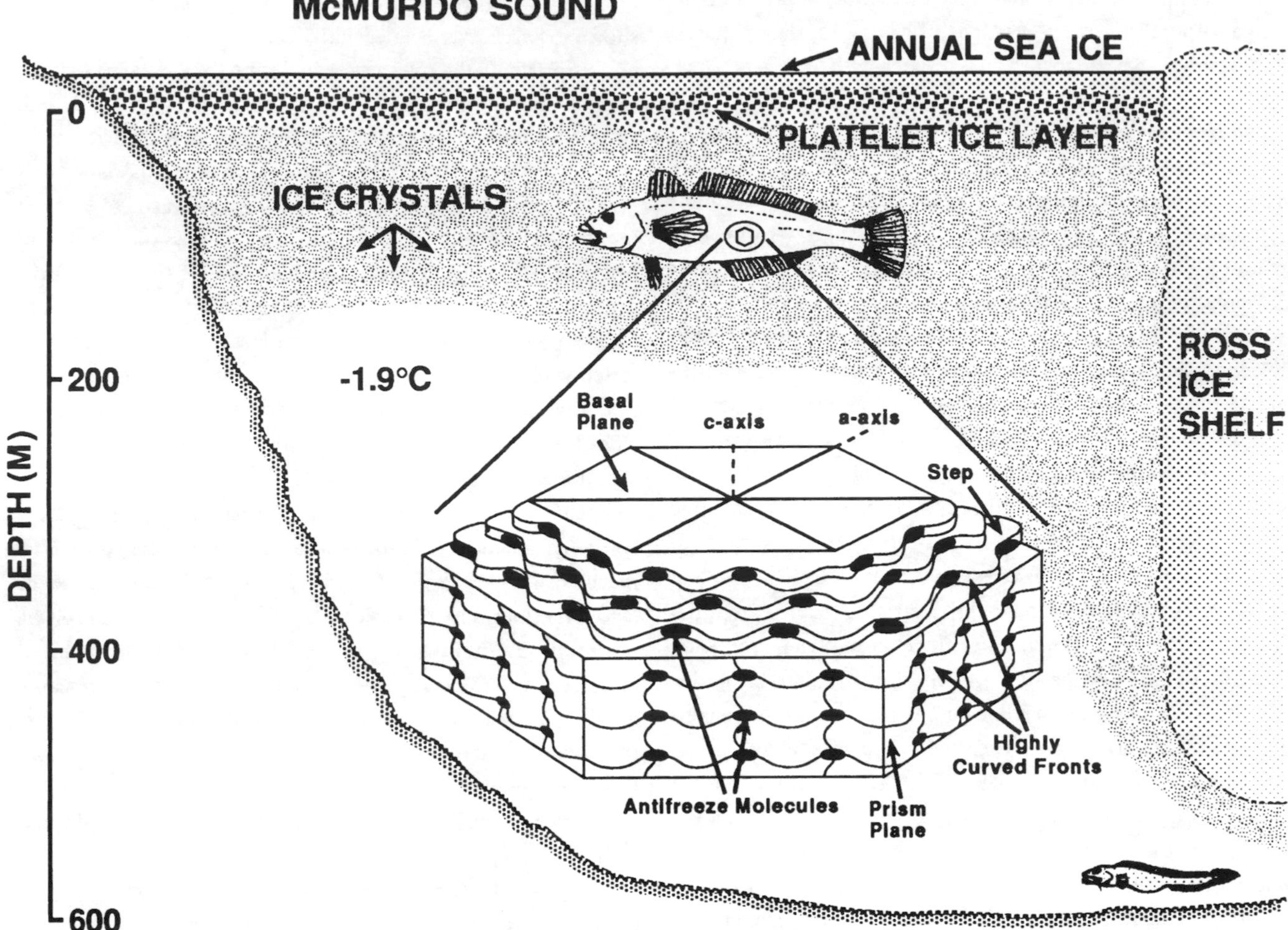

Fig. 27.3. An Antarctic nototheniid fish swimming through water containing small ice crystals in McMurdo Sound. Deep-water ice is formed as a result of the upwelling of supercooled water orginating beneath the Ross Ice Shelf. These ice crystals occasionally enter the fish but their growth is inhibited by antifreeze proteins. Antifreeze proteins adsorb to specific crystal planes and inhibit the growth by causing a 'roughening' of the fast growth prism planes as shown by the highly curved fronts on the hexagonal ice crystal. The bottom-dwelling eel pout (*Rhigophila dearborni*) inhabits deep water where ice does not form.

ever been identified in the spleen of the deep-water fish, suggesting that they rarely see ice or perhaps not at all. Despite this paradox it must be kept in mind though, that in the past ice shelves may have been much thicker and thus all but the deepest water may have been ice laden. Deep-water fishes certainly would have required AFs to survive and the high levels may in part be the result of an evolutionary response to past environmental conditions. Certainly, Antarctic fishes inhabiting the shallow waters near the islands at the Antarctic Convergence have relatively high levels of AFs, although the water temperature there is several degrees above the freezing point.

Correlation of severity of environment with blood antifreeze content

Although ice cover and low water temperature prevail through out most of the year near thick ice shelves (Foldvid & Kvinge 1977), ice cover and annual temperatures vary considerably during the year in regions distant from large ice shelves. The waters near the Antarctic Peninsula have mean annual water temperatures well above freezing (+1 °C) and are ice covered for only a few winter months (Shabica *et al.* 1977). By contrast, McMurdo Sound is often ice covered year round and has a mean annual water temperature of −1.81 °C, very close to its freezing point (Littlepage 1965). Despite apparently similar winter freezing conditions, shallow-water peninsula fishes such as *Notothenia coriiceps* have less circulating AFGPs as indicated by a blood freezing point of −2.2 °C (measured in summer fish), than their McMurdo Sound counterparts whose AFGP levels are the highest among all Antarctic fishes (DeVries & Lin 1977b). Shallow water *Trematomus hansoni* in McMurdo Sound have a blood freezing point of −2.7 °C (Ahlgren & DeVries 1984). This difference in freezing points is a result of differing amounts of circulating AFGPs and is likely to be related to the occurrence of extreme ice conditions and temperatures in the water column in McMurdo Sound associated with its proximity to the Ross Ice Shelf and the influence of shelf water as described above, and the absence of similar events in the peninsula waters. The upward excursion of supercooled shelf water not only potentially exposes deep-water fishes to ice, but also subjects them to temperatures as low as −2.3 °C, which is substantially below the surface- water freezing point (−1.93 °C). Thus, it appears that the high levels of AFGPs in McMurdo

Sound shallow-water fishes are necessitated not only by their year- round constant freezing environment but also by their periodic exposure to extremes in water temperature and to ice in the water column at depths well below the surface. In contrast, some of the Antarctic Peninsula fishes, although exposed to freezing seawater immediately beneath the ice cover during the winter, are unlikely to be exposed to substantial amounts of ice in deep water nor to temperature extremes because of the absence of nearby ice shelves. Thus, the levels of AFGP in *N. coriiceps*, although not as high as that in its McMurdo Sound counterparts, appear to be adequate to prevent freezing in their shallow-water freezing habitat where ice formation only occurs a few metres below the ice–water interface. Other Peninsula fishes, such as *Chaenocephalus aceratus,* are caught in the same waters as *N. coriiceps* during the summer; they have freezing points of −1.6 °C, and therefore much lower levels of circulating AFGPs. Obviously, these fishes have insufficient AFGP to survive during the winter and must either migrate to warmer water or to deeper water where the surface ice does not penetrate and remain super-cooled. It is not known to what depths ice crystals occur in the peninsula water column, and thus the extent of ice exposure the fishes face. The extent of interplay between AFGP level, temperature and ice conditions, and freezing avoidance in the Peninsula fishes has yet to be elucidated.

Genomic bases of AF heterogeneity and high-level production

A striking feature of AFs is their marked molecular heterogeneity. Each fish species produces not just one, but a heterogeneous population of AF molecules. The AFGPs are compositionally near identical but size variant, while the AFPs are compositionally variant but very similar in size. The discovery of the 14 KDa AFP of the Antarctic eel pout *L. dearborni,* which is twice as large as all other known eel pout AFPs, represents the only exception to the latter. It has been known for some time now that AFs are coded by multigene families. In the northern fishes, each gene encodes a single AFP, and thus AFP heterogeneity arises from a family of related genes (Gourlie *et al.* 1984, Scott *et al.* 1988, Hew *et al.* 1988). In the Antarctic eel pouts the heterogeneity also arises from a family of related genes.

In the case of AFGPs from Antarctic cods, the presence of unique DNA sequences for the proline-containing AFGPs 6,7 and 8 indicates that each size variant is also encoded separately, except they all occur within a large polyprotein gene (Hsiao *et al.* 1990). The large AFGPs are also encoded within a gene with multiple copies in tandem repeat followed by several copies of AFGPs 7 and 8. The frequency of occurrence of genes and the number of copies within a gene for a particular size of AFGP appear to correlate with the abundance of the mature protein, with there being many more copies of the small AFGPs than the large ones.

Besides forming the basis for protein heterogeneity, multiple AFGP hybrid polyprotein genes may also be a key factor in providing the gene dosage necessary for the production of constantly high levels (3 %W/V) of mature protein. The single AFGP 8 polyprotein gene of *D. mawsoni* that has been characterized is equivalent to 41 single-copy genes. From genomic library screening statistics there appear to be at least 30 genes in the AFGP gene family of *D. mawsoni*. This multigene, multiple copy per gene organization creates a huge gene dosage, which correlates with the high blood AFGP level of the McMurdo Sound nototheniids, the highest among all antifreeze-bearing fishes.

ACKNOWLEDGEMENTS

This research was supported by the Office of Polar Programs, National Science Foundation.

REFERENCES

Ahlgren, J. A. & DeVries, A. L. 1984. Comparison of antifreeze glycopeptides from several Antarctic fishes. *Polar Biology*, **3,** 93–97.

Ahlgren, J. A., Cheng, C. C., Schrag, J. D. & DeVries, A. L. 1988. Freezing avoidance and the distribution of antifreeze glycopeptides in body fluids and tissues of Antarctic fish. *Journal of Experimental Biology*, **137,** 549–563.

Andriashev, A. P. 1970. Cryopelagic fishes in the Arctic and Antarctic and their significance in polar ecosystems. In Holdgate, M. W., ed. *Antarctic ecology*, Vol. 1. London: Academic Press, 297–304.

Black, V. S. 1951. Some aspects of the physiology of fish. II. Osmotic regulation in teleost fishes. *University of Toronto Studies in Biology Series 59*, **71,** 53–89.

Cheng, C. C. & DeVries, A. L. 1989. Structures of antifreeze peptides from the Antarctic eel pout, *Austrolycichthys brachycephalus. Biochimica et Biophysica Acta*, **997,** 55–64.

Cheng, C. C. & DeVries, A. L. 1991. The role of antifreeze glycopeptides and peptides in freezing avoidance of cold-water fishes. In di Prisco, G., ed. *Life under extreme conditions*. Berlin, Heidelberg: Springer-Verlag, 1–14.

Davies, P. L. & Hew, C. L. 1990. Biochemistry of fish antifreeze proteins. *FASEB Journal*, **4,** 2460–2468.

Dayton, P. K., Robilliard, G. A. & DeVries, A. L. 1969. Anchor ice formation in McMurdo Sound, Antarctica and its biological effects. *Science*, **163,** 273–274.

DeVries, A. L. 1971. Glycoproteins as biological antifreeze agents in Antarctic fishes. *Science*, **172,** 1152–1155.

DeVries, A. L. 1974. Survival at freezing temperatures. In Sargent, J. S. & Mallins, D. W., eds. *Biochemical and biophysical perspectives in marine biology*, Vol. 1. London: Academic Press, 289–230.

DeVries, A. L. 1982. Biological antifreeze agents in coldwater fishes. *Comparative Biochemistry and Physiology*, **73A,** 627–640.

DeVries, A. L. 1984. Role of glycoptides and peptides in inhibition of crystallization of water in polar fishes. *Philosophical Transactions of the Royal Society, London*, **B304,** 575–588.

DeVries, A. L. 1986. Antifreeze glycopeptides and peptides: interactions with ice and water. *Methods in Enzymology*, **127,** 293–303.

DeVries, A. L. & Lin, Y. 1977a. Structure of a peptide antifreeze and mechanism of adsorption to ice. *Biochimica et Biophysica Acta*, **495,** 388–392.

DeVries, A. L. & Lin, Y. 1977b. The role of glycoprotein antifreezes in the survival of Antarctic fishes. In Llano, G. A., ed. *Adaptations within Antarctic ecosystems*. Washington DC: Smithsonian Institution, 439–458.

DeVries, A. L., Komatsu, S. K. & Feeney, R. E. 1970. Chemical and physical properties of freezing point-depressing glycoproteins from Antarctic fishes. *Journal of Biological Chemistry*, **245,** 2901–2908.

Dieckmann, G., Rohardt, G., Hellmer, H. & Kipfstuhl, J. 1986. The occurrence of ice platelets at 250 m depth near the Filchner Ice Shelf and its significance for sea ice biology. *Deep-Sea Research*, **33,** 141–148.

Dobbs, G. H., Lin, Y. & DeVries, A. L. 1974. Aglomerularism in Antarctic fish. *Science*, **85,** 793–794.

Eastman, J. T., DeVries, A. L., Coalson, R. E., Nordquist, R. E. & Boyd, R. B. 1979. Renal conservation of antifreeze peptide in Antarctic eel pout, *Rhigophila dearborni. Nature*, **282,** 217–218.

Foldvid, A. & Kvinge, T. 1977. Thermohaline convection in the vicinity of an ice shelf. In Dunbar, M. J., ed. *Polar oceans.* Montreal:Arctic Institute of North America, 247–255.

Gourlie, B., Lin, Y., Price, J., DeVries, A. L., Powers, D. & Huang, R. C. C. 1984. Winter flounder antifreeze proteins: a multigene family. *Journal of Biological Chemistry*, **264,** 11 313–11 316.

Hew, C. L., Wang, N-C., Joshi, S., Fletcher, G. L., Scott, G. K., Hayes, P. H., Buettner, B. & Davies, P. L. 1988. Multiple genes provide the basis for antifreeze protein diversity and dosage in the ocean pout, *Macrozoarces americanus*. *Journal of Biological Chemistry*, **263,** 12 049–12 055.

Hobbs, P. V. 1974. *Ice physics*. Oxford: Clarendon Press, 583–589.

Hsiao, K-C., Cheng, C. C., Fernandes, I. E., Detrich, H. W. & DeVries, A. L. 1990. An antifreeze glycopeptide gene from the Antarctic cod *Notothenia coriiceps neglecta* encodes a polyprotein of high peptide copy number. *Proceedings of the National Academy of Sciences, Washington, DC*, **87,** 9265–9269.

Hudson, A. P., DeVries, A. L. & Hashchemeyer, A. E. V. 1979. Antifreeze glycoprotein biosynthesis in Antarctic fishes. *Comparative Biochemistry & Physiology*, **62B,** 179–183.

Knight, C. A., Cheng, C. C. & DeVries, A. L. 1991. Adsorption of -helical antifreeze peptides on specific ice crystal surface planes. *Biophysical Journal*, **59,** 409–418.

Knight, C. A., Driggers, E. & DeVries, A. L. 1993. Adsorption of antifreeze glycopeptides 7 & 8 to ice. *Biophysical Journal*, **64,** 252–259.

Komstsu, S. K., DeVries, A. L. & Feeney, R. E. 1970. Studies of the structure of the freezing point-depressing glycoproteins from an Antarctic fish. *Journal of Biological Chemistry*, **245,** 2901–2908.

Lin, Y., Duman, J. G. & DeVries, A. L. 1972. Studies on the structure and activity of low molecular weight glycoproteins from an Antarctic fish. *Biochimica et Biophysica Research Communications*, **46,** 87–92.

Littlepage, J. L. 1965. Oceanographic investigations in McMurdo Sound, Antarctica. *Antarctic Research Series*, **5,** 1–37.

Morris, H. R., Thompson, M. R., Osuga, D. T., Ahmed, A. I., Chan, S. M., Vandenheede, J. R. & Feeney, R. E. 1978. Antifreeze glycoproteins from the blood of an Antarctic fish. *Journal of Biological Chemistry*, **253,** 5155–5162.

O'Grady, S. M., Clarke, A. & DeVries, A. L. 1982a. Characterization of glycoprotein antifreeze biosynthesis in isolated hepatocytes from *Pagothenia borchgrevinki*. *Journal of Experimental Zoology*, **220,** 179–189.

O'Grady, S. M., Ellory, J. C. & DeVries, A. L. 1983. The role of low molecular weight antifreeze glycopeptides in the bile and intestinal fluid of Antarctic fishes. *Journal of Experimental Biology*, **104,** 149–162.

O'Grady, S. M., Schrag, J. D., Raymond, J. A. & DeVries, A. L. 1982b. Comparison of antifreeze glycopeptides from Arctic and Antarctic fishes. *Journal of Experimental Zoology*, **224,** 177–185.

Prosser, C. L. 1973. Water: osmotic balance; hormonal regulation. In Prosser, C. L., ed. *Comparative animal physiology*. Philadelphia: Saunders, 1–78.

Raymond, J. A. & DeVries, A. L. 1977. Adsorption inhibition as a mechanism of freezing resistance in polar fishes. *Proceedings of the National Academy of Sciences, Washington DC*, **74,** 2589–2593.

Raymond, J. A., Wilson, P. W. & DeVries, A. L. 1989. Inhibition of growth of non-basal planes in ice by fish antifreezes. *Proceedings of the National Academy of Sciences, Washington DC.*, **86,** 881–885.

Scholander, P. F., Vandam, L., Kanwisher, J. W., Hammel, H. T. & Gordon, M. S. 1957. Supercooling and osmoregulation in arctic fish. *Journal of Cell Comparative Physiology*, **49,** 5–24.

Scott, C. K., Hayes, P. H., Fletcher, G. L. & Davies, P. L. 1988. Wolffish antifreeze protein genes are primarily organized as tandem repeats that each contain two genes in inverted orientation. *Molecular and Cell Biology*, **8,** 3670–3675.

Shabica, S. V., Hedgpeth, J. W. & Park, P. K. 1977. Dissolved oxygen and pH increases by primary production in the surface water of Arthur Harbor, Antarctica, 1970–1971. In Llano, G. A., ed. *Adaptations within Antarctic ecosystems*. Washington DC: Smithsonian Institution, 83–97.

Shier, W. T., Lin, Y. & DeVries, A. L. 1972. Structure and mode of action of glycoproteins from an Antarctic fish. *Biochimica et Biophysica Acta*, **263,** 406–413.

Shier, W. T., Lin, Y. & DeVries, A. L. 1975. Structure of the carbohydrate of antifreeze glycoproteins from an Antarctic fish. *FEBS Letters*, **54,** 135–138.

Sonnichsen, F. D., Sykes, B. D., Chao, C. & Davies, P. L. 1993. The nonhelical structure of antifreeze protein type III. *Science*, **259,** 1154–1157.

Turner, J. D., Schrag, J. D. & DeVries, A. L. 1985. Ocular freezing avoidance in Antarctic fish. *Journal of Experimental Biology*, **118,** 121–131.

Valerio, P. F., Ming, H.Kao & Fletcher, G. L. 1992. Fish skin: an effective barrier to ice crystal propagation. *Journal of Experimental Biology*, **164,** 135–151.

VanVoorhies, W. V., Raymond, J. A. & DeVries, A. L. 1978. Glycoproteins as biological antifreeze agents in the cod, *Gadus ogac*. *Physiological Zoology*, **51,** 347–353.

Wöhrmann, A. P. A., 1977. Freezing resistance in Antarctic fish. In Battaglia, B., Valencia, J. & Walton, D.W.H., eds. *Antarctic communities: species, structure and survival.* Cambridge: Cambridge University Press, 209–216.

28 Freezing resistance in Antarctic fish

ANDREAS P. A. WÖHRMANN
Institut für Polarökologie, Universität Kiel, Wischhofstraße 1-3, Geb. 12, D-24148 Kiel, Germany

ABSTRACT

Antifreeze glycopeptides and peptides have been isolated from 37 species of Antarctic fish representing the families Nototheniidae, Artedidraconidae, Bathydraconidae, Channichthyidae, Muraenolepididae, Liparididae, Zoarcidae and Myctophidae. Amino acid and carbohydrate analysis as well as antifreeze activity indicate that all investigated notothenioids contain antifreeze glycopeptides (AFGP). Pleuragramma antarcticum, Lepidonotothen kempi, Bathydraco marri *and* Dolloidraco longedorsalis *synthesize additional antifreeze molecules. The non-notothenioid species possess antifreeze peptides,* Muraenolepis marmoratus *possesses an antifreeze glycopeptide like the AFGP. A novel glycopeptide comprised of the carbohydrate residue N-acetyl-glucosamine and the amino acids asparagine, glutamine, glycine, alanine, and traces of arginine, valine and threonine were isolated and characterized from* P. antarcticum. *A maximal thermal hysteresis of 3.4 °C was measured, while the hysteresis decreases with increasing amounts of ice in the sample and decreasing peptide concentration. The molecular weight of this* Pleuragramma-*antifreeze glycopeptide PAGP is approximately 146 kDa. Surprisingly, antifreeze molecules are not specific to a given taxa, as was previously thought, but are found throughout the phylogenetic tree of fishes, even between highly dissimilar taxa. The species studied here are representative of most of the habitats of the Antarctic Ocean.*

Key words: antifreeze, AFGP, cold adaptation, notothenioidei, thermal hysteresis, *Pleuragramma.*

INTRODUCTION

Many regions of the ocean surrounding Antarctica remain near the freezing point of seawater (−1.86 °C) throughout the year (Gordon *et al.* 1978). Fish living in these waters avoid freezing by synthesizing a unique family of glycopeptides which act to depress the freezing point of body fluids. These antifreeze molecules are built up of a tripeptide repeating unit $(\text{Ala–Ala–Thr})_n$, to which the disaccharide ß-D-galactosyl-(1,3)-α-N-acetyl-D-galactosamine is linked by the hydroxyl oxygen of the threonyl residue (Feeney & Yeh 1978). In most Antarctic fishes and in some northern fishes these solutes are relatively large glycopeptides, whereas in some other polar fishes they are peptides. The glycopeptides appear in several sizes with molecular weights ranging between 2600 and 33 700 Dalton (DeVries *et al.* 1970, Feeney & Yeh 1978), whereas the peptides occur in fewer size classes, ranging between 3200 and 14 000 Dalton molecular weight depending on the species (Duman & DeVries 1976, Slaughter *et al.* 1981, DeVries 1988).

On a mass basis both the glycopeptides and peptides are nearly as effective as sodium chloride in depressing the freezing point of water. On a molar basis, however, they depress the freezing point by 200–300 times more than expected on the basis of colligative relations alone (DeVries 1971a, b). These glycopeptides and peptides appear to lower the freezing point only in a non-colligative manner but show the expected colligative effect on the melting point of the solid phase. This antifreeze activity, also termed thermal hysteresis, of the blood is characteristic for fishes belonging to the Antarctic families of the suborder Notothenioidei, and for many different families in the northern and southern hemisphere (Cheng & DeVries 1991).

If ice were not present, fishes could theoretically exist in a supercooled state and survive without antifreeze. Supercooling is, however, risky under Antarctic conditions. Sooner or later fish without antifreeze will probably freeze, especially if seeded by ice crystals. Since there is a good chance of contacting ice in most nearshore waters, antifreezes were necessary for the evolutionary adaptation of notothenioids to the full spectrum of ice-laden habitats. Surprisingly, few Antarctic notothenioids and non-notothenioids, e.g. *Pleuragramma antarcticum*, *Pagetopsis macropterus*, *Lepidonotothen kempi* and *Pogonophryne scotti*, have been reported to lack antifreeze, on the basis of an analysis for high alanine ratios and antifreeze activity of blood serum (DeVries & Lin 1977, Haschemeyer & Jannasch 1983, Eastman 1993).

In this study I tried to gain more information about the distribution, chemical composition and function of antifreeze peptides and glycopeptides purified from several members of Notothenioidei, and members of the Antarctic fish families Muraenolepididae, Liparididae, Zoarcidae and Myctophidae. These fishes occupy diverse ecological habitats in the Weddell Sea and the Lazarev Sea.

MATERIAL AND METHODS

Fish were collected in 1989 and 1991 during the expeditions 'ANT VII/4' and 'ANT IX/3' of the German research vessel RV *Polarstern* in the Weddell Sea and the Lazarev Sea between 69° S and 76° S latitude. The fish specimens were caught at twelve stations in water depths between 120 and 1400 m with a commercial 140 feet (42 m) bottom trawl and an Agassiz trawl. Water temperatures were around −1.8 °C. Detailed information about the cruise and the scientific programmes are given by Rankin *et al.* (1990) and Wöhrmann & Zimmermann (1992). Species, standard length (SL) and fresh weight (FW) were determined on board the ship. Species were identified according to the FAO identification sheets (Fischer & Hureau 1985). Samples were deep frozen at −80 °C.

Antifreeze peptides were isolated as described previously (Wöhrmann 1993). Samples were passed over Bio-Gel TSK DEAE 5PW (diethylaminoethyl cellulose; BIO-RAD) ion exchange resin (column 75×7.5 mm I.D.), equilibrated with 20 mM Tris-HCl, pH 9.5, and a salt-concentration gradient (0.8 M NaCl in 20 mM Tris-HCl, pH 9.5). Collected fractions were further purified by HPLC on an Vydac C4 reverse-phase column (5 μm, 250×4.6 mm). A linear acetonitrile gradient was used to elute the peptides and glycopeptides, which were detected by absorbance at 215 and 280 nm (Wöhrmann & Haselbeck 1992).

Composition and structure of amino acids and carbohydrates, molecular weight, secondary structure and thermal hysteresis were determined. A detailed description of all methods is given by Wöhrmann (1993).

Samples of reverse-phase HPLC were used for plasma-desorption mass spectrometry (PD-MS) analysis. Measurements were performed on a 252Cf plasma-desorption time-of-flight mass spectrometer (Applied Biosystems, Foster City, CA, USA) with a flight tube length of 15 cm. Acceleration voltages were 15 kV in both the positive ion and negative ion modes. Spectra measured in the positive and negative ionization modes were accumulated for ten million fission events each.

The molecular weight was obtained at 25 °C by coupling on-line HPSEC, a multi-angle laser light scattering photometer (MALLS detector, a DAWN-F fitted with a K5 flow cell and a He–Ne laser (λ=632.8 nm) from Wyatt Technology Corporation, Santa Barbara, CA) and a refractometer. Proteins were chromatographed on a TSK-gel G3000SW column with an elution buffer 50 mM KH_2PO_4, 150 mM NaCl, pH 7.0 and a flow rate of 1.0 ml min^{-1}. Weight average MW and root-mean-square radii were established with ASTRETTE and EASY software (v. 3.04) (Wyatt 1994).

RESULTS

Utilizing DEAE ion exchange chromatography the isolation of antifreeze glycopeptides is simple and rapid. Most serum peptides adsorb to the resin while the antifreeze glycopeptides (AFGP) elute at the void volume. Other peptides or glycopeptides are then eluted by increasing the ionic strength of the Tris-HCl buffer, pH 9.5 to 0.8 M NaCl.

As shown in Table 28.1, all investigated notothenioids possess AFGP but in different concentrations. The highest AFGP content was found in *Trematomus pennellii*, which lives in ice-laden shallow waters. The lowest content could be determined in the benthopelagic channichthyid *Neopagetopsis ionah*, which lives in warmer (−0.5 °C) and deeper water. The maximal thermal hysteresis (antifreeze activity) of AFGP, measured at concentrations of 20 mg ml^{-1}, differ from 0.52 °C in *N. ionah* to 1.20 °C in *Pleuragramma antarcticum* (Fig. 28.1). *P. antarcticum*, *Lepidonotothen kempi*, *Bathydraco marri* and *Dolloidraco longedorsalis* possess further antifreeze peptides at the same concentrations as AFGP. However, the antifreeze activity is lower (0.41 °C to 0.84 °C at peptide concentration of 40 mg ml^{-1}). So far, these peptides could not be further characterized.

Antifreeze peptides could be isolated from non-notothenioid species, as well. *Muraenolepis marmoratus*, *Macrourus holotrachys* and *Paraliparis somovi* synthezise antifreeze substances at very low concentrations; there is no thermal hysteretic effect in the blood serum. At peptide concentrations of 40 mg ml^{-1} antifreeze activity could be measured (0.13 °C in *M. holotrachys*, up to 0.56 °C in *M. marmoratus*). The antifreeze peptide of *M. marmoratus* is glycosylated, probably like the AFGP (Table 28.2).

In contrast to earlier investigations on *Pleuragramma antarcticum* (Haschemeyer & Jannasch 1983) antifreeze glycopeptides could be isolated from this species caught in the southeastern Weddell Sea. The concentration of AFGP is about 80% less than in other notothenioids from the same region. The results of the amino acid analysis (Table 28.2) show that the

Table 28.1. *Antifreeze peptides (AFP) and glycopeptides (AFGP) of investigated notothenioids and non-notothenioids of the Weddell Sea and the Lazarev Sea. AF=antifreeze substance; % FRG=antifreeze substance % fresh weight; TH=thermal hysteresis of AFGP (20 mg ml^{-1}) and other antifreeze substances ([1]50 mg ml^{-1})*

Species	AF	% FRG	TH (°C)	Region of catch	Distribution (m)	Depth (m)
Nototheniidae						
Gobionotothen gibberifrons	AFGP	0.0270	0.67	Elephant Isl.	5–750	200
Lepidonotothen kempi	AFGP	0.0636	0.52	Lazarev Sea	100–900	560
	AF II	0.0598	0.73[1]			
Aethotaxis mitopteryx	AFGP	0.0356	0.89	Lazarev Sea	100–850	702
Pleuragramma antarcticum	AFGP	0.0267	1.20	Weddell Sea	0–900	450–630
	AF II	0.0032	0.45[1]			
	PAGP	0.0143	3.21[1]			
Dissostichus mawsoni	AFGP	0.1053	1.10	Lazarev Sea	80–1600	626
Trematomus bernacchii	AFGP	0.1021	1.01	Weddell Sea	100–700	626
Trematomus eulepidotus	AFGP	0.1989	1.02	Weddell Sea	70–550	467
Trematomus lepidorhinus	AFGP	0.1351	0.97	Weddell Sea	200–900	617
Trematomus loennbergii	AFGP	0.1204	0.95	Weddell Sea	60–830	574
Trematomus pennellii	AFGP	0.3337	1.06	Lazarev Sea	0–730	405
Artedidraconidae						
Artedidraco loennbergi	AFGP	0.0977	0.85	Weddell Sea	230–600	343
Dolloidraco longedorsalis	AFGP	0.0879	0.81	Weddell Sea	200–2250	626
	AF II	0.1322	0.72[1]			
Pogonophryne marmorata	AFGP	0.1595	0.87	Weddell Sea	140–1400	626
Pogonophryne scotti	AFGP	0.1627	0.89	Weddell Sea	110–1200	830
Pogonophryne barsukovi	AFGP	0.1544	0.86	Lazarev Sea	220–1120	800
Pogonophryne permitini	AFGP	0.1509	0.87	Lazarev Sea	430–1120	830
Pogonophryne macropogon	AFGP	0.1487	0.85	Lazarev Sea	570–840	830
Bathydraconidae						
Bathydraco marri	AFGP	0.0279	0.85	Weddell Sea	300–1250	623
	AF II	0.0651	0.84[1]			
Bathydraco macrolepis	AFGP	0.0219	0.84	Lazarev Sea	450–2100	1400
Bathydraco antarcticus	AFGP	0.0231	0.85	Lazarev Sea	320–2250	1400
Gymnodraco acuticeps	AFGP	0.1973	0.90	Weddell Sea	0–550	197
Cygnodraco mawsoni	AFGP	0.1794	0.89	Weddell Sea	110–300	197
Gerlachea australis	AFGP	0.1643	0.84	Weddell Sea	200–670	407
Racovitzia glacialis	AFGP	0.1147	0.84	Weddell Sea	220–610	574
Channichthyidae						
Chaenodraco wilsoni	AFGP	0.2835	0.57	Weddell Sea	200–800	509
Chionodraco hamatus	AFGP	0.2576	0.80	Weddell Sea	0–600	407
Chionodraco myersi	AFGP	0.1544	0.89	Weddell Sea	200–800	623
Cryodraco antarcticus	AFGP	0.0920	0.65	Weddell Sea	200–800	623
Dacodraco hunteri	AFGP	0.0809	1.00	Lazarev Sea	300–800	623
Neopagetopsis ionah	AFGP	0.0621	0.52	Weddell Sea	20–900	799
Pagetopsis maculatus	AFGP	0.1796	0.94	Weddell Sea	200–800	453
Pagetopsis macropterus	AFGP	0.2498	0.97	Weddell Sea	0–650	506
Muraenolepididae						
Muraenolepis marmoratus	(AFGP)	0.0076	0.56	Lazarev Sea	20–1600	830
Macrouridae						
Macrourus holotrachys	(AFGP)	0.0031	0.13[1]	Lazarev Sea	150–1100	742
Liparididae						
Paraliparis somovi	AFP	0.0103	0.54[1]	Lazarev Sea	400–850	623
Zoarcidae						
Lycenchelys hureaui	AFP	0.0041	0.18[1]	Lazarev Sea	560–940	830
Myctophidae						
Gymnoscopelus opisthopterus	AFP	0.0070	≈0.1[1]	Lazarev Sea	≥500	742

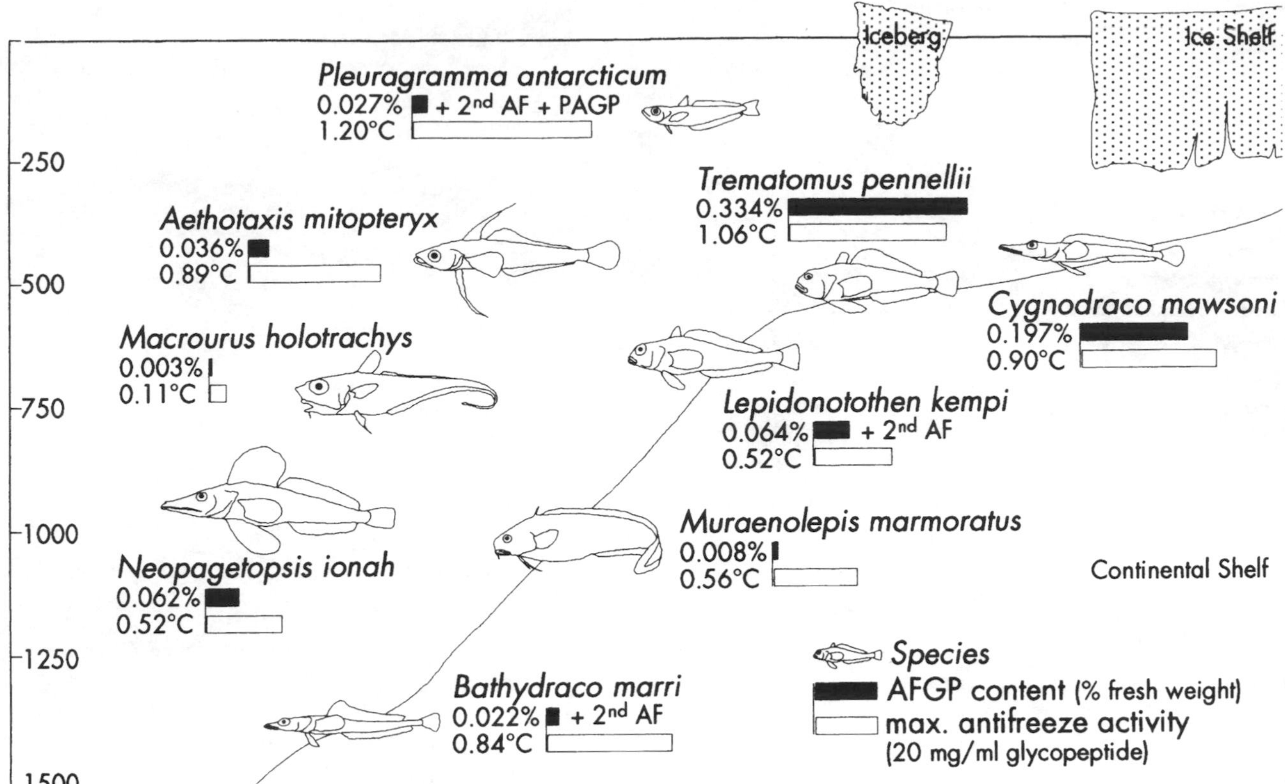

Fig. 28.1. Antifreeze glycopeptides (AFGP) of some investigated Antarctic fish species in relation to water depth and mode of life. Species shown here are representative of most of the ecological habitats of the Weddell Sea and the Lazarev Sea. The antifreeze activity (°C) was measured at a protein concentration of 20 mg ml^{-1} with differential scanning calorimetry.

larger glycopeptides are composed solely of the amino acid residues alanine and threonine in a two to one ratio. Analysis revealed only alanine and threonine and some proline in the lower molecular weight glycopeptide fractions. Carbohydrate analysis by HPAEC-PAD also revealed the presence of N-acetylgalactosamine (GalNAc) and galactose (Gal) after acid hydrolysis (trifluoroacetic acid) and of Gal-ß(1,3)-Gal-N-Ac after ß-elimination ($NaOH/NaBH_4$) and treatment with endo-α-N-acetylgalactosaminidase. It seems apparent that the molecular weights for the antifreeze glycopeptides may vary in the range 35 000 to 2600 Dalton from species to species. The molecular weight of AFGP 7 and 8 were determined with the plasma desorption-mass spectrometry at 3277 Da and 2669 Da.

An additional glycopeptide, occuring in similar concentrations as AFGPs, was isolated (Table 28.2). Amino acid analysis of this novel glycopeptide, called PAGP (*Pleuragramma*- antifreeze glycopeptide), shows glycine (23.9 Mol%), alanine (20.9 Mol%), aspartic acid (17.7 Mol%), glutamic acid (15.8 Mol%) and traces of valine, leucine, arginine and threonine as the amino acids. Sugar analysis by HPAEC-PAD indicates N-acetylglucosamine as the carbohydrate residue. The molecular weight of PAGP is about 146 kDa, and the root mean square radius is 57.3 nm, determined by laser-light scattering. The analysis of circular dichroism for PAGP obtained at temperatures of 20 °C shows ß-sheet (56%), α-helical (19%) and random chain (25%) characteristics. Both AFGP and PAGP are expanded in secondary structure. The antifreeze compounds total 2.46 mg ml^{-1} blood serum in adult specimens (SL 22 cm) of *P. antarcticum*.

There is a variation in the content of both AFGP and PAGP in relation to the age of *P. antarcticum* (Fig. 28.2). Early post-larvae and maturing adults, abundant near the ice shelf in the southeastern Weddell Sea, possess higher antifreeze concentrations (AFGP: 0.283-0.295% FW; PAGP: 0.188–0.219% FW) than juvenile fish feeding on krill in the East Wind Drift (AFGP: 0.279% FW; PAGP: 0.139% FW). Moreover, the adult specimens possess the highest concentrations of high molecular weight AFGPs (0.194% FW).

DISCUSSION

The presence of antifreeze glycopeptides in fish is an important adaptation which permits survival in freezing seawater. Because of this adaptation Antarctic fishes are found occupying most ecological niches of the Antarctic Ocean including the surface and midwaters which are often rich in food but ice-laden year round (for review see DeVries 1988, Cheng & DeVries 1991, Eastman 1993).

Studies of other notothenioids, including the deeper living bathydraconids, channichthyids and non-notothenioid species, reveal that antifreeze glycopeptides (AFGP) are common in all high-Antarctic notothenioids, and probably in the gadiform *M. marmoratus*. It is striking that the amino acid and sugar composition of the isolated antifreeze compounds are similar.

Table 28.2. *Characteristics of antifreeze glycopeptides of PAGP and AFGP of* P. antarcticum *(Pa),* A. mitopteryx *(Am),* P. scotti *(Ps),* B. marri *(Bm),* L. kempi *(Lk),* P. macropterus *(Pm) and* Muraenolepis marmoratus *(Mm). Amino acid and carbohydrate compositions were determined on the HPLC purified antifreeze components. Molecular weight was determined by SDS polyacrylamide gel electrophoresis, plasma desorption-mass spectrometry and laser light scattering. Thermal hysteresis (TH) was measured by differential scanning calorimetry (DSC) at a scan rate of 1 °C min.$^{-1}$ and the secondary structure was measured by circular dichroism*

Species	Pa	Pa	Pa	Am	Lk	Bm	Ps	Pm	Mm
Antifreeze	PAGP	AFGP1–5	AFGP6–8	AFGP	AFGP	AFGP	AFGP	AFGP	AFGP
Amino acid									
Aspartic acid	17.7%								11.0%
Glutamic acid	15.8%								15.7%
Serine									13.7%
Glycine	23.9%								
Arginine	5.1%								6.4%
Threonine	3.8%	38.8%	34.1%	33.5%	33.8%	32.7%	33.4%	34.1%	18.0%
Alanine	20.9%	61.2%	62.4%	63.6%	63.1%	64.4%	63.5%	62.6%	18.4%
Proline			3.5%	2.9%	3.1%	2.9%	3.1%	3.3%	10.2%
Valine	6.4%								
Leucine	6.2%								
Carbohydrate									
Galactose		50%	50%	50%	50%	50%	50%	50%	50%
GalNAc		50%	50%	50%	50%	50%	50%	50%	50%
GlcNAc	100%								
Mol.wt.	146 000[1]	33 700	–2667[2]						
Secondary structure		expanded in all cases							
TH (20 mg ml^{-1})	0.16 °C	1.20 °C	0.92 °C	0.89 °C	0.52 °C	0.85 °C	0.89 °C	0.97 °C	0.56 °C

Notes:

[1] determined by laster light scattering, [2] determined by plasma desorption-mass spectrometry.

As was already shown in earlier investigations (review DeVries 1988), the content of AFGPs depend upon the fish's habitats. Species such as the nototheniid *Trematomus pennellii*, living in shallow waters which may come in contact with anchor ice, possess large amounts of AFGP, which is also true for the channichthyid *Pagetopsis macropterus* and the bathydraconid *Cygnodraco acuticeps*.

The AFGP content in *B. marri* and *D. longedorsalis* is relatively low compared with other notothenioids. However, they possess further antifreeze compounds. Bathydraconids and artedidraconids are usually referred to as the 'typical high-Antarctic' species (DeWitt 1970) and bathydraconids clearly prefer deep and cold areas (Ekau *et al.* 1986, Schwarzbach 1988, Ekau 1990). Bathydraconids are commonly found in the Filchner Depression, which is a deep trench running from the Filchner ice shelf to the continental slope. The prevailing water body is the ice-shelf-water (ISW) with temperatures as low as −2.2 °C and salinities of 34.6–34.7‰ (Hellmer & Bersch 1985). The relatively high abundance of bathydraconids in Gould Bay, which is adjacent to the Filchner Depression, could be due to the extremely low temperatures of −2.0 to −2.2 °C, as proposed by Ekau (1990), or to the high pressure or a combination of both and the possession of additional antifreeze compounds behind the AFGP.

Aethotaxis mitopteryx and *Pleuragramma antarcticum* belong to one tribe (Pleuragrammini; together with *Cryothenia peninsulae* and *Gvozdarus svetovidovi*), which is the most advanced amongst Nototheniidae and which has developed fairly recently, less than ten million years ago (Andersen 1984). All four species are more or less confined to a pelagic/benthopelagic mode of life. Some special adaptations (e.g. neutral buoyancy, antifreeze, blood characteristics) may be of relatively recent origin (Andersen 1984), and could be assigned to recent changes in lifestyle. Apparently, this unique mode of life for fishes, i.e. pelagic and sluggish, seems to be an energy-saving adaptation, providing advantages for fish life in the pelagial or at least does not have any obvious disadvantages (Kunzmann 1991, Kunzmann & Zimmermann 1992). In contrast to *A. mitopteryx*, *P. antarcticum* has a second antifreeze glycopeptide and an additional antifreeze peptide in lower amounts. This is particularly interesting, because of *P. antarcticum's* seasonal migrations (Hubold 1985). Different water masses are probably crossed during migrations and functionally different antifreeze compounds could be helpful when environmental temperature varies.

An attempt to correlate antifreeze type with the assumed phylogenetic relationships between fishes and their zoogeographical distribution leads to many anomalous situations. The five major antifreeze types characterized to date are distributed over ten families, spanning at least five suborders (Fig. 28.3).

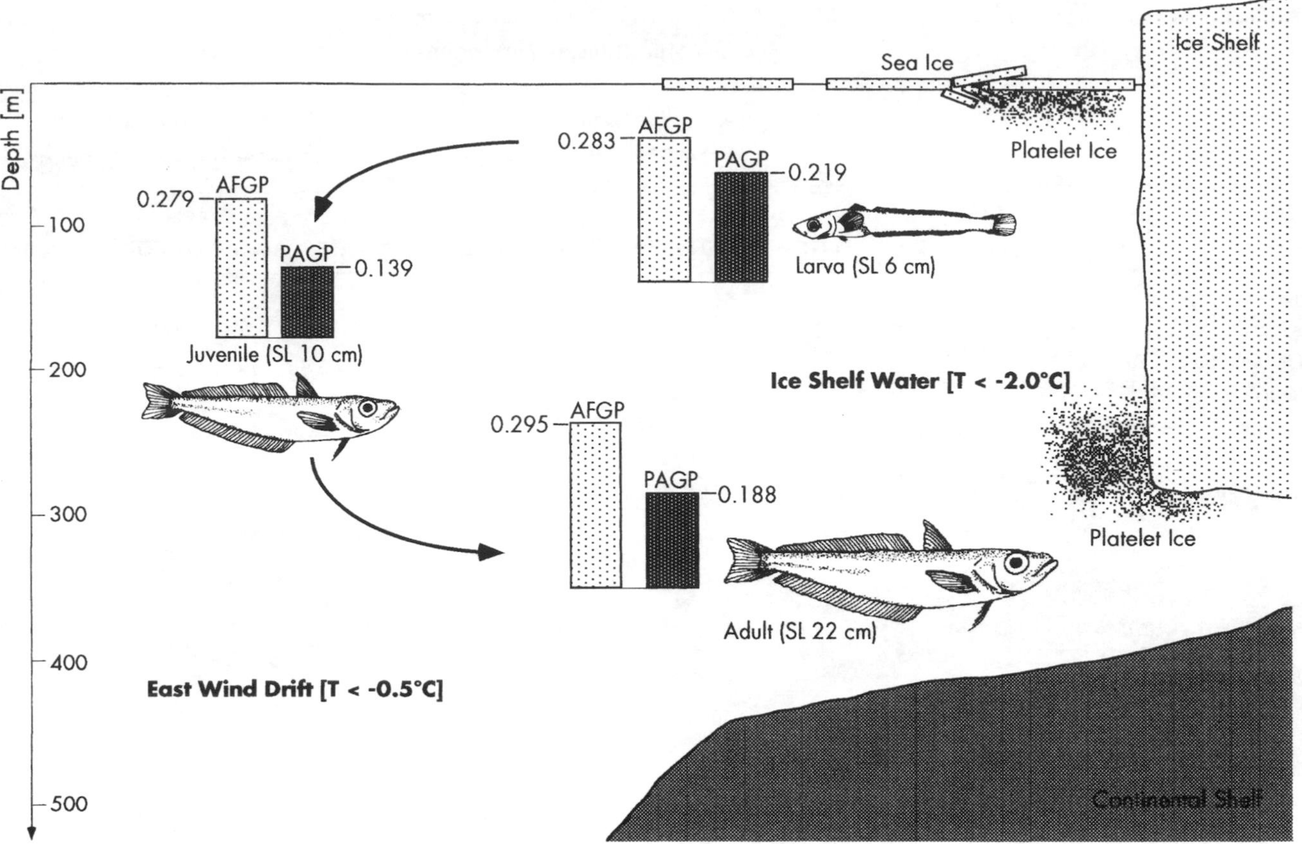

Fig. 28.2. Antifreeze glycopeptides in *Pleuragramma antarcticum* in relation to the age of fish. AFGP=antifreeze glycopeptide (AFGP 1–5, high molecular weight fraction; AFGP 6–8, low molecular weight fraction); PAGP=*Pleuragramma*-antifreeze glycopeptide. The ages of the fishes were calculated indirectly using the standard length after Hubold & Tomo (1989).

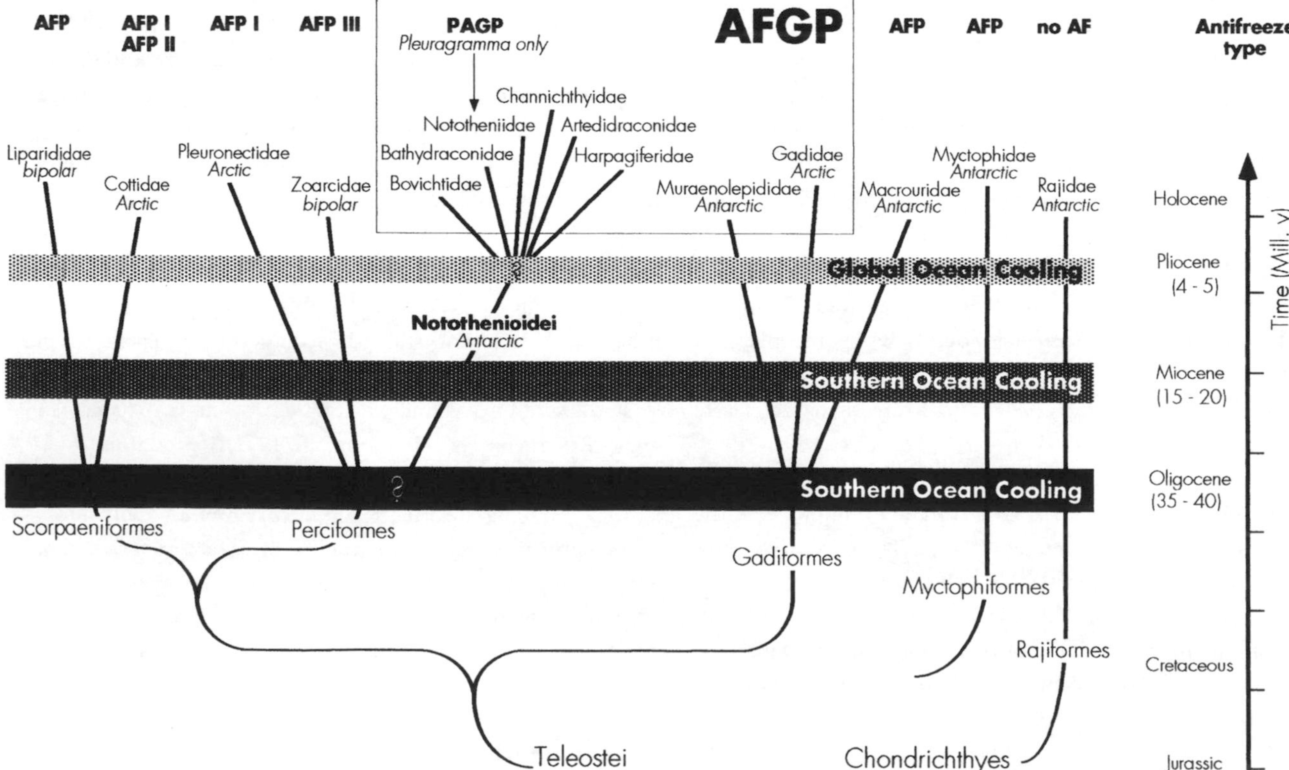

Fig. 28.3. Schematic outline showing the phylogenetic relationships of present-day fish that possess antifreeze peptides or glycopeptides (changed after Scott *et al.* 1986). The broad bands at the Oligocene, Miocene and Pliocene indicate the onsets of Antarctic and Arctic glaciation (Kennett 1977, Clarke & Crame 1992).

Some closely related fishes inhabiting the same environment produce very different antifreezes, while others thought to be of different orders and inhabiting opposite polar waters produce essentially identical antifreezes. Scott *et al.* (1986) suggested that the first antifreeze proteins of marine teleosts were established during the dramatic Cenozoic cooling events initiated approximately 36 million years ago at the Eocene–Oligocene boundary (Clarke & Crame 1992) and that the identical AFGPs of Arctic gadiforms such as *Boreogadus saida* and Antarctic notothenioids imply a close relationship. Eastman (1993) states that AFGPs evolved independently in gadiforms and notothenioids, and that notothenioid AFGPs appeared during the past 10–15 million years, possibly even later. At the conclusion of the Eocene/Oligocene cooling event at 36 Ma, waters were simply too warm (5–7 °C) to require the presence of AFGPs in a benthic fish stock. And, if gadiforms evolved in the Southern Hemisphere and possessed AFGPs early in their history, we would expect them to be more extensively represented in the coldest shelf waters of modern Antarctica. However, the present investigation indicates that Antarctic gadiforms also possess AFGPs. I suggest that before the continental drift occured precursor proteins (e.g. blood proteins or lectins) to the present antifreeze proteins existed, which have evolved during the Cenozoic cooling events initiated approximately 36 Ma, 16 Ma and 3 Ma (Clarke 1990, Kennett 1977) into the various antifreeze peptides and glycopeptides of marine teleosts to be found today.

CONCLUSIONS

Cold adaptation and freezing resistance are important adaptations which all Antarctic fishes have successfully developed. All high-Antarctic notothenioids possess antifreeze glycopeptides (AFGPs), the non-notothenioid species possess non-glycosylated peptides, with the exception of *Muraenolepis marmoratus*. The ambient water temperature near the freezing point is not a selective mechanism for the distribution of Antarctic fish. However, the pelagic *Pleuragramma antarcticum* was confronted with various new problems which the species managed to overcome. To avoid freezing in the presence of frazil ice, *P. antarcticum* synthesize different kinds of antifreeze glycopeptides. *P. antarcticum* is a prime example of the complexity of adaptations necessary to thrive in the widely unoccupied pelagic niche in Antarctic shelf waters.

ACKNOWLEDGEMENTS

I wish to express my thank to A. Haselbeck and M. Wozny (Boehringer Mannheim GmbH, R & D Biotechnology) for their helpful discussions and advice. The excellent technical assistance of A. Roos, S. Schneid and P. Kratzsch is gratefully acknowledged. C. Zimmermann assisted in the graphic presentation. I would like to thank the crew of the RV *Polarstern* for their skilful support during the cruise. Part of this work was supported by the Deutsche Forschungs Gemeinschaft (DFG) and the European Science Foundation (ESF).

REFERENCES

Andersen, N. C. 1984. Genera and subfamilies of the family Nototheniidae (Pisces, Perciformes) from the Antarctic and Subantarctic. *Steenstrupia*, **10,** 1–34.

Cheng, C. C. & DeVries, A. L. 1991. The role of antifreeze glycopeptides and peptides in the freezing avoidance of cold-water fish. In di Prisco, G., ed. *Life under extreme conditions. Biochemical adaptation.* Berlin: Springer-Verlag, 1–15.

Clarke, A. 1990. Temperature and evolution: Southern Ocean cooling and the Antarctic marine fauna. In Kerry, K. R. & Hempel, G., eds. *Antarctic ecosystems: change and conservation.* Berlin: Springer-Verlag, 9–22.

Clarke, A. & Crame, J. A. 1992. The origin of the Southern Ocean marine fauna. In Crame, J. A., ed. *Origins and evolution of the Antarctic biota. Special Publication of the Geological Society of London*, No. 47, 253–268.

DeVries, A. L. 1971a. Glycoproteins as biological antifreeze agents in Antarctic fishes. *Science*, **172,** 1152–1155.

DeVries, A. L. 1971b. Freezing resistance in fishes. In Hoar, W. S. & Randall, D. J., eds. *Fish physiology*, Vol. VI. New York: Academic Press, 157–190.

DeVries, A. L. 1988. The role of antifreeze glycopeptides and peptides in the freezing avoidance of Antarctic fishes. *Comparative Biochemistry and Physiology,* **90B,** 611–621.

DeVries, A. L. & Lin, Y. 1977. The role of glycoprotein antifreezes in the survival of Antarctic fishes. In Llano, G. A., ed. *Adaptations within Antarctic ecosystems.* Washington DC: Smithsonian Institution, 439–458.

DeVries, A. L., Komatsu, S. K. & Feeney, R. E. 1970. Chemical and physical properties of freezing point-depression glycoproteins from Antarctic fishes. *Journal of Biological Chemistry*, **245,** 2901–2913.

DeWitt, H. H. 1970. The character of the midwater fish fauna of the Ross Sea, Antarctica. In Holdgate, M. W., ed. *Antarctic ecology*, Vol. 1. London: Academic Press, 305–314.

Duman, J.G. & DeVries, A.L. 1976. Isolation, characterization and physical properties of protein antifreezes from the winter flounder, *Pseudopleuronectes americanus. Comparative Biochemistry and Physiology*, **54B**, 375–380.

Eastman, J. T. 1993. *Antarctic fish biology.* San Diego: Academic Press, 322 pp.

Ekau, W. 1990. Demersal fish fauna of the Weddell Sea, Antarctica. *Antarctic Science*, **2,** 129–137.

Ekau, W., Hubold, G. & Wöhrmann, A. P. A. 1986. Fish and fish larvae. *Berichte zur Polarforschung*, **39,** 210–218.

Feeney, R. E. & Yeh, Y. 1978. Antifreeze proteins from fish bloods. *Advances in Protein Chemistry*, **32,** 191–282.

Fischer, W. & Hureau, J. C. 1985. *FAO species identification sheets for fishery purposes. Southern Ocean*. Rome: FAO, 471 pp.

Gordon, A. L., Molinelli, E. J. & Baker, T. N. 1978. Large-scale dynamic topography of the Southern Ocean. *Journal of Geophysical Research*, **83,** 3023–3032.

Haschemeyer, A. E. V. & Jannasch, H. W. 1983. Antifreeze glycopeptides of Antarctic fishes. *Comparative Biochemistry and Physiology*, **76B,** 545–548.

Hellmer, H. H. & Bersch, M. 1985. The Southern Ocean. A survey of oceanographic and marine meteorological research work. *Berichte zur Polarforschung*, **26,** 1–115.

Hubold, G. 1985. The early life history of the high-Antarctic silverfish, *Pleuragramma antarcticum.* In Siegfried, W. R., Condy, P. R. & Laws, R. M., eds. *Antarctic nutrient cycles and food webs.* Berlin: Springer-Verlag, 445-451.

Hubold, G. & Tomo, A. P. 1989. Age and growth of Antarctic silverfish *Pleuragramma antarcticum* Boulenger, 1902, from the southern Weddell Sea and Antarctic Peninsula. *Polar Biology*, **9,** 205–212.

Kennett, J. P. 1977. Cenozoic evolution of Antarctic glaciation, the circumAntarctic ocean and their impact on global paleoceanography. *Journal of Geophysical Research*, **82,** 3843–3876.

Kunzmann, A. 1991. Blood physiology and ecological consequences in Weddell Sea fishes (Antarctica). *Berichte zur Polarforschung*, **91,** 1–79.

Kunzmann, A. & Zimmermann, C. 1992. *Aethotaxis mitopteryx*, a high-Antarctic fish with benthopelagic mode of life. *Marine Ecology Progress Series*, **88,** 33–40.

Rankin, J. C., Johnson, T., Kunzmann, A. & Wöhrmann, A. P. A. 1990. Physiological studies on teleost fish. *Berichte zur Polarforschung*, **68,** 144–152.

Schwarzbach, W. 1988. Die Fischfauna des östlichen und südlichen Weddellmeeres: geographische Verbreitung, Nahrung und trophische Stellung der Fischarten. [The demersal fish fauna of the eastern and southern Weddell Sea: geographical distribution, feeding of fishes and their trophic position in the food web.] *Berichte zur Polarforschung*, **54,** 1–94.

Scott, G. K., Fletcher, G. L. & Davies, P. L. 1986. Fish antifreeze proteins: recent gene evolution. *Canadian Journal of Fisheries and Aquatic Science*, **43,** 1028–1034.

Slaughter, D., Fletcher, G. L., Ananthanarayanan, V. S. & Hew, C. L. 1981. Antifreeze proteins from the Sea Raven, *Hemitripterus americanus*. Further evidence for diversity among fish polypeptide antifreezes. *Journal of Biological Chemistry*, **256,** 2022–2026.

Wöhrmann, A. P. A. 1993. Gefrierschutz bei Fischen der Polarmeere. [Freezing resistance in Antarctic and Arctic fishes.] *Berichte zur Polarforschung*, **119,** 1–99.

Wöhrmann, A. P. A. & Haselbeck, A. 1992. Characterization of antifreeze glycoproteins of *Pleuragramma Antarcticum* (Pisces: Notothenioidei). *Biological Chemistry Hoppe-Seyler*, **373,** 854.

Wöhrmann, A. P. A. & Zimmermann, C. 1992. Comparative investigations on fishes of the Weddell Sea and the Lazarev Sea. *Berichte zur Polarforschung*, **100,** 208–222.

Wyatt, P. J. 1994. *DAWN-F, ASTRETTE, EASY Instruction manual*. Santa Barbara: Wyatt Technology Corporation.

29 An ontogenetic shift in the use of visual and non-visual senses in Antarctic nototheniod fishes

JOHN C. MONTGOMERY
School of Biological Sciences, University of Auckland, Private Bag 92019, Auckland, New Zealand

ABSTRACT

This paper proposes the hypothesis that there will be an ontogenetic shift in the relative importance of visual and non-visual senses in Antarctic fishes. The majority of Antarctic fish species hatch in spring or early summer, such that early larval life occurs at a time when zooplankton production is high, and light conditions are appropriate for visual feeding. Winter darkness will represent a critical period during which this ontogenetic shift is likely to occur.

Support for the hypothesis is based on our current understanding of the sensory biology of Antarctic fish species and consideration of the general capabilities and limitations of visual and non-visual senses in fish.

In Pagothenia borchgrevinki *visual acuity is sharpest in fish of about 100 mm total length and deteriorates as the fish grow. Rod density is also maximal at about this body length, and decreases with increasing body size. These developmental patterns produce an adult retina that, as in other notothenioids, is devoid of extreme specializations, so in low-light feeding is thought to be mediated by non-visual systems. If larvae feed visually in their first summer, and adults depend more on non-visual systems in winter, it is reasonable to propose an ontogenetic shift in the use of visual and non-visual senses coinciding with the first winter.*

Key words: Nototheniidae, fish vision, lateral lines.

INTRODUCTION

The majority of Antarctic fish species hatch in spring or early summer (Kock & Kellermann 1991). So their early larval life occurs at a time when zooplankton production is high, and light conditions are appropriate for visual feeding. Some species have a short pelagic phase and settlement occurs at the onset of the first winter, whereas other species extend the pelagic phase over several seasons (Kock & Kellermann 1991). In either of these life history scenarios the arrival of winter darkness will represent a critical period during which it is postulated that there will be an ontogenetic shift in the relative importance of visual and non-visual senses. This hypothesis is based on the following:

(a) that notothenioid larvae feed visually during their first summer,

(b) that they continue to feed during winter darkness,

(c) that as adults non-visual senses make an important contribution to feeding capabilities during periods of low light.

The ontogenetic shift hypothesis essentially follows as a corollary of these premises. Current evidence for the premises will be presented and evaluated followed by a further discussion of the hypothesis itself.

Visual feeding in fish larvae

The great majority of fish larvae are visual feeders. Currently, only half a dozen or so species have been shown to continue feeding in the dark (Blaxter 1986), and in a few species this ability has been shown to be mediated by the lateral line system (Vischer 1989, Jones & Janssen 1992). The almost universal use of vision for first feeding implies a significant advantage of this sense over others at this early stage of development. And indeed, the early pelagic phase of most larval teleosts can be interpreted as a strategy that will put the larvae in conditions suitable for visual feeding.

Most of the exceptions to the rule of hatching in spring and early summer are species like the channichthyids with larger larvae that can take up to 3 months to absorb their yolk-sac (North 1991). So for the majority of species, including some of those that hatch in winter, the critical period of first feeding will occur under light conditions suitable for visual feeding.

Notothenioid larvae are pelagic, and occur mainly in the upper 250 m of the water column (North 1991). In McMurdo sound the larvae are often found in the area immediately below the fast-ice (personal observation). So not only the time of hatching, but also the depth distribution, will put the larvae in suitable conditions for visual feeding. Since light controls productivity, if larval feeding and growth occur at the times and depth that provide the best conditions for vision, they also coincide with the times and areas of greatest food abundance.

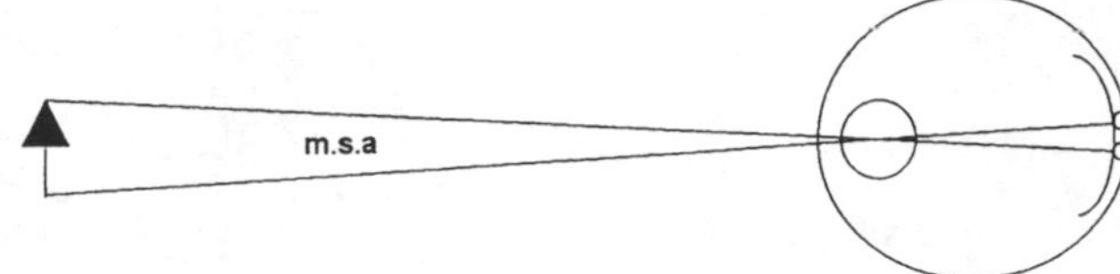

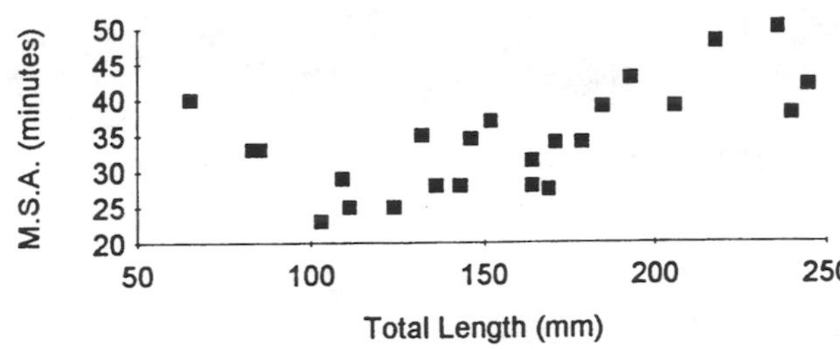

Fig. 29.1. Minimum separable angle (minutes of arc) as a function of fish length for the cryopelagic fish *P. borchgrevinki* (from Pankhurst & Montgomery 1990). The illustration above the graph summarises the geometry used to calculate the minimum separable angle based on optics and receptor packing. *P. borchgrevinki* attains maximum resolving power at about 100 mm.

Feeding during the first winter

If Antarctic fish larvae were purely visual feeders they would not be expected to feed during winter and growth would be expected to cease, or even reverse. Published growth rates in larval fishes only cover the periods of spring to early autumn (North 1991). However, there is no suggestion in the literature that growth ceases completely in winter, only that increased seasonality of food availability and day length towards the poles depresses the annual growth rate in Antarctic fish larvae (Clarke & North 1991). The microstructure of otoliths and scales show a compression rather than a cessation of microincrements over the space of the winter check (White 1991).

In essence, non-visual feeding over the first winter is not essential to the ontogenetic shift hypothesis itself, but only constrains the timing of the shift.

Non-visual feeding in adult Antarctic fishes

Ontogeny of the visual system has been studied in the cryopelagic fish *Pagothenia borchgrevinki* (Pankhurst & Montgomery 1990). Visual acuity is a measure of the resolving power of the visual system. Theoretical acuity is a function of optics and density of packing of retinal receptors (Fig. 29.1) and can be determined from anatomical measurements. However, functional or behavioural acuity includes other 'downstream' processes such as receptor convergence onto ganglion cells and CNS processing and is generally about 2–3 times less than theoretical acuity (Pankhurst *et al.* 1993). In general, there is a trade-off between acuity and sensitivity. Larger cones can operate at lower light levels, but require more space when packed in the retina. Low values of convergence from receptors to ganglion cells increase acuity, but at the expense of sensitivity. For *P. borchgrevinki* visual acuity is sharpest in fish of about 100 mm total length, and progressively deteriorates as the fish get larger (Fig. 29.1). Rod density is also maximal at about this body length, and decreases with increasing body size. These developmental patterns produce an adult retina that is generally similar to those of other coastal fishes. The adult retina in Antarctic fishes is devoid of the extreme specializations seen in other species, such as deep-sea fishes, which also live in low light environments (Eastman 1988, Pankhurst & Montgomery 1989). For comparison Fig. 29.2 plots rod density versus fish size for *P. borchgrevinki* and the orange roughy (*Hoplostethus atlanticus*). Similar rod densities to *P. borchgrevinki* are found in larval orange roughy, but in this deep-water species rod density continues to increase with increasing size (Pankhurst 1987). The fact that spatial resolution in *P. borchgrevinki* reaches a maximum at about 100 mm body length, and the relative lack of retinal specializations in the adult fish, can be taken as indications that high visual performance is not critical to their mode of feeding when as adult fish they find themselves in low light conditions. It appears that coinciding with the end of the juvenile period there is a relaxation in selective pressure driving further anatomical specialization of the retina.

Montgomery *et al.* (1989) have tried to assess the limitations of visual feeding in the cryopelagic habitat. The reduction in light levels under snow and ice, and the vertical distribution of prey items, mean that visually mediated feeding is likely to be marginal even under the most favourable of conditions. Accumulations of snow, discoloration of the ice by ice-algae, and reduced light levels during the winter months further limit visibility. Under these conditions non-visual systems become important. Planktivorous midwater fishes such as *Pleurogramma antarcticum*, and benthic fish feeding on zooplankton or benthic animals hidden in the substrate would also have problems detecting prey visually (Foster & Montgomery 1993).

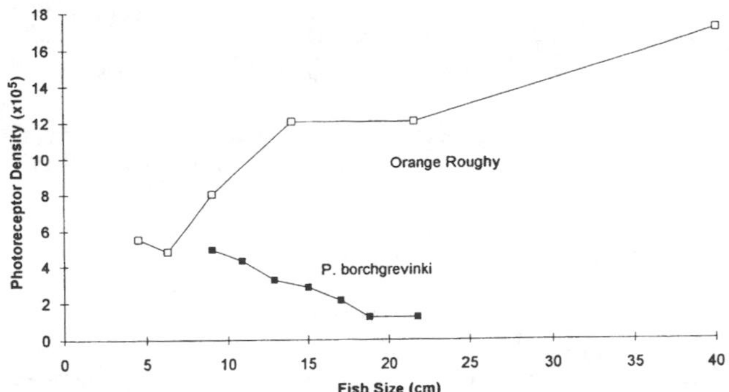

Fig. 29.2. Photoreceptor density as a function of fish size for *P. borchgrevinki* and the mesopelagic orange roughy (*Hoplostethus atlanticus*). This shows that in some fish photoreceptor density can continue to increase into adulthood, whereas in *P. borchgrevinki* it decreases.

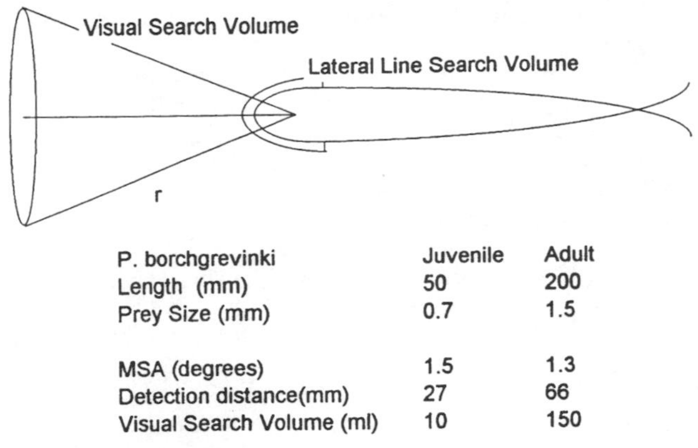

P. borchgrevinki	Juvenile	Adult
Length (mm)	50	200
Prey Size (mm)	0.7	1.5
MSA (degrees)	1.5	1.3
Detection distance(mm)	27	66
Visual Search Volume (ml)	10	150
Approx Head Area (mm2)	300	4800
Reaction Distance (mm)	0.5	5
Lat Line Search Volume (ml)	0.15	24

Fig. 29.3. A comparison of visual and lateral line search volumes in juvenile and adult *P. borchgrevinki*. In adult fish the lateral line search volume increases disproportionately compared with visual search volume.

The mechanosensory lateral line has been shown to be appropriate for the detection of planktonic prey in *P. borchgrevinki* (Montgomery & Macdonald 1987) and lateral line and tactile mediated feeding behaviour have been described in laboratory studies of blinded benthic notothenioids (Janssen *et al.* 1990, Janssen 1992), and in field observations (Janssen *et al.* 1991).

Discussion of the ontogenetic shift hypothesis

One way of thinking about the relative contribution of the senses to feeding behaviour is to think in terms of the relative search areas that the senses can provide. Fig. 29.3 presents some indicative values for visual and lateral line search volumes in juvenile and adult *P. borchgrevinki*. These values are based on the following assumptions and observations. The volume in which prey can be detected visually is conical with a cone angle of about 45° (O'Brien *et al.* 1990) creating a visual search volume of approximately $\pi r^3/6$. The behavioural minimum separable angle is twice the theoretical values based on retinal histology (Pankhurst *et al.* 1993). So a larval *P. borchgrevinki* (total body length 50 mm) feeding on *Oithonia* (length about 0.7 mm) (Hoshiai & Tanimura 1981) should be able to detect them at about 27 mm providing a maximal search volume of about 10 ml. An adult feeding on *Limacina* would have a search volume of about 150 ml. In contrast, the reaction distance of responses mediated by the lateral line of larval fish to prey of about 0.7 mm is only 0.5 mm (Jones & Janssen 1992). Assuming that lateral line sense organs on the head are sufficiently close that the effective search volume is equal to the head area times the reaction distance, then the lateral line search volume for the juvenile is about 0.15 ml. For an adult feeding on larger prey the reaction distance increases to about 5 mm (Montgomery & Milton 1993) creating a search volume of 24 ml. Given the number of assumptions involved and the uncertainties associated with them, these values should not be treated as anything but indicative. However, taken at face value they do show that in the juvenile, visual search volume is 67 times larger than the lateral line search volume, whereas in the adult it is only six times bigger. Visual search volumes are calculated on a 'best case scenario', so visual conditions in the adult would not have to deteriorate much before visual search volume fell to a comparable level to that of the lateral line system. The very much larger visual than lateral line search volumes in the juvenile may help explain the premium for visual feeding in larval fishes.

Current thinking is that larval notothenioid fishes are visual feeders, and that adults can and do feed non-visually under conditions of low light. Thus, it seems likely that there will be an ontogenetic shift in the use of visual and non-visual senses in notothenioid fishes. Further information is required on the relative timing of the development of peripheral and central components of the visual and lateral line systems, the sensory biology and behaviour of larval fishes, and feeding and growth over the first winter period to test this hypothesis adequately.

ACKNOWLEDGEMENTS

I wish to thank Matthew Halstead and Dr John Macdonald for critical reading of the manuscript.

REFERENCES

Blaxter, J. H. S. 1986. Development of sense organs and behavior of teleost larvae with special reference to feeding and predator avoidance. *Transactions of the American Fisheries Society*, **115**, 98–114.

Clarke, A. & North, A. W. 1991. Is the growth of polar fish limited by temperature? In di Prisco, G., Maresca, B. & Tota, B. , eds. *Biology of Antarctic fish*. Berlin: Springer-Verlag, 54–69.

Eastman, J. T. 1988. Ocular morphology in Antarctic Notothenioid fishes. *Journal of Morphology*, **196**, 282–306.

Foster, B. A. & Montgomery, J. C. 1993. Planktivory in benthic nototheniid fish in McMurdo Sound, Antarctica. *Environmental Biology of Fishes*, **36**, 313–318.

Hoshiai, T. & Tanimura, A. 1981. Copepods in the stomach of a nototheniid fish *Trematomus borchgrevinki* fry at Syowa Station, Antarctica. *Memoirs of the National Institute of Polar Research*, Ser.E, **34**, 44–48.

Janssen, J. 1992. Responses of Antarctic fishes to tactile stimuli. *Antarctic Journal of the United States*, **27**, 142–143.

Janssen, J., Coombs, S., Montgomery, J. & Sideleva, V. 1990. Comparisons in the use of the lateral line for detecting prey by

Notothenioids and sculpins. *Antarctic Journal of the United States*, **25**, 214–215.

Janssen, J., Sideleva, V. & Montgomery, J. 1991. Under-ice observations of fish behavior at McMurdo Sound. *Antarctic Journal of the United States*, **26**, 174–175.

Jones, W. R. & Janssen, J. 1992. Lateral line development and feeding behavior in the mottled scuplin, *Cottus bairdi* (Scorpaeniformes: Cottidae). *Copeia*, 1992, 485–492.

Kock, K. H. & Kellermann, A. 1991. Reproduction in Antarctic Notothenioid fish. *Antarctic Science*, **3**, 125–150.

Montgomery, J. C. & Macdonald, J. A. 1987. Sensory tuning of lateral line receptors in Antarctic fish to the movements of planktonic prey. *Science*, **235**, 195–196.

Montgomery, J. C. & Milton, R. C. 1993. Use of the lateral line for feeding in the torrentfish (*Cheimarrichthys fosteri*). *New Zealand Journal of Zoology*, **20**, 121–125.

Montgomery, J. C., Pankhurst, N. W. & Foster, B. A. 1989. Limitations on visual feeding in the planktivorous Antarctic fish, *Pagothenia borchgrevinki*. *Experientia*, **45**, 395–397.

North, A. W. 1991. Review of the early life history of Antarctic Notothenioid fish. In di Prisco, G., Maresca, B. & Tota, B., eds. *Biology of Antarctic fish*. Berlin: Springer-Verlag, 70–86.

O'Brien, W. J., Browman, H. I. & Evans, B. I. 1990. Search strategies of foraging animals. *American Scientist*, **78**, 152–160.

Pankhurst, N. W. 1987. Intra- and interspecific changes in retinal morphology among mesopelagic and demersal teleosts from the slope waters of New Zealand. *Environmental Biology of Fishes*, **19**, 269–280.

Pankhurst, N. W. & Montgomery, J. C. 1989. Visual function in four Antarctic nototheniid fishes. *Journal of Experimental Biology*, **142**, 311–324.

Pankhurst, N. W. & Montgomery, J. C. 1990. Ontogeny of vision in the Antarctic fish *Pagothenia borchgrevinki* (Nototheniidae). *Polar Biology*, **10**, 419–422.

Pankhurst, P. M., Pankhurst, N. W. & Montgomery, J. C. 1993. Comparison of behavioural and morphological measures of visual acuity during ontogeny in a teleost fish *Fosterygion varium*, *Tripterygiidae* (Foster, 1801). *Brain Behaviour and Evolution*, **42**, 178–188.

Vischer, H. A. 1989. The development of lateral-line receptors in *Eigenmannia* (Teleostei, Gymnotiformes). I. The mechanosensory lateral line system. *Brain Behavior and Evolution*, **33**, 205–222.

White, M. G. 1991. Age determination in Antarctic fish. In Di Prisco, G., Maresca, B. & Tota, B., eds. *Biology of Antarctic fish*. Berlin: Springer-Verlag, 87–100.

30 Hydration-related spatial and temporal variation of photosynthetic activity in Antarctic lichens

BURKHARD SCHROETER[1], FLORIAN SCHULZ[2] AND LUDGER KAPPEN[1,2]

[1]Botanisches Institut, Universität Kiel, Olshausenstr. 40, D-24098 Kiel, Germany. [2]Institut für Polarökologie, Universität Kiel, Wischhofstr. 1-3, Geb. 12, D-24148 Kiel, Germany

ABSTRACT

Hydration-dependent metabolic activity in Antarctic lichens is analysed by using the chlorophyll a *fluorescence response of the lichen photobionts. Variations in the dehydration pattern within one thallus of the crustose species* Buellia frigida *was detected as a function of the microtopography of its substratum. In* Placopsis contortuplicata *measurements of chlorophyll* a *fluorescence reveal that within the same thallus the green algal photobiont can be activated by water vapour if a dry thallus is exposed to high air humidity whilst the cyanobacterial photobiont needs liquid water for metabolic activation. In long-term measurements of microclimatic parameters of* Usnea aurantiaco-atra, *chlorophyll* a *fluorescence measurements provide an important tool for an unattended and automatic recording of the time periods of metabolic activity. From the recorded data, and a CO_2 gas exchange response model, estimates of primary production of lichens in their Antarctic habitat are possible.*

Key words: lichen, chlorophyll *a* fluorescence, photosynthesis, Antarctica, primary production.

INTRODUCTION

In the terrestrial ecosystems of Antarctica lichens form a major element of the vegetation. Because of the poikilohydrous nature of lichens the ambient water relations are a key factor that determines production and growth and consequently the distribution of the thalli. Therefore, any investigation of the photosynthetic carbon production of lichens has to take into account the metabolic activity of the mycobiont and the photosynthetic production of the green algal or cyanobacterial photobiont as functions affecting thallus hydration. CO_2 gas exchange methods have proved to be successful for measuring net photosynthesis and dark respiration in Antarctic lichens in the laboratory as well as in the field and even *in situ* (Kappen *et al.* 1987, 1988, 1990, Kappen & Breuer 1991, Schroeter 1991, Schroeter *et al.* 1991, 1994). However, these methods are not capable of detecting small-scale differences in photosynthetic activity within one thallus (Schroeter *et al.* 1992). CO_2 gas exchange methods are also not applicable for unattended long- term *in situ* measurements of photosynthetic activity (Schroeter *et al.* 1991). For such cases, methods and protocols of recently developed chlorophyll *a* fluorescence measurements can overcome the limitations of the CO_2 exchange methods (Schroeter *et al.* 1991, 1992, 1996, Schroeter 1994). Using chlorophyll *a* fluorescence measurements we are able to show how water relations influence the photosynthetic activity within one thallus during water uptake and dehydration. We also demonstrate how primary production in lichens in Antarctica depends primarily on the periods of hydration of the lichen over a 12 month period.

MATERIAL AND METHODS

The lichen species

Buellia frigida Darb., a crustose lichen, grows on the surface of rocks and boulders mainly in continental Antarctica where it is one of the most prominent lichen species. Its upper cortical layer is darkly pigmented and the photobiont belongs to the chlorophycean alga genus *Trebouxia*. The crustose lichen *Placopsis contortuplicata* Lamb is widely distributed in the maritime Antarctic where it is often found in hygrophilous lichen communities in meltwater-influenced sites. The green photobiont of the thallus belongs to the genus *Trebouxia*. External cephalodia in the central region of the circular thallus, which can reach diameters of more than 10 mm, consist of cyanobacteria of the Nostoc-type (Lamb 1947). *Usnea aurantiaco-atra* (Jacq.) Bory is one of the most common fruticose lichens in the maritime Antarctic (Redon 1985). It grows on rocks and boulders and often forms extensive lichen 'heath'-like formations (Lindsay 1971, Kappen 1993).

Measurements of chlorophyll *a* fluorescence

Chlorophyll *a* fluorescence was measured by using a pulse-amplitude-modulation system (PAM 2000, Walz, Germany) as described by Schreiber (1986). The pulse–amplitude–modulation technique allows measurements of the 'fluorescence yield' ΔF/Fm′(Genty *et al.* 1989) and, if multiplied with the photosynthetic photon flux density (PPFD) incident at the sample surface, provides an estimate of the electron transport rate through photosystem II (ETR=ΔF/Fm′×PPFD; see Schroeter *et al.* 1992, Winter & Lesch 1992) under ambient light conditions and without any previous dark adaption of the lichen. For unattended automatic long-term measurements in the field a modified fluorometer (Fl-1, BBE, Germany) was used (Schroeter *et al.* 1991). The relationship of chlorophyll *a* fluorescence and the degree of hydration in a lichen thallus was demonstrated by Schroeter (1991) and Schroeter *et al.* (1991) in combined chlorophyll *a* fluorescence, CO_2 gas exchange and weight measurements. Further technical and theoretical details can be found in Bolhàr-Nordenkampf *et al.* (1989), Genty *et al.* (1989), Krause & Weis (1991) and Schroeter *et al.* (1992). Besides the chlorophyll *a* fluorescence measurements, the microclimatic conditions at an *Usnea aurantiaco-atra* thallus site, such as temperature (minithermistor probes, Grant, UK), photosynthetic photon flux density (SKP 215, Skye, UK) and air humidity (data not shown here) were automatically recorded at 5 min. intervals and averaged every 30 min. by a datalogger (SQ1259, Grant, UK). A detailed description and analysis of the long-term microclimatic recordings will be published elsewhere.

RESULTS

Non-homogenous water relations within one thallus

Stages of photosynthetic activity during water loss in an originally fully soaked thallus of *Buellia frigida* are shown in Fig. 30.1. The images show that the chlorophyll *a* fluorescence yield (ΔF/Fm′) as a measure of photosynthetic activity varies in space as a function of a non- homogenous water loss within the thallus area. During the desiccation period a detailed grid pattern of the fluorescence parameter ΔF/Fm′ was measured at certain time intervals over the whole thallus, which covered an area of 15.3 cm². The small area (0.3–0.5 cm²) investigated with each measurement allowed a very precise location of the photosynthetic status of the thallus, with more than 70 individual measurements taken for each image. When completely wetted (5 min.) the ΔF/Fm' value was homogenously distributed over the whole thallus area. During dehydration, while photon flux density (PAR: 150 μmol m^{-2} s^{-1}) and temperature (18 °C) conditions were constant, a pattern of decreasing ΔF/Fm′ values was non-homogenously distributed over the thallus. Thallus dehydration apparently first affected chlorophyll *a* fluorescence at the upper margins, where the thallus was spread over a 30° inclined surface of a small granite boulder. During the further course of dehydration a zone of photosynthetic inactivation extended from an initial section to 20% of the thallus area within 120 min. Within small depressions of the rock surface the dehydration process was apparently retarded, and these parts of the thallus remained photosynthetically active after 240 minutes. After 300 min. the whole thallus was inactive as it became dehydrated to the ambient humidity.

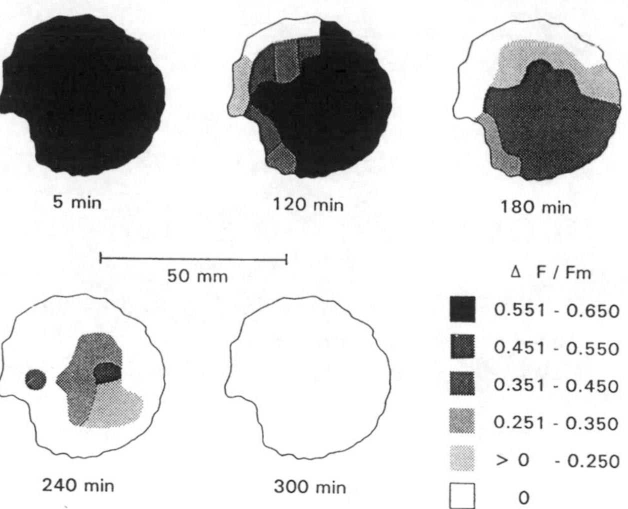

Fig. 30.1. Distribution of classes of ΔF/Fm' over a thallus of *Buellia frigida* during dehydration after artificial water saturation. The times given are the time elapsed after initial water saturation (after Schroeter *et al.* 1992).

In the case of *Placopsis contortuplicata* the green algae and the cyanobacteria react independently from each other to the same moisture conditions within the same thallus (Fig. 30.2). If the whole dry thallus is exposed to an atmosphere with high water vapour pressure for more than 150 h, the green algal photobionts develop considerable photosynthetic activity (shown by the electron transport rate through photosystem II, ETR), while the cyanobacteria in the cephalodium stayed photosynthetically inactive. The latter were activated only by spraying the thallus with liquid water. Subsequent to moistening the thallus with liquid water the photosynthetic activity of the green algal photobionts was higher than that produced by high water vapour pressure in the atmosphere. Moreover, the green algal photobionts could sustain higher electron transport rates through photosystem II than the cyanobacteria at any photon

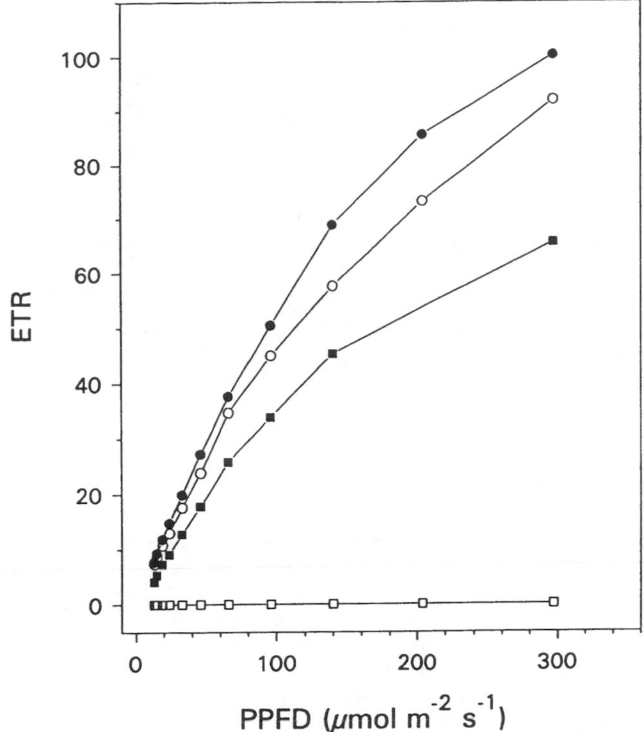

Fig. 30.2. Light response curve of relative rate of PS II electron transport (ETR) in the green (circles) and the cyanobacterial (squares) photobiont within the same thallus of *Placopsis contortuplicata* after exposure to air humidity of 94% (open symbols) and after activation by liquid water (black symbols) (after Schroeter 1994).

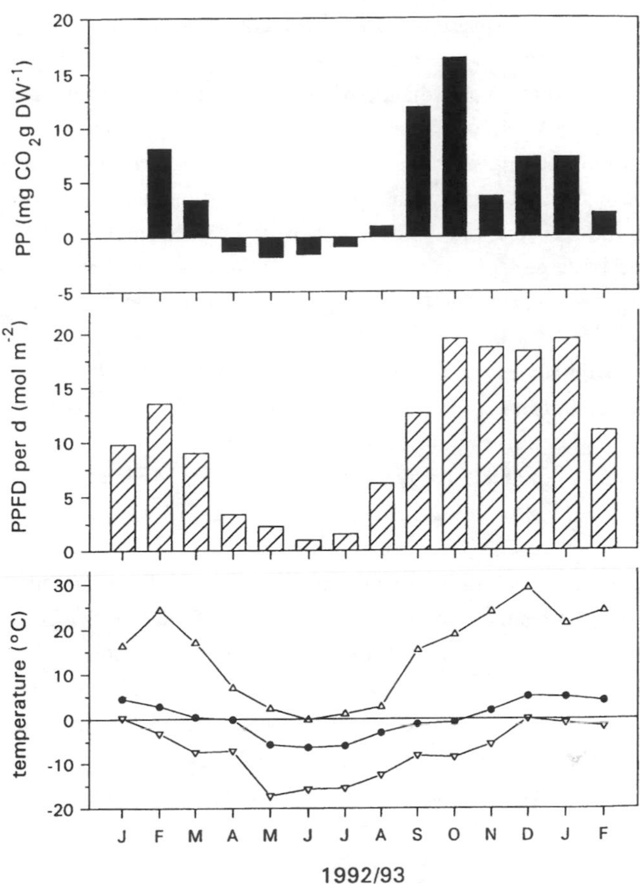

Fig. 30.3. (Top) Annual course of the potential, maximum realizable primary production (PP) calculated for *Usnea aurantiaco-atra* in Livingston Island, South Shetland Islands, Antarctica. (Middle) Photosynthetic photon flux densities (PPFD), shown as the monthly mean sum per day. (Bottom) Maximum and minimum monthly thallus temperatures (open triangles) and mean monthly thallus temperatures (closed circles).

flux density, even when the algal photobiont did not receive liquid water.

Whole thallus activity during long-term measurements

Results of the long-term measurements of the microclimatic conditions in *Usnea aurantiaco-atra* growing on rocks in Livingston Island, maritime Antarctica, are shown in Fig. 30.3. The course of the monthly sum of photosynthetic photon flux density (PPFD) and the monthly average of thallus temperatures reveal a pronounced seasonal trend. A maximum thallus temperature of +30.3 °C was recorded on 25 December 1992, while a minimum thallus temperature of −16.1 °C was reached in May 1992. During the whole winter period between May and August the maximum thallus temperatures frequently exceeded 0 °C. The continuously recorded values of chlorophyll *a* fluorescence indicated potential photosynthetic activity if the signal was different from zero. For all time periods with a signal of metabolic activity photosynthetic productivity and respiration were calculated as a function of PPFD and temperature conditions using a CO_2 gas exchange response model for *Usnea aurantiaco-atra* that was established from CO_2 exchange measurements in the field under the varying environmental conditions in January and February 1993. The construction of the model follows Schroeter (1991) and Schroeter *et al.* (1995) and details will be given elsewhere.

The monthly CO_2 balance, shown in Fig. 30.3, gives an estimate of the photosynthetic production with respect to CO_2 uptake during the day and CO_2 loss during metabolic active periods in darkness. The calculation of the CO_2 balance over time reveals that *Usnea aurantiaco- atra* was metabolically active throughout the winter, even though the carbon balance was negative from May to July. The most productive period in the annual cycle, presented here, was in the spring months September and October, where the temperature and water availability met the metabolic requirements and PPFD greatly exceeded the light compensation point for net photosynthesis.

DISCUSSION

It is well known that chlorophyll *a* fluorescence from photosystem II can provide an important tool for the analysis of a great variety of stress effects on photosynthesis (Schreiber *et al.* 1988). Lichens react sensitively to all stresses which directly or indirectly affect photosystem II, such as air pollution (Scheidegger & Schroeter 1994), high PPFD levels (Demmig-Adams *et al.* 1990) or increased UV-B (chapter 31 this volume). These stresses can be monitored by means of chlorophyll *a* fluorescence measurements. In the poikilohydrous lichens drought stress occurs regularly as a consequence of drying and wetting cycles due to variations in the ambient moisture conditions. Laboratory measurements of thallus hydration and related photosynthetic performance of lichens from various habitats including Antarctica have been reported by various authors, based on CO_2 gas exchange, as well as chlorophyll *a* fluorescence measurements. Such studies demonstrated differences between green algal and cyanobacterial lichens at different hydration levels (Lange *et al.* 1989, Green *et al.* 1993). This is confirmed by our measurements on the crustose lichen species *Placopsis contortuplicata,* which also show that this difference in physiological performance between the green algal and the cyanobacterial photobionts is maintained even though they are in close contact within the same thallus. The need of the cyanobacterial photobiont for liquid water may explain the pronounced preference of *Placopsis contortuplicata* for wet sites, such as meltwater streams in the maritime Antarctic. Our results for *Buellia frigida* demonstrate small-scale differences in water-dependent photosynthetic activity in response to water status changes. Both the examples presented illustrate that small-scale chlorophyll *a* fluorescence measurements allow a detailed determination of the location of metabolic activity in a lichen thallus, while other methods such as CO_2 measurements mostly integrate the overall response of the lichen. They further demonstrate a novel way of investigating metabolic activity of cryptogams *in situ* under natural environmental conditions.

For long-term measurements and calculations of primary production the continuous recording of the hydration status of photosynthetically active organisms is the crucial factor as far as poikilohydrous organisms such as lichens and mosses are concerned. Again, in this context, the chlorophyll *a* fluorescence measurements provide a reliable tool for recording periods of metabolic activity automatically and *in situ* and also provide direct information on photosynthetic potential ($\Delta F/Fm'$) or rates (ETR) of photosynthetic activity. In lichens, chlorophyll *a* fluorescence measurements only relate to the photosynthetically productive symbiont while the respiratory activity of the mycobiont cannot be directly recorded. However, respiratory activity of the thallus can be evaluated because of the close relation in the hydration status between both symbionts in the lichen thallus (Scheidegger & Schroeter unpublished results). It is possible to calculate primary production in lichens from chlorophyll *a* fluorescence measurements using microclimatic measurements of PPFD and thallus temperature if a CO_2 gas exchange response model which considers the dark respiration of the photobiont as well as the respiratory activity of the mycobiont is computed (Schroeter 1991, Schroeter *et al.* 1991). These long-term data, as shown here, may serve for thallus age estimations and calculations of primary production on a community level as well as a basis for the analysis of possible effects of global atmospheric influences on the primary production of the terrestrial ecosystems in Antarctica.

ACKNOWLEDGEMENTS

The authors thank Dr T. G. A. Green, New Zealand, Dr R. D. Seppelt, Australia, and M. Sommerkorn, Kiel, for companionship in the field and many fruitful discussions. Special thanks are due to the members of the Spanish Antarctic expedition at BAE Juan Carlos I, Livingston Island, in 1992–1994 and in particular to Prof. Dr J. Castellví Piulachs, Dr A. Castejon and Dr L. G. Sancho, Madrid. M. Mempel and H. Timm, Kiel, are thanked for their respective contributions. Financial support by the Deutsche Forschungsgemeinschaft is gratefully acknowledged.

REFERENCES

Bolhàr-Nordenkampf, H. R., Long, S. P., Baker, N. R., Öquist, G., Schreiber, U. & Lechner, E. G. 1989. Chlorophyll fluorescence as a probe of the photosynthetic competence of leaves in the field: a review of current instrumentation. *Functional Ecolology*, **3,** 497–514.

Demmig-Adams, B., Maguas, C., Adams III, W. W., Meyer, A., Kilian, E. & Lange, O. L. 1990. Effect of high light on the efficiency of photochemical energy conversion in a variety of lichen species with green and blue-green phycobionts. *Planta*, **180,** 400–409.

Genty, B., Briantais, J-M. & Baker, N. 1989. The relationship between the quantum yield of photosynthetic electron transport and quenching of chlorophyll fluorescence. *Biochimica et Biophysica Acta*, **990,** 87–92.

Green, T. G. A., Büdel, B., Heber, U., Meyer, A., Zellner, H. & Lange, O.L. 1993. Differences in photosynthetic performance between cyanobacterial and green algal components of lichen photosymbiodemes measured in the field. *New Phytologist*, **125,** 723–731.

Jackson, A. E. & Seppelt, R. D. 1997. Physiological adaptations to freezing and UV radiation exposure in *Prasiola crispa*, an Antarctic terrestrial alga. In: Battaglia, B., Valencia, J. & Walton, D. W. H., eds. *Antarctic communities: species, strategy and survival.* Cambridge: Cambridge University Press, 226–233.

Kappen, L. 1993. Lichens in the Antarctic region. In Friedmann, E.I., ed. *Antarctic microbiology*. New York: Wiley-Liss, 433–490.

Kappen, L. & Breuer, M. 1991. Ecological and physiological investigations in continental antarctic cryptogams. II. Moisture relations and photosynthesis of lichens near Casey Station, Wilkes Land. *Antarctic Science*, **3,** 273–278.

Kappen. L., Bölter, M. & Kühn, A. 1987. Photosynthetic activity of lichens in natural habitats in the maritime Antarctic. *Bibliotheca Lichenologica*, **25,** 297–312

Kappen, L., Meyer, M. & Bölter, M. 1988. Photosynthetic production of the lichen *Ramalina terebrata* Hook. f. et Tayl., in the maritime Antarctic. *Polarforschung*, **58,** 181–188.

Kappen, L., Schroeter, B. & Sancho, L. G. 1990. Carbon dioxide exchange of Antarctic crustose lichens *in situ* measured with a CO_2/H_2O porometer. *Oecologia*, **82,** 311–316.

Krause, G. H. & Weis, E. 1991. Chlorophyll fluorescence and photosynthesis: the basics. *Annual Review of Plant Physiology and Plant Molecular Biology*, **43,** 313–349.

Lamb, I. M. 1947. A monograph of the lichen genus *Placopsis* Nyl. *Lilloa*, **13,** 151–288.

Lange, O. L., Bilger, W., Rimke, S. & Schreiber, U. 1989. Chlorophyll fluorescence of lichens containing green and blue-green algae during hydration by water vapor uptake and by addition of liquid water. *Botanica Acta*, **102,** 306–313.

Lindsay, D. C. 1971. Vegetation of the South Shetland Islands. *Bulletin British Antarctic Survey*, No. 25, 59–83.

Redon, J. 1985. *Liquenes antárticos*. Santiago de Chile: INACH, 123 pp.

Scheidegger, C. & Schroeter, B. 1995. Effects of ozone fumigation on epiphytic macrolichens: ultrastructure, CO_2-gas exchange and chlorophyll fluorescence. *Environmental Pollution*, **88,** 345–354.

Schreiber, U. 1986. Detection of rapid induction kinetics with a new type of high-frequency modulated chlorophyll-fluorometer. *Photosynthesis Research*, **9,** 261–272

Schreiber, U., Bilger, W., Klughammer, C. & Neubauer, C. 1988. Application of the PAM fluorometer in stress detection. In Lichtenthaler, K. H., ed. *Applications of chlorophyll fluorescence.* Dordrecht: Kluwer Academic Publications, 151–155.

Schroeter, B. 1991. *Untersuchungen zu Primärproduktion und Wasserhaushalt von Flechten der maritimen Antarktis unter besonderer Berücksichtigung von* Usnea antarctica *Du Rietz.* Dissertation, Kiel: Universität Kiel, 148 pp.

Schroeter, B. 1994. *In situ* photosynthetic differentiation of the green algal and the cyanobacterial photobiont in the crustose lichen *Placopsis contortuplicata*. *Oecologia*, **98,** 212–220.

Schroeter, B., Green, T. G. A., Kappen, L. & Seppelt, R. D. 1994. Carbon dioxide exchange at subzero temperatures. Field measurements on *Umbilicaria aprina* in Antarctica. *Cryptogamic Botany*, **4,** 233–241.

Schroeter, B., Green, T. G. A., Seppelt, R. D. & Kappen, L. 1992. Monitoring photosynthetic activity of crustose lichens using a PAM-2000 fluorescence system. *Oecologia*, **92,** 457–462.

Schroeter, B., Kappen, L. & Moldaenke, C. 1991. Continuous *in situ* recording of the photosynthetic activity of Antarctic lichens – established methods and a new approach. *Lichenologist*, **23,** 253–265.

Schroeter, B., Kappen, L. & Schulz, F. 1996. Long-term measurements of microclimatic conditions in the fruticose lichen *Usnea aurantiaco-atra* in the maritime Antarctic. *Actas del V. Simposio Antártico (Barcelona).*

Schroeter, B., Olech, M., Kappen, L. & Heitland, W. 1995. Ecophysiological investigations of *Usnea antarctica* in the maritime Antarctic. I. Annual microclimatic conditions and potential primary production. *Antarctic Science*, **7,** 251–260.

Winter, K. & Lesch, M. 1992. Diurnal changes in chlorophyll *a* fluorescence and carotenoid composition in *Opuntia ficus-indica*, a CAM plant, and in three C_3 species in Portugal during summer. *Oecologia*, **91,** 505–510.

31 Physiological adaptations to freezing and UV radiation exposure in *Prasiola crispa*, an Antarctic terrestrial alga

A E. JACKSON AND R. D. SEPPELT
Australian Antarctic Division, Channel Highway, Kingston, 7050 Tasmania, Australia

ABSTRACT

In the Antarctic environment, the alga Prasiola crispa *is exposed to extreme conditions such as high levels of incident radiation and freezing temperatures. This study investigated physiological adaptations to these factors. Samples were collected throughout a 13 month period and analysed for free amino acid composition. There was a marked increase in the levels of proline during April, concurrent with a decrease in other amino acids. In January, proline constituted 2.8 ($\pm$0.2)% of the total free amino acid pool (1.2$\pm$0.1 mol (g dry weight)$^{-1}$) whereas by mid-April it was the major component at 48.8 ($\pm$2.9)% and 28.4$\pm$2.9 mol (g dry weight)$^{-1}$. Proline declined to summer levels if algae were kept in culture. Amino acids and other solutes are involved in the preservation of photosynthetic activities during freezing and it is possible that proline is involved in cryoprotection in this species. The effect of natural UV radiation levels on photosynthesis and the production of water-soluble UV absorbing compounds were also studied. After one month of experimental screening, ratios of variable to maximal fluorescence were measured at 2 h intervals over a 24 h period using a PAM-2000 fluorometer. All treatments followed a similar trend over the diurnal course of measurement. Fv/Fm ratios were higher at 2100 h than at 1500 h local time due to photoinhibition of photosynthesis. At lower light levels, a higher Fv/Fm ratio became apparent in screened, as opposed to unscreened, algae. Fluorescence measurements were also taken 11 months later, on screened and unscreened algae, at midday and midnight. The Fv/Fm ratio was again higher in screened algae. Levels of UV absorbing compounds were significantly higher in unscreened algae. These pigments remained present in* P. crispa *throughout the year.*

Key words: amino acids, green alga, UV effects, cryoprotection.

INTRODUCTION

The supralittoral, green alga *Prasiola crispa* forms sheet-like thalli on the surface of wet ground, and is thus exposed to potentially damaging levels of UV radiation, as well as photo-inhibitory photosynthetic photon flux densities (PPFD), during the Antarctic summer. In winter the algae are permanently frozen and undergo repeated freeze/thaw cycles in the spring and autumn. *P. crispa* is also subject to considerable fluctuations in salinity due to influx of freshwater from snow melt, exposure to salt spray and nutrient enrichment from birds. In spite of these factors, *P. crispa* has successfully colonized large areas of Antarctica. Indeed, in terms of biomass, it is probably the most abundant alga in the Windmill Islands region (centred at 66°22' S, 110°30' E), which is where this study took place. We examined physiological adaptations to UV radiation and freezing in *P.*

crispa that are likely to be important to the survival of this species in Antarctica.

The accumulation of organic solutes in plants under water stress, induced either by dehydration or hyperosmotic conditions, has been widely reported in the literature (e.g. Stewart & Larher 1980, Edwards *et al.* 1988, Vance & Zaerr 1990), and this includes studies on Antarctic marine algae (Karsten *et al.* 1991). Although the effects of variations in salinity on organic solute concentrations have been studied in *P. crispa* in laboratory culture (Jacob *et al.* 1991), the effect of the reduction in available water that occurs with freezing has not been investigated. Davey (1989) has already suggested, on the basis of evidence that photosynthesis in *P. crispa* continued down to −15 °C (Becker 1982), that this alga might contain large quantities of cryoprotectants. In our study, the concentrations of free amino acids in *P. crispa* were analysed during the year to ascertain whether they are involved in cryoprotection.

The seasonal depletion in stratospheric ozone, which is particularly pronounced over Antarctica, has led to concern over the effects of UV radiation on the Antarctic biosphere. Enhanced UV-B radiation has been reported to depress photosynthesis in plants (Caldwell 1981). However, its effects have been studied primarily on agricultural species in growth chambers and only a few studies have been done in the field (Tevini 1993). The potential for damage rendered by UV-A light has been less studied. In this study, we relied on screening of natural UV radiation to determine the effects of UV-A and UV-B radiation on photosynthesis in *P. crispa*. UV-B levels measured at Casey during a clear day in February 1989 reached more than 0.7 Wm^{-2} in the middle of the day (Wood 1989).

We also investigated the possible role of UV absorbing compounds in the protection of *P. crispa* against UV radiation. These substances occur in a diverse range of organisms, including *P. crispa* (Post *et al.* 1992) and other Antarctic algae (Karentz *et al.* 1991), and are often thought to be produced for screening out potentially damaging radiation.

MATERIALS AND METHODS

Study site

Experiments were done on a large, uniform patch of *P. crispa* growing on a flat, moist, exposed area in the proximity of penguin colonies on Shirley Island in the Windmill Island group. Due to large amounts of blowing snow in May, the area became covered by a drift tail that remained until January 1994.

The effect of UV radiation on photosynthesis and pigment production

UV radiation was screened from plots of algae using 0.1 mm thick Mylar (UV-B only) and polycarbonate (UV-A and B) screens (30 cm×30 cm). A 0.1 mm layer of Mylar is opaque to radiation of wavelengths shorter than 320 nm (Lorenzen 1979). Unscreened frames delineated patches of algae used as controls. Screens were raised off the ground with small rocks to ensure a throughflow of air. Thallus temperatures in screened and unscreened algae were measured with Grant miniature thermistor probes (one probe per algal patch) and logged every hour on a Grant Squirrel data logger. Photosynthetic photon flux densities (PPFD) were recorded with quantum sensors attached to an electronic data collecting device (Li-1000, Li-Cor, Lincoln, NB, USA).

When the screens had been in place for a month (March 1993), ratios of variable (Fv) to maximal (Fm) fluorescence were measured at 2 h intervals over a 24 h period using a PAM-2000 chlorophyll fluorometer (Walz, Effeltrich, Germany). Samples were dark adapted for 10 min. prior to Fv/Fm measurements.

Extraction of photosynthetic and UV absorbing pigments was done under conditions of low light and temperature. Algal samples were homogenized in 10 ml of methanol and centrifuged at 1500*g* for 5 min. The pellet was re-extracted in fresh solvent and the supernatants were combined and made up to volume. Visible-UV absorption spectra of the methanol extracts were recorded with an Hitachi model U-3210 scanning spectrophotometer. The efficiency for methanol extraction of UV absorbing compounds was calculated using the method of Dunlap *et al.* (1989). Eight samples of algae were extracted four times each with 10 ml volumes and the absorbances at 325 and 666 nm (UV absorbing pigments and chlorophyll respectively) were measured.

Screens were left in place and the UV absorbing pigments in unscreened and polycarbonate screened algae were measured again at the end of January 1994.

Differences in pigment concentrations and fluorescence measurements between treatments were tested for by one-way ANOVA.

UV absorbing compounds were extracted from triplicate field samples, taken from outside the screens, at least once a month throughout the year to determine seasonal changes.

Free amino acid analysis by HPLC

Samples of algae were collected, at least once a month throughout the year, from undisturbed patches in the vicinity of, but outside the screens. These were either extracted immediately, or stored frozen at −70 °C. Freezing had no effect on the amino acid profiles. Triplicate samples of algae were homogenized in 80% methanol and extracted overnight at 4 °C. Solid material in the extract was removed by centrifugation and the supernatant was deproteinized by filtration through a Millipore PLGC membrane (10 000 molecular weight cut-off). Free amino acids were analysed using the Waters PicoTag method which involved precolumn derivatization with phenylisothiocyanate (PITC). Aliquots of 30 μl of algal extract were vacuum dried with 10 nmoles of methionine sulphone as an internal standard. Prior to PITC derivatization, the dried samples were resuspended in 20 μl of redrying solution (methanol:water:triethylamine, 2:2:1 v:v:v) and redried. To derivatize, 20 μl of a solution of methanol:water:triethylamine:phenyliso-thiocyanate (7:1:1:1 by volume) were added and the samples were left to react for 20 min. at room temperature before vacuum drying. When dry, the samples were resuspended in 90 μl of Waters sample diluent and 10 μl of methanol, placed in a WISP autosampler and applied to the column in 30 μl aliquots.

The amino acid derivatives were separated by reversed-phase high-performance liquid chromatography, at 46 °C, on a Waters C_{18} column (4 μm, 3.9 mm×300 mm) with an in-line precolumn filter. The gradient conditions were 0 to 9% Waters Eluent 2 (E2) for 30 min., 9 to 34% E2 for 20 min., 34 % E2 for 12 min., 34 to 100 % E2 in 0.5 min., 100 % E2 for 4 min. and 100 to 0% E2 in 0.5 min. Equilibration was with 100 % Waters Eluent 1 for 8 min. The flow rate was 1.0 ml $min.^{-1}$. PTC derivatives were detected on a Waters 440 UV detector at 254 nm. Amino acids were identified by co-chromatography with individual standard amino acids.

Algae were oven dried for at least 24 h and dry weights determined on a Sartorius precision electronic balance.

Recovery of photosynthetic activity after freezing

The recovery of photosynthetic activity after freezing was monitored in *P. crispa* brought in from the field in July. Algae were thawed in petri dishes held over water in three replicate desiccators under low light at 5 °C (three petri dishes per desiccator). Photosynthetic yield values (calculated as ΔF/Fm′) were measured using the PAM-2000 chlorophyll fluorometer over a 48 h period. At least three measurements were taken on each petri dish of algae. Data from each time point were pooled and averaged.

RESULTS

UV and photosynthesis

In the UV screening experiment (February to March 1993), Fv/Fm ratios in all treatments followed a similar trend over the diurnal course of measurement (Fig. 31.1). Values were low during the day when light levels were highest, reflecting photoinhibition. Recovery occurred when solar radiation dropped below 600 mol m^{-2} s^{-1} after 1500 h (Fig. 31.2a). Fv/Fm reached a minimum in all treatments at 0700 h (Fig. 31.1). This was probably the result of tissue freezing as the ambient temperature fell below the freezing point of *P. crispa* (−7 °C) during the previous 2 hour period (Fig. 31.2b). The presence of screens appeared to have only a minor effect on PPFD and temperature (Fig. 31.2a,b). Measured PPFD was reduced slightly by the presence of a polycarbonate screen. Screening did not have a significant effect on algal temperature, as evidenced by the small standard errors around the mean when the temperatures of all treatments and controls were pooled for each time point (Fig. 31.2b).

At lower light levels, there was a small but significant ($P<0.001$) difference in the Fv/Fm ratios of polycarbonate screened, as opposed to unscreened, algae (0.63±0.02 polycarbonate screened; 0.54±0.01 unscreened) indicating that there may have been damage to PSII caused by UV radiation. There was no significant difference in screened and unscreened algae

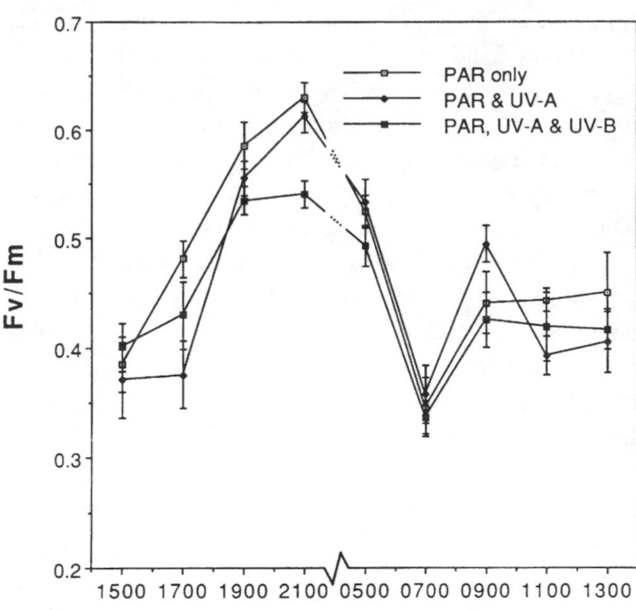

Fig. 31.1. Diurnal Fv/Fm ratios in screened and unscreened algae (±SE). PAR = photosynthetically active radiation.

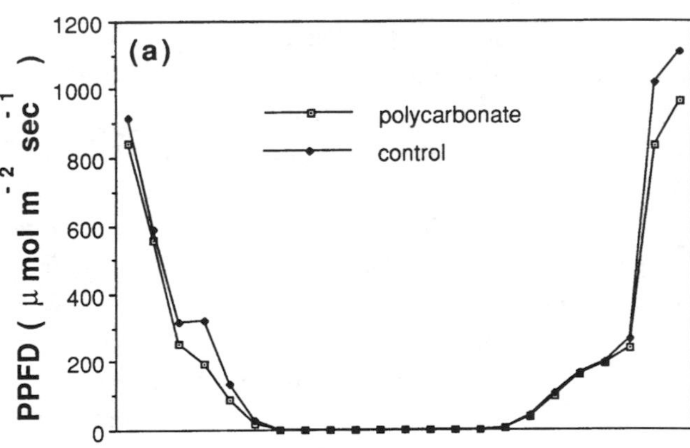

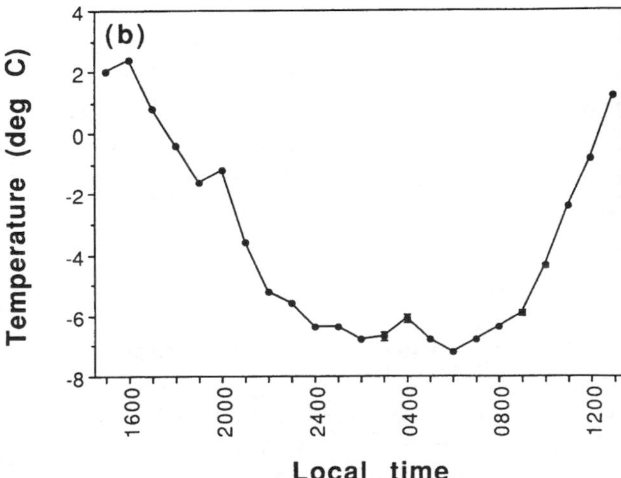

Fig. 31.2. Microclimate variables over the diurnal course of the UV experiment. (a) The photosynthetic photon flux density PPFD reaching polycarbonate screened and unscreened algae. (b) Averages of temperatures recorded from all treatments and controls.

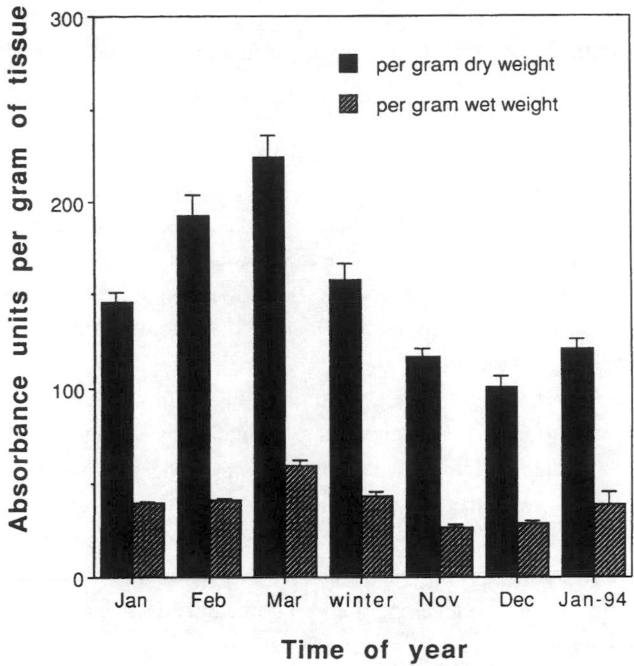

Fig. 31.3. UV absorbance of compounds extracted from *Prasiola crispa* throughout the year (±SE).

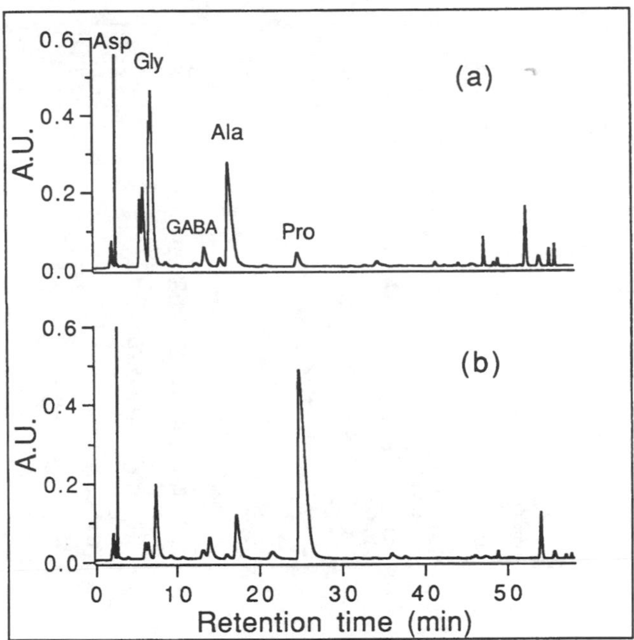

Fig. 31.4. Amino acid profiles of algae collected in (a) January and (b) April. Asp=aspartine, Gly=glycine, GABA=gamma aminobutyric acid, Ala=alanine, Pro=proline.

when Fv/Fm was depressed by high light levels during the day (mean of all treatments 0.43±0.02 in March 1993). In January 1994, a small difference (P=0.02) was apparent in the midday Fv/Fm values of polycarbonate screened and unscreened algae (0.48±0.02 and 0.42±0.01 respectively). Fv/Fm values measured at night in January 1994, on polycarbonate screened algae (0.64±0.01), were significantly higher (P<0.001) than those of unscreened algae (0.58±0.01). There was no difference between Mylar and polycarbonate screened algae and hence no evidence of a UV-A effect on photosynthetic competence in *P. crispa*.

UV absorbing compounds

Methanol extracted UV absorbing pigments from *P. crispa* with an efficiency of 72%. This was taken into account when calculating total pigments. Levels of UV absorbing pigments were significantly higher in unscreened algae (P<0.001). In March 1993 after the screens had been in place for 1 month, 39±3, 33±3 and 59±3 absorbance units per gram wet weight (AU g^{-1}) were extracted from polycarbonate, Mylar and unscreened algae respectively. In 1994 the overall pigment levels were lower but there was still a significant difference (P<0.05) between polycarbonate screened algae and the controls (14±1 and 39±7 AU g^{-1} respectively). Seasonal samples were also collected and assayed for UV absorbing pigments (Fig. 31.3). UV absorbing pigment levels showed no immediate decline after the summer period, and were still high in April (57±3 AU (g wet weight)$^{-1}$, 157±13 AU (g dry weight)$^{-1}$). Extracts of algae from culture also absorbed a substantial amount of UV radiation (41±2 AU g^{-1}).

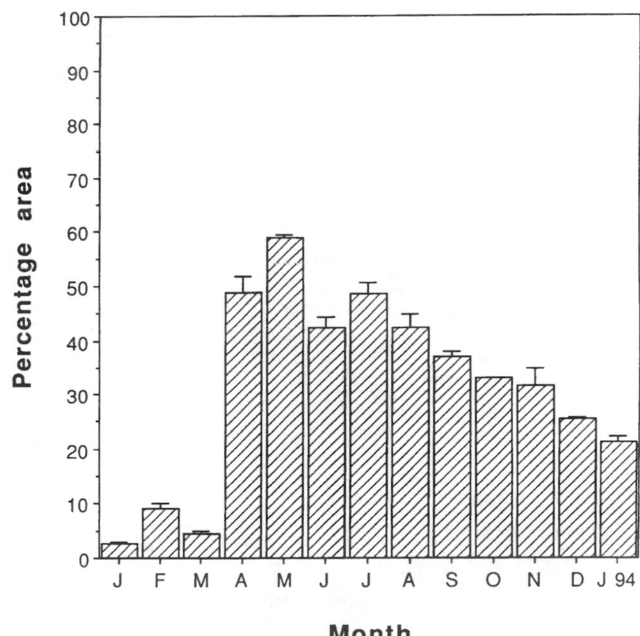

Fig. 31.5. Proline as a percentage of total amino acids during the year (±SE).

Amino acids and recovery of photosynthetic activity

The amino acid profiles of algae collected in January and April are shown in Fig. 31.4. There was a marked increase in the levels of proline between March and April, concurrent with a decrease in other amino acids (Figs. 31.5 and 31.6). Proline as a percentage of total amino acids was calculated in terms of units of area under the peaks in HPLC chromatograms. In January, proline constituted 2.8 (±0.2)% of the total free amino acid pool (1.2±0.1 μmol (g dry weight)$^{-1}$), whereas by mid-April it was the major component at 48.8 (±2.9)% and 28.4±2.9 μmol (g dry

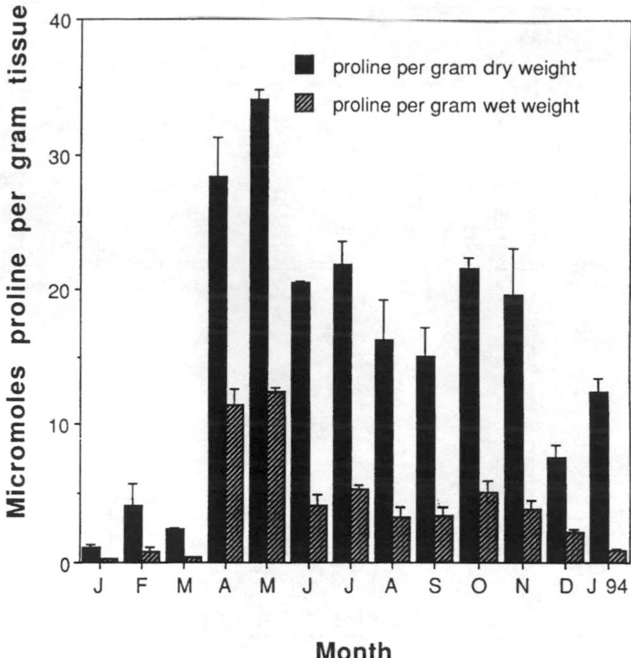

Fig. 31.6. Seasonal changes in the proline content of *Prasiola crispa* (±SE)

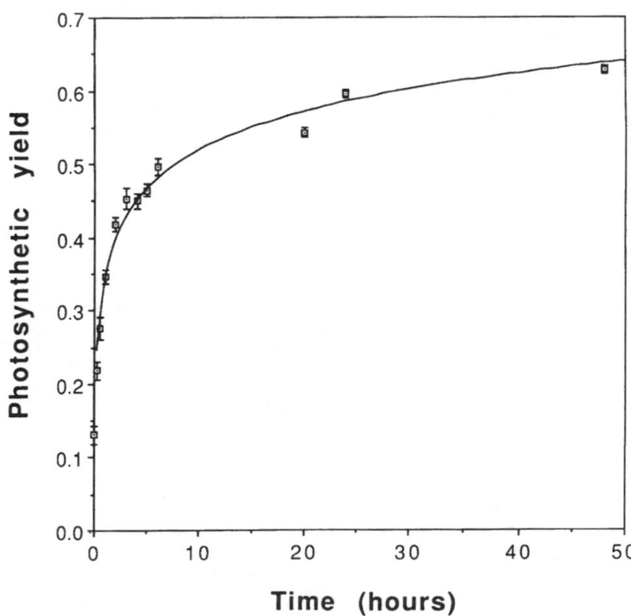

Fig. 31.7. Recovery of photosynthetic yield after prolonged freezing (±SE). Equation of best fit: $y=0.34619+0.7267 \times \log x$, $r^2=0.987$.

weight)$^{-1}$. Proline declined to summer levels if algae were kept in culture. Algae collected in winter recovered photosynthetic activity logarithmically over time of thawing (Fig. 31.7). There was rapid initial recovery to yield values of around 0.4 after 2 h. This slowed thereafter, with full recovery taking at least 48 h.

DISCUSSION

UV-B had a detrimental effect on PSII reaction centres in *P. crispa,* which became more apparent when ambient light was low. This concurs with evidence gained from laboratory studies under artificial UV-B irradiance (Post *et al.* 1992). However, the inhibition of photosynthesis by UV radiation that occurs in *P. crispa* is small relative to the effect on phytoplankton for example, in which photosynthesis has been reported to increase by 200–300% with UV screening.

Fv/Fm values decreased at high PPFD in both screened and unscreened algae, indicating that in *P. crispa* the midday depression in photosynthesis is not due to UV wavelengths as has been suggested for phytoplankton (Maske 1984). It may be that UV absorbing pigments in *P. crispa* protect PSII from some UV radiation induced damage.

Screening UV radiation from algae in summer reduced the concentrations of UV absorbing pigments by around 40%, which indicates that these compounds afford the alga UV protection; hence, production declined when UV radiation was reduced. The production of protective pigments is a common survival strategy in algae exposed to high levels of UV radiation. However, pigment levels showed no immediate decline after the summer period, and were still high in April (Fig. 31.3). Also, extracts of algae from culture absorbed a substantial amount of UV radiation. Thus, although pigment levels in this alga are influenced by UV radiation regime, consistent with their proposed role in UV protection, it is noteworthy that large quantities of this pigment remain in algae for months after exposure.

UV absorbing pigments continued to decline after the winter period and the levels were lower in the summer of 1993/94 than in 1992/93 (Fig. 31.3). The summer of 1993/94 was markedly colder than the previous one. Average temperatures for January 1993 were 3.8 °C (maximum) and −1.5 °C (minimum), whereas for January 1994 they were 1.4 °C (maximum) and −3.9 °C (minimum). The algae remained covered by a persistent snow drift until early January 1994. This would have reduced their UV exposure and may account for the lower pigment levels.

There was no evidence from this study that UV-A wavelengths depressed photosynthesis over and above the reduction in Fv/Fm observed in the UV-B exposed algae. The effect of UV-A on growth and photosynthesis in algae varies between species. For example, Jokiel & York (1984), found that growth rates in the diatom *Phaeodactylum tricornutum* were inhibited by UV- B but not UV-A radiation, whereas radiation in the UV-A region has been reported to be responsible for over 50% of the total inhibition of photosynthesis due to UV radiation in some natural phytoplankton assemblages (Helbling *et al.* 1992).

In field measurements of photosynthetic yield Fv/Fm reached a minimum in all treatments after the temperature fell to −7 °C, the measured freezing point of the tissue. This is in accord with the previous measurements of photosynthesis by gas-exchange made by Davey (1989). He reported that *P. crispa* continued to photosynthesize at sub-zero temperatures in the laboratory, ceasing at around −7 °C. However, Becker (1982) measured $^{14}CO_2$ fixation in *P. crispa* down to −15 °C.

P. crispa recovers photosynthetic activity when brought in from the field in the middle of winter, despite having been frozen for months (Fig. 31.7). Experiments with isolated chloroplast membranes have established that amino acids and other solutes are involved in the preservation of photosynthetic activities during freezing. In view of the accumulation of proline in *P. crispa*, coincident with the onset of winter, it is possible that proline is involved in cryoprotection in this species.

There are a number of reports of cross-protection in the tolerance mechanisms of plants to dehydration, frost and salt stress (e.g. Schmidt *et al.* 1986, Duncan & Widholm 1987). This commonality is assumed to be related to decreases in the availability of free water which result from these different conditions. Proline accumulation is one example. It has been reported to protect against damage induced by dehydration (Rudolph *et al.* 1986), freezing (Santarius 1992) and high salinity (Stewart & Lee 1974).

Jacob *et al.* (1991), analysed for various organic substances, including total amino acids and proline (ninhydrin method), in *P. crispa* cultured in artificial seawater at various salinities. These authors found that although sucrose and sorbitol increased in hypersaline media, there was no accumulation of proline. Preliminary analyses which we have undertaken, of amino acids in *P. crispa* collected from shallow pools of water with different degrees of exposure to salt spray and guano, revealed no differences in the proportion of proline. Thus, although proline may accumulate as a tolerance mechanism to freezing in *P. crispa*, there is no evidence that it also accumulates as an adaptation to salt stress.

It is important to note that Jacob *et al.* (1991) found high levels of proline even in algae grown at low salinity (30% of the total amino acids in algae cultured in freshwater for 14 days), and that in control plants the amount of proline exceeded that of all other amino acids combined. It is possible that these algae were already stressed in some way from being in culture, since proline has been shown to ameliorate deleterious effects from a range of factors such as heat, pH, salt and chemicals (Stewart 1989).

Davey (1989) compared the susceptibility of the algae *P. crispa* and *Phormidium* to freezing and desiccation. He noted different degrees of susceptibility to each stress, and suggested that the nature of the cell's resistance to freezing might not be the same as that for dehydration.

It is conceivable that in *P. crispa* the nature of the damage due to freezing is different to that caused by dehydration. Some of the freezing injury which occurs may be due to intracellular ice-formation, as distinct from extracellular ice-formation, which removes water from the cell.

The mechanism of cryoprotection is not completely understood. Compatible solutes such as proline may accumulate in water-stressed plants in order to stabilize membranes and macromolecules, or to restore osmotic balance.

Freezing can damage biomembranes either mechanically or by solute action due to the increase in solute concentration that occurs as the tissue freezes (Heber *et al.* 1979). In membrane stabilization, cryoprotective solutes like sucrose and glycerol seem to interact directly with the polar head groups of lipids. The effect of proline may depend on hydrophobic interactions (Santarius 1992) but also possibly on its intercalation between phospholipid head groups (Rudolph *et al.* 1986).

Timasheff (1982) concluded that protein stabilizing compounds are preferentially excluded from contact with the surface of the protein. This explanation is different from the previous postulate of Schobert & Tschesche (1978) that proline increases the hydrophilic area on the surface of proteins by some specific hydrophobic interaction.

Although the accumulation of proline may assist in osmotic adjustment, balancing concentration differences between the cytoplasm and the central vacuole (Field 1976), in some plants it does not accumulate in sufficient quantities (Dix & Pearce 1981). Whilst vacuoles are lacking in *P. crispa* grown at salinities between 0.35 and 35‰, they developed under hyperosmotic conditions (Jacob *et al.* 1992) and possibly acted as deposition sites for inorganic ions. It has also been suggested (Stewart 1989) that proline may act to detoxify free radicals which can cause damage in water-stressed plants.

The increase in proline in *P. crispa* was sharp and occurred between March and April (Figs. 31. 5 and 31.6). It was in this period that the average maximum temperature dropped below the freezing point of *P. crispa* tissue (Fig. 31.8). A sharp increase in proline at a critical level of water stress has been observed in other plants (Vance & Zaerr 1990).

The increase in proline in *P. crispa* between January and April is of a similar magnitude to that reported in some drought-sensitive resurrection plants after water deprivation (Tymms & Gaff 1979). However, from literature reports, there is no clear indication of whether the accumulation of proline during water stress is associated with drought-resistant plant varieties or vice versa (Singh *et al.* 1972, Hanson *et al.* 1979). There are similar

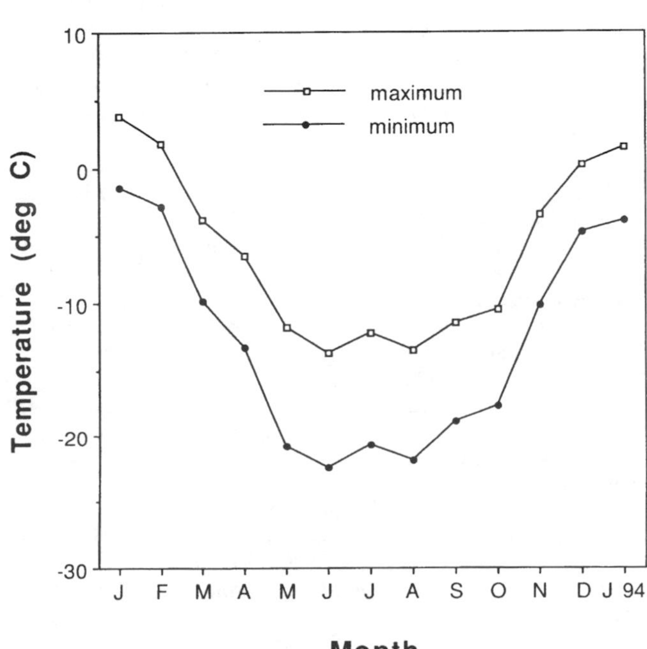

Fig. 31.8. The mean maximum and minimum monthly thallus temperatures for 1993.

discrepancies reported to occur with NaCl-sensitive versus resistant strains (Dix & Pearce 1981, Pandey & Ganapathy 1985). It is yet to be determined whether or not there are intraspecific variations in susceptibility to freezing-damage and proline accumulation in *P. crispa*.

Although there was a gradual decline in proline with the approach of summer 1993/94, the concentrations were still higher than in 1992/93 (Figs. 31.5 and 31.6). As has already been discussed, the temperatures during the second summer were markedly colder than those the previous year, and this may have resulted in higher levels of proline.

The accumulation of substances such as UV absorbing compounds and proline in *P. crispa*, under conditions of UV exposure and freezing respectively, probably enables this alga to survive successfully in the Antarctic environment. *P. crispa* provides a good model for studying the effects of such factors in the Antarctic terrestrial ecosystem and further work is needed to elucidate the synergistic effects of various stresses.

ACKNOWLEDGEMENTS

The Australian Antarctic Division provided logistic support for this project. The authors would like to thank the members of the 1993–94 ANARE, particularly Andrew Croke and Sean Johnson for field assistance and David Melick for informative discussions. Mean monthly temperatures were supplied by the Australian Bureau of Meteorology. We are grateful to Drs B. Schroeter and A. H. L. Huiskes for helpful comments on the original manuscript.

REFERENCES

Becker, E. W. 1982. Physiological studies on antarctic *Prasiola crispa* and *Nostoc commune* at low temperatures. *Polar Biology*, **1**, 99–104.

Caldwell, M. M. 1981. Plant response to solar ultraviolet radiation. In Lange, O. L., Noble, P. S., Osmond, C. B. & Ziegler, H. eds. *Physiological plant ecology. I. Responses to physical environment. Encyclopaedia of plant physiology*, Vol. 12A. Berlin: Springer-Verlag, 169–197.

Davey, M. C. 1989. The effects of freezing and desiccation on photosynthesis and survival of terrestrial Antarctic algae and cyanobacteria. *Polar Biology*, **10**, 29–36.

Dix, P .J. & Pearce, R. S. 1981. Proline accumulation in NaCl-resistant and sensitive cell lines of *Nicotiana sylvestris*. *Zeitschrift für Pflanzenphysiologie*, **102**, 243–248.

Duncan, D. R. & Widholm, J. M. 1987. Proline accumulation and its implication in cold tolerance of regenerable maize callus. *Plant Physiology*, **83**, 703–708.

Dunlap, W. C., Williams, D. McB., Chalker, B. E. & Banaszak, A. T. 1989. Biochemical photoadaptation in vision: UV-absorbing pigments in fish eye tissues. *Comparative Biochemistry and Physiology*, **93B**, 601–607.

Edwards, D. M., Reed, R. H. & Stewart, W. D. P. 1988. Osmoacclimation in *Enteromorpha intestinalis*: long-term effects of osmotic stress on organic solute accumulation. *Marine Biology*, **98**, 467–476.

Field, C. D. 1976. Salt tolerance in halophytes. *Nature*, **264**, 510–511.

Hanson, A. D., Nelson, C. E., Pedersen, A. R. & Everson, E. M. 1979. Capacity for proline accumulation during water stress in barley and its implications for breeding for drought resistance. *Crop Science*, **19**, 489–493.

Heber, U., Volger, H., Overbeck, V. & Santarius, K. A. 1979. Membrane damage and protection during freezing. In Fennema, O., ed. *Advances in Chemistry Series*, No. 180, 159–189.

Helbling, E. W., Villafañe, V., Ferrario, M. & Holm-Hansen, O. 1992. Impact of natural ultraviolet radiation on rates of photosynthesis and on specific marine phytoplankton species. *Marine Ecology Progress Series*, **80**, 89–100.

Jacob, A., Kirst, G. O., Wiencke, C. & Lehmann, H. 1991. Physiological responses of the Antarctic green alga *Prasiola crispa* ssp. *antarctica* to salinity stress. *Plant Physiology*, **139**, 57–62.

Jacob, A., Lehmann, H., Kirst, G. O. & Wiencke, C. 1992. Changes in the ultrastructure of *Prasiola crispa* ssp. *antarctica* under salinity stress. *Botanica Acta*, **105**, 41–46.

Jokiel, P. L. & York Jr, R. H. 1984. Importance of ultraviolet radiation in photoinhibition of microalgal growth. *Limnology and Oceanography*, **29**, 192–199.

Karentz, D., McEuen, F. S., Land, M. C. & Dunlap, W. C. 1991. Survey of mycosporin-like amino acid compounds in Antarctic marine organisms: potential protection from ultraviolet exposure. *Marine Biology*, **108**, 157–166.

Karsten, L. F., Wiencke, C. & Kirst, G. O. 1991. The effect of salinity changes upon the physiology of eulittoral green macroalgae from Antarctica and Southern Chile. II. Intracellular inorganic ions and organic compounds. *Journal of Experimental Botany*, **42**, 1533–1539.

Lorenzen, C. J. 1979. Ultraviolet radiation and phytoplankton photosynthesis. *Limnology and Oceanography*, **24**, 1117–1120.

Maske, H. 1984. Daylight ultraviolet radiation and the photoinhibition of phytoplankton carbon uptake. *Journal of Plankton Research*, **6**, 351–357.

Pandey, R. & Ganapathy, P. S. 1985. The proline enigma: NaCl-tolerant and NaCl-sensitive callus lines of *Cicer arietinum*. *Plant Science*, **40**, 13–17.

Post, A., Gentle, S. & Larkum, A. W. D. 1992. Algal photosynthesis: inhibition by UV- radiation, recovery and UV-absorbing pigments. In Murata, N., ed. *Research in photosynthesis*, Vol IV. Dordrecht: Kluwer Academic Publishers, 847–850.

Rudolph, A. S., Crowe, J. H. & Crowe, L. M. 1986. Effects of three stabilising agents – proline, betaine and trehalose on membrane phospholipids. *Archives of Biochemistry and Biophysics*, **245**, 134–143.

Santarius, K. A. 1992. Freezing of isolated thylakoid membranes in complex media. VIII. Differential cryoprotection by sucrose, proline and glycerol. *Physiologia Plantarum*, **84**, 87–93.

Schmidt, J. E., Schmidt, J. M., Kaiser, W. M. & Hincha, D. K. 1986. Salt treatment induces frost hardiness in leaves and isolated thylakoids from spinach. *Planta*, **168**, 50–55.

Schobert, B. & Tschesche, H. 1978. Unusual solution properties of proline and its interaction with proteins. *Biochimica et Biophysica Acta*, **541**, 270–277.

Singh, T. N., Aspinall, D. & Paleg, L. G. 1972. Proline accumulation and varietal adaptability to drought in barley; a potential metabolic measure of drought resistance. *Nature*, **236**, 188–190.

Stewart, G. R. 1989. Desiccation injury, anhydrobiosis and survival. In Jones, H. G., Flowers, T. J. & Jones, M. B., eds. *Plants under stress: biochemistry, physiology and ecology and their application to plant improvement*. Cambridge: Cambridge University Press, 115–130.

Stewart, G. R. & Larher, F. 1980. Accumulation of amino acids and related compounds in relation to environmental stress. In Stumpf, P. K. & Conn, E. E., eds. *The biochemistry of plants*, Vol. 5. London: Academic Press, 609–635.

Stewart, G. R. & Lee, J. A. 1974. The role of proline accumulation in halophytes. *Planta*, **120**, 279–289.

Tevini, M. 1993. Effects of enhanced UV-B radiation on terrestrial plants. In Tevini, M., ed. *UV-B radiation and ozone depletion: effects on humans, animals, plants, microorganisms, and materials*. Boca Raton, FL: Lewis Publishers, 125–153.

Timasheff, S. N. 1982. Preferential interactions in protein–water–cosolvent systems. In Franks, F., ed. *Biophysics of water*. London: John Wiley, 70–72.

Tymms, M. J. & Gaff, D. F. 1979. Proline accumulation during water stress in resurrection plants. *Journal of Experimental Botany*, **30**, 165–168.

Vance, N. C. & Zaerr, J. B. 1990. Analysis by high-performance liquid chromatography of free amino acids extracted from needles of drought-stressed and shaded *Pinus ponderosa* seedlings. *Physiologia Plantarum*, **79**, 23–30.

Wood, W. F. 1989. The effects of ultraviolet radiation on the marine biota. *Transactions of the Menzies Foundation*, **15**, 179–185.

32 The delicate stability of lichen symbiosis: comparative studies on the photosynthesis of the lichen *Mastodia tesselata* and its free-living phycobiont, the alga *Prasiola crispa*

A. H. L. HUISKES, N .J. M. GREMMEN AND J. W. FRANCKE

Netherlands Institute of Ecology, Centre for Estuarine and Coastal Ecology, (NIOO-CEMO), Vierstraat 28, 4401 EA Yerseke, The Netherlands

ABSTRACT

Photosynthesis of the lichen species Mastodia tesselata, *its phycobiont* Prasiola crispa *ssp.* antarctica, *also occurring as free living alga in the same habitat (Antarctic rocky shores), and their intermediate form was studied in relation to experimentally prolonged thallus hydration.*

Lichen thalli, kept hydrated and illuminated for increasing time periods, showed a significant decrease in rates of photosynthesis. Neither in the free-living algal thalli, nor in the thalli of the intermediate form between the lichen and the alga, did this decrease in photosynthetic rates occur. The rates of dark respiration did not show any effect of prolonged hydration.

It is assumed that prolonged thallus hydration allows the phycobiont to photosynthesize for longer periods, resulting initially in a decrease in the rate of photosynthesis, caused by accumulation of photosynthates in the lichen thallus. Subsequently, the phycobiont reacts with an increase in cell production, resulting in an intermediate form between the lichen and the alga, thus destabilizing the symbiosis. The ecological implications with respect to habitat differences between M. tesselata *and* P. crispa *are discussed.*

Key words: lichen symbiosis, thallus hydration, photosynthesis.

INTRODUCTION

The lichen *Mastodia tesselata* (Hook.f. et Harv.) Hook.f. et Harv. occurs in the supralittoral of the maritime Antarctic on exposed rock surfaces, often in places with a moderate to strong influence of birds and sea spray, resulting in elevated levels of nutrients and salt. Moisture conditions in this habitat are such that hydration of the thallus is intermittent (Gremmen *et al.* 1994).

The phycobiont of *M. tesselata* is a green alga of the genus *Prasiola* (Henssen & Jahns 1974, Brodo 1976). It is likely that the phycobiont is the same species as the free-living *Prasiola crispa* ssp. *antarctica* (Kütz.) Knebel, which occurs also in the supralittoral of the same area but mainly in rock crevices where meltwater may accumulate, resulting in a wet, nutrient-rich habitat (Davey 1989, Gremmen *et al.* 1994).

Although a number of authors (Printz 1964, Brodo 1976, Redon 1985, Kappen *et al.* 1987) regard *M. tesselata* as a lichen, other publications question the status of the taxon. Ahmadjian (1967) regards the association between the alga and the fungus to be 'lichen-like' and not a lichen at all. Henssen & Jahns (1974) describe *M. tesselata* as a primitive lichen species, a first step in the evolution of lichenization. They suggest, in suboptimal (drier) habitats for *P. crispa*, a process involving the fungus *Guignardia* sp. growing into the algal thallus and breaking it up into isolated clusters of algal cells, resulting in a morphological entity of fungus and alga, called *M. tesselata*, with a niche differing from that of the free-living alga. Other authors (Dughi 1939, Jaag 1945) assume that for some lichen species the mycobiont cannot exist in wet places and disappears when the circumstances become unfavourable for the fungus.

Both possibilities suggest the existence of a transitory process from the alga *P. crispa* to the lichen *M. tesselata* (or vice versa) which implies the occurrence of intermediate forms, maybe as a result of the less well evolved symbiosis.

During field studies in the coastal area of the Argentine Islands (between 65°6′ S and 65°20′ S, 64°16′ W), we actually found intermediate forms between *M. tesselata* and *P. crispa*. Intermediate forms are also reported from Signy Island (60°43′ S, 46°16′ W) (R. I. Lewis Smith, pers. commun.)

The occurrence of *M. tesselata* and *P. crispa* in habitats in the supralittoral with differing moisture conditions suggests different responses of the free-living alga and its lichenized form, *M. tesselata*, with respect to this environmental factor. The aim of this study is to compare the relation between photosynthesis and thallus moisture of the alga *P. crispa*, the lichen *M. tesselata* and their intermediate form. The results might explain their different habitat preferences, and consequently point towards ecological implications of the symbiosis of alga and fungus.

MATERIALS AND METHODS

Experiments on fresh samples of the alga *Prasiola crispa*, the lichen *Mastodia tesselata* and their intermediate forms were performed under controlled conditions at the British Antarctic Survey Station Faraday, Argentine Islands, Antarctic Peninsula (65°14′ S, 64°16′ W).

The samples were collected in the vicinity of Faraday, mainly on Galindez Island. As dry samples proved to be too brittle to collect properly, *M. tesselata*, a black-brown foliose lichen, and the intermediate form between *M. tesselata* and *P. crispa*, morphologically similar to *M. tesselata* but with an olive-green to dark-green thallus, were sampled after spraying them with tap water, using a hand spray. The thus moistened samples were taken off the rocks using a putty knife. *P. crispa*, with a bright green foliose thallus, could be collected from wet places without any pretreatment.

Subsamples of the material were placed in petri dishes of 3.5 cm in diameter and of known weight. Care was taken that the samples covered the complete area of the petri dishes without disturbing their natural three-dimensional structure. The filled petri dishes were weighed again, in order to obtain the wet weight of the sample.

Net photosynthesis and dark respiratory rates were measured by placing two petri dishes in a perspex incubator with an internal volume of 150 cm^3. The incubator contained a light sensor, sensitive in the range of photosynthetic active radiation (PAR, 400–700 nm), and a temperature sensor. The incubator had a bolted-on lid; sealing was effected by a neoprene gasket and silicon vacuum grease. The temperature inside the incubator was kept at +2 °C by immersing the incubator in a thermostatic bath containing ethylene glycol. Illumination, kept at 300 mol m^{-2} s^{-1}, was achieved by using a slide projector and a mirror reflecting the light beam onto the incubator.

To measure the CO_2 flux in the incubator, outside air was pumped through, drawn in from 25 m above ground level via a pvc tube and an air pump. Air samples were drawn from the outside air and from the air leaving the incubator by the two internal sample pumps of an ADC 225 Mk III Infra Red Gas Analyser. The air flow in both air streams was 250 ml min^{-1}. The pressure of the air streams was kept at 2 mbar above ambient air pressure to prevent air or ethylene glycol leaking into the system. Before entering the analyser the air samples were led through a cold trap in order to prevent the interference of water vapour with the measurements and through dust filters. CO_2 concentration in the air flows was measured, using differential measurement. A more elaborate description of the measuring system may be found in Long & Hällgren (1993).

Temperature and light intensity in the incubator and the difference in CO_2 concentration between the two air streams (i.e. the CO_2 flux of the plant samples) were recorded on a paper chart recorder (Laumann Minikas 12).

After the experiment the contents of one of the two petri dishes were dried at 70 °C for 36 hours in order to determine their dry weight.

The contents of the other petri dish were deep frozen (−20 °C) and shipped back to the Netherlands where the chlorophyll content of the thallus was determined using high performance liquid chromatography (HPLC) after grinding it by hand in liquid nitrogen followed by two extractions with acetone (5 ml) for 10 minutes each.

Rates of net photosynthesis and dark respiration were related to the concentration of chlorophyll *a* in the sample and expressed in g CO_2 μg chl a^{-1} h^{-1}.

Moisture loss of the samples was measured by repetitive weighing. For this the petri dishes with the samples were taken out of the incubator and weighed on an electronic balance (Sartorius 1475) with a precision of 1 mg.

Three different kinds of experiment were performed.

1. Rates of photosynthesis of samples were measured immediately after collection ('short experiments').
2. Samples were sprayed liberally with tap water after being put into the petri dishes and left to soak for one hour in order to obtain complete moisture saturation of the thallus. Subsequently adhering water was removed by blotting the thalli with filter paper. The rates of photosynthesis were measured continuously while the samples were allowed to desiccate ('long experiments').
3. Samples were regularly sprayed with tap water during a

period of 72 hours in order to obtain prolonged periods of saturated thallus moisture. They were kept under natural light, temperature and air humidity conditions. After this period the samples were left to dehydrate for different periods of time (2, 4, 6, 8 or 10 h). Subsequently, the rates of photosynthesis were measured after removal of any adhering water if necessary ('very long experiments').

RESULTS

In order to explain the relationship between moisture content and net photosynthesis and dark respiration in the lichen samples we used a simple regression model. Details of this model will be published elsewhere. The model is based on the generally accepted assumption that the rates of net photosynthesis increase with increasing thallus moisture content to a certain optimum, after which the rates decline again (Lange & Tenhunen 1981, Snelgar *et al.* 1981, Kappen 1985, Kershaw 1985, Matthes-Sears & Nash 1986).

The regression model used is:

$$NP=a\,[m_1(m_1<m_{opt})-m_0]+\{a\,(m_{opt}-m_0)-b[m_2\,(m_2>m_{opt})-m_{opt}]\}$$

in which NP=the rate of net photosynthesis; a and b are constants; m_1 and m_2 are thallus moisture contents in the ascending and the descending leg of the regression line, respectively; m_{opt} is the thallus moisture content at which the rate of net photosynthesis is maximal; and m_0 is the thallus moisture content at which the rate of net photosynthesis is zero. The same regression model was used to describe the dark respiration rates.

The statistical analysis was executed using the SYSTAT program. At first, the regression curves were calculated for the different experiments separately. If the 95% confidence limits of the regression coefficient overlapped by less than 10%, the results of the experiments were regarded to differ significantly from each other. If not, the results were combined and a regression analysis on the combined results was performed.

Fig. 32.1 shows that, at the same thallus moisture content, the rates of net photosynthesis are higher in the short experiments than in the long experiments; the very long experiments show even lower rates. The rates of dark respiration are also based on the chlorophyll *a* concentrations of the thallus (which did not differ from each other significantly in the three experiments).

Fig. 32.2 shows the results obtained from samples of *P. crispa*

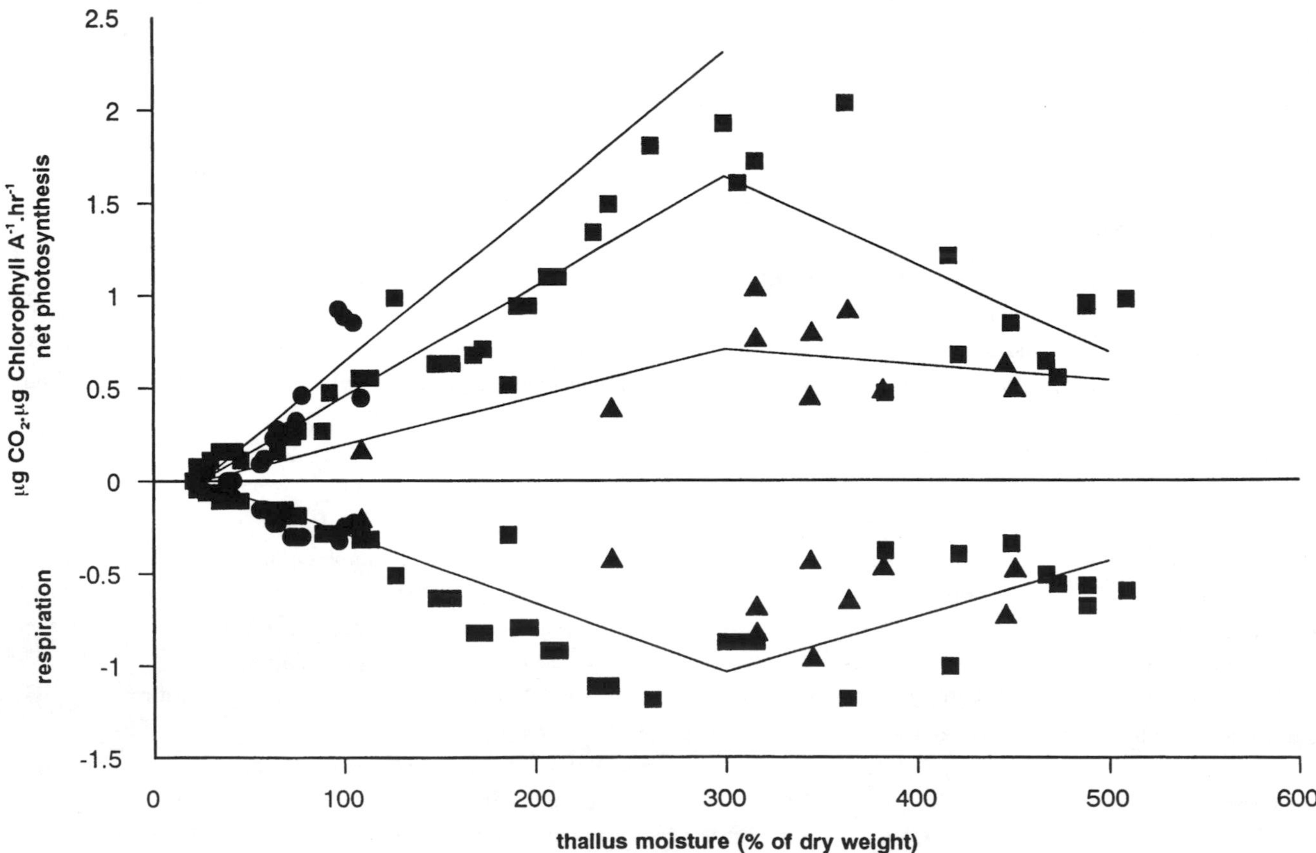

Fig. 32.1. Rates of net photosynthesis (at 300 mol $m^{-2}\,h^{-1}$ PAR) and of dark respiration at +2 °C (based on the chlorophyll *a* concentration of the thallus), of samples of *Mastodia tesselata*, collected in the Argentine Islands, Antarctica. ●=photosynthesis measured directly after collection (the 'short experiments', three replicates); photosynthetic photon exposure during thallus hydration <100 mol m^{-2} PAR; r^2 of the regression line=0.872. ■=samples sprayed with tap water and allowed to desiccate while photosynthesis was measured (the 'long experiments', two replicates); photosynthetic photon exposure between 100 and 2000 mol m^{-2} PAR; r^2 of the regression line=0.830. ▲=samples kept under natural light conditions while hydration of the thallus was maintained for 72 hours followed by different periods of dehydration (2, 4, 6, 8 or 10 hours) under natural circumstances; photosynthesis was measured subsequently (the 'very long experiments', two replicates); photosynthetic photon exposure >>2000 mol m^{-2} PAR; r^2 of the regression=0.942. r^2 of the regression on the rates of dark respiration=0.865.

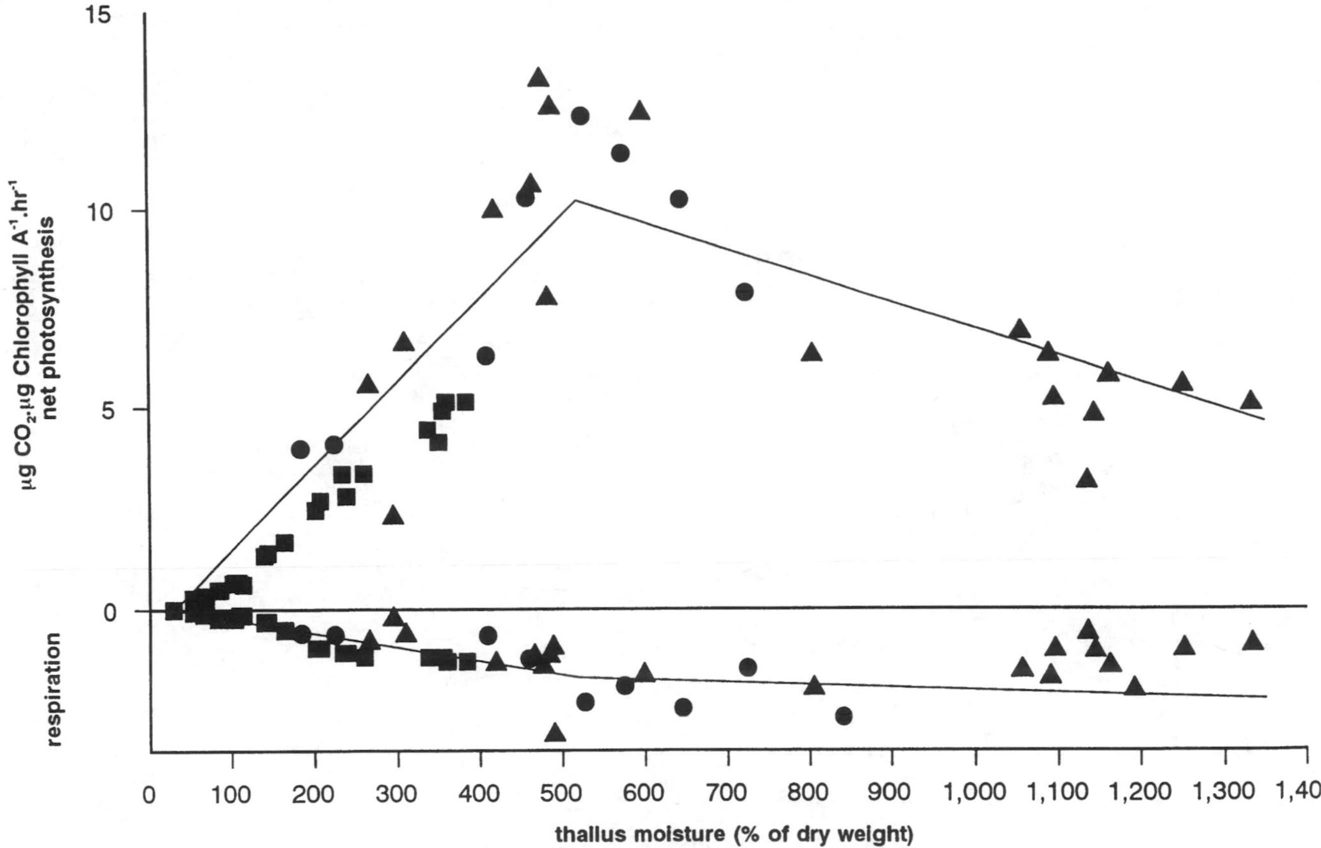

Fig. 32.2. Rates of net photosynthesis (at 300 mol m^{-2} h^{-1} PAR) and rates of dark respiration at +2 °C and based on the chlorophyll *a* concentration of the thallus, of samples of *Prasiola crispa* collected in the Argentine Islands, Antarctica. Results were obtained in the same experiments as performed on *M. tesselata* (Fig. 32.1). See Fig. 32.1 for definitions of symbols. r^2 of the regression on the rates of net photosynthesis=0.915. r^2 of the regression on the rates of dark respiration=0.930.

in the same three kinds of moisture regime, as performed on *M. tesselata* (Fig. 33.1). No significant differences were found in the rates of net photosynthesis or the rates of dark respiration between the three experiments.

The relation between the thallus moisture content and the rates of net photosynthesis and the rates of dark respiration of the intermediate form in the three moisture regimes are shown in Fig. 32.3. Again no significant differences were found between the rates in the three experiments.

Although based on the concentration of chlorophyll *a* in the thallus, the level of the rates of net photosynthesis in *P. crispa* was found to be about four to five times higher than in *M. tesselata* or the intermediate forms.

DISCUSSION

The relationship between thallus moisture content and rates of net photosynthesis is generally accepted to be triphasic: an increase in the rates of net photosynthesis, ascribed to an increase in biochemical activity with increasing hydration (Matthes-Sears & Nash 1986), towards an optimum moisture content where the CO_2 resistance is lowest, followed by a decrease in the rates of net photosynthesis, ascribed to an increase in CO_2 diffusion resistance (Lange & Tenhunen 1981). The course of the CO_2 resistance in the thalli may differ between different lichen species, depending on the anatomy of the thallus, especially that of the lower cortex (Snelgar *et al.* 1981). Our results are comparable with the course of the CO_2 exchange rates of lichen species with a poorly developed or no tomentum (i.e. a felt-like lower cortex), growing in exposed areas; *M. tesselata* fits this description.

Also in the rates of dark respiration a similar triphasic relationship with thallus moisture exists: an increase in the rates of dark respiration to the optimum thallus moisture content and beyond that point a levelling off to an almost constant rate or even a decrease (in stress situations the rates should decrease even further).

The model adopted also fits the relationship between the rates of net photosynthesis and thallus moisture content in *P. crispa*. Jacob (1992) shows a similar relationship.

The differences between the rates of net photosynthesis at corresponding thallus moisture contents in the short, long and very long experiments with *M. tesselata* (Fig. 32.1) may be explained by the total amount of photosynthetic photon exposure received by the samples in the period that the thallus was in a hydrated state and therefore able to photosynthesize. Comparing the total amount of photons received at e.g. 110% thallus moisture, the samples received less than 100 mol m^{-2} PAR in the short experiment, about 2000 mol m^{-2} in the long experiment and over 10000 mol m^{-2} in the very long experiment.

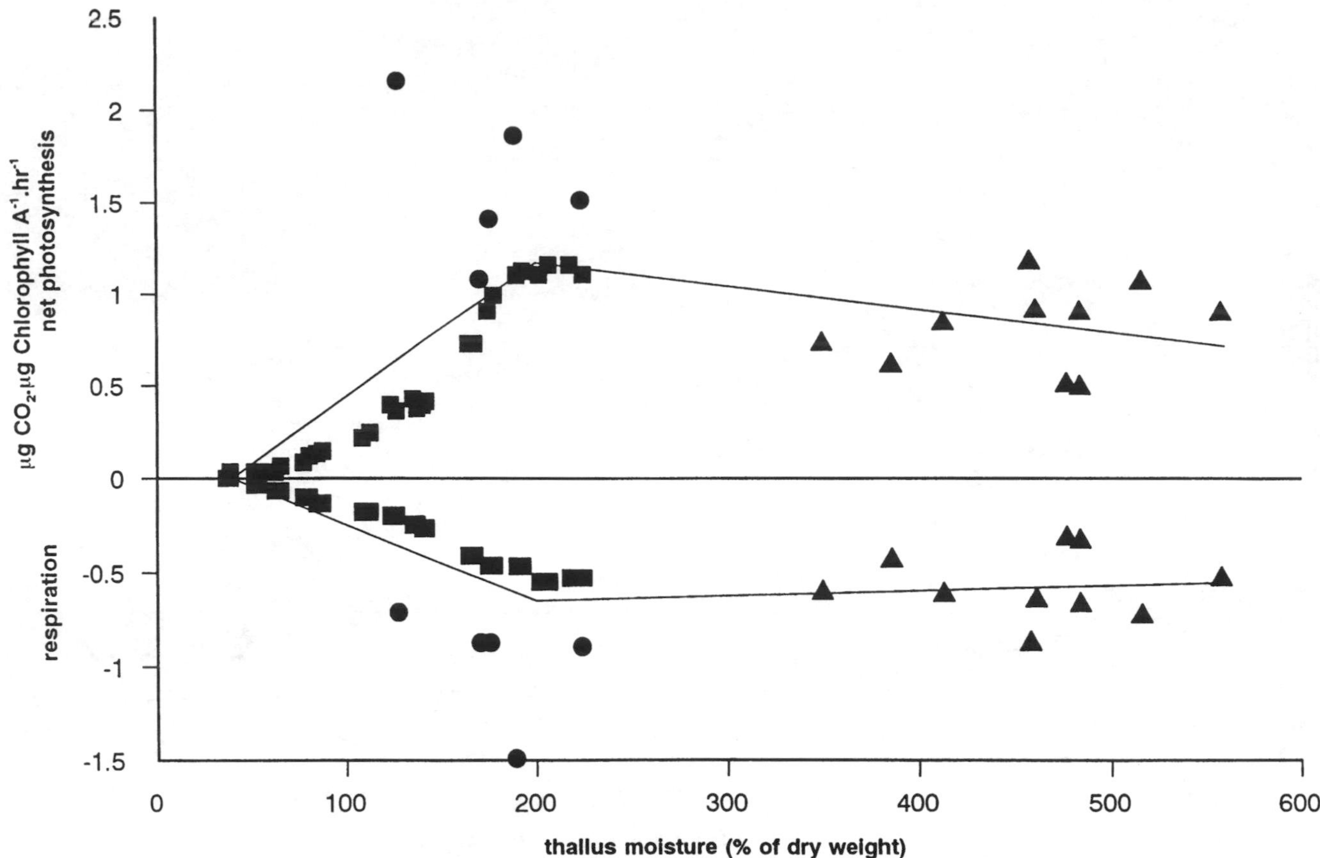

Fig. 32.3. Rates of net photosynthesis (at 300 µmol m^{-2} h^{-1} PAR) and rates of dark respiration at +2 °C and based on the chlorophyll *a* concentration of the thallus, of samples of the intermediate form between *M. tesselata* and *P. crispa*, collected in the Argentine Islands, Antarctica. Results were obtained in the same experiments as performed on *M. tesselata* (Fig. 32.1). r^2 of the regression on the rates of net photosynthesis=0.850. r^2 of the regression on the rates of dark respiration=0.846.

As the rates of dark respiration did not differ in the three experiments, the decrease in rates of net photosynthesis was not the result of an increased rate of respiration, pointing towards a stress response in the mycobiont.

As the differences in photosynthetic photon exposure do not cause differences in the rates of net photosynthesis in *P. crispa* or in the intermediate between *P. crispa* and *M. tesselata*, photosynthetic adaptation of the phycobiont to the light climate (Kershaw & MacFarlane 1980) is not likely to be the cause of the decrease in the rates of net photosynthesis.

An explanation might be found in a possible accumulation of photosynthates in the algal cells. Lichens photosynthesize when the thallus has a sufficient moisture content (>25–40% of thallus dry weight, Kappen 1985). Under natural circumstances, this situation may exist for only part of the day (Lange *et al.* 1970, Kappen *et al.* 1987, 1988), except under wet atmospheric conditions. The amount of photosynthates produced during a relatively short time period is apparently sufficient for succesful functioning of the phycobiont as well as of the mycobiont of the lichen. Whether the role of the mycobiont in the transfer of photosynthates from the phycobiont is active or passive is unclear (Richardson 1973).

When the hydrated state of the lichen thallus is experimentally prolonged while exposed to photosynthetic active radiation, more than the usual amount of photosynthates may be produced. The decrease in the rates of photosynthesis in the long and very long experiments with *M. tesselata* may indicate an accumulation of photosynthates (Kurssanov 1933). This accumulation may be caused by a slow transport rate of carbohydrates from the phycobiont to the mycobiont. But as the carbohydrates in the phycobiont differ from the soluble carbohydrates in the mycobiont, the rate of conversion may also be an inhibiting factor (Richardson 1973). The algal cells might react initially by a lowering of the rate of photosynthesis, but if this situation persists, the phycobiont may utilize the photosynthates for increased growth resulting in intermediate forms between the lichen and the alga. Support for this hypothesis may be found in the fact that the 6C-carbohydrate concentration in the lichen is increased by a factor of two over 72 hours when thallus moisture is experimentally prolonged. In the intermediate form the carbohydrate concentration remains stable over the same period and is about four to five times lower than in the lichen (Huiskes & De Klerk unpublished results).

If the photosynthetic period of the lichen is relatively short, accumulation of photosynthates can occur but after that the concentration of carbohydrates decreases subsequently to the original level at the start of the photosynthetic period (Huiskes & De Klerk unpublished results).

In the free-living *P. crispa*, the prolonged thallus hydration does not cause a decrease in the rates of photosynthesis nor does it cause accumulation of carbohydrates. We assume that in the free-living alga the photosynthates are used directly for the maintenance of existing cells and the production of more algal cells (growth and reproduction).

Under continuous imbibition in a climate room, *M. tesselata* thalli were observed to be 'greening up'. Keeping these thalli subsequently under comparatively dry circumstances reverted the process. Microscopic observations on the lichen samples kept for some time under continuous imbibition show that the clusters of algal cells are larger than those in the dark black-brown lichen thalli.

This 'greening up' is known from other lichen species as well (Dughi 1939, Jaag 1945). Jaag (1945) explained this by assuming that the mycobiont cannot exist in continously wet circumstances.

Based on the findings presented in this paper we propose the following hypothesis with respect to the dispersal and ecology of *M. tesselata* and *P. crispa*.

Under natural circumstances *P. crispa*, growing vigorously under favourable (moisture and light) conditions, is able to produce large amounts of akinetes (asexual diaspores developed by metamorphosis of vegetative cells; production of gametes by *P. crispa* is unconfirmed, Smith 1955). These akinetes will, after dispersal, develop into small *P. crispa* thalli. Young *P. crispa* thalli have been found on dry exposed rock faces, but their development is stunted (Kovacik, pers. commun.). Under these circumstances they may form an easy target for fungal attack, e.g. by *Guignardia* sp., and the algal thallus is fragmented by the fungal hyphae into clusters of algal cells, forcing a three-dimensional morphology, called *M. tesselata* (Henssen & Jahns 1974). The fungus may provide conditions that enable the algal cells to survive in an otherwise unsuitable environment and to produce sufficient photosynthates for the growth of both the algal clusters and the fungal hyphae under fluctuating hydration regimes.

If the environment becomes more moist again, e.g. by meltwater flowing over the thallus, the alga may react by increased production of carbohydrates which accumulate in the phycobiont and are subsequently utilized for increased growth of the phycobiont at the expense of the mycobiont. Intermediate forms between the lichen and the alga may then occur.

The results presented in this paper may support the argument that, in the case of *M. tesselata*, the symbiotic relationship between an alga and a fungus is dependent on the environmental moisture conditions.

ACKNOWLEDGEMENTS

The research reported in this paper was funded by the Netherlands Antarctic Research Programme 1989–1994, administered by The Netherlands Foundation for Marine Research (since 1994: Netherlands Geosciences Foundation). This is publication number 745 of the Netherlands Institute of Ecology, Centre for Estuarine and Coastal Ecology, Yerseke, The Netherlands.

The authors wish to thank the British Antarctic Survey (Cambridge) for logistic support and hospitality at Faraday Station. Special thanks are due to the inhabitants of Faraday for their assistance, especially to Mr Russell Manning, boatman in the 1990/91 summer season and as such responsible for our transport in the Argentine Islands archipelago.

We thank Messrs Almekinders, Hoekman and Haazen (NIOO-CEMO), for the construction of the incubator we used and for further technical assistance. Daniëlla De Klerk, trainee research technician, performed a number of experiments on the carbohydrate concentration of the species.

We acknowledge the inspiring discussions following presentation of our poster at the VIth SCAR Biology Symposium, Venice, especially with Dr Lubomír Kovacik, and the useful comments by Dr Marten Hemminga and Prof. Ludger Kappen on an early draft of the paper which improved the discussion section of this paper significantly.

REFERENCES

Ahmadjian, V. 1967. A guide to the algae occurring as lichen symbionts: isolation, culture, cultural physiology, and identification. *Phycologia*, **6**, 127–160.

Brodo, I. M. 1976. Lichenes canadenses exsiccati: fascicle II. *The Bryologist*, **79**, 385–405.

Davey, M. C. 1989. The effects of freezing and desiccation on photosynthesis and survival of terrestrial Antarctic algae and cyanobacteria. *Polar Biology*, **10**, 29–36.

Dughi, R. 1939. Domaine de la stabilité de la symbiose lichénique. L' énantiohygrie. *Comptes rendus hebdomadaires des séances de l' Académie des Sciences*, **208**, 2017–2019.

Gremmen, N. J. M., Huiskes, A. H. L. & Francke, J. W. 1994. Epilithic lichen vegetation of the Argentine Islands area, Antarctica. *Antarctic Science*, **6**, 463–471.

Henssen, A. & Jahns, H. M. 1974. *Lichenes. Eine Einfuhrung in die Flechtenkunde*. Stuttgart: Thieme Verlag, 467 pp.

Jaag, O. 1945. Untersuchungen über die Vegetation und Biologie der Algen des Nackten Gesteins in den Alpen, im Jura und im Schweizerischen Mittelland. *Beiträge zur Krypto gamenflora der Schweiz*, **9**, 1–560.

Jacob, A. 1992. Physiologie und Ultrastruktur der Antarktischen Grünalge *Prasiola crispa* ssp. *antarctica* unter Osmotischem Streß und Austrocknung. *Berichte zur Polarforschung*, **102**, 7–143.

Kappen, L. 1985. Water relations and net photosynthesis of *Usnea*. A comparison between *Usnea fasciata* (maritime Antarctic) and *Usnea sulphurea* (continental Antarctic) In Brown, D. H., ed. *Lichen physiology and cell biology*. New York: Plenum Publishing Company, 41–56.

Kappen, L., Bölter M. & Kühn, A. 1987. Photosynthetic activity of lichens in natural habitats in the maritime Antarctic. Progress and problems in lichenology in the eighties. *Bibliotheca Lichenologia*, **25**, 297–312.

Kappen, L., Meyer, M. & Bölter, M. 1988. Photosynthetic production of the lichen *Ramalina terebrata* Hook. f. et Tayl., in the maritime Antarctic. *Polarforschung*, **58**, 181–188.

Kershaw, K. A. 1985. *Physiological ecology of lichens*. Cambridge: Cambridge University Press, 286 pp.

Kershaw, K. A. & MacFarlane, J. D. 1980. Physiological–environmental interactions in lichens. X. Light as an ecological factor. *New Phytologist*, **84**, 687–702.

Kurssanov, A. L. 1933. Über den Einflusz der Kohlenhydrate auf den Tagesverlauf der Photosynthese. *Planta*, **20**, 535–548.

Lange, O. L. & Tenhunen, J. D. 1981. Moisture content and CO_2 –

exchange of lichens. II. Depression of net photosynthesis in *Ramalina maciformis* at high water content is caused by increased thallus carbon dioxide diffusion resistance. *Oecologia*, **51,** 426–429.

Lange, O. L., Schulze, E. D. & Koch, W. 1970. Experimentell-Ökologische Untersuchungen an Flechten der Negev-Wüste. III. CO_2 Gaswechsel und Wasserhaushalt von Krüsten- und Blattflechten am naturlichen Standort während der sommerliche Trockenperiode. *Flora*, **159,** 525–538.

Long, S. P. & Hällgren, J-E. 1993. Measurement of CO_2 assimilation by plants in the field and the laboratory. In Hall, D. O., Scurlock J. M. O., Bolhàr-Nordenkampf, H. R., Leegood, R. C. & Long, S. P., eds. *Photosynthesis and production in a changing environment: a field and laboratory manual*. London: Chapman and Hall, 129–167.

Matthes-Sears, U. & Nash III, T. H. 1986. A mathematical description of the net photosynthetic response to thallus water content in the lichen *Ramalina menziesii*. *Photosynthetica*, **20,** 377–284.

Printz, H. 1964. Die Chaetophoralen der binnengewasser. *Hydrobiologia*, **24,** 1–376.

Redon, J. 1985. *Liquenes Antarticos*. Santiago: Instituto Antartico Chileno, 123 pp.

Richardson, D. H. S. 1973. Photosynthesis and carbohydrate movement. In Ahmadjian, V. & Hale, M. S., eds. *The lichens*. New York: Academic Press, 249–288.

Smith, G. M. 1955. *Cryptogamic botany. I. Algae and fungi*, 2nd edn. New York: McGraw- Hill, 546 pp.

Snelgar, W. P., Green, T. G. A. & Wilkins, A. L. 1981. Carbon dioxide exchange in lichens: resistances to CO_2 uptake at different thallus water contents. *New Phytologist*, **88,** 353–361.

33 Crabeater seal cohort variation: demographic signal or statistical noise?

PETER L. BOVENG AND JOHN L. BENGTSON
National Marine Mammal Laboratory, Alaska Fisheries Science Center, National Marine Fisheries Service, 7600 Sand Point Way N.E., Bldg 4, Seattle, Washington 98115, USA

ABSTRACT

*This study considered whether the fluctuations in strengths of annual cohorts of crabeater seals (*Lobodon carcinophagus*) are genuine demographic phenomena ('signal') or artifacts ('noise') of sampling or analysis. The evidence was then examined for support of a periodic interpretation of the fluctuations.*

Age estimates were obtained from 2852 crabeater seals collected near the Antarctic Peninsula between 1964 and 1990. These 'catch-at-age' data were analysed by a maximum likelihood technique, to produce a time series of relative cohort strengths for the 1945–1988 cohorts. Monte Carlo techniques were used to assess the uncertainty of the cohort strength estimates and the power of the sampling scheme to detect true fluctuations. It was found that, if certain assumptions are correct, the relative cohort strengths are well determined by the data and therefore the fluctuations in cohorts are likely to contain signals from genuine demographic events. Time series modelling of the data, however, indicated that there is little support for interpreting the fluctuations in cohort strength as periodic. Several models, with differing implications for strength and frequency of periodicity, fit the data equally well. It remains, however, that the cohort strengths are strongly autocorrelated. This feature, common among ecological time series, poses difficulties for judging the significance of relationships between long-term data sets.

Key words: cohort analysis, *Lobodon carcinophagus*, autocorrelation, population cycles, Monte Carlo simulation.

INTRODUCTION

Crabeater seals (*Lobodon carcinophagus*), because of their great abundance, wide distribution and specialized trophic position, are useful subjects for studying variability in the Antarctic marine ecosystem. Previous studies have drawn attention to variability in the age structure of the population of crabeater seals inhabiting the sea ice and waters surrounding the Antarctic Peninsula (Laws 1984, Bengtson & Laws 1985, Testa *et al.* 1991). This variability occurs in the form of apparent fluctuations – termed periodic or quasi-cyclic – in the relative strengths of representation by the cohorts (year classes) born between the mid-1940s and the mid-1980s. Fluctuations in these cohort strengths have been compared with interannual fluctuations in crabeater seal reproductive parameters and historic whale catches (Bengtson & Laws 1985), and interannual fluctuations in Weddell seal (*Leptonychotes weddellii*) pupping rates, leopard seal (*Hydrurga leptonyx*) sightings and the El Niño/Southern Oscillation Index (Testa *et al.* 1991).

An adequate explanation of the causes of crabeater seal cohort variation has yet to emerge from comparisons with variation in these other data. Either the data used in these comparisons bear insufficient relation to the crabeater seal cohorts or the apparent fluctuations in cohorts reflect variability caused by factors such as random sampling events or analytical artifacts that should not be expected to relate to other biological or physical parameters of the ecosystem.

In the present study, new estimates of crabeater seal cohort strength are derived from the age data used in the previous studies plus additional samples obtained between 1988 and 1990. To assess better the prospects for relating this new series of cohort strengths to environmental data that might help to explain the cohort variation, tests are devised to evaluate the extent to which the variation derives from genuine demographic variation (signal), rather than variation caused by sampling or analytical effects (noise). Specifically, these tests address the statistical significance of the patterns observed in the cohort strengths, and the statistical power of the sampling and analytical procedures to detect genuine differences among cohort strengths. To evaluate whether the fluctuations are, in fact, periodic, time-series models are applied to the series of cohort strengths. Finally, the time-series characteristics of the crabeater seal cohort strengths are used to demonstrate certain difficulties in comparing series of ecological data to derive explanations of their variability.

DATA COLLECTION AND ANALYSIS

Estimation of cohort strengths

Age estimates (Fig. 33.1) were obtained by counting annuli in cementum and/or dentine of teeth (Laws 1958, 1962, McCann 1993) from 2852 crabeater seals collected near the Antarctic Peninsula and South Orkney Islands between 1964 and 1990 (Øritsland 1970, Laws 1977, 1984, Bengtson & Laws 1985, Bengtson unpublished data). Relative strengths of cohorts were estimated from the age distribution of the seals by a maximum likelihood technique (Testa *et al.* 1991, Boveng 1993). A likelihood function for the sample age distribution was derived assuming that (1) the age samples were obtained at random with respect to the population age structure (i.e. age samples follow multinomial distributions); (2) age-specific survival rates of seals older than one year were equal for all cohorts and followed a five-parameter competing risks model (Siler 1979, Barlow & Boveng 1991); (3) ages were known with certainty; and (4) there was no trend in crabeater seal population size. The resulting multinomial likelihood function was maximized – using the simplex method (Nelder & Mead 1965, Press *et al.* 1992) – over the age frequency distribution of seals greater than one year of age, because of a possibility that the first age class is under-represented in the samples (Siniff *et al.* 1979, Testa *et al.* 1991). The new estimates of cohort strength (Fig. 33.2) for the 1945–1988 cohorts are hereafter called the 'observed' cohort strengths.

Statistical reliability of cohort strengths

Tests to evaluate the statistical reliability (power and significance) of the observed cohort strengths were conducted by Monte Carlo techniques, sampling repeatedly from hypothetical, model age distributions and estimating cohort strengths from each sample. To assess the statistical power of the sampling scheme, a model age distribution was constructed under the hypothesis that the true pattern in cohort strengths was the same as in the observed series. This hypothesis, coupled with an assumption that survivorship of all cohorts followed the Siler model (estimated in association with the observed series), defined the age distribution of this hypothetical population for each year of the sampling period (1964–1990). This hypothetical population was then sampled 500 times, simulating as nearly as possible the actual sampling scheme used to obtain the original age data. Cohort strengths were estimated from each of the 500 samples, using the same maximum likelihood method that produced the observed series (Fig. 33.3).

The zone formed by the upper and lower bounds of the central 95% of Monte Carlo outcomes (Fig. 33.3) defined a narrow band around the observed series of cohort strengths; any other series of cohort strengths chosen within that zone would recapitulate the main pattern of the observed series (with the exception of the strength of the 1988 cohort, which is highly variable because it is estimated by only one age class in one sample year and that year, by random chance, may not have been included in some Monte Carlo samples). That the Monte Carlo outcomes followed the same pattern as the observed series indicates that if the assumptions are correct, the sampling scheme and analytical procedures were sufficiently powerful to detect cohort fluctuations of the type in the observed series.

The second aspect of statistical reliability – significance, or whether the cohort variation could have arisen as sampling artifacts – was addressed with two additional Monte Carlo tests. In the first, a hypothetical population age structure was constructed assuming there was no variation in true cohort strengths and that survivorship again followed the Siler model. This age structure was resampled 500 times, estimating cohort strengths from each sample as described above (Fig. 33.4). If the central 95% band for Monte Carlo cohort strengths from this scenario is assumed to represent a joint 95% confidence interval for the cohort strengths, 5% or about two to three, of the 44 estimated cohort strengths would be expected to fall outside the band by chance alone, under a null hypothesis of constant true cohort strengths. In fact, 18 values in the observed series of cohort strengths fell outside the band. The probability – from the binomial distribution – of 18 or more values outside the band is approximately 10^{-12}. Even if the central 95% of the Monte Carlo cohort strengths overestimated the true confidence interval by as much as 5%, the probability of 18 or more values outside the band would be only 10^{-8}. Thus, the observed variability was much greater than would be expected under constant cohort strengths. Because the true cohort strength and the survivorship were constant in this simulation, the variability indicated by the confidence bands is due entirely to the sampling process and serves to illustrate the amount of the variation in cohort strength estimates that could be due solely to the sampling scheme. Therefore, from these results, it is unlikely that fluctuations as large as those in the observed series of cohort

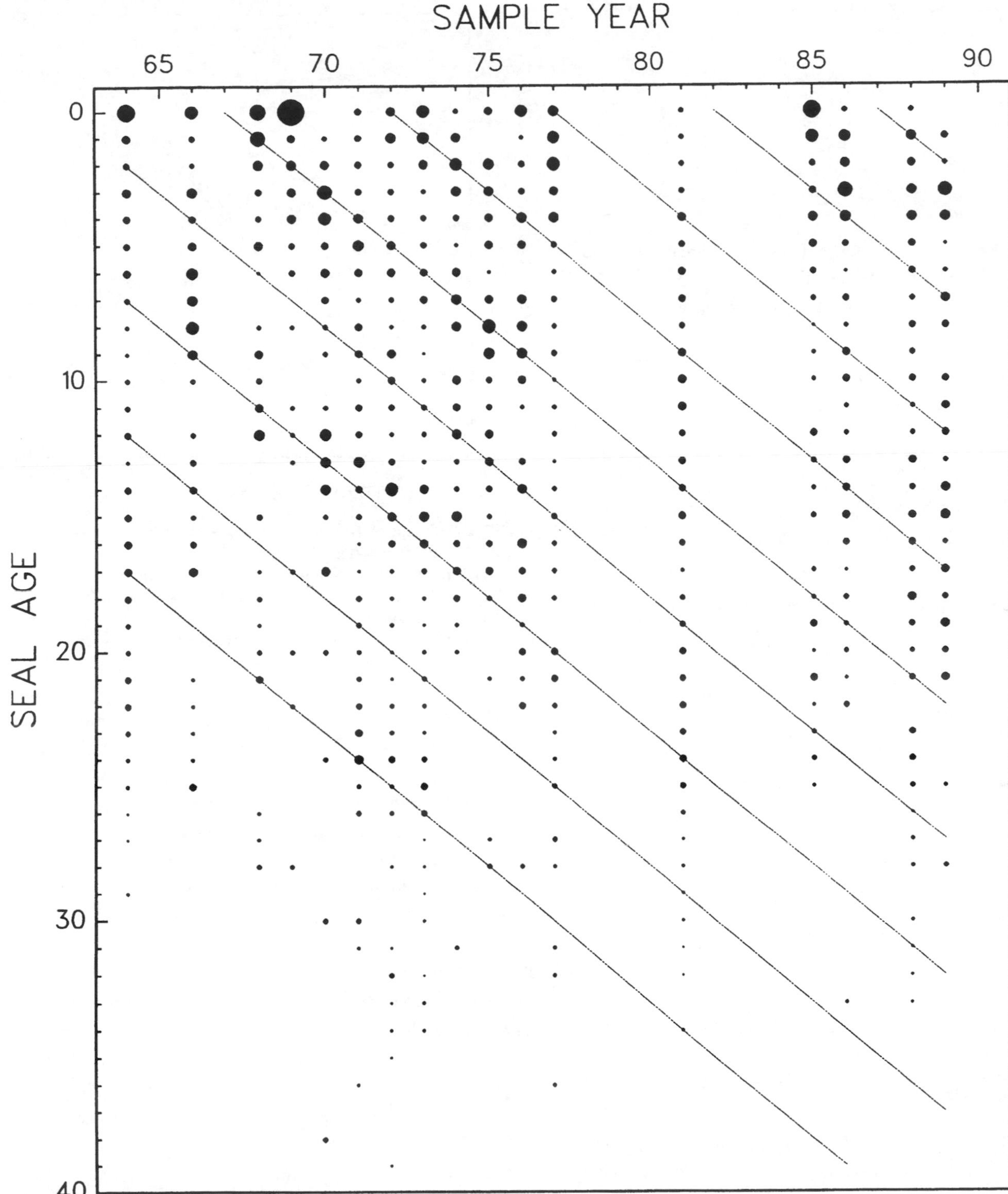

Fig. 33.1. Distributions of age estimates obtained from 2852 crabeater seals sampled between 1964 and 1990. Ages and sample years were assigned as if all collections were made on 1 November of each year. The area of the circle at each coordinate is proportional to the number of seals of a particular age, sampled in a particular year, normalized by the total number sampled in that year. Diagonal lines are added to assist the eye in following cohorts through time.

strengths could have resulted from a population with constant cohort strength.

The second Monte Carlo test for significance of the cohort fluctuations employed independent cohort strengths drawn from a log-normal distribution with variance equal to the variance among the observed cohort strengths. Cohort series drawn at random in this scenario would, on average, exhibit no autocorrelation among the cohorts. Each series drawn in this fashion (again, coupled with a survivorship model) defined a hypothetical population that could be resampled in the manner previously described. Five-hundred of these hypothetical populations were constructed and one simulated sample of the

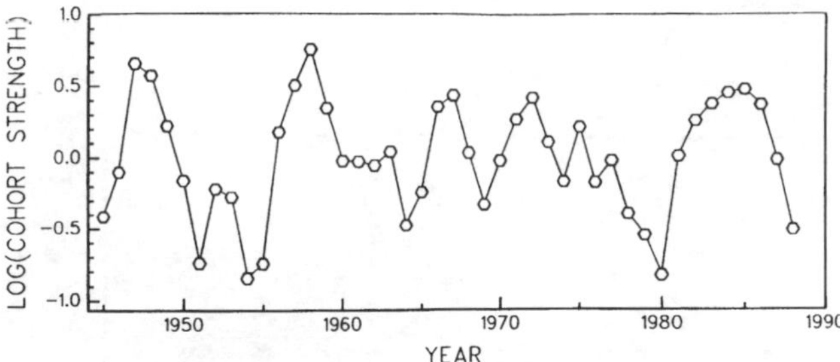

Fig. 33.2. Estimates of the relative strengths of crabeater seal cohorts born in the Antarctic Peninsula region (the 'observed' cohort series) for 1945–1988.

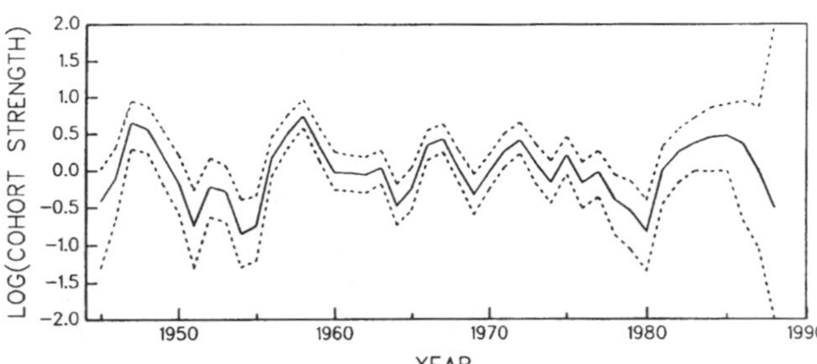

Fig. 33.3. The observed cohort strengths, with the central 95% of 500 Monte Carlo estimates of cohort strength (dashed lines) based on samples drawn from a hypothetical population with cohorts equal to the observed series and age-specific survivorship of all cohorts following a five-parameter Siler (1979) model. The area between the two dashed lines defines the zone within which cohort fluctuations could be expected to occur solely by chance.

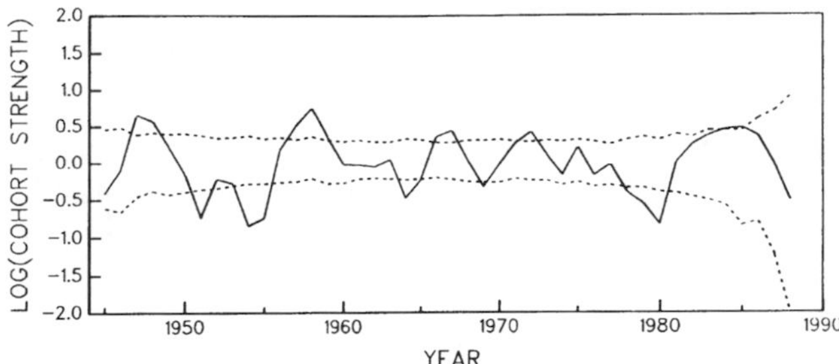

Fig. 33.4. The observed cohort series, with the central 95% of Monte Carlo estimates of cohort strength (dashed lines) from a hypothetical population with constant cohorts and age-specific survivorship following the Siler model.

age distribution was drawn from each. From each hypothetical cohort series, the 'true' first-order (1-year lag) autocorrelation and the first-order autocorrelation of the cohort strength estimates were computed. The first-order autocorrelations were chosen because the peak in the autocorrelation function of the observed cohort series was at a 1-year lag. Comparison of the distributions of autocorrelations in the 'true' and estimated series allowed testing of whether the first-order autocorrelation of the observed series could have arisen strictly from the sampling and/or estimation procedure.

The first-order autocorrelations of the 'true' and estimated cohort series were compared graphically by the quantile–quantile (Q–Q) method (Chambers *et al.* 1983). The nearly one-to-one relationship between the quantiles (Fig. 33.5) indicates that the

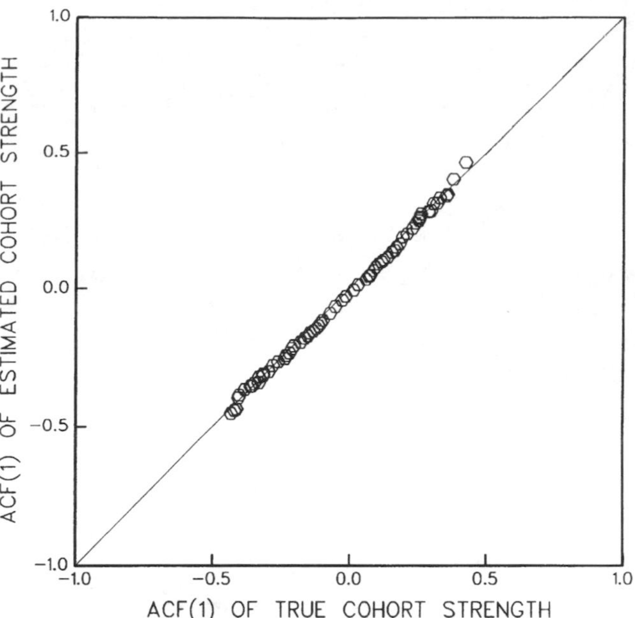

Fig. 33.5. Q–Q plot of 1-year-lag autocorrelations in 'true' versus estimated series of cohort strengths. The 'true' autocorrelations were measured in hypothetical populations defined by independently normally distributed cohort strengths. The estimated autocorrelations were measured in series obtained by sampling from the hypothetical populations, then forming maximum likelihood estimates of cohort strength and computing the autocorrelation coefficients.

model sampling and estimation procedures did not modify the distribution of first-order autocorrelations. The first-order autocorrelation of the observed series was 0.57; this value was greater than any of the 500 values obtained by Monte Carlo simulations of cohort series from an underlying model with zero autocorrelation, indicating that the autocorrelation in the observed cohort series is unlikely to have arisen by chance or as an artifact of the analytical procedure for estimating cohort strengths.

Periodicity in cohort strengths

The time-series characteristics of the cohort series were investigated by the Box–Jenkins ARIMA (autoregressive integrated moving average) model identification procedures (Box & Jenkins 1970), which make use of the autocorrelation and partial autocorrelation functions. ARIMA model coefficients were estimated using the Gaussian maximum likelihood routine (arima.mle) in S-Plus®† (Statistical Sciences, Inc. 1993). Alternative model formulations were compared using Akaike's Information Criterion (AIC) (Akaike 1974). A pure autoregressive model of order 7 (AR[7]) provided the best fit (i.e. minimum AIC) within a broad class of ARIMA models, but the fit was not significantly better than those from an AR[2] and an AR[3] model. The one-step-ahead predictions from the AR[7] model were mostly similar in pattern to the observed series of cohort strengths (Fig. 33.6).

To evaluate the strength and frequency of a periodic pattern in the observed cohort series, the power spectra (e.g. Shumway

† Reference to trade names does not imply endorsement by the National Marine Fisheries Service, NOAA.

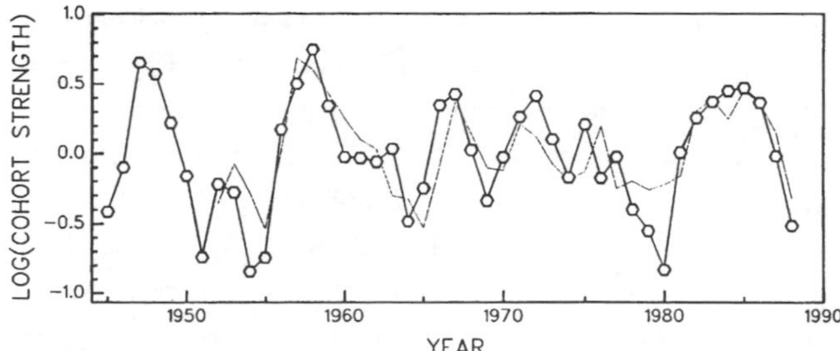

Fig. 33.6. One-step-ahead predictions of cohort strength from an autoregressive model of order 7 (dotted line), superimposed on the observed series of cohort strengths.

1988) of the three autoregressive models were investigated. The power spectrum of the AR[7] model had strong peaks at periods of 11 and 4.9 years. The AR[2] and AR[3] models had only a single, rounded peak at about 7 years. Thus, any particular periodic characteristic of the cohort series was not well supported. It remains, however, that a primary feature of the variation in the observed cohort series is its autocorrelation.

Implications of autocorrelation

To illustrate a potential problem in studies using correlation to compare two or more long-term series of ecological data (e.g. a demographic and an environmental series), a simulation was conducted. The AR[7] model of crabeater seal cohort strengths (Fig. 33.6) was used to generate 1000 random series, each consisting of 10 annual cohort strengths. Correlations (zero-lag) were formed between each of these series and 10 values from another time series that was chosen so as to have no mechanistic relationship to the crabeater seal cohorts. The time series chosen for this comparison was one based on market demand for double-knit polyester fabric (Montgomery & Johnson 1976; p. 269). Upon visual inspection, this series appeared to have fluctuations of about the same length as fluctuations in the crabeater seal cohorts and it was therefore chosen in much the same manner as one might (somewhat naïvely) choose an environmental parameter as a potential explanatory variable.

Using the standard test for significance of a correlation coefficient (e.g. Snedecor & Cochran 1980; p. 477), one would expect about 50 of the 1000 simulation trials to yield correlations greater in absolute magnitude than 0.632, if the two data sets are unrelated. In this simulation, however, 378 of the simulated correlations between crabeater seal cohort strengths and demand for double-knit polyester fabric exceeded 0.632 in absolute value, implying a greatly increased risk of falsely rejecting an hypothesis of no relationship between the series. This result is largely a consequence of the autocorrelation present in each of the two data sets (e.g. Yamaguchi 1986, Bence 1995).

DISCUSSION

Statistical reliability of cohort strengths

The series of cohort strengths estimated in this study was very similar to that estimated previously by Testa *et al.* (1991), despite a sample size that was approximately 25% larger than the one used by those authors. This similarity of results suggests that the estimates are robust, which is further supported by the results of the Monte Carlo tests concerning the statistical reliability of the cohort strength estimates.

Results of the Monte Carlo tests were consistent with a hypothesis that the variation in the observed cohort strengths reflects genuine demographic variation (signal) in the crabeater seal population rather than random sampling or analytical artifacts (noise). That is, the first Monte Carlo test indicated that the estimates were well determined by the sample age structure. The second Monte Carlo test indicated that the magnitude of variation in the cohort strength estimates was greater than would be expected from a population with no variation in cohort strength. The third Monte Carlo test indicated that although substantial variation could be due to random sampling events, there was a coherence among adjacent year classes (autocorrelation) that is unlikely to be due to chance or to the estimation procedure. Therefore, it seems highly unlikely that the features interpreted by this and previous studies as variation in crabeater seal cohort strengths could have arisen by chance from a stable (age) population or as artifacts of analysis, provided the assumptions are not violated.

One assumption, implicit in the maximum likelihood technique used to obtain estimates of cohort strength, is that ages of the crabeater seals are known with certainty. In reality, of course, there are precision errors (and possibly bias errors) associated with the age estimates. Precision errors in age estimates are known to reduce variability in cohort strength estimates and may introduce autocorrelation (Rivard 1989, Bradford 1991). Boveng (1993) investigated the effects of, and partially corrected for, precision errors in the crabeater seal age estimates. Further analyses are needed, however, to assess the extent to which autocorrelation in the crabeater seal cohort series may be due to errors in age estimation. Those analyses notwithstanding, the results of the present study support the conclusion that the observed series of cohort strengths contains signals from genuine demographic variability and that further investigation of the causes and characteristics of fluctuations in crabeater seal cohorts is warranted.

Periodicity in cohort strengths

Much effort has been devoted to the study of cycles in vertebrate populations, most notably several species of mammals and birds in the North American tundra and boreal forest (reviewed in Akçakaya 1989, Sinclair *et al.* 1993). Those studies have been based upon relatively long time series exhibiting many cycles of about 10 years duration, so there can be little doubt that the fluctuations are nearly periodic. There has, however, been considerable debate about the period of the cycles and the causes. The periodicity in those population time series has strongly influenced efforts to explain the variation and to relate it to changes in other ecological parameters.

Variability in crabeater seal cohorts has also been suggested to be cyclic (Laws 1984, Bengtson & Laws 1985, Testa *et al.* 1991), with a period of approximately 4–5 years. It must be noted, however, that the crabeater seal time series is relatively short and that an apparent cycle of this duration could be expected to arise by chance from non-cyclic variation much more readily than say, a 10 year cycle. Several time-series models, with strikingly different implications for strength and periodicity of cycles, fit the cohort series equally well. Thus, despite the indications from the Monte Carlo tests that there is genuine variability in crabeater seal cohorts, the data do not presently support interpreting these fluctuations as being truly periodic. Therefore, in attempting to explain the fluctuations in crabeater seal cohorts, no particular emphasis should be given a priori to hypotheses or data that exhibit cyclic behavior. It is, however, important to take account of the autocorrelation among cohort strengths.

As noted previously, estimates of cohort strength may exhibit autocorrelation as an artifact of imprecise age estimation, but autocorrelation may also be a genuine feature of a cohort series. Environmental events that affect production and survival of the young-of-the-year class may also affect survival of other age groups, especially juveniles, leading to autocorrelated cohort strengths. Testa (1990) illustrated this phenomenon – wherein the effect of a poor (or a favourable) year is spread over several adjacent cohorts – with an age-structured, density-dependent population model for crabeater seals. Although the phenomenon has not been observed directly in crabeater seals, observations from other pinnipeds (e.g. Trillmich *et al.* 1991) and from seabirds (e.g. Wooller *et al.* 1992) indicate it is reasonable to expect that crabeater seals and perhaps most higher predator species will exhibit autocorrelated cohort strengths in a fluctuating environment such as the Southern Ocean.

Implications of autocorrelation

As the availability of long-term biological data from Antarctica increases, so will efforts to establish links between the biological and the environmental components of the ecosystem (e.g. Rounsevell 1988, Testa *et al.* 1991, Fraser *et al.* 1992, Chastel *et al.* 1993). These long-term biological data will usually exhibit autocorrelation, in some cases because of strong links to autocorrelated environmental features, but in other cases because of mechanisms largely unrelated to the environment. As the example of correlations between crabeater seal cohorts and fabric demand shows, applying standard tests for significance of correlations between autocorrelated series can lead to false impressions of relationships (for another example, see Yamaguchi 1986). The effective number of independent observations is smaller than the actual number in an autocorrelated series (Bayley & Hammersley 1946), and therefore the risk of incorrectly identifying significant correlations decreases as the number of fluctuations or 'events' in the data series increases. Thus, if a parameter typically goes through fluctuations of say, 3 to 4 years duration, a series 10 years in length will be insufficient to gauge reliably correlations with other parameters (using standard tests), but a series of several decades may suffice. These properties of autocorrelated time series have been widely recognized in the fields of statistics and econometrics, but much less so in ecology (Hurlbert 1984, Jassby & Powell 1990, Bence 1995). Several strategies have emerged for mitigating the effects of autocorrelation on tests of relationships between two series. These include adjusting the variance or degrees of freedom of an estimator (e.g. Bartlett 1966), fitting ARIMA models to the series and then comparing the (independent) residual series (e.g. Haugh 1976, Yamaguchi 1986), and using likelihood estimates that explicitly account for intra-series correlation (Bence 1995). However, even these techniques will often lead to overconfidence in the strength of relationships between data sets (Bence 1995). In general, such comparisons should be approached with caution, and with due consideration for the appropriate models of temporal variability and the implications of the models' properties for statistical tests.

ACKNOWLEDGEMENTS

We gratefully acknowledge the dependence of this work upon data generously shared by R. M. Laws, T. Øritsland, D. B. Siniff and T. J. Härkönen and upon the many persons involved in field collections and laboratory analysis of the tooth samples. M. K. Schwartz provided valuable database assistance. J. W. Testa, D. P. DeMaster and an anonymous reviewer provided helpful comments. This work was supported by NOAA's National Marine Fisheries Service as part of its Antarctic Marine Living Resources (AMLR) Program, by the National Science Foundation's Office of Polar Programs (grants OPP75-17719, DPP80-20087 and DPP84-20851), and by collaborations with the British Antarctic Survey and the Swedish Antarctic Research Program.

REFERENCES

Akaike, H. 1974. A new look at statistical model identification. *IEEE Transactions in Automation and Control*, **AC-19,** 716–723.

Akçakaya, H. R. 1989. *Population cycles of mammals: theory and evidence*. PhD thesis, State University of New York, 176 pp.

Barlow, J. & Boveng, P. 1991. Modeling age-specific mortality for marine mammal populations. *Marine Mammal Science*, **7,** 50–65.

Bartlett, M. S. 1966. *An introduction to stochastic processes*, 2nd edn. Cambridge: Cambridge University Press, 362 pp.

Bayley, G. V. & Hammersley, J. M. 1946. The effective number of independent observations in an autocorrelated time series. *Journal of the Royal Statistical Society*, **B8,** 184–197.

Bence, J. R. 1995. Analysis of short time series: correcting for autocorrelation. *Ecology*, **76,** 628–639.

Bengtson, J. L. & Laws, R. M. 1985. Trends in crabeater seal age at maturity: an insight into Antarctic marine interactions. In Siegfried, W. R., Condy, P. R. &. Laws, R. M., eds. *Antarctic nutrient cycles and food webs*. Berlin & Heidelberg: Springer-Verlag, 669–675.

Boveng, P. L. 1993. *Variability in a crabeater seal population and the marine ecosystem near the Antarctic Peninsula*. PhD thesis, Bozeman, Montana State University, 110 pp.

Box, G. E. P. & Jenkins, G. M. 1970. *Time series analysis, forecasting, and control*. San Francisco: Holden Day, 553 pp.

Bradford, M. J. 1991. Effects of ageing errors on recruitment time series estimated from sequential population analysis. *Canadian Journal of Fisheries and Aquatic Sciences*, **48,** 555–558.

Chambers, J.M., Cleveland, W.S., Kleiner, B. & Tukey, P.A. 1983. *Graphical methods for data analysis*. Pacific Grove, CA: Wadsworth and Brooks/Cole Publishing, 395 pp.

Chastel, O., Weimerskirch, H. & Jouventin, P. 1993. High annual variability in reproductive success and survival of an Antarctic seabird, the snow petrel *Pagodroma nivea*: a 27 year study. *Oecologia*, **94,** 278–285.

Fraser, W. R., Trivelpiece,W. Z., Ainley, D. G. & Trivelpiece, S. G. 1992. Increases in Antarctic penguin populations: reduced competition with whales or a loss of sea ice due to environmental warming? *Polar Biology*, **11,** 525–531.

Haugh, L. D. 1976. Checking the independence of two covariance-stationary time series: a univariate residual cross-correlation approach. *Journal of the American Statistical Association*, **71,** 378–385.

Hurlbert, S. H. 1984. Pseudoreplication and the design of ecological field experiments. *Ecological Monographs*, **54,** 187–211.

Jassby, A. D. & Powell, T. H. 1990. Detecting changes in ecological time series. *Ecology*, **71,** 2044–2052.

Laws, R. M. 1958. Growth rates and ages of crabeater seals, *Lobodon carcinophagus* Jacquinot and Pucheran. *Proceedings of the Zoological Society of London*, **130,** 275–288.

Laws, R. M. 1962. Age determination of pinnipeds with special reference to growth layers in the teeth. *Zeitschrift für Säugetierkunde*, **27,** 129–146.

Laws, R. M. 1977. The significance of vertebrates in the Antarctic marine ecosystem. In Llano, G. A., ed. *Adaptations within Antarctic ecosystems.* Washington, DC: Smithsonian Institution, 411–438.

Laws, R. M. 1984. Seals. In Laws, R. M., ed. *Antarctic ecology*, Vol. 2. London: Academic Press, 621–715.

McCann, T. S. 1993. Age determination. In Laws, R. M., ed. *Antarctic seals: research methods and techniques*. Cambridge: Cambridge University Press, 199–227.

Montgomery, D. C. & Johnson, L. A. 1976. *Forecasting and time series analysis*. New York: McGraw-Hill, 299 pp.

Nelder, J. A. & Mead, R. 1965. A simplex method for function minimization. *Computer Journal*, **7,** 308–313.

Øritsland, T. 1970. Sealing and seal research in the south-west Atlantic pack ice, Sept.-Oct. 1964. In Holdgate, M. W., ed. *Antarctic ecology*, Vol. 1. London: Academic Press, 367–376.

Press, W. H., Teukolsky, S. A., Vetterling, W. T. & Flannery, B. P. 1992. *Numerical recipes*, 2nd edn. New York: Cambridge University Press, 963 pp.

Rivard, D. 1989. Overview of the systematic, structural, and sampling errors in cohort analysis. In Edwards, E. F. & Megrey, B., eds. *Mathematical analysis of fish stock dynamics. American Fisheries Society Symposium*, No. 6. Bethesda, MD: American Fisheries Society, 49–65.

Rounsevell, D. 1988. Periodic irruptions of itinerant leopard seals within the Australasian sector of the Southern Ocean, 1976–86. *Papers and Proceedings of the Royal Society of Tasmania*, **122,** 189–191.

Shumway, R. H. 1988. *Applied statistical time series analysis*. Englewood Cliffs, NJ: Prentice Hall, 379 pp.

Siler, W. 1979. A competing risk model for animal mortality. *Ecology*, 60, 750–757.

Sinclair, A. R. E., Gosline, J. M., Holdsworth, G., Krebs, C. J., Boutin, S., Smith, J. N. M., Boonstra, R. & Dale, M. 1993. Can the solar cycle and climate synchronize the snowshoe hare cycle in Canada? Evidence from tree rings and ice cores. *American Naturalist*, **141,** 173–198.

Siniff, D. B., Stirling, I., Bengtson, J. L. & Reichle, R. A. 1979. Social and reproductive behavior of crabeater seals (*Lobodon carcinophagus*) during the austral spring. *Canadian Journal of Zoology*, **57,** 2243–2255.

Snedecor, G. W. & Cochran, W. G. 1980. *Statistical methods*, 7th edn. Ames, IA: Iowa State University Press, 507 pp.

Statistical Sciences Inc. 1993. *S-Plus for Windows reference manual, version 3.1.* Seattle, WA: Statistical Sciences, Inc.

Testa, J. W. 1990. A simulation of the age structure of crabeater seals in a fluctuating environment. In Kerry, K. R. & Hempel, G., eds. *Antarctic ecosystems: ecological change and conservation*. Berlin & Heidelberg: Springer-Verlag, 246–252.

Testa, J. W., Oehlert, G., Ainley, D. G., Bengtson, J. L., Siniff, D. B., Laws, R. M. & Rounsevell, D. 1991. Temporal variability in Antarctic marine ecosystems: periodic fluctuations in the phocid seals. *Canadian Journal of Fisheries and Aquatic Sciences*, **48,** 631–639.

Trillmich, F., Ono, K. A., Costa, D. P., DeLong, R. L., Feldkamp, S. D., Francis, J. M., Gentry, R. L., Heath, C. B., LeBoeuf, B. J. ,Majluf, P. & York, A. E. 1991. The effects of el Niño on pinniped populations in the eastern Pacific. In Trillmich, F. & Ono, K. A., eds. *Pinnipeds and El Niño: responses to environmental stress*. Berlin & Heidelberg: Springer-Verlag, 248–270.

Wooller, R. D., Bradley, J. S. & Croxall, J. P. 1992. Long-term population studies of seabirds. *Trends in Ecology and Evolution*, **7,** 111–114.

Yamaguchi, D. K. 1986. Interpretation of cross correlation between tree-ring series. *Tree-ring Bulletin*, **46,** 47–53.

IV Adaptative mechanisms

INTRODUCTION

Today, the classic statement: 'organisms survive because they are adapted' appears tautological. Nevertheless, the underlying relationship between survival and adaptation is quite clear: adaptation is the ability of organisms to cope with environmental features and their changes. The greater the level of this adaptation, the greater the potential to survive and reproduce.

In one of the previous Symposia (Llano 1977), adaptations were classified as physiological and ecological. Nowadays, a more satisfactory classification, supported by our improved understanding of the mechanisms involved, might be one based on the distinction between physiological and genetic adaptations. The former are those which work through phenotypic stability and plasticity; the latter are those based on mechanisms which operate by replacing, through selection, genotypes conferring a lower degree of fitness with others conferring a higher degree of fitness. In any case, the adaptive nature of a character must always be demonstrated rather than presumed.

The theme of adaptive strategies in Antarctica is perhaps the central one of Antarctic biology. Present studies encompass further developments of existing techniques as well as the rapidly expanding application of new techniques such as molecular biology.

The marine environment is cold and stable, promoting a very different suite of adaptive features to those found in terrestrial organisms (Clarke 1990). Fish are the most extensively investigated marine group and di Prisco (Chapter 34), reviewing the biochemical and physiological specializations, pays special attention to oxygen transport mechanisms and to respiration processes.

The problem of the mechanisms utilized to compensate for the decrease in enzyme catalytic rates at low temperatures has been tackled by Ciardiello *et al.* (Chapter 35). In the blood of Antarctic fish, the enzyme G6PD seems to have acquired particular features related to metabolic rate compensation. The authors describe the properties and adaptive significance of this enzyme from channichthyid and nototheniid fish. The properties of another enzyme, superoxide dismutase from *Pagothenia bernacchii,* were studied by Cassini *et al.* (Chapter 36). Whilst protein structure here appears generally conservative substitutions in amino acid sequences, which suggest a high evolutionary rate, have also been observed.

Antarctic fish are also the subject of a study by Coscia & Oreste (Chapter 37) on the molecular structure and antibody specificity of immunoglobulins (Ig). The results contribute also to the clarification of the role of immunoglobulins in the immune response to parasitic nematodes. Another paper has referred to the ecological significance of fat accumulation in young fish. Hubold & Hagen (Chapter 38), in a very detailed study of the diet of *Pleuragramma*, conclude that fat is needed in this species, which has no swim bladder, to provide buoyancy as compensation for the gain in weight with age. Fat acquisition is finely tuned to the zooplankton cycles of abundance, showing how complex adaptive features can be in Antarctic fish.

The question of lipid acquisition and accumulation is also treated by Mayzaud (Chapter 39) in his study of krill populations. Biochemical analysis of lipid composition and the discussion of possible uses of the different kinds of lipids is then related to their supposed function in development and reproduction, and to energy metabolism. The author concludes that the evidence at present is still insufficient to elucidate the adaptive value of the different lipid types available.

Krill has been the subject of a study by Vetter *et al.* (Chapter 40) on certain specific enzymes involved in the cellular supply of energy. In particular, the functional properties of citrate-synthetase (CS) from Antarctic krill have been compared with those of the same enzyme in other crustaceans from temperate and Arctic regions. The differences between species can be seen as adaptations to the ambient temperature changes with regard to the eury- and stenothermic tolerance of these crustaceans. It has also been suggested that other enzymes, such as NAGase, are less influenced by temperature and more by moulting hormones and feeding habits.

Adaptations can be found at the levels of cell organelles. Microtubular structures develop in the ciliate *Euplotes* for swimming, reproducing and preserving the body architecture. Miceli *et al.* (Chapter 41), in a thorough study of the structure of the tubulin genes in *E. focardii*, could show the existence of four distinct isoforms in the expressed genomes of the somatic nucleus (macronucleus). In a detailed analysis of sequence and transcription mechanisms the tubulin sequence shows structural traits related in particular to temperature.

A different approach, using the perspective of chemical ecology, was taken by Capasso *et al.* (Chapter 42) in work on *Sterechinus neumayeri.* The authors describe the biochemical properties of zinc-binding polypeptides and metallothioneines, and the possible adaptive value of these compounds as detoxifying agents and in the control of the Zn and Cu metabolism in this sea urchin. These polypeptides were also found in other Antarctic organisms, implying that this may be a generalized mechanism for the metabolism of these metals. On the other hand, Slattery & McClintock (Chapter 43) studied the ecological adaptive implications of the biosynthesis of toxic compounds in three species of soft corals. Their findings indicate that predation pressure can be neutralized by these chemical defenses and suggest that they may also function as antifoulants against bacteria and diatoms.

On a very different scale, Block (Chapter 44) gives a very thorough analysis of the adaptive features of two species of terrestrial invertebrates with special emphasis on the micro-environmental variability of temperature and moisture. In the case of the mite *Alaskozetes* desiccation stress induces biosynthesis of glycerol, which functions as an antifreeze and allows supercooling of the body fluids, whilst the springtail *Cryptopygus* produces a multi-component cryoprotectant to enhance adult survival during winter. Block suggests that these species will be able to colonize new recently deglaciated areas because of these ecophysiological adaptations. In a comparison of life-cycle strategies in arthropods and bryophytes Convey (Chapter 45) finds that a key feature of reproductive success is flexibility in responding to positive environmental periods whilst maintaining the ability to survive unfavourable seasons without wasting scarce resources on unsuccessful reproduction. Both of these papers underline the increasing evidence that survival in terrestrial Antarctic habitats does not require any special adaptations but depends on the exploitation in different ways of widely held physiological, biochemical and behavioural characteristics.

The adaptive value of the ontogenetic changes in the diving behaviour of Weddell seal pups is discussed by Moss & Testa (Chapter 46). Within three months, the pups of this species are able to gradually improve their diving performance in duration, depth and frequency. These diving abilities are tied to the diel cycles of vertical migration of the prey. The ontogentic physiological changes described allow the pups to obtain their food in sufficient quantities to face the high energy demand of their metabolism and the seasonal variation of food availability.

Motivated by the recent declines in numbers of southern elephant seals, Burton *et al.* (Chapter 47) compiled data of weaning mass of pups from five sub-Antarctic islands. They postulate that the weaning mass of pups is a function of the fat reserves of the mother and propose that the resulting differences may be attributed to food availability for the mothers.

REFERENCES

Clarke, A. 1990. Temperature and evolution: Southern Ocean cooling and the Antarctic marine fauna. In Kerry, K. R. & Hempel, G., eds. *Antarctic ecosystems.* Berlin: Springer-Verlag. pp. 9–22 .

Llano, G. A., ed. 1977. *Adaptations within Antarctic ecosystems.* Washington DC: Smithsonian Institution.

34 Physiological and biochemical adaptations in fish to a cold marine environment

GUIDO DI PRISCO

Institute of Protein Biochemistry and Enzymology, CNR, Via Marconi 10, 80125 Naples, Italy

ABSTRACT

The progressive cooling of Antarctic seawater gradually induced in the fish fauna physiological and biochemical adaptations, which are at the extremes found among the vertebrates. Some adaptations essentially characterize Antarctic fish as absolutely unique (e.g. freezing avoidance; efficient protein polymerization and enzymatic catalysis at low temperature). Fish haematological features have also become highly specialized. The temperature-induced increase in blood viscosity is counteracted by drastic reduction of erythrocytes and haemoglobin (Hb); 'icefishes', grouped in one of the six families of the highly endemic suborder Notothenioidei, which dominates the fish fauna of the shelf and upper slope, are totally devoid of this haemoprotein. Some of the adaptive specializations of Antarctic fish, i.e. neutral buoyancy, freezing avoidance, polymerization of tubulins and actins, enzyme catalysis, haematological parameters and the oxygen transport system, are reviewed. The latter topic, dealing with basic tools for studying the physiology and biochemistry of a vital biological function such as respiration, has been analysed in some detail. It is suggested that the search of correlations, based on molecular data, between the various forms of adaptation and fish lifestyle will be a very important development of future biological research.

Key words: Antarctic coastal zone, fish, Notothenioidei, adaptation, oxygen transport, haemoglobin.

INTRODUCTION

The supercontinent Gondwana, of which Antarctica was the central portion, existed since the late Precambrian (590 Ma) and remained intact for almost 400 Ma, during the Paleozoic and part of the Mesozoic, through the Jurassic; then fragmentation began and continued during the Cretaceous. The continental drift took Antarctica to its present position about 65 Ma, at the beginning of the Cenozoic. Final separation of East Antarctica from Australia occurred around 38 Ma, in the Eocene–Oligocene transition; that of West Antarctica from South America occurred 22–25 Ma, in the Oligocene–Miocene transition. The opening of the Drake passage gave rise to the Antarctic Convergence (or Antarctic Polar Front). Although sea ice may have already been present at the end of the Eocene (40 Ma), it has been suggested that extensive ice sheets were formed, with periodic occurrence every 1–3 Ma, only after the middle Miocene (14 Ma), and that the latest expansion of the ice sheet, with progressive cooling leading to the present climatic conditions, began 2.5 Ma, in the Pliocene.

Paleogeographic events are the necessary background for understanding the evolutionary history and the adaptations of fish to the current environmental conditions of the Antarctic Ocean. The following references will provide comprehensive information on the palaeogeography of Gondwana (King 1958, Adie 1970, Craddock 1970, Frakes & Crowell 1970, Kennett 1977, Harwood 1987, Pickard *et al.* 1988, Webb 1990, Parrish 1990, Eastman 1991, 1993). The fossil records reveal that these variations produced many diversified forms of terrestrial and aquatic life (see Eastman 1991).

Table 34.1. *Numbers of species in the families of Notothenioidei*

Family	Antarctic species	Non-Antarctic species	Total
Bovichtidae	1	10	11
Nototheniidae	34	15	49
Harpagiferidae	6	0	6
Artedidraconidae	24	0	24
Bathydraconidae	15	0	15
Channichthyidae	15	0	15
	95	25	120

THE ANTARCTIC FISH FAUNA

The origin of the first teleosts can be traced in the Jurassic, approximately 200 Ma. These fish continued to evolve through the Cretaceous and, until the late Eocene, were quite cosmopolitan. In contrast, the modern Antarctic fish fauna (exclusively marine) is largely endemic and, unlike the populations of the other continental shelves, is dominated by a single group: the suborder Notothenioidei, which comprises 120 species in total. Of the 174 species living on the shelf or upper slope of the Antarctic continent, 95 (55%) are notothenioids. Fifty-three of the remaining 79 belong to families originating in the northern part of the Pacific Ocean (Liparididae and Zoarcidae); other fish typical of the boreal hemisphere are virtually absent (Gon & Heemstra 1990).

No fossil record of Notothenioidei is as yet available, leaving a void of 38 million years from the Eocene to the present. There is a lack of information about their site of origin, on the existence of a transition fauna and on the time of their radiation in the Antarctic. However, indirect indications suggest that notothenioids appeared in the early Tertiary, filling the ecological void on the shelf left by most of the other fish fauna (which experienced local extinction during maximal glaciation), and began to diversify in the middle Tertiary. Reduced competition and increasing isolation favoured speciation. Notothenioids fill a varied range of ecological niches normally occupied by taxonomically diverse fish communities in temperate waters. The suborder comprises six families (Table 34.1). Only one of 11 species of Bovichtidae, the most primitive notothenioid family, lives south of the Antarctic Polar Front; 15 of 49 Nototheniidae species are non-Antarctic. Harpagiferidae, Artedidraconidae, Bathydraconidae and Channichthyidae are all Antarctic (Gon & Heemstra 1990). Notothenioids are red-blooded, with the remarkable exception of Channichthyidae, the only know vertebrates whose pale whitish blood is devoid of haemoglobin (Hb) and has a greatly reduced number of erythrocyte-like cells (Ruud 1954, Hureau *et al.* 1977).

The evolutionary adaptation of Antarctic fish included physiological and biochemical specializations, some of which characterize these organisms as unique. The ensemble of these specializations was developed in the last 20–30 Ma, during increasing isolation in the cooling southern seas. These fish are finely adjusted to the environment and intolerant of warmer temperatures, thus an increase in temperature of only a few degrees centigrade has lethal effect (Somero & DeVries 1967). They live in isolation south of the Antarctic Polar Front, a natural barrier to migration in both directions and thus a key factor for fish evolution. In this habitat, fish had to cope with specific conditions, e.g. temperatures below the freezing point of the body fluids and high oxygen pressures. Some of the adaptive specializations developed in order to overcome the limiting effects of low temperature are: neutral buoyancy; freezing avoidance; aglomerular kidneys; preference for some metabolic pathways for selecting metabolic fuels (unsaturated lipids rather than carbohydrates); optimization of the haematological, enzymatic, muscular, membrane and nervous systems (although there are alternative explanations for some of these adaptations, see Eastman 1993, p.278). In addition to the low and constant temperature, seasonality is another factor of primary importance, which may influence biological processes such as feeding, growth and reproduction (Clarke & North 1991). It does not apply so much to temperature, but other variables (e.g. light and productivity) show marked seasonal fluctuations. In most species adaptation has established an energy balance between high costs in some processes (e.g. antifreeze biosynthesis and reproduction) and savings in others (e.g. activity and growth).

A comprehensive review of the broad field of fish adaptations and life strategies would be overlong. Also, the constraints of the Antarctic marine environment have produced evolutionary responses at all levels (organismal, organic, cellular, molecular) of life organization. Therefore only a few examples of adaptation have been selected for consideration. Furthermore, in molecular terms, although all macromolecules fulfil a key role, only specializations involving the structure and function of proteins will be discussed. A number of comprehensive sources of basic concepts and detailed information will integrate the reader's knowledge (Clarke 1983, Hochachka & Somero 1984, Macdonald *et al.* 1987, di Prisco *et al.* 1988a, 1991a, di Prisco 1991, Gon & Heemstra 1990, Eastman 1993).

SOME ADAPTIVE SPECIALIZATIONS

Neutral buoyancy

All Antarctic fish lack the swim bladder, which would produce neutral buoyancy and allow energy saving during locomotion and displacement in the water column since a neutrally buoyant fish is weightless. In notothenioids (a primarily benthic group, although some species have become seasonally or even permanently adapted to pelagic life) evolutionary compensation for the lack of a swim bladder induced modifications in a variety of body systems, facilitating rapid vertical migration with maximal energy conservation. This was attained through a combination of strategies of high efficiency in the Antarctic marine habitat (DeVries & Eastman 1978, Eastman & DeVries 1982, Eastman 1993), i.e. reduction of bone and scale mineralization (actually, many species are scaleless); substitution of bone with cartilage (less dense); storage of lipid (providing static lift); and production of a subdermal layer of watery, gelatinous tissue. Two pelagic notothenioids (*Pleuragramma antarcticum* and *Dissostichus mawsoni*), and possibly a third one (*Aethotaxis mitopteryx*), are neutrally buoyant. In general, benthic and cryopelagic notothenioids are lighter than expected and are able to perform easy vertical displacement in the water column.

Freezing avoidance

Depending on depth, the water temperature in the coastal part of the Antarctic Ocean varies between −1 °C and −2 °C. The average year-round value is −1.87 °C, the equilibrium temperature of ice and seawater, well below the freezing temperature (−0.8 °C) of a typical marine teleost hyposmotic to seawater (Black 1951). A few species avoid freezing by supercooling but the supercooled state is metastable and requires absence of contact with ice. Although in deeper water ice formation is impaired by the hydrostatic pressure effect (Littlepage 1965), currents may nevertheless carry ice crystals, making supercooling a dangerous strategy. It should be recalled that freezing, even partial, invariably causes death (Scholander *et al.* 1957).

In Notothenioidei, freezing is avoided by lowering the freezing point of blood and other tissue fluids. About one half of the freezing point depression is mostly due to NaCl; the other half is provided by solutes in the colloidal fraction of the fluid, which exert their effect non-colligatively. In most Antarctic fish species these solutes are glycopeptides with molecular mass ranging from 2600 to 33 700 Da, which contain a repeating unit of three amino acid residues in the sequence $[\text{Ala-Ala-Thr}]_n$; a disaccharide is linked with each Thr. In the lighter glycopeptides, Pro periodically substitutes Ala at position 1 of the tripeptide. Antifreeze glycopeptides are synthesised year-round in the liver, secreted into the circulatory system and then distributed into the extracellular fluids, where their concentration approaches 3.5%.

Some non-notothenioid species possess other peptides with antifreeze activity. In the Arctic, some fish species have antifreeze peptides whilst others have glycopeptides similar in structure to the Antarctic ones, posing interesting questions on molecular evolution. In the Arctic Ocean, isolation is much less stringent and the synthesis of antifreeze molecules very often occurs only during winter.

Most of our knowledge on the antifreeze molecules (identification, tissue distribution, mechanism of action, molecular structure, identification of the gene, relationships with kidney morphology, phyletic distribution) comes from the studies of Arthur DeVries (DeVries 1971, 1980, 1988, DeVries *et al.* 1970, 1971, Raymond & DeVries 1977, Eastman & DeVries 1986, Ahlgren *et al.* 1988, Hsiao *et al.* 1990, Cheng & DeVries 1991, DeVries chapter 27, this volume, see also Eastman 1993).

Cytoskeletal polymers: tubulins and actins

Tubulins are proteins which, together with microtubule-associated proteins (MAPs), assemble and form subcellular structures (microtubules). These are a major component of the cytoskeleton of eukaryotic cells and participate in many processes, e.g. mitosis, nerve growth, intracellular transport of organelles. The *in vitro* assembly of microtubules is temperature sensitive. In temperate fish, mammals and birds, tubulins associate near 37 °C, but disassemble at temperatures as low as 4 °C. In notothenioids, however, adaptive changes in the tubulin molecular structure have made these proteins able to stay polymerized at temperatures as low as −2 °C. Thermodynamic analysis of polymerization showed large positive enthalpy and entropy changes and indicated that the reaction is entropically driven (Detrich 1991a,b) and that, in organisms with widely different evolutionary histories, the entropic control increases with decreasing average body temperature. Interspecific differences in polymerization thermodynamics have an important adaptive consequence: the critical concentration for microtubule assembly is conserved within a narrow range of tubuline concentrations, and each organism can efficiently assemble microtubules within the limits of its body temperature. Polymerization of Antarctic fish tubulins relies on entropy-generating interactions, and, more specifically, on an increase in hydrophobic interactions (rather than in exothermic electrostatic bonds) between the molecular domains involved in the process (Detrich & Overton 1986, 1988, Detrich *et al.* 1987).

Skeletal muscle actins are another interesting example of protein polymerization (Sweezey & Somero 1982, Hochachka & Somero 1984, Somero 1991). Subunit self-assembly is temperature dependent. Unlike tubulins, in 14 vertebrate species, the standard entropy and enthalpy of the globular to filamentous transformation of actins increase as average body temperature increases; stabilization of actin filaments in cold-adapted fish is suggested to depend on exothermic polar bonds, rather than on endothermic hydrophobic interactions. The low enthalpy change is interpreted as adaptive, since a lower heat input is needed to drive polymerization.

Thus, cytoskeletal polymer assembly at low temperatures is regulated by at least two adaptive strategies (Detrich 1991b): (i) modification of the bond types at subunit contacts gives an entropy of association sufficient to overcome unfavourable enthalpy changes; (ii) a preponderant role of bonds making negative contributions to the overall enthalpy of polymerization limits the destabilizing enthalpy changes.

In considering contractile proteins, reference should be made to the work of Ian Johnston and collaborators on white muscle biochemistry, physiology and mechanics. Space does not allow to review the extensive research of this group and the reader is encouraged to consult two recent reviews (Johnston & Altringham 1988, Johnston *et al.* 1991).

Enzyme catalysis

The relationship between cold adaptation and enzymatic activity is directly and/or indirectly related to the concept of 'metabolic cold adaptation' (MCA) (Wohlschlag 1964). Experiments on oxygen consumption and routine metabolic rate have led to controversial conclusions and to alternative definitions of MCA

(Holeton 1974, Smith & Haschemeyer 1980, Clarke 1983, 1991, Wells 1987, Macdonald *et al.* 1987, Sidell 1991). In this context, MCA and the disagreements that have flourished will not be discussed.

The effect of temperature on enzyme stability and activity (the former factor interferes with the latter; this is likely to have an impact on reaction rates) opens the door to thermodynamic analysis. The data in the literature are not suggestive of unique patterns. For instance, some Antarctic fish enzymes are more labile than the corresponding mammalian ones (see Clarke 1983, and Macdonald *et al.* 1987). In others, the differences in heat denaturation are either non-existent or very small, and obviously of no (or very indirect) physiological relevance (Genicot *et al.* 1988, Ciardiello *et al.* chapter 35, this volume, and unpublished), and suggest high conservation of the molecular structure during evolution.

At least two types of adaptation can account for high catalytic rates at low temperatures (Hochachka & Somero 1984, Somero 1991): a higher intracellular enzyme concentration, and a higher inherent catalytic activity per active site. In the first mechanism, an increased number of catalytic sites compensates for the temperature-induced lower rate per site. For example, in the skeletal muscle of non-Antarctic green sunfish, acclimation from 25 °C to 5 °C increased cytochrome *c* over 50%, with a concomitant increase in its steady-state concentration (Sidell 1977). In the second mechanism, higher activity is achieved by means of fewer molecules of a more efficient enzyme: myofibrillar ATPase from cold-adapted fish has greater activity than the enzyme from temperate species, with decreased energy barrier to the reaction and enthalpic contribution (Johnston & Walesby 1977).

Reports on other enzymes suggest, however, that both mechanisms function at the same time in the majority of cases. Although the temperature dependence of the kinetic parameters K_m (a measure of the substrate or coenzyme affinity), k_{cat} (the catalytic efficiency), k_{cat} / K_m (the physiological efficiency), often offers meaningful indications, and is at the basis of thermodynamic analysis, this criterion seldom permits unequivocal conclusions. To mention one example, discussed in more detail by Ciardiello *et al.* (1995, 1997), thermodynamic analysis gives hints that glucose-6-phosphate dehydrogenase (G6PD) from the blood of two notothenioids is a 'better enzyme' than G6PD from temperate fish. However, the highly increased amount of G6PD in the few cells of the Hb-less blood of the channichthyid *Chionodraco hamatus* strongly suggests temperature compensation also via the synthesis of a larger number of G6PD molecules.

Somero (1991) has discussed some general criteria for defining thermal optima for biochemical functions, stressing that an optimal situation under *in vitro* conditions need not be optimal in a physiological sense. Thus, it would be incorrect to identify enzyme thermal optima with the temperatures at which a given property (e.g. reaction rate, or substrate affinity) is highest. The strong conservation among species of K_m values under physiological conditions of temperature, hydrostatic pressure, osmotic concentration and pH, which all strongly interfere with enzyme–ligand interactions, was interpreted by Hochachka & Somero (1984) as a mechanism which enzymes adopt to retain optimal regulatory capacity (e.g. maximal responsiveness to changes in substrate concentration and to allosteric effectors) in response to variations of the above-mentioned physico-chemical factors. Thus, thermal optima are defined in a physiologically appropriate way in terms of temperatures at which values for enzymatic properties are held within the range strongly conserved among the species (Somero 1991). The conservation of K_m in the range of physiological temperatures (generally accompanied by a sharp increase above this range) in many enzymes of cold-adapted and temperate species supports this view (Somero 1991). Acetylcholinesterase of warm-acclimated *Pagothenia borchgrevinki*, which shows a very sharp increase in K_m at 3–4 °C (Baldwin & Hochachka 1970; Baldwin 1971), is an intriguing example. This enzyme is critical for synaptic transmission. The release of acetylcholine increases very sharply in this notothenioid with increasing temperature (Macdonald & Montgomery 1982), and the enzyme may be unable to bind the substrate. Under these conditions, accumulation of large amounts of acetylcholine at synapses can disrupt synaptic transmission. This control breakdown could partially account for the heat death of notothenioids at temperatures below 10 °C (Somero & DeVries 1967, Somero 1991).

Blood and the oxygen-transport system

The haematological features of many Antarctic Notothenioidei have been extensively investigated in the past decades (Everson & Ralph 1968, Hureau *et al.* 1977, Wells *et al.* 1980; Kunzmann 1991). Correlations with fish activity patterns have been proposed, although the dramatic effect that stress invariably exerts on the haematological parameters (Macdonald *et al.* 1987, Wells *et al.* 1990) calls for extreme caution.

A clear difference between the fish below the Antarctic Polar Front and temperate and tropical species is the reduction of erythrocyte number and Hb concentration in the blood. The subzero seawater temperature would greatly increase the viscosity of blood, but the potentially negative physiological effects caused by this increase (e.g. higher cardiac work) are counterbalanced, as previous studies have pointed out (Hemmingsen & Douglas 1970, 1977, Macdonald *et al.* 1987, Wells *et al.* 1990, Macdonald & Wells 1991), by reducing or eliminating erythrocytes and Hb, an adaptive mechanism aimed at reducing the energy required for circulation.

Channichthyidae, the most phyletically derived of notothenioids (Iwami 1985, Eastman 1993) have attained the extreme of this trend. Their colourless blood totally lacks Hb (Ruud 1954). It has been reported to have a small number of erythrocyte-like cells, 2–3 orders of magnitude less than temperate fish and 1–2 orders of magnitude less than temperate fish and red-blooded notothenioids (Hureau *et al.* 1977, Kunzmann 1991). These residual cells contain enzymes with key metabolic functions, e.g. G6PD (di Prisco & D'Avino 1989, Ciardiello *et al.*, chapter 35, this volume), which offers an explanation for their physiological significance. Hb has not been replaced by another oxygen carrier; oxygen is physically dissolved in the plasma, and the oxygen-carrying capacity of channichthyid blood is only 10% of that of red-blooded fish. However, it is beyond doubt that these fish are not disadvantaged by lack of Hb. The physiological adaptations which enable channichthyids to prosper without Hb have been discussed in several papers (Hemmingsen & Douglas 1970, 1977, Holeton 1970, Macdonald *et al.* 1987). They include: low metabolic rate; large, well-perfused gills; large blood volume; large heart and stroke volume; large capillary diameter; and cutaneous respiration (channichthyids are scaleless). In addition to reducing the overall metabolic demand for oxygen, low temperatures increase its solubility in the plasma, so that more oxygen can be carried in physical solution and less needs to be bound to Hb. The co-existence of Hb-less and naturally cytopenic red-blooded species suggests that the need for an oxygen-carrier in a stable, cold environment is reduced in both groups. Functional incapacitation of Hb in *Pagothenia bernacchii* (di Prisco *et al.* 1992) had no discernible effect on the vital function under routine metabolic conditions. The same was observed after induced reduction of the haematocrit to 1–2% (Wells *et al.* 1990). Like channichthyids, red-blooded Antarctic fish can carry routinely needed oxygen dissolved in plasma. There is also evidence of some extent of cutaneous respiration in scaled nototheniids (Wells 1987).

In channichthyids (Hamoir 1988), muscles also lack myoglobin. The evolutionary loss of the respiratory pigments is a highly specialized condition which raises several questions. If the Hb-less state is adaptive for one family, why not also for the others living in the same habitat? No unequivocal answer has yet been found, and Wells (1990) has suggested that the Hb-less state may be non-adaptive. Another questions is: what happened to the α- and the ß-globin genes? Relieved of selective pressure of expression, they may have diverged from those of red-blooded notothenioids, or been lost altogether. The characterization of globin DNA sequences from several species of red-blooded and Hb-less notothenioids has indicated that three channichthyids, spanning the clade from primitive to advanced genera, share retention of α-globin-related DNA sequences in their genomes and apparent loss, or rapid mutation, of ß-globin genes (Cocca *et al.* 1995). This common pattern suggests that loss of globin-gene expression is a primitive character, established in the common ancestral channichthyid approximately 25 Ma (Eastman 1993), prior to diversification within the clade. Deletion of the ß-globin locus of the ancestor may have been the primary event leading to the Hb-less phenotype. The α-globin gene(s), no longer under selective pressure for expression, would then have accumulated mutations leading to loss of function without, as yet, complete loss of sequence information.

Another common feature of endemic Antarctic fish is the markedly reduced Hb multiplicity. This is hardly unexpected, since multiplicity is linked with variability in the environment (Riggs 1970), and the Antarctic waters are a stable habitat. Among notothenioids, 34 sedentary bottom dwellers (*A. mitopteryx* and *D. mawsoni,* are benthopelagic but the former is very sluggish and the latter very moderately active) have a single major Hb (Hb 1) and often a second, minor component (Hb 2, about 5% of the total, usually having the ß-chain in common with Hb 1) (Table 34.2); in both Hbs, oxygen binding is generally strongly regulated by pH and endogenous organophosphates (di Prisco 1988, D'Avino & di Prisco 1988, 1989, di Prisco *et al.* 1988b, 1991b, D'Avino *et al.* 1989, 1991, Kunzmann 1991, Kunzmann *et al.* 1991, Caruso *et al.*, 1991, 1992, Camardella *et al.* 1992, di Prisco *et al.* 1994, see Dickerson & Geis 1983). Another component (Hb C) is present at less than 1% in all species. On the other hand, Antarctic Zoarcidae (suborder Zoarcoidei, with all-latitude distribution) possess four to five

Table 34.2. *Hbs in the blood of Notothenioidei*

Family	Species	Hb components
Bovichtidae	*Pseudaphritis urvillii*[1]	Hb 1 (95%), Hb 2 (5%)
Nototheniidae	*Notothenia coriiceps*	Hb 1 (95%), Hb 2 (5%)
	Notothenia rossii	Hb 1 (95%), Hb 2 (5%)
	Notothenops angustata[1]	Hb 1 (95%), Hb 2 (5%)
	Nototheniops nudifrons	Hb 1 (95%), Hb 2 (5%)
	Notothenia larseni	Hb 1 (95%), Hb 2 (5%)
	Gobionotothen gibberifrons	Hb 1 (90%), Hb 2 (10%)
	Pagothenia hansonii	Hb 1 (95%), Hb 2 (5%)
	Pagothenia bernacchii	Hb 1 (98%), (Hb 2?)
	Dissostichus mawsoni	Hb 1 (98%), (Hb 2?)
	Aethotaxis mitopteryx	One Hb (99%)
	Trematomus nicolai	Hb 1 (95%), Hb 2 (5%)
	Trematomus pennellii	Hb 1 (95%), Hb 2 (5%)
	Trematomus loennbergi	Hb 1 (95%), Hb 2 (5%)
	Trematomus eulepidotus	Hb 1 (95%), Hb 2 (5%)
	Trematomus lepidorhinus	Hb 1 (95%), Hb 2 (5%)
	Trematomus scotti	Hb 1 (95%), Hb 2 (5%)
Bathydraconidae	*Cygnodraco mawsoni*	Hb 1 (97%), Hb 2 (5%)
	Racovitzia glacialis	Hb 1 (90%), Hb 2 (10%)
	Parachaenichthys charcoti	One Hb (99%)
	Gymnodraco acuticeps	One Hb (99%)
	Bathydraco marris	One Hb (99%)
	Bathydraco macrolepis	One Hb (99%)
	Akarotaxis nudiceps	One Hb (99%)
	Gerlachea australis	One Hb (99%)
Artedidraconidae	*Artedidraco skottsbergi*	One Hb (99%)
	Artedidraco orianae	One Hb (99%)
	Artedidraco shackletoni	One Hb (99%)
	Histiodraco velifer	One Hb (99%)
	P. scotti	One Hb (99%)
	Pogonophryne sp.1	One Hb (99%)
	Pogonophryne sp.2	One Hb (99%)
	Pogonophryne sp.3	One Hb (99%)
Harpagiferidae	*Harpagifer antarcticus*	One Hb (99%)

Notes:
The blood of all species contains traces (less than 1%) of Hb C.
[1] These two species non-Antarctic, all others Antarctic.

functionally distinct major Hbs (di Prisco & D'Avino 1989, di Prisco *et al.* 1990. Also, two species of non-endemic Macrouridae and Anotopteridae have three and four major Hbs, respectively (Kunzmann 1991).

Many papers have been published in the last decade on the molecular structure and biological function of the Hbs of many Antarctic species (D'Avino *et al.* 1991, 1992, 1994; see di Prisco *et al.* 1991b). The effect of pH and endogenous organophosphate effectors on oxygen equilibria and saturation (the Bohr and Root effects; Root 1931; see Brittain 1987 and Riggs 1988) has been investigated in 30 (plus an additional one) of the 34 notothenioids of Table 34.2 (Table 34.3). The Hb system has been examined in a non-Antarctic nototheniid (Fago *et al.* 1992) and bovichtid (D'Avino *et al.*, 1997). The haematocrit, erythrocyte number, Hb concentration and MCHC of non-cold-adapted *Notothenia angustata* are higher than those of Antarctic nototheniids (Tetens *et al.* 1984, Macdonald & Wells 1991), as expected in a fish of lower latitudes. But Hb multiplicity and structural/functional features closely resemble those of the Antarctic species of the same family and suborder (Fago *et al.* 1992). The most striking similarity concerns the primary structure, with a degree of identity with Antarctic *N. coriiceps* Hbs of 93% in the ß chains and 99% in the α chains of Hb 1; the α chains of Hb2 are identical. This identity between a cold-adapted and a non-cold-adapted species of the same family is the highest ever found among notothenioids Hbs. Did separation of *N. angustata* from Antarctic notothenioids occur before or after development of cold adaptation? If the two *Notothenia* diverged evolutionary prior to the establishment of the Antarctic Polar Front, as suggested by Andersen (1984) and Balushkin (1988), cooling exerted no pressure in determining the amino acid sequence. Molecular evolution of Hb took place prior to notothenioid cold adaptation, since Antarctic coastal waters cooled to approximately 0 °C only 10 Ma (Clarke 1983). The sequence similarity would simply reflect an ancestral condition and the common phylogenetic origin of these species (Fago *et al.* 1992).

The most recent data reveal that three notothenioids do not follow the pattern of low multiplicity. *Trematomus newnesi, P. antarcticum* and *P. borchgrevinki*, all Nototheniidae, have three to five components and these multiple Hbs are functionally distinct (Table 34.4). These fish, unlike the 34 species listed in Tables 34.2 and 34.3, are not sluggish bottom dwellers, and display peculiar lifestyles.

T. newnesi actively swims and feeds near the surface (Eastman 1988). It is the only species (D'Avino *et al.* 1994) in which Hb C is not present in traces, but reaches 20–25%. Unlike the other notothenioids, the oxygen binding of Hb 1 and Hb 2 is not regulated by pH and organophosphates; on the other hand, Hb C displays effector-enhanced Bohr and Root effects. Thus, *T. newnesi* is the only known notothenioid having two functionally distinct major Hbs, which may be required by this more active fish to ensure oxygen binding at the gills, and oxygen delivery to tissues in conditions of acidosis. In order to balance the lack of regulation in Hb 1 and Hb 2 by protons and other physiological effectors, evolution may have preserved the expression of high Hb C levels, conceivably redundant in the other notothenioids.

P. antarcticum is the most abundant species of high-Antarctic shelf areas and has great importance in the pelagic system. Among notothenioids, it has the highest multiplicity of major components (three). These Hbs display effector-enhanced Bohr

Table 34.3. *Regulation by pH and physiological effectors of oxygen binding of Hbs of Antarctic and non-Antarctic (*N. angustata *and* P. urvillii*) Notothenioidei*

Family	Species	Bohr & Root effects; effect of organophosphates
Bovichtidae	*Pseudaphritis urvillii*	Strong in Hb 1, Hb 2
Nototheniidae	*Notothenia coriiceps*	Strong in Hb 1, Hb 2
	Notothenia rossii	Strong in Hb 1, Hb 2
	Notothenia angustata	Strong in Hb 1, Hb 2
	Gobionotothen gibberifrons	Strong in Hb 1, Hb 2
	Pagothenia hansoni	Strong in Hb 1, Hb 2
	Pagothenia bernacchii	Strong in Hb 1
	Dissostichus mawsoni	Strong in Hb 1
	Aethotaxis mitopteryx	Root, absent; Bohr, weak
	Trematomus nicolai	Strong in Hb 1, Hb 2
	Trematomus pennellii	Strong in Hb 1, Hb 2
	Trematomus loennbergi	Strong in Hb 1, Hb 2
	Trematomus eulepidotus	Strong in Hb 1, Hb 2
	Trematomus lepidorhinus	Strong in Hb 1, Hb 2
	Trematomus scotti [1]	Strong in Hb 1, Hb 2
Bathydraconidae	*Cygnodraco mawsoni*	Strong in Hb 1, Hb 2
	Racovitzia glacialis	Strong (in haemolysate)
	Parachaenichthys charcoti	Strong
	Gymnodraco acuticeps	Absent
	Bathydraco marri	Strong
	Bathydraco macrolepis[1]	Strong, only with ATP
	Akarotaxis nudiceps[1]	Strong
	Gerlachea australis[1]	Strong
Artedidraconidae	*Artedidraco orianae*	Weak (Root only with ATP)
	Artedidraco shackletoni[1]	Weak, only with ATP
	Histiodraco velifer	Weak
	Pogonophryne scotti	Weak (Root only with ATP)
	Dolloidraco longedorsalis[1]	Weak (in haemolysate)
	Pogonophryne sp.1	Weak (Root only with ATP)
	Pogonophryne sp.2	Weak (Root only with ATP)
	Pogonophryne sp.3	Weak (Root only with ATP)

Notes:
[1] The Bohr effect was not measured.

Table 34.4. *Antarctic Notothenioidei (family Nototheniidae) with higher Hb multiplicity and with functionally distinct components*

Species	Hb components	Bohr effect	Root effect	Effect of organophosphates
Trematomus newnesi	Hb C (20%)	Strong	Strong	Strong
(2 major Hbs)	Hb 1 (75%)	Weak	Absent	Absent
	Hb 2 (5%)	Weak	Absent	Absent
Pleuragramma antarcticum[1]	Hb C (traces)	Strong	Strong	Strong
(3 major Hbs)	Hb 1 (30%)	Strong	Strong	Strong
	Hb 2 (20%)	Strong	Strong	Strong
	Hb 3 (50%)	Strong	Strong	Strong
Pagothenia borchgrevinki	Hb C (traces)	Strong	Strong	Strong
(1 major Hb)	Hb 0 (10%)	Strong	Strong	Strong
	Hb 1 (70%)	Weak	Weak	Absent
	Hb 2 (10%)	Weak	Weak	Weak
	Hb 3 (10%)	Weak	Weak	Weak

Note:

[1] The Hbs of *P. antarcticum* differ thermodynamically (they have different values of heat of oxygenation).

and Root effects, but differ thermodynamically in the heats of oxygenation (Tamburrini *et al.* 1994, 1996). Temperature-regulated oxygen affinity may reflect a highly refined molecular adaptation to the pelagic lifestyle of this fish, allowing migration with optimal energy savings across water regions with significant temperature differences and fluctuations.

P. borchgrevinki is an active cryopelagic species. It has five Hbs (Tamburrini & di Prisco, unpublished), Hb 1 accounting for 70–80% of the total. Besides Hb C, present in traces, another component (Hb 0) displays strong, effector-enhanced Bohr and Root effects. Hb 1, Hb 2 and Hb 3 bind oxygen with weak pH-dependence and weak or no influence of organophosphates. In view of this high multiplicity of functionally distinct Hbs, the oxygen-transport system of *P. borchgrevinki* is the most specialized among notothenioids.

CONCLUDING REMARKS

The adaptive evolution of Antarctic fish has the advantage of being staged within a simplified framework (a reduced number of variables in a stable environment, dominated by a taxonomically uniform group). Although correlations of molecular data with physiological and biochemical adaptations and with ecology and lifestyle are difficult to identify, the primary importance of this objective is increasingly attracting the interest of scientists. Haematology (a source of basic information on the biochemistry and physiology of a vital function such as respiration) and the key constituent Hb (a protein whose molecular features have been conserved throughout evolution) have been evaluated as potential tools. Investigations on a highly representative number of species (38 out of 80 red-blooded Antarctic – and two non-Antarctic – notothenioids) allow us to reach two conclusions of general bearing: (i), the more phyletically derived notothenioid families have lower erythrocyte numbers and Hb concentration and multiplicity; (ii), a correlation often appears between lifestyle and Hb multiplicity and functional features. Bottom dwellers have a single major Hb with a higher affinity for oxygen, which in the most inactive species binds to the protein with no (or weak) regulation by pH and effectors (di Prisco *et al.* 1991b, Kunzmann 1991, Kunzmann *et al.* 1992, D'Avino *et al.* 1992, Tamburrini *et al.* 1992). The constant physico-chemical conditions of the ocean may have reduced the need for multiple Hbs. The observation that three pelagic and active nototheniids have multiple, functionally distinct Hbs supports this interpretation. In these cases, a link with adaptation to habitat and with lifestyle becomes possible (di Prisco & Tamburrini 1992), since the selective advantage of multiple Hb genes is clear. In most other cases, it is not clear whether multiple genes are surviving selectively neutral gene duplication, with no obvious correlation with environmental conditions. In fact 'adaptive mutations are much less frequent than selectively neutral or nearly neutral substitutions caused by random drift' (Kimura 1983).

Perutz (1987) has analysed the possible relationship between structure and function of Hb and adaptation. Functional adaptation may have been produced by the gradual accumulation of minor mutations or by substitutions in key positions, and the two alternatives may not be mutually exclusive. The amino acid sequences of Antarctic fish Hbs will thus be a useful tool in the molecular approach to adaptive evolution. In the construction of phylogenetic trees, this study logically implies further extensions to non-Antarctic notothenioids of the non-endemic families Bovichtidae and Nototheniidae, the most ancient of the suborder. Because of its importance in evolution history, the structure of the globin genes in Hb-less Channichthyidae is also well worth analysis.

ACKNOWLEDGEMENTS

This paper includes work sponsored by the Italian National Programme for Antarctic Research and carried out thanks to the enthusiast participation of L. Camardella, V. Carratore, C. Caruso, M. A. Ciardiello, E. Cocca, R. D'Avino, A. Fago, M. Romano, A. Riccio, M. Tamburrini and the late B. Rutigliano. The helpful comments of two reviewers, A. Clarke and G. Hubold, are gratefully acknowledged.

REFERENCES

Adie, R. J. 1970. Past environment and climates of Antarctica. In Holdgate, M.W., ed. *Antarctic ecology*, Vol.1, London: Academic Press, 7–14.

Ahlgren, J. A. Cheng, C. C., Schrag, J D. & deVries, A. L. 1988. Freezing avoidance and the distribution of antifreeze glycopeptides in body fluids and tissues of Antarctic fish. *Journal of Experimental Biology,* **137,** 549–563.

Andersen, N. C. 1984. Genera and subfamilies of the family Nototheniidae (Pisces, Perciformes) from the Antarctic and Subantarctic. *Steenstrupia,* **10,** 1–34.

Baldwin, J. 1971. Adaptation of enzymes to temperature: acetylcholinesterase in the central nervous system of fishes. *Comparative Biochemistry and Physiology,* **40B,** 181–187.

Baldwin, J. & Hochachka, P. W. 1970. Functional significance of iso-enzymes in thermal acclimatization: acetylcholinestease from trout brain. *Biochemical Journal,* **116,** 883–887.

Balushkin, A. V. 1988. Suborder Notothenioidei. In McAllister, D. E., ed. *A working list of fishes of the world.* Ottawa: National Museum of Canada, 1118–1126.

Black, V. S. 1951. Some aspects of the physiology of fish. II. Osmotic regulation in teleost fishes. *University of Toronto Studies in Biology Series 59,* **71,** 53–89.

Brittain, T. 1987. The Root effect. *Comparative Biochemistry and Physiology,* **86B,** 473–481.

Camardella, L., Caruso, C., D'Avino, R., di Prisco, G., Rutigliano, B., Tamburrini, M., Fermi, G,. & Perutz, M. F. 1992. Haemoglobin of the Antarctic fish *Pagothenia bernacchii.* Amino acid sequence, oxygen equilibria and crystal structure of its carbonmonoxy derivative. *Journal of Molecular Biology,* **224,** 449–460.

Caruso, C., Rutigliano, B., Riccio , A., Kunzmann, A. & di Prisco, G. 1992. The amino acid sequence of the single hemoglobin of the high-Antarctic fish *Bathydraco marri* Norman. *Comparative Biochemistry and Physiology,* **102B,** 941–946.

Caruso, C., Rutigliano, B., Romano, M., di Prisco, G. 1991. The hemoglobins of the cold-adapted Antarctic teleost *Cygnodraco mawsoni. Biochimica et Biophysica Acta,* **1078,** 273–282.

Cheng, C. C., & DeVries, A. L. 1991.The role of antifreeze glycopeptides and peptides in the freezing avoidance of cold-water fish. In di Prisco, G., ed. *Life under extreme conditions. Biochemical adaptations.* Berlin: Springer-Verlag, 1–14.

Ciardiello, M. A., Camardella, L. & di Prisco, G. 1995. Glucose-6-phosphate dehydrogenase from the blood cells of two Antarctic teleosts: correlation with cold adaptation. *Biochimica et Biophysica Acta,* **1250,** 76–82.

Ciardiello, M. A., Camardella, L. & di Prisco, G. 1997. Enzymes in cold-adapted Antarctic fish: glucose-6-phosphate dehydrogenase. In Battaglia, B., Valencia, J. & Walton, D. W. H., eds. *Antarctic communities: species, structure and survival.* Cambridge: Cambridge University Press, 261–265.

Clarke, A. 1983. Life in cold water: the physiological ecology of polar marine ectotherms. *Annual Review of Oceanography and Marine Biology,* **21,** 341–453.

Clarke, A. 1991. What is cold adaptation and how should we measure it? *American Zoology,* **21,** 81–92.

Clarke, A. & North, A. W. 1991. Is the growth of polar fish limited by temperature? In di Prisco, G., Maresca, B. & Tota, B., eds. *Biology of Antarctic fish.* Berlin: Springer-Verlag, 54–69.

Cocca, E., Ratnayake-Lecamwasam, M., Parker, S. K., Camardella, L., Ciaramella, M., di Prisco, G. & Detrich, H. W., III. 1995. Genomic remnants of α-globin genes in the hemoglobinless antartic icefishes. *Proceedings of the National Academy of Science, USA,* **92,** 1817–1821.

Craddock, C. 1970. Antarctic geology and Gondwanaland. *Antarctic Journal of the United States,* **5**(3), 53–57.

D'Avino, R. & di Prisco, G. 1988. Antarctic fish hemoglobin: an outline of the molecular structure and oxygen binding properties. 1. Molecular structure. *Comparative Biochemistry and Physiolgy,* **90B,** 579–584.

D'Avino, R. & di Prisco, G. 1989. Hemoglobin from the Antarctic fish *Notothenia coriiceps neglecta.* 1. Purification and characterisation. *European Journal of Biochemistry,* **179,** 699–705.

D'Avino, R. & di Prisco, G. 1997. The hemoglobin system of Antarctic and non-Antarctic notothenioid fishes. *Comparative Biochemistry and Physiology,* in press.

D'Avino, R., Caruso, C., Romano, M., Camardella, L., Rutigliano, B. & di Prisco, G. 1989. Hemoglobin from the Antarctic fish *Notothenia coriiceps neglecta.* 2. Amino acid sequence of the α chain of Hb 1. *European Journal of Biochemistry,* **179,** 707–713.

D'Avino, R., Caruso, C., Camardella, L., Schininà, M. E., Rutigliano, B., Romano, M., Carratore, V., Barra, D. & di Prisco, G. 1991. An overview of the molecular structure and functional properties of the hemoglobins of a cold-adapted Antarctic teleost. In di Prisco, G., ed. *Life under extreme conditions. Biochemical adaptations.* Berlin: Springer-Verlag, 15–33.

D'Avino, R., Fago, A., Kunzmann, A. & di Prisco, G. 1992. The primary structure and oxygen-binding properties of the high-Antarctic fish *Aethotaxis mitopteryx* DeWitt. *Polar Biology,* **12,** 135–140.

D'Avino, R., Caruso, C., Tamburrini, M., Romano, M., Rutigliano, B., Polverino de Laureto, P., Camardella, L., Carratore, V. & di Prisco, G. 1994. Molecular characterization of the functionally distinct hemoglobins of the Antarctic fish *Trematomus newnesi. Journal of Biological Chemistry,* **269,** 9675–9681.

Detrich, H. W., III. 1991a. Cold-stable microtubules from Antarctic fish. In di Prisco, G., ed. *Life under extreme conditions. Biochemical adaptations.* Berlin: Springer-Verlag, 35–49.

Detrich, H. W., III. 1991b. Polymerization of microtubule proteins from Antarctic fish. In di Prisco, G., Maresca, B. & Tota, B., eds. *Biology of Antarctic fish.* Berlin: Springer-Verlag, 248–262.

Detrich, H. W., III & Overton, S. A. 1986. Heterogeneity and structure of brain tubulins from cold-adapted Antarctic fishes: comparison to brain tubulins from a temperate fish and a mammal. *Journal of Biological Chemistry,* **261,** 10922–10930.

Detrich, H. W., III & Overton, S.A. 1988. Antarctic fish tubulins: heterogeneity, structure, amino acid compositions, and charge. *Comparative Biochemistry and Physiology,* **90B,** 593–600.

Detrich, H. W., III, Prasad, V. & Ludueña, R. F. 1987. Cold-stable micro-tubules from Antarctic fishes contain unique α-tubulins. *Journal of Biological Chemistry,* **262,** 8360–8366.

DeVries, A. L. 1971. Glycoproteins as biological antifreeze agents in Antarctic fishes. *Science,* **172,** 1152–1155.

DeVries, A. L. 1980. Biological antifreezes and survival in freezing environments. In Gilles, R., ed. *Animals and environmental fitness.* Oxford: Pergamon Press, 583–607.

DeVries, A. L. 1988. The role of antifreeze glycopeptides and peptides in the freezing avoidance of Antarctic fishes. *Comparative Biochemistry and Physiology,* **90B,** 611–621.

DeVries, A. L. 1997. The role of antifreeze proteins in survival of Antarctic fishes in freezing environments. In Battaglia, B., Valencia, J. & Walton, D. W. H., eds. *Antarctic communities: species, structure and survival.* Cambridge: Cambridge University Press, 202–208.

DeVries, A. L. & Eastman, J. T. 1978. Lipid sacs as a buoyancy adaptation in an Antarctic fish. *Nature,* **271,** 352–353.

DeVries, A. L., Komatsu, S. K. & Feeney, R. E. 1970. Chemical and physical properties of freezing-point depressing glycoproteins from Antarctic fishes. *Journal of Biological Chemistry,* **245,** 2901–2908.

DeVries, A. L., Vandeheede, J. & Feeney, R. E. 1971. Primary structure of freezing-point depressing glycoproteins. *Journal of Biological Chemistry,* **246,** 305–308.

Dickerson, R. E. & Geis, I. 1983. *Hemoglobin: structure, function, evolution and pathology.* Menlo Park, CA: Benjamin/Cummings Publishing Co., 176 pp.

di Prisco, G. 1988. A study of hemoglobin in Antarctic fishes: purification and characterisation of hemoglobins from four species. *Comparative Biochemistry and Physiology*, **90B**, 631–637.

di Prisco, G., ed. 1991. *Life under extreme conditions. Biochemical adaptations*. Berlin: Springer-Verlag, 144 pp.

di Prisco, G. & D'Avino, R. 1989. Molecular adaptation of the blood of Antarctic teleosts to environmental conditions. *Antarctic Science*, **1**, 119–124.

di Prisco, G. & Tamburrini, M. 1992. The hemoglobins of marine and freshwater fish: the search for correlations with physiological adaptation. *Comparative Biochemistry and Physiology*, **102B**, 661–671.

di Prisco, G., Maresca, B. & Tota, B., eds. 1988a. Marine biology of Antarctic organisms. *Comparative Biochemistry and Physiology*, **90B**, 459–637.

di Prisco, G., Giardina, B., D'Avino, R., Condo', S. G., Bellelli, A. & Brunori, M. 1988b. Antarctic fish hemoglobin: an outline of the molecular structure and oxygen binding properties II. Oxygen binding properties. *Comparative Biochemistry and Physiology*, **90B**, 585–591.

di Prisco, G., D'Avino, R., Camardella, L., Caruso, C., Romano, M. & Rutigliano, B. 1990. Structure and function of hemoglobin in Antarctic fishes and evolutionary implications. *Polar Biology*, **10**, 269–274.

di Prisco, G., Maresca, B. & Tota, B., eds. 1991a. *Biology of Antarctic fish*. Berlin: Springer-Verlag, 292 pp.

di Prisco, G., D'Avino, R., Caruso, C., Tamburrini, M., Camardella, L., Rutigliano, B., Carratore, V. & Romano, M. 1991b. The biochemistry of oxygen transport in red-blooded Antarctic fish. In di Prisco, G., Maresca, B. & Tota, B., eds. *Biology of Antarctic fish*. Berlin: Springer-Verlag, 263–281.

di Prisco, G., Macdonald, J. A. & Brunori, M. 1992. Antarctic fishes survive exposure to carbon monoxide. *Experientia*, **48**, 473–475.

di Prisco, G., Camardella, L., Carratore, V., Caruso, C., Ciardiello, M. A., D'Avino, R., Fago, A., Riccio, A., Romano, M., Rutigliano, B. & Tamburrini, M. 1994. Structure and function of hemoglobins, enzymes and other proteins from Antarctic marine and terrestrial organisms. In *Proceedings of the 2nd Meeting 'Biology in Antarctica'*. Padova: Edizioni Universitarie Patavine, 157–177.

Eastman, J. T. 1988. Ocular morphology in Antarctic notothenioid fishes. *Journal of Morphology*, **196**, 283–306.

Eastman, J. T. 1991. The fossil and modern fish faunas of Antarctica: evolution and diversity. In di Prisco, G., Maresca, B. & Tota B., eds. *Biology of Antarctic fish*. Berlin: Springer- Verlag, 116–130.

Eastman, J. T. 1993. *Antarctic fish biology. Evolution in a unique environment*. San Diego, CA: Academic Press, 322 pp.

Eastman, J. T. & DeVries, A. L. 1982. Buoyancy studies of notothenioid fishes in McMurdo sound, Antarctica. *Copeia*, 1982 (2), 385–393.

Eastman, J. T. & DeVries, A. L. 1986. Renal glomerular evolution in Antarctic notothenioid fishes. *Journal of Fish Biology*, **29**, 649–662.

Everson, I. & Ralph, R. 1968. Blood analyses of some Antarctic fish. *British Antarctic Survey Bulletin*, No. 15, 59–62.

Fago, A., D'Avino, R & di Prisco, G. 1992. The hemoglobins of *Notothenia angustata*, a temperate fish belonging to a family largely endemic to the Antarctic Ocean. *European Journal of Biochemistry*, **210**, 963–970.

Frakes, L. A. & Crowell, J. C. 1970. Geologic evidence for the place of Antarctica in Gondwanaland. *Antarctic Journal of the United States*, **5** (3), 67–69.

Genicot, S., Feller, G. & Gerday, C. 1988. Trypsin from Antarctic fish (*Paranotothenia magellanica* Forster) as compared with trout (*Salmo gairdneri*) trypsin. *Comparative Biochemistry and Physiology*, **90B**, 601–609.

Gon, O. & Heemstra, P. C., eds. 1990. *Fishes of the Southern Ocean*. Grahamstown: JLB Smith Institute of Ichthyology, 462 pp.

Hamoir, G. 1988. Biochemical adaptation of the muscles of the Channichthyidae to their lack of hemoglobin and myoglobin. *Comparative Biochemistry and Physiology*, **90B**, 557–559.

Harwood, D. M. 1987. Diatom biostratigraphy and paleoecology with a Cenozoic history of Antarctic ice sheets. *Dissertation Abstracts International*, **47B**, 3276–3277.

Hemmingsen, E. A. & Douglas, E. L. 1970. Respiratory characteristics of the hemoglobin-free fish *Chaenocephalus aceratus*. *Comparative Biochemistry and Physiology*, **33**, 733–744.

Hemmingsen, E. A. & Douglas, E. L. 1977. Respiratory and circulatory adaptations to the absence of hemoglobin in Chaenichthyid fishes. In Llano, G. A., ed. *Adaptations within Antarctic ecosystems*. Washington DC: Smithsonian Institution, 479–487.

Hochachka, P. W. & Somero, G. N. 1984. *Biochemical adaptation*. Princeton, NJ: Princeton University Press, 537 pp.

Holeton, G. F. 1970. Oxygen uptake and circulation by a hemoglobinless Antarctic fish (*Chaenocephalus aceratus* Lönnberg) compared with three red-blooded Antarctic fish. *Comparative Biochemistry and Physiology*, **34**, 457–471.

Holeton, G. F. 1974. Metabolic cold adaptation of polar fish: fact or artefact? *Physiological Zoology*, **73A**, 137–152.

Hsiao, K., Cheng, C. C., Fernandez, I. E., Detrich, H. W. & DeVries, A. L. 1990. An antifreeze glycopeptide gene from the Antarctic cod *Notothenia coriiceps neglecta* encodes a protein of high peptide copy number. *Proceedings of the National Academy of Science, USA*, **87**, 9265–9269.

Hureau, J. C., Petit, D., Fine, J. M. & Marneux, M. 1977. New cytological, biochemical and physiological data on the colorless blood of the Channichthyidae (Pisces, Teleosteans, Perciformes). In Llano, G. A., ed. *Adaptations within Antarctic ecosystems*. Washington DC: Smithsonian Institution, 459–477.

Iwami, T. 1985. Osteology and relationships of the family Channichthyidae. *Memoir of the Institute of Polar Research, Tokyo, Series E*, **36**, 1–69.

Johnston, I. A. & Altringham, J. D. 1988. Muscle contraction in polar fishes. *Comparative Biochemistry and Physiology*, **90B**, 547–555.

Johnston, I. A. & Walesby, N. J. 1977. Molecular mechanisms of temperature adaptation in fish myofibrillar adenosine triphosphatases. *Journal of Comparative Physiology*, **119**, 195–206.

Johnston, I. A., Johnson, T. P. & Battram, J. C. 1991. Low temperature limits burst swimming performance in Antarctic fish. In di Prisco, G., Maresca, B & Tota, B., eds. *Biology of Antarctic fish*. Berlin: Springer-Verlag, 179–190.

Kennett, J. P. 1977. Cenozoic evolution of the Antarctic glaciation, the circum-Antarctic ocean and their impact on global paleoceanography. *Journal of Geophysical Research*, **82**, 3843–3876

Kimura, M. 1983. The neutral theory of molecular evolution. In Nei, M. & Kohen, R. K., eds. *Evolution of genes and proteins*. Sunderland, MA: Sinauer Associates, 208–233.

King, L. C. 1958. Basic palaeogeography of Gondwanaland during the late Paleozoic and Mesozoic eras. *Quarterly Journal of the Geological Society, London*, **114**, 47–77.

Kunzmann, A. 1991. Blood physiology and ecological consequences in Weddell Sea fishes (Antarctica), *Berichte für Polarforschung*, **91**, 79 pp.

Kunzmann, A., Caruso, C. & di Prisco, G. 1991. Haematological studies on a high-Antarctic fish: *Bathydraco marri* Norman. *Journal of Experimental Marine Biology and Ecology*, **152**, 243–255.

Kunzmann, A., Fago, A., D'Avino, R. & di Prisco, G. 1992. Haematological studies on *Aethotaxis mitopteryx* DeWitt, a high-Antarctic fish with a single haemoglobin. *Polar Biology*, **12**, 141–145.

Littlepage, J. L. 1965. Oceanographic investigations in McMurdo Sound, Antarctica. *Antarctic Research Series*, **5**, 1–37.

Macdonald, J. A. & Montgomery, J. C. 1982. Thermal limits of neuromuscular function in an Antarctic fish. *Journal of Comparative Physiology*, **147**, 237–250.

Macdonald, J. A. & Wells, R. M. G. 1991. Viscosity of body fluids from Antarctic notothenioid fish. In di Prisco, G., Maresca, B. & Tota B., eds. *Biology of Antarctic fish*. Berlin: Springer-Verlag, 163–178.

Macdonald, J. A. & Montgomery, J. C. & Wells R. M. G. 1987. Comparative physiology of Antarctic fishes. *Advances in Marine Biology*, **24**, 321–388.

Parrish, J. T. 1990. Gondwanan paleogeography and paleoclimatology. In Taylor, T. N. & Taylor, E. L., eds. *Antarctic paleobiology: its role in the reconstruction of Gondwana*. New York: Springer-Verlag, 15–26.

Perutz, M. F. 1987. Species adaptation in a protein molecule. *Advances in Protein Chemistry,* **36,** 213–244.

Pickard, J., Adamson, D. A., Harwood, D. M., Quilty, P. G. & Dell, R. K. 1988. Early Pliocene marine sediments, coastline and climate of East Antarctica. *Geology,* **16,** 158–161.

Raymond, J. A. & DeVries, A. L. 1977. Adsorption inhibition as a mechanism of freezing resistance in polar fishes. *Proceeding of the National Academy of Science, USA,* **74,** 2589–2593.

Riggs, A. 1970. Properties of fish hemoglobins. In Hoar, W. S. & Randall, D. J., eds. *Fish physiology*, Vol. 4. New York: Academic Press, 209–252.

Riggs, A. 1988. The Bohr effect. *Annual Review of Physiology*, **50,** 181–204.

Root, R. W. 1931. The respiratory function of blood in marine organisms. *Biological Bulletin,* **61,** 427–456.

Ruud, J. T. 1954. Vertebrates without erythrocytes and blood pigment. *Nature,* **173,** 848–850.

Scholander, P. F., Flaff, W., Hock, R. J. & Irving, L. 1957. Studies on the physiology of frozen plants and animals in the Arctic. *Journal of Cellular and Comparative Physiology,* **42,** suppl. 1, 1–56.

Sidell, B. D. 1977. Turnover of cytochrome *c* in skeletal muscle of green sunfish (*Lepomis cyanellus,* R.) during thermal acclimation. *Journal of Experimental Zoology,* **199,** 233–250.

Sidell, B. D. 1991. Physiological roles of high lipid content in tissues of Antarctic fish species. In di Prisco, G., Maresca, B & Tota B., eds. *Biology of Antarctic fish*. Berlin: Springer-Verlag, 220–231.

Smith, M. A. K. & Haschemeyer, A. E. V. 1980. Protein metabolism and cold adaptation in Antarctic fishes. *Physiological Zoology,* **53,** 373–382.

Somero, G. N. 1991. Biochemical mechanisms of cold adaptation and stenothermality in Antarctic fish. In di Prisco, G., Maresca, B. & Tota B., eds. *Biology of Antarctic fish*. Berlin: Springer-Verlag, 232–247.

Somero, G. N. & DeVries, A. L. 1967. Temperature tolerance of some Antarctic fishes. *Science,* **156,** 257–258.

Sweezey, R. R. & Somero, G. N. 1982 Polymerization thermodynamics and structural stabilities of skeletal muscle actins from vertebrates adapted to different temperatures and hydrostatic pressures. *Biochemistry,* **21,** 4496–4503.

Tamburrini, M., Brancaccio, A., Ippoliti, R. & di Prisco, G. 1992. The amino acid sequence and oxygen-binding properties of the single hemoglobin of the cold adapted Antarctic teleost *Gymnodraco acuticeps. Archives of Biochemistry and Biophysics,* **292,** 295–302.

Tamburrini, M., D'Avino, R., Fago, A., Carratore, V., Kunzmann, A. & di Prisco, G. 1994. The unique hemoglobin system of *Pleuragramma antarcticum,* a high-Antarctic fish with holopelagic mode of life. *SCAR 6th Biology Symposium, Poster Abstracts,* 261.

Tamburrini, M., D'Avino, R., Fago, A., Carratore, V., Kunzmann, A. & di Prisco, G. 1996. The unique hemoglobin system of *Pleuragramma antarcticum*, an Antarctic migratory teleost. *Journal of Biological Chemistry*, **271**, 23 780–23 785.

Tetens, V., Well, R. M. G. & DeVries, A. L. 1984. Antarctic fish blood: respiratory properties and the effect of thermal acclimation. *Journal of Experimental Biology,* **109,** 265–279.

Webb, P. N. 1990. The Cenozoic history of Antarctica and its global impact. *Antarctic Science,* **2,** 3–21.

Wells, R. M. G. 1987. Respiration of Antarctic fish from McMurdo sound. *Comparative Biochemistry and Physiology.* **88A,** 417–424.

Wells, R. M. G. 1990. Hemoglobin physiology in vertebrate animals: a cautionary approach to adaptionist thinking. In Boutilier, R. G., ed. *Advances in Comparative and Environmental Physiology*, Vol. 6. Berlin: Springer-Verlag, 143–161.

Wells, R. M. G., Ashby, M. D., Duncan, S. J. & Macdonald, J. A. 1980. Comparative studies of the erythrocytes and haemoglobins in nototheniid fishes from Antarctica. *Journal of Fish Biology,* **17,** 517–527.

Wells, R. M. G., Macdonald, J. A. & di Prisco, G. 1990. Thin-blooded Antarctic fishes: a rheological comparison of the haemoglobin-free icefishes, *Chionodraco kathleenae* and *Cryodraco antarcticus,* with a red-blooded nototheniid, *Pagothenia bernacchii. Journal of Fish Biology,* **36,** 595–609.

Wohlschlag, D. E. 1964. Respiratory metabolism and ecological characteristics of some fishes in McMurdo Sound, Antarctica. *Antarctic Research Series,* **1,** 33–62.

35 Enzymes in cold-adapted Antarctic fish: glucose-6-phosphate dehydrogenase

M. A. CIARDIELLO, L. CAMARDELLA AND G. DI PRISCO
Insititute of Protein Biochemistry and Enzymology, CNR, Via Marconi 10, 80125 Naples, Italy

ABSTRACT

Glucose-6-phosphate dehydrogenase (G6PD) in the blood of nototheniid and channichthyid fish has been investigated. Assays carried out on freshly drawn blood indicate that (i) G6PD activity in channichthyid (icefish) blood is about 20% of that present in the same volume of nototheniid blood; (ii) the few blood cells of the icefish contain two- to four-fold more G6PD than nototheniid blood cells; (iii) the total blood G6PD activity is similar on a weight basis in fish of both families, since the blood volume of the icefish is about four-fold higher.

Biochemical characterization of G6PD from erythrocytes from the nototheniid Dissostichus mawsoni *shows that some properties such as optimal pH and thermostability are similar to those of G6PD from temperate organisms. Other properties such as temperature dependence of activity and the kinetic parameters, analysed in comparison with the G6PD from sea bass (*Dicentrarchus labrax*) liver, show some differences which might be related to adaptation to a low temperature environment.*

Key words: G6PD, enzyme, Antarctica, fish, kinetics, adaptation.

INTRODUCTION

Antarctic fish have developed special anatomical, physiological and biochemical characteristics permitting life and reproduction at very low temperatures. This fauna offers an excellent opportunity to understand the biochemical mechanisms underlying cold adaptation. A comparative study of structural and functional properties of enzymes purified from Antarctic organisms and their counterparts from temperate regions and, when possible, thermophilic organisms, may shed light on the molecular basis sustaining 'cold adaptation' and, in general, the temperature effect on enzyme function.

The enzymes from Antarctic organisms so far investigated show some properties that might be correlated with the environmental temperatures. In some cases they represent adaptive changes for improved function at low temperature (Kobori *et al.* 1984, Genicot *et al.* 1988, Feller *et al.* 1990, 1992); in other cases, they are neutral consequences of other differences. Several enzymes from Antarctic fish are more active at low temperature than the corresponding temperate ones, whereas some other enzymes show little or no functional difference related to low temperature (for a review, see McDonald *et al.* 1987). Thus, it is difficult to define a general behaviour; it seems that different behaviours could reflect the relative importance of specific metabolic pathways. In Antarctic fish the glycolytic pathway seems poorly developed and the pentose-phosphate pathway might assume a quantitatively more important role in carbohydrate metabolism (Somero *el al.* 1968). This pathway is closely linked to lipid biosynthesis and enzymes from Antarctic fish of fatty acid metabolism have higher catalytic efficiencies than those from temperate fish (Johnston & Harrison 1985).

In order to study Antarctic fish enzymes of special metabolic significance, we have purified glucose-6-phosphate dehydrogenase (G6PD) from some nototheniid and channichthyid (icefish) species (Ciardiello *et al.* 1995). This enzyme is present in all cells, where it plays a key role in the metabolism of glucose. It catalyses the first reaction of the pentose-phosphate pathway (oxidation of D-glucose-6-phosphate to D-glucono-δ-lactone-6-phosphate by simultaneous reduction of the coenzyme NADP). G6PD is thought to control the rate of this pathway, although the detailed mechanism of regulation has not yet been understood. The main function of this metabolic pathway is production of NADPH (used in lipid biosynthesis and in the protection of the cell from oxidative stress) and of 5-phosphoribosyl pyrophosphate (used in RNA synthesis). This pathway is responsible for 10% of total glucose consumption in all tissues where biosynthesis of lipids and steroids is important. In erythrocytes, where glucose is the primary energy source, G6PD is involved both in the production of metabolites which can enter the glycolytic pathway and in the production of NADPH, which is necessary as coenzyme of glutathione reductase and which has a role in the maintenance of the active form of the catalase (Kirkman *et al.* 1987).

MATERIALS AND METHODS

Specimens of *Notothenia coriiceps, Gobionotothen gibberifrons* (Nototheniidae) and *Chaenocephalus aceratus, Pseudochaenichthys georgianus* and *Champsocephalus gunnari* (Channichthyidae) were collected by bottom trawling in Dallman Bay and Low Island, Antarctic Peninsula. Specimens of *Chionodraco hamatus* (Channichthyidae) were collected by gill nets in Terra Nova Bay, Ross Sea. Fish were kept in aquaria supplied with running seawater at 0–1.5 °C. *Dissostichus mawsoni* was available thanks to the courtesy of A. DeVries.

Determination of the G6PD content in the blood of nototheniid and channichthyid specimens was performed on freshly drawn blood and on separated cells.

Blood was drawn from the caudal vein of unanaesthetized fish, by means of heparinized syringes; EDTA was normally used as anticoagulant. The cells were immediately separated by low-speed centrifugation in the cold. The packed erythrocytes from the nototheniid blood were washed with a solution of 1.7% NaCl in 1 mM Tris-HCl, pH 8.1 (D'Avino & di Prisco 1988). The small white sediment of the channichthyid blood was used without prior saline washing, in view of the reported fragility of the erythrocyte-like cells. (Hureau *et al.* 1977); this observation was confirmed by the detection of a substantial amount of G6PD activity in the supernatant after every cell washing attempt.

2′,5′ ADP-Sepharose 4B was purchased from Pharmacia; Tris, NADP and G6P from SIGMA; cellulose acetate from Gelman Int. Ltd; molecular weight calibration proteins, protein assay reagent and reagents for polyacrylamide gel electrophoresis from Bio-Rad. All other chemicals were of the highest purity commercially available.

Protein was determined with the Bio-Rad method (Bradford 1976). For enzymes spectrophotometric assays were made at 20 °C by recording the reduction of NADP, at 340 nm, with a Varian DMS 300 spectrophotometer. The assays for the saturation kinetics were made on a Jasco FP-777 spectrofluorometer by recording the emission at 460 nm with excitation at 340 nm. The standard assay mixture contained 0.1 M Tris-HCl, pH 7.6, 0.2 mM NADP and 1 mM G6P.

Cellulose acetate electrophoresis was carried out in Gelman High Resolution Buffer (35.1% w/w Tris, 13.7% w/w Barbital, 54.2% w/w sodium Barbital; pH 8.8) at 250 V for 30–40 min. at room temperature. Staining was carried out in a solution of 0.2% Ponceau S in 3% trichloroacetic acid, followed, after 3–5 min., by rinsing in 3% v/v acetic acid. Enzymatic activity was revealed by the staining procedure described by Kahn & Dreyfus (1974).

To determine thermostability the enzyme was incubated at various temperatures, in a mixture containing 50 mM potassium phosphate buffer, pH 7.5, 25 mM NaCl, 1 mM EDTA, 0.2% ß-mercaptoethanol. Aliquots were assayed in a standard conditions assay at 20 °C. The energy of activation (E_a) was calculated from the Arrhenius plot (slope $= -E_a/R$).

In order to perform a more extensive characterization of G6PD from Antarctic fish blood, the enzyme was partially purified. Packed erythrocytes from nototheniid blood and the white sediment of the channichthyid blood were analysed in an equal volume of 1 mM Tris-HCl, pH 8.1 (D'Avino & di Prisco 1988); removal of stromas and of high molecular weight material was carried out by centrifugation at 100 000*g*; all manipulations were carried out at 0 °C. The purification method involves affinity chromatography on 2′,5′ ADP-Sepharose; the enzyme was recovered from the affinity column by means of stepwise elution with the coenzyme NADP (Descalzi-Cancedda *et al.* 1984). Removal of minor, low molecular weight contaminating proteins was achieved by means of gel filtration on a Sephadex G100 column; this step was, in most cases, omitted due to the extreme paucity of the enzyme (1–4 μg ml^{-1} of blood). The fractions containing the enzyme were pooled and concentrated either by precipitation with ammonium sulphate at 65% saturation or by centrifugation in Amicon Centricon 30.

RESULTS AND DISCUSSION

G6PD content in the blood of nototheniids and channichthyids

In 23 experiments with individual specimens of the channichthyid *C. aceratus,* the total G6PD activity in the blood averaged 0.225 IU ml^{-1}, with values ranging from 0.128 to 0.305; values of 1.125 IU ml^{-1} were found in pooled blood of *N. coriiceps* and of *G. gibberifrons*. Therefore, the amount of G6PD in the icefish blood is about 20% of that found in an equal volume of the nototheniid blood. Considering that the number of cells in icefish blood is 5–10% of the number of nototheniid blood cells, it appears that a blood cell of the icefish contains two- to four-fold more G6PD than a nototheniid blood cell. The increased icefish blood volume (about four-fold) accounts for a similar level of G6PD activity on a weight basis in the blood of Antarctic species from both families.

In channichthyid blood, G6PD activity was found distributed in the plasma and the cells in approximately equal amounts (47% and 53%, respectively, with individual variations in the range 40–60%, (Table 35.1). Following centrifugation at 150 000*g*, the plasma activity was totally recovered in the supernatant, indicating that it is not particle bound. In contrast, less than 1% of the blood activity was detected in the plasma of nototheniids. Plasma and cell G6PD from three channichthyid species, *C. aceratus, P. georgianus* and *C. gunnari* have undergone comparative partial characterization which suggested that the electrophoretic mobility, (activity as a function of pH and apparent K_m for substrates of the two enzyme fractions) was identical (data not shown). This evidence does not support the existence of a soluble plasma isoenzyme in fish of this family and the discovery of G6PD in the plasma can be attributed to leakage due to the fragility of the icefish blood cells.

The purification of G6PD obtained from three nototheniid and two channichthyid species is outlined in Table 35.2. The higher purification factor observed with the enzyme from nototheniid erythrocytes is due to the removal of haemoglobin, the major component of erythrocyte extracts, which is totally absent in channichthyid blood cells. In all preparations the electrophoretic analysis on cellulose acetate, followed by activity staining, indicated the presence of a single activity containing band (thus excluding the presence of isoenzymes). All bands had very similar electrophoretic mobility.

Table 35.1. *Distribution of G6PD activity in the plasma and cells of channichthyid and nototheniid blood*

	Plasma(IU ml^{-1})		Cells[1](IU ml^{-1})	
Species	Mean	Range	Mean	Range
C. aceratus (23 individuals)	0.110	0.09–0.13	0.12	0.09–0.13
N. coriiceps (pool of 25)	0.007	—	1.00	—
G. gibberifrons (pool of 18)	0.007	—	0.90	—

Note:
[1] 150 000*g* supernatants of lysates of cells of *C. aceratus* and erythrocytes of nototheniids.

Characterization of *Dissostichus mawsoni* G6PD

Further biochemical characterization was performed on G6PD from *D. mawsoni,* a large-size nototheniid with circumpolar distribution. The response of activity to pH variations measured in the range 6.0–10.0 showed that the enzyme reached optimal activity at pH 8.0 (Fig. 35.1), indicating the presence of an ionizable group in the active site of the enzyme. Similar to other G6PD (Camardella *et al.* 1988), the presence of an essential

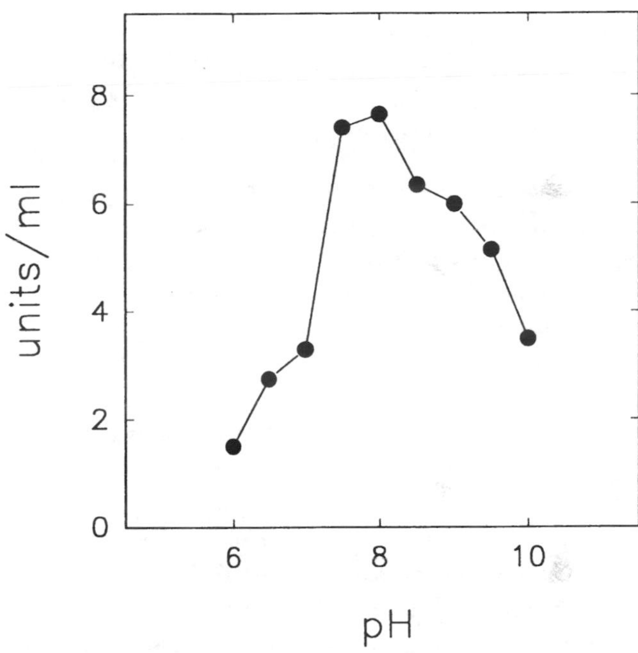

Fig. 35.1. G6PD activity as a function of pH. Assays were in 0.1 M Tris-HCl (above pH 7.0) and Bistris-HCl (below pH 7.0) buffers.

Table 35.2. *G6PD purification from nototheniid and channichthyid blood cells*

Source[1]	Step	Volume (ml)	Total activity (IU)	Protein (mg ml^{-1})	Specific activity (IU mg^{-1})	Purif. factor
N. coriiceps	extract	75	60.3	136.40	0.006	
	ADP-Seph.	0.07	16.9	1.16	209	34 833
G. gibberifrons	extract	70	69.8	87.30	0.011	
	ADP-Seph.	0.01	25.7	1.20	217	19 727
D. mawsoni[2]	extract	55	55.7	26.20	0.039	
	ADP-Seph.	1	25.6	0.17	150.6	3860
C. aceratus	extract	70	20.2	2.90	0.099	
	ADP-Seph.	0.2	1.3	0.18	38.6	390
C. hamatus[2]	extract	9.5	4.0	4.70	0.089	
	ADP-Seph.	0.2	1.4	0.06	117	1315

Note:
[1] Initial blood volume was 110 ml (*N. coriiceps*), 70 ml (*N. gibberifrons*)), 31 ml (*D. mawsoni*), 220 ml (*C. aceratus*,), 100 ml (*C. hamatus*). [2] From frozen cells.

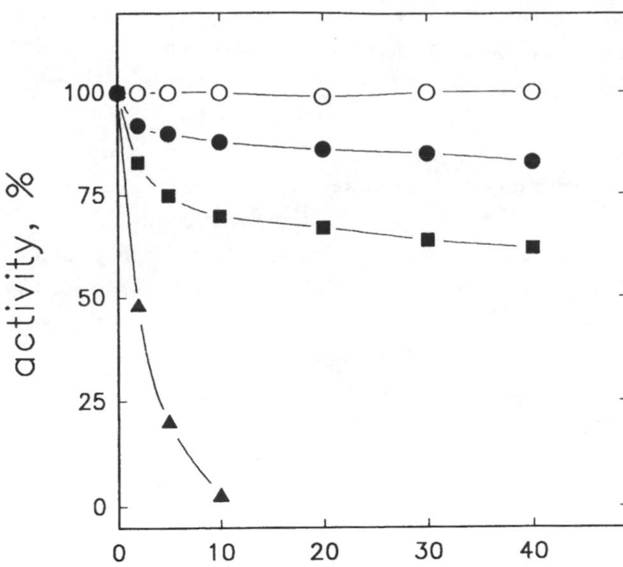

Fig. 35.2. Time course of heat inactivation at 40 °C (○), 45 °C (●), 50 °C (■) and 55 °C (▲). Protein concentrated was 55 μg ml^{-1}.

lysine residue in the active site was also revealed by reacting the enzyme with pyridoxal 5′-phosphate, a reagent having some structural similarity with the substrate G6P, and capable of forming covalent bonds with the ϵ-amino group of lysine.

Investigation of protein thermostability showed that the enzyme retained its activity after 30 min at 40 °C; at 50 °C it was totally inactivated in 20 min, while at 60 °C very fast inactivation occurred (Fig. 35.2). This behaviour, which resembles that of the enzyme isolated from human erythrocytes (Ciardiello *et al.* unpublished), indicates that the thermostability of Antarctic G6PD does not differ significantly from that of the enzyme isolated from non-cold-adapted organisms. It has been already noted that even if thermal stability of proteins correlates with adaptation temperature, it is a poor index of the temperature effects on the functional properties of a protein, mainly because thermal denaturation of proteins occurs many degrees above the physiological temperature of the organisms (Somero 1991).

The kinetic parameters which characterize the enzyme efficiency at two temperatures are shown in Table 35.3. The comparison with sea bass (*Dicentrarchus labrax*) liver G6PD (Bautista *et al.* 1989) indicates that the Antarctic enzyme has higher specific activity and lower K_m values for both the substrate G6P and the coenzyme NADP at both temperatures. Therefore, the physiological efficiency (the ratio between the specific activity and K_m for substrate and coenzyme) of the Antarctic enzyme is significantly higher. The observed differences could indeed be related to the existence of different isoenzymes with specific tissue expression; however, no clear evidence exists in the literature about the presence of G6PD isoenzymes in fish or other organisms.

The temperature dependence of the catalytic rate was investigated from 0 to 55 °C; above this temperature fast inactivation occurred during the assay, preventing reliable activity measurements. When plotting the values according to the Arrhenius equation (Fig. 35.3) two slopes were obtained, accounting for two different activation energies and two different Q_{10}: between 5 °C and 35 °C, E_a=9.3 kcal mol^{-1} and Q_{10}=1.8; and between 35 °C and 55 °C, E_a=4.8 kcal mol^{-1} and Q_{10}=1.3. These results suggest that the enzyme exists in two active forms, which has been suggested for G6PD from sea bass liver, in which the transition between the two forms occurs at 30 °C (Bautista *et al.* 1989). The reported E_a and Q_{10} values for the latter enzyme are lower in the low temperature range (E_a–2.2 kcal mol^{-1}, Q_{10}=1.1). This result indicates that the enzyme in Antarctic fish (which experiences only a very limited temperature range) is more sensitive to temperature variation in the low range.

The obvious effect of decreasing temperature is a decrease in enzyme catalytic rates. It appears that organisms living at different temperatures have evolved mechanisms to compensate for this effect and maintain comparable metabolic velocities. Hochachka & Somero (1984) have suggested two possible mechanisms for achieving this metabolic rate compensation in the enzymatic system: (i), an increase in the number of enzyme molecules; and (ii), an increase in the inherent catalytic activity per enzyme molecule. In the blood of Antarctic fish, G6PD seems to have acquired

Table 35.3. *Kinetic parameters for G6PD from* D. mawsoni *and* D. labrax

Species	Temperature (°C)	Specific activity (IU mg^{-1})	K_m G6P (μM)	K_m NADP (μM)
D. mawsoni	10	200	13	1.7
(erythrocyte)	20	300	10	4.5
D.labrax[1]	10	106	114	21.3
(liver)	25	202	96	30.0

Note:
[1] Bautista *et al.* (1989).

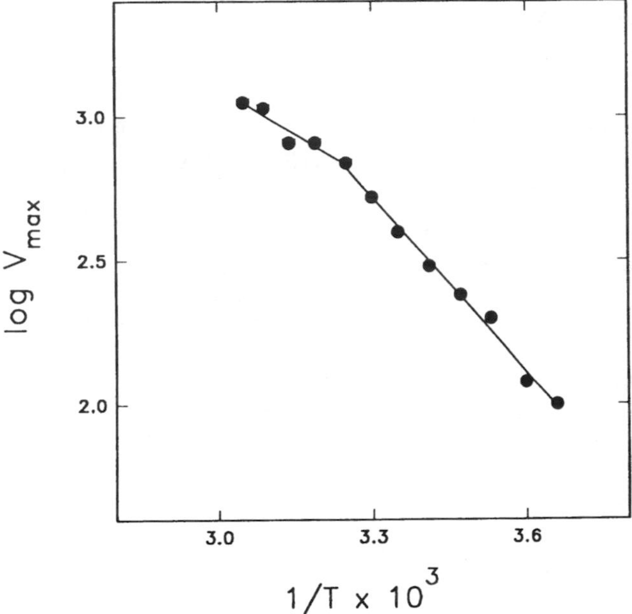

Fig. 35.3. Effect of temperature on V_{max}.

some characteristics related to metabolic rate compensation. As far as Channichthyidae are concerned, the amount in circulating blood cells is maintained, even in the absence of normal erythrocyte cells. In *D. mawsoni* G6PD, high catalytic efficiency ensures the functioning of the pentose-phosphate pathway, which is of primary importance in blood cell metabolism.

ACKNOWLEDGEMENTS

This study is part of the Italian National Programme for Antarctic Research. Thanks are due to the reviewers, G. Gerday and B. Sidell, for useful suggestions.

REFERENCES

Bautista, J. M., Soler, G. & Garrido, A. 1989. The regulation of glucose 6-phosphate dehydrogenase from *Dicentrarchus labrax* (bass) liver. *International Journal of Biochemistry*, **21**, 783–789.

Bradford, M. M. 1976. A rapid and sensitive method for the quantitation of microgram quantities of protein utilizing the principle of protein-dye binding. *Analytical Biochemistry*, **72**, 248–254.

Camardella, L., Caruso, C., Rutigliano, B., Romano, M., di Prisco, G. & Descalzi-Cancedda, F. 1988. Human erythrocyte glucose-6-phosphate dehydrogenase. Identification of a reactive lysyl residue labelled with pyridoxal 5′-phosphate. *European Journal of Biochemistry*, **171**, 485–489.

Ciardiello, M. A., Camardella, L. & di Prisco, G. 1995. Glucose-6-phosphate dehydrogenase from the blood cells of two Antarctic teleosts: correlation with cold adaptation. *Biochemica et Biophysica Acta*, **1250**, 76–82.

D'Avino, R. & di Prisco, G. 1988. Antarctic fish hemoglobin: an outline of the molecular structure and oxygen binding properties. 1. Molecular structure. *Comparative Biochemistry and Physiology*, **90B**, 579–584.

Descalzi-Cancedda, F., Caruso, C., Romano, M., di Prisco, G. & Camardella, L. 1984. Amino acid sequence of the carboxy-terminal end of human erythrocyte glucose-6-phosphate dehydrogenase. *Biochemical and Biophysical Research Communications*, **118**, 332–338.

Feller, G., Thiry, M., Arpigny, J. L, Mergeay, M . & Gerday, C. 1990. Lipases from psychrotrophic Antarctic bacteria. *FEMS Microbiology Letters*, **66**, 239–234.

Feller, G., Lonhienne, T., Deroanne, C., Libioulle, C., Van Beeumen, J . & Gerday, C. 1992. Purification, characterization, and nucleotide sequence of the thermolabile α-amylase from the Antarctic psychrotroph *(Alteromonas haloplanctis)* A23. *Journal of Biological Chemistry*, **267**, 5217–5221.

Genicot, S., Feller, G. & Gerday, C. 1988. Trypsin from Antarctic fish *(Paranotothenia magellanica* Forster) as compared with trout (*Salmo gairdneri)* trypsin. *Comparative Biochemistry and Physiology*, **90B**, 601– 609.

Hochachka, P. W. & Somero, G. N. 1984 . *Biochemical adaptation*. Princeton: Princeton University Press. 537 pp.

Hureau, J. C., Petit, D., Fine, J. M. & Marneux, M. 1977. New cytological, biochemical and physiological data on the colorless blood of the Channichthyidae (Pisces, Teleosteans, Perciformes). In Llano, G. A., ed. *Adaptations within Antarctic ecosystems*. Washington DC: Smithsonian Institution, 459–477.

Johnston, I. A. & Harrison P. 1985. Contractile and metabolic characteristics of muscle fibres from Antarctic fish. *Journal of Experimental Biology*, **116**, 223–236.

Kahn, A. & Dreyfus J. C. 1974. Purification of glucose-6-phosphate dehydrogenase from red blood cells and from human leukocytes. *Biochimica et Biophysica Acta*, **334**, 257–265.

Kirkman, H. N., Galiano, S. & Gaetani, G. F. 1987. The function of catalase-bound NADPH. *Journal of Biological Chemistry*, **262**, 660–666.

Kobori, H., Sullivan, C. W. & Shizuya, H. 1984. Heat-labile phosphatase from Antarctic bacteria: rapid 5′ end-labeling of nucleic acids. *Proceedings of the National Academy of Science, USA*, **81**, 6691–6695.

McDonald, J. A. Montgomery, J. C. & Wells, R. M. G. 1987. Comparative physiology of Antarctic fishes. *Advances in Marine Biology*, **24**, 321–387.

Somero, G. N., 1991. Biochemical mechanisms of cold adaptation and stenothermality in Antarctic fish. In di Prisco, G., Maresca, B. & Tota, B., eds. *Biology of Antarctic fish*. Berlin, Heidelberg: Springer-Verlag, 232–247.

Somero, G. N., Giese, A. C. & Wolschlag, D. E. 1968. Cold adaptation of Antarctic fish *Trematomus bernacchii*. *Comparative Biochemistry and Physiology*, **26**, 223–233.

36 Cu–Zn superoxide dismutase from *Pagothenia bernacchii*: catalytic and molecular properties

A. CASSINI[1], M. FAVERO[1], P. POLVERINO DE LAURETO[2] AND V. ALBERGON[1]
[1]Department of Biology and [2]CRIBI Biotechnology Centre, University of Padua, Via Trieste 75, 35121 Padova, Italy

ABSTRACT

Copper–zinc superoxide dismutase, the ubiquitous antioxidant enzyme, is considered a very conservative class of proteins. The purpose of this study was to investigate the catalytic and molecular properties of this enzyme from an Antarctic teleost fish, to evaluate the conservative aspect of superoxide dismutase in an organism which has been isolated for a long time. Cu–Zn superoxide dismutase was purified from livers of Pagothenia bernacchii *collected near the Italian Scientific Station in Terra Nova Bay. The specific activity (3800 units mg protein^{-1}), molecular weight (16 000 subunit^{-1}), isoelectric point (5.0), metal content (3.8 mg mg protein^{-1}), amino acid composition and N-terminal amino acid sequence of the enzyme were all characterized. A phylogenetic tree based on the protein parsimony algorithm was constructed from the sequence. Results confirm the conservative aspect of this class of proteins, but also reveal a number of substitutions in the N-terminal amino acid sequence, which indicates a high rate of evolution among these fish.*

Key words: Cu–Zn superoxide dismutase (Cu–Zn SOD), Antarctic teleost, protein evolution.

Cu–Zn SUPEROXIDE DISMUTASE AND THE ANTARCTIC ENVIRONMENT

Copper–zinc superoxide dismutase (Cu–Zn SOD), the ubiquitous cytosolic enzyme which catalyzes the dismutation of superoxide radicals to oxygen and hydrogen peroxide, represents the first line of defence against oxygen toxicity (Fridovich 1989) and is a highly conserved class of enzymes, at least as regards its catalytic properties.

A comparative study of Cu–Zn SODs from ox, sheep, pig and yeast showed only minor variations in their catalytic parameters (O'Neill *et al.* 1988). Comparison of six proteins with great differences in net protein charge in the area surrounding the active sites displayed an identical electrostatic potential distribution, which has been conserved in the evolution of this protein family (Desideri *et al.* 1992).

An interesting situation for SOD study seems to be the Antarctic environment, where adaptation to extreme conditions has given rise to many functional modifications. Among these, of particular importance is the lack of both haemoglobin and myoglobin (Ruud 1954) with reduced oxygen carrying capacity (Holeton 1970) in the family of Channichthyidae, the icefish. In these conditions, according to the theory of superoxide mediated oxygen toxicity, which indicates the dependance of SOD induction on oxygen supplied to the tissues (Fridovich 1975), it might be expected that SOD activity should be lower in the haemoglobinless icefish, and some studies have in fact found lower SOD activity in tissues of white-blooded fish than in those of red-blooded and temperate ones (Witas *et al.* 1984, Cassini *et al.* 1993).

Considering that these adaptations may involve modifications in biochemical systems related to oxygen metabolism, Natoli *et al.* (1990) investigated the SOD of the Antarctic icefish *Chaenocephalus aceratus,* but found that the isolated protein exhibited several structural and functional properties of the erythrocyte bovine enzyme considered typical of this class of enzymes.

To verify if the conservation of enzyme properties is also a common feature of organisms living in extreme conditions, we examined the biochemical and structural properties of Cu–Zn SOD from another Antarctic teleost, *Pagothenia bernacchii,* in which adaptation to the extreme environment has given rise to fewer modifications than those of icefish, such as decreases in Hb content and erythrocyte number.

EVOLUTION OF Cu–Zn SOD

Besides living in extreme conditions, Antarctic fish are characterized by long-term isolation, so that effects produced by casual mutations, associated with selective pressure or not, may be studied in them.

In this respect, Cu–Zn SOD is a particularly interesting protein since, in contrast to Mn and Fe SODs, which appear to have evolved at a relatively constant rate over the entire history of eukaryotes, its evolution has been erratic, having been rapid in recent times (Lee *et al.* 1985). Mammalian, frog, insect and plant Cu–Zn SOD sequences differ more than expected, whereas some earlier divergences, as in fungi and plants, exhibit unexpectedly slow rates of change (Smith & Doolittle 1992).

A comparative study on the evolutionary aspects of Cu–Zn SOD showed that alignment of 19 Cu–Zn SOD amino acid sequences from organisms including bacteria and mammals revealed 21 invariant and 19 nearly invariant (18/19 sequences) amino acid sequence positions and also that the Cu–Zn SOD sequence of all sources contains invariant and variable regions (Bannister *et al.* 1991). The former regions are those directly involved as metal ligands or responsible for maintaining the structure and function of the active site, dimer contacts and β-barrel folds and have many invariant residues, even in alignments from very distant species, so that this enzyme may be considered as one of the most stable globular proteins characterized so far (Getzoff *et al.* 1989). However, the variable regions permit evaluation of divergences among various groups of the animal kingdom. As one of these regions is a broad zone (from position 13 to 36 approximately) in the N-terminal portion, we aimed at deeper knowledge of the Cu–Zn SOD evolution rate by determining the N-terminal amino acid sequence from an organism that has been isolated from the early Tertiary onwards.

STUDY OF Cu–Zn SUPEROXIDE DISMUTASE FROM *PAGOTHENIA BERNACCHII*

Enzyme purification

Specimens of *P. bernacchii* were collected in the Ross Sea (74°41′ S, 164°7′ E), during Italian expeditions to Terra Nova Bay. Fish were frozen at −80 °C until the time of analysis.

Pooled *Pagothenia* livers were washed in cold physiological saline and homogenized in a Polytron homogenizer with five volumes of cooled 0.1 M Tris-HCl buffer, pH 7.5. The homogenate was clarified by centrifugation at 27 000*g* for 30 min. at 4 °C. The supernatant was dialysed against 20 mM Tris-HCl buffer, pH 7.8, and filtered through 0.22 μm filters.

The sample was chromatographed on a Sephadex G-75 column (26×850 mm), equilibrated and eluted with 20 mM Tris-HCl buffer, pH 7.8. This allowed the separation of two SOD activity peaks, fraction 1 containing manganese SOD and fraction 2 containing Cu–Zn SOD. The latter was exhaustively dialysed against 5 mM Tris-HCl buffer, pH 8.0, and chromatographed on a DEAE Sephadex A-50 anion exchange column (8×150 mm), previously equilibrated with the same buffer. The adsorbed sample was eluted with 150 ml of a linear gradient of 0-200 mM NaCl. Polyacrylamide gel electrophoresis revealed that, at this stage of purification, some proteins had still not been completely removed from the SOD preparation.

Further purification was then accomplished by chromatography on a strong ion exchanger. Fractions having activity in excess of 50 units of SOD per mg of protein were pooled and equilibrated with 10 mM Tris-HCl buffer, pH 7.5, and then applied to a Mono Q HR 5/5 anion exchange column, previously equilibrated with the same buffer, and eluted with a linear gradient of 0–400 mM NaCl, pH 7.5. This step was repeated, equilibrating the samples at pH 7.8 and eluting the column with a linear gradient of 0–200 mM NaCl. Fractions having SOD activity were pooled, and a part was used for physico-chemical characterization. The other was concentrated, resuspended with 0.1% TFA in Milli-Q water, and applied to a ProRPC HR 5/2 reverse phase column on an HPLC system. The sample was eluted with a gradient of 10–90% of 0.1% TFA in 60% acetonitrile.

Biochemical assay

SOD activity was assayed by the method of Beauchamp & Fridovich (1971), which is based on the inhibition of the reduction of nitroblue tetrazolium (NBT) by O_2 produced via riboflavin photoreduction. A unit of SOD activity was defined as the amount of sample enzyme required for 50% inhibition of NBT conversion. Protein concentration was determined by the Lowry method (Lowry *et al.* 1951).

Native molecular mass was determined by gel exclusion chromatography on Sephadex G-75 column using the following molecular weight standards: bovine serum albumin (MW 67.000), ovalbumin (MW 43.000), chymotrypsinogen A (MW 25.000) and cytochrome *c* (MW 12.300).

Electrophoresis on 7.5% polyacrylamide gels was performed as described by Davis (1964), stained for activity with the Beauchamp & Fridovich method (1971) and for protein with Coomassie Brillant Blue (Diezel *et al.* 1972).

Subunit molecular weight was estimated by SDS/PAGE, using the PhastGel gradient 10–15 in the presence of SDS (1%) and β-mercaptoethanol (0.1 M), according to Schägger & Von Jagow (1987). The gel was silver stained (Nielsen & Brown 1984). Molecular weight standards used were: phosphorylase (MW 94 000), bovine serum albumin (MW 67 000), ovalbumin (MW 43 000), carbonic anhydrase (MW 30 000), trypsin inhibitor (MW 20 100) and a-lactoalbumin (MW 14 000).

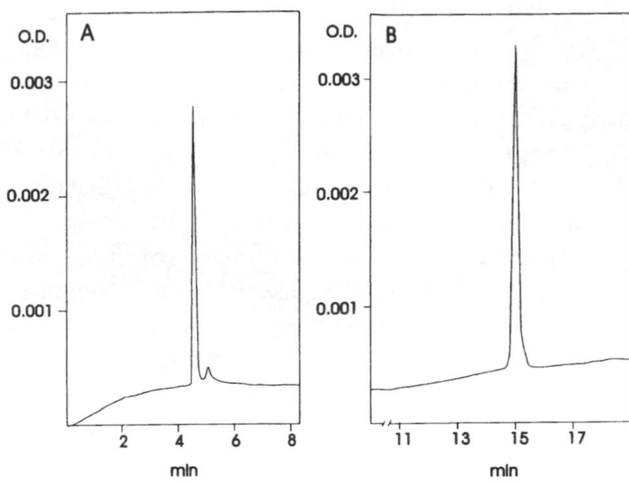

Fig. 36.1. HPCE electropherograms of *P. bernacchii* liver Cu–Zn superoxide dismutase. A: acidic conditions; B: alkaline conditions.

Capillary electrophoresis under acidic (0.1 M sodium phosphate, pH 2.5) and alkaline conditions (0.1 M sodium phosphate, pH 8.0) was used to assess sample purity before sequence analysis. The isoelectric point was determined by isoelectric focusing on a gradient gel in the pH range 3–9, as suggested by Baumann & Chrambach (1976). Copper and zinc contents were measured by atomic absorption spectroscopy, using a Perkin Elmer spectrophotometer with graphite furnace.

Amino acid analysis was performed on a Carlo Erba amino acid analyser. Samples were hydrolyzed at 110 °C in 6 M HCl for 24 h. Half cysteine was evaluated as cysteic acid according to Hirs (1967).

The N-terminal amino acid sequence was determined by an Automated Edman degradation of protein on an Applied Biosystems sequencer equipped with an on-line phenylthiohydantoin (PTH) amino acid analyser. Sequence analysis was conducted according to the standard program of Applied Biosystems. Polybrene was used as a carrier for protein sample.

Table 36.1. *Purification of Cu–Zn superoxide dismutase from* Pagothenia bernacchii *liver*

Procedure	Protein ($mg\ ml^{-1}$)	Activity ($units\ ml^{-1}$)	Specific activity ($units\ mg^{-1}$)	Purification factor
Crude extract	12.4	15	1.2	1
Sephadex G-75	2.0	340	170	142
DEAE A-50	1.2	303	253	211
I Mono Q	0.1	147	1470	1225
II Mono Q	0.03	114	3800	3167

Table 36.2. *Cu–Zn SOD specific activity in fish, mammalian and yeast*

Source	Specific activity[1] ($units\ mg^{-1}$)	References
P. bernacchii	3800	this work
Swordfish	3000	Bannister *et al.*, 1977
Common carp	2100	Vig *et al.*, 1989
White marlin	2720	Martin & Fridovich, 1981
Speckled trout	4210	Martin & Fridovich, 1981
Pony fish	3500	Martin & Fridovich, 1981
Blue shark	3300	Galtieri *et al.*, 1986
Sheep	2800	Schinninà *et al.*, 1986
Horse	3300	Albergoni & Cassini, 1974
Rabbit	3600	Reinecke *et al.*, 1989
Rat	3600	Asayama & Burr, 1985
Bovine	3300	McCord & Fridovich, 1969
Pig	3000	Hering *et al*,. 1985
Yeast	3330	Goscin & Fridovich, 1972

Notes:

[1] Determined by various spectrophotometric methods at 25 °C and pH 7.8.

Data analysis

Eukaryotic sequences were retrieved from the SwissProt database and considered as the 42 N-terminal amino acid sequences. Sequences were aligned using the CLUSTAL V program (Pearson & Lipman 1988). Phylogenetic trees were constructed with the Phyllip program which employs the Protpars function based on the protein parsimony algorithm (Dayhoff 1978).

CATALYTIC AND STRUCTURAL PROPERTIES

The purity of the Cu–Zn superoxide dismutase preparation from *P. bernacchii* was indicated by a single peak by Mono Q, tested by polyacrylamide electrophoresis and confirmed by acidic and alkaline analysis in HPCE (Fig. 36.1 reports HPC electropherograms). Table 36.1 summarizes the isolation procedure and Table 36.2 indicates that specific activity was within the range observed for this enzyme from a variety of sources.

The native enzyme molecular weight, determined by gel filtration, was approximately 31 500. Polyacrylamide gel electrophoresis, after exposure to sodium dodecyl sulphate in the presence of ß-mercaptoethanol, gave an estimated molecular weight of about 16 000 to be estimated for the enzyme subunits, confirming that *P. bernacchii* Cu–Zn SOD also consists of two subunits of equal size. Atomic absorption analysis for metals indicated two atoms of copper and two of zinc per enzyme molecule. Iron and manganese were below detection limits. Isoelectric focusing revealed a single protein fraction with an isoelectric point at pH 5.

The amino acid composition of Cu–Zn SOD is shown in Table 36.3. For comparison we also report in this table the values for the corresponding enzymes of the icefish *C. aceratus* and two other fish whose amino acid sequence is known, the teleost *Xiphias gladius* and the elasmobranch *Prionace glauca*. The range of values of amino acid composition resulting from the comparative study of Martin & Fridovich (1981) are also reported.

Table 36.3. *Amino acid composition of* Pagothenia bernacchii *Cu–Zn superoxide dismutase. Values for corresponding enzymes of icefish,* Chaenocephalus aceratus *(Natoli* et al. *1990), of swordfish,* Xiphias gladius *(Rocha* et al. *1984) and of shark,* Prionace glauca *(Calabrese* et al. *1989) are reported for comparison. The range of values for corresponding enzymes of six other teleosts are also reported (Martin & Fridovich 1981). Values are expressed as mols of residues/mols of native enzymes*

	P. bernacchii	*C. aceratus*	*X. gladius*	*P. glauca*	Value range of other fish
Asp	34	38	36	36	36–44
Thr	20	22	22	18	16–25
Ser	16	12	12	14	16–25
Glu	30	26	20	28	26–39
Pro	14	10	10	10	10–16
Gly	48	46	52	52	41–52
Ala	26	32	26	20	25–33
Cys	6	8	6	6	—
Val	22	18	22	24	14–26
Met	4	4	2	2	3–6
Ile	14	16	20	14	12–15
Leu	18	18	18	20	18–21
Tyr	4	4	4	2	3–5
Phe	10	12	8	10	7–11
Trp	—	—	—	—	0–2
His	14	16	14	16	13–14
Lys	18	20	20	18	18–22
Arg	8	6	10	12	6–11
Total	306	308	302	302	

The amino acid composition of Cu–Zn SOD from *P. bernacchii* is similar to that of *C. aceratus* and, for all amino acids except aspartic acid, falls within the range of Cu–Zn SOD values from teleosts reported by Martin & Fridovich (1981). The absence of tryptophan is also common to several Cu–Zn enzymes.

N-TERMINAL AMINO ACID SEQUENCE

Fig. 36.2 shows the N-terminal amino acid sequence of Cu–Zn SOD from *P. bernacchii* and 18 other organisms (mammals, amphibians, fish, arthropods and fungi) using an alignment which maximizes the number of invariant residues.

This alignment indicates that eight out of 42 residues are invariant in all organisms. Lys 3, Cys 6, Val 29, Leu 38 and Gly 41 are substitued only in fungi. Thus, excluding these organisms, 13 residues out of 42 are invariant. Again excluding fungi and noting that Gly 33 is substituted only in swordfish, Leu 8, Val 18 and Ile 35 are substituted by similar amino acids, and Gly 12 finds a gap in fruit fly, the invariant or nearly invariant residues being 18 out of 42.

The N-terminal amino acid is Val in all eukaryotes except mammals, in which it is constantly Ala. The same occurs for Val 18, which in mammals is Ile. Mammals also have invariant residues in Asp 11, Pro 13 and Val 14.

Conservative aspects

Like other vertebrate enzymes, Cu–Zn SOD from *P. bernacchii* is a homodimer protein with a molecular weight of 16 000 per subunit, each subunit containing one Cu and one Zn ion. The isoelectric point of about 5 is similar to that of most Cu–Zn SODs, although there are many proteins with isoelectric points close to neutral or slightly alkaline pH (Martin & Fridovich

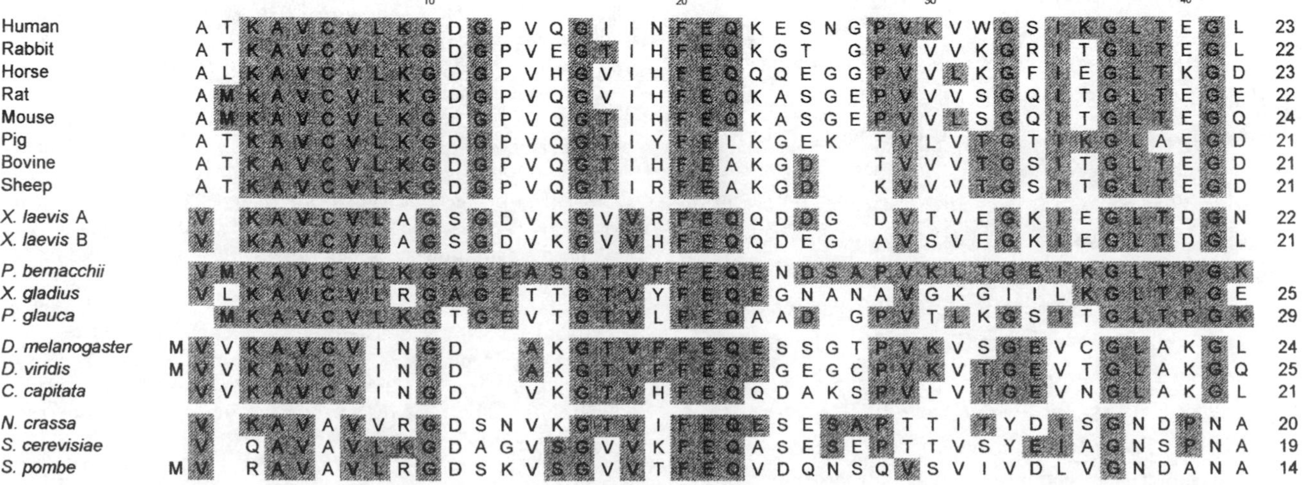

Fig. 36.2. Amino acid sequences of the N-terminal regions of Cu–Zn superoxide dismutase from *P. bernacchii* and other 18 organisms: human, rabbit, horse, rat, mouse, pig, bovine, sheep, frog (*Xenopus laevis* A and B), swordfish (*Xiphias gladius*), blue shark (*Prionace glauca*), fruit fly (*Drosophila melanogaster*, *D. virilis*, *Ceratitis capitata*) and fungi (*Neurospora crassa*, *Saccharomyces cerevisiae*, *Schizosaccharomyces pombe*). The sequences were aligned using FASTA program and corrected by visual inspection to maximize the number of invariant residues. The position number of the residues is based on the sequence of Cu–Zn SOD from *P. bernacchii*. The amino acid residues shared by *Pagothenia* are shaded. Numbers in the column on the right indicate invariant residues as related to *P. bernacchii* Cu–Zn SOD sequence.

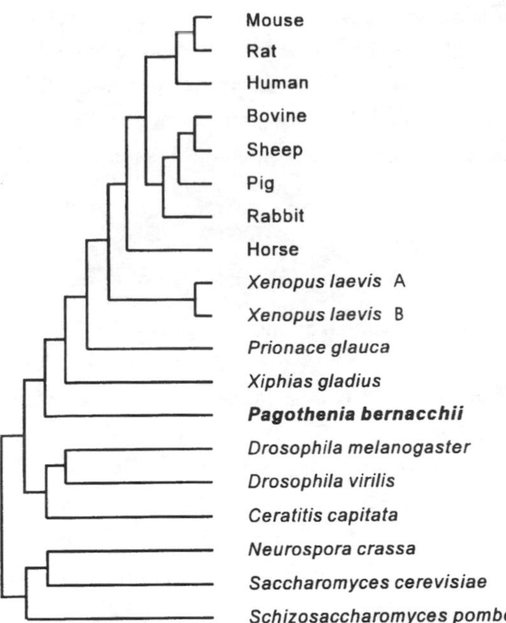

Fig. 36.3. Phylogenetic tree of eukaryotic animal Cu–Zn superoxide dismutase sequences derived from the alignment shown in Fig. 36.2.

1981). The UV spectrum did not show any peculiar characteristics.

The amino acid composition of Cu–Zn SOD from *P. bernacchii* is closely related to that of corresponding enzymes, particularly those of fish (Table 36.3). Thus, as regards structural and catalytic properties, our data confirm that Cu–Zn SODs are a conserved class of enzymes. As observed in *C. aceratus* (Natoli *et al.* 1990), extreme conditions, such as low temperature and high partial pressure of oxygen, do not cause functional modifications to the properties of Cu–Zn SOD.

EVOLUTIONARY CONSIDERATIONS

The sequence alignment reported in Fig. 36.2 also partially confirms the conservative aspect of the enzyme, which displays a homogeneous initial region, at least in animals. However, in the *P. bernacchii* Cu–Zn SOD sequence, 13 amino acids are substituted with respect to blue shark and 17 with respect to swordfish. As regards other animals, the number of substitutions varies from 17 to 22, indicating a high rate of SOD evolution among fish, higher than would be expected on the basis of the rest of the sequences.

The evolutionary tree obtained from the alignment of the N-terminal amino acid sequence (Fig. 36.3) is similar to those based on complete amino acid sequences (Smith & Doolittle, 1992; Wolf *et al.* 1993, Bordo *et al.* 1994, Fitch & Ayala 1994), indicating that the N-terminal region usually plays a crucial role in the elaboration of the tree. This relation still remains to be verified, but, if it is true, it suggests an interesting observation: the reported tree shows that *P. bernacchii* Cu–Zn SOD evolved independently not only from a distantly related fish such as *Prionace glauca* (Chondrichthyes), but also from a close one such as *Xiphias gladius* (Osteichthyes). Whatever the importance of this observation, the high heterogeneity of the *P. bernacchii* Cu–Zn SOD N-terminal portion is evident. It is probably only due to casual mutations which occurred during its long isolation and not to selective pressure, since the latter would also produce functional modifications. This high heterogeneity confirms the erratic rate of Cu–Zn SOD evolution and agrees with the theory of the 'unclocklike' regulation of Cu–Zn SOD evolution, particularly with the possibility that different subsets of amino acids evolve at different rates (Fitch & Ayala 1994).

ACKNOWLEDGEMENTS

We are indebted to Drs A. Valbonesi and M. Nigro for sample collection. This work was supported by the Programma Nazionale di Ricerche in Antartide (PNRA).

REFERENCES

Albergoni, V. & Cassini, A. 1974. A cupro-zinc protein with superoxide dismutase activity from horse liver. Isolation and properties. *Comparative Biochemistry and Physiology,* **47B,** 767–777.

Asayama, K. & Burr, I. M. 1985. Rat superoxide dismutases. Purification, labeling, immunoassay, and tissue concentration. *Journal of Biological Chemistry,* **260,** 2212–2217.

Bannister, J. V., Anastasi, A. & Bannister, W. H. 1977. Cytosol superoxide dismutase from swordfish (*Xiphias gladius L.*) liver. *Comparative Biochemistry and Physiology,* **56B,** 235–238.

Bannister, W. H., Bannister, J. V., Barra, D., Bond, J. & Bossa, F. 1991. Evolutionary aspects of superoxide dismutase: the copper/zinc enzyme. *Free Radical Research Communications*, **5,** 349–361.

Baumann, G. & Chrambach, A. 1976. A highly crosslinked, transparent polyacrylamide gel with improved mechanical stability for use in isoelectric focusing and isotachophoresis. *Analytical Biochemistry*, **70,** 32–38.

Beauchamp, C. & Fridovich, I. 1971. Superoxide dismutase: improved assays and an assay applicable to acrilamide gels. *Analytical Biochemistry*, **44,** 276–281.

Bordo, D., Djinovic', K. & Bolognesi, M. 1994. Conserved patterns in the Cu,Zn superoxide dismutase family. *Journal of Molecular Biology*, **238,** 366–386.

Calabrese, L., Polticelli, F., O'Neill, P., Galtieri, A., Barra, D., Schinninà, E. & Bossa, F. 1989. Substitution of arginine for lysine 134 alters electrostatic parameters of the active site in shark Cu,Zn superoxide dismutase. *FEBS Letters*, **250,** 49–52.

Cassini, A., Favero, M. & Albergoni, V. 1993. Comparative studies of antioxidant enzymes in red-blooded and white-blooded Antarctic teleost fish *Pagothenia bernacchii* and *Chinodraco hamatus. Comparative Biochemistry and Physiology,* **106C,** 333–336.

Davis, B. J. 1964. Disc electrophoresis (II). Method and application to human serum proteins. *Annals of New York Academy of Science, USA*, **121,** 404–427.

Dayhoff, M. O. 1978. *Atlas of protein sequence and structure*, Vol. 5. Washington, DC: National Biochemical Research Foundation.

Desideri, A., Falconi, M., Ponticelli, F., Bolognesi M., Djinovic, K. & Rotilio, G. 1992. Evolutionary conservativeness of electric field in the Cu,Zn superoxide dismutase active site. *Journal of Molecular Biology,* **223,** 337–342.

Diezel, W., Kopperschlager, G. & Hofmann, E. 1972. An improved procedure for protein in polyacrylamide gels with a new type of coomassie brillant blue. *Analytical Biochemistry*, **48,** 617–620.

Fitch, W. M. & Ayala, F. J. 1994. The superoxide dismutase molecular clock revisited. *Proceedings of the National Academy of Sciences, USA*, **91,** 6802 6807.

Fridovich, I. 1975. Superoxide dismutases. *Annual Review of Biochemistry*, **44,** 147–159.

Fridovich, I. 1989. Superoxide dismutase. An adaptation to a paramagnetic gas. *Journal of Biological Chemistry*, **267,** 7761–7764.

Galtieri, A., Natoli, G., Lania, A. & Calabrese, L. 1986. Isolation and characterization of Cu,Zn superoxide dismutase of the sark *Prionance glauca*. *Comparative Biochemistry and Physiology*, **83B,** 555–559.

Getzoff, E. D., Tainer, J. A., Stempien, M. M., Graeme, I. B. & Hallewell, R. A. 1989. Evolution of CuZn superoxide dismutase and the β-barrel structural motif. *Proteins*, **5,** 322–336.

Goscin, S. A. & Fridovich, I. 1972. The purification properties of superoxide dismutase from *Saccharomyces cerevisiae*. *Biochimica et Biophysica Acta*, **289,** 276–283.

Hering, K., Kim, S. M., Michelson, A. M., Otting, F., Puget, K., Steffens, G. J. & Flohe, L. 1985. The primary structure of porcine of Cu,Zn superoxide dismutase. Evidence for allotypes of superoxide dismutase in pigs. *Biological Chemistry Hoppe-Seyler*, **366,** 435–445.

Hirs, C. H. W. 1967. Determination of cystine as cysteic acid. *Methods in Enzymology*, **11,** 59–62.

Holeton, G. F. 1970. Oxygen uptake and circolation by a haemoglobinless Antarctic fish compared with tree red-blooded Antarctic fish. *Comparative Biochemistry and Physiology*, **34,** 457–465.

Lee, J. M., Friedman, D. J. & Ayala, F. J. 1985. Superoxide dismutase: an evolutionary puzzle. *Proceedings National Academy of Sciences, USA*, **82,** 834–828.

Lowry, O. H., Rosebrough, N. J., Farr, A. L. & Randall, R. J. 1951. Protein measurement with the Folin phenol reagent. *Journal of Biological Chemistry*, **193**, 265–275.

Martin, J. P. & Fridovich, I. 1981. Evidence for a natural gene transfert from the Ponifish to its bioluminescent bacterial symbiont *Photobacter leiognathi*. *Journal of Biological Chemistry*, **256,** 6080–6089.

McCord, J. M. & Fridovich, I. 1969. Superoxide dismutase. An enzymic function for erythrocuprein (hemocuprein). *Journal of Biological Chemistry*, **244,** 6049–6055.

Natoli, G., Calabrese, L., Capo, C., O'Neill, P. & Di Prisco, G. 1990. Icefish (*Chaenocephalus aceratus*) Cu,Zn superoxide dismutase. Conservation of the enzyme properties in extreme adaptation. *Comparative Biochemistry and Physiology*, **95B,** 29–33.

Nielsen, B. L. & Brown, L. R. 1984. The basic for colored silver–protein complex formation in stained polyacrylamide gels. *Analytical Biochemistry*, **141,** 311–315.

O'Neill, P., Davies, S., Fielden, E. M., Calabrese, L., Capo, C., Marmocchi, F., Natoli, G. & Di Prisco, G. 1988. The effects of pH and various salts upon the activity of a series of superoxide dismutases. *Biochemistry Journal*, **251,** 41–46.

Pearson, W. R. & Lipman, D. J. 1988. Improved tools for biological sequence comparison. *Proceedings National Academy of Sciences, USA*, **85,** 2444–2448.

Reinecke, K., Wolf, B., Michelson, A. M., Puget, K., Steffens, G. J. & Flohe, L. 1989. The amino-acid sequence of rabbit Cu,Zn superoxide dismutase. *Hoppe-Seyler's Zeitschrift für Physiologische Chemie*, **369,** 715–725.

Rocha, H. A., Bannister, W. H. & Bannister, J. V. 1984. The amino acid sequence of copper/zinc superoxide dismutase from swordfish liver. Comparison of copper/zinc superoxide dismutase sequences. *European Journal of Biochemistry*, **145,** 477–484.

Ruud, J. T. 1954. Vertebrates without erythrocytes and blood pigments. *Nature*, **173,** 848–850.

Schägger, H. & Von Jagow, G. 1987. Tricine-sodium dodecyl sulfate-polyacrylamide gel electrophoresis for the separation of proteins in range from 1 to 100 Kda. *Analytical Biochemistry*, **166**, 368–379.

Schininà, M. E., Barra, D., Gentilomo, S., Bossa, F., Capo, C. & Rotilio, G. 1986. Primary structure of a cationic of Cu,Zn superoxide dismutase. The sheep enzyme. *FEBS Letters*, **207,** 7–10.

Smith, M. V. & Doolittle, R. F. 1992. A comparison of evolutionary rates of the two major kinds of superoxide dismutase. *Journal of Molecular Evolution*, **34,** 175–184.

Vig, E., Gabryelak, T., Leyko, W., Nemcsok, J. & Matkovics, B. 1989. Purification and characterization of Cu,Zn-superoxide dismutase from common carp liver. *Comparative Biochemistry and Physiology*, **94B,** 395–397.

Witas, H., Gabryelak, T. & Matkovics, B. 1984. Comparative studies on superoxide dismutase and catalase activities in livers of fish and other Antarctic vertebrates. *Comparative Biochemistry and Physiology*, **74C,** 409–411.

Wolf, B., Reinecke, K., Autmann, K. D., Brigelius-Flohe', R. & Flohe', L. 1993. Taxonomical classification of the guinea pig based on its Cu/Zn-superoxide dismutase sequence. *Biological Chemistry Hoppe-Seyler*, **334,** 641–649.

37 Structure and antibody specificity of immunoglobulins from plasma of Antarctic fishes

M. R. COSCIA AND U. ORESTE

Institute of Protein Biochemistry and Enzymology, CNR, Via G. Marconi, 10, 80125 Naples, Italy

ABSTRACT

Plasma was collected from different species of Antarctic fish: Chionodraco hamatus, Pagothenia bernacchii, Notothenia coriiceps, Notothenia gibberifrons, Chaenocephalus aceratus, Pseudochaenichthys georgianus *and* Bathyraja maccaini. C. hamatus *and* P. bernacchii *immunoglobulins (Igs) were purified by ammonium sulphate precipitation and DE-52 ionic-exchange chromatography. It produced at least two peaks (*c. *190 kDa and 750 kDa) by SE-FPLC, and two bands of 27 kDa (L chain) and 75 kDa (H chain) by SDS-PAGE. Antibody specificity was also investigated. Nematode proteins, recovered from* Contracaecum osculatum *parasites, precipitated antibodies from the plasma of* C. hamatus, P. bernacchii, N. coriiceps, N. gibberifrons, *and* C. aceratus *by immunodiffusion. In addition, a 52 kDa nematode antigen was identified by immunoblotting, and specific antibodies were isolated by affinity chromatography.* C. hamatus *and* P. bernacchii *Ig precipitated anti-mammalian Ig antisera by immunodiffusion. Moreover, anti-fish plasma antisera raised in rabbit recognized mammalian Ig. Our results extend the knowledge of the molecular structure of Antarctic fish Ig, and suggest its role in the immune response to parasites.*

Key words: fish immunoglobulins, assembled immunoglobulins, Antarctic fish, *Contracaecum*, parasite antigens, anti-Ig antisera.

PLASMA IMMUNOGLOBULINS OF ANTARCTIC FISH

Immunoglobulins (Ig) are mediators of the vertebrate immune response. They are complex heterodimers of two multidomain polypeptides: the heavy (H) and the light (L) chains.

IgM is considered the most ancient immunoglobulin from the evolutionary standpoint (Du Pasquier 1989). Its heavy chain (chain) is heavily glycosylated, and consists of one variable (V_H) and four constant (C_H) domains in all species; C_H4, corresponding to the transmembrane region, is the most conserved domain. IgM is present in the serum and secretions of many vertebrate species, including fishes. Its molecular weight varies from species to species because of the varying degree of polymerization of its basic structure H_2L_2; in fact, pentameric IgM is present in the serum of mammalians, avians and cartilaginous fish and tetrameric and hexameric IgMs are found in the serum of bony fishes and amphibians, respectively. A monomeric Ig has also been found in the serum and cutaneous mucus of some fish species. In addition, Igs containing the heavy chain of the non-M isotype, consisting of four domains, have been detected in lower vertebrates; they can be considered G-like by their molecular weight (Kobayashi *et al.* 1984). Differences in the amino acid sequence from Teleostei onward suggest the presence of two classes of L chains (κ and λ) (Lobb *et al.* 1984). Ig gene organization in fish consists of hundreds of clusters of V, D, J and C elements (Ghaffari & Lobb 1992). Rearrangement seems to occur exclusively within one cluster, with a consequently reduced diversity. Biological data agree with the low antibody repertoire of fish species: fishes, immunized with either low-molecular-weight or bacterial antigens, produce a low-affinity low-heterogeneous antibody response varying

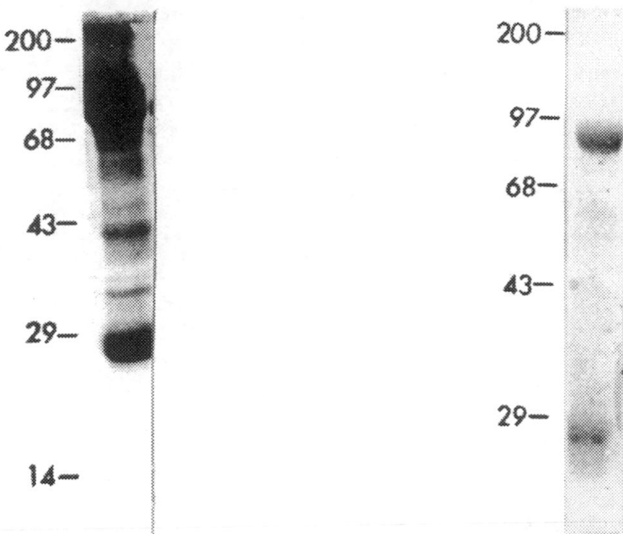

Fig. 37.1. Electrophoretic analysis of *P. bernacchii* plasma pool (left) and *P. bernacchii* Ig fraction (right) by SDS-PAGE under reducing conditions. A 5–10% (w/v) polyacrylamide gradient slab gel was used in plasma analysis, and a 10–15% gradient in Ig fraction analysis.

slightly among different individuals (Cossarini-Dunier *et al.* 1986).

Ig structure has been extensively investigated in many species of fish, such as *Carassius aurata, Ictaulurus punctulatus, Archosargus probactocephalus*, *Ginglymyostoma cirratum, Raja konojei, Heterodontus francisci, Anguilla anguilla* and *Cyprinus carpio.* We focused our attention on Antarctic species. Plasma samples were collected from *Pagothenia bernacchii* and *Chionodraco hamatus* caught in the Terra Nova Bay area, and from *Notothenia coriiceps, Notothenia gibberifrons, Chaenocephalus aceratus, Pseudochaenichthys georgianus* and *Bathyraja maccaini* caught at Palmer Station. All plasma were stored at −20 °C.

Five individual plasma samples from *P. bernacchii* and as many from *C. hamatus* were pooled. The electrophoretic heterogeneity of *P. bernacchii* plasma pool proteins (37.5 mg ml^{-1}) is shown in Fig. 37.1.

The Ig fractions of *P. bernacchii* and *C. hamatus* plasma pools were precipitated by adding a 45% saturated solution of ammonium sulphate. Precipitated proteins (about 55%) from both plasma pools were dialysed and chromatographed on a DE-52 column: each chromatographic step separated one excluded and one retained fraction. The excluded fractions, eluted as symmetrical peaks by 5mM Tris-HCl, pH 8, accounted for about 49% of the loaded proteins (27% of total plasma proteins); the retained ones were eluted, in several peaks, by adding 100 mM NaCl to the buffer. All the fractions were analysed by SDS-PAGE under reducing conditions: the excluded fractions showed a pattern of two bands, of 25 kDa and 75 kDa, which were absent in the retained ones and could be identified as L and H Ig chains (Fig.37.1).

The molecular heterogeneity of *P. bernacchii* and *C. hamatus* Ig fractions (DE-52 excluded fractions) was analysed by size-exclusion FPLC. The elution profile of the Ig fraction showed three major peaks (about 750, 380 and 190 kDa) in *P. bernacchii* (Fig. 37.2a), and a major peak of about 750 kDa, as well as another of about 190 kDa, in *C. hamatus*. The 750 kDa *P. bernacchii* peak was analysed by SDS-PAGE under both native and reducing conditions (Fig.37.2b). Under native conditions, a single high molecular weight band was revealed; under reducing conditions, two bands of 75 and 27 kDa were stained, clearly indicating that the 750 kDa peak is a polymeric form of the 190 kDa one.

These results show that, besides the currently represented monomeric form of Ig, multimeric forms, presumably dimeric and tetrameric, are present in the plasma of *P. bernacchii;* in the plasma of *C. hamatus*, instead, a monomeric and a presumably tetrameric Ig could be identified in comparable amounts. Our results agree with the data reported in the literature on the tetrameric structure of fish Igs. As regards monomeric Igs, observed also by other investigators (Kobayshi *et al.* 1984, Buchmann *et al.* 1992), our data did not allow us to detect whether their heavy chains are indistinguishable from those of the tetrameric form, or are isotype specific; the latter hypothesis suggests a class heterogeneity of fish Igs. The dimeric form, found in *P. bernacchii* but not in *C. hamatus*, could be either an intermediate in the assembling of the tetrameric form, or a product of degradation.

Antibody specificity

Nematode parasites in many fish species has been well documented. The life cycle of the anisakid nematode *Contracaecum* has been studied in the marine Antarctic ecosystem, and fishes and seals have been reported to serve as paratenic hosts for

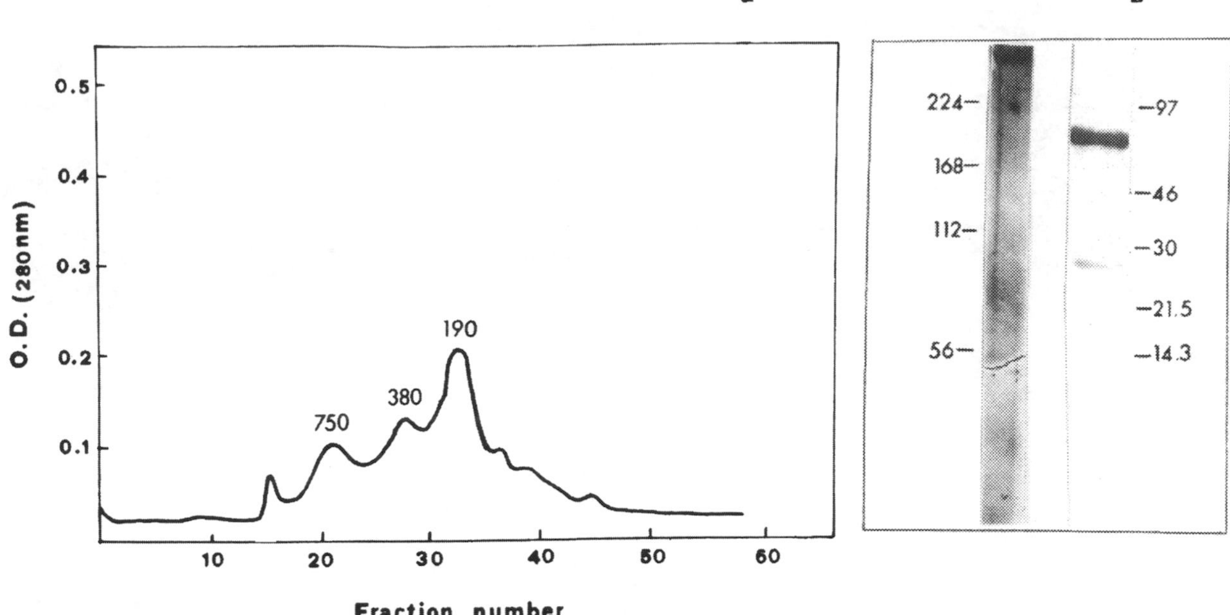

Fig. 37.2. (a) Size-Exclusion FPLC analysis of *P. bernacchii* Ig fraction on a Superose 6HR 10/30 column. Elution with a phosphate buffer, pH 7.0, containing 150 mM NaCl at a flow rate of 0.5 ml min^{-1}. (b) PAGE analysis of the 750 kDa peak from SE-HPLC of *P. bernacchii* Ig fraction, under native (left) and reducing (right) conditions.

the infective third larval stage. Two types of larva have been distinguished in *Contracaecum*, the short and the long type (Kloser *et al.* 1992). The long type, morphologically corresponding to *C. osculatum*, is more abundant in benthic fish species, whereas the short type, assigned to the species *C. radiatum*, is mostly present in pelagic fish. A positive relationship between eutrophication and fish parasitism has also been suggested. In fact, the parasite fauna is the result of the feeding habitat and the trophic level of the host (Moser & Cowen 1991). So far, the anti-parasite immune response of Antarctic fish has not yet been investigated.

The presence of anti-parasite antibodies in the plasma of Antarctic fish was tested. Nematodes, collected from the liver of a *C. hamatus* fish, were identified as larvae of the *Contracaecum osculatum* complex (Nematoda, Ascaridida, Ascaridoidea) at the Institute of Parasitology, 'La Sapienza' University, Rome. This identification did not include the genotypic analysis of the *C. osculatum* A,B,C subspecies (Nascetti *et al.* 1993).

After washing in 100 mM Phosphate buffer pH 8.0, the larvae were boiled for 10 min. in the same buffer containing 2% SDS, 10% ß-mercaptoethanol; the recovered proteins (*Co* proteins) were dialysed against water and analysed by SDS-PAGE under reducing conditions (Fig. 37.3).

Plasma of individual fishes of seven Antarctic species were tested by double immunodiffusion (IDD) with *Co* proteins. Immunoprecipitates were observed with several plasma samples from *C. hamatus, P. bernacchii, C. aceratus, N. coriiceps* and *N. gibberifrons*; none of the tested *P. georgianus* and *B. maccaini* plasma samples gave immunoprecipitation. In particular, among the species whose individual plasma samples were available in a larger number, IDD positivity was found in six out of the 10 plasma samples from *C. hamatus* and two out of the 10 from *P. bernacchii.* Fig. 37.3b shows the immunodiffusion of a *C. aceratus* and a *B. maccaini* plasma sample against *Co* proteins.

Immunoblotting was utilized to identify antigenic *Co* protein(s) accounting for the immunoprecipitation of fish antibodies in IDD. *Co* proteins were separated by SDS-PAGE, electroblotted onto nitrocellulose sheets and incubated with *P. bernacchii* plasma. The binding of individual *Co* antigenic bands to *P. bernacchii* Ig was revealed by radiolabelled rabbit IgGs, specific for *P. bernacchii* Ig. These IgGs were recovered from an anti-*P. bernacchii* plasma antiserum by MabTrap G affinity chromatography, and were labelled with ^{125}I. The ability of these rabbit IgGs to bind *P. bernacchii* Igs had been previously verified (data not shown). Autoradiography showed the presence of a single *C. osculatum* antigen of 52 kDa molecular weight (Fig. 37.3a).

The anti-*Co* proteins antibodies were purified by affinity chromatography from an IDD-positive *P. bernacchii* plasma sample: 2.7 mg of *Co* proteins dissolved in 1 ml of 200 mM $NaHCO_3$, pH 8.3 containing 500 mM NaCl (coupling buffer) were immobilized on a HiTrap NHS-activated 1 ml column (Pharmacia). *P. bernacchii* plasma (0.75 ml), dialysed against the coupling buffer, was loaded on the column and eluted with the same buffer. The bound fraction was recovered with 200 mM Glycine-HCl, pH 2.6, and analysed by SDS-PAGE under reducing conditions. The presence of only two bands (75 and 27 kDa) confirmed the successful purification of the plasma specific antibodies (data not shown).

The method followed to recover *Co* proteins did not allow us to distinguish them on the basis of their origin from the different tissues. In addition, parasite antigens, released *in vivo* in the host body, could not be detected, although they are presumably

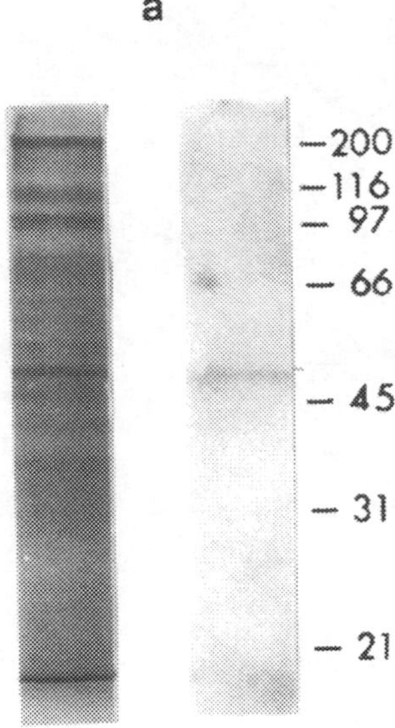

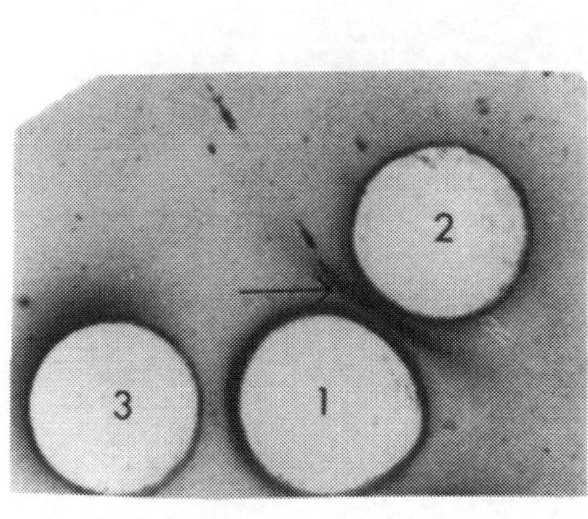

Fig. 37.3. (a) SDS-PAGE analysis of *Co* proteins using an 8% (w/v) polyacrylamide slab gel: on the left Coomassie staining, on the right immunoblotting. (b) IDD analysis: well 1 *Co* proteins, well 2 *C. aceratus* plasma, well 3 *B. maccaini* plasma.

involved in the host–parasite immune relationship. To get a deeper insight into this matter, a larger number of live nematodes needs to be available in future. The lack of anti-parasite-specific antibodies in some of the individual plasma samples tested may reflect the inadequacy of the experimental method used (IDD) to reveal non-precipitating antibodies or lower titres of the precipitating ones. Obviously, the absence of anti-parasite antibodies in individual fishes might also be related to the absence of *C. osculatum* larvae in their bodies. We do not know whether all the IDD positive plasma were collected from fishes with ongoing *C. osculatum* infection; the finding of antibodies in non-infected fishes would suggest a protective role. All the tested plasma from fishes of the *B. maccaini* and *P. georgianus* species were found negative; however, the number of samples was insufficient to predict a species-specific resistance to the infection.

The anti-parasite antibodies purified by us can be very useful for isolating, by affinity chromatography, the 52 kDa *C. osculatum* antigen to be characterized and localized at tissue level. In addition, a comparison between the structural features of anti-parasite antibodies and those of natural fish Igs would be interesting in order to establish whether they belong to a single Ig class.

Antigenic crossreactivity

Surprisingly, Igs from ancestrally divergent species have been found to be crossreactive, supporting the idea that these defence molecules form a strongly conserved family (Marchalonis *et al.* 1992).

Antigenic crossreactivity of Antarctic fish Igs with mammalian Igs was investigated by testing the ability of *P. bernacchii* and *C. hamatus* Igs to recognise antibodies in antisera produced against Igs of different mammalian species. Moreover, we tested the recognition of mammalian Igs by antisera produced by immunizing rabbits with *P. bernacchii* and *C. hamatus* plasma proteins.

Igs from *P. bernacchii* and *C. hamatus* plasma were analysed by IDD with a panel of rabbit antisera (or purified Ig fractions) anti-human Ig, anti-mouse Ig and anti-goat Ig, specific for ζ, μ, λ and κ chains. Rabbit non-immune sera were used as controls. Immunoprecipitates were obtained when *P. bernacchii* or *C. hamatus* Ig diffused against anti-human IgG (F_c specific) antiserum, anti-human λ chain antiserum, anti-goat IgG antiserum and affinity purified antibody fraction of anti-mouse IgG antiserum. Immunoprecipitation bands were not obtained when *P. bernacchii* or *C. hamatus* Igs were tested with non-immune sera. These results are partially shown in Fig. 37.4.

Moreover, we immunized rabbits with *P. bernacchii* and *C. hamatus* plasma proteins. The antisera produced contained antibodies against many plasma proteins, including Igs, as revealed by immunoelectrophoresis and IDD (data not shown). These antisera, together with non-immune rabbit sera used as controls, were tested by IDD with human, mouse and goat IgG. A neat precipitation was obtained with goat IgG (Fig. 37.4); doubtful results were obtained with human and mouse IgG, while no precipitation was obtained with mammalian Igs and non-immune sera.

These results need verifying by alternative approaches, including inhibition tests of the Ag/Ab binding. However, they indicate that Igs of very distant species share common epitopes.

ACKNOWLEDGEMENTS

We thank Dr G. di Prisco and Dr L. Camardella for providing us with fish plasma and nematodes and Dr S. Mattiucci, Institute of Parasitology, University of Rome, for identification of the nematode larvae.

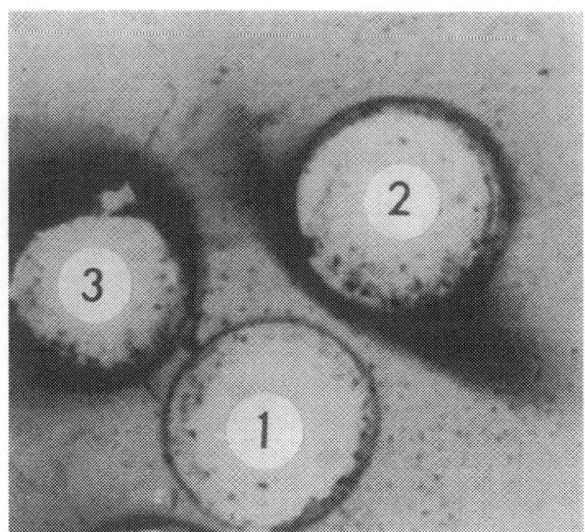

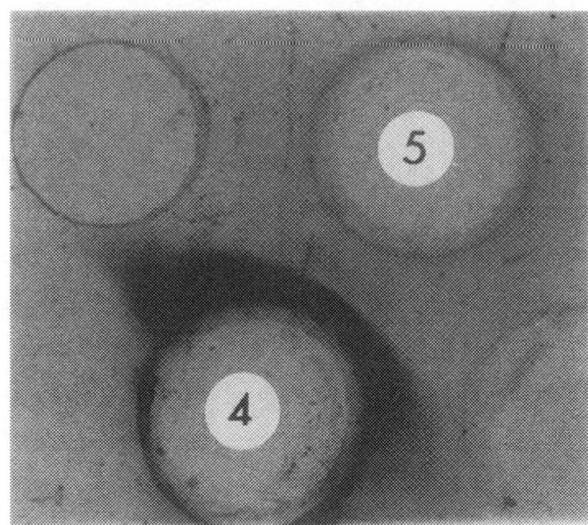

Fig. 37.4. IDD analysis of antigenic crossreactivity of *P. bernacchii* and mammalian Igs. Wells: 1 *P. bernacchii* plasma, 2 rabbit anti-mouse IgG antibodies, 3 rabbit non immune serum used as a control, 4 goat IgG, 5 rabbit anti-*C. hamatus* plasma proteins antiserum.

REFERENCES

Buchmann, K., Ostergaard, L. & Glamann, J. 1992. Affinity purification of antigen-specific serum immunoglobulin from the European eel (*Anguilla anguilla*). *Scandinavian Journal of Immunology*, **36,** 89–97.

Cossarini-Dunier, M., Desvaux, F. X. & Dorson, M. 1986. Variability in humoral responses to DPN-KLM of rainbow trout. *Developmental & Comparative Immunology*, **10,** 207–217

Du Pasquier, L. 1989. *Evolution of the immune system. W.E. Paul fundamental immunology*. New York: Raven Press, 139–165.

Ghaffari, S. H. & Lobb, C. J. 1992. Organisation of immunoglobulin heavy chain constant and joining region genes in the channel catfish. *Molecular Immunology*, **29,** 151–159.

Klöser, H., Plotz, J., Palm, H., Bartsh, A. & Hubold, G. 1992. Adjustment of anisakid nematode life cycles to the high antarctic food web as shown by *Contracaecum radiatum* and *Contracaecum osculatum* in the Weddell Sea. *Antarctic Science*, **4,** 171–178.

Kobayashi, K., Tomogama, S. & Kajii, T. 1984. A second class of immunoglobulin other than IgM present in the serum of a cartilaginous fish, the skate, *Raja kenojei*: isolation and characterisation. *Molecular Immunology*, **21,** 397–404.

Lobb, C. J., Olson, M. O. J. & Clem, L. W. 1984. Immunoglobulin light chain classes in a teleost fish. *Journal of Immunology*, **132,** 1917–1923.

Marchalonis, J., Schluter, S., Yang, H., Hohman, V., McGee, K. & Yeaton, L. 1992. Antigenic crossreactivity among immunoglobulin of diverse vertebrates (Elasmobranchs to man) detected using xenoantisera. *Comparative Biochemistry and Physiology*, **101,** 675–687

Moser, M. & Cowen, R. K. 1991. The effects of periodic eutrophication on parasitism and stock identification of *Trematomus bernacchii* (Pisces: Nothotheniidae) in McMurdo Sound, Antarctica. *Journal of Parasitology*, **77,** 551–556.

Nascetti, G., Cianchi, R., Mattiucci, S., D'Amelio, S., Orecchia, P., Paggi, L., Brattey, J. Berland, B., Smith, J. W. & Bullini, L. 1993. Three sibling species within *Contracaecum osculatum* (Nematoda, Ascaridida, Ascaridoidea) from the Atlantic arctic-boreal region: reproductive isolation and host preferences. *International Journal of Parasitology*, **23,** 105–120.

38 Seasonality of feeding and lipid content in juvenile *Pleuragramma antarcticum* (Pisces: Nototheniidae) from the southern Weddell Sea

G. HUBOLD[1] AND W. HAGEN[2]

[1]*Bundesforschungsanstalt für Fischerei, Palmaille 9, D-22767 Hamburg, Germany,* [2]*Institut für Polarökologie, Universität Kiel, Wischhofstr. 1-3, Gebäude 12, D-24148 Kiel, Germany*

ABSTRACT

During late winter/early spring (October/November 1986) pelagic fishes were collected under seasonal sea ice in the southeastern Weddell Sea for investigations on their feeding ecology and energy reserves (lipids). The pelagic ichthyofauna of the upper 500 m water column consisted almost entirely of juvenile Pleuragramma antarcticum *ranging in size from 29 to 115 mm standard length (SL). Food was found in 148 of the 155 juvenile* Pleuragramma antarcticum *investigated. The average number of prey items per gut was 29 (0–220). Except for a negligible number of juvenile* Euphausia crystallorophias *and chaetognaths, the diet consisted entirely of copepods with various developmental stages of 14 species and a size range of 0.3 to 9.2 mm. By numbers, the cyclopoid copepods* Oncaea *spp. (62%) and* Oithona *spp. (15%) dominated the diet of the smaller* P. antarcticum *(<50 mm SL). In this size group* Calanoides acutus *was more important in terms of biomass with about 40%, whereas* Oncaea *spp. contributed only 24%. The food of the larger* P. antarcticum *(>60 mm SL) was clearly dominated by* Calanus propinquus *(mainly copepodite stage V, CV), both in terms of abundance (68%) and biomass (70%).* Rhincalanus gigas *ranked second with 11% and 17%, respectively. Biochemical analyses of* Pleuragramma antarcticum *revealed total lipid contents of 15% to 41% of dry weight (%DW) depending on the size of these specimens, which exhibit a pronounced ontogenetic lipid accumulation. These values compare well with lipid levels of* P. antarcticum *specimens of similar size from late summer. Hence,* P. antarcticum *thrives in Antarctic continental shelf waters without notable seasonal effects on its activity and condition.*

Key words: life history, seasonality effects, copepods, diet composition, lipid storage.

INTRODUCTION

High latitude pelagic ecosystems are governed by a pronounced seasonality in light regime and ice cover. In the southern Weddell Sea at 70–78° S, phytoplankton growth is mainly confined to the months December through March (Fukuchi *et al.* 1985, Gieskes *et al.* 1987). Life cycles of herbivorous zooplankton are closely linked to the seasonal phytoplankton production. Many herbivores reproduce and feed in epipelagic waters in spring/summer, but overwinter at depth (Foxton 1956, Voronina 1970), although there are exceptions from this generalized pattern (Schnack-Schiel & Hagen 1994).

Surface macroplankton and micronekton collected by pelagic trawls in late winter/early spring (October/November 1986) was composed of euphausiids as well as large amphipods, gelatinous carnivores (ctenophores) and juvenile fishes. The smaller mesoplankton was represented by the omnivorous copepods *Oncaea* spp. and *Oithona* spp. at maximum densities of 1000 ind. m^{-3} (Fransz 1988). Typically, in the southeastern Weddell Sea late winter abundances of the dominant larger mesoplankton species in the epipelagial, e.g. *Calanoides acutus* and *Calanus propinquus*, were only 10–25% of summer values (Hubold & Hempel 1987, Schnack-Schiel & Hagen 1994). In spring, females of these species ascended in high densities to the surface layer to spawn (Schnack–Schiel *et al.* 1991). During summer, different stages of calanoid species dominate the mesoplankton biomass (Boysen-Ennen *et al.* 1991). Mesoplankton abundance in the epipelagial peaks in December and January, and from January to February the portion of adult stages decreases by 50%, while the numbers of young stages increase (Schnack-Schiel & Hagen 1994). By this time older stages of e.g. *Calanoides acutus* have accumulated extensive lipid stores (Hagen 1988) and return to deeper layers to overwinter. By March/April, the surface plankton of the Coastal Current is mainly composed of omnivorous and carnivorous species (Hosie *et al.* 1988) and these seem to remain dominant throughout the winter period, albeit at low abundances.

The pelagic fish fauna of the southeastern Weddell Sea is composed of larvae and juveniles of about 25 species, mainly notothenioids. *Pleuragramma antarcticum* is overwhelmingly dominant with generally 90–99% of the larvae, juveniles and adult fishes during summer (Hubold 1990). The larvae hatch in early November and feed on the rapidly growing populations of cyclopoid copepods. Throughout their first summer larval condition is generally excellent (von Dorrien 1989) and daily growth rates were found to be comparable to those of boreal pelagic species, e.g. herring *Clupea harengus* (Hubold 1985). By the end of summer, *P. antarcticum* young of the year have grown from <10 mm to about 30 mm. Their lipid contents increase continuously throughout the summer from 15 to 25% DW at the end of the productive season (Hagen 1988).

The fate of juvenile *Pleuragramma antarcticum* during their first winter has been unknown so far. An obvious hypothesis would be that the juveniles follow the copepod migration and survive on overwintering copepods in deeper layers. Alternatively, the fishes could rely on their lipid reserves accumulated during the summer feeding season. In both cases, the first winter may be a critical phase in the survival of *P. antarcticum*. To answer these questions, the pelagic fish fauna of the southeastern Weddell Sea was investigated with respect to life-cycle strategies and adaptations to the seasonal pattern of plankton productivity.

MATERIAL AND METHODS

The fishes were collected by RV *Polarstern* during the first German winter expedition to the ice-covered southeastern Weddell Sea in October/November 1986. The expedition report includes detailed accounts on sampling procedures (Schnack-Schiel 1987).

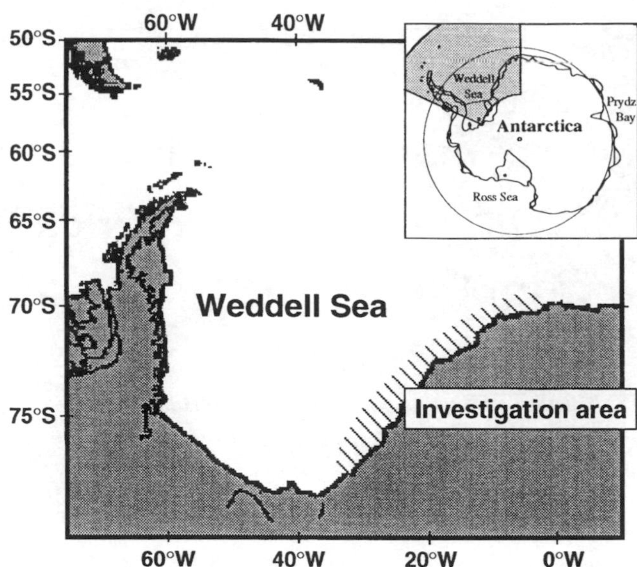

Fig. 38.1. Area investigated in the southeastern Weddell Sea during the winter/spring expedition 'ANT V/3' in October/November 1986.

A Rectangular Midwater Trawl RMT 1+8 (325 μm/4500 μm mesh) and a pelagic krill net (10×10 m^2 opening, mesh size of 10 mm in the cod end) were employed in the pack-ice area of the southeastern Weddell Sea (Fig. 38.1). The nets were hauled through the upper 500 m. Sampling was hampered by low air temperatures and severe sea-ice conditions, which often did not allow for quantitative estimates of tow path and filtered volumes. Quantitative estimates on pelagic fish abundance and biomass were therefore not attempted. A number of specimens, especially of *Pleuragramma antarcticum*, were damaged by the difficult hauling procedures and could not be used for further analyses.

The fishes were identified to species level, measured (standard length SL, rounded to the lower mm) and either immediately deep-frozen at −80 °C for lipid analysis or preserved in 4% buffered formaldehyde for diet analyses. In the home laboratory the prey items from the fish guts were identified to the lowest taxon (or stage) possible under a dissecting microscope. Clusters of copepod eggs and single eggs were recorded but excluded from the quantitative data. The length of food particles was derived from plankton data of the same cruise published by Mizdalski (1988) and a typical mean value was attributed to each prey item. Dry weight measurements of different plankton species and stages from this cruise were also available from Mizdalski (1988). These data were used for the biomass calculations.

The specimens for the lipid analyses were lyophilized to constant weight (48 h), reweighed for dry weight (±0.1 mg) and stored at −80°C for the lipid extractions, which followed the gravimetric method of Folch *et al.* (1957). The samples were homogenized at 1200 rpm in a chloroform : methanol solution (2:1, v:v) at 0 °C, with 0.01% butylhydroxytoluene added as antioxidant. Non-lipid contaminants were removed by shaking the combined solvents with one quarter their total volume of 0.88%

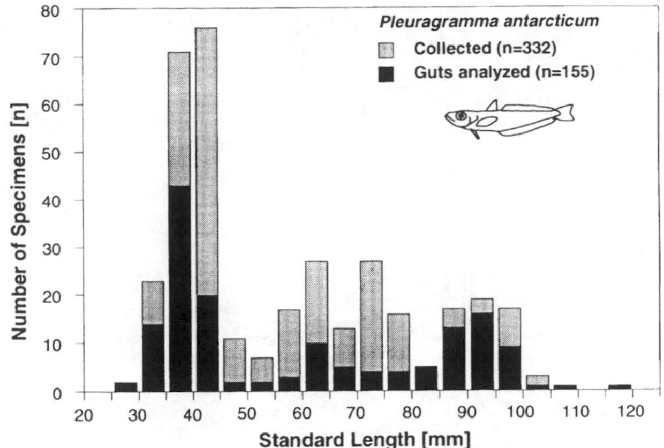

Fig. 38.2. Length/frequency distributions of all undamaged *Pleuragramma antarcticum* specimens collected (shaded columns) and of those investigated for the gut content analyses (black columns).

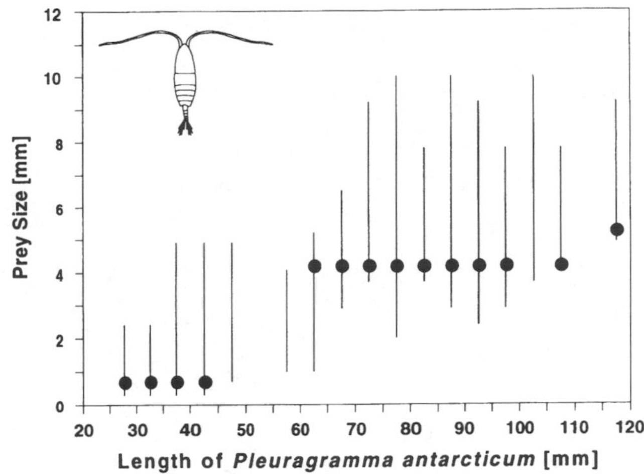

Fig. 38.3. Size range (lines) and modal groups (filled circles) of prey items versus standard length of *Pleuragramma antarcticum*.

potassium chloride solution. After centrifuging (2612 g, 30 min., 5 °C), the aquatic phase was discarded and the lower phase, which contained the extracted lipid, was transferred to a pre-weighed glass vial. The lipid extracts were evaporated under a flow of nitrogen to near dryness, dried in a vacuum desiccator (30 min.) and weighed (±0.1 mg). Lipid contents are expressed as a percentage of dry weight (%DW). Altogether, 24 specimens of *Pleuragramma antarcticum* with a standard length between 29 to 72 mm were analysed.

RESULTS

The RMT and krill nets deployed in the upper 500 m yielded a total of 484 juvenile fishes, of which 469 (97%) belonged to age classes 1 and older of *Pleuragramma antarcticum*. Second in abundance were specimens of the notothenioid family Channichthyidae. Starting in early November, freshly hatched yolk-sac larvae of *P. antarcticum* and channichthyids occurred in bongo hauls; these were not considered for the present study.

Of 332 undamaged *Pleuragramma antarcticum* 155 specimens were selected for the gut content analyses. The fish covered a size spectrum between 29 mm and 115 mm SL (Fig. 38.2). Lengths of 35–45 mm can be attributed to one-year-old fishes born in November 1985. The subsequent sizes of 60–75 mm and 90–95 mm may represent ages 2 and 3, respectively (Hubold & Tomo 1989). The length distribution shows that during winter several year classes of juvenile *P. antarcticum* remain in the upper water layers.

Of the fishes investigated 148 specimens (96%) contained food. The number of food particles per gut ranged from 0 to 220 (mean of 29). The diet consisted of 42 distinguishable items including 14 species of copepods represented by various developmental stages (Table 38.1). Prey sizes were between 0.1 and >10 mm.

In all guts investigated, the cyclopoid copepods *Oncaea* spp. dominated by numbers and comprised 49%. *Calanus propinquus* (copepodite stage V) and *Oithona* spp. each contributed 12%. *Microcalanus* spp. and *Ctenocalanus* spp. accounted for 6% and 4% of the food items, respectively. Smaller and larger food particles, e.g. copepod eggs of 0.1 mm (in clusters) and remains of chaetognaths and euphausiids (≥10 mm), occurred only occasionally.

Prey sizes in *Pleuragramma antarcticum* guts were clearly related to fish lengths. Fishes from <30 mm to 50 mm contained prey items between 0.1 and 5 mm long. In this size range the prey length mode was 0.7 mm (Fig. 38.3). Fishes >60 mm switched to larger prey sizes between 2 mm and 10 mm with a distinctive size mode at 4.2 mm. From the prey size distribution, two feeding types can thus be distinguished related to fish sizes <50 mm and >60 mm, respectively.

The prey size mode at 0.7 mm in the smaller *Pleuragramma antarcticum* juveniles was due to the large number of cyclopoid copepods *Oncaea* spp. in the guts. They represented more than 60% of all prey items consumed by the <50 mm *P. antarcticum* size group (Fig.38.4A). Second in abundance was *Oithona* spp. of 0.9 mm (15%). By biomass, *Oncaea* spp. represented 24% (Fig. 38.4B). The much larger *Calanoides acutus* (mainly females, 4.9 mm) were frequently found in guts of *P. antarcticum* from 35 mm SL onwards. Although they occurred in low numbers, in terms of biomass this prey made up more than a third of the total.

In the >60 mm size group, the most numerous prey item was *Calanus propinquus* (mainly copepodite stage V), which represented two thirds of the food items (Fig. 38.4C) and 70% of the food biomass (Fig. 38.4D). Other calanoid copepods such as *Rhincalanus gigas* (11%) and *Calanoides acutus* (8%) were of secondary importance.

Total lipid contents were rather stable in the *Pleuragramma antarcticum* size group <50 mm (year class 1, $n=15$) and ranged between 15–22% DW with a mean of 18% DW. However, there was a clear increase in lipid content discernible in the larger specimens between 55–72 mm with a mean of 37% DW. Lipid accumulation appeared to be correlated with length and lipid levels increased steadily from 29% DW in a 55 mm specimen to

Table 38.1. *Diet composition (shown as number of prey items) of size classes of* Pleuragramma antarcticum *(Stage – C: copepodid, M: male, F: female)*

Prey items	Stage	Size (mm)	*Pleuragramma* size interval (mm)																		
			<30	<35	<40	<45	<50	<55	<60	<65	<70	<75	<80	<85	<90	<95	<100	<105	<110	<115	<120
Eggs		0.1	—	45	78	16	—	—	—	—	—	—	—	—	—	—	—	—	—	—	—
Nauplii indet.		0.3	1	3	12	2	—	—	—	—	—	—	—	—	—	—	—	—	—	—	—
Copepodids indet.		?	2	5	45	19	3	8	2	21	9	—	—	1	3	1	2	1	—	—	4
Microcalanus sp.		0.5	—	79	150	34	—	—	—	—	—	—	—	—	—	—	—	—	—	—	—
Oncaea sp.		0.7	74	348	1134	604	2	—	—	—	—	—	—	—	—	—	—	—	—	—	—
Oithona sp.		0.9	8	47	305	147	2	—	—	—	—	—	—	—	—	—	—	—	—	—	—
Metridia gerlachei	CII	1.0	—	—	3	—	—	—	—	—	—	—	—	—	—	—	—	—	—	—	—
Scolecitricella sp.		1.0	—	2	50	8	1	—	—	1	—	—	—	—	—	—	—	—	—	—	—
Stephos longipes		1.0	—	—	9	—	—	—	—	—	—	—	—	—	—	—	—	—	—	—	—
Ctenocalanus sp.		1.0	—	22	124	33	4	—	1	—	—	—	—	—	—	—	—	—	—	—	—
Euchaeta sp.	C?	1.1	—	—	2	14	—	—	—	—	—	—	—	—	—	—	—	—	—	—	—
Euchaeta sp.	CI	1.1	—	—	—	2	—	—	—	—	—	—	—	—	—	—	—	—	—	—	—
Metridia gerlachei	CIII	1.2	1	2	17	8	—	—	—	—	—	—	—	—	—	—	—	—	—	—	—
Spinocalanus sp.		1.3	—	—	—	—	—	—	—	1	—	—	—	—	—	—	—	—	—	—	—
Calanoides acutus	CII	1.4	—	—	1	—	—	—	—	—	—	—	—	—	—	—	—	—	—	—	—
Calanus propinquus	C?	1.5	—	—	2	—	—	—	—	—	—	—	—	—	—	—	—	—	—	—	—
Metridia gerlachei	CIV	1.6	2	4	16	5	—	—	—	—	—	—	—	—	—	—	—	—	—	—	—
Scaphocalanus sp.		1.7	—	—	—	—	—	—	—	1	—	—	—	—	—	—	—	—	—	—	—
Euchaeta sp.	CII	1.7	—	—	1	1	—	—	—	—	—	—	—	—	—	—	—	—	—	—	—
Calanoides propinquus	CIII	2.0	—	—	6	3	1	—	—	1	—	—	1	—	—	—	—	—	—	—	—
Calanoides acutus	CIII	2.4	—	—	11	3	3	—	—	—	—	—	—	—	—	2	—	—	—	—	—
Metridia gerlachei	CV	2.4	1	1	6	4	—	—	—	—	—	—	—	—	—	—	—	—	—	—	—
Metridia gerlachei	M	2.4	—	—	3	1	—	—	—	—	—	—	—	—	—	—	—	—	—	—	—
Euchaeta sp.	CIII	2.6	—	—	1	11	—	—	—	1	—	—	—	—	—	1	—	—	—	—	—
Calanus propinquus	CIV	2.9	—	—	1	1	—	—	—	1	1	—	—	—	1	3	1	—	—	—	—
Calanoides acutus	CIV	3.3	—	—	8	8	3	—	1	3	—	—	—	—	—	2	—	—	—	—	—
Heterorhabdus minor		3.7	—	—	—	—	—	—	—	1	—	1	—	2	1	—	3	—	—	—	—
Metridia gerlachei	F	3.7	—	—	2	1	—	—	1	4	1	1	2	1	4	15	1	1	—	—	—
Rhincalanus gigas	CIII	3.7	—	—	—	1	—	—	—	—	—	—	—	—	—	—	—	—	—	—	—
Euchaeta sp.	CIV	4.0	—	—	—	—	—	—	—	—	—	—	—	—	—	1	—	—	—	—	—
Calanoides acutus	CV	4.1	—	—	4	1	3	—	3	2	1	3	1	2	7	10	—	—	1	—	—
Calanus propinquus	CV	4.2	—	—	1	1	—	—	—	10	9	34	43	56	108	134	44	—	8	—	—
Calanoides acutus	F	4.9	—	—	9	14	1	—	—	3	—	2	7	3	8	2	2	—	1	—	1
Calanus propinquus	M	5.0	—	—	—	—	—	—	—	1	—	1	—	1	1	—	4	—	1	—	—
Rhincalanus gigas	CIV	5.0	—	—	—	—	—	—	—	—	—	—	—	—	—	1	1	—	—	—	—
Calanus propinquus	F	5.2	—	—	—	—	—	—	—	4	—	3	3	7	10	16	3	—	1	—	2
Euchaeta sp.	CV	6.0	—	—	—	—	—	—	—	—	—	—	—	2	1	—	—	—	—	—	—
Rhincalanus gigas	CV	6.5	—	—	—	—	—	—	—	—	2	3	8	2	4	6	3	—	2	—	—
Rhincalanus gigas	F	7.8	—	—	—	—	—	—	—	—	—	4	5	3	15	15	11	—	1	—	—
Euchaeta antarctica	F	9.2	—	—	—	—	—	—	—	—	—	1	—	—	1	2	—	—	—	—	1
Chaetognatha indet.		10.0	—	—	—	—	—	—	—	—	—	—	1	—	1	—	—	—	—	—	—
Euphausia crystallorophias	juv.	10.0	—	—	—	—	—	—	—	—	—	—	—	—	—	—	—	1	—	—	—
Total (excl. eggs)			89	513	1923	926	23	8	8	55	23	53	71	80	165	211	75	3	15	0	8

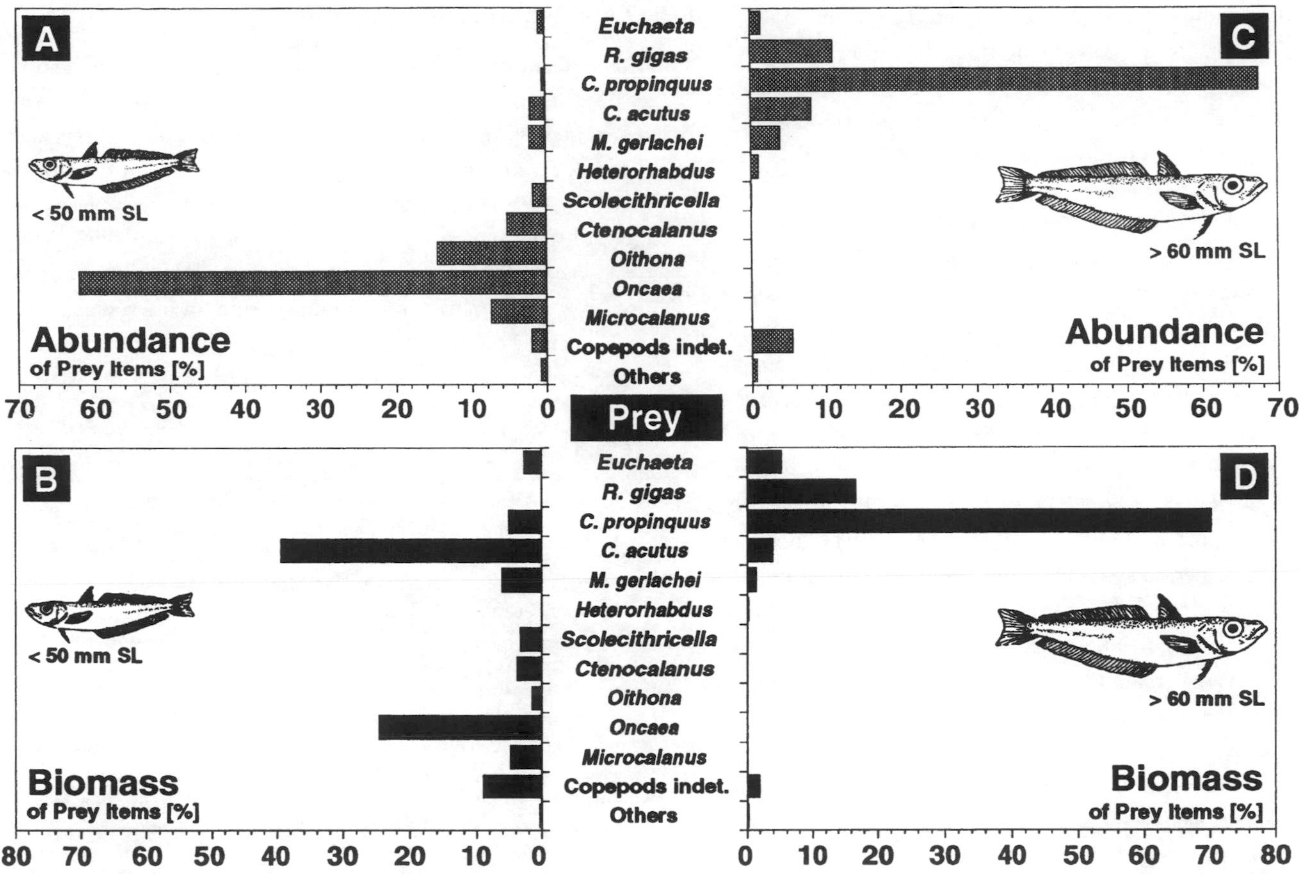

Fig. 38.4. A–D. Abundance and biomass of prey items (in decreasing order of prey size, except for the last two items) in *Pleuragramma antarcticum* <50 mm (A+B) and *Pleuragramma antarcticum* >60 mm (C+D).

41% DW in a 72 mm specimen. These winter lipid data agree well with late summer data for similar fish sizes from the same area (Fig. 38.5).

DISCUSSION

Summer investigations revealed a distinct vertical separation of the different age classes of *Pleuragramma antarcticum* with a clear positive relationship between water depth and fish size. This phenomenon was interpreted as an adaptation to avoid cannibalistic feeding on larvae and juveniles by adults (Hubold 1985, 1992). The winter/spring study shows that during their first two to three years juvenile *P. antarcticum* inhabit the surface layers of the southern Weddell Sea during all seasons. The fishes hence maintain a year-round vertical separation of age classes, in spite of the pronounced seasonal vertical migrations of some of their major prey items.

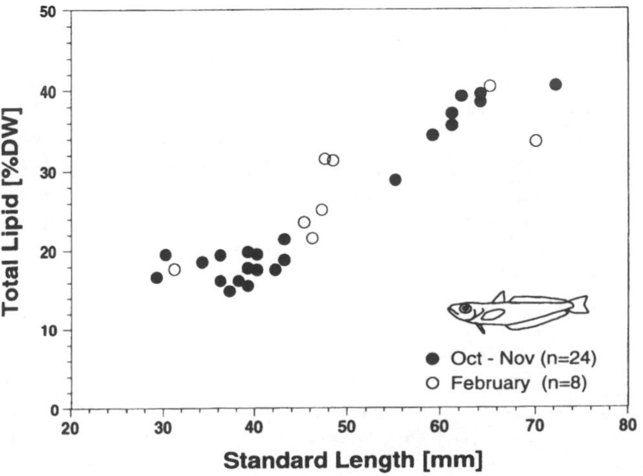

Fig. 38.5. Lipid content versus standard length of *Pleuragramma antarcticum* specimens (filled circles: late winter data from October/November 1986; open circles: summer data from February 1985 after Hagen, 1988).

In summer, food of juvenile *Pleuragramma antarcticum* (40–60 mm) consisted mainly of the calanoid copepods *Calanoides acutus* and *Metridia gerlachei* and of larval *Euphausia crystallorophias* (calyptopis I–III) in the size range 1.5–4.9 mm. The calyptopis stages I and II (<3 mm) made up about half of these prey items (Hubold & Ekau 1990). The larger *P. antarcticum* specimens between 60 mm and 140 mm contained *C. acutus, M. gerlachei* and *C. propinquus* as well as subadult and adult *E. crystallorophias* (prey size 3–30 mm). The dominant prey size (by numbers) of these larger fishes was between 4 mm and 5 mm. The overall prey spectrum of *P. antarcticum* in summer consisted of 12 species including various developmental stages. The mean number of prey items per gut was 20 (Hubold & Ekau 1990).

In comparison with these summer data the diet composition in late winter was characterized by a reduced number of calanoid copepods (*Calanoides acutus, Metridia gerlachei*) and an almost complete absence of euphausiids. On the other hand, *Calanus propinquus* was a major prey species in the larger juvenile *Pleuragramma antarcticum* (>60 mm) during winter. This reflects the seasonal changes in zooplankton composition and indicates the opportunistic feeding of *P. antarcticum*. The diet of the smaller *Pleuragramma* specimens (<50 mm) in late winter was dominated by cyclopoid copepods. This may be due to the absence of other suitable prey for these smaller fishes, such as early developmental stages of calanoid copepods and euphausiids, at that time of the year. It is also a continuation of the feeding behaviour determined in postlarval *P. antarcticum* (13–25 mm) during the first summer when three quarters of their diet consisted of the cyclopoids *Oithona* and *Oncaea* (von Dorrien 1989). Hence, prey selection reflects seasonal as well as ontogenetic influences. During its early life history *P. antarcticum* seems to rely almost completely on cyclopoid copepods. This period spans from the initial feeding during the first summer and the following winter to the second summer, when the fishes switch to larger prey at sizes of >60 mm onwards. The results show that in late winter juvenile *Pleuragramma antarcticum* is able to feed sufficiently in the upper water layers, although overall zooplankton density is strongly reduced (Hubold & Hempel 1987, Schnack-Schiel *et al.* 1991, Schnack-Schiel & Hagen 1994).

This foraging activity of *Pleuragramma antarcticum* in October/November was apparently preceded by similar behaviour during the winter period, as can be inferred from the chemical composition of *P. antarcticum*. The analyses revealed total lipid contents of 15% to 41% DW depending on the size of the specimens. These values compare well with similar lipid levels of *P. antarcticum* specimens collected in the same area during late summer 1985 (Hagen 1988). Seasonal variability in zooplankton occurrence does not therefore seem to affect the physiological condition of juvenile *P. antarcticum*. The species is an opportunistic predator well-adapted to feed on a wide spectrum of zooplankton depending on seasonal availability. High summer concentrations of lipid-rich copepods are apparently not utilized to accumulate large energy reserves for a winter season of low food supply. Instead, the low zooplankton levels available in winter are sufficient to support the population. The life cycle of juvenile *P. antarcticum* seems to be uncoupled from the highly seasonal plankton production and *P. antarcticum* specimens maintain a rather stable amount of lipid according to size. The strong increase in lipid content with size leads to extremely high lipid levels of up to 58% DW in the adult specimens (Friedrich & Hagen 1994). This indicates that lipid deposits are in fact mainly needed as buoyancy aid to compensate for the increasing weight with age in this swimbladderless species, rather than an energy reserve for the winter season (see also DeVries & Eastman 1978, Clarke *et al.* 1984, Eastman & DeVries 1989, Friedrich & Hagen 1994). It shows furthermore that the energetic basis for the juvenile *Pleuragramma* population is provided by the low amount of winter food, rather than by the elevated plankton production during the short summer period.

ACKNOWLEDGEMENTS

This study would not have been possible without the cooperation of Elke Mizdalski during the gut content analyses and her expertise in identifying the prey species and stages. We acknowledge the support and patience of the chief scientist, Prof. Gotthilf Hempel, during the cruise. We are grateful to the captain and crew of RV *Polarstern* for the professional help during the deployment of the nets under extremely difficult sea-ice conditions. We thank Christopher Zimmermann for assistance in the graphic presentation. The manuscript benefited from the critical suggestions provided by the anonymous reviewers.

REFERENCES

Boysen-Ennen, E., Hagen, W., Hubold, G. & Piatkowski, U. 1991. Zooplankton biomass in the ice-covered Weddell Sea, Antarctica. *Marine Biology*, **111**, 227–235.

Clarke, A., Doherty, N., DeVries, A. L. & Eastman, J. T. 1984. Lipid content and composition of three species of Antarctic fish in relation to buoyancy. *Polar Biology*, **3**, 77–83.

DeVries, A. L. & Eastman, J. T. 1978. Lipid sacs as a buoyancy adaptation in an Antarctic fish. *Nature*, **271**, 352–353.

Dorrien, C. von. 1989. *Ichthyoplankton in Abhängigkeit von Hydrographie und Zooplankton im Weddellmeer*. [Ichthyoplankton in relation to hydrography and zooplankton in the Weddell Sea.] MSc thesis, Kiel University, 66 pp. [Unpublished.]

Eastman, J. T. & DeVries, A. L. 1989. Ultrastructure of the lipid sac wall in the Antarctic notothenioid fish *Pleuragramma antarcticum*. *Polar Biology*, **9**, 333–335.

Folch, J., Lees, M. & Sloane-Stanley, G. H. 1957. A simple method for the isolation and purification of total lipids from animal tissues. *Journal of Biological Chemistry*, **226**, 497–509.

Foxton, P. 1956. The distribution of the standing crop of zooplankton in the Southern Ocean. *Discovery Report*, **28**, 191–236.

Fransz, H. G. 1988. Vernal abundance, structure and development of epipelagic copepod populations of the eastern Weddell Sea (Antarctica). *Polar Biology*, **9**, 107–114.

Friedrich, C. & Hagen, W. 1994. Lipid contents of five species of notothenioid fish from high-Antarctic waters and ecological implications. *Polar Biology*, **14**, 359–369.

Fukuchi, M., Tanimura, A. & Ohtsuka, H. 1985. Zooplankton community conditions under sea ice near Syowa Station, Antarctica. *Bulletin of Marine Science*, **37**, 518–528.

Gieskes, W. W. C., Veth, C., Wöhrmann, A. & Gräfe, M. 1987. Secchi disc visibility world record shattered. *Eos*, **68**, 123.

Hagen, W. 1988. Zur Bedeutung der Lipide im Antarktischen Zooplankton. [On the significance of lipids in Antarctic zooplankton.] *Berichte zur Polarforschung*, **49**, 1–129. [English Version: *Canadian Translation of Fisheries and Aquatic Sciences*, **5458**, 149 pp.]

Hosie, G.W., Ikeda, T. & Stolp, M. 1988. Distribution, abundance and population structure of the Antarctic krill (*Euphausia superba* Dana) in the Prydz Bay region, Antarctica. *Polar Biology*, **8**, 213–224.

Hubold, G. 1985. The early life-history of the high-Antarctic silverfish *Pleuragramma antarcticum*. In Siegfried, W. R., Condy, P. R. & Laws, R. M., eds. *Antarctic nutrient cycles and food webs*. Berlin: Springer-Verlag, 445–451.

Hubold, G. 1990. Seasonal patterns of ichthyoplankton distribution and abundance in the Southern Weddell Sea. In Kerry, K. R. & Hempel, G., eds. *Antarctic ecosystems. Ecological change and conservation*. Berlin: Springer-Verlag, 149–158.

Hubold, G. 1992. Zur Ökologie der Fische im Weddellmeer. [Ecology of Weddell Sea fishes.] *Berichte zur Polarforschung*, **103**, 1–157.

Hubold, G. & Ekau, W. 1990. Feeding patterns of postlarval and juvenile notothenioids in the southern Weddell Sea (Antarctica). *Polar Biology*, **10**, 255–260.

Hubold, G. & Hempel, I. 1987. Seasonal variability of zooplankton in the Weddell Sea. *Meeresforschung*, **31**, 185–192.

Hubold, G. & Tomo, A. 1989. Age and growth of Antarctic silverfish *Pleuragramma antarcticum* Boulenger, 1902, from the southern Weddell Sea and Antarctic Peninsula. *Polar Biology*, **9**, 205–212.

Metz, C. 1995. Seasonal variation in the distribution and abundance of *Oithona* and *Oncaea* species (Copepoda, Crustacea) in the southeastern Weddell Sea, Antarctica. *Polar Biology*, **15**, 3, 187–194.

Mizdalski, E. 1988. Weight and length data of zooplankton in the Weddell Sea in austral spring 1986 (ANT V/3). *Berichte zur Polarforschung*, **55**, 1–72.

Schnack-Schiel, S. B. 1987. The Winter Expedition of RV P*olarstern* to the Antarctic (ANT V/1–3). *Berichte zur Polarforschung*, **39**, 1–259.

Schnack-Schiel, S. B. & Hagen, W. 1994. Life cycle strategies and seasonal variations in distribution and population structure of four dominant calanoid copepod species in the eastern Weddell Sea, Antarctica. *Journal of Plankton Research*, **16**, 1543–1566.

Schnack-Schiel, S. B., Hagen, W. & Mizdalski, E. 1991. Seasonal comparison of *Calanoides acutus* and *Calanus propinquus* (Copepoda: Calanoida) in the southeastern Weddell Sea, Antarctica. *Marine Ecology Progress Series*, **70**, 17–27.

Voronina, N. M. 1970. Seasonal cycles of some common Antarctic copepod species. In Holdgate, M. V., ed. *Antarctic ecology*, Vol. 1. London: Academic Press, 162–172.

39 Spatial and life-cycle changes in lipid and fatty acid structure of the Antarctic euphausiid *Euphausia superba*

PATRICK MAYZAUD

Laboratoire d'Océanographie Biochimique et d'Ecologie, URA-CNRS 2077. Observatoire Océanologique, BP 28, 06230 Villefranche sur mer, France

ABSTRACT

Lipid composition of summer (January–February) Antarctic krill was studied in relation to feeding ecology (spatial variability) and internal requirements (developmental stages) in the south Indian sector (FIBEX and SIBEX cruises). Quantification of the lipid classes showed that triglycerides dominated the neutral lipid fraction. Polar lipids were dominated by phosphatidyl-choline (Ph-choline), with phosphatidyl-ethanolamine (Ph-ethanolamine) and an unknown glycolipid present in lesser proportions. Comparison between growth and sexual maturity stages showed that total lipid concentrations was maximal in gravid females (and ovaries), intermediate in spent females and subadults, and minimum in males. Both polar lipids and triglycerides showed a significant linear relationship with total lipid, suggesting that in E. superba *Ph-choline was used as depot fat contrary to Ph-ethanolamine and glycolipids, which appeared to be true structural components. Changes in growth or maturity stages was also reflected in the fatty-acid composition of the different lipid classes with the highest levels of unsaturation in Ph-choline and glycolipids. Ph-ethanolamine showed a dominance of saturates and monoenes with maximum percentages of polyunsaturates in spent females. In glycolipids, polyunsaturated fatty acid proportions were maximum in males, and decreased steadily from spent females to gravid female muscle to ovaries.*

During the FIBEX cruise, a comparison of the spatial variability in lipid content of krill populations was carried out using mixed-stage samples and stage-specific samples (mostly males). Parallel changes were observed between mixed stages and males suggesting a major influence of local trophic conditions. Spatial variability in fatty-acid composition of mixed populations was analysed by factorial correspondence analysis for both triglycerides and glycolipids. As anticipated, the influence of phytoplankton fatty acids largely explained the variance in the fatty-acid composition of E. superba *triglycerides. On the contrary, the spatial changes in the fatty-acid composition of glycolipids appeared related to variations both in stages of sexual maturity and in synthetic pathways (e.g. the activity of Δ^9 desaturase). The possible role of glycolipids in the reproduction cycle is discussed.*

Key words: lipids, fatty acids, *Euphausia superba*, reproduction.

INTRODUCTION

The current knowledge of the chemical composition of the Antarctic euphausiid *Euphausia superba* has expanded significantly over the past two decades. The early analyses by Raymont *et al.* (1971), Van der Veen *et al.* (1971), Bottino (1974, 1975), and Tsuyuki & Itoh (1976) were made on samples obtained from different locations of the Antarctic Ocean and illustrated the wide variability in values of major biochemical components, as well as in lipid and fatty acid composition. Further studies by Clarke (1980, 1984), Saether *et al.* (1986), Hagen (1988) and Virtue *et al.* (1993) have illustrated the confounding effect of life history and krill physiology on the lipid composition reported.

The central role of lipid storage in both the survival and reproductive strategies of *Euphausia superba* is now well established, but the information remains limited to populations from coastal locations (South Georgia, Bransfield Strait, Antarctic Peninsula or Prydz Bay) and little is known on the biochemical characteristics of open-ocean krill outside of the Weddell Sea area. In addition, the proposal by Ellingsen (1982, cited by Saether *et al.* 1986) that phospholipids may also serve as an energy reserve brings a renewed interest in the seasonal or spatial changes of their composition and fatty-acid structure.

This paper presents the results of lipid analyses of krill from the South Indian area of the Antarctic Ocean, with a special emphasis on the spatial variability at the population level.

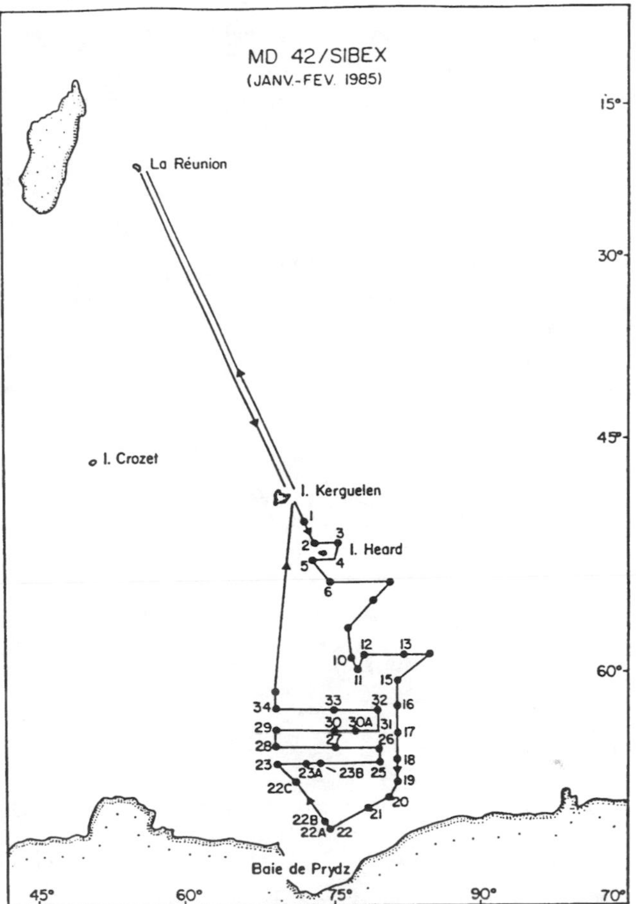

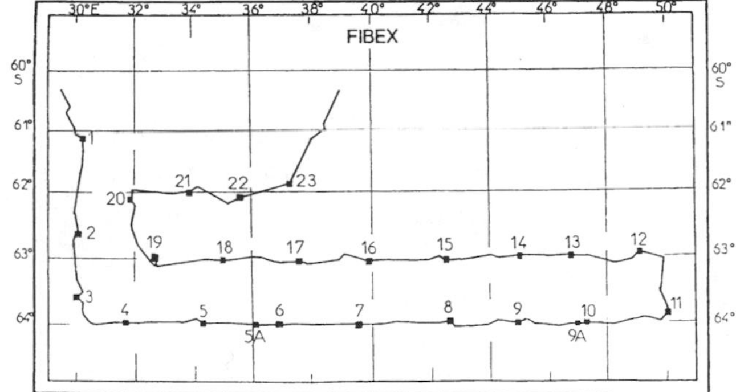

Fig. 39.1. Cruise tracks and stations operated as part of FIBEX (February 1981) and SIBEX (January–February 1985).

MATERIALS AND METHODS

Samples of *Euphausia superba* were obtained from RMT 8 oblique tows made to a depth of 100 m during two cruises of the RV *Marion Dufresne* in February 1981 (FIBEX) and January–February 1985 (SIBEX). Details of the sampling stations and their position are given in Fig. 39.1. The individuals were separated immediately after capture in the net, rinsed with distilled water and deep frozen (−80 °C). At most stations, 15 to 20 individuals were sorted to developmental stages according to Makarov & Denys (1981) before deep-freezing. Samples were stored at −70 °C under nitrogen and transported to France within four months.

Frozen krill was placed on crushed ice and brought to 0 °C. Size was estimated to the nearest mm and individuals were weighed on a Mettler balance (±0.05 mg) after draining excess water on sharkskin paper. Specimens were dried at 60 °C until constant weight to determine the dry weight:fresh weight ratios. Frozen gravid females were dissected to detach cephalothorax from abdomen. A razor blade was used and the individuals maintained on an ice-cold metal block to avoid any loss of internal fluid.

Lipid extraction was carried out according to the method of Bligh & Dyer (1959). Extraction was made upon arrival in the laboratory and the lipids stored in sealed vials under argon at −70 °C. Further analyses were carried out within the next 18 months. Total lipids from bulk samples were weighed in tarred vials and the concentration measured using a microbalance (±0.001 mg). Total lipid content from each specimen was estimated using the spectrophotometric method of Barnes & Blackstock (1973). Comparison of both methods yielded similar results.

Lipid classes were quantified after chromatographic separation coupled with FID detection on a Iatroscan Mark III TH 10 (Ackman 1981). Total lipids were applied to chromarod SIII using microcapillaries (1 μl) and analysed in duplicate. Neutral lipids were separated using a double development procedure with the following solvent systems: n-hexane:benzene:formic acid 80:20:0.5 (by volume) followed by n-hexane:diethyl ether:formic acid 97:3:0.5 (v/v). Glycolipids were separated

Table 39.1. *Lipid composition (% wet weight) and lipid class composition (% total lipids) of* Euphausia superba, *SIBEX:station 23B, 1985*

	Juvenile	Male	Gravid Female			Spent female
			Whole	Cephalothorax	Abdomen	
Total lipid content						
(% wet weight (±SD))	2.93	2.29	4.23	3.94	0.59	2.97
	(1.02)	(0.75)	(1.22)	(1.12)	(0.07)	(0.98)
Lipid classes						
(% total lipids (±SD))						
Phosphatidyl choline	29.3	28.0	24.2	30.3	17.8	27.0
	(1.5)	(0.6)	(1.5)	(0.6)	(4.4)	(2.3)
Phosphatidyl ethanolamine	14.7	17.6	11.5	7.9	11.1	18.6
	(0.3)	(0.9)	(0.8)	(0.2)	(0.1)	(2.8)
Acetone mobile glycolipid	11.7	17.0	15.9	11.2	11.3	19.6
	(0.6)	(0.2)	(0.6)	(0.8)	(0.2)	(1.1)
Triglycerides	31.0	26.4	40.4	47.2	59.7	30.6
	(0.9)	(0.4)	(1.2)	(1.9)	(17.6)	(1.5)
Free fatty acids	ND	Tr	ND	ND	Tr	ND
Cholesterol	13.3	11.1	1.2	0.8	1.0	4.2
	(0.9)	(1.0)	(0.06)	(0.03)	(0.05)	(0.8)
Sterol ester	ND	ND	5.6	2.6	Tr	Tr
			(0.03)	(0.08)		

Notes:
ND=not detected; Tr=traces not detected by the integrator (<0.5%). Analyses on five replicates.

according to Hirayama & Morita (1980) with chloroform:ethyl acetate:acetone: methanol:acetic acid:H_2O 60:12:15:16:3:3 (v/v). Phospholipids were separated with chloroform: methanol:H_2O 65:35:4 (v/v).

Neutral and polar lipids were isolated on a preparative scale by column chromatography on silica gel (Bio-Sil HA, minus 325 mesh). The neutral lipid fraction was eluted with 6 column volumes of chloroform, the acetone-mobile compounds were eluted with 4 volumes of acetone and the phospholipids were eluted with 6 volumes of methanol. All operations took place under nitrogen. Each fraction collected was further separated by thin-layer chromatography (TLC) on precoated silica gel plates (Analtech, Uniplate) and developed with hexane:diethyl ether:acetic acid 80:20:1.5 (v/v) for neutral lipids, or chloroform:methanol:aqueous ammonia 85: 30: 1 (v/v) for glycolipids, and chloroform:methanol:acetic acid:H_2O 25:15:4:2 (v/v) for polar lipids. Lipid classes were visualized using dichlorofluorescein and identification was achieved by comparison with standard mixtures. Specific detection of glycolipids on TLC was made with diphenylamine (Stahl 1969).

Fatty acid methyl esters (FAME) of total and individual lipid classes were prepared with 7% boron trifluoride in methanol (Morrison & Smith 1964). Gas liquid chromatography (GLC) of all esters were carried out on a 46 mm length×0.25 mm ID quartz capillary column coated with Silar 5CP in a Perkin-Elmer 8320 gas chromatograph equipped with a flame ionization detector. The column was operated isothermally at 165 °C. Helium was used a carrier gas at 65 psi. Detector and injector temperatures were maintained at 250 °C. In addition to the examination of esters recovered, part of all ester samples was completely hydrogenated and the product examined quantitatively and qualitatively by GLC. The quantitative results are given to two decimal places, but this does not imply this level of accuracy. Major components (>10%) should be accurate to ± 5%, moderate sized components (1 to 9%) to ±10% and minor components (<1%) to up to ±30%.

RESULTS

Lipid content and lipid classes composition.

The lipid composition of subadult and adult *Euphausia superba* is given in Table 39.1. Total lipid content, on a wet weight basis, was maximum for gravid females, with most lipid accumulated in the cephalothorax (mostly ovaries). Subadults and spent females showed intermediate concentrations, whereas minimum levels were observed in males. The composition was similar in all stages with polar lipids made up mostly of phosphatidylcholine, phosphatidyl-ethanolamine and an unknown glycolipid. Triglycerides dominated the neutral fractions with cholesterol, and sterol esters as minor constituents.

Lipids in gravid females were dominated by triglycerides (40% of total lipids), with the largest percentage (almost 60%) in the abdomen muscles. Phosphatidyl-choline was the second constituent in importance with similar levels for all developmental stages. Phosphatidyl-ethanolamine concentrations ranged from 8 to 19%, with maximum values in spent females and males and minimum values in ovaries. Cholesterol and sterol esters were always in small concentrations (1–5%), with sterol esters located essentially in the ovaries. Subadults and spent females displayed similar levels of triglycerides (30%), while males showed minimal accumulation (26%). Concentrations of acetone mobile glycolipids were lowest in juveniles (12%) and highest in spent females (20%).

Spatial changes in lipid content and lipid classes composition. Krill was sampled at 13 stations during the FIBEX cruise. On a wet weight basis, lipid content displayed high levels of variation with sexual maturity stage within a station, and between stations (Fig. 39.2). The four stages were not always present in the various samples, but a comparison between lipid content of the krill mixed population and of adult males showed that lipid accumulation followed a similar pattern over space (r=0.562, N=13), despite differences in the proportion of the population structure sampled at each station (Table 39.2). This suggests that spatial changes are not solely related to the proportion of growth stages but are also influenced by local conditions. Despite a bimodal east–west distribution of the populations and food resources over the sampling area (Mayzaud *et al.* 1985,

Table 39.2. *Developmental stage structure of the total population of* Euphausia superba *sampled during the FIBEX cruise*

Station	Developmental stage (% total)		
	Juveniles	Males	Females
2	96	2	2
3	37	36	27
5	70	10	20
5A	13	51	36
6	0	40	60
9	6	72	22
9A	16	37	47
10	10	45	45
11	5	68	27
17	25	34	41
18	90	7	3
19	49	37	14
23	56	24	20

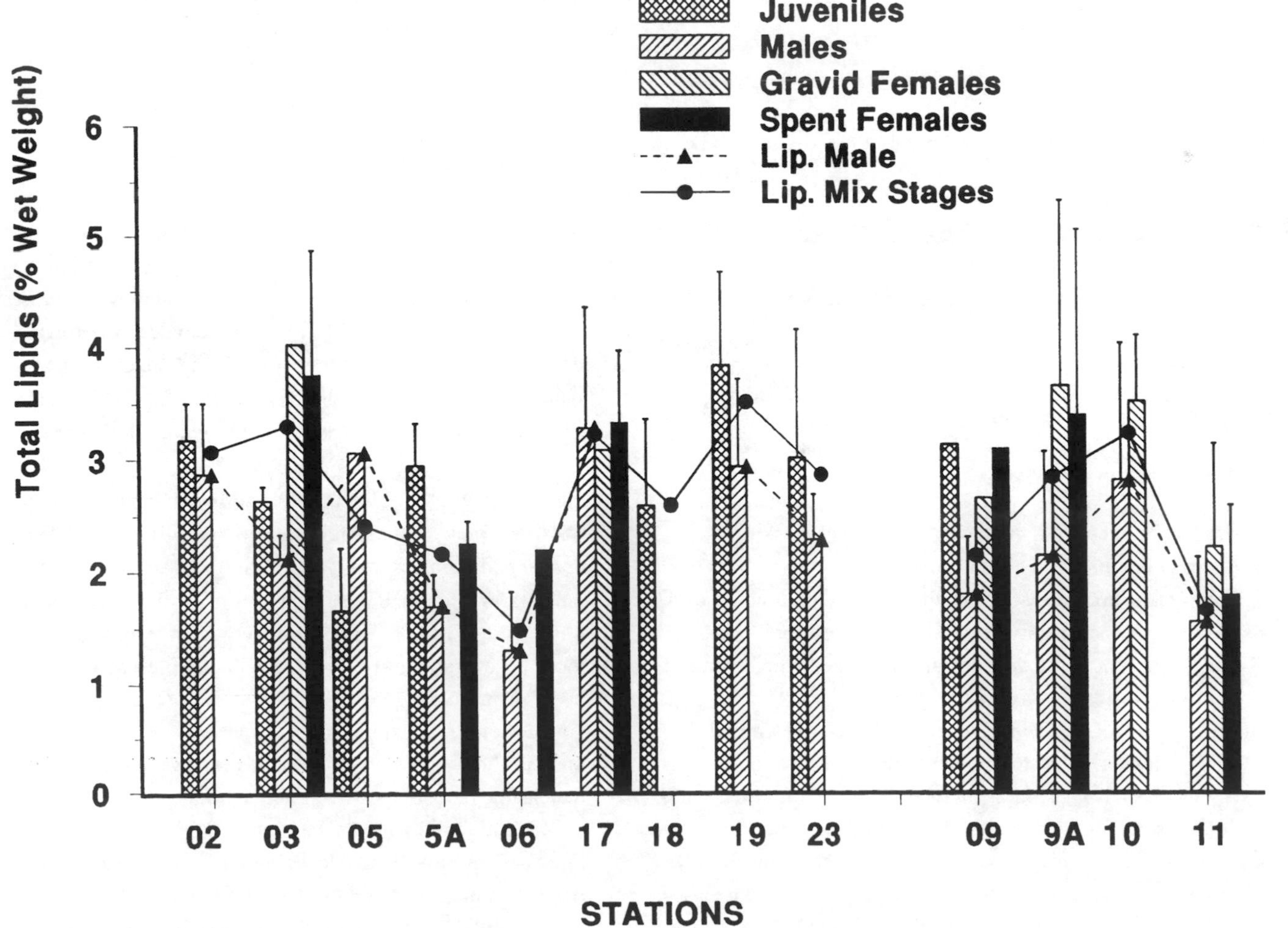

Fig. 39.2. Spatial changes in lipid content for the different developmental stages sampled during the FIBEX cruise. Vertical bars=standard deviations.

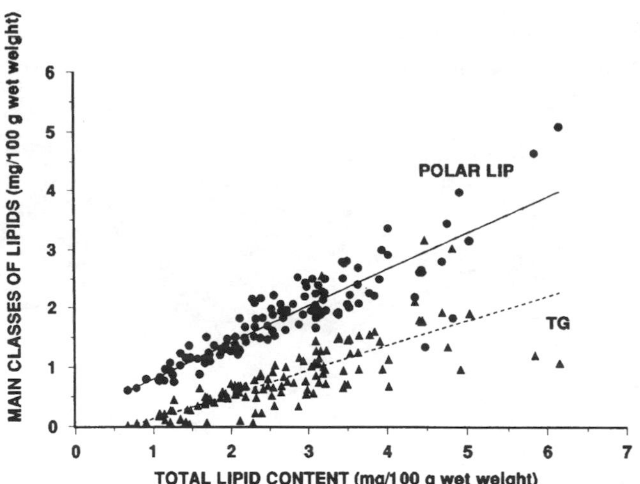

Fig. 39.3. Relationships between total lipid content and concentration of the two main lipid classes in *Euphausia superba* sampled during FIBEX. All developmental stages sampled at the various location were combined. Regression equations are: Polar lipids (Polar lip)=0.18+0.615 total lipids (R^2=0.793, n=124). Triglycerides (TG) =0.28+0.410 total lipids (R^2=0.602, n=124).

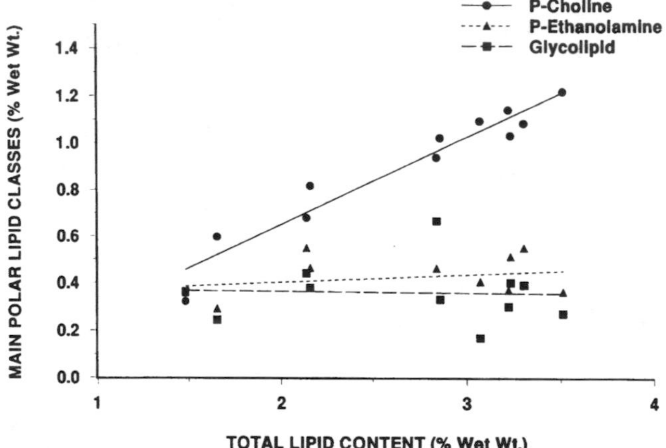

Fig. 39.4. Relationships between total lipid content and concentrations of the main polar lipids in extracts of total populations of *Euphausia superba* sampled during SIBEX. Regression equations are: Phosphatidyl-choline=−0.10+0.372 total lipids (R^2=0.930, n=11). Phosphatidyl-ethanolamine=0.34+0.030 total lipids (R^2=0.062, n=11). Glycolipids=0.38+−0.011 total lipids (R^2=0.001, n=11).

Farber-Lorda 1986), no significant differences could be established either north–south or east–west.

Changes in lipid class composition were followed at each station for both specific growth stages and total populations. For all stages combined, the concentrations of polar lipids and of triglycerides displayed linear relationships with total lipid concentration (Fig. 39.3). Regressions were significantly different with a higher slope for polar lipids than for triglycerides ($F_{1,244}$=24, p>0.001). Separation of polar lipids into their main constituents was carried out on the extract from the total population and showed clearly that unlike phosphatidyl-ethanolamine and glycolipids, phosphatidyl-choline increased linearly with total lipid concentration (Fig. 39.4). A cluster analysis of the distance matrix 1-R (Pearson correlation coefficient), computed from the spatial changes in lipid classes

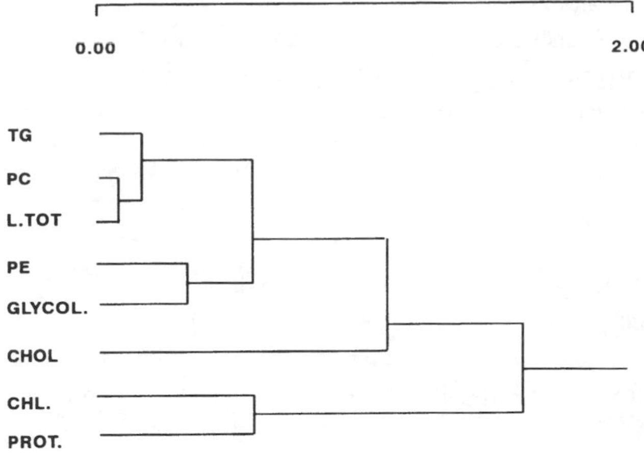

Fig. 39.5. Single linkage grouping of total lipid (L.TOT) classes (as μg animal^{-1}) for *Euphausia superba* and particulate chlorophyll (CHL) and protein (PROT) sampled during FIBEX. Grouping is based on the distance 1-R (Pearson). TG=triglycerides, PE=phosphatidyl-ethanolamine, PC=phosphatidyl-choline, GLYCOL=glycolipids, CHOL=choline.

over the FIBEX sampling area, confirmed the strong similarity between total lipids, triglycerides and phosphatidyl-choline on the one hand, and phosphatidyl- ethanolamine and glycolipids on the other hand (Fig. 39.5). No correlation could be established between lipid content and a local food supply index such as chlorophyll (phytoplankton) or protein (phytoplankton and microzooplankton).

Fatty acid composition of krill lipid classes and developmental stages.

The fatty acid composition of the main classes of polar and neutral lipids of *Euphausia superba* was analysed according to sex and stage of sexual development, and arranged by degree of unsaturation (Figs. 39.6 and 39.7). Gravid females were further divided into cephalothorax (for ovaries and digestive gland lipids) and digested gland (for lipids in muscles) to establish the respective composition of ovaries, digestive glands and muscles.

- Phosphatidyl-ethanolamine showed intermediate levels of unsaturation with maximal proportions of 20:5 and 22:6 in non-gravid females (42% of total fatty-acids). Ovaries, muscles and male lipids showed similar levels of polyunsaturates (18–22%) and relatively high percentages of monounsaturated fatty acids (38–41%), represented mostly by vaccenic acid.
- The glycolipid fraction was characterized by high levels of unsaturation with minimum values for ovaries (26%) and increasing percentages in gravid female muscle (33.6%), non-gravid females (46.2%) and males (57.6%). Total saturates and monoenes displayed similar compositions with the dominance of 16:0, 14:0, 18:1 (n-9) and 18:1(n-7).
- Phosphatidyl-choline showed higher levels of unsaturation than phosphatidyl-ethanolamine with maximum values for male individuals (52%) and intermediate percentages (39%)

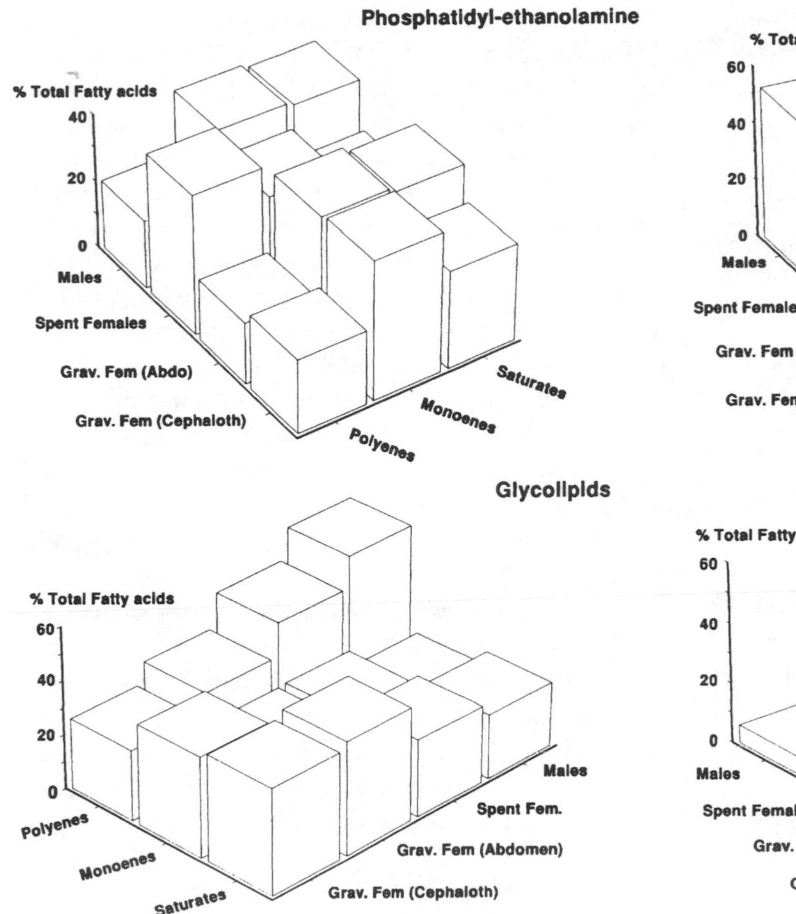

Fig. 39.6. Influence of the sexual development on the degree of unsaturation of fatty acids per class in phosphatidyl-ethanolamine and glycolipids of *E. superba* from SIBEX samples. Symbols : Grav. Fem=gravid females; Abdo=abdomen; Cephaloth=cephalothorax.

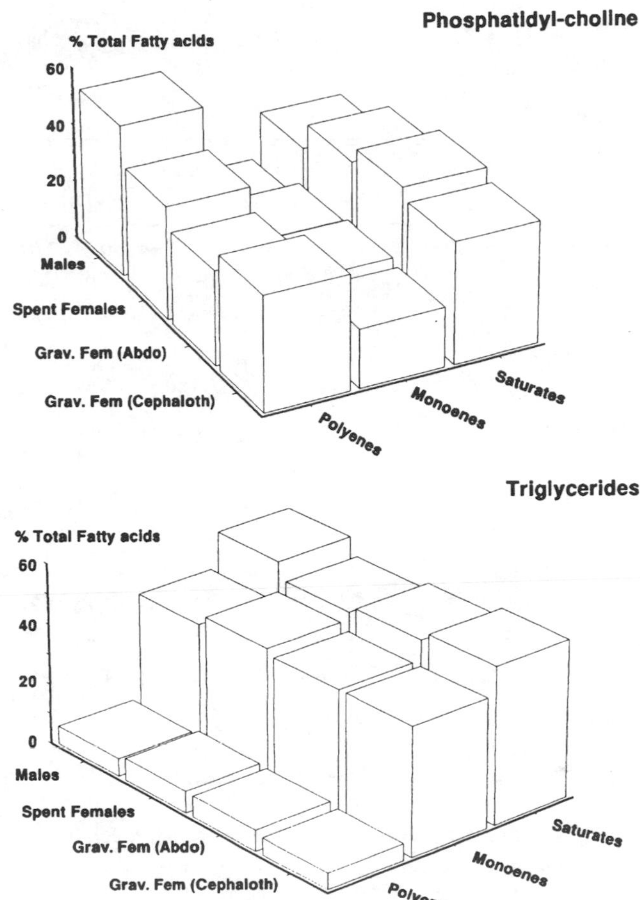

Fig. 39.7. Influence of the sexual development on the degree of unsaturation of fatty acids per class in phosphatidyl-choline and triglycerides of *E. superba* from SIBEX samples. Abbreviations : see Fig.39.6

for gravid female ovaries and spent females. Total monoenes and saturated acids followed similar patterns with greater levels of 16:0, 18:1(*n*-9) and 18:1(*n*-7) in gravid female ovaries and muscles.

- Triglycerides were always considerably less unsaturated than polar lipids, regardless of sex or fraction considered, due to smaller percentages in 20:5*n*-3 and 22:6*n*-3 and the occurrence of high levels of 14:0, 16:0, 16:1(*n*-7) and 18:1(*n*-9).

The isomer ratio 16:1(*n*-7)/16:0 displayed values <1 for all stages considered and showed increasing values from polar lipids to triglycerides in all cases (Fig. 39.8). An opposite trend was observed for 18:1(*n*-7)/18:1(*n*-9), with values ranging from 0.4 (triglycerides) and 11 (phosphatidyl-ethanolamine) but it appears to represent different pathways of fatty acid synthesis and/or incorporation (Fig. 39.8). Glycolipid and triglyceride ratios illustrated a high concentration of 18:1(*n*-9), while phosphatidyl-choline and phosphatidyl-ethanolamine ratios showed a higher proportion of 18:1(*n*-7). The ratio of *n*-3/*n*-6 acids indicated constitutive differences between triglycerides and polar fractions with respective values ranging from 1.6 (high proportion of 18:2(*n*-6)) to >10 (dominance of 20:5 and 22:6). Interestingly, the increasing value of the ratio from neutral to glycolipids and polar lipids (Fig. 39.8) observed in both gravid and spent females was related to the changes in the proportion of 20:5(*n*-3) and not to the decreased concentration of *n*-6 unsaturated acids.

Spatial changes in fatty acid composition of krill lipid classes

During the FIBEX cruise, fatty acid composition for mixed stages of krill was estimated at 11 to 13 stations for both polar lipid classes and triglycerides. The levels of spatial variability were moderate and varied with the lipid class and the group of fatty acid considered. Indeed, phosphatidyl-choline and phosphatidyl-ethanolamine content in total saturated, monoenoic and polyunsaturated acids appeared moderately conservative over space (Table 39.3), although the proportions of polyunsaturated acids in phosphatidyl-ethanolamine displayed a greater variability than in phosphatidyl-choline. In the glycolipid fraction, the distribution of fatty acids in saturates, monoenes and polyenes with 6 double bonds (Table 39.3) appeared less conservative than in the previous two polar lipids. The total saturates and monoenes of triglycerides were quite uniform, unlike the polyunsaturates with 5 and 6 double bonds, which displayed the highest variability for this group (Table 39.3).

Table 39.3. *Mean fatty acid composition of* Euphausia superba *lipid classes sampled over the FIBEX grid of stations and grouped by degree of unsaturation*

Σ fatty acids	*p*-Choline (±SD)	*p*-Ethanolamine (±SD)	Glycolipids (±SD)	Triglycerides (±SD)
Saturates	38.68±8.66	25.05±6.35	23.01±7.64	50.74±3.04
% variability	(22.4)	(25.3)	(33.2)	(6.0)
Monoenes	14.51±2.66	24.17±4.38	19.61±5.92	37.94±2.77
% variability	(18.3)	(18.1)	(30.2)	(7.3)
Dienes	3.16±0.60	1.47±0.56	2.56±0.52	2.42±0.52
% variability	(18.9)	(38.1)	(20.3)	(21.5)
Poly 3	1.68±0.40	0.30±0.31	1.56±0.56	1.11±0.52
% variability	(23.8)	(103)	(35.9)	(46.8)
Poly 4	3.61±0.74	1.88±1.13	4.71±1.29	2.58±1.06
% variability	(20.4)	(60.1)	(27.3)	(41.1)
Poly 5	25.38±3.87	19.13±4.15	28.49±6.34	2.17±1.21
% variability	(15.2)	(21.6)	(22.2)	(55.8)
Poly 6	12.26±2.11	25.86±6.54	18.64±5.52	1.49±1.14
% variability	(17.2)	(25.9)	(29.6)	(76.5)
	n=14	*n*=12	*n*=11	*n*=13

Notes:
Symbols: SD=Standard deviation; *n*=number of samples; Poly *n*=polyunsaturated acids with *n* double bonds.

Since the covariation among a host of descriptors is always difficult to describe in simple terms of relative abundance, analyses of the fatty acid interrelationships in glycolipids and triglycerides were carried out using factorial correspondence analysis. The correlation coefficients between the first two factorial axes which explain 64% of the total variance, the variables (fatty acids) and the observations (stations) are presented in Table 39.4. The first axis discriminates between the highly polyunsaturated fatty acids (>4 double-bonds) and the rest of the fatty acids, and separated stations 03 and 09 with highly unsaturated fatty acid profiles (but with widely different profiles of saturates and monoenes) from the least unsaturated station, 17. The second axis differentiated between palmitic acid (16:0) on the one hand and palmitoleic and oleic on the other hand and singled out those stations characterized by a high level of 16:1(*n*-7) and 18:1(*n*-9) (i.e. high activity of Δ^9 desaturase), such as stations 08, 17 and 03. The relationship between the degree of unsaturation of the glycolipid fraction and sex suggests that the first axis could illustrate the variability of sex and maturity stages in the population sampled, while the second axis represents differences in the level of *de-novo* synthesis of monoenoic acids. The composition of the population sampled (see Table 39.2) confirmed the interpretation of the first axis, with station 9 largely male dominated, as opposed to station 17 dominated by females and juveniles. Not surprisingly, a different composi-

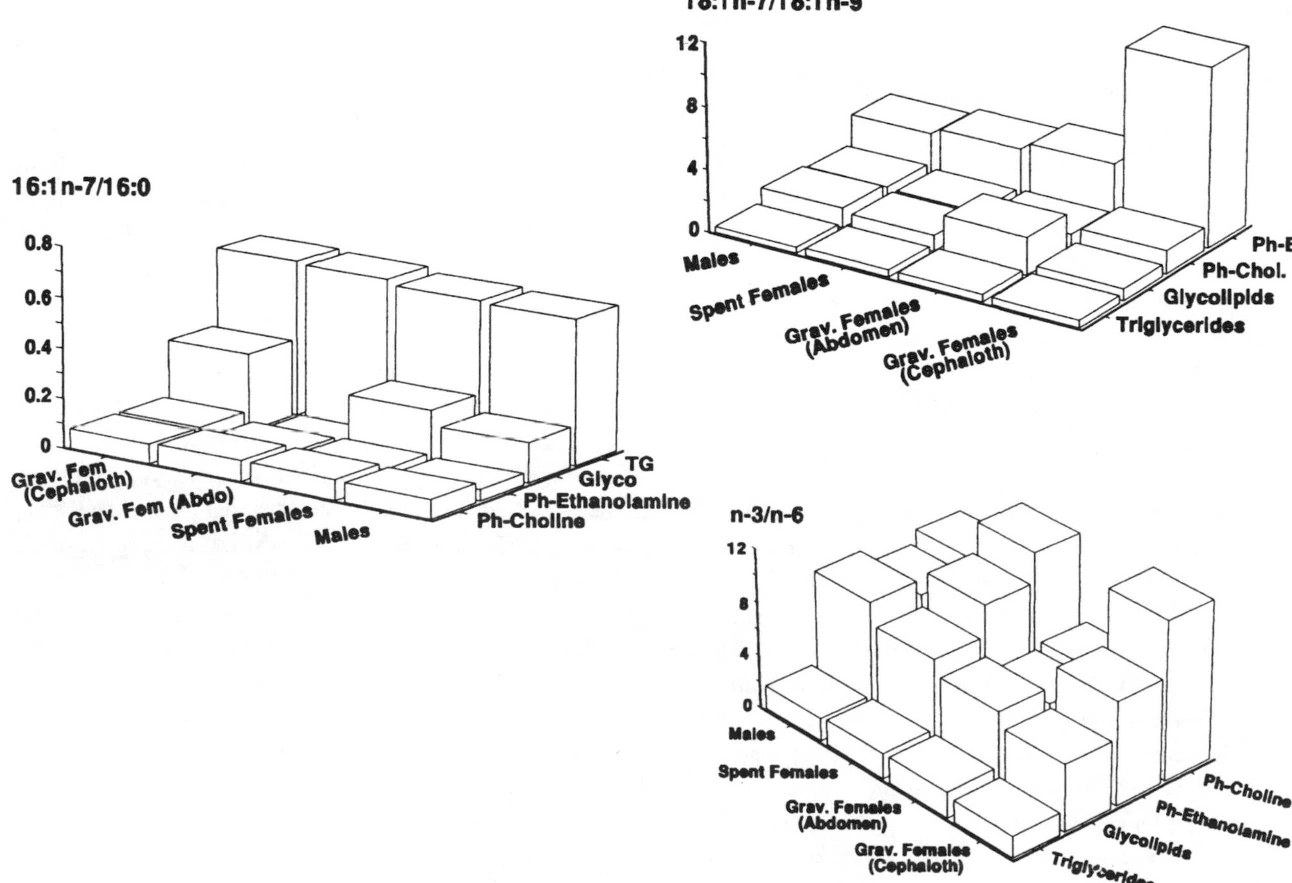

Fig. 39.8. Isomer ratios in the lipid classes of the different sexual development stages of *Euphausia superba* from SIBEX samples. Abbreviations: see Fig. 39.6.

Table 39.4. *Factorial correspondance analysis of the fatty acid spatial changes in* E. superba *triglycerides. Correlations are between axes and variables or stations. Percentages are the percentage of variance explained by each factorial axis*

Fatty acids	Axis 1 52.6%	Fatty acids	Axis 2 23.4%	Stations	Axis 1	Stations	Axis 2
14:0	−0.543	14:0	−0.149	02	−0.349	02	−0.121
Iso16:0	−0.316	16:0	−0.752	5A	−0.130	07	−0.352
18:0	−0.506	18:1(*n*−9)	−0.225	07	−0.313	08	−0.101
16:1(*n*−7)	−0.521	20:1(*n*−9)	−0.361	9A	−0.157	17	−0.131
18:1(*n*−9)	−0.462	18:1(*n*−7)	0.774	17	−0.415	19	−0.044
Iso15:0	0.615	18:1(*n*−5)	0.725	19	−0.445	23	−0.343
15:0	0.615	20:1(*n*−7)	0.705	23	−0.530	03	0.012
17:0+Phyt.	0.711	18:3(*n*−3)	0.923	03	0.007	5A	0.140
14:1(*n*−9)	0.831	16:4(*n*−1)	0.869	06	0.008	06	0.222
16:1(*n*−5)	0.907	18:4(*n*−3)	0.483	08	0.899	09	0.616
18:2(*n*−6)	0.735	20:4(*n*−3)	0.597	09	0.076	9A	0.251
16:3(*n*−4)	0.824	18:5(*n*−3)	0.826	10	0.002	10	0.133
16:3(*n*−3)	0.671			11	0.165	11	0.684
16:4(*n*−3)	0.554						
20:5(*n*−3)	0.859						
22:6(*n*−3)	0.974						

Note:
Phyt = phytanic acid.

Table 39.5. *Factorial correspondance analysis of the fatty acid spatial changes in* E. superba *glycolipids. Correlations are between axes and variables or stations. Percentages are the percentage of variance accounted for by the factorial axes*

Fatty acids	Axis 1 39.3%	Fatty acids	Axis 2 24.6%	Stations	Axis 1	Stations	Axis 2
14:0	−0.893	16:0	−0.476	02	−0.100	02	−0.645
16:0	−0.441	20:4(*n*−9)	−0.296	08	−0.010	06	−0.245
18:0	−0.399	17:0+Phyt.	0.235	17	−0.693	09	−0.113
16:1(*n*−9)	−0.375	16:1(*n*−7)	0.754	23	−0.277	23	−0.089
16:1(*n*−7)	−0.220	16:1(*n*−5)	0.376	03	0.598	03	0.294
16:1(*n*−5)	−0.202	18:1(*n*−9)	0.530	5A	0.080	5A	0.018
18:1(*n*−9)	−0.345	18:2(*n*−6)	0.338	06	0.067	07	0.004
18:1(*n*−7)	−0.227			07	0.019	08	0.455
20:1(*n*−9)	−0.285			09	0.692	17	0.152
18:2(*n*−6)	−0.582						
18:3(*n*−3)	−0.364						
17:3+Phyt.	0.320						
16:4(*n*−6)	0.627						
20:5(*n*−3)	0.784						
22:6(*n*−3)	0.798						

Note:
Phyt = phytanic acid.

tion was observed at station 03, suggesting that other factors are also involved in the control of glycolipid unsaturation.

The results of a similar analysis with the fatty acids constitutive of the triglycerides is presented in Table 39.5. The first two axes accounted for 76% of the total variance. From the levels of correlation between variables and axes, it can be seen that the first axes plots 14:0, 18:0 and 16:1(*n*-7) against the polyunsaturated acids and 14:1(*n*-9), 16:1(*n*-5) and distinguishes the northeastern stations (17, 19, 23) from the southern ones (08). The second axis separates the phytoplankton descriptors (16:4(*n*-1), 18:5(*n*-3), 18:3(*n*-3)) from the saturates and some of the monoenoic acids. It singles out the southern area (stations 09, 11) with higher levels of phytoplankton fatty acid markers.

DISCUSSION

The lipid content of oceanic populations of *Euphausia superba* was relatively low compared with lipid-rich zooplankton from high latitudes (Lee 1975, Lee *et al.* 1971, Sargent & Henderson 1986) but in agreement with the range of values reported for euphausiids in general (Ackman *et al.* 1970, Sargent & Falk-Petersen 1981, Saether *et al.* 1986) and for summer populations of *E. superba* from the Antarctic peninsula area (Raymont *et al.* 1971, Bottino 1975, Clarke 1980, 1984, Reinhart & Van Vleet 1986, Hagen 1988). The lipid class composition found here, for summer populations, is broadly similar to those found by Bottino (1975), Clarke (1980) and Hagen (1988), with a dominance of polar lipids. Phosphatidyl-choline constituted the major phospholipid fraction, with a lower content of phosphatidyl-ethanolamine and traces of lyso-forms and phosphatidyl-inositol (Mayzaud, unpubl.). The main differences were in the relatively high proportion of an acetone-mobile glycolipid, which ranged from 10 to 20% of the total lipids, and in the levels of triglycerides, which not only dominated the neutral lipids but also were the most abundant single lipid constituent. Clarke (1984) and Hagen (1988) have shown that summer accumulation of lipid reserves was a dominant feature of the seasonal cycle of *E. superba* biochemical composition.

Somatic growth and sexual maturity have been shown to control to a large extent the lipid accumulation. Clarke (1980) reported that maturing females contained most lipids, and spent females and males the least, while immature individuals displayed intermediate amounts. The differences were clearly related to triacylglycerol accumulation with no major changes in the proportions of phosphatidylcholine and phosphatidyl-ethanolamine. Our results confirmed that the degree of triglyceride accumulation was maximum in mature females, intermediate in subadults and spent females and minimum in males, but also showed that the levels of phosphatidyl-ethanolamine and glycolipids were also stage dependent with maximum values in males and spent females and minimum ones in the ovarics. Spatial variability observed for given maturity stages strongly suggests that besides growth and reproduction, lipid accumulation was also related to local conditions. Hagen (1988) reached the same conclusions with organisms from different seasons in the Antarctic Peninsula and the Weddell Sea.

In contrast to many other zooplankton, euphausiids of the genus *Euphausia* and *Thysanoessa* apparently use part of their phospholipids as fat deposits in addition to triglycerides and/or wax-esters (Ellingsen 1982, cited by Saether *et al.* 1986). This hypothesis was supported by the results of Hagen (1988), which showed that only lecithin (phosphatidyl-choline) was stored in significant amounts. Both studies reported steeper or equal slopes for the linear increases of triglycerides (or wax-esters) with total lipids rather than with phospholipids. Moreover, Hagen (1988) reported that lipid accumulation in euphausiids occurred independently of sex, season or geographical origin. Our results confirmed these conclusions but suggest that in open-ocean populations the accumulation of phosphatidyl-choline proceeds at a faster rate than that of triglycerides. To some extent, this difference could have been related to sexual development, as females displayed a steeper slope of polar lipid accumulation than males and subadults (Mayzaud unpublished data), but could it also be related to local conditions since higher rate of triglyceride accumulation was never observed in any of the growth stages considered. Why *Euphausia* and *Thysanoessa* species store phosphatidyl-choline when their northern counterpart *Meganyctiphanes norvegica* does not (Saether *et al.* 1986) remains unclear. Adaptation to low temperature have been proposed by Hagen (1988) on the basis that phosphatidyl-choline could be more easily made available to specific organs such as muscles than triglycerides or wax-esters, but the evidences remained mostly speculative. Obviously, further work is needed to understand the respective role of deposits of phospholipids and neutral lipids.

The same comment could apply to the relatively high content of acetone-mobile glycolipids. The presence of several, minor, uncharacterized glycolipids has been reported by Clark (1980) but not quantified. Virtue *et al.* (1993) also observed a glycolipid fraction which was not quantified but analysed for fatty acid composition. In the present study, the acetone-mobile compounds accounted for 10 to 20% of the total lipids, a value far higher than any reported up to now. Since the levels of phospholipid constituents found in the present study are similar to those reported previously (i.e. 35–44% of the total lipids) and all major compounds are accounted for (Van der Veen *et al.* 1971, Clark 1980, Hagen 1988, Reinhart & Van Vleet 1986, Virtue *et al.* 1993), the possibility of a cross-reaction with some unknown phospholipid is unlikely. The dependence of glycolipid levels on growth stage (lowest in subadults and highest in spent females and males), and on sexual development, suggests some coupling with the reproductive cycle and possibly the development of the fat body described by Cuzin-Roudy (1993) in males and maturing females. The close association with phosphatidyl-ethanolamine shown by the cluster analysis certainly suggests a structural role rather than a metabolic one. The difference with previous reports could then be related either to local (open-ocean populations) and/or seasonal factors.

The lipids of the krill *Euphausia superba* had an overall fatty acid composition broadly similar to the reported composition for different growth stages, sexual maturity and locations (Van der Veen *et al.* 1971, Bottino 1975, Clarke 1980, Fricke *et al.* 1984, Virtue *et al.* 1993). The fatty acid composition reported in the present study shows the classic pattern for marine crustaceans in that triglycerides are more variable and less unsaturated than polar lipids. Influence of dietary input in triglyceride composition can be derived from the different levels of diatom-related fatty acids such as 16:4(*n*-1) (Ackman *et al.* 1968) or dinoflagellate, prymnesiophyte fatty acid such as 18:5(*n*-3) or 18:3(*n*-3) (Mayzaud *et al.* 1976, Volkman *et al.* 1981).

The phospholipid constituents displayed relatively different fatty acid composition, but contrary to the findings of Van der Veen *et al.* (1971) and Clarke (1980), lower content in monoenoic acids and higher unsaturation were generally observed in phosphatidyl-choline compared with phosphatidyl-ethanolamine. With the exception of the spent females where levels of unsaturation are similar, this difference was related to lower proportions of 20:5 in phosphatidyl-ethanolamine. It is difficult to evaluate if these discrepancies are related to differences in methodologies or to local conditions : Van der Veen *et al.* (1971) and Clarke (1980) used a chromatographic system which did not separate glycolipids from phospholipids, whereas Brockerhoff & Ackman (1968) and Ackman *et al.* (1970) suggested that since phospholipids draw certain fatty acids from the same dietary pool as triglycerides a certain degree of variability should be expected.

Major differences were also observed in the fatty acid structure of the glycolipid fraction which, contrary to the findings of Virtue *et al.* (1993) on sexually immature krill, showed a relatively high level of unsaturation with increasing levels of 22:6 from gravid females to spent females to males. In six instances, 15:0, 16:4(*n*-1), 16 PUFA, 16:1(*n*-7), 18:1(*n*-9), 20:5(*n*-3), Virtue *et al.* (1993) illustrated significant changes in the glycolipid fatty acid composition with food supply (diatom versus prymnesiophyte), which suggests a strong relationship with local food conditions. Interestingly, this structural component was the fraction most influenced by the succession of experimental diets, well above the triglycerides which showed significant changes in only one instance (16:1(*n*-7)). The disparity in unsaturation could then be related to differences in both developmental stages and local trophic conditions.

CONCLUSIONS

Despite the numerous studies on krill biochemistry, we are still far from a complete understanding of the complex interrelationships between internal requirements and the trophic environment which regulate lipid accumulation and use. The classic view of lipid dynamics entirely driven by neutral lipids, as opposed to constant structural polar lipids, is made more complex in *Euphausia superba* with the possibility that polar lipids may serve as fat deposits. The actual role of phosphatidylcholine in the overall metabolism needs to be clarified further but appears indirectly linked with development and reproduction. Similarly, the presence of significant levels of an unknown glycolipid possibly associated with sexual development but with a fatty acid composition that appears to be influenced by the nature of the diet, needs to be clarified further as to both its structure and its role. To what extent these features are characteristics of high-polar euphausiids and/or associated to specific adaptations is worth further consideration.

ACKNOWLEDGEMENTS

The author is grateful to Dr J. C. Hureau, chief scientist, and the crew of the RV *Marion Dufresne* for their help and assistance throughout the sea operations. I also thank Dr J. Farber-Lorda for his help in staging the krill from the FIBEX cruise and Dr J. Carlos for staging the krill from the SIBEX cruise. I am also grateful to M. C. Corre for assistance in the field and J. Chiaverini, P. Drouin, M. Gay and G. Ouellet for technical assistance in the laboratory. This work was supported by the TAAF scientific mission and CNRS-INSU.

REFERENCES

Ackman, R. G. 1981. Application of flame ionisation detector to thin layer chromatography on coated quartz rods. *Methods in Enzymology*, **72**, 205–252.

Ackman, R. G., Eaton, C. A., Sipos, J. C., Hooper, S. N. & Castell, J. D. 1970. Lipids and fatty acids of two species of North Atlantic krill (*Meganyctiphanes norvegica* and *Thysanoessa inermis*) and their role in the aquatic food web. *Journal of Fisheries Research Board of Canada*, **27**, 513–533.

Ackman, R. G., Tocher, C. S. & McLachlan, 1968. Marine phytoplankters fatty acids. *Journal of Fisheries Research Board of Canada*, **25**, 1603–1620.

Barnes, H. & Blackstock, J. 1973. Estimation of lipids in marine animals and tissues : detailed investigation of the sulphovanilin method for total lipids. *Journal of Experimental Marine Biology and Ecology*, **12**, 103–118.

Bligh, E. G. & Dyer, W. J. 1959. A rapid method of total lipid extraction and purification. *Canadian Journal of Biochemistry and Physiology*, **37**, 911–917.

Bottino, N. R. 1974. The fatty acids of Antarctic phytoplankton and euphausiids. Fatty acid exchange among trophic levels of the Ross Sea. *Marine Biology*, **27**, 197–204.

Bottino, N. R. 1975. Lipid composition of two species of Antarctic krill: *Euphausia superba* and *E. crystallorophias*. *Comparative Biochemistry and Physiology*, **50B**, 479–484.

Brockerhoff, H. & Ackman, R. G. 1968. Positional distribution of isomers of monoenoic fatty acids in animal glycerolipids. *Journal of Lipid Research*, **8**, 661–666.

Clarke, A. 1980. The biochemical composition of krill *Euphausia superba* Dana, from South Georgia. *Journal of Experimental Marine Biology and Ecology*, **43**, 221–236.

Clarke, A. 1984. Lipid content and composition of Antarctic krill, *Euphausia superba* Dana. *Journal of Crustacean Biology*, **4**, 285–294.

Cuzin-Roudy, J. 1993. Reproductive strategies of the mediterranean krill, *Meganyctiphanes norvegica* and the Antarctic krill *Euphausia superba* (Crustacea: Euphausiacea). *Invertebrate Reproduction and Development*, **23**, 105–114.

Ellingsen, T. E. 1982. *Biokjemiske studier over antarkisk krill*. Dr. Ing. thesis, University of Trondheim, 382 pp.

Farber-Lorda, J. 1986. *Etudes biologiques, énergétiques et biochimiques du krill antarctique* Euphausia superba *et* Thysanoessa macrura, *récolté au cours de la campagne FIBEX*. Doctorat, Université Aix-Marseille II. Faculté des Sciences de Luminy, 214 pp.

Fricke, H., Gerken, G., Schreiber, W. & Oehlenschkager, J. 1984. Lipid, sterol and fatty acid composition of Antarctic krill (*Euphausia superba*, Dana). *Lipids*, **19**, 821–827.

Hagen, W. 1988. On the significance of lipids in Antarctic zooplankton. *Berichte zur Polarforschung*, **49**, 92–117. [Canadian translation of Fisheries and Aquatic Sciences # 5458.]

Hirayama, O. & Morita, K. 1980. A simple and sensitive method for the quantitative analysis of chloroplast lipids by use of thin layer chromatography and flame ionization detector. *Agricultural Biological Chemistry*, **44**, 2217–2219.

Lee, R. F. 1975. Lipids of arctic zooplankton. *Comparative Physiology and Biochemistry*, **51B**, 263–266.

Lee, R. F., Hirota, J. & Barnett, A. M. 1971. Distribution and importance of wax esters in marine copepods and other zooplankton. *Deep-Sea Research*, **18**, 1147–1165.

Makarov, R. R. & Denys, C. J. 1981. Stages of sexual maturity of *Euphausia superba* Dana. *BIOMASS handbook no. 11*. SCAR/SCOR/IABO/ACMRR.

Mayzaud, P., Eaton, C. A. & Ackman, R. G. 1976. The occurrence and distribution of octadecapentaenoic acid in a natural plankton population. A possible food chain index. *Lipids*, **11**, 858–862.

Mayzaud, P., Farber-Lorda, J. & Corre, M. C. 1985. Aspect of the nutritional metabolism of two Antarctic euphausiids: *Euphausia superba* and *Thysanoessa macrura*. In Siegfried, W. R., Condy, P. R. & Laws, R. M., eds. *Antarctic nutrient cycles and food webs*. Berlin: Springer-Verlag, 330–338.

Morrison, W. R. & Smith, L. M. 1964. Preparation of fatty acid methyl esters and dimethyl acetals from lipids with boron fluoride-methanol. *Journal of Lipid Research*, **5**, 600–608.

Raymont, J.E.G., Srinivasagam, R.T. & Raymont, J.K.B. 1971. Biochemical studies on marine zooplankton. IX. The biochemical composition of *Euphausia superba*. *Journal of the Marine Biology Association of UK*, **51**, 581–588.

Reinhardt, S. B. & Van Vleet, E. S. 1986. Lipid composition of twenty two species of Antarctic midwater zooplankton and fish. *Marine Biology*, **91**, 149–159.

Saether, O., Ellingsen, T. E. & Mohr, V. 1986. Lipids of North Atlantic. *Journal of Lipid Research*, **27**, 274–285.

Sargent, J. R. & Falk-Petersen, S. 1981. Ecological investigations on the zooplankton community in Balsfjorden, northern Norway: lipids and fatty acids in *Meganyctiphanes norvegica*, *Thysanoessa rashii* and *T. Inermis* during mid-winter. *Marine Biology*, **62**, 131–137.

Sargent, J. R. & Henderson, R. J. 1986. Lipids. In Corner, E. D. S. & O'Hara, S. C. M., eds. *The biological chemistry of copepods*. Oxford: Clarendon Press, 59–164.

Stahl, E. 1969. *Thin-layer chromatography. A laboratory handbook*. Berlin: Springer-Verlag.

Tsuyuki, T. & Itoh, S. 1976. Fatty acid component of lipid of *Euphausia superba*. *Scientific Reports of the Whales Research Institute*, **28**, 167–174.

Van der Veen, J., Medwadowski, R. & Olcott, H. S. 1971. The lipid of krill (*Euphausia superba*) and red crab (*Pleurocondes planipes*). *Lipids*, **6**, 481–485.

Virtue, P., Nichols, P. D., Nicol, S., McMinn, A. & Sikes, E. L. 1993. The lipid composition of *Euphausia superba* Dana in relation to the nutritional value of *Phaeocystis pouchetii* (Hariot) Lagerheim. *Antarctic Science*, **5**, 169–177.

Volkman, J. K., Smith, D. J., Eglinton, G., Forsberg, T. E. V. & Corner, E. D. S. 1981. Sterol and fatty acid composition of four marine haptophycean algae. *Journal of the Marine Biology Association of UK*, **61**, 509–527.

40 Temperature adaptation and regulation of citrate synthase in the Antarctic krill compared with other crustaceans from different climatic zones

R-A.H. VETTER[1], R. SABOROWSKI[1], G. PETERS[2] AND F. BUCHHOLZ[1]

[1]*Biologische Anstalt Helgoland, Marine Station, Department of Marine Zoology, D-27483 Helgoland, Germany,* [2]*Institut für Meereskunde, Department of Marine Zoology, Düsternbrooker Weg 20, D-24105 Kiel, Germany*

ABSTRACT

The high physiological performance and capability of Antarctic krill, Euphausia superba, *should be reflected in specific properties of various enzymes. Citrate synthase (CS), one of the regulating enzymes of intermediary metabolism was studied in order to differentiate between the effects of temperature and those of the specific life-style (pelagic versus benthic) of the species investigated. Accordingly, CS in crustaceans from different habitats and climatic zones was assayed: the pelagic euphausiids* Euphausia superba *from the Antarctic, and* Meganyctiphanes norvegica *from the Scandinavian Kattegat and the Mediterranean, were compared with the benthic isopods* Serolis polita *(Antarctica) and* Idotea baltica *from the Baltic. In addition, the non-metabolic chitin-degrading enzyme N-acetyl-ß-D-glucosaminidase (NAGase) was investigated in the same four species to compare the results of the two enzymes with different functions.*

CS is regulated in various ways depending on the temperature tolerance, the climatic zone and the life-style of the crustaceans investigated. The different adaptive mechanisms include varying the inhibiting effects of ATP, reducing the activation energy, and adapting specific enzyme activity to ambient temperatures. In contrast, NAGase showed less pronounced adaptations to ambient temperature regimes. This enzyme and particularly its isoforms are mainly influenced by feeding or by hormones of the moulting cycle.

Key words: citrate synthase, N-acetyl-ß-D-glucosaminidase, enzyme regulation, temperature adaptation, krill, crustacea.

INTRODUCTION

The Antarctic krill, *Euphausia superba*, is distinguished from other polar invertebrates by its high physiological efficiency, particularly high swimming speed and potential fast growth (Kils 1981, Clarke & Morris 1983, Buchholz *et al.* 1989, Buchholz 1991). Physiological performance and capability should be reflected in the properties of enzymes, since enzyme activity is involved at all levels of metabolism. Changes in enzyme kinetics or synthesis are involved in acclimation and adaptation to ambient temperatures (Clarke 1983). In this study, enzymatic temperature adaptation has been examined by comparing partially purified enzymes of different functions in crustaceans of different life-style, and from different climatic zones. Accordingly, the constantly active, pelagic euphausiids *Euphausia superba* from the Antarctic and *Meganyctiphanes norvegica* from the Scandinavian Kattegat and the Mediterranean were compared with the less active, benthic

Table 40.1. *Characteristics of citrate synthase from different crustaceans*

Species	*E. superba*	*M. norvegica*	*M. norvegica* (med.)	*S. polita*	*I. baltica*
Ambient temp. (°C)	−2 to +2	0 to 15	11 to 13	−2 to +2	0 to 20
Temp. optimum (°C)	40	40	37	45	43
E_a (kJ mol^{-1})	10.9±1.21[1]	45.1±1.5	43.3±3.3	41.8±2.4	43.7±2.8
K_i [ATP] (mmol l^{-1})	0.779±0.097	0.927±0.234	1.036±0.399	2.867±0.594[1]	0.594±0.114[1]

Notes:
Data given as x±SD, n=3. [1] significant differences (p≤0.05). med.=Mediterranean.

isopods *Serolis polita* (Antarctica) and *Idotea baltica* from the Baltic. The contrasting temperature regimes can be characterized as polar: cold with little variation; boreal: strong vertical and seasonal differences; mediterranean: warm and constant.

The central object of the present study was the lyase citrate synthase (CS, EC 4.1.3.7). CS, is one of the important enzymes of the intermediary metabolism, involved in the cellular supply of energy. The enzyme plays an essential part in controlling the citric acid cycle (e.g. Krebs & Johnson 1937, Wiskich 1980). In order to gain insights into more general regulating components some further comparisons were drawn with the hydrolase N-acetyl-ß-D-glucosaminidase (NAGase, EC 3.2.1.30), a catabolic, chitin-degrading enzyme involved in digestion (Arnould & Jeuniaux 1982, Buchholz 1989) and moulting (Jeuniaux 1963, Buchholz & Buchholz 1989).

Temperature-dependent characteristics of the two enzymes were investigated to evaluate mechanisms of thermal regulation, such as temperature optima, activation energies and kinetic parameters. In the case of NAGase immunotechniques were used for further characterization of the existing isoforms (Buchholz & Vetter 1993). Additionally, kinetic parameters of both enzymes were estimated in euphausiids maintained at different temperatures to specify the eco-physiological relevance of short time acclimation.

MATERIALS AND METHODS

E. superba (euphausiid) was caught in Admiralty Bay, King George Island, *M. norvegica* (euphausiid) in the Scandinavian Kattegat and the Ligurian Sea (Mediterranean) using a 1 m^2 ring trawl. *S. polita* (isopod) was sampled by divers in Admiralty Bay (*c.* 5 m depth). *I. baltica* (isopod) was dredged in Kiel Bay between 5 and 10 m (for details see Buchholz & Vetter 1993).

Additionally, *E. superba* and *M. norvegica* were maintained at different constant temperatures for 11 days (5 specimens in 5 l glass jars, n=100 specimens, fed with natural phytoplankton, with seawater exchanged at 48 h intervals).

Specimens were shock-frozen and stored at −80 °C. Extracts were prepared from whole animals by Ultra-Turrax homogenizer (Janke & Kunkel) in appropriate buffers, on ice (Buchholz & Vetter, 1993). The enzymes were stable at 4 °C. Addition of anti-oxidants was not necessary.

Enzymes were partially purified using fast protein liquid chromatography (FPLC) with a pre-packed anion-exchange column HiLoad 16/10 Q-Sepharose HP (Pharmacia). NAGase was eluted in imidazole buffer 0.01 mol l^{-1} (pH 6.8) using a linear gradient of NaCl (0–0.85 mol l^{-1}) and CS was chromatographed with the same NaCl gradient using Tris/HCl 0.04 mol l^{-1} (pH 8.0) containing 20 mmol l^{-1} KCl and 4 mmol l^{-1} $MgSO_4$. All fractions (4.3 ml each) containing more than 10% of the maximum enzyme activity were pooled for further characterization. Protein concentrations of the pools were determined according to Bradford (1976).

Enzyme assays of NAGase were performed after Spindler & Buchholz (1988), assays of CS were modified after Stitt (1984) using DTNBA as co-substrate (Dawson *et al.* 1986). Kinetic parameters (K_M^{app}, K_i and V_{max}) were calculated from measurements with variable substrate concentrations after Wilkinson (1961). Each substrate of CS was varied between 1.8 and 363.6 mmol l^{-1} (oxaloacetate and acetyl-CoA, Boehringer). Additional kinetic measurements were carried out to quantify the influence of 0.2 to 1.8 mmol l^{-1} ATP (Boehringer) on CS. Kinetics of NAGase were performed using p-nitrophenyl-N-acetyl-ß-D-glucosaminide (Sigma) from 0.05 to 3.33 mmol l^{-1}.

A specific antibody for NAGase was isolated from rabbit-serum, using standard procedures. Two different isoforms of NAGase were detected by FPLC after Buchholz & Vetter (1993), and their activity in extracts quantified by immunotitration after Mentlein *et al.* (1985) using a polyclonal antibody.

Statistics: Overall differences between species were tested by ANOVA followed by Student-Newman–Keuls multiple comparison test (Sachs 1984). Slopes of Arrhenius plots were compared by ANCOVA after Sokal & Rohlf (1981).

RESULTS

FPLC and electrophoretic separations demonstrated the existence of only one form of CS. Distinct temperature optima of CS of all species investigated were found to be between 37 °C and 45 °C. These were far outside of the natural ambient temperatures ranges of the animals (Table 40.1). Additionally, activation energies (E_a) were calculated from the slopes of Arrhenius plots. With the exception of *E. superba*, which had an outstandingly low value of 10.9 kJ mol^{-1}, the resulting values of 41.8 to 45.1 kJ mol^{-1} of the other species (Table 40.1) were not significantly different (ANCOVA p≤0.01).

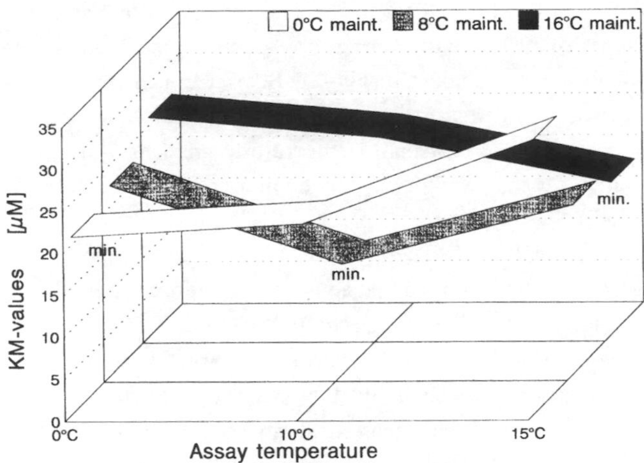

Fig. 40.1. Influence of maintenance temperature on the K_M^{app} values of *Meganyctiphanes norvegica*. Apparent K_M values of acetyl-CoA were determined in *M. norvegica* maintained at different temperatures (0, 8, 16 °C). The resulting data are plotted against the incubation temperature of the enzyme assay. Standard deviations were always below 15% of means (n=3). min.=lowest K_M value of the respective maintenance series.

The influence of the CS-inhibitor ATP was estimated by calculating the K_i values of competitive inhibition from kinetic measurements with different ATP concentrations. In the euphausiids the differences between the K_i values were not significant (ANOVA, t test, $p \leq 0.05$), but the values of the isopods *S. polita* and *I. baltica* were significantly higher or lower, respectively, than those of the euphausiids.

Nordic and Antarctic krill were maintained at different temperatures (*M. norvegica*: 0, 8, 16 °C; *E. superba*: −1, +2, +8 °C) to estimate the influence of acclimation temperatures on kinetic properties of CS. In the case of *E. superba*, CS from aquaria-maintained specimens showed no differences in apparent K_M-values (K_M^{app}) and specific activity (V_m mg^{-1} protein) compared with animals taken directly from the open water. In contrast, maintenance of *M. norvegica* resulted in variation of enzyme–substrate affinity (K_M^{app}) and specific activity. Highest enzyme–substrate affinity, i.e. minima of K_M^{app}, were always found at the respective maintenance temperatures (Fig. 40.1). Additionally, changes in specific activity were found. At each assay temperature-specific activities of the coldest maintained specimens was always highest and specimens maintained at 16 °C had lowest values (Fig. 40.2). Comparing only those data of specific V_{max}, at incubation temperatures (enzyme assay) which corresponded to the respective maintenance temperature, the values remained constant in the range from 0 °C to 10 °C (Fig. 40.2). Further increase of temperature to 15 °C led to decreasing specific activity, due to the very low values in the specimens maintained at 16 °C.

The chitinolytic NAGase was also investigated with regard to temperature adaptation. In contrast to CS, which had no isoenzymes, NAGase could be separated into two isoforms. However, these two isoforms did not contribute to temperature adaptation to the same extent as was demonstrated in CS. Only a limited capacity of regulation of K_m values in relation to climatic adaptation was shown in NAGase (Buchholz & Vetter 1993).

Table 40.2. *Distribution of NAGase isoforms in stomach and integument of* M. norvegica. *Data resulting from immunotitration are given as percentage of total NAGase activity in the specified organs*

	Stomach	Integument
NAGase B	8.5%	90%
NAGase C	91.5%	8%

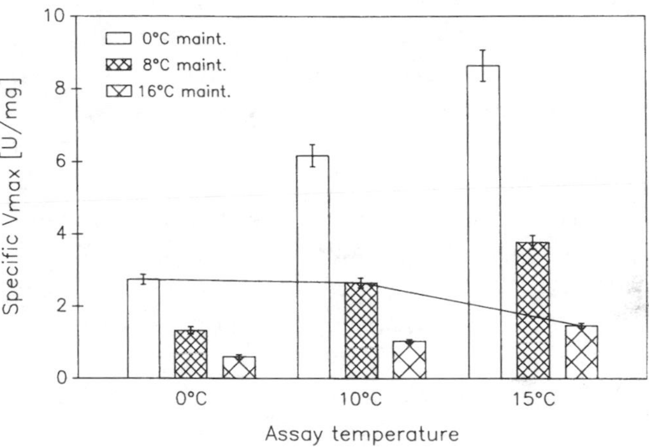

Fig. 40.2. Temperature dependence of specific CS activity of *Meganyctiphanes norvegica*. maintained at different temperatues. Specific activity (V_{max} per mg protein) of maintained *M. norvegica* (0, 8, 16 °C) is plotted against incubation temperature of the enzyme assay. The thin line represents those values where the maintenance temperature corresponds to the appropriate assay temperature. n=3.

Nevertheless, using polyclonal antibodies it was found that the isoforms were localized in different organs with different functions (Table 40.2). Isoform B is present in the exuvial cleft of the integument. Isoform C is an enzyme-variant present in the digestive tract. Furthermore, immunotitration showed no temperature-related differences of activity in either variant in the acclimated specimens (maintenance experiments, see above).

DISCUSSION

To date, mechanisms of temperature adaptation in poikilotherms have been controversial (e.g. Hochachka & Somero 1973, Hazel & Prosser 1974, Clarke 1983). In fact, the lack of clear ecophysiological patterns complicated the current discussion. Most of the previous investigations on enzymatic temperature adaptation were carried out with crude extracts only, and with a limited set of experimental temperatures. In this study, an attempt was made to purify enzymes partially by chromatography and investigate temperature-dependent parameters directly. In addition, maintenance experiments had been performed to demonstrate the effects of short-time adaptations to temperature.

A conspicuous characteristic found in CS was the similarity of the temperature profiles and optima. Apparently, there is no shift of optima into the ambient temperature regime of the specific animal (Table 40.1). This was also found in NAGase (Buchholz & Vetter 1993). However, differences in activation energies occurred in CS, but not in NAGase.

The remarkably low E_a value of CS of *E. superba*, which is only 25% of that of the other crustaceans (Table 40.1), indicates a possible adaptation to polar temperatures. Although E_a of the polar isopod *S. polita* is not reduced, this mechanism can be related to low-temperature adaptation. The missing reduction of E_a in *S. polita* may be due to its benthic, i.e. less active, life- style.

Instead of a reduction of E_a, the K_i-value of ATP is increased in the polar isopod as a consequence of low ATP inhibition. The low inhibitory effect of ATP in the Antarctic *S. polita* leads to a decreased regulatory capability of CS. However, this effect ensures sufficient enzyme activity at low temperature. The disadvantage of minimized ATP regulation is offset by a low basal metabolism of the benthic Antarctic isopod (Luxmore 1982, Clarke 1983). The boreal *I. baltica* does not show an increase in K_i. Due to higher temperatures in the Baltic there is no need for *I. baltica* to reduce the inhibitory effect of ATP.

Accordingly, two different mechanisms were found in CS to regulate enzyme activity depending on life-style and climatic zone. The pelagic, i.e. constantly swimming, euphausiids show an essential requirement for ATP-regulation. The polar *E. superba* is additionally characterized by a decreased value of activation energy, and thus increased enzyme efficiency, which helps to compensate for the low-temperature effects. In contrast, the benthic, polar *S. polita* is little affected by ATP regulation, resulting in generally higher CS activities and presumably relatively constant levels of ATP. In the slow moving, benthic isopods, a reduction of the energy of activation does not seem to be necessary, irrespective of the climatic zone. In any case, low-temperature adaptation depends on the variation of either E_a or K_i.

Mechanisms of short-term adaptations were investigated in laboratory-maintained euphausiids. Differences between *E. superba* and *M. norvegica* were obvious. Whereas maintenance temperature had no effects on the stenothermic *E. superba*, the eurythermic *M. norvegica* was influenced in two ways: variation of K_M^{app} and specific V_{max} of CS. Minima of K_M^{app} were well related to the respective maintenance temperature in each case (Fig. 40.1). Accordingly, the enzyme–substrate affinity for acetyl-CoA was highest at ambient temperatures. In addition, cold-maintained *M. norvegica* had higher specific CS activities than warm-maintained specimens (Fig. 40.2). As a result, the specific activity was almost constant when comparing the values of corresponding assay and maintenance temperature (Fig. 40.2). These variations of K_M^{app} and specific V_{max} indicate a connection with temperature changes in the natural habitat. Specific activity and enzyme substrate affinity were increased in *M. norvegica* at low temperatures to compensate the loss of activity with decreasing temperature. This eurythermic euphausiid has to regulate its enzyme activity as temperature changes up to 10 °C daily during vertical migration (Buchholz *et al.* 1995) and seasonally (Buchholz & Boysen-Ennen 1988). This regulation is not necessary in the stenothermic Antarctic *E. superba*, because seasonal temperature changes within a maximal range of only 4 °C are less important in the natural habitat.

Before analysing the temperature control of NAGase in the digestive tract, we verified that activities are endogenous and not of bacterial origin (Donachie *et al.* 1995). None of the described mechanisms of biochemical temperature adaptation were found in endogenous NAGase in the same crustacean species. However, apparent Michaelis constants showed minima which corresponded to ambient water temperatures. Here, a certain degree of thermal fine-tuning of enzyme activities in relation to the climatic regime seemed possible (Buchholz & Vetter 1993). The immunological investigations showed that there is no temperature-related production of two existing isoforms of NAGase. However, a clear compartmentation of the two forms exists: they have different functions in the integument and the digestive tract. Their activities are controlled by synthesis according to physiological necessities:

- Isoform B is regulated by the hormones of the moult cycle (Buchholz 1991) and is responsible for reabsorbtion of chitin from the cuticle. The activities of isoform B increase three-fold during the moult-stage immediately prior to ecdysis. During this phase the degradation of the old cuticle is at its maximum (Buchholz & Buchholz 1989).
- Isoform C is an enzyme variant located in the digestive tract, and is directly inducable by chitin containing food (Buchholz 1989, Saborowski & Buchholz 1991).

Apparently, temperature does not influence the regulative processes of the NAGase variants to a significant extent. This might be typical for the function of hydrolases in general.

In conclusion, different enzymes exert thermal control of activities in very different ways. Further investigations seem necessary to systematize the interrelation of function, compartmentation and adaptive mechanisms of different classes of enzymes in invertebrates.

ACKNOWLEDGEMENTS

Special thanks are due to Sonja Böhm, Marion Ziebarth and Gerrit Sahling for skilful assistance. We are also grateful to Prof. Dr D. Adelung for generous supply of laboratory space and equipment.

REFERENCES

Arnould, C. & Jeuniaux, C. 1982. Les enzymes hydrolytiques du système digestif chez les Crustacés Pagurides. *Cahiers de Biologie Marine*, **23**, 89–103.

Bradford, M. M. 1976. A rapid and sensitive method for the quantitation of microgram quantities of protein utilizing the principle of protein-dye binding. *Analytical Biochemistry*, **72**, 248–254.

Buchholz, F. 1989. Moult cycle and seasonal activities of chitinolytic enzymes in the integument and digestive tract of the Antarctic krill, *Euphausia superba*. *Polar Biology*, **9,** 311–317.

Buchholz, F. 1991. Moult cycle and growth of Antarctic krill, *Euphausia superba*, in the laboratory. *Marine Ecology Progress Series*, **69,** 217–229.

Buchholz, F. & Boysen-Ennen, E. 1988. *Meganyctiphanes norvegica* in the Kattegat: studies on the horizontal distribution in relation to hydrography and zooplankton. *Ophelia*, **29,** 71–82.

Buchholz, C. & Buchholz, F. 1989. Ultrastructure of the integument of a pelagic crustacean: moult cycle related studies on the Antarctic krill, *Euphausia superba*. *Marine Biology*, **101,** 355–365.

Buchholz, F. & Vetter, R. A. H. 1993. Enzyme kinetics in cold water: characteristics of N-acetyl-ß-D-glucosaminidase activity in the Antarctic krill, *Euphausia superba*, compared with other crustacean species. *Journal of Comparative Physiology*, **163B,** 28–37.

Buchholz, F., Buchholz, C., Fischer, J. & Reppin, J. 1995. Diel vertical migrations of *Meganyctiphanes norvegica* in the Kattegat: comparison of net catches and measurement with Acoustic Doppler Current Profilers. *Helgoländer Meeresuntersuchungen*, **49,** 1–4, 849–864.

Buchholz, F., Morris, D. J. & Watkins, J. L. 1989. Analyses of field moult data: prediction of intermoult period and assessment of seasonal growth in Antarctic krill, *Euphausia superba* Dana. *Antarctic Science*, **1,** 301–306.

Clarke, A. 1983. Life in cold water: the physiological ecology of polar marine ectotherms. *Oceanography and Marine Biology Annual Review*, **21,** 341–453.

Clarke, A. & Morris, D. J. 1983. Towards an energy budget for krill: the physiology and biochemistry of *Euphausia superba* Dana. *Polar Biology*, **2,** 69–86.

Dawson, R. M., Elliot, D. C., Elliot, W. H. & Jones, K. M. 1986. *Data for biochemical research*. Oxford: Clarendon Press, 580 pp.

Donachie, S. P., Saborowski, R., Peters, G. & Buchholz, F. 1995. Bacterial digestive enzyme activity in the stomach and hepatopancreas of *Meganyctiphanes norvegica* (M. Sars, 1857). *Journal of Experimental Marine Biology and Ecology*, **188,** 151–165.

Hazel, J. R. & Prosser, C. L. 1974. Molecular mechanisms of temperature compensation. *Physiological Reviews*, **54,** 620–670.

Hochachka, P. W. & Somero, G. N. 1973. *Strategies of biochemical adaptation*. Philadelphia: Saunders, 403 pp.

Jeuniaux, C. 1963. *Chitine et chitinolyse*. Paris: Masson, 181 pp.

Kils, U. 1981. The swimming behaviour, swimming performance and energy balance of Antarctic krill, *Euphausia superba*. *BIOMASS Science Series*, **3,** 1–122.

Krebs, H. A. & Johnson, W. A. 1937. The role of citric acid in intermediate metabolism in animal tissues. *Enzymologia*, **4,** 148–156.

Luxmore, R. A. 1982. Moulting and growth of serolid isopods. *Journal of Experimental Marine Biology and Ecology*, **56,** 63–85.

Mentlein, R., Berge, R. K. & Heymann, E. 1985. Identity of purified monoacylglycerol lipase, palmitoyl-CoA hydrolase and aspirin metabolizing carboxyesterase from rat liver microsomal fractions. *Biochemistry Journal*, **232,** 479–483.

Saborowski, R. & Buchholz, F. 1991. Induction of enzymes in the digestive tract of the Antarctic krill, *Euphausia superba*. *Verhandlungen der Deutschen Zoologischen Gesellschaft*, **84,** 422.

Sachs, L. 1984. *Angewandte Statistik*. Berlin: Springer-Verlag, 552 pp.

Sokal, R. R. & Rohlf, F. J. 1981. *Biometry*. San Francisco: Freeman, 454–560.

Spindler, K-D. & Buchholz, F. 1988. Partial characterization of chitin degrading enzymes from two euphausiids, *Euphausia superba* and *Meganyctiphanes norvegica*. *Polar Biology*, **9,** 115–122.

Stitt, M. 1984. Citrate synthase (condensing enzyme). In Bergmeyer, H.U., ed. *Methods of enzymatic analysis*, Vol. IV. Weinheim: Verlag Chemie, 353–358.

Wilkinson, G. N. 1961. Statistical estimations in enzyme kinetics. *Biochemistry Journal*, **80,** 324–332.

Wiskich, J. T. 1980. Control of the Krebs Cycle. In Davies, D. D., ed. *The biochemistry of plants*, Vol. 2. *Metabolism and respiration*. New York: Academic Press, 243–278.

41 The β-tubulin gene family of the Antarctic ciliate *Euplotes focardii*: determination of the complete sequence of the *β-T1* gene

CRISTINA MICELI, S. PUCCIARELLI, P. BALLARINI, A. VALBONESI AND P. LUPORINI
Dipartimento di Biologia Molecolare, Cellulare e Animale, Università di Camerino, 62032, Camerino (MC), Italy

ABSTRACT

In the Antarctic ciliate Euplotes focardii *three β-tubulin genes have been identified. The complete sequence of one of them, denoted as* ß-T1 *and amplified to* 5×10^3 *copies in the expressed cell somatic nucleus, has been determined. It shows some salient distinctive features. One is related to the eccentric use of the genetic code in ciliates: a TAG codon apparently specifies one Trp residue in the position 21 of the predicted amino acidic sequence. The others are mainly concerned with (i) a concentration of four unique amino acid substitutions in the candidate region for binding GTP, and (ii) a decrease of the acidity and extent of polyglutamilation of the molecule.*

Key words: ciliated Protozoa, *Euplotes*, cytoskeleton, tubulin structure.

INTRODUCTION

A rich variety of protozoa inhabit Antarctic waters. Amongst them, ciliates are predominant in the number of species which can be easily collected and grown in cultures capable of multiplying true to type under controlled laboratory conditions. As individual cells directly exposed to natural selection, they provide unique opportunities to investigate cellular mechanisms and biological phenomena underlying elementary adaptive strategies for living in Antarctica. In line with this perspective, we approached the problem of the differential microtubule stability to cold in one of these 'domesticated' Antarctic ciliates, *Euplotes focardii* Valbonesi & Luporini (Valbonesi & Luporini 1990, 1993). The vast array and abundance of microtubular structures developed by this species for swimming, reproducing, and preserving the body architecture are well illustrated in Fig. 41.1.

IDENTIFICATION OF THE TUBULIN GENES IN THE *E. FOCARDII* MACRONUCLEUS

Like other hypotrich ciliates, *E. focardii* contains two structurally and functionally distinct genomes. One, lying in the cell germinal nucleus (micronucleus), is not expressed and consists of chromosomes. The other, lying in the cell somatic nucleus (macronucleus), is expressed and consists of short DNA molecules, 400–25 000 base pairs (bp) long, that represent and function as individual 'free' genes. Each one appears to be amplified to thousands of copies, which comprise a central coding region flanked by two non-translated 3′ and 5′ sequences terminating in telomeres and usually not interrupted by introns (Klobutcher & Prescott 1986). These macronuclear genes are derived by a process of chromosome diminution taking place in one of the mitotic products of the synkaryon that the cell develops as a result of a sexual process.

In *E. focardii* macronuclear DNA, one ß-tubulin gene of 1800 bp and three ß-tubulin genes of 1600, 1900, and 2150 bp (denoted *β-T1*, *β-T2*, and *β-T3*, respectively) have been identified (Miceli *et al.* 1994), after hybridization with tubulin genes cloned from another cosmopolitan *Euplotes* species, *E. crassus* (Harper & Jahn 1989). By Southern blot analyses carried out with a procedure described in detail for *E. raikovi* pheromone genes (La Terza *et al.* unpublished), they were shown to be amplified to 5×10^3, 7×10^3, and 25×10^3 copies per cell macronucleus, respectively, i. e. equivalent to five- 20-fold the average copy numbers of tubulin genes of other *Euplotes* species (Baird & Klobutcher 1991).

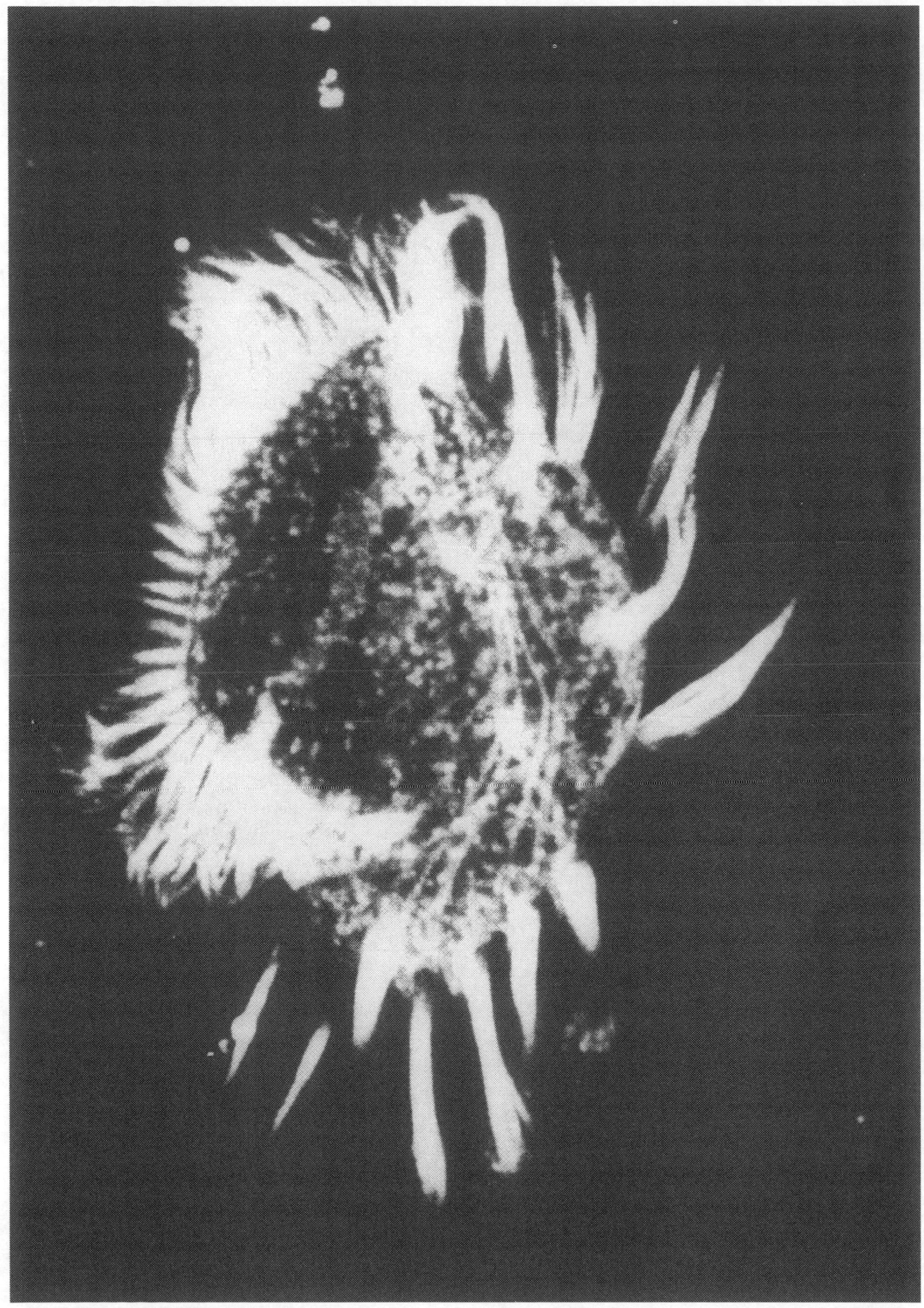

Fig.41.1. A confocal microscopy projection of *E. focardii* showing the microtubular structures associated with the compound ciliary organelles (i.e. the membranelles of the adoral zone and the cirri of the cell ventral surface). The cell was incubated with an anti-a-tubulin antibody and anti-mouse IgG conjugated to fluorescein isothiocyanate. Magnification: ×750.

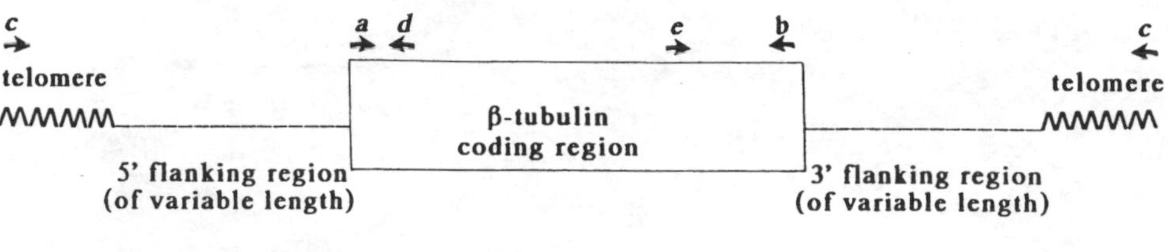

Fig. 41.2. Schematic representation of *E. focardii* ß-tubulin gene organization, with the location of primers used in PCR strategies.

Most effort has so far been directed to defining the structure of the three ß-tubulin genes, using different polymerase chain reaction (PCR) strategies, as shown in Fig. 41.2. The amplification of the coding regions of all the three genes was produced by a single reaction involving, as sense and antisense primers, two oligonucleotides (designated *a* and *b* in Fig. 41.2) corresponding to ß-tubulin sequences conserved in other organisms (reviewed by Burns 1991). On the other hand, the characterization of the 5′ and 3′ non-translated regions required two distinct reactions for each gene. The oligonucleotide $(C_4A_4)_4$ (*c* in Fig. 41.2), corresponding to the telomeric sequence strictly conserved in different species of *Euplotes* (reviewed by Jahn 1991), was used in both, while the other two primers were specifically synthesized for each gene. One (*d* in Fig. 41.2) corresponded to the sequence located 50 nucleotides downstream from primer a ; the other (*e* in Fig. 41.2) to the sequence located 150 nucleotides upstream from primer b.

To date, the only reconstruction of an entire nucleotide sequence to be completed is that of the *β-T1* gene. As shown in Fig. 41.3, it comprises 1541 nucleotides: of these, 1332 constitute the coding region, 113 the 3′ region and 96 the 5′ region.

ECCENTRIC USE OF ONE TAG CODON IN THE *β-T1* GENE SEQUENCE

One in-frame TAG codon appeared in the position of residue 21 of the nucleotide sequence of the *β-T1* coding region. The possibility that this codon interrupts translation, hence that the *β-T1* gene behaves as a pseudogene, was excluded by observing that cDNA clone sequences exactly correspond to the *β-T1* gene sequence. Thus, most likely, this TAG codon specifies Trp, which is conserved in position 21 of every α-, ß-, and γ-tubulin sequence so far determined in different organisms (Burns 1991 as a review). However, as we have not yet determined the *E. focardii* ß-tubulin sequence by direct chemical analysis, the possibility that it specifies Gln, as occurs in *Paramecium*, *Tetrahymena* and *Stylonychia* (Martindale 1989), cannot be excluded.

A number of deviations in the use of the genetic code have been detected in ciliate species branching from the same ancestor, thus suggesting that this eccentric genetic behaviour is a recent ciliate acquisition (Harper & Jahn 1989, Miceli *et al.* 1989). The identification of one TAG codon in the coding frame of the *E. focardii β-T1* gene strengthens this view and, consistent with the finding that the same codon may also terminate gene translation in *E. crassus* (Jahn & Klobutcher, pers. commum.), more generally suggests that most deviations from the universal use of the genetic code are species-specific events in ciliates.

PREDICTED AMINO ACID SEQUENCE OF THE *β-T1* GENE

The analysis of the predicted amino acid sequence of the *E. focardii β-T1* gene shows it to be markedly different from the ß-tubulin sequences of any other organism. There are eight unique amino acid substitutions, plus a number of unusual ones. Seven of the eight unique substitutions occur in the amino terminal domain of the molecule, whose association with the ß-tubulin carboxy-terminal domain is apparently weakened in cold-sensitive microtubules, which disassemble into α/ß-tubulin heterodimers at a low temperature (Kirchner & Mandelkow 1985). Four, namely Arg-57, Ile-59-Cys-60 and Met-63, are closely clustered together in a candidate region for binding GTP (Mandelkow *et al.* 1985, Burns 1991), and hence are involved in the regulation of microtubule stability (Davis *et al.* 1994). Another two, namely Pro-217 and Ile-268, lie in regions of the molecule which are exposed on the microtubule surface (Burns 1991). Ile-268, in addition, is adjacent to substitution Ile-269, which has so far been detected only in the ß-tubulin sequence of one other Antarctic organism, the fish *Notothenia coriiceps neglecta* (Detrich & Parker 1993).

There are at least two other overall structural features of *E. focardii* ß-tubulin that find counterparts in the tubulin of cold-poikilotherm fishes (Detrich & Overton 1988, Detrich *et al.* 1990) and deserve attention for their potential relationship with the activity of the molecule. One is a net overall decrease in the molecule negative charge and acidity, due to the introduction of one basic amino acid, Arg-57, in place of Gly, as well as the substitutions of two Glu residues by Lys-158 and Gln-378. The other feature is the relative abundance of Asp with respect to Glu residues in the acidic carboxy-terminal region, which probably causes a decrease in the extent of molecule polyglutamilation. This process is usually regarded as a post-translational tubulin modification important for microtubule stability (Alexander *et al.* 1991), as it favours microtubule interactions with its associated proteins (MAPS) (Wolff *et al.* 1992). It thus

$(C_4A_4)_4$TTTTTAGGAATTTATTCAAATTTTATTCCTATTCGCCTATAATAATATTTTAAAAAGCAAGGTA 96

```
Met Arg Glu Ile Val His Ile Gln Ala Gly Gln Cys Gly Asn Gln Ile Gly Ala   18
ATG AGA GAG ATC GTA CAT ATC CAA GCA GGT CAA TGC GGT AAC CAG ATT GGT GCC  150

Lys Phe Trp Glu Val Ile Ser Asp Glu His Gly Val Asp Pro Thr Gly Thr Tyr   36
AAA TTT TAG GAA GTC ATT TCT GAT GAA CAT GGA GTT GAT CCA ACC GGT ACT TAC  204

His Gly Asp Ser Asp Leu Gln Leu Glu Arg Ile Asn Val Tyr Phe Asn Glu Ala   54
CAC GGA GAT TCC GAT CTT CAA TTG GAG AGA ATC AAC GTT TAT CTT AAC GAA GCA  258

         *       *   *               *
Thr Gly Arg Lys Ile Cys Pro Arg Ala Met Leu Met Asp Leu Glu Pro Gly Thr   72
ACT GGC AGG AAG ATA TGT CCA AGA GCT ATG CTC ATG GAT CTT GAA CCA GGC ACT  312

Met Asp Ser Val Arg Ala Gly Pro Phe Gly Gln Leu Phe Arg Pro Asp Asn Phe   90
ATG GAC TCA GTC AGA GCT GGA CCA TTT GGA CAA CTT TTC AGA CCA GAC AAC TTT  366

Val Phe Gly Gln Ser Gly Ala Gly Asn Asn Trp Ala Lys Gly His Tyr Thr Glu  108
GTC TTT GGT CAA AGT GGA GCC GGT AAT AAT TGG GCT AAG GGT CAC TAT ACC GAG  420

Gly Ala Glu Leu Ile Asp Ser Val Leu Asp Val Val Arg Lys Glu Ala Glu Gly  126
GGT GCT GAG CTT ATC GAC TCT GTA CTT GAT GTC GTA AGA AAG GAA GCT GAA GGA  474

Cys Asp Cys Leu Gln Gly Phe Gln Ile Thr His Ser Leu Gly Gly Gly Thr Gly  144
TGT GAT TGC CTC CAA GGA TTC CAG ATT ACT CAC TCT TTA GGT GGT GGT ACT GGT  528

                                                     *
Ser Gly Met Gly Thr Leu Leu Ile Ser Lys Val Arg Glu Lys Tyr Pro Asp Arg  162
TCA GGA ATG GGA ACC CTC TTG ATC TCC AAG GTC AGA GAA AAG TAC CCA GAC AGA  582

Ile Met Ala Thr Phe Ser Val Val Pro Ser Pro Lys Val Ser Asp Thr Val Val  180
ATC ATG GCT ACT TTC TCA GTC GTC CCA TCA CCA AAG GTC TCA GAT ACC GTC GTC  646

Glu Pro Tyr Asn Ala Thr Leu Ser Val His Gln Leu Val Glu Asn Ala Asp Glu  198
GAG CCA TAC AAC GCC ACC CTA TCA GTC CAT CAA CTC GTC GAA AAC GCT GAT GAG  690

Val Met Cys Ile Asp Asn Glu Ala Leu Tyr Asp Ile Cys Phe Arg Thr Leu Lys  216
GTT ATG TGT ATC GAT AAC GAA GCC CTC TAC GAT ATC TGC TTC AGA ACC TTA AAG  744

 *
Pro Thr Thr Pro Thr Tyr Gly Asp Leu Asn His Leu Val Ser Ala Val Ile Ser  234
CCG ACC ACC CCA ACT TAT GGT GAT TTA AAC CAT TTG GTA TCC GCC GTT ATC TCA  798

Gly Val Thr Ser Cys Leu Arg Phe Pro Gly Gln Leu Asn Ser Asp Leu Arg Lys  252
GGA GTT ACT TCA TGC CTC AGA TTC CCA GGT CAA TTA AAC TCT GAT TTG AGA AAA  852

                                                             *
Leu Ala Val Asn Leu Ile Pro Phe Pro Arg Leu His Phe Phe Met Ile Gly Phe  270
TTA GCT GTG AAC TTA ATT CCA TTC CCA AGA CTC CAT TTC TTC ATG ATC GGT TTC  916

Ala Pro Leu Thr Ser Arg Gly Ser Gln Gln Tyr Arg Ala Leu Thr Val Pro Glu  288
GCC CCA TTA ACC TCC AGA GGA TCT CAA CAA TAC AGA GCT TTG ACC GTC CCA GAA  960

Leu Thr Gln Gln Met Phe Asp Ala Lys Asn Met Met Cys Ala Ser Asp Pro Arg  306
CTT ACC CAA CAA ATG TTC GAT GCT AAA AAC ATG ATG TGT GCT TCC GAT CCA AGA 1014

His Gly Arg Tyr Leu Thr ALa Ser Ala Met Phe Arg Gly Arg Met Ser Thr Lys  324
CAC GGA AGA TAC CTT ACA GCC TCA GCT ATG TTC AGA GGA AGA ATG TCA ACT AAA 1068

Glu Val Asp Glu Gln Met Leu Asn Val Gln Asn Lys Asn Ser Ser Tyr Phe Val  342
GAA GTT GAT GAA CAA ATG CTC AAT GTC CAG AAC AAG AAC TCA TCA TAC TTT GTT 1122

Glu Trp Ile Pro Asn Asn Ile Lys Ser Ser Val Cys Asp Ile Pro Pro Lys Gly  360
GAA TGG ATT CCA AAC AAC ATT AAA TCA TCT GTC TGT GAT ATT CCA CCA AAG GGA 1176

                                                             *
Leu Lys Met Ser Ser Thr Phe Ile Gly Asn Ser Thr Ala Ile Gln Glu Met Phe  378
CTT AAG ATG TCT TCT ACC TTT ATT GGT AAC TCT ACC GCC ATC CAG GAG ATG TTC 1230

Lys Arg Val Ala Glu Gln Phe Thr Ala Met Phe Arg Arg Lys Ala Phe Leu His  396
AAG AGA GTA GCT GAA CAA TTT ACT GCT ATG TTC AGA AGA AAA GCT TTC TTG CAT 1284

Trp Tyr Thr Gly Glu Gly Met Asp Glu Met Glu Phe Thr Glu Ala Glu Ser Asn  414
TGG TAC ACT GGA GAA GGT ATG GAT GAG ATG GAG TTC ACT GAA GCC GAG TCT AAC 1338

Met Asn Asp Leu Val Ser Glu Tyr Gln Gln Tyr Gln Asp Ala Thr Ala Glu Glu  432
ATG AAC GAT CTC GTT TCT GAA TAC CAA CAA TAT CAA GAT GCC ACC GCC GAA GAA 1392

Glu Gly Glu Phe Asp Asp Glu Glu Glu Met Asp Val
GAA GGA GAA TTC GAC GAT GAA GAA GAG ATG GAT GTT TAAATTTAATCTTGGATGAAATA 1451
```

ATAGAATTTAGAACAATTATTAGAACATTCTTTTTTTTCAATAGGATTTAATAAAAAT$(G_4T_4)_4$ 1541

Fig. 41.3. Nucleotide and predicted amino-acid sequence of the complete *E. focardii β-T1* gene. The TAG codon located in the coding frame is boxed. The amino acid substitutions which differ from ß-tubulin sequences of other organisms are underlined; eight of them, which are unique to the ß-tubulin sequence of *E. focardii*, are additionally marked by an asterisk.

seems likely that *E. focardii* evolved ß-tubulin post-translational modifications, alternative or additional to polyglutamilation, to maintain more stable interactions with MAPS in the cold.

EXPRESSION OF THE *ß-T1* GENE

The nucleotide sequences flanking the coding region deserve attention due to the presence of genetic elements with presumptive properties for the regulation of the expression of the *ß-T1* gene which, like every 'free' macronuclear gene of ciliates, is apparently autonomous for this function. The 3′ region contains the putative polyadenylation signal AATAAT located at nucleotide 1448 as the only relevant element. This variant of the canonical signal AATAAA has been previously detected in a number of other eukaryotic genes. On the other hand, in the 5′ region there is a sequence rich in AT that resembles the TATA-box element known to control transcription in higher eukaryotes. However, this sequence seems to be positionally disadvantaged for functioning as a transcription promoter, as it is located 27 nucleotides upstream of the initiation site of translation. It is thus plausible that promoter functions in the *ß-T1* gene are played by the terminal telomeric sequences, as suggested for other *Euplotes* genes (Ghosh *et al.* 1994), although this assumption is weakened by the observation that the three *E. focardii* ß-tubulin genes, despite their identical telomeres, are transcribed to remarkably varied extents.

ACKNOWLEDGEMENTS

Financial support was provided by the Italian National Program of Antarctic Researches (PNRA).

REFERENCES

Alexander, J .E., Hunt, D. F., Lee, M. K., Shabanowitz, J., Michel, H., Berlin, S.C., Macdonald, T. L., Sundberg, R. J., Rebhun, L. I. & Frankfurter, A. 1991. Characterization of posttranslational modifications in neuron-specific class III ß-tubulin by mass spectrometry. *Proceedings of the National Academy of Sciences of the USA*, **88**, 4685–4689.

Baird, S. E. & Klobutcher, L. A. 1991. Differential DNA amplification and copy number control in the hypotrichous ciliate *Euplotes crassus*. *Journal of Protozoology*, **38**, 136–140.

Burns, R. G. 1991. α-, ß-, and γ-tubulins: sequence comparisons and structural constrains. *Cell Motility and Cytoskeleton*, **20**, 181–189.

Davis, A., Sage, C. S., Dougherty, C. A. & Farrell, K. W. 1994. Microtubule dynamics modulated by guanosine triphosphate hydrolysis activity of ß-tubulin. *Science*, **264**, 839–841.

Detrich, H. W. & Overton, S. A. 1988. Antartic fish tubulins: heterogeneity, structure, amino acid composition and charge. *Comparative Biochemistry and Physiology*, **90**, 593–600.

Detrich, H. W. & Parker, S. K. 1993. Divergent neural ß-tubulin from the antarctic fish *Notothenia coriiceps neglecta*: potential sequence contributions to cold adaptation of microtubule assembly. *Cell Motility and Cytoskeleton*, **24**, 156–166.

Detrich, H. W., Neighbors, B. W., Sloboda, R. D. & Williams, R. C. J. 1990. Microtubules-associated proteins from Antarctic fishes. *Cell Motility and Cytoskeleton*, **17**, 174–186.

Ghosh, S., Jaraczewski, J. W., Klobutcher, L. A. & Jahn, C. L. 1994. Characterization of transcription initiation, translation initiation, and poly(A) addition sites in the gene-sized macronuclear DNA molecules of *Euplotes*. *Nucleic Acid Research*, **22**, 214–221.

Harper, D. S. & Jahn, C. L. 1989. Inconstancy of the genetic code among ciliated protozoa. *Proceedings of the National Academy of Sciences of the USA*, **86**, 3052–3056.

Jahn, C. L. 1991. The nuclear genomes of hypotrichous ciliates: maintaining the maximum and the minimum of information. *Journal of Protozoology*, **38**, 252–258.

Kirchner, K. & Mandelkow, E. V. 1985. Tubulin domains responsible for assembly of dimers and protofilaments. *EMBO Journal*, **4**, 2397–2402.

Klobutcher, L. A. & Prescott, D. M. 1986. The special case of the hypotrichs. In Gall, J. G., ed. *The molecular biology of ciliated protozoa*. Orlando, FL: Academic Press, 111–144.

Mandelkow, E. M., Herrmann, M. & Ruhl, U. 1985. Tubulin domains probed by limited proteolysis and subunit specific antibodies. *Journal of Molecular Biology*, **185**, 311–327.

Martindale, D. W. 1989. Codon usage in *Tetrahymena* and other ciliates. *Journal of Protozoology*, **36**, 29–34.

Miceli, C., La Terza, A. & Melli, M. 1989. Isolation and structural characterization of cDNA clones encoding the mating pheromone E*r*-1 secreted by the ciliate *Euplotes raikovi*. *Proceedings of the National Academy of Sciences of the USA*, **86**, 3016–3020.

Miceli, C., Di Giuseppe, G., Ballarini, P., Valbonesi, A. & Luporini, P. 1994. Identification of the tubulin gene family and sequence determination of one ß-tubulin gene in a cold-poikilotherm protozoan, *Euplotes focardii*. *Journal of Eukaryotic Microbiology*, **41**, 420–427.

Valbonesi, A. & Luporini, P. 1990. *Euplotes focardii*, a new marine species of *Euplotes* (Ciliophora, Hypotrichida) from Antarctica. *Bulletin of the British Museum (Natural History) Zoology*, **56**, 57–61.

Valbonesi, A. & Luporini, P. 1993. Biology of *Euplotes focardii*, an Antarctic ciliate. *Polar Biology*, **13**, 489–493.

Wolff, A., de Néchaud, B., Chillet, D., Mazarguil, H., Desbruyères, E., Audebert, S., Eddé, B., Gros, F. & Denoulet, P. 1992. Distribution of glutamylated and ß-tubulin in mouse tissues using a specific monoclonal antibody, GT335. *European Journal of Cell Biology*, **59**, 425–432.

42 Purification and characterization of atypical zinc-binding polypeptides from the Antarctic sea urchin *Sterechinus neumayeri*

CLEMENTE CAPASSO[1], R. SCUDIERO[2], A. CAPASSO[1], R. D'AVINO[1], L. CAMARDELLA[1], G. DI PRISCO[1] AND E. PARISI[1]

[1]*CNR Institute of Protein Biochemistry and Enzymology, via Marconi 10, 80125 Napoli, Italy,* [2]*Department of Evolutionary and Comparative Biology, University of Napoli, Italy*

ABSTRACT

The majority of the cellular zinc in the gonads of the Antarctic sea urchin Sterechinus neumayeri *is bound to small, metal-containing polypeptides which cannot be classified as typical metallothioneins. These peptides have been purified by a combination of gel-permeation and anion-exchange chromatography. In the presence of mercaptoethanol, these peptides dissociated into smaller forms. They were shown to be particularly rich in histidine, and also contained high amounts of glycine, cysteine, aspartate and glutamate.*

Key words: heavy metals, zinc-binding proteins, metallothionein, sea urchin, Antarctica.

INTRODUCTION

In most organisms, zinc is bound to low molecular mass polypeptides termed metallothioneins (MTs) (Margoshes & Vallee 1954). MTs are cysteine-rich proteins which do not contain aromatic residues, arginine and histidine (Kägi & Schäffer 1988); they are ubiquitously distributed in a large variety of species, including vertebrates, invertebrates, plants and microorganisms (Hamer 1986). It has been inferred that these proteins are involved in the transport and homeostasis of physiological oligoelements (mostly Zn and Cu) (Bremner & Beattie 1990), as well as in the detoxification of a number of non-essential trace elements such as Cd and Hg (Dallinger 1993).

It is generally accepted that MTs play a major role as zinc donors to cope with the high demand of metabolic zinc; however, evidence exists that some tissues are deficient in MTs and contain metal-binding polypeptides of similar molecular mass, but very different amino acid composition (Waalkes & Goering 1990). In general, these proteins are characterized by a lower cysteine content than MTs, by the presence of arginine, histidine and aromatic amino acids, and by an unusually high content of acidic residues (Stone & Overnell 1985).

In a previous study, we found that the liver of the Antarctic icefish *Chionodraco hamatus* does not contain detectable amounts of MT; instead, it contains a zinc-binding protein unlike MT with characteristics similar to those described above (Scudiero *et al.* 1992). For this reason, we decided to investigate metal-binding proteins in other Antarctic species. The present study reports the unusual properties of heat-stable low molecular mass zinc-containing polypeptides isolated from the ovaries of the Antarctic sea urchin *Sterechinus neumayeri*.

MATERIAL AND METHODS

Biological material

Specimens of *S. neumayeri* were collected in the proximity of the Terra Nova Bay Station (Italy) and immediately frozen at −80 °C. Ovaries were quickly removed from the frozen animals and homogenized in 5 volumes of acetone prechilled at −20 °C. The homogenate was filtered on a Whatman 3MM paper filter and the resulting residue was extracted two more times in the same way. The final residue was dried and stored at 4 °C.

Determination of MT content

MT content in sea urchin extracts was determined by silver saturation assay as described by Scheuhammer & Cherian (1991). Five hundred mg of acetone powder were homogenized in 2 ml of 0.5 M sucrose. The homogenate was centrifuged at 12 000*g* for 20 min. Four hundred μl of this extract were used for the assay.

Protein isolation

Ten grams of acetone powder were homogenized in 50 mM Tris-HCl pH 8.6, 2 mM dithiotreitol (DTT) and 0.1 mM phenyl-methyl-sulfonyl-fluoride (PMSF). The homogenate was centrifuged at 10 000*g* for 30 min. The supernatant was heated for 10 min. at 90 °C and centrifuged again at 80 000*g* for 1 h. The final supernatant was chromatographed on Sephadex G-75 (2.8×74 cm column). The two major zinc-containing protein peaks were individually pooled and chromatographed on DEAE-Sephadex A-25 (1.5×13 cm column). The zinc-containing fractions eluted by a 0–0.4 M NaCl gradient in 10 mM Tris-HCl pH 8.6, 2 mM DTT were pooled, concentrated by lyophilization and further purified by reverse-phase HPLC (0.4×25 cm Super Pac Pep-S column, Pharmacia) using as eluent a 36 min. linear gradient of 0–60% acetonitrile in 0.1% trifluoroacetic acid. Flow rate was adjusted to 0.8 ml min.$^{-1}$, and the absorbance was monitored at 280 nm. The UV-absorbing peaks obtained by repeated runs were pooled, lyophilized and reconstituted in a solution containing 1M 1-mercaptoethanol and 0.3% trifluoroacetic acid. After incubation under N_2 at room temperature for 3 h, this material was subjected to HPLC (0.39×30 cm μBondapack C-18 reverse-phase column, Waters) using as eluent a 37 min. linear gradient of 0–20% acetonitrile in 0.3% trifluoroacetic acid. Flow rate was 0.6 ml min. $^{-1}$.

Analytical procedures

Zinc and silver were determined by atomic absorption spectrometry using a model 5100 Perkin Elmer apparatus equipped with a Zeeman graphite furnace. Amino acid analysis was performed on a Carlo Erba 3A30 automatic analyser. Samples were oxidized with performic acid (Böhlen & Schroeder 1982) and hydrolized in 6 N HCl at 150 °C for 1 h under N_2. Norleucine was used as an internal standard to check recovery.

RESULTS AND DISCUSSION

A search for MT in *S. neumayeri* was initially carried out on the whole extract by means of the silver saturation assay. This method was found to be very suitable for the measurement of MT level in sea urchins (Scudiero *et al.* 1994). The results show that the amount of MT present in *Sterechinus* was well below the detection limit of the method (0.01), but the addition of horse MT to the extract resulted in a significant increase in the amount

Table 42.1. *Progress of purification of* S. neumayeri *Zn-binding proteins*

Step	Protein (mg)	Zn (μg)	μg Zn mg protein^{-1}
Heated extract	120.0	360.0	3.0
Gel filtration (Peak I)	16.2	125.0	7.7
(Peak II)	35.7	200.0	5.6
DEAE (I)	0.86	21.7	25.3
(II)	1.17	27.5	23.5

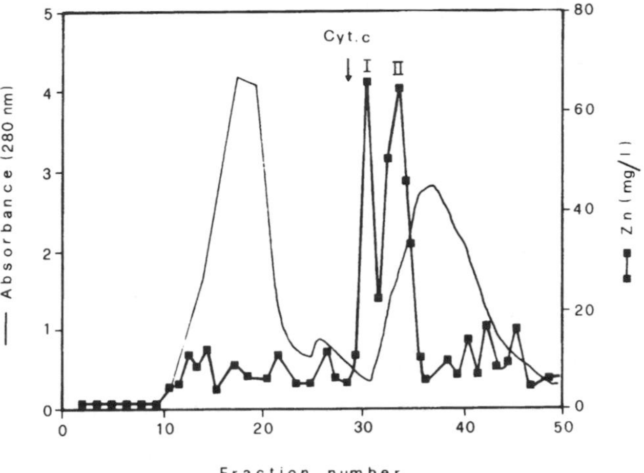

Fig. 42.1. Gel filtration of an extract from ovaries of *S. neumayeri*. An extract prepared as described in the text was applied on a Sephadex G-75 column which was eluted with 10 mM Tris-HCl pH 8.6, 2 mM DTT at 4 °C. Five millilitres were collected per tube at a flow rate of 60 ml h^{-1}.

of bound Ag (2.8±0.15). Hence, unlike the Mediterranean sea urchin *Paracentrotus lividus* in which a Zn-protein similar to MT was identified (Scudiero *et al.* 1995), no such protein is present in *S. neumayeri*.

In order to establish whether a non-MT Zn-binding protein could be detected in *Sterechinus*, an ovarian extract was fractionated as described above. The elution profile of a gel-permeation chromatography is shown in Fig. 42.1. Most of the applied zinc (>90%) was eluted as two peaks (forms I and II in Fig. 42.1). Each form was further purified by anion-exchange chromatography, following the procedure described above. The progress of purification is summarized in Table 42.1. The material from the DEAE-cellulose step was analyzed by HPLC. The elution profiles of HPLC depicted in Fig. 42.2a,b show the presence of a single peak with absorbance at 226 and 280 nm for both form I and form II. However, as shown in Fig. 42.2c,d these two peaks could be resolved in two fractions following incubation in 1-mercaptoethanol and rechromatography on a C-18 column. Amino acid analyses carried out on the four isolated isoforms are reported in Table 42.2. It is evident that the most striking peculiarity of the *S. neumayeri* proteins is their unusual composition, as they show cysteine contents lower than that of vertebrate MTs, but higher than that reported for the non-MT proteins isolated from vertebrate

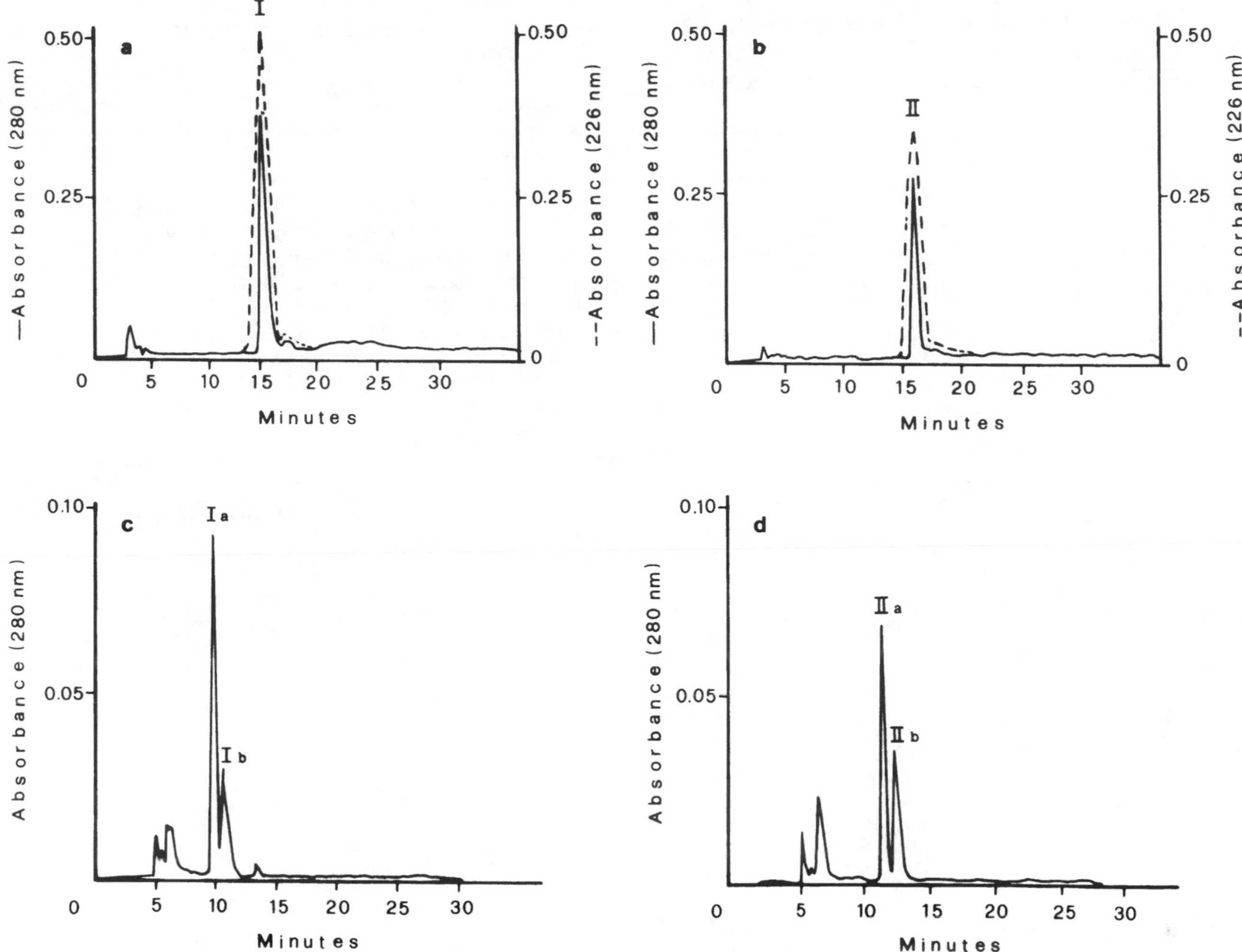

Fig. 42.2. Reverse-phase HPLC elution profile of ovarian Zn-binding proteins. The upper panels represent the elution profile of forms I and II separated by gel-permeation chromatography and purified by anion-exchange chromatography. The bottom panels show the rechromatography of peaks I and II after incubation with mercaptoethanol.

gonads (Waalkes *et al.* 1988, Baer & Thomas 1991). In addition, the *Sterechinus* proteins have an unusually high content of histidine and an abundance of aspartate and glutamate. The mercaptoethanol-induced dissociation of forms I and II into distinct isoforms suggests the existence of disulphide-linked eterodimers. The presence of high amounts of histidine, aspartate and glutamate suggests that these residues may be putative metal-binding sites, together with cysteine. In particular, Zn^{++} displays a strong preference to be in the plane of the carboxilate, and this optimizes the interaction with the oxigen sp^2 lone electron pair (Christanson 1991). Clearly, more work is needed to elucidate the nature of the binding groups, and how these may affect the storage and homeostasis of essential oligoelements.

ACKNOWLEDGEMENTS

This research was supported by the Italian National Programme for Antarctic Research. We are grateful to the Atomic Spectrometry Service of the CNR Research Area of Naples for the use of the atomic absorption spectrometer.

REFERENCES

Baer, K. N. & Thomas, P. 1991. Isolation of novel metal-binding proteins distinct from metallothionein from spotted seatrout (*Cynoscion nebulosus*) and Atlantic croaker (*Micropogonias undulatus*) ovaries. *Marine Biology*, **108**, 31–37.

Bremner, I. & Beattie, J. H. 1990. Metallothionein and the trace minerals. *Annual Review of Nutrition*, **10**, 63–83.

Böhlen, P. & Schroeder, R. 1982. High-sensitivity amino acid analysis methodology for determination of amino acid compositions with less than 100 pmoles of peptides. *Analytical Biochemistry*, **22**, 151–164.

Christianson, D. W. 1991. Structural biology of zinc. *Advances in Protein Chemistry*, **42**, 281–355.

Dallinger, R. 1993. Strategies of metal detoxification in terrestrial invertebrates. In Dallinger, R. & Rainbow, P. S., eds. *Ecotoxicology of metals in invertebrates*. Boca Raton, FL: Lewis Publishers, 245–289.

Hamer, D. H. 1986. Metallothionein. *Annual Review of Biochemistry*, **27**, 913–951.

Kägi, J. H. R. & Schäffer, A. 1988. Biochemisty of metallothionein. *Biochemisty*, **27**, 8509–8515.

Margoshes, M. & Vallee, B. L. 1954. A cadmium protein from equine kidney cortex. *Journal of the American Chemistry Society*, **79**, 4813–4818.

Scheuhammer, A. M. & Cherian, M. G. 1991. Quantification of metallothionein by silver saturation. *Methods in Enzymology*, **205**, 78–83.

Table 42.2. *Amino acid composition of* S. neumayeri *Zn-binding proteins (in moles %)*

	Form			
Residue	Ia	Ib	IIa	Iib
cys	10.0	18.0	9.1	9.7
asx	9.4	7.2	6.7	7.6
thr	4.0	6.4	2.6	3.0
ser	4.8	8.0	5.5	6.2
glx	10.5	5.1	9.4	13.0
pro	—	2.5	1.6	—
gly	7.1	9.9	8.8	12.4
ala	4.8	5.6	3.0	4.6
val	4.0	5.3	2.3	3.0
met	—	—	0.4	0.4
ile	1.8	3.4	1.1	1.7
leu	1.8	5.0	1.0	2.0
tyr	—	1.6	—	—
phe	—	—	—	—
his	33.0	10.6	42.5	31.7
lys	2.5	3.9	1.8	1.8
trp	n.d.	n.d.	n.d.	n.d.
arg	5.8	6.6	3.5	2.6

Note:
n.d.=not determined.

Scudiero, R., Capasso, C., De Prisco, P. P., Capasso, A., Filosa, S. & Parisi, E. 1994. Metal-binding proteins in eggs of various sea urchin species. *Cell Biology International,* **18**, 47–53.

Scudiero, R., Capasso, C., Del Vecchio-Blanco, F., Savino, G., Capasso, A., Parente, A. & Parisi, E. 1995. Isolation and primary structure determination of a metallothionein from *Paracentrotus lividus* (Echinodermata, Echinoidea). *C.B.P. Biochemistry and Molecular Biology,* **111**, 2, pp. 329–336.

Scudiero, R., De Prisco, P. P., Camardella, L., D'Avino, R., di Prisco, G. & Parisi, E. 1992. Apparent deficiency of metallothionein in the liver of the antarctic icefish *Chionodraco hamatus*. Identification of a zinc-binding protein unlike metallothionein. *Comparative Biochemistry and Physiology*, **103B**, 201–207

Stone, H. & Overnell, J. 1985. Non-metallothionein cadmium binding proteins. *Comparative Biochemistry and Physiology*, **80C**, 9–14.

Waalkes, M. P. & Goering, P. L. 1990. Metallothioneins and other cadmium-binding proteins: recent developments. *Toxicology*, **3**, 281–288.

Waalkes, M. P., Perantoni, A. & Palmer, A. E. 1988. Isolation and partial characterization of the low-molecular mass zinc/cadmium-binding protein from the testes of the patas monkey (*Erythrocebus patas*). *Biochemical Journal*, **256**, 131–137.

43 An overview of the population biology and chemical ecology of three species of Antarctic soft corals

MARC SLATTERY AND JAMES B. MCCLINTOCK
Department of Biology, University of Alabama at Birmingham, Birmingham, Alabama 35294-1170, USA

ABSTRACT

The Antarctic soft corals Alcyonium paessleri *and* Clavularia frankliniana *are conspicuous abundant members of the shallow (12–33 m depth) hard-bottom communities of eastern McMurdo Sound. In contrast, the larger, comparatively rare,* Gersemia antarctica *occurs below 18 m on soft-bottom habitats of western McMurdo Sound. All three soft corals contain bioactive compounds that deter potential sea star and fish predators, while only* A. paessleri *and* G. antarctica *possess toxic metabolites that may serve as antifoulants.* Alcyonium paessleri *and* C. frankliniana *both brood embryos that are released as planula larvae, while asexual propagation via fission is particularily prevalent in* C. frankliniana. *Laboratory-measured growth rates during the austral spring and summer were 0.8 mm per month in* A. paessleri *and 1.2 cm per month in* C. frankliniana, *while field growth measured over an entire year in* G. antarctica *indicated a growth rate of 0.7 cm per month. These rates are similar to those measured in soft corals from temperate habitats. A field test of contact-mediated competition indicated that the Antarctic sponge* Mycale acerata *suffers significant tissue necrosis when contacted by the soft coral* A. paessleri, *suggesting that defensive chemicals may assist in preventing overgrowth by this rapidly growing, space dominating, sponge.*

Key words: soft coral, Antarctica, chemical defense, population biology, benthic ecology.

INTRODUCTION

Soft corals are a diverse group of conspicuous fleshy marine invertebrates that often comprise significant proportions of total biomass in hard-bottom communities (Preston & Preston 1975, Benayahu & Loya 1981, Tursch & Tursch 1982, Dinesen 1983). They are especially well represented in the tropical Pacific and Atlantic, but their geographic distribution extends into temperate and polar ecosystems (Bayer 1981 and references within). The basis of the evolutionary success of soft corals (Sammarco & Coll 1990), which generally lack morphological defenses, has been debated (Tursch *et al.* 1978, Bakus *et al.* 1986, Sammarco & Coll 1988). Clearly soft coral species diversity has been influenced by numerous physical and biological factors including geologic events, notably the closing of the Isthmus of Panama (Rosen 1978, Stanley 1984, Veron 1986), glaciation and oceanic cooling (Dana 1975, Frost 1977, Stanley 1979), reproductive patterns (Alino & Coll 1989, Achituv & Benayahu 1990, Lasker 1990, Brazeau & Lasker 1992) and dispersal (Sebens 1983a, b, Benayahu & Loya 1985, Walker & Bull 1983). However, recent work has demonstrated the importance of diverse secondary metabolites, sequestered within soft coral tissues, that may serve as defensive agents against predation, competition and fouling (Bakus *et al.* 1986, Sammarco & Coll 1990, Paul 1992).

Chemical defenses in sessile or slow-moving tropical marine invertebrates, including soft corals, appear to have evolved in response to predation pressure from a myriad of browsing fish (Bakus & Green 1974, Vermeij 1978). Decreased predation by browsing fish at higher latitudes (Nuedecker 1979, Palmer 1979), and the potential for impaired synthesis of noxious metabolites at low temperature (Yamanouchi 1955), led earlier researchers to suggest that evolutionary selection for defensive compounds may be reduced in high-latitude communities (Bakus & Green 1974, Bakus 1981). This hypothesis is also applicable to soft corals along a latitudinal gradient across the Great Barrier Reef, Australia (Coll *et al.* 1982). However, recent evidence of chemical defenses in Antarctic benthic marine invertebrates (McClintock *et al.* 1990, 1994) suggests that more information is needed before such biogeographic relationships can be adequately evaluated. The presence of bioactivity in many Antarctic benthic marine invertebrates probably reflects the prevalence of biological interactions such as predation and competition which structure both tropical (Porter 1972, Green 1977, Nuedecker 1979) and Antarctic (Dayton *et al.* 1974) marine systems. In Antarctica, sea stars have replaced fish as the dominant browsing predators of sessile and sluggish marine invertebrates (McClintock 1994), and selective pressures have apparently been sufficient over an extensive geological history to drive the evolution of defensive metabolites (McClintock 1987, McClintock *et al.* 1990). Chemical defense may also be related to the slow growth rates and long lifespans which typify the Antarctic marine invertebrate fauna (Clarke 1983, Pearse *et al.* 1991).

The shallow hard-bottom communities of McMurdo Sound, Antarctica can be divided into three discrete zones based on depth and the degree of physical disturbance (Dayton *et al.* 1970). At depths less than 15 m, sessile biota are virtually absent due to annual ice disturbance events including anchor ice and iceberg scouring (Dayton 1989, Dayton *et al.* 1969, 1970, 1974). At depths greater than 33 m, the benthos is characterized by extreme physical stability and sponges dominate community biomass (Littlepage 1965, Dayton *et al.* 1969, 1974, Barry & Dayton 1988, Pearse *et al.* 1991, Dayton *et al.* 1994). The transition depths between these two zones (15–33 m depth) are comprised of a fauna dominated by cnidarians, including sea anemones, soft corals and hydroids (Dayton *et al.* 1974, Slattery 1994). The purpose of this paper is to review recently published information on the distribution, abundance and chemical defense mechanisms of the Antarctic soft corals *Alcyonium paessleri, Clavularia antarctica* and *Gersemia antarctica* at sites on both the eastern and western sides of McMurdo Sound. Moreover, new unpublished information is presented on reproductive modes, recruitment and growth, and contact mediated competition.

Table 43.1. *Summary of the population dynamics of the Antarctic soft corals* Alcyonium paessleri, Clavularia frankliniana, *and* Gersemia antarctica *at two sites in McMurdo Sound; Arrival Heights (AH) and Explorer's Cove (EC)*

	Alcyonium paessleri	*Clavularia frankliniana*	*Gersemia antarctica*
Distribution (depth m)			
AH	12–30	12–30	—
EC	1	1	18–33
Abundance (corals m²)			
AH	7.3	1337.3	—
EC	0.02	0.2	0.04
Mean biomass per individual[2] (g dry wt)	1.81	0.008	45.0
Energetic content (kJ per individual)	15.9	17.3	14.5
Energetic content (kJ m²)			
AH	210.1	185.1	—
EC	0.58	0.03	26.1

Notes:
[1] Rare, dependant on availability of hard substrata.
[2] Based on mean-sized individual (Slattery and McClintock 1995).
Source: Data are summarized from Slattery & McClintock 1995)

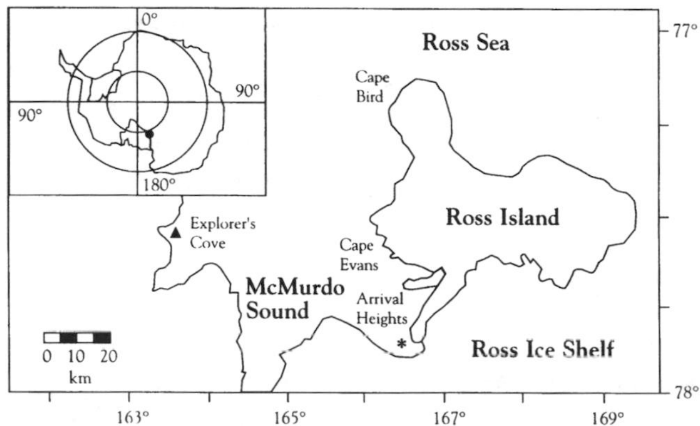

Fig. 43.1. Map of McMurdo Sound, Antarctica showing the location of the study sites at Arrival Heights (*) and Explorer's Cove (▲). Insert shows the location of McMurdo Sound (●) in relation to the Antarctic continent.

OBSERVATIONS

Field sampling and collections of *Alcyonium paessleri, Clavularia frankliniana* and *Gersemia antarctica* were undertaken at Arrival Heights (77°51' S, 166°39' E) and Explorer's Cove (77°34' S, 163°35' E), McMurdo Sound, Antarctica (Fig. 43.1). Populations of *A. paessleri* and *C. frankliniana* were monitored in November and December 1989, October, November and December 1992, January, September, October, November and December 1993 and January and February 1994, along permanent transects or within marked quadrats. Populations of *G. antarctica* were censused using a belt transect and marked individually with stakes driven into the sediments. Observations were made in November, 1989, 1992, 1993 and February 1994. Depth distribution, abundance, recruitment and height of individual colonies were measured (Table 43.1; Slattery 1994,

Table 43.2. *Summary of chemically mediated defenses in the Antarctic soft corals* Alcyonium paessleri, Clavularia frankliniana, *and* Gersemia antarctica

Bioassay	*Alcyonium paessleri*	*Clavularia frankliniana*	*Gersemia antarctica*
Feeding deterrence			
Seastar tube-foot retraction	+	+	+
Ichthy-emetic response	+	+	+
Toxicity			
Sea urchin			
Sperm motility	+	—	+
Embryo growth	+	—	+
Adult survival	+	—	+
Antifouling activity			
Microbial			
Attachment	+	—	+
Growth	+	—	+
Algal			
Settlement	+	—	+
Growth	+	—	+
Competition			
Contact			
Sponges	+	ns	*
Soft corals	ns	ns	*
Sea anemone	—	ns	*

Notes:
+, at least one bioactive extract from an organism; —, no bioactivity; *, the assay was not conducted; and ns, in contact experiments, indicates no significant difference between experimentals and controls. All data except competition contact experiments are summarized from Slattery and McClintock (1995) and Slattery *et al.* (1995).

Slattery & McClintock 1995). Organic extracts of each soft coral, obtained by partitioning whole colony tissues with organic solvents of increasing polarity (hexane, chloroform, methanol, and aqueous methanol), were assayed for their ability to deter feeding by predatory sea stars and fish (Slattery & McClintock 1995), induced mortality in sea urchin gametes (Slattery & McClintock 1995) and inhibition of microbial and algal growth (Slattery *et al.* 1995). In addition, *in situ* manipulative experiments between soft corals and other common benthic macroinvertebrates (notably sponges and a sea anemone) were performed to test for contact-mediated competitive interactions (Table 43.2; Slattery 1994).

Soft coral populations remained remarkably stable between 1989 and 1992 for all species and at both study sites (Slattery 1994, Slattery & McClintock 1995). *Alcyonium paessleri* and *C. frankliniana* occurred between 12 and 33 m depths on rocky substrates at Arrival Heights and were limited to the patchy hard substrata at Explorer's Cove, an area comprised mostly of soft sediments. *Gersemia antarctica* was not found on the eastern side of McMurdo Sound (i.e. Arrival Heights) but was able to colonize soft-bottom habitats below 18 m depth and at least to 33 m depth throughout Explorer's Cove by anchoring to mollusc shells and echinoid tests, or by production of a 'clay anchor' (i.e. a compressed ball of hard clay) within the basal disc. Mean population densities ($x\pm1$ SD) of *A. paessleri* and *C. frankliniana* were 7.3 ± 0.8 and 0.02 ± 0.03 colonies m^{-2} and 1337.3 ± 223.1 and 0.18 ± 1.4 polyps m^{-2}, at Arrival Heights and Explorer's Cove, respectively (Slattery & McClintock 1995). The mean density of *G. antarctica* at Explorer's Cove was much lower with a density of 0.04 ± 0.01 colonies m^{-2}. In the austral spring of 1993, a significant reduction in density of *A. paessleri* at the Arrival Heights population was noted (a reduction to <1 colony m^{-2}); however, a similiar decline was not observed in the population of *C. frankliniana* at this site (Slattery & McClintock 1995).

Recruitment patterns among these three soft corals were variable both temporally and spatially, and probably represent differences in reproductive modes and habitat selection (Slattery 1994). *Alcyonium paessleri* colonies held in the laboratory and observed in the field brood planula larvae in their basal solenia (Slattery personal observations). The shed planula appear to settle rapidly, often within a few centimetres of a parent colony, although colonies observed on field settling plates at locations isolated from adult populations suggest that long-range dispersal may occur (Slattery personal observations). Planula shedding in *A. paessleri* was observed in the laboratory in May 1993 and in the field in October 1993. This suggests that reproductive activity is prolonged, spanning at least five months (Slattery 1994). In addition to sexual propagation, *A. paessleri* colonies also undergo asexual fission. Based on observations of 351 and 200 individuals during the austral summers of 1989 and 1992, respectively, 7 and 11% of the populations had individuals undergoing fission (Slattery 1994). Presumably, this is an adaptation to take advantage of suitable habitat within the area colonized by the parent colony.

Clavularia frankliniana, which has an extensive sub-benthic network of stolons, may produce new polyps continuously throughout the austral spring–summer period (mean 29.4 ± 37.6 new polyps $colony^{-1}$ $month^{-1}$; determined from 14 counts of 31 tagged colonies between September 1993 and February 1994). Colonies held in the laboratory brooded embryos within basal stolons that are released as planula larvae once fully developed (Slattery personal observations). Recruits of *C. frankliniana* may suffer high mortality rates. For example, although 300 recruits were counted and tagged in November1989, none were observed at this same site in December 1989. *Clavularia frankliniana* is more dependent upon clonal propagation, via stolon extension and overgrowth, than gamete production as a means of colonizing local available substrata.

Juvenile *Gersemia antarctica* did not recruit at Explorer's Cove on the western side of McMurdo Sound between 1989 and 1992, suggesting this species may exhibit only episodic spawning events or suffer intense mortality soon after recruitment. In December 1993, 30 *G. antarctica* recruits were observed in Explorer's Cove attached to empty shells of the scallop *Adamusium colbecki*. Each was tagged individually with

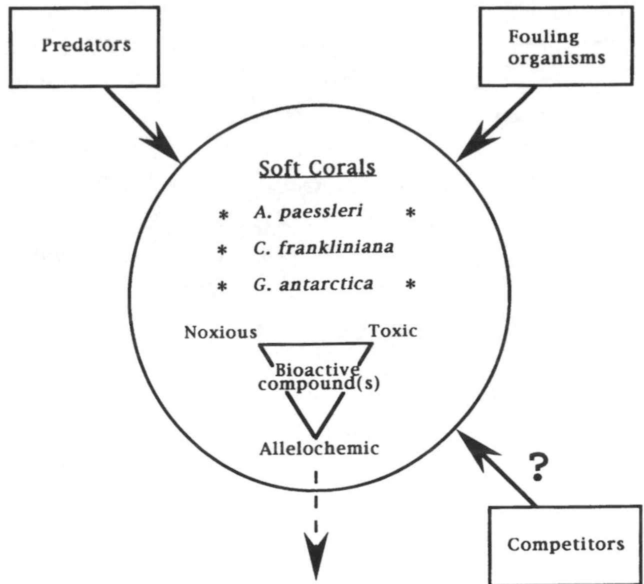

Fig. 43. 2. Summary of proposed ecological roles of bioactive compounds from the Antarctic soft corals *Alcyonium paessleri*, *Clavularia frankliniana*, and *Gersemia antarctica*. An asterisk indicates bioactivity of a noxious or toxic nature.

a numbered tie-wrap. However, by February 1994 no tagged colonies remained, indicating that either biotic or abiotic disturbances can have a significant impact on recruitment in this soft coral.

Field observations indicate that *A. paessleri* and *G. antarctica* colonies can survive for at least 4.5 years. Individual *C. frankliniana* polyps disappear within 1.5 years (Slattery personal observations); however, clones of these individuals may continue to survive over several seasons. Growth rates ($x \pm 1$ SD) of 41 and 100 laboratory-reared juvenile *A. paessleri* and *C. frankliniana* were 0.8 ± 0.6 mm month^{-1} and 1.2 ± 0.4 cm month^{-1}, averaged from weekly measurements made over 8 and 5 months, respectively. Field-marked juveniles of *G. antarctica* grew 8 ± 7.2 cm y^{-1} based on an initial measurement of three individuals in November 1993 and subsequent measurements of these individuals in December 1993 and February 1994. These growth rates are similar to those measured in temperate soft corals, including at least one congenor (Sebens 1983b, Benayahu & Loya 1985). Growth rates slow markedly once mean colony size is attained (Slattery 1994). Adult *A. paessleri* colonies (n=25), marked during the 1989 austral spring, showed no measurable growth over 4 years. A similiar lack of measurable adult growth was noted for *G. antarctica* colonies (n=5) marked in October 1992 and remeasured after one year.

The ecological interactions that are potentially mediated by noxious and toxic bioactive compounds in *Alcyonium paessleri*, *Clavularia frankliniana*, and *Gersemia antarctica* are summarized in Fig. 43.2. Predator deterrence was examined in two seperate bioassays targeted at sympatric sea stars and fish (Slattery & McClintock 1995). A retraction response of the chemo-sensory terminal tube-feet of the Antarctic seastars *Odontaster validus* and *Perknaster fuscus* was utilized as one indicator of bioactivity in organic extracts of *A. paessleri*, *C. frankliniana* and *G. antarctica*. Significant tube-foot retraction responses were recorded in response to extracts of all three species of soft corals. In addition, pieces of whole colony tissue and crude aqueous extracts embedded in 5% agar pellets containing a feeding stimulant were consistently rejected when offered to the Antarctic fish *Pseudotrematomus bernacchii* and *Pagothenia borchgrevinki*. Feeding deterrence was detected across a gradient of increasingly polar solvent extractions, indicating that a suite of noxious compounds are probably present in *A. paessleri*, *C. frankliniana* and *G. antarctica*.

Toxicity of soft coral extracts was examined in three life history stages of the common Antarctic sea urchin *Sterechinus neumayeri* (Slattery & McClintock 1995). Sea urchin sperm were incubated in the presence of crude aqueous extracts of each soft coral at tissue level concentrations, and flagellar motility, or the lack thereof, was recorded. Similiarly, crude aqueous soft coral extracts were injected via the oral opening into the gut of adult sea urchins to determine if intestinal absorption of the compounds was lethal. Embryological development rates were assayed in the presence of four organic extracts, from each species of soft coral, in order to determine effects of the toxic metabolites. Among these three sea urchin bioassays, only extracts from *A. paessleri* and *G. antarctica* exhibited results consistent with the presence of toxic metabolites (Slattery & McClintock in press). Furthermore, two distinct groups of toxins, one soluble in chloroform and the second in aqueous methanol, were discerned. These findings suggest that both noxious and toxic compounds exist in these three Antarctic soft corals, and that they may act in a synergistic manner.

Antifouling activity of soft coral extracts was tested in field and laboratory assays that focused on the inhibition of sympatric bacterial and microalgal settlement and growth (Slattery 1994; Slattery *et al.* 1995). Differential growth inhibition was noted in three Antarctic marine bacteria exposed to soft coral extracts; however, in settlement and growth assays, only extracts from *A. paessleri* and *G. antarctica* were bioactive. Moreover, the bioactivity was detected in both chloroform and aqueous methanol fractions, suggesting that antifouling activity is probably the result of toxic metabolites, perhaps sequestered within the epithelial surface of these soft corals (Slattery *et al.* 1995). Microalgal *in situ* settlement and laboratory growth assays exhibited similiar trends in species specificity and extract bioactivity. In month-long field assays conducted at Palmer Station during October–November 1991, aqueous methanol extract of *Alcyonium paessleri* caused inhibition of fouling invertebrates as well as of microalgae (Slattery 1994). Low densities of fouling invertebrates on settling plates retrieved from McMurdo Sound precluded statistical analyses.

Contact-mediated competitive interactions of *Alcyonium paessleri* and *Clavularia frankliniana* were assayed *in situ* over a one-year period (Fig. 43.3; Slattery 1994). Each soft coral (n=4 replicates per treatment) was placed in contact with the other soft coral species, a sea anemone (*Urticinopsis antarctica*) and

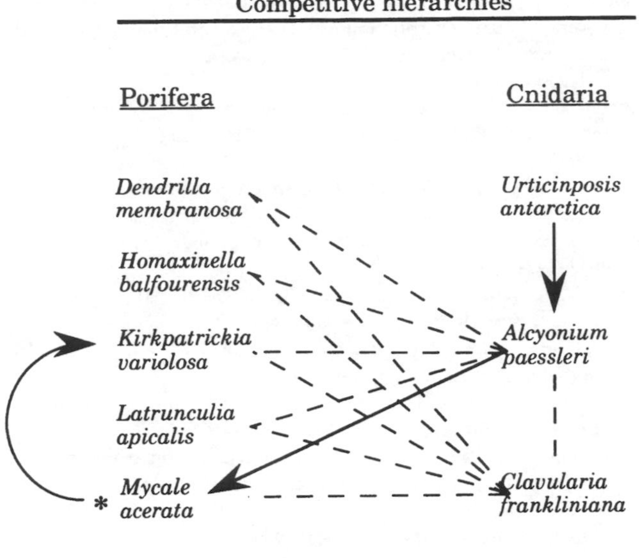

Fig. 43.3. Proposed competitive hierarchy for Antarctic shallow-water invertebrates. All interactions are based on contact-mediated interactions. Solid lines with arrows indicate significant competitive hierarchies that have been experimentally documented, dashed lines indicate potential interactions. The competitive interaction between the sponges *Mycale acerata* and *Kirkpatrickia variolosa* was documented by Dayton *et al.* (1974).

five common species of sponge (*Dendrilla membranosa, Homaxinella balfourensis, Kirkpatrickia variolosa, Latrunculia apicalis* and *Mycale acerata*). Controls consisted of conspecific contacts and soft corals treated in a manner similar to experimentals, but never placed in contact with another organism. After one year the samples were retrieved and examined for signs of tissue necrosis at the zone of tissue contact. The sponge *Mycale acerata* exhibited significantly larger zones of tissue necrosis at the site of contact with *A. paessleri* (P^2=0.05; Student's *t*-test; *n*=4) than controls (necrotic area=10.8±4.1 cm^2). *Alcyonium paessleri* suffered significant mortality when placed in contact with the sea anemone *Urticinopsis antarctica* (P^2=0.05; Fisher's exact test; *n*=4). Although all colonies contacting *U. antarctica* died, there was no evidence of tissue necrosis. *Clavularia frankliniana* did not cause tissue necrosis when placed in contact with either cnidarians or sponges, or exhibit signs of contact-mediated necrosis of its own tissues. Similarly, there was no evidence of a contact effect in pairings of *A. paessleri* and *C. frankliniana*, or among solitary controls.

DISCUSSION

The Antarctic soft corals *Alcyonium paessleri, Clavularia frankliniana*, and *Gersemia antarctica* are abundant and energy-rich members of their respective communities (Slattery 1994, Slattery & McClintock 1995). The distribution of *A. paessleri* and *C. frankliniana* on the eastern side of McMurdo Sound is coincident with the lower depth limits of anchor ice formation and the upper depth limits of a sponge spicule mat. Previous research has indicated the importance of biological interactions such as predation and competition in structuring Antarctic benthic communities below 33 m depth (Dayton *et al.* 1974). Competition may, in part, explain the inability of soft corals to colonize depths below 33 m. Our observations suggest that physical disturbance may be more important in structuring the cnidarian-dominated 'transition' communities (15–33 m depths) of eastern McMurdo Sound. Burial by glacial flour and localized scour by icebergs and volcanic rubble can have significant negative effects on these soft corals (Slattery, personal observations). On the western side of McMurdo Sound, *A. paessleri* and *C. frankliniana* are rare and occur only on patchy hard substrata. Populations of *G. antarctica*, only found on the western side of McMurdo Sound, may be structured more by biotic factors. For instance, lower overall levels of productivity in Explorer's Cove (Dayton & Oliver 1977, Dayton *et al.* 1986) may limit population densities and/or sustainable biomass. Moreover, pycnogonids, which prey on soft corals, are at least an order of magnitude more abundant on the western than the eastern side of McMurdo Sound (Slattery 1994, Slattery & McClintock 1995).

Reproductive patterns may be important in governing the ability of *Alcyonium paessleri, Clavularia frankliniana* and *Gersemia antarctica* to colonize available substrate (*sensu* Thorson 1950, Schick *et al.* 1979). However, once established these soft corals appear to rely on their chemical defenses as one means of sustaining their population numbers within the community (Slattery 1994, Slattery & McClintock 1995, Slattery *et al.* 1995). Despite the potential energetic resource represented by these soft corals, predation pressure is effectively neutralized through the biosynthesis of defensive compounds which are noxious to common predators in McMurdo Sound. The production of toxic metabolites suggests that bioactive compounds may also function as antifoulants against benthic bacteria and diatoms in *A. paessleri* and *G. antarctica* (Slattery *et al.* 1995). The lack of production of toxic compounds by *C. frankliniana* suggests that its significant sub-benthic biomass may reduce the need for antifoulants or indicate a high energetic cost of secondary metabolite production (see Herms & Mattson 1992). Preliminary evidence also suggests that contact-mediated competition, possibly chemically based, plays a role in structuring the distribution of the cnidarians *Urticinopsis antarctica* and *Alcyonium paessleri* and the common sponge *Mycale acerata* in shallow benthic communities of McMurdo Sound. This is particulary noteworthy as these species have growth rates among the fastest recorded for benthic marine invertebrates in McMurdo Sound (Dayton *et al.* 1974, Slattery 1994), and are therefore likely to compete for space.

ACKNOWLEDGEMENTS

We wish to thank Dale Andersen, Larry Basch, Dan Bockus, Sid Bosch, Patrick Bryan, Mark Hamann, John Heine, Jim Mastro and Jim Weston for assistance during field studies. Mark

Hamann and Wes Yoshida provided valuable insights on chemical extraction techniques and Patrick Bryan, Deneb Karentz and Mary Voytek offered suggestions on microbial assays. The Antarctic Support Associates the National Science Foundation, and the US Naval Antarctic Support Force at McMurdo and Palmer Stations provided logistical support for which we are grateful. Various aspects of this research were supported by NSF grants awarded to J. B. McClintock (OPP-8815959 & OPP-9118864) and D. Karentz (OPP-9017664).

REFERENCES

Achituv, Y. & Benayahu, Y. 1990. Polyp dimorphism and functional, sequential hermaphroditism in the soft coral *Heteroxenia fuscescens* (Octocorallia). *Marine Ecology Progress Series*, **64**, 263–269.

Alino, P. M. & Coll, J. C. 1989. Observations of the synchronized mass spawning and post- settlement activity of octocorals on the Great Barrier Reef, Australia: biological aspects. *Bulletin of Marine Science*, **45**, 697–707.

Bakus, G. J. 1981. Chemical defense mechanisms and fish feeding behaviour on the Great Barrier Reef, Australia. *Science*, **211**, 497–499.

Bakus, G. J. & Green, G. 1974. Toxicity in sponges and holothurians: a geographic pattern. *Science*, **185**, 497–498.

Bakus, G. J., Targett, N. M. & Schulte, B. 1986. Chemical ecology of marine organisms: an overview. *Journal of Chemical Ecology*, **12**, 951–987.

Barry, J. P. & Dayton, P. K. 1988. Current patterns in McMurdo Sound, Antarctica and their relationship to local biotic communities. *Polar Biology*, **8**, 367–376.

Bayer, F. M. 1981. Key to the genera of Octocorallia exclusive of Pennatulacea (Coelenterata: Anthozoa), with diagnoses of new taxa. *Proceedings of the Biological Society of Washington*, **94**, 902–947.

Benayahu, Y. & Loya, Y. 1981. Competition for space among coral reef sessile organisms at Eilat, Red Sea. *Bulletin of Marine Science*, **31**, 514–522.

Benayahu, Y. & Loya, Y. 1985. Settlement and recruitment of a soft coral: why is *Xenia macrospiculata* a successful colonizer. *Bulletin of Marine Science*, **36**, 177–188.

Brazeau, D. A. & Lasker, H. R. 1992. Reproductive success in the Caribbean gorgonian *Briareum asbestinum*. *Marine Biology*, **114**, 157–163.

Clarke, A. 1983. Life in cold water: the physiological ecology of polar marine ectotherms. *Oceanography and Marine Biology, Annual Review*, **21**, 341–453.

Coll, J. C., La Barre, S., Sammarco, P. W., Williams, W. T. & Bakus, G. J. 1982. Chemical defense in soft corals (Coelentera: Octocorallia) of the Great Barrier Reef: a study of comparative toxicities. *Marine Ecology Progress Series*, **8**, 271–278.

Dana, T. F. 1975. Development of contemporary eastern Pacific coral reefs. *Marine Biology*, **33**, 355–374.

Dayton, P. K. 1989. Interdecadel variation in an Antarctic sponge and its predators from oceanographic climate shifts. *Science*, **245**, 1484–1486.

Dayton, P. K. & Oliver, J. S. 1977. Antarctic soft-bottom benthos in oligotrophic and eutrophic environments. *Science*, **197**, 55–58.

Dayton, P. K., Mordida, B. J. & Bacon, F. 1994. Polar marine communities. *American Zoologist*, **34**, 90–99.

Dayton, P. K., Robilliard, G. A. & DeVries, A.L. 1969. Anchor ice formation in McMurdo Sound, Antarctica, and its biological effects. *Science*, **163**, 273–274.

Dayton, P. K., Robiliard, G. A. & Paine, R. T. 1970. Benthic faunal zonation as a result of anchor ice at McMurdo Sound, Antarctica. In Holdgate, M. W., ed. *Antarctic ecology*, Vol. 1. London: Academic Press, 244–258.

Dayton, P. K., Robilliard, G. A., Paine, R. T. & Dayton, L. B. 1974. Biological accomodation in the benthic community at McMurdo Sound, Antarctica. *Ecological Monographs*, **44**, 105–128.

Dayton, P. K., Watson, D., Palmisano, A., Barry, J. P., Oliver, J. S. & Rivera, D. 1986. Distribution patterns of benthic microalgal standing stock at McMurdo Sound, Antarctica. *Polar Biology*, **6**, 207–213.

Dinesen, Z. D. 1983. Patterns in the distribution of soft corals across the central Great Barrier Reef. *Coral Reefs*, **1**, 229–236.

Frost, S. H. 1977. Miocene to Holocene evolution of Caribbean Province reef-building corals. *Proceedings of the Third International Coral Reef Symposium, Miami*, **2**, 353–359.

Green, G. 1977. Ecology of toxicity in marine sponges. *Marine Biology*, **40**, 207–215.

Herms, D. A. & Mattson, W. J. 1992. The dilema of plants: to grow or defend. *Quarterly Review of Biology*, **67**, 283–335.

Lasker, H. R. 1990. Clonal propagation and population dynamics of a gorgonian coral. *Ecology*, **71**, 1578–1589.

Littlepage, J. L. 1965. Oceanographic investigations in McMurdo Sound, Antarctica. *Antarctic Research Series*, **5**, 1–37.

McClintock, J. B. 1987. Investigation of the relationship between invertebrate predation and biochemical composition, energy content, spicule armament and toxicity of benthic sponges at McMurdo Sound, Antarctica. *Marine Biology*, **94**, 479–487.

McClintock, J. B. 1994. The trophic biology of Antarctic echinoderms. *Marine Ecology Progress Series*, **111**, 191–202.

McClintock, J. B., Baker, B., Slattery, M., Hamann, M., Kopitzke, B. & Heine, J. 1994. Chemotactic tube-foot responses of the spongivorous sea star *Perknaster fuscuc* to organic extracts from Antarctic sponges. *Journal of Chemical Ecology*, **20**, 859–870.

McClintock, J. B., Heine, J., Slattery, M. & Weston, J. 1990. Chemical bioactivity in common shallow-water Antarctic marine invertebrates. *Antarctic Journal of the United States*, **25**(5), 204–206.

Nuedecker, S. 1979. Effects of grazing and browsing fish on the zonation of corals in Guam. *Ecology*, **60**, 666–672.

Palmer, A. R. 1979. Fish predation and the evolution of gastropod shell sculpture: experimental and geographic evidence. *Evolution*, **33**, 697–713.

Paul, V. J. 1992. *Ecological roles of marine natural products*. Ithaca: Cornell University Press, 245 pp.

Pearse, J. S., McClintock, J. B. & Bosch, I. 1991. Reproduction of Antarctic benthic invertebrates: tempos, modes, and timing. *American Zoologist*, **31**, 65–80.

Porter, J. W. 1972. Predation by *Acanthaster* and its effect on coral species diversity. *American Naturalist*, **106**, 487–492.

Preston, E. M. & Preston, J. L. 1975. Ecological structure in a West Indian gorgonian fauna. *Bulletin of Marine Science*, **25**, 248–258.

Rosen, D. E. 1978. Vicariance patterns and historical explanation in biogeography. *Systematic Zoology*, **27**, 159–188.

Sammarco, P. W. & Coll, J. C. 1988. The chemical ecology of alcyonarian corals (Coeleneterata: Octocorallia). In Scheuer, P. J., ed. *Bioorganic marine chemistry*, Vol. 2. Berlin: Springer-Verlag, 87–116.

Sammarco, P. W. & Coll, J. C. 1990. Lack of predictability in terpenoid function: multiple roles and integration with related adaptations in soft corals. *Journal of Chemical Ecology*, **16**, 273–289.

Schick, J. M., Hoffmann, R. J. & Lamb, A. N. 1979. Asexual reproduction, population structure, and genotype–environment interactions in sea anemones. *American Zoologist*, **19**, 699–713.

Sebens, K. P. 1983a. The larval and juvenile ecology of the temperate octocoral *Alcyonium siderium* Verrill. I. Substratum selection by benthic larvae. *Journal of Experimental Marine Biology and Ecology*, **71**, 73–89.

Sebens, K. P. 1983b. The larval and juvenile ecology of the temperate octocoral *Alcyonium siderium* Verrill. II. Fecundity, survival, and juvenile growth. *Journal of Experimental Marine Biology and Ecology*, **72**, 263–285.

Slattery, M. 1994. *A comparative study of population structure and chemical defenses in the soft corals* Alcyonium paessleri *May,* Clavularia frankliniana *Rouel, and* Gersemia Antarctica *Kukenthal in McMurdo Sound, Antarctica*. PhD thesis, Birmingham: University of Alabama, 109 pp.

Slattery, M. & McClintock, J. B. 1995. Population structure and defensive strategies of three soft coral species in McMurdo Sound, Antarctica. *Marine Biology*, **122**, 461–470.

Slattery, M., McClintock, J. B. & Heine, J. N. 1995. Chemical defenses in Antarctic soft corals: evidence for antifouling compounds. *Journal of Experimental Marine Biology and Ecology*, **190**, 61–77.

Stanley, S. M. 1979. *Macroevolution: pattern and process*. San Francisco: W. H. Freeman, 301 pp.

Stanley, S. M. 1984. Mass extinctions in the ocean. *Scientific American*, **250**, 46–54.

Thorson, G. 1950. Reproduction and larval ecology of marine bottom invertebrates. *Biological Review*, **25**, 1–45.

Tursch, B. & Tursch, A. 1982. The soft coral community on a sheltered reef quadrat at Laing Island (Papua, New Guinea). *Marine Biology*, **68**, 321–332.

Tursch, B., Braekman, J. C., Daloze, D. & Kaisin, M. 1978. Terpenoids from coelenterates. In Scheuer, P. J., ed. *Marine natural products: chemical and biological perspectives*, Vol. 2. New York: Academic Press, 247–296.

Vermeij, G. J. 1978. *Biogeography and adaptation*. Boston: Harvard University Press, 332 pp.

Veron, J. E. N. 1986. *Corals of Australia and the Indo-Pacific*. Sydney: Angus and Robertson, 656 pp.

Walker, T. A. & Bull, G. D. 1983. A newly discovered method of reproduction in gorgonian corals. *Marine Ecology Progress Series*, **12**, 137–143.

Yamanouchi, T. 1955. On the poisonous substances contained in holothurians. *Publications of Seto Marine Biology Lab*, **4**, 183–213.

44 Ecophysiological strategies of terrestrial arthropods in the maritime Antarctic

WILLIAM BLOCK

British Antarctic Survey, Natural Environment Research Council, High Cross, Madingley Road, Cambridge CB3 0ET, UK

ABSTRACT

The environmental constraints acting on terrestrial microarthropods in maritime Antarctic habitats are listed. Six ecophysiological features have been identified as the basis of the ecophysiological strategies of two common species: Alaskozetes antarcticus *(Acari, Cryptostigmata) and* Cryptopygus antarcticus *(Collembola, Isotomidae). The latter species is considered to be pre- adapted for responding to polar warming.*

Key words: environmental constraints, ecophysiology, strategies, climate change, *Cryptopygus antarcticus*, *Alaskozetes antarcticus*.

INTRODUCTION

Antarctic terrestrial invertebrates inhabit some of the most extreme environments in the world, particularly in respect of temperature and moisture. A range of strategies is adopted to survive sub-zero temperatures and often drought conditions during both summer and winter. Among the most successful species are those arthropods belonging to the Acari (mites, especially Cryptostigmata and Prostigmata) and the Collembola (wingless insects termed springtails). Successful species, both resident and colonist, have to optimise their performance during short periods in the austral summer when relatively warm and moist conditions prevail.

This paper emphasizes the environmental constraints which affect the two common species of terrestrial microarthropods in the maritime Antarctic, and proposes an ecophysiological strategy for their success under present environmental conditions.

The oribatid mite *Alaskozetes antarcticus* (Acari, Podacaridae) is common throughout Antarctica in the maritime and sub-Antarctic zones and is especially abundant close to areas fertilized by birds and seals (Block & Convey 1995). The springtail *Cryptopygus antarcticus* (Collembola, Isotomidae) is ubiquitous in the maritime Antarctic, occurring wherever there is vegetation (Block 1984); it is found also in the continental and sub-Antarctic zones.

Both species of microarthropod occupy a wide range of terrestrial habitats. In the maritime Antarctic they are found in recently deglaciated areas and fellfields, well-developed moss banks and algal-dominated habitats close to the shore and near penguin rookeries, seal wallows and elsewhere.

Environmental constraints

There are seven major constraints which influence the ecology of such terrestrial species, some of which are closely inter-related. Together, they largely determine the ecophysiology of the individuals and their populations. These are:

- low temperature, often with freezing conditions,
- freeze-thaw events,
- drought and dryness,
- wet–dry cycles,
- snow cover,
- seasonal light climate,
- increased UV-B in certain areas.

Low environmental temperatures exist for much of the year in Antarctic terrestrial habitats and there is a constant risk of freezing conditions occurring on any day of the year in maritime and continental Antarctica (Walton 1984). Variation in both snow depth and its compaction influences winter temperatures of the underlying substrate. Davey *et al.* (1992), in their study of

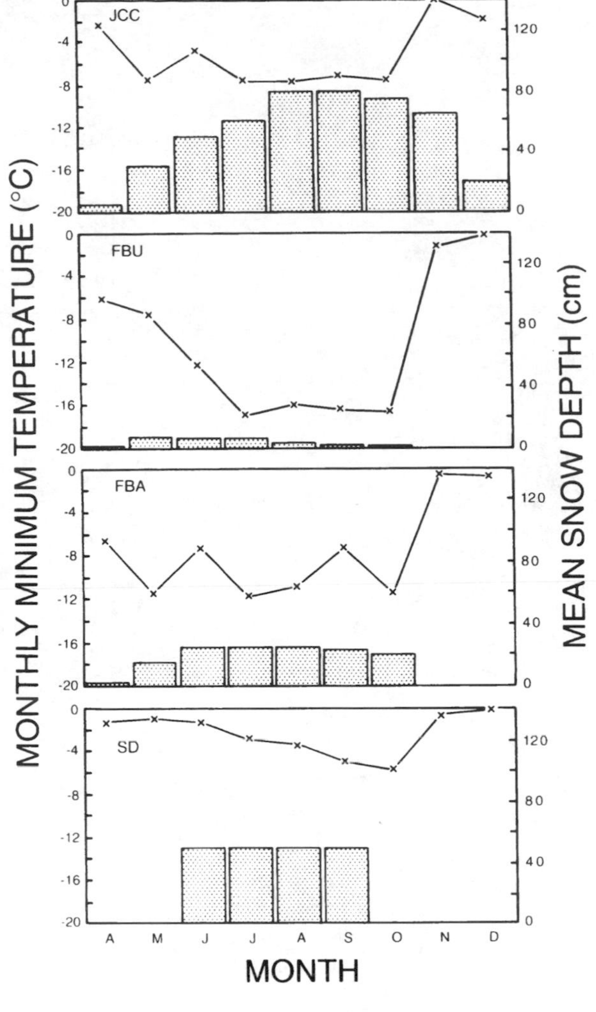

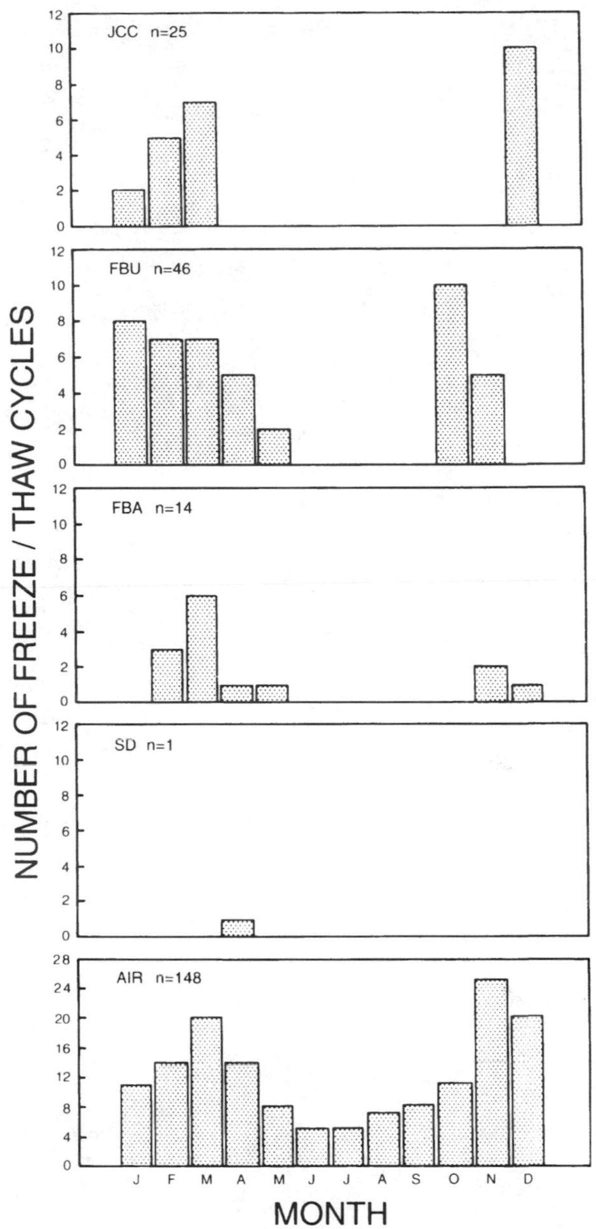

Fig. 44.1. (left) Monthly minimum soil temperatures and snow depth; (right) frequency of freeze–thaw events at four terrestrial sites on Signy Island during 1987 (after Davey *et al.* 1992). JCC: Jane Col polygon centre; FBU: Factory Bluffs *Usnea* (lichen); FBA: Factory Bluffs *Andreaea* (moss); SD: Station site *Drepanocladus* (moss); AIR: meteorological screen.

four fellfield habitats over one year at Signy Island, demonstrated that temperature minima in the upper soil layers were lower (*c.* −16 °C) in areas with <10 cm depth of snow than in other areas (*c.* −8 °C) with deeper (>70 cm) snow cover (Fig. 44.1). The number and frequency of freeze–thaw events (when the recorded temperatures show a transition through 0 °C (0.5 to −0.5 °C or vice versa) in the soil were also very different at the four sites and were fewer in number compared with those in the air at screen height. Freeze–thaw events in soil habitats were frequent during spring and autumn with up to 10 such events per month in contrast to >20 freeze–thaw events per month being recorded at screen height (Fig. 44.1).

Data on wet–dry cycles are sparse for Antarctic soil habitats, but microarthropod distribution and habitat occupancy appears to be linked to both their dehydration resistance and the moisture conditions of different microsites (Worland & Block 1986). A biogeographical synthesis undertaken by Kennedy (1993) concluded that water plays a primary role in determining the distribution and abundance of arthropods and other terrestrial organisms in Antarctica at various spatial scales (continent to microsite). He considered that many of the limiting effects previously attributed to severe (i.e. low) temperatures may operate through organism water balance. In addition, seasonal changes in light levels, light quality and intensity, as well as in UV-B, may be critical for different life stages of such species, and these constraints may be most important during the spring and summer periods in the Antarctic.

These constraints result in a short active season with variable food supply for most terrestrial microarthropods, but there

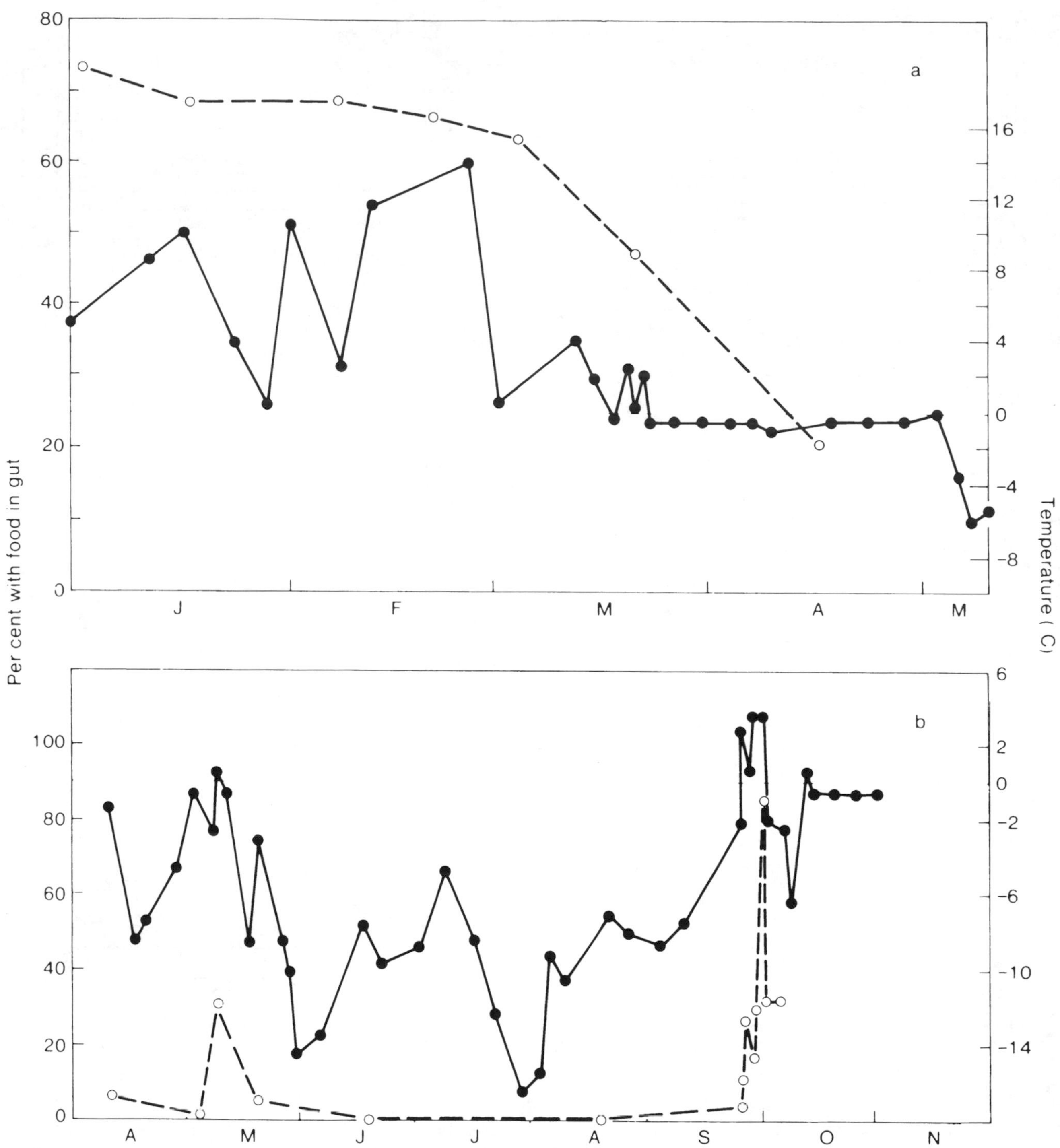

Fig. 44.2. Feeding activity of *Cryptopygus* on Signy Island in (a) summer and autumn in moss turf, and (b) winter and spring in *Prasiola* alga. Soil temperatures at 1 cm depth (●) are compared with the proportion of the population with observable gut contents (○) (Burn 1982).

appears to be little interspecific competition due to low species diversity of the fauna (Usher *et al.* 1989).

Ecophysiological features

1. Microarthropod activity is limited by low temperature and snow cover: A study of the activity pattern of *Cryptopygus* in the field at Signy Island, maritime Antarctic, showed a positive relationship between activity and habitat temperature and a diurnal pattern, both of which were less apparent during periods of snow cover (Burn & Lister 1988). Feeding of this species was determined largely by soil temperature, especially during summer and early winter (Fig. 44.2) (Burn 1982). *Alaskozetes* feeds mainly during summer and at occasional winter thaws.
2. Metabolism responds rapidly to small temperature increments: Q_{10}s for *Alaskozetes* and *Cryptopygus* over their summer field temperature range suggest a rapid increase of metabolism between 0 and 5 °C, especially for some juveniles and adult *Alaskozetes* (Table 44.1).

Table 44.1. *Temperature coefficients (Q_{10}s) calculated from rates of oxygen consumption measured by Cartesian Diver micro-respirometry for juveniles and adults of two species of arthropods from the maritime Antarctic*

	Temperature range (°C)	
Life stage/size class	0–5	5–10
Alaskozetes antarcticus		
Larva	5.2	1.8
Protonymph	4.7	3.1
Deutonymph	2.1	2.1
Tritonymph	2.0	2.9
Adult male	7.2	1.8
Adult female	4.8	2.4
Cryptopygus antarcticus		
Size class I	4.4	2.0
Size class II	3.6	1.3
Size class III	3.1	1.7
Size class IV	2.7	2.2
Size class V	2.5	2.6

Source: From Block & Tilbrook (1975) and Block (1977).

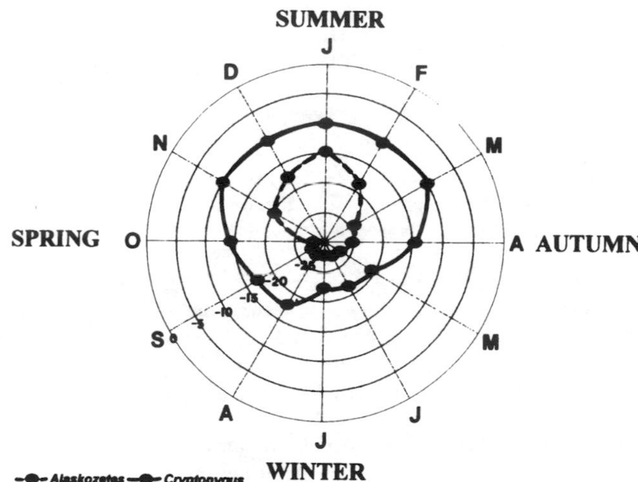

Fig. 44.3. Annual supercooling profiles for *Cryptopygus* and *Alaskozetes* averaged over 8 years at Signy Island depicted as a radar graph plot. Mean monthly supercooling points are shown with >500 data points per month.

3. Freeze avoidance is by extensive supercooling of body fluids: Neither species can tolerate freezing and die at their supercooling points. All life stages of *Alaskozetes* have the ability for consistent deep supercooling throughout the year, whilst supercooling in *Cryptopygus* is shallow and more variable (Fig. 44.3).
4. Winter survival is by both juveniles and adults (rarely as eggs): Experimental enclosures show winter survival rates better than 60% for four life stages of *Alaskozetes* under maritime Antarctic conditions (Convey 1994a). Similar field experiments suggest that *Cryptopygus* has a more variable overwinter survival but with >20% for adults and immatures (Convey 1994b). The cause of overwinter mortality may be cold injury or spontaneous freezing during extensive periods when individuals are supercooled.
5. Water balance exhibits a seasonal cycle in some species: The body water content of *Cryptopygus* is maximal in February and September with a minimum in July (Block & Harrisson 1995), whereas that of *Alaskozetes* is maintained at *c*.65% of its live weight for most of the year (Block 1986).
6. Antifreezes are produced in response to low temperatures and dehydration: Four compounds are synthesized by *Alaskozetes*; glycerol is quantitatively dominant and increases during autumn and remains high throughout winter (Fig. 44.4). Desiccation has been shown experimentally to stimulate glycerol production in this mite (Young & Block 1980). A multi-component cryoprotectant system is elaborated by *Cryptopygus* (Fig. 44.4).

CONCLUSIONS

- *Cryptopygus* is a labile species with the ability to respond more rapidly to fluctuations in environmental conditions than *Alaskozetes*.
- Both species utilize a freeze avoidance overwintering strategy in all their post- embryonic stages.
- The annual supercooling profile of *Cryptopygus* correlates with the seasonal cycle of water balance, individuals being more cold resistant when partially dehydrated in winter (Cannon *et al.* 1985).
- Desiccation stress promotes glycerol synthesis by *Alaskozetes* at the onset of winter.
- Both species utilize existing mechanisms to overcome the physiological challenges of cold and drought; no unique adaptations have been found.
- Both forms are 'successful' in the environment of the maritime Antarctic. *Cryptopygus* is an efficient colonizer of new areas by its high mobility and dispersal rate, whereas *Alaskozetes* is more resistant to physical stresses (e.g. salt-water immersion).
- *Cryptopygus* is better able to capitalize, both biologically and ecologically, from warming of its maritime Antarctic habitats than *Alaskozetes*.
- It is predicted that there would be a rapid spread of species such as *Cryptopygus* from established metapopulations during polar warming, provided food resources were available.

ACKNOWLEDGEMENTS

The support of the British Antarctic Survey for my research over several years is much appreciated. I thank Dr P. Convey for helpful discussions, and Professors L. Sømme and J. S. Bale for their constructive comments on the manuscript.

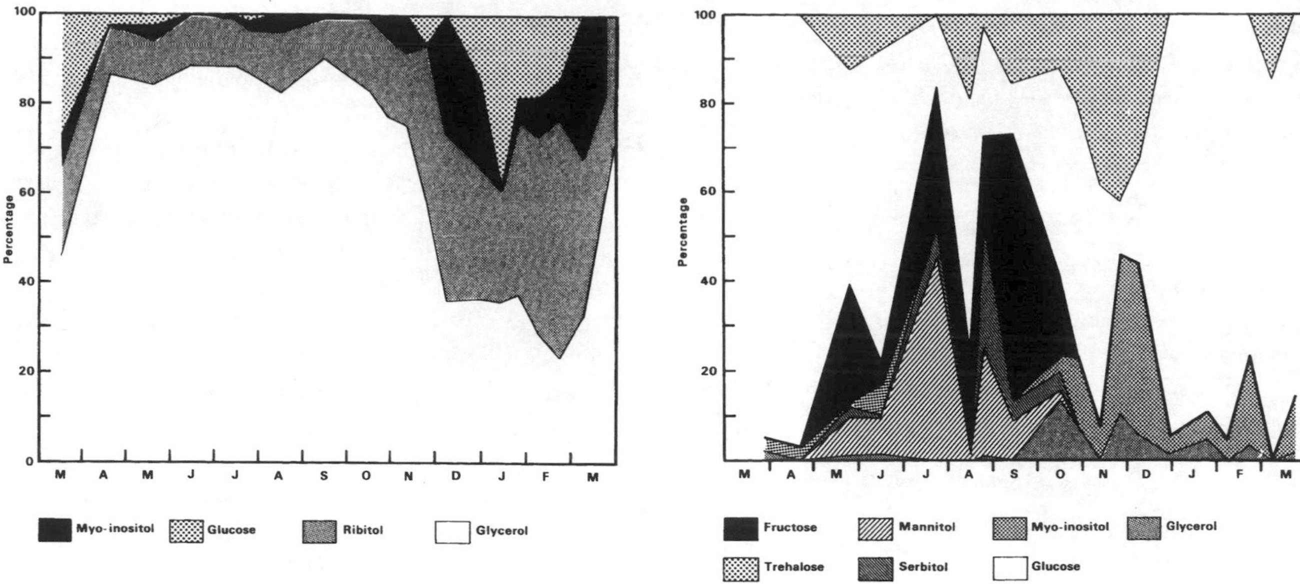

Fig. 44.4. Annual cycle of antifreeze production by (left) adult *Alaskozetes* and (right) adults and juveniles of *Cryptopygus* at Signy Island. Each compound is expressed as a percentage of the total polyols and sugars detected (after Sømme & Block 1982; after Cannon & Block 1988).

REFERENCES

Block, W. 1977. Oxygen consumption of the terrestrial mite *Alaskozetes antarcticus* (Acari: Cryptostigmata). *Journal of Experimental Biology,* **68,** 69–87.

Block, W. 1984. Terrestrial microbiology, invertebrates and ecosystems. In Laws, R.M., ed. *Antarctic ecology*, Vol. 1. London: Academic Press, 163–236.

Block, W. 1986. Temperature and drought effects on antarctic land arthropods. In *Proceedings of the 3rd European Congress of Entomology, Amsterdam, 1986*. Amsterdam: Nederlandse Entomologische Vereniging, 247–250.

Block, W. & Convey, P. 1995. The biology, life cycle and ecophysiology of the antarctic mite *Alaskozetes antarcticus* (Michael). *Journal of Zoology*, **236,** 431–449.

Block, W. & Harrisson, P. M., 1995. Collembolan water relations and environmental change in the Antarctic. *Global Change Biology*, **1,** 347–359.

Block, W. & Tilbrook, P. J. 1975. Respiration studies on the antarctic collembolan *Cryptopygus antarcticus. Oikos,* **26,** 15–25.

Burn, A. J. 1982. Effects of temperature on the feeding activity of *Cryptopygus antarcticus. Comité National Français des Recherches Antarctiques,* **51,** 209–217.

Burn, A. J. & Lister, A. 1988. Activity patterns in an antarctic arthropod community. *British Antarctic Survey Bulletin,* No. 78, 43–48.

Cannon, R. J. C. & Block, W. 1988. Cold tolerance of microarthropods. *Biological Reviews*, **63,** 23–77.

Cannon, R. J. C., Block, W. & Collett, G. D. 1985. Loss of supercooling ability in *Cryptopygus antarcticus* (Collembola: Isotomidae) associated with water uptake. *Cryo- Letters*, **6,** 73–80.

Convey, P. 1994a. Growth and survival strategy of an antarctic mite *Alaskozetes antarcticus. Ecography,* **17,** 97–197.

Convey, P. 1994b. The use of field exclosures to measure growth and mortality rates in an antarctic collembolan. *Acta Zoologica Fennica*, **195,** 18–22.

Davey, M. C., Pickup, J. & Block, W. 1992. Temperature variation and its biological significance in fellfield habitats on a maritime Antarctic island. *Antarctic Science,* **4,** 383–388.

Kennedy, A. D. 1993. Water as a limiting factor in the Antarctic terrestrial environment: a biogeographical synthesis. *Arctic & Alpine Research,* **25,** 308–315.

Sømme, L. & Block, W. 1982. Cold hardiness of Collembola at Signy Island, maritime Antarctic. *Oikos*, **38,** 168–176.

Usher, M. B., Block, W. & Jumeau, P. J. A. M. 1989. Predation by arthropods in an antarctic terrestrial community. In Heywood, R. B., ed. *University research in Antarctica, 1989–1992* Cambridge: British Antarctic Survey, 123–129.

Walton, D. W. H. 1984. The terrestrial environment. In Laws, R. M., ed. *Antarctic ecology,* Vol. 1. London: Academic Press, 1–60.

Worland, M. R. & Block, W. 1986. Survival and water loss in some antarctic arthropods. *Journal of Insect Physiology,* **32,** 579–584.

Young, S. R. & Block, W. 1980. Experimental studies on the cold tolerance of *Alaskozetes antarcticus. Journal of Insect Physiology,* **26,** 189–200.

45 Comparative studies of Antarctic arthropod and bryophyte life cycles: do they have a common strategy?

PETER CONVEY
British Antarctic Survey, Natural Environment Research Council, High Cross, Madingley Road, Cambridge CB3 0ET, UK

ABSTRACT

Influences that the extreme environmental characteristics experienced in Antarctic terrestrial habitats have on various aspects of the life history, physiology and metabolism of the dominant plant and animal groups (bryophytes and micro-arthropods) are compared qualitatively with the predictions of two general life history models, the C–S–R triangle of plant strategies and the similar r–K–*A formulation. A number of the features of the life cycles of both cryptogamic plants and invertebrates (e.g. long lifespan, low reproductive output, survival adaptations), agree with the predictions of S- (stress) or A- (adversity) selection. However, these are often primitive features of the respective taxonomic groups, with little evidence of novel adaptation. Antarctic species, most of which have probably colonized within the last 15 000 years, are likely to be derived from groups that already possess appropriate 'pre-adaptations', rather than sharing evolved common responses to new environmental challenges. Identification of the selective forces leading to these pre-adaptations requires study of the habitats of the parent groups of such recent colonists, in addition to Antarctic populations directly. Specific novel adaptations to life at low temperatures are, however, present in some species (particularly invertebrates) in the form of maintenance of growth and production rates, increased metabolic rates and low enzyme activation energies, and increased investment in cryoprotectants (some novel) relative to temperate species.*

Key words: adversity/stress selection, arthropod, bryophyte, life history, pre-adaptation.

INTRODUCTION

The extreme environmental conditions of Antarctica present the terrestrial biota with ecological and physiological challenges which must be overcome for them to survive, maintain populations or colonize new areas. Even the most complex Antarctic terrestrial communities are formed by, at most, two phanerogams, a limited number of cryptogams (mosses, liverworts, lichens), micro-arthropods (mites, springtails), microscopic invertebrates (e.g. nematodes, tardigrades) and microbiota (algae, fungi, cyanobacteria, bacteria) (Lewis Smith 1984, Block 1984). Does this limited biota show any common response to the stresses presented by the extreme environment?

The concept that certain environmental factors may drive the evolution of aspects of life history, or select for organisms with pre-existing characters, has led to the development of a range of life history models. One productive approach to the study of life history strategies is to search for general responses to environmental stimuli over a wide range of taxonomic groups. Models such as the *r–K*–A formulation developed by Southwood (1977) and Greenslade (1983), and the C–S–R triangle of plant strategies of Grime (1974, 1977, 1988), have been applied widely and show much overlap (Southwood 1988). They lead to the

Table 45.1. *Major life history correlates of the three principal ecological determinants in the general models developed by Grime (1974, 1977, 1988) (C–S–R), Southwood (1977) and Greenslade (1983) (*r–K–*A)*

Determinant	Correlate
C/*K* 'competition'	Delayed reproduction Iteroparity High parental investment in a small number of young Lower reproductive investment overall Efficiency of resource capture Competitive ability
S/A 'stress' or 'adversity'	Low fecundity Ability to delay reproduction Low growth potential Low competitive ability Survival adaptations Cold tolerance Herbivory/predator avoidance Conservation of resources Long life history
R/*r* 'ruderal'	Asexual reproduction Large reproductive effort Many progeny No parental care Rapid growth rates Early maturity Dispersal ability

prediction of a range of likely correlates of major ecological determinants (Table 45.1).

Few environments are likely to provide examples of a single predominating determinant. One such, experiencing a wide range of potential stresses (Table 45.2), is that occupied by Antarctic terrestrial biota. Using features of the biology of the dominant Antarctic terrestrial plant and animal groups as a baseline for comparison, this short review considers whether environmental stresses have led to the development of comparable life history features in taxonomically widely separated groups, and whether any features observed give support to the predictions of general life history models.

PROPERTIES OF ANTARCTIC SPECIES

Lifespan

Long life-cycles, often described as 'extended', have been reported in many polar and alpine invertebrates (Danks 1981, 1992) including some Antarctic species (Burn 1984, Sømme 1986, Convey 1994a). In contrast, typical temperate Collembola often complete several generations per year (e.g. Christiansen 1964, Joosse & Veltkamp 1970), although some members of groups such as cryptostigmatid mites may have relatively long life-cycles (1–2 year) wherever they occur (Mitchell 1977, Norton 1994). It is plausible that life-cycle elongation is merely a consequence of a limited heat budget, and thus cannot be considered a specific adaptation, although being consistent with an A-/S-selected life history. However, life-cycle extension does require associated adaptive development of overwintering capabilities, beyond those present in temperate source populations.

Table 45.2. *Environmental stresses which may be experienced in Antarctic terrestrial habitats*

Stress	Comment
Temperature	Importance of extreme maxima and minima, daily and seasonal range, and low mean temperature
Freeze–thaw events	Possible occurrence at any time of year
Water availability	Limited due to low precipitation, ice formation or desiccation
Light availability	Highly seasonal at high latitudes; may reach limiting levels for photosynthesis in summer; unproven effects of ozone depletion
Nutrient stress	Particularly important for plants
Growing season length	Short at high latitudes; may be reduced by late snow cover
Unpredictability	Whilst the magnitude of each stress is largely predictable, the timing and duration are not

The concept of life-cycle length is not particularly useful when considering bryophytes, with the exception of annuals or biennials. Although an annual strategy is common among temperate bryophytes (During 1979), this strategy is very rare in the Antarctic, despite the apparent abundance of suitable transient or short-lived habitats, suggesting that an annual life history is not viable. Ephemeral plant species are expected to have low resistance to predation and mechanical or climatic damage (Grime 1988). The latter, in particular, may provide a selective barrier to their colonization of and survival in Antarctic sites.

Reproduction

Increased occurrence of asexual reproduction is often associated with ruderal (R-/*r*-selected) life history strategies, allowing a rapid increase in population. Its importance may also increase in stressful environments, where the chance of sexual encounters is limited at low population densities. Both asexual and sexual modes of reproduction occur in all the major invertebrate and plant groups present in Antarctica. However, with the exception of sub-Antarctic dipterans (Duckhouse 1985), there is insufficient evidence to indicate that parthenogenesis is more frequent in Antarctic than temperate arthropods. Various mechanisms of asexual reproduction predominate in Antarctic bryophytes, with the incidence of successful sexual reproduction dropping from 80–90% of tropical and temperate species, to 25% in the maritime and 10% in the continental Antarctic (Webb 1973,

Longton 1988, Convey & Lewis Smith 1993). Any increased reliance on mechanisms of asexual reproduction in Antarctic species is likely to be a preadapted feature, although there may also be selection against sexual reproduction both by genetic costs (loss of specific successful adaptations) and by simple non-viability of parts of the sexual cycle (e.g. different temperature tolerances of the two sexes of moss plants – Longton 1988).

Reproductive output

Of the few data available for reproductive output of polar micro-invertebrates, only those for the cryptostigmatid mites are comparable with temperate species. Members of this group typically produce small numbers of eggs (e.g. Luxton 1981, Norton 1994), with Antarctic species at the bottom of the range observed (Convey 1994b, Block & Convey 1995). It is also likely that females of Antarctic species will only mature a single batch of eggs in their lifetime, unlike many temperate species which have the opportunity to mature several batches (Convey 1994a). Antarctic Collembola are similarly likely to show reduced fecundity relative to temperate species, as egg production depends on moulting frequency, which itself increases with temperature (Joosse *et al.* 1973). Low lifetime fecundity and rate of egg production are consistent with the predictions of A- or S-selected life histories.

The measurement of reproductive effort in bryophytes is complicated (Convey & Lewis Smith 1993) and no comparative data for species over a wide latitudinal range are available. A study comparing sub- and maritime Antarctic samples of several species revealed no overall pattern, although finding evidence of reduced spore output by some species at the more extreme site (Convey, 1994d). Insufficient information is available to justify discussion of possible adaptations in bryophytes.

Dispersal

The presence of several species of mite and springtail at virtually any suitable site, despite isolation by many kilometres of ice or sea, implies that dispersal must occur. Wind is likely to be the most important non-biological agent (Wynn-Williams 1991), although few species have sufficient desiccation resistance to survive extended airborne transport (Worland & Block 1986, Harrisson *et al.* 1990). Likewise, transport by agents such as birds has not been proven. Some oribatid mites can survive extended immersion in seawater (Strong 1967) but, with this possible exception, there is no evidence for dispersal adaptations in either juvenile or adult stages of Antarctic micro-arthropods.

In contrast, bryophytes have clear morphological features associated with dispersal. Spores produced as a consequence of sexual reproduction are capable of both local and long-range dispersal (During 1979, Miles & Longton 1992), whilst various forms of asexual propagule are generally capable only of local dispersal. Spore viability is largely unknown and probably low, but has been demonstrated both in the field (Lewis Smith 1972, Miles & Longton 1990) and after long-distance high altitude aerial transport (van Zanten 1978, van Zanten & Gradstein 1987). As mentioned above, the occurrence of sexual reproduction decreases in Antarctic regions. Although this failure has a number of possible causes, an important contributory factor is the problem of gamete fusion in dioecious species (those with separate male and female plants). The majority of Antarctic species which fruit successfully are monoecious, a derived evolutionary feature (During 1979, Longton & Schuster 1983). Although monoecism is not limited to, or better represented in, Antarctic mosses, it can be considered a pre-adaptation which is likely to increase the chance of successful sexual reproduction in extreme environments.

Both micro-arthropods and bryophytes generally show a lack of, or reduction in, dispersal mechanisms, which is consistent with the predictions of A/S-selection. However, it is unreasonable to view the lack of a particular feature as an adaptation in itself.

Growth rates

Given the long life cycles considered above, annual growth rates of Antarctic invertebrates are inevitably much lower than those of temperate relatives. This gives a false impression, as growth only occurs during the short summer period. The information available on the growth of Antarctic species is limited to oribatid mites (Burn 1986, Convey 1994a,b,c) and indicates that rates comparable to those of temperate species (e.g. Lebrun *et al.* 1991, Norton 1994) may be achieved during the short Antarctic summer, at temperatures 10–20 °C lower than those experienced in temperate microhabitats.

In favourable habitats, polar bryophytes, similarly, may show net productivity comparable to many temperate sites (Collins 1977, Davis 1983, Ino 1992), although it is clear that mosses in many Antarctic sites do not achieve such high values. It is a feature of the biology of many temperate bryophytes that, although optimum temperatures for growth are in the range 15–25 °C, relative growth rates are reduced by less than 50% at low temperatures (5 °C) (Furness & Grime 1982). In addition, net photosynthesis at sub-zero temperatures has been measured in several Antarctic plants (Montiel & Cowan 1993). Although not directly comparable with productivity, the relationship between net photosynthesis and temperature found in the few Antarctic mosses examined shows little difference from that of temperate species (see Longton 1988; Dilks & Proctor 1975, Proctor 1982 and Convey 1993 give examples of temperate, boreal and Antarctic species). However, Arctic mosses are reported to have lower temperature optima for photosynthesis and respiration (Oechel & Collins 1976).

The ability of some populations or species in both groups to maintain rates of growth comparable to those in temperate species during the short austral summer is suggestive of underlying metabolic adaptations allowing compensation for the low summer temperatures experienced. This feature suggests a more 'ruderal' response (R-/*r*-selected), where growth is maximized during the short opportunity provided.

Competition

Competition is generally thought to have limited significance in Antarctic terrestrial habitats (Lewis Smith 1972, Block 1985, Burn 1986), although few explicit studies exist (but see Tréhen *et al.* 1985, Davies 1987, Usher *et al.* 1989). The majority of micro-arthropods are generalist primary consumers or detritivores, usually with an apparent superabundance of food available. Lister *et al.* (1988) considered the predatory mite *Gamasellus racovitzai* (often the only arthropod predator in maritime Antarctic invertebrate communities) unlikely to be limited by the availability of its major springtail prey. General evidence for competition between plant species exists in the form of directional (long term) succession (Longton 1988, Lewis Smith 1993), although the pattern of succession (if any) in many communities is poorly defined and may consist of a single stage of autosuccession (Muller 1952). Bryophytes appear to be under a phylogenetic limitation such that they do not include true 'competitors', and lie along the stress-tolerant–ruderal axis of the C–S–R triangle (Grime *et al.* 1990).

Low levels of competition are a precursor for the development of both stress/adversity and ruderal life histories; however, their existence does not indicate low competitive ability *per se.* Although it is tempting to suggest that extreme environments, as found in Antarctica, provide 'refuges' for species of low competitive ability, there is little direct evidence to support such a view.

Cold tolerance

Mechanisms of cold tolerance of polar and alpine arthropods have been well-studied (Cannon & Block 1988, Block 1990). Polar species use a range of strategies (e.g. behavioural avoidance, supercooling, production/exclusion of ice nucleating agents, antifreezes, thermal hysteresis proteins) to either avoid freezing altogether or restrict it to extracellular fluid, thereby avoiding cell damage. Much of the biochemical basis of cold tolerance, with the exception of thermal hysteresis proteins, relies on ancestral features of arthropod physiology (Pullin 1994), and may be closely related to adaptations required for desiccation resistance (Pullin 1994, Ring & Danks 1994). Indeed, many temperate and tropical arthropods produce sufficient cryoprotectant chemicals to allow survival over a range of temperatures that they never experience, although experimental studies indicate that death may occur well above the supercooling point in some (e.g. Bale 1987). In the context of survival of both low temperatures and desiccation in polar environments, the production of cryoprotectants involves significant diversion of resources (Cannon & Block 1988, Block 1990, Ring & Danks 1994). The extent of investment in survival adaptations, although often an extension of primitive synthetic abilities, is likely to be an adaptive feature of the life histories of Antarctic micro-invertebrates, consistent with the predictions of A/S-selection.

Mechanisms of cryoprotection in plants generally, and bryophytes specifically, are unclear. There is evidence that survival of a wide range of stresses including cold tolerance and desiccation resistance may be reduced to, or at least correlated with, tolerance of the major underlying stress of nutrient limitation (Grime 1988, MacGillivray & Grime 1995). Low molecular weight sugars and polyols known to have a cryoprotectant role in other organisms are present in Antarctic algae and bryophytes, and leach into the environment during freeze–thaw cycles (Tearle 1987, Hawes 1990, Melick & Seppelt 1992, Montiel & Cowan 1993). Studies of temperate bryophytes have identified seasonal variation in cold-hardiness, but have failed to demonstrate a clear connection between this and specific cryoprotectants, although suggesting that sugars such as sucrose may have an important role (e.g. Rütten & Santarius 1992, 1993).

Respiration and metabolism

Detectable respiration at sub-zero temperatures has been reported in both Antarctic micro-arthropods and bryophytes (Block 1987, Young & Block 1980, Montiel & Cowan 1993). However, to be of significance to life history, respiratory or metabolic adaptations must be demonstrable over the range of summer temperatures experienced by the organism, when virtually all activity occurs.

The ability of Antarctic micro-invertebrates and bryophytes to show growth rates comparable with temperate species at lower field temperatures is strongly suggestive of underlying metabolic adaptations. Identification of low optimum temperatures for invertebrate feeding and growth (Young 1979, Burn 1986, Convey 1994b) likewise implies temperature specialization of enzyme systems. However, there are few explicit demonstrations of such adaptations. Two significant lines of evidence have been proposed for Antarctic micro-arthropods. First, rates of standard metabolism may be elevated compared with those of temperate species over the same temperature range (Block & Young 1978, Sømme *et al.* 1989). Debate continues over the importance of this effect (Clarke 1991, 1993, Chown 1997). Second, Young (1979) and Block (1982) have demonstrated the existence of low enzyme activation energies in Antarctic mites and sub-Antarctic beetles. These adaptations minimize the effect of low temperatures on poikilotherm physiology.

Bryophytes may not be under significant metabolic limitation at low temperatures during Antarctic summer conditions. As discussed above, temperature response curves of photosynthesis show little difference from those of many temperate species, which are often characterized by wide, flat curves allowing significant rates to be achieved at low temperatures. Additionally, bryophyte photosynthesis is often maximized at relatively low light intensities (e.g. under snow cover), and inhibited by high intensities (Kappen *et al.* 1989, Post *et al.* 1990). These general characteristics of bryophytes lead to the possibility of significant levels of assimilation occurring at temperatures and light intensities well below 'optimum' (Longton 1988).

Table 45.3. *Comparisons of life history traits of Antarctic terrestrial micro-arthropods and bryophytes in relation to temperate species*

Trait	Arthropods	Bryophytes
Lifespan	Often increased, more life stages may overwinter	Annual life history not viable
Reproduction	Increased occurrence of parthenogenesis, at least in sub-Antarctic	Sexual reproduction often not possible, asexual (vegetative) propagules more important
Reproductive output (total and rate)	Often lower	Possibly lower, little evidence
Dispersal	Very limited, no specific dispersal stage	Long-range dispersal often less significant due to failure of sexual reproduction
Growth rates	Similar at lower temperatures (over short periods)	Dry mass production may be similar at lower temperatures, little data
Competition	Less	Possibly less (little evidence)
Cold tolerance	Greater utilization of cryoprotectants, also related to desiccation resistance	Not known
Metabolism	Temperature-specific respiration rates may be increased; lower enzyme activation energies	Achieve photosynthesis and respiration rates similar to many temperate species, although may have lower temperature optima. Maintain significant rates at sub-zero and low positive temperatures

CONCLUSIONS

The main conclusions of this brief review are presented in Table 45.3. Antarctic micro-arthropods and bryophytes show a number of features consistent with the predictions of both A-/S- and R-/r- selected life histories. However, many of these are characteristic of the taxonomic group concerned, or a direct consequence of the low energy budget of the Antarctic environment, rather than adaptations to particular environmental stresses. There are also features of the biology of Antarctic arthropods, in particular, which probably represent derived adaptations. These include the increased utilization of cryoprotectant compounds (some novel) representing significant diversion of resources, and the (unidentified) biochemical adaptations which must underlie the ability to maintain growth rates at low temperature, decrease enzyme activation energies and allow increased metabolic activity at low temperature. Danks (1981) drew similar conclusions, suggesting that Arctic arthropods, while having a suite of general life history characters contributing to their success, possess few specific adaptations.

It is likely that the great majority of Antarctic biota are post-glacial colonists, as suitable terrestrial habitats became available only as major icecaps and glaciers retreated at the end of the Quaternary glaciation, 10–15 000 years ago. Some micro-arthropods (e.g. Wallwork 1967), lichens and endolithic microbes could have survived on nunataks, although these species form a small, truly endemic and little-studied element of the Antarctic biota. However, for the majority of colonizers, the period available for development of novel adaptations is short in evolutionary terms, particularly allowing for the long life-cycles observed. Thus, the species currently inhabiting Antarctica are likely to be from groups whose primitive features include characteristics which are similar to many of the predictions of A-/S-selection, and are therefore 'pre-adapted' to the extreme environment. Such characteristics as extended lifespan, high incidence of asexual reproduction, reduced investment in reproduction, limited dispersal ability, and ability to survive in nutrient-limited habitats, generally found in Antarctic terrestrial invertebrates and/or plants, may be considered pre-adaptations.

The concept of A-/S-selection, therefore, provides a good description of the pre-adapted characteristics which are a pre-requisite for colonizing 'jumps' to be successful from largely unknown source populations to current Antarctic terrestrial habitats. The ecological conditions and processes leading to the evolution of these characteristics in stressful habitats at lower latitudes are therefore of great interest, as are the mechanics of long- and short-distance colonization of Antarctic sites. These requirements limit the pool of potential colonists and this, combined with the geographical isolation of sites, both within the Antarctic and from other islands and continents as sources of propagules, contributes to the low species diversity found in Antarctic terrestrial habitats.

ACKNOWLEDGEMENTS

This paper has benefitted from many discussions with colleagues in the Terrestrial & Freshwater and Marine Life Sciences Divisions of the British Antarctic Survey. I thank particularly Drs W. Block, R. I. Lewis Smith, D. W. H. Walton and Professor A. Clarke for their input, and Professor J. P. Grime for his constructive criticism.

REFERENCES

Bale, J. S. 1987. Insect cold hardiness: freezing and supercooling – an ecological perspective. *Journal of Insect Physiology*, **33,** 899–908.

Block, W. 1982. Respiration studies on some South Georgian Coleoptera. *CNFRA*, **51,** 183–192.

Block, W. 1984. Terrestrial microbiology, invertebrates and ecosystems. In Laws R. M., ed. *Antarctic ecology*, Vol. 1. London: Academic Press, 163–236.

Block, W. 1985. Arthropod interactions in an Antarctic terrestrial community. In Siegfried, W. R., Condy, P. R. & Laws, R. M., eds. *Antarctic nutrient cycles and food webs.* Berlin: Springer-Verlag, 614–619.

Block, W. 1987. Ecophysiology of terrestrial arthropods. *CNFRA*, **58,** 99–106.

Block, W. 1990. Cold tolerance of insects and other arthropods. *Philosophical Transactions of the Royal Society of London,* **B326,** 613–633.

Block, W. & Convey, P. 1995. The biology, life cycle and ecophysiology of the Antarctic mite *Alaskozetes antarcticus* (Michael). *Journal of Zoology*, **236,** 431-449.

Block, W. & Young, S. R. 1978. Metabolic adaptations of Antarctic terrestrial micro-arthropods. *Comparative Biochemistry and Physiology*, **61A,** 363–368.

Burn, A. J. 1984. Life cycle strategies in two Antarctic Collembola. *Oecologia*, **64,** 223–229.

Burn, A. J. 1986. Feeding rates of the cryptostigmatid mite *Alaskozetes antarcticus* (Michael). *British Antarctic Survey Bulletin*, No. 71, 11–17.

Cannon, R. J. C. & Block, W. 1988. Cold tolerance of microarthropods. *Biological Reviews*, **63,** 23–77.

Chown, S. L. 1997. Sub-antarctic weevil assemblages: species, structure and survival. In Battaglia, B., Valencia, J. & Walton, D. W. H., eds. *Antarctic communities: species, structure and survival.* Cambridge: Cambridge University Press, 152–161

Christiansen, K. 1964. Bionomics of Collembola. *Annual Review of Entomology*, **9,** 147–178.

Clarke, A. 1991. What is cold adaptation and how should we measure it? *American Zoologist*, **31,** 81–92.

Clarke, A. 1993. Seasonal acclimatization and latitudinal compensation in metabolism: do they exist? *Functional Ecology*, **7,** 139–149.

Collins, N. J. 1977. The growth of mosses in two contrasting communities in the maritime Antarctic: measurement and prediction of net annual production. In Llano, G. A., ed. *Adaptations within Antarctic ecosystems*. Washington DC: Smithsonian Institution, 921–933.

Convey, P. 1993. Photosynthesis and dark respiration in Antarctic mosses – an initial comparative study. *Polar Biology*, **14,** 65–69.

Convey, P. 1994a. Growth and survival strategy of the Antarctic mite *Alaskozetes antarcticus*. *Ecography*, **17,** 97–107.

Convey, P. 1994b. Sex ratio, oviposition and early development of the Antarctic oribatid mite *Alaskozetes antarcticus* (Acari: Cryptostigmata) with observations on other oribatids. *Pedobiologia*, **38,** 161–168.

Convey, P. 1994c. The influence of temperature on individual growth rates of the Antarctic mite *Alaskozetes antarcticus*. *Acta Oecologica*, **15,** 43–53.

Convey, P. 1994d. Modelling reproductive effort in sub- and maritime Antarctic mosses. *Oecologia*, **100,** 45–53.

Convey, P. & Lewis Smith, R. I. 1993. Investment in sexual reproduction by Antarctic mosses. *Oikos*, **68,** 293–302.

Danks, H. V. 1981. *Arctic arthropods. A review of systematics and ecology with particular reference to the North American fauna.* Ottawa: Entomological Society of Canada, 608 pp.

Danks, H. V. 1992. Long life cycles in insects. *Canadian Entomologist*, **124,** 167–187.

Davies, L. 1987. Long adult life, low reproduction and competition in two sub-Antarctic carabid beetles. *Ecological Entomology*, 12, 149–162.

Davis, R. C. 1983. Prediction of net primary production in two Antarctic mosses by two models of net CO_2 fixation. *British Antarctic Survey Bulletin*, No. 59, 47–61.

Dilks, T. J. K. & Proctor, M. C. F. 1975. Comparative experiments on temperature responses of bryophytes: assimilation, respiration and freezing damage. *Journal of Bryology*, **8,** 317–336.

Duckhouse, D. A. 1985. Psychodidae (Diptera, Nematocera) of the sub-antarctic islands with observations on the incidence of parthenogenesis. *International Journal of Entomology*, **27,** 173–184.

During, H. J. 1979. Life strategies of bryophytes: a preliminary review. *Lindbergia*, **5,** 2–18.

Furness, S. B. & Grime, J. P. 1982. Growth rate and temperature responses in bryophytes II. A comparative study of species of contrasted ecology. *Journal of Ecology*, **70,** 525–536.

Greenslade, P. J. M. 1983. Adversity selection and the habitat templet. *American Naturalist*, **122,** 352–365.

Grime, J. P. 1974. Vegetation classification by reference to strategies. *Nature*, **250,** 26–31.

Grime, J. P. 1977. Evidence for the existence of three primary strategies in plants and its relevance to ecological and evolutionary theory. *American Naturalist*, **111,** 1169–1194.

Grime, J. P. 1988. The C–S–R model of primary plant strategies – origins, implications and tests. In Gottlieb, L. D. & Jain, S. K., eds. *Plant evolutionary biology*. London: Chapman & Hall, 371–393.

Grime, J. P., Rincon, E. R. & Wickerson, B. E. 1990. Bryophytes and plant strategy theory. *Botanical Journal of the Linnean Society*, **104,** 175–186.

Harrisson, P. M., Block, W. & Worland, M. R. 1990. Moisture and temperature dependent changes in the cuticular permeability of the Antarctic springtail *Parisotoma octooculata* (Willem). *Revue d'Écologie et de Biologie du Sol*, **27,** 435–448.

Hawes, I. 1990. Effects of freezing and thawing on a species of *Zygnema* (Chlorophyta) from the Antarctic. *Phycologia*, **29,** 326–331.

Ino, Y. 1992. Estimation of the net production of moss community at Langhovde, East Antarctica. *Nankyoku Shiryô* [*Antarctic Record*], **36,** 49–59.

Joosse, E. N. G. & Veltkamp, E. 1970. Some aspects of growth, moulting and reproduction in five species of surface dwelling Collembola. *Netherlands Journal of Zoology*, **20,** 315–328.

Joosse, E. N. G., Brugman, F. A. & Veld, C. J. 1973. The effects of constant and fluctuating temperatures on the production of spermatophores and eggs in populations of *Orchesella cincta* (Linné), (Collembola, Entomobryidae). *Netherlands Journal of Zoology*, **23,** 488–502.

Kappen, L., Lewis Smith, R. I. & Meyer, M. 1989. Carbon dioxide exchange of two ecodemes of *Schistidium antarctici* in continental Antarctica. *Polar Biology*, **9,** 415–422.

Lebrun, P. H., van Impe, G., de Saint Georges-Gridelet, D., Wauthy, G. & André, H. M. 1991. The life strategies of mites. In Schuster, R. & Murphy, P. M., eds. *The Acari: reproduction, development and life-history strategies*. London: Chapman & Hall, 3–22.

Lewis Smith, R. I. 1972. Vegetation of the South Orkney Islands, with particular reference to Signy Island. *British Antarctic Survey Scientific Reports*, No. 68, 124 pp.

Lewis Smith, R. I. 1984. Terrestrial plant biology of the sub-Antarctic and Antarctic. In Laws R. M., ed. *Antarctic ecology*. London: Academic Press, 61–162.

Lewis Smith, R.I. 1993. Dry coastal ecosystems of Antarctica. In van der Maarel, E., ed. *Ecosystems of the world*, Vol. 2A. *Dry coastal ecosystems, polar regions and Europe*. Amsterdam: Elsevier, 51–71.

Lister, A., Block, W. & Usher, M. B. 1988. Arthropod predation in an Antarctic terrestrial community. *Journal of Animal Ecology*, **57,** 957–971.

Longton, R. E. 1988. *Biology of polar bryophytes and lichens*. Cambridge: Cambridge University Press, 391 pp.

Longton, R. E. & Schuster, R. M. 1983. Reproductive biology. In Schuster, R. M., ed. *New manual of bryology*. Nichinan: Hattori Botanical Laboratory, 386–462.

Luxton, M. 1981. Studies on the oribatid mites of a Danish beechwood soil IV. Developmental biology. *Pedobiologia*, **21,** 312–340.

MacGillivray, C. W. & Grime, J. P. 1995. Testing predictions of the resistance and resilience of vegetation subjected to extreme events. *Functional Ecology*, **9,** 640–649.

Melick, D. R. & Seppelt, R. D. 1992. Loss of soluble carbohydrates and changes in freezing point of Antarctic bryophytes after leaching and repeated freeze–thaw cycles. *Antarctic Science*, **4**, 399–404.

Miles, C. J. & Longton, R. E. 1990. The role of spores in reproduction in mosses. *Botanical Journal of the Linnean Society*, **104**, 149–173.

Miles, C. J. & Longton, R. E. 1992. Deposition of moss spores in relation to distance from parent gametophytes. *Journal of Bryology*, **17**, 355–368.

Mitchell, M. J. 1977. Life history strategies of oribatid mites. In Dindal, D. L., ed. *Biology of oribatid mites*. New York: State University of New York, 65–69.

Montiel, P. O. & Cowan, D. A. 1993. The possible role of soluble carbohydrates and polyols as cryoprotectants in Antarctic plants. In Heywood, R. B., ed. *University research in Antarctica 1989–1992*. Cambridge: British Antarctic Survey, 119–126.

Muller, C. H. 1952. Plant succession in arctic heath and tundra in northern Scandinavia. *Bulletin of the Torrey Botanical Club*, **79**, 296–309.

Norton, R. A. 1994. Evolutionary aspects of oribatid mite life histories and consequences for the origin of the Astigmata. In Houck, M., ed. *Ecological and evolutionary analyses of life- history patterns*. New York: Chapman & Hall, 99–135.

Oechel, W. C. & Collins, N. J. 1976. Comparative CO_2 exchange patterns in mosses from two tundra habitats at Barrow, Alaska. *Canadian Journal of Botany*, **54**, 1355–1369.

Post, A., Adamson, E. & Adamson, H. 1990. Photoinhibition and recovery of photosynthesis in antarctic bryophytes under field conditions. In Baltscheffsky, M., ed. *Current research in photosynthesis*, Vol. IV. Dordrecht: Kluwer Academic Publishers, 635–638.

Proctor, M. C. F. 1982. Physiological ecology: water relations, light and temperature responses, carbon balance. In Smith, A. J. E., ed. *Bryophyte ecology*. London: Chapman & Hall, 333–381.

Pullin, A. S. 1994. Evolution of cold hardiness strategies in insects. *Cryo-Letters*, **15**, 6–7.

Ring, R. A. & Danks, H. V. 1994. Desiccation and cryoprotection: overlapping adaptations. *Cryo-Letters*, **15**, 181–190.

Rütten, D. & Santarius, K. A. 1992. Relationship between frost tolerance and sugar concentration of various bryophytes in summer and winter. *Oecologia*, **91**, 260–265.

Rütten, D. & Santarius, K. A. 1993. Seasonal variation in frost tolerance and sugar content of two *Plagiomnium* species. *Bryologist*, **96**, 564–568.

Sømme, L. 1986. Ecology of *Cryptopygus sverdrupi* (Insecta: Collembola) from Dronning Maud Land, Antarctica. *Polar Biology*, **6**, 179–184.

Sømme, L., Ring, R. A., Block, W. & Worland, M. R. 1989. Respiratory metabolism of *Hydromedion sparsutum* and *Perimylops antarcticus* (Col., Perimylopidae) from South Georgia. *Polar Biology*, **10**, 135–139.

Southwood, T. R. E. 1977. Habitat, the templet for ecological strategies. *Journal of Animal Ecology*, **46**, 337–365.

Southwood, T. R. E. 1988. Tactics, strategies and templets. *Oikos*, **52**, 3–18.

Strong, J. 1967. Ecology of terrestrial arthropods at Palmer Station, Antarctic Peninsula. *Antarctic Research Series*, **10**, 357–371.

Tearle, P. V. 1987. Cryptogamic carbohydrate release and microbial response during spring freeze–thaw cycles in Antarctic fellfield fines. *Soil Biology and Biochemistry*, **19**, 381–390.

Tréhen, P., Bouché, M., Vernon, P. H. & Frenot, Y. 1985. Organization and dynamics of Oligochaeta and Diptera on Possession Island. In Siegried, W. R., Condy, P. R. & Laws, R. M., eds. *Antarctic nutrient cycles and food webs*. Berlin: Springer-Verlag, 606–613.

Usher, M. B., Block, W. & Jumeau, P. J. A. M. 1989. Predation by arthropods in an Antarctic terrestrial community. In Heywood, R. B., ed. *University research in Antarctica, 1989–1992*. Cambridge: British Antarctic Survey, 123–129.

Wallwork, J. A. 1967. Cryptostigmata (oribatid mites). *Antarctic Research Series*, **10**, 105–122.

Webb, R. 1973. Reproductive behaviour of mosses on Signy Island, South Orkney Islands. *British Antarctic Survey Bulletin*, No. 36, 61–77.

Worland, M. R. & Block, W. 1986. Survival and water loss in some Antarctic arthropods. *Journal of Insect Physiology*, **32**, 579–584.

Wynn-Williams, D. D. 1991. Aerobiology and colonization in Antarctica – the BIOTAS programme. *Grana*, **30**, 380–393.

Young, S. R. 1979. Respiratory metabolism of *Alaskozetes antarcticus*. *Journal of Insect Physiology*, **25**, 361–369.

Young, S. R. & Block, W. 1980. Some factors affecting metabolic rate in an Antarctic mite. *Oikos*, **34**, 178–185.

van Zanten, B. O. 1978. Experimental studies on trans-oceanic long-range dispersal of moss spores in the Southern Hemisphere. *Journal of the Hattori Botanical Laboratory*, **44**, 455–482.

van Zanten, B. O. & Gradstein, S. R. 1987. Feasibility of long-distance transport in Colombian hepatics. *Symposia Biologica Hungarica*, **35A**, 315–322.

46 Developmental changes and diurnal and seasonal influences on the diving behaviour of Weddell seal (*Leptonychotes weddellii*) pups

JENNIFER M. BURNS[1] *AND J. WARD TESTA*[2]

[1]*Institute of Marine Science, University of Alaska Fairbanks, Fairbanks, Alaska, 99775-7220, USA,* [2]*Alaska Department of Fish and Game, 333 Raspberry Road, Anchorage, Alaska, 99502, USA*

ABSTRACT

The development of diving behaviour of Weddell seal pups in McMurdo Sound, Antarctica, was monitored with time depth recorders (TDRs) during the austral springs and summers of 1992 and 1993. Pups (n=17, 20) carried TDRs for periods of several days every second week between the ages of 2 and 13 weeks. To track diving behaviour throughout the autumn and winter, satellite-linked time depth recorders (SLTDRs) were deployed in January, 1993, on seven of the pups that had carried TDRs during 1992.

Pups began to dive within two weeks of birth. The number of dives per day, and the mean depth and duration of dives increased significantly over the first 13 weeks. During this period, dive behaviour was determined primarily by pup age, although diel effects were apparent in dive frequency. The SLTDR records from pups between the ages of 11 and 32 weeks revealed diel effects in all measured parameters. The longest and deepest dives occurred during the afternoon period, night dives were short and shallow, and the morning and evening dives were intermediary. Dive frequency was highest during the night, and lowest during the afternoon. These effects predominated once the pups were older than three months. Combined SLTDR and TDR data suggest possible seasonal trends in dive behaviour, although the pattern was confounded by pup age. Mean dive depth and duration increased throughout the study only during the afternoon period, the hours of maximum light. In all other periods, dive depth and duration declined or remained constant. In combination, depth, duration, and frequency data suggest that dives were more shallow and less frequent later in the season in periods when light was limited. The diel and seasonal pattern in dive behaviour is consistent with the hypothesis that pups are foraging throughout the day on vertically migrating prey species.

Key words: diving, development, diel variation, Weddell seal, pups, foraging.

INTRODUCTION

Many studies have documented the diving behaviour of pinnipeds. Variations in dive behaviour within and between species have been attributed to differences in foraging behaviour, age, size, sex, season and time of day (Kooyman 1975, Kooyman *et al.* 1983, Croxall *et al.* 1985, Testa *et al.* 1985, LeBoeuf *et al.* 1986, Feldkamp *et al.* 1989, Boyd *et al.* 1991, Castellini *et al.* 1991, Thorson & LeBoeuf 1994). Most studies have focused on the dive behaviour of adults and juveniles;

little work has been performed on young pups due to difficulties of instrument deployment on small individuals, and with recapture (Lyderson *et al.* 1992, Thorson & LeBoeuf 1994). However, because Weddell seal (*Leptonychotes weddellii* Lesson) pups are relatively large, and easy to approach and recapture both before and after weaning, they offer a unique opportunity to study the development of pup diving ability. In addition, the diving behaviour of adult, juvenile and a few weaned Weddell seal pups has been documented (Kooyman 1968, 1981, Kooyman *et al.* 1983, Castellini *et al.* 1991), and can be compared with that of the pups in this study. Understanding the development of pup diving ability is important for several reasons: differences in diving behaviour at an early age may be linked to differential survival, diving behaviour can provide clues about diet and foraging habits, and diving development might yield information on developmental physiology.

In McMurdo Sound, Antarctica, Weddell seals are born in October and November. Traditional colonies are generally in fast ice near tidal cracks which, in addition to breathing holes maintained by the seals, provide access to water (Siniff *et al.* 1975, Stirling 1977, Kooyman 1981, Testa *et al.* 1989). Pups weigh approximately 25 kg at birth, and at the end of the six week lactation period can exceed 100 kg (Bryden *et al.* 1984, Hill *et al.* 1986, Castellini *et al.* 1991). By two weeks of age, pups have begun to enter the water and they spend increasing amounts of time swimming and diving as lactation progresses (Tedman & Bryden 1979, Hill 1987). By the end of December pups are weaned and the fast ice has begun to thin and break up. By February much of the Sound is open water and the majority of the seals have dispersed (Castellini *et al.* 1991). The maximum depth of McMurdo Sound is approximately 700 m, with depths greater than 200 m within 1 km of shore, making it unlikely that pup diving behaviour is constrained by bathymetry while they remain in the Sound. The short lactation period, rapid growth of the pups, and disappearance of their haul out substrate, suggests that the development of pup diving behaviour is rapid.

Daylight patterns in the McMurdo Sound area are extreme. From October 21 until February 23, the sun never sets, and between April 23 and August 24, the sun never rises. While there is an absence of a typical light–dark cycle for eight months of the year, during the summer there is variation in the incident radiation due to the sun's angle, and during the winter the moon provides enough light to see distant features. There is no sunlight only during June and July (Castellini *et al.* 1991). During the spring and summer, adult Weddell seal dive behaviour varies with time of day. Seals are generally on the surface during the day, and dives are concentrated in the twilight and night periods. The little information on diving patterns in the mid-winter months suggests that this pattern breaks down once the light levels stop fluctuating (Kooyman 1975, Thomas & DeMaster 1983, Castellini *et al.* 1991, Testa 1994). The role of diel and seasonal variation in the dives of pups or juveniles has not previously been studied. These effects, and those of age, were examined to determine the role of diurnal and seasonal variation in the development of pup diving behaviour throughout lactation, and for several months post weaning.

METHODS

This study was conducted in McMurdo Sound (77°45' S, 166°30' E), during the austral summers of 1992 and 1993. Long-term tagging studies aided in adult and pup identification. Pup ages were known to within two days. Both male and female pups were studied (1992, n=6,11; 1993, n=13,7). Time depth recorders (TDRs) (Wildlife Computers, Woodinville, WA, USA, Mark 5 or 6) were deployed on individual pups for several days every two weeks to monitor the development of their diving behaviour between the ages of 2–13 weeks. Seventeen pups carried TDRs during 1992 for a total of 60 deployments, and TDRs were deployed a total of 54 times on 20 pups in 1993. On average, pups carried TDRs four times in 1992 and three times in 1993. Until pups moulted their lanugo pelage TDRs were carried on anklets consisting of a 4 cm wide, 3 mm thick neoprene rubber band fastened around the ankle with small corrosive bolts. The TDRs were attached to the anklet with Velcro and nylon ties. After the moult, TDRs were glued to the mid-dorsal pelage with an epoxy adhesive. Data collected from TDRs deployed in a similar fashion on four yearlings in 1992 is reported for comparison. TDRs monitored depth every 10 s when wet, had a depth resolution of 2 m, and a maximum depth of 500 m. Data from the TDRs consisted of a continuous time-line of depths.

To track dive behaviour throughout the autumn and winter, satellite-linked time depth recorders (SLTDRs) were deployed in January, 1993, on seven of the pups that had carried TDRs earlier in the season. The half-watt SLTDRs (Wildlife Computers, SDR-T6, Service Argos PTTs) were attached to the mid-dorsal pelage with epoxy and netting. Like the TDRs, the SLTDRs sampled depth every 10 s during dives, had a depth resolution of 2 m, and a maximum depth of 500 m. However, due to transmission limitations, the SLTDRs placed all dives into one of six depth and duration bins (80 m and 4 min. intervals) and transmitted the number of dives in each bin, rather than the actual depths and durations of dives. The data were transmitted to a polar orbiting satellite, and relayed to the University of Alaska via Service Argos (Argos 1984, Fancy *et al.* 1988). For analysis, all SLTDR recorded dives were given depth and duration values. The values assigned were determined by simultaneously collecting TDR and SLTDR dive data from individual weaned pups and yearlings, and calculating the mean depth and duration of those dives which fell within each histogram bin range. Additional information transmitted via Service Argos included status messages that reported the depth of the deepest dive in the previous 24 h. The longest dive duration was not reported.

Data from both the TDRs and SLTDRs were processed with software from Wildlife Computers. The minimum depth for a dive to be analysed was 12 m, and minimum duration was 30 s. In this analysis, the mean dive depth is the deepest depth of each dive averaged over all dives for each seal at each age (in weeks, rounded down) rather than the mean of the average depth of each dive. Average dive duration is the duration of each dive, averaged over all dives for each seal at each age. Similarly, average dive frequency is the average number of dives per day grouped by individual and age. Weekly maximum dive depths are an average of the depths of the deepest dives made by each seal each day.

Diel effects were searched for by dividing the day into four equal periods: night (2100–0259), morning (0300–0859), afternoon (0900–1459), and evening (1500–2059). In part, this grouping of dive data by period was mandated by the need to compress data for transmission by the SLTDRs, but the times of the four periods were chosen to take into account the fluctuating daylight in McMurdo Sound and the known behaviour of the seals. The average depth, duration and frequency of dives was then calculated for each seal by period and deployment, and these values used in all analyses.

All comparisons were performed using multivariate linear regression, with age as a continuous variable, and year and period as categorical variables (STATISTIX software). Non-linear curve fitting was performed, but in all cases linear regression yielded the best fit. The null hypothesis assumed no difference in mean dive depth, duration or frequency, by pup age, year, period or month. Significance for rejection of the null hypothesis was set at $p<0.05$. Only equations from significant regressions are presented. Reported rates of increase in average dive depth, duration or frequency are the slope coefficient from the best fit regression equation for the appropriate subset of data. Differences in rates were considered significant if the pertinent categorical variable was significant in the overall regression. All figures show the overall average dive depth, duration or frequency, grouped first by individual, then by period or year, as indicated.

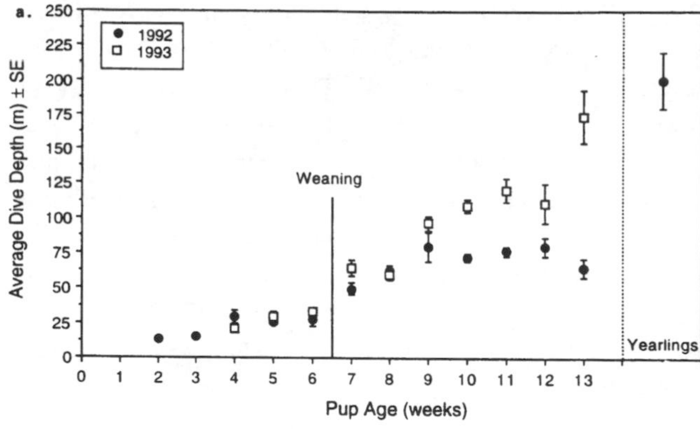

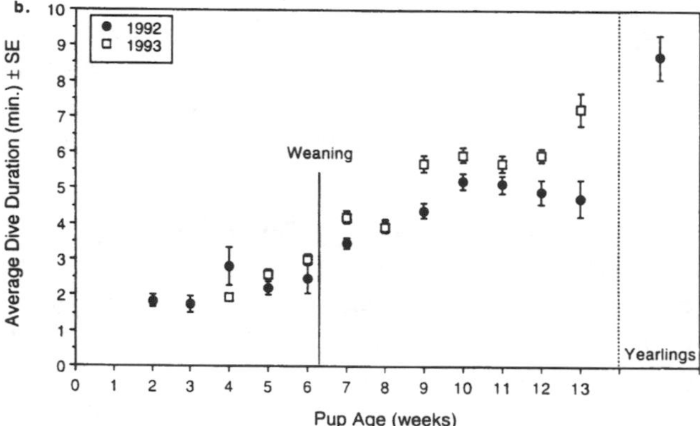

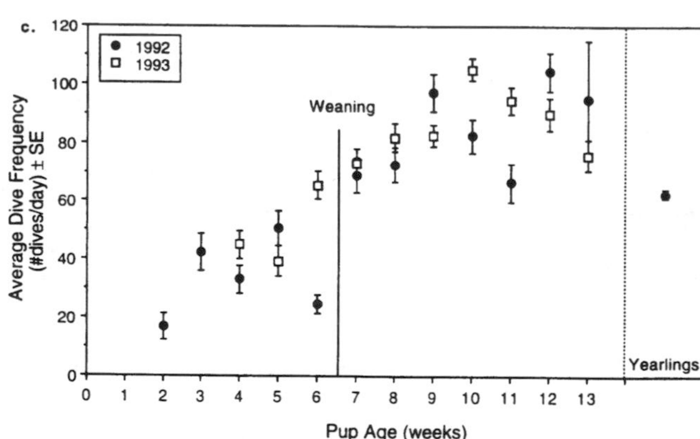

Fig. 46.1. Average dive depth, duration and frequency in 1992 and 1993 as recorded by TDR. (a) Average maximum dive depth, (b) average dive duration, (c) average dive frequency. $n=119$. Yearling values are shown for comparison. Error bars show standard error.

RESULTS

Data from 58 500 dives were analysed. In 1992, 10 188 pup dives and 1810 yearling dives came from TDR records, and 29 920 from SLTDR records. In 1993, all 16 582 dives were collected by TDR. Dives were collected from TDRs between November and February, and from SLTDRs from January through May. No dives made by pups or yearlings ever exceeded the 500 m maximum resolution.

Mean depth, duration and frequency of dives made by pups between the ages of 2 and 13 weeks were mainly determined by the age of the pup, which accounted for 50–72% of the variation in dive behaviour. However, the average depth and duration of dives was significantly greater in 1993 than in 1992, as were the weekly rates of increase (depth: 7.52 versus 16.27 m week^{-1}; duration: 0.37 versus 0.59 min. week^{-1}). As a result, 1992 and 1993 dive depth and duration data were analysed separately. There was no difference in the average dive frequency between 1992 and 1993, and these were combined for subsequent analysis (Fig. 46.1). Because no significant difference in dive behaviour existed between male and female pups in either 1992 or 1993, the sexes were combined.

Once the pups were older than three months, pup age (in weeks) accounted for little of the variation in dive behaviour (data recorded by SLTDRs). There was no significant increase in the average dive depth and dive duration with increasing pup age from 3–8 months, but average dive frequency was positively correlated with pup age. However, this correlation accounted for only 7% of the variation in dive frequency (dive frequency$=57.87+4.46\times$ pup age, $r^2=0.07$ $n=333$, $p<0.001$).

In 1992, maximum dive depth increased with the age of pups across both the TDR and SLTDR records, but the rate of increase was significantly greater in the TDR record (TDR record: maximum depth$=-46.56+21.15 \times$ pup age, $r^2=0.67$, $n=52$, $p<0.001$; SLTDR record: maximum depth$=160.21+7.25\times$pup age, $r^2=0.18$, $n=56$, $p<0.001$).

To determine if, in addition to pup age, dive behaviour was influenced by time of day, period of day was added into the model as a categorical value. In the TDR records from both 1992 and 1993 for pups between the ages of 2 and 13 weeks there were no significant differences in average dive depth or duration by period, so all periods were combined. Dive frequency (average number of dives per period) did not vary by period in either 1992 or 1993. In all case, the absence of significant differences between periods in average dive depth, duration and frequency, could be attributed to the large amount of individual variation in the diving behaviour of pups younger than three months. As a result there was little evidence of any consistent diurnal variation in the diving behaviour of pups younger than three months. For these pups, age, rather than time of day, appears to be the primary determinant of diving behaviour.

In contrast, once the pups were older than three months, the period of day had a significant effect on average dive depth, duration, and frequency as recorded by SLTDRs, and age had less influence. The average depth of dives was significantly different in each of the four periods. The deepest dives occurred in the afternoon, followed by the evening, morning, and night periods (Fig. 46.2). The average depth of dives increased significantly with pup age only during the afternoon period (average depth$=-11.7+7.59\times$pup age, $r^2=0.35$, $n=70$, $p<0.001$), and was not correlated with pup age in the morning, afternoon or evening periods. Maximum dive depths could not be analysed by period because the time of the dive was not reported by the SLTDR.

Average dive duration of pups older than three months also varied by period, but not all periods were significantly different from each other. Dives in the night were shortest, followed by dives in the morning. The longest dives occurred in the afternoon and evening periods, which were grouped because dive durations were not significantly different (Fig. 46.2). Pup age was only a significant factor in the combined afternoon and evening period when dive duration increased slightly with pup age (average duration$=4.05+0.07\times$pup age, $r^2=0.05$, $n=151$, $p=0.004$).

Dive frequency showed the least variation by period. There were significantly more dives in the night and morning periods than in the afternoon or evening periods (Fig. 46.2). Pup age was a significant factor in the combined night and morning period, when dive frequency increased slightly as pups aged (average frequency$=27.48+0.77\times$pup age, $r^2=0.03$, $n=157$, $p=0.017$). There was no significant relationship between pup age and dive frequency in the afternoon period. The greatest increase in dive frequency with age occurred during the evening period (average frequency$=-11.15+2.19\times$pup age, $r^2=0.37$, $n=69$, $p<0.001$).

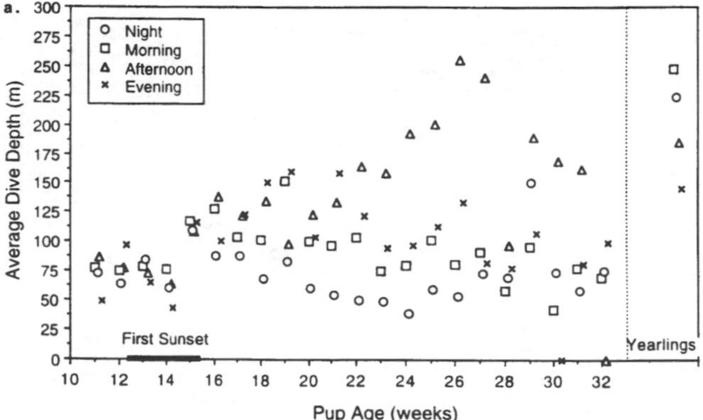

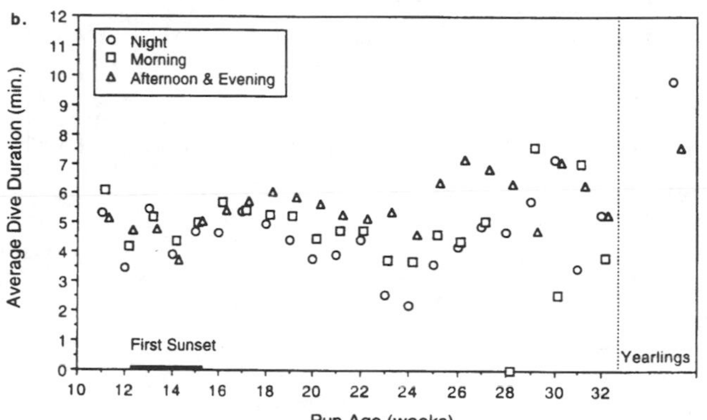

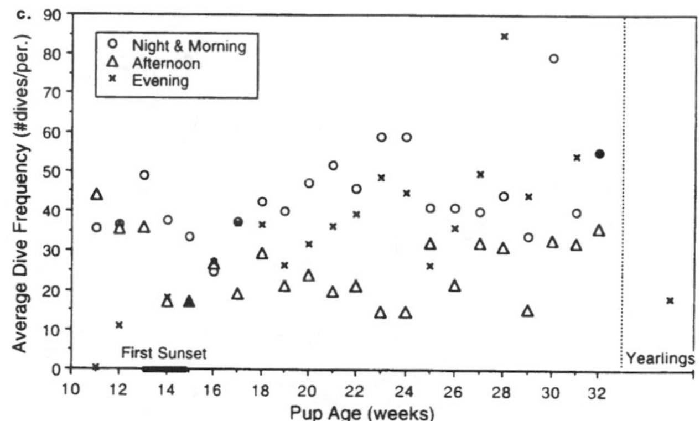

Fig. 46.2. Average dive depth, duration, and frequency by period, for those periods which differed significantly. (a) Average maximum dive depth, (b) average dive duration, (c) average dive frequency. Data collected from SLTDRs deployed on 7 pups in 1993. Yearling values are shown for comparison.

Analyses of seasonal trends were confounded by the overlap between calendar month and pup age. When month was treated as a categorical variable, the correlation between month and pup age was too large to separate the two effects. As a result, seasonal effects could only be searched for by using pup age as a proxy, and examining each of the periods separately. This technique revealed some interesting patterns. Once the pups were older than three months, average dive depths and durations were uncorrelated with pup age or month during morning, evening and night periods (see Fig. 46.2). In contrast, both average depth and duration increased with pup age throughout the TDR and SLTDR records during the afternoon period. Despite the

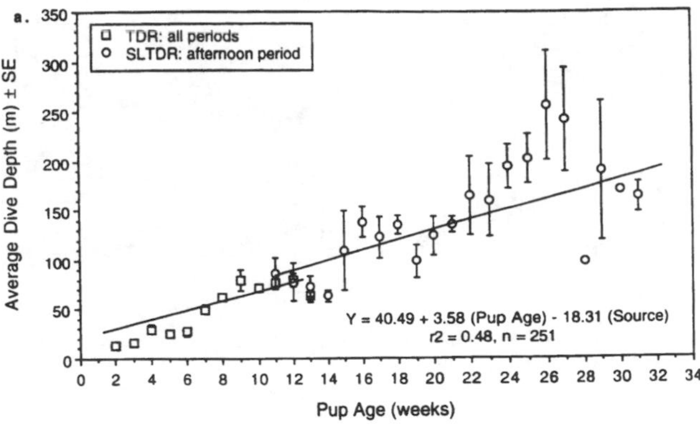

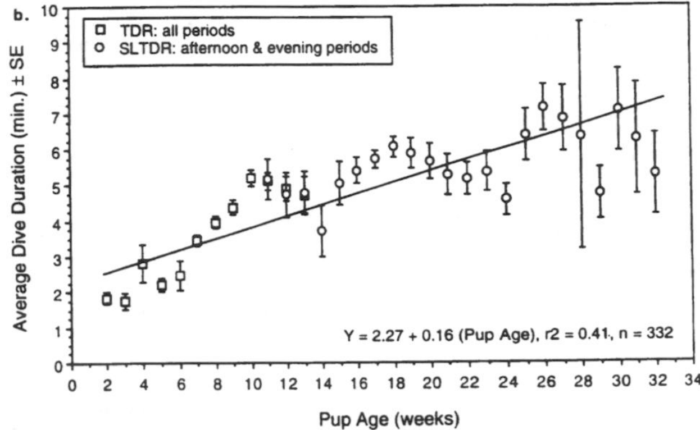

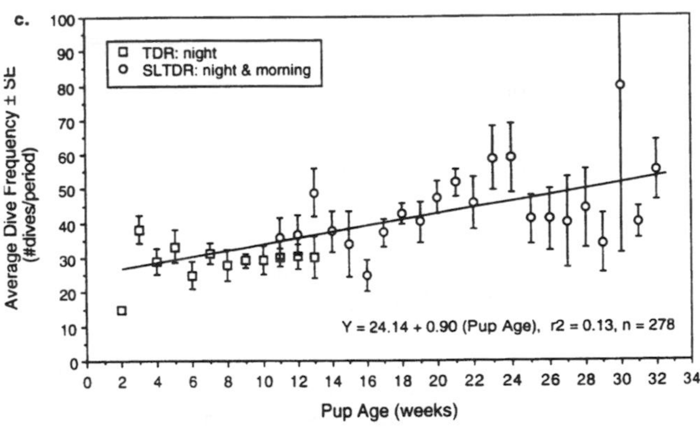

Fig. 46.3. Average dive depth, duration, and frequency with pup age throughout the TDR and SLTDR records from 1992. (a) Average dive depth, (b) average dive duration, (c) average dive frequency. The least squared regression line fit for all the data is shown. Source=1 for TDR data and 0 for SLTDR data. Error bars show standard error.

unusual light regime in McMurdo Sound, the afternoon period did receive some solar insolation during all of the SLTDR record. During the afternoon period, the rate of increase in average dive depth and duration as recorded by SLTDRs was not statistically different from the TDR records. Similarly, dive frequency continued to increase at a constant rate throughout the record only during the night (Fig. 46.3). During these periods, pup age accounted for greater than 40% of the variation in dive behaviour.

In both the TDR and SLTDR recorded data, there were strong correlations between average dive depth and average dive duration. This correlation was greatest once the pups were older than three months, and was present in all periods of the SLTDR data. Table 46.1 shows the partial correlation values for average dive depth and duration, controlled for pup age. Because seal

Table 46.1. *Partial correlation between average dive depth and duration*

Source	Year	Period	*N*	Partial correlation
TDR	1992	All combined	52	0.36
TDR	1993	All combined	67	0.50
SLTDR	1992	All combined	92	0.68
SLTDR	1992	Night (0)	73	0.68
SLTDR	1992	Morning (1)	69	0.41
SLTDR	1992	Afternoon (2)	61	0.62
SLTDR	1992	Evening (3)	64	0.74

pups throughout the study spent a large proportion of their time on the surface, there was no clear correlation between dive duration and dive frequency.

DISCUSSION

The diving ability of Weddell seal pups developed rapidly during their first three months of life. During this period, pup age was the most important factor in determining the average depth and duration of dives. By the time pups were 13 weeks old their diving performance was still less than that of yearlings, although they were completely weaned and foraging on their own (Bryden *et al.* 1984, Hill *et al.* 1986). The differences in the rate of increase in average dive depth and duration by pup age between 1992 and 1993 could be a reflection of differences between the two years in pup growth rates or in ecological factors such as prey distribution. However, as dive frequency did not differ between the two years, it is more likely that the differences were tied to ecological rather than physiological causes. If 1992 pups were less capable divers than 1993 pups, then all three parameters should have been lower in 1992, which was not the case. One such ecological cause could be differences in yearly prey distributions. Unfortunately, little is known about such variation in McMurdo Sound.

Time of day clearly played an important role in determining the dive behaviour of Weddell seal pups. However, it did not appear as a significant factor in dive depth and duration until after the pups were older than three months, foraging on their own, and the sun had started to set. In these older pups, trends in all measured variables followed a similar pattern: dives were short, shallow and frequent during the night period, when light levels were lowest; and longest, deepest and least frequent during the afternoon period, when insolation was highest. The morning and evening periods when light levels changed rapidly, showed intermediate characteristics. The strength and continuity of this pattern throughout the study was somewhat unexpected, for McMurdo Sound does not have a typical light–dark cycle. However, even in May when the observations ceased, the afternoon period received some light, suggesting that dive patterns were influenced by insolation throughout the study.

The dive behaviour of many pinniped species has been linked

to the distribution of their prey (Croxall *et al.* 1985, LeBoeuf *et al.* 1986, Boyd *et al.* 1991), which supports the hypothesis that Weddell seal dives track their prey distribution. Unfortunately, while there is good information on the diet and dive behaviour of adult and juvenile Weddell seals, there is little to no information on the ecology of their prey items in McMurdo Sound. Adult Weddell seals prey primarily on the Antarctic silverfish (*Pleuragramma antarcticum* Boulanger), and secondarily on several fish species, squid, octopods and crustaceans (Dearborn 1965, Clarke & MacLeod 1982, Testa *et al.* 1985, Plötz 1986, Green & Burton 1987, Castellini *et al.* 1991). The few pup scats collected in McMurdo Sound contained otoliths from *Pleuragramma*, squid beaks and crustacean parts, suggesting that pups forage on the same species as do adults (Green & Burton 1987). This is in contrast to early reports (Lindsey 1937, Bertram 1940), which suggest that weaned Weddell seal pups in other regions of Antarctica undergo a transitional period during which they prey primarily on crustaceans.

Information on the ecology of these prey items comes from ice-free Antarctic waters outside of McMurdo Sound, where the majority of these species occur midwater, make diurnal vertical migrations, or are separated in size by depth (Eastman 1985, Hubold & Ekau 1985, Kellerman 1986, White & Piatkowski 1993). However, little information is available regarding their distribution or movements under ice or in the winter. Thus, while both the diet and dive information suggest that Weddell seal adults, juveniles, and pups forage on similar prey, and that these species show some degree of vertical migration throughout the year, there is no direct evidence for such behaviour.

In this study, the influence of season on Weddell seal dive behaviour was difficult to determine. The correlation between pup age and season made it impossible to separate the two effects in all attempted statistical analyses. Although there was a steady increase in dive depth and duration throughout the TDR record, and during the afternoon period in the SLTDR record, after the pups were older than three months of age there was little significant change in dive behaviour with age (or month) in any other period. Yet the lack of correlation between pup age and average dive depth, duration and frequency in the morning, evening and night periods might itself be a result of seasonal trends in light patterns. As light levels decline, vertically migrating prey might move closer to the surface. Seals foraging on these species could then capture their prey with shorter and shallower dives. As (until mid-May) the afternoon period has the highest light levels, seals may have to continue to dive deeply in this period in order to obtain sufficient food.

Increases throughout the record in dive depth and duration in the afternoon period, and in dive frequency during the night, suggest that pup diving ability continues to develop through the first year. This is confirmed by comparing average afternoon dive depths and durations of pups at the end of the study (116±13 m, 5.6±0.5 min.) to the overall averages for yearlings (200±35 m, 8.69±0.64 min.), and adults (144±83 m, 10.4±2.9 min.) (Castellini *et al.* 1991, this study). Additionally, the steady change in maximum dive depth by pups in this study indicates that both the frequency and depth of deep dives continues to increase. Clearly, pups less than nine months of age have yet to develop the skills of yearling and adults. That pup dive depths are closer to adult values than are their dive durations suggests that the development of breath hold ability (oxygen stores) is slower than the development of the ability to reach great depths. Physiologically, this is to be expected. Smaller, younger seals have higher mass specific metabolic rates than adults and require more oxygen per unit mass for a given dive duration (Lavigne *et al.* 1986). Even juvenile Weddell seals (between 2–4 years old) have lower aerobic dive limits (ADL) than adults, (10–13 min. versus 16–20 min., Kooyman *et al.* 1983), indicating that increases in average dive duration continue for several years.

How, or if, the increase in aerobic capacity is developed after the pups are weaned has not yet been determined. Weaned pups are of approximately the same mass as yearlings, but may have a lower proportion of muscle mass (Rea & Costa 1992). This, along with differences in haematocrit, haemoglobin and myoglobin content due to age, also might account for age-specific differences in the aerobic capacity. Assuming that pups swim at 1.5 m s^{-1} (an average swim speed for many pinnipeds), a 400 m dive requires nine 'minutes' of oxygen (Davis *et al.* 1985, Williams & Kooyman, 1985). Pups can dive for this duration, but to make such a deep dive energetically feasible may require more time at that depth than they are capable of spending during an aerobic dive. Thus, while there are no physiological reasons that pups should be unable to dive to 'adult' depths, they may not have sufficient oxygen stores to make such deep dives worth the effort. The strong correlation between depth and duration in both the TDR and SLTDR records corroborates this hypothesis.

The results of this study support several conclusions: that the development of diving behaviour in Weddell seal pups is rapid, and that age is the primary determinant during the first three months. During this period of rapid development in diving ability, average maximum dive depth increased from 13 to 185 m, average dive duration from 1.7 to 7.6 min. and average daily dive frequency from 5 to >500 dives day^{-1}. After the pups were three months old and foraging on their own, diving ability continued to improve, and diurnal influences on diving patterns became evident. While the role of seasonal effects is unclear, the pattern suggested agrees with that indicated by diurnal variation: shorter and shallower dives during periods and seasons when light is limited. In combination, these findings suggest that Weddell seals in McMurdo Sound forage on prey species that make diurnal vertical migrations under the ice. In addition, as similar variations in dive behaviour are seen in adults, juveniles and pups, it is likely that all ages forage on similar species.

ACKNOWLEDGEMENTS

This work was funded by grant DPP-9119885 from the National Science Foundation (USA) to J. W. Testa and M. A. Castellini. Logistical Support was provided through the Division of Polar

Programs at NSF, by Antarctic Support Associates, and the US Navy and Coast Guard. For help in the field, we thank Amal Ajmi, Michael Castellini, Brian Fadely, Janey Fadely, Kelly Hastings, Lorrie Rea, Brad Scotton, Kate Wynne, and all the volunteers. Wildlife Computers' staff were generous in their help. Support from the University of Alaska Fairbanks through a Chancellors fellowship to J. M. Burns is also gratefully acknowledged. We are grateful for the helpful comments from Drs Ian Boyd and Mark Hindell.

REFERENCES

Argos. 1984. *Location and data collection satellite system user's guide*. Landover, MD: Service Argos, Inc., 36 pp.

Bertram, G. C. L. 1940. The biology of the Weddell and crabeater seals: with a study of the comparative behaviour of the Pinnipedia. *British Graham Land Expedition 1934–1937, Reports*, **1**, 1–139.

Boyd, I. L., Lunn, N.J . & Barton, T. 1991. Time budgets and foraging characteristics of lactating Antarctic fur seals. *Journal of Animal Ecology*, **60**, 577–592.

Bryden, M. M., Smith, M. S. R., Tedman, R. A. & Fetherstone, D. W. 1984. Growth of the Weddell seal, *Leptonychotes weddellii* (Pinnipedia). *Australian Journal of Zoology*, **32**, 33–41.

Castellini, M. A., Davis, R. W. & Kooyman, G. L. 1991. Annual cycles of diving behavior and ecology of the Weddell seal. *Bulletin of the Scripps Institute of Oceanography*, **28**, 1–54.

Clarke, M. R. & MacLeod, N. 1982. Cephalopod remains in the stomachs of eight Weddell seals. *British Antarctic Survey Bulletin*, No. 57, 33–40.

Croxall, J. P., Everson, I., Kooyman, G. L., Ricketts, C. & Davis, R. W. 1985. Fur seal diving behaviour in relation to vertical distribution of krill. *Journal of Animal Ecology*, **54**, 1–8.

Davis, R. W., Williams, T. M. & Kooyman, G. L. 1985. Swimming metabolism of yearling and adult harbor seals *Phoca vitulina*. *Physiological Zoology*, **58**, 590–596.

Dearborn, J. H. 1965. Food of Weddell seals at McMurdo Sound, Antarctica. *Journal of Mammalogy*, **46**, 37–43.

Eastman, J. T. 1985. *Pleuragramma antarcticum* (Pisces, Nototheniidae) as food for other fishes in McMurdo Sound, Antarctica. *Polar Biology*, **4**, 155–160.

Fancy, S. G., Pank, L. F., Douglas, D. C., Curby, C. H., Garner, G. W., Amstrup, S. C. & Regelin, W. L. 1988. Satellite telemetry: a new tool for wildlife research and management. *U.S. Fish and Wildlife Resource Publication*, No. 172, 54 pp.

Feldkamp, S.D., DeLong, R.L. & Antonellis, G.A. 1989. Diving patterns of California sea lions, *Zalophus californianus*. *Canadian Journal of Zoology*, **67**, 872–883.

Green, K. & Burton, H.R. 1987. Seasonal and geographical variation in the food of Weddell Seals, *Leptonychotes weddellii*, in Antarctica. *Australian Wildlife Research*, **14**, 475–489.

Hill, S. 1987. *Reproductive ecology of Weddell seals* (Leptonychotes weddellii) *in McMurdo Sound, Antarctica*. PhD thesis, University of Minnesota, 106 pp.

Hill, S., Vacca, M., Bartsh, S., Reichle, R., Testa, J.W. & Siniff, D. 1986. Patterns of weight gain by Weddell seal pups. *Antarctic Journal of the United States*, **21**, 209–210.

Hubold, G. & Ekau, W. 1985. Midwater fish fauna of the Weddell Sea, Antarctica. In *Proceedings of the 5th Congress of European Ichthyologists*. Stockholm: Swedish Museum of Natural History, 391–396.

Kellerman, A. 1986. Geographical distribution and abundance of post-larval and juvenile *Pleuragramma antarcticum* (Pisces, Notothenioidei) off the Antarctic Peninsula. *Polar Biology*, **6**, 111–119.

Kooyman, G. L. 1968. An analysis of some behavioral and physiological characteristics related to diving in the Weddell seal. *Antarctic Research Series*, **2**, 227–261.

Kooyman, G. L. 1975. A comparison between day and night diving in the Weddell seal. *Journal of Mammalogy*, **56**, 563–574.

Kooyman, G. L. 1981. *Weddell seal: consummate diver*. Cambridge: Cambridge University Press. 135 pp.

Kooyman, G. L., Castellini, M. A., Davis, R. W. & Maue, R. A. 1983. Aerobic diving limits of immature Weddell seals. *Journal of Comparative Physiology B*, **151**, 171–174.

Lavigne, D. M., Innes, S., Worthy, G. A. J., Kovacs, K. M., Schmitz, O. J. & Hickle, J. P. 1986. Metabolic rates of seals and whales. *Canadian Journal of Zoology*, **64**, 279–284.

LeBoeuf, B. J., Costa, D. P., Huntley, A. C., Kooyman, G. L. & Davis, R. W. 1986. Pattern and depth of dives in northern elephant seals, *Mirounga, angustirostris*. *Journal of Zoology*, **208**, 1–7.

Lindsey, A. A. 1937. The Weddell seal in the Bay of Whales, Antarctica. *Journal of Mammalogy*, **18**, 127–144.

Lydersen, C., Ryg, M. S., Hammill, M. O. & O'Brien, P. J. 1992. Oxygen stores and aerobic dive limit of ringed seals (*Phoca hispida*). *Canadian Journal of Zoology*, **70**, 458–461.

Plötz, J. 1986. Summer diet of Weddell seals *Leptonychotes weddellii* in the Eastern and Southern Weddell Sea, Antarctica. *Polar Biology*, **6**, 97–102.

Rea, L. D. & Costa, D. P. 1992. Changes in standard metabolism during long-term fasting in Northern elephant seal pups (*Mirounga angustirostris*). *Physiological Zoology*, **65**, 97–111.

Siniff, D., Reichle, R., Hofman, R. & Kuehn, D. 1975. Movements of Weddell seals in McMurdo Sound, Antarctica, as monitored by telemetry. *Rapports et Procès-verbaux des Reunions. International Journal du Conseil International pour l'Exploration de la Mer*, **169**, 387–393.

Stirling, I. 1977. Adaptations of Weddell and ringed seals to exploit the polar fast ice habitat in the absence or presence of surface predators. In Llano, G. A., ed. *Proceedings of the Third SCAR Symposium on Antarctic Biology*. Washington, DC: Smithsonian Institution, 741–748.

Tedman, R. A. & Bryden, M. M. 1979. Cow–pup behaviour of the Weddell seal, *Leptonychotes weddellii*, in McMurdo Sound, Antarctica. *Australian Wildlife Research*, **6**, 19–37.

Testa, J. W. 1994. Overwinter movements and diving behaviour of female Weddell seals (*Leptonychotes weddellii*) in the South West Ross Sea, Antarctica. *Canadian Journal of Zoology*, **72**, 1700–1710.

Testa, J. W., Siniff, D. B., Ross, M. J. & Winter J. D. 1985. Weddell seal–Antarctic cod interactions in McMurdo Sound, Antarctica. In Siegfried, W. R., Condy, P. R. & Laws, R. M., eds. *Antarctic nutrient cycles and food webs*. Berlin, Heildelberg: Springer-Verlag, 561–565.

Testa, J. W., Hill, S. E. B. & Siniff, D. B. 1989. Diving behavior and maternal investment in Weddell seals (*Leptonychotes weddellii*). *Marine Mammal Science*, **5**, 399–405.

Thomas, J. A. &. DeMaster, D. P. 1983. Diel haul-out patterns of Weddell seal (*Leptonychotes weddellii*) females and their pups. *Canadian Journal of Zoology*, **61**, 2084–2086.

Thorson, P. H. & LeBoeuf, B. J. 1994. Developmental aspects of diving in northern elephant seal pups. In LeBoeuf, B. J. & Laws, R. M., eds. *Elephant seals: population ecology, behavior, and physiology*. Berkeley: University of California Press, 271–289.

White, M. G. & Piatkowski,U. 1993. Abundance, horizontal and vertical distribution of fish in eastern Weddell Sea micronekton. *Polar Biology*, **13**, 41–53.

Williams, T. M. & Kooyman, G. L. 1985. Swimming performance and hydrodynamic characteristics of harbor seals *Phoca vitulina*. *Physiological Zoology*, **58**, 576–589.

47 Significant differences in weaning mass of southern elephant seals from five sub-Antarctic islands in relation to population declines

H.R. BURTON[1], T. ARNBOM[2], I.L. BOYD[3], M. BESTER[4], D. VERGANI[5] AND I. WILKINSON[4]

[1]*Australian Antarctic Division, Channel Highway, Kingston, Tasmania 7050, Australia,* [2]*Department of Zoology, University of Stockholm, 106 91 Stockholm, Sweden,* [3]*British Antarctic Survey, Natural Environment Research Council, High Cross, Madingley Road, Cambridge CB3 0ET, UK,* [4]*Mammal Research Institute, University of Pretoria, Pretoria 0002, South Africa,* [5]*Instituto Antarctico Argentino, CERLAP, Calle 8N: 1467, 1900 La Plata, Argentina.*

ABSTRACT

The weaning mass of southern elephant seal pups has been shown to be determined mainly by the energy reserves of the mother when she gives birth. Recent declines in the number of elephant seals across parts of their range could be a response to changes in their food supply. Since food availability may be reflected in a change in the energy reserves of mothers then this will also affect the mass of their pups at weaning. The mass of southern elephant seal pups (n=3292) was measured at weaning and compared across the major populations. Weaning mass was found to vary significantly between locations. Weaning mass was greatest in the Atlantic sector of the Southern Ocean where populations have been stable and it was smallest in the Pacific and Indian Ocean sectors where populations have been in decline. Within the Indian Ocean sector, the weaning masses at Marion and Heard islands, where populations are continuing to decline rapidly (4–5% per annum), were significantly less than at Macquarie Island, where populations have declined less (<1% per annum) in recent years. Female elephant seals at Marion Island deliver up to 30% less energy to their pups during lactation than females at King George Island. It would, however, be premature to conclude that these differences in weaning mass show that nutritional conditions for adult females differ in the areas of declining populations compared with the regions where populations are stable. They may also be the result of natural selection due to differences in the importance of weaning mass for subsequent survival.

Key words: southern elephant seal, *Mirounga leonina*, weaning, birth, mass, sub-Antarctic.

INTRODUCTION

Populations of southern elephant seals at Macquarie, Marion and Heard islands as well as at Iles Kerguelen have declined significantly over the last 30 years at rates of 1.6–5.8% per annum (Guinet *et al.*1992, Hindell & Burton 1987, SCAR 1991, Wilkinson 1992, Hindell *et al.* 1994). At other locations, mainly in the region of South Georgia, the Antarctic Peninsula and at Peninsula Valdez in Patagonia, populations appear to have been relatively stable over the same period of time (McCann & Rothery 1988, SCAR 1991). Potential explanations for this difference between regions include differences in the history of exploitation (Hindell 1991) and differences in food availability, possibly mediated by long-term ecosystem changes that differ

Table 47.1. *Sample sizes and mean mass of male and female southern elephant seal pups weighed at birth and weaning at different breeding sites. No measurements were made of birth mass at Heard Island*

	Female			Male			Combined	
Island	*n*	Mean mass (kg)	SD	*n*	Mean mass (kg)	SD	Mean mass (kg)	SD
(a) Birth								
King George	162	43.9	6.2	145	48.6	6.6	46.1	6.9
South Georgia	127	40.1	5.9	114	45.7	6.0	42.7	7.0
Macquarie	191	37.5	4.8	141	40.4	5.1	38.7	6.3
Marion	54	34.1	4.5	66	41.1	6.8	37.9	6.3
(b) Weaning								
King George	388	152.6	28.4	402	165.5	29.5	159.1	29.8
South Georgia	206	123.0	24.7	168	130.0	25.4	126.5	25.1
Macquarie	463	116.8	19.6	477	122.6	20.0	118.7	20.7
Marion	172	109.4	19.7	195	119.3	21.2	114.7	21.4
Heard	411	111.3	19.3	409	116.5	20.7	114.1	20.6

between the regions. Redistribution of animals is thought to be an unlikely explanation because no tagged or branded individuals have been observed to move away from their natal population and Hoelzel *et al.* (1993) have shown that southern elephant seal breeding populations tend to be genetically distinct.

Arnbom *et al.* (1993) have shown that 55% of the variance in southern elephant seal pup weaning mass is determined by the mass of the mother at parturition. Since mothers arriving at the pupping grounds to give birth have all the energy reserves necessary for lactation then mass is a good indicator of the energy reserves available to be passed on to the pup. It then follows that the growth of the pup up to weaning depends on the energy transferred from mother to pup and the ultimate weaning mass of pups is, therefore, a function of maternal energy reserves. Although Fedak *et al.* (1996) have shown that these investment patterns are not a linear function of maternal energy reserves they nevertheless provide a potentially useful indication of maternal energy reserves. In this study, we use the mass of pups at weaning to compare energy investment in reproduction by female southern elephant seals between different breeding populations. The aim was to (1) test the hypothesis that weaning mass differed between populations and (2) to use the patterns of maternal investment in pup growth during lactation to examine the possibility that the rates of change in the different populations are caused by different nutritional conditions.

METHODS

A total of 972 pups were weighed (precision 1 kg) on the day of birth at South Georgia, Marion, King George (South Shetland Islands) and Macquarie islands and were tagged with an individually numbered plastic tag in a rear flipper web. Harems selected for study were checked daily. Newly born pups were identified and marked by tagging or with a spot of paint or dye on the back. The marked pups were identified and reweighed on the day of weaning, defined as the time after birth when the pup left the harem. A further 2320 pups from all of the populations included in the study of birth mass were also weighed at weaning. This included 820 weaned pups from Heard Island. When appropriate, birth date was recorded and for all pups in the study weaning date and sex were also recorded. Weights were obtained between 1985 and 1991 inclusive but not in all years from all sites.

Statistical analyses mainly involved multiple linear regression using the GLM procedure in the SAS Statistical Package (SAS Institute, Carey, NC, USA), which was particularly suited to the unbalanced nature of the sample sizes from each site and across years.

RESULTS

Birth mass varied significantly across breeding sites (ANOVA, $F_{3,533}=11.62$, $P<0.001$) and ranged from 37.9 kg at Marion Island to 46.1 kg at King George Island (Table 47.1a). The birth mass differed significantly ($P<0.001$) between all sites except between Macquarie and Marion islands ($t=1.19$; df=450; $P>0.05$). Birth date or year of birth had no significant effect on birth mass. Male pups were significantly heavier at birth than female pups at all sites (Student's *t*-test, $P<0.001$ in each case; Table 47.1a). Overall, 26% of variation in birth mass was due to the sex of the pup whereas only 0.9% was due to interannual differences (Table 47.2).

Weaning mass also varied significantly across breeding sites (ANOVA, $F_{3,533}=12.22$, $P<0.001$) and ranged from 109.4 kg at Marion Island to 152.6 kg at King George Island (Table 47.1b). There was no significant difference in the weaning mass at Marion and Heard islands ($t=0.46$; df=1185; $P>0.05$) but weaning mass was significantly different between all the other breeding sites ($P<0.01$). There was a positive and significant relationship between birth mass and weaning mass in individuals for which there were data about both parameters (females, $F_{1,405}=347$, $P<0.001$; males, $F_{1,350}=382$, $P<0.001$). When

Table 47.2. *The percentage variance in pup birth and weaning mass attributable to the sex of pups and to year for southern elephant seals*

	Birth			Weaning		
Island	*n*	Sex	Year	*n*	Sex	Year
King George	307	20.0	2.6	791	8.6	11.5
South Georgia	241	27.4	0.0	374	3.2	2.9
Macquarie	332	10.0	0.0	940	1.6	15.7
Marion	120	46.4	0.9	367	9.6	3.9
Heard	—	—	—	820	2.6	—
Combined	1000	26.0	0.9	3292	5.1	14.3

examined as the only independent variable, birth mass explained 46% and 52% of the variation in weaning mass for females and males respectively.

The overall difference between birth mass and weaning mass at each site is a measure of the relative investment made in lactation by mothers. Comparing the mean values shows that at King George Island, pups gained an average of 108.7 kg during lactation, whereas at Marion Island, this was reduced to 75.3 kg. Therefore, on average, mothers at Marion Island invested only 69% of the amount being invested in the growth of pups during lactation by mothers at King George Island.

When variation in weaning mass due to sex and location were controlled for in a multiple regression ANOVA in which year was nested within location, interannual variation accounted for 14.3% of the variation in weaning mass (excluding Heard Island because measurement were from only one year). In contrast to the case for birth mass, where sex accounted for a substantial proportion of the variation in weaning mass (Table 47.2), pup sex accounted for a much smaller amount of the variation in weaning mass, for weaning mass was more sensitive than birth mass to interannual variation (Table 47.2).

DISCUSSION

Most variation in weaning mass is the direct result of variation in maternal mass (Arnbom *et al.* 1993). Moreover, Arnbom *et al.* (1993) also showed that the variation in weaning mass between the sexes disappears after variation in maternal mass is taken into account and it appears that mass differences between the sexes of pinniped pups may be caused by differnces in the body composition of pups rather than differences in energy delivered to the pup by the mother (Arnould *et al.* 1996). Therefore, maternal mass is the over-riding consideration when examining variation in weaning mass and most of the variations observed in this study are likely to reflect variations in maternal mass rather than fundamental differences in the patterns of growth or strategies of energy transfer to pups used by mothers.

Superficially, therefore, it appears that adult female southern elephant seals from Macquarie, Marion and Heard islands, where there have been recent declines in numbers, are likely to be smaller than at the other sites where the numbers have remained stable through the same period of time (SCAR 1991). An implication of this is that, on average, mothers provide up to 30% less nutrition to their pups up to the time they become independent in the regions of decline than in the regions of stable population size. Although there are few data from southern elephant seals which allow us to relate mass at weaning to subsequent survival, no relationship was found for northern elephant seals (LeBoeuf *et al.* 1994). However, the specific circumstances of the northern elephant seal study, where there is potentially high predation of juvenile seal by sharks, may invalidate it as a comparison with the situation in southern elephant seals. Nevertheless, the variability in weaning mass between breeding sites may correlate with juvenile survival.

Although the ultimate cause of the observed variability in weaning mass is likely to be the condition or fatness of mothers, there are several potential proximate causes of this.

(1) Mothers may be phenotypically and congenitally smaller at Macquarie, Marion and Heard islands than at South Georgia and King George islands. Carrick *et al.* (1962) and Bryden (1968) found that female elephant seals from Macquarie Island were smaller than those from South Georgia and it seems likely that this difference still exists. No evidence exists to suggest that there is an exchange of individuals between these two sets of breeding sites and, even within the sets, there may be little or no significant interchange between sites.This could have tended to produce different morphological types of seals if the selection pressures on size differ between the two sets of breeding locations. While a high weaning mass may provide newly weaned pups with a longer period to learn to feed or find an appropriate feeding location, it will also have costs associated with the additional costs of swimming with a thick buoyant blubber layer. Depending on the specific selective pressures that exist at each of the breeding sites there may have been no particular advantage for mothers to have attained a large size and, therefore, to produce large pups at Macquarie, Heard and Marion islands compared with South Georgia and King George island. Perhaps conditions at Macquarie, Heard and Marion islands are similar to those found for the northern elephant seal (LeBoeuf *et al.* 1994), where large weaning mass made no difference to subsequent survival, but at South Georgia and King George Island there may be a survival advantage associated with large size at weaning.

(2) The small size of mothers, indicated by low weaning mass in this study, is caused by reduced food availability in the Indian Ocean sector of the Southern Ocean when compared with the Atlantic Sector. This hypothesis was suggested by McCann (1985) but there are no other data to support the view that food availability differs between the Atlantic and Indian Ocean sectors. The diet of southern elephant seals is poorly understood and it is not possible to relate trends in abundance and weaning mass to specific measurements of changes in prey abundance.

(3) Increased adult female mortality could have resulted in a

decline in numbers and, as a consequence, a decline in the average size of adults, which has influenced the weaning mass of pups. The logic underlying this is that female elephant seals continue to grow throughout life (Bryden 1968) and any reduction in the survival rate would reduce the average age of the population together with the average body size of females. Populations of different age structure would, therefore, tend to produce pups with different weaning masses.

It is not possible to dismiss any of these potential proximate causes of the observed variability in weaning mass. However, an ultimate cause of the population decline at Macquarie Island appears to have been juvenile survival. Throughout the 1950s the average survival rate to age 1 year was between 0.4 and 0.5, but between 1960 and 1965 this declined to <0.2 (Hindell 1991). Unfortunately, there are no concurrent data about the weaning mass of pups through this period of change so it is not possible to conclude that the change in survival was related to the resources that pups obtained from mothers during lactation.

Although this study has demonstrated significant variation in weaning mass between elephant seals at different breeding sites and a broad relationship between the regional variation in weaning mass and trends in the abundance of elephant seals in those regions, it may be premature to conclude that these differences are indicative of food depletion as the proximate cause of the recent declines in elephant seal populations. If weaning mass was a sensitive indicator of such changes, we would expect to observe density-dependent changes in weaning mass. In those populations which have declined most, weaning mass might have been expected to have begun to return to levels similar to those observed in the region where no population decline has been observed. This is particularly true for Macquarie Island, where the population appears to be approaching stability (Hindell *et al.* 1994). The significantly higher weaning mass at Macquarie Island than at Heard and Marion islands, where declines have continued, is suggestive of a potential density-dependent effect, but the lack of a sufficient time-series of data from Macquarie Island means that it would be premature to infer that this difference was related to the rate of change in the population.

Perhaps the most significant feature of this study is that it supports the view that there are at least two distinct stocks of southern elephant seals; one mainly in the southern Indian Ocean basin and the other in the South Atlantic Ocean basin. They may be subjected to conditions that are sufficiently distinct, but internally consistent and possibly characteristic of each ecosystem, that natural selection or environmentally induced differences in development have produced populations of animals with subtle differences in behaviour, morphology and/or demography. This study has been a first step towards the development of a better understanding of the comparative dynamics of the populations from each region.

ACKNOWLEDGEMENTS

Many individuals have assisted with the collection of the information used in this study and all the authors are grateful for the help they have received. This study was undertaken partly under the auspices of the SCAR Group of Specialists on Seals, which provided the forum in which this collaborative study was devised. Early drafts of this paper were prepared by H. R. Burton but the final version was written by I. L. Boyd.

REFERENCES

Arnbom, T. A., Fedak, M. A., Boyd, I. L. & McConnell, B. J. 1993. Variation in weaning mass of pups in relation to maternal mass, post weaning fast duration and weaned pup behaviour in southern elephant seals (*Mirounga leonina*) at South Georgia. *Canadian Journal of Zoology,* **71,** 1722–1781.

Arnould, J. P. Y., Boyd, I. L. & Socha, D. G. 1996. Milk consumption and growth efficiency in Antarctic fur seal (*Arctocephalus gazella*) pups. *Canadian Journal of Zoology,* **74,** 254–266.

Bryden, M. M. 1968. Control of growth in two populations of elephant seals. *Nature,* **217,** 1106–1108.

Carrick, R., Csordas, S. E. & Ingham, S. E. 1962. Studies on the southern elephant seal, *Mirounga leonina* (L.). IV. Breeding and development. *CSIRO Wildlife Research,* **7,** 161–197.

Fedak, M. A., Arnbom, T. & Boyd, I. L. 1996. The influence of the mother's size on the transfer of mass, energy and materials to southern elephant seal pups. *Physiological Zoology*, **69,** 887–911.

Guinet, C., Jouventin, P. & Weimerskirch, H. 1992. Population changes, movements of southern elephant seals on Crozet and Kerguelen Archipelagos in the last decades. *Polar Biology*, **12,** No. 3–4, 349–356.

Hindell, M. A. 1991. Some life-history parameters of a declining population of southern elephant seals, *Mirounga leonina*. *Journal of Animal Ecology,* **60,** 119–134.

Hindell, M. A. & Burton, H. 1987. Past and present status of the southern elephant seal (*Mirounga leonina*) at Macquarie Island. *Journal of Zoology, London,* **213,** 365–380.

Hindell, M. A., Slip, D. J. & Burton, H. R. 1994. Possible causes of the decline of southern elephant seal populations in the southern Pacific and southern Indian Oceans. In LeBoeuf, B .J. & Laws, R. M., eds. *Elephant seals: population ecology, behavior and physiology*. Berkeley: University of California Press, 66–84.

Hoelzel, A. R., Halley, J., O'Brien, S. J., Campagna, C., Arnbom, T. R., LeBoeuf, B. J., Ralls, K. & Dover, G. A. 1993. Elephant seal genetic variation and the use of simulation models to investigate historical population bottlenecks. *Journal of Heredity,* **84,** 443–449.

Le Boeuf, B. J., Morris, P. & Reiter, J. 1994. Juvenile survivorship of northern elephant seals. In Le Boeuf, B. J. & Laws, T. M., eds. *Elephant Seals: population ecology, behavior and physiology*. Berkeley: University of California Press, 121–136.

McCann, T .S. 1985 Size, status and demography of southern elephant seal (*Mirounga leonina*) populations. In Ling, J. K. & Bryden, M.M., eds. *Sea mammals of south latitudes. Proceedings Symposium ANZAAS Congress,* **52,** 1–17.

McCann, T. S. & Rothery, P. 1988. Population size and status of the southern elephant seal (*Mirounga leonina*) at South Georgia, 1951–1985. *Polar Biology*, **8,** 153–160.

SCAR. 1991. *Report of the workshop on southern elephant seals, Monterey, California 22–23 May 1991.* Cambridge: Scientific Committee for Antarctic Research, 29 pp.

Wilkinson, I. S. 1992. *Factors affecting reproductive success of southern elephant seals,* Mirounga leonina, *at Marion Island.* PhD thesis, University of Pretoria, 185 pp.

V Human impact and environmental change

INTRODUCTION

Environmental change

The world is changing, as it has done for millennia. The difference now is that the change might be due, at least in part, to human activities rather than being only naturally induced. Global change has become an important, even fashionable, research area with great public and political interest in the resolution of cause and effect.

A significant part of the international effort is devoted to developing and refining models of climate change at a global level (GCM) to provide both predictive ability and tests for sensitivity to mitigating measures. The predictions of these models have been used, along with many other data, by the Intergovernmental Panel on Climate Change (Houghton *et al.* 1995) to assess the likely trends in world climate for the next few decades. All the data and models suggest that there will be further increases in global temperature and changes in patterns and amounts of precipitation, with significant differences between regions. The polar regions are expected to be especially affected by the warming. The separate but linked depletion of stratospheric ozone with the resulting increase in UV is also a matter of major Antarctic concern.

The extent and importance of global or even regional warming is a matter of considerable dispute at present. Karlén (chapter 48) introduces a wide variety of data sources, many non-biological, to illustrate the historical variability of climate in both the temperate and polar regions. Whilst he concludes that the changes so far reported fall within those observed during the last 100 years, he says little about the present rate of change.

Analysis of the historical record may provide useful indications of the way in which the climate system responded in the past. Studies by Berkman (Chapter 49) on Holocene Antarctic macrofossils from raised beaches indicate that there was a cooling of coastal seawater temperatures of nearly 2 °C during the last 6000 years, with a sudden cold interval around 500 years ago. His data are especially interesting in two ways. First, they suggest that there have been significant changes in penguin distribution in the Ross Sea during the Holocene, which may have been induced by changes in sea-ice persistence. Second, his estimated seawater palaeo-temperatures do not agree very closely with those for atmospheric temperatures deduced from ice cores, providing a salutary warning that ecologically important palaeoenvironmental indicators cannot necessarily be deduced from ice-core data without considerable qualification.

On much shorter timescales it has proved difficult to obtain unequivocal evidence for global warming from traditional measurements of meteorological variables. The signal, at a decadal level, is largely lost in the noise. Perhaps this is the more difficult way to prove the case. Smith (1994) has already made a persuasive case for the use of plants as bioindicators of regional warming in the Antarctic Peninsula region, suggesting that they are far more sensitive indicators of climate change than our physical measurements and, by integrating very small improvements over a growing season, can provide early warning of change. Frenot *et al.* (Chapter 50), in examining primary colonization of recently deglaciated areas on Kerguelen, note the connection between the differential survival in the two *Poa* species and the local changes in both temperature and precipitation.

The effects of enhanced UV

The original report by Farman *et al.* (1985) on ozone destruction above the Antarctic has stimulated major research activity both inside and outside the Antarctic. From a biological viewpoint the most important implication of ozone loss is the enhancement in UV levels in spring. A wide variety of effects at both the species and community levels were predicted and gradually we are accumulating evidence from field measurements and experiments to test the hypotheses. Marchant (Chapter 51) reviews the existing information for all groups. He concludes that whilst Antarctic birds and mammals are unlikely to be affected directly, terrestrial cryptogams are stimulated to produce protective pigments to minimize damage from UV-B. In the sea, marine phytoplankton also produce protective pigments but damage still reduces primary productivity in the marginal ice zone by 6–12% and may in turn cause changes in community composition.

In direct contrast to this, Holm-Hansen *et al.* (Chapter 52) conclude that from their experiments with phytoplankton enhanced UV-B will not significantly affect primary production although UV-A does show important inhibitory effects on photosynthesis. In addition, their measurements showed all effects of impact had stopped by 12 m depth, although the euphotic zone extended to 50 m. If this was true perhaps evidence existed in sediment cores to show that the UV had caused no change to pre-existing community composition. McMinn *et al.* (Chapter 53) tested this out on high-resolution cores from fjords at the Vestfold Hills. At least as far as the diatom component of the phytoplankton is concerned they found no evidence of change attributable to enhanced UV, but it is perhaps important to note that this was only with respect to diatoms with silica frustules which may provide important protection against UV.

In examining the response of the terrestrial microbiota to UV Vincent & Quesada (Chapter 54) take a more general approach to assessing vulnerability. Their conclusion that enhanced UV-B is likely to affect community structure, especially during colonization, is based on a model of coupling between genotypic specialization and niche utilization. In concluding that none of the three strategies described is characteristic of the polar zones they reflect what other workers have already documented – that flexibility in performance and stability in genotype are characteristics of colonizing species anywhere.

Human impacts on the environment

The rapid progress towards the implementation of the Protocol for the Protection of the Antarctic Environment has had a number of important effects on Antarctic science. It has changed the organisation of research and logistics by requiring environmental impact assessments of all activities, by increasing the requirements for environmental monitoring and by highlighting the lack of knowledge in some key areas for implementation. For example, there was previously little agreement about how to deal with oil spills, what was the fate of a wide variety of manmade materials when released into the Antarctic environment, how resilient terrestrial communities were to disturbance etc. In addition the rapid development of tourism has raised other issues about potential damage to the Antarctic environment, flora and fauna. These under-worked areas are now attracting an increasing amount of research attention, as this symposium clearly demonstrated.

The establishment of stations on the limited areas of snow-free ground has always been a matter for concern. In competing for this especially scarce resource with the native flora and fauna it is essential that the station activities cause minimum impact on the surrounding environment. Ellis-Evans *et al.* (Chapter 55) have provided a clear case study of the impacts of station water usage on a single lake. Whilst the increased sediment load from vehicle activity in the catchment could have been predicted the changes in the lake communities due to recycling of hot water almost certainly could not. It is unclear from their data if the lake could revert to its original state now that the damage has been done.

Toxic compounds are produced by all stations – in waste water, in vehicle and generator exhausts, and from incinerators. How biologically significant the effects of these compounds are in the local environment has yet to be resolved. One approach is to identify key species and, by monitoring some aspect of performance against uptake of a key compound, determine the potential impact at the species and community level. Focardi *et al.* (Chapter 56) have assessed the value of using a key physiological parameter – in this case the mixed function oxidase enzyme – as an index of pollution. Nigro *et al.* (Chapter 57) on the other hand have used a key group – molluscs – to assess the potential for accumulation of heavy metals in the benthic ecosystem. Their data provide an object lesson in the possible pitfalls – natural sources of the metals rather than human pollution, identification of seasonal patterns in availability as well as in effects, determination of accumulation sites within organisms. One major drawback of monitoring in this way is that the animals have to be killed for the analysis, thus destroying any possibility of characterising individual response with time. The technique using preen gland oil or blood from birds described by van den Brink (Chapter 58) goes a long way towards meeting this objection as well as providing a more ethically acceptable protocol for dealing with larger animals.

Since the Agreed Measures for the Conservation of the Antarctic Flora and Fauna were accepted at the Antarctic Treaty Consultative Meeting in 1964 there has been continuing concern about the potential biological disaster ensuing from the introduction of alien species to the Antarctic. Whilst this concern had also been shared for the sub-Antarctic islands (Walton 1986) it came a little late for islands which already had a significant range of deliberate and accidental introductions (Dingwall 1995). Whilst there already exist a number of papers on the effects of grazing by introduced herbivores on the sub-Antarctic islands, effects on the native species of other introductions have been little studied. Gremmen (Chapter 59) utilizes surveys spread over 29 years to assess how aggressive alien plant species have proved to be on Marion Island. Success appears to be linked to seed production and, in at least some of the nine plant associations in which aliens reach dominance, there is a consequent reduction in the number of native plant species present. He suggests that strict measures are essential to avoid further introductions, a point emphasized by Chevrier *et al.* (Chapter 60) in their paper on alien insects on Kerguelen.

It would seem important to be able to predict the potential impacts of an introduction on the native communities. Chevrier *et al.* illustrate how essential it is to understand the key life-cycle features of the principal species involved and how their activities interact at a community level. In this instance it appears that the decline of the native fly *Anatalanta aptera* is due to increased predation by the introduced beetle *Oopterus soledadinus* when resource competition from the introduced fly *Calliphora vicina* forces *Anatalanta* larvae out of its preferred niche. It seems unlikely that this interaction could have been predicted!

More obvious perhaps was the effect of increasing populations of fur seals on Antarctic vegetation. In an earlier paper

Smith (1988) had described the destruction of the cryptogamic communities at Signy Island by the seals. In this present paper Smith (Chapter 61) reports on the potential for re-establishing the original communities by excluding the seals with fencing. He concludes that the prospects are poor, because of both substrate removal and the toxic levels of nutrients from faeces.

Finally, the problems of direct impact from human activities are a subject of increasing concern. The Protocol requires a high level of preparedness to cope with hydrocarbon spills, a disaster illustrated by the sinking of *Bahia Paraiso*. In providing a case history of this event Penhale *et al.* (Chapter 62) have underlined the value of international assistance in the true spirit of the Antarctic Treaty and highlighted the need for a better approach to oil spill contingency planning. More information on nutrient pathways would also help in deciding how, where and when the principal effects of an oil spill will be visible.

With increasing numbers of tourists visiting the Antarctic there has been a great deal of speculation on the impacts that they are creating and many demands for increased legislation to control tourism. Little hard evidence has been available to support any of the more extreme effects predicted but equally the negative – no effects – could also not be proved. In their paper on penguins and tourist disturbance Fraser & Paterson (Chapter 63) provide the most extensive data yet and conclude that adverse effects due to tourism or research are well within the effects imposed by natural environmental change. Whilst this will not necessarily apply to all sites and certainly not to all birds it does at least indicate that proving adverse effects unequivocally may well prove an almost impossible task.

REFERENCES

Dingwall, P., ed. 1995. *Progress in conservation of the subantarctic islands. Proceedings of the SCAR/IUCN Workshop on protection, research and management of the subantarctic islands, Paimpont, France, 27–29 April 1992*. Gland: IUCN. 225 pp.

Farman, J. C., Gardiner, B. C. & Shanklin, J. D. 1985. Large losses of total ozone in Antarctica reveal seasonal ClO_x/NO_x interaction. *Nature*, **315**, 207–210.

Houghton, J. T., Meira Filho, L. G., Callander, B. A. Harris, N., Kattenberg, A. & Markell, K. , eds. 1995. *Climate change 1995. The science of climate change*. Cambridge: Cambridge University Press. 572 pp.

Smith, R. I. L. 1988. Destruction of Antarctic terrestrial ecosystems by a rapidly increasing fur seal population. *Biological Conservation*, **45**, 55–72.

Smith, R. I. L. 1994. Vascular plants as bioindicators of regional warming in Antarctica. *Oecologia*, **99**, 322–328.

Walton, D. W. H., rapporteur. 1986. *The biological basis for the conservation of the subantarctic islands*. Cambridge:SCAR/IUCN. 37 pp.

48 Climatic change and the recent climatic record

WIBJÖRN KARLÉN
Department of Physical Geography, Stockholm University, 106 91 Stockholm, Sweden

ABSTRACT

Information about climatic change available from studies of icecores from Greenland and Antarctica, fluctuations of alpine glaciers, changes in the treeline altitude and variations in tree-ring width are reviewed. Large-amplitude, high-frequency temperature fluctuations in Greenland (7 °C shifts over a period of several decades) during the last ice age and also possibly during the previous warm interval around 130 000 years BP have been recognized in both ice cores from Greenland and peat deposits in central Europe. Variations in tree-ring width indicate that temperature fluctuations of about 1.5 °C have occurred during periods as short as a few decades during the last 1000 years. The paleoclimatic record shows that changes in temperature during the last 100 years are not unique and are well within the limits of natural variability. Changes in the climate lasting several hundred years appear to be recognized over most of the globe, while fluctuations lasting decades may be local.

Key words: Antarctica, climatic change, Holocene, temperature records.

INTRODUCTION

Much research has been concentrated on studying climate and the effect climate has on ecosystems (Houghton *et al.* 1990). If the anthropogenic release of greenhouse gases is continued at the same rate as at present, and if current predictions are correct, global temperature may increase to such an extent that it will cause environmental problems within a century (Swedish Environmental Protection Agency 1992). Present predictions indicate that climatic trends could be rapid in comparison with changes during the last 1000 years; it has been suggested that many species may not be able to adapt or migrate fast enough to survive (Swedish Environmental Protection Agency 1992).

Climate has always fluctuated. Major changes in climate, such as the ice ages, have changed vegetational distribution. Certain species have become extinct when migration to suitable localities has not been possible. However, many species presently alive have survived dramatic changes in climate on a timescale represented by the ice ages as well as on one represented by decades.

PRE-HOLOCENE CLIMATIC FLUCTUATIONS

Much information about the climate during the last 250 000 years has been obtained from ice cores retrieved from the Greenland and Antarctic ice sheets. The isotopic composition of the snow that forms these ice sheets yields information about temperature. Analysis of the air trapped between the ice crystals indicates the composition of the atmosphere some time after the deposition of the snow. Impurities in the snow give information about volcanic eruptions, atmospheric dust content, changes in solar irradiation and, possibly, also about the algae in surrounding seas (Lorius 1988).

Cores from both Greenland and Antarctica show that after a short warm interval (Eemian, 130 000 y BP), when the temperature was about 2 °C warmer than at present, the climate cooled down. Ice-age conditions were reached about 110 000 yr BP and marked by a series of partial climatic recoveries until the last glacial maximum *c.*20 000 years ago (Alley *et al.* 1993). In Antarctica the climate fluctuated only slightly during this period of about 100 000 y, but the Greenland record shows that large-amplitude, high-frequency fluctuations were superimposed on low-frequency trends (Johnsen *et al.* 1992, Jouzel *et al.* 1993,

Dansgaard *et al.* 1993, GRIP members 1993). Although the temperature during the warmer phases of the last glaciation was about 7 °C warmer (in Greenland) than that of the cold periods, it was still 5 °C colder than the present temperature. The mild periods, which lasted 500–2000 years, started with a rapid temperature increase, with the maximum temperature being reached within a few decades, possibly even less than a decade for the most recent warming at the end of the Younger Dryas. Afterwards the climate ramped more gradually back to cold conditions (Stauffer 1993, Lehman 1993).

Dramatic changes from a warm to cold climate and vice versa have also been observed in several other records. Data from the Camp Century ice core shows distinct variations (Dansgaard *et al.* 1982). In France, Woillard & Mook (1982) studied the pollen assemblage in a peatbog at Grand Pile, NE France, where sediments had accumulated from before the Eemian. They found that forest vegetation had been replaced by tundra vegetation on repeated occasions at the end of the Eemian and early Weichselian (Weichsel = the name used for the last glaciation in northern Europe). The number of rapid climatic changes observed is smaller than in the Greenland ice cores. This is not surprising since the slow rate of accumulation at Grand Pile makes it much more difficult to find evidence of short-term events. Recently the pattern of climatic change found at Grand Pile has been confirmed; similar dramatic changes in the climate have been recognized at a second locality in France (Guiot *et al.* 1989, Thouveny *et al.* 1994). However, the vegetational changes in southwestern North America and in Europe may not be directly correlated with the high-frequency climatic changes known from the Greenland ice sheet.

HOLOCENE

The climate of the Holocene appears to have been relatively stable, when compared with that of the previous hundreds of thousands of years for which detailed records are available. However, even during this period changes in climate have occurred. These changes are best known from studies of glacier fluctuations, changes in the altitude of the tree limit, dendrochronology, ice-cores and palynology.

The warming at the end of the last glaciation was abrupt. The warming in Greenland after the cold event called 'Younger Dryas', which occurred around 11 600 y BP (based on the counting of annual layers in ice cores), may have been over in 50 years (Johnsen *et al.* 1992, Alley *et al.* 1993). The authors conclude that the temperature increased by about 7 °C and the snow accumulation may have doubled in a period as short as two or three years.

The record from Summit, central Greenland (Dansgaard *et al.* 1993), does not indicate any long-term (century–millenium) change in temperature during the Holocene. However, records from Camp Century (Dansgaard *et al.* 1984) and Devon Island (Paterson *et al.* 1977) show a slight tendency for a cooler climate during early and late Holocene (before and after the climatic optimum). This pattern is also known from a number of other areas, such as the Caribbean (Hodell *et al.* 1991) and Scandinavia. While some Norwegian studies indicate a distinct period between 8000 and 5000 years BP that was warm (Nesje & Kvamme 1991), studies from northern Scandinavia show only a slight Holocene warming during the same period (Karlén 1982, 1991, 1993).

Evidence of Holocene temperature fluctuations that lasted several hundreds of years is known from a number of areas with alpine glaciers (Röthlisberger 1986, Grove 1988). Glacier advances and tree-limit variations in the Alps and in northern Sweden show that short cold events also occurred during the period often referred to as the Hypsithermal, which is supposed to have been a warm period. During the early Holocene the cold periods appear to have been shorter and less cold than during the last 3000 years. An Antarctic ice core covering the last 9000 years shows fluctuations in temperature very similar to the climatic changes dated in Scandinavia (Kameda *et al.* 1990).

Röthlisberger (1986) has concluded that most major shifts in climate have occurred at approximately the same time in the majority of areas where glacier fluctuations have been studied, whereas Grove has maintained that only the last climatic change, the so-called 'Little Ice Age', has been so dated; however, at present it seems unlikely to have been a globally simultaneous change. Wigley (1988) has discussed the problem and believes that major climatic changes occurred in both hemispheres simultaneously.

Probably, the best series of dates on glacier advances in the Antarctic region has been obtained from ^{14}C dating of moss beds preserved at Signy Island (Smith 1990). These dates indicate glacier expansions around AD 500, 900, the mid-1200s, mid-1300s, late 1400s, late 1500s, late 1700s and mid-1800s. These dates are remarkably similar to dates on cold events dated in Scandinavia (Briffa *et al.* 1992, Karlén 1993). Less precise dates obtained on lacustrine sediments, as well as a ^{14}C date of a moraine and lichenometric data from the area around Hope Bay on the Antarctic Peninsula (Zale & Karlén 1989), all indicate glacier expansions around AD 650 (maximum date) and after AD 1150. Using the estimate of the lichen growth rate on Signy Island of 5 mm per century (Smith 1990), measurements of lichens from Hope Bay indicate glacier expansions around AD 800, 1100 and 1300. The margin of error of these estimated ages is large. In addition, on the South Shetland Islands Sugden & Clapperton (1986) have dated a glacier advance to between AD 1250–1550 and Rabassa (1983) has dated a moraine on James Ross Island to around AD 1250. Even considering the large margin of error, these dates indicate some event like 'The Little Ice Age' also in the Antarctic region.

THE LITTLE ICE AGE

The last few hundred years have sometimes been called 'The Little Ice Age' because glacier advances are known to have been extensive at this time in areas such as the European Alps, Iceland

and Scandinavia. Further studies have revealed that in many areas local moraines fronting glaciers were most likely formed at approximately the same time.

Ice-core data from Greenland and Devon Island show that the climate became cooler around AD 1300 (Dansgaard 1975, Paterson *et al.* 1977). Ice-core data from Qualqua, Peru (Thompson *et al.* 1986, Thompson 1992), as well as several long tree-ring records (LaMarche 1978), support the glacier record. A number of palaeotemperature records from various parts of the world and covering the last 500–2000 years have been summarized by Morgan (1985) and show a cool period beginning between AD 1100 and AD 1400. The Little Ice Age cold period apparently begins a few hundred years earlier in some regions of the world than in Europe.

Tree-ring width is closely correlated with climate (Fritts 1976). In areas close to the northern boreal, as well as the alpine tree limit, summer temperature is frequently the most important factor determining the width of the tree rings, while precipitation is the limiting factor in dry areas. Temperature reconstructions based on tree-ring width usually yield reliable information about annual and short-term fluctuations (variability in the climate). However, long-term changes may be less easily detected because of the statistical method used for smoothing the chronologies. The dendrochronological technique permits dating with good precision. The error is frequently reported to be less than a few percent. A large number of dendroclimatological records covering the last few hundred years have been published. In addition, a few records are available covering the period before which temperature observations were made. Dendrochronologies for the southern hemisphere are rare, but long chronologies for southern Argentina, New Zealand and Tasmania has been published (Cook *et al.* 1991, Bradley & Jones 1993).

A temperature reconstruction covering the last 1500 years has been published for northern Fennoscandia (Briffa *et al.* 1992). This record shows that the temperature was low in AD 500–700, 800–900, around 1150, 1200–1400, 1580–1720 and during the last decades of the 1800s. Particularly warm periods occurred around AD 900–1100 and 1400–1500. During cold periods the 10-year-average summer temperature often was 0.5 °C below the present temperature. During warm periods it was up to 0.7 °C above the present temperature. Temperature changes of up to 1.5 °C have occurred over periods as short as a few decades. Dendrochronologies from Argentina (37° S–39° S) show that the summer temperature was low during the 1600s, late 1700s and around 1850 (Bradley & Jones 1993). A distinct, short cold period during early 1900s is documented for Argentina and Tasmania as well as being known from many other records.

TEMPERATURE OBSERVATIONS

The first temperature observations were initiated in England in 1659 (Manley 1974), and a few more observation series were begun around 1750. The methods of observation varied in the beginning, and the number of stations was small until the mid 1800s.

Continental observations show a temperature increase of 0.3–0.6 °C after AD 1850 (Houghton *et al.* 1990). Before corrections are added, the maritime observations show that the temperature over the oceans was relatively warm until about AD 1900 (Oort *et al.* 1987). After a period of low temperature from 1900–1940, the temperature has increased. Since the 1950s the air temperature over the oceans has been approximately as high as it was during the late 1800s. The maritime temperatures have been corrected (Bottomley *et al.* 1990) for an assumed error and the marine temperature used for calculating the global mean temperature differ, after this correction, considerably from the observations but are similar to the continental temperature.

Even if the temperature records are short they are important for our understanding of regional and temporal variations in temperature. Only for small regions (e.g. a country) are seasonal and annual temperatures well correlated. However, the mean temperature over 10–30 years seems to be similar for much of the area of a continent. Note that the temperature increase of the 1920s and 1930s was not worldwide. For example, changes in air circulation and changes in the radiation balance caused a temperature decrease in central Siberia and parts of Antarctica, at the same time as the temperature increased in most of the rest of the world. Available records indicate that the temperature increase in the 1920s and 1930s was more distinct in the Arctic regions than in the middle and low latitudes. In the region around the South Orkney Islands (the only Antarctic area with an adequate record) the temperature decreased by about *c.* 0.5 °C during this period (Pittock 1978).

The temperature around the Antarctic Peninsula appears to be out-of-phase with the global temperature. Observations from the South Orkney Islands (Smith 1990) and South Georgia (Schwerdtfeger 1984) (Fig. 48.1), island stations with long records, indicate low temperature in the 1930s, when the 'global mean temperature' was high. A temperature maximum was reached on these islands first in the 1950s, e.g. 20 years later than the temperature rose in much of the world. Also, isotope studies of ice cores from Filchner-Ronne Ice Shelf and Siple Station indicate that the temperature in the region of the Antarctic Peninsula is out of phase with the typical global trend (Mosley-Thompson 1992, p. 580). It appears that this is a result of large-scale changes in mid-latitude circulation; when zonal westerlies are strong, temperatures are anomalously warm in the Peninsula area (e.g. higher at Faraday and Bellinghausen than at Halley and Orcadas in the 1970s; Mosley-Thompson 1992, p. 584, Fig. 29.4). Data from New Zealand shows that the temperature increase here also came first in the 1950s and not in the 1930s (Salinger & Gunn 1975).

There is evidence for a distinct recent warming of the Antarctic climate. Summer temperature observations at Laurie Island and Signy Island, South Orkney Islands, show an almost continuous increase in mean summer temperature since about

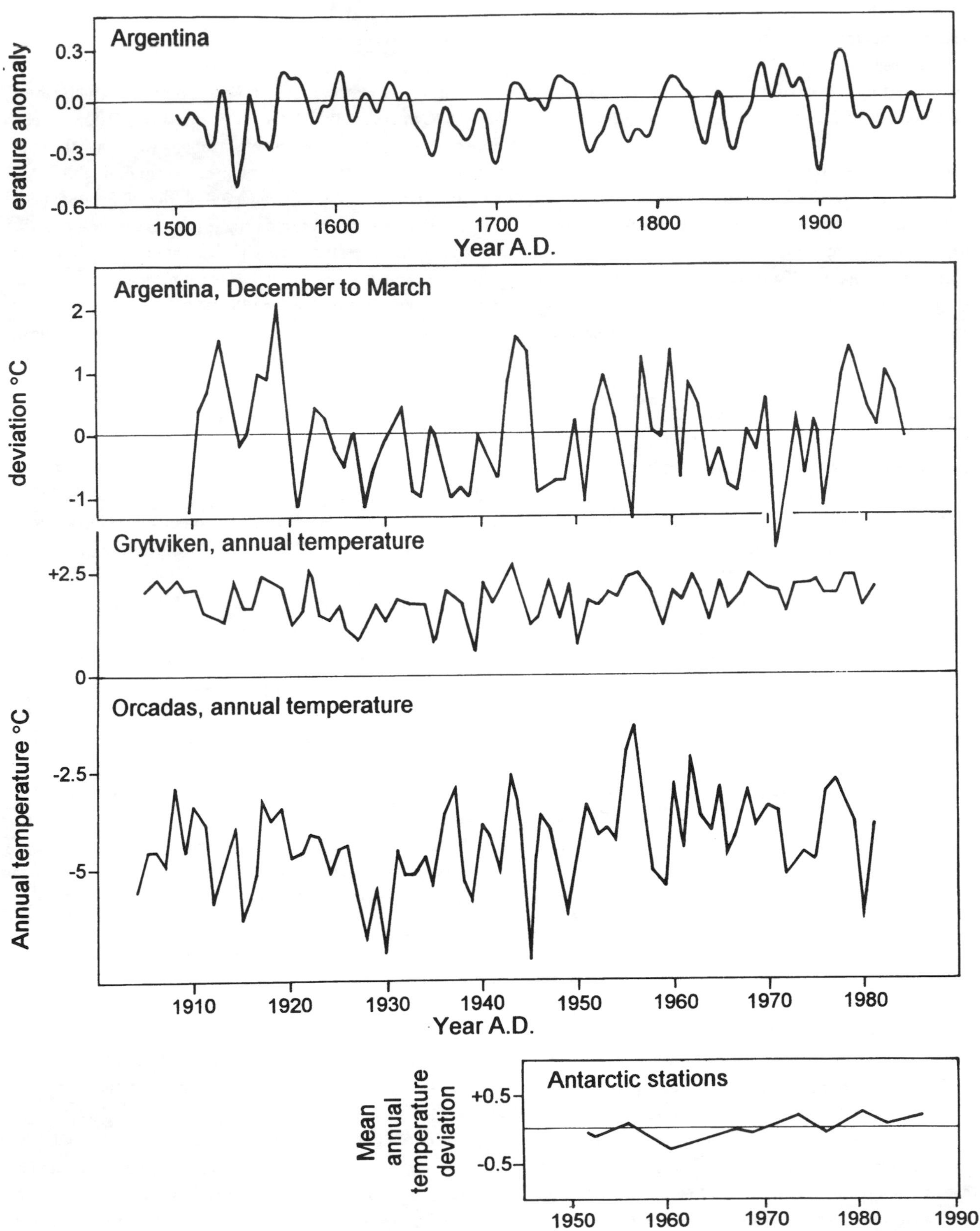

Fig. 48.1. Reconstructed summer temperature anomalies for South America (37–39 °S) based on dendrochronology (redrawn from Bradley & Jones 1993), South American summer (December to March) temperature departure computed as the regional average of Collunco, Bariloche, Mascardi Esquel and Sarmiento (redrawn from Lara & Villalba 1993); annual temperatures at Orcades and Grytviken (redrawn from Schwerdtfeger 1984); an Antarctic temperature compilation of mean annual temperatures from the Antarctic stations with long, continuous records as well as from Macquarie Island. The Antarctic data has been smoothed by a gaussian filter (bandwidth 3 years). The scale of the original figure (Morgan *et al.* 1991) has been recalculated to be similar to scales used for Grytviken and Orcadas. In South America, South Georgia (Grytviken) and the South Shetland Islands (Orcadas) the temperature was relatively low during the 1930s.

1950 (Smith 1990). The same trend is apparent in a compilation of mean annual temperatures from Antarctic stations with long, continuous records and from Maquarie Island (where there is information on the mean temperature since 1950) (Morgan *et al.* 1991). However, this recent trend is not seen in annual temperature records from the South Orkney Islands and South Georgia, where observations started around 1905 or Ushuaya (Schwerdtfeger 1984). Neither do ice core data from Dronning Maud Land (130 year long record) show a distinct recent trend (Isaksson *et al.*, submitted).

Although temperature records were made at a few stations even earlier than the mid-1800s, the number and geographical distribution of stations is such that it is not possible to calculate a global mean temperature from them; information is not available for large areas. Information from several of these long temperature records was recently published by Bradley & Jones (1992). These records, which are mostly from northern Europe and eastern USA, show that the late 1800s was a very cold period (Bradley & Jones 1992, fig. 13.1c). For example, the June–August, 5-year mean temperature for Stockholm was 2.1 °C colder in the 1860s than in the 1930s (Liljeqvist 1949). Even before the late 1800s the temperature fluctuated around an average that appears to have been lower than the temperature has been since about the 1930s. A satellite study of the global temperature from 1979–1993 shows no trend for this short period (Christy & McNider 1994), a period during which the temperature increase has been rapid, according to the IPCC report (Houghton *et al.* 1990, p. xxix).

PREDICTION OF FUTURE TEMPERATURE

Predictions concerning the climate in the future are based on a calculated atmospheric response to a predicted increase in the atmospheric concentration of greenhouse gases such as CO_2. Although several different types of gases actually are important, the effect of these gases is frequently calculated as CO_2 equivalents and their combined effect is used in calculations. Commonly, the temperature increase at a doubling of the concentration of these gases ($2\times CO_2$) is calculated and the temperature increase at this greenhouse-gas concentration is called 'sensitivity'. Most studies indicate a sensitivity of between 1.5 and 4.5 °C (Houghton *et al.* 1990).

Climatic changes occurred even before anthropogenic CO_2 was released. It has been proposed that variations in the content of volcanic dust and aerosols in the atmosphere (Hammer *et al.* 1980, Bradley 1988, Minnis *et al.* 1993), or variations in solar irradiance (Denton & Karlén 1973, Friis-Christensen & Larssen 1991, Mayewski *et al.* 1993, Zielinski *et al.* 1994), have caused these fluctuations, but no accepted theory has up till now been presented. It is possible that the warming that took place during the 1920s and 1930s, a period with a relatively limited release of greenhouse gases, was to some extent caused by natural processes (Balling 1993).

Recently Kelly & Wigley (1992) tried to model the temperature increase of the last 100 years, using varying greenhouse sensitivity indexes and a model for solar irradiance variations based on changes in the sunspot cycle length (Friis-Christensen & Larssen 1991, Balling 1993). The best correlation with the observed temperature was found for a model based only on variations in solar irradiance. However, this result was disregarded because the authors considered the greenhouse effect too well-established to be omitted from their models.

Prognoses are based on equilibrium models, which commonly do not consider the dampening effects caused by direct backscattering and by an increased cloud albedo owing to the anthropogenic release of SO_2 (Houghton *et al.* 1992, p. 137; compare p. 118). These dampening effects are calculated to be of the same order of magnitude as the total greenhouse effect (Kiehl & Briegleb 1993, Pearce 1994, Taylor & Penner 1994).

ACKNOWLEDGEMENTS

Dr D. W. H. Walton, British Antarctic Survey, has been very helpful and his comments have greatly improved the manuscript.

REFERENCES

Alley, R. B., Meese, D. A., Shuman, C. A., Gow, A. J., Taylor, K. C., Grootes, P. M., White, J. W. C., Ram, M., Waddington, P. A., Mayevski, P. A. & Zielinski, G. A. 1993. Abrupt increase in Greenland snow accumulation at the end of the Younger Dryas event. *Nature*, **362,** 527–29.

Balling, Jr, R. C. 1993. The global temperature data. *National Geographic, Research & Exploration*, spring 1993, 201–207.

Bottomley, M., Folland, C. K., Husing, J., Newell, R. E. & Parker, D. E. 1990. *Global ocean surface temperature atlas*. Meteorological Office and Massachusetts Institute of Technology, 20 pp+313 plates.

Bradley, R. 1988. The explosive volcanic eruption signal in northern hemisphere continental temperature records. *Climatic Change*, **12,** 221–243.

Bradley, R. S. & Jones, P. D. 1992. Climatic variations in the longest instrumental records. In Bradley, R. S. & Jones, P. D., eds. *Climate since A.D. 1500.* London: Routledge, 246–268.

Bradley, R. S. & Jones, P. D. 1993. 'Little Ice Age' summer temperature variations: their nature and relevance to recent global warming trends. *The Holocene*, **3,** 367–376.

Briffa, K. R., Jones, P. D., Bartholin, T. S., Eckstein, D., Schweingruber, F. H., Karlén, W., Zetterberg, P. & Eronen, M. 1992. Fennoscandian summers from A. D. 500: Temperature changes on short and long timescales. *Climatic Dynamics*, **7,** 111–119.

Christy, J. R. & McNider, R. T. 1994. Satellite greenhouse signal. *Nature*, **367,** 325.

Cook, E., Bird, T., Peterson, M., Barbethi, M., Buckley, B., D'Arrigo, R., Francey, R. & Tans, P. 1991. Climatic change in Tasmania inferred from a 1089-year tree-ring chronology of Huron pine. *Science*, **253,** 1266–1268.

Dansgaard, W. 1975. *Klimatsvängningar och människoöden*. Stockholm: Ymer 1975, 197–215.

Dansgaard, W., Clausen, H. B., Gundestrup, N., Hammer, C. U., Johnsen, S. F., Kristinsdottir, P. M. & Reeh, N. 1982. A new Greenland deep ice core. *Science*, **218**, 1273–1277.

Dansgaard, W., Johnsen, S. J., Clausen, H. B., Dahl-Jensen, D., Gundestrup, N., Hammer, C. U. & Oeschger, H. 1984. North Atlantic climatic oscillations revealed by Greenland ice cores. *Climate Processes and Climate Sensitivity*, Geophysical Monograph 29, Maurice Ewing Volume 5, 288–298.

Dansgaard, W., Johnsen, S. J., Clausen, H. B., Dahl-Jensen, D., Gundestrup, N. S., Hammer, C. U., Hvidberg, C. S., Steffensen, J. P., Svelnbjörnsdottir, A. E., Jouzel, J. & Bond, G. 1993. Evidence for general instability of past climate from a 250-kyr ice-core record. *Nature*, **364,** 218–220.

Denton, G. H. & Karlén, W. 1973. Holocene climatic variations – their pattern and possible cause. *Quaternary Research*, **3**, 155–205.

Friis-Christensen, E. & Larssen, K. 1991. Length of the solar cycle: an indicator of solar activity closely associated with climate. *Science*, **245**, 698–700.

Fritts, H. C. 1976. *Tree rings and climate*. New York: Academic Press, 567 pp.

GRIP members (Greenland Ice-core Project). 1993. Climate instability during the last interglacial period recorded in the GRIP ice core. *Nature*, **364**, 203–207.

Grove, J. M. 1988. *The Little Ice Age*. London: Methuen, 498 pp.

Guiot, J., Pons, A., de Beaulieu, J. L. & Reille, M. 1989. A 140,000-year continental climate reconstruction from two European pollen records. *Nature*, **338**, 309–313.

Hammer, C. U., Clausen, H. B. & Dansgaard, W. 1980. Greenland ice sheet evidence of post-glacial volcanism and its climatic impact. *Nature*, **288**, 230–235.

Hodell, D. A., Curtis, J. H., Jones, G. A., Higuera-Gundy, A., Brenner, M., Binford, M. W. & Dorsey, K. T. 1991. Reconstruction of Caribbean climate change over the past 10,500 years. *Nature*, **352**, 790–793.

Houghton, J. T., Jenkins, G. J. & Ephraums, J. J. 1990. *Climatic change. The IPCC scientific assessment.* World Meteorological Organization/United Nations Environment Programme. Cambridge: Cambridge University Press, 365 pp.

Houghton, J. T., Callander, B. A. & Varney, S. K. 1992. *Climatic change. The supplementary report to the IPCC scientific assessment*. World Meteorological Organization/United Nations Environment Programme. Cambridge: Cambridge University Press, 200 pp.

Johnsen, S. J., Clausen, H. B., Dansgaard, W., Fuhrer, K., Gunnestrup, N., Hammer, C. U., Iversen, P., Jouzel, J., Stauffer, B. & Steffensen, J.P. 1992. Irregular glacial interstadials recorded in a new Greenland ice core. *Nature*, **359**, 311–313.

Jouzel, J., Barkov, N. I., Barnola, J. M., Bender, M., Chappellaz, J., Genthon, C., Kotlyakov, V. M., Lipenkov, V., Lorius, C., Petit, J. R., Raynaud, D., Raisbeck, G., Ritz, C., Sowers, T., Stievenard, M., Yiou, F. & Yiou, P. 1993. Extending the Vostok ice-core record of palaeoclimate to the penultimate glacial period. *Nature*, **364**, 407–412.

Kameda, T., Nakawo, M., Mae, S., Watanabe, O. & Naruse, R. 1990. Thinning of the ice sheet estimated from total content of ice cores in Mizuho Plateau, East Antarctica. *Journal of Glaciology*, 131–135.

Karlén, W. 1982. Holocene glacier fluctuations in Scandinavia. *Strie*, 18, 26–34.

Karlén, W. 1991. Glacier fluctuations in Scandinavia during the last 9000 years. In Starkel, L., Gregory, K. J. & Thornes, J. B., eds. *Temperate palaeohydrology*. London: John Wiley, 395–412.

Karlén, W. 1993. Glaciological, sedimentological and paleobotanical data indicating Holocene climatic change in Northern Fennoscandia. In Frenzel, B., ed. *Oscillations of alpine and polar tree limits in the Holocene*. Gustav Fischer Verlag, 69–83.

Kelly, P. M. & Wigley, T. M. L. 1992. Solar cycle length, greenhouse forcing and global climate. *Nature*, **360**, 328–330.

LaMarche, Jr, V. C. 1978. Tree-ring evidence of past climatic variability. *Nature*, **276**, 334–338.

Lehman, S. 1993. Ice sheets, wayward winds and sea change. *Nature*, **365**, 108–109.

Liljeqvist, G. 1949. On fluctuations of the summer mean temperature in Sweden. *Geografiska Annaler*, 1949(1-2), 157–178.

Lorius, C. 1988. Polar ice cores: A record of climatic and environmental changes. In Bradley, ed. *Global changes of the past*. Boulder, CO: UCAR/Office for Interdiciplinary Earth Studies, 261–294.

Manley, G. 1974. Central England temperatures: monthly means 1659 to 1973. *Quaternary Journal of Royal Meteorological Society*, **100**, 389–405.

Mayewski, P. A., Meeker, L. D., Whitlow, S., Twickler, M. S., Morrison, M. C., Alley, R. B., Bloomfield, P. & Taylor, K. 1993. The atmosphere during the Younger Dryas. *Science*, **261**, 195–197.

Minnis, P., Harrison, E. F., Stowe, L. L., Gibson, G. G., Denn, F. M., Doelling, D. R. & Smith Jr, W. L. 1993. Radiative climate forcing by the Mount Pinatubo eruption. *Science*, **259**, 1411–1415.

Morgan, V. I. 1985. An oxygen isotope – climate record from the Law Dome, Antarctica. *Climatic Change*, **7**, 415–426.

Morgan, V. I., Goodwin, I. D., Etheridge, D. M. & Wookey, C. W. 1991. Evidence from Antarctic ice cores for recent increases in snow accumulation. *Nature*, **354**, 58–60.

Mosley-Thompson, E. 1992. Paleoenvironmental conditions in Antarctica since A.D. 1500: ice core evidence. In Bradley, R. S. & Jones, P. D., eds. *Climate since A.D. 1500*. London: Routledge, 572–591.

Nesje, A. & Kvamme, M. 1991. Holocene glacier and climate variations in western Norway: Evidence for early Holocene glacier demise and multiple Neoglacial events. *Geology*, **19**, 610–612.

Oort, A. H., Pan, Y. H., Reynolds, R. W. & Ropelewski, C. F. 1987. Historical trends in the surface temperature over the oceans based on the CODAS. *Climate Dynamics*, **2**, 9–38.

Paterson, W. S. B., Koener, R. M., Fisher, D., Johnson, S. J., Clausen, H. B., Dansgaard, W., Bucher, P. & Oeschger, H. 1977. An oxygen-isotope climatic record from Devon Island ice cap, Arctic Canada. *Nature*, **266**, 508–511.

Pearce, F. 1994. Not warming but cooling. *New Scientist*, July 1994, 37–41.

Pittock, A. B. 1978. A critical look at long-term sun–weather relationships. *Reviews of Geophysics and Space Physics*, **16**, 400–420.

Rabassa, J. 1983. Stratigraphy of the glacienic deposits in Northern James Ross Island, Antarctic Peninsula. In Evenson, E., Schluchter, C. & Rabassa, J., eds. *Tills and related deposits*. Rotterdam: Balkema, 329–340.

Röthlishberger, F. 1986. *10,000 Jahre Gletschergeschichte der Erde*. Aarau: Verlag Sauerländer, 416 pp.

Salinger, M. J. & Gunn, J. M. 1975. Recent climatic warming around New Zealand. *Nature*, **256**, 396–398.

Schwerdtfeger, W. 1984. *Weather and climate of the Antarctic*. Amsterdam: Elsevier, 261 pp.

Smith, R. I. L. 1990. Signy Island as a paradigm of biological and environmental change in Antarctic terrestrial ecosystems. In Kerry, K. R. & Hempel, G., eds. *Antarctic ecosystems, ecological change and conservation*. Berlin: Springer-Verlag, 32–50.

Stauffer, B. 1993. The Greenland ice core. *Science*, **260**, 1766–1767.

Sugden, D. E. & Clapperton, C. M. 1986. Glacial history of the Antarctic Peninsula and South Georgia. *South African Journal of Science*, **82**, 508–509.

Swedish Environmental Protection Agency. 1992. Åtgärder mot klimatförändringar. *Naturvårdsverket Rapport 4120. Nordstedts tryckeri AB*, 1–292.

Taylor, K. E. & Penner, J. E. 1994. Response of the climate system to atmospheric aerosols and greenhouse gases. *Nature*, **369**, 734–737.

Thompson, L. G. 1992. Ice core evidence from Peru and China. In Bradley, R. S. & Jones, P. D., eds. *Climate since A.D. 1500*. London: Routledge, 517–548.

Thompson, L. G., Mosley-Thompson, E., Dansgaard, W. & Grootes, P. M. 1986. The Little Ice Age as recordes in the stratigraphy of the Tropical Quelccaya Ice Cap. *Science*, **234**, 361–364.

Thouveny, N., de Beaulieu, J-L., Bonifay, E., Creer, K. M., Guiot, J., Icole, M., Johnsen, S., Jouzel, J., Reille, M., Williams, T. & Williamson, D. 1994. Climatic variations in Europe over the past 140 kyr deduced from rock magnetism. *Nature*, **371**, 503–506.

Wigley, T. M. L. 1988. Climate variability on the 10–100 year time scale: observations and possible causes. In Bradley, ed. *Global changes of the past*. Boulder, CO: UCAR/Office for Interdiciplinary Earth Studies, 83–101.

Woillard, G. M. & Mook, W. G. 1982. Carbon-14 dates at Grand Pile: correlation of land and sea chronologies. *Science*, **215**, 159–161.

Zale, R. & Karlén, W. 1989. Lake sediment cores from the Antarctic Peninsula and surrounding islands. *Geografiska Annaler*, **71A**, 211–220.

Zielinski, G. A., Mayewski, P. A., Meeker, L. D., Whitlow, S., Twickler, M. S., Morrison, M., Meese, D. A., Gow, A. J. & Alley, R. B. 1994. Record of volcanism since 7000 B.C. from the GISP2 Greenland ice core and implications for the volcano-climate system. *Science*, **264**, 948–951.

49 Ecological variability in Antarctic coastal environments: past and present

PAUL ARTHUR BERKMAN
Byrd Polar Research Center, Ohio State University, Columbus, Ohio 43210, USA

ABSTRACT

Holocene macrofossils in emerged beaches can be used to interpret environmental variability in the coastal zone around Antarctica which has been directly impacted by ice-sheet advance and retreat during the last 10 000 years. Radiocarbon dating was used to determine fossil ages based on reservoir corrections derived from the average pre-bomb ^{14}C-ages of different Antarctic species: seals (1424±200 years), penguins (1130±134 years) and molluscs (1300±100 years). In the modern marine environment, the oxygen isotope composition of mollusc shells varies seasonally and across nearshore depth gradients associated with glacial meltwater input. In the fossil record, scallop shells in emerged beaches along the Victoria Land coast have oxygen isotope ratios that reflect a relatively long warm period around 6000 years before present (y BP) and a brief cold period 500 y BP which may be associated with the Little Ice Age. The persistence of scallop and Adélie penguin populations along the Victoria Land coast since the middle Holocene, as well as their localized migrations during this period, can be attributed unambiguously to natural phenomena in the absence of human activities. Viewed in a circumpolar perspective, these long-term dynamics of coastal populations can be used to interpret the transitory nature of anthropogenic impacts in the Antarctic marine ecosystem. Coordinated among nations, this interdisciplinary research in Antarctic coastal areas will enhance our understanding of the Earth's climate system.

Key words: climate, Holocene, marine, meltwater, oxygen isotope, radiocarbon.

INTRODUCTION

The purpose of this paper is to identify long-term datasets that can be developed to distinguish natural and anthropogenic impacts in the Antarctic marine ecosystem. Such long-term datasets, which extend over the lifetime of a population, can be created by monitoring experiments into the future or by assessing fossil phenomena into the past. The principal advantage of assessing fossil phenomena is that the dynamics of populations can be interpreted unambiguously in relation to natural habitat variations, in the absence of human impacts over seasonal, annual, decadal, century and millennium timescales.

Antarctic fossil phenomena during the Holocene, which represent the current climate regime over the last 10 000 years, can be directly related to modern events. Antarctic species have well-developed populations throughout this period, but only a few have fossil assemblages with high abundances and broad distributions that can be analysed to resolve their temporal variability. These Holocene fossils can be found in marine sediments or adjacent terrestrial areas in the circumpolar coastal zone, which is influenced directly by the advance and retreat of the Antarctic ice sheets.

Sediment cores provide continuous records for interpreting the dynamics of marine microflora and microfauna under varying environmental conditions around Antarctica during the Holocene (Domack 1988). The composition and abundances of these algal and foraminiferan species have been used to assess water column productivity (DeMaster *et al.* 1993), sea-ice extent

(Leventer & Dunbar 1988), circulation patterns (Dunbar *et al.* 1985) and climatic events (Leventer *et al.* 1993). However, because of low Holocene sedimentation rates, which range from 3.4 mm y^{-1} to as low as 0.05 mm y^{-1} (Ledford-Hoffman *et al.* 1986, Krissek 1988, Domack *et al.* 1989), there are difficulties with extracting seasonal to decadal information from these biogenic deposits.

In contrast, Antarctic beaches often have Holocene marine fossils with seasonal to decadal environmental records in their growth morphologies (Fig. 49.1). Antarctic beach fossils have been identified among 37 algal species, 45 foraminifera species, and nearly 70 species from 10 phyla of macrofauna (Table 49.1). The emerged macrofossils include species that are restricted to

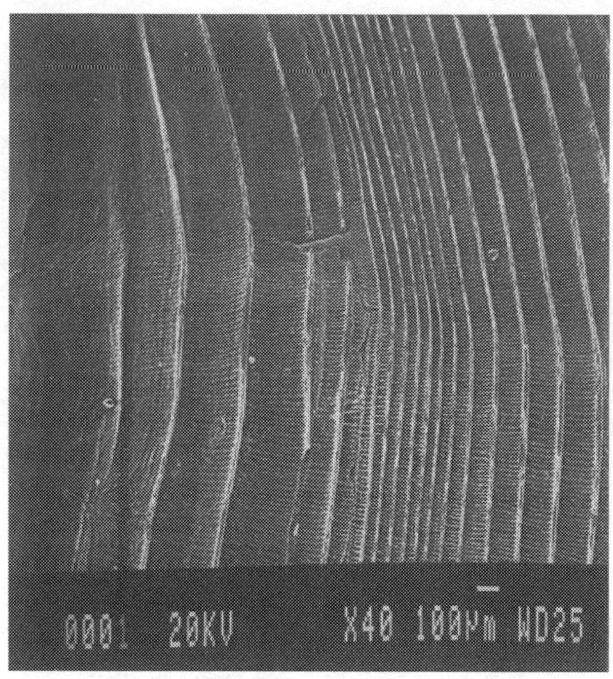

Fig. 49.1. External summer (wide) and winter (narrow) shell growth bands of the Antarctic scallop (*Adamussium colbecki*), which were imaged with a JEOL JSM-820 scanning electron microscope at 40× magnification with a scale bar of 100 μm (Berkman 1991). Similar seasonal growth patterns have been identified in the shells of other mollusc species (Picken 1980), the otoliths of fish (Kock 1992), and the teeth of seals (Laws 1952).

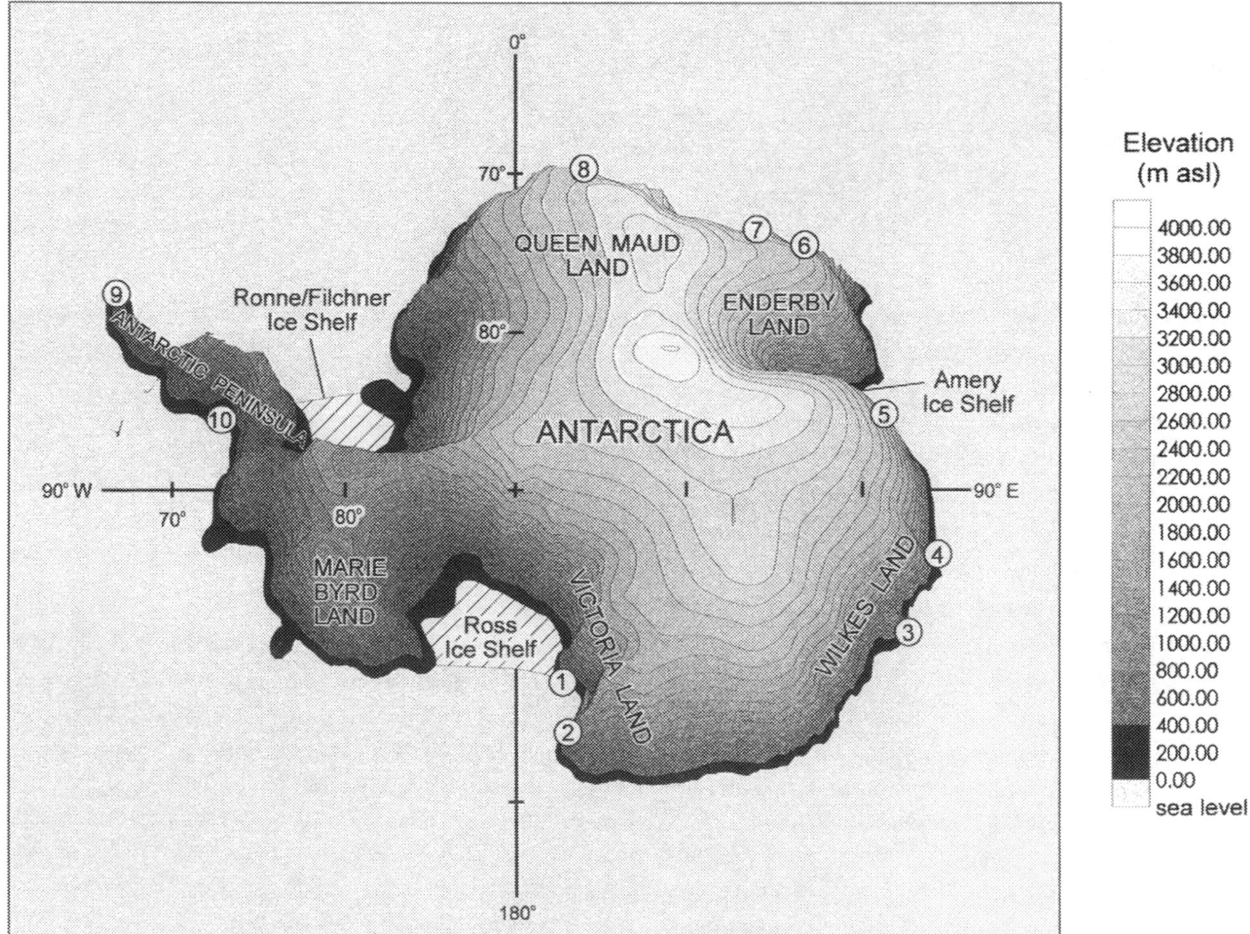

Fig. 49.2. The circumpolar distribution of Quaternary beaches and oases in Antarctic coastal areas (Korotkevich 1971, Pickard 1986, Kirk 1991, Adamson & Colhoun 1992, Lopez-Martinez *et al.* 1992, Goodwin 1993, Hayashi & Yoshida 1994) in relation to icesheet elevation around the continent. (1) McMurdo Sound, (2) Terra Nova Bay, (3) Windmill Hills, (4) Bunger Hills, (50 Vestfold Hills, (6) Thala Hills, (7) Soya Coast, (8) Schirmacher Ponds, (9) South Shetland Islands, (10) Ablation Point. The three-dimensional contours were derived from a digital database taken from Drewry (1983).

Table 49.1. *Number of marine fossil taxa identified in Holocene beaches around Antarctica*

Fossil phyla	0°–60° W	60°–120° W	120°–180° W	180°–120° E	120°–60° E	60°–0° E
Algae	—	—	—	—	36	1
Protozoa	—	—	—	14	18	22
Bryozoa	—	—	—	2	—	—
Porifera	—	—	—	2	4	—
Annelida	—	—	—	4	7	2
Mollusca	3	—	—	21	15	3
Echinodermata	—	—	—	5	—	1
Arthropoda	—	1	—	7	—	1
Pisces	—	—	—	1	—	—
Aves	—	—	—	2	3	—
Mammalia	3	—	—	3	4	2

Note:
Based on data compiled in Berkman (1994a).

the marine environment during their lifetime, such as molluscs and whales, as well as those that can move onto the land, such as penguins and seals. Even though less than 5% of the coastline is exposed (Drewry *et al.* 1982), almost 150 of these emerged Holocene beaches have been studied around Antarctica (Fig. 49.2).

Despite their abundance and relevance for interpreting long-term environmental changes around the continent during the Holocene, marine macrofossils in Antarctic beaches are virtually unstudied. The following discussion will illustrate the interdisciplinary capability for analysing the habitat ecology of emerged marine macrofauna (Fig. 49.3) in relation to natural environmental variability in Antarctic coastal areas during the Holocene.

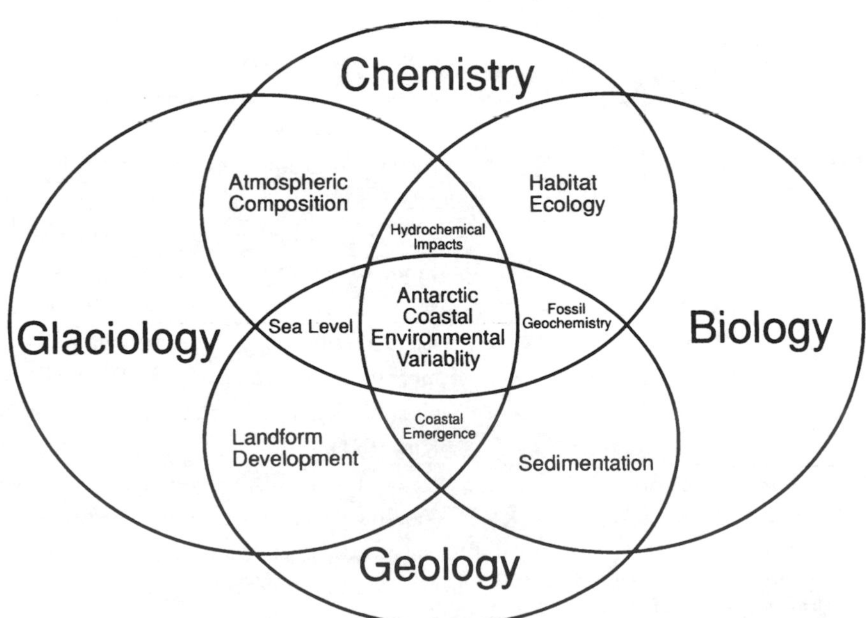

Fig. 49.3. The rosette of interdisciplinary research showing the integration between biology, chemistry, geology and glaciology to assess Antarctic coastal environmental variability during the Holocene.

THE RADIOCARBON RESERVOIR IN THE ANTARCTIC MARINE ECOSYSTEM

Accurate age estimates are critical for understanding environmental variability. In the modern ocean, radiocarbon ages of marine species can vary from several hundred to several thousand years at different times and places in the Antarctic marine ecosystem (Omoto 1983, Gordon & Harkness 1992). This radiocarbon age variability depends on the ^{14}C content of oceanic deep waters (Stuiver *et al.* 1983) and the atmosphere (Bard *et al.* 1990), dilution from melted Antarctic continental ice (Domack *et al.* 1989, Melles *et al.* 1994), vital effects among species (Gordon & Harkness 1992) and global ^{14}C-enrichment associated with nuclear explosions (Broecker & Peng 1982, Berkman & Forman 1996). Because of the above factors, there is no accepted convention on the use of radiocarbon reservoir corrections for Antarctic marine species.

Over the last 6000 years, which includes most of the Holocene fossils which have been radiocarbon dated from Antarctica (Stuiver & Braziunas 1985, Berkman 1992a), the marine radiocarbon reservoir essentially has been constant (Bard *et al.* 1990). During this period, the major change in the radiocarbon reservoir can be attributed to nuclear explosions this century. To resolve the radiocarbon variability associated with these nuclear activities it is necessary to assess the $\Delta^{14}C$ (Broecker & Peng 1982) in Antarctic marine species and seawater both before and after the atmospheric explosions began in 1945 (Berkman & Forman 1996).

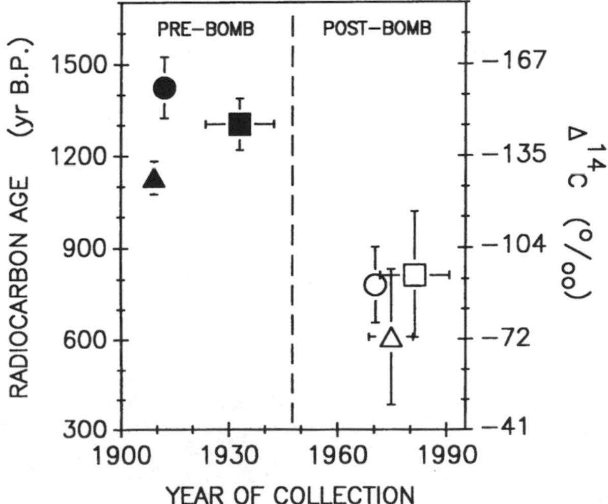

Fig. 49.4. The relation between the radiocarbon age and $\Delta^{14}C$ content of selected taxa in the Antarctic marine ecosystem before and after nuclear explosions began in 1945, based on compiled data (Gordon & Harkness 1992) for seals (circles) and penguins (triangles), and recent measurements (Berkman & Forman 1996) for molluscs (squares).

Post-bomb radiocarbon ages of Antarctic marine species as well as pre-bomb ages of seals and penguins which were killed during the Heroic Era of Antarctic exploration have been compiled (Gordon & Harkness 1992). Pre-bomb Antarctic molluscs, which were obtained from the Smithsonian Institution, also have been radiocarbon-dated (Berkman & Forman 1996). The radiocarbon relationships between these pre-bomb and post-bomb seals, penguins and molluscs from south of the Antarctic Convergence (55° S) are shown in Fig. 49.4.

Differences between the pre-bomb and post-bomb $\Delta^{14}C$ of all three species are around 50‰ (Fig. 49.4). Associated with the latitudinal pattern of evaporation and precipitation (Pickard & Emery 1982), seawater radiocarbon variability south of the Antarctic Convergence before and after 1945 also is around 50‰ (Broecker & Peng 1982). The close correspondence between the $\Delta^{14}C$ variation in the marine species and seawater indicates that they have a conservative relationship which reflects the general distribution of radiocarbon in the Antarctic marine ecosystem. Based on the pre-bomb radiocarbon ages of the different taxa (Fig. 49.4), the average radiocarbon reservoir corrections for Antarctic seals, penguins and molluscs are 1424±200 years, 1130±134 years, and 1300±100 years, respectively (Berkman & Forman 1996).

HOLOCENE ENVIRONMENTAL RECORDS FROM EMERGED BEACHES ALONG THE VICTORIA LAND COAST

Grounded ice sheet recession in the Ross Sea based on glacial geomorphology

Since the Last Glacial Maximum, 18 000 years before present (y BP), the north–south-trending Victoria Land Coast from 72° S to 78° S latitude has been influenced by the southward retreat of ice sheets which were grounded near the shelf break in the Ross Sea (Kellogg *et al.* 1979, Anderson *et al.* 1992). Soil morphologies, radiocarbon ages and glacial drift deposits indicate that massive recession of the grounded ice was underway in the western Ross Embayment by 13 000 y BP (Denton *et al.* 1989). From 10 000 to 8000 y BP, the grounded ice sheet lingered in McMurdo Sound and by 6000 y BP it had already reached Black Island (Denton *et al.* 1989, Kellogg *et al.* 1990). Today, there are more than 50 Holocene beaches which have been identified along the Victoria Land Coast (Stuiver *et al.* 1981, Mabin 1986, Kirk 1991).

Temporal and spatial patterns of ice-sheet retreat along the Victoria Land Coast are matched by the progressive isostatic emergence of Holocene beaches. Marine limits decrease with increasing latitude from 30 m at Terra Nova Bay to less than 5 m at Explorers Cove (Fig. 49.5a). Similarly, corrected radiocarbon-ages of marine mollusc fossils indicate that the maximum ages of raised beaches at Terra Nova Bay were greater than 6005 y BP and 4850 y BP in the vicinity of Explorers Cove (Fig. 49.5b). Relative ages and elevations of these Holocene beaches provide the geologic context for interpreting the habitat ecology of emerged marine fossils along the Victoria Land coast (Fig. 49.3).

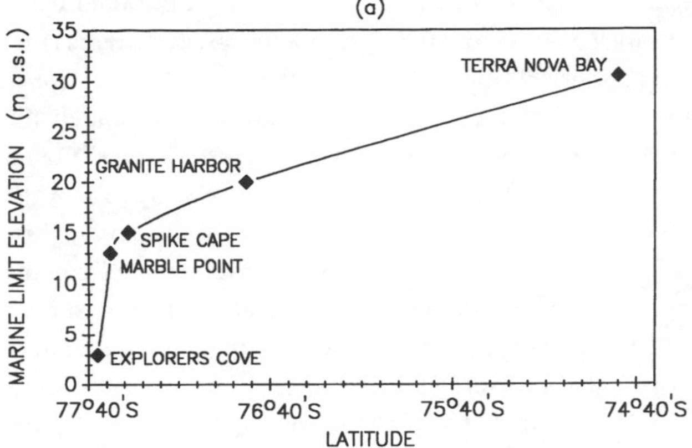

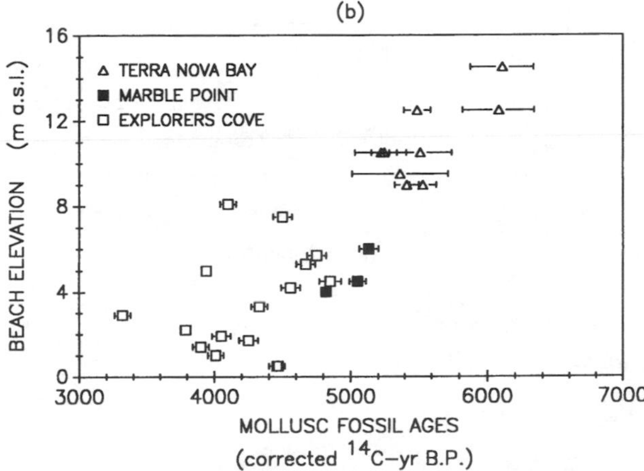

Fig. 49.5. Characteristics of the Holocene beaches along the Victoria Land Coast. (a) Marine limit elevations in metres above sea level (m asl) based on the data of Stuiver *et al.* (1981); (b) relative emergence of the raised beaches in the Terra Nova Bay and Marble Point regions based on the elevations and radiocarbon ages of nearshore marine bivalve mollusc fossils from Terra Nova Bay (Baroni & Orombelli 1991) and the McMurdo Sound area (Stuiver *et al.* 1981). The fossil mollusc radiocarbon ages were corrected by a factor of 1300 years (Fig. 49.4).

Oxygen isotope geochemistry of modern macrofauna: seasonal meltwater and temperature estimates

Oxygen isotopic ratios in marine carbonates are commonly used for assessing continental ice-volume and seawater-temperature changes (Shackleton 1967, Rye & Sommer 1980). To interpret the oxygen isotope geochemistry of Holocene fossils around Antarctica, isotopic signals first should be experimentally assessed in their living analogues. As an illustration, oxygen isotope measurements from shells of the circumpolar Antarctic scallop (*Adamussium colbecki*) will be discussed.

Live *Adamussium* shells were collected along a depth gradient (from 3 to 27 m) adjacent to glacial streams at Explorers Cove (77°38′ S, 166°25′ E) during the 1986–87 austral summer (Berkman 1990, Berkman *et al.* 1992). Across the growth surfaces of these *Adamussium* shells (Fig. 49.1), oxygen isotope ratios indicate that there were seasonal changes in their hydrochemical environment (Fig. 49.6a). Based only on the outermost shell-growth bands, oxygen isotope ratios also demonstrate that there were significant hydrochemical changes with depth during the summer (Fig. 49.6b).

The oxygen isotopic ratios in Fig. 49.6a, b, which overlapped with those from previous *Adamussium* shell measurements (Barerra *et al.* 1990), produced palaeotemperature estimates which exceeded the –0.2 °C maximum and 1.6 °C range in McMurdo Sound (Littlepage 1965, Barry 1988). Together with the complementary measurements of shell trace element concentrations, calcite unit cell volumes and epizoic species (Berkman 1994a, b, Berkman *et al.* 1992), the oxygen isotope data indicate that the nearshore *Adamussium* at Explorers Cove were being impacted by glacial meltwater during the summer. Oxygen isotope signatures of meltwater have been described in mollusc shells from the Antarctic Peninsula region (Marshall *et al.* 1993).

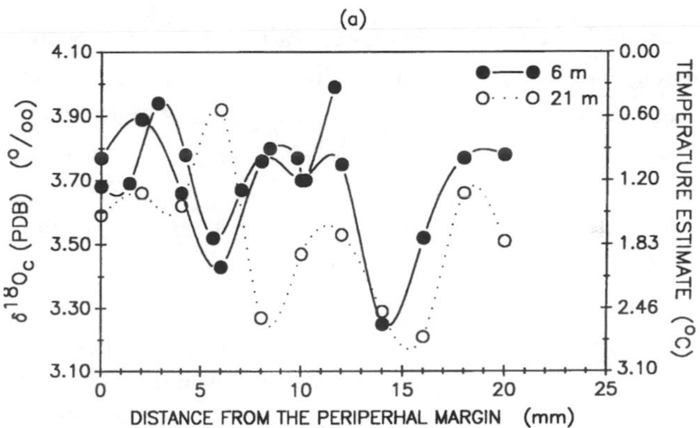

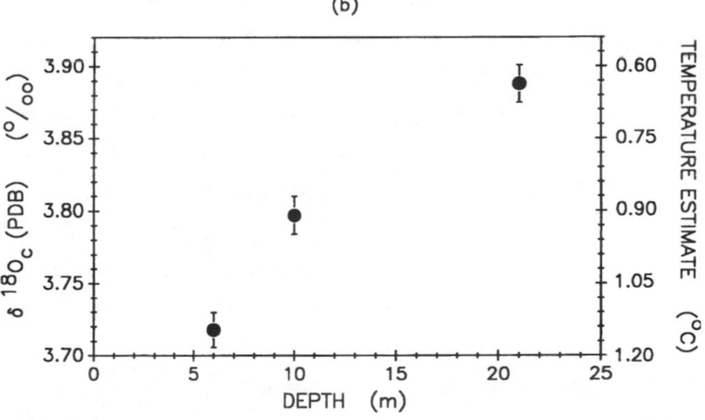

Fig. 49.6. Oxygen isotope ratios in *Adamussium* shells which were collected from 3 to 27 m depth adjacent to meltwater streams in Explorers Cove during the 1986–87 austral summer (Berkman 1994a). Accompanying palaeotemperature estimates were derived according to Rye & Sommer (1980). (a) The cyclic seasonal pattern of oxygen isotope variation across the surfaces of *Adamussium* shells from 6 m and 21 m depths. (b) Oxygen isotope measurements from the outermost shell growth bands which were produced during that summer decreased significantly from 6 to 21 m depth (df 2, 33; F=7.82; p<0.002).

Using a two-point mixing equation, the volume of meltwater entering the ocean from the glacial meltwater streams at Explorers Cove during the summer can be estimated:

$$V_{melt} O_{melt} + V_{sea} O_{sea} = (V_{melt} + V_{sea})\, O_{mix} \quad (49.1)$$

where the volumes (V) and oxygen isotopic characteristics (O) of the meltwater streams (melt) and surface seawater above 20 m depth in the Ross Sea (sea) are mixed (mix). Average $\delta^{18}O$ values of the glacial meltwater streams (Berkman 1994a) and the Ross Sea above 20 m (Jacobs *et al.* 1985) were −29.28‰ and −0.39‰, respectively. Summer $\delta^{18}O$ values in Explorers Cove were measured (−1.04‰ to −0.79‰) from the seawater as well as predicted (−0.84‰ to −0.67‰) from the oxygen isotopic composition of the *Adamissium* shells (Fig. 49.6b). Considering their close agreement, the predicted oxygen isotope values of the mixed seawater were used in the meltwater volume calculation. Lastly, the study area at Explorers Cove was estimated to be 2 km wide, 1 km across and 30 m deep with a volume of 0.06 km^3. The impact of sea-ice melting or freezing on seawater oxygen isotopic characteristics was considered to be negligible (Redfield & Friedman 1969).

Using the above input parameters in Equation 49.1, the volume of glacial meltwater runoff into Explorers Cove during the summer was estimated to range from 0.57×10^6 m^3 to 0.93×10^6 m^3. The reliability of this first approximation can be seen in relation to the annual volume transport of the Onyx River in the adjacent Dry Valleys, which is the largest meltwater stream in Antarctica. Summer data from 1968 to 1988 indicate that the Onyx River had a maximum flow of 14.98×10^6 m^3, a minimum of 0 m^3 and a mean of $3.78 \pm 3.49 \times 10^6$ m^3 during this 20 year period (Chinn 1993). Moreover, flow measurements of the Lost Seal Stream (which drains the Commonwealth Glacier landward into Lake Fryxell rather than seaward into Explorers Cove) revealed total summer discharges of 0.53×10^6 m^3 in 1990/91 and 0.47×10^6 m^3 in 1991/92 (H. R. House, pers. commun. 1995). These stream gauge measurements indicate that glacial meltwater volume estimates (Equation 49.1) for Explorers Cove are reasonable and can be derived from the oxygen isotopic composition of nearshore mollusc shells. Therefore, analysing the oxygen isotope composition of fossil scallop shells should provide information for interpreting the glaciological and hydrochemical variability (Fig. 49.3) along the Victoria Land coast during the Holocene.

Oxygen isotope geochemistry of fossil macrofauna: Holocene meltwater and temperature estimates

In contrast to the peripheral shell-growth bands which reflect seasonal seawater composition only, entire shells represent the integrated oxygen isotopic composition of the ambient seawater throughout the decadal lifespan (Berkman 1990) of *Adamussium*. Based on the bulk oxygen isotopic composition of fossil *Adamussium* from Terra Nova Bay (Baroni *et al.* 1991) and modern shells from Terra Nova Bay (Berkmen, unpublished), palaeotemperature estimates suggest that average coastal seawater temperatures along the Victoria Land coast have varied by more than 2 °C during the last 6000 years (Fig. 49.7). High-resolution climate data from Antarctic ice cores, however, indicate that atmospheric temperatures around the continent varied by less than 1 °C during the last 6000 years (Lorius *et al.* 1979, Cias *et al.* 1992). This discrepancy in the ranges of estimated palaeotemperatures could be resolved by accounting for more meltwater or precipitation along the Victoria Land coast during the middle Holocene. Interestingly, elevated temperatures and increased precipitation are thought to have influenced mid-Holocene glacier advances in coastal areas along East Antarctica (Domack *et al.* 1991) and the Antarctic Peninsula (Ingolfsson *et al.* 1992).

The oxygen isotope data in Fig. 49.7 also suggest that there was a relatively cold interval along the Victoria Land coast around 500 y BP. This cold interval may coincide with the Little Ice Age, which persisted for several hundred years after the thirteenth century (Porter 1986) and is recognized as the most important global climate fluctuation during the current millennium (Grove 1988). It is noteworthy that the Little Ice Age also has been inferred from marine sedimentary deposits (Leventer *et al.* 1993) and penguin rookeries along the Victoria Land coast (Baroni & Orombelli 1994) as well as from ice cores around Antarctica (Benoist *et al.* 1982, Mosley-Thompson *et al.* 1990).

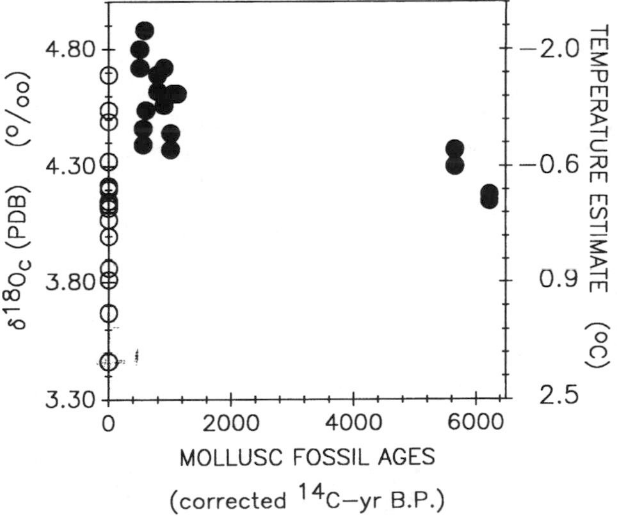

Fig. 49.7. Oxygen isotopic composition of fossil *Adamussium* from emerged Holocene beaches (Baroni *et al.* 1991) and modern *Adamussium* that were collected alive in 1995 (Berkman unpublished) in Terra Nova Bay. The oxygen isotope analyses (see Prentice *et al.* 1993 for the methodology) were conducted individually on entitre *Adamussium* shells that were pulverized into homogenous powders. The fossil radiocarbon ages were corrected by a factor of 1300 years (Fig. 49.4).

Open water conditions based on fossil penguin rookeries

Palaeoenvironmental data from penguin rookeries are ideal for assessing long-term changes that can occur naturally over the lifetime of a population. Abandoned Adélie penguin (*Pygoscelis adelie*) rookeries have been identified along the Budd Coast in

East Antarctica (Goodwin 1993) as well as along the Victoria Land coast, where 17 abandoned rookeries have been identified which extend from 13 000 y BP to the present (Baroni & Orombelli 1994).

Based on the latitudinal distribution of the abandoned rookeries along the Victoria Land coast (Fig. 49.8), which range from Cape Adare south to McMurdo Sound, it can be seen that Adélie populations have persisted continuously since the middle Holocene only in the vicinity of Terra Nova Bay at 74°50′ S. The geographic stability of the Adélie penguin rookeries in Terra Nova Bay may reflect the persistence of the Drygalski Tongue as a dominant environmental feature along the Victoria Land coast during the Holocene. From 3000–4000 y BP, during a period which has been termed the 'penguin optimum' (Baroni & Orombelli 1994), 10 of the 17 rookeries along the Victoria Land coast were occupied. Another phase of rookery occupation existed for several hundred years after 1000 y BP (Baroni & Orombelli 1994). This latter period may have coincided with the Little Ice Age that has been characterized by open-water diatom assemblages along the Victoria Land coast (Leventer *et al.* 1993).

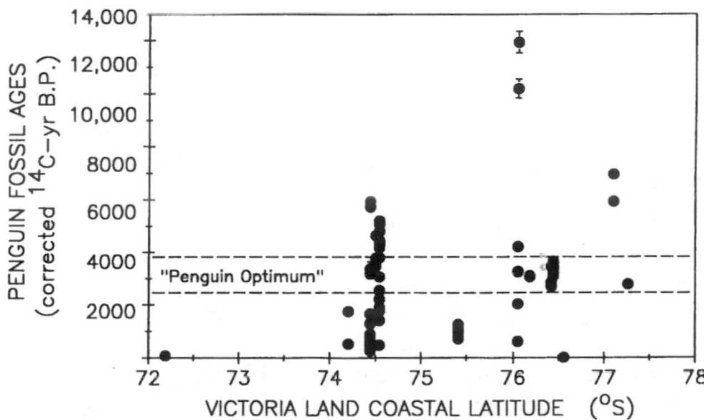

Fig. 49.8. Abandoned Adélie penguin rookeries which have been radiocarbon-dated along the Victoria Land coast (Baroni & Orombelli 1994). Locations of these rookeries have been redrawn in relation to their latitudes. The 'penguin optimum', as defined by Baroni & Orombelli (1994), was a favourable environmental period when 10 of the 17 now-abandoned rookeries were occupied. For comparison, today there are 21 Adélie penguin rookeries along the Victoria Land coast, all of which are north of Terra Nova Bay (Taylor *et al.* 1990). The fossil penguin radiocarbon ages were corrected by a factor of 1130 years (Fig. 49.4).

Occupation of Adélie penguin rookeries along the Victoria Land coast may be strongly influenced by environmental factors which affect penguin access to the sea-ice edge (Baroni & Orombelli 1994). Throughout the Holocene, the greatest diffusion of Adélie penguin rookeries along the Victoria Land coast has been south of Terra Nova Bay (Fig. 49.8). However, today the predominance of Adélie penguins are northward along the Victoria Land coast near Cape Adare (Taylor *et al.* 1990). This change in the distribution of occupied and abandoned Adélie penguin rookeries along the Victoria Land coast may reflect a northward progression of the sea-ice edge in the Ross Sea during the last several hundred years.

CONCLUSIONS

As indicated by the fossil scallop (Fig. 49.5b) and penguin (Fig. 49.8) populations, marine species have persisted along the Victoria Land coast but with localized variations that have been occurring over century and millennium time scales throughout the Holocene. These localized variations in the Antarctic marine ecosystem can be attributed unambiguously to natural phenomena. Within these periods, the morphology and geochemistry of the fossils (Fig. 49.3) also can be analysed to interpret seasonal to decadal environmental variability (Figs. 49.1 and 49.6a). Natural environmental variations during the Holocene are further reflected in adjacent terrestrial populations (R. I. L. Smith 1984, Smith & Friedman 1993).

Viewed in a circumpolar perspective, the long-term dynamics of coastal populations will provide time constraints for judging the nature of transitory impacts associated with the rational use of Antarctic marine resources (Berkman 1992b). Coordinated among nations (Weller & Lorius 1989, Weller 1992), interdisciplinary research in Antarctic coastal areas (Fig. 49.3) will enhance our understanding of the Earth's climate system.

ACKNOWLEDGEMENTS

Preparation of this manuscript was supported by a National Science Foundation grant to P.A.B. (OPP 92-21784). This paper is Contribution 941 from the Byrd Polar Research Center at The Ohio State University.

REFERENCES

Adamson, D. A. & Colhoun, E. A. 1992. Late Quaternary glaciation and deglaciation of the Bunger Hills, Antarctica. *Antarctic Science*, **4,** 435–446.

Anderson, J. B., Shipp, S. S., Bartek, L. R. & Reid, D. E. 1992. Evidence for a grounded ice sheet on the Ross Sea continental shelf during the late Pleistocene and preliminary paleodrainage reconstruction. *Antarctic Research Series*, **57,** 39–62.

Bard, E., Hemelin, B., Fairbanks, R. G. & Zindler, A. 1990. Calibration of the ^{14}C timescale over the past 30,000 years using mass spectrometric U–Th ages from Barbados corals. *Nature*, **345,** 405–410.

Baroni, C. & Orombelli, G. 1991. Holocene raised beaches at Terra Nova Bay, Victoria Land, Antarctica. *Quaternary Research*, **36,** 157–177.

Baroni, C. & Orombelli, G. 1994. Abandoned penguin rookeries as Holocene paleoclimatic indicators in Antarctica. *Geology*, **22,** 23–26.

Baroni, C., Stenni, B. & Longinelli, A. 1991. Isotopic composition of Holocene shells from raised beaches and ice shelves of Terra Nova Bay (Victoria Land, Antarctica). *Memoirie Della Societa Geologica Italiana*, **46,** 93–102.

Barrera, E., Tevesz, J. S. & Carter, J. G. 1990. Variations in oxygen and carbon isotopic compositions and microstructure of the shell of *Adamussium colbecki* (Bivalvia). *Palaios*, **5,** 149–159.

Barry, J. P. 1988. Hydrographic patterns in McMurdo Sound, Antarctica, and their relationship to local benthic communities. *Polar Biology*, **8,** 377–391.

Benoist, J. P., Jouzel, J., Lorius, C., Merlivat, L. & Pourchet, M. 1982. Isotope climatic record over the last 2.5 KA from Dome C, Antarctica. *Annals of Glaciology*, **3,** 17–22.

Berkman, P. A. 1990. The population biology of the Antarctic scallop, *Adamussium colbecki* (Smith, 1902) at New Harbor, Ross Sea. In Kerry, K. R. & Hempel, G., eds. *Antarctic ecosystems: ecological change and conservation*. Berlin & Heidelberg: Springer-Verlag, 281–288.

Berkman, P. A. 1991. Holocene meltwater variations recorded in Antarctic coastal marine benthic assemblages. In Weller, G., Wilson, C. L. & Severin, B. A. B., eds. *Proceedings of the International Conference on the Role of the Polar Regions in Global Change*. Fairbanks: University of Alaska, 440–449.

Berkman, P. A. 1992a. Circumpolar distribution of Holocene marine fossils in Antarctic beaches. *Quaternary Research*, **37**, 256–260.

Berkman, P. A. 1992b. The Antarctic marine ecosystem and humankind. *Reviews in Aquatic Sciences*, **6**, 295–333.

Berkman, P. A. 1994a. Geochemical signatures of meltwater in mollusc shells from areas during the Holocene. In Berkman, P. A. & Yoshida, Y., eds. *Holocene environmental changes in Antarctic coastal areas. Memoirs of the National Institute of Polar Research, Special Issue*, No. 50, 11–33.

Berkman, P. A. 1994b. Epizoic zonation on growing scallop shells in McMurdo Sound, Antarctica. *Journal of Experimental Marine Biology and Ecology*, **179**, 49–67.

Berkman, P. A. & Forman, S. L. 1996. Pre-bomb radiocarbon and the reservoir correction for calcareous marine species in the Southern Ocean. *Geophysical Research Letters*, **23**, 363–366.

Berkman, P. A., Foreman, D. W., Mitchell, J. C. & Liptak, R. J. 1992. Scallop shell mineralogy and crystalline characteristics: proxy records for interpreting Antarctic nearshore marine hydrochemical variability. *Antarctic Research Series*, **57**, 27–38.

Broecker, W. S. & Peng, T-H. 1982. *Tracers in the Sea*. Palisades: Eldigio Press.

Cias, P., Petit, J. R., Jouzel, J., Lorius, C., Barkov, N. I., Lipenkov, V. & Nicolaiev, V. 1992. Evidence for an early Holocene climatic optimum in the Antarctic deep ice-core record. *Climate Dynamics*, **6**, 169–177.

Chinn, T. 1993. Physical hydrology of the Dry Valley lakes. *Antarctic Research Series*, **59**, 1–52.

DeMaster, D. J. Dunbar, R. B., Gordon, L. I., Leventer, A. R., Morrison, J. M., Nelson, D. M., Nittrouer, C. A. & Smith, W. O. 1993. The cycling and accumulation of biogenic silica and organic matter in high latitude sediments: the Ross Sea. *Oceanography*, **5**, 146–153.

Denton, G. H., Bockheim, J. G., Wilson, S. C. & Stuiver, M. 1989. Late Wisconsin and early Holocene glacial history, inner Ross embayment, Antarctica. *Quaternary Research*, **31**, 151–182.

Domack, E. W. 1988. Biogenic facies in the Antarctic glacimarine environment: basis for a polar glacimarine summary. *Palaeogeography, Palaeoclimatology, Palaeoecology*, **63**, 357–372.

Domack, E. W., Jull, A. J. T., Anderson, J. B., Linick, T. W. & Williams, C. R. 1989. Application of Tandem Accelerator Mass-Spectrometer dating to Late Pleistocene–Holocene sediments of the East Antarctic continental shelf. *Quaternary Research*, **3**, 277–287.

Domack, E. W., Jull, A. J. T. & Nakao, S. 1991. Advance of East Antarctic outlet glaciers during the Hypsithermal: implications for the volume state of the Antarctic ice sheet under global warming. *Geology*, **19**, 1059–1062.

Drewry, D. J. 1983. The surface of the Antarctic Ice Sheet. In Drewry, D. J., ed. *Glaciological and Geophysical Folio*. Sheet 2. Cambridge: Scott Polar Research Institute.

Drewry, D. J., Jordan, S. R. & Jankowski, E. 1982. Measured properties of the Antarctic ice sheet: surface configuration, ice thickness, volume and bedrock characteristics. *Annals of Glaciology*, **3**, 83–91.

Dunbar, R. B., Anderson, J. B., Domack, E. W. & Jacobs, S. S. 1985. Oceanographic influences on sedimentation along the Antarctic Continental Shelf. *Antarctic Research Series*, **43**, 291–312.

Goodwin, I. D. 1993. Holocene deglaciation, sea-level change, and the emergence of the Windmill Islands, Budd Coast, Antarctica. *Quaternary Research*, **40**, 70–80.

Gordon, J. E. & Harkness, D. D. 1992. Magnitude and geographic variation of the radiocarbon content in Antarctic marine life: implications of reservoir corrections in radiocarbon dating. *Quaternary Science Reviews*, **11**, 697–708.

Grove, J. M. 1988. *The Little Ice Age*. London: Methuen, 498 pp.

Hayashi, M. & Yoshida, Y. 1994. Holocene raised beaches in the Lutzhow–Holm Bay region, East Antarctica. In Berkman, P. A. & Yoshida, Y., eds. *Holocene environmental changes in Antarctic coastal areas. Memoirs of the National Institute of Polar Research, Special Issue, No. 50*, 67–71.

Ingolfsson, O., Hjort, C., Bjorck, S. & Smith, R. I. L. 1992. Late Pleistocene and Holocene glacial history of James Ross Island, Antarctic Peninsula. *Boreas*, **21**, 209–222.

Jacobs, S. S., Fairbanks, R. G. & Horibe, Y. 1985. Origin and evolution of water masses near the Antarctic continental margin: evidence from $H_2{}^{18}O/H_2{}^{16}O$ ratios in seawater. *Antarctic Research Series*, **43**, 59–86.

Kellogg, T. B., Kellogg, D. E. & Stuiver, M. 1990. Late Quaternary history of the southwestern Ross Sea: evidence from debris bands on the McMurdo Ice Shelf. *Antarctic Research Series*, **50**, 24–56.

Kellogg, T. B. Truesdale, R. S. & Osterman, L. E. 1979. Late Quaternary extent of the West Antarctic Ice Sheet: new evidence from Ross Sea cores. *Geology*, **7**, 249–253.

Kirk, R. M. 1991. Raised beaches, Late Quaternary sea levels and deglacial sequences on the Victoria Land Coast, Antarctica. In Gillieson, D. S. & Fitzsimons, D., eds. *Quaternary research in Australian Antarctica: future directions*. Canberra: Australian Defense Force Academy, 85–105.

Kock, K. H. 1992. *Antarctic fish and fisheries*. Cambridge: Cambridge University Press, 359 pp.

Korotkevich, Ye S. 1971. Quaternary marine deposits and terraces in Antarctica. *Soviet Antarctic Expedition Information Bulletin*, **8**, 185–190.

Krissek, L. A. 1988. Sedimentation history of the Terra Nova Bay region, Ross Sea, Antarctica. *Antarctic Journal of the United States*, **23**, 104-105.

Laws, R. M. 1952. A new method for age determination of mammals. *Nature*, **169**, 972.

Ledford-Hoffman, P. A., DeMaster, D. J. & Nittrouer, C. A. 1986. Biogenic silica accumulation in the Ross Sea and the importance of Antarctic continental shelf deposits in the marine silica budget. *Geochemica Cosmochimica Acta*, **50**, 2099–2110.

Leventer, A. & Dunbar, R. B. 1988. Recent diatom record of McMurdo Sound, Antarctica: implications for history of sea ice extent. *Paleoceanography*, **3**, 259–274.

Leventer, A., Dunbar, R. B. & DeMaster, D. J. 1993. Diatom evidence for Late Holocene climatic events in Granite Harbor, Antarctica. *Paleoceanography*, **8**, 373–386.

Littlepage, J. L. 1965. Oceanographic investigations in McMurdo Sound, Ross Sea, Antarctica. *Antarctic Research Series*, **5**, 1–37.

Lopez-Martinez, J., Martinez De Pison, E. & Arche, A. 1992. Geomorphology of Hurd Peninsula, Livingston Island, South Shetland Islands. In Yoshida, Y., Kaminuma, K. & Shiraishi, K., eds. *Recent Progress in Antarctic Earth Science*. Tokyo: Terrapub., 751–756.

Lorius, C., Merlivat, L., Jouzel, J. & Pourchet, M. 1979. A 30,000-yr climatic record from Antarctic ice. *Nature*, **280**, 642–648.

Mabin, M. 1986. The Ross Sea section of the Antarctic ice sheet at 18,000 yr. BP: evidence from Holocene sea-level changes along the Victoria Land coast. *South African Journal of Science*, **82**, 506–508.

Marshall, J. D., Pirrie, D. & Nolan, C. 1993. Oxygen stable isotope composition of skeletal carbonates from living Antarctic marine invertebrates. In Heywood, R. B., ed. *University Research in Antarctica, 1989–1992*. Cambridge: British Antarctic Survey, 81–84.

Melles, M., Verkulich, S. R. & Hermichen, W. D. 1994. Radiocarbon dating of lacustrine and marine sediments from the Bunger Hills, East Antarctica. *Antarctic Science*, **6**, 375–378.

Mosley-Thompson, E., Thompson, L.G., Grootes, P. M. & Gundestrup, N. 1990. Little Ice Age (Neoglacial) paleoenvironmental conditions at Siple Station, Antarctica. *Annals of Glaciology*, **14**, 199–204.

Omoto, K. 1983. The problem and significance of radiocarbon geochronology in Antarctica. In Oliver, R. L., James, P. R. & Jago, J. B., eds. *Antarctic earth science*. Cambridge: Cambridge University Press, 450–452.

Pickard, J., ed. 1986. *Antarctic oasis: terrestrial environments and history of the Vestfold Hills*. New York: Academic Press, 367 pp.

Pickard, G. L. & Emery, W. J. 1982. *Descriptive physical oceanography, an introduction*, 4th edn. New York: Pergamon Press.

Picken, G. B. 1980. The distribution, growth, and reproduction of the Antarctic limpet *Nacella (Patinigera) concinna* (Strebel, 1908). *Journal of Experimental Marine Biology and Ecology*, **42,** 71–85.

Porter, S. C. 1986. Pattern and forcing of northern hemisphere glacier variations during the last millennium. *Quaternary Research*, **26,** 27–48.

Prentice, M. L., Bockheim, J. G., Wilson, S. C., Burckle, L. H., Hodell, D. A., Schuchter, C. & Kellogg, D. E. 1993. Late Neogene Antarctic glacial history: evidence from Central Wright Valley. *Antarctic Research Series*, **60,** 207–249.

Redfield, A. C. & Friedman, I. 1969. The effect of meteoric water, meltwater and brine on the composition of polar sea water and of the deep water of the ocean. *Deep-Sea Research*, **16,** 197–214.

Rye, D. M. & Sommer, M. A. 1980. Reconstructing paleotemperature and paleosalinity regimes with oxygen isotopes. In Rhoads, D. C. & Lutz, R. A., eds. *Skeletal growth of aquatic organisms: biological records of environmental change*. New York: Plenum Press, 169–202.

Shackleton, N. 1967. Oxygen isotope analysis and Pleistocene temperature re-assessed. *Nature*, **215,** 5547–5548.

Smith, G. I. & Friedman, I. 1993. Lithology and paleoclimatic implications of lacustrine deposits around Lake Vanda and Don Juan Pond, Antarctica. *Antarctic Research Series*, **59,** 83–94.

Smith, R. I. L. 1984. Terrestrial plant biology of the sub-Antarctic and Antarctic. In Laws, R. M., ed. *Antarctic Ecology*, Vol. 1. New York: Academic Press, 61–162.

Stuiver, M. & Braziunas, T. F. 1985. Compilation of isotopic dates from Antarctica. *Radiocarbon*, **27,** 117–304.

Stuiver, M., Denton, G. H., Hughes, T. J. & Fastook, G. L. 1981. History of the marine ice sheet in West Antarctica during the last glaciation: a working hypothesis. In Denton, G. H. & Hughes, T. J., eds. *The last great ice sheets*. New York: Wiley, 319–436.

Stuiver, M., Quay, P. D. & Ostlund, H. G. 1983. Abyssal water carbon-14 distribution and the age of the world oceans. *Science*, **219,** 849–851.

Taylor, R. H., Wilson, P. R. & Thomas, B. W. 1990. Status and trends of Adelie penguin populations in the Ross Sea. *Polar Record*, **26,** 293–304.

Weller, G. 1992. Antarctica and the detection of environmental change. *Transactions of the Royal Society of London*, **B338,** 201–208.

Weller, G. & Lorius, C. 1989. *The Role of Antarctica in global change: scientific priorities for the International Geosphere–Biosphere Programme (IGBP)*. Cambridge: ICSU Press/SCAR, Scott Polar Research Institute.

50 Climate change in Kerguelen Islands and colonization of recently deglaciated areas by *Poa kerguelensis* and *P. annua*

YVES FRENOT[1], JEAN-CLAUDE GLOAGUEN[2] AND PAUL TRÉHEN[1]

[1]Université de Rennes I, UMR 6553, Station Biologique, F-35380 Paimpont, France, [2]Université de Rennes I, URA 1853 CNRS, Laboratoire d'Ecologie Végétale, Campus de Beaulieu, F-35042 Rennes Cedex, France

ABSTRACT

This paper considers the potential effects of the recent climate changes on the vegetation of Kerguelen. The changes in temperature, precipitation, sunshine and atmospheric humidity on Kerguelen are analysed by year and month between 1951 and 1993: the range of annual mean temperature is high (1.9 °C) over the whole period but the trend of increase is low (about 0.1 °C). The current temperature (from 1982 to the present) remains at a high level. Precipitation drastically decreased in the early 1960s and its current level is lower than in the 1950s by about 500 mm. These climatic changes have accelerated the retreat of most of the glacier snouts since 1970. P. kerguelensis *(native) and* P. annua *(alien) are used as an example of changes in the primary succession on the recently deglaciated areas. Seed arrival and germination occur even in the most recent glacier forelands. The inflorescence biomass (%) is more important in* P. annua *than in the native species, whereas seed production remains low.* P. annua *seeds germinate immediately after their production, but in* P. kerguelensis *overwintering seems necessary to induce germination. This disadvantage is compensated by a deep root architectural pattern. The disappearance of* P. annua *on the oldest sites can be related to this root pattern and to the accessibility of soil nutrients. These results suggest that the colonization dynamics and the interaction processes between the two* Poa *species on the Ampère Glacier forelands are essentially governed by the soil development, in terms of stabilization, cryoturbation, and nutrient and water accessibility. The current climatic modifications* per se *play probably a minor (or indirect) role in the primary succession processes in the Ampère Valley. However, this conclusion cannot be generalized to the whole Kerguelen archipelago and other examples demonstrate the ecological impact of the current warming and water deficiency.*

Key words: sub-Antarctic, primary succession, glacier forelands, introduced species, climate change, seed plants.

INTRODUCTION

In the past decades there has been a substantial increase in studies on the relationship between global climate change and atmospheric chemistry (Pearman 1991). Climatologists at present expect that the greatest warming will occur in the middle and high latitudes (Houghton & Woodwell 1989). Most of the sub-Antarctic glaciers have retreated during the last 30 years on South Georgia (Gordon & Timmis 1992), Heard (Allison & Keage 1986) and on Kerguelen (Frenot *et al.* 1993). On this

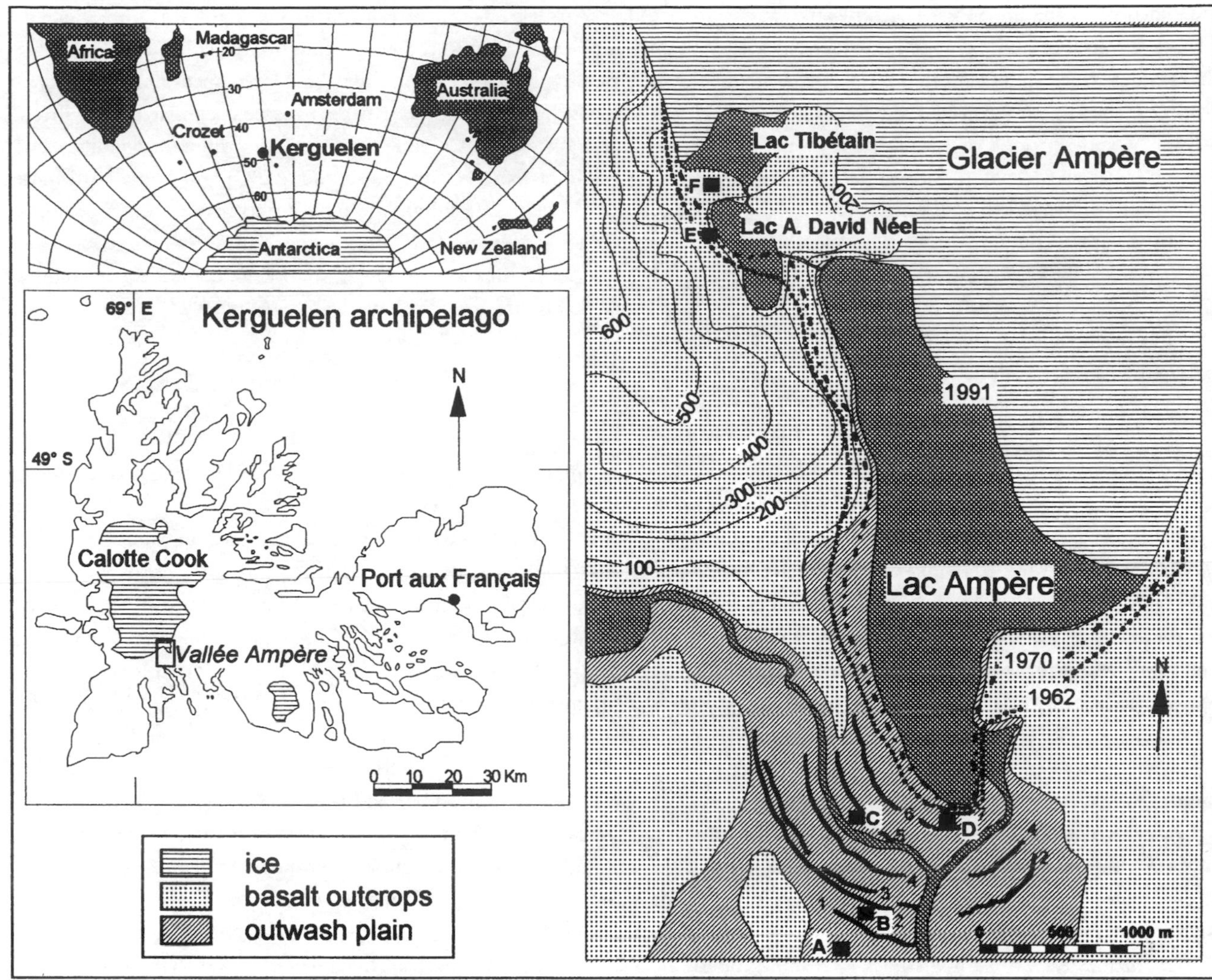

Fig. 50.1. Map of the study area within its regional context. Dotted and dashed lines show the boundaries of Ampère Glacier in 1962 and 1970, respectively. Morainic arcs (continuous lines) are numbered from 1 to 7. Black squares indicate locations of the study sites (A to F) along the chronosequence.

archipelago, a slow retreat of the Ampère Glacier snout began 200 years ago (1 km retreat between 1800 and 1970) but has accelerated since the early 1970s (3 km retreat). This recent trend in glacier behaviour, as in the northern hemisphere (see review in Matthews 1993), suggests that the current climate is changing, resulting in the strongly negative glacier mass budget. Smith (1990) has emphasized that the low biodiversity and the small range of environmental variables characteristic of the circum-Antarctic islands should contribute to their high sensitivity to even small fluctuations in climate. In consequence, it is essential to pay careful attention to the potential responses of these sensitive ecosystems to the recent climate change, taking into account the demographic traits and plasticity of the species and their interspecific relationships.

This paper analyses the climate change recorded at Port-aux-Français Station between 1951 and 1993 and discusses its influence on the plant communities. As the amelioration in climate might be expected to benefit the alien species in particular (Walton 1975), we present the results obtained on two *Poa* species, one long-lived, native to Kerguelen and with distribution restricted to this archipelago and Heard Island (*P. kerguelensis* Steud.), and the other short-lived, introduced and with widespread distribution (*P. annua* L.). Their abundance and colonizing strategies in terms of production of seeds, germination and reproductive investment, are compared in the primary succession stages on the Ampère Glacier forelands (Fig. 50.1), where only seven vascular species are involved in the colonization processes (Frenot *et al.* 1993). Finally, we provide some other examples which demonstrate the reality of the changes in the communities on Kerguelen.

METHODS

Climate

Air temperature, precipitation, hours of sunshine and atmospheric humidity recorded at Port-aux-Français meteorological station (Meteo France) over 1951–1993 period were analysed by year and month using regression analysis. For each parameter, the 43-year period was subdivided into sub-periods according to the ability to adjust successive linear models to the annual data.

The significance of the slopes was tested by analysis of variance in respect of variance homogeneity and stochastic distribution of the residues. When significant trends were detected, the slopes of the linear models were considered as the rate of change per year over the sub-period. The mean monthly values were similarly analysed over each of these sub-periods. The three-year running means shown in Fig. 50.2 were not used in these calculations, only the annual data.

Study area

Six sites were investigated on the sandur along the chronosequence in the Ampère Valley (Fig. 50.1) according to the age (in 1992) of their deglaciation (Frenot *et al.* 1993). The maximum extent of Ampère Glacier was reached at the end of the Little Ice Age (site A, about 200 y BP). The age of site B is estimated at 180 y BP and that of site C at 40–50 y old. Aerial photographs allowed the precise determination of the age of the following sites: sites D, E and F are 30, 15 and two years old, respectively.

It is worth noting that the climate in the Ampère Valley is colder (about 4 °C, estimation from some fragmentary records during summer months) and moister than at Port-aux-Français. However, we assume that the trends recorded by the meteorological station also occurred on the west coast of Kerguelen.

Vegetation

Poa kerguelensis and *P. annua* individuals were counted in 100 contiguous 1 m^2 quadrats (10 m×10 m area) in each site of the chronosequence in February 1992. The two species were counted again on site F in 1993 and 1994. The confidence intervals given in the text are standard errors. Above-ground and below-ground biomass was estimated using 100 individuals of each species (covering the natural range of diameters) collected on site D, where their maximum abundance was observed. Roots were carefully hand-sorted and washed. Observations on the root architecture (length, diameter, depth) were recorded during the sampling. The above-ground biomass of the fertile individuals was subdivided into green material biomass and inflorescence biomass. Samples were dried at 60 °C and weighed.

Inflorescences of *P. kerguelensis* and *P. annua* were collected in site D on 14 February 1992 (36 plants harvested) and 4 March 1992 (84 plants) respectively, when seeds became mature. The number of inflorescences per plant, the number of spikelets per inflorescence and the number of flowers per spikelet were determined. In addition, seeds of *P. annua* and *P. kerguelensis* were examined with a binocular microscope to estimate the percentage of viable seeds (Frenot & Gloaguen 1994) on the basis of their morphological development (presence or absence of a well developed caryopsis between the lemma and palea). Finally, the mean number of seeds produced per fertile plant was calculated to give an account of the fertility of the studied population. Comparisons of means were statistically performed using the t test (in respect of the normality of the distribution and the homogeneity of variances) or the non-parametric Mann Whitney test.

Air-dried seeds were taken to the laboratory in France at ambient temperature and germination tests began about three months after their collection. Well-developed and undamaged seeds were placed in Petri dishes (25 seeds per dish) on two sheets of moistened filter paper and stored in different conditions of constant temperature (5, 10 and 20 °C) and light (24 h dark (dark sets) or 16 h light – 8 h dark photoperiod (light sets)). A cold pretreatment (5 °C during one month) was also conducted on sets of the two species of seed. Details of the germination tests used are given in Frenot & Gloaguen (1994).

RESULTS

Trends in climate

Mean annual temperature over the 1951–1993 period shows three distinct sub-periods (Fig. 50.2a). The first one (1951–1964) was characterized by a drastic decrease of about 1.15 °C ($p<0.005$). Despite the general trend in the decrease of all the monthly values during this period, only the mean August temperature decrease of 2.14 °C was significant ($p<0.001$, Fig. 50.2b). The second period (1964–1982) showed a smaller but highly significant increase of about 1.27 °C ($p<0.001$) due to the significant warming trend of the summer months (January, February, March, $p<0.01$) and of three inter-seasonal months (September, November and May, $p<0.05$) (Fig. 50.2c). No significant change has been observed since 1982 in the annual or monthly means (Fig. 50.2d). On the basis of these calculations and despite the wide range of annual means over the 43- year period (varying from 3.5 °C in 1964 to 5.4 °C in 1982), the increase in the annual mean temperature was slight (0.12 °C).

Total annual precipitation shows also three distinct sub-periods (Fig. 50.2e). No change was observed during the 1950s. The precipitation remained high during this decade (about 1120 mm y^{-1}) and only the April mean value (Fig. 50.2f) significantly decreased ($p<0.01$). The 1960–1965 period was characterized by an important decrease in the precipitation ($p<0.01$) at about 132.5 mm y^{-1}. Nearly all monthly values showed a similar decrease (Fig. 50.2g) but only January and September precipitation showed significant change ($p<0.05$). A slight but significant increase ($p<0.01$) of about 267 mm was then recorded from 1965 to the present, principally during the winter months (May, June, July) and March ($p<0.05$) (Fig. 50.2h). The summer precipitation remained at a very low level. Despite the recent upward trend in the annual precipitation, the present values are still about 500 mm lower than during the 1950s.

For sunshine four periods may be distinguished (Fig. 50.2i): 1951–1957 was characterized by a significant increase ($p<0.02$) of about 292 h, whereas from 1957 to 1963 the hours of sunshine decreased significantly ($p<0.01$), by about 540 h. No monthly trends have been observed during these two periods. A significant increase of about 324 hours was then recorded between 1963 and 1976 ($p<0.001$). The June, August and November values showed a similar trend during this period ($p<0.02$). No significant change in the yearly hours of sunshine can be

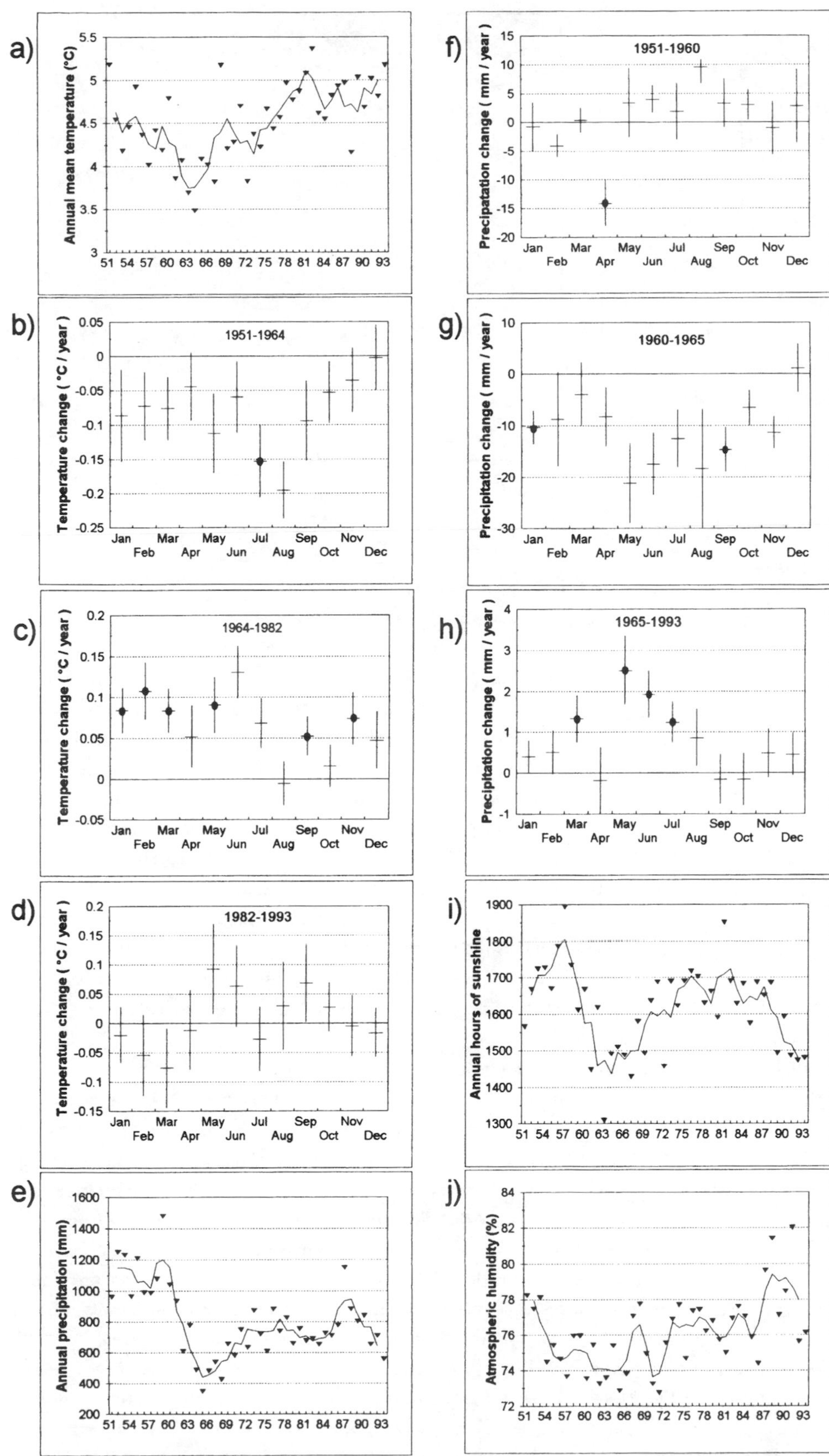

Fig. 50.2. Climate change on Kerguelen between 1951 and 1993 (Port aux Français Station, Meteo France data):(a) annual mean temperature ; (b) to (d) rate of monthly mean temperature change over the three main periods of annual mean temperature change (means±SD) ; (e) total annual precipitation ; (f) to (h) rate of monthly mean precipitation change over the three main periods of annual precipitation change; (i) annual hours of sunshine ; (j) atmospheric humidity. Continuous lines in (a), (e), (i) and (j) are three-year running means. Points in (b) to (d) and (f) to (h) indicate the significant trends of change ($p<0.05$).

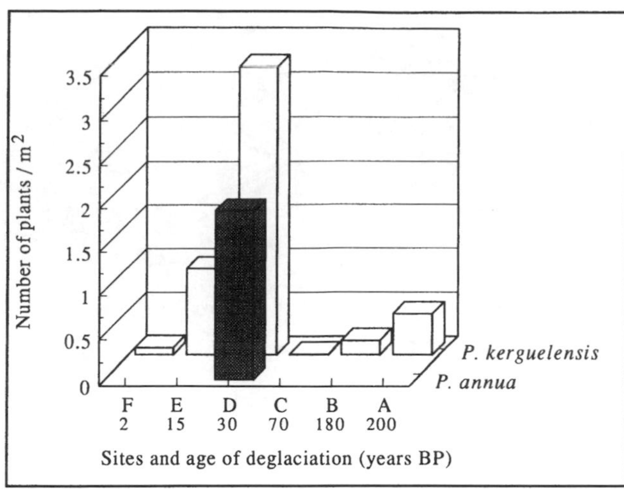

Fig. 50.3. Abundance of *Poa annua* and *P. kerguelensis* along the chronosequence (in February 1992) in relation to the age of deglaciation.

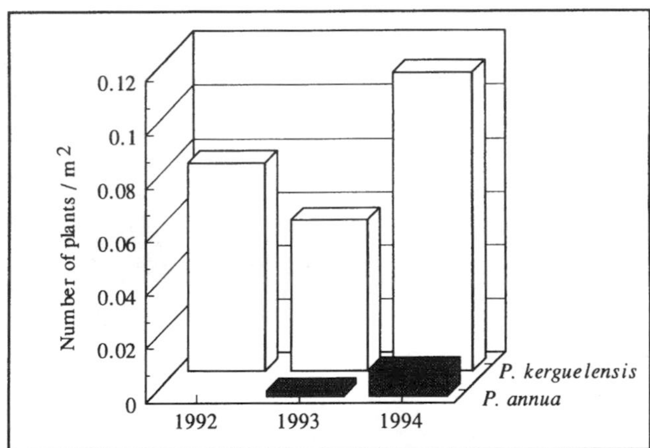

Fig. 50.4. Change in the densities of *Poa annua* and *P. kerguelensis* on the youngest site (F) from 1992 to 1994.

detected during the last period (1976–1993), despite the low values recorded recently. These overall fluctuations have not resulted in any major change in the present annual hours of sunshine compared with the early 1950s.

No clear periods nor significant trends in atmospheric humidity can be distinguished over the 1951–1993 period (Fig. 50.2j). However, the interannual changes seem to be greater during recent years.

Poa kerguelensis and *P. annua* status along the chronosequence

In February 1992, *P. kerguelensis* was present in all the study sites of the chronosequence (Fig. 50.3). Its density increased regularly from site F to reach its maximum on the 30-year-old site D (3.26±0.48 plants m^{-2}, 412±94 mg m^{-2}). It was virtually absent from site C but was present on the older sites (B and A). *P. annua* occurred only on site D (1.91±0.21 plants m^{-2}, 78± 8 mg m^{-2}). The first individual was observed on the unstabilized site F the next year and the number of individuals of *P. annua* increased in 1994 (Fig. 50.4). Conversely, the density of *P. kerguelensis* fluctuated on the site during these three years.

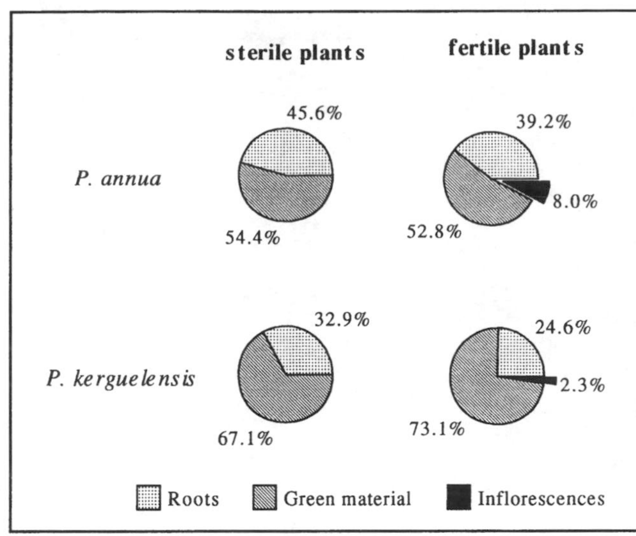

Fig. 50.5. Percentage of root, green material and inflorescence biomass in *Poa annua* and *P. kerguelensis* on the 30-year-old site (D).

Investment and root architecture

P. annua was characterized by its very superficial root system which colonized the upper 3–5 cm of the soil and extended *c.* 7 cm out (maximum 14 cm). The root biomass reached more than 45% of the total biomass in the sterile plants (Fig. 50.5). The roots of *P. kerguelensis* were deeper (>10 cm deep) and longer (*c.* 20 cm long, maximum 50 cm). Despite this greater development, the root weight did not exceed 33% of the total biomass in the sterile plants. The reproductive investment was indicated by the biomass of the inflorescences, which reached 8.0% in *P. annua* and only 2.3% in *P. kerguelensis*. The presence of several large fertile clumps in this perennial species (up to 9 cm in diameter) resulted in an increase in the percentage of green material when plants became fertile. This change in the root/green material ratio did not occur in the shorter-lived *P. annua*.

Reproductive capacities

P. annua and *P. kerguelensis* produced a small mean number of inflorescences per fertile plant (Fig. 50.6), 1.50 and 5.25 respectively. However, some *P. kerguelensis* individuals can produce up to 36 inflorescences. The two *Poa* species show a similar pattern of spikelet distribution in the inflorescences, varying from three to 15. The number of flowers per spikelet in *P. kerguelensis* is significantly lower than in *P. annua* ($p<0.001$, t-test). In some inflorescences of both species, a few spikelets have no flowers. Visual examination of the seeds revealed that 58% of *P. annua* and 75% of *P. kerguelensis* flowers produce fertile seeds. These results allow an estimation of the mean seed production per fertile plant of 32.3 for *P. kerguelensis* and 11.2 for *P. annua*.

Without cold pre-treatment, *P. kerguelensis* had almost no success in germination (only two seedlings from 300 seeds). The storage of seeds for one month at 5 °C increased germination to 84% at 10 °C in the light and 76% at 20 °C in the light. Conversely, *P. annua* germinated easily at all temperatures without cold pre-treatment. The highest percentages occurred at

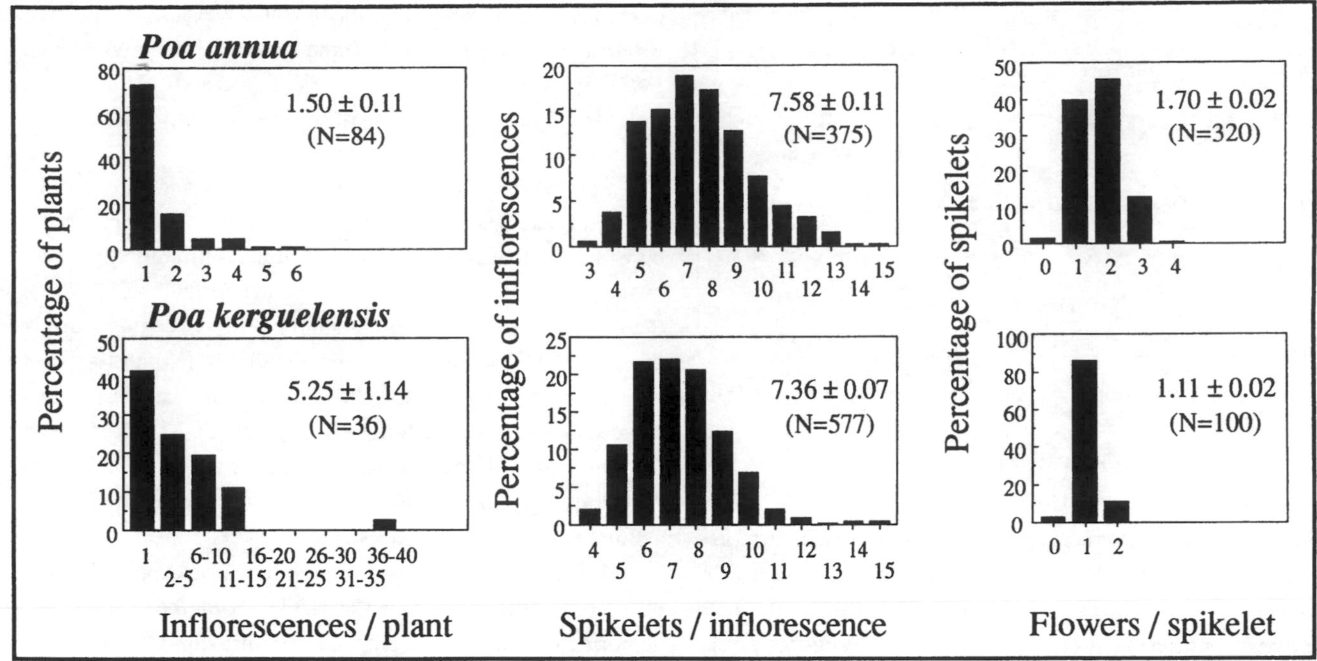

Fig. 50.6. Distribution of the main reproductive characteristics used to estimate the production of seeds on the Ampère Valley sandur. Mean, SE and number of replicates are given in each diagram.

10 °C in the light (89%) or in the dark (80%). The germination success was also high at 5 °C in the light (64%) whereas it was lower at 20 °C (32%). The cold pre-treatment did not significantly modify the percentage of germination at 10 °C, but it markedly increased the number of seedlings at 20 °C. More details on these germination experiments are given in Frenot & Gloaguen (1994).

DISCUSSION

The trends in temperature on Kerguelen are in accordance with other observations in the sub-Antarctic province. Warming occurred from the late 1960s to early 1980s at Marion Island (Smith & Steenkamp 1990) and at Macquarie Island (Adamson *et al.* 1988). Gordon & Timmis (1992) also reported rising trends in mean annual temperature over the last 30 years on South Georgia. However, some differences exist between the islands: at Marion Island, for example, surface air temperature was fairly constant between 1951 and 1964, whereas it significantly decreased on Kerguelen. These divergent observations result in the noticeable difference between the range of warming on those two islands between 1951 and 1988, estimated to be about 1.1 °C on Marion (Smith & Steenkamp 1990) and 0.1 °C on Kerguelen.

The changes in annual totals of precipitation since 1951 differ markedly between the islands. Smith & Steenkamp (1990) noted that the changes between air temperature and precipitation on Marion (late 1960s to mid 1980) were in opposite directions whereas the two parameters changed in the same way on Kerguelen. However, it seems that no precipitation deficiency is currently observed on Marion. Another difference between the changes in climate on Marion and Kerguelen is found in the annual hours of sunshine which increased significantly by about 4 h per year over 1951–1988 period on Marion, but overall no change was observed on Kerguelen despite slight oscillations.

Thus, the trends in climate change on Kerguelen cannot be generalized to the whole sub-Antarctic province and differences in the responses of the terrestrial plant communities are to be expected. The results for the two *Poa* species on the glacier forelands contribute to characterizing these responses on Kerguelen.

It appears that very few studies have been carried out on *Poa kerguelensis* (Cour 1958, Chastain 1958, Hennion 1992, Frenot & Gloaguen 1994), but there is an extensive literature on *Poa annua* because of its cosmopolitan status as a weed and turfgrass. In the sub-Antarctic region *P. annua* is one of the most widespread and successful introduced species (Walton 1975), but few data on its reproductive capacity in this biogeographical province are available.

Poa kerguelensis and *P. annua* are obviously pioneer species (Grubb 1987) on the Kerguelen glacier forelands because they occurred in the youngest unstabilized sites only a few months after the retreat of the glacier. This suggests both good seed dispersion and germination abilities. Wind could account for dispersal to the newly deglaciated areas. In laboratory studies, we have shown that *P. annua* seeds germinate well over a wide range of temperatures (5–20 °C), and the optimum, in terms of rate and maximum germination, is reached at a lower temperature (10 °C) than for the native species. Moreover, the *P. kerguelensis* seeds showed a clear post- harvest dormancy. In the Ampère Valley communities, no dormancy in *P. annua* seeds have been observed. This characteristic is common in primary successional colonizers (Chapin 1993). For these two reasons, the local temperature conditions could favour the germination of *P. annua* seeds.

P. annua generally produces abundant seeds. For instance,

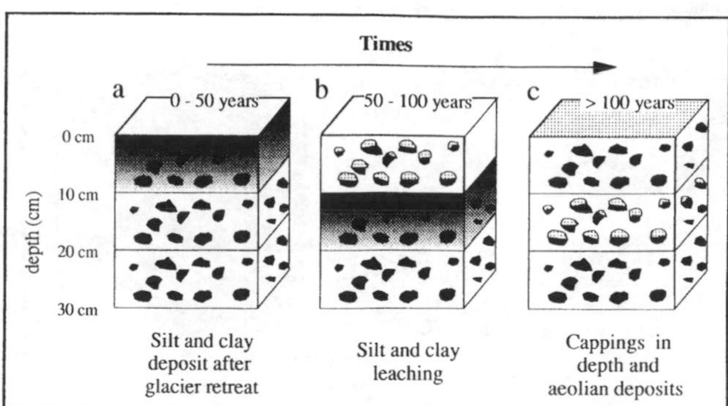

Fig. 50.7. Soil development in the Ampère Glacier forelands. See text for explanations.

Law *et al.* (1977) reported 450 inflorescences per fertile plant in a temperate pasture. This is in agreement with the usual high production of seeds in annual or biennial species and with their *r*-strategist status (Harper 1977). Despite a relatively high reproductive investment (Fig. 50.5), *P. annua* produces a very low number of seeds per plant on Kerguelen glacier forelands. This is probably related to the climatic and edaphic constraints: low temperatures, cryogenic perturbations and low nutrient availability. In consequence, the number of seeds produced by a *P. annua* fertile plant is lower than in the perennial native species, *P. kerguelensis*.

The environmental conditions of the Ampère Valley allow the development of both *P. annua* and *P. kerguelensis* populations in the early stages (0–30 years) of the primary succession. However, the harshness of these conditions seems to decrease the colonizing capacities of the alien *P. annua* and may consequently explain the greatest abundance of the native *P. kerguelensis*. This is accordance with the assumption that pioneers at sites poor in resources are typically long-lived and that perennial species are dominant (Crocker & Major 1955, Grubb 1987, Gray 1993).

The following stages of succession are characterized by the complete disappearance of *P. annua* and the strong decrease in the abundance of *P. kerguelensis* (Fig. 50.3). This phenomenon can be related to the major changes in the soil characteristics with time (Frenot *et al.* 1993). Two stages of soil development (Fig. 50.7a) occur during the early stages of succession (0–30 years).

(1) the first stage (sites F and E) is characterized by eutrophic conditions in the soil, instability of the substrate, importance of freeze–thaw cycles in relation to particle size and by a constant high humidity. (2) the second stage (site D) is characterized by a decrease in the meltwater run off, the absence of fresh loamy sediments and the stabilization of the coarse elements of the pavement at the surface after deflation of the smallest particles. As a result of the glacier retreat, the distance from the snout increases and frost is less efficient. Large amounts of fine particles, silt and clay occur in the first horizon, protected by the pavement. This stage contains all the requirements of soil fertility allowing the persistence of the two *Poa* species. Its onset is estimated at 10–15 years after the retreat of the glacier. But it is a very transient stage, since leaching (Paton 1978, *in* Matthews, 1993) quickly decreases the nutrient availability for plants. Colonization and establishment of the two *Poa* species takes place during these two soil development stages.

Later, the increase in the leaching processes induces a third stage, characterized by an impoverishment in fine particles in the upper horizon. Silt and clay are only preserved in this upper part of the profiles as cappings on the pebbles (Fig. 50.7b). However, these cappings have a low stability and they progressively decrease in number and in volume. Finally, more than 100 years after the glacier retreat, the silty cappings are transferred in a coarser facies to deeper horizons (Fig. 50.7c). The decrease in the abundance of *Poa* species coincides with the onset of this third soil stage (Site C) and marks the end of the potential interaction between the two species.

The presence of numerous fine roots in the cappings, easily observed in the field, and their absence in the soil matrix, demonstrate that the cappings are the principal nutrient and water resources for plants in the late stages. The root architecture of the species explains why the shallow-rooting *P. annua* disappears, because it cannot use these resources when they are deep in the soil. Conversely, the more-developed root system of *P. kerguelensis* allows the survival of plants. However, the disappearance of fine particles in the top soil results in the lack of microsites suitable for new seedling establishment and limits the recruitment to the plant population (Eriksson & Ehrlén 1992).

A thin layer of fresh aeolian clay and silt admixture develops under the pavement of the oldest sites (B and A), entrapped by the nearly continuous lichen cover (Fig. 50.7c). This leads, about 200 years after the retreat of the glacier, to an incipient A0/A1 horizon development. These new fine particle deposits allow the establishment of *P. kerguelensis* seedlings again (Fig. 50.3). The absence of *P. annua* in these sites could be related to the poor quality or quantity of nutrients in these aeolian deposits.

Smith (1990) considered that climate is undoubtedly the prime force responsible for determining ecological change and fluctuations. In Ampère Valley, the consequences of the recent climate changes on the colonization processes are not obvious. The temperature increase on Kerguelen between 1951 and 1993 (0.1 °C) is small and unlikely to influence the species composition of the plant communities, as assumed by Smith & Steenkamp (1990) at Marion Island, nor the reproductive performance of the two *Poa* species in the Ampère Valley. Conversely, the current low level of precipitation and the lack of fine particles in the soil matrix of the oldest sites may cause water stress. However, the colonization dynamics and the interaction between the two *Poa* species on the Ampère Glacier forelands seem essentially governed by the soil development, in terms of stabilization, cryoturbation, and nutrient and water accessibility. In conclusion, the current climatic modifications *per se* probably play only a minor (or indirect) role in the primary succession processes in the Ampère Valley.

However, this conclusion cannot even be generalized to the whole Kerguelen archipelago. Numerous field observations demonstrate the consequences of the low precipitation level on other species. For instance, the death of *Azorella selago* Hook. f. cushions on several islands in the Golfe du Morbihan (east of Kerguelen) may be related to the annual drought observed during the last few years. Similarly, this drought has dried up small ponds on the Péninsule Courbet in summer leading to the disappearance of water plant communities. On the other hand, *Acaena magellanica* Vahl, which is a species with very wide ecological tolerance (Walton 1976), currently appears to be becoming more widespread on Kerguelen and has an invasive behaviour in some eastern localities, suggesting a good drought resistance. A similar change in its distribution since the 1950s in the nearby Heard Island is reported by Scott (1990). There is no evidence that the current climatic conditions benefit the other native vascular plants.

These climate changes could also affect the colonization dynamics of some introduced plant parasites. *Albugo candida* (Pers.) Ktze, a fungal parasite of Brassicaceae, was observed on numerous *Pringlea antiscorbutica* Brown individuals and aphids were exceptionally abundant during the warm summer of 1994. However, these phenomena are more related to annual climatic conditions than to a recent climatic trend. Conversely, the successful establishments of the lepidopteran *Plutella xylostella* L. at Marion Island (Crafford & Chown 1987, Chown & Avenant 1991) and of the dipteran *Calliphora vicina* Robineau-Desvoidy at Kerguelen (Chevrier *et al.* 1997) have probably been facilitated by the high temperatures during the last decade.

Greene & Longton (1970) considered it unlikely that shortage of water was the most serious factor influencing vegetation diversity on the sub-Antarctic islands, considering that temperature was more important. The present study shows that water deficiency combined with high temperature may play an important role in biological changes : whilst the low level of precipitation during the early 1960s associated with low temperatures probably had only a slight impact on the vegetation, the combination of drought and high temperatures occurring today on Kerguelen could result in serious consequences. A monitoring of the responses of the native and introduced species to the climate changes is necessary, taking into account the biological and adaptive characteristics of the species, to answer questions on both biodiversity and community succession.

ACKNOWLEDGEMENTS

This study was supported by the Institut Français pour la Recherche et la Technologie Polaires, the CNRS and the French Environment Ministry. We are grateful to Gabriel Picot, Frank Foesser and Thierry Coulée for their assistance in the field and in the laboratory. We thank Drs P. M. Selkirk and R. I. Lewis Smith for their helpful comments on the manuscript.

REFERENCES

Adamson, D. A., Whetton, P. & Selkirk, P. M. 1988. An analysis of air temperature records for Macquarie Island: decadal warming, ENSO cooling and southern hemisphere circulation patterns. *Proceedings of the Royal Society of Tasmania*, **122,** 107–112.

Allison, I. F. & Keage, P. L. 1986. Recent changes in the glaciers of Heard Island. *Polar Record*, **23,** 255-271.

Chapin, F. S. 1993. Physiological controls over plant establishment in primary succession. In Miles J. & Walton, D. W. H., eds. *Primary succession on land.* Oxford: Blackwell Scientific Publications, 161–178.

Chastain, A. 1958. La flore et la végétation des îles de Kerguélen. *Mémoire du Museum National d'Histoire Naturelle, série B Botanique*, **11,** 1–136.

Chevrier, M., Vernon, P. & Frenot, Y. 1997. Potential effects of two alien insects on a sub-Antarctic wingless fly in the Kerguelen islands. In Battaglia, B., Valencia, J. & Walton, D. W. H., eds. *Antarctic communities: species, structure and survival.* Cambridge: Cambridge University Press, 424–431.

Chown, S. L. & Avenant, N. 1991. Status of *Plutella xylostella* at Marion Island six years after its colonisation. *South African Journal of Antarctic Research*, **22,** 37–40.

Cour, P. 1958. A propos de la flore de l'archipel de Kerguelen. *Terres Australes et Antarctiques Françaises, Paris*, **4-5,** 10–32.

Crafford, J. E. & Chown, S. L. 1987. *Plutella xylostella* L. (Lepidoptera: Plutellidae) on Marion Island. *Journal of the Entomological Society of South Africa*, **50,** 259–260.

Crocker, R. L. & Major, J. 1955. Soil development in relation to vegetation and surface age at Glacier Bay, Alaska. *Journal of Ecology*, **43,** 427–448.

Eriksson, O. & Ehrlén, J. 1992. Seed and microsite limitation of recruitment in plant populations. *Oecologia*, **91,** 360–364.

Frenot, Y. & Gloaguen, J. C. 1994. Reproductive performance of native and alien colonizing phanerogams on a glacier foreland, Îles Kerguelen. *Polar Biology*, **14,** 473–481.

Frenot, Y., Gloaguen, J. C., Picot, G., Bougère, J. & Benjamin, D. 1993. *Azorella selago* Hook. used to estimate glacier fluctuations and climatic history in the Kerguelen Islands over the last two centuries. *Oecologia*, **95,** 140–144.

Gordon, J.E. & Timmis, R.J. 1992. Glacier fluctuations on South Georgia during the 1970s and early 1980s. *Antarctic Science*, **4,** 215–226.

Gray, A. J. 1993. The vascular plant pioneers of primary successions: persistence and phenotypic plasticity. In Miles, J. & Walton, D. W. H., eds. *Primary succession on land.* Oxford: Blackwell Scientific Publications, 179–191.

Greene, S. W. & Longton, R. E. 1970. The effects of climate on Antarctic plants. In Holdgate M. W., ed. *Antarctic ecology.* London: Academic Press, 786–800.

Grubb, P. J. 1987. Some generalizing ideas about colonization and succession in green plants and fungi. In Gray A. J., Crawley, M. J. & Edwards, P. J., eds. *Colonization, succession and stability.* Oxford: Blackwell Scientific Publication, 81–102.

Harper, J. L. 1977. *Population biology of plants.* London: Academic Press, 892 pp.

Hennion, F. 1992. *Etude des caractéristiques biologiques et génétiques de la flore endémique des îles Kerguelen.* Thèse du Muséum National d'Histoire Naturelle, Paris. 264 pp.

Houghton, R. A. & Woodwell, G. M. 1989. Global climatic change. *Scientific American*, **260,** 2–10.

Law, R., Bradshaw, A. D. & Putwain, P. D. 1977. Life history variation in *Poa annua*. *Evolution*, **31,** 233–246.

Matthews, J. A. 1993. *The ecology of recently deglaciated terrain. A geoecological approach to glacier forelands and primary succession.* Cambridge: Cambridge University Press, 386 pp.

Pearman, G. I. 1991. Changes in atmospheric chemistry and the greenhouse effect – a southern hemisphere perspective. *Climatic Change*, **18,** 131–146.

Scott, J. J. 1990. Changes in vegetation on Heard Island 1947–1987. In

Kerry, K. R. & Hempel, G., eds. *Antarctic ecosystems: ecological change and conservation*. Berlin: Springer-Verlag, 61–76.

Smith, R. I. L. 1990. Signy Island as a paradigm of biological and environmental change in antarctic terrestrial ecosystems. In Kerry, K. R. & Hempel, G., eds. *Antarctic ecosystems: ecological change and conservation.* Berlin: Springer-Verlag, 32–50.

Smith, V. R. & Steenkamp, M. 1990. Climatic change and its ecological implications at a sub- Antarctic island. *Oecologia*, **85,** 14–24.

Walton, D. W. H. 1975. European weeds and other alien species in the sub-Antarctic. *Weed Research,* **15,** 271–282.

Walton, D. W. H. 1976. Dry matter production in *Acaena* (Rosaceae) on a sub-Antarctic island. *Journal of Ecology*, **64,** 399–415.

51 Impacts of ozone depletion on Antarctic organisms

HARVEY J. MARCHANT

Australian Antarctic Division, Channel Highway, Kingston, Tasmania 7050, Australia

ABSTRACT

Springtime stratospheric ozone depletion over Antarctica leads to a period in October and November when incident solar UV-B radiation (280–320 nm) is at least as high as at the summer solstice. That solar UV-B radiation is damaging to a wide range of organisms, with the shorter wavelengths generally having the greatest impact, has been well documented. The terrestrial plants of Greater Antarctica (algae, lichens and bryophytes) that have been examined possess UV absorbing compounds and differ in their responses to UV exposure. Ozone depletion leads to an alteration in the underwater light climate, which alters the balance between spectrally specific processes including photosynthesis, photoinhibition and photoprotection, and as a consequence primary productivity in the Antarctic marginal ice- edge zone is reduced by at least 6 to 12%. Some Antarctic marine phytoplankton synthesize UV-B photoprotective compounds and most, if not all, have repair mechanisms. Interspecific differences in productivity and growth in response to enhanced UV-B exposure may lead, or have lead, to changes in phytoplankton species composition. Antarctic marine bacteria may also be stressed by present levels of UV-B exposure, as laboratory investigations indicate that their viability and extracellular enzymic activity are markedly impaired by UV-B exposure. Antarctic birds and mammals are unlikely to be directly effected by UV-B. Whether they are subject to indirect effects resulting from perturbations of the microbial components of the food web has yet to be determined. Other than finding that some possess UV screening compounds, there has been no work to determine whether UV affects invertebrates and fish. Photolysis has been proposed as the rate limiting step for the degradation of biologically refractory dissolved organic material and increased UV flux is likely to increase this rate. Investigations on the photochemistry of organic compounds in Antarctic seawater are required.

Key words: Antarctica, marine, organisms, ozone depletion, Southern Ocean, terrestrial, UV-B.

INTRODUCTION

Since the mid-1970s there has been a marked decrease in the concentration of stratospheric ozone over Antarctica during springtime from around 300 Dobson Units (DU) to frequently less than 150 DU and on occasion to less than 100 DU (Farman *et al.* 1985, Madronich 1993)(Fig. 51.1). Only some 7% of solar radiation reaching the surface of the Earth is in the UV region of the spectrum, <400 nm. Ozone in the atmosphere is the principal agent for the complete absorption of UV-C (200–280 nm) and partial absorption of UV-B (280–320 nm). UV-A (320–400 nm), similar to 'visible light' or photosynthetically active radiation (PAR), with a spectral range of 400–700 nm, is not significantly absorbed by atmospheric ozone. A consequence of ozone depletion is the atmosphere becoming more transparent to shorter wavelengths of UV-B. The springtime UV irradiance at Antarctic coastal sites has been found to be as high or higher than at the summer solstice (Frederick & Snell 1988).

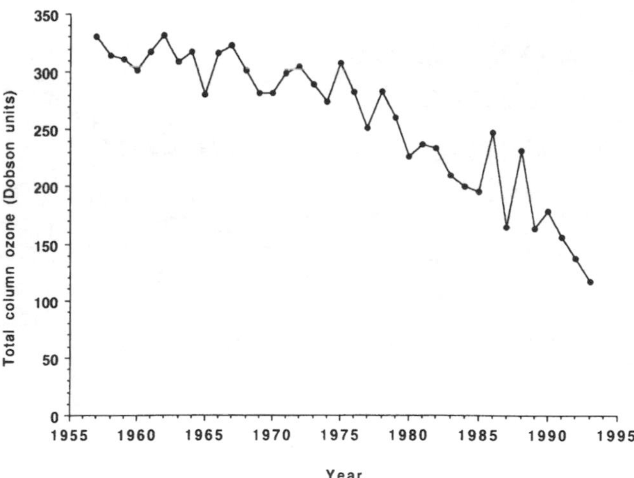

Fig 51.1. Mean total ozone concentrations during October over Halley Bay, Antarctica. NB 1993 value is preliminary. (Redrawn from Farman *et al.* 1985 with recent data kindly provided by Brian Gardiner, British Antarctic Survey.)

UV-A exposure has both harmful and beneficial effects on organisms while UV-B is strongly absorbed by various cellular components, notably nucleic acids, inhibiting processes of growth and reproduction, with the damaging effects being substantially greatest at shorter wavelengths (Tevini 1993a). As small a change in wavelength as 7 nm in the UV-B spectrum can produce an order of magnitude increase in effect. The spectral effects of UV on organisms are usually described as action spectra or spectral biological weighting functions, which reflect the net effect of the damage to organisms and processes of repair.

Not surprisingly, the effects of enhanced UV exposure have attracted considerable attention (see Tevini 1993b for a comprehensive review). The potential direct effects of enhanced UV-B radiation on the Antarctic biota as a consequence of the 'ozone hole' have stimulated much of this research. Weiler & Penhale (1994) provide a comprehensive review of this work. Here, I summarize the impacts of increased UV irradiance on Antarctic terrestrial and marine organisms and the responses they employ to minimize these impacts. I also briefly consider the potential indirect biological effects of UV by photolysis of organic compounds in aquatic environments and indicate directions for future research.

BIOLOGICAL IMPACTS AND RESPONSES

Terrestrial systems

The principal components of the terrestrial flora of continental Antarctica are cyanobacteria, algae, lichens and mosses. Other than during the winter or when not covered by snow these organisms are directly exposed to solar UV radiation. As yet unidentified UV absorbing compounds were evident in 85% of a survey of 35 Antarctic lichens but no UV absorption was found in the four species of moss examined (Karentz *et al.* 1991). The Antarctic mosses *Grimmia antarctici* and *Bryum pseudotriquetrum* contain compounds which absorb around 323 nm. These compounds apparently function as effective UV filters, as field experiments indicate that present ambient UV radiation does not contribute to photosynthetic stress in these bryophytes. The concentration of UV absorbing pigments in both of these mosses was found to be high in early summer, when UV irradiance is high, and declined over summer (Adamson & Adamson 1992). The terrestrial Antarctic green alga *Prasiola* and the cyanobacterium *Nostoc* grow in regions usually with a northerly aspect and enriched with nutrients derived from penguin rookeries or the nesting sites of other birds. *Prasiola* contains low concentrations of compounds which absorb at around the same wavelengths as the mosses *Grimmia* and *Bryum*. Chlorophyll concentration and photosynthesis in this alga are depressed by ambient UV exposure, with no apparent change in the ratio of the concentrations of UV-B absorbing compounds to chlorophyll (Post & Larkum 1993). Preliminary investigations indicate that photosynthesis in the heavily pigmented lichen *Usnea* is not affected by ambient UV exposure (Adamson & Adamson 1992). Vincent & Quesada (1994) have shown that Antarctic cyanobacteria differ substantially in their ability to cope with UV radiation. They point out that during exposure of these organisms to UV-B before or early in the growing season, when they are frozen, biosynthetic protective and repair mechanisms are unlikely to be active. Thus, UV-B induced damage could remain unchecked until the recommencement of active metabolism in the summer. As Wynn-Williams (1994) concludes, because most of the Antarctic terrestrial flora are cosmopolitan they are expected to retain the potential for UV screening and/or repair and acclimatizing to increasing levels of UV radiation. The extent to which they can continue to do this is uncertain. In addition, as yet there has been no detailed investigations of the consequential effects of UV-B enhancement on the cryptic invertebrate communities. Nematodes, tardigrades, rotifers and arthropods which colonize the terrestrial vegetation would be expected to be effected by any diminution of its growth.

Lakes

Surprisingly little has been reported on the impacts of UV-B on the organisms that inhabit Antarctic lakes considering the level of interest and our present understanding of the biology of these environments on sub-Antarctic islands as well as on the continent itself. Eklund (1992) reported reduction in the motility and survival of *Cryptomonas* sp. from a pond on South Georgia but no apparent effect of UV-B on the growth of the chlorophytes *Botryococcus* and *Stauastrum* or the cyanophyte, *Lyngbya*. *Cryptomonas* was collected from a pond with a high concentration of humic material, whereas the lake from which the other organisms came had a lower concentration of these compounds. The high transparency of many of the lakes on the Antarctic continent suggests that UV-B penetration is likely to be significant and that growth of benthic as well as planktonic organisms may be effected. The few lakes that have been investigated are characterized by low species diversity. As will be discussed in the next section, marine phytoplankton exhibit marked interspecific differences in their responses to UV-B exposure, especially productivity and growth rates. It is likely therefore that the community structure of species depauperate lakes could be effected by increased UV-B radiation.

Marine phytoplankton and sea-ice algae

Because of their role as primary producers, marine phytoplankton have been the principal focus of research on the impact of ozone depletion in Antarctica. Besides phytoplankton, the marine environment contains a substantial biomass of bacteria and protozoa and there are complex interactions between them (Azam *et al.* 1991, Garrison 1991, Marchant & Murphy 1994). Little attention has been given to the impacts of UV-B irradiation on these other microbial components of marine ecosystems.

There is a considerable body of evidence indicating that UV exposure affects phytoplankton, including inhibition of primary production (eg Smith *et al.* 1992), growth and reproduction (Davidson & Marchant 1994), nitrogen metabolism (Döhler 1990, Döhler *et al.* 1987), and motility and phototactic orientation (Häder & Worrest 1991). That Antarctic marine phytoplankton are UV stressed was first shown by El-Sayed *et al.* (1990). In their experiments mixed phytoplankton assemblages were exposed to four treatments – enhanced UV, ambient UV, reduced UV-B, and excluded UV-A and UV-B. They found a marked reduction of primary productivity in all treatments other than that screened from UV-B. In another pioneering study to establish the influence of depth penetration of different wavelengths of UV on the productivity of Antarctic phytoplankton, Holm-Hansen *et al.* (1989) demonstrated that incident solar UV radiation significantly depressed photosynthesis to a depth of 10–15 m. The radiation in the 305–350 nm spectral range accounted for some 75% of the inhibition of photosynthesis. Following these early studies there have been numerous investigations to elucidate the underwater radiation climate and the impacts and responses of Antarctic marine phytoplankton to UV exposure (see Bidigare 1989, Karentz 1991, Voytek 1990, Smith *et al.* 1992 for reviews). Despite these investigations, much more work is required to ascertain the species-specific responses necessary to develop a quantitative assessment of the impacts of UV enhancement.

Inhibition of photosynthesis by phytoplankton, either in Antarctica or more temperate waters, can result from exposure to UV-A and PAR as well as UV-B (Smith *et al.* 1992, Helbling *et al.* 1994, Neale *et al.* 1994, Prézelin *et al.* 1994). Helbling *et al.* (1992) found that UV-A was responsible for over 50% of the photoinhibition, with wavelengths shorter than 305 nm accounting for 15 to 20% of the inhibition. Neale *et al.* (1994) showed that the action spectra, or biological weighting function, of UV inhibition of photosynthesis varied with both absolute amount and ratio of UV-B:UV-A and PAR but not UV:PAR. They also found that the larger microplanktonic organisms were more inhibited than the smaller nanoplanktonic forms, with the diatom *Chaetoceros* stimulated to produce resting spores and, together with *Thalassiosira*, exhibiting greatest morphological change. Tropical phytoplankton exhibited markedly higher resistance to UV exposure than Antarctic organisms. Phytoplankton from below the mixed layer in Antarctic waters appeared to be more tolerant of solar radiation than organisms from below the mixed layer in tropical waters. These authors conclude that this is likely to be due to the relative instability and deeper mixing of Southern Ocean waters and the photoadaptive processes of those organisms from surface waters. The role of vertical mixing in mitigating photoinhibition and UV-B exposure has been addressed by several authors (Kullenberg 1982, Smith & Baker 1982, Cullen & Lesser 1991, Helbling *et al.* 1994). Cullen & Lesser (1991) found inhibition of photosynthesis of the marine diatom *Thallasiosira pseudonana* by UV-B to be a function of irradiance as well as of dose, so that for equal doses a short, high level exposure is more inhibitory than a longer exposure of lower irradiance. They conclude that conventional incubation times of 4–24 hours for the measurement of primary productivity will be representative only if the water column is relatively stable over the same timespan.

Despite high concentrations of macronutrients, for most of the year the standing crop of phytoplankton and primary productivity of the Southern Ocean are low. Productivity is light limited during the winter when the sun angle is low and the sea is ice and snow covered. In summer the high wind forcing leads to deep mixing and low production. Overall, the annual productivity of the Southern Ocean is similar to that of oligotrophic oceanic waters. The most productive part of the Southern Ocean is the marginal ice zone (MIZ) associated with the southward retreating sea ice in spring (Smith & Nelson 1986). The release of freshwater from the melting ice produces a mixed zone some 10–20 m deep in which phytoplankton reach high numbers in the high light, nutrient-rich environment (Smith 1987). The major components of the phytoplankton in the MIZ are diatoms and the colonial stage in the life cycle of the prymnesiophyte *Phaeocystis* (Fryxell & Kendrick 1988, Davidson & Marchant 1992). Grazing and sedimentation are apparently the principal fates of much of this ice-edge production (Riebesell *et al.* 1991, Karl 1993). This bloom in the MIZ coincides with the springtime ozone hole and the subsequent period of high UV-B radiation over the Southern Ocean.

Measurements of UV-B attenuation in Antarctic waters have shown penetration of biologically damaging wavelengths to at least 10 m depth (Prézelin *et al.* 1994). Transmission often reached 20–30 m under conditions similar to those existing in the MIZ (Karentz & Lutze 1990, Gieskes & Kraay 1990). Evidence of photoinhibition has been detected to depths of 25 m in the MIZ (Smith *et al.* 1992). Estimates of mixing times from the surface layer of the world's oceans to a depth of 10 m range from 30 minutes to hundreds of hours (Denman & Gargett 1983). The depth of the pycnocline in the MIZ can be ≤10 m for periods of up to 6 days (Veth 1991). Wind forcing can greatly reduce this time, but the high biological activity in the MIZ results from its shallow mixing. Organisms in this surface layer will have little opportunity to avoid UV exposure. Thus, productivity incubations over a timespan of hours to tens of hours will be representative of the natural conditions.

The effect of the nutritional status of phytoplankton on their UV tolerance has only been explored in a few studies, none apparently in Antarctica. Cullen & Lesser (1991) found that primary productivity of nitrate-limited cultures of the diatom *Thallassiosira pseudonana* were some nine times more sensitive to UV-B irradiation than nutrient-replete cultures. Behrenfeld *et al.* (1994) report a UV-B induced reduction of 20% in photosynthetic rates in nutrient-replete cultures of another diatom, *Phaeodactylum tricornutum,* and 36% reduction in nutrient-depleted cultures of this alga. However, in this study the specific growth rate and biomass remained essentially unchanged in nutrient-limited cultures but were reduced by 2 to 16% in those that were nutrient replete. This decrease in carbon fixation in nutrient-depleted cultures but without a corresponding decrease in growth rate and biomass implies an inconsistency in the carbon mass balance. Behrenfeld *et al.* (1994) point out that phytoplanktonic photosynthetic production is often greater than nutrient-limited organisms' utilization rate and this 'excess' photosynthate is secreted. Thus, if the amount of carbon utilized by the organisms was determined by a limiting nutrient, the UV-B induced decrease in carbon fixation would result in a reduction in the excretion of photosynthate. These authors suggest that the growth of phytoplankton in nutrient-rich areas of the ocean may be more susceptible to UV-B inhibition than those in regions where nutrients limit growth. Their findings have obvious direct relevance to the phytoplankton of Antarctic waters, where macronutrient (nitrate, phosphate and silicate) limitation of growth is only rarely encountered (Lizotte & Sullivan 1992).

It has become clear that different organisms use different strategies to protect themselves from UV exposure and to repair UV induced damage. In a survey of 57 species of intertidal, subtidal and planktonic Antarctic marine organisms, comprising one fish, 48 invertebrates and eight algae, Karentz *et al.* (1991) found that most contained mycosporine-like amino acids (MAAs), a group of water-soluble compounds which absorb strongly between 310 and 360 nm. Pigmentation in the Antarctic marine macrophyte *Palmaria* changes through the year, the highest concentration of pigments being found in summer (Post & Larkum 1993). These authors suggest that these species have some degree of natural biochemical protection from UV exposure. Although this may well be true, there is clear evidence that just because an organism has UV absorbing compounds it is not necessarily better equipped to survive UV exposure than organisms with low concentrations or which lack these compounds. Antarctic colonial *Phaeocystis* has high concentrations of UV-B absorbing compounds, while the motile stage in the life cycle of this alga and many Antarctic diatoms either lack or have low levels of UV-B absorbing compounds. The motile cells of *Phaeocystis* are much more susceptible to UV damage than the colonies of this alga, but diatoms which have low concentrations of UV-B absorbing compounds are more UV-B tolerant than Antarctic colonial *Phaeocystis* (Davidson & Marchant 1994). Thus diatoms apparently use mechanisms other than screening to mitigate UV damage. It is clear that there is substantial interspecific variation in the response of phytoplankton to UV-B exposure (Davidson & Marchant 1994, Karentz 1994, Vernet *et al.* 1994). As a consequence, ozone depletion is likely to cause subtle changes in species composition and interactions. Understanding the nature and magnitude of these changes requires elucidation of the UV photobiology of key species.

Cullen *et al.* (1992) showed that UV-A significantly reduces primary productivity of a non-Antarctic marine diatom and a dinoflagellate and that the effect of UV-B was even more profound. On the basis of their model, they predict a reduction of near-surface primary production by 12 to 15% due to the Antarctic 'ozone hole'. Based on work conducted in the MIZ, Smith *et al.* (1992) estimated a minimum of 6 to 12% reduction in phytoplanktonic production associated with the 'ozone hole'. This estimate is now considered to be too conservative (Prézelin *et al.* 1994). Such assessments of inhibition of productivity should be viewed in the context of an estimated interannual variation of around ±25% in primary productivity in the MIZ (Smith *et al.* 1988).

Most of the investigations on the impacts of UV-B on Antarctic algae have concentrated on the phytoplankton, with relatively little attention being given to sea-ice organisms. At its maximum extent the sea ice around Antarctica covers an area of some 17×10^6 km^2 (Zwally *et al.* 1983). Sea ice is colonized by a microbial community comprising microalgae, protozoa and bacteria. The algal biomass and productivity are small relative to the phytoplankton but cannot be ignored (Rivkin *et al.* 1989). Sea ice during spring is relatively transparent to UV-B (Trodahl & Buckley 1989), transmitting up to 10% of incident radiation (Ryan 1992). Productivity of algae isolated from sea ice was 5% lower in treatments exposed to UV-B than in screened controls (Ryan & Beaglehole 1994). These authors conclude that the present elevated levels of UV-B have little effect on the bottom sea-ice algae.

The long-term impacts of UV-B exposure on organisms and on ecosystems is very difficult to assess. Experimental investigations clearly indicate that marine phytoplankton exhibit substantial interspecific differences in their responses to UV-B exposure. As a consequence, changes in species composition favouring those organisms that are more tolerant of UV would be expected. Since Antarctic springtime ozone depletion has been increasing since the mid-1970s a change in species composition may already have happened and the apparent robustness of present day sea-ice algae and phytoplankton to UV-B exposure reflects that. There is no clear evidence reported of a change in phytoplanktonic species composition over the last two decades, which unless it was dramatic would be extremely difficult to detect anyway. Detailed pre-ozone hole quantitative investigations of species composition are scarce and variations in temporal and spatial distribution as well as interannual variability makes identifying differences in species composition extremely difficult. The sedimentary record, however, provides some evidence of a change in species composition. Examination of diatom assemblages in cores, spanning the last 20 years, taken from anoxic basins in Antarctic fjords which usually remain ice-covered for most of the summer have revealed that although there has been no significant change in the accumulation and species composition of phytoplanktonic diatoms there has been a decline in the sedimentation of some sea-ice diatoms. Species showing some evidence of decline, including *Entomoneis kjellmannii*, *Nitzschia stellata* and *Berkeleya rutilans*, were all from the bottom-ice community (McMinn *et al.* 1994). This study is important for two reasons; it is the first evidence of a shift in species composition that correlates with stratospheric ozone depletion and it indicates that UV induced changes to the sea-ice community may be more profound than previously thought.

Marine bacteria

Over the last decade our understanding of the role of bacteria in marine processes has increased dramatically. It is now recognized that these abundant organisms, which can consume up to 50% of marine primary production, play a pivotal role in biogeochemical pathways, trophodynamics and such physical processes as light scattering and trace metal binding (Cho & Azam 1990, Simon *et, al.* 1992). Their importance in the Southern Ocean is no less than in other parts of the world's oceans (Cota *et al.* 1990, Sullivan *et al.* 1990, Fiala & Delille 1992).

Compared with the work on phytoplankton there is relatively little published on the effect of solar UV on bacteria in marine environments (cf. Bailey *et al.* 1983, Sieracki & Sieburth 1986). Much of the impact of sunlight on bacteria has been ascribed to UV-A. However, Herndl *et al.* (1993) reported a 40% reduction of bacterial activity in the top 5 m in the Adriatic Sea resulting from a 30 minute solar UV-B exposure. In addition, they found that after four hours exposure to surface solar irradiance, bacterial extracellular enzymic activity was reduced by about 67%. However, enzymic activity in bacteria-free seawater was reduced by 50 to 60% indicating that both enzyme production is inhibited and the enzymes themselves are broken down by UV-B photolysis. Because of the importance of bacteria in marine biogeochemical cycles they concluded that UV-B irradiance may have considerable influence on marine processes involving them.

Evidence of the effect of UV-B on mixed populations of Antarctic marine microorganisms (bacteria, algae and protozoa) comes from the investigation of Vosjan *et al.* (1990), who reported an average decrease of around 75% in ATP concentration of these organisms collected from depths to 30 m following five hours exposure to UV-B at 1.35 W m^{-2}. This result and the investigation by Herndl *et al.* (1993) prompted us to conduct a preliminary study on the UV tolerance of Antarctic marine bacteria. Three strains of Antarctic marine bacteria were incubated in an artificial UV-B illumination gradient approximately equivalent to 50 to 600% of surface mid-summer UV-B irradiance measured at 69° S (Davidson & Marchant 1994). The viability of these strains was reduced by about 30% following a 30 minute exposure at 12 °C (their optimum growth temperature) to 1.0 W m^{-2} (Fig. 51.2). As bacterial protein was lost during this exposure it appeared that the cells were lysing. The activity of extracellular alkaline phosphatase was also depressed following exposure of the bacterial to UV-B. However, it was not ascertained whether this diminution in activity was due to inhibition of the production of the extracellular enzyme by the bacteria or photolysis of it in solution. Despite the obvious inherent uncertainties of extrapolating laboratory experimental data to the field situation, our laboratory findings mirror the field studies of Herndl *et al.* (1993) and indicate that Antarctic marine bacteria may be under stress from ambient UV-B exposure. The extent to which this is the case and the consequences are yet to be determined.

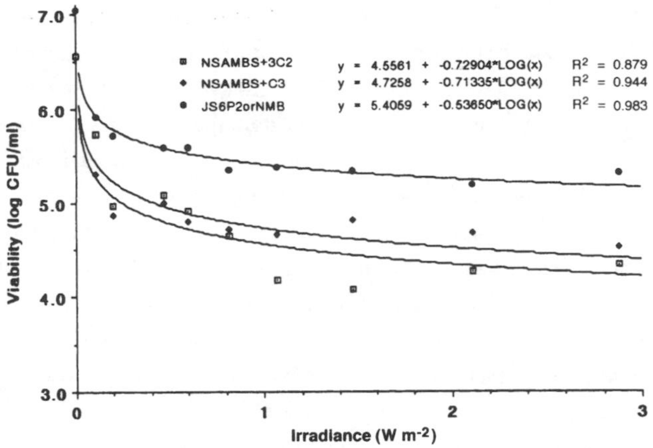

Fig. 51.2. Viability (colony-forming units ml^{-1}) of three strains of Antarctic marine bacteria following 30 min. exposure at 12 °C to various irradiances of UV-B. (See text for details of methodology.)

Higher trophic levels

Of the planktonic invertebrates screened for MAAs, highest concentrations were found in the pteropod *Limacina helicina*, the copepod *Calanus propinquus* and the euphausiid *Euphausia superba*, organisms which live high in the water column. However, as Karentz *et al.* (1991) stress, correlations of MAA concentration with habitat are inconsistent as not all planktonic organisms possess high MAA concentrations. Although the planktonic ctenophores and salps lacked MAAs they contained methanol-soluble UV absorbing compounds.

Many Antarctic seals and penguins spend a substantial amount of time out of the water in spring and early summer when breeding. Crabeater and Weddell seals pup on the sea ice in October and November. Adélie and emperor penguins are also rearing chicks at this time. These animals tend to favour localities with a northerly exposure, presumably to shelter from prevailing southerly winds and to gain warmth from the sun. In such locations they are exposed to maximum solar UV irradiation. As UV is known to promote the incidence of cataract and various carcinoma of ocular and surrounding tissue (Charman 1990), concern has been expressed about the impact of increased UV on the eyes of these animals. Hemmingsen & Douglas (1970) report that the threshold for UV damage to the corneas of Antarctic birds is higher than for those from temperate regions. As previously discussed (Marchant 1994), the tears of southern elephant seals absorb strongly at wavelengths shorter than 300 nm. Seals lack a naso-lachrymal duct and as a result tears run over the face of these animals. Whether the tears of other species of seal absorb at these wavelengths and whether these tears provide UV protection to the eyes and surrounding hairless tissue remains to be ascertained.

Seasonal variation in serum vitamin D in humans and sheep has been interpreted as a consequence of the role of sunlight on the precursors of this vitamin. An investigation of serum vitamin D concentration in crabeater and Weddell seals and Adélie and chinstrap penguins revealed that the seals have large interspecific differences in vitamin concentrations and penguin sera contained only low concentrations with no evidence of prolonged exposure to solar radiation (Griffiths & Fairney 1988). Thus, it appears that direct impacts of UV on birds and mammals are minimal and any adverse effects are likely to be indirect, resulting from perturbations to the food web.

Photolysis of dissolved organic material

Most of the work on the impacts of UV-B on natural systems has concentrated on living organisms. Only few studies report UV photolysis of dissolved organic material (DOM) and dissolved inorganic material (DIM) and none, to my knowledge, in Antarctic waters, despite there being good reasons to believe there are likely to be significant effects. Oceanic dissolved organic carbon (DOC) is one of the largest reservoirs of carbon on Earth yet the processes regulating the cycling of this material are not well understood. DOC in the oceans at depths below 500 m is apparently principally composed of biologically refractory compounds and is some thousands of years old, indicating that its turnover is slow (Williams & Druffel 1987). Photochemical degradation by sunlight of this material to biologically labile and/or volatile compounds has been suggested as the rate-limiting step for its removal (Mopper *et al.* 1991). They conclude that this rate will increase with increasing solar UV-B flux. Concentrations of DOM can be extremely high in Antarctic surface waters during the early summer bloom of phytoplankton, especially if *Phaeocystis* is the major component of the bloom (Davidson & Marchant 1992). The effect of UV-B on the dynamics of the cycling of this material has yet to be elucidated.

Dimethyl sulphide (DMS) ventilated from the sea is estimated to constitute about half of the total biogenic input of sulphur to the atmosphere. DMS is important as it apparently is a precursor of tropospheric aerosols and cloud condensation nuclei thereby influencing global radiative balance and therefore climate (Bates *et al.* 1987, Charlston *et al.* 1987). The precursor of DMS is dimethylsulphoniopropionate (DMSP), which is produced by phytoplankton, especially some prymnesiophytes and dinoflagellates. In Antarctic waters *Phaeocystis* has been estimated to be the source of around 10% of the global production of DMS (Gibson *et al.* 1989). As well as ventilation to the atmosphere, DMS is removed from seawater by various physical, chemical and biological pathways. Kiene & Bates (1990) showed that DMS removal in tropical waters is dominated by biological processes, with microbial consumption of DMS being more than ten times faster than atmospheric ventilation. These authors stress that DMS concentrations are intimately linked to both the structure and level of activity of the microbial food web. In light of the evidence outlined here of phytoplanktonic and bacterial activities being substantially inhibited by present levels of UV radiation, attention should be given to the impact of UV-B not only on the organisms that synthesize the precursor of DMS but also on those organisms involved directly or indirectly in its utilization. In addition, another fate of DMS in seawater is photochemical oxidation (Brimblecombe & Shooter 1986). Just as increasing flux of UV-B is thought to increase the rate of photochemical degradation of DOC (Mopper *et al.* 1991), so too is the rate of degradation of dissolved DMS.

Conclusions and priorities for future work

The physics, chemistry and climatology that give rise to the springtime diminution of stratospheric ozone over Antarctica are becoming well understood. This research provides a robust scientific backdrop for international agreements and national policies to limit the production and use of halogenated compounds that contribute to stratospheric ozone depletion. In contrast, our understanding of the effects of increased UV-B irradiation on biological systems is not as well developed. There is clear evidence from temperate waters that, in addition to phytoplankton and bacteria, the survival and growth of marine metazoa including zooplankton (Damaeker & Dey 1983) and larval fish (Hunter *et al.* 1981), are affected by UV-B exposure. At present there are no reports of the effects of UV-B on such organisms from Antarctic waters. There is also good evidence from temperate and tropical waters that biologically refractory DOM is photochemically degraded to biologically labile and volatile compounds. Thus, in addition to investigations of the impacts of UV-B on organisms, there is the need to understand better the impact of UV exposure on DOM in Antarctic waters.

There is a pressing need for a better knowledge of UV-B induced impacts on biological systems for us to be able to predict confidently the consequences, especially in concert with other global changes such as the increasing concentration of atmospheric CO_2 and warming. This has been stressed in a number of workshops including UNEP (1991), SCOPE (1993) and in the SCAR document *The role of the Antarctic in global change: an international plan for a regional research program* (Weller 1993). There is no internationally coordinated program to investigate UV impacts in Antarctica. Rather, because of the diversity of organisms and systems effected by UV-B exposure and the inadvisability of establishing another international program overlapping existing ones, it is logical that investigations on UV impacts be incorporated into, or conducted as an adjunct to, existing programs such as SCAR Global Change in Antarctica (GLOCHANT), SCAR Ecology of the Antarctic Sea-Ice Zone (SCAR-EASIZ), Southern Ocean Joint Global Ocean Flux Study (SO-JGOFS) and Global Ocean Ecosystems Dynamics Research (SO-GLOBEC) and Biological Investigations of Terrestrial Antarctic Systems (BIOTAS).

ACKNOWLEDGEMENTS

I am most grateful to Drs Barbara Prézelin and Mike Behrenfeld for providing me with their unpublished manuscripts and to Dr Patrick Quilty and Andrew Davidson for their comments on the manuscript. This review was largely written while I was an exchange scientist with the 35th Japanese Antarctic Research Expedition. I thank the leader and members of JARE 35 and the captain, officers and crew of the icebreaker *Shirase* for their hospitality and companionship.

REFERENCES

Adamson, H. & Adamson, E. 1992 . Possible effects of global climate change on Antarctic terrestrial vegetation. In Quilty, P., ed. *Impacts of climate change on Antarctica*. Canberra: Department of the Arts, Sport, the Environment, Tourism and Territories, 52–62.

Azam, F., Smith, D. C. & Hollibaugh, J. T. 1991. The role of the microbial loop in Antarctic pelagic ecosystems. *Polar Research*, **10**, 239–243.

Bailey, C. A., Neihof, R. A. & Tabor, P. S. 1983. Inhibitory effect of solar radiation on amino acid uptake in Chesapeake Bay bacteria. *Applied and Environmental Microbiology*, **46**, 44–69.

Bates, T. S., Charlson, R. J. & Gammond, R. H.1987. Evidence for the climatic role of marine sulphur. *Nature*, **329**, 319–321.

Behrenfeld, M. J., Lee, H. II & Small, L. F. 1994. Interactions between nutritional status and long-term responses to ultraviolet-B radiation stress in a marine diatom. *Marine Biology*, **118**, 523–530.

Bidigare, R. R. 1989. Potential effects of UV-B radiation on marine organisms of the Southern Ocean: distributions of phytoplankton and krill during austral spring. *Photochemistry and Photobiology*, **50**, 469–477.

Brimblecombe, P. & Shooter, D. 1986. Photo-oxidation of dimethylsulphide in aqueous solution. *Marine Chemistry*, **19**, 343–353.

Charlston, R. L. Lovelock, J. E. Andreae, M. O. & Warren, S. G. 1987. Oceanic phytoplankton, atmospheric sulphur, cloud albedo and climate. *Nature*, **326**, 655–661.

Charman, W. N. 1990. Ocular hazards arising from depletion of the natural atmospheric ozone layer: a review. *Opthalmology and Physiolgical Optics*, **10**, 333–341.

Cho, B. C. & Azam, F. 1990. Biogeochemical significance of bacterial biomass in the ocean's euphotic zone. *Marine Ecology Progress Series*, **63**, 253–259.

Cota, G. F., Kottmeier, S. T., Robinson, D. H., Smith, W.O. Jr & Sullivan, C. W. 1990. Bacterioplankton in the marginal ice zone of the Weddell Sea: biomass, production and metabolic activities during austral autumn. *Deep-Sea Research*, **37**, 1145–1167.

Cullen, J. J. & Lesser, M. P. 1991. Inhibition of photosynthesis by ultraviolet radiation as a function of dose and dosage rate: results for a marine diatom. *Marine Biology*, **111**, 183–190.

Cullen, J. J., Neale, P. J. & Lesser, M. P. 1992. Biological weighting function for the inhibition of phytoplankton photosynthesis by ultraviolet radiation. *Science*, **258**, 646–650.

Damaeker, D. M. & Dey, D. B. 1983. UV damage and photoreactivation potentials of larval shrimp *Pandalus platyceros* and adult euphausiids *Thysanoessa raschii*. *Oecologia*, **60**, 169–175.

Davidson, A. T. & Marchant, H. J. 1992. Protist interactions and carbon dynamics of a *Phaeocystis*-dominated bloom at an Antarctic coastal site. *Polar Biology*, **12**, 387–395.

Davidson, A. T. & Marchant, H. J. 1994. The impact of UV radiation on *Phaeocystis* and selected species of Antarctic marine diatoms. *Antarctic Research Series*, **62**, 187–205.

Denman, K. L. & Gargett, A. E. 1983. Time and space scales of vertical mixing and advection of phytoplankton in the upper ocean. *Limnology and Oceanography*, **28**, 801–815.

Döhler, G. 1990. Impact of UV-B radiation on uptake of ^{15}N-ammonia and ^{15}N-nitrate by phytoplankton of the Wadden Sea. *Marine Biology*, **112**, 485–489.

Döhler, G., Worrest, R. C., Biermann, L. & Zink, J. 1987. Photosynthetic $^{14}CO_2$ fixation and ^{15}N-ammonia assimilation during UV-B radiation of *Lithodesmium variabile*. *Physiologia Plantarum*, **70**, 511–515.

Eklund, N. G. A. 1992. Studies on the effects of UV-B radiation on phytoplankton of sub-Antarctic lakes and ponds. *Polar Biology*, **12**, 533–537.

El-Sayed, S. Z., Stephens, F. C., Bidigare, R. A. & Ondrusek, M. E. 1990. Effect of ultraviolet radiation on Antarctic marine phytoplankton. In Kerry, K. R. & Hempel, G., eds. *Antarctic ecological change and conservation*. Berlin: Springer-Verlag, 379–385.

Farman, J. C., Gardiner, B. G. & Shanklin, J. D. 1985. Large losses of total ozone in Antarctica reveal seasonal ClOx/NOx interaction. *Nature*, **315**, 207–210.

Fiala, M. & Delille, D. 1992. Variability and interactions of phytoplankton and bacterioplankton in the Antarctic neritic area. *Marine Ecology Progress Series*, **89**, 135–146.

Frederick, J. E. & Snell, H. E. 1988. Ultraviolet radiation levels during the Antarctic spring. *Science*, **241**, 438–440.

Fryxell, G. A. & Kendrick, G. A. 1988. Austral spring microalgae across the Weddell Sea ice edge: spatial relationships found along a northward transect during AMEREZ 83. *Deep-Sea Research*, **35**, 1–20.

Garrison, D. L. 1991. An overview of the abundance and role of protozooplankton in Antarctic waters. *Journal of Marine Systems*, **2**, 317–331.

Gibson, J. A. E., Garrick, R. C., Burton, H. R. & McTaggart, A. R. 1989. Dimethylsulfide and the alga *Phaeocystis pouchetii* in Antarctic coastal waters. *Marine Biology*, **104**, 339–346.

Gieskes, W. W. C. & Kraay, G. W. 1990. Transmission of ultraviolet light in the Weddell Sea: report of the first measurements made in the Antarctic. *BIOMASS Newsletter*, **12**, 12–14.

Griffiths, P. & Fairney, A. 1988. Vitamin D metabolism in polar vertebrates. *Comparative Biochemistry and Physiology*, **91** B3, 511–516.

Häder, P. & Worrest, R. C. 1991. Effects of enhanced solar ultraviolet radiation on aquatic ecosystems. *Photochemistry and Photobiology*, **53**, 717–725.

Helbling, E. W., Villafañe, V., Ferrario, M. & Holm-Hansen, O. 1992. Impact of natural ultraviolet radiation on rates of photosynthesis and on specific marine phytoplankton species. *Marine Ecology Progress Series*, **80**, 89–100.

Helbling, E. W., Villafañe, V. & Holm-Hansen, O. 1994. Effects of ultraviolet radiation on Antarctic marine phytoplankton photosynthesis with particular attention to the influence of mixing. *Antarctic Research Series*, **62**, 202–227.

Hemmingsen, E. A. & Douglas, E. L. 1970. Ultraviolet radiation thresholds for corneal injury in Antarctic and temperate-zone animals. *Comparative Biochemistry and Physiology*, **32**, 593–600.

Herndl, G. J., Müller-Niklas, G. & Frick, J. 1993. Major role of ultraviolet-B in controlling bacterioplankton growth in the surface layer of the ocean. *Nature*, **361**, 717–719.

Holm-Hansen, O., Mitchell, B. G. & Vernet, M. 1989. Ultraviolet radiation in Antarctic waters: effects on rates of primary production. *Antarctic Journal of the United States*, **24**, 177–178.

Hunter, J. R., Kaup, S. E. & Taylor, J. H. 1981. Effects of solar and artificial ultraviolet-B radiation on larval Northern Anchovy, *Engraulis mordax*. *Photochemistry and Photobiology*, **34**, 477–486.

Karentz, D. 1991. Ecological considerations of Antarctic ozone depletion. *Antarctic Science*, **3**, 3–11.

Karentz, D. 1994. Ultraviolet tolerance mechanisms in Antarctic marine organisms. *Antarctic Research Series*, **62**, 93–110.

Karentz, D. & Lutze, L. H. 1990. Evaluation of biologically harmful ultraviolet radiation in Antarctica with a biological dosimeter designed for aquatic environments. *Limnology and Oceanography*, **35**, 548–561.

Karentz, D., McEuen, F. S., Land, M. C. & Dunlap, W. C. 1991. Survey of mycosporine-like amino acid compounds in Antarctic marine organisms: potential protection from ultraviolet exposure. *Marine Biology*, **108**, 157–166.

Karl, D. M. 1993. Microbial process in the southern oceans. In Friedmann, E. I., ed. *Antarctic microbiology*. New York: Wiley-Liss, 1–63.

Kiene, R. P. & Bates, T. S. 1990. Biological removal of dimethyl sulphide from sea water. *Nature*, **345**, 702–705.

Kullenberg, G. 1982. Note on the role of vertical mixing in relation to effects of UV radiation on the marine environment. In Calkins, J., ed. *The role of solar ultraviolet radiation in marine ecosystems*. New York: Plenum, 283–292.

Lizotte, M. P. & Sullivan, C. W. 1992. Biochemical composition and photosynthate distribution in sea ice microalgae of McMurdo Sound, Antarctica: evidence for nutrient stress during the spring bloom. *Antarctic Science*, **4**, 23–30.

Madronich, S. 1993. UV radiation in the natural and perturbed atmosphere. In Tevini, M., ed. *UV-B radiation and ozone depletion*. Boca Raton, FL: Lewis Publishers, 17–69.

Marchant, H. J. 1994. Biological impacts of seasonal ozone depletion. In Hempel, G., ed. *Antarctic science: global concerns*. Springer-Verlag, 95–109.

Marchant, H. J. & Murphy, E. 1994. Interactions at the base of the Antarctic food web. In El-Sayed, S. Z., ed. *Southern Ocean ecology; the BIOMASS perspective*. Cambridge: Cambridge University Press, 267–285.

McMinn, A., Heijnis, H. & Hodson, D. 1994. Minimal effects of UVB radiation on Antarctic diatoms over the past 20 years. *Nature*, **370**, 547–549.

Mopper, K., Zhou, R. J., Kieber, D. J. Sikorski, R. J. & Jones, R. D. 1991. Photochemical degradation of dissolved organic carbon and its impact on the oceanic carbon cycle. *Nature*, **353**, 60–62.

Neale, P. J., Lesser, M. P. & Cullen, J. J. 1994. Effects of ultraviolet radiation on photosynthesis of phytoplankton in the vicinity of McMurdo Station, Antarctica. *Antarctic Research Series*, **62**, 125–142.

Post, A. & Larkum, A. W. D. 1993. UV-absorbing pigments, photosynthesis and UV exposure in Antarctica: comparison of terrestrial and marine algae. *Aquatic Botany*, **45**, 231–243.

Prézelin, B. B., Boucher, N. P. & Smith, R. C. 1994. Marine primary production under the influence of the Antarctic ozone hole: Icecolors '90. *Antarctic Research Series*, **62**, 159–186.

Riebesell, U. Schloss, I. & Smetacek, V. 1991. Aggregation of algae released from melting sea ice: implications for seeding and sedimentation. *Polar Biology*, **11**, 239–248.

Rivkin, R. B., Putt, M., Alexander, S. P., Meritt, D. & Gaudet, L. 1989. Biomass and production in polar planktonic and sea ice microbial communities: a comparative study. *Marine Biology*, **101**, 273–283.

Ryan, K. G. 1992. UV radiation and photosynthetic production in Antarctic sea ice microalgae. *Journal of Photochemistry and Photobiology*, **13**, 235-240.

Ryan, K. G. & Beaglehole, D. 1994. Ultraviolet radiation and bottom-ice algae: laboratory and field studies from McMurdo Sound, Antarctica. *Antarctic Research Series*, **62**, 229–242.

SCOPE. 1993. *Effects of increased ultraviolet radiation on global ecosystems*. Paris: Scientific Committee on Problems of the Environment, 47 pp.

Sieracki, M. E. & Sieburth, J. M. 1986. Sunlight induced delay of planktonic marine bacteria in filtered seawater. *Marine Ecology Progress Series*, **33**, 19–27.

Simon, M., Cho, B. C. & Azam, F. 1992. Significance of bacterial biomass in lakes and the ocean: comparison to phytoplankton biomass and biogeochemical implications. *Marine Ecology Progress Series*, **86**, 103–110.

Smith, R. C. & Baker, K. S. 1982. Assessment of the influence of enhanced UV-B on marine primary productivity. In Calkins, J., ed. *The role of solar ultraviolet radiation in marine ecosystems*. New York: Plenum, 509–537.

Smith, R. C., Prézelin, B. B., Baker, K. S., Bidigare, R. R., Boucher, N. P., Coley, T., Karentz, D., MacIntyre, S., Matlick, H. A., Menzies, D., Ondrusek, M., Wan, Z. & Waters, K. J. 1992. Ozone depletion: ultraviolet radiation and phytoplankton biology in Antarctic waters. *Science*, **255**, 952–959.

Smith, W. O., Jr 1987. Phytoplankton dynamics in marginal ice zones. *Oceanography and Marine Biology Annual Review*, **25**, 11–38.

Smith, W. O., Jr & Nelson, D. M. 1986. Importance of ice edge phytoplankton production in the Southern Ocean. *BioScience*, **36**, 251–257.

Smith, W. O., Jr, Keene, N. K. & Comiso. J. C. 1988. Interannual variability in estimated primary productivity of the Antarctic marginal ice zone. In Sahrhage, D., ed. *Antarctic Ocean and resources variability*. Berlin: Springer-Verlag, 131–139.

Sullivan, C. W., Cota, G. F., Krempin, D. W. & Smith, W. O, Jr 1990. Distribution and activity of bacterioplankton in the marginal ice zone of the Weddell–Scotia seas during austral spring. *Marine Ecology Progress Series*, **63**, 239–252.

Tevini, M. 1993a. Molecular biological effects of ultraviolet radiation. In Tevini, M., ed. *UV-B radiation and ozone depletion*. Boca Raton, FL: Lewis Publishers, 1–15.

Tevini, M., ed. 1993b. *UV-B radiation and ozone depletion*. Boca Raton, FL: Lewis Publishers, 248 pp.

Trodahl, H. J. & Buckley, R. G. 1989. Ultraviolet levels under sea ice during the Antarctic spring. *Science*, **245**, 194–195.

UNEP 1991. *UNEP report on the environmental effects of ozone depletion*. New York: United Nations Environment Program.

Vernet, M., Brody, E. A., Holm-Hansen, O. & Mitchell, B. G. 1994. The response of Antarctic phytoplankton to ultraviolet radiation: absorption, photosynthesis and taxonomic composition. *Antarctic Research Series*, **62**, 143–158.

Veth, C. 1991. The evolution of the upper water layer in the marginal ice zone, austral spring 1988, Scotia – Weddell Sea. *Journal of Marine Systems*, **2**, 451–464.

Vincent, W. F. & Quesada, A. 1994. Ultraviolet radiation effects on cyanobacteria: implications for Antarctic microbial ecosystems. *Antarctic Research Series*, **62**, 111–124.

Vosjan, J. H., Döhler, G. & Nieuwland, G. 1990. Effect of UV-B irradiance on the ATP content of microorganisms of the Weddell Sea (Antarctica). *Netherlands Journal of Sea Research*, **25**, 391–393.

Voytek, M. A. 1990. Addressing the biological effects of decreased ozone on the Antarctic environment. *Ambio*, **19**, 52–61.

Weiler, C. S. & Penhale, P., eds. 1994. Ultraviolet radiation in Antarctica: measurements and biological effects. *Antarctic Research Series*, **62**, 1–257.

Weller, G. E. 1993. *The role of the Antarctic in global change: an international plan for a regional research program*. Cambridge: SCAR, 54 pp.

Williams, P. M. & Druffel, E. R. M. 1987. Radiocarbon in dissolved organic matter in the central North Pacific Ocean. *Nature*, **330**, 246–248.

Wynn-Williams, D. D. 1994. Potential effects of ultraviolet radiation on Antarctic primary terrestrial colonizers: cyanobacteria, algae, and cryptogams. *Antarctic Research Series*, **62**, 243–257.

Zwally, H. J., Parkinson, C. L. & Comiso, J. C. 1983. Variability of Antarctic sea ice and changes in carbon dioxide. *Science*, **220**, 1005–1012.

52 Effects of solar ultraviolet radiation on primary production in Antarctic waters

OSMUND HOLM-HANSEN, VIRGINIA E. VILLAFAÑE AND E. WALTER HELBLING
Polar Research Program, Scripps Institution of Oceanography, University of California, San Diego, La Jolla, CA 92093-0202, USA

ABSTRACT

The effects of solar ultraviolet radiation on inhibition of rates of photosynthesis in Antarctic phytoplankton were studied at Palmer Station, Antarctica, from early October to the end of December, 1993. A well-developed ozone hole, with column ozone values as low as 140 Dobson Units, was present in October and part of November, before column ozone values returned to normal in late November and December. It was thus possible to estimate the magnitude of photosynthetic inhibition due to UV-B radiation on days with or without significant enhancement of the shorter UV-B wavelengths due to ozone depletion. Daily temperature-controlled incubator experiments with natural phytoplankton assemblages from Arthur Harbor showed that UV-A radiation was responsible for approximately twice as much inhibition of photosynthesis as that resulting from UV-B radiation, even on days with low column ozone concentrations. Results from in situ *incubations also demonstrated that UV-A radiation was responsible for much more inhibition of photosynthesis than UV-B radiation, and that no inhibition by either UV-B or UV-A radiation could be detected below 10–12 m in a water column where the euphotic zone extended to approximately 50 m. These data suggest that the impact of enhanced UV-B radiation resulting from ozone depletion will not seriously lessen the rate of primary production in Antarctic waters.*

Key words: phytoplankton, ultraviolet radiation, ozone hole, primary production.

INTRODUCTION

Previous studies of ours have shown that both UV-B (280 to 320 nm) and UV-A radiation (320 to 400 nm) can cause significant decreases in photosynthetic rates of Antarctic phytoplankton, but that the impact of UV-A is generally considerably greater than that of UV-B (Helbling *et al.* 1992, Holm-Hansen *et al.* 1993). The results of *in situ* incubations of Smith *et al.* (1992), who worked with ice-edge populations of phytoplankton, also show that UV-A radiation is responsible for a greater loss of primary production than the loss caused by UV-B radiation. Some of our studies referred to above were done in the time period of October to December, 1988 and 1989, but as the column ozone levels were >300 Dobson Units (DU) during our experimental periods, we had no direct measurement of the impact of the enhanced UV-B radiation that results from ozone depletion in the stratosphere. Spectral enhancement of UV-B radiation resulting from decreased ozone concentrations is essentially zero at 320 nm, and increases exponentially toward the shorter UV-B wavelengths (Holm-Hansen *et al.* 1993). There has been much speculation that such enhancement of the shorter UV-B wavelengths might be particularly damaging to cells due to direct absorption of the high-energy UV photons by deoxyribonucleic acid (DNA).

During the austral spring of 1993 we investigated the effects of the ozone hole on photosynthetic rates of Antarctic phytoplankton and compared the magnitude of these effects at the time of the ozone hole with that of those later in the season when atmospheric ozone concentrations were again within the normal range. The data presented in this paper report the effects of both enhanced UV-B radiation and 'normal' UVR when there was no ozone hole, on natural phytoplankton assemblages sampled from Arthur Harbor, close to Anvers Island.

MATERIALS AND METHODS

Sampling protocol

All work was done at Palmer Station (64.7° S, 64.1° W) from October 10 to December 31, 1993. Part of the experimental procedure was to obtain samples of natural phytoplankton assemblages each day from Arthur Harbor and to determine the impact of incident solar ultraviolet radiation (UVR) on photosynthetic rates in short-term incubations (<one day). During the time that Arthur Harbor was covered with sea ice (from the time of our arrival on Station to mid-November), a 5.0 litre Go-flo bottle was lowered through a hole in the ice about 2 m from the shoreline and water samples were taken at approximately 1 m depth. When there was sufficient ice-free water in Arthur Harbor, the Go-flo bottle was used from a boat (*Zodiac*) so that samples could be taken from various depths at a station to the north of Bonaparte Point.

Incubation of samples

For the studies done in the deck incubators where many replicate samples were often used in any one experiment (up to 48 tubes), water samples from any one depth were poured into a 16 l jug with a bottom spigot, so that all experimental vessels (50 ml quartz or Pyrex tubes with Teflon-lined screw-caps) could be filled within a few minutes. For the *in situ* studies, the water sample was poured into a 1.0 l bottle, which was used to fill the six tubes to be incubated at any one depth. Precautions were taken not to expose the samples to any light shock prior to start of the incubation period. After addition of radiocarbon (0.1 ml of ^{14}C-bicarbonate containing 5 μCi) to each experimental tube, the samples were either placed in an outdoor incubator with flowing seawater for temperature control and exposed to solar radiation, or they were placed on anodized aluminum frames and incubated *in situ* at the same depths from which the water samples had been obtained. Either duplicate or triplicate samples were used for each of the three different treatments, which included (i) quartz tubes with no filters, so that the phytoplankton were exposed to both UV-B and UV-A radiation, in addition to visible radiation (PAR) from 400 to 700 nm, (ii) pyrex tubes wrapped with one layer of Mylar, so that the cells would be exposed to just UV-A+PAR, and (iii) pyrex tubes covered with Plexiglas UF-3, so that the cells would be exposed to just PAR and no UVR. Incubations lasted for 8–10 hours, centered approximately at local apparent noon.

Measurement of irradiance

A PUV-510 unit (Biospherical Instruments, Inc.) with sensors for recording of spectral UVR (four channels, at wavelengths of 305, 320, 340, and 380 nm) and PAR (400 to 700 nm) was mounted in a shade-free location above the outdoor incubators. Data from each channel were recorded in a computer once per minute during our entire stay at Palmer Station. A submersible unit (PUV-500) with the same sensors (plus sensors for depth and temperature) was used in conjunction with all *in situ* incubations so that the magnitude of inhibition of photosynthesis could be related to the fluence of UVR and PAR at any depth. In addition to the above measurements, we were also provided with the spectral UVR and visible radiation recorded by the UV-spectroradiometer operated at Palmer Station by the US National Science Foundation (Booth *et al.* 1994). Data from this instrument were used to estimate column ozone concentrations expressed in Dobson Units (DU).

RESULTS AND DISCUSSION

During the experimental period at Palmer Station there was a well-developed ozone hole from early October to late November, with column ozone values fluctuating considerably on a daily basis, but reaching values as low as 140 DU on Oct. 14 (Fig. 52.1A). Normal ozone values (>300 DU) were not re-established until sometime during the fourth week of November. These fluctuations in ozone values are reflected in the noontime UV-B irradiance, which showed maximal values at the end of October (Fig. 52.1B). Irradiance of UV-A or PAR should increase gradually to maximal values close to the summer solstice (December 21), but variable cloud cover caused great fluctuations in incident radiation as shown in Fig. 52.1C. The highest UV-A irradiance (57 W m^{-2}) was recorded on December 31. The mean UV-A irradiance during the month of December was considerably higher than during the time period of the ozone hole, in contrast to UV-B irradiance. The result of the ozone hole is thus to increase not only the UV-B irradiance, but to increase the ratio of UV-B to UV-A irradiance, as shown by the ratio of irradiances at 305 nm to 380 nm (Fig. 52.1D).

Data in Fig. 52.2 indicate the magnitude of inhibition of photosynthetic rates caused by UV-B (Fig. 52.2A) and UV-B+UV-A radiation (Fig. 52.2B) as expressed as percentage enhancement over the controls in quartz tubes. It is seen that the percentage enhancement when all UV-B has been removed with a pre-filter of Mylar increases with increasing UV-B radiation, and that the enhancement during maximal development of the ozone hole (when UV-B is over 3 W m^{-2}) is in the range of 12–30%. When all UVR has been removed with a pre-filter of UF-3, the percentage enhancement is in the range of 60–120%. This strongly suggests that UV-A radiation has a significantly greater impact on decreasing rates of primary production in Antarctic waters than does UV-B radiation, including those days when there is a deep ozone hole with its associated

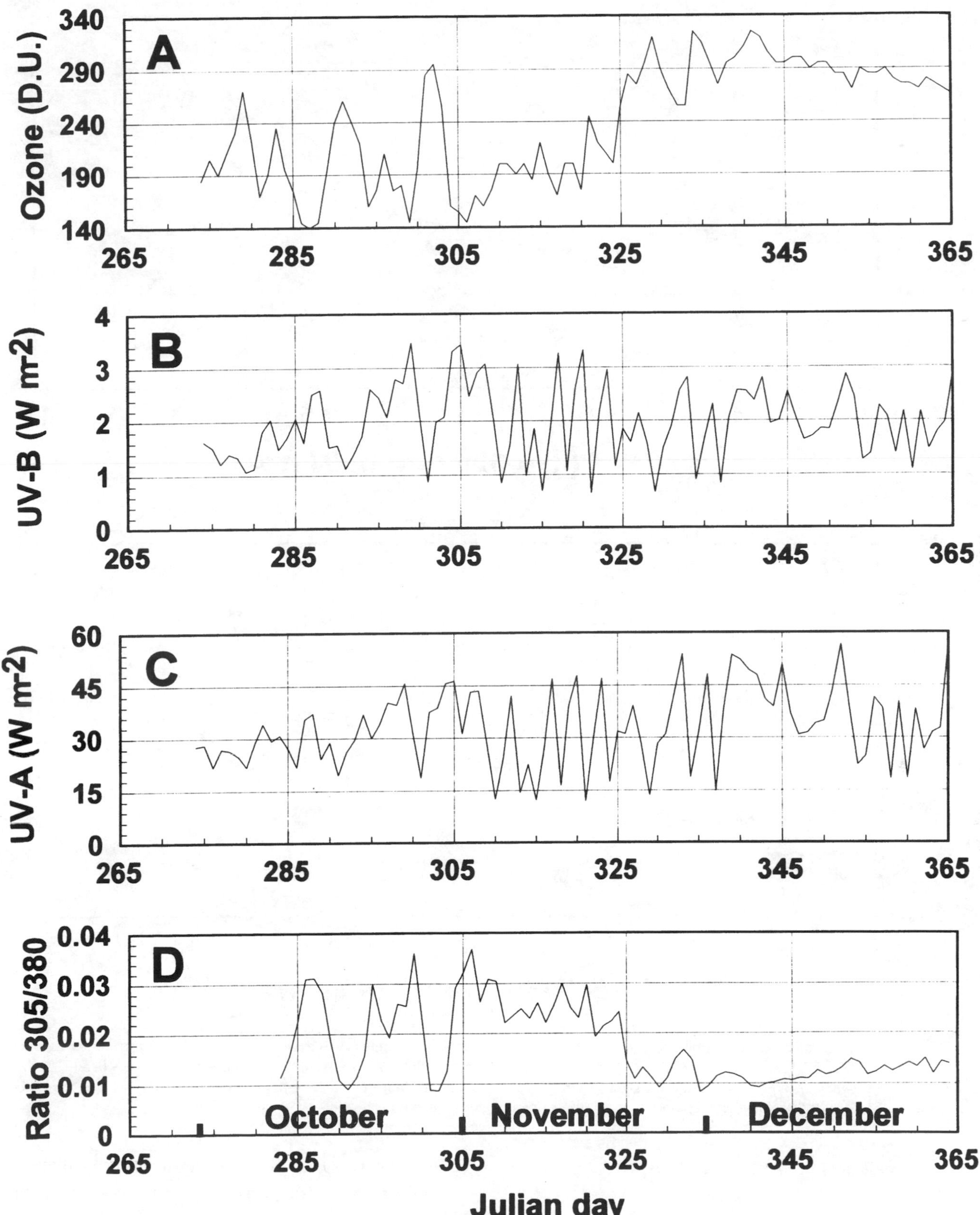

Fig. 52.1. Daily fluctuations at Palmer Station in column ozone concentrations and in incident ultraviolet radiation from October 1 (Julian day 274) to December 31 (Julian day 365), 1993. All values derive from the spectral scan closest to local apparent noon. A, column ozone concentrations expressed in Dobson Units. B, UV-B irradiance (290 to 320 nm) in Watts per m^2. C, UV-A irradiance (320 to 400 nm) in Watts per m^2. D, ratio of the incident energy at 305 nm divided by the energy at 380 nm.

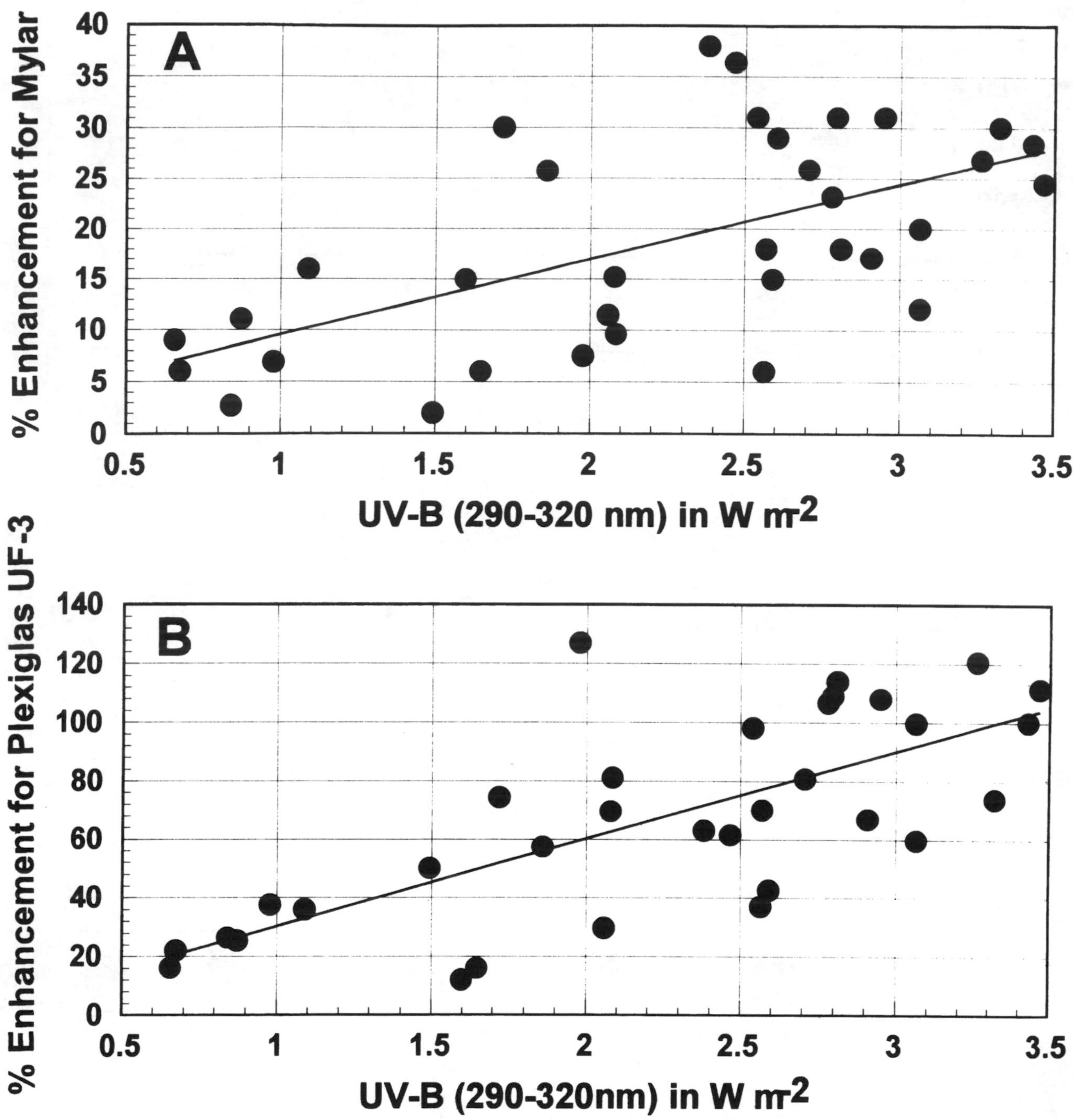

Fig. 52.2. The percentage enhancement in photosynthetic rates as a function of the noontime UV-B irradiance during the incubation period when spectral regions of solar UVR are removed by filters covering the phytoplankton samples. A, percentage enhancement in cultures where UV-B has been removed with a Mylar filter as compared with controls in quartz tubes. B, percentage enhancement when both UV-A and UV-B have been removed with a Plexiglas filter. Note that the same abscissa is used in both figures. Lines indicate the least square fit of the data.

enhanced irradiances of the shorter UV-B wavelengths. The considerable variability seen in the biological response at any specific UV-B irradiance (Fig. 52.2) is thought to be due primarily to: (i) changes in species composition of the phytoplankton assemblages, which showed large changes in relative numbers of diatoms, small flagellates and cryptophytes during the three-month study, and (ii) variability in photoadaptational state of phytoplankton, as some samples came from ice-covered waters, while other samples came from ice-free waters in Arthur Harbor. Previous studies have shown that there is considerable variability in sensitivity to UVR by different species of Antarctic phytoplankton (Karentz *et al.* 1991, El-Sayed *et al.* 1990).

By the time that there was open water in Arthur Harbor and *in situ* incubations were possible (mid to late November), column ozone concentrations were returning to normal (Fig. 52.1A). The results of one typical *in situ* deployment are shown in Fig. 52.3. It is seen that the magnitude of inhibition of photosynthesis by UV-A radiation is much greater than that caused by UV-B radiation, and that it also extends to greater depth.

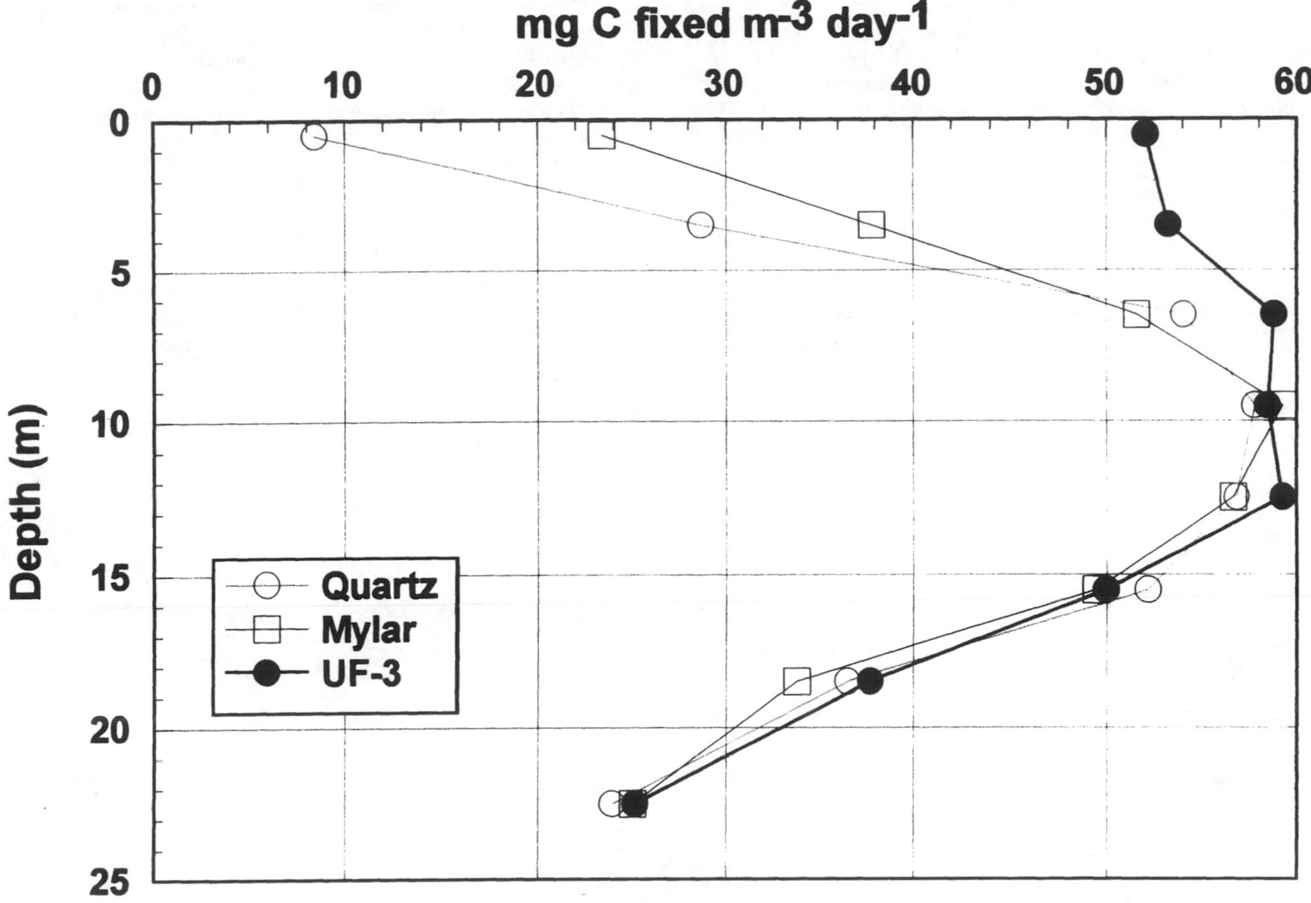

Fig. 52.3. Effect of UV-B and UV-A radiation on photosynthetic rates of Antarctic phytoplankton when incubated *in situ* (Arthur Harbor, December 11, 1993). The area between the open circles and squares indicates the magnitude of loss of primary production due to UV-B radiation, while the area between squares and the solid circles indicates the integrated loss of primary production due to UV-A radiation.

Although the 1% light depth (PAR) was at approximately 50 m, there was no significant inhibition of photosynthesis below 10 m. The mean irradiance of UV-B at 10 m was <0.3 W m^{-2}, which is considerably lower than the lowest irradiance shown in Fig. 52.2.

The experimental data discussed above, which include studies of the magnitude of inhibition by UVR of primary production under the deepest ozone hole yet recorded at Palmer Station, lend support to the views previously expressed (Holm-Hansen *et al.* 1993) that the enhanced UV-B radiation resulting from seasonal ozone depletion in the Antarctic will not have dramatic impacts on seasonal primary production of the Southern Ocean ecosystem as was first suggested by El-Sayed (1988). It must be noticed, however, that our paper deals only with short-term effects of UVR on rates of primary production. It is possible that other metabolic processes (e.g. DNA replication) may be more vulnerable to enhanced UV-B than are photosynthetic rates. On the other hand, our data may indicate the 'worst case scenario' in that the ozone hole occurs before the period of maximal rates of primary production in December to February, and also that long-term effects of UVR may not have the same deleterious effects on primary production as short-term experiments indicate (Bothwell *et al.* 1994, Kim & Watanabe 1994).

ACKNOWLEDGEMENTS

This work was supported by the US National Science Foundation (Grant DPP92-20150) and by the Alternative Fluorocarbon Environmental Acceptability Study (AFEAS). We extend our gratitude to personnel from Antarctic Support Associates (ASA) and to all personnel at Palmer Stations for their generous help and support.

REFERENCES

Booth, C. R., Lucas, T. B., Morrow, J. H., Weiler, C. S. & Penhale, P. A. 1994. The United States National Science Foundation's polar network for monitoring ultraviolet radiation. *Antarctic Research Series*, **62**, 17–37.

Bothwell, M. L., Sherbort, D. M. J. & Pollock, C. M. 1994. Ecosystem response to solar ultraviolet-B radiation: influence of trophic level interactions. *Science*, **265**, 97–100.

El-Sayed, S. Z. 1988. Fragile life under the ozone hole. *Natural History*, **97**, 72–80.

El-Sayed, S. Z., Stephens, F. C., Bidigare, R. R. & Ondrusek, M. E. 1990. Effects of ultraviolet radiation on Antarctic marine phytoplankton. In Kerry, K. R. & Hempel, G., eds. *Antarctic ecosystems: ecological change and conservation*. Berlin & Heidelberg: Springer-Verlag, 379–385.

Helbling, E. W., Villafañe, V., Ferrario, M. & Holm-Hansen, O. 1992. Impact of natural ultraviolet radiation on rates of photosynthesis and on specific marine phytoplankton species. *Marine Ecology Progress Series*, **80**, 89–100.

Holm-Hansen, O., Helbling, E. W. & Lubin, D. 1993. Ultraviolet radiation in Antarctica: inhibition of primary production. *Photochemistry and Photobiology*, **58**, 567–570.

Karentz, D., Cleaver, J. E. & Mitchell, D. L. 1991. Cell survival characteristics and molecular responses of Antarctic phytoplankton to ultraviolet-B radiation. *Journal of Phycology*, **27**, 326–341.

Kim, D. S. & Watanabe, Y. 1994. Inhibition of growth and photosynthesis of freshwater phytoplankton by ultraviolet A (UVA) radiation and subsequent recovery from stress. *Journal of Plankton Research*, **16**, 1645–1654.

Smith, R. C., Prezelin, B. B., Baker, K. S., Bidigare, R. R., Boucher, N. P., Coley, T., Karentz, D., MacIntyre, S., Matlick, H. A., Menzies, D., Ondrusek, M., Man, Z. & Waters, K. J. 1992. Ozone depletion: ultraviolet radiation and phytoplankton biology in Antarctic waters. *Science*, **255**, 952–959.

53 Preliminary sediment core evidence against short-term UV-B induced changes in Antarctic coastal diatom communities

A. MCMINN[1], H. HEIJNIS[2] AND D. HODGSON[3]

[1]Antarctic CRC and Institute of Antarctic and Southern Ocean Studies, University of Tasmania, Box 252C, Hobart 7001, Tasmania, Australia, [2]Environmental Radiochemistry Laboratory, Australian Nuclear Science and Technology Organisation, Private Mailbag 1, Menai, NSW 2234, Australia, [3]Dept of Plant Science, University of Tasmania

ABSTRACT

Springtime UV-B levels have been increasing in Antarctic marine ecosystems since the 1970s. Effects on natural phytoplankton and sea-ice algal communities, however, remain unresolved. At the Marginal Ice Edge Zone, enhanced springtime UV-B levels coincide with a shallow, stratified water column and a major phytoplankton bloom. In these areas it is possible that phytoplankton growth and survival is adversely impacted by enhanced UV-B. In coastal areas, however, the sea ice, which attenuates most of the UV-B before it reaches the water column, remains until December/January, by which time UV-B levels have returned to long-term seasonal averages. Phytoplankton from these areas are unlikely to show long-term changes resulting from the hole in the ozone layer.

Fjords of the Vestfold Hills, eastern Antarctica, have anoxic basins which contain high-resolution, unbioturbated sedimentary sequences. Diatom assemblages from these sequences reflect the diatom component of the phytoplankton and sea-ice algal assemblages at the time of deposition. Twenty-year records from these sequences show no consistent record of change in species composition, diversity or species richness. Six-hundred-year records from the same area also show changes in species abundance greater than those seen in the last 20 years. From these records it can be seen that recent changes in diatom abundances generally fall within the limits of natural variability and there is little evidence of recent changes that might be associated with UV-B-induced change.

Key words: Antarctic, phytoplankton, sea ice, diatoms, UV-B.

INTRODUCTION

Annual springtime stratospheric ozone depletion, with resultant elevated UV-B irradiance, has occurred over the Antarctic continent and Southern Ocean since the 1970s. The effects of increased UV-B on Antarctic phytoplankton communities is, however, still uncertain. El Sayed *et al.* (1990) documented a reduction in phytoplankton growth rates at ambient Antarctic spring/summer irradiances at Palmer on the Antarctic Peninsula. Investigations conducted in the Bellingshausen Sea further demonstrated a reduction of at least 6 to 12% in primary productivity due to UV-B; fluctations that are within the limits of interannual variability (Smith *et al.*, 1992). Laboratory results have similarly shown that most species in culture are susceptible to high UV-B irradiances, although the level at which either growth or photosynthesis is inhibited is variable (Karentz *et al.* 1991a, Marchant & Davidson 1991, Davidson *et al.* 1994). Karentz (1991) and Marchant & Davidson (1991) considered this variable response would lead to a shift in phytoplankton species composition. Holm-Hansen *et al.* (1989), however, maintain that physical factors such as cloud cover, ice cover and deep mixing of the water column combine to minimize the time phytoplankton cells remain exposed to the potential lethal doses of UV-B and consequently little change is likely to occur in natural phytoplankton communities.

Modern phytoplankton communities have already been exposed to approximately two decades of elevated UV-B levels. Antarctic phytoplankton generation times are mostly between two and five days (Fiala & Oriol 1990) and therefore ample time has elapsed since the ozone hole first appeared for the more competitive species under higher UV-B levels to out compete those that are less well adapted. If changes in phytoplankton productivity and species composition are going to occur they should already be apparent (Karentz 1991).

Southern Ocean phytoplankton communities are usually dominated by an abundant and diverse diatom component (Medlin & Priddle 1990, Kang & Fryxell 1991), which is readily fossilized. *Phaeocystis pouchetii*, a prymnesiophyte, is one of the few non-diatom phytoplankton species that frequently dominates communities both in ice-edge and coastal blooms. However, as it leaves no fossilized remains, its past abundance cannot be readily estimated from sediment analysis.

In most marine environments biological mixing of sediments prevents interpretation of sedimentary sequences at high resolution (ie $<$100 y). In locations where the bottom waters are anoxic and where there is consequently no bioturbation it is possible to resolve the sedimentary record on a much finer, even annual scale. The fjords of the Vestfold Hills, eastern Antarctica, contain numerous such anoxic basins. These basins have open connections with the sea and so have a phytoplankton community characteristic of the Antarctic coastal zone (McMinn & Hodgson 1993). Diatom preservation in these basins is outstanding, as most of the small, lightly silicified taxa that are normally destroyed in bioturbated sediments are well preserved (McMinn 1995). Preliminary results of this work have been summarized in McMinn *et al.* (1994). In this paper results of a multivariate analysis, cell size changes and *Chaetoceros* resting spore ratios are presented.

MATERIALS AND METHODS

Six short, 20–40 cm, sediment cores were taken with a Glew Corer (Glew 1989) from the fjords of the Vestfold Hills in November 1992. Of these, three were later selected for diatom analysis on the basis of the consistency of their ^{210}Pb profiles. Cores were subsectioned into 0.5 or 1 cm intervals depending on sediment consistency. Subsamples for diatom analysis were prepared by washing in distilled water, soaking in H_2O_2 for three days, washing again and mounting in Naphrax. Prepared slides were examined on a Zeiss Axioskop microscope using a 100×, oil immersion lens with Nomaski illumination; 1200 specimens were counted from each sample and the remainder of the slide scanned for rare species.

Total ^{210}Pb (=supported ^{210}Pb) activity was determined by measuring its alpha-emitting grand-daughter ^{210}Po, which was assumed to be in equilibrium with ^{210}Pb. Measurements were made using a Po-coated silver disk. This silver source was counted in an alpha spectrometer using the 5.30 MeV peak for ^{210}Po and the ^{209}Po (4.88 MeV) peak for yield tracing. The supported ^{210}Pb was determined by measuring the ^{226}Ra activity using alpha spectrometry (4.78 MeV peak). Chemical recovery was calculated using ^{133}Ba peak (356 KeV) of the gamma spectrum of the same radium/barium-sulphate colloidal precipitate source. The excess ^{210}Pb (=unsupported ^{210}Pb) was calculated by subtracting the ^{226}Ra activity. Sedimentation rates were calculated using the CIC model (Robbins & Edlington 1975, Goldberg *et al.* 1977). Here sedimentation rate=the slope of the regression line of depth versus ln excess ^{210}Pb, i.e. $t_z=(1/k)\times\ln(A_0/A_2)$; where t_z=age at depth z, k=decay constant for ^{210}Pb, A_0=initial $^{210}Pb_{act}$ and A_2=unsupported $^{210}Pb_{act}$ at depth z. Profiles of excess ^{210}Pb against depth for the three sediment cores are presented in Fig. 53.1. ^{137}Cs and ^{241}Am levels were too low to be of use in dating.

Statistical analysis involved the use of the ordination technique, non-metric multidimensional scaling (NMDS), which was carried out on the samples using the statistical program BIOΣTAT 2 (Pimental, R.A. & Smith, J.D. 1985, Sigma Soft, Placentia, California). The NMDS starting configuration was generated by principal coordinates analysis of the dataset and performed over 100 iterations. NMDS, which provides a view of sample similarities in two, three or four dimensions, is considered one of the most robust ordination techniques available (Gray *et al.* 1988, Pimentel 1993)

RESULTS

Diatom abundance

McMinn *et al.* (1994) based the interpretation of the effects of enhanced UV-B on the diatom community over the last 20 years on changes in single species abundance, species richness and assemblage diversity in the sediment cores. Here, results of a multivariate nonlinear multidimensional scaling analysis (NMDS) and other ecological parameters are also included. Most diatom taxa in the sediment cores show considerable fluctuations in abundance over relatively short time periods. An examination of the relative abundance of four of the most common phytoplankton taxa in Ellis Fjord (i.e. *Chaetoceros* spp. (vegetative cells), *Fragilariopsis cylindrus*, *Fragilariopsis curta* and *Thalassiosira dichotomica*, which are abundant in most of the cores) shows that no consistent change in abundance has occurred across the Vestfold Hills over the last 20 years (Fig.53.1a–d, McMinn *et al.* 1994). These species are also widespread in Antarctic springtime oceanic, coastal and sea-ice communities (Medlin & Priddle 1990, Scott *et al.* 1994).

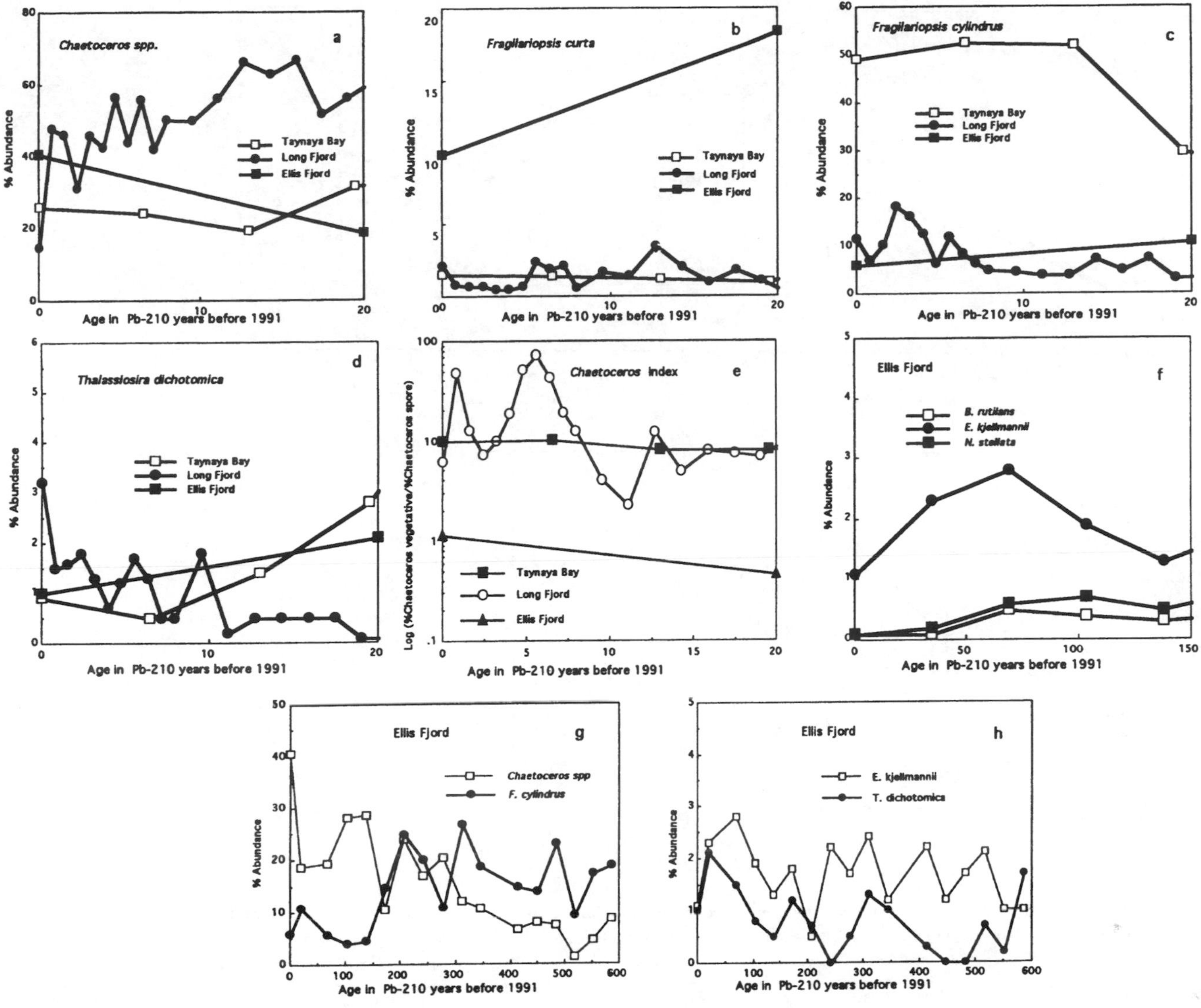

Fig. 53.1. Relative (%) species abundance of diatoms in cores from the Vestfold Hills. (a) *Chaetoceros* spp. (vegetative cells); (b) *Fragilariopsis curta*; (c) *Fragilariopsis cylindrus*; (d) *Thalassiosira dichotomica*; (e) *Chaetoceros* Index, log of ratio of *Chaetoceros* vegetative cells to resting spores; (f) Abundance of sea-ice taxa in Ellis Fjord; (g) 600 y record of *Chaetoceros* spp. (vegetative cells) and *F. cylindrus* from Ellis Fjord; (h) 600 y record of *E. kjellmannii* and *T. dichotomica* from Ellis Fjord.

Multivariate ordination analysis of the entire diatom data set using NMDS identifies both changes at the community level and temporal trends in the data. A line joining sequential samples, a time track, has been placed through the data to demonstrate either the presence or absence of community change over the last 20 years. If the diatom community were responding significantly to a new environmental stress by changes in species composition then the later samples should be spatially segregated from earlier samples and/or the direction of the time track should change. In the NMDS analyses from the Vestfold Hills there is little clear evidence for either a separation of younger samples or a change in direction of the time track. Samples from the last 20 years in the profile from Ellis Fjord (Fig. 53.2a) show a continuation of a much longer-term change. The Long Fjord profile shows a moderately consistent shift in species composition over the last 20 years but as the trend commenced before significant ozone depletion it may not be a response to increasing UV-B irradiance (Fig. 53.2b). The samples from the last 20 years in Taynaya Bay show no separation from earlier samples (Fig. 53.2c). Future sampling at all sites should determine whether subsequent annual diatom fluxes are responding to changes in UV-B irradiances.

The abundance of *Chaetoceros* resting spores is thought to be associated with either environmental stress or elevated productivity, i.e. nutrient stress (Fryxell & Medlin 1981). In the Taynaya Bay and Ellis Fjord cores, the ratio of *Chaetoceros* vegetative cells to resting spores changes little over the last 20 years (Fig. 53.1e). The ratio fluctuates markedly in the Long Fjord core but does not reflect a sequential increase in stress and possibly indicates interannual variations in ice cover. The long-term record from Ellis Fjord shows a rise in the vegetative cell/resting spore ratio that parallels a consistent rise in the abundance of vegetative cells (Fig. 53.1e). Average cell size, another possible indicator of environmental stress, fluctuates within a relatively narrow range but shows no systematic change within the last 20 years (Fig. 53.3).

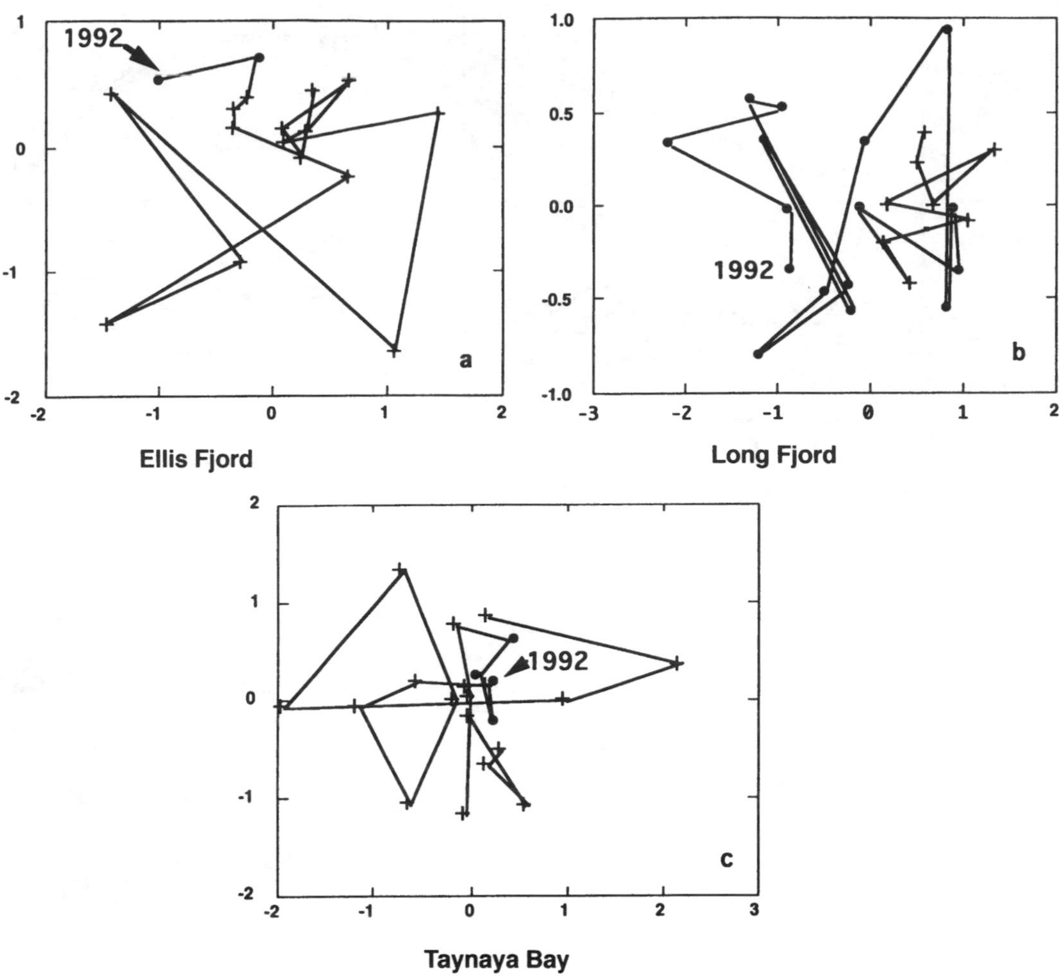

Fig. 53.2. Nonmetric multidimensional scaling analysis of sediment core diatom abundance data from the Vestfold Hills. Units and scale on both axes are arbitrary. Points indicated with a '●' represent those younger than 20 years; those with a '+' are older than 20 years. The line joining the points is sequential from oldest to youngest. The youngest point is indicated by an arrow pointing to 1992.

Diatom assemblages from the fjords of the Vestfold Hills show little evidence of changes in species richness or diversity over the last 20 years (McMinn *et al.* 1994). The species richness in each core fluctuates within a relatively narrow range and no significant decline is indicated. The Shannon diversity index likewise shows no appreciable variation over this time interval (McMinn *et al.* 1994). Certainly, there is no evidence that any species has become extinct in the fjords over the last 20 years. Investigation of changes in the abundance of these species over longer time periods shows fluctuations of a greater magnitude than are seen within the last 20 years (Fig. 53.1f–h). This implies that the recent changes in abundance fall within the limits of natural variability and that factors other than UV-B are responsible for most of the changes in the phytoplankton community (McMinn *et al.* 1994).

^{210}Pb Analysis

The use of $^{210}Pb/^{226}Ra$ as a tool for measuring sedimentation rates is more difficult in Antarctic environments than elsewhere. The supply of ^{210}Pb from atmospheric fallout around Antarctica is extremely low (Appleby & Oldfield 1992). Therefore, the excess ^{210}Pb contributes very little to the total ^{210}Pb in the sample. Furthermore, the anoxic basins act as a sink for Uranium, as U^{6+} is reduced by bacteria to the less soluble U^{4+} and is easily scavenged to the bottom. It then contributes to a longer ^{210}Pb *in situ* production of ^{226}Ra and subsequently ^{210}Pb. Thus, the production of ^{226}Ra and ^{210}Pb in the water column is reduced and the *in situ* production of ^{210}Pb is high. These factors contribute to the very low initial excess ^{210}Pb concentrations at depth (Fig. 53.4). The possibility of analytical error was eliminated by re-analysing the samples and cross checking with analyses from other areas. The ^{210}Pb profile from Long Fjord is less problematical than those from Taynaya Bay and Ellis Fjord, which had negligible amounts of excess ^{210}Pb below 10 cm, and can be applied with confidence throughout its entire length. In all three cores the record of the last 20 years lies well above the negligible ^{210}Pb values.

DISCUSSION

Phytoplankton composition, productivity and competitiveness is closely associated with local climatic conditions. In the Vestfold Hills there is no evidence for consistent, long-term changes in average surface temperature, cloud cover, wind speed,

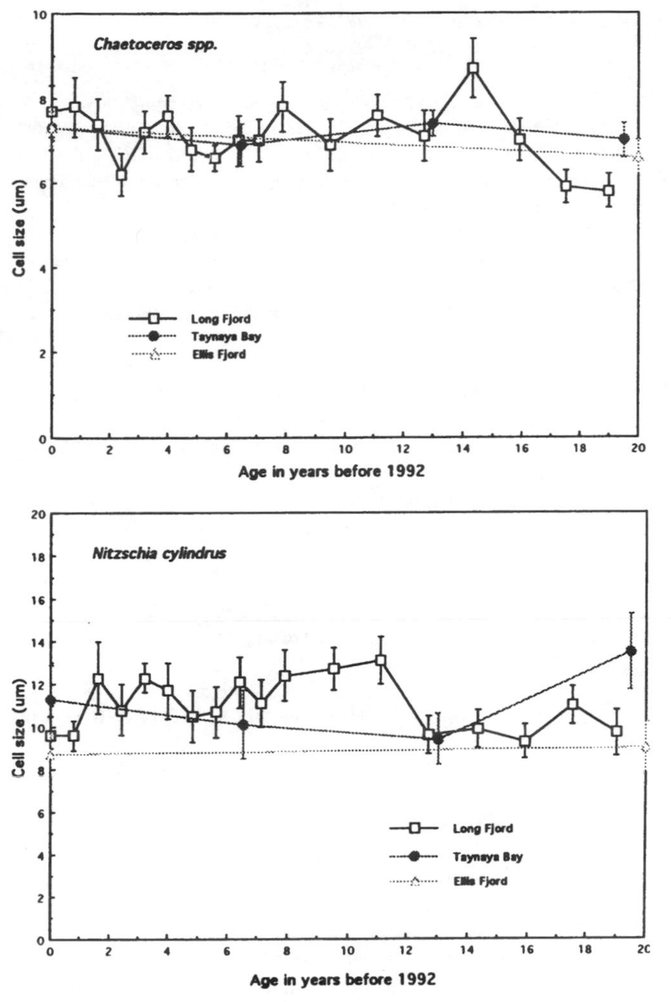

Fig. 53.3. Changes in cell size of *Chaetoceros* spp. and *Nitzschia cylindrus* over the last 20 years.

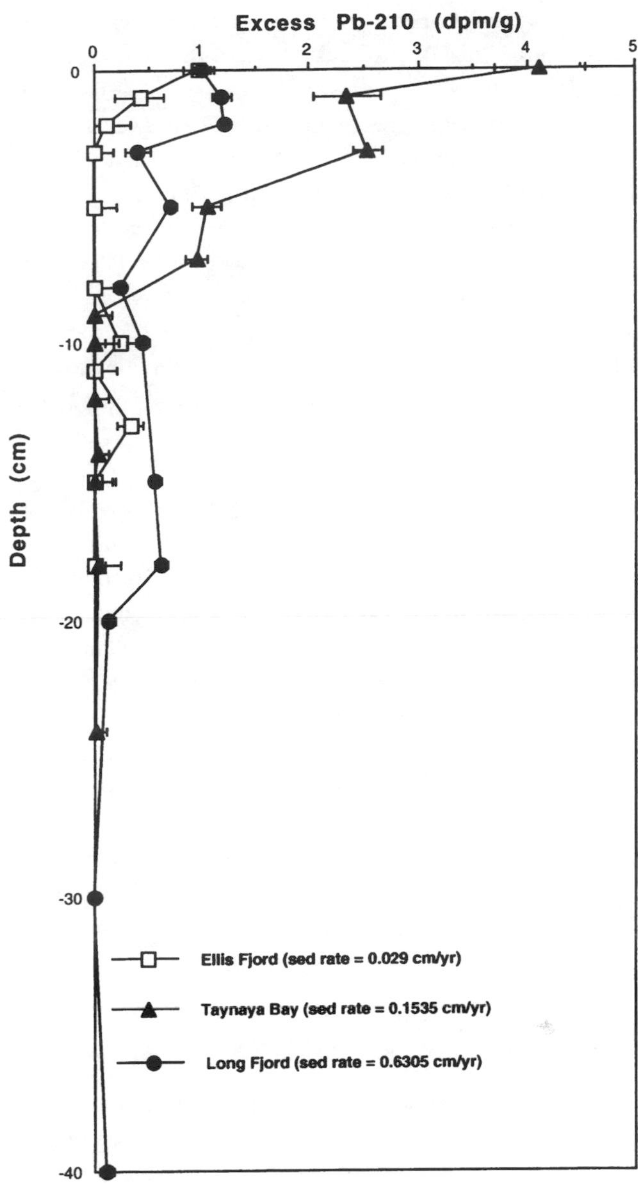

Fig. 53.4. ^{210}Pb profiles of cores from the Vestfold Hills.

water temperature or ice cover over the last 20 years (Jacka *et al.* 1984, Russell-Head & Simmonds 1993, Allison unpublished data). There are no data available on variations in non-climatic controls on productivity such as nutrient levels or predation, but these are unlikely to have changed significantly over this period. Graneli *et al.* (1993) recently demonstrated experimentally that under certain conditions grazers could influence Southern Ocean algal successions. The only data on the UV-B tolerance of these grazers, however, focused on the presence of elevated levels of mycosporine-like amino acids (MAA), which may offer some protection from UV-B-induced damage (Karentz *et al.* 1991b). Although UV-B irradiances have changed significantly, there appears to have been little change in the diatom community and so it is reasonable to suggest that the change in the UV-B irradiance has been insufficient to influence diatom growth and competitiveness.

Minimum ozone levels are usually recorded in October and return to normal in late November or early December (Stolarski *et al.* 1990). At the Marginal Ice Edge Zone in spring, the sea-ice cover has dissipated and water column stratification, high primary productivity (Smith & Nelson 1986) and enhanced UV-B coincide (Bridigare 1989). The phytoplankton in this area is thus potentially susceptible to the harmful effects of UV-B. At most Antarctic coastal sites, sea ice is usually still present until December/January. While snow-free fast ice transmits approximately 10% of UV-B in October (Ryan 1992), this declines to only 0.25% in late November (Trodahl & Buckeley, 1989) and snow cover will further substantially reduce transmittance. By the time the sea ice has dissipated, UV-B levels have returned to seasonal, pre-ozone hole levels and the major coastal bloom, which typically develops in January (Perrin *et al.*, 1987), is no longer experiencing elevated UV-B.

Sea-ice algae are potentially threatened by an increase in ambient UV-B but as yet there has been comparatively little investigation of their susceptibility. They bloom in October when the effects of the ozone hole are at a maximum and they are retained close to the surface on the underside of the ice. Trodahl & Buckeley (1989) have estimated a 20-fold increase in the under-ice UV-B irradiance in October due to ozone depletion in a period of relatively high transmittance. A recent report of a 5% reduction in primary productivity by some sea-ice algal species exposed to UV-B in culture supports this suggestion (Ryan 1992). Sea-ice diatoms were mostly present at only low levels in the sediment cores of the Vestfold Hills but a small decline in the abundance of *E. kjellmannii*, *N. stellata* and *B. rutilans* can be seen in Ellis Fjord (McMinn *et al.* 1994). These species characterize the bottom strand community, which has a non homogeneous spatial distribution (Watanabe 1988), and this may also contribute to the temporal fluctuations in the cores. Currently, the only experimental growth or survival data for these species under enhanced UV-B conditions are those of Ryan & Beaglehole (1994), which indicated that elevated UV-B levels had little effect on bottom-dwelling sea-ice algae. If, however, the decline noted in the sediment cores is reflecting real stress in the fast-ice algal community then this community may possibly be experiencing a genuine decline in productivity.

If UV-B is having a significant impact on diatom communities, then changes in the community composition should have become apparent over the last 20 years. Analysis of sedimentary diatom records from the last 20 years reveals that while fluctuations in abundance have occurred, they are inconsistent between sites and generally fall within the longer-term natural variability in abundance of these species. UV-B is the only environmental factor with both a consistent temporal gradient over the last 20 years and also the potential to impact the phytoplankton. The lack of change in the sediment core diatom assemblages suggests that UV-B is having a minimal effect on the diatom component of the phytoplankton.

ACKNOWLEDGEMENTS

We are grateful for the support of Australian National Antarctic Research Expeditions (ANARE) for logistical and field support. Financial support was recieved from an Antarctic Scientific Advisory Committee (ASAC) grant and assistance with ^{210}Pb analyses was recieved from an Australian Institute of Nuclear Science and Engineering (AINSE) grant. Anonymous reviewers helped us improve the manuscript considerably.

REFERENCES

Appleby, P. G. & Oldfield, R. 1992. Application of Lead ^{210} to sedimentation studies. In Ivanovich, M. & Harmon, R. S., eds. *Uranium series disequilibrium, applicants to earth, marine and environmental sciences*, 2nd edn. Oxford: Oxford University Press.

Bridigare, R. R. 1989. Potential effects of UV-B radiation on marine organisms of the Southern Ocean: distributions of phytoplankton and krill during austral spring. *Photochemistry and Photobiology*, **50**, 469–477.

Davidson, A. T., Bramich, D., Marchant, H. J. & McMinn, A. 1994. Effects of UV-B irradiation on growth and survival of Antarctic marine diatoms. *Marine Biology*, **119**, 507–515.

El Sayed, S. Z., Stephens, F. C., Bridigare, R. R. & Ondrusek, M. E. 1990. Effect of ultraviolet radiation on Antarctic marine phytoplankton. In Kerry, K. R. & Hempel, G., eds. *Antarctic ecosystems. Ecological change and conservation*. Berlin: Springer-Verlag, 379–385.

Fiala, M. & Oriol, L. 1990. Light–temperature interactions on the growth of Antarctic diatoms. *Polar Biology*, **10**, 629–636.

Fryxell, G. A. & Medlin, L. K. 1981. Chain forming diatoms: evidence of parallel evolution in *Chaetoceros*. *Cryptogamie Algologie*, **2**, 3–29.

Glew, J. R. 1989. A new type of mechanism for sediment samples. *Journal of Palaeolimnology*, **2**, 241–243.

Goldberg, E. D., Gambles, E., Griffin, J.J . & Koide, M. 1977. Pollution history in Narangansett Bay as recorded in its sediment. *Estuarine Coastal and Marine Science*, **5**, 549–561.

Graneli, E., Graneli, W., Rabbani, M. M., Daugbjerg, N. Fransz, G., Cuzin-Roudy, J. & Alder, V. A. 1993. The influence of copepod and krill grazing on the species composition of phytoplankton communities from the Scotia–Weddell Sea. *Polar Biology*, **13**, 201–213.

Gray, J. S., Aschan, M., Carr, M. R., Clarke, K. R., Green, R. H., Pearson, T. H., Rosenberg, R. & Warwick, R. M. 1988. Analysis of community attributes of the benthic macro-fauna of Frierfjord/Langesundfjord and in a mesocosm experiment. *Marine Ecology Progress Series*, **46**, 151–165.

Holm-Hansen, O., Mitkchell, B. G. & Vernet, M. 1989. Ultraviolet radiation in Antarctic waters: effects on rates of primary productivity. *Antarctic Journal of the United States*, **24**, 177–178.

Jacka, T. H., Christou, L. & Cook, B. J. 1984. A data bank of mean monthly and annual surface temperatures for Antarctica, the Southern Ocean and South Pacific Ocean. *ANARE Research Notes*, No. 22, 1–97.

Kang, S. H. & Fryxell, G. A. 1991. Most abundant diatom species in water column assemblages from five ODP Leg 119 drill sites in Prydz Bay, Antarctica: distributional patterns. *Proceedings of the Ocean Drilling Program: Scientific Results*, **119**, 645–666.

Karentz, D. 1991. Ecological considerations of Antarctic ozone depletion. *Antarctic Science*, **3**, 3–11.

Karentz, D., Cleaver, J. E. & Mitchell, D. 1991a. Cell survival characteristics and molecular responses of Antarctic phytoplankton to ultraviolet-B radiation. *Journal of Phycology*, **27**, 326–341.

Karentz, D., McEuen, F. S., Land, M. C. & Dunlap, W. C. 1991b. Survey of mycosporine-like amino acid compounds in Antarctic marine organisms: potential protection from ultraviolet exposure. *Marine Biology*, **108**, *157–166*.

Marchant, H. J. & Davidson, A. T. 1991. Possible impacts of ozone depletion on trophic interactions and biogenic vertical carbon flux in the Southern Ocean. In Weller, G. *et al.*, eds. *Proceedings of the International Conference on the Role of Polar Regions in Global Change*. Fairbanks: Geophysical Institute, 397–400.

McMinn, A. 1995. Comparison of diatom preservation between oxic and anoxic basins in Ellis Fjord, Antarctica. *Diatom Research*, **10**, 145–151.

McMinn, A. & Hodson, D. 1993. Seasonal phytoplankton succession in Ellis Fjord, eastern Antarctica. *Journal of Plankton Research*, **15**, 925–938.

McMinn, A., Heijnis, H., Hodgson, D. 1994. Minimal effects of UV-B on Antarctic diatoms over the past 20 yr. *Nature*, **370**, 547–549.

Medlin, L. K. & Priddle, J. 1990. *Polar marine diatoms*. Cambridge: British Antarctic Survey, 214 pp.

Perrin, R. A., Lu, P. & Marchant, H. J. 1987. Seasonal variation in marine phytoplankton and ice algae at a shallow Antarctic coastal site. *Hydrobiologia*, **146**, 33–46.

Pimental, R. A. 1993. *BIOiTAT II A multivariate statistical toolbox. Tutorial manual*, 3rd edn. San Luis Obispo, CA: Sigma Soft, 297 pp.

Robbins, J. A. & Edlington, D. N. 1975. Determination of recent sedimentation rates in Lake Michigan using ^{210}Pb and ^{137}Cs. *Geochemica Cosmochemica Acta*, **39**, 285–304.

Russell-Head, D. & Simmonds, I. 1993. Temporal structure of surface weather parameters at Casey, Davis, Mawson and Macquarie Island. *University of Melbourne, School of Earth Sciences Publication*, No. 35, 1–181.

Ryan, K. G. 1992. UV radiation and photosynthetic production in Antarctic sea-ice microalgae. *Journal of Photochemistry and Photobiology*, **13**, 235–240.

Ryan, K. G. & Beaglehole, D. 1994. Ultraviolet radiation and bottom-ice algae: laboratory and field studies from McMurdo Sound,

Antarctica. *Antarctic Research Series,* **62,** 229–242.

Scott, P., McMinn, A. & Hosie G. 1994. Physical parameters influencing diatom community structure in eastern Antarctic sea ice. *Polar Biology*, **14,** 507–517.

Smith, W. O. & Nelson, D. M. 1986. The importance of ice edge phytoplankton production in the Southern Ocean. *BioScience*, **36,** 251–257.

Smith, R. C., Prézelin, B. B., Baker, K. S., Bridigare, R. R., Boucher, N. P., Coley, T., Karentz, D., Macintyre, S., Matlick, H. A., Menzies, D., Ondrusek, M., Wan, Z. & Waters, K. J. 1992. Ozone depletion: ultraviolet radiation and phytoplankton biology in Antarctic waters. *Science*, **255,** 952–959.

Stolarski, R. S., Schoeberl, M. R., Newman, P. A., McPeters, R. D. & Krueger, A. J. 1990. The 1989 Antarctic ozone hole as observed by TOMS. *Geophysical Research Letters*, **17,** 1269–1270.

Trodahl, H. J. & Buckeley, R. G. 1989. Ultraviolet levels under the sea ice during the Antarctic spring. *Science*, **245,** 194–195.

Watanabe, K. 1988. Sub-ice micro algal strands in the Antarctic coastal fast ice area near Syowa Station. *Sorui (Japanese Journal of Phycology)*, **36,** 221–229.

54 Microbial niches in the polar environment and the escape from UV radiation in non-marine habitats

WARWICK F. VINCENT[1] AND ANTONIO QUESADA[2]

[1]*Département de biologie et Centre d'études nordiques, Université Laval, Sainte-Foy, Québec G1K 7P4, Canada,*
[2]*Departemento de biologia, Universidad Autonoma de Madrid, 28049 Madrid, Spain*

ABSTRACT

The non-marine environments of Antarctica contain a physically and chemically diverse variety of habitats for microbial colonization and growth. Three potential niche strategies are evaluated: specialized genotypes that occupy narrow niches in which they outperform other colonizing species; generalist genotypes that grow sub-optimally, but survive because of their tolerance to environmental extremes; and generalists that experience broad, variable conditions including periods (or patches) of optimal and suboptimal growth. Ecological success may ultimately depend on flexibility, and the speed of response to changing conditions. No single strategy appears to be characteristic of the polar zones, and it is likely that all three often operate in the same species in different niche dimensions. This combination of traits may influence the response of microbiota to the changing ultraviolet-B radiation flux over the polar zones. For example, repair of UV-B damage by biosynthetic processes may be inefficient given the lack of cold temperature adaptation by many genotypes and the additional stresses (e.g. osmotic) imposed by the polar environment. There are large differences between species in their ability to escape UV damage, and organisms may be especially vulnerable during the colonization phase.

Key words: algae, Antarctic, Arctic, bacteria, competition, cyanobacteria, microbial ecology, niche, ultraviolet radiation, UV-B.

INTRODUCTION

Although the islands and continental margins of Antarctica lack the higher plant and animal communities which characterize the north polar zone, they offer a physically and geochemically diverse range of habitats for microbial life-forms. These non-marine environments include freshwater and saline lakes, streams, ice-shelf meltwaters, rock, soils, glacier ice and snow. As in the Arctic, microbial communities appear to flourish in many of these types of environment despite the exposure to extreme seasonal conditions. The microorganisms include bacteria, microalgae, fungi and protozoa, and these assemblages often control, and sometimes exclusively dominate, the biological flux of carbon and energy. Antarctic microbiota must now contend with another seasonal extreme, the springtime exposure to increasingly high dosages of ultraviolet-B radiation (UV-B) (e.g. Smith *et al.* 1992). Similar conditions are also developing in high latitude regions of the Northern Hemisphere (e.g. Madronich 1992).

In this review we first examine the diversity of microbial strategies in the non-marine polar environment, with emphasis on Antarctica. We consider how niche space may be partitioned between species, and evaluate the hypothesis that polar organisms occupy broad realized as well as fundamental niches (*sensu* Hutchinson as given in Giller 1984). We then identify the niche strategies which might allow microorganisms to escape the damaging effects of UV radiation, and consider how the continued rise in ambient UV-B may impact on polar microbial ecosystems.

Niche breadth in the polar environment

Microbial community structure is a function of immigration rates during the primary colonization phase, the competition for available resources (e.g. habitat space), the growth and survival response curves over the environmental range, and loss processes. The latter include biological sources of mortality such as pathogenic attack, grazing and predation, as well as abiotic mechanisms such as desiccation and mechanical disruption (e.g. by ice crystal formation). Three potential strategies may allow microorganisms to achieve ecological success during and subsequent to the colonization phase (Fig. 54.1):

1. Specialized genotypes occupy narrow niches in which they outperform other colonizing species.
2. Generalist genotypes grow suboptimally, but survive because of their tolerance to environmental extremes.
3. Generalist genotypes experience broad, variable conditions including periods (or patches) of optimal and suboptimal growth. Niche space changes in different parts of the habitat, or over the course of a season.

Although the first and third of these strategies differ with respect to niche breadth, they share the characteristic of an actual or realized niche that largely corresponds to the potential or fundamental niche; the organism expresses its optimal fitness (measured for example as maximum growth rate) within the environmental range of the habitat. The second strategy implies that tolerance is more important than competitive ability; the realized niche of the microorganism is confined to a small part of its fundamental niche, and optimal fitness is rarely achieved.

Niche dimensions

The enormous spatial and temporal variability which characterizes the natural world means that perfect fitness of a microorganism to its environment in all niche dimensions is neither likely, nor a strategy for long-term ecological and evolutionary survival. Furthermore, an organism may differ greatly in its goodness-of-fit with respect to different niche dimensions. In the following sections we evaluate to what extent the three microbial strategies defined above operate along specific dimensions of niche space within the polar environment.

Temperature

Low temperature is a persistent feature of Arctic and Antarctic habitats, and a type 1 strategy (optimal growth at low temperatures) would appear to be highly advantageous. Contrary to expectation, true psychrophily (defined by Morita 1975 as growth within (or below) the range 0 to 20 °C, with an optimal growth at or below 15 °C) seems rare in the polar environment relative to psychrotrophy (as defined by growth at 5 °C or less but with a temperature optimum above 15 °C). Even in the stable cold environment of the Antarctic Ocean many of the bacteria appear to be psychrotrophic, although Atlas & Morita (1986) report that the proportion of psychrophilic species appears to be higher than in the Arctic Ocean. Psychrotrophy implies a type 2 strategy (suboptimal growth under natural conditions). It should be noted, however, that there is evidence of low temperature adaptation at the level of membrane transport in certain polar psychrotrophs (Nedwell & Rutter 1994 and references therein).

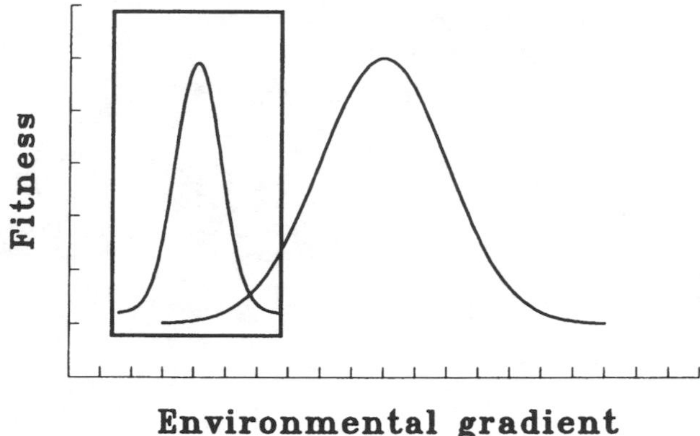

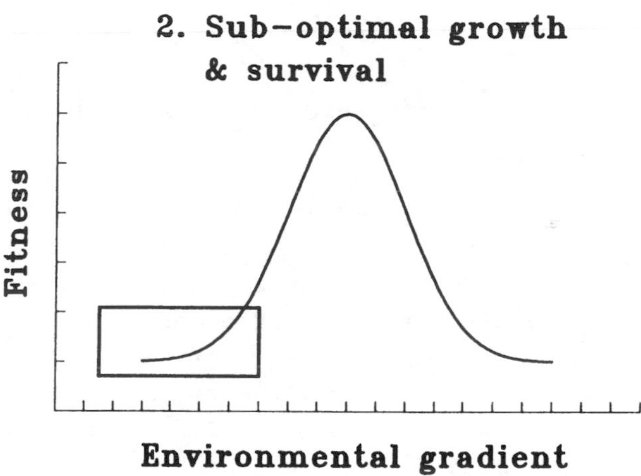

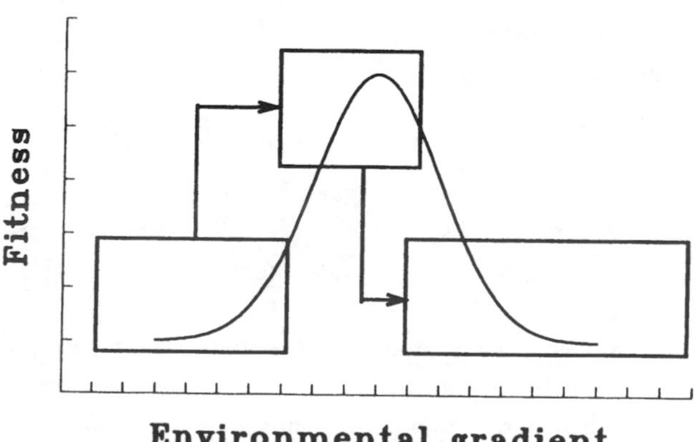

Fig. 54.1. Three microbial strategies in the polar environment. The bell-shaped curves represent the fundamental niche, and the boxes represent available niche space.

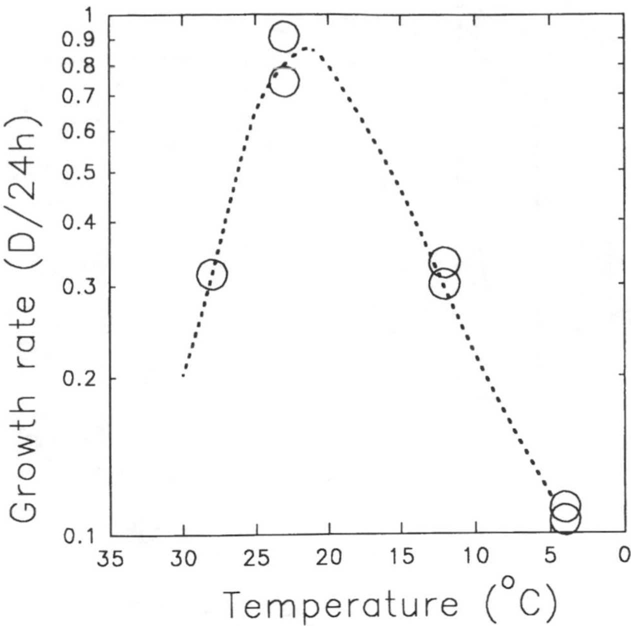

Fig. 54.2. The growth versus temperature curve for *Oscillatoria priestleyi* from Salt Pond on the McMurdo Ice Shelf (D=doublings). Redrawn from Castenholz & Schneider (1994).

The disparity between environmental temperatures and the temperature optima for growth is especially manifest in terrestrial and freshwater polar environments. Many of the heterotrophic bacteria brought into culture from Antarctic soils have optima 10 to 20 °C above the upper environmental limit (e.g. Miller *et al.* 1983, Chauhan & Shivaji 1994). Cyanobacteria, which are ubiquitously distributed throughout the Arctic (e.g. Hamilton & Edlund 1994) and Antarctic (e.g. Wynn-Williams 1994), show maximal rates of photosynthesis, respiration and growth at temperatures well above 10 °C, while their natural habitats rarely warm above 5 °C. For example, the growth optimum for the mat-forming species *Oscillatoria priestleyi* isolated from the McMurdo Ice Shelf is in the range 21 to 25 °C, with long-term growth at temperatures as high as 28 °C. Growth was extremely slow (<0.2 doublings d^{-1}) at 0 to 10 °C, the temperature range found in its native pondwater environment (Fig. 54.2; Castenholz & Schneider 1994). These types of response show that fast growth at low temperature is not a prerequisite for success in the polar zones, and temperature may be a less important niche dimension relative to other constraints in the polar environment.

In some terrestrial environments the ability to tolerate and adapt to fluctuations in temperature (a modified type 3 response) may be more important than genetic adaptation to a restricted subset of conditions. Rutter & Nedwell (1994) addressed this question using two species of bacteria isolated from Antarctic lake sediments. When mixed cultures containing both isolates were exposed to fluctuations in temperature, the competitive success of each species depended in part upon their temperature-dependent growth parameters (especially K_s), but also upon their speed of adjustment to the changing conditions, which differed markedly between the two species. There is a need for polar ecologists to extend this work to studies of other microbial groups exposed to non-equilibrium conditions at a variety of timescales.

Light

Arctic and Antarctic phototrophs experience extremes of irradiance, from continuous sunlight in summer to continuous winter darkness. Many species escape the effects of bright summer UV and PAR via their selection of (or by) habitat, and throughout the growing season they live under extremely low photon fluence rates. Such conditions are likely to favour strong competition for light as a limiting resource, thereby selecting for Type 1 strategies (as defined above). Evidence of this level of niche specialization comes from studies of photosynthesis (P) as a function of irradiance (I) in several types of Antarctic environment, in particular the slope of the initial P versus I curve (alpha), and the irradiance which marks the onset of saturation (I_k). In the McMurdo Dry Valley lakes, for example, photosynthetic carbon incorporation as well as inorganic nitrogen uptake saturate at irradiances less than 1% of ambient (Priscu *et al.* 1987, Priscu 1989 Lizotte & Priscu 1992). Similarly low light saturation values have been observed for phytoplankton in maritime Antarctic lakes (Hawes 1985).

Shade species of *Prasiola* have been found in Antarctic glacier-fed streams (Broady 1989), and more recently in equivalent stream habitats from the high Arctic (Hamilton & Edlund 1994). Mat-forming cyanobacteria in both polar zones contain most of the photosynthetically active biomass within a deep chlorophyll maximum where the cells experience extreme shade, and the communities generally show high alpha values (Vincent *et al.* 1993).

Terrestrial algal species which live beneath translucent rocks (subliths, as described in Broady 1981), within the soil, within rock fissures (chasmoendoliths) and in the spaces between rock crystals (cryptoendoliths) are known to occur widely in both polar zones, and are likely to be strongly shade-adapted or acclimated. For example, the Antarctic green alga *Hemichloris antarctica* is found within certain rocks below a lichen-dominated community, and its irradiance regime during summer is restricted to the range 0.05 to 10 μmol m^{-2} s^{-1}. This selection of habitat space may reflect competition with the higher-light-requiring-lichens (Nienow & Friedmann, 1993), and it again implies that specialization towards low light may be an important requirement for survival and growth in Antarctica.

Nutrients

Nutrient supply may be an important factor influencing plant biomass and diversity in the Arctic tundra and associated freshwaters (Shaver *et al.* 1992). The importance of nutrient supply as a microbial niche dimension requires much closer attention in the Antarctic environment. There is recent evidence from Signy Island (maritime Antarctic) fellfield soils that microbial populations are limited by nitrogen availability, and that nitrogen supply may also influence their resilience to freeze-up (Davey &

Rothery 1992). Phosphorus and-or nitrogen limit planktonic algal biomass in many Arctic and Antarctic lakes (Priddle *et al.* 1986). However, nutrients appear to be present in excess in the mats of cyanobacteria in the Antarctic lakes and streams studied to date (Vincent *et al.* 1993). Thus nutrient supply may be a niche dimension selecting for a type 1 strategy in many but not all polar environments.

Grazer resistance

Herbivore activity is an important feature influencing niche breadth and diversity in some environments. For example, in Hudson Bay (Canadian sub-Arctic) grazing pressure by the lesser snow goose can either increase or decrease plant species diversity depending on whether the birds feed on above-ground vegetation or below-ground rhizomes (Hunter *et al.* 1992). Exclusion experiments in mountain streams at temperate latitudes have shown that grazer activity can also influence microbial diversity: microbial mats were overgrown by felts of diatoms when grazing fish and invertebrates were excluded (Power cited in Hunter *et al.* 1992). Mats of cyanobacteria are common throughout aquatic environments in the Arctic where high populations of grazing arthropods can develop, as well as in the Antarctic where the benthic grazer communities are generally restricted to small organisms (nematodes, rotifers, tardigrades and protozoa) that are unlikely to have a major impact on the total standing stock of periphyton (e.g. Hawes 1989). Many of the terrestrial micro-organisms in Antarctica inhabit environments that are too severe for larger invertebrates, and grazer resistance thus seems an unlikely influence on niche breadth and diversity.

For the plankton within polar lakes, top-down grazer controls may operate both within and above the microbial loop. In Antarctica only the more northerly lakes (e.g. lakes on Signy Island) contain substantial populations of larger zooplankton that might exert a significant effect (Priddle *et al.* 1986). Crustacean zooplankton occur in lakes throughout the Arctic, but the evidence to date points to a greater importance of bottom-up controls on the microbial food web, mediated by nutrient and light effects on phytoplankton production (Rublee 1992).

Disturbance, tolerance and resilience

The ability to tolerate and recover from disruptions to the physical and chemical environment appears to be an important factor influencing the structure of higher plant communities in certain temperate latitude habitats (Grime 1979) and may play a role in Arctic and Antarctic microbial ecosystems. Some microbial communities inhabit physically stable environments, for example in permanently ice-covered lakes (Priscu *et al.* 1987), that are conducive to type 1 strategies. Elsewhere, the potential disturbances could be severe, for example by freezing and ice-crystal formation, by wind-scouring of soils and other exposed habitats, and by the scouring effects of flowing waters. Specific examples include the exfoliation of rocks containing cryptoendoliths (Nienow & Friedmann 1993) and the scouring of benthic stream communities by sediment-laden flowing water (Howard-

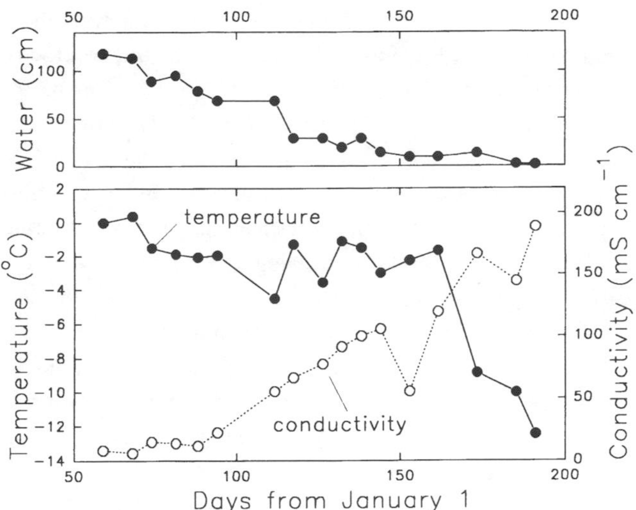

Fig. 54.3. Changes in salinity and temperature during freeze-up in a coastal pond on Ross Island. Redrawn from Schmidt *et al.* (1991).

Williams *et al.* 1986). These habitats may select for broad-niche, type 3 species capable of adjusting to major temporal and spatial changes in the environment.

The biota in many types of polar environment also experience a major chemical disturbance each year, the extreme dehydration and rise in osmolarity during freeze-up. Evidence of the potential magnitude of this effect comes from a study of two small ponds near the coast of Ross Island (Schmidt *et al.* 1991). These waters were relatively fresh during open water conditions in summer, but during the freeze-up process there was a gradual concentration of salts in the remaining water (Fig. 54.3). In mid-winter the benthic microbial communities were exposed to a thin layer of concentrated brine, with salinities more than six times greater than seawater and liquid water temperatures of −12 °C. Salinity may therefore be a dimension of niche space that selects for broadly tolerant, type 3 genotypes. Antarctic microbial mats appear to be highly resistant of desiccation, although there are differences in the degree of tolerance between different species assemblages (Davey 1989, Hawes *et al.* 1992).

Escape from UV radiation effects

The increasing dosages of UV-B radiation now experienced in Antarctica as a result of stratospheric ozone depletion suggest that this high-energy component of the solar spectrum may increasingly play a role in restricting or modifying the realized niches of the biota. However, many groups of organisms have four lines of defense against the damaging effects of UV-B radiation: avoidance, screening, quenching and repair. The following sections briefly consider to what extent these may operate in polar microbial environments. Additional, complementary details are given in Wynn-Williams (1994) and Vincent & Quesada (1994).

Avoidance

Many microbial communities in Antarctica live in sheltered habitats that are protected from freezing and bright light during the summer growing season. This same protection may provide an important shield against UV radiation. In the cryptoendolith environments, for example, short wavelengths appear to be absorbed and scattered differentially resulting in a dim light regime that is enriched in wavelengths towards the red end of the spectrum (Nienow *et al.* 1988). A similar orange–red light regime lacking in high-energy wavelengths characterizes the growth environment for the deep-living (and photosynthetically most active) component of Antarctic microbial mats (Vincent *et al.* 1993). The primary colonization phase, however, may remain vulnerable to UV-B effects, and should be a high priority area of microbial research in both the Antarctic and Arctic. One escape mechanism that may be much less available in the south relative to north polar zone is the screening of aquatic communities by dissolved humic materials. These humic materials derived from terrestrial vegetation are effective absorbers of UV wavelengths (Vincent & Roy 1993), and are unlikely to occur at high concentration in the streams draining continental Antarctic catchments, where higher plants are usually absent and the vegetation often sparsely distributed. However, dissolved organic carbon values up to 9 mg l^{-1} have been recorded for maritime Antarctic lakes where extensive cryptogamic communities occur in the catchments (Ellis-Evans 1981).

Cellular motility may also provide an escape mechanism from UV radiation effects. Species of cyanobacteria from the McMurdo Ice Shelf, for example, can rapidly adjust their position within the microbial mat environment by trichome gliding in response to bright or dim light (Vincent *et al.* 1993). This ability does not appear to be impaired by low to moderate levels of UV radiation, and these short wavelengths appear to induce a self-shading clustering response (Fig. 54.4). Motility is also a feature of many of the pennate diatoms which inhabit benthic habitats, as well as of the flagellates within Antarctic lake and pond phytoplankton assemblages.

Screening

A broad range of UV-absorbing pigments are produced by polar micro-organisms. For example the black or dark gold pigmentation associated with microbial mats and films is often due to the cyanobacterial pigment scytonemin (Fig. 54.5). These darkly pigmented communities are common in the shallow waters of Arctic lakes and ponds (Fig. 54.5), Arctic and subarctic tundra streams (Sheath *et al.* 1995), exposed Arctic rock faces (Konhauser *et al.* 1994) and equivalent habitats in Antarctica (Vincent & Quesada 1994). Scytonemin absorbs maximally in the UV-A end of the spectrum, and is produced in higher concentrations in response to UV or bright light (Garcia-Pichel & Castenholz 1992). Additional water soluble pigments that absorb at lower UV wavelengths are also known to occur in some polar species (Fig. 54.5). These compounds may be mycosporine-like amino acids which also can be induced to accumulate to higher concentration by incubating cultures in bright light or UV (Garcia-Pichel *et al.* 1993). Flavonoids may also play a similar role in certain assemblages in the Antarctic terrestrial environment (Wynn-Williams 1994). Antarctic lichens often contain a dark pigmentation which is thought to act as a UV-screen and which increases in concentration in communities exposed to bright light (see fig. 11.18 in Kappen 1993).

Quenching

A potentially severe problem associated with solar UV radiation is the production of high energy photooxidants such as free radicals and hydrogen peroxide (Vincent & Roy 1993). Many Antarctic microbial species, heterotrophs as well as phototrophs, are rich in carotenoids, compounds that are known effectively to quench these damaging photochemical products as well as stabilize the photosynthetic reaction centres. Antarctic lichens are often rich in cellular carotenoids (Kappen 1993). Stream and pond mats of cyanobacteria in the Arctic as well as Antarctica often contain a surface layer that is highly enriched in carotenoids such as myxoxanthophyll and canthaxanthin. Such compounds may also be transferred to grazers and thereby confer a similar protection to certain animal populations; for example, HPLC chromatograms of acetone extracts of the rotifer *Philodina gregaria* from ponds in the McMurdo Sound region show the accumulation of myxoxanthophyll and other carotenoids that are characteristic of the cyanobacterial mats on which they feed (M.T. Downes & W.F. Vincent, unpublished). It is probable, however, that some of the carotenoids play additional roles, for example in light-harvesting by photosynthetic organisms, and in membrane stability across the environmental temperature range (e.g. Chauhan & Shivaji 1994).

Repair

Most microorganisms have at least some capacity to identify and repair the damage associated with UV exposure. The recovery mechanisms include DNA excision and repair, and the resynthesis and replacement of photosynthetic proteins. The efficacity of these processes in polar terrestrial and freshwater environment is a major unknown at present, but it is likely that, as with the other escape mechanisms, there are large differences between species. The net cellular damage may represent the balance between damage and repair (Lesser *et al.* 1994); if the latter processes are enzymatic and thus temperature-sensitive, high latitude communities may be much more sensitive to UV than their temperate counterparts, particularly given their lack of temperature adaptation as noted above. The temperature-dependence of UV effects will be an important focus for future research in the high latitude zones of both hemispheres. The repair processes may also have other complex requirements. For example, our recent work with cyanobacteria from the McMurdo Ice Shelf indicates that the spectral balance across the UV waveband has a major influence on growth responses; increasing levels of UV-A can offset the inhibitory effects of low UV-B dosages (Quesada *et al.* 1995).

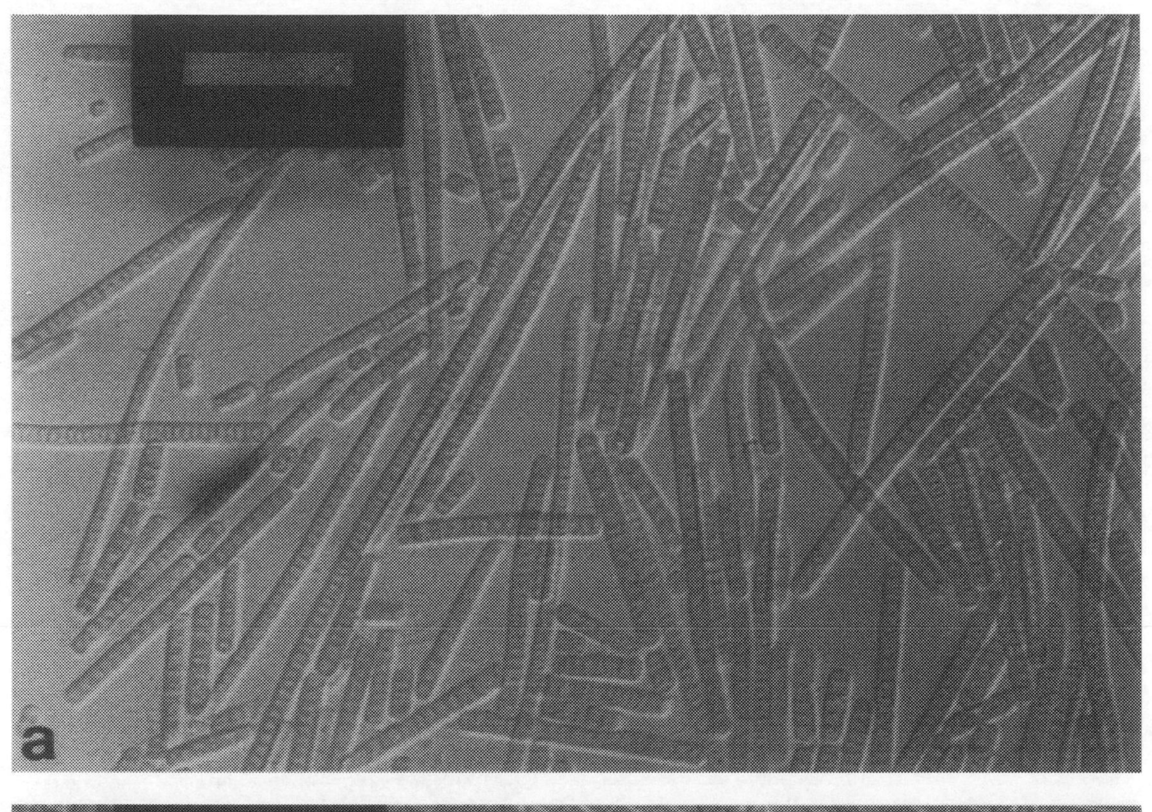

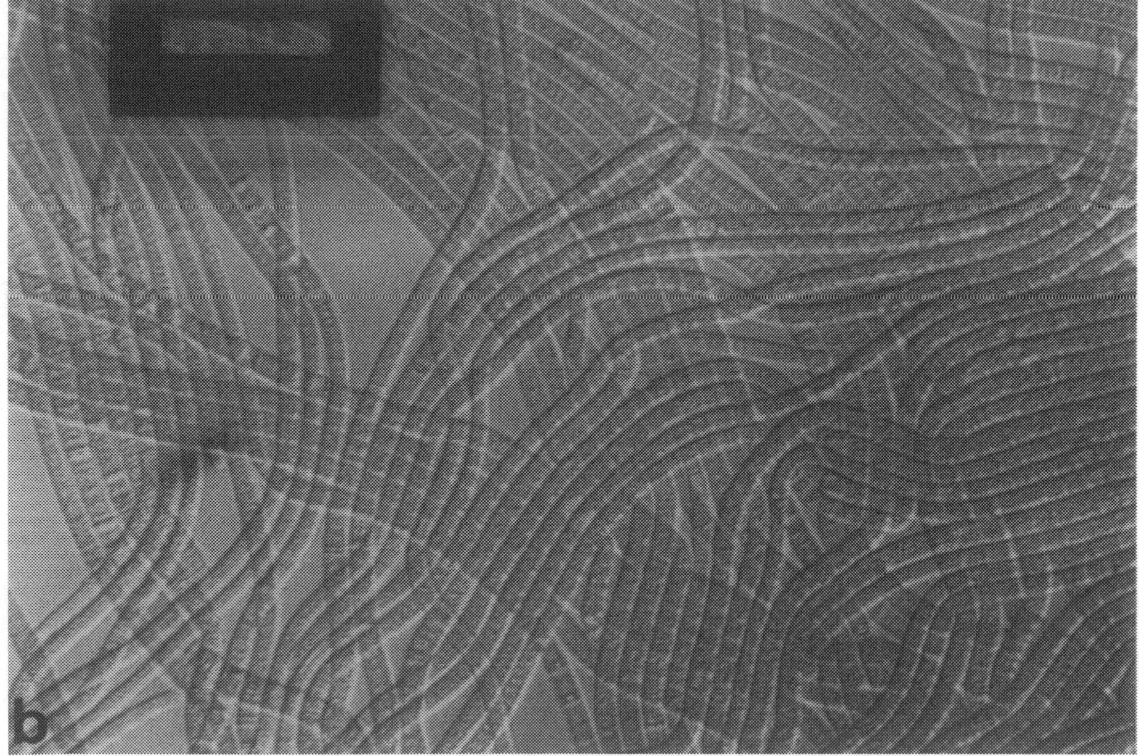

Fig. 54.4. *Oscillatoria priestleyi* from Salt Pond grows as randomly dispersed filaments (a) under white light (120 μmol m^{-2} s^{-1}), but low dosage rates of UV-B (100 μW cm^{-2}over 6 h; the lamp spectrum is given in Quesada *et al.* 1995) induce a self-shading response via trichome gliding and a re-alignment of filaments (b). The scale bar (inner open region of the scale mask) is 20 μm.

CONCLUSIONS

Polar environments contain a great variety of habitats for microbial colonization and growth. The hypothesis that polar microbes inhabit broad realized niches that correspond to their fundamental niche is only partially supported. Certain dimensions of niche space have selected for highly specialised type 1 genotypes that compete for low levels of certain resources, for example the photosynthetic species forced into shade conditions because of the severe environment and instability of exposed

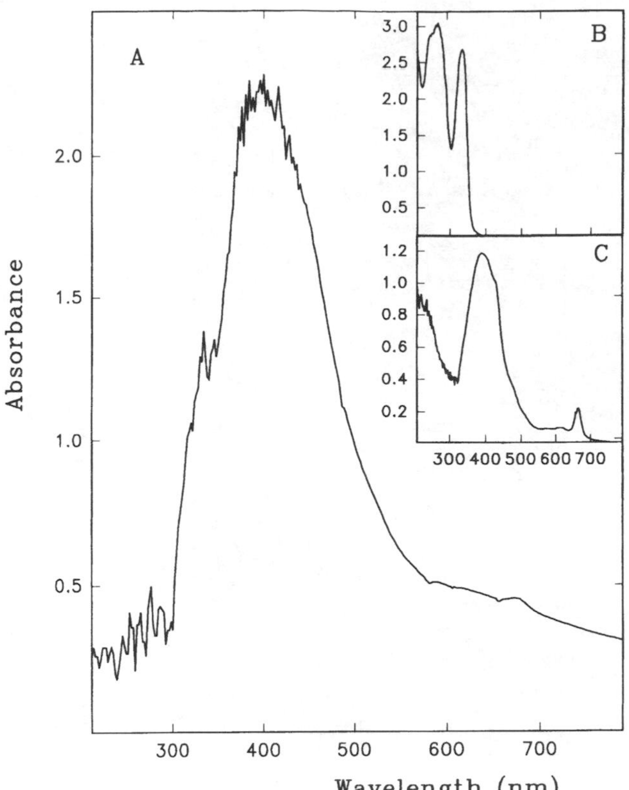

Fig. 54.5. UV-screening by *Nostoc commune* from Toolik Lake, Alaska. Curve A is an *in vivo* absorbance spectrum obtained with an Hewlett-Packard diode array spectrophotometer fitted with an integrating sphere. The right-hand curves are spectrophotometric scans of extracts of this material with either 90% methanol–water (B), or 90% acetone–water (C).

habitats. Persistent conditions of salinity or nutrients may have favoured certain genotypes in particular environments such as the halophilic species *Planococcus* in Dry Valley soils (Miller *et al.* 1983), and picoplanktonic organisms in ultra-oligotrophic lakes. Other niche dimensions, such as temperature, seem to have played a much lesser selective role, and the organisms often show a type 2/tolerance–suboptimal growth response. In the physically or chemically unstable parts of Antarctica microbial success is dependent upon broad, flexible type 3 tolerances to environmental conditions, for example, salinity. The ability to adjust rapidly to major fluctuations in the environment may be an important determinant of competitive success. No single strategy appears to dominate in the polar environment, and many species may exhibit all three depending on the niche dimension.

This same combination of niche traits may strongly influence the response of the polar biota to changes in ambient UV-B. For example, the lack of optimal growth at cold ambient temperatures may mean that such microbes are more vulnerable to UV exposure. The preference of many communities, however, for well-protected shade habitats will greatly lessen the impact of solar UV. Polar microbes possess a broad range of UV escape mechanisms, but there are major variations between species, and the continued long-term exposure to enhanced UV-B is likely to affect community structure, particularly in unstable habitats during the colonization phase.

ACKNOWLEDGEMENTS

The preparation of this paper was supported by the Natural Sciences and Engineering Research Council of Canada, and the Ministry of Education, Spain. We thank R. W. Castenholz and two anonymous reviewers for their comments.

REFERENCES

Atlas, R. M. & Morita, R. Y. 1986. Bacterial communities in nearshore Arctic and Antarctic marine ecosystems. In Megusar, F. & Gantar, M., eds. *Perspectives in microbial ecology*. Ljubljana: Slovene Society for Microbiology, 185–190.

Broady, P. A. 1981. The ecology of sublithic terrestrial algae at the Vestfold Hills, Antarctica. *British Phycological Journal*, **16**, 231–40.

Broady, P. A. 1989. The distribution of *Prasiola calophylla* (Carmich.) Menegh. (*Chlorophyta*) in Antarctic freshwater and terrestrial habitats. *Antarctic Science*, **1**, 109–118.

Castenholz, R. W. & Schneider, A. J. 1994. Cyanobacterial dominance at high and low temperatures: optimal conditions or precarious existence? In Guerrero, R. & Pedors-Alio, C., eds. *Trends in microbial ecology*. Barcelona: Spanish Society for Microbiology, 19–24.

Chauhan, S. & Shivaji, S. 1994. Growth and pigmentation of *Sphingobacterium antarcticus*, a psychrotrophic bacterium from Antarctica. *Polar Biology*, **14**, 31–36.

Davey, M. C. 1989. The effects of freezing and desiccation on photosynthesis and survival of terrestrial Antarctic algae and cyanobacteria. *Polar Biology*, **10**, 29–36.

Davey, M. C. & Rothery, P. 1992. Factors causing the limitation of growth of terrestrial algae in maritime Antarctica during late summer. *Polar Biology*, **12**, 595–601.

Ellis-Evans, J. C. 1981. Freshwater microbiology in the Antarctic I & II. *British Antarctic Survey Bulletin*, No. 54, 85–121.

Garcia-Pichel, F. & Castenholz, R. W. 1992. Characterization and biological implications of scytonemin, a cyanobacterial sheath pigment. *Journal of Phycology*, **27**, 395–409.

Garcia-Pichel, F., Wingard, C. E. & Castenholz, R. W. 1993. Evidence regarding the UV sunscreen role of a mycosporine-like compound in the cyanobacterium *Gloeocapsa* sp. *Applied and Environmental Microbiology*, **59**, 170–176.

Giller, P. S. 1984. *Community structure and the niche*. London: Chapman & Hall, 176 pp.

Grime, J. P. 1979. *Plant strategies and vegetation processes*. London: John Wiley, 222 pp.

Hamilton, P. B. & Edlund, S. A. 1994. Occurrence of *Prasiola fluviatilis* (Chlorophyta) on Ellesmere Island in the Canadian Arctic. *Journal of Phycology*, **30**, 217–221.

Hawes, I. 1985. Light climate and phytoplankton photosynthesis in maritime Antarctic lakes. *Hydrobiologia*, **123**, 69–79.

Hawes, I. 1989. Filamentous green algae in freshwater streams on Signy Island, Antarctica. *Hydrobiologia*, **172**, 1–18.

Hawes, I., Howard-Williams, C. & Vincent, W. F. 1992. Desiccation and recovery of Antarctic cyanobacterial mats. *Polar Biology*, **12**, 587–594.

Howard-Williams, C., Vincent, C. L., Broady, P. A., & Vincent, W. F. 1986. Antarctic stream ecosystems: variability in environmental properties and algal community structure. *Internationale Revue der gesamten Hydrobiologie*, **71**, 511–544.

Hunter, M. D., Ohgushi, T. & Price, P. W. 1992. *Effects of resource distribution on animal–plant interactions*. San Diego: Academic Press, 505 pp.

Kappen, L. 1993. Lichens in the Antarctic region. In Friedmann, E. I., ed. *Antarctic microbiology*. New York: Wiley-Liss, 433–490.

Konhauser, K. O., Fyfe, W. S., Schultze-Lam, S., Ferris, F. G. & Beveridge, T. J. 1994. Iron phosphate precipitation by epilithic microbial films in Arctic Canada. *Canadian Journal of Earth Sciences*, **31**, 1320-1324.

Lesser, M. P, Cullen, J. J. & Neale, P. J. 1994. Carbon uptake in a marine diatom during acute exposure to ultraviolet B radiation: relative importance of damage and repair. *Journal of Phycology*, **30**, 183–192.

Lizotte, M. P. & Priscu, J. C. 1992. Photosynthesis irradiance relationships in phytoplankton from a perennial ice-covered lake (Lake Bonney). *Journal of Phycology*, **28**, 179–185.

Madronich, S. 1992. Implications of recent total atmospheric ozone measurements for biologically active ultraviolet radiation reaching the Earth's surface. *Geophysical Research Letters*, **19**, 391–395.

Miller, K. L., Leschine, S. B. & Huguenin, R. L. 1983. Halotolerance of micro-organisms isolated from saline antarctic dry valley soils. *Antarctic Journal of the United States*, **18**, 222–223.

Morita, R. Y. 1975. Marine psychrophilic bacteria. *Bacteriological Reviews*, **39**, 144–167.

Nedwell, D. B. & Rutter, M. 1994. Influence of temperature on growth rate and competition between two psychrotolerant Antarctic bacteria: low temperature diminishes affinity for substrate uptake. *Applied & Environmental Microbiology*, **60**, 1984–1992.

Nienow, J. A. & Friedmann, E. I. 1993. Terrestrial lithophytic (rock) communities. In Friedmann, E. I., ed. *Antarctic microbiology*. New York: Wiley-Liss, 343–412.

Nienow, J. A., McKay, C. P. & Friedmann, E. I. 1988. The cryptoendolithic microbial environment in the Ross Desert of Antarctica: light in the photosynthetically active region. *Microbial Ecology*, **16**, 271–289.

Priddle, J., Hawes, I., Ellis-Evans, J.,C. & Smith, T. J. 1986. Antarctic aquatic ecosystems as habitats for phytoplankton. *Biological Reviews*, **61**, 199–238.

Priscu, J. C. 1989. Photon dependence of inorganic nitrogen transport by phytoplankton in perennially ice-covered Antarctic lakes. *Hydrobiologia*, **172**, 173–82.

Priscu, J. C., Priscu, L. R., Howard-Williams, C. & Vincent, W. F. 1987. Photosynthate distribution by microplankton in permanently ice-covered Antarctic desert lakes. *Limnology & Oceanography*, **21**, 260–70.

Quesada, A., Mouget, J-L. & Vincent, W. F. 1995. Growth of Antarctic cyanobacteria under ultraviolet radiation: UV-A counteracts UV-B inhibition. *Journal of Phycology*, **31**, 242–248.

Rublee, P. A. 1992. Community structure and bottom-up regulation of heterotrophic microplankton in Arctic LTER lakes. *Hydrobiologia*, **240**, 133–141.

Rutter, M. & Nedwell, D. B. 1994. Influence of changing temperature on growth rate and competition between two psychrotolerant Antarctic bacteria: competition and survival in non-steady-state temperature environments. *Applied & Environmental Microbiology*, **60**, 1993–2002.

Schmidt, S., Moskal, W., De Mora, S. J., Howard-Williams, C. & Vincent, W. F. 1991. Limnological properties of Antarctic ponds during winter freezing. *Antarctic Science*, **3**, 379–388.

Shaver, G. R., Billings, W. D., Chapin III, F. S., Giblin, A. E., Nadelhoffer, K. J., Oechel, W. L. & Rastetter, E. B. 1992. Global change and the the carbon balance of arctic ecosystems. *BioScience*, **42**, 433–441.

Sheath, R. G., Vis, M. L., Hambrook, J. A. & Cole, K. M. 1995. Tundra stream macroalgae of North America: composition, distribution and physiological adaptations. In Kristiansen, J., ed. *Biogeography of freshwater algae*. Dordrecht: Kluwer Academic Publishers.

Smith, H. G. 1992. Distribution and ecology of testate rhizopod fauna of the continental Antarctic zone. *Polar Biology*, **12**, 629–634.

Smith, R. C., Prézelin, B. B., Baker, K. S., Bidigare, R. R., Boucher, N. P., Coley, T., Karentz, D., Macintyre, S., Matlick, H. A., Menzies, D., Ondrusek, M., Wan, Z. & Waters, K. J. 1992. Ozone depletion: ultraviolet radiation and phytoplankton biology in Antarctic waters. *Science*, **255**, 952–959.

Vincent, W. F. & Quesada, A. 1994. Ultraviolet radiation effects on cyanobacteria: implications for Antarctic microbial ecosystems. *Antarctic Research Series*, **62**, 111–124.

Vincent, W. F. & Roy, S. 1993. Solar ultraviolet-B radiation and aquatic primary production: damage, protection and recovery. *Environmental Reviews*, **1**, 1–12.

Vincent, W. F., Castenholz, R. W., Downes, M. T. & Howard-Williams, C. 1993 . Antarctic cyanobacteria: light, nutrients and photosynthesis in the microbial mat environment. *Journal of Phycology*, **29**, 745–755.

Wynn-Williams, D. D. 1994. Potential effects of ultraviolet radiation on Antarctic primary terrestrial colonizers: cyanobacteria, algae and cryptogams. *Antarctic Research Series*, **62**, 243–257.

55 Human impact on an oligotrophic lake in the Larsemann Hills

J. CYNAN ELLIS-EVANS[1], JOHANNA LAYBOURN-PARRY[2,3], PETER R. BAYLISS[1] AND STEPHEN T. PERRISS[2]

[1]British Antarctic Survey, Natural Environment Research Council, High Cross, Madingley Road, Cambridge CB3 0ET, UK, [2]Department of Zoology, La Trobe University, Bundoora, Victoria 3083, Australia, [3] Present address: Department of Physiology and Environmental Science, University of Nottingham, Sutton Bonington, Loughborough, Leics. LE12 5RD, UK.

ABSTRACT

No Worry Lake (unofficial name) is immediately downslope of the CHINARE station, Zhongshan, which was established on Mirror Peninsula, Larsemann Hills in 1989. Studies pre-1989 indicated a slightly brackish, clear, shallow lake with a gravel catchment and virtually no terrestrial vegetation, comparable to other lakes in the area. Sampling during summer 1993 indicated that the lake, now surrounded by buildings and vehicles, had planktonic microbial communities markedly different to any other lakes in the Larsemann Hills. There was evidence of recent increased rates of sediment input, a significant presence of silt particles in the previously sandy catchment soils and high levels of nutrients in the soils immediately downslope from the main station building. Wet terrestrial areas were covered with filamentous green algae (otherwise rare in the Hills) and microelectrode profiles indicated totally anoxic conditions 1 mm beneath the surface of wet areas. The lake sediments were sulphide-rich and dominated by anaerobes. The water column was turbid from the high particulate content, much of it organic. Very large numbers of ciliate protozoa were feeding on the abundant microbial flora indicative of very rapid nutrient cycling. Human activities in the catchment are causing changes in meltwater and sediment input patterns. Recycling of lake water to facilitate cooling of the station generators has resulted in bottom-water warming, which appears to promote heterotrophic microbial activity even under winter ice cover.

Key words: Antarctic, thermal environment, environmental damage.

INTRODUCTION

Any station in Antarctica has to consider water requirements, both for drinking water and for logistic requirements. Whilst seawater can be used in both cases to provide a secure supply of freshwater, it does involve complex and expensive technology which can be difficult to establish and maintain. Drinking water can also be imported but this is expensive. It has therefore been common for stations to be established whenever possible near freshwater lakes to provide a direct source. Given the limited number of suitable ice-free areas and the frequency of lakes and pools in these same areas (even if the lake itself is not required as an exploitable source), there is a high probability of locating a station within a lake catchment.

Antarctic water bodies are located in poorly developed rocky catchments and are characterized biotically by very short food chains, with microbial components playing a predominant role (Vincent 1988). Antarctic freshwater lakes have a low chemical buffering capacity, with the biota largely comprised of psychrotrophic microorganisms capable of rapid response to

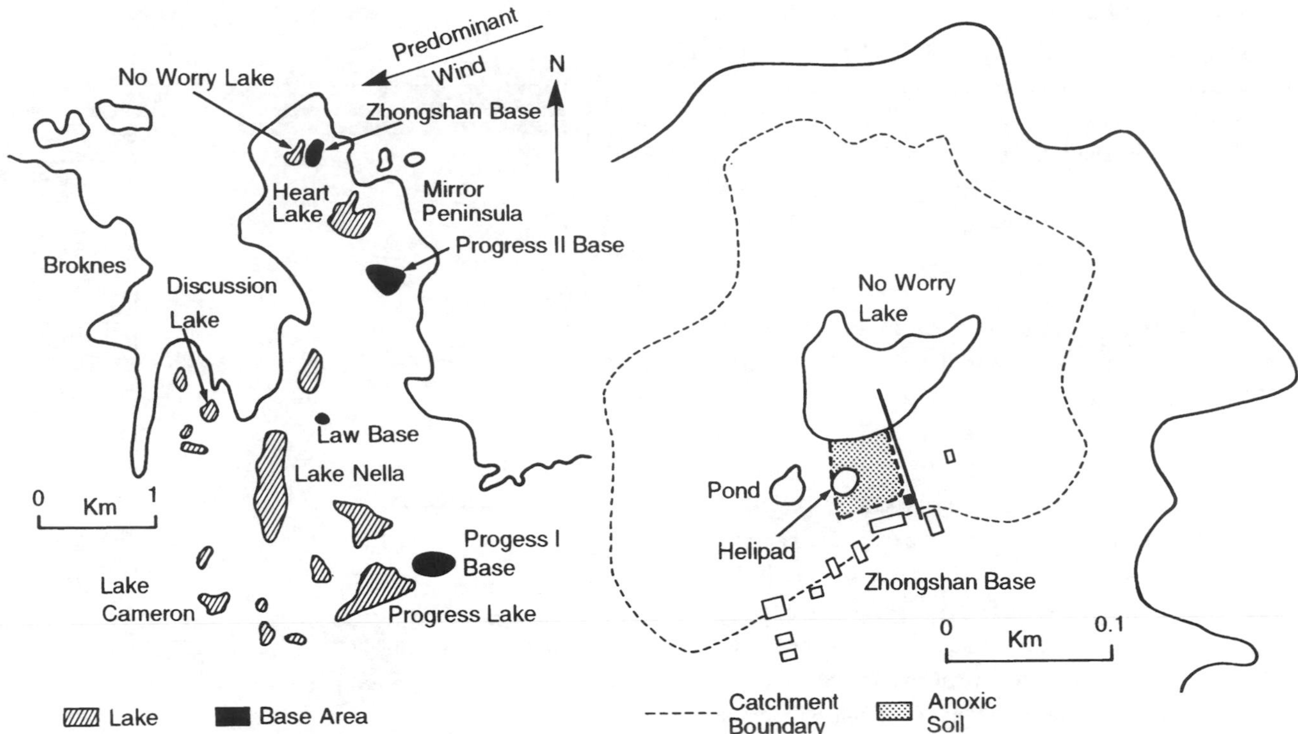

Fig. 55.1. (left) Line diagram of Mirror Peninsula showing stations and lake locations. (right) Line diagram of the seaward end of Mirror Peninsula, Larsemann Hills, showing the location of Zhongshan Station in relation to the catchment (dotted line) of No Worry Lake and location of the anoxic area.

environmental change even at low temperatures (Herbert & Bell 1977, Ellis-Evans & Wynn-Williams 1985, Nedwell & Rutter 1994). Grazing pressure is often low in continental Antarctic locations with very few grazers present (Laybourn-Parry *et al.* 1991), but can be more substantial in enriched maritime Antarctic systems (Priddle *et al* 1986). It follows that the presence of any significant pertubation in the lake or its catchment is likely to produce a significant effect on these fragile ecosystems.

There has been some effort in recent years to undertake impact assessment studies of the terrestrial environment in the vicinity of stations (e.g. effects of cement dust at Casey Station – Adamson & Seppelt 1989). Very little attention until now has been paid to the lake and stream systems of Antarctica, which effectively integrate catchment changes and therefore constitute a potentially useful assessment tool. There are plans by New Zealand scientists to examine the possible impact of waste (grey water, hydrocarbons, heavy metals) from Vanda Station (southern Victoria Land Dry Valleys) on ultra-oligotrophic Lake Vanda over the next three years (pers. comm., C. Howard-Williams).

This paper summarizes a range of observations made in and around No Worry Lake (unofficial name), which is located in close proximity to the CHINARE Station, Zhongshan, in the Larsemann Hills, and represents the first attempt to examine holistically human impact on an Antarctic lake. Publication constraints limit this account to a preliminary examination of human impact on a single lake, but a more more wide-ranging dissertation on the biology and chemistry of Larsemann Hills lakes is in preparation.

LOCATION AND SITE CHARACTERISTICS

Larsemann Hills

The Larsemann Hills (69°24′ S, 76°20′ E) are a series of rocky peninsulas and islets in Prydz Bay, situated on the Ingrid Christensen Coast of Princess Elizabeth Land, between the Amery Ice Shelf and the Sorsdal Glacier. The two largest peninsulas are Stornes and Broknes, with the Mirror Peninsula being a subdivision of the latter. There are in excess of 150 freshwater lakes throughout the Hills (Gillieson *et al.* 1990), including supra-glacial and proglacial lake types together with open and closed rock basins and colluvium dammed ponds. Meltwater streams are infrequent, except on Stornes, the most westerly peninsula, where they are generally well developed.

Mirror Peninsula lies at the eastern end of the Larsemann Hills and is the location for all four research stations (Australia's Law Base, the Russian Progress I and Progress II Stations and the Chinese Zhongshan Station) established in the Hills (Fig. 55.1). The area shows a far greater degree of weathering activity than Stornes at the western end and indeed there is a clear gradient of decreasing weathering activity westward towards the milder environmental conditions imposed by the Amery Ice Shelf. The peninsula is deeply dissected below a planation surface situated roughly 60 m above sea level. Virtually all exposed rock in the area, comprising chiefly gneisses and granites, has a thick weathered crust and extensive surficial deposits occur. Many outcrops are deeply pitted with weathering hollows (tafoni) and snowpack gravel and sand fans provide substantial mineral sediment inputs to the lakes and pools. Vegetation is extremely sparse.

No Worry Lake and Zhongshan Station

During summer 1989, the Chinese National Antarctic Research Expedition established the wintering station of Zhongshan, at the head of the Mirror Peninsula, on a series of small ridges (Fig. 55.1) bordering the catchment of No Worry Lake (unofficial local name, also recorded as LH69 by Gillieson *et al.* 1990). The station has continued to expand over the past five years and offers accomodation for around 60 personnel in summer and 20–30 over winter.

During 1986/87, when Australian scientists undertook a limnological reconnaisance of over 70 lakes in the Larsemann Hills, LH69 (referred to here as No Worry Lake) was a small, optically clear, slightly brackish lake (electrical conductivity, 1000 μS cm^{-1}) with reasonable sediment deposits and surrounded by a small (25 hectares) gravel and sand catchment. The northern end where the lake is situated is particularly exposed to the daily katabatic winds, which results in extensive aeolian and salt weathering and considerable slope instability. All the lakes in the area show an ionic balance influenced by marine salt spray but, particularly in the closed basin catchments, there is also evidence of the influence of evaporation and inputs of catchment weathering products (Gillieson *et al.* 1990).

Prior to the construction of the research station, the catchment was a barren area, relieved only by the presence of snowbanks which melted in summer and drained into the lake. There was no evidence that the area had ever been frequented by large seabirds or seals and, from general observations throughout the Peninsula area, bird populations would have consisted of a couple of Macormick skua pairs and small colonies of snow petrels and Wilson's storm petrels (Wang & Norman 1993). Since the station was established storage of food in open drums has attracted an increasing number of skuas, whose faeces constitute an increased nutrient input into the catchment of No Worry Lake.

CATCHMENT

There is considerable variation in the status of lithosols in the catchment of No Worry Lake. Away from the station buildings the soils are aerobic and contain little organic matter, but in the area below the main buildings there is a large seepage/rivulet area where nutrient levels increase markedly and anoxic conditions prevail subsurface as indicated by black subsurface sulphide deposits and the measured redox, dissolved oxygen and sulphate reduction profiles (Fig. 55.2). This seepage area receives snowmelt channelled by the station pathways, together with inorganic material resulting from snow-clearing activity. There is a significant sediment fan formed towards the bottom of the slope to the lake and particle size distribution measurements reveal a substantial size fraction of small sand or silt grades which are not seen elsewhere in the Larsemann Hills.

A particularly significant component of this seepage area is the presence of large amounts of the green filamentous algae *Mougeotia* on the soil surface, overlying a thin layer (2 mm) of a

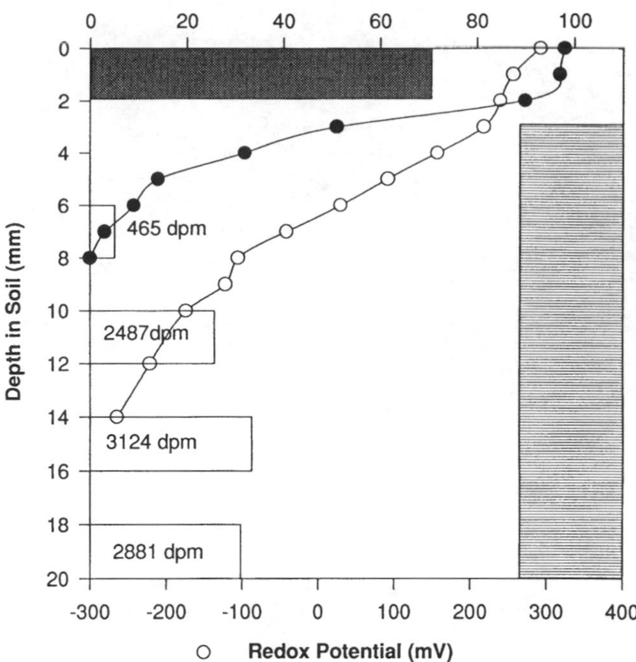

Fig. 55.2. Vertical profiles of dissolved oxygen (●) and redox potential (○) in enriched soil from the anoxic area by No Worry Lake. Rates of sulphate reduction (open bars, data expressed as disintegrations per minute, dpm). Zone of sulphide deposits (filled vertical bar) and distribution of cyanobacterial mat (filled horizontal bar) are also shown.

Phormidium-dominated mucilaginous cyanobacterial/diatom assemblage (Fig. 55.2). Whilst the cyanobacterial mats occurred in many other seepage areas green filamentous algae occurred very rarely, other than in lake shallows.

Closer examination of the seepage area revealed greasy deposits, on the water surface and on the algae/cyanobacteria, which was traceable to leaking drums of waste food outside the Station kitchen, providing the basis for enhanced microbial activity. Further microbial processing within the seepage area was evidenced by the presence of substantial quantities of mucilage, bacterial cell counts two orders of magnitude higher than in other lithosol sites in the catchment and a far greater diversity of physiologically distinct microbial groups, including sulphate- reducing and denitrifying bacteria.

For anoxia to develop in sand/gravel lithosols, it is necessary for the movement of oxygenated water through the soil profile to be limited and for the oxygen present to be consumed. The presence of fine (silt) particles and the mucilage would have restricted water flow, and the presence of organic and inorganic nutrient sources clearly promoted the development of larger populations of microbes, which would have then removed oxygen. Solar warming of the surface mats resulted in temperatures as high as 18 °C, which would also have enhanced microbial activity. Inorganic particles washed downslope into the lake accumulated an organic coating *en route* and thus provided microsites for further microbial development within the lake, whilst organic aggregates from the nutrient-enriched soil community would have also contributed to the lake carbon pool.

PHYSICAL AND CHEMICAL CHARACTERISTICS OF THE LAKE

The lake was sampled for nutrients (methods after Mackereth *et al.* 1978) and these and temperature were measured in both summer 1993 and 1994. Photosynthetically active radiation (PAR) profiles were undertaken only in summer 1993 using cosine-corrected PAR sensors (Skye Instruments). All values measured in ice-covered 'typical' lakes were corrected for potential error due to reflectance from the lake sediments and the under-ice surface, but in practice this was negligible in the PAR range.

A preliminary bathymetry was established from a series of transects, using a plumb line. The lake has a maximum depth of 3.8 m but over 85% of the lake area is less than 2 m deep and at least 65% is less than 1 m deep. Ice cover of 1 m was reported in November 1993, which was approximately 20 cm less than ice thickness measurements made for a wide range of other shallow lakes (<5 m) in the area during December 1992 – January 1993. Assuming 1 m of ice, only about 37 % of the summer lake volume would have remained unfrozen each winter.

Temperature profiling in Antarctic ice-covered freshwater lakes generally indicate an inverse temperature profile with a temperature of 0.1–0.3 °C immediately under the ice and bottom temperatures of 2–3 °C (Vincent 1988). No Worry Lake was profiled in November, when the lake was entirely ice covered and revealed surface temperatures of 1.2 °C under the ice cover and a range of 2.3–5.0 °C in the remainder of the profile. Summer water temperatures were typically 8–10 °C in the main water body, substantially higher than in other comparable systems (typically 4–7 °C). This can in part be attributed to black-body effects associated with lake turbidity and indeed, temperatures as high as 15 °C were recorded in 20 cm of water at the lake edge. Nearby Heart Lake, which has a similar maximum depth to the study lake, had an under-ice gradient of 0-3 °C and summer open-water temperatures of 5–6 °C, following the general trend for Larsemann Hills lakes.

Water is extracted year round from No Worry Lake via a pipeline that runs to the deepest part of the lake via a jetty. This water is used for generator cooling and subsequently returned, obviously somewhat warmer than previously, to the lake. A proportion of the water circulated through the pipe is removed to provide shower facilities but this is subsequently run to waste and does not return to the lake. The vertical temperature profile in the lake suggests that the piped circulation system causes active under-ice mixing of the lake water beyond the usual diffusion processes and that the high heat removal capacity of the ice cover cannot entirely compensate for the heat input, hence the observed thinner ice sheet.

Conductivity measurements of No Worry Lake in 1987 indicated values around 1000 μS cm^{-1}, but in February 1993 values were 500 μS cm^{-1} just after ice-out. Near the sediment surface

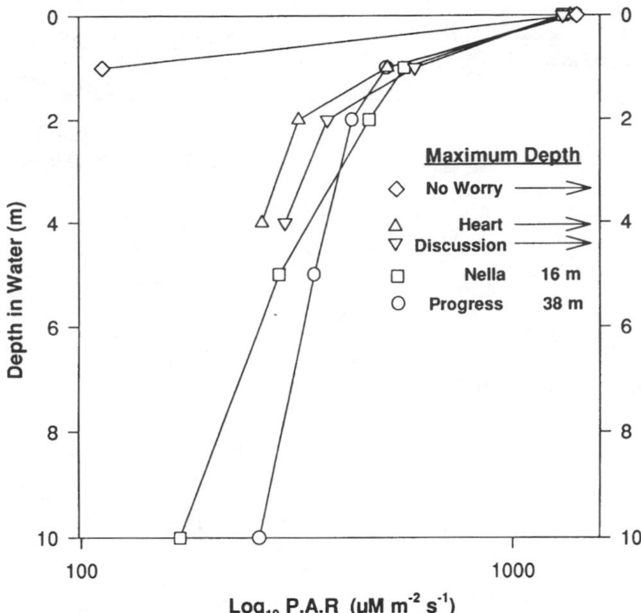

Fig. 55.3. Penetration of photosynthetically active radiation (PAR) through the water column of five oligotrophic lakes of varying depth, including No Worry Lake, located in the Larsemannn Hills. Maximum depth of shallow lakes is shown by arrows.

values were closer to 1200 μS and average values of 1500 μS were recorded by CHINARE staff the previous winter (He Jianfeng, pers. commun.) when the concentration effect of ice formation and winter anoxia would have markedly increased conductivity in this shallow lake. This evidence would suggest that the salinity of the lake is not increasing markedly at present but this takes no account of possible human impact on catchment hydrology.

Evaporation has been a feature of the lake over long time periods, as evidenced by old shorelines over a metre above the existing lake level. This is a feature of a number of open and closed basin lakes in the area and is clearly a function of long-term variations in precipitation and snow accumulation. The small catchment areas associated with these lakes, the low summer precipitation rates and the absence, in most cases, of links to permanent icefields, makes these water bodies particularly sensitive to changes in catchment hydrological balance. There has been concern amongst Station staff that the lake has been evaporating more quickly in recent years as a result of the water extraction process and this has prompted the building of earthworks to deepen and enlarge the small pond about 30 m south of the lake (Fig. 55.1). Snow meltwater is directed there, by digging small channels or snow-ploughing snow from the area over to the pond so that it is not lost to the system. Water has subsequently been transferred each summer from the holding pond to top up lake water levels, a practice which would clearly alter the seasonal hydrological patterns.

The most obvious feature of this lake is its high turbidity (Fig. 55.3), which is in marked contrast to every other substantial lake in the Larsemann Hills, including many of comparable depth and wind exposure. The turbidity is evident, even

Table 55.1. *Cell counts of microbial loop components, chlorophyll* a *and inorganic nutrients in No Worry Lake, Lake Cameron and Heart Lake*

Sampling site and date	Bacteria ($\times10^{-8}\,l^{-1}$)	PNAN ($\times10^{-5}\,l^{-1}$)	HNAN ($\times10^{-5}\,l^{-1}$)	Large flagellates ($\times10^{-3}\,l^{-1}$)	Ciliates ($\times10^{-2}\,l^{-1}$)	Chlorophyll-*a* ($\mu g\,l^{-1}$)	Ammonium ($\mu g\,l^{-1}$)	0-reactive phosphate ($\mu g\,l^{-1}$)	Soluble reactive Si ($\mu g\,l^{-1}$)
No Worry Lake 8 Jan. 1993	23.00	21.60	15.00	—	249.99	3.96	15.9	0.4	66.2
No Worry Lake 15 Nov. 1994	14.07	0.82	7.02	2.73	29.40	0.85	34.3	1.57	28.8
No Worry Lake 30 Nov. 1994	10.54	1.26	1.94	5.30	51.00	0.59	23.8	2.9	08.3
Heart Lake 15 Nov. 1994	9.68	ND	2.87	0.80	1.24	0.43	2.7	7.3	01.9
Cameron Lake 16 Jan. 1993	2.79	0.38	84.0	—	1.52	0.28	65.0	84.7	31.2
Lake Cameron 15 Nov. 1994	8.45	ND	3.31	0	9.13	0.54	15.1	2.7	36.7
No Worry Lake Catchment	—	—	—	—	—	—	25.0	18.2	24.8

Notes:
ND None detected
— Not done

under ice cover, though the amount of suspended material is reduced in these circumstances. The under-ice turbidity is further evidence of active mixing and resuspension processes in addition to simple diffusion and can be attributed to the piped circulation.

The water column in most lakes is very transparent (vertical attenuation coefficients typically 0.1–0.22 m^{-1}) and Fig. 55.3 illustrates the contrast in light penetration between a number of typical lakes of varying depth and the turbid No Worry Lake water column (vertical attenuation coefficient of 2.52 m^{-1}). Whilst substantial suspension could be attributed to the daily katabatic winds, nearby Heart Lake, with similar depth and wind exposure, shows no indication of this.

Examination of the shallow water (<1 m) sediments has revealed evidence of silt-sized particles (9% by weight), which are otherwise virtually absent from Larsemann Hills lakes (Gillieson *et al.* 1990, Ellis-Evans unpublished data). As previously mentioned, silt occurred predominantly in the area between the main buildings and the lake edge and was virtually absent from the lithosols on the far side of the lake where human activity is comparitively minimal. Motorized tracked vehicles regularly move around the station buildings within the lake catchment area as a great deal of snow-ploughing is necessary to keep pathways clear and to transfer snow to the holding pond. These movements, coupled with the original construction programme, have clearly impacted on the areas between the base buildings and the lake shore, probably mechanically crushing the sand/gravel particles. The action of snow-clearing also has the effect of transferring large amounts of lithosol material around the catchment and generating 'dirty' snow, which melts more quickly, and alters the temporal and spatial patterns of drainage within the catchment.

Inorganic nutrient concentrations in this lake are not significantly enhanced and indeed are comparable to values found in truly oligotrophic lakes elsewhere in the Hills (Table 55.1). In fact, certain oligotrophic systems such as Lake Cameron, which lies only 600 m from the edge of the polar plateau, have summer phosphorus concentrations substantially (up to 20 times) greater than that recorded in No Worry Lake. The high P concentrations found in Cameron during January open water appear to be derived from weathering products flushed into the lake by meltwater as phosphate concentrations of 60–94 g PO_4-P l^{-1} were recorded at the same time in meltwater emerging from lithosols overlain by snowpack. Catchment concentrations of ammonium and phosphorus varied considerably from lake to lake (Ellis-Evans *et al.* unpublished), but given the substantial weathering activity throughout the area it seems likely that all the catchments contribute quantities of phosphorus to the lakes during the summer melt period.

Comparison of November under-ice nutrient concentrations with those in open water on January (Table 55.1) indicates significant apparent changes in nutrient concentrations within No Worry Lake. These can be largely attributed to the dilution effects of the melting of the metre- thick ice cover and the influx of snow meltwater. However, it is clear that there has been removal of ammonium and phosphate during the transition from ice cover to open water as catchment water contained significant amounts of phosphate and ammonium (Table 55.1).

The most likely explanation for the observed nutrient removal would seem to lie with the increased chlorophyll concentrations (3.96 μg l^{-1}) observed in summer open water, which are substantial when compared with other lakes of the Larsemann Hills (Table 55.1) despite evidence of considerable grazing pressure.

LAKE PLANKTON

Antarctic lakes (continental lakes in particular) are characterized by very short food chains and the dominance of microbial components. The largest animal in these systems is the cladoceran *Daphniopsis studerii*, which occurs, in low numbers, in most Larsemann Hills lakes including No Worry Lake. However, the important components of the plankton are the bacteria, the nanoplankton (both photosynthetic autotrophs, PNAN, and heterotrophs, HNAN), ciliates, large heterotrophic flagellates and, sometimes, rotifers. These comprise the elements of the microbial loop (Laybourn-Parry 1992) which rapidly recycles carbon and major elements within the water column.

Table 55.1 presents mean counts of microbial loop components for the study lake and two representative lakes, Heart Lake and Cameron Lake, and these data indicate substantial differences with respect to No Worry Lake. No Worry Lake has consistently higher bacterial and ciliate numbers and the latter are equivalent to those normally reported for temperate mesotrophic systems (Laybourn-Parry 1992). PNAN numbers are high in summer open water but are very low under ice, as might be expected for a system where turbidity is significant under ice and the light regime therefore particularly unfavourable for primary producers. The other lakes are typical oligotrophic systems with low nutrient concentrations and a sparse plankton. Of particular interest is the considerable temporal variation in PNAN numbers in No Worry Lake. Counts for November indicate insufficient primary production to support the heterotrophic community, which must therefore be primarily utilizing allochthonous carbon. The particles suspended in the water column are inorganic and these all have an organic coating of mucilages and various micro-organisms. The remainder of the particles are purely organic aggregates, often surrounded by ciliates. Comparison of 1.0 μm filtered versus unfiltered water indicates that much of the microbial community is attached to particulate material, rather than free living, and that attached material is the source of carbon to the under-ice community.

With the onset of open water and access to potentially higher light levels and nutrients, chlorophyll *a* and PNAN counts increase in No Worry Lake (Table 55.1) and these increases are matched by the heterotrophic components. The increase in heterotrophs may reflect higher summer temperatures and the input of fresh allochthonous carbon from the soils below the station buildings rather than increased availability of photosynthetic assimilation products.

Comparison of the plankton in the three lakes also indicates differences in species composition, particularly amongst the ciliates and large flagellates, which reflect different patterns of food

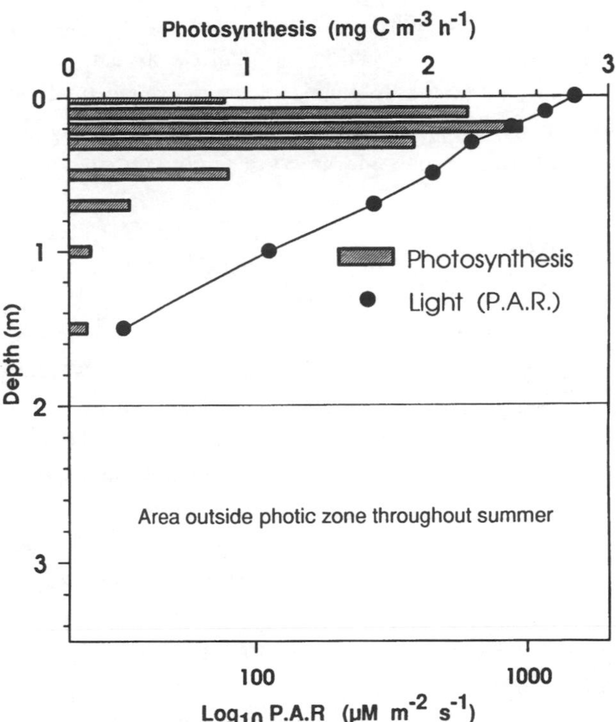

Fig. 55.4. Vertical profile of PAR in No Worry Lake and the distribution of photosynthetic activity ($^{14}CO_2$-uptake). The surface PAR value was measured 5 cm below the water surface.

availability. Lake Cameron has only very small ciliates, mostly oligotrichs and scuticociliates, whereas large ciliates occur in both Heart and No Worry. In Heart Lake an aplastidic oligotrich dominated, whilst a large hypotrich (*c*. 300 μm long), a large heterotrich (*c*. 200 μm long) and two rotifer species were relatively common. The large flagellate, probably a euglenid, was also seen in some numbers. Two oligotrichs dominated No Worry Lake – a small (30 μm) aplastidic form and one that may be *Strombidium viride*. These two ciliates were present in substantial numbers. Other ciliates occurred in lower numbers, together with two large coloured flagellates (a quadriflagellate prasinomonadid and a biflagellate euglenid). The HNAN community was much the same between lakes with *Paraphysomonas* dominant at all three sites.

Measurements of photosynthetic activity (Fig. 55.4) confirm the conclusions from the light-penetration data that the water column is very turbid and the photic zone restricted to the top 1 m. Peak activity in moored bottles was observed at 0.2–0.3 m depth and no activity was observed below 1 m. Photo-inhibition was evident in the top two samples. In summer, wind mixing would circulate the phytoplankton in and out of the photic zone, so the curve presented in Fig. 55.4 is potentially misleading. However, as the katabatic winds only blew between *c*. 2100 and 1000 h, active mixing was actually restricted to periods when light levels ranged from 0–65% of maximum light levels. At times of high radiation, the virtual absence of wind mixing would have favoured the establishment of vertical patterns of photosynthetic organisms that broadly followed the pattern of photosynthetic activity seen in Fig. 55.4. Given the limited degree of light penetration, the band of photosynthetic activity in these circumstances would be very narrow so that grazing pressures would be locally significant, when compared with a wind-mixed situation where the phytoplankton were more dispersed and predator–prey encounters therefore reduced.

BENTHIC DATA

Examination of sediment cores from several locations in the lake indicated that the cyanobacterial mats present broadly complied with the general pattern seen in Larsemann Hills lakes which appears to relate to conductivity and lake depth (Ellis-Evans *et al.* unpublished data). In the shallows (<1 m) of those lakes with a conductivity of 500–1000 μS and maximum depth <5 m, a thin, but discrete mat of red-brown cyanobacterial filaments (dominated by a non-heterocystous oscillatoreacean with a filament width of 1–2 m) overlies gravel/stone deposits on a rock base. Within this filament structure a number of different coccoid chlorophytes and, occasional *Nostoc* assemblages, are embedded. Below 2 m the mat structure changes to a spongy blue-green mat (comprising oscillatoracean filaments and coccoid chlorophytes) which forms characteristic broadly spherical plates (diameter 10–25 cm, depth 1–2 cm) that cover the lake bottom, piled haphazardly to a depth of approximately 1 m. These deep-water mats develop anoxia in winter, but there is no evidence that this extends to the mat surface. Certainly in summer the mats are well oxygenated in the upper layers (Fig. 55.5). Heart Lake has the 'discus' mat composition described above and we suggest this was probably the situation in No Worry Lake before 1987.

In No Worry Lake at present, the shallows are currently characterized by sand/gravel sediments which include a significant silt component. The cyanobacterial mat is unconsolidated and ill-defined because sedimentation and resuspension occurs throughout the summer growing season. Such mats clearly have to migrate upwards constantly as sedimentation and resuspension steadily bury or disrupt them. *In situ* oxygen and redox microprobe profiles (1 mm resolution) of the shallow-water sediments (after methods described in Ellis-Evans & Bayliss 1993) are presented in Fig. 55.5. Whilst the overlying water is oxygenated, a sharp transition towards reduced/anoxic conditions occurs approximately 1 mm above the sediment surface and oxygen penetrates less than 10 mm into the sediment. Redox measurements indicate conditions suitable for sulphate reduction in the top 20 mm and this observation is supported by the black colour (pyrite) of subsurface sediments and measurable reduction of ^{35}S-sulphate in the 10–20 mm section (data unpublished). From the profiles it is apparent that the shallow site sediments in this lake are very active microbiologically and the high water temperatures of 8–15 °C recorded in 10–20 cm of water at the lake edge would have contributed to rapid turnover of organic carbon.

Examination of deep-water cores (3.5 m water depth) revealed a largely inorganic base, 8 cm below the sediment surface, overlain by greenish cyanobacterial mat material (1 cm thick), a black layer of mat debris 5–6 cm thick and a 1 cm thick layer of yellow/grey mat debris/ inorganic particles. The core profiles are yet to be dated, so interpretation of the core is still speculative. Visual examination indicated that the core profile from 1–8 cm is structurally cyanobacterial mat material, suggesting that the top 1 cm would seem to represent sedimentation and resuspension over the past 5 years. At least part of this resuspension must be directly attributable to jetty construction.

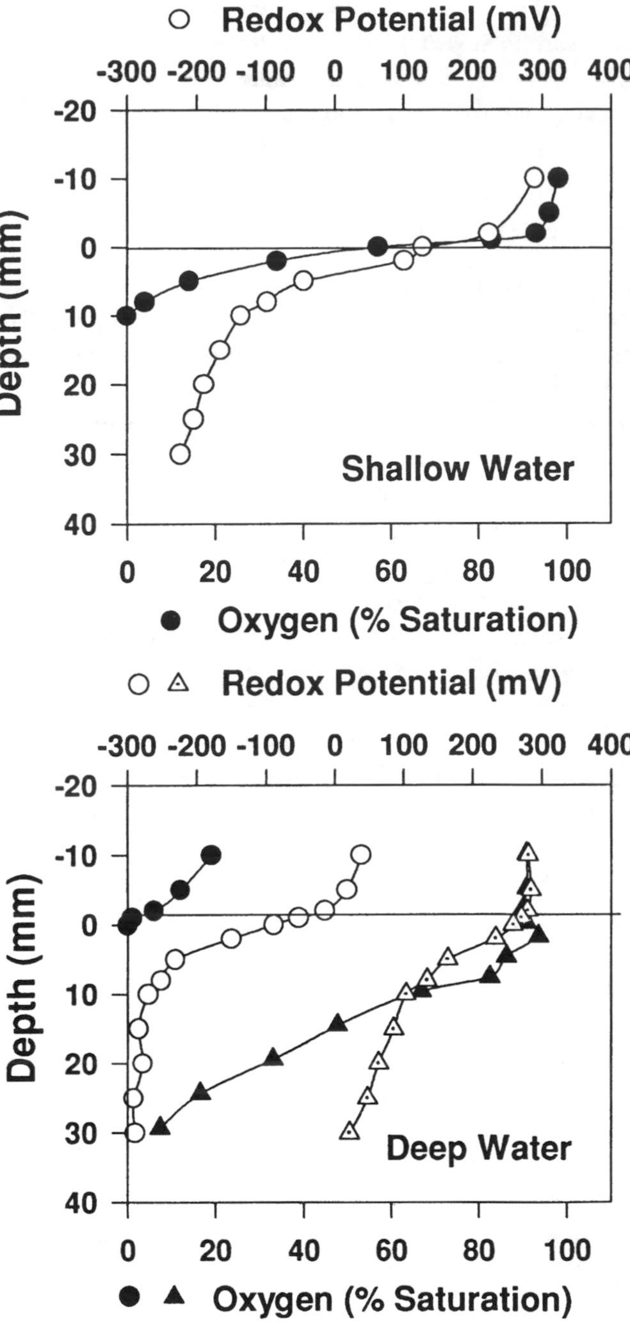

Fig. 55.5. Vertical profiles of dissolved oxygen and Eh in (top) shallow-water sediments and (bottom) deep-water sediments of No Worry Lake (circles) and of Heart Lake (triangles).

The 1–7 cm section lost its black colour if left in air as the sulphides were oxidised, and revealed a mat structure of the 'discus' form seen in Heart and Discussion Lake, but the mat was partly decomposed, indicating substantial microbial activity under highly reduced conditions. Very little oxygen (Fig. 55.5) was found in the overlying 20 mm water layer and none was detected at the sediment surface. Redox profiling gave a range of −120 mV at the sediment surface, falling to −290 mV subsurface. Such reduced conditions would be highly appropriate for sulphate reduction, accounting for the sulphide deposits overlying the mat debris. Experiments with ^{35}S sulphate confirmed that there was measureable reduction occurring between 0.5 cm and 5 cm depth (data unpublished). Whilst such reducing conditions are not unusual in shallow lakes and pools elsewhere in the Antarctic (Ross Ice Shelf pools, Signy Island lakes – Vincent 1988) such conditions are rare in Larsemann Hills lakes (excepting the brackish lakes) and certainly the 'discus' mat form is not associated with anoxia on the scale currently seen in No Worry Lake. The evidence suggests a transformation from a clear, benthic mat-dominated lake in comparatively recent times which accords with observations made in the original survey (Gillieson *et al.* 1990).

In winter it might be anticipated that much of the water body would become deoxygenated under ice cover and that anaerobic heterotrophic processes would dominate carbon cycling. The evidence of a previously well-established mat pre-1987 that does not tolerate gross anoxia suggests that the lake water body remained oxygenated as in Heart and Discussion Lake. It does not seem likely that a cyanobacterial mat would have developed at the deep station to the extent evident from examination of No Worry Lake cores if the lake had been as turbid as it is now.

The circulation induced in the bottom waters of the lake by the generator cooling system clearly maintains large amounts of material in suspension and resuspends sediment material. The presence of substantial amounts of organic matter, and elevated winter temperatures, undoubtedly enhances carbon turnover even where much of the organic material comprises slowly metabolized molecules such as cellulose and proteins.

CONCLUSIONS

Whilst a number of national groups have undertaken limnological research in the Antarctic and much evidence to demonstrate the simplicity and presumed fragility of these systems has been presented (see Vincent 1988, Friedmann 1993), there has, as yet, been little attempt to study the effects of change in such systems. The work of Bayly (1986) and Burke & Burton (1988) on variations in zooplankton and photosynthetic bacteria caused by the irregular influxes of sea water to Burton Lake, Vestfold Hills and a recent study of the impact, on lake chemistry, of fur seal invasion of a lake catchment over a 10 year period (Ellis-Evans 1992) are isolated examples.

Evidence from the sediment record broadly indicates that many Antarctic lakes have experienced substantial climate change events which have influenced catchment status and that both lake chemistry and lake biota have responded to these changes (Jones *et al.* 1994, Björck *et al.* 1996). It has also been demonstrated that the impact of marine animals on both oligotrophic and eutrophic lakes at Signy Island, South Orkney Islands will not always result in a simple eutrophication response (Ellis-Evans 1992). Antarctic ice-free areas are currently being subjected to a wide range of impacts, including, in some areas, rapid deglaciation of adjoining ice caps and reduced snow fall, changing patterns of land-based activity by marine animals, increasing presence of research stations and, of course, the growing presence of tourists. Lakes offer a well preserved record of past events spanning thousands of years but also provide a sensitive indication of changes in catchment characteristics on timescales of years (Appleby *et al.* 1995).

No Worry Lake has responded very quickly to human activities in its catchment, despite comparatively low temperatures and relatively low levels of nutrient enrichment by temperate standards. Zhongshan Station appears to have had relatively little impact on the local area, but the lake holds evidence that indicates real changes in the dynamics of the catchment. These strongly seasonal, microbially dominated lakes and their catchments have considerable merit as sites for investigating community processes and the impact of environmental pertubation. This preliminary study suggests a need for further studies of lakes where catchment changes have either occurred, or are developing, to increase our understanding of the dynamics of unperturbed systems.

POSTSCRIPT

During the 1993/94 summer season, further observations at Zhongshan Station revealed that the waste food was no longer being stored outside the building in 45 gallon drums and that feeding of the skuas at the kitchen door had also ceased. As a result, the large 'batchelor' groups of skuas had disappeared and nutrient additions to the slope leading to the lake would have been reduced. Water continues to be extracted from the lake and replaced by water transferred from the pond.

ACKNOWLEDGEMENTS

The authors are grateful to the Australian Antarctic Division and numerous Division staff for their enthusiastic logistic support. The authors are also grateful to the CHINARE staff of Zhongshan Station for providing station facilities and kind hospitality. JCEE would like to thank Dr James Burgess and Andy Spate for the invitation to work on their programme in the Larsemann Hills and for being such exceptional field companions. JCEE is also grateful to British Antarctic Survey for the opportunity to undertake this work. JLP, PB and STP were funded by Natural Environment Research Council (PB), the Leverhulme Foundation (JLP) and the Australian Antarctic Science Advisory Committee (STP). Special thanks are due to Dr Clive Howard-Williams and Prof. Max Tilzer for constructive criticisms of an earlier draft.

REFERENCES

Adamson, E. & Seppelt, R. 1989. Alkaline pollution damage to lichens growing downwind of the concrete batching site at Casey. *ANARE News*, June, 16–17.

Appleby, P. R., Jones, V. J. & Ellis-Evans, J. C. 1995. Radiometric dating of lake sediments from Signy Island (maritime Antarctic): evidence of recent climatic change. *Journal of Palaeolimnology*, **13**, 179–191.

Bayly, I. A. E. 1986. Ecology of the zooplankton of a meromictic Antarctic lagoon with special reference to *Drepanopus bispinosus* (Copepoda: Calanoida). *Hydrobiologia*, **140**, 173–181.

Björck, S., Olsson, S., Ellis-Evans, J. C., Häkansson, H., Humlum, O. & De Lirio, J. M. 1996. Late Holocene palaeoclimatic records from lake sediments on James Ross Island, Antarctica. *Palaeogeography, Palaeoclimatology, Palaeoecology*, **121**, 195–220.

Burke, C. M. & Burton, H. R. 1988. The ecology of photosynthetic bacteria in Burton Lake, Vestfold Hills, Antarctica. In Ferris, J. M., Burton, H. R., Johnstone, G. W. & Bayly, I. A. E., eds. *Biology of the Vestfold Hills, Antarctica. Developments in Hydrobiology*, **34**, 1–12.

Ellis-Evans, J. C. 1992. Evidence for change in the chemistry of nutrient-enriched Heywood Lake. In Kerry, K. R. & Hempel, G., eds. *Antarctic ecosystems: ecological change and conservation*. Berlin: Springer-Verlag, 77–82.

Ellis-Evans, J. C. & Bayliss, P. R. 1993. Biologically-mediated activity gradients in Antarctic lakes and streams. *Verhandlungen der Internationalen Vereinigung für Theoretische und Angewandte Limnologie*, **25**, 948–952.

Ellis-Evans, J. C. & Wynn-Williams, D. D. 1985. The interaction of soil and lake microflora at Signy Island. In Seigfried, W. R., Condy, P. R. & Laws, R. M., eds. *Antarctic nutrient cycles and food webs*. Springer-Verlag, Berlin, 662–668.

Friedmann, E. I., ed. 1993. *Antarctic microbiology*. New York: Wiley-Liss, 634 pp.

Gillieson, D., Burgess, J. S., Spate, A. P. & Cochrane, A. 1990. An atlas of the lakes of the Larsemann Hills. *ANARE Research Notes*, No. 74, 1–173.

Herbert, R. A. & Bell, C. R. 1977. Growth characteristics of an obligately psychrophilic *Vibrio*. *Archiv für Microbiologie*, **113**, 215–220.

Jones, V. J., Juggins, S. & Ellis-Evans, J. C. 1994. The relationship between water chemistry and sediment diatom assemblages in maritime Antarctic lakes. *Antarctic Science*, **5**, 339–348.

Laybourn-Parry, J. 1992. *Protozoan plankton ecology*. London & New York: Chapman & Hall, 231 pp.

Laybourn-Parry, J., Marchant, H. J. & Brown, P. 1991. The plankton of a large oligotrophic Antarctic lake. *Journal of Plankton Research*, **13**, 1137–1149.

Mackereth, J. F. H., Heron, J. & Talling, J. F. 1978. *Water analysis: some revised methods for limnologists*. Kendal: Titus Wilson.

Nedwell, D. B. & Rutter, M. 1994. Influence of temperature on growth rate and competition between two psychrotolerant Antarctic bacteria: low temperature diminishes affinity for substrate uptake. *Applied and Environmental Microbiology*, **60**, 1984–1992.

Priddle, J., Heywood, R. B. & Theriot, E. 1986. Some environmental factors influencing phytoplankton in the Southern Ocean around South Georgia. *Polar Biology*, **5**, 65–79.

Vincent, W. F. 1988. *Microbial ecosystems of Antarctica*. Cambridge: Cambridge University Press, 304 pp.

Wang, Z. & Norman, F. I. 1993. Timing of breeding, breeding success and chick growth in south polar skuas (*Catharacta maccormicki*) in the eastern Larsemann Hills, Princess Elizabeth Land, east Antarctica. *Notoris*, **40**, 1–15.

56 Investigation of mixed function oxidase activity in Antarctic organisms

S. FOCARDI, M. C. FOSSI, L. LARI, S. CASINI AND R. BARGAGLI
Dipartimento di Biologia Ambientale, Via delle Cerchia 3, 53100 Siena, Italy

ABSTRACT

*Potential hazards to Antarctic fish and birds in relation to exposure to pollutants were investigated by: 1. determination of mixed function oxidase (MFO) activity in the south polar skua (*Catharacta maccormicki*) and Adélie penguin (*Pygoscelis adeliae*); 2. evaluation of MFO activity in the fish* Pagothenia bernacchii *and* Chionodraco hamatus*; 3. study of MFO induction in fish experimentally treated with xenobiotics.*

Key words: MFO activity, Antarctic organisms, xenobiotic induction, biomarker.

INTRODUCTION

The mixed function oxidase (MFO) system is an important and highly versatile group of enzymes that metabolize a wide variety of lipophilic compounds (Kappas & Alvares 1975, Coon & Persson 1980, De Matteis 1984). It may be used as a biomarker of such compounds in the environment and as an index of the susceptibility of organisms to xenobiotics. One of the main properties of this enzyme system is its substrate inducibility (Payne 1984). In this study we report the results of pilot studies for a programme to assess potential and actual hazard to Antarctic wildlife in relation to exposure to pollutants. Although this environment may be regarded as pristine because of its remoteness from man-made sources of contamination, recent research shows that certain persistent pollutants have reached the Antarctic food chain by atmospheric transport. Hence, it is important to investigate the potential tolerance of these organisms and their capacity to detoxify persistent xenobiotics such as chlorinated hydrocarbons.

MATERIALS AND METHODS

Specimens of south polar skua (*Catharacta maccormicki*) and Adélie penguin (*Pygoscelis adeliae*) were collected at the northern rookery, Cape Bird, Ross Island. Liver samples obtained from five injured skuas (fatally wounded in intraspecific combat) and eight penguins (killed by leopard seals) within 30 min. of death were fresh-frozen and stored in liquid nitrogen until analysis of MFO activities (aldrin epoxidation and 7-ethoxyresorufin dealkylation, EROD). Several specimens of two species of the suborder Notothenioidei (*Chionodraco hamatus* and *Pagothenia bernacchii*) were captured during the Antarctic summers of 1987/88, 1989/90 and 1991/92 near the Italian Scientific Station 'Baia Terra Nova' (Ross Sea). At the station the specimens were dissected and the liver placed in liquid nitrogen until analysis. For the induction experiments the specimens were placed in an aquarium containing seawater, kept at 0 °C by a cooling system. Some were sacrificed and the livers stored in liquid nitrogen; others were acclimatized for five days in the aquarium and then induced.

Induction experiment

(a) After acclimatization 40 samples of each notothenioid species (*P. bernacchii* and *C. hamatus*) were divided into four groups, three experimental and one control. In the three experimental groups, 85 mg kg^{-1} phenobarbital (PB), 15 mg kg^{-1} 3-methylchol-anthrene (3-MC) and 130 mg kg^{-1} Arochlor 1260 (PCBs) in corn oil were injected into the caudal vein. The controls were injected with corn oil alone. Twenty-four hours after injection the fish were sacrificed. The livers were placed in liquid nitrogen and brought to Italy for analysis of benzo(α)pyrene

Table 56.1. *MFO activities and organochlorine residues in Antarctic birds (mean and (SD))*

Species	Aldrin epoxidase	EROD[1]	NADPH-cyt *c* red	PCBs[2]	DDTs[2]
Adélie penguin *n*=8	0.074 (0.064)	0.003 (0.002)	27.31 (12.08)	0.619 (0.346)	0.029 (0.020)
South polar skua *n*=5	0.468 (0.082)	0.110 (0.050)	40.54 (8.54)	2.546 (0.168)	0.314 (0.249)

Notes:
[1] nmol sub. mg protein $^{-1}$ min^{-1}; [2] mg kg^{-1} dw.

Table 56.2. *MFO and reductase Activities in Antarctic fish (mean and (SD))*

Species	Aldrin epoxidase[1]	BPMO[2]	NADPH-cyt *c* red[3]	NADPH-cyt *c* red[3]	NADH-ferrired
C. hamatus *n*=9	0.133 (0.055)	0.031 (0.016)	28.90 (3.60)	6.17 (1.51)	529.8 (55.7)
P. bernacchii *n*=9	0.040 (0.009)	0.023 (0.017)	7.90 (5.20)	12.7 (7.80)	939.0 (153.7)

Notes:
[1] pmol mg protein^{-1} min^{-1}; [2] AU mg protein^{-1} min^{-1}; [3] nmol sub. mg protein^{-1} min^{-1}

monooxygenase, EROD, benzyloxyresorufin dealkylation (BROD), NADPH-cytochrome *c* reductase (NADPH-CYTCRED), NADH-cytochrome *c* reductase (NADH-CYTCRED) and NADH-ferricyanide reductase (NADH-FERRIRED). All of these were taken as measures of detoxifying activity.

(b) After acclimatization 30 specimens of *P. bernacchii* were divided into three groups: one group was treated with 10 mg kg^{-1} benzo(α)pyrene, one with 50 mg kg^{-1} Aroclor 1260 (PCBs) and one was left as a control. The treatments were repeated 24 hours later. After two, four and ten days three controls and five treated animals were sacrificed. The livers were placed in liquid nitrogen until MFO activities (Benzo(α)pyrene monooxygenase, EROD and BROD) were determined. Kinetic studies using different incubation periods and temperatures were also performed with both species with the same three substrates.

Biochemical analysis

For the determination of MFO activity, the livers were homogenized in buffer at pH 7.5 (0.25 M sucrose) and centrifuged for 20 min. at 9000*g* to remove the cell debris; the supernatant was then centrifuged for 90 min. at 100 000*g* to obtain the microsomal fraction. Before enzyme analysis the microsomal fraction was resuspended with buffer at pH 7.5 (1.15% KCl). Benzo(α)pyrene monooxygenase activity (BPMO) was determined by the method of Kurelec *et al.*(1977); aldrin epoxidase was determined by the method of Krieger and Wilkinson (1969); EROD and BROD activity were determined by the method of Lubet *et al.* (1985); NADPH-cytochrome *c* reductase (NADPH-CYTCRED), NADH-cytochrome *c* reductase (NADH-CYTCRED) and ferricyanide reductase (NADH-CYTCRED) activities were determined by the method of Livingstone & Farrar (1984).

In the bird specimens, enzyme activities were determined at body temperature (42 °C) ; in fish (living at −1.8 °C) all activities were determined at +20 °C in order to amplify the signal and reduce experimental error.

RESULTS AND DISCUSSION

MFO activity in Antarctic birds

Striking differences in MFO activity were found in the two species (Table 56.1). The values found in skuas were similar to those recorded in larids of the northern hemisphere (Walker *et al.* 1984, Fossi *et al.* 1986, Focardi et al. 1992a). These interspecies differences are probably due to the different body sizes (Walker *et al.* 1984) and migratory habits of the birds (the species had radically different organochlorine levels, see Table 56.1). The interspecies differences in MFO activity are also related to feeding habits, as previously reported by other authors (Ronis & Walker 1985). In fact fish-eating birds usually have lower activity than omnivorous scavenging species.

MFO activity in Antarctic fish

In *C. hamatus* and *P. bernacchii* (Table 56.2), interspecific differences in activity were found for most of the enzymes tested. The values of the BPMO and NADPH-cytochrome c reductase were lower in both species than in fish of temperate seas (Payne 1984, Stegeman 1989, Focardi *et al.* 1992b). This data is of ecotoxicological interest as a measure of the MFO detoxication system in marine organisms distant from sources of man-made contamination.

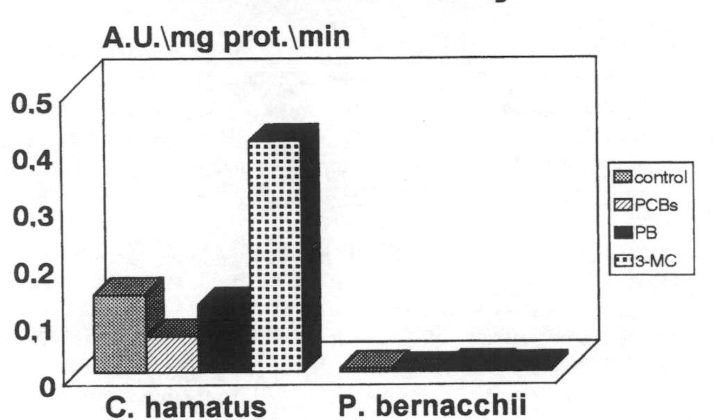

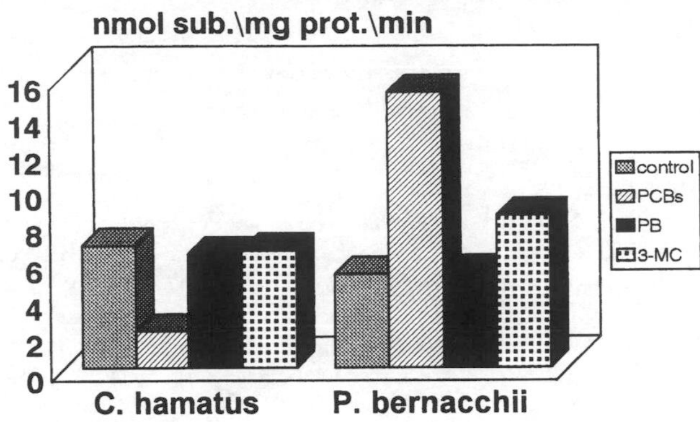

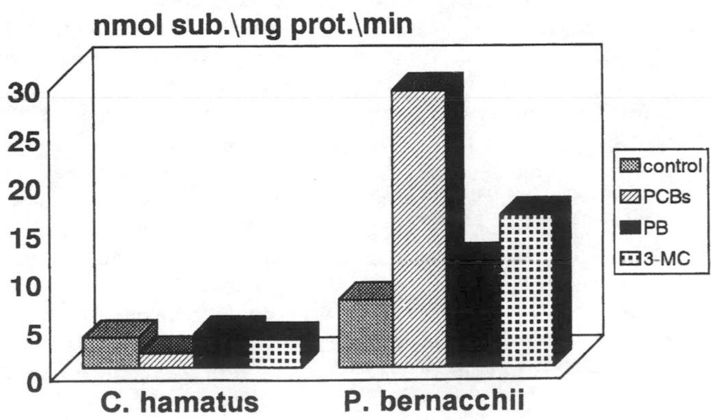

Fig. 56.1. Microsomal enzyme activities measured at 20 °C in *Pagothenia bernacchii* and *Chionodraco hamatus* treated experimentally with PB, 3-MC and PCBs.

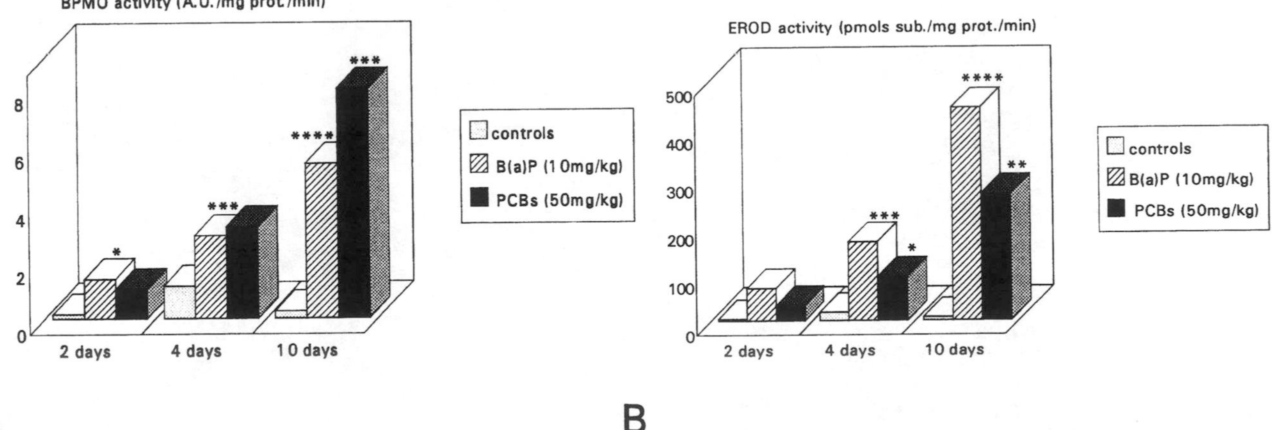

Fig. 56.2. Changes in enzyme activities with time in *Pagothenia bernacchii* after treatment with benzo(α)pyrene and PCBs.

MFO activity in Antarctic fish induced experimentally with xenobiotic compounds

Fig. 56.1 shows significant differences in BPMO activities in the two species of fish experimentally induced with PB, 3-MC and PCBs, the three classical inducers of the MFO system. The chemicals did not cause induction of MFO activity in *P. bernacchii* after 24 h. On the other hand, in *C. hamatus*, a statistically significant induction was detected only for 3-MC. This data confirms the absence of PB-type induction in fish as previously reported by other authors (Stegeman 1989). When we investigated the response of the MFO system in *P. bernacchii* after treatment with benzo()pyrene and PCBs, BPMO, EROD and BROD (benzyloxyresorufin-O- deethylase) activities were found to be strongly induced for longer experimental periods, especially beyond 10 days (Fig. 56.2). The long time required for the detoxifying system of these fish to respond to xenobiotics is probably related to the low temperature of their environment, which makes their metabolism very slow. This factor was probably also responsible for the lower MFO activity with respect to temperate sea fish (Melancon *et al.* 1987, Stegeman 1989, Focardi *et al.* 1992b). Moreover, the low basal MFO activities found in fish suggest that they could be susceptible to toxic stress and to accumulation of liposoluble contaminants.

REFERENCES

Coon, M. J. & Persson, A. V. 1980. Microsomal cytochrome P-450: a central catalyst in detoxication reactions. In Jakoby, W. B., ed. *Enzymatic basis of detoxication*, Vol. 1. New York: Academic Press, 117–130.

De Matteis, F. 1984. Metabolismo epatico di sostanze esogene, influenze sulla tossicita' e interazioni metaboliche. *47° Congresso della societa' di medicina del lavoro e igiene industriale. Assisi 17–20, Ottobre 1984.*

Focardi, S., Fossi, M. C., Leonzio, C., Lari, L., Marsili, L., Court, G. S. & Davis, L. S. 1992a. Mixed function oxidase and chlorinated hydrocarbons residues in Antarctic seabirds: south polar skua (*Catharacta maccormicki*) and Adélie penguin (*Pygoscelis adeliae*). *Marine Environmental Research,* **34,** 201–205.

Focardi, S., Fossi, M. C., Lari, L., Marsili, L., Leonzio, C. & Casini, S. 1992b. Induction of mixed function oxidase (MFO) system in two species of Antarctic fish from Terra Nova Bay (Ross Sea). *Polar Biology,* **12,** 721–725.

Fossi, M. C., Leonzio, C. & Focardi, S. 1986. Mixed function oxidase activity and cytochrome P-450 forms in black-headed gulls feeding in different areas. *Marine Pollution Bulletin,* **17,** 546–548.

Kappas, A. & Alvares, A. P. 1975. How liver metabolizes foreign substances. *Scientific American,* **232,** 22–31.

Krieger, R. L. & Wilkinson, C. F. 1969. Microsomal mixed function oxidases in insects. 1: Localisation and properties of an enzyme system affecting aldrin epoxidation in larvae of the southern armyworm (*Prodenia eridania*). *Biochemical Pharmacology,* **19,** 1403–1415.

Kurelec, B., Britvic, S., Rijavec, M., Muller, W. E. G. & Zahn, R. K. 1977. Benzo(α)pyrene monooxigenase induction in marine fish-molecular response to oil pollution. *Marine Biology,* **44,** 211–216.

Livingstone, D.R. & Farrar, S. 1984. Tissue and subcellular distribution of enzyme activity of mixed-function and benzo(α)pyrene metabolism in the common mussell *Mytilus edulis. Science of the Total Environment,* **39,** 209–235.

Lubet, R. A., Nims, R. W., Mayer, R. T., Cameron, J. W. & Schechtman, L. M. 1985. Measurement of cytochrome P450 dependent dealkylation of alkoxyphenoxazones in hepatic S9s and hepatocyte homogenates: effects of dicumarol. *Mutation Research,* **142,** 188–197.

Melancon, M. J., Yeo, S. E. & Lech, J. J. 1987. Induction of hepatic microsomal monooxigenase activity in fish by exposure to river water. *Enviromental Toxicology and Chemistry,* **6,** 127–135.

Payne, J. F. 1984. Mixed fuction oxygenase in biological monitoring programs: review of potential usage in different phyla of aquatic animals. In Persoone, G., Jaspers, E. & Clause, C. eds., *Ecotoxicological testing for the marine environment,* Vol.1, Ghent: State University of Ghent, 625–655.

Ronis, M. J. J. & Walker, C. H. 1985. Species variations in the metabolism of liposoluble organochlorine compounds by hepatic microsomal monooxygenase: comparative kinetics in four vertebrate species. *Comparative Biochemical Physiology,* **82C,** 445–449.

Stegeman, J. J. 1989. Cytochrome P-450 forms in fish: catalytic, immunological and sequence similarities. *Xenobiotica,* **19,** 1093–1110.

Walker, C. H., Knight, G. C., Chipman, J. K. & Ronis, M. J. J. 1984. Hepatic microsomal monooxygenase of sea birds. *Marine Environmental Research,* **14,** 416–419.

57 Heavy metals in Antarctic molluscs

MARCO NIGRO, F. REGOLI, R. ROCCHI AND E. ORLANDO
Dipartimento di Biomedicina Sperimentale Infettiva e Pubblica, sez. Biologia e Genetica, Via A. Volta, 4, 56126 Pisa, Italy

ABSTRACT

Concentrations of Mn, Fe, Ni, Cu, Zn, As, Se, Cd and Pb were determined in the gills, digestive gland and kidney of the Antarctic bivalves Adamussium colbecki, Laternula elliptica *and* Yoldia eightsi *and in the gastropod* Neobuccinum eatoni *collected at Terra Nova Bay (Ross Sea). The molluscs always showed high levels of cadmium, indicating a high bioavailability of this metal in the Antarctic marine environment, which is probably related to the upwelling of Cd-enriched deep waters. Moreover, concentrations up to 400 $\mu g\ g^{-1}$ dry weight were found in the organs of* Y. eightsi. *This accumulation has been discussed with reference to the feeding strategies of this species. Concentrations of Mn, Fe, Cu and Zn in the digestive gland of* A. colbecki *were shown to decrease significantly during the austral summer, confirming the importance of considering seasonal variations for correct interpretation of the data.*

Key words: heavy metals (Mn, Fe, Ni, Cu, Zn, As, Se, Cd, Pb), molluscs, Antarctica, biomonitors.

INTRODUCTION

Antarctic, as well as temperate, molluscs could be profitably used as biomonitors for interpreting environmental variability at different levels: 1. on a geological timescale, through the study of fossils (Baroni & Orombelli, 1994; Berkman, 1992; Berkman, Chapter 49, this volume); 2. for monitoring the anthropogenic input of metals, through the study of bioaccumulation (Mauri *et al.* 1990, Berkman & Nigro 1992); and 3. to assess the biological impact of disturbance through the study of opportune biomarkers. However, reliable baseline data and a better knowledge of the variability related to various natural processes are needed for a correct use of molluscs as experimental organisms for the interpretation of environmental changes (Berkman *et al.*, 1992). The present study concerns the concentrations of nine elements (seven heavy metals and two metalloids) in four species of Antarctic molluscs, and gives some preliminary insight regarding their variability during the austral summer.

MATERIALS AND METHODS

The bivalves *Adamussium colbecki, Laternula elliptica* and *Yoldia eightsi* and the gastropod *Neobuccinum eatoni* were collected by SCUBA diving or dredging at Terra Nova Bay (74°5′ S 164°30′ E), in the vicinity of the Italian Antarctic Base, during the 1990/91 and 1991/92 expeditions. Variations of heavy metal concentrations during the summer were investigated in the scallop *A. colbecki*, collected before and after sea-ice melting at a depth of 12 m. After sampling, the molluscs were maintained for 1–2 days in aquaria to eliminate any sediment from their digestive tract. Specimens were dissected and the organs (gills, digestive gland and kidney) isolated and freeze dried in the laboratory at the base. Dry samples were digested with concentrated nitric acid (Merck, Suprapur), first overnight, at room temperature, and then for 8 h under pressure in a teflon bomb. For quality assurance and control, a blank was processed with samples at each digestion (1 blank+7 samples), and standard reference material (lobster hepatopancreas, provided by the National Research Council, Canada) was digested using the same procedure. The metals were analysed by atomic absorption spectrophotometry (IL mod. S11 equipped with a deuterium background corrector or Varian SpectrAA-300 Zeeman). Cu, Fe, Mn, Zn and Cd were determined by flame atomization and Pb, Se, As, Ni were analysed by electrothermal atomization. In the latter case, appropriate chemical modifiers (such as

Table 57.1. *Concentrations (mean ±SD µg g^{-1} dry weight) of heavy metals in the Antarctic molluscs from Terra Nova Bay*

Species	Organ	Mn	Fe	Ni	Cu	Zn	As	Se	Cd	Pb
A. colbecki	D G	3.4±1.2[1]	292±63.0[1]	8.5±2.3	12.6±3.3[1]	74.9±25.4[1]	11.2±2.8	29.0±5.5	142±57[1]	0.18±0.05
65–80 mm	Gills	5.7±1.5[1]	119±45.0[1]	3.9±2.7	6.5±2.2[1]	114±33.0[1]	14.7±7.2	6.5±1.3	6.8±1.1[1]	0.25±0.18
n=25	Kidney	16.3±8.9[1]	69.2±31.7[1]	6.9±3.4	4.0±1.7[1]	199±89.0[1]	6.6±2.3	10.9±4.9	11.6±3.3[1]	0.46±0.30
L. elliptica	D G	4.5±0.4	145±57.7	8.3±2.3	33.3±5.1	119±10.9	37.2±16.9	47.8±20.2	48.2±15.6	0.24±0.09
65–90 mm	Gills	5.1±0.7	178±56.3	2.2±0.8	7.8±0.8	207±63.6	10.1±1.4	18.2±2.3	12.5±3.8	0.05±0.02
n=20	Kidney	26.5±15	277±156.4	21.5±12.3	8.0±1.2	3300±1400	12.6±3.2	15.2±4.2	360±150	5.37±2.56
Y. eightsi	D G	31.5±9.6	2330±787	7.0±1.1	39.1±10.7	105±8.7	283±32.0	7.6±1.2	31.9±8.2	1.11±0.42
20–30 mm	Gills	4.5±3.3	1590±137	18.9±7.5	32.8±12.3	96.6±13.7	394±112	85.6±30.9	111±32.2	9.55±1.60
n=20	Kidney	—	—	—	—	—	—	—	—	—
N. eatoni	D G	5.5±1.4	721±288	6.8±2.6	21.4±38.2	803±1141	31.4±21.0	25.7±10.7	211±226	0.17±0.04
50–70 mm	Gills	—	—	—	—	—	—	—	—	—
n=20	Kidney	217±42.1	—	—	15.0±10.0	450±200	—	—	107±130	—

Note:
DG: Digestive Gland.
Source: [1]=from Mauri *et al.* (1990).

palladium solution plus a reducing agent or orthophosphoric acid for lead) were added to samples, blanks and standards (see Regoli & Orlando 1994 for details on analytical procedure). The concentrations obtained for the standard reference material were always within the 95% confidence interval of certified values. Differences in metal concentrations among different organs and species were assessed by ANOVA followed by the Scheffe test.

RESULTS AND DISCUSSION

The metal concentrations in the digestive gland, gills and kidney of the four Antarctic molluscs are shown in Table 57.1. Concentrations and organ distribution showed marked differences according to the metal and species considered. Metal levels in the kidney and digestive gland were usually significantly higher than in the gills. Honda *et al.* (1987) reported the levels of heavy metals in several organisms, including *A. colbecki*, collected around Syowa Station. Their values were similar to those obtained in the present study with the exception of lead and nickel, which were respectively one order of magnitude higher and five-fold lower than in specimens from Terra Nova Bay. Compared with temperate molluscs, the Antarctic species always showed relatively high Cd concentrations. This finding is in accordance with data reported by Honda *et al.* (1987) and with our previous observations (Mauri *et al.* 1990, Berkman & Nigro 1992). It is most likely that the marked bioaccumulation of Cd in Antarctic organisms reflects a high bioavailability of this metal in the Antarctic marine environment. Although studies on metal levels in Antarctic waters are scarce, Cd concentrations higher than in the central gyres of both Atlantic and Pacific oceans have been reported (Mart *et al.* 1982, Orren & Monteiro 1985, Honda *et al.* 1987). Cadmium shows a nutrient-like distribution in oceanic waters, i.e. depleted near the surface and increasing with depth (Bruland, 1980). Orren & Monteiro (1985) confirmed this pattern for the Antarctic waters and related the Cd enrichment in surface waters to the upwelling of deep waters into the euphotic zone. Alternatively, the Cd levels in Antarctic molluscs might be due to a high potential of these organisms for metal bioaccumulation. This hypothesis is supported by the detection of Cd-metallothioneins in *A. colbecki* (Viarengo *et al.* 1993), and by the results of experimental exposure to heavy metals, which demonstrated that Cd is rapidly accumulated but slowly excreted by this species (Nigro *et al.* 1994).

A remarkably high accumulation of As was observed in *Y. eightsi*, with mean values one order of magnitude higher than those observed in the other species investigated (Table 57.1). Moreover, this bivalve also showed the highest concentrations of Pb, Fe and Cu. Considering that all species investigated were collected in the same area, spatial differences in metal bioavailability can be excluded. A wide variability in As concentrations among molluscs has been well documented (Phillips 1990 for review). The highest As levels are reported for the giant clam *Tridacna derasa* (454–1025 µg g^{-1} dry weight) and *T. maxima* (953–1004 µg g^{-1} dry weight) from the Great Barrier Reef (Benson & Summons 1981). Alternative explanations for the tendency of *Y. eightsi* to accumulate metals can be suggested, for example, the feeding behaviour of this infaunal bivalve, which mainly feeds on the subsurface layer of sediments (Davenport 1988). In this respect, it should be noted that sediments are considered a significant source of arsenic, at least in some species (Langston 1980).

Metal concentrations varied during the austral summer. Table 57.2 shows the concentrations of metals in the digestive gland of *A. colbecki* collected in December and January, respectively before and after sea-ice melting. The levels of Mn, Fe, Cu and Zn decreased significantly from December to

Table 57.2. A. colbecki. *Variations of metal concentrations in digestive gland (mean±SD μg g^{-1} dry weight) and condition index (CI=digestive gland wet weight/shell height×10) during the austral summer*

Month	Mn	Fe	Cu	Zn	Cd	CI
December	5.5±0.6	478.7±78.2	18.5±3.5	69.5±7.8	127.2±27.2	2.96±0.42
January	3.7±0.5	232.1±41.1	14.0±3.5	50.7±8.0	138.0±20.8	4.62±0.47
Student test	*	*	*	*	ns	*

Note:
n=12; *=P<0.01; ns=not significant.

January while Cd concentrations remained almost constant. During this period a massive removal of sea ice and an intense bloom occurred in Terra Nova Bay. The consequent increase in food availability caused a marked growth of the digestive gland biomass, as indicated by the increase of the condition index (i.e. digestive gland wet weight/shell height×10 of scallops (Table 57.2). The decrease of metals is supposed to be the consequence of a 'dilution effect' of the increase of biomass, mainly due to the accumulation of lipid reserve material (characterized by a low metal content). Bryan (1973) first reported seasonal variations in tissue metal levels in pectinids, suggesting that this fluctuation may be related to food availability, as concentrations were highest in the autumn and winter months, when phytoplankton productivity was low. The cadmium level in the digestive gland was not affected by the dilution process, its concentrations remaining unchanged as the organ increased its biomass. This fact may be explained by assuming that cadmium uptake from phytoplankton is very efficient whilst excretion is negligible, as previously demonstrated by experimental exposure to the metal (Nigro *et al.* 1994).

CONCLUSIONS

The present study confirms the utility of molluscs, especially *A. colbecki* and *L. elliptica*, as 'sentinels' for monitoring the Antarctic marine environment. These species have a wide distribution around Antarctica (Dell 1972), a body size suitable for dissection and single organ analyses, and their growth and population parameters are well known (Ralph & Maxwell 1977, Stockton 1984, Berkman 1990, Nigro 1993). On the other hand, the present study also showed marked differences in metal concentrations related to the species investigated. Better knowledge of those variables influencing metal accumulation in Antarctic molluscs is necessary before these organisms can be effectively used as biomonitors of environmental changes.

ACKNOWLEDGEMENTS

Financially supported by Ente Nazionale Energie Alternative (ENEA) in the framework of the Italian Antarctic Research Program (PNRA).

REFERENCES

Baroni, C. & Orombelli, G. 1994. Holocene glacier variations in the Terra Nova Bay area (Victoria Land, Antarctica). *Antarctic Science,* **6**, 497–506.

Benson, A. A. & Summons, R. E. 1981. Arsenic accumulation in Great Barrier Reef invertebrates. *Science*, **211**, 482–483.

Berkman, P. A. 1990. The population biology of the Antarctic scallop *Adamussium colbecki* (Smith, 1902) at New Harbor, Ross Sea. In Kerry, K. R. and Hempel, G., eds. *Antarctic ecosystems: ecological change and conservation*. Berlin: Springer-Verlag, 281–288.

Berkman, P. A. 1992. Circumpolar distribution of Holocene marine fossils in Antarctic beaches. *Quaternary Research*, **37**, 256–260

Berkman, P. A. 1997. Ecological variability in Antarctic coastal environments: past and present. In Battaglia, B., Valencia, J. & Walton, D. W. H., eds. *Antarctic communities: species, structure and survival.* Cambridge: Cambridge University Press, 349–357.

Berkman, P. A. & Nigro, M. 1992. Trace metal concentrations in scallops around Antarctica: extending the mussel watch programme to the southern ocean. *Marine Pollution Bulletin*, **24**, 322–323.

Berkman, P. A., Foreman, D. W., Mitchell, J. C., Liptak, R. J. 1992. Scallop shell mineralogy and crystalline characteristics: proxy records for interpreting Antarctic nearshore marine hydrochemical variability. *Antarctic Research Series*, **57**, 27–38.

Bruland, K. W. 1980. Oceanographic distribution of cadmium, zinc, nickel and copper in the north Pacific. *Earth and Planetary Science Letters*, **47**, 176–198.

Bryan, G. W. 1973. The occurrence and seasonal variation of trace metals in the scallops *Pecten maximus* (L.) and *Clamys opercularis* (L.). *Journal of the Marine Biology Association UK,* **53**, 145–166.

Davenport, J. 1988. The feeding mechanism of *Yoldia* (=*Aequiyoldia*) *eightsi* (Courthouy). *Proceedings Royal Society of London*, **B232**, 431–442.

Dell, R. K. 1972. Antarctic benthos. *Advances in Marine Biology*, **10**, 1–216.

Honda, K., Yamamoto, Y., Tatsukawa, R. 1987. Distribution of heavy metals in Antarctic marine ecosystem. *Proceedings NIPR Symposium Polar Biology*, **1**, 184–197.

Langston, W. J. 1980. Arsenic in U.K. estuarine sediments and its availability to benthic organisms. *Journal of the Marine Biology Association*, **60**, 869–881.

Mauri, M., Orlando, E., Nigro, M., Regoli, F. 1990. Heavy metals in the Antarctic molluscs *Adamussium colbecki*. *Marine Ecology Progress Series*, **67**, 27–33.

Mart, L., Rutzel, H, Klahre, P., Sipos, L., Platzek, U., Valenta, P., Nurnberg, H. W. 1982. Comparative studies on the distribution of heavy metals in the ocean and coastal waters. *Science of the Total Environment*, **26**, 1–17.

Nigro, M. 1993. Nearshore population characteristics of the circumpolar Antarctic scallop *Adamussium colbecki* (Smith, 1902) at Terra Nova Bay (Ross Sea). *Antarctic Science,* **5**, 377–378.

Nigro, M., Mauri, M., Regoli, F. & Orlando, E. 1994. Ecology and metal distribution in Antarctic molluscs. In Battaglia, B., Bisol, P. M. & Varotto, V., eds. *Proceedings 2nd Meeting on 'Antarctic Biology', Padova, 26–28 February 1992*. Padova: Edizioni Universitarie Patavine Padova, 235–256.

Orren, M. J. & Monteiro, P. M. S. 1985. Trace element geochemistry in the Southern Ocean. In Siegfried, R. W., Condy, P. R. & Laws, R. M., eds. *Antarctic nutrient cycles and food webs.* Berlin: Springer-Verlag, 30–37.

Phillips, D. J. H. 1990. Arsenic in aquatic organisms: a review, emphasizing chemical speciation. *Aquatic Toxicology*, **16**, 151–186.

Ralph, R. & Maxwell, J. G. H. 1977. Growth of two Antarctic lamellibranchs: *Adamussium colbecki* and *Laternula elliptica*. *Marine Biology*, **42**, 171–175.

Regoli, F., & Orlando, E. 1994. Seasonal variations of trace metals concentrations in the digestive gland of the mediterranan mussel *Mytilus galloprovincialis*: comparison between a polluted and a non-polluted site. *Archives of Environmental Contamination and Toxicology*, **27**, 36–43.

Stockton, W. L. 1984. The biology and ecology of the epifaunal scallop *Adamussium colbecki* on the west side of McMurdo Sound, Antarctica. *Marine Biology*, **78**, 171–178.

Viarengo, A., Canesi, L., Mazzuccotelli, A., Ponzano, E. & Orunesu, M. 1993. Cu, Zn, Cd content in different tissues of the Antarctic scallop *Adamussium colbecki* (Smith, 1902): role of metallothionein in the homeostasis and detoxication of heavy metals . *Marine Environmental Research,* **35**, 216–217.

58 Preengland oil and blood samples: non-destructive methods for monitoring organochlorine levels in Antarctic top predators

NICO W. VAN DEN BRINK
Institute for Forestry and Nature Research (IBN-DLO), Department for Aquatic Ecology, P.O. Box 23, NL-6700, Wageningen, The Netherlands

ABSTRACT

*Different tissues of adult cape petrels (*Daption capense*) were analysed for correlations between concentration of Σ-CB, HCB and Σ-DDT. Linear regression analyses indicate positive relationships between concentrations in fat, muscle, liver, kidney and preengland and the concentrations in either blood or preengland oil. Although the number of birds used in this study is small, the results illustrate the potential use of non-destructive sampling of blood and preengland oil to assess concentrations of organochlorines in the body. Beside ethical concerns, another important advantage of these methods is the opportunity they afford to study trends by repeated sampling of the same individuals over a long period of time.*

Key words: organochlorine pollutants, monitoring, non-destructive sampling, fulmarine petrels, cape petrel.

INTRODUCTION

Organochlorine pollutants were detected some time ago in Antarctica (see Bacci *et al.* 1986, Focardi *et al.* 1992, Larsson *et al.* 1992, Risebrough *et al.* 1976, Subramanian *et al.* 1983, Schneider *et al.* 1985, Tanabe *et al.* 1983). Because of their lipophilic and persistent properties and their toxicity, organochlorines are everywhere a major environmental threat and are the subject of monitoring on local scales (e.g. Peakall & Fox 1987). The remote Antarctic continent offers excellent opportunities for monitoring of global background levels of organochlorine pollutants (Walton & Shears 1994).

The evolution of ethical standards increasingly requires consideration of non-destructive sampling methods in long-term monitoring projects on animals. Collection of blood and preengland oil is non-destructive and relatively easy. Blood samples have been proposed as monitors for concentrations of organochlorine pollutants in other tissues of the birds (Friend *et al.* 1979, Henny & Meeker 1981). This has also been suggested for preengland oil (Larsson & Lindegren 1987). Nevertheless, compartmentation of the body of an animal can mean that one tissue does not always reflect the concentrations of organochlorine pollutants in another (Aguilar 1985). In order to assess the use of blood and preengland oil collected from Antarctic birds as indicators of global background contamination, different tissue samples of cape petrels (*Daption capense*) were analysed to establish the relationship between organs for organochlorine pollutants.

MATERIALS AND METHODS

Three failed breeding pairs of cape petrels were collected in January 1991, at Demay Point, King George Island, by members of the Dutch Antarctic Research Expediton. The birds were dissected in a laminar-flow cabinet under clean lab conditions. The following tissues were obtained: subcutaneous fat, intestinal fat, liver, breast muscle, kidney, brain, preengland tissue, preengland oil from the feather boundle on the gland (1–3 mg) and blood

Table 58.1. *Concentrations of some organochlorine pollutants in cape petrels (geometric mean (n=6) and ranges in μg g⁻¹ fat)*

	Σ-CB		HCB		Σ-DDT	
	Geom. mean	Range	Geom. mean	Range	Geom. mean	Range
Blood	2.04	1.21–5.50	0.57	0.84–1.65	2.18	0.74–13.3
Preen oil	1.58	0.75–4.71	2.25	0.73–10.4	2.64	0.56–14.5
Intest. fat	1.79	0.98–5.90	1.01	0.35–4.09	5.23	1.36–17.3
Subcut. fat	2.09	1.23–7.06	1.31	0.70–4.94	3.29	1.04–13.8
Liver	0.83	0.43–3.01	0.66	0.38–2.43	5.49	2.78–23.5
Kidney	0.63	0.41–2.44	0.62	0.21–3.45	6.01	1.50–39.4
Muscle	1.44	0.88–4.70	1.10	0.40–5.27	2.43	0.83–11.8
Preen gland	0.60	0.32–2.17	0.65	0.31–2.72	0.68	0.29–2.14
Brains	2.38	0.49–7.54	1.16	0.72–11.9	Na	Na

Note:
Na=not analysed.

from the heart (post-mortem, 1–5 ml). The samples were extracted with Soxhlet extraction using n-hexane, after drying with sodium sulphate. Organochlorine pollutants were analysed by gas-chromatography with a Ni63 electron capture detector (J&W, DB-5 column, 0.25 mm ID, 0.25 μm film, 30 metre program: chlorinated biphenyl (CB) fraction: 60 °C injection, 30 °C min.⁻¹ to 140 °C, 10 °C min.⁻¹ to 200 °C, 5 minutes stationary, 4 °C min.⁻¹ to 280 °C, 17 minutes stationary; pesticide fraction: 60 °C injection, 1 minute stationary, 30 °C min .⁻¹ to 140 °C, 6 °C min.⁻¹ to 200 °C, 20 °C min.⁻¹ to 280 °C, 6 minutes stationary, carrier gas helium). A mixture of specific compounds was used for calibration of the instrument, analyses of recovery by these methods and confirmation of peaks. Σ-CB consists of congeners with the following IUPAC-number: 31, 28, 45, 52, 44, 66, 101/90, 97, 87, 77/110, 151, 107/108, 149, 118, 146, 153, 132/105, 141, 179, 138/163/164, 187/182, 183, 128, 174, 177, 180, 170, 196, 194, 206. To conform to De Boer & Wester (1991) congener 077/110 is determined as 110, 138/163/164 as 138 and 187/182 as 182. Congeners 101/90 and 107/108 and 132/105 are determined as 101, 107 and 105 respectively, all bearing in mind that this has not been exactly confirmed. Σ-DDT is defined as *op*-DDT+1.1((pp-DDE+*pp*- DDD+*pp*-DDD).

Data concerning low concentrations of pollutants are generally not normally distributed. Hence, prior to statistical analysis Σ-CB, HCB and Σ-DDT concentrations have been log-transformed. Linear regression on transformed data was used to analyse relations between concentrations in different tissues. R^2_{adj} is a measure of the percentage of the variance accounted for by the model in the regression analysis ($R^2_{adj}=100\times(1-MS_{res}/MS_{tot})$ with MS_{res}=mean squares of the residuals and MS_{tot}=total mean squares) (Genstat 1987).

RESULTS AND DISCUSSION

The geometric mean and ranges of Σ-CB, hexachlorobenzene (HCB) and Σ-DDT are given in Table 58.1. Traces of dieldrin, oxychlordane, trans-nonachlor, endrin, pentachlorobenzene and mirex, in concentrations in the order of magnitude of the detection limits, were also detected in most samples. No calculations have been performed on these latter compounds.

Most Antarctic bird species show low levels of contamination (e.g. De Boer & Wester 1991, Focardi *et al.* 1992, Luke *et al.* 1989, Lukowski 1978, Subramanian *et al.* 1986) compared with seabirds in other regions (Walker 1992). The levels found in this study are comparable to the ones reported for cape petrels in other studies (Luke *et al.* 1989, Lukowski 1978) and are intermediate to levels found in true Antarctic species such as Adélie penguins (*Pygoscelis adeliae*) and more migratory species like the Wilson's storm petrel (*Oceanites oceanicus*) and south polar skua (*Catharacta maccormicki*).

Blood is considered to be a major compartment and distribution medium and its organochlorine levels are expected to be proportional to levels in other tissues (Norstrom *et al.* 1986). Preen oil, however, is not in contact with the rest of the internal body once it is excreted. This implies that its concentrations of organochlorine pollutants are expected to reflect concentrations in the body at the time of production. In this approach preen oil has a lag time in reacting to changes in body concentrations. This lag time is a function of the rate of production of oil and the rate of use by the bird, but is probably not significant compared to half life times of organochlorines (Peakall & Fox 1987).

The significance and the R^2_{adj} of the regressions between the concentrations of the major organochlorine pollutants (Σ-CB, HCB and Σ-DDT) in different tissues, and blood of preen oil as regressors are given in Table 58.2. Concentrations of Σ-CB and Σ-DDT in different tissues, except brain tissue, are proportional to concentrations found in blood or preen oil. This implies an equilibrium between the compartments and a small lag-time for preen oil. Comparison of relations to other studies shows similarities. Douthwaite *et al.* (1992) report a linear relation between log concentrations of Σ-DDT in subcutaneous fat and liver of reed cormorants ($\log_{fat}=0.98\times\log_{liver}+0.71$). We find a comparable relation: $\log_{fat}=1.02\times\log_{liver}-0.54$. Larsson & Lindegren (1987) report relations (on non-transformed data) for

Table 58.2. *Significance of* F-*statistic and* R^2_{adj} *of linear regressions*

	Σ-CB				HCB				Σ-DDT			
	Blood		Preen oil		Blood		Preen oil		Blood		Preen oil	
Regressor	Sign.	R^2_{adj}	Sign.	R^2_{adj}	Sign	R^2_{adj}	Sign.	R^2_{adj}	Sign.	R^2_{adj}	Sign.	R^2_{adj}
Preen oil	**	65			*	58			**	57		
Intest. fat	**	94	*	55	**	66	**	94	**	85	*	55
Subcut.fat	**	88	**	75	—	<0	**	59	**	84	**	89
Liver	**	85	**	72	0	11	**	78	**	84	**	63
Kidney	**	92	**	71	*	52	**	97	**	83	**	89
Muscle	**	87	**	75	*	53	**	97	**	93	**	80
Preen gland	**	89	**	87	*	40	**	96	*	48	**	95
Brains	—	38	—	16	—	38	*	42	na	na	na	na

Notes:
na = not analysed.
Significance: *$P \leq 0.05$; **$0.05 < P < 0.1$.

CB's in the preengland and muscle of goosander (*Mergus merganser*) ($CB_{gland} = 0.66 \times CB_{muscle} - 5.8$), for feathers and muscles ($CB_{feathers} = 0.57 \times CB_{muscle} + 20.5$) and for DDE in preengland and muscle ($DDE_{gland} = 0.97 \times DDE_{muscle} - 0.006$). Organochlorines on feathers are considered to be mainly derived from preengland oil. In this study we found respectively (for non-transformed data): $CB_{gland} = 0.48 \times CB_{muscle} - 0.077$, $CB_{oil} = 0.97 \times CB_{muscle} \pm 0.237$ and $DDE_{gland} = 1.22\ DDE_{muscle} - 0.13$. The coefficients are comparable, and differences can probably be related to the small number of samples, analytical differences and taxonomic differences. However, there is a marked difference between the study of Larsson & Lindegren (1987) and this study: we find higher concentrations in the preen oil than in the preengland itself (paired *t*-test, two-tailed $t = 3.63$, $P < 0.05$), contrary to Larsson & Lindegren. Apparently with cape petrels, the organochlorine pollutants have a higher affinity to the preen oil as a matrix than to the preengland tissue. Differences in composition of preen oil can probably explain interspecies differences (Jacob & Hoerschmann 1982).

HCB levels in preen oil are proportional to levels in other tissues except brain tissue, but such regressions are less or not significant for HCB in blood. A possible explanation for this phenomena is that HCB is relatively volatile. Bidleman *et al.* (1993) report recovery rates for HCB in their analyses of 65%. Recovery rates for HCB found in this study are >80%, but with this small number of samples it is possible that small variations in recoveries between blood samples do cause this low significance.

Levels of Σ-CB and HCB in brain tissue of cape petrels in this study were not proportional to levels found in blood or preen oil. Lambeck *et al.* (1991) found a linear relation between CB concentrations in breast muscle and brains of oystercatchers. They found variation in the regression coefficient due to the age of the individual. At older ages a relatively larger proportion of the body burden was located in the brains. We do not know the exact age of the birds, so this might be a source of variation. Moreover, transport of organochlorine pollutants to the brain tissue can be altered due to the haematoencephalic barrier and to the differences in molecular structure of the fat in the brain (Aguilar 1985).

Although the number of birds used in this study is relatively small, it is concluded that blood and preengland oil are adequate indicators of body concentrations of organochlorine pollutants in live Antarctic petrels, except for brain tissue. Collecting samples from live birds has, apart from the positive ethical aspect, the important advantage that repeated samples can be collected from the same bird. This will allow questions on seasonal fluctuations, accumulation and metabolism rates to be addressed in future and the impact of fluctuations in the body condition of birds on organochlorine pollutants.

ACKNOWLEDGEMENTS

Research was supported by the Geosciences Foundation (GOA) of the Netherlands Organisation for Scientific Research (NWO) and is part of a project conducted in cooperation with the Australian Antarctic Division. I am grateful to the members of the Dutch Antarctic Expedition of 1990/91 who collected the cape petrels for this study. I thank Wim Wolff, Rob Gast, Elze de Ruiter-Dijkman and Jan Andries van Franeker for critically reading the manuscript.

REFERENCES

Aguilar, A. 1985. Compartmentation and reliability of sampling procedures in organochlorine pollution surveys of cetaceans. *Residue Reviews*, **95**, 91–114.

Bacci, E., Calamari, D., Gaggi, C., Fanelli, R., Focardi, S. & Morosini, M. 1986. Chlorinated hydrocarbons in lichen and moss samples from the Antarctic Peninsula. *Chemosphere*, **15**, 747–754.

Bidleman, T. F., Walla, M. D., Roura, R., Carr, E. & Schmidt, S. 1993. Organochlorine pesticides in the atmosphere of the Southern Ocean and Antarctica, January–March, 1990. *Marine Pollution Bulletin*, **26**, 258–262.

De Boer, J. & Wester, P. 1991. Chlorobiphenyls and organochlorine pesticides in various sub-Antarctic organisms. *Marine Pollution Bulletin*, **22**, 441–447.

Douthwaite, R. J., Hustler, C. W., Kruger, J. & Renzoni, A. 1992. DDT residues and mercury levels in reed cormorants on Lake Kariba: a hazard assessment. *Ostrich*, **63**, 123–127.

Focardi, S., Fossi, M. C., Leonzio, C., Lari, L., Marsili, L., Court, G. S. & Davis, L. S. 1992. Mixed function oxidase activity and chlorinated hydrocarbon residues in Antarctic sea birds: south polar skua (*Catharacta maccormicki*) and Adélie penguin (*Pygoscelis adeliae*). *Marine Environmental Research*, **34**, 201–205.

Friend, M., Haegle, M., Meeker, D., Hudson, R. & Baer, C. 1979. Correlations between residues of dichlorodiphenylethane, polychlorinated biphenyl and dieldrin in the serum and tissues of mallard ducks (*Anas platyrhynchos*). In *Animals as monitors of environmental pollutants*. Washington, DC: National Academy of Sciences, 319–326.

Genstat. 1987. *GENSTAT 5 Reference Manual*. GENSTAT 5 Committee. Oxford: Clarendon Press.

Henny, C. & Meeker, D. 1981. An evaluation of blood plasma for monitoring DDE in birds of prey. *Environmental Pollution, series A*, **25**, 291–304.

Jacob, J. & Hoerschmann, H. 1982. Chemotaxonomische Untersuchungen zur Systematik de Röhrennasen (Procellariiformes). *Journal für Ornithologie*, **123**, 63–84.

Lambeck, R. H. D., Niewenhuize, J. & van Liere, J. M. 1991. Polychlorinated biphenyls in Oystercatchers (*Haematopus ostralegus*) from the Oosterschelde (Dutch Delta area) and the Western Wadden Sea, that died from starvation during severe winter weather. *Environmental Pollution*, **71**, 1–16

Larsson, P. & Lindegren, A. 1987. Animals need not be killed to reveal their body-burdens of chlorinated hydrocarbons. *Environmental Pollution*, **45**, 73–78.

Larsson, P., Järnmark, C. & Södergen, A. 1992. PCB's and chlorinated pesticides in the atmosphere and aquatic organisms of Ross Island, Antarctica. *Marine Pollution Bulletin*, **25**, 281–287.

Luke, B. G., Johnstone, G. W. & Woehler, E. J. 1989. Organochlorine pesticides, PCB's and mercury in Antarctic and sub-Antarctic seabirds. *Chemosphere*, **19**, 2007–2021.

Lukowski, A. B. 1978. DDT and its metabolites in Antarctic birds. *Polski Archiwum Hydrobiologii*, **25**, 729–737.

Norstrom, R. J., Clark, T. P., Jeffrey, D. A. & Won, H. T. 1986. Dynamics of organochlorine compounds in Herring Gulls (*Larus argentatus*): 1. Distribution and clearance of [^{14}C]DDE in free-living herring gulls (*Larus argentatus*). *Environmental Toxicology and Chemistry*, **5**, 41–48.

Peakall, D. B. & Fox, G. A. 1987. Toxicological investigations of pollutant-related effects in Great Lake gulls. *Environmental Health Perspectives*, **71**, 187–193.

Risebrough, R., Walker, H., Scmidt, T., Lappe, B. & Connors, C. 1976. Transfer of chlorinated biphenyls to Antarctica. *Nature*, **264**, 738–739.

Schneider, R., Steinhagen-Schneider, G. & Drescher, H. 1985. Organochlorines and heavy metals in seals and birds from the Weddell Sea. In Siegfried, W. R., Condy, P. R. & Laws, R. M., eds. *Antarctic nutrient cycles and food webs*. Berlin: Springer Verlag, 652–655.

Subramanian, B., Tanabe, S., Hidaka, H. & Tatsukawa, R. 1983. DDT's and PCB isomers and congeners in Antarctic fish. *Archives of Environmental Contamination and Toxicology*, **12**, 621–626.

Subramanian, B., Tanabe, S., Hidaka, H. & Tatsukawa, R. 1986 Bioaccumulation of organochlorines (PCBs and p,p'-DDE) in Antarctic Adélie Penguins *Pygoscelis adeliae* collected during a breeding season. *Environmental Pollution, series A*, **40**, 173–189.

Tanabe, S., Hidaka, H. & Tatsukawa, R. 1983. PCB's and chlorinated hydrocarbon pesticides in Antarctic atmosphere and hydrosphere. *Chemosphere*, 12, 277–288.

Walker, C. H. 1992. The ecotoxicology of persistent pollutants in marine fish-eating birds. In Walker, C.H. & Livingstone, D.R., eds. *Persistent pollutants in marine ecosystems*. Oxford: Pergamon Press.

Walton, D. W. H. & Shears, J. 1994. The need for environmental monitoring in Antarctica: baselines, environmental impact assessments, accidents and footprints. *International Journal of Environmental Chemistry*, **55**, 77–90.

59 Changes in the vegetation of sub-Antarctic Marion Island resulting from introduced vascular plants

NIEK J. M. GREMMEN

Netherlands Institute of Ecology – Centre for Estuarine and Coastal Ecology, Vierstraat 28, 4401 EA Yerseke, The Netherlands

ABSTRACT

Of 42 species of vascular plants recently found on Marion Island, 18 are a result of human introduction. Six of the aliens have disappeared again. Others, however, have been spreading rapidly.

Alien plants have been observed in 21 of the 23 plant associations recognized on the island. They may reach dominance in nine associations, mostly in relatively nutrient-rich habitats. Associated with dominance by alien plants is a marked reduction in the number of native species in the vegetation. In Acaena*-dominated communities of streambanks and drainage lines the invasion and subsequent dominance by alien grasses has led to a 50% decrease in the number of native species in the vegetation. Characteristics of the reproductive strategies of the introduced species appear to determine the success of their introduction on the island. All aliens which produce seed on the island have become widespread. Of the species which do not produce seed, some are able to persist by their ability to spread laterally by vegetative means, but species lacking this ability have disappeared again.*

The increase in Poa annua *and* Cerastium fontanum *is believed to be related to the increase in temperature at Marion Island during the past two decades.*

A regular botanical survey to detect any new introductions and any changes in the distribution and phenology of alien species with at present a restricted distribution on the island is necessary. To avoid further introductions it is important to adhere strictly to the quarantine measures that are presently in force, and to ensure that all personnel involved in the islands are aware of and appreciate the reason for these measures.

Key words: sub-Antarctic, Marion Island, introduced plants, colonization, vegetation change, climate change, biodiversity.

INTRODUCTION

Compared with most other oceanic islands, sub-Antarctic islands have been little affected by human activity. This makes them of considerable interest from a scientific as well as from a biological and conservation viewpoint (Clark & Dingwall 1985, Smith 1987, Smith & Smith 1987). The most important factor affecting the ecology of Marion Island with respect to human activities has been the breakdown of its isolation by the immigration of people, and importation of building materials, food, etc., allowing species of plants and animals which are not able to cross large distances of ocean unaided to reach the islands (Smith & Smith 1987). Consequently, there has been a continuous interest in the occurrence of alien species in the islands (see Watkins & Cooper 1986 and Cooper & Condy 1988 for references). The present amelioration of the island's climate (Smith 1992) may lead to a change in the competitive abilities of introduced and native species, and may allow a larger

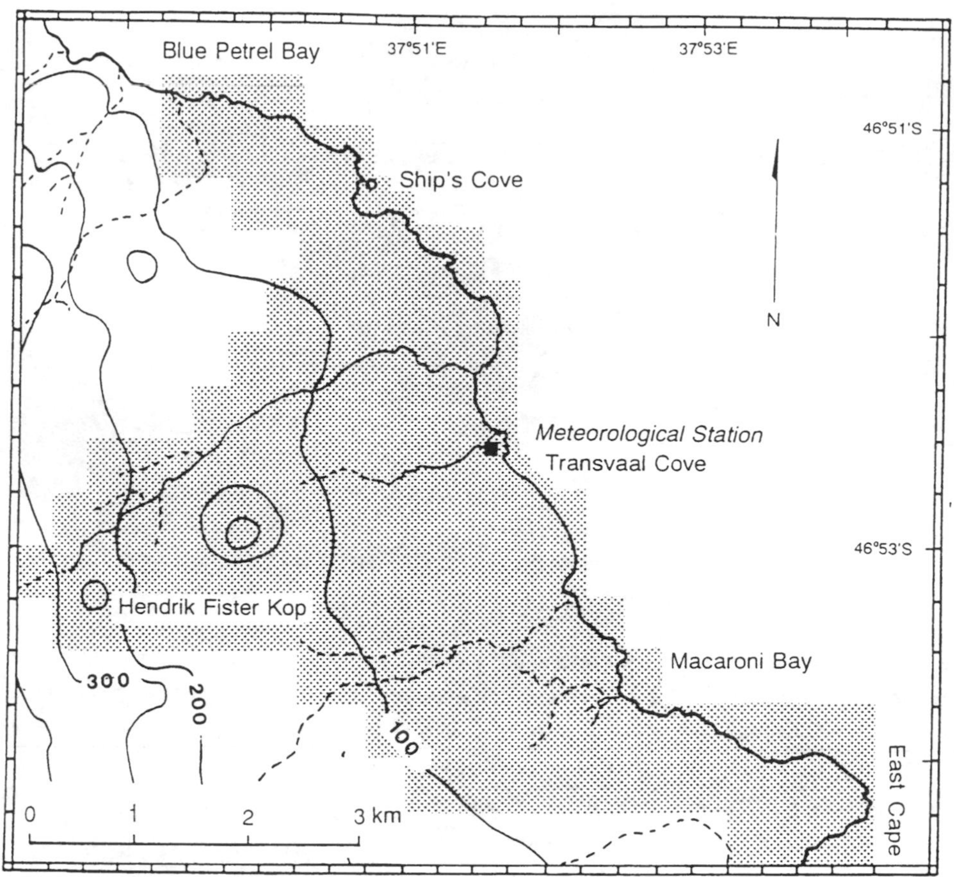

Fig. 59.1. Map of the study area in the northeastern part of Marion Island. Thin lines=contour lines with height in m; solid/dashed lines=permanent/intermittent stream.

proportion of introduced species to succeed in becoming established (cf. Holdgate 1986).

In the present study the success of alien vascular plants in colonizing Marion Island, and the changes they caused in the natural vegetation were investigated. An attempt is made to explain the success of the alien species in relation to their dispersal strategies.

Marion Island (46°50′ S, 37°50′ E) is a relatively young, volcanic island, 290 km^2 in area, situated in the Southern Indian Ocean, just north of the Antarctic Convergence. The island has a cool, extremely oceanic climate, characterized by a mean annual temperature of *c.* 5 °C, with little diurnal and seasonal variation, high precipitation (>2500 mm annually, mainly as rainfall), and high wind speeds (Schulze 1971). Since the 1970s the annual mean temperature has increased by *c.* 1 °C (Smith 1992). Marion and its neighbour Prince Edward Island, 22 km to the north, are 950 km from the nearest island group (Iles Crozet), and 1800 km from the nearest continent (Africa). The island's environment and biota are described by Van Zinderen Bakker *et al.* (1971) and Smith (1987). The island's flora is relatively poor in species. Forty-two species of vascular plants have been recorded, 18 of which are considered introduced. Some 85 species of mosses, 35 hepatics and 60 lichens are known. The vegetation of the island has been classified into 23 associations, which are grouped into six community complexes (Gremmen 1981; see Table 59.2).

The island was possibly first sighted in 1663, but the date usually cited for its discovery is 1772 (Van Zinderen Bakker 1971). Sealers had established camps on the island by 1802, and were intermittently active throughout the nineteenth and early twentieth century (Cooper & Avery 1986). The last sealing expedition to visit the island was by the sealer *Kildalkey* in 1930. In 1948 a meteorological station was established at Transvaal Cove on the island's northeastern coast, and has been occupied continuously since then. Until the early 1970s human activities were more or less confined to this site, but since then huts have been erected at a number of localities and human activities, associated with research and management (e.g. the cat eradication programme), have spread into previously unaffected areas.

METHODS

Survey of species distribution

In 1993/94 the distribution of alien vascular plants in the northeastern coastal plain of Marion Island was investigated. The study area was bounded by East Cape, Ship's Cove and Hendrik Fister Kop, but included the coastal area between Ship's Cove and Blue Petrel Bay (Fig. 59.1). The area was divided into grid cells *c.* 320×475 m. Each grid cell was examined carefully at least once, usually twice, and the presence, abundance and phenology of alien plants was noted.

Data from surveys in 1965/66 (Huntley 1971), 1975

Table 59.1. *Data on Marion Island alien vascular plants, from own observations and from the literature (Cooper & Avery 1986, Gremmen 1975, 1981, 1982; Gremmen & Smith 1981, Grime* et al. *1988).* Sonchus *sp. and* Senecio *sp., which were eradicated, are excluded*

Code	Species	dis	inc	dom	rec	si	spr	spe	wgt	fer	veg
Av	*Avena sativa* L.	1	–	–	1965	M	–	–	5	–	
Hl	*Holcus lanatus* L.	1	–	–	1953	M	–	+	3	si	+
Hr	*Hypochoeris radicata* L.	1	–	–	1953	M	–	–	4	si	–
Pl	*Plantago lanceolata* L.	1	–/ex	–	1965	M	–	+	5	si	–
Aa	*Alopecurus australis* Nees.	1	=	+	1965	O	–	?	2	sc	++
Ar	*Agropyron repens* (L.) Beauv.	1	=	+	1965	O	–	+	5	si	++
Fr	*Festuca rubra* L.	2	=	+	1966	O	–	–	4	si	++
Ra	*Rumex acetosella* L.	3	=	+	1953	M,O	–?	+	3	d	++
Ag	*Agrostis gigantea* Roth	1		+	1994	M	?	+	1	?	++
Ac	*Agrostis castellana* Boiss. & Reuter	w	?	+	1975	M	+	+	1	?	++
As	*Agrostis stolonifera* L.	w	++	+	1965	M	+	+	1	?	++
Cf	*Cerastium fontanum* Baumg.	w	+	–	1873	O	++	+	2	sc	–
Pa	*Poa annua* L.	w	+	+	1948	O	++	+	2	sc;c	+
Pp	*Poa pratensis* L.	w	++	+	1965	M	+	+?	2	a	+
Sp	*Sagina procumbens* L.[1]	w	++	+	1965	M	++	+	1	sc	+
Sm	*Stellaria media* (L.) Vill.	w	=	–	1873	O	++	+	3	sc;c	–

Notes:

Abbreviation used in Table 59.1. dis=distribution pattern: 1–3=number of sites; w=widespread. inc=rate of increase: –=disappeared; ==stable distribution; +=increasing; ++=rapid increase; ex=exterminated. dom=dominance: +=locally dominant; –=never dominant. rec=date of first record. si=site of introduction: M=meteorological station; O=other. spr=seed production at Marion Island: ++=abundant; +=not abundant; –=usually no seed produced. spe=seed persistence: +=persistent seedbank; ±=short-lived seedbank; –=no persistent seedbank. wgt=seed weight: 1=<0.1 mg; 2=0.1–0.3 mg; 3=0.3–0.6 mg; 4=0.6–1.2 mg; 5=>1.2 mg. fer=fertilization: sc=self-compatible; si=self-incompatible; d=dioecious; a=apoximis; c=cleistogamy. veg=capacity for vegetative spread: ++=high; +=intermediate; –=low or absent.

[1]in previous surveys erroneously listed as *S. apetala* Ard.

(Gremmen 1975, Gremmen & Smith 1981) and 1981 (Gremmen 1982) were entered into the same grid cell system. The 1965/66 data were taken from the published map (Huntley 1971, p. 107), but more precise information on the locality of some species was derived from herbarium labels. Data from the 1975 and 1981 surveys were taken directly from field maps and field notes made by the author. Bergstrom & Smith (1990) surveyed the distribution of alien vascular plants on Marion and Prince Edward Islands in 1989. The scale of their maps, however, does not yield the resolution needed for a detailed quantitative comparison with the present survey.

Vegetation survey

Whenever alien plants were discovered a note was made on the plant communities in which they occurred, and whether they reached dominance or not. More detailed information on the position of the alien species in the vegetation was collected by the analysis of the species composition of the vegetation in 200 2×2m sample plots, using the methods of Braun-Blanquet (Westhoff & Van der Maarel 1973, Gremmen 1981). The species composition of plots containing alien plants was compared with plots containing only native species (Gremmen 1981 and unpublished data). For each plant community the number of species was compared between plots without aliens, plots with aliens covering between 1 and 50% of the area and plots with more than 50% cover by alien vascular plants.

RESULTS AND DISCUSSION

Alien vascular flora

The *Challenger* expedition of 1873 recorded the first alien vascular plants on the island (*Cerastium fontanum* and *Stellaria media*; Moseley 1874, Oliver 1874). During the 1965/66 biological expedition 14 alien vascular plants were collected (Huntley 1971, Gremmen & Smith 1981; Table 59.1). Four of these had disappeared by 1973 (Gremmen 1975). Two new introductions occurred during the 1980s, at the site of the Meteorological Station: *Sonchus* sp. in 1983, and *Senecio* sp. in 1988 (Bergstrom & Smith 1990, Watkins & Cooper 1986). In 1988 *Plantago lanceolata* was observed again, signifying a reintroduction of this species (Bergstrom & Smith 1990). All plants of *Sonchus, Senecio*, and *Plantago* were removed after their discovery (Bergstrom & Smith 1990).

In 1993/94 12 alien species of vascular plants were observed, two of which were not recorded during previous surveys (Table 59.1). One of these, *Agrostis castellana*, was subsequently found in collections from 1975, but was then not recognized as different from *A. stolonifera*. The second, *Agrostis gigantea*, is

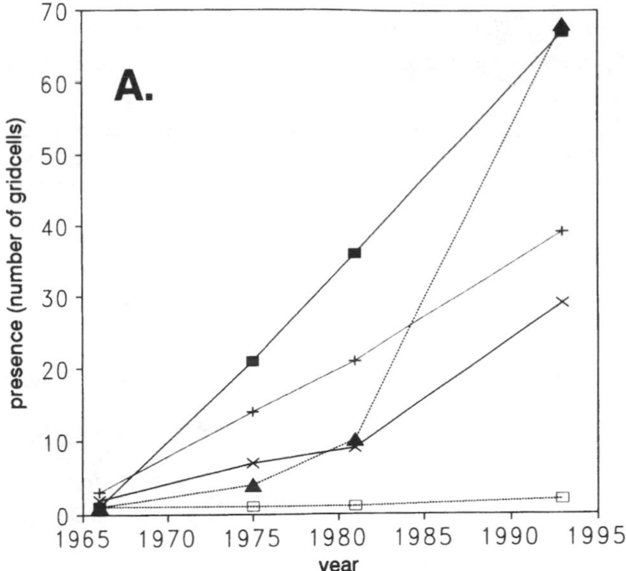

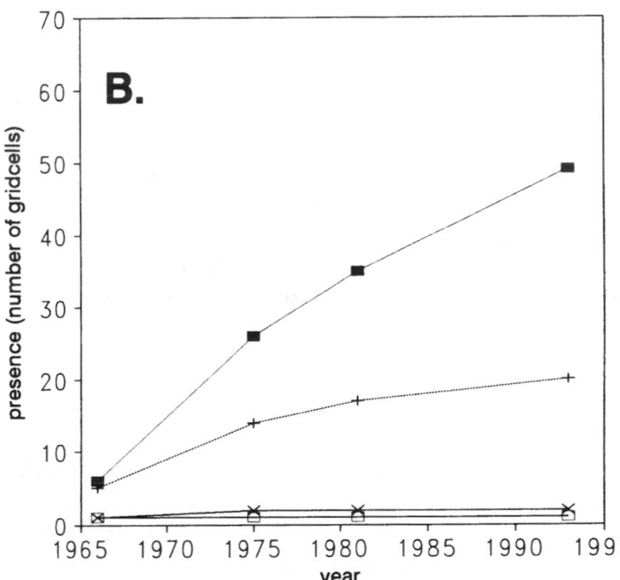

Fig. 59.2. The abundance of alien vascular plants in the northeastern part of Marion Island. **A.** ■——■ = *Agrostis stolonifera*, +·····+ = *Cerastium fontanum*, x——x = *Poa pratensis*, ▲-----▲ = *Sagina procumbens*, □-----□ = *Rumex acetosella*. **B.** ■·····■ = *Poa annua*, +-----+ = *Stellaria media*, x——x = *Festuca rubra*, □——□ = *Agropyron repens*.

considered a very recent introduction. It was found only at the Meteorological Station. Its size, conspicuously larger than *A. stolonifera* and *A. castellana*, makes it unlikely that it was overlooked during previous surveys. Data on the distribution, first record and presumed site of introduction of the alien vascular species are summarized in Table 59.1. Quantitative data on the spread of the aliens in the northeastern part of the island are given in Fig. 59.2.

The distribution pattern and date of first record of seven species point to an introduction at sites distant from the Meteorological Station, prior to 1948 (cf. Gremmen 1975, 1982). Of these seven species *Alopecurus australis, Agropyron repens, Festuca rubra* and *Rumex acetosella* occur at three or less sites (Table 59.1). The other three (*Cerastium fontanum, Poa annua* and *Stellaria media*) are widespread. *Stellaria* has reached a stable distribution, but both *P. annua* and *C. fontanum* are still increasing in abundance (Fig. 59.2). Of ten species introduced at the weather station *Avena sativa, Holcus lanatus, Hypochoeris radicata* and *Plantago lanceolata* have disappeared again (Gremmen 1975, Bergstrom & Smith 1990). *Agrostis stolonifera, Poa pratensis* and *Sagina procumbens* have spread rapidly from the Meteorological Station (Fig. 59.2). Insufficient data are available on the distribution of *Agrostis castellana*, while *A. gigantea* is well established at the weather station, but has not yet spread from there.

For the species introduced at the Meteorological Station the time since their introduction largely coincides with the period of climatic amelioration, and our data do not allow a distinction between the effects of the temperature increase and the time available for colonization on the species' area. Data from Huntley (1971) and Gremmen (1975), however, suggest that the widespread aliens introduced at earlier times, i.e. *Poa annua, Cerastium fontanum* and *Stellaria media*, had colonized the whole of the island at the start of the observation period. The observed increase of *P. annua* and *C. fontanum* in the study area is not due to invasion of areas not previously reached by the species, but to an increase in the number of colonies, resulting in a larger number of grid cells occupied by these species. It is suggested that this increase in abundance is related to the recent temperature increase at the island (Smith 1992), resulting in a longer growing season, which in turn leads to an increased seed production.

Based on their increase in area during the past three decades, further spread of *Agrostis stolonifera, Poa pratensis* and *Sagina procumbens* may be expected. *Agrostis* and *Sagina* have been reported from parts of the island outside the northeastern coastal plain, their occurrence apparently resulting from increased pedestrian activity and helicopter transport between the Meteorological Station and outlying huts (Bergstrom & Smith 1990). It is predicted that *A. stolonifera* and *S. procumbens* will become common and widespread throughout the island, within a period of some 50–100 years. *Poa pratensis* has a slower dispersal rate, but is also expected to spread eventually over all the island.

IMPACT OF ALIEN PLANTS ON THE NATIVE VEGETATION

Huntley (1971) listed alien plants in seven of the 13 plant communities which he recognized on the island, but stated that only *Poa annua* competed with undisturbed natural vegetation. In 1993/94 alien plants were observed in 21 of the 23 plant associations described by Gremmen (1981), and reach dominance in nine associations, mostly belonging to the *Callitriche antarctica – Poa cookii* and the *Acaena magellanica – Brachythecium* complex (Table 59.2). These complexes are characterized by a relatively abundant nutrient supply compared to the oligotrophic *Juncus – Blepharidophyllum,*

Table 59.2. *Occurrence of alien vascular plants in Marion Island plant communities. The communities are described in Gremmen (1981)*

	Introduced species									
Communities	Ar	As	Aa	Cf	Fr	Pa	Pp	Ra	Sp	Sm
Crassula complex (salt spray vegetation)										
1.1. *Cotulo plumosae – Crassuletum moschatae*		d				*			d	
1.2. *Crassulo moschatae – Clasmatocoleetum vermicularis*		*							*	
1.3. *Crassulo moschatae – Azorelletum selaginis*		*							*	*
Callitriche – Poa complex (biotically influenced vegetation)										
2.1. *Montio fontanae – Callitrichetum antarcticae*		d				*			*	
2.2. *Callitricho antarcticae – Poetum annuae*		*	d			d			*	
2.3. *Poo cookii – Cotuleum plumosae*		*				d			*	
2.4. *Montio fontanae – Clasmatocoleetum vermicularis*		*				*		d	d	
2.5. *Leptodontio proliferi – Poetum cookii*		*		*		*			*	*
Acaena – Brachythecium complex (springs, flushes, drainage lines)										
3.1. *Brachythecietum subplicati*		d				*				
3.2. *Acaeno magellanicae – Drepanocadetum uncinati*		d				*			*	*
3.3. *Acaeno magellanicae – Brachythecietum rutabuli*	d	d		*	d	*	d		*	*
3.4. *Acaeno magellanicae – Agrostietum stoloniferae*		d		*		*				*
Juncus – Blepharidophyllum complex (oligotrophic mires)										
4.1. community of *Juncus scheuchzerioides*										
4.2. *Junco scheuchzerioidis – Drepanocladetum uncinati*		d							*	
4.3. *Distichophylletum fasciculati*										
4.4. *Blepharidophyllo densifolii – Clasmatocoleetum humilis*							*		d	
4.5. *Uncinio compactae – Ptychomnietum ringiani*		*								
4.6. *Lycopodio magellanici – Jamesonielletum coloratae*		*							*	
4.7. *Jamesonielletum grandiflorae*		d		*			*		*	
4.8. *Bryo laevigati – Breutelietum integrifoliae*						*			*	
Blechnum complex (well-drained slopes)										
5.1. *Isopterigio pulchelli – Blechnetum penna-marinae*		*		*		*			*	
Andreaea – Racomitrium complex (feldmark vegetation)										
6.1. *Jungermannio coniflorae – Racomitrietum crispuli*		*		*		*			*	
6.2. *Andreaeo acutifoliae – Racomitrietum crispuli*		*		*					*	

Notes:

*=species found in this community, but not dominant; d = species occurs as dominant in this community. See Table 59.1 for a key to the plant name codes.

Blechnum, and *Andreaea– Racomitrium* complex (Gremmen 1981). The areas of *Crassula* saltspray vegetation dominated by aliens also tend to be influenced by animal excreta. Most naturalized aliens are able to reach dominance in the vegetation. Only *Cerastium fontanum* and *Stellaria media* never reached high cover values (Table 59.2).

S. procumbens and *P. annua*, which predominantly spread by seed, are reaching high cover values mostly in communities where bare ground is available for colonization. *S. procumbens*, however, also seems able to germinate in compact bryophyte mats, e.g. in mire vegetation. Once it becomes established it may rapidly reach high cover values, overgrowing the bryophyte stratum in the vegetation.

Where aliens have invaded the vegetation but have not become dominant, the number of native species appears unaffected. Where they reached dominance, however, the number of native vascular and bryophyte species decreased markedly (Table 59.3). In communities of the *Acaena* complex dominated by alien grasses total cover of the bryophyte stratum is generally strongly reduced (Table 59.3). This may be due to the persistent cover of dead grass during winter, compared with the deciduous leaves of *Acaena*, which fall in autumn and quickly decompose, leaving the bryophyte stratum exposed to ambient light during winter. Table 59.3 shows average species numbers per 2×2 m sample plot. The total number of species in several plots per community gave similar results. For example, in ten plots from different stands of *Acaeno–Brachythecietum rutabuli* drainage line vegetation without alien plants a total of 14 native species occurred. However, ten plots in stands of this community dominated by *Poa pratensis* yielded a total of only seven native species. Similarly, plots in three stands of *Acaeno–Drepanocladetum* unaffected by aliens contained 16

Table 59.3. *Cover of herb and bryophyte strata, and average numbers of species in vegetation without alien plants and in vegetation dominated by aliens (averages per 2×2m sample plot)*

Community	Dominant alien	*n*	hc	ac	bc	#nat	#vasc	#bryo	#lich
Cotulo–Crassuletum moschatae	—	29	90	0	5	4.8	4.1	0.4	0.3
Cotulo–Crassuletum moschatae	*Sagina procumbens*	6	90	80	<1	4.5	4.0	0.2	0.3
Poo cookii–Cotuletum plumosae	—	12	100	0	<1	4.0	3.6	0.4	—
Poo cookii–Cotuletum plumosae	*Poa annua*	3	95	90	4	2.7	2.3	0.3	—
Acaeno–Drepanocladetum uncinati	—	3	100	<1	85	9.0	5.3	3.6	—
Acaeno–Drepanocladetum uncinati	*Agrostis stolonifera*	4	100	90	60	4.0	2.0	2.0	—
Acaeno–Brachythecietum rutabuli	—	17	100	<1	90	6.2	3.6	2.6	—
Acaeno–Brachythecietum rutabuli	*Poa pratensis*	12	100	90	5	2.7	1.8	1.0	—
Acaeno–Brachythecietum rutabuli	*Agrostis stolonifera*	10	95	90	20	4.1	2.1	2.0	—

Notes:
n=number of sample plots. hc=cover of herb layer as % of area. ac=cover of alien vascular species as %. bc=cover of bryophyte stratum as %. #nat=number of native species. #vasc=number of native vascular plants. #bryo=number of bryophyte species. #lich=number of lichen species.

native species, while four plots in stands of this community dominated by *Agrostis stolonifera* possessed only eight native species. This signifies a 50% reduction in biodiversity in these communities.

In the study area in the northeastern coastal plain the total area affected by alien plants is estimated to be less than 5% of the total vegetated surface. The area in which aliens dominate the vegetation is less than 1%. As aliens, however, are concentrated in specific communities and their associated habitat, a large proportion of these habitat types may be affected. In the study area over 50% of the area of *Acaena* communities is colonized by *Agrostis stolonifera*. All drainage line vegetation on coastal slopes and most *Acaena* vegetation of stream banks are invaded by this grass. Only inland drainage line vegetation, away from the apparent invasion routes along the coast and streams, appears to have escaped invasion. Outside the northeastern coastal area the proportion of vegetation invaded by aliens is considerably less. As the most successful aliens are spreading into these areas, however, the situation will become similar to that presently found in the study area. As a result of colonization by *A. stolonifera* and *P. pratensis*, the *Acaeno magellanicae – Brachythecietum rutabuli* community of drainage lines and stream banks (Gremmen 1981) will be considerably reduced in abundance, and may even eventually disappear.

Species characteristics

Characteristics related to reproductive strategies of the alien plant species are listed in Table 59.1. Significantly, all of the transient and most of the naturalized aliens with a localized distribution are self-incompatible. This would make establishment of a viable population from a single colonizing individual impossible. All transient species have been observed to flower, but no seed was observed in the available herbarium specimens. The localized naturalized species produce seed rarely, if at all. This greatly reduces the chance of dispersal over any appreciable distance. Their ability to persist, and spread locally, appears to result from vegetative extension by rhizomes, stolons, etc. However, this restricts the spread of individual plants to less than about 1 m per year.

All widespread species seem to produce seed abundantly, although in the case of the *Agrostis* species and *Poa pratensis* probably only in favourable years. With the exception of *Stellaria media*, they have very light seeds (less than 0.3 mg), which may be easily transported by wind. The seeds are able to form persistent seedbanks (Table 59.1). It seems significant that all naturalized species which are able to spread by vegetative means are able to become dominant in the vegetation. Of the three species which do not spread laterally by vegetative means only one (*Poa annua*) may attain dominance. The latter species dominates locally in biotically influenced habitats, where it forms dense swards of perennial individuals. This perennial form has some capacity for vegetative lateral spread (Grime 1979, p. 107).

From the above information, predictions can be made on the outcome of new introductions on the island. Newly introduced species which are able to produce viable seed can be expected to become widespread. Species which do not produce seed will remain localized, but may persist by their ability to spread vegetatively. Species which do not produce seed, nor have the potential for vegetative spread, are expected to disappear again. An increase in temperature at Marion Island (Smith 1992) may allow some of the species presently precluded by low temperatures from producing viable seed to reproduce generatively, and thus to become widespread on the island.

MANAGEMENT IMPLICATIONS

Invasion by aggressive alien vascular plants poses an unnatural impact on the native flora and vegetation of Marion Island by imposing a new level of competition which the latter may be unable to tolerate. As a management response to such aliens it is desirable to remove these species. There is little experience with removal of introduced plants from Southern Ocean islands (Watkins & Cooper 1986). It should be possible to remove the localized aliens, but at present these do not pose a serious threat to the native ecosystems. They dominate the vegetation at the sites where they have become established, but their impact is restricted to a minute part of the island, and under present environmental conditions the risk of them becoming more widely spread is small. A further amelioration of the island's climate, however, may possibly allow these species to produce viable seeds, and to disperse into other parts of the island.

The removal of widespread species is impossible without severe and possibly disastrous damage to the native ecosystems of the island. Therefore, it is recommended that new arrivals should be eradicated as soon as they are discovered, as was done with *Plantago lanceolata, Senecio* sp. and *Sonchus* sp. in the 1980s (Bergstrom & Smith 1990). This necessitates a regular botanical survey of the area of the Meteorological Station, other areas where personnel or equipment are landed, and along regularly used pedestrian routes (cf. Wace 1986). This survey should include the monitoring of the distribution and phenology of the naturalized aliens with a restricted distribution. When these species appear to be spreading or producing seed, eradication should be seriously considered.

The best strategy, obviously, is to prevent propagules from reaching the island. Despite measures to avoid introduction of organisms new to the island several introductions have taken place during the past decade. It is therefore important to strictly adhere to the quarantine measures that are presently in force, and to ensure that *all* personnel involved in the island, those actually visiting the island as well as those responsible for the procurement and transport of goods to the island, are aware of and appreciate the reason for these measures. This is even more important with respect to neighbouring Prince Edward Island, where at present only two alien vascular plants occur, *Poa annua* and *Cerastium fontanum* (Bergstrom & Smith 1990), which do not seriously affect the native vegetation.

ACKNOWLEDGEMENTS

I am grateful to Prof V. R. Smith, University of Orange Free State, for his invitation to visit the island, for his help during the fieldwork, and for constructive discussions and comments. Logistic facilities provided by the South African Department of Environment Affairs are gratefully acknowledged. This work was made possible by financial assistance from the Netherlands Antarctic Research Programme. Thanks are due to Dr R. van der Meijden, Rijksherbarium, Leiden, for his confirmation of the identification of *Sagina procumbens*, and to Dr R. I. L. Smith, Dr A. H. L. Huiskes and Prof V. R. Smith, whose comments on the manuscript were greatly appreciated. Publication no. 2003 of the Netherlands Institute of Ecology – Centre for Estuarine and Coastal Ecology, Vierstraat 28, 4401 EA Yerseke, The Netherlands.

REFERENCES

Bergstrom, D. M. & Smith, V. R. 1990. Alien vascular flora of Marion and Prince Edward Islands: new species, present distribution and status. *Antarctic Science*, **2,** 301–308.

Clark, M. R. & Dingwall, P. R. 1985. *Conservation of islands in the Southern Ocean: a review of the protected areas of Insulantarctica.* Gland: IUCN, 188 pp.

Cooper, J. & Avery, G., eds. 1986. Historical sites at the Prince Edward Islands. *South African National Scientific Programmes Report*, No. 128, 82 pp.

Cooper, J. & Condy, P. R. 1988. Environmental conservation at the sub-Antarctic Prince Edward Islands: a review and recommendations. *Environmental Conservation*, **15,** 317–326.

Gremmen, N. J. M. 1975. The distribution of alien vascular plants on Marion and Prince Edward Islands. *South African Journal of Antarctic Research*, **5,** 25–30.

Gremmen, N. J. M. 1981. *The vegetation of the subantarctic islands Marion and Prince Edward.* The Hague: Junk Publishers, 145 pp.

Gremmen, N. J. M. 1982. Alien vascular plants on Marion Island (subantarctic). *CNFRA*, **51,** 315–323.

Gremmen, N. J. M. & Smith, V. R. 1981. *Agrostis stolonifera* L. on Marion Island (subantarctic). *South African Journal of Antarctic Research*, **10/11,** 33–34.

Grime, J. P. 1979. *Plant strategies and vegetation processes.* Chichester: Wiley, 222 pp.

Grime, J .P., Hodgson, J. G. & Hunt, R. 1988. *Comparative plant ecology. A functional approach to common British species.* London: Unwin Hyman, 742 pp.

Holdgate, M. W. 1986. Summary and conclusions: characteristics and consequences of biological invasions. *Philosophical Transactions of the Royal Society of London*, **B314,** 733–742.

Huntley, B. J. 1971. Vegetation. In Van Zinderen Bakker, E. M., Winterbottom, J. M. & Dyer, R. A., eds. *Marion and Prince Edward Islands.* Cape Town: Balkema, 98–160.

Moseley, H. N. 1874. On the botany of Marion Island, Kerguelen's Land and Young Island of the Heard group. *Journal of the Linnean Society (Botany)*, **14,** 387–388.

Oliver, D. 1874. List of plants collected by H. N. Moseley, M A., on Kerguelen's Land, Marion Island and Young Island. *Journal of the Linnean Society (Botany)*, **14,** 389–390.

Schulze, B. R. 1971. The climate of Marion Island. In Van Zinderen Bakker, E. M., Winterbottom, J. M. & Dyer, R. A., eds. *Marion and Prince Edward Islands.* Cape Town: Balkema, 16–31.

Smith, V. R. 1987. The environment and biota of Marion Island. *South African Journal of Science*, **83,** 211–220.

Smith, V. R. 1992. Surface air temperatures at Marion Island, sub-Antarctic. *South African Journal of Science*, **88,** 575578.

Smith, V. R. & Smith, R. I. L. 1987. The biota and conservation status of sub-Antarctic islands. *Environment International*, **13,** 95–104.

Van Zinderen Bakker, E. M. 1971. Introduction. In Van Zinderen Bakker, E. M., Winterbottom, J. M. & Dyer, R. A., eds. *Marion and Prince Edward Islands.* Cape Town: Balkema, 1–15.

Van Zinderen Bakker, E. M., Winterbottom, J. M. & Dyer, R. A., eds. 1971. *Marion and Prince Edward Islands.* Cape Town: Balkema, 427 pp.

Wace, N. 1986. The arrival, establishment and control of alien plants on Gough Island. *South African Journal of Antarctic Research*, **16,** 95–101.

Watkins, B. P. & Cooper, J. 1986. Introduction, present status and control of alien species at the Prince Edward islands, sub-Antarctic. *South African Journal of Antarctic Research*, **16,** 86–94.

Westhoff, V. & van der Maarel, E. 1973. The Braun-Blanquet approach. In Whittaker, R. H., ed. *Ordination and classification of communities. Handbook of vegetation science 5.* The Hague: Junk Publishers, 617–726.

60 Potential effects of two alien insects on a sub-Antarctic wingless fly in the Kerguelen islands

M. CHEVRIER, P. VERNON AND Y. FRENOT

Université de Rennes I, UMR 6553, Station Biologique, F-35380 Paimpont, France

ABSTRACT

On sub-Antarctic islands, human settlements have increased the number of introduced species. Although the climatic conditions might be considered to be harsh, the characteristics of the native community facilitate the success of invading species. Several introductions are accidental, especially those of the invertebrate fauna. In Îles Kerguelen, Oopterus soledadinus *(Coleoptera, Carabidae) and* Calliphora vicina *(Diptera, Calliphoridae) were first observed in 1939 and 1978 respectively. Flightless* O. soledadinus *is a cold temperate predator species while* C. vicina *is a worldwide necrophagous blowfly. Both species interact with native invertebrates and particularly with a wingless fly* Anatalanta aptera *(Diptera, Sphaeroceridae). The introduced species present spatial and temporal distributions related to their own life history parameters and have distinct impacts on the native populations of* A. aptera. O. soledadinus *directly threatens the survival of populations in the littoral zones. The presence of* C. vicina *could lead to intraspecific competition between* A. aptera *larvae for resource use. The co-occurrence of both introduced species could increase their impact on* A. aptera. *Since the species richness of the native fauna on sub-Antarctic islands is poor, the role played by even a few introduced species is likely to be important.*

Key words: Sub-Antarctic, insects, alien species, biogeography, predation, competition.

INTRODUCTION

Biological introductions or invasions are a frequent process and constitute an important research area for the understanding of community structure. Not all introductions are necessarily associated with drastic changes in species composition of native communities (Niemelä & Spence 1991). The ecological impacts of exotic species may be considered in two main categories: direct interactions with native populations and habitat modifications. The interactions between alien and native species include competition, diet, niche overlap, predation, parasitism, disease and facilitation. Besides these phenomena, the complexity and stability of the ecosystem, the role of keystone species and the influence of history on community structure are of great importance for the study of community assembly (Vitousek *et al.* 1987, Niemelä & Spence 1991, Doube *et al.* 1991, Atkinson & Cameron 1993, Lodge 1993). When considering the effects of introductions there are two principal factors: the biological characteristics and colonizing abilities of the introduced species, and the characteristics of the target community (Pimm 1987, Niemela & Spence 1991, Lodge 1993). Indeed, Pimm (1987) characterizes the main rules for a successful introduction as: 1. introduce species into places where predators or competitors are few or absent; 2. introduce highly polyphagous species, and 3. introduce species into relatively simple communities.

Thus, it can be seen that oceanic islands, such as sub-Antarctic ones, are likely to be very sensitive to species introductions. The native communities of sub-Antarctic islands are species poor and very simply structured (Chapuis *et al.* 1991). The native terrestrial ecosystems are devoid of carnivorous and

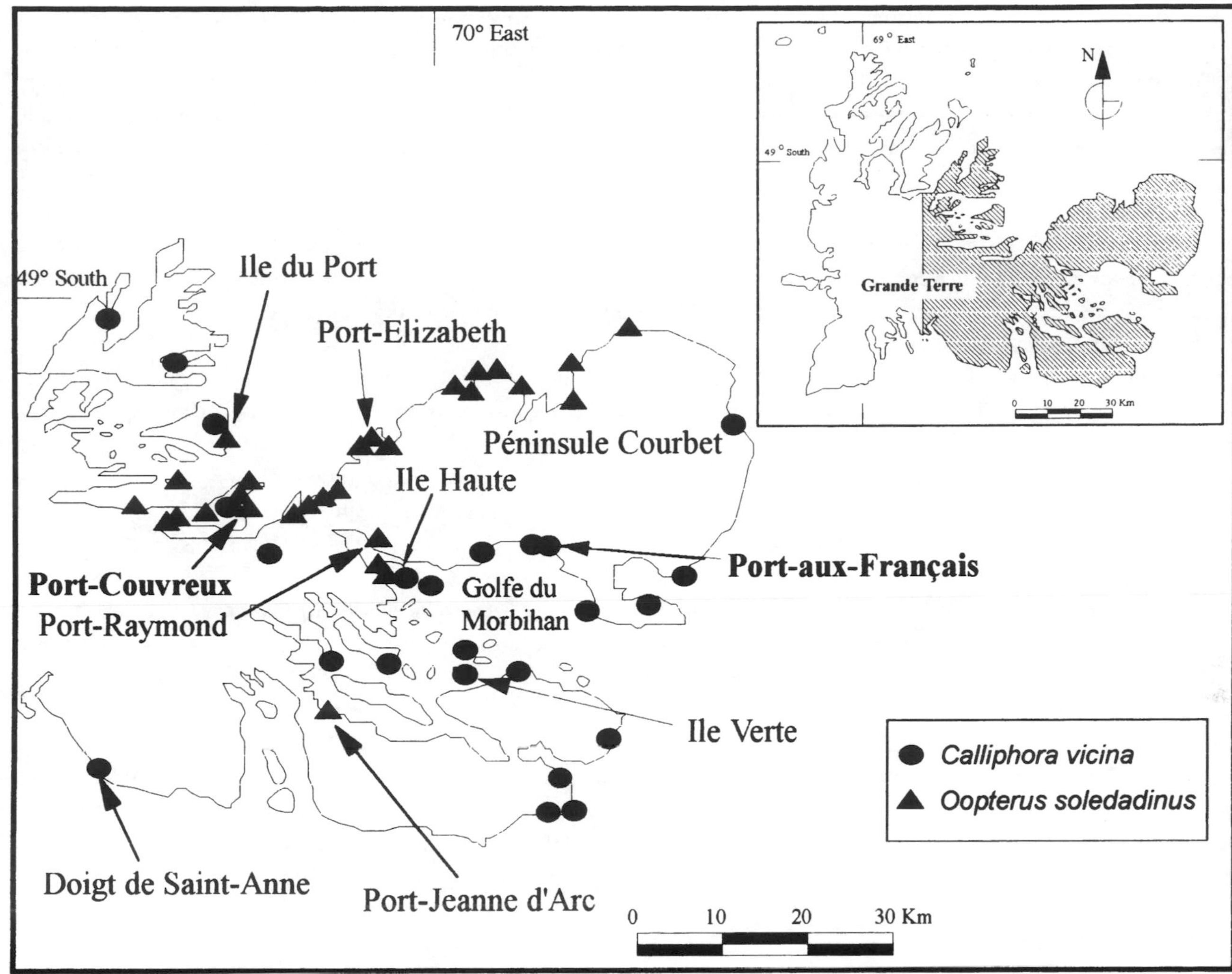

Fig. 60.1. Location of the study areas and spatial distribution of introduced species in the east part of the Kerguelen archipelago.

herbivorous mammals, and consist mainly of producer and decomposer oriented communities. With few native taxa there is ample opportunity for successful colonization by alien species (Block 1984, Crafford 1986, Crafford & Chown 1990, Chapuis *et al.* 1994). The risk of accidental establishment of exotic species is reduced by the remoteness, harsh climate and environmental constraints of these islands and most successful establishments are attributable to human influence (settlements, transport).

In Îles Kerguelen several introductions have occurred since the commencement of human activity in the archipelago, in the nineteenth century. Some of them were intentional (herbivorous mammals) and others were accidental (arthropod introductions) (Clark & Dingwall 1985, Chapuis *et al.* 1994). The under-utilization of some niches could have facilitated the establishment of alien species, depending on their colonizing abilities.

Calliphora vicina Robineau-Desvoidy (Diptera, Calliphoridae), a necrophagous blowfly, and *Oopterus soledadinus* Guérin (Coleoptera, Trechidae), a predatory carabid, could have taken advantage of the presence of available niches not fully utilized by invertebrate native species. These two recently introduced species interact with native invertebrates and particularly with the flightless fly *Anatalanta aptera* Eaton (Diptera, Sphaeroceridae).

The necrophagous *Calliphora vicina* is cosmopolitan. It occurs from tropical to Arctic environments (Nuorteva 1967, Shewell 1987). This blowfly was first recorded in Îles Kerguelen in 1978 by L. Davies at Port-aux-Français (Fig. 60.1), and is found in almost all biotopes: littoral, herbfield, seabird colonies, fell-field. Adults show a diurnal activity.

The distribution of *Oopterus soledadinus* is restricted to the southern cold temperate zone (Johns 1974). Its populations are found in Patagonia (Darlington 1970), on the Falkland Islands, South Georgia and Îles Kerguelen. On the last islands (Fig. 60.1) the beetle was observed for the first time in February 1939 at Port-Couvreux (Jeannel 1940). This species is likely to have been introduced with sheep from the Falkland Islands, where it is common. Sheep rearing occurred at Port-Couvreux during the early part of this century and stopped in 1932. Jeannel concluded that *O. soledadinus* was already present at Kerguelen in 1932. It is active during the night and may be found during day time particularly beneath pieces of stones or kelp belts (Ottesen 1990).

The sub-Antarctic flightless fly *Anatalanta aptera* occurs in

Crozet, Heard and Kerguelen Islands (Séguy 1940). On Kerguelen, it has been observed throughout the whole archipelago from sea level up to 600 m asl and in almost all habitats. Larvae and adults are saprophagous, the flies feeding on various types of decayed organic matter, and are especially abundant at littoral sites (Vernon 1981).

A. aptera is active all year long (mainly at night) even in sites where food supply is scarce and generally seasonal. Laboratory experiments showed that adult fasting performances are very high (up to five months at 5 °C) and that adults have also a long mean life-span of 265 days at 5 °C (Tréhen 1982, Vernon 1986a, b). The winter and summer fecundity is equivalent (60 eggs for an oviposition cycle) but the winter egg is larger than the summer one (Chevrier 1992).

This native fly is saprophagous and it is very often found on carrion, potentially competing with the necrophagous alien blowfly for resource utilization at the larval stage. *A. aptera* itself also constitutes potential prey for the carabid *O. soledadinus*. The native species could therefore be threatened by both recently introduced species. An interesting point could also be noted: *O. soledadinus* is a cold southern temperate species while *C. vicina* is a true alien species. This paper presents the spatial and temporal distribution of the introduced species followed by a preliminary review of the interaction type and potential impact of both introduced species on *A. aptera* populations.

MATERIALS AND METHODS

Spatial distribution of introduced species

Data available from the general invertebrate surveys on Kerguelen in 1983, 1991, 1992 and 1994 are used in this study to describe the spatial distribution of the introduced species. During field work, various collecting methods were used : pitfall-traps (circular box: 9 cm diameter and 4 cm high), yellow traps (20×20 cm, 5 cm high), soil samples (20×20 cm, 5 cm high) and visual observations. In addition, five sites were chosen for field experiments in the eastern part of the archipelago : Port-aux-Français (scientific base), Ile Verte and Ile Haute for the interaction between flies, Ile Haute, Port-Couvreux and Port-Elizabeth for predation study (Fig. 60.1).

Temporal distribution of the three species

The seasonal activity of *C. vicina* and *A. aptera* adults was studied in 1992 on Ile Verte (Golfe du Morbihan, Fig. 60.1) in five sites with no or little organic matter: upper beach, herbfields with various vegetation cover and fell-field (altitude range of 2 to 30 m asl). One yellow trap and three pitfall traps were used for five days each month at each site. The yellow trap was expected to attract flying insects (*C. vicina*) and the pitfall trap to catch the flightless ones (*A. aptera*).

The temporal distribution of *O. soledadinus* was assessed in field observations during 1991, 1992, 1993 and summer 1994.

The flying activity of *C. vicina* was studied using yellow bait traps in two sites at Port-aux- Français in 1993 and 1994 (one yellow bait trap per site). The first site was near the accommodation buildings, the other beside the scientific buildings. Traps were opened twice each month for five days each. The bait, which consisted of pieces of meat placed in a box inside the yellow trap, was replaced at the beginning of each catching period. Maximum temperature, mean and maximum wind speed, mean relative humidity, rainfall, sunshine duration and nebulosity were recorded daily during the experiment (Meteo France data, Port-aux-Français station).

The attractiveness of the bait to *C. vicina* adults was studied from the same experiment by assuming that the decomposition of carrion was linear over the five days of carrion exposure: each day characterized a decay state. For each decay state the number of *C. vicina* adults was recorded. The results were compared with a null hypothesis corresponding to an equivalent number of catchings of *C. vicina* for each five decay states.

Interaction between introduced insects and autochtonous fly

An interaction experiment between *A. aptera* and *C. vicina* was conducted on the field at Ile Haute (Fig. 60.1) during summer 1993/94. The experiment was performed in containers 13 cm high and 20 cm in diameter. Gauze formed the centre of the lid (mesh 0.315 mm) and of the bottom (mesh 0.5 mm) for ventilation. The containers were filled with a layer of mineral soil 5 cm thick. Half of a dead rabbit was placed on the soil surface in the cage. The containers were placed close to a mouflon carcass on a littoral site (2 m from kelp zone) which is protected from west winds. One of the containers was opened for two days and the other for five days. This experiment began on 12 February 1994 and ended on 8 March 1994. The larvae of each species in the litter and in the carrion were counted.

The impact of the introduced carabid on the *A. aptera* populations was studied by comparing the field observations conducted by Jeannel in 1939 with some recent observations from 1983 and 1994.

The catchability of *A. aptera* eggs and first instar larvae by *O. soledadinus* adults was observed during field investigations.

RESULTS

Spatial distribution of introduced species

Since it was first recorded in 1978, *C. vicina* has spread out in all directions and colonized sites up to a distance of 66 km from the station (Doigt de Sainte Anne in 1983) and up to an altitude of 600 m asl. *C. vicina* has also colonized several islands of Golfe du Morbihan (Fig. 60.1). These islands are inhabited by numerous seabirds and several by introduced vertebrate species (rabbits, mice, sheep, mouflons). Carcasses of these vertebrates constitute a significant resource. Prevailing westerly winds do not seem to have stopped the alien fly from colonizing the western part of the archipelago. Nevertheless, during invertebrate surveys *C. vicina* adults were observed more often than larvae. For sites where larvae were recorded, it can be said

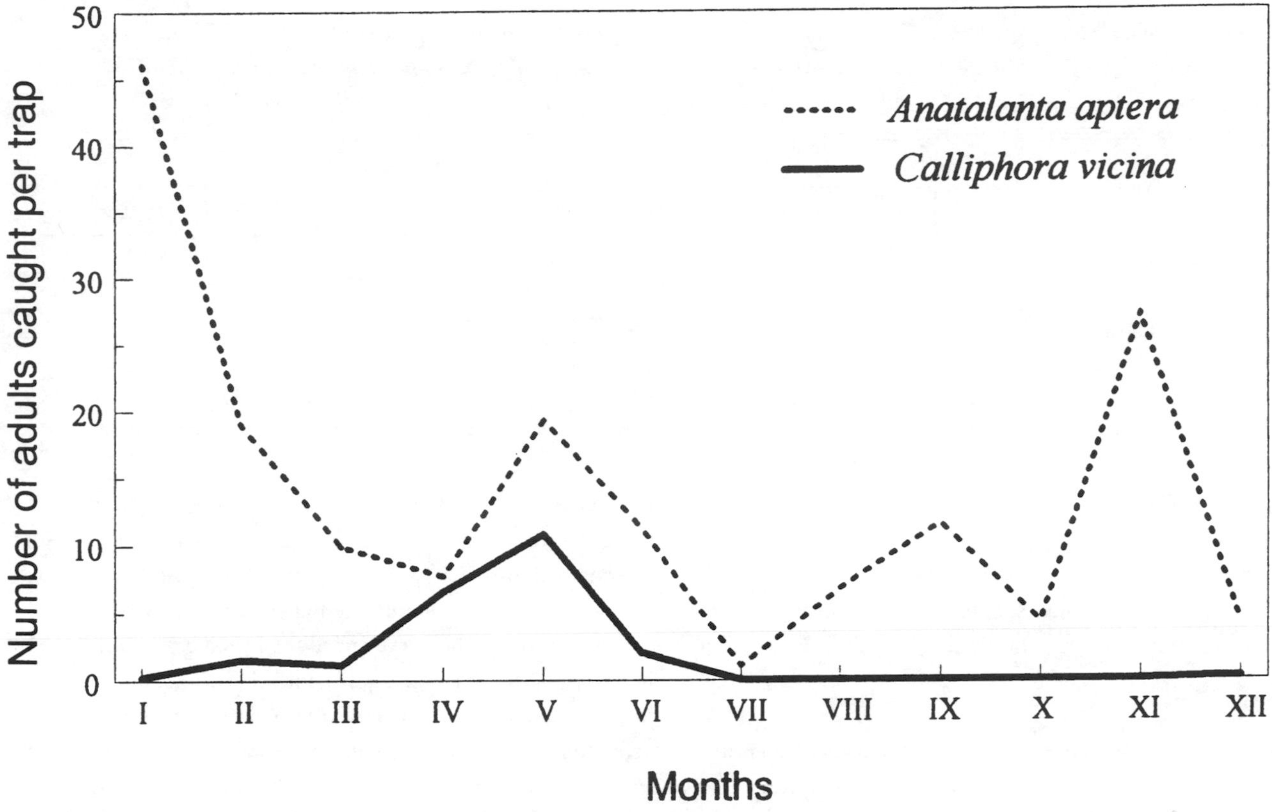

Fig 60.2. Temporal distribution of *A. aptera* and *C. vicina* at Ile Verte during 1992.

that there is a true population of *C. vicina* from which next generation of adults could disperse to other sites.

From its introduction on the site of Port-Couvreux, *O. soledadinus* has now colonized a significant part of Grande Terre, mainly the north coast of the Péninsule Courbet (Fig. 60.1) (Tréhen & Voisin 1984, Dreux *et al.* 1992). Its habitats are essentially coastal, frequently in herbfield. No *O. soledadinus* have been found above 110 m asl.

In the past few years *O. soledadinus* has colonized Port-Raymond, Ile Haute and Port- Jeanne d'Arc. How it dispersed to those sites is not yet known. On Ile Haute (8.5 km^2), it was first recorded in October 1992 at the north west tip of the island beneath mouflon carcasses. Individuals were observed in March 1994 on coasts of the island at 4 km (south coast) and 4.4 km (north coast) from the first record site. From the proximity between the northwest tip of the island and Grande Terre, it seems reasonable to assume that *O. soledadinus* invaded Ile Haute via the northwest tip. On this island, its spread could be roughly estimated at 3 km per year (it has covered 4 km in 16 months from October 1992 to February 1994). This rapid colonization could be related to the considerable numbers of dead mouflon whose carcasses provide food for several invertebrate species which are potential prey for *O. soledadinus*. Moreover, at those sites which are far from its original point of introduction, larvae populations were found, making these potential new dispersal sites.

Temporal distribution of the three species

The native fly is active throughout the year and presents peaks of abundance at the beginning of spring and during summer (Fig. 60.2). Conversely, *C. vicina* shows no winter activity, with its development pausing at different stages, as in temperate environments. In the case of adults, this phenomenon is probably related to a quiescence process. Few adults can be observed during winter in shelters or in yellow traps. *C. vicina* adults show a well-marked period of activity with a peak of abundance at the end of summer.

Active adults of *O. soledadinus* have been observed in all seasons during the surveys of 1991, 1992 and 1993. Also, some adults were hand caught in May, August, October and November. Soil samples taken during February and March 1994 at several sites (Ile Haute, Port-Couvreux and Port-Elizabeth) indicated that all development stages were present : larvae, pupae, teneral and mature adults. Moreover, one teneral male was observed in winter period just after emerging (August 1992). Thus, *O. soledadinus* does not seem to have as well marked a period of activity during the year as *A. aptera*.

The flying activity of *C. vicina* is directly related to daily climatic conditions. Using the data corresponding to the activity period of *C. vicina* (January to May 1993, November 93, December 1993, January and February 1994) Spearman's rank correlation coefficients were calculated between each climatic factor and number of adults of *C. vicina* caught each day (n=223 adults for 82 catching days). Of the seven parameters, only four were correlated with the flying activity of *C. vicina* : sunshine duration (r=0.393, p<0.01), nebulosity (r=−0.374, p<0.01), mean wind speed (r=−0.346, p<0.01) and maximum temperature (r=0.301, p<0.01).

The bait attractiveness to *C. vicina* was studied for the same

Table 60.1. *Distribution of* A. aptera *and* C. vicina *in the soil litter and carrion in containers opened for two and for five days. Some* A. aptera *females were observed in the containers: two in the container opened for two days and eight in that opened for five days. One young larva of* O. soledadinus *was found in the container opened for two days and one adult in that opened for five days*

		Container opened 2 days			Container opened 5 days			
		Soil litter	Carrion	Total	Soil litter	Carrion	Total	Larvae total
Anatalanta aptera	Larvae III	189	173		2373	290		
	Larvae II		9	371	13	17	2693	3064
Calliphora vicina	Pupae	88	2		10			
	Larvae III	290	8	388	364	1	375	763
Total of flies, larvae and pupae				759			3068	

time period as for the flying activity study. Assuming that trapping had no impact on the population size, *C. vicina* adults were more likely to be attracted to fresh than older carrion ($\chi^2=17.87$, $df=4$, $p<0.01$): 77% of adults were attracted during the first three days of trap opening (among the 216 adults captured in total).

Interaction between introduced insects and the autochtonous fly

The preliminary results of rearing experiments conducted on Ile Haute are summarized in Table 60.1. Second and third instar *A. aptera* larvae occurred in the containers after one month. At the same time, third instar larvae and pupae of *C. vicina* were observed. Otherwise, the oviposition of *A. aptera* occurred only during the open days for the containers even if some females were caught in the cages.

Considering the larvae of both species, 759 and 3068 larvae were observed in containers opened for two and five days respectively. This difference could be explained by the number of days open. The longer the bait is accessible to females of *A. aptera*, the more they deposit eggs (Table 60.1). In contrast, the number of *C. vicina* larvae remained almost constant (388 in two days versus 375 in five days). Two hypotheses could explain *C. vicina* colonization of carrion: either (i) the females of *C. vicina* are attracted only by fresh carrion, or (ii) the climatic conditions during the five open days favoured *C. vicina* oviposition for the first two opening days.

In total, 3064 *A. aptera* larvae and 763 *C. vicina* larvae were recorded (Table 60.1). A female of *A. aptera* could deposit 60 eggs during an oviposition cycle while a *C. vicina* may deposit 100–300 eggs in temperate areas (Wardle 1930). It seems in the present case that *A. aptera* females are more favoured at this site and *C. vicina* females were either very few or less fecund than the temperate ones.

When the containers were examined, *C. vicina* larvae were mostly at the end of their development and were found in the soil litter, while *A. aptera* maggots were early third larval instars. The distribution of *C. vicina* maggots was similar in both containers and *A. aptera* maggots showed no substrate preference in the two days container, but preferred the soil litter in the five days container (Table 60.1).

C. vicina has a short development time compared to *A. aptera*: less than two months at Kerguelen (Chapuis *et al.* 1991). After one month, *C. vicina* larvae are at third instars and pupal stage, and have migrated into the soil beneath the carrion to seek a place to pupate. They have ended their food consumption and they store up reserves. As equivalent proportions of larvae are observed in the soil litter and in carrion for the two cages, it seems that they have not really been influenced by the larval density in the containers. The mean weight of migrating larvae of *C. vicina* is 67.5 mg (Putman 1977), so the equivalent biomass of *C. vicina* prepupae is about 20 g (for a total larval biomass of 23 g) and 25 g (for a total larval biomass of 45 g) for the two days and five days containers respectively. Thus, since *C. vicina* larvae appears to represent the major part of the larval biomass in the two containers, they may have used the greater part of the potential food.

A. aptera has a slower development – more than two months. After one month, third instar larvae have been observed, with a mean weight of 7.8 mg (Chapuis *et al.* 1991). The distribution of larvae in the two cages was very different: 52% of larvae III in the litter and 48% in the carrion for the two days container and 89% in the litter and 11% in the carrion for the 5 days container. This could be explained by the amount of suitable resource available for the larval density observed. In this case, the advanced state of decomposition of the carrion in the 5 days container allows a great density of larvae III to migrate into the soil beneath the carrion before they consumed all their necessary resources. However, *A. aptera* is saprophagous, as are other Sphaeroceridae. Its larvae may be able to use nutrients leaching from the carcass into the soil.

The impact of the predaceous *O. soledadinus* on the *A. aptera* populations could be described initially by comparing the distribution of both species at different dates. In 1939, *O. soledadinus* was present at Port-Couvreux (Jeannel 1940). Its populations then colonized two other sites surveyed in 1939 : Port-Elizabeth and Ile du Port (Table 60.2). *A. aptera* populations had disappeared at Port-Couvreux and Port-Elizabeth by 1983 at the latest. The population at Ile du Port was still extant in 1991.

Table 60.2. *Changes in presence (+)/absence (−) of* Anatalanta aptera *and* Oopterus soledadinus *at three stations on Kerguelen since the introduction of* Oopterus soledadinus

		Anatalanta aptera	*Oopterus soledadinus*
Port-Couvreux	1939	+	+
	1983	−	+
	1991	−	+
	1994	−	+
Port-Elizabeth	1939	+	−
	1983	−	+
	1994	−	+
Ile du Port	1939	+	−
	1991	+	+

DISCUSSION

The two alien species *O. soledadinus* and *C. vicina*, observed for the first time at settlements on Kerguelen in 1939 and 1978 respectively, are now permanently established. The current distribution of these species is not completely known because of incomplete field surveys of the western part of the archipelago. Fig. 60.1 shows that spatial distribution differs between the species, due to specific differences in dispersal abilities.

On South Georgia *O. soledadinus* is restricted in its distribution to the immediate vicinity of the whaling station at Husvik (Ernsting 1993) and the area around Grytviken, where it is found from sea level to 150 m asl, as at Kerguelen (Darlington 1970). Although this species is present in similar habitats on the two island groups, it has a wider distribution on Kerguelen. Moreover, at South Georgia the presence of another predatory carabid, *Trechisibus antarcticus*, which is abundant in many types of vegetation, may probably hinder the dispersal of *O. soledadinus* (Ernsting 1993).

The climatic conditions of the two islands are also different. South Georgia has a harsher climate than Kerguelen (annual mean temperatures of 1.8 °C and 4.5 °C at South Georgia and Kerguelen, respectively). The duration of winter is longer on South Georgia (Block & Somme 1983), which generally has permanent snow cover from May to October. On the Kerguelen coastal area, the mean number of days with snow cover is three for July, the coldest month (mean data for 1951–1990). The climatic conditions in the Kerguelen archipelago could therefore be considered more favourable than on South Georgia, accelerating the activity, the development cycle and the dispersal of *O. soledadinus*. Moreover, from observations of the ovaries, it appears that this species may have one breeding season, occurring during the summer, on South Georgia (Ernsting 1993). The overlap in generations observed at Kerguelen in February 1939 (Jeannel 1940) and in February and March 1994 could suggest that on this archipelago the breeding activity of *O. soledadinus* is not restricted to summer only or that its development cycle is faster.

There are several factors to be considered in assessing the degree of interaction between introduced and native species. The *season* plays an important role in the activity of all species. The native species, although active during winter, shows a slower development cycle and activity than in summer. *O. soledadinus* is supposed to be active during winter but *C. vicina* is, with few exceptions, inactive from June to October. Most interactions are therefore likely to occur during the summer, especially if the adults of *C. vicina* colonize carrion mainly under the influence of the sunshine, as found in sub-Arctic conditions (Nuorteva 1965). The *locality* specifies the kind of interaction. The three species involved potentially co-occur on vegetated coast, while at higher altitudes (fell-field) only *A. aptera* and *C. vicina* have been observed.

The *resource quality* corresponds to the decomposition state of carrion which could attract the flies and, consequently, the carabid. A study where rabbit carrion was placed under cages at Kerguelen in 1993 showed that *A. aptera* deposited eggs up to two months after the beginning of the experiment. However, because of its nocturnal activity, it could not colonize carrion during day time. *C. vicina* is attracted by carrion during the first three days of decomposition. The daily climatic conditions, especially sunshine duration, also significantly influence the flying activity of *C. vicina*. However, in a temperate environment, no oviposition has been observed after 48 h of decay of mice in the field (Putman 1977). Thus, it seems that the 'attractiveness' of carrion is the determinant factor for the colonization of a resource.

The *resource quantity* and distribution could be also an important determinant factor for interactions. Whereas carrion are patchily distributed on fell-field, carcasses can be abundant in the littoral zones or on small islands, where most vertebrates occur. A single carcass can support a limited density of consumers. Putman (1977) shows that the number of eggs of *C. vicina* was not correlated to carcass weight but could be limited by the restriction of suitable sites for oviposition. At Kerguelen, consumption of carrion is often attributed to scavenging birds such as skuas (*Stercorarius skua lönnbergi*) and giant petrels (*Macronestes giganteus* and *M. halli*), thereby reducing the available resource for invertebrates.

The interaction between *A. aptera* and *C. vicina* might be at the larval diet stage. From the preliminary results, it does not seem that there is a direct interaction between the two species which may lead to exclusion. However, it might be hypothesized that the presence of *C. vicina* larvae in carrion could lead to intraspecific competition for *A. aptera* larvae when the suitable resource is less abundant, i.e. after its utilization by *C. vicina* larvae. The *A. aptera* larvae III would then have to exploit a secondary trophic niche corresponding to the soil litter just beneath the carcass.

The two introduced species may have a different impact on the *A. aptera* populations. Its dispersal abilities mean that *O. soledadinus* colonizes sites slower but more permanently than *C. vicina*. Instead, *C. vicina* population establishment depends mainly on the trophic resource distribution. When an *O. soledadinus* population is established in a site, it has the potential to become one of the dominant species. This is probably because this species has a cold temperate origin and belongs to a high trophic level which is not well represented on Kerguelen. *A. aptera*, one of the carabid's potential prey, has few native predators (apart from the spider *Myro kerguelensis*) and has disappeared from at least two sites some decades after their colonization by *O. soledadinus*. When the three species are present (e.g. Ile Haute), the occurrence of *C. vicina* larvae could lead to a greater density of *A. aptera* third instars in the soil litter beneath carrion. These *A. aptera* larvae III could be then more vulnerable to the impact of *O. soledadinus*. Indeed, preliminary predation experiments in the laboratory suggest that *O. soledadinus* adults eat all stages of *A. aptera* from eggs to third instar larvae.

A more extensive knowledge of life history characteristics of introduced species in the sub- Antarctic is of paramount importance to assess their current and future impact on native invertebrates. It is also necessary to take into account the climatic changes observed on Kerguelen since the 1950s (Frenot *et al.*, 1997) as these newly introduced species could be favoured in their colonization strategies by climatic amelioration.

The species richness of the native fauna in the sub-Antarctic islands is poor and the role played by introduced species is likely to increase. Species introductions do not necessarily cause the extinction of native ones. Scenarios which lead to an increase of species diversity cannot be excluded. These studies are especially important in the context of global concerns about conservation issues on remote oceanic islands.

ACKNOWLEDGEMENTS

This study was supported by IFRTP (Programmes 136 and 238), and CNRS (GDRE 1069 and URA 1853). We are indebted to T. Coulee, F. Foesser and G. Picot for their assistance in the field, to P. Trehen and Y. Delettre for comments, and to G. Ernsting and an anonymous referee for valuable criticism.

REFERENCES

Atkinson, A. E. & Cameron, E. K. 1993. Human influence on the terrestrial biota and biotic communities of New Zealand. *Trends in Ecology and Evolution*, **8**, 447–451.

Block, W. 1984. Terrestrial microbiology, invertebrate and ecosystems. In Laws R. M., ed. *Antarctic ecology*, Vol. 1. London: Academic Press, 163–236.

Block, W. & Somme, L. 1983. Low temperature adaptations in beetles from the sub-Antarctic island of South Georgia. *Polar Biology*, **2**, 109–114.

Chapuis, J. L., Bousses, P. & Barnaud, G. 1994. Alien mammals, impact and man management in the French sub-Antarctic islands. *Biological Conservation*, **67**, 97–104.

Chapuis, J. L., Vernon, P. & Frenot, Y. 1991. Fragilité des peuplements insulaires: exemple des îles Kerguelen, archipel subantarctique. In *Réactions des êtres vivants aux changements de l'environnement*. Paris: PIREN, CNRS, 235–248.

Chevrier, M. 1992. *Croissance et reproduction chez un Diptère subantarctique: interaction thermique et trophique. Rapport D.E.A.* Université de Rennes I, 1–28. [Unpublished]

Clark, M. R. & Dingwall, P. R. 1985. *Conservation of islands in the Southern Ocean: a review of the protected areas of Insulantarctica.* Gland: IUCN, 188 pp.

Crafford, J. E. 1986. A case study of an alien invertebrate (*Limnophyes pusillus*, Diptera, Chironomidae) introduced on Marion Island: selective advantages. *South African Journal of Antarctic Research*, **16**, 115–117.

Crafford, J. E. & Chown, S. L. 1990. The introduction and establishment of the diamondback moth (*Plutella xylostella* L., Plutellidae) on Marion Island. In Kerry, K. R. & Hempel, G., eds. *Antarctic ecosystems: ecological change and conservation*. Berlin & Heidelberg: Springer- Verlag, 354–358.

Darlington, P. J. 1970. Coleoptera: Carabidae of South Georgia. *Pacific Insects Monograph*, **23**, 234.

Doube, B. M., Macqueen, A., Ridsdill-Smith, T. J. & Weir T. A. 1991. Native and introduced dung beetles in Australia. In Hanski, I. & Cambefort, Y., eds. *Dung beetle ecology*. Princeton, NJ: Princetown University Press, 255–278.

Dreux, P., Galiana, D. & Voisin, J. F. 1992. Acclimatation de *Merizodus soledadinus* Guérin dans l'archipel de Kerguelen (Coleoptera, Trechidae). *Bulletin de la Société Entomologique de France*, **97**, 219–221.

Ernsting, G. 1993. Observations on life cycle and feeding ecology of two recently introduced predatory beetle species at South Georgia, sub-Antarctic. *Polar Biology*, **13**, 423–428.

Frenot, Y., Gloaguen, J-C. & Tréhen, P. 1997. Climate change in Kerguelen Islands and colonization of recently deglaciated areas by *Poa kerguelensis* and *P. annua*. In Battaglia, B., Valencia, J. & Walton, D.W.H., eds. *Antarctic communities: species, structure and survival*. Cambridge: Cambridge University Press, 358–366.

Jeannel, R. 1940. Croisière du Bougainville aux îles Australes Françaises. III. Coléoptères. *Mémoires Museum National d'Histoire Naturelle, serie A*, **14**, 63–202.

Johns, P. M. 1974. Arthropoda of the Sub-Antarctic Islands of New Zealand. I. Coleoptera: Carabidae. Southern New Zealand, Patagonian and Falkland Islands insular Carabidae. *Journal of the Royal Society of New Zealand*, **4**, 283–302.

Lodge, D. M. 1993. Biological invasions: lessons for ecology. *Trends in Ecology and Evolution*, **8**, 133–137.

Niemelä, J. & Spence, J. R. 1991. Distribution and abundance of an exotic ground-beetle (Carabidae) – a test of community impact. *Oikos*, **62**, 351–359.

Nuorteva, P. 1965. The flying activities of blowflies (Dipt., Calliphoridae) in subarctic conditions. *Annales Entomologici Fennici*, **31**, 242–245.

Nuorteva, P. 1967. Observation on the blowflies (Dipt., Calliphoridae) from Sptisbergen. *Annales Entomologici Fennici*, **33**, 62–64.

Ottesen, P. S. 1990. Diel activity patterns of Carabidae, Staphylinidae and Perimylopidae (Coleoptera) at South Georgia, sub-Antarctic. *Polar Biology*, **10**, 515–519.

Pimm, S. L. 1987. Determining the effects of introduced species. *Trends in Ecology and Evolution*, **2**, 106–108.

Putman, R. J. 1977. Dynamics of the blowfly, *Calliphora erythrocephala*, within carrion. *Journal of Animal Ecology*, **46**, 853–866.

Séguy, E. 1940. Croisière du Bougainville aux îles Australes Françaises. IV. Diptères. *Mémoires Museum National d'Histoire Naturelle, serie A*, **14**, 203–268.

Shewell, G. E. 1987. Calliphoridae. In McAlpine , J. F., ed. *Manual of Nearctic Diptera*, Vol. 2. Research Branch Agriculture, Canada, 1133–1145.

Tréhen, P. 1982. Cycle du développement et stratégies de la reproduction chez quelques espèces de Diptères des îles subantarctiques. *CNFRA*, **51**, 149–156.

Tréhen, P. & Voisin, J. F. 1984. Sur la présence de *Merizodus soledadinus* Guerin à Kerguelen (Coléoptère Trechidae). *L'Entomologiste*, **40**, 53–54.

Vernon, P. 1981. *Peuplement diptèrologique des substrats enrichis en milieu insulaire subantarctique (Îles Crozet)*. Etude des

Sphaeroceridae du genre *Anatalanta*. Université de Rennes I, 1–115. [Unpublished]

Vernon, P. 1986a. Evolution des réserves lipidiques en fonction de l'état physiologique des adultes dans une population expérimentale d'un diptère subantarctique: *Anatalanta aptera* Eaton. *Bulletin de la Société d'Ecophysiologie*, **11**, 95–116.

Vernon, P. 1986b. Ecological aspects of adult starvation in a sub-Antarctic wingless Dipteran: *Anatalanta aptera* Eaton (Sphaeroceridae). In Darvas, B. & Papp, L., eds. *First International Congress of Dipterology*, Abstract Volume. Budapest, Hungary, 247 pp.

Vitousek, P. M., Loope, L. L. & Stone, C. P. 1987. Introduced species in Hawaii: biological effects and opportunities for ecological research. *Trends in Ecology and Evolution*, **2**, 224–227.

Wardle, R. A. 1930. Significant variables in the blowfly environment. *Annals of Applied Biology*, **17**, 554–574.

61 Impact of an increasing fur seal population on Antarctic plant communities: resilience and recovery

R. I. LEWIS SMITH

British Antarctic Survey, Natural Environment Research Council, High Cross, Madingley Road, Cambridge CB3 0ET, UK

ABSTRACT

The impact of an increasing summer population of non-breeding male fur seals on the vegetation of Signy Island, where they have caused widespread devastation in most lowland terrestrial and freshwater ecosystems, has been monitored since 1985. Fenced exclosures were erected in stands representative of six different cryptogamic communities in areas occupied by seals. Changes in plant species composition and cover abundance were recorded after four and seven years in permanent quadrats within each exclosure and in adjacent unprotected control plots. Chemical analyses of the soils were also undertaken. Seal impact has been severe at five of the six sites and in each instance physical disturbance and nutrient toxicity in the control plots have resulted in severe or complete destruction. Within the protected plots, all experienced an initial recovery by some moss species but most eventually deteriorated because of lateral leaching of toxic levels of nitrogenous or other compounds. Concurrent with the removal of bryophytes and lichens has been the rapid colonization of these perturbed areas by the nitrophilous alga Prasiola crispa. *Even if the complete removal of fur seals were possible, recovery of these sensitive communities to their former state of ecological equilibrium and complexity is expected to take many decades or even centuries.*

Key words: fur seals, monitoring, vegetation destruction, soils, resilience, recovery.

INTRODUCTION

Antarctic fur seals (*Arctocephalus gazella* Peters) have been increasing dramatically at their main breeding location, South Georgia, since around 1970 (Payne 1977, Boyd 1993). During the past two decades increasing numbers of non-breeding males from South Georgia haul out in summer on ice-free coastlines throughout the northern maritime Antarctic (Aguayo 1978, Furse 1986, Smith 1988, Vergani & Coria 1989, Bengtson *et al.* 1990, W. Fraser pers. commun.). On the west coast of the Antarctic Peninsula in recent years several thousands come ashore on Brabant Island, Melchior Islands and the Arthur Harbour area of southern Anvers Island, and up to 1000 in northern Marguerite Bay where there is also a small breeding population. There is no record of these animals formerly occurring in the Antarctic Peninsula region. In the South Orkney Islands the annual summer immigration amounts to 35 000–40 000 individuals, causing widespread devastation to large tracts of lowland vegetation up to 50 m above sea level and to 500 m inland. This problem has been monitored at Signy Island (Smith 1988, 1990) where peak numbers (February) have increased from only a few animals each year in the 1950s and 1960s to a few dozens in the early 1970s, but then to 2000–3000 between 1976 and 1983, 8000–17 000 between 1984 and 1993 and, most recently, 20 500 in 1994 and 1995.

The fragile Antarctic terrestrial ecosystem dominated by cryptogams is extremely sensitive to the physical impact of these gregarious and highly mobile animals, and is quickly destroyed if seal presence persists. On Signy Island, in recent years, some of the freshwater lakes have undergone considerable eutrophica-

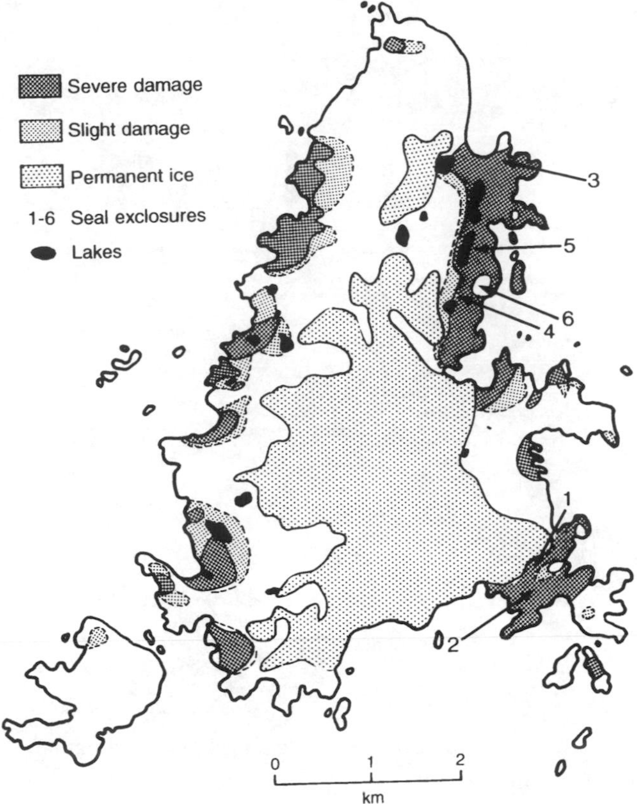

Fig. 61.1. Signy Island showing extent and degree of fur seal impact on vegetation in 1992, and location of monitoring sites (1–6).

tion, biological change and perturbation of their 5000–7000-year-old sediments. Several giant petrel (*Macronectes giganteus* Gmelin) and Antarctic tern (*Sterna vittata* Gmelin) colonies have been destroyed. Extensive bryophyte and lichen communities have been eradicated or have suffered some damage by disruption and removal of the living vegetation (Fig. 61.1). The active layer above the permafrost of several 3000–5000-year-old moss banks has been seriously eroded. Elsewhere, the plant and underlying organic accumulations are being either compressed (altering the hydrology of lake catchments), or eroded and blown into lakes or the sea. Seal excrement and 'fur-wash' causes elemental toxicity which quickly kills the vegetation. The influence of this major natural phenomenon has allowed a critical assessment of the capacity of individual plant species and of communities to tolerate such impact and to recover if the cause is removed.

METHODS

To follow changes in the vegetation and soil caused by fur seals, fenced exclosures were erected on the east side of Signy Island (60°43′ S, 45°38′ W) in February 1985 at six sites (see Fig. 61.1) representing different plant communities in various stages of degradation. Sites 1–4 and 6 were 25 m^2 and Site 5 was 50 m^2. Five 0.25 m^2 permanent quadrats were marked within each exclosure, and a further five in an adjacent unfenced control plot. In 1985, 1989 and 1992 the abundance of all living plant species in each quadrat was visually estimated as percentage cover and averaged for each site. In this way the change in status of each species was determined. Also, in 1985 and 1992, three samples of soil from 1–3 cm below the surface were collected within each exclosure and control plot for chemical analysis.

SITES

Site 1. Sloping closed deep moss turf dominated by *Chorisodontium aciphyllum* Hook. f. et Wils., *Polytrichum alpestre* Hoppe and several fruticose lichens (Tilbrook Hill, 50 m asl; five bryophyte, nine lichen species).

Site 2. Level closed deep moss turf dominated by *Chorisodontium aciphyllum* (Hillier Moss, 20 m asl; three bryophyte, two lichen species).

Site 3. Level open community dominated by *Ceratodon purpureus* (Hedw.) Brid., *Chorisodontium aciphyllum* and *Drepanocladus uncinatus* (Hedw.) Warnst. (Tern Cove, 18 m asl; seven bryophyte, four lichen species).

Site 4. Gently sloping open calcareous fellfield in an area of limestone outcrops dominated by *Schistidium antarctici* (Card.) Savicz. et Smirn., with several other calcicolous mosses, the liverwort *Marchantia berteroana* Lehm. & Lindenb. and the terricolous lichen *Catapyrenium lachneoides* O. Breuss also prominent (Marble Knolls, 30 m asl; 12 bryophyte, seven lichen species).

Site 5. Closed soligenous mire dominated by *Calliergon sarmentosum* (Wahlenb.) Kindb. and *Drepanocladus uncinatus* (Mirounga Flats, 15 m asl; five bryophyte, no lichen species).

The sixth site (Waterpipe Beach, 35 m asl; eight bryophyte, 15 lichen species), erected in acidic fellfield dominated by species of *Usnea* and *Andreaea* has, so far, not been invaded by fur seals and has undergone no change. It is not included here.

RESULTS

Changes in vegetation

The change in abundance of the principal species in the five sites is illustrated in Fig. 61.2.

When selected, Site 1 was free of seals but it soon became populated and the macrolichens were quickly detached by the seals and removed by wind. Within three years the control plot was totally destroyed, although the moss turf became extensively colonized by the nitrophilous foliose alga *Prasiola crispa* (Lightf.) Menegh. and filamentous algae. Inside the exclosure little change was experienced at first, but after several years the uphill margins became progressively 'scorched' by leaching of nitrogenous products from the seals from outside the fence. This caused the death and detachment of the macrolichens and browning of the moss, which became progressively more extensive.

Site 2 was already quite badly damaged when the exclosure

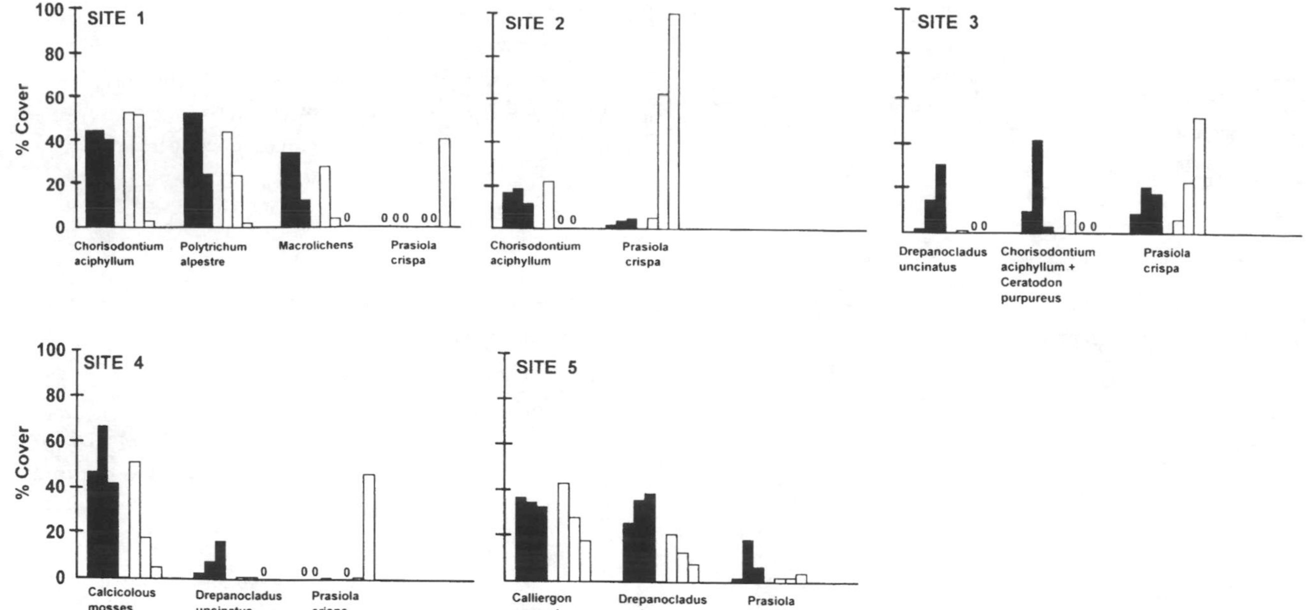

Fig. 61.2. Change in mean percentage cover of dominant plant species or groups in 0.25 m^2 plots (n=5) at five fur-seal-damaged sites on Signy Island. Solid bars: exclosures, open bars: control plots. In each block, left bar: 1985, centre bar: 1989, right bar: 1992. O: Not present.

was erected. After three years the control plot was totally destroyed but there was some early recovery by *Chorisodontium* within the exclosure. Marginal leaching was minimal as the site was level, but the long-term effect was death of the moss and very slight increase of algae. In the control plot the moribund moss became completely overgrown by *Prasiola* and filamentous algae.

At Site 3 there had been minor damage when the exclosure was established. While all mosses declined in abundance in the control, within the fence *Ceratodon* and *Chorisodontium* made an initial recovery before declining, while *Drepanocladus* showed a moderate degree of recovery. The muscicolous crustose lichens *Lepraria* spp. and *Ochrolechia frigida* (Swartz) Lynge also increased slightly over the seven years. *Prasiola* increased considerably in the control.

Site 4 had the greatest diversity of species and had undergone almost no disturbance when the exclosure was erected. As the seals encroached into the site the loosely attached *Schistidium antarctici* and *Tortula saxicola* Card. were gradually removed from the control plot. Several small turf- or mat-forming bryophytes either resisted seal trampling or, being ephemeral species (e.g. *Bryum argenteum* Hedw., *Encalypta patagonica* Broth., *Pottia austro-georgica* Card., *P. heimii* (Hedw.) Hamp.), were able to regenerate to some extent each summer from vegetative propagules or spores. Within the exclosure *Schistidium* and *Drepanocladus* were largely unaffected. The latter, having a relatively rapid growth rate, increased at the expense of *Schistidium*. Again, *Prasiola* became dominant in the control, although it was not abundant, being somewhat calcifuge.

The gently sloping Site 5 was in the early stages of perturbation when the exclosure was erected. However, it deteriorated rapidly as the number of seals increased. The unprotected wet moss carpet was reduced to a morasse of dead and detached living moss and its underlying shallow peat. The predominant mosses (*Calliergon* and *Drepanocladus*) within the fence were unaffected and grew vigorously, being tolerant of the high levels of nitrogenous and other nutrients seeping through the carpet. The near-absence of *Prasiola* here may have been caused by the very high levels of calcium and/or phosphorus (see below).

Changes in soil chemistry

The soil or peat at most sites showed marked and sometimes highly significant changes in concentrations of common elements over the period of this study (Fig. 61.3); other elements (K, Mn, Al, Fe) showed little change. Sites 1, 2 and 3, with acidic substrata, showed slight increases in pH within the exclosures but substantial increases in the control plots where the seal impact persisted, particularly in the moss peat at Sites 1 and 2. However, the calcareous fellfield at Site 4 and soligenous mire at Site 5 had reduced pHs after seven years, especially in the controls.

Sodium and magnesium followed a similar trend to that of pH, with considerable increases in the control plots at Sites 1–3. The concentration of Mg doubled in the acid peat of the control plots at Sites 1 and 2. The source of the massive input of Na and Mg is probably from seawater and sweat washed out of the seals' hair by rain, and to a lesser extent from their moulted hair and excrement. There was virtually no change in the low levels of these elements at the calcareous fellfield. Calcium levels increased in the moss peat of the controls at Sites 1 and 2, but showed little change in the two mineral soil sites (Sites 3 and 4). Although the highly calcareous mineral soil of Site 4 had a low Ca content, the high absorptive capacity of the mire peat was responsible for the exceptionally high Ca levels at Site 5 (which lies in a catchment including several limestone outcrops). Here, the action of the seals caused considerable reduction in calcium.

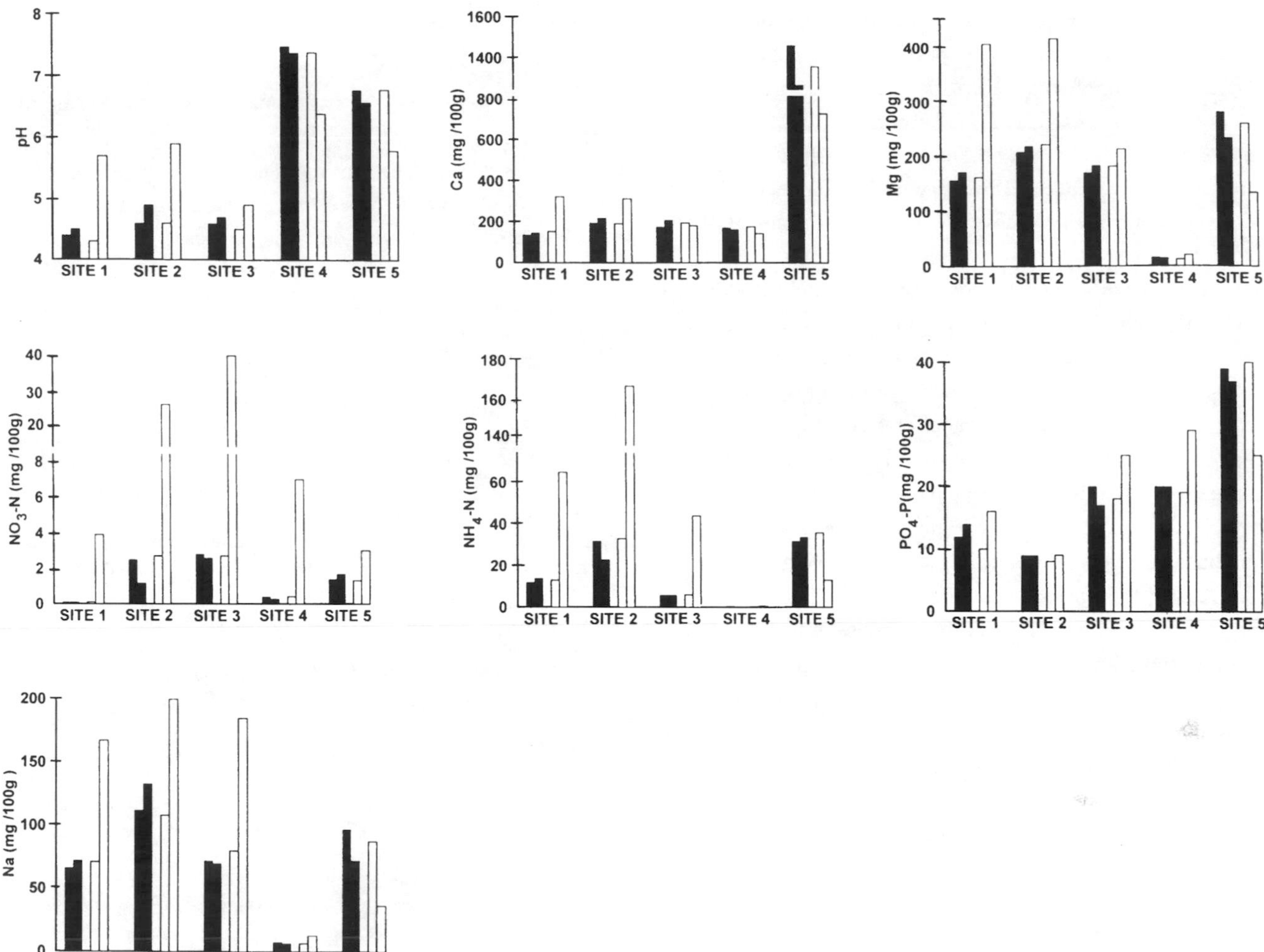

Fig. 61.3. Change in mean (n=3) soil pH and nutrient concentrations at five fur-seal-damaged sites on Signy Island. Solid bars: exclosures, open bars: control plots. In each block, left bar: 1985, right bar: 1992.

Phosphorus (as phosphate-P) appeared to be the least affected element; all the drier sites showed increases in the controls, the originally damaged Site 2 showed no change, while P at the mire site declined, especially in the control.

As anticipated there were some dramatic changes in the concentrations of nitrogen (as nitrate-N and ammonium-N), the major source of which is seal urine and excrement. The peat at Sites 1 and 2, and the mineral soils at Sites 3 and 4, had massive increases in NO_3-N in the control plots, although within the exclosures there was little change. The mire site (5) experienced, proportionally, only small increases. NH_4-N followed the same general trend as NO_3-N, although there was no change in the calcareous fellfield, which had very low levels, and those in the disrupted moss and peat in the mire control declined nearly threefold.

CONCLUSIONS

Monitoring the impact of the huge annual summer invasion of fur seals on the state of the ecologically sensitive cryptogamic vegetation at five sites on Signy Island has confirmed the observational prediction that, in areas of high seal concentrations, the damage caused by these animals is probably irreversible (Smith 1988). Once the vegetation cover, particularly that formed by mosses, has been destroyed and removed soil erosion ensues. While some mosses (notably those with a broad ecological amplitude and therefore tolerant of a wide range of environmental conditions (e.g. *Ceratodon purpureus*, *Drepanocladus uncinatus*) are able to withstand a small degree of physical disturbance and increased concentrations of nutrients, none can tolerate continued perturbation or excessive levels of some elements. Elsewhere on the island the grass *Deschampsia antarctica* Desv. has shown moderate resilience to the physical and nutrient effects of the seals, but it too is killed by persistent seal presence.

This study has shown that when Antarctic vegetation, which has already undergone some degradation, is protected from seal activity some species can recover, but the effect seems to be rather transient as, at each site, their general health or abundance declined after a few years. Even where seal impact commenced after the erection of an exclosure (Site 1), marginal leaching of toxic levels of certain nutrients into the fenced plot (but not actually detected in the soil analyses) is believed to be responsible for the progressive death of the vegetation towards the centre of the exclosure. In four of the five sites, as the natural community became increasingly destroyed and the bryophytes and lichens removed, the soil or peat surface quickly became colonized by cyanobacterial crusts, the foliose nitrophilous green alga *Prasiola crispa*, and one or more species of green filamentous algae. Some areas of Signy Island, which were once extensively dominated by a diversity of bryophyte and lichen communities, are are now covered by many hectares of these algae and little else.

There is no historical evidence of such seal impact at Signy Island in the past. Even with the complete removal of fur seals the edaphic conditions have been altered so much that recovery of these plant communities, either by recolonization or by regeneration, may take many decades or even centuries to attain their former state of ecological equilibrium and complexity. It therefore seems that the terrestrial and freshwater ecosystems of Signy Island are experiencing an irreversible change. Because the island has the longest continuous record of biological research in Antarctica, the fur seal phenomenon here offers an unique opportunity for studying further the response of terrestrial and freshwater biota and communities to a major natural environmental perturbation, and to allow a better understanding of species and ecosystem resilience to and possibly recovery from such disturbance. This monitoring study has also provided information useful in assessing aspects of environmental impact associated with human activity in Antarctica.

ACKNOWLEDGEMENTS

I am indebted to the late W. N. Bonner for help in constructing the seal exclosures, to G. Collett, D. Wright, H. MacAlister, M. Smithers and M. Chalmers for their subsequent maintenance, and to the Institute of Terrestrial Ecology, Merlewood Research Station, for undertaking the soil chemical analyses. I also thank M. R. Worland for preparing the histograms.

REFERENCES

Aguayo, A. L. 1978. The present status of the Antarctic fur seal *Arctocephalus gazella* at the South Shetland Islands. *Polar Record*, **19**, 167–173.

Bengtson, J. L., Ferm, L. M., Härkönen, T J. & Stewart, B. S. 1990. Abundance of Antarctic fur seals in the South Shetland Islands, Antarctica, during the 1986/87 austral summer. In Kerry, K. R. & Hempel, G., eds. *Antarctic ecosystems: ecological change and conservation*. Berlin: Springer-Verlag, 265–270.

Boyd, I. L. 1993. Pup production and distribution of breeding Antarctic fur seals (*Arctocephalus gazella*) at South Georgia. *Antarctic Science*, **5**, 17–24.

Furse, C. 1986. *Antarctic year. Brabant Island Expedition*. London: Croom Helm.

Payne, M. R. 1977. Growth of a fur seal population. *Philosophical Transactions of the Royal Society, London*, **B279**, 67–79.

Smith, R. I. L. 1988. Destruction of Antarctic terrestrial ecosystems by a rapidly increasing fur seal population. *Biological Conservation*, **45**, 55–72.

Smith, R. I. L. 1990. Signy Island as a paradigm of biological and environmental change in Antarctic terrestrial ecosystems. In Kerry, K. R. & Hempel, G., eds. *Antarctic ecosystems: ecological change and conservation*. Berlin: Springer-Verlag, 32–50.

Vergani, D. F. & Coria, N. R. 1989. Increase in numbers of male fur seals *Arctocephalus gazella* during the summer–autumn period at Mossman Peninsula (Laurie Island). *Polar Biology*, **9**, 487–488.

62 The *Bahia Paraiso*: a case study in environmental impact, remediation and monitoring

POLLY A. PENHALE[1], JON COOSEN[2,3] AND ENRIQUE R. MARSCHOFF[4]

[1]*Office of Polar Programs, National Science Foundation, 4200 Wilson Boulevard, Arlington, VA 22230 US,* [2]*National Institute for Coastal and Marine Management/RIKZ, P.O. Box 8039, 4330 EA Middelburg, The Netherlands,* [3]*current address: Directorate-General for Public Works and Water Management, Directorate Zeeland, P.O. BOX 5014, 4330 EA Middleburg, The Netherlands,* [4]*Instituto Antartico Argentino, Cerrito 1248, 1010 Buenos Aires, Argentina*

ABSTRACT

The sinking of the Argentine resupply ship, Bahia Paraiso, *in the vicinity of Palmer Station (United States), Antarctica in January, 1989 resulted in a diesel fuel spill of about 600 000 litres. After an initial international effort directed toward rescue of personnel and damage containment, the United States National Science Foundation (US NSF) organized a quick-response team of scientists from the United States, Argentina and Chile to assess the environmental impacts. The affected area consisted of nearly 20 small islands within a 2 km radius of the station. These islands, with well-developed intertidal and subtidal communities, are breeding sites for several species of seabirds and a principal habitat for seals. The islands have been the site of research for more than 20 years. Initial results showed an immediate negative impact to the rocky intertidal community and to certain seabird species. United States scientists have continued a long-term monitoring effort which has shown recovery by certain species and long-term negative effects on others. An estimated 400 000 litres of fuel was believed to remain in the ship. Continuing fuel leakage from the* Bahia Paraiso *prompted The Netherlands to propose a cooperative program with Argentina to recover the fuel remaining in the submerged wreck. This resulted in a formal bilateral agreement, The preparation of an Initial Environmental Evaluation (IEE), and the subsequent joint fuel recovery effort conducted by The Netherlands and Argentina between December 2, 1992 and January 2, 1993. The application of modern hot tap salvage techniques for the penetration of the submerged tanks and for fuel pumping proved environmentally sound; no disturbance of wildlife was observed. Due to damage and cracks in the hull of the ship, a great deal of fuel escaped prior to the recovery operation. The more harmful lubricants which remained in the ship were also removed. The example of the* Bahia Paraiso *provides a case study for international cooperation in environmental clean-up activities in Antarctica.*

Key words: oil-spill, monitoring, environmental impact, oil-salvage, *Bahia Paraiso*.

INTRODUCTION

On January 28, 1989, an Argentine Navy resupply ship, *Bahia Paraiso*, ran aground several kilometres from Palmer Station (Fig. 62.1), a US National Science Foundation (NSF) research facility located on Anvers Island, Antarctic Peninsula (64°46′ S 64°03′ W). The ship, *en route* to resupply the Argentine stations in the vicinity, had called at Palmer Station to allow the tourists on board an opportunity to visit. Although no passengers or crew members were injured as a result of the accident, diesel fuel

Fig. 62.1. The *Bahia Paraiso* on January 29, 1989, shortly after running aground. DeLaca Island is in the distance. NSF photo by Ted DeLecca.

Fig. 62.2. Deck view of the *Bahia Paraiso*, after it drifted and rolled on January 31, 1989, with fuel drums and gas bottles lashed on deck. NSF photo by Ted DeLecca.

and other petroleum products began leaking from the hull. This presented a threat to the environment and to the on-going scientific research programs in the local area. The objective of this paper is to describe the event, to provide an overview of the initial spill containment and subsequent salvage efforts, and to discuss the initial assessment of the impact on the environment. A long-term commitment by the US NSF to study the effects of the spill presented a unique opportunity to evaluate the resilience and recovery of a polar marine ecosystem. On February 18, 1992, representatives of the Netherlands and Argentina signed a Memorandum of Understanding, which resulted in a joint salvage effort between the two countries (Coosen & de Jonge 1993). This effort, which resulted in the removal of most of the fuel, was preceded by an Initial Environment Evaluation (IEE), according to the Protocol on Environmental Protection to the Antarctic Treaty. The impact assessment and salvage operation provides lessons for future cooperation in Antarctica.

The grounding, oil spill and initial containment effort

Shortly after the grounding of the ship, the 26 Palmer Station personnel responded to a request for emergency assistance by the captain of the *Bahia Paraiso*. Using inflatable boats, station personnel transported passengers and crew from the ship to Palmer Station. The US Antarctic Program contacted two tour ships in the vicinity, the *Society Explorer* and the *Illyria*, which immediately altered course to Palmer Station, where they evacuated 202 passengers and crew from the *Bahia Paraiso* to South America. The remaining 114 crew members spent several days camped at the station before being transported by Chilean, Spanish and Argentine vessels to either King George Island or South America. No loss of life occurred, but considerable logistical and scientific research resources from a number of nations and tour operators were diverted as a response to the accident.

The *Bahia Paraiso* carried various petroleum products such as diesel fuel arctic (DFA), jet fuel (JP-1), light marine diesel fuel, lubricating oil, as well as compressed gas bottles (Fig. 62.2).

Fig. 62.3. A containment boom, placed by the Chileans, extends around the stern of the *Bahia Paraiso*. NSF photo by Thomas Forhan.

When the ship initially grounded, a 10 m long tear in the hull allowed fuel to seep out. Initially, a 1 km^2 fuel slick formed near the disabled ship. After about 4 hours, the slick began to spread into the vicinity of the station. The spill remained fairly concentrated for the next 24 to 36 hours. On January 31, a combination of wind, currents, and tides freed the *Bahia Paraiso* from its original grounding site. The ship drifted several hundred metres; it then sank, leaving only 20% of the ship above the surface of the ocean.

At this point, the fuel slick spread throughout the 30 km^2 area around the station (Kennicutt *et al.* 1991).

In early February, Argentine and Chilean navy ships surveyed the site, collected water samples, and deployed the initial containment booms (Fig. 62.3). By February 7, a rapid-response team of oil spill response experts from the US arrived with 52 tons of equipment, including a small boat with a skimmer-boom system and various other oil spill clean-up equipment (Anonymous 1989a). After Chilean divers surveyed the wreck and the US crew began skimming operations to collect fuel on the surface, Argentine divers worked to patch the

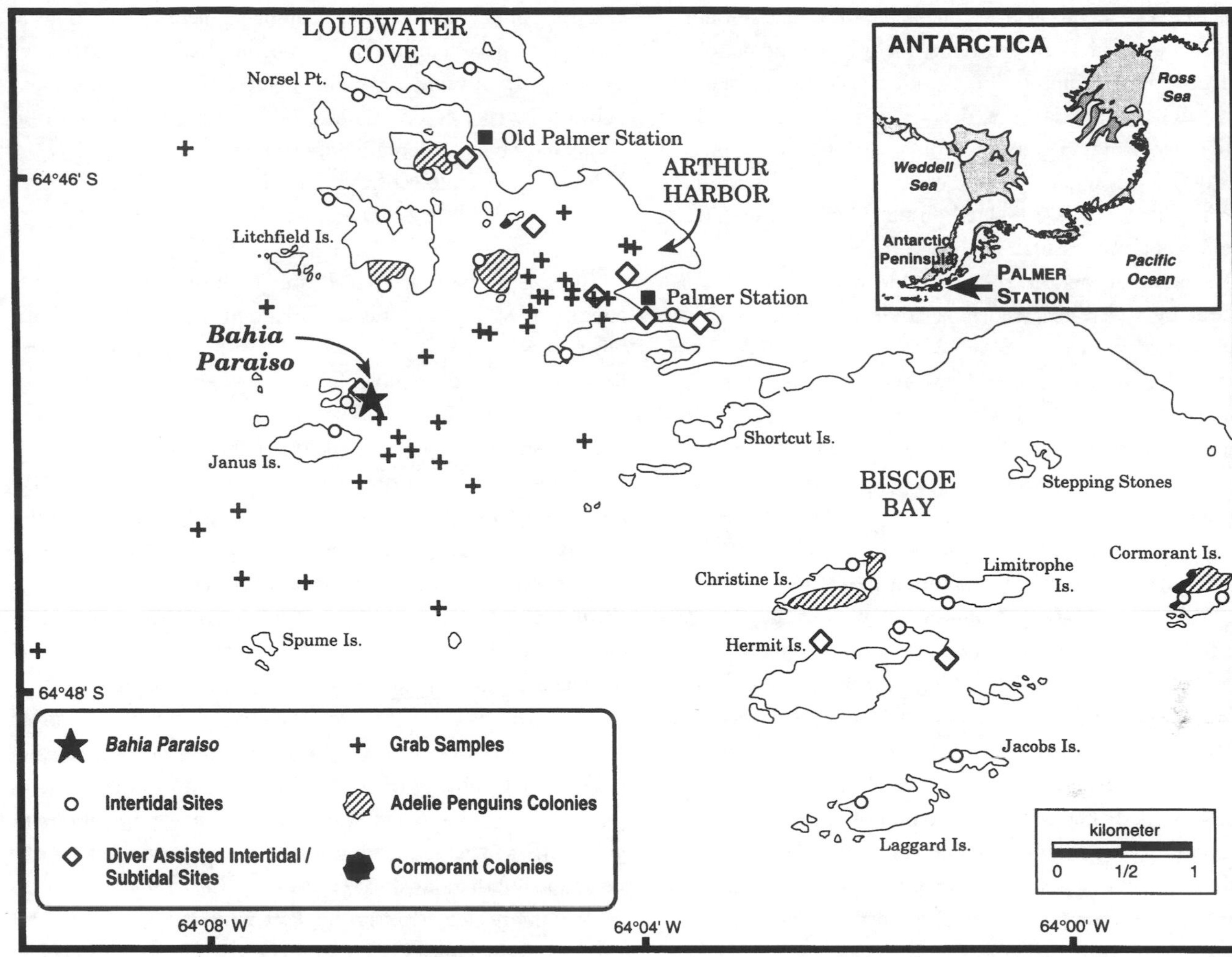

Fig. 62.4. Sampling sites occupied during the *Bahia Paraiso* study. Adélie penguin (*Pygoscelie adeliae*) and cormorant (*Phalacrocorax atricepts*) colonies are noted; all other bird species occur on all islands in the vicinity. See text for details and references.

ship where possible. Experts from the US and Argentina agreed that, within the first week, much of the spilled fuel had dissipated by natural factors, including evaporation, dissolution, photooxidation and dispersion. By mid-March approximately 65 000 l of fuel had been removed from the ship, and US and Argentine clean-up and salvage experts completed their efforts. An estimated 600 000 l of petroleum was believed to have been lost, with approximately 400 000 l estimated to remain in the tanks at that time. The major environmental threats at that time were considered to be a possible, but unlikely, break-up of the ship and the continued leakage of fuel from the ship's breached tanks (Anonymous 1989b).

The environmental impact of the spill

The study of the environmental effects of the oil spill on the marine environment in the vicinity of Palmer Station consisted of an initial sampling effort begun within days of the grounding of the *Bahia Paraiso* (Penhale 1989). During the month following the spill, on-site scientists and station personnel collected samples of water, intertidal macroalgae and invertebrates, beach sediments and morbid birds at various sites within the 30 km^2 area around Palmer Station (Fig. 62.4) in order to examine the extent of the oil spill and its impact. This was followed by an international, interdisciplinary scientific research team, organized by the US NSF. Fifteen scientists, from nine institutions in the US, Chile and Argentina, arrived within six weeks of the accident to begin a long-term research programme aimed at understanding the environmental impact of the accident. The objectives were to identify and determine the fate of hydrocarbons in the environment, to study the microbial degradation of petroleum hydrocarbons, and to observe and record the ecological effects of the spill on the plants and animals in the area. After this initial study, research has continued through efforts of the US scientists. The establishment of a Long-term Ecological Research site at Palmer Station and vicinity in 1990 ensures that data will continue to be collected (Quetin & Ross 1992).

The early observations showed that by four days after the Bahia ran aground, a 30 km^2 area surrounding Palmer Station was covered by an oil slick (Kennicutt 1990). Because of differences in tides, winds and currents, the quantity of petroleum products and the duration of exposure varied from island to island. Rocky intertidal sites, particularly those that had been continuously exposed to the spill for 3 to 4 days, showed the first signs of ecological damage. By February 4, a 50% mortality of intertidal limpets was estimated and algal mats in the littoral zone also appeared dead (Penhale 1989).

The initial impact on seabirds in the area varied. Although few dead birds were observed, most adult birds appeared to have been exposed to fuel oil (Fraser & Patterson, Chapter 63, this volume). Initial surveys of all local Adélie penguin (*Pygoscelis adeliae*) colonies indicated that more than 80 % of the birds had been exposed to the spill. The exposure of the various species of seabirds in the area depended on their behaviour. Adults were exposed primarily through feeding on krill and fish present in the area. As the spill occurred during the period of peak chick growth for many species, chicks were exposed to oiled parents and oiled food. Adélie penguin chicks fledged several weeks after the spill, and were thus exposed again at that time.

Initial observations on other ecosystem components suggested negligible effects (Kennicutt 1990). Marine mammals (which were primarily absent from the area during February and March) and benthic fish appeared not to be affected by the spill. Initial results of microbial degradation of hydrocarbons indicated low levels of activity. Preliminary examination of the subtidal community suggested little impact. Most of the results and interpretations required further laboratory and data analyses or long-term observations before conclusions could be drawn on the environmental impact of the spill.

Research on various components of the marine ecosystem indicate variations in persistence and level of impact. The distribution and fate of the petroleum hydrocarbons was evaluated by analysing various components of the ecosystem (Kennicutt *et al.* 1991). Analysis of hydrocarbon contamination of macroalgae (*Monostroma* sp. and *Leptosomia simplex*), limpets (*Nacella concinna*), birds (*Phalacrocorax atricepts*, *Catharacta maccormicki*, *Larus dominicanus* and *Pygoscelis adeliae*), clams (*Laternula elliptica*), bottom- feeding fish (*Notothenia coriiceps neglecta* and *Harpagifer anarcticus*) and sediments collected over several months after the spill showed contamination to varying degrees. Kennicutt *et al.* (1991) concluded that the relatively small size of the volatile fuel in a high-energy environment resulted in limited toxic effects. Samples taken a year following the spill showed some contamination, primarily in sediments in the vicinity of the *Bahia Paraiso*, due to low-level release from the ship. Studies conducted two years after the spill (Kennicutt *et al.* 1992a, b) showed little contamination in limpets and subtidal sediments, but periodic releases of small quantities of fuel from the ship resulted in continued contamination of some intertidal beaches.

Studies of the subtidal communities showed little effect. A study of benthic macrofauna assemblages near the oil spill site and a comparable control site showed no significant differences between the oil spill area and control sites in numbers of individuals, species or families, nor any major difference in dominant fauna or community composition (Hyland *et al.* 1994). These results correspond with hydrocarbon analyses which showed little contamination of subtidal organisms (Kennicutt *et al.* 1991).

Conclusions from laboratory studies of oxygen evolution by microalgal intertidal communities dominated by the chlorophyte *Urospora penicilliformis* suggested that differences between island sites most likely stemmed from sources of variability other than the hydrocarbons released during and after the spill (Ferreyra & Alder 1990). Laboratory studies were conducted on the short-term (3–7 days) effects of diesel fuel arctic on sediment microbial processes in the Palmer Station vicinity; the acute effects appeared negligible (Karl 1992). Longer-term (120 days) exposure showed either no effect or a slight stimulatory effect on metabolic activity and production. Very low rates of microbial hydrocarbon oxidation potential were detected in both the oil spill area and control regions.

The effect of the oil spill on the seabird community varied with time and with species; effects appeared both direct and indirect. Both during the time of the spill and in subsequent years, results showed that the adult breeding populations of giant petrels (*Macronectes giganteus*), brown skuas (*Catharacta skua*) and South Polar skuas (*Catharacta maccormicki*) were not adversely affected (Fraser & Patterson, Chapter 63, this volume). At the time of the oil spill, however, a population-wide mortality of South Polar skua chicks occurred.

Alternative hypotheses exist as to whether this mortality was due to the oil spill, weather conditions or a regional food limitation, or a combination of these factors (Eppley 1992, Eppley & Rubega 1990, Trivelpiece *et al.* 1990).

Comparison of changes in Adélie penguin (*Pygoscelis adeliae*) populations in the vicinity of Palmer Station with those receiving no exposure to oil suggests that colonies exposed to oil lost an additional 16% of their numbers, when the loss due to natural variability was less than 3% (Fraser & Patterson, Chapter 63, this volume). Populations exposed to oil showed no significant differences compared with control sites in subsequent years. Cormorants (*Phalacrocorax atriceps*) showed a near 100% mortality of chicks following the spill; in subsequent years, the number of active nests has decreased by approximately 85% (Fraser & Patterson, Chapter 63, this volume). This is most likely due to exposure to the fuel spill in areas of bathing and foraging used by cormorants.

An example of indirect effects has been suggested for the kelp gull (*Larus dominicanus*) (Fraser & Patterson, Chapter 63, this volume). While the adult populations did not appear affected at the time of the spill, the number of active nest sites in the area exposed to the oil spill has shown a steady decline since 1989. During the breeding season, males defend intertidal feeding territories to obtain limpets for themselves and their mates. Fraser & Patterson proposed that the damage to intertidal limpet populations accelerated the depletion of this resource in an area where resource limitation was severe, thus driving limpet densities below the threshhold levels needed to maintain gull reproduction.

Events leading to the fuel removal

On February 7, 1991, a joint US/Argentine oil spill response exercise was conducted at the site of the *Bahia Paraiso* (Strickland 1991). The purpose of the exercise, conducted by Palmer Station personnel and Argentine Navy personnel operating from the *Irigoyen*, was to test the contingency plans for dealing with a spill that might emanate from the wreck. The exercise was conducted during weather conditions of light, variable wind, calm seas and moderate temperature. Although the participants were well-prepared and enthusiastic, the exercise showed how difficult spill response and clean-up actions could be, even in reasonably good weather conditions. The drill also showed the need for more comprehensive response plans, more effective equipment (booms, mooring systems, skimmer systems), additional training and more frequent drills.

Following this exercise, the US developed an Oil Spill Contingency Plan for the Palmer Station area (Anonymous 1995), which addressed overall plans for the station as well as specific plans related to the *Bahia Paraiso*.

The presence of fuel remaining in the tanks and the hull of the *Bahia Paraiso* remained a source of concern. The slow leakage from the wreck and the possibility of a larger fuel release due to a possible hull breakage presented a continued and future threat to the local environment. This situation prompted the Netherlands Government to offer to collaborate with the Argentine Government to remove the remaining fuel. The well-known expertise of the Netherlands in salvage operations in the North Sea, and elsewhere in the world, together with logistical support from Argentina, combined to make the operation possible (Coosen & de Jonge 1993).

Following the adoption of the Protocol on Environmental Protection to the Antarctic Treaty by the Antarctic Treaty Consultative Parties in Madrid, Spain on October 4, 1991, discussions between the Governments of the Argentine Republic and the Kingdom of the Netherlands resulted in a Memorandum of Understanding (MOU), signed on February 18, 1992. The MOU presented the scope of work for the fuel removal project as follows:

'The two Parties will undertake a joint project to further study, prepare and possibly implement measures to remove from the Argentine vessel *Bahia Paraiso*, sunken on January 28, 1989, off the coast of Antarctica (Arthur Harbour), any remaining oil and other products, which, if released from the ship, would be harmful to the Antarctic environment.'

The MOU also stated that: 'In order to fully protect the ecosystem of the area, an environmental impact assessment of the planned measures will be part of the project'. Following the principles and requirements of the new Protocol to the Antarctic Treaty, a draft Initial Environmental Evaluation (IEE) was prepared by the Rijkswaterstaat and the Argentine Antarctic Institute. The environmental aspects were discussed by specialists during the SCAR XXII meeting in Bariloche, Argentina, in June 1992. Comments provided by experts from other Antarctic Treaty nations were used in the production the final IEE (Coosen *et al.* 1992), which was presented to all Delegates at the XVII Antarctic Treaty Consultative Meeting in Venice, Italy, in November 1992.

The assessment of environmental impacts of the fuel removal was considered to be critical to the process of decision-making.

Initial options considered were (1) no action, (2) salvaging the fuel remaining in the hull, and (3) removing the hull, either by cutting or refloating. The operation proposed by the Netherlands and Argentina was option (2); option (3) was eliminated due to the prohibitive costs and possible adverse environmental impact. The IEE was limited to considering the possible impact of option (2), compared with option (1).

The objective of the IEE was to evaluate the impacts expected from the salvaging of the fuel and oil remaining in the hull of the *Bahia Paraiso* in order to establish if the proposed activity might have significant negative impacts to the environment. To assess the impacts, a matrix was developed (Coosen *et al.* 1992) which included 14 possible operational actions versus 28 ecosystem components (based on data from research conducted at Palmer Station and vicinity, summarized in Quetin & Ross 1989). A separate category, termed PAISA (Protection of Antarctic Integrated Study Areas), was developed to account for those elements related to scientific research.

Identified interactions were classified as direct, indirect and long term. When the proposed actions themselves affected the environmental components, the effects were considered direct; if an affected component affected another ecosystem component, this effect was considered indirect. After identification of possible impacts, each impact was categorized, taking into account the spatial extent, duration and resilience of the system. Impacts were assessed as nil or insignificant, low, medium and high.

Additionally, several alternatives to the operations were analysed; these included using a bigger workboat, postponing the operation several months or until the following year, and alternative methods of waste storage. Highly vulnerable components, either environmental or scientific, were identified.

The IEE identified no impacts above the low or low/medium level for the preferred option, fuel removal, under normal field conditions. This was in part due to the transient nature of the impacts expected because of the expected short time span (three to four weeks) of the operation. The most significant impact was considered to be a possible spill of 4000 l; this was categorized as low/medium impact. The most probable impacts were identified as those produced by the operation, such as human presence, noise and wastes; these impacts were categorized as negligible or low. The final conclusion, after reviewing the options, was that removing the oil under controlled conditions would best protect the environment and the scientific research against further adverse impacts. To prevent such impacts, the IEE recommended a series of mitigating measures.

The joint operation to remove the remaining fuel from the wreck

Argentina provided the navy ship Canal Beagle, which carried all the equipment necessary for the operation. The Netherlands provided the main items of the salvage equipment. These included two working launches, diesel generators, hot-tap oil removal equipment with hydraulic-powered drilling machines and pumps, oily water separator with recorder, diving equipment, oil containment booms, and skimmers, absorbents, and steam cleaners. In order to reduce the pollution risk, vegetable oil was used on all hydraulic tools. The recovery team consisted of 17 people.

After arrival at the site on December 6, 1992 and consultation with the Operations Manager and scientific staff of Palmer Station, the Canal Beagle anchored in Arthur Harbour about 1 km from the wreck. The Argentine Navy then positioned an oil boom around the wreck. The two work launches were assembled to support recovery operations, being used as a platform for divers and for pumping and ferrying the recovered oil and oily water mixtures to the tanker. Removal operations began on December 9, after a delay due to a heavy storm on December 8. A initial diving inspection was performed and the first penetration of tanks using hot-tap connections were then made. At this time, ecological monitoring started with collection of water surface and sediment samples.

Between December 9 and 21, 1992 all tanks of the wreck were perforated, examined and all fuel discovered was pumped out. Operations were frequently stopped by swells exceeding 1.5 m or by large amounts of ice. On December 21, the cleaning of the engine room began but poor weather conditions caused further delays. After repositioning the oil containment booms and towing heavy amounts of ice from the wreck, the first attempt to enter the engine room was made on December 24. This required cutting several holes into the ship's structure before divers could access the engine room on December 30. Very little oil was pumped out, as it had already escaped from cracks in the hull. The oil recovery operation was completed on January 2, 1993. Prior to the operation, approximately 400 000 l of petroleum products were estimated to remain in the tanks and engine room of the ship. This estimate proved high; 148 500 l were recovered. The *Bahia Paraiso* was considered empty of fuel, and no further environmental damage due to oil leakage was expected.

Environmental considerations

The IEE contained a number of general recommendations to ensure protection of the environment. These included:

(1) No operations under marginal weather conditions. The operation was delayed numerous times due to two major storms, in combination with large pieces of ice nearby, and periods of high swell.

(2) No helicopter flights.

(3) Special contingency measures (including booms) to prevent the fouling of special sites. After a small spill of approximately 20 l occurred in Arthur Harbour, a containment boom was placed to contain the spill; no other spills occurred.

(4) Noise generation should be reduced. The launches were outfitted with silencers and reduced-sound generators and no disturbance to the breeding Southern Giant Petrels at nearby islands was observed.

(5) No unauthorized visits to islands or Palmer Station. A spirit of cooperation developed between the joint Argentine/Netherlands crew and the personnel on station. Several planned visits to the station and US research sites on nearby islands for the purpose of scientific cooperation were made.

(6) The presence of tourist ships during the joint operation should be discouraged. Scheduled visits to Palmer Station by the research ships *James Clark Ross* (UK) and *Polar Duke* (US) were made, but no tourist ships visited. A Chilean navy tugboat and an Argentine navy ship visited the area. None of the ships interfered with the salvage operation.

(7) No discharges of any kind on floating ice. There was no ice within the containment boom during pumping operations; thus, no discharges occurred.

(8) Any substantial spilling of oil (>100 l) should be reported. No such spill occurred during salvage operations.

(9) Cutting of the engine room should be carried out in calm weather conditions; this was achieved.

(10) Oil boom anchoring should not use rock bolts on DeLaca Island, nearest the site. After consultation with Palmer Station scientists about possible impact, existing rock bolts were used.

(11) If spills occur, their fate should be monitored to reduce impacts on penguin colonies under study. No significant spills occurred.

(12) Ice should be prevented from entering into the containment boom areas. While it was impossible to prevent larger pieces of ice from entering the site of the wreck, trapped ice was removed after stopping operations and opening the boom.

(13) Weather and ice forecasts should be available on-site during the whole operation. Argentine as well as US meteorological services were employed.

(14) Waste disposal (non food waste as well as food wastes), should be stored in containers on board until disposal at port facilities in Argentina. Contrary to what was stated in the IEE, sewage produced on board of *Canal Beagle* was

treated in a sewage treatment plant. The solid residue was stored on board and the treated effluent was discharged to the sea. Total production of solid waste was estimated at 20–25 kg per day.

Short- and long-term monitoring

Short-term environmental monitoring included: (1) visual inspection (including video and photographs) of the waters around the ships (e.g. the wreck, the work boats and the support vessel) during the operation; (2) observations of the breeding giant petrels (*Macronectes giganteus*) on islands nearest the wreck; (3) water sampling at three locations in the vicinity of the wreck near the bottom, mid-water and surface; these were stored for analysis of hydrocarbon content, if necessary; (4) sampling the discharge from the oily water separator, on a continuous basis; (5) recording of the volume and fate of any spillage.

To establish a baseline so that any hydrocarbon contamination as a consequence of the operation could be detected, samples were taken at the beginning, during and at the end of the operation. These included samples of sediment in the vicinity of the wreck and intertidal limpets (*Nucella concinna*) and macroalgae (*Monostroma* sp. and *Leptosomia simplex*) at the nearby DeLaca and Janus Islands. These samples remain frozen, to be analysed, if necessary, for hydrocarbons. Since no signs of contamination were observed during sampling, the stored samples have not been analysed. Intertidal communities on the nearby islands did not differ from those at islands at a larger distance from the wreck.

The long-term monitoring of those elements of the marine ecosystem components which proved most affected by the spill continues to be conducted by US scientists. Additionally, the establishment of a Long-term Ecological Research site at Palmer Station and the vicinity ensures continued assessment of the entire system (Quetin & Ross 1992).

SUMMARY

The *Bahia Paraiso*, beginning with its grounding and subsequent oil spill near Palmer Station and the associated environmental impacts, and continuing through the removal of the fuel and continued long-term monitoring of the fates and effects of the spill, provides many examples and lessons. Despite ice-strengthened ships and improved technology for navigation and ice detection, Antarctic operations remain hazardous and constant caution is required. When the accident occurred, international cooperation happened immediately to ensure that personnel were rescued and oil was contained and removed. This was followed by an international research effort. The fuel recovery operation again showed the great value and importance of international collaboration and cooperation in Antarctica.

Despite all precautions, similar accidents may happen in the future. To ensure the greatest protection to the environment, oil spill contingency plans need to be prepared for station and ship operations in Antarctica. These plans should include not only detailed response procedures, but alsoprotocols for any scientific research needed to detect the fate and effects of any pollution.

ACKNOWLEDGEMENTS

The authors would like to thank their governments and all those who participated in the operational response and review, fuel removal activities, and scientific study related to the grounding of the *Bahia Paraiso* and subsequent fuel spill. The views expressed in this paper are those of the authors and not necessarily those of the US National Science Foundation. The detailed comments of two anonymous reviewers were greatly appreciated.

REFERENCES

Anon. 1989a. NSF launches oil-spill containment team. *Antarctic Journal of the United States*, **24** (2), 8–9.

Anon. 1989b. Argentine ship sinks near Palmer Station. *Antarctic Journal of the United States*, **24** (2), 3–7.

Anon. 1995. *Palmer Station area oil spill contingency plan*. Prepared for the National Science Foundation by Jamestown Marine Service, Inc., 144 pp.

Coosen, J. & de Jonge, A. M. 1993. *Oil removal from* Bahia Paraiso. *Final Report DGW-93.028*. Rijkswaterstaat, The Netherlands: Ministry of Transport, Public Works & Water Management, 142 pp.

Coosen, J., Acero, J. M., Agraz, J. L., Aguirre, C. A. & Marschoff, E. R. 1992. *Initial environmental evaluation associated with the salvage operation of the remaining oil of the* Bahia Paraiso. Rijkswaterstaat, The Netherlands: Ministry of Transport, 60 pp.

Eppley, Z. A. 1992. Assessing the indirect effects of oil in the presence of natural variation: the problem of reproductive failure of south polar skuas during the *Bahia Paraiso* oil spill. *Marine Pollution Bulletin*, **25**, 307–312.

Eppley, Z. A. & Rubega, M. A. 1990. Indirect effects of an oil spill: reproductive failure in a population of south polar skuas following the '*Bahia Paraiso*' oil spill in Antarctica. *Marine Ecology Progress Series*, **67**, 1–6.

Ferreyra, G. A. & Alder, V. A. 1990. The evolution of oxygen in microalgal intertidal communities near Palmer Station. *Antarctic Journal of the United States*, **25** (5), 183–184.

Fraser, W. R. & Patterson, D. L. 1997. Human disturbance and long-term changes in Adélie Penguin populations: a natural experiment at Palmer Station, Antarctic Peninsula. In Battaglia, B., Valencia, J. & Walton, D. W. H., eds. *Antarctic communities: species, structure and survival*. Cambridge: Cambridge University Press, 445–452.

Hyland, J., Laur, D., Jones, J., Shrake, J., Cadian, D. & Harris, L. 1994. Effects of an oil spill on the soft-bottom macrofauna of Arthur Harbour, Antarctica compared with long-term natural change. *Antarctic Science*, **6**, 37–44.

Karl, D. M. 1992. The grounding of the *Bahia Paraiso*: microbial ecology of the 1989 Antarctic oil spill. *Microbiology Ecology*, **24**, 77–89.

Kennicutt II, M. C. 1990. Oil spillage in Antarctica: initial report of the National Science Foundation-sponsored quick-response team on the grounding of the *Bahia Paraiso*. *Environmental Science & Technology*, **24**, 620–624.

Kennicutt II, M. C. & Sweet, S. T. 1992. Hydrocarbon contamination on the Antarctic Peninsula. III. The *Bahia Paraiso* – two years after the spill. *Marine Pollution Bulletin*, **25**, 303–306.

Kennicutt II, M. C., Sweet, S. T., Fraser, W. R., Stockton, W. L. & Culver, M. 1991. Grounding of the *Bahia Paraiso*, Arthur Harbour, Antarctica. I. Distribution and fate of oil spill related hydrocarbons. *Environmental Science and Technology*, **25**, 509–518.

Kennicutt II, M. C., McDonald, T. J., Denoux, G. J. & McDonald, S. J. 1992a. Hydrocarbon contamination on the Antarctic Peninsula. I. Arthur Harbour – subtidal sediments. *Marine Pollution Bulletin*, **24**, 499–506.

Kennicutt II, M. C., McDonald, T. J., Denoux, G. J. & McDonald, S. J. 1992b. Hydrocarbon contamination on the Antarctic Peninsula. II. Arthur Harbour – inter and subtidal limpets (*Nacella concinna*). *Marine Pollution Bulletin*, **24**, 506–511.

Penhale, P. A. 1989. Research team focuses on environmental impact of oil spill. *Antarctic Journal of the United States*, **24** (2), 9–12.

Quetin, L. B. & Ross, R. M. 1992. A long-term ecological research strategy for polar environmental research. *Marine Pollution Bulletin*, **25**, 9–12.

Stickland, E. 1991. *Report on joint US Palmer Station/Argentine Navy oil spill response exercise*. Report to the US National Science Foundation by Jamestown Marine Services, Inc., 10 pp.

Trivelpiece, W. Z., Ainley, D. G., Fraser, W. R. & Trivelpiece, S. G. 1990. Skua survival. *Nature*, **345**, 211–212.

63 Human disturbance and long-term changes in Adélie penguin populations: a natural experiment at Palmer Station, Antarctic Peninsula

WILLIAM R. FRASER AND DONNA L. PATTERSON
Polar Oceans Research Group, Department of Biology, Montana State University, Bozeman, Montana 59717, USA

ABSTRACT

Human activities (tourism and research) near Palmer Station, Anvers Island, Antarctic Peninsula, has increased significantly since 1975. Although these activities were focused on the large, easily accessible populations of Adélie penguins on Litchfield and Torgersen islands, Litchfield Island became a Specially Protected Area (SPA) in 1978. This ended tourism on the island and reduced research-related activity to negligible levels. Despite SPA status, the total breeding population of Adélie penguins on Litchfield Island decreased by 43% between 1975 and 1992. In contrast, on Torgersen Island, where tourism and research-related activities continued to increase over the same time period, the decrease in these populations was only 19%. There is increasing concern that tourism and other human activities may adversely impact Antarctic wildlife populations. Although this concern may be justified for some types of human activity, our data suggest that the potentially adverse effects of tourism and research may be negligible relative to the effects imposed by long-term changes in other environmental variables.

Key words: Antarctica, Adélie penguin, tourism, research, human activities, disturbance, population trends.

INTRODUCTION

Human disturbance due to tourism or research has been implicated in the decline of Adélie penguin (*Pygoscelis adeliae*) populations at many localities in Antarctica (Thompson 1977, Muller-Schwartze 1984, Wilson *et al.* 1990, Young 1990, Acero & Aguirre 1994, Woehler *et al.* 1994). Close inspection of some of the long-term demographic data that underpin these implications, however, reveals certain patterns that are inconsistent with the view that human disturbance is the only factor linked to changes in these Adélie penguin populations. These patterns include sharp increases in Adélie penguin populations following the construction of research stations (Young 1990, Acero & Aguirre 1994); asynchrony in the chronology of population change relative to the inception, expansion and abandonment of research stations (Wilson *et al.* 1990); and, most notably, increases and decreases in populations disturbed by humans coinciding with similar changes in undisturbed populations over broad geographic areas and spanning several decades (Stonehouse 1965, Taylor *et al.* 1990, Wilson *et al.* 1990).

Although these patterns do not completely rule out human disturbance as a possible reason for the decline of some Adélie penguin populations (see Wilson *et al.* 1990, Woehler *et al.* 1994), they agree with hypotheses presented by Blackburn *et al.* (1991), Fraser *et al.* (1992) and Taylor & Wilson (1990) that suggest that more complex environmental factors ultimately force changes in Antarctic penguin populations. In this study, we examine long-term population data on Adélie penguins breeding in the vicinity of Palmer Station, Antarctic Peninsula, relative to both exposure to human activity and the implications

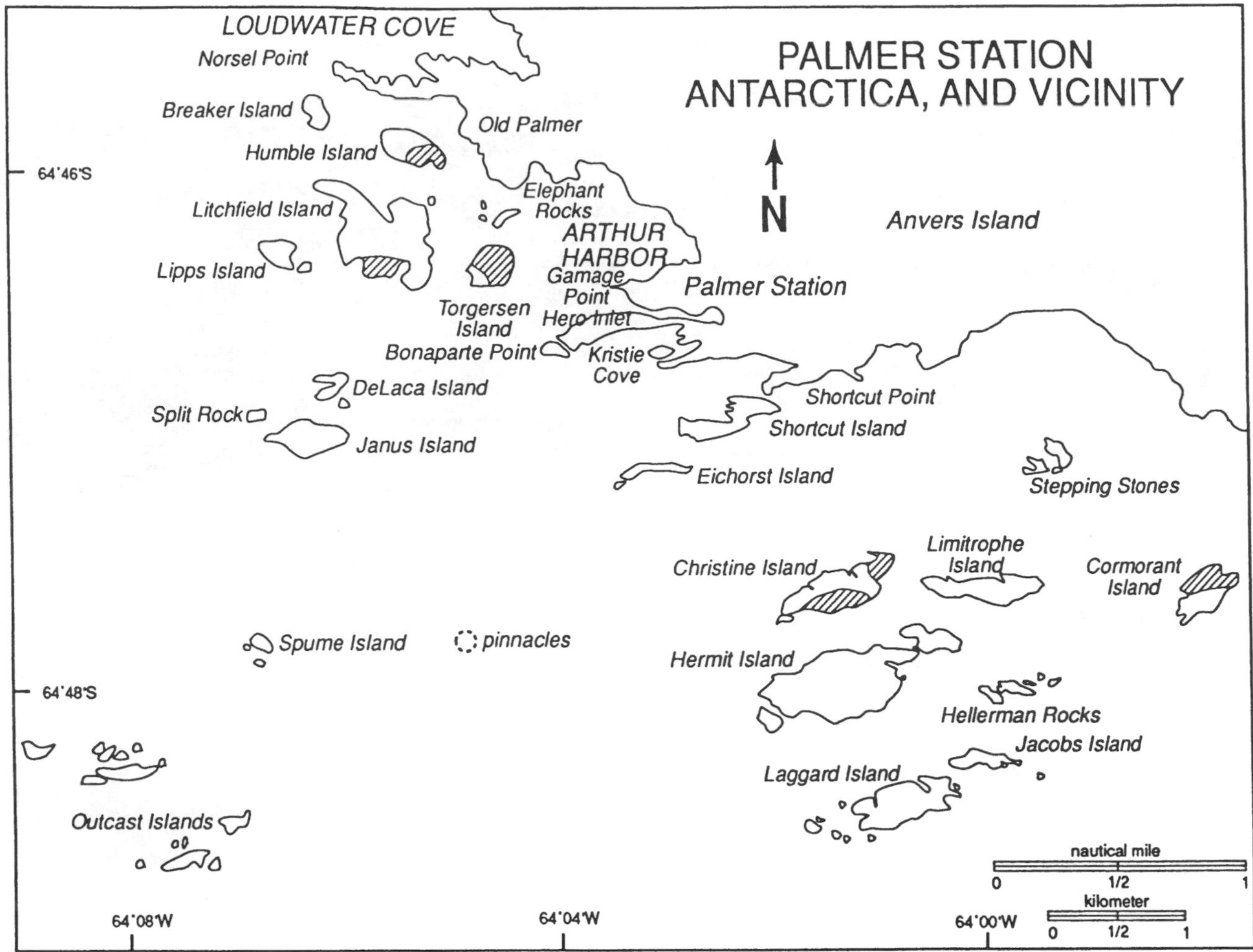

Fig. 63.1. Palmer Station, Antarctica, and vicinity. Adélie penguin colony sites on each island are identified by shading.

inherent in these hypotheses. Three factors make Palmer Station an ideal site for examining the issue of human disturbance. First, Adélie penguin populations near Palmer Station have been exposed to concurrent tourist and research-related activities for more than 20 years. Second, because all colonies occur on islands away from the station, construction activities never altered the availability of breeding habitat. Third, the designation of Litchfield Island (Fig. 63.1) as a Specially Protected Area (SPA No. 17) in 1978, combined with tourist management policies initiated by the US National Science Foundation (NSF) in 1986 and 1990, greatly reduced human access to penguin colonies, thus setting up a natural experiment in which the effects of human activity on penguin populations could be monitored at visited (experimental) and non-visited (control) colonies.

METHODS

Human activity

Data on the number of shipborne tourists visiting the Palmer Station area were obtained from unpublished field records (WRF, 1975–1977) and from information maintained by the US National Science Foundation (1980–1992). Data on the number of support and science personnel at Palmer Station were obtained from unpublished journal notes (WRF, 1975–1977, 1987–1992) and from science and logistics summaries published in the *Antarctic Journal of the United States*, 1975–1992. Additional information on the frequency of landings on the various islands by Palmer Station personnel was obtained from logs maintained at the station for October 1991 – March 1993 on behalf of the NSF when Multiple Use Planning Area (MPA) guidelines were implemented in 1990 (see Results, below).

Adélie Penguin censuses

Data were obtained from published censuses in Parmelee & Parmelee (1987), Heimark & Heimark (1988), Ainley & Sanders (1988), Fraser (1992) and Fraser *et al.* (1993): data for 1975 are based on unpublished records (WRF). In selecting census data, the following criteria were applied: (1) only N_1 counts were considered. These counts denote active (egg or chick present) nest counts with an accuracy of ±5% (see Croxall & Kirkwood 1979). Exceptions were made (1) for large colonies (>1100 pairs) on Torgersen and Humble islands where accuracy was estimated at ±10%; (2) where more than one census was done in a season, the census closest to the 7–25 November period (see item 4 below) was used; (3) no censuses were considered if they

Table 63.1. *Years during which islands in the Palmer Station area occupied by Adélie penguins were open to various human visitors*

Island	Use		
	Tourism	Research	Recreation by station personnel
Litchfield	1975–1978	1975–1992	1975–1978
Humble[1]	1975–1989	1975–1992	1975–1989
Christine[1]	1975–1989	1975–1992	1975–1989
Cormorant[1]	1975–1989	1975–1992	1975–1989
Torgersen	1975–1992	1975–1992	1975–1992

Note:

[1] Islands where tourism was restricted by agreement between the National Science Foundation and tour operators after the Litchfield Island SPA was established in 1978. All tourism and recreation by station personnel ended on these islands in 1989.

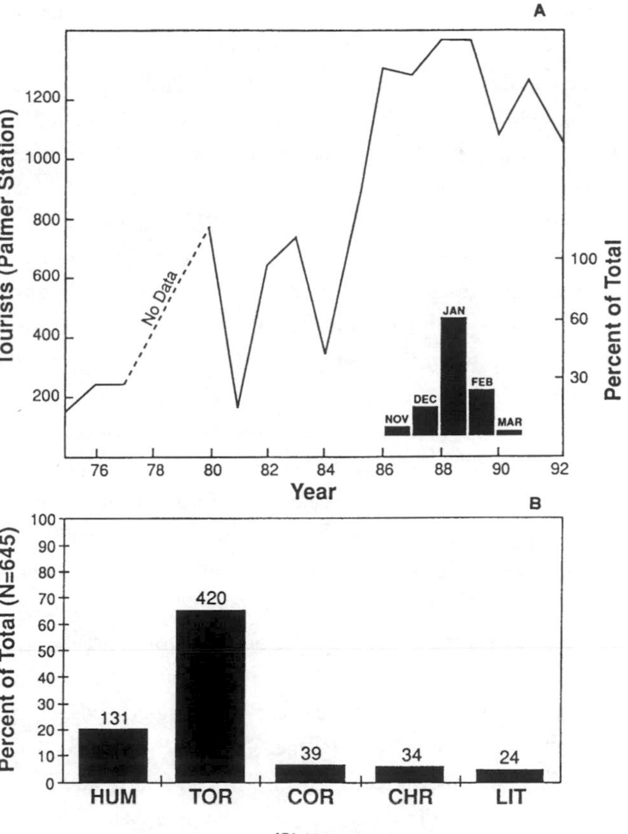

Fig. 63.2. (A) Trends in shipborne tourist landings on Torgersen Island, Palmer Station. Landings by month are shown as a percentage of the total for all years. (B) Landings by island for science and non-science Palmer Station personnel based on data obtained between October 1991 and March 1993. Island codes are HUM (Humble), TOR (Torgersen), COR (Cormorant), CHR (Christine) and LIT (Litchfield).

occurred after 5 January to avoid the possibility that the creche period had begun; (4) published censuses were used without correcting the data to account for variations in census date and the progressive loss of nests that occurs during the breeding season. Calculating such corrections requires concominant data on annual variability in breeding chronology and mortality, which are not yet well developed for the Palmer Station area, as is true for most of the Antarctic. However, based on the data that are available (censuses and reports by WRF and others in 1975, 1991 and 1992), peak egg laying occurs between 12–21 November and peak hatching between 13–22 December. By peak hatching, 14–16% of the nests in the Palmer area have, on average, been lost. Virtually every census conducted in the area, aside from those by WRF, fall between peak egg laying and peak hatching based on the WRF 1975, 1991 and 1992 data. This would suggest that true breeding populations based on censuses conducted after peak egg laying are underestimated by 14–16%. Censuses in 1991 and 1992 by WRF and others were made during the peak egg laying period and require no corrections, meaning overall estimates of population decrease through 1992 are conservative relative to censuses done in prior years.

RESULTS

Patterns of human activity

Human access to islands populated by Adélie penguins (Fig. 63.1) has been progressively curtailed since 1975, Torgersen Island being the only exception (Table 63.1). Three factors were involved. First, the Litchfield Island SPA was established in 1978, which officially ended most human activity on the island. Second, an unofficial agreement between tour operators and the NSF to protect research sites was reached coincident with implementation of the SPA in 1978. This effectively ended nearly all tourist access to Humble, Cormorant and Christine islands at about the same time. Third, in 1990, guidelines were implemented in accordance with Multiple-Use Planning Area (MPA) management objectives that restricted personnel without permits from landing on islands with large seabird populations during the nesting season (1 October – 1 March).

Trends in the number of shipborne tourists to the area are shown in Fig. 63.2A. During 1975–1977, Litchfield and Torgersen islands were each visited with equal frequency by the same number of tourists (WRF unpublished data). Tourist landings doubled after 1985, and by 1989 only Torgersen Island still remained accessible due to restrictions placed on other sites (Table 63.1). In 1986, the NSF began to limit the number of tourists permitted to visit the area to no more than 1300 annually due to the demands placed on station resources and personnel. A summary of tourist landings by month during 1975–1992 is shown in Fig. 63.2A. In all years, landings occurred primarily during the December–February period, with 60% taking place in January. This period coincides precisely with the peak hatching, growth and fledging stages of Adélie penguins (Fraser *et al.* 1993). Ships typically carried 80–120 passengers, landing groups of 20–40 on Torgersen Island for up to 1 h per group. Visitation frequency increased from one ship every 9–14 days before 1985 to one every 4–6 days beginning in 1986.

There is no historical record comparable to the one presented above for tourist activity that details the frequency with which Palmer Station personnel visited the islands with Adélie penguin populations during the October–March breeding season. In the case of Litchfield Island, however, where access was regulated by permit beginning in 1978 (Table 63.1), it appears that fewer than 35 people entered the site between 1978–1992 (unpublished data based on logistics reviews and the NSF permit offices). Moreover, there is no evidence that any science project working with Adélie penguins on Litchfield Island ever engaged in anything more than population censuses during 1975–1992, suggesting the island was probably visited no more than one to three times a season (see Parmelee & Parmelee 1987, Fraser *et al.* 1993). Use patterns for the island before 1978 cannot be ascertained. However, Palmer Station is small, able to house no more than 43 people. Before 1987, the number of science and support personnel present at the station during the October–February Adélie penguin breeding season rarely exceeded 22 people. Thus, their use of Litchfield Island during 1975–1977 may have been inconsequential relative to the activities of tourists (Fig. 63.2A).

Torgersen Island is currently the most frequently visited of the islands by personnel from Palmer Station (Fig. 63.2B). This pattern was obviously affected by implementation of the MPA guidelines in 1990 (Table 63.1); however, there is reason to believe that such a use pattern would have existed prior to enforcement of the MPA because it reflects two key factors that have not changed since 1975. The first is that the island is close to the station (Fig. 63.1) and holds the area's largest penguin colonies (*c.* 8000 pairs; Fraser *et al.* 1993): what draws tourists to the island also draws station personnel. The second, and perhaps most pertinent factor relative to this study, is that during nearly 18 years between 1975–1992, one to three projects involving two to ten researchers per season have used the Adélie penguin colonies for a variety of studies investigating physiology, behaviour, foraging ecology, demography, chick development and reproductive biology (unpublished NSF records). Because virtually every study has involved serial sampling, the island has been the most intensively used by researchers in the area. In contrast to Torgersen Island, Cormorant and Christine islands are remote and accessible only during good weather, making them poor choices for research sites and of limited use for recreation by personnel in general; the use patterns shown in Fig. 63.2B reflect these qualities and probably also apply to the pre-1975 period. The use pattern shown for Humble Island (Fig. 63.2B) reflects its status as a long-term Adélie penguin monitoring site since 1987 (Fraser & Ainley 1988). Historical use patterns for this island are unknown, but prior to implementation of the MPA it probably received more visits from station personnel in general than it does at present.

Long-term changes in Adélie Penguin populations

As shown in Fig. 63.3, Adélie penguin populations near Palmer Station were at an 18-year high in 1975, and have been decreasing steadily since that time, a pattern noted in other sectors of

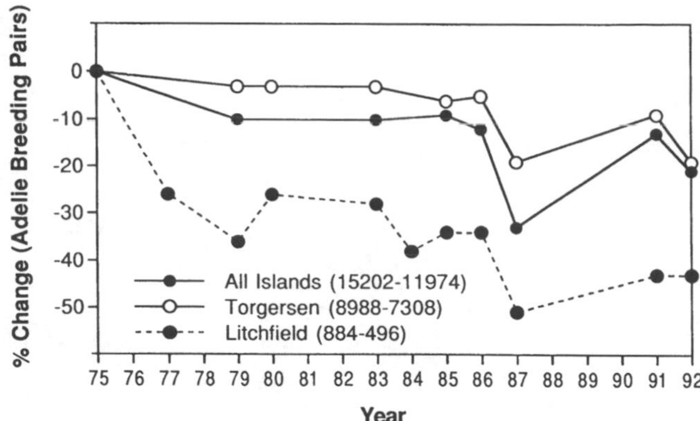

Fig. 63.3. Long-term change in Adélie penguin populations on Torgersen and Litchfield islands, compared with data from all islands, 1975–1992. Numbers in parentheses represent the number of breeding pairs present at the beginning (1975) and end (1992) of the census period. Change is shown relative to the number of breeding pairs present in 1975.

the Antarctic Peninsula as well (Poncet & Poncet 1987). The large decrease in 1987 corresponds with the disruption of food webs in many parts of the world as a result of altered climate patterns (Croxall *et al.* 1988, Ainley & Boekelheide 1990); it is not yet clear if the large decreases evident in 1992 can be similarly explained. It is notable that despite closure of Litchfield Island in 1978 to virtually all human activity (Table 63.1), penguin populations continued to decrease. By 1992, this decrease represented 388 breeding pairs, or approximately 43% of the pairs present in 1975. On Torgersen Island, in contrast, where the volume of human activity increased dramatically over the same time period (Fig. 63.2A), annual decreases in penguin populations were, on average, of smaller magnitude than the changes for the area as a whole, and quite different than the trends exhibited on Litchfield Island.

Factors associated with long-term change in Adélie Penguin populations

During 1975–1992, the percentage decrease in Adélie penguins on Litchfield Island was greater than on any of the other islands in the area. Litchfield Island also has the largest number of extinct colonies in the area (Fig. 63.4A), indicating not only that higher numbers of penguins previously bred on the island (Ainley & Sanders 1988), but also suggesting that the decrease in populations started earlier than 1975. Extinct colonies occur on each of the islands currently occupied by Adélie penguins except Humble Island. The most obvious factor characterizing these extinct colonies when compared with currently active colonies is that 86% of them (18 out of 21 total) are located on the southwest side of topographic features >5 m in height (Fig. 63.4B). In contrast, only 17% of active colonies exhibit this aspect. For the Palmer Station area, where the predominant wind direction during storms is from the northeast (Fraser *et al.* 1993), this means that most extinct colonies are situated downwind from, and in the lee side of, high topographic features. On Litchfield Island, the entire penguin rookery (active and extinct colonies) is

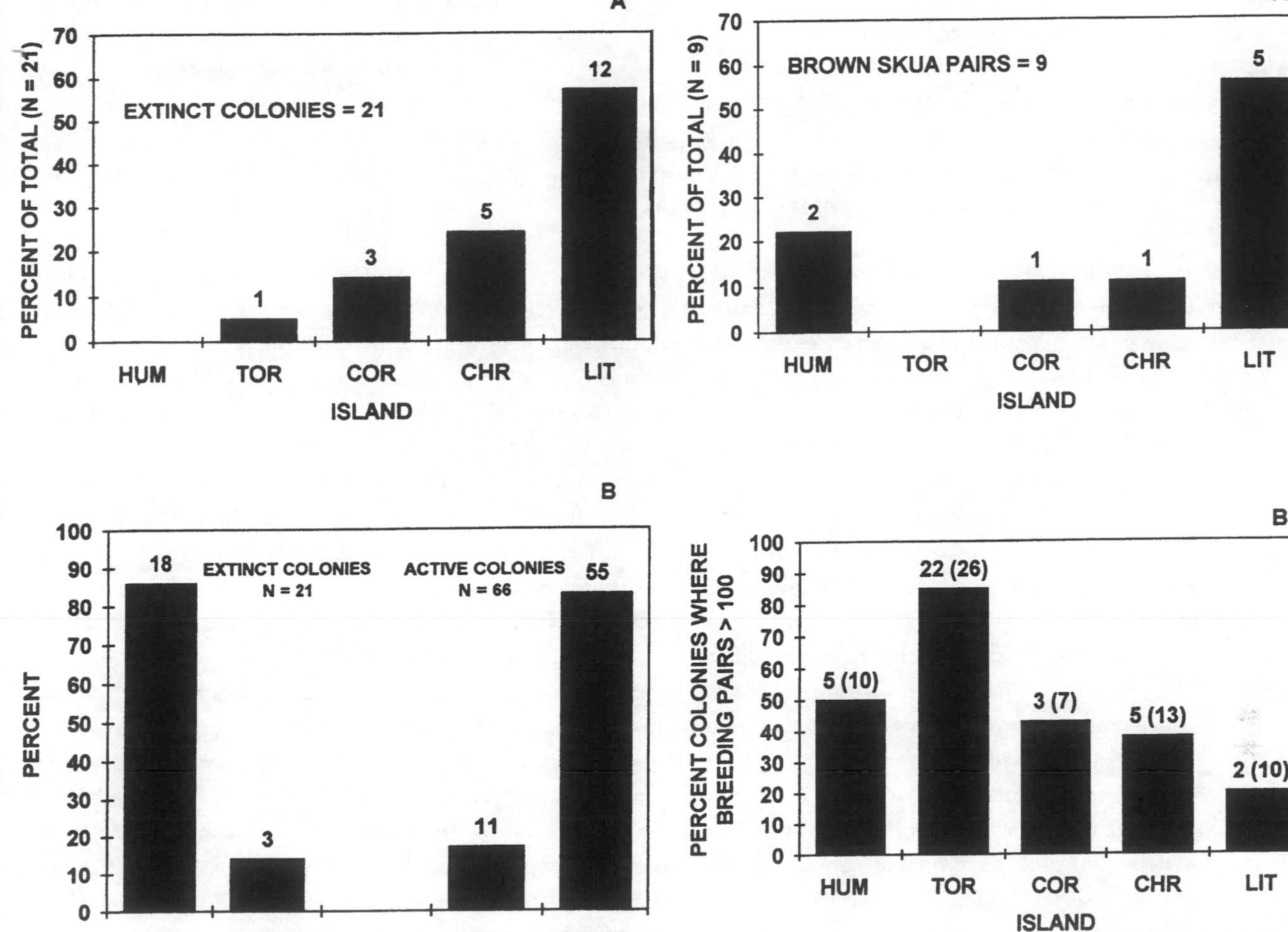

Fig. 63.4. (A) Extinct colonies by island in the Palmer Station area. (B) Aspect (SW=southwest) relative to colony status (active or extinct). Numbers above each bar in both figures are the actual number of colonies present for each category.

Fig. 63.5. (A) Breeding brown skua pairs by island in the Palmer Station area. Numbers above each bar represent the actual number of breeding pairs. (B) Large colonies (>100 breeding pairs) of Adélie penguins by island in the Palmer Station area. Percentages are calculated for each island. The number before the parentheses is the actual number of large colonies; the number in parentheses is the total number of active colonies on each island.

located on the southwest side of a series of high (60–70 m) rock outcrops. Indeed, nine of the 11 active colonies with a southwest aspect occurring in the Palmer Station area (Fig. 63.4B) are also part of the Litchfield Island rookery. Torgersen Island, in contrast, has no similar topographic features, and colonies are directly exposed to the predominant wind direction. This is also true of most of the other active colonies on the islands (Fig. 63.4B).

A second factor associated with Adélie penguin population decrease on Litchfield Island is the presence of brown skuas (*Catharacta lonnbergi*), which during the breeding season prey exclusively on penguin eggs and chicks (Neilson 1983). As shown in Fig. 63.5A, 56% of the nine brown skua pairs that occur in the area are found on Litchfield Island, a ratio that has persisted since at least 1974 (Neilson 1983), and probably longer. Since 1987, two penguin colonies have become extinct on Litchfield Island (Fraser & Ainley1988). These were small colonies (<30 pairs) within brown skua feeding territories where skuas took every egg and/or chick every season (Fraser *et al.* 1993). Litchfield Island has the smallest percentage (two out of nine total) of large (>100 pairs) penguin colonies of any of the islands (Fig. 63.5B) and the remaining 7 colonies average only 37 pairs (n=7–83 pairs). According to Trivelpiece *et al.* (1980), brown skua feeding territories encompass 90–2011 (x=1028) breeding pairs of Adélie penguins, suggesting that predation pressure from brown skuas on the Litchfield Island colonies was potentially high even as early as 1975 (884 breeding pairs of penguins, Fig. 63.3).

DISCUSSION

Questions related to the real and potential consequences of tourism and the activities of researchers on Adélie penguin populations are not new to Antarctica (Thompson 1977, Muller-Schwartze 1984). However, the rapid proliferation of these activities since 1970, particularly on the Antarctic Peninsula (Harris 1991, Enzenbacher 1992), has not only forced an extension of these questions to wildlife populations in general, but also coloured them with a sense of urgency and controversy that has polarized the opinions of private industry, scientists, government organizations and environmental groups (Tangley 1988).

In Adélie penguins, the clearest relationships between human activity and potential disturbance have been demonstrated in seasonal short-term studies that have correlated exposure to different types of human activity with changes in physiological, behavioural and reproductive parameters of individual birds (Wilson *et al.* 1989, 1991, Culik *et* al. 1990, Fraser & Trivelpiece 1994). The results of these studies have in general concluded that human activity has potentially negative consequences to Adélie penguins (see Fraser & Trivelpiece 1994). However, support for these conclusions based on data in which human activity has been examined relative to long-term demographic changes has, with rare exception (see Woehler *et al.* 1994), not been forthcoming. In these studies, environmental variability rather than human disturbance has been implicated as being the key factor forcing change in penguin populations (Taylor & Wilson 1990, Blackburn *et al.* 1991).

Different scales of measurement thus appear to offer different conclusions regarding the relative effects of human disturbance on Adélie penguin populations. What may seem obvious over the short-term based on the responses of individual birds does not necessarily correspond with what is observed over the long-term at the colony and population levels. The results presented in this study add further support to this observation. The data shown in Fig. 63.3 suggest a lack of correspondence between island-specific human activity patterns and long-term change in Adélie penguin populations. Despite a six-fold increase in tourist-related activities alone, and the continued long-term use of Torgersen Island colonies for research purposes and recreation (Fig. 63.2A), trends in the Adélie penguin population remained characteristic of changes exhibited by the area as a whole. This was not the case for Litchfield Island, which by 1992 had lost 43% of its penguin population despite a 15-year absence of human activity. Resolving the apparent contradictions in conclusions between short-term and long-term studies may rest on the idea that in order for human disturbance to affect colonies and populations, disturbance must first become pathological (i.e. lead to death or long-term reproductive failure) to the individuals that ultimately constitute these demographic groups. These were the joint observations of a recent workshop that noted and addressed this issue (see Fraser & Trivelpiece 1994, p. 10). Pathological responses of the type that would lead to population declines have not been documented in short-term studies. This suggests that the changes in parameters being measured in response to 'disturbance' may fall within the adaptive range of an individual's ability to deal with environmental stress in general, and not specifically with human activity. For Torgersen Island, this would imply that the types of human activity associated with tourism, research and recreation have not been incompatible with the long-term fitness of Adélie penguins.

The Adélie penguin population decrease on Litchfield Island between 1975 and 1992 (Fig. 63.3) in the absence of human activity suggests that other processes have been involved. Here we propose that the key agents forcing this decline are related to long-term change in the patterns of snow accumulation on the island, combined with the effects of predation by brown skuas. As shown in Fig. 63.4B, 18 of the 21 extinct colonies in the area have a southwest aspect. At Palmer Station, where the predominant wind direction during storms is from the northeast (Fraser *et al.* 1993), snow accumulates on the southwest side of all topographic features. Colony aspect and status (active or extinct) thus do not appear to be the result of random processes. This is further supported by the fact that 55 of the 66 active colonies in the area have aspects other than southwest (Fig. 63.4B). As shown in Fig. 63.1, the entire Litchfield Island rookery is located on the southwest side of the island. Its location is also directly below the island's highest topographic feature (see Results, above). The height of this feature obviously determines how much snow accumulates below it, and, in turn, combined with the effects of temperature, the rate at which the accumulation melts during the season.

Adélie penguins nest only where mounds or ridges provide ground where neither snow nor meltwater accumulate (Wilson *et al.* 1990). The high number of extinct colonies on Litchfield Island (Fig. 63.4A) thus suggests that at some time in the past the availability of these nesting areas coincided with the breeding chronology of the penguins. Fraser *et al.* (1992) have suggested that a warming trend on the Antarctic Peninsula in the last four decades has forced a decrease in the number of cold years with extensive winter sea-ice cover, a pattern recently confirmed by Stammerjohn (1993). Because sea ice blocks the exchange of water vapour with the atmosphere, colder winters with extensive sea ice cover would have potentially resulted in diminished snowfall during the past (Barry 1982, Foster 1989), conditions that would have promoted the availability and use of currently extinct colonies. We therefore suggest that an increase in the frequency of years with open water (Fraser *et al.* 1992) has resulted in a gradual increase in mean annual snowfall in the area, and because of the prevailing wind direction, accumulations are amplified at breeding sites with a southwest aspect. Although a corresponding increase in temperature has occurred, melt rates associated with these sites may simply be inadequate to accommodate the temporal requirements of Adélie penguin breeding chronology. An interesting analogue to the process herein proposed is described in Wilson *et al.* (1990). When the joint US–NZ base was built at Cape Hallett in 1956, snowdrifts that developed down-wind from the buildings covered several small Adélie Penguin colonies. These colonies, initially abandoned, were subsequently recolonized three decades later after the buildings were taken down and natural patterns of wind flow and snow deposition restored.

The decrease in the Litchfield Island Adélie penguin population since 1975 has been characterized by the extinction of small colonies and a reduction in the size of larger colonies (Fraser & Ainley 1988, Fraser *et al.*1993). The latter has involved the coincident processes of habitat loss around colony perimeters due to meltwater accumulation and/or the persistence of snow, and a decrease in the number of breeding pairs. The 'final step' in the extinction process, however, appears to be aided by predation from brown skuas, which are effectively able to remove every egg or chick from smaller colonies. Based on the two extinctions recorded on the island since 1987 (see Results, above), the vulnerable colony size is approximately 25–30 pairs. In both cases, the colonies became extinct (i.e. no adults returned to breed) after two consecutive seasons of complete egg and chick losses. This suggests there is some critical minimum density of breeding adults required to maintain colony viability in relation to predation from brown skuas, although habitat changes associated with other environmental factors ultimately mediate the consequences of predation by brown skuas on Adélie penguins.

ACKNOWLEDGEMENTS

This research was funded by grants from the National Science Foundation, Office of Polar Programs, and by the National Marine Fisheries Service. During 1975–1992, these grants were awarded to one or more of the following individuals: David F. Parmelee, University of Minnesota; David G. Ainley, Point Reyes Bird Observatory; William R. Fraser and Wayne Z. Trivelpiece, Old Dominion University and Montana State University. We thank D. Neilson and B. Showers for help in the acquisition of early data, and support personnel of the US Antarctic Program for logistical support. We are grateful to B. Houston for his commitment to the research during 1991 and 1992, and to E. Stephens and D. Keller for help in data acquisition. Comments by J. Croxall, G. Robertson and E. Woehler on earlier drafts of this paper greatly improved the final manuscript.

REFERENCES

Acero, J. M. & Aguirre, C. A. 1994. Adélie penguin breeding site selection and its relation to human presence. In Fraser, W. R. & Trivelpiece, W. Z., eds. *Report: workshop on researcher–seabird interactions.* Washington, DC: Joint Oceanographic Institutions, 57 pp.

Ainley, D. G. & Boekelheide, R. J. 1990. *Seabirds of the Farallon Islands: ecology, structure and dynamics of an upwelling-system community.* Palo Alto, CA: Stanford University Press.

Ainley, D. G. & Sanders, S. R. 1988. *The status of seabirds in the Arthur Harbour/Biscoe Bay area, Antarctica, 1987–88.* Report to the US National Science Foundation, Washington, DC.

Barry, R. G. 1982. Snow and ice indicators of possible climatic effects of increasing atmospheric carbon dioxide. In Beatty, N. B., ed. *Carbon dioxide effects and assessment program, Proceedings of the DOE Workshop on First Detection of CO_2 Effects, Harpers Ferry, West Virginia, June 1981.* Washington, DC: Office of Energy Research, 207–236.

Blackburn, N.,Taylor, R. H. & Wilson, P. R. 1991. An interpretation of the growth of the Adélie penguin rookery at Cape Royds, 1955–1990. *New Zealand Journal of Ecology*, **15,** 117–121.

Croxall, J. P. & Kirkwood, E. D. 1979. *The distribution of penguins on the Antarctic Peninsula and islands of the Scotia Sea.* Cambridge: British Antarctic Survey, 179 pp.

Croxall, J. P., McCann, T. S., Prince, P. A. & Rothery, P. 1988. Variation in reproductive performance of seabirds and seals at South Georgia, 1976–1986, and its implication for Southern Ocean monitoring studies. In Sahrhage, D., ed. *Antarctic ocean and resources variability.* Berlin & Heidelberg: Springer-Verlag, 261–285.

Culik, B. M., Adelung, D., Woakes, A. J. 1990. The effects of disturbance on the heart rate and behaviour of Adélie penguins (*Pygoscelis adeliae*) during the breeding season. In Kerry, K. R. & Hempel, G., eds. *Antarctic ecosystems: ecological change and conservation.* Berlin: Springer-Verlag, 177–182.

Enzenbacher, D. J. 1992. Antarctic tourism and environmental concerns. *Marine Pollution Bulletin*, **25,** 258–265.

Foster, J. L. 1989. The significance of the date of snow disappearance on the Arctic tundra as a possible indicator of climate change. *Arctic and Alpine Research*, **21,** 60–70.

Fraser, W. R. 1992. *US seabird research undertaken as part of the CCAMLR Ecosystem Monitoring Program at Palmer Station 1991–1992. Annual Report.* La Jolla, CA: National Marine Fisheries Service, 1–28.

Fraser, W. R. & Ainley, D. G. 1988. *US seabird research undertaken as part of the CCAMLR Ecosystem Monitoring Program at Palmer Station 1987–1988. Annual Report.* La Jolla, CA: National Marine Fisheries Service, 1–31.

Fraser, W. R. & Trivelpiece, W. Z. 1994. *Report: workshop on researcher–seabird interactions.* Washington, DC: Joint Oceanographic Institutions, 57 pp.

Fraser, W. R., Trivelpiece, W. Z., Ainley, D. G. & Trivelpiece, S. G. 1992. Increases in Antarctic penguin populations: reduced competition with whales or a loss of sea-ice due to global warming? *Polar Biology*, **11,** 525–531.

Fraser, W. R., Trivelpiece, W. Z., Houston, B. & Patterson, D. L. 1993. *US seabird research undertaken as part of the CCAMLR Ecosystem Monitoring Program at Palmer Station 1992–1993. Annual Report.* La Jolla, CA: National Marine Fisheries Service, 1–33.

Harris, C. M. 1991. Environmental effects of human activities on King George Island, South Shetland Islands, Antarctica. *Polar Record*, **27,** 193–204.

Heimark, G. M. & Heimark, R. I. 1988. Observations of birds and marine mammals at Palmer Station November 1985 to November 1986. *Antarctic Journal of the United States*, **23,** 3–8.

Muller-Schwarze, D. 1984. Possible human impact on penguin populations in the Antarctic Peninsula area. *Antarctic Journal of the United States*, **19,** 158–159.

Neilson, D. R. 1983. *Ecological and behavioural aspects of sympatric breeding south polar skua (*Catharacta maccormicki*) and the brown skua (*Catharacta lonnbergi*) near the Antarctic Peninsula.* Masters Thesis, University of Minnesota, Minneapolis, 1–79.

Parmelee, D. F. & Parmelee, J. M. 1987. Revised penguin numbers and distribution for Anvers Island, Antarctica. *British Antarctic Survey Bulletin*, No. 76, 65–73.

Poncet, S. & Poncet, J. 1987. Censuses of penguin populations of the Antarctic Peninsula, 1983–87. *British Antarctic Survey Bulletin*, No. **77,** 109–129.

Stammerjohn, S.E. 1993. *Spatial and temporal variability in southern ocean sea-ice coverage.* Masters Thesis, University of California, Santa Barbara, 1–111.

Stonehouse, B. 1965. Counting Antarctic animals. *New Scientist*, **29,** 273–276.

Tangley, L. 1988. Who's polluting Antarctica? *BioScience*, **38,** 590–594.

Taylor, R. H. & Wilson, P. R. 1990. Recent increase and southern expansion of Adélie penguin populations in the Ross Sea, Antarctica, related to climate warming. *New Zealand Journal of Ecology*, **14,** 25–29.

Taylor, R. H., Wilson, P. R. & Thomas, B. W. 1990. Status and trends of Adélie penguin populations in the Ross Sea region. *Polar Record*, **26,** 293–304.

Thompson, R. B. 1977. Effects of human disturbance on an Adélie penguin rookery and measures of control. In Llano, G. A., ed. *Adaptations within Antarctic ecosystems.* Washington DC: Smithsonian Institution, 1177–1180.

Trivelpiece, W. Z., Butler, R. G. & Volkman, N. J. 1980. Feeding territories of brown skuas (*Catharacta lonnbergi*). *The Auk*, **97**, 669–676.

Wilson, K. J., Taylor, R. H. & Barton, K. J. 1990. The impact of man on Adélie penguins at Cape Hallett, Antarctica. In Kerry, K. R. & Hempel, G., eds. *Antarctic ecosystems: ecological change and conservation*. Berlin & Heidelberg: Springer-Verlag, 183–190.

Wilson, R. P., Coria, N. R., Spairani, H. J., Adelung, D. & Culik, B. 1989. Human-induced behaviour in Adélie penguins (*Pygoscelis adeliae*). *Polar Biology*, **10**, 7–80.

Wilson, R. P., Culik, B., Danefeld, R. & Adelung, D. 1991. People in Antarctica: how much do penguins care? *Polar Biology*, **11**, 363–370.

Woehler, E. J., Penney, R. L., Creet, S. M. & Burton, R. H. 1994. Impacts of human visitors on breeding success and long-term population trends in Adélie penguins at Casey, Antarctica. *Polar Biology*, **14**, 269–274.

Young, E. C. 1990. Long-term stability and human impact in Antarctic skuas and Adélie penguins. In Kerry, K. R. & Hempel, G., eds. *Antarctic ecosystems: ecological change and conservation.* Berlin & Heidelberg: Springer-Verlag, 231–236.

VI Postscript

FUTURE OPPORTUNITIES

In looking forward to suggest what may be the scientific objectives for Antarctic biologists over the next ten years the editors claim no special knowledge. However, the organization of such a major symposium as the meeting in Venice and the selection and editing of many of the contributions has provided us with an incentive to look more closely at what could be the major themes relevant to studies of communities.

Barring some major new discovery like the ozone hole, the inertia built into Antarctic research by detailed planning systems and complex logistics suggests that there will be no revolutionary changes but that there will be significant shifts in emphasis and direction. These will reflect to some extent the priority fields identified in the national research plans of Consultative Parties but they are also likely to underline an existing trend towards more team research and more international collaboration. Global agreements may also play a part in forcing the pace in some fields.

For instance, the Biodiversity Convention requires assessment of biodiversity in several ways and for all regions of the world. This has highlighted the existing inadequacies in taxonomy for many terrestrial and marine groups, with particular difficulties for community ecologists. There are already plans to produce continent-wide floras for lichens and bryophytes, and agreement seems likely on the taxonomy of freshwater diatoms. The continuing series of US and German publications on benthos will provide a sound basis for the identification of many key groups, providing much better support for the studies of marine communities in the future. However, there remain several major groups of marine organisms which require taxonomic revision, a situation common for marine ecosystems world wide. The continuing lack of resources devoted to basic taxonomy sits oddly beside the international interest in biodiversity and appears unlikely to change.

At the level of the species there is already evidence of the application of the latest techniques to Antarctic problems. The facilities provided at the Crary Laboratory at McMurdo have produced a major step forward in what can already be undertaken in the Antarctic. With other countries planning to provide upgraded scientific facilities for biology on both ships and stations there seems likely to be much greater opportunities in the next few years for state-of-the art research into biochemistry and physiology. Of major interest are the effects of increased UV. There seems certain to be a further expansion of existing studies to research in more detail the effects at the cellular level, whether they are concerned with mitigation through the production of screening compounds, or repair of damage to organelles or transcription systems. Other aspects of climate change – temperature, water availability and carbon dioxide – also seem likely to be included in experimental investigations of the effects of single and interacting stresses at both the organism and community levels.

A major area of growth seems likely to be molecular biology. This symposium has already seen a range of studies on evolution and population characterization, whilst there is a growing body of work on the role of specific gene sequences in controlling stress responses and behavioural patterns. The importance of this general field has been recognised by the formation of a new subcommittee of the SCAR Working Group on Biology to coordinate activities in evolutionary biology. Already there has been some work on identifying molecular characteristics of populations to characterize genetic diversity in widespread populations. The studies in this volume on moss populations are important in establishing the potential of the tool for terrestrial species. There have already been papers describing its application to fish stock assessment where clear population distinctions are essential to provide a more scientific base for evaluating the application of catch quotas. The need for this seems certain to increase rapidly over the next few years. A second focus, already under rapid development, is the use of molecular tools to answer phylogenetic questions. Studies on birds and bacteria have already demonstrated the value of this, especially with respect to endemism. A third field, so far hardly touched, is the identification of the source of immigrants to the Antarctic, be they terrestrial or aquatic.

It is not simply the present communities that are of interest. In unravelling the relationships between present and past communities it is essential to analyse whatever historical data can be obtained. The lake sediments can tell us a great deal about the major environmental changes of the past 5–10 000 years. Whilst there are already detailed sequences available from the lakes in the South Orkney and South Shetland archipelagoes, only recently have there been data from East Antarctica and the Dry Valleys. A major challenge will be to synthesize the sediment data from widely separated lakes to decide if there is a continent-wide synchronized chronology for climate change. Such a chronology could then be compared with snow core data for further validation. An as yet unexplored resource is the sediments from sub-Antarctic lakes.

For some people the Protocol is simply another piece of bureaucracy designed to make life difficult for the Antarctic scientist. Looked at more dispassionately, it can be seen as both a protection for the future use of the Antarctic for scientific purposes and as a framework pointing up the inadequacies of our present understanding of the Antarctic ecosystems. At the most basic level we need better taxonomic tools. Even with these there are some major tasks as yet inadequately addressed. For example, what are the assemblage rules that govern the make up of communities? Where are the niche descriptions that characterize the structure of communities and allow some assessment of resilience and recovery? Whilst there is already some research on immigration and colonization in both terrestrial and aquatic communities a great deal more needs to be known if we are to make well-based predictions on the rate of change of communities by the introduction of new species and the possibilities for micro-evolution in disjunct populations. In terms of most forms of impact – physical, chemical or biological – we have little experimental data on which to base our predictions. A great deal of fundamental research is needed to ensure the application of environmental impact assessments.

If the proposed 'Report on the state of the Antarctic environment' is undertaken by SCAR it is certain that this will expose major inadequacies in a wide variety of fields. There are few data on ecotoxicology and on baseline levels of pollutants for the region as a whole. Chemical speciation under low temperatures and its relevance to toxicity has hardly been touched. A great deal still remains to be done to characterize adequately the implications of natural change, e.g. El Niño–Southern Oscillation, on communities in order to extract a clearer signal of the effects of anthropogenic activities.

Above all, the next decade should see a growth in the development and application of models. One of the greatest strengths of Antarctic biology over the past 30 years has been the recognition of the close coupling between terrestrial and aquatic ecosystems. The growth of Antarctic data now needs to be utilized effectively and that requires data management tools (such as GIS), an holistic approach to ecosystems and greater investment in modelling. This will obviously apply at the levels of process, species and communities.

Antarctic biology is alive and well. The development of closer links with IGBP, the enthusiastic launch of major new SCAR biology programmes – EASIZ and APIS – and the continuing participation of Antarctic scientists in the global programmes of GLOBEC, WOCE, JGOFS, etc. have increased both the relevance and the visibility of Antarctic science world wide. The future looks full of new opportunities!

Index